General Chemistry

Tenth Edition

William R. Robinson Purdue University

Jerome D. Odom University of South Carolina

Henry F. Holtzclaw, Jr. Professor Emeritus,
University of Nebraska, Lincoln
Contributing Editor

Houghton Mifflin Company   Boston   New York

Senior Sponsoring Editor: Richard Stratton
Senior Associate Sponsoring Editor: Susan Warne
Development Editor: Betty Hoskins
Senior Project Editor: Maria Morelli
Editorial Assistant: Christian Zabriskie
Senior Production/Design Coordinator: Jennifer Waddell
Senior Manufacturing Coordinator: Lisa Merrill
Executive Marketing Manager: Karen Natale

Cover Design: Harold Burch, Harold Burch Design, New York City.

Cover Image: Vitamin B_{12} Crystals, Peter Arnold Inc. © Manfred Kage

Advanced Placement test questions selected from Advanced Placement Examination: Chemistry © 1995 College Entrance Examination Board. Reprinted by permission of Educational Testing Service and the College Entrance Examination Board, the copyright owners of the test questions. Disclaimer: Permission to reprint AP materials does not constitute review or endorsement by Educational Testing Service or the College Board of this publication as a whole or any testing information it may contain.

For permission to use photographs, grateful acknowledgment is made to the copyright holders listed under Photo Credits at the back of the book, hereby considered an extension of this copyright page.

Printed in the U.S.A.

Library of Congress Catalog Card Numbers:
    General Chemistry with Qualitative Analysis: 96-76955
    General Chemistry: 96-76954
    Essentials of General Chemistry: 96-76953

International Standard Book Numbers:
    General Chemistry with Qualitative Analysis: 0-669-35482-1
        Exam Copy: 0-669-41864-1
    General Chemistry: 0-669-35483-X
        Exam Copy: 0-669-41861-7
    Essentials of General Chemistry: 0-669-35484-8
        Exam Copy: 0-669-41860-9

1 2 3 4 5 6 7 8 9—VH—00 99 98 97 96

Atomic Masses of the Elements (based on carbon-12 with the uncertainties in parentheses)

Name	Symbol	Atomic Number	Atomic Mass	Name	Symbol	Atomic Number	Atomic Mass
Actinium*	Ac	89	227.028	Molybdenum	Mo	42	95.94(1)
Aluminum	Al	13	26.981539(5)	Neilsbohrium*	Ns	107	(262)
Americium*	Am	95	(243)^a	Neodymium	Nd^g	60	144.24(3)
Antimony	Sb	51	121.75(3)	Neon	Ne^{g,m}	10	20.1797(6)
Argon	Ar^{g,r}	18	39.948(1)	Neptunium*	Np	93	237.048
Arsenic	As	33	74.92159(2)	Nickel	Ni	28	58.6934(2)
Astatine*	At	85	(210)^a	Niobium	Nb	41	92.90638(2)
Barium	Ba	56	137.327(7)	Nitrogen	N^{g,r}	7	14.00674(7)
Berkelium*	Bk	97	(247)^a	Nobelium	No	102	(259)^a
Beryllium	Be	4	9.012182(3)	Osmium	Os^g	76	190.23(3)
Bismuth	Bi	83	208.98037(3)	Oxygen	O^{g,r}	8	15.9994(3)
Boron	B^{g,m,r}	5	10.811(5)	Palladium	Pd^g	46	106.42(1)
Bromine	Br	35	79.904(1)	Phosphorus	P	15	30.973762(4)
Cadmium	Cd^g	48	112.411(8)	Platinum	Pt	78	195.08(3)
Calcium	Ca^g	20	40.078(4)	Plutonium*	Pu	94	(244)^a
Californium*	Cf	98	(251)^a	Polonium*	Po	84	(209)^a
Carbon	C	6	12.011(1)	Potassium (kalium)	K	19	39.0983(1)
Cerium	Ce^g	58	140.115(4)	Praseodymium	Pr	59	140.90765(3)
Cesium	Cs	55	132.90543(5)	Promethium*	Pm	61	(145)^a
Chlorine	Cl	17	35.4527(9)	Protactinium*	Pa	91	231.03588(2)
Chromium	Cr	24	51.9961(6)	Radium*	Ra	88	226.025
Cobalt	Co	27	58.93320(1)	Radon*	Rn	86	(222)^a
Copper	Cu^r	29	63.546(3)	Rhenium	Re	75	186.207(1)
Curium*	Cm	96	(247)^a	Rhodium	Rh	45	102.90550(3)
Dysprosium	Dy^g	66	162.50(3)	Rubidium	Rb^g	37	85.4678(3)
Einsteinium*	Es	99	(252)^a	Ruthenium	Ru^g	44	101.07(2)
Erbium	Er^g	68	167.26(3)	Rutherfordium*	Rf	104	(261)
Europium	Eu^g	63	151.965(9)	Samarium	Sm^g	62	150.36(3)
Fermium*	Fm	100	(257)^a	Scandium	Sc	21	44.955910(9)
Fluorine	F	9	18.9984032(9)	Seaborgium*	Sg	106	(263)
Francium*	Fr	87	(223)^a	Selenium	Se	34	78.96(3)
Gadolinium	Gd^g	64	157.25(3)	Silicon	Si^r	14	28.0855(2)
Gallium	Ga	31	69.723(1)	Silver	Ag	47	107.8682(2)
Germanium	Ge	32	72.61(2)	Sodium (natrium)	Na	11	22.989768(6)
Gold	Au	79	196.96654(3)	Strontium	Sr	38	87.62(1)
Hafnium	Hf	72	178.49(2)	Sulfur	S	16	32.066(6)
Hahnium*	Ha	105	(262)	Tantalum	Ta	73	180.9479(1)
Hassium*	Hs	108	(265)	Technetium*	Tc	43	(98)^a
Helium	He^{g,r}	2	4.002602(2)	Tellurium	Te^g	52	127.60(3)
Holmium	Ho	67	164.93032(3)	Terbium	Tb	65	158.92534(3)
Hydrogen	H^{g,m,r}	1	1.00794(7)	Thallium	Tl	81	204.3833(2)
Indium	In	49	114.818(3)	Thorium*	Th^{g,Z}	90	232.0381(1)
Iodine	I	53	126.90447(3)	Thulium	Tm	69	168.93421(3)
Iridium	Ir	77	192.217(3)	Tin	Sn^g	50	118.710(7)
Iron	Fe	26	55.845(2)	Titanium	Ti	22	47.88(3)
Krypton	Kr^{g,m}	36	83.80(1)	Tungsten (wolfram)	W	74	183.85(3)
Lanthanum	La^g	57	138.9055(2)	Ununbium*	Uub	112	(277)
Lawrencium*	Lr	103	(260)^a	Ununnilium*	Uun	110	(269)
Lead	Pb^{g,r}	82	207.2(1)	Unununium*	Uuu	111	(272)
Lithium	Li^{g,m,r}	3	6.941(2)	Uranium*	U^{g,m,Z}	92	238.0289(1)
Lutetium	Lu^g	71	174.967(1)	Vanadium	V	23	50.9415(1)
Magnesium	Mg	12	24.3050(6)	Xenon	Xe^{g,m}	54	131.29(2)
Manganese	Mn	25	54.93805(1)	Ytterbium	Yb^g	70	173.04(3)
Meitnerium*	Mt	109	(266)	Yttrium	Y	39	88.90585(2)
Mendelevium*	Md	101	(258)^a	Zinc	Zn	30	65.39(2)
Mercury	Hg	80	200.59(3)	Zirconium	Zr^g	40	91.224(2)

^{a}Atomic mass values in parentheses are used for radioactive elements the atomic masses of which are not knows precisely or that cannot be quoted precisely without knowledge of the origin of the elements; the value given is the atomic mass number of the isotope of that element that has the longest known half-life.

^{g}Geological specimens are known in which the element has an isotopic composition outside the limits for normal material. The difference between the atomic mass of the element in such specimens and that given in this table may exceed the implied uncertainty.

^{m}Modified isotopic compositions may be found in commercially available material because it has been subjected to an undisclosed or inadvertent isotopic separation. Substantial deviations in atomic mass of the element from that given in the table can occur.

^{r}Range in isotopic composition of normal terrestrial material prevents a more precise mass being given; the tabulated mass value should be applicable to any normal material.

^{Z}An element, without stable nuclide(s), exhibiting a range of characteristic terrestrial compositions of long-lived radionuclide(s) such that a meaningful atomic mass can be given.

*Element has no stable nuclides.

Adapted from *Pure and Applied Chemistry,* Vol. 66, pages 2426-2427 (1994).

Brief Contents

1. Some Fundamental Concepts   1
2. The Language of Chemistry   33
3. Chemical Stoichiometry   70
4. Thermochemistry   116
5. Atoms: Their Structure, Spectra, and Properties   143
6. Chemical Bonding: General Concepts   183
7. Molecular Structures and Models of Covalent Bonds   216
8. Chemical Reactions and the Periodic Table   257
9. Structure and Bonding at Carbon   293
10. Gases   325
11. Intermolecular Forces, Liquids, and Solids   371
12. Solutions and Colloids   413
13. Chemical Kinetics   460
14. An Introduction to Chemical Equilibrium   507
15. Acids and Bases   553
16. Ionic Equilibria of Weak Electrolytes   582
17. The Formation and Dissolution of Precipitates   638
18. Chemical Thermodynamics   668
19. Electrochemistry and Oxidation-Reduction   709
20. Nuclear Chemistry   756
21. The Representative Metals   786
22. The Semi-Metals and the Nonmetals   807
23. The Transition Elements and Coordination Compounds   864
24. The Chemistry of Materials   894
25. Biochemistry   926

Appendixes   A-1
Photo Credits   B-1
Glossary   C-1
Answers to Odd-Numbered Exercises   C-12
Index   C-35

Contents

Preface   xix

1 Some Fundamental Concepts   1

The Basics   3

1.1   Doing Chemistry   3
1.2   Matter   4
1.3   Energy   5
1.4   Chemical and Physical Properties   7
1.5   Classifying Matter   8
1.6   Atoms and Molecules   10
1.7   Classifying Elements: The Periodic Table   11
1.8   The Domains of Chemistry   13

Measurement in Chemistry   14

1.9   Units and Measurement   14
1.10 SI Base Units   16
1.11 Derived SI Units   17
1.12 Uncertainty in Measurements and Significant Figures   18
1.13 Conversion of Units   21
1.14 Conversion of Temperature Units   24

For Review   26
Summary   26   Key Terms and Concepts   27   Exercises   27

2 The Language of Chemistry   33

2.1   Dalton's Atomic Theory   33
2.2   The Composition of Atoms   35
2.3   Symbols and Formulas   37
2.4   Isotopes   41
2.5   The Periodic Table   44
2.6   Formation of Ions   46
2.7   Ionic and Covalent Bonds   49
2.8   Chemical Equations   52
2.9   Ionic Equations   55
2.10 Classification of Chemical Reactions   56
2.11 Naming Inorganic Compounds   59

For Review   63
Summary   63   Key Terms and Concepts   64   Exercises   64

3 Chemical Stoichiometry   70

3.1   Atomic and Molecular Masses   71
3.2   Moles of Atoms; Avogadro's Number   73
3.3   Moles of Molecules   77
3.4   Percent Composition from Formulas   80
3.5   Derivation of Empirical Formulas   81
3.6   Derivation of Molecular Formulas   84
3.7   Solutions   85
3.8   Dilution of Solutions   87

Stoichiometry and Chemical Change   90
3.9   Mole Relationships and Chemical Equations   90
3.10 Material Balances   92
3.11 Theoretical Yield, Actual Yield, and Percent Yield   98
3.12 Limiting Reactant   99

Quantitative Analysis   101
3.13 Titration   101
3.14 Gravimetric Analysis   102
3.15 Combustion Analysis   104

For Review   106
Summary   106   Key Terms and Concepts   107   Exercises   107

4 Thermochemistry   116

4.1   Thermal Energy, Heat, and Temperature   118
4.2   Measurement of Heat   119
4.3   Calorimetry   120
4.4   Thermochemistry and Thermodynamics   123
4.5   Enthalpy Changes   124
4.6   Hess's Law   131
4.7   Fuel and Food   134

For Review   136
Summary   136   Key Terms and Concepts   137   Exercises   137

5 Atoms: Their Structure, Spectra, and Properties   143

Historical Development of the Atomic Theory   145
5.1   Atomic Architecture: The Nuclear Atom   146
5.2   Light, Atomic Spectra and Atomic Structure   147
5.3   The Bohr Model of the Atom   151

Quantum Mechanics   155
5.4   Behavior in the Microscopic World   155
5.5   The Quantum-Mechanical Model of the Atom   157
5.6   Results of the Quantum-Mechanical Model of the Atom   158
5.7   Orbital Energies and Atomic Structure   163

The Periodic Table: Electron Configuration and Atomic Properties   164
5.8   The Aufbau Process   164
5.9   Electron Configurations in the Periodic Table   169
5.10   Variation of Atomic Properties within Periods and Groups   171

For Review   177
Summary   177   Key Terms and Concepts   178   Exercises   178

6   Chemical Bonding: General Concepts   183

6.1   Ionic Bonding: Electrostatic Attractions Between Atoms   184
6.2   The Electronic Structures of Ions   185
6.3   The Lattice Energies of Ionic Crystals   187
6.4   Covalent Bonds: Chemical Bonding by Electron Sharing   188
6.5   Polar Covalent Bonds: Electronegativity   189
6.6   Lewis Electron-Dot Symbols and Formulas   192
6.7   Writing Lewis Structures: Covalently Bonded Atoms with a
       Noble Gas Configuration   194
6.8   Writing Lewis Structures: Covalently Bonded Atoms
       without a Noble Gas Configuration   197
6.9   Formal Change   200
6.10   Oxidation State   203
6.11   Resonance   205
6.12   The Strength of Covalent Bonds   206

For Review   209
Summary   209   Key Terms and Concepts   210   Exercises   210

7   Molecular Structures and Models of Covalent Bonds   216

Valence Shell Electron-Pair Repulsion Theory   217
7.1   Predicting Molecular Structures   217
7.2   Rules for Predicting Electron-Pair Geometry and
       Molecular Geometry   222
7.3   Molecular Polarity and Dipole Moment   227

Valence Bond Theory: Hybridization of Atomic Orbitals   229
7.4   Valence Bond Theory   230
7.5   Hybridization of Atomic Orbitals   232
7.6 sp Hybridization   233
7.7 sp^2 Hybridization   235

7.8	sp^3 Hybridization	235
7.9	sp^3d and sp^3d^2 Hybridization	237
7.10	Assignment of Hybrid Orbitals to Central Atoms	238
7.11	Hybridization Involving Double and Triple Bonds	240

Molecular Orbital Theory   243

7.12	Molecular Orbitals	243
7.13	Molecular Orbital Energy Diagrams	245
7.14	Bond Order	247
7.15	H_2 and He_2 Molecules	247
7.16	Diatomic Molecules of the Second Period	248

For Review   251
Summary   251   Key Terms and Concepts   252   Exercises   252

8 Chemical Reactions and the Periodic Table   257

8.1	Chemical Behavior and the Periodic Table	258
8.2	Metals, Nonmetals, and Semi-Metals	259
8.3	Oxidation States	261
8.4	Periodic Variations in Oxidation States	265
8.5	Chemical Reactions	267
8.6	Variation in Metallic and Nonmetallic Behavior of the Representative Elements	272
8.7	The Activity States	273
8.8	Prediction of Reaction Products	274
8.9	Chemical Properties of Some Important Industrial Chemicals	279

For Review   286
Summary   286   Key Terms and Concepts   286   Exercises   287

9 Structure and Bonding at Carbon   293

Elemental Carbon and its Inorganic Compounds   294

9.1	Elemental Carbon	294
9.2	Inorganic Compounds of Carbon	297

Organic Compounds of Carbon   299

9.3	Alkanes	300
9.4	The Basics of Organic Nomenclature	302
9.5	Alkenes	304
9.6	Alkynes	306
9.7	Aromatic Hydrocarbons	307

Derivatives of Hydrocarbons   308

9.8	The Behavior of Hydrocarbon Derivatives	308
9.9	Alcohols	309

9.10   Ethers   310
9.11   Amines   310
9.12   Aldehydes and Ketones   311
9.13   Carboxylic Acids and Esters   312
9.14   Natural Products   314
9.15   Polymers   315

For Review   317
Summary   317   Key Terms and Concepts   318   Exercises   318

10 Gases   325

The Macroscopic Behavior of Gases   326

10.1   The Pressure of a Gas   326
10.2   Volume and Pressure: Boyle's Law   329
10.3   Volume and Temperature: Charles's Law   333
10.4   The Kelvin Temperature Scale   336
10.5   Moles of Gas and Volume: Avogadro's Law   337
10.6   The Ideal Gas Equation   338
10.7   Standard Conditions of Temperature and Pressure 341
10.8   Densities and Molecular Masses of Gases   343
10.9   The Pressure of a Mixture of Gases: Dalton's Law 345
10.10   Diffusion and Effusion of Gases: Graham's Law   348
10.11   Stoichiometry Involving Gases   349
10.12   The Atmosphere   352

The Microscopic Behavior of Gases   356

10.13   The Kinetic-Molecular Theory   356
10.14   Molecular Velocities and Kinetic Energy   357
10.15   Relationship of the Behavior of Gases to
        the Kinetic–Molecular Theory   359
10.16   Derivation of the Ideal Gas Equation from
        Kinetic–Molecular Theory   361
10.17   Deviations from Ideal Gas Behavior:
        Descriptions of Real Gases   362

For Review   364
Summary   364   Key Terms and Concepts   364   Exercises   365

11 Intermolecular Forces, Liquids, and Solids   371

11.1   Kinetic-Molecular Theory: Liquids and Solids   372

Forces Between Molecules   374

11.2   Dipole-Dipole Attractions   374
11.3   Dispersion Forces   375
11.4   Hydrogen Bonding   376

Properties of Liquids and Solids   378

11.5   Evaporation of Liquids and Solids   378
11.6   Boiling of Liquids   379
11.7   Distillation   380
11.8   Melting of Solids   382
11.9   Critical Temperature and Pressure   383
11.10 Phase Diagrams   384
11.11 Cohesive Forces and Adhesive Forces   387

The Structures of Crystalline Solids   389

11.12 Types of Solids   389
11.13 Crystal Defects   391
11.14 The Structures of Metals   391
11.15 The Structures of Ionic Crystals   394
11.16 The Radius Ration Rule   396
11.17 Unit Cells   397
11.18 Calculation of Ionic Radii   402
11.19 The Born-Haber Cycle   403
11.20 X-Ray Diffraction   404

For Review   406
Summary 406   Key Terms and Concepts 406   Exercises 407

12

Solutions and Colloids   413

12.1   The Nature of Solutions   414

The Process of Dissolution   416

12.2   The Formation of Solutions   416
12.3   Dissolution of Ionic Compounds   419
12.4   Dissolution of Molecular Electrolytes   421

Macroscopic Properties of Solutions   422

12.5   Solutions of Gases in Liquids   422
12.6   Solutions of Liquids in Liquids   424
12.7   The Effect of Temperature on the Solubility of Solids
       in Water   425
12.8   Solid Solutions   426

Expressing Concentration   427

12.9   Percent Composition   428
12.10 Molarity   429
12.11 Molality   430
12.12 Mole Fraction   431

Colligative Properties of Solutions   435

12.13 Lowering of the Vapor Pressure of a Solvent   435
12.14 Elevation of the Boiling Point of a Solvent   437
12.15 Distillation of Solutions   439

12.16   Depression of the Freezing Point of a Solvent   440
12.17   Phase Diagram for an Aqueous Solution of a Nonelectrolyte   443
12.18   Osmosis and Osmotic Pressure of Solutions   444
12.19   Determination of Molecular Masses   445
12.20   Colligative Properties of Electrolytes   447

Colloid Chemistry   448

12.21   Colloids   448
12.22   Preparation of Colloidal Systems   450
12.23   Soaps and Detergents   450
12.24   Electrical Properties of Colloidal Particles   452
12.25   Gels   453

For Review   453
Summary   453   Key Terms and Concepts   454   Exercises   454

13   Chemical Kinetics   460

Rates of Chemical Reactions   461

13.1   Rate of Reaction   461
13.2   Relative Rates of Reaction and Reaction Velocity   464
13.3   Factors that Affect Reaction Rates   468
13.4   Rate Laws   470
13.5   Integrated Rate Laws   477
13.6   The Half-Life of a Reaction   483

The Microscopic Explanation of Reaction Rates   486

13.7   Collision Theory of Reaction Rates   486
13.8   Activation Energy and the Arrhenius Equation   488
13.9   Elementary Reactions   492
13.10   Reaction Mechanisms   495
13.11   Catalysts   497

For Review   499
Summary   499   Key Terms and Concepts   500   Exercises   500

14   An Introduction to Chemical Equilibrium   507

An Introduction to Equilibrium   508

14.1   The State of Equilibrium   508
14.2   Reaction Quotients and Equilibrium Constants   510
14.3   Homogeneous and Heterogeneous Equilibria   513
14.4   Changes in Concentrations at Equilibrium:
        Le Châtelier's Principle   517
14.5   Predicting the Direction of a Reversible Reaction   521

Kinetics and Equilibria   523

14.6   The Relationship of Reaction Rates and Equilibria   523
14.7   Reaction Mechanisms Involving Equilibria   525

Equilibrium Calculations   527

14.8   Concentration and Pressure Changes   527
14.9   Calculations Involving Equilibrium Concentrations   529
14.10 Calculation of Changes in Concentration   532
14.11 Techniques for Solving Equilibrium Problems   535

For Review   543
Summary   543   Key Terms and Concepts   544   Exercises   544

15 Acids and Bases   553

The Brønsted-Lowry Concept of Acids and Bases   554

15.1   The Protonic Concept of Acids and Bases   554
15.2   Amphiprotic Species   558
15.3   The Strength of Brønsted Acids and Bases   559
15.4   Acid–Base Neutralization   563
15.5   The Relative Strength of Strong Acids and Bases   564
15.6   Properties of Brønsted Acids in Aqueous Solutions   566
15.7   Preparation of Brønsted Acids   566
15.8   Monoprotic, Diprotic, and Triprotic Acids   567
15.9   Properties of Brønsted Bases in Aqueous Solutions   568
15.10 Preparation of Hydroxide Bases   568
15.11 Quantitative Reactions of Acids and Bases   569
15.12 Equivalents of Acids and Bases   570

The Lewis Concept of Acids and Bases   572

15.13 Definitions and Examples   572

For Review   574
Summary   574   Key Terms and Concepts   575   Exercises   575

16 Ionic Equilibria of Weak Electrolytes   582

16.1   Ion Concentrations in Solutions of Strong Electrolytes   583
16.2   pH and pOH   585
16.3   The Ionization of Weak Acids and Weak Bases   590
16.4   Concentrations in Solutions of Weak Acids or Weak Bases   596
16.5   Acid–Base Properties of Solutions of Salts   602
16.6   Mixtures of Acids or Bases; Polyprotic Acids   604
16.7   The Ionization of Very Weak Acids   609
16.8   The Salt Effect and the Common Ion Effect   613
16.9   Buffer Solutions   617

16.10   The Ionization of Hydrated Metal ions   623
16.11   Acid–Base Indicators   625
16.12   Titration Curves   626

For Review   629
Summary 629   Key Terms and Concepts 630   Exercises 630

17 The Formation and Dissolution of Precipitates   638

17.1   The Solubility Product   639

Precipitation and Dissolution   642

17.2   Solubilities and Solubility Products   642
17.3   Calculation of Solubilities from Solubility Products   644
17.4   The Precipitation of Slightly Soluble Solids   646
17.5   Concentrations Necessary to Form a Precipitate   648
17.6   Concentrations Following Precipitation   649
17.7   Fractional Precipitation   650

Multiple Equilibria Involving Solubility   651

17.8   Dissolution by Formation of a Weak Electrolyte   652
17.9   Dissolution by Formation of a Complex Ion   656

For Review   661
Summary 661   Key Terms and Concepts 661   Exercises 661

18 Chemical Thermodynamics   668

Work, Heat, and Changes in Internal Energy   669

18.1   Systems and Surroundings   670
18.2   Internal Energy, Heat, and Work   671
18.3   The First Law of Thermodynamics   674
18.4   State Functions   676

Enthalpy, Entropy, and Spontaneous Processes   677

18.5   Spontaneous Chemical Reactions   678
18.6   Minimization of Energy, Increase in Disorder, and
       Spontaneous Change   680
18.7   Changes in Energy: Enthalpy Changes   681
18.8   Changes in Disorder: Entropy and Entropy Changes   684
18.9   Calculation of Entropy Changes   686
18.10 The Third Law of Thermodynamics   688

Free Energy Changes   689

18.11 Free Energy Changes, Enthalpy Changes, and
       Entropy Changes   689
18.12 Free Energy of Formation   692

18.13   The Second Law of Thermodynamics   693
18.14   Free Energy Changes and Nonstandard States   694
18.15   The Relationship Between Free Energy Changes and
         Equilibrium Constants   696

For Review   699
Summary   699   Key Terms and Concepts   699   Exercises   700

19 Electrochemistry and Oxidation–Reduction   709

Galvanic Cells and Cell Potentials   710

19.1   Galvanic Cells   710
19.2   Cell Potentials   714
19.3   Standard Electrode Potentials   716
19.4   Circulation of Cell Potentials   721
19.5   Cell Potential, Electrical Work, and Free Energy   723
19.6   The Effect of Concentration on Cell Potential:
         The Nernst Equation   725
19.7   Relationship of the Cell Potential and the
         Equilibrium Constant   727

Batteries   729

19.8   Primary Cells   730
19.9   Secondary Cells   730
19.10   Fuel Cells   732
19.11   Corrosion   733

Electrolytic Cells   734

19.12   The Electrolysis of Molten Sodium Chloride   734
19.13   The Electrolysis of Aqueous Solutions   736
19.14   Electrolytic Disposition of Metals   738
19.15   Faraday's Law of Electrolysis   740

Oxidation–Reduction Reactions   743

19.16   Balancing Redox Equations   743

For Review   747
Summary   747   Key Terms and Concepts   748   Exercises   748

20 Nuclear Chemistry   756

The Stability of Nuclei   757

20.1   The Nucleus   757
20.2   Nuclear Binding Energy   758
20.3   Nuclear Stability   761
20.4   The Half-Life of Radioactive Materials   762

Nuclear Reactions   763

20.5   Equations for Nuclear Reactions   763
20.6   Radioactive Decay   765
20.7   Radioactive Dating   767
20.8   Synthesis of Nuclides   769

Nuclear Energy and Other Applications   771

20.9   Nuclear Fission and Nuclear Power Reactors   771
20.10   Nuclear Fusion and Fusion Reactors   776
20.11   Uses of Radioisotopes   777
20.12   Interaction of Radiation with Matter   779

For Review   781
Summary   781   Key Terms and Concepts   782   Exercises   782

21 The Representative Metals   786

The Elemental Representative Metals   788

21.1   Periodic Relationships Among Groups   788
21.2   Preparation of Representative Metals   791

Compounds of the Representative Metals   795

21.3   Compounds with Oxygen   796
21.4   Hydroxides   798
21.5   Carbonates and Hydrogen Carbonates   799
21.6   Salts   803

For Review   00
Summary   803   Key Terms and Concepts   803   Exercises   804

22 The Semi-Metals and the Nonmetals   807

The Semi-Metals   809

22.1   The Chemical Behavior of the Semi-Metals   809
22.2   Structures of the Semi-Metals   810
22.3   Occurrence and Preparation of Boron and Silicon   811
22.4   Boron and Silicon Hydrides   812
22.5   Boron and Silicon Halides   815
22.6   Boron and Silicon Oxides and Derivatives   816

The Nonmetals   818

22.7   Periodic Trends and the General Behavior of the Nonmetals   818
22.8   Structures of the Nonmetals   821
22.9   Occurrence, Preparation, and Uses of the Nonmetals   824
22.10   Hydrogen   825

22.11   Oxygen   827
22.12   Nitrogen   829
22.13   Phosphorus   829
22.14   Sulfur   830
22.15   The Halogens   831
22.16   The Noble Gases   833

Properties of the Nonmetals   834

22.17   Hydrogen   834
22.18   Oxygen   835
22.19   Nitrogen   837
22.20   Phosphorus   837
22.21   Sulfur   838
22.22   The Halogens   839
22.23   The Noble Gases   842

Compounds of Selected Nonmetals   843

22.24   Hydrogen Compounds   844
22.25   Nonmetal Halides   848
22.26   Nonmetal Oxides   850
22.27   Nonmetal Oxyacids and Their Salts   853

For Review   859
Summary 859   Key Terms and Concepts 861   Exercises 861

23   The Transition Elements and Coordination Compounds   864

Periodic Relationships of Transition Elements   865

23.1   The Transition Elements   865
23.2   Properties of the Transition Elements   866
23.3   Preparation of the Transition Elements   868
23.4   Compounds of the Transition Elements   872
23.5   Copper Oxide Superconductors   875

Coordination Compounds   876

23.6   Basic Concepts   877
23.7   The Naming of Complexes   878
23.8   The Structures of Complexes   879
23.9   Isomerism in Complexes   880
23.10   Uses of Complexes   881

Bonding and Electron Behavior in Coordination Compounds   882

23.11   Valence Bond Theory   882
23.12   Crystal Field Theory   884
23.13   Magnetic Moments of Molecules and Ions   887
23.14   Colors of Transition Metal Complexes   888

For Review   890
Summary 890   Key Terms and Concepts 890   Exercises 890

24 The Chemistry of Materials 894

Polymers 895

24.1 Factors That Affect Properties of Polymers 896
24.2 Polymer Properties 899
24.3 Elastomers, Thermoplastics, and Thermosetting Polymers 901
24.4 Recycling of Polymers 901
24.5 The Future of Polymers 903

Metals, Insulators, and Semiconductors 904

24.6 Band Theory 904
24.7 Metals 906
24.8 Semiconductors and Insulators 906
24.9 The Solar Cell 907

Ceramics 909

24.10 Catalytic Converters and Space Shuttle Tiles 910
24.11 Glass 910
24.12 Superconducting Ceramics 913

Thin Films 916

24.13 Films on Winows and Lenses 916
24.14 Diamond Thin Films 917
24.15 Synthesis of Thin Films 918

Materials in Medicine 919

24.16 Artificial Hips 919
24.17 Nitinol 920

For Review 922
Summary 922   Key Terms and Concepts 922   Exercises 923

25 Biochemistry 926

The Cell 927

25.1 Components of a Cell 927

Metabolism 929

25.2 Energy Sources and Utilization 929
25.3 ATP 931

Proteins 931

25.4 Amino Acids 932
25.5 Peptides and Proteins 935
25.6 Classes and Structures of Proteins 936
25.7 Enzymes 940

Carbohydrates   940

25.8   Monosaccharides   941
25.9   Disaccharides and Polysaccharides   942

Lipids   944

25.10   Classes of Lipids   944

Nucleic Acids and Genetic Code   947

25.11   Nucleic Acids   947
25.12   The Genetic Code and Protein Synthesis   950
25.13   Recombinant DNA Technology   955
25.14   Gene Mutation   955

For Review   956

Summary   956   Key Terms and Concepts   957   Exercises   958

Appendixes

A.   Chemical Arithmetic   A1
B.   Units and Conversion Factors   A5
C.   General Physical Constants   A6
D.   Solubility Products   A6
E.   Formation Constants for Complex Ions   A8
F.   Ionization Constants of Weak Acids   A8
G.   Ionization Constants of Weak Bases   A9
H.   Standard Electrode (Reduction) Potentials   A10
I.   Standard Enthalpies of Formation, Standard Free Energies
     of Formation, and Absolute Standard Entropies [298.15K
     (25°C), 1 atm]   A11
J.   Composition of Commercial Acids and Bases   A19
K.   Half-Life Times for Several Radioactive Isotopes   A19

Photo Credits   B-1

Glossary   C-1

Answers to Odd-Numbered Exercises   C-12

Index   C-35

Preface

A general chemistry textbook is more than a guide to learning chemistry; it is also a book about chemistry. As we planned and wrote this book, we spent time listening to general chemistry instructors, reading the chemical education literature, listening to users of previous editions of this book and users of our short book, *Concepts and Models,* and reading reviewers comments in order to determine appropriate coverage and depth as well as what proves useful to the student. This input has helped us write a text that students should find clear, helpful, and interesting to read. It presents chemistry at a breadth and depth appropriate for chemistry majors, other science majors, engineering majors, and students in preprofessional programs. It is also written for students with varying backgrounds, math skills, and learning styles.

Major Features of this Edition

- This text emphasizes the idea that the foundation of modern chemistry is the use of **microscopic behavior to explain observations in the macroscopic world.** Consequently, concepts of atomic and molecular structure are covered early in the book and play an important role in many of the subsequent chapters.

- **Important changes in topical coverage and organization** have been made throughout the tenth edition. Chapter 8, "Chemical Equations and the Periodic Table," presents a general introduction to chemical reactivity based on the periodic table, an approach that had proved popular in earlier editions of this text. Chapter 9, "Structure and Bonding at Carbon," offers the choice of introducing organic chemistry as a logical follow-up and reinforcement to topics in chemical bonding and molecular geometry or of deferring it to the more traditional location at the end of the course. Chapters 21, 22, and 23 present a condensed descriptive chemistry section that spotlights the more general kinds of chemical behavior. Chapter 24, "The Chemistry of Materials," is new. It presents exciting information on materials chemistry from cutting-edge research to applications in everyday life.

- **An innovative approach to the introduction of conceptual and quantitative problem-solving** has been used throughout the text. Quantitative concepts are introduced fully early on in the book, but when one appears in an example in a later chapter we ask the students to work through the parts of the example involving the concept introduced earlier. This reinforces the idea that skills learned in earlier chapters will be continually reused.

- **Conceptual questions** are included throughout the book to make students think about why they are doing their numerical calculations.

- The text has been rewritten in a less formal style, one that is **less intimidating for students yet retains the clarity** of previous editions. We paid very close attention to the precise use of words because it is easy for inexperienced students to get lost in imprecise language.

- **The concepts and models of chemical behavior are introduced clearly before applying them to real-world situations** where additional complications can easily mask the principles involved. Although our introductions draw on familiar substances or behaviors as examples, we have been careful to develop the concepts before extending them to more complicated situations. This is much like learning to drive on an empty parking lot; one finds the controls before worrying about traffic.

- **Multimedia support for the instructor** is provided through the Chemistry Interactive CD-ROM. This Windows and Macintosh product includes animations, video clips, and molecular models that illustrate key concepts in the text. It also includes drawings and tables from text as well as all overhead transparencies along with a simple-to-use presentation program to enhance multimedia classroom presentations of these materials.

In-Text Learning Aids

It is our conviction that the textbook is an important tool that helps students succeed in chemistry. With that in mind, we provide a number of important learning aids in the text.

- The text includes **233 worked-out examples,** which proceed from simple straightforward problems to more complex ones. Many of the exercises and examples illustrate the relevance of chemistry to many other disciplines.

- **Solutions in examples are presented in four different ways** (with text, graphics, outlines, and equations) in order to accommodate as many different learning styles as possible. Moreover, the text and outlines give students models of how to talk about problems instead of just talking about what to do with numbers.

- Worked-out examples are followed by analogous problems with answers called **Check Your Learning,** which give students the opportunity for immediate feedback on their understanding.

- **Quantitative concepts are introduced with fully worked out examples, but discussed in increasingly briefer form in examples in subsequent chapters.** For example, gram-to-gram stoichiometry calculations are introduced in complete examples in Chapter 3. In subsequent examples in that chapter (say, in a percent yield calculation) the gram-to-gram conversion is not explained as fully. In subsequent chapters the discussion of gram-to-gram calculations gets briefer and eventually we indicate that such a conversion is required and then simply give the result. This technique points out the similarities in various problems and compels students to remember previously introduced material as they move on to a new chapter.

- A total of over 2000 exercises are divided into **exercises grouped by topic**—to help students find practice material easily—and **Applications and Additional Exercises**—to give students more challenging exercises, some of which invoke topics from previous chapters.

- We have included **questions that probe conceptual and qualitative understanding** to help students improve their understanding of why they are manipulating numbers and formulas. Many students have difficulties learning chemistry because they are trained to do calculations but not to understand why they are doing them.

- **Quantitative exercises that require a discussion of the problem solution** are also included in many of the chapters that involve numerical concepts. These ask students to describe how they solve the problem as well as to actually perform the calculations.

- We are careful to provide **answers for every odd-numbered question or exercise,** not just answers for numerical exercises.

Organization and Topical Coverage

In preparing the tenth edition, we have aimed to provide a solid foundation in chemistry for science and engineering majors and at the same time offer instructors flexibility and support in conveying the course content in a manner most appropriate for their students.

Our experience and much research literature tells us that many students come to general chemistry with serious misconceptions about the nature of matter. Rather than assuming that students understand these fundamentals, instructors can use Chapter 1 to challenge these misconceptions. One particular area of difficulty is the distinction between the microscopic world of atoms and molecules and the macroscopic world of events that can be seen. To complicate the problem, the same symbols are used for both worlds. Chapter 1 includes an introduction to the microscopic and macroscopic worlds and a discussion of how the same symbols are used for both. (We continue to address the distinction between the two domains throughout the text.)

Chapter 1 also introduces uncertainty in measurements and describes the use of significant figures as one approximation for propagating uncertainty through calculations. We also have included a brief introductory section on using ratios to make unit conversions. We then point out how unit conversion factors arise while solving a ratio and continue to use conversion factors in the remainder of the text. We include ratios because many students have a basic understanding of ratios. We find that these students rapidly move to conversion factors; however, in the process students develop a better understanding of what a conversion factor is doing.

Chapter 2, "The Language of Chemistry," has been shortened. It now focuses on the basic language of chemistry, including a review of the postulates of Dalton's atomic theory and the meaning of formulas and chemical equations. Your students can use it as an introduction or a review, depending on their background.

Chapter 3, on stoichiometry chapter, continues to appear early so that students have the calculation skills necessary for laboratory work. We have added an optional section describing the three traditional methods of chemical analysis, gravimetric analysis, titration, and combustion analysis.

The introduction to the structure of the atom and its physical properties has been condensed into a single chapter—Chapter 5, "Atoms: Their Structure, Spectra, and Properties." Because most students have seen the history of the development of the atom in previous physical science and chemistry courses, we simply outline the chronology of this development. We retained our discussion of the Bohr model of the atom because it serves to introduce energy levels, quantum numbers, and spectra using a mathematical model that students at this level can understand. Moreover, when we move to the quantum mechanical model the differences between it and the Bohr model are clear because the Bohr model is fresh in a student's mind.

A discussion of chemical bonding follows the introduction of atomic theory and periodicity. However, the treatment of molecular orbital theory has been reduced.

Chapter 8, "Chemical Reactions and the Periodic Table," presents the discussion of the periodic table, chemical reactions, and predicting chemical reactions that is based on concepts of structure and bonding and that has proved popular in earlier editions.

Chapter 9, "Structure and Bonding at Carbon," is in a new location in this edition.

This optional introduction to carbon chemistry includes a look at the element and a few of its inorganic compounds as well as an introduction to organic functional groups and polymers. The chapter provides a chance to review Lewis structures, molecular geometry, and aspects of hybridization.

Chapters 14, 16, and 17, which introduce equilibria, start with an introduction to the various types of equilibrium systems and their qualitative behavior. Problem solving follows. The common features in the various problems are highlighted in order to emphasize the idea that the chemical species may be different but the arithmetic and logic of many equilibrium problems is the same.

Descriptive chemistry has been retained but the material has been condensed to focus on more characteristic and general types of reactions and on more important chemistry. Chapters 21, 22, and 23 treat active metals, semimetals and nonmetals, and transition metals, respectively. The result is a descriptive chemistry section that is more like a text and less like an encyclopedia.

Finally, a new chapter describing materials chemistry, Chapter 24, has been added and Chapter 25, the biochemistry chapter, has been retained.

Alternate Versions of the Text

This edition is available in three versions so instructors have the flexibility to support the content of their courses with a book that is not excessively long. *General Chemistry with Qualitative Analysis,* tenth edition, provides comprehensive coverage of concepts and principles as well as a nine-chapter introduction to qualitative analysis. *General Chemistry,* tenth edition, is available for instructors who do not require qualitative analysis. It omits the qualitative analysis chapters, but retains all other material including descriptive chemistry, materials, and biochemistry. A third version of the text, *Essentials of General Chemistry,* tenth edition, provides streamlined coverage by omitting the chapters on materials chemistry and biochemistry.

Supplements

An extensive learning and teaching package has been designed to make this book more useful to both student and instructor.

For the Student

Study Guide by Norman Griswold   Directs and assists the student in the study of the text by providing an overview of the chapter, performance goals and suggestions for study, words frequently mispronounced, and a self-help test with answers.

Solutions Guide by John Meiser   Provides detailed solutions to odd numbered end-of-chapter exercises.

Basic Laboratory Studies by Grace Hered   Presents 45 laboratory-tested exercises which include thought questions and a new section on waste disposal. This manual includes a detailed discussion of safety issues and laboratory skills and devices.

For the Instructor

Instructors Guide by Jack Breazeale   A resource which lists prerequisite chapters, student goals, and a detailed chapter overview broken down by section.

Complete Solutions Guide by John Meiser   Provides detailed solutions to all end-of-chapter exercises.

Instructors Guide for Basic Laboratory Studies by Grace Hered   Includes helpful suggestions for performing the experiments in class, safety precautions, suitable unknowns, and answers to the questions in the student's manual.

Test Item File by Stacey Lowery Bretz, Richard L. Bretz, and Christian Clausen   Approximately 2,000 test questions, with roughly 25% new to this edition, that are keyed to the appropriate text section. The test items consist of multiple-choice, true/false, short answer, matching, and fill-in-the-blank questions. A computerized version is also available in Mac, Windows, and DOS formats.

Transparencies   A set of 125 full color transparencies, 45 more than the last edition.

Houghton Mifflin Chemistry Lecture Demonstration: Videotape Series A with Guide, Series B with Guide, Series C with Guide, and Series D   A set of 133 chemistry lecture demonstrations in VHS format performed by John Luoma, Cleveland State University; John J. Fortman and Rubin Battino, Wright State University; Patricia L. Samuel, Boston University; and Paul Kelter, University of Nebraska, Lincoln.

Houghton Mifflin Chemistry Videodisc   Contains video clips of lecture demonstrations, animations of important chemical processes and concepts, still images from the text that can be used in classroom presentations, as well as a companion guide.

Chemistry in Motion software   Macintosh software developed by Leonard J. Soltzberg of Simmons College provide animation of concepts and processes that are best understood when viewed in motion and with dimensionality; for example, Le Châtelier's principle, crystal structure, states of matter, molecular motion, and entropy.

Chemistry Interactive, Instructors Edition   A CD-ROM containing animation, videos, and molecular models that support major topics in general chemistry. In addition, drawings and tables from the text and all overhead transparencies are included, along with a simple-to-use classroom presentation program.

Power Presentation Manager   Includes on floppy disks the classroom presentation tool from the Chemistry Interactive CD-ROM along with still images and tables from the text. Available in both Macintosh and Windows versions.

Acknowledgments

No text is written in isolation; many people play important roles in its development. We wish to thank the following reviewers who read all or part of the manuscript and acknowledge their many thoughtful suggestions:

Susan Andolfi, *Oakton Community College*
William H. Breazeale, Jr., *Francis Marion University*
Frank Brimelow, *Mountain Empire Community College*
Larry Brown, *Texas A&M University*
Albert W. Burgstahler, *The University of Kansas*
Loren Carter, *Boise State University*
Wayne B. Counts, *Georgia Southwestern College*
Patricia A. Cunniff, *Prince George's Community College*

Michael I. Davis, *The University of Texas at El Paso*
David C. Easter, *Southwest Texas State University*
Anina K. El-Ashawy, *Collin County Community College*
Donald Frantz, *Hypercube, Inc.*
Patrick M. Garvey, *Des Moines Area Community College*
Steve Heninger, *Allegheny Community College*
Philip M. Jaffe, *Oakton Community College*
Sidney A. Katz, *Rutgers University*
Howard C. Knachel, *University of Dayton*
Larry K. Krannich, *The University of Alabama at Birmingham*
Peter J. Krieger, *Palm Beach Community College*
John D. Lewis, *St. Edward's University*
Edwin F. Meyer, *DePaul University*
Terry L. Morris, *Southwest Virginia Community College*
David J. Morrissey, *Michigan State University*
Bruce R. Osterby, *University of Wisconsin—La Crosse*
Colonel Daniel Y. Pharr, *Virginia Military Institute*
Earl R. Poore, *Jacksonville State University*
Henry D. Schreiber, *Virginia Military Institute*
Jane Joseph Sheppard, *Wake Forest University*
Al Shina, *San Diego City College*
Mary Jane Shultz, *Tufts University*
Wesley D. Smith, *Ricks College*
Frank Stark, *Boise State University*
Alexander Svager, *Central State University*
Jesse Yeh, *South Plains College*

In particular, we would like to thank Beverly Foote, and Catherine Keenan of Chaffey College for their careful work in checking the accuracy of the final manuscript and subsequent proof.

We acknowledge and appreciate the valuable contribution of Professor Brian J. Johnson to the addition of a materials chemistry chapter and of Professor James R. Hunsley in the revision of the biochemistry chapter. Brian and Jim, thanks for your help.

When two publishing companies merge, as happened during the preparation of this manuscript, authors lose old colleagues and gain new ones. We send our thanks to those with whom we worked at D.C. Heath and Company. Our ties there were close and congenial. The new production staff at Houghton Mifflin picked up a project that was already underway and caught up with it quickly. We thank them for their assistance and we look forward to continuing to work with them.

In closing we wish to acknowledge a long standing and particularly valuable colleague and contributor. After guiding this textbook through eight editions, Henry Holtzclaw has retired from active authorship and assumed the role of Contributing Editor. As such, he continues his most valuable contributions to this text; specifically, the traditions of clear discussion and concern for the student that characterize the years during which he lead our authoring team. Henry, thank you for sharing your experience and for offering your guidance.

William R. Robinson
Jerome D. Odom

CHAPTER OUTLINE

The Basics

1.1 Doing Chemistry
1.2 Matter
1.3 Energy
1.4 Chemical and Physical Properties
1.5 Classifying Matter
1.6 Atoms and Molecules
1.7 Classifying Elements: The Periodic Table
1.8 The Domains of Chemistry

Measurement in Chemistry

1.9 Units and Measurement
1.10 SI Base Units
1.11 Derived SI Units
1.12 Uncertainty in Measurements and Significant Figures
1.13 Conversion of Units
1.14 Conversion of Temperature Units

1

Some Fundamental Concepts

Chemical plants and power plants both convert one form of matter into another.

There are many reasons that we could cite for studying chemistry. Several of them are summarized in the following letter to the *Journal of Chemical Education*. It was written by a professor of chemistry to remind other chemistry teachers of the importance of our science. It was written for instructors, but the message is universal.

On a Friday in April, I watched as my nine-year-old son won a foot race by a good 20 or 30 feet and thought nothing of it. After all he always wins races, he's very fast. Twenty nights later, I watched my son walk across my bedroom, hesitantly and wobbly, but I had never been prouder of him. The intervening 20 days gave me a new perspective of the importance of chemistry and physics in our lives and I would like to share this perspective with you.

Less than 30 minutes after winning the race my son told me that he could not move his left hand. For the next two days he continued to lose more and more of his left side body function and finally on Saturday night, he could no longer urinate. We took him to the hospital where the doctor immediately ordered a set of cerebral spine X-rays and a CT scan of the head and spine. These scans showed no abnormality so an MRI scan was ordered for the next morning. The MRI scan showed a significant amount of blood accumulated around the front of the spinal cord and an early diagnosis of an arterial venous malformation in the spinal cord was made. Intravenous steroids were administered to reduce the swelling around the cord. Later on Sunday the local neurosurgeon told us that there was nothing that could be done for him locally and suggested that we move him to a large research hospital.

Monday morning found us at another hospital where the pediatric neurosurgery group began studying the MRI scans. Monday afternoon the neurosurgeons performed an angiogram in which a plastic catheter was inserted into one of my son's veins near the groin and snaked up to the neck area. The progress of the catheter was followed using a series of X-ray scans that essentially gave a TV picture of the catheter moving through the veins and arteries. Once the neck region was entered, an iodine-containing dye was inserted into the catheter and the flow of blood in the neck area could be monitored on the TV screens. This revealed a hole in one of the arteries or veins in front of the spinal cord. This particular problem is called a fistula. Next came the discussions of what to do and how to do it.

On Wednesday the decision was made to perform another angiogram but this time its purpose was to shut off the flow of blood to this artery. Thursday morning, for the next two hours, a slow blockage of the artery was performed in the following manner. First the catheter was inserted from the groin to the neck area and positioned just in front of the fistula. Then roughly one-inch-long silk threads, that had been soaked in a polyvinyl alcohol solution, were inserted into the catheter and lodged in the artery just in front of the fistula. As these threads formed a blockage in the artery, the polyvinyl alcohol caused the blood to clot and seal off the artery past the point of the blockage. As the artery was sealed off and blood clotting continued, the artery that had caused the paralysis was effectively rendered useless. This decreased the pressure on the spinal cord so that osmotic forces could begin removing the blood from the spinal cord area.

Friday morning the physical therapist arrived and for the first time in a week my son walked a little on his own. On Monday a third angiogram showed that the blockage was complete and discharge from the hospital followed on Tuesday. Two weeks after the initial event, my son returned home. By Tuesday he was back in school full time. He walked slowly and hesitantly to class, but he walked on his own.

Every time that I have told this story, the response has been that it is a miracle of modern medicine, which it certainly is. But there is another miracle here—the miracle of some old, middle-aged, and modern chemistry and physics. Not that many years ago my son would have been written-off as unsalvageable because we could not see what his problem was nor get to the area to fix it. Now with X-rays, CT scans, MRI scans, and angiograms, seeing the problem is a much simpler process. When I hear CT scan, MRI and angiogram, the names Roentgen, discoverer of X-rays in 1895, Bloch and Purcell, discoverers of NMR in 1946, and Shockley, Brattain and Bardeen, discoverers of the transistor in 1948, rush to my thoughts. The steroids that were administered to relieve the pressure in my son's spinal cord were probably made from a process that was initiated by Woodward as he determined how to synthesize steroids in 1951. The catheter and the polyvinyl alcohol, that were so instrumental in blocking the artery, are the direct descendants of the pioneering work on polymers done by Baekeland in 1907 and Carothers in 1928.

What this experience has brought to my attention is that neurosurgeons are the hands of modern medicine and the miracles that it can make. The spinal cord of modern medicine is chemistry and physics. If the blood supply to the spinal cord is choked off, as we fear will happen in this era of lack of interest in the physical sciences, then the ultimate result will be paralysis of modern medicine. That is why it is important that we teach chemistry and physics and do a good job at it. Our students must realize that chemists and physicists are every bit as important to the recovery of my son's health as the neurosurgeons that applied the fruits of the labor of chemists and physicists.

Who knows what the miracles of modern chemistry and physics will give us tomorrow. Charles H. Atwood, University of Georgia; Athens, GA 31207.

The Basics

1.1 Doing Chemistry

Even before the beginning of recorded time, people tried to convert matter into more useful forms. Initial attempts involved changing the shape of a substance without changing the substance itself. Our stone-age ancestors chipped pieces of flint into useful tools, cut skins into garments, and carved wood into statues and toys. As their knowledge increased, they began to change the composition of the substances—clay was converted into pottery, copper ores into copper, and grain into bread.

One can argue that humans began to practice chemistry at the time they learned to control fire and use it to cook, make pottery, and smelt metals. Subsequently, they began to examine specific components of matter. A variety of drugs were isolated, primarily from plants. Dyes were isolated from plant and animal matter. Metals were combined to form alloys, and more elaborate smelting techniques produced iron. Alkalis were extracted from ashes, and soaps were prepared by combining these alkalis with fats. Alcohol was produced by fermentation and distillation.

Attempts to explain the behavior of matter extend back over 2500 years. As early as the sixth century B.C., Greek philosophers discussed a system in which water was the basis of all things. Many of us are aware of the Greek postulate that matter consists of four elements: earth, air, fire, and water. Subsequently, an amalgamation of chemical technologies and philosophical speculations spread from Egypt, China, and/or the eastern Mediterranean area in the form of alchemy, a practical and theoretical science that attempted to explain and carry out the transformation of one form of matter into another.

Thus medical practice (with its isolation of drugs from natural sources); technology, metallurgy, and the dye industry; and alchemy all contributed to the development of modern chemistry—a science that still aspires to understand and control the behavior of matter. Ironically, this effort to achieve understanding has been so successful that many people do not realize how pervasive chemistry is. Chemistry and the language of chemists play a vital role in biology, soil science, materials science, pharmacy and medicine, food science, civil engineering, plant and animal sciences, environmental studies, waste management, and a vast variety of other areas.

Oil and nylon, gasoline and water, salt and sugar, and iron and gold are familiar types of matter that differ from each other because of differences in their composition and structure. These and most other types of matter can be converted into different types

(A)

(B)

Figure 1.1

Chemical changes. (A) Copper and nitric acid undergo a chemical change to form copper nitrate and brown gaseous nitrogen dioxide. (B) During the combustion of a match, cellulose and oxygen (from the air) undergo a chemical change to form carbon dioxide and water vapor.

(Fig. 1.1). Indeed, our very existence depends on such changes: Digestion of food and its assimilation by our bodies are changes in matter that are essential to our life processes.

As we proceed through this book, we will discover many different examples of changes in the composition and structure of matter. We will also learn how to classify these changes and how they occur, and we will study their causes, the changes in energy that accompany them, and the principles and laws involved. As you discover these things, you will be doing **chemistry,** *the investigation of the composition, structure, and properties of matter and of the reactions by which one form of matter may be produced from or converted into other forms.* The practice of chemistry is not limited to chemistry books or laboratories; it happens whenever someone is interested in changes in matter or in conditions that may lead to such changes.

Chemistry is a science that is based on experiment. The **laws** we encounter are statements or equations that summarize a vast number of experimental observations but that provide no explanation for the observations. **Theories** are explanations that people invent to explain laws or large bodies of experimental data. Theories are accepted because they appear to be satisfactory explanations, but they can be modified if new data become available.

Doing chemistry involves attempting to answer questions and explain observations in terms of the laws and theories of chemistry, using procedures that are accepted by the scientific community. There is no single approach to answering questions or explaining observations, but all have one thing in common. Each uses knowledge that is based on experiments that can be reproduced to verify the results. Some approaches involve a **hypothesis,** a tentative explanation of observations or a plan that acts as a guide for gathering and checking information. We test a hypothesis by experiment, calculation, and/or comparison with the experiments of others, and then we refine it as the results dictate.

Some hypotheses are attempts to explain the behavior that is summarized in laws. If such a hypothesis turns out to be capable of explaining a large body of experimental data, it can reach the status of a theory. The path of discovery that leads from question and observation to law or from hypothesis to theory, combined with experimental verification of the hypothesis and any necessary modification of the theory, is sometimes called the **scientific method.**

1.2 Matter

Matter is anything that occupies space and has mass. Solids and liquids are obviously matter; we can feel that they take up space, and their weight tells us they have mass. Gases are also matter; if gases did not take up space, balloons filled with gas would collapse.

Solids, liquids, and gases are the three forms, or states, of matter that are commonly found on the earth (Fig. 1.2 on page 5). **Solids** are rigid and possess a definite shape. **Liquids** flow and take the shape of a container, except that they form a flat upper surface. (However, in space in zero gravity, unconfined liquids assume a spherical shape.) Both liquid and solid samples have volumes that are very nearly independent of pressure. **Gases** take both the shape and the volume of their container.

A fourth type of matter, **plasma,** is found in the interior of stars and in some other high-temperature situations. A plasma is a high-temperature, ionized gas composed of electrons (Section 2.2) and positive ions (Section 2.6) in such numbers that the gas is essentially electrically neutral.

Figure 1.2

The three states of matter as illustrated by water: solid water (ice), liquid water, and gaseous water. The clouds and mist form when gaseous water, which is invisible, condenses into very small drops of liquid water.

Some samples of matter appear to have properties of solids, liquids, and/or gases at the same time. This can occur when the sample is composed of many small pieces or is a mixture. We can pour sand as though it were a liquid because it is composed of many small grains of solid sand. Clouds appear to behave somewhat like gases, but they are mixtures of air and tiny drops of liquid water.

The **mass** of an object is a measure of the amount of matter in it. One way to measure an object's mass is to measure the amount of energy it takes to accelerate the object. We use much more energy pushing a car up to a speed of 5 miles per hour than pushing a bicycle up to 5 miles per hour because the car has much more mass. A more common way to determine the mass of an object is to use a balance to compare its mass with a standard mass.

Although weight is related to mass, it is not the same thing. **Weight** is the force that gravity exerts on an object. This force increases as the mass of the object increases. The weight of an object changes as the force of gravity changes, but its mass does not. Astronauts have the same mass on the earth and on the moon; however, their weight on the moon is only one-sixth their earth weight because the force of gravity on the moon is one-sixth that on earth.

Many observations are summarized in the **law of conservation of matter:** *There is no detectable change in the total quantity of matter present when matter converts from one type to another (a chemical change) or changes among solid, liquid, and gaseous forms (a physical change).* A flash bulb provides an example of conservation of mass (Fig. 1.3). It contains magnesium wire that burns in the atmosphere of pure oxygen inside the bulb and produces magnesium oxide when the bulb is fired. The mass of the bulb is the same before and after this chemical change, because the magnesium oxide has the same mass as the magnesium metal and oxygen from which it formed.

Figure 1.3

The mass of a flash bulb is the same before and after it is used, because the mass of the magnesium oxide produced when the bulb flashes equals the mass of the magnesium metal and oxygen gas that reacted.

1.3   Energy

Energy can be defined as the capacity for doing work. One type of **work** is the process of causing matter to move against an opposing force. For example, we do work when we pump up a bicycle tire—we move matter (the air in the pump) against the opposing force of the air already in the tire.

Like matter, energy comes in different types. One scheme classifies energy into two types: **kinetic energy,** the energy an object possesses because of its motion, and **potential energy,** the energy an object has because of its position, composition, or condition (Fig. 1.4). Falling water has kinetic energy that can be used to do work and produce electricity in a hydroelectric plant. Water at the top of a waterfall has potential energy because of its position; if it falls in a hydroelectric plant, it can do work. A battery has potential energy because the chemicals within it can produce electricity that can do work.

Energy can be converted from one form to another, but all of the energy present before a change occurs always exists in some form after the change is completed. This observation is expressed in the **law of conservation of energy:** *During a chemical change, energy can be neither created nor destroyed, although it can be changed in form.*

There is always a conversion of one form of energy into another when one kind of matter is converted into another. Usually heat is released or absorbed, but sometimes the conversion also involves light, electrical energy, or some other form of energy. When magnesium metal combines with oxygen in a flash bulb, potential energy is converted into light and heat.

During chemical changes, the laws of conservation of matter and energy hold very well. However, these laws do not describe nuclear reactions, where conversion of matter into energy or energy into matter can occur. To encompass both chemical and nuclear changes we combine the two laws of conservation into one statement: *The total quantity of matter and energy in the universe is fixed.*

1.4 Chemical and Physical Properties

The characteristics that enable us to distinguish one substance from another are called properties. The capacity of one type of matter to change into another type (or its inability to do so) is a **chemical property.** We observe a chemical property of iron when it combines with oxygen in the presence of water to form rust. We observe a chemical property of chromium when we determine that it does not rust (Fig. 1.5).

In order to identify a chemical property we look for a chemical change. A **chemical change** always produces one or more types of matter that differ from what was present before the change. The formation of rust is a chemical change because rust is a different kind of matter from the iron, oxygen, and water present before it formed. The formation of magnesium oxide in a flash bulb (Fig. 1.3) is also a chemical change because magnesium oxide is a different kind of matter from magnesium and oxygen.

A **physical property** is a characteristic that does not involve a change in the composition of matter. Familiar examples of physical properties include mass, volume, length, color, hardness, the temperature at which a substance melts or boils, and electrical conductivity. Iron, for example, melts at 1535°C; chromium melts at a different temperature, 1900°C. As they melt, the metals change form (from solids to liquids), but their composition does not change: They remain iron or chromium.

We can observe some physical properties, such as mass and color, without changing the form of matter observed. Other physical properties, such as the melting temperature of iron and the freezing temperature of water, can be observed only as matter undergoes a physical change. A **physical change** is one that does not involve the change of one kind of matter into another with a different composition. We observe a physical change when wax melts, when water freezes, and when steam condenses to liquid water. In each of these events, there is a change in the form of the substance but no change in its chemical composition.

Mass, volume, and melting temperature are examples of properties that can be measured. Such properties fall into one of two categories. If the property depends on the amount of matter present, it is an **extensive property.** If the property of a sample of matter does not depend on the amount of matter present, it is an **intensive property.** The mass and volume of a substance are examples of extensive properties. For example, the more water we have, the larger the mass and volume of the water. Temperature is an example of an intensive property. If we combine two separate cups of water, each at 20°C (room temperature), the temperature remains 20°C.

Figure 1.5

One of the chemical properties of iron is that it rusts; one of the chemical properties of chromium is that it does not.

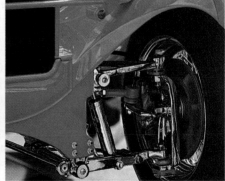

Figure 1.6

The classification of matter.

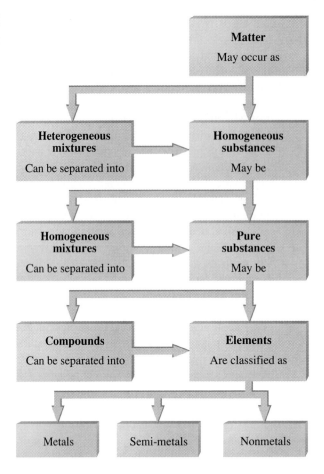

1.5   Classifying Matter

We can classify matter into several categories (Fig. 1.6). Two broad categories are heterogeneous mixtures and homogeneous substances. A **mixture** is composed of two or more types of matter that can be present in varying amounts and that can be separated by physical changes, such as evaporation. A mixture with a composition that varies from point to point is called a **heterogeneous mixture.** Italian dressing (Fig. 1.7) is an example of a mixture that is heterogeneous. Its composition varies when we make it from varying amounts of oil, vinegar, and herbs. It is not the same from point to point—one drop may be mostly vinegar, whereas a different drop may be mostly oil or herbs because the oil and vinegar separate, and the herbs settle out. Other heterogeneous mixtures include chocolate chip cookies (we can see the separate bits of chocolate, nuts, and cookie dough), frozen orange juice, sand in water, and concrete (cement, sand, and gravel).

A **homogeneous substance** is a substance that is the same from point to point. One type of homogeneous substance is a **homogeneous mixture,** often called a **solution.** A second type of homogeneous substance is called a **pure substance.** All specimens of a pure substance are identical in composition as well as having identical chemical and physical properties.

Figure 1.7

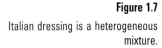

Italian dressing is a heterogeneous mixture.

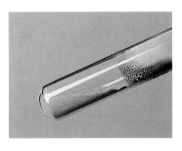

Figure 1.8

Mercury oxide is a compound. When heated, it decomposes into silvery drops of mercury and invisible oxygen gas.

An unshaken carbonated soft drink is a homogeneous mixture, typically of water, sugar, coloring, flavoring agents, and carbon dioxide. Each drop in a particular soft drink tastes the same because each drop contains the same amount of water, sugar, and other components. If we shake the drink so it foams, we make a heterogeneous mixture that consists of bubbles of carbon dioxide and liquid. Other examples of homogeneous mixtures are syrup, air, gasoline, and a solution of salt in water.

A carbonated drink is not a pure substance by our definition, because it is a mixture—its composition can vary. The amount of carbon dioxide in the drink decreases after the bottle is opened and the drink begins to lose its fizz. A sample of carbon dioxide alone, however, is a pure substance. Any sample of carbon dioxide that weighs 44 grams can always be converted into 12 grams of carbon and 32 grams of oxygen. Every sample of carbon dioxide also has the same melting temperature, color, and other properties, no matter what brand of beverage or other source it is isolated from.

We can divide pure substances into two classes: elements and compounds. Pure substances that cannot be separated into any simpler substances by chemical or physical changes are called **elements.** Iron, silver, gold, aluminum, sulfur, oxygen, and copper are familiar examples of elements. At the present time 112 elements are known; a list of these is printed on the inside front cover of this book. Of these elements, 90 occur naturally on the earth, and the other 22 have been created in laboratories (the most recent in 1996). We classify the elements as metals, semi-metals, and nonmetals, as described in Section 1.7.

Pure substances that can be decomposed by chemical changes are called **compounds.** This decomposition may produce either elements or other compounds, or both. Mercury oxide can be broken down by heat into the elements mercury and oxygen (Fig. 1.8). When heated in the absence of air, the compound sucrose decomposes into the element carbon and the compound water (Fig. 1.9).

The properties of combined elements are different from those in the free, or uncombined, state. For example, white crystalline sugar (sucrose) is a compound resulting from the chemical combination of the element carbon, which is a black solid in one of its uncombined forms, and the two elements hydrogen and oxygen, which are colorless gases when uncombined. Free sodium, an element that is a soft, shiny, metallic solid,

Figure 1.9

(A) Sucrose (table sugar) before it is heated. (B) Upon heating, sucrose decomposes into elemental carbon and water.

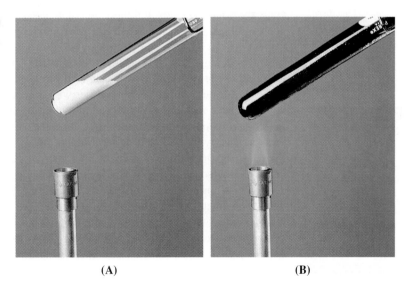

(A)   (B)

Table 1.1

Percentages of Elements in the
Earth's Crust and Atmosphere,
by Mass

Oxygen	49.20%	Chlorine	0.19%
Silicon	25.67	Phosphorus	0.11
Aluminum	7.50	Manganese	0.09
Iron	4.71	Carbon	0.08
Calcium	3.39	Sulfur	0.06
Sodium	2.63	Barium	0.04
Potassium	2.40	Nitrogen	0.03
Magnesium	1.93	Fluorine	0.03
Hydrogen	0.87	Strontium	0.02
Titanium	0.58	All others	0.47

and free chlorine, an element that is a yellow-green gas, combine to form sodium chloride, a compound that is a white crystalline solid.

Although there are only 112 known elements, several million chemical compounds result from different combinations of these elements. Each compound has a specific composition and exhibits definite chemical and physical properties by which we can distinguish it from all other compounds.

Eleven elements make up about 99% of the earth's crust and atmosphere (Table 1.1). Oxygen constitutes nearly one-half, and silicon about one-fourth, of the total quantity of these elements. Almost all of the elements on the earth are found in chemical combinations with other elements; about one-fourth of the elements are also found in the free state.

1.6   Atoms and Molecules

An **atom** is the smallest particle of an element that can enter into a chemical combination. Silicon is an element. If we could take a piece of silicon (from which computer chips are made) and look at it more and more closely, we would eventually see a single atom of silicon (Fig. 1.10). This atom would no longer be silicon if it were divided any further.

The first suggestion that matter is composed of atoms is attributed to early Greek philosophers. However, it was not until the early nineteenth century that John Dalton (1766–1844), an English school teacher with a keen interest in science, presented the hypothesis that matter was composed of atoms *and* supported it with quantitative measurements (Section 2.1). Since that time, repeated experiments have confirmed many aspects of this hypothesis, and it has become one of the central theories of chemistry. Many aspects of Dalton's atomic theory are still used with only minor revisions.

An atom is so small that its size is difficult to imagine. One of the smallest things we can see with our unaided eye is a single thread of a spider web; these threads are about 0.0001 centimeter (1/25,000 inch) in diameter. The carbon atoms in a spider web are much smaller, with a diameter of about 0.000000015 centimeter. It would take about 7000 carbon atoms to make a line the same length as the diameter of a single thread of a spider web. If we could take a pair of scissors and cut one such thread into two pieces, the flat end of each piece would be so small that most of us would have trouble seeing

Figure 1.10

A scanning-tunneling microscope (STM) can generate images of the surfaces of crystals, such as this image of the silicon atoms on a crystal of silicon. Each of the spheres is an image of an atom.

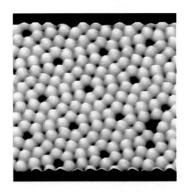

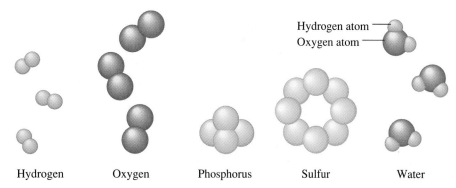

Figure 1.11

Representations of molecules of the elements hydrogen, oxygen, phosphorus, and sulfur and of the compound water. Each water molecule is composed of one oxygen atom and two hydrogen atoms.

Hydrogen atom
Oxygen atom

Hydrogen   Oxygen   Phosphorus   Sulfur   Water

the thread "end on." If each carbon atom in a cut thread suddenly expanded to the size of a dime, the flat end of a cut thread would expand so much that the end would be a little larger than a football field. Imagine the number of dimes it would take to cover a football field. It takes that many atoms to cover the cut end of the thread of a spider web.

An atom is so light that its mass is difficult to imagine. A billion lead atoms (1,000,000,000 atoms) weigh about 3×10^{-13} grams, a mass that is far too light to be weighed on the world's most sensitive balance. It would require about 300,000,000,000,000,000 lead atoms to be weighed, and they would weigh only 0.0001 gram (0.0000004 ounce).

It is rare to find collections of individual atoms. Only a few elements, such as the gases helium, neon, and argon, consist of a collection of individual atoms that move about independently of one another. Other elements, such as the gases hydrogen, nitrogen, oxygen, and chlorine, are composed of units that consist of two atoms (Fig. 1.11). The element phosphorus consists of units composed of four phosphorus atoms; sulfur, of units composed of eight sulfur atoms. These units are called molecules. A **molecule** consists of two or more atoms joined by strong forces called chemical bonds. The atoms in a molecule move around as a unit, much like the cans of soda in a sixpack. A molecule may consist of two or more identical atoms, as in the molecules found in the elements hydrogen, oxygen, and sulfur, or it may consist of two or more different atoms, as in the molecules found in water. Each water molecule is a unit that contains two hydrogen atoms and one oxygen atom (Fig. 1.11). Subdivision of the molecules in water by an electric current results in the formation of the gases hydrogen and oxygen (Fig. 1.12) which are composed, respectively, of hydrogen and oxygen molecules.

Like atoms, molecules are incredibly small and light. If a glass of water were enlarged to the size of the earth, the water molecules would be about the size of golf balls.

Figure 1.12

The battery provides an electric current that decomposes water into the gases hydrogen (on the right) and oxygen (on the left). Although the gases hydrogen and oxygen both look like air, they have properties that are very different from those of air.

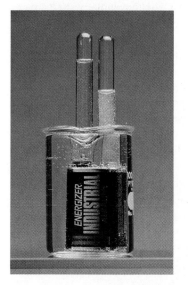

1.7 Classifying Elements: The Periodic Table

Many elements differ dramatically in their chemical and physical properties, but other elements are similar in their behavior. Thus we can identify sets of elements that exhibit common properties. For example, many elements conduct heat and electricity well, whereas others are poor conductors. These properties can be used to sort the elements into two classes: **metals** (elements that conduct well) and **nonmetals** (elements that conduct poorly). Some chemists classify a few elements as **semi-metals.** These elements conduct heat and electricity, but not very well, and possess some of the other properties of metals as well as some properties of nonmetals.

The elements can be subdivided further by the composition of the compounds they form. For example, some nonmetals form compounds composed of molecules that consist of one atom of the element and one hydrogen atom; we classify these elements into a group. Other nonmetals form compounds composed of molecules that consist of one atom of the element and two atoms of hydrogen; we classify these elements into a different group.

The **periodic table** is a table of the elements that places elements with similar properties close together. Figure 1.13 is a simplified version of the periodic table. The letters are symbols for the names of the elements, and the elements are arranged in increasing order of their atomic number, a property that will be discussed in Chapter 2. A list of the elements and their symbols is located inside the front cover of this book, along with a more comprehensive version of the periodic table.

All elements shaded in green in the upper righthand portion of the table in Figure 1.13 exhibit nonmetallic properties. Those elements shaded in red to the left of the table exhibit metallic properties. The elements shaded in blue are the borderline semi-metallic elements.

A periodic table consists of 7 horizontal rows of elements, called **periods** or **series,** and 32 vertical columns, called **groups.** In order that the table fit on a single page, parts

Figure 1.13

A periodic table. Each block contains the atomic number of an element and the symbol of the element. Blocks shaded red indicate elements that are classified as metals, blue blocks indicate semi-metals, and green blocks indicate nonmetals.

1 1A	2 2A	3 3B	4 4B	5 5B	6 6B	7 7B	8	9 8B	10	11 1B	12 2B	13 3A	14 4A	15 5A	16 6A	17 7A	18 8A
1 H																	2 He
3 Li	4 Be											5 B	6 C	7 N	8 O	9 F	10 Ne
11 Na	12 Mg											13 Al	14 Si	15 P	16 S	17 Cl	18 Ar
19 K	20 Ca	21 Sc	22 Ti	23 V	24 Cr	25 Mn	26 Fe	27 Co	28 Ni	29 Cu	30 Zn	31 Ga	32 Ge	33 As	34 Se	35 Br	36 Kr
37 Rb	38 Sr	39 Y	40 Zr	41 Nb	42 Mo	43 Tc	44 Ru	45 Rh	46 Pd	47 Ag	48 Cd	49 In	50 Sn	51 Sb	52 Te	53 I	54 Xe
55 Cs	56 Ba	57 La*	72 Hf	73 Ta	74 W	75 Re	76 Os	77 Ir	78 Pt	79 Au	80 Hg	81 Tl	82 Pb	83 Bi	84 Po	85 At	86 Rn
87 Fr	88 Ra	89 Ac†	104 Rf	105 Ha	106 Sg	107 Ns	108 Hs	109 Mt	110 Uun	111 Uuu	112 Uub						

*Lanthanides	58 Ce	59 Pr	60 Nd	61 Pm	62 Sm	63 Eu	64 Gd	65 Tb	66 Dy	67 Ho	68 Er	69 Tm	70 Yb	71 Lu
†Actinides	90 Th	91 Pa	92 U	93 Np	94 Pu	95 Am	96 Cm	97 Bk	98 Cf	99 Es	100 Fm	101 Md	102 No	103 Lr

of two of the rows, a total of 14 columns, have been written below the main body of the table. The series called *lanthanides* fits between elements 57 and 72, and the series called *actinides* fits between elements 89 and 104.

With the exception of the lanthanides and the actinides, groups are labeled at the top of each column. In the United States, the labels traditionally have been numerals with capital letters. However, the International Union of Pure and Applied Chemistry (IUPAC) recommends that the numbers 1 through 18 be used, and these labels are becoming more common.

We can predict many of the properties of an element from its location in the periodic table. For example, we can determine whether an element is a metal because an element in the red region of the table in Figure 1.13 is metallic. Thus, even though you may be unfamiliar with the element osmium (Os, element 76) you can predict that it will conduct electricity because it is classified as a metal. (Your prediction would be correct.)

As we will see in subsequent chapters, we also can use the periodic table to predict the composition of compounds. Elements in the same group exhibit similar chemical properties. For example, nonmetals in Group 7A, the column headed by fluorine (F), form compounds composed of molecules that consist of one atom of the element and one hydrogen atom. Nonmetals in Group 6A, the column headed by oxygen (O), form compounds composed of molecules that consist of one atom of the element and two atoms of hydrogen. Elements in Group 2A, the column headed by beryllium (Be), form compounds with the element chlorine that contain two atoms of chlorine for each atom of the element in the compound.

1.8   The Domains of Chemistry

We study and describe the behavior of matter and energy in three different domains: a macroscopic domain, a microscopic domain, and a symbolic domain. These domains are different ways of considering and describing chemical behavior.

Macro comes from Greek and means "large." The **macroscopic domain** is familiar to us: It is the domain of everyday things that are large enough to see and measure. This domain encompasses the chemistry of both everyday life and the laboratory. The part of chemistry that observes, detects, and classifies physical and chemical properties or changes is part of the macroscopic domain. This is the domain of metal wires, salt crystals, color changes, growth and decay, heat and cold. Measurements of mass, volume, and boiling temperature are part of the macroscopic domain.

Micro comes from Greek and means "small." The **microscopic domain** of chemistry is almost always visited in the imagination. Except for occasional snapshots, such as the STM image in Fig. 1.10, the subjects of the microscopic domain of chemistry generally are too small to be seen and must be pictured in the mind. However, things we know only in the imagination can also be real. Most of us must imagine what it is like to be in Australia because we have never been there, but Australia does exist.

Atoms and molecules are part of the chemist's microscopic domain. Other components of the microscopic domain include ions and electrons, protons and neutrons, and chemical bonds; all of these are far, far too small to see. This domain includes the individual metal atoms in a wire, the ions that compose a salt crystal, the alterations in individual molecules that result in a color change, the conversion of nutrient molecules into tissue and energy, and the evolution or absorption of heat as the bonds that hold atoms together form and break.

We interpret the physical and chemical behavior that can be observed in the macroscopic domain in terms of the much smaller components of the microscopic domain.

One of the things that makes chemistry fascinating is the use of a domain that must be imagined to explain behavior in a domain that can be observed.

The **symbolic domain** contains the specialized language used to represent components of the macroscopic and microscopic domains. Chemical symbols (such as those used in the periodic table in Fig. 1.13), chemical formulas, and chemical equations are part of the symbolic domain, as are graphs and drawings. Calculations also can be considered to be part of the symbolic domain. The components of the symbolic domain play an important role in chemistry because they provide clues about how to interpret the behavior of the macroscopic domain in terms of the components of the microscopic domain.

Part of the difficulty some students have with chemistry is that they do not recognize that the same symbols can represent different things in the macroscopic and microscopic domains. Many of you have studied the behavior of water in previous science classes, so we will use water to illustrate the three domains. (If some of the following ideas are not completely clear, do not worry about it now. We will address them throughout this book.)

We observe the chemical change of water into the gases hydrogen and oxygen (Fig. 1.12) in the macroscopic domain. The macroscopic properties of water include the observations that water is a liquid that boils at 100°C and freezes at 0°C.

In the microscopic domain we describe water as composed of molecules that consist of two hydrogen atoms and one oxygen atom (Fig. 1.11). Its decomposition is described as a rearrangement of the atoms in its molecules into molecules composed of two hydrogen atoms and molecules composed of two oxygen atoms. We can explain the liquid behavior of water and its boiling and freezing temperatures as resulting from weak bonding forces between the molecules. Even though no one has seen any of these atoms or molecules, experimental evidence is so compelling that chemists accept their existence as firmly as we accept the existence of the wind or the songs of birds.

Almost everyone is familiar with a portion of the symbolic domain of chemistry: Most people would expect a bottle with the label H_2O to contain water. In this case H_2O is used as a symbol for macroscopic amounts of water. The same symbol, H_2O, is used in the microscopic domain to indicate that one molecule of water is composed of two hydrogen atoms and one oxygen atom.

Measurement in Chemistry

Measurements provide the macroscopic information that is the basis for most of the hypotheses, theories, and laws that describe the behavior of matter and energy in both the macroscopic and the microscopic domains of chemistry. Every measurement provides three kinds of information: the size or magnitude of the measurement (a number), an indication of the uncertainty in the measurement (Section 1.12), and a standard of comparison for the measurement, a unit (Section 1.9)

1.9   Units and Measurement

Units, such as liters and pounds, are standards of comparison for measurements. When we buy a 2-liter bottle of a soft drink, we expect that the volume of the drink was measured so that it is two times larger than the volume everyone agrees to be 1 liter. Simi-

Table 1.2

The Seven Base Units
of the SI System

Property Measured	Name of Unit	Symbol
Length	meter	m
Mass	kilogram	kg
Time	second	s
Temperature	kelvin	K
Electric current	ampere	A
Amount of substance	mole	mol
Luminous intensity	candela	cd

larly, the meat used to prepare a quarter-pound hamburger is measured so that it weighs one-fourth as much as a pound.

We usually report the results of scientific measurements in SI units, an updated version of the metric system that employs the units listed in Table 1.2. Other units can be derived from these base units. The standards for these units are fixed by international agreement and they make up the **International System of Units,** or **SI units** (from the French, *Le Système International d'Unités*). SI units have been used by the United States National Institute of Standards and Technology (formerly the NBS, the National Bureau of Standards) since 1964.

Sometimes we use units that are fractions or multiples of some base unit. Ice cream is sold in quarts, pints (half a quart), or gallons (4 quarts). We also use fractions or multiples of units in the SI system, but these fractions or multiples are always multiples of 10 times a base unit. Fractional or multiple SI units are named by using a prefix and the name of the base unit. For example, a length of 1000 meters is also called a kilometer because the prefix *kilo* means 10^3 (1 kilometer = 1000 meters = 10^3 meters). The prefixes used and the powers to which 10 is raised are listed in Table 1.3.

Table 1.3

Common Prefixes Used with SI Units

Prefix	Symbol	Factor	Example
pico	p	10^{-12}	1 picometer (pm) = 1×10^{-12} m (0.000000000001 m)
nano	n	10^{-9}	1 nanogram (ng) = 1×10^{-9} g (0.000000001 g)
micro	μ*	10^{-6}	1 microliter (μL) = 1×10^{-6} L (0.000001 L)
milli	m	10^{-3}	2 milliseconds (ms) = 2×10^{-3} s (0.002 s)
centi	c	10^{-2}	5 centimeters (cm) = 5×10^{-2} m (0.05 m)
deci	d	10^{-1}	1 deciliter (dL) = 1×10^{-1} L (0.1 L)
kilo	k	10^3	1 kilometer (km) = 1×10^3 m (1000 m)
mega	M	10^6	3 megagrams (Mg) = 3×10^6 g (3,000,000 g)
giga	G	10^9	5 gigameters (Gm) = 5×10^9 m (5,000,000,000 m)
tera	T	10^{12}	1 teraliter (TL) = 1×10^{12} L (1,000,000,000,000 L)

*Greek letter mu.

Figure 1.14

Relative lengths of (A) 1 m, (B) 1 yd, (C) 1 cm, and (D) 1 in. (not actual size) and a comparison of 2.54 cm and 1 in. (actual size) and of 1 m and 1.059 yd.

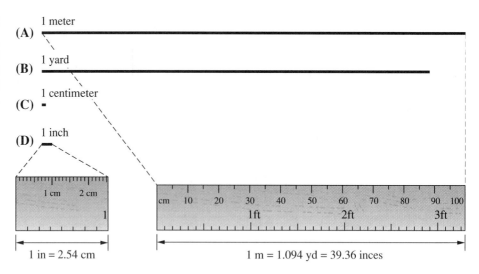

Figure 1.14

Relative lengths of (A) 1 m, (B) 1 yd, (C) 1 cm, and (D) 1 in. (not actual size) and a comparison of 2.54 cm and 1 in. (actual size) and of 1 m and 1.059 yd.

1.10   SI Base Units

The initial units of the metric system, which eventually evolved into the SI system, were established in France during the French Revolution. The original standards for the meter and the kilogram were adopted in 1799, and other countries slowly accepted them. This section introduces four of the SI base units commonly used in chemistry. Other SI units will be introduced as needed in subsequent chapters.

Length.   The base unit of length in both the SI and the original metric systems is the **meter (m).** A meter was originally defined as 1/10,000,000 of the distance from the north pole to the equator. It is now defined in terms of the wavelength of a certain color of light. In familiar terms, a meter is about 3 inches longer than a yard (Fig. 1.14); 1 meter is about 39.37 inches, or 1.094 yards. Longer distances are often reported in kilometers (1 km = 1000 m = 10^3 m). Shorter distances (Fig. 1.14) can be reported in centimeters (1 cm = 0.01 m = 10^{-2} m) or millimeters (1 mm = 0.001 m = 10^{-3} m).

Figure 1.15

The standard kilogram.

Mass.   The base unit of mass in the SI system is the **kilogram (kg).** A kilogram was originally defined as the mass of a liter of water (a cube of water with an edge length of exactly 0.1 meter). It is now defined as the mass of a certain cylinder of platinum-iridium alloy, which is kept in France (Fig. 1.15). Any object with the same mass as this cylinder is said to have a mass of 1 kilogram. One kilogram is about 2.2 pounds. The gram (g) is exactly equal to 0.001 times the mass of the kilogram (10^{-3} kg).

Temperature.   Temperature is an intensive property. The SI unit of temperature is the **kelvin (K).** (Note that *degree* is not used in the name, nor is ° used in the symbol.) The **degree Celsius (°C)** is also allowed in the SI system. (*Degree* is used with Celsius, and ° is used in the symbol.) Water freezes at 273.15 K (0°C) and boils at 373.15 K (100°C). Body temperature is about 310 K (37°C). Converting between these two units and the Fahrenheit scale will be discussed later in this chapter.

Time.   The SI base unit of time is the **second (s).** Large time intervals can be expressed with the appropriate prefixes; for example, 10 megaseconds = 10,000,000 s. Alternatively, hours, days, and years can be used.

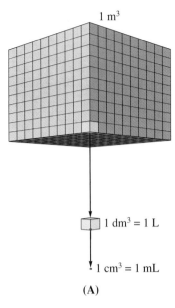

1 m³

1 dm³ = 1 L

1 cm³ = 1 mL

(A)

├─ 1.8 cm ─┤   1 cubic centimeter
Dime   = 1 milliliter

(B)

Figure 1.16

(A) Relative volumes of 1 m³ (the largest cube), 1 dm³ (1 L), and 1 cm³ (1 mL). These cubes are not actual size. (B) Comparison between a dime (actual size) and a volume of 1 cm³ (1 mL).

1.11 Derived SI Units

We can derive many more units from the seven SI base units. For example, we can use the base unit of length to define a unit of volume, and the base units of mass and length can be used to define a unit of density.

Volume.   The standard SI unit of volume is defined in terms of the base unit of length. The standard volume is a **cubic meter (m³),** a cube with an edge length of exactly 1 meter. In order to dispense a cubic meter of water, we could build a cubic box with an inner edge length of exactly 1 meter. This box would hold a cubic meter of water or of any other substance.

A more convenient unit of volume (Fig. 1.16 A) is derived from the decimeter (0.1 m, or 10 cm). A cube with an edge length of exactly 1 decimeter contains a volume of 1 cubic decimeter (dm³). A **liter (L)** is the more common name for the cubic decimeter. A volume of 1 liter holds about 1.06 quarts.

A **cubic centimeter (cm³)** is the volume of a cube with an edge length of exactly 1 centimeter (Fig. 1.16 B). A cubic centimeter is also called a **milliliter (mL)** and is 0.001 liter.

Density.   We use the mass and volume of a substance to determine its density. Thus the units of density are defined by the base units of mass and length.

The **density** of a substance is the ratio of the mass of a sample of the substance to its volume. The SI unit for density is the kilogram per cubic meter (kg/m³); however, many chemists regard this as an inconvenient unit and use units of grams per cubic centimeter (g/cm³) for the densities of solids and liquids. Units of grams per liter (g/L) are used for gases. Although there are some exceptions, most liquids and solids have densities that range from 0.9 g/cm³ (the density of ice) to 11.3 g/cm³ (the density of lead). The density of air is about 1.2 g/L. Many students find it helpful to think of density as the mass of a cubic centimeter of a solid or a liquid or the mass of a liter of a gas.

One way to determine the density of a sample is to measure its mass and volume and then to divide the mass of the sample by its volume.

$$\text{Density} = \frac{\text{mass}}{\text{volume}}$$

Example 1.1   *Calculation of Density*

For centuries people have thought about filling the centers of hollow gold bricks with lead to fool buyers into thinking that the entire brick is gold. It does not work: Lead is a dense substance, but its density is not as great as that of gold, 19.3 g/cm³. What is the density of lead if a cube of lead has an edge length of 2.00 cm and a mass of 90.7 g?

Solution: The density of a substance can be calculated by dividing its mass by its volume. The volume of a cube is calculated by cubing the edge length. Thus

$$\text{Volume of lead cube} = 2.00 \text{ cm} \times 2.00 \text{ cm} \times 2.00 \text{ cm} = 8.00 \text{ cm}^3$$

$$\text{Density} = \frac{90.7 \text{ g}}{8.00 \text{ cm}^3} = \frac{11.3 \text{ g}}{1 \text{ cm}^3} = 11.3 \text{ g/cm}^3$$

(We will discuss the reason for reporting only one decimal place in the next section.)

Check your learning: **(a)** To three decimal places, what is the volume of a cube with an edge length of 0.843 cm? **(b)** If the cube in part (a) is copper and has a mass of 5.34 g, what is the density of copper to two decimal places?

Answer: **(a)** 0.599 cm^3 **(b)** 8.91 g/cm^3

1.12   Uncertainty in Measurements and Significant Figures

All measurements have some degree of **uncertainty,** and an indication of its uncertainty should accompany the report of every measurement. In this section we will consider one common approximation that is used to indicate the uncertainty in a measurement. Before we do, however, we should recognize that there are some quantities that are exact and have no uncertainty.

A counted quantity is exact, provided that the objects counted do not change while they are being counted. If we count the eggs in a carton, we know exactly how many eggs the carton contains. Defined quantities are also exact. By definition, 1 foot is exactly 12 inches, and 1 kilogram is exactly 1000 grams. Finally, the exponential part in exponential notation (Appendix A) is exact. The 10^3 in the quantity 3.49×10^3 grams is exact; it simply tells us where the decimal point belongs.

Measured quantities are not exact. Suppose we weigh a dime and find that its mass is closer to 2.3 grams than to either 2.2 or 2.4 grams. We report its mass as 2.3 grams even though it is only *about* 2.3 grams. The uncertainty in the measurement is 0.1 gram, because the actual mass of the dime is somewhere between 2.25 and 2.35 grams. If we weigh the dime on a more sensitive balance, we may find its mass is 2.294 grams. This means its mass lies between 2.2935 and 2.2945 grams, an uncertainty of 0.001 gram. Note that even with a better measuring device, we still have some uncertainty in our measurement. In fact, **any measurement has an uncertainty of at least one unit in the last digit of the reported value.**

The mass of 2.3 grams has an uncertainty of 0.1 gram; 2.294 grams has an uncertainty of 0.001 gram; a measured volume of 25.2 milliliters has an uncertainty of at least 0.1 milliliter; and a measured length of 125 meters has an uncertainty of at least 1 meter. *We will use the approximation that any measurement has an uncertainty of one unit in the last digit of the reported value.*

All of the measured digits in a determination, including the uncertain last digit, are called **significant figures.** We can use the following rules to determine the number of significant figures in a measurement.

1.   Starting with the first nonzero digit on the left, count this digit and all remaining digits to the right. This is the number of significant figures in the measurement unless the last digit is a zero lying to the left of the decimal point. (This case is covered in Rule 2.)

First nonzero digit on the left

<u>1</u>267 m

Four significant figures

First nonzero digit on the left

55.0 g

Three significant figures

First nonzero digit on the left

70.607 mL

Five significant figures

First nonzero digit on the left

0.00832407 s

Six significant figures

The first three zeros in the last example above are not significant. They merely tell us where the decimal point is located. We could use exponential notation (as described in Appendix A) and express the number as 8.32407×10^{-3}; then the number 8.32407 contains all of the significant figures, and 10^{-3} locates the decimal point. (Recall, numbers that indicate the location of the decimal point are exact.)

2. The number of significant figures is uncertain in a number that ends with a zero to the left of the place where the decimal point would fall. The zeros in the measurement 1300 grams could be significant *or* they could simply indicate where the decimal point is located. The ambiguity can be avoided with the use of exponential notation: 1.3×10^3 (two significant figures), 1.30×10^3 (three significant figures), or 1.300×10^3 (four significant figures). For convenience in this text, *we assume that all the zeros are significant in a measurement that ends with a zero to the left of a decimal point.* A measurement like 1300 grams has four significant figures *in this text.*

A second important observation concerning uncertainty is that **results calculated from a measurement are as uncertain as the measurement itself.** Thus we must take the uncertainty in our measurements into account in order not to misrepresent the uncertainty in calculated results. One way to do this involves using the correct number of significant figures. The number of significant figures that should be reported from a calculation involving uncertain quantities is governed by the following three rules for **rounding numbers.**

1. When adding or subtracting numbers, we round the result to the same number of decimal places as the number with the least number of decimal places.

Example 1.2   Addition and Subtraction with Significant Digits

(a) Add 1.0023 g and 4.383 g. (b) Subtract 421.23 g from 486 g.

Solution: (a)
```
  1.0023 g
 +4.383  g
 +5.3853 g = 5.385 g  {3 decimal places}
```
(b)
```
  486    g
 -421.23 g
  64.77 g = 65 g  {No decimal places}
```

Check your learning: (a) Add 2.334 mL and 0.31 mL. (b) Subtract 55.8752 m from 56.533 m.

Answer: (a) 2.64 mL (b) 0.658 m

2. When multiplying or dividing numbers, we round the result to the same number of digits as the number with the least number of significant digits.

Example 1.3   Multiplication and Division with Significant Digits

(a) Multiply 0.6238 cm by 6.6 cm. (b) Divide 421.23 g by 486 mL.

Solution: (a) 0.6238 cm $\times$ 6.6 cm   = 4.1 cm^2
                    4 significant   2 significant   2 significant
                    digits   digits   digits

            (b) 421.23 g $\div$ 486 mL   = 0.867 g/mL
                    5 significant   3 significant   3 significant
                    digits   digits   digits

Check your learning: (a) Multiply 2.334 cm by 0.320 cm. (b) Divide 55.8752 m by 56.53 s.

Answer: (a) 0.747 cm^2 (b) 0.9884 m/s

---

3. When rounding numbers:
    a. If the leftmost digit to be dropped is less than 5, we do not change the remaining digits. (For two significant digits, 3.4456 rounds to 3.4.)
    b. If the leftmost digit to be dropped is greater than 5 or is a 5 followed by other nonzero digits, we increase the last remaining digit by 1. (For three significant digits, 23.387 and 23.3511 both round to 23.4.)
    c. If the digit to be dropped is a 5 or a 5 followed *only* by zeros, look at the last remaining digit. We increase it by 1 if it is odd and leave it unchanged if it is even. (For three significant digits, 3.415 and 3.425 both round to 3.42.)

Example 1.4   Calculation with Significant Digits

One common bathtub is 13.44 dm long, 5.920 dm wide, and 2.54 dm deep. Assume that the tub is rectangular and calculate its approximate volume in liters.

Solution: $V = l \times w \times d$
            = 13.44 dm $\times$ 5.920 dm $\times$ 2.54 dm
            = 202.09459 dm^3 = 202 dm^3 (or 202 L)
                On calculator   three significant
                                     digits

Check your learning: What is the density (Section 1.11) of a liquid if its mass is 31.1415 g and its volume is 30.13 cm^3?

Answer: 1.034 g/cm^3

Table 1.4

Common Conversion Factors (to four significant figures)

Length	
1 meter = 1.094 yards	1 inch = 2.540 centimeters
Volume	
1 liter = 1.057 quarts	1 cubic foot = 28.32 liters
Mass	
1 kilogram = 2.205 pounds	1 ounce = 28.35 grams

1.13 Conversion of Units

SI units are used in the scientific community, but many different units are used in other disciplines. For example, we may see units of mass as grams and kilograms in science, pounds and tons in engineering, and grains and drams in medicine. In this section we will consider how to convert measurements from one system of units to another.

Relationships between some common units are given in Table 1.4, in Appendix B, and inside the back cover of this book. These **conversion factors** indicate the size of the same quantity in two different units. For example, a length of exactly 1 meter is the same as a length of 1.094 yards (Fig. 1.14), and a mass of exactly 1 ounce is the same as 28.35 grams. Conversion factors can be used to convert units by at least two methods— using ratios and using unit conversion factors.

Using Ratios to Convert Units. The **ratio** of two quantities is the fraction formed by dividing one quantity by the other. In many applications we simplify the ratio by dividing the numerator (the top term) by the denominator (the bottom term), thus reducing the denominator to 1. For example, a density calculation (Section 1.11) gives us the mass of 1 milliliter of a sample because, in its simplified form, a ratio of quantities gives us the amount of the top quantity in one unit of the bottom quantity. A sample with a mass of 49 g and a volume of 70.0 mL has the ratio 49 g/70.0 mL (this is read "49 grams per 70.0 mL"), which simplifies to 0.70 g/mL ("0.70 grams per milliliter") or, as is useful in conversion problems, 0.70 g/1 mL.

Ratios can be set up from conversion factors. For example, we can use the conversion factor 1 in. = 2.540 cm (Table 1.4) to set up the ratios 2.540 cm/1 in. or 1 in./2.540 cm, which reduces to 0.3937 in./1 cm. We can see in Table 1.5 that the ratios of the

Table 1.5

The Ratios of Several Lengths in Centimeters to the Lengths in Inches

Unit	Length, cm	Length, in.	Ratio, cm/in.	Ratio, in./cm
An inch	2.540 cm	1 in.	$\dfrac{2.540 \text{ cm}}{1 \text{ in.}} = \dfrac{2.540 \text{ cm}}{1 \text{ in.}}$	$\dfrac{1 \text{ in.}}{2.540 \text{ cm}} = \dfrac{0.3937 \text{ in.}}{1 \text{ cm}}$
A foot	30.48 cm	12 in.	$\dfrac{30.48 \text{ cm}}{12 \text{ in.}} = \dfrac{2.540 \text{ cm}}{1 \text{ in.}}$	$\dfrac{12 \text{ in.}}{30.48 \text{ cm}} = \dfrac{0.3937 \text{ in.}}{1 \text{ cm}}$
A yard	91.44 cm	36 in.	$\dfrac{91.44 \text{ cm}}{36 \text{ in.}} = \dfrac{2.540 \text{ cm}}{1 \text{ in.}}$	$\dfrac{36 \text{ in.}}{91.44 \text{ cm}} = \dfrac{0.3937 \text{ in.}}{1 \text{ cm}}$
One author's height	190.1 cm	74.85 in.	$\dfrac{190.1 \text{ cm}}{74.85 \text{ in.}} = \dfrac{2.540 \text{ cm}}{1 \text{ in.}}$	$\dfrac{74.85 \text{ in.}}{190.1 \text{ cm}} = \dfrac{0.3937 \text{ in.}}{1 \text{ cm}}$

lengths of any distance measured both in centimeters and in inches are always equal to 2.540 cm/1 in. or to 0.3937 in./1 cm (1 in./2.540 cm). We can use these ratios to convert measurements between inches and centimeters.

Example 1.5   Unit Conversion Using Ratios

How large is a 21-inch television screen in centimeters?

Solution: If we call the length we are trying to find x cm, we can write the ratio of this length to 21 inches as x cm/21 in. The conversion factor between centimeters and inches is 2.540 cm = 1 in. (Table 1.4), and the ratio of these quantities is 2.540 cm/1 in. The ratio x cm/21 in. and the ratio 2.540 cm/1 in. are equal when the unit *centimeters* is in the numerator on both sides of the equals sign and the unit *inches* is in the denominator.

$$\frac{x \text{ cm}}{21 \text{ in.}} = \frac{2.540 \text{ cm}}{1 \text{ in.}}$$

Now we solve the equation in order to find the length in centimeters. If we multiply both sides by 21 in., we do not change the fact that both sides are equal, because we are multiplying both sides by the same quantity. Like quantities in the numerator and denominator can be canceled; this includes both numbers and units.

$$21 \text{ in.} \times \frac{x \text{ cm}}{21 \text{ in.}} = 21 \text{ in.} \times \frac{2.540 \text{ cm}}{1 \text{ in.}}$$

Now we can calculate x.

$$x \text{ cm} = 21 \text{ in.} \times \frac{2.540 \text{ cm}}{1 \text{ in.}}$$

$$= 53 \text{ cm}   \text{(two significant digits)}$$

A length of 21 inches corresponds to 53 centimeters.

Check your learning: Use a ratio from the conversion factor 1 L = 1.057 qt to convert a volume of 0.77 qt to liters.

Answer: 0.73 L

In summary, the steps used in the example to convert a quantity from one set of units to another set of units using ratios are

1. Set up a ratio containing the quantity to be found (the unknown quantity) and the quantity given. Some students find the calculation simpler if the ratios are written so that the unknown quantity is in the numerator.

2. Set up the ratio determined from the conversion factor between the two units involved.

3. Set the two ratios equal to each other, but be certain that the same units are in the numerator (on top) in both ratios and that the second set of units are in the denominator (on the bottom).

4. Solve the resulting equation for the value of the unknown quantity.

Using Unit Conversion Factors to Convert Units.

A ratio involving conversion factors can be used as a unit conversion factor for the conversion of units. A **unit conversion factor** is the ratio between two units; for example, 2.540 cm/1 in. (2.540 cm = 1 in.), 91.44 cm/3 ft (91.44 cm = 3 ft), 0.9463 L/32 oz (0.9463 L = 32 oz [liquid ounces]). When we multiply a quantity (such as a measured distance) in one set of units by an appropriate unit conversion factor, we convert the quantity to a value with different units, but the quantity (in this case, the distance) is not changed.

$$\frac{\text{Number of}}{\text{units of B}} = \frac{\text{number of}}{\text{units of A}} \times \text{unit conversion factor} \qquad (1)$$

We can determine that a unit conversion factor is set up correctly if all unwanted units cancel.

Example 1.6 — Using a Unit Conversion Factor

The mass of a competition Frisbee is 125 g. Convert its mass to ounces using the unit conversion factor derived from the relationship 1 oz = 28.35 g (Table 1.4).

Solution: If we have the conversion factor, we can determine the mass in kilograms using an equation similar to Equation 1.

$$x \text{ oz} = 125 \text{ g} \times \text{unit conversion factor}$$

We write the unit conversion factor in its two forms: 1 oz/28.35 g and 28.35 g/1 oz. The correct unit conversion factor is the ratio that cancels the units of grams and leaves ounces.

$$x \text{ lb} = 125 \,\cancel{g} \times \frac{1 \text{ oz}}{28.35 \,\cancel{g}}$$

$$= 4.41 \text{ oz}   \text{(three significant digits)}$$

Note that using Equation 1 to apply a unit conversion factor is simply the second step in converting units by using ratios.

Check your learning: Use a unit conversion factor to convert a volume of 9.345 qt to liters.

Answer: 8.841 L

Remember that a ratio in its simplified form gives us the amount of the top quantity in one unit of the bottom quantity. Thus a simplified unit conversion factor gives the number of units in the numerator in one of the units in the denominator. For example, the ratio 1 oz/28.35 g simplifies to 0.03527 oz/1 g and shows that there is 0.03527 ounce in 1 gram. Thus when we use Equation 1, we are in effect multiplying the number of units of A by the number of units of B in one unit of A.

Example 1.7 — Unit Conversion Factors with Multiple Conversions

What is the density of the common antifreeze ethylene glycol in units of g/mL? A 4.00-qt (1-gal) sample of ethylene glycol weighs 9.26 lb.

Solution: To determine the density in grams per milliliter, we need the mass in grams and the volume in milliliters. This requires four conversions: (1) quarts to liters, (2) liters to milliliters, (3) pounds to kilograms, and (4) kilograms to grams. The necessary conversion factors are given in Table 1.4 and Section 1.10: 1 L = 1.057 qt, 1 kg = 2.205 lb, 1 L = 1000 mL, and 1 kg = 1000 g.

$$\frac{\text{Number of}}{\text{units of B}} = \frac{\text{number of}}{\text{units of A}} \times \text{unit conversion factor}$$

(1) x kg $= 9.26 \text{ lb} \times \dfrac{1 \text{ kg}}{2.205 \text{ lb}} = 4.20 \text{ kg}$

(2) x g $= 4.20 \text{ kg} \times \dfrac{1000 \text{ g}}{1 \text{ kg}} = 4.20 \times 10^3 \text{ g}$

(3) x L $= 4.00 \text{ qt} \times \dfrac{1 \text{ L}}{1.057 \text{ qt}} = 3.78 \text{ L}$

(4) x mL $= 3.78 \text{ L} \times \dfrac{1000 \text{ mL}}{1 \text{ L}} = 3.78 \times 10^3 \text{ mL}$

$$\text{Density} = \frac{\text{mass}}{\text{volume}} = \frac{4.20 \times 10^3 \text{ g}}{3.78 \times 10^3 \text{ mL}}$$

$$= 1.11 \text{ g/mL}$$

Alternatively, the calculation could be set up in a way that uses the four unit conversion factors sequentially as follows:

$$\frac{9.26 \text{ lb}}{4.00 \text{ qt}} \times \frac{1 \text{ kg}}{2.205 \text{ lb}} \times \frac{1000 \text{ g}}{1 \text{ kg}} \times \frac{1.057 \text{ qt}}{1 \text{ L}} \times \frac{1 \text{ L}}{1000 \text{ mL}} = 1.11 \text{ g/mL}$$

Check your learning: What is the volume in liters of 1.000 oz given that 1 L = 1.057 qt and 1 qt = 32 oz (exactly)?

Answer: 2.956×10^{-2} L

1.14   Conversion of Temperature Units

We use the word **temperature** to refer to the hotness or coldness of a body. One way we measure a change in temperature is to use the fact that most substances expand when their temperatures increase and contract when their temperatures decrease. The mercury or alcohol in a common glass thermometer changes its volume as the temperature changes. Because the volume of the liquid changes more than the volume of the glass, we can see the liquid expand when it gets warmer and contract when it gets cooler.

To mark a scale on a thermometer, we need a set of reference values; two of the most commonly used are the freezing and boiling temperatures of water at a pressure of 1 atmosphere. On the **Celsius scale,** 0°C is defined as the freezing temperature of water and 100°C as the boiling temperature of water. The heights of a mercury column in a thermometer at these two temperatures determine the 0° and 100° points on a Celsius ther-

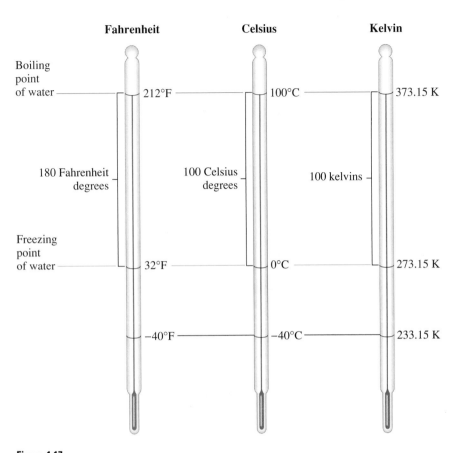

Figure 1.17

Comparison of the Fahrenheit, Celsius, and Kelvin temperature scales.

mometer. The space between the two temperatures is divided into 100 equal intervals, which we call degrees. On the **Fahrenheit scale,** the freezing point of water is taken as 32° and the boiling temperature as 212°. The space between these two points on a Fahrenheit thermometer is divided into 180 equal parts (degrees).

A Fahrenheit degree is 100/180, or 5/9, of a Celsius degree (Fig. 1.17). This gives us the relationships

$$°C = \frac{5}{9}(°F - 32) \qquad \text{and} \qquad °F = \frac{9}{5}°C + 32$$

The readings below 0° on either scale are treated as negative.

The SI unit of temperature is the kelvin (K). The freezing temperature of water on the **Kelvin scale** is 273.15 K and its boiling temperature 373.15 K. The sizes of the units on the Kelvin scale and on the Celsius scale are identical. We can use the following equations to convert between the two scales:

$$K = 273.15 + °C \qquad \text{and} \qquad °C = K - 273.15$$

Figure 1.17 shows the relationship among the three temperature scales. Recall that we do not use the degree sign with temperatures on the Kelvin scale.

Example 1.8   Conversion from Celsius

Normal body temperature is 37°C. What is this temperature on the Kelvin scale and on the Fahrenheit scale?

Solution: $K = °C + 273.15 = 37 + 273.15 = 310\ K$

$$°F = \frac{9}{5}°C + 32 = \left(\frac{9}{5} \times 37\right) + 32 = 67 + 32 = 99°F$$

Check your learning: Convert 80.92°C to K and °F.

Answer: 354.07 K, 177.7°F

Example 1.9   Conversion from Fahrenheit

A favorite recipe calls for an oven temperature of 325°F. If you are in Europe, and your oven thermometer uses the Celsius scale, what is the proper setting? What is the corresponding Kelvin temperature?

Solution: $°C = \frac{5}{9}(°F - 32) = \frac{5}{9}(325 - 32) = \frac{5}{9} \times 293 = 163°C$

$$K = °C + 273.15 = 163 + 273.15 = 436\ K$$

Check your learning: Convert 50°F to °C and K.

Answer: 10°C, 283 K

For Review   Summary

Chemistry deals with the composition, structure, and properties of matter and the ways in which various forms of matter may be interconverted. Thus it occupies a central place in the study and practice of science and technology.

Matter is anything that occupies space and has **mass.** The basic building block of matter is the **atom,** the smallest unit of an **element** that can enter into combination with atoms of the same element or other elements. In many substances atoms are combined into **molecules.** On earth, matter commonly exists in three states: **solids,** of fixed shape and volume; **liquids,** of variable shape but fixed volume; and **gases,** of variable shape and volume. Under high-temperature conditions, matter also can exist as a **plasma.** Most matter is a **mixture;** it is composed of two or more types of matter that can be present in varying amounts and can be separated by physical means. **Heterogeneous mixtures** vary in composition from point to point; **homogeneous mixtures** have the same composition from point to point. Pure substances are also homogeneous and consist of only one type of matter. A **pure substance** can be an **element,** which consists of only one type of atom and cannot be decomposed by a chemical change, or a **compound,** which consists of two or more types of atoms. Its position in the **periodic table** indicates whether an element is a metal, semi-metal, or nonmetal. The conversion of matter from one type to another is always accompanied by a conversion of one form of **energy** into another.

A pure substance may be identified by its characteristic **chemical and physical properties.** A property of a substance that is independent of the amount of substance is called an **intensive property.** A property that depends on the amount of substance is called an **extensive property.** We study and describe the behavior of matter and energy in three different domains: a **macroscopic domain,** the domain of everyday things that are large enough to see and measure; a **microscopic domain,** which must be visited in the imagination because its subjects are too small to be seen; and a **symbolic domain,** which consists of the specialized language used to represent components of the macroscopic and microscopic domains. Part of the difficulty some students have with chemistry is that they do not recognize that the same symbols can represent different things in the macroscopic and microscopic domains.

Chemistry is a science based on laws and theories that are derived from data yielded by experiments that can be reproduced to verify the results. The path of discovery that leads from question and observation to law or hypothesis to theory, combined with experimental verification, is called the **scientific method.** Quantitative measurements used to gather data utilize the metric system, generally with **SI units.** A measurement that is expressed in one unit (for example, units of A) can be expressed in an alternative unit (for example, units of B) by using a ratio of the two units derived from their conversion factor in an equation that involves ratios:

$$\frac{\text{Unknown number of units of B}}{\text{number of units of A}} = \frac{\text{number of units of B}}{\text{(in) number of units of A}}$$

(ratio from conversion factor)

or by using an equation that involves **unit conversion factors:**

$$\frac{\text{Unknown number}}{\text{of units of B}} = \frac{\text{number of}}{\text{units of A}} \times \text{unit conversion factor}$$

The unit conversion factor is the ratio determined from the conversion factor.

Key Terms and Concepts

Note: The section number in parentheses indicates where each term is introduced and defined.

atom   (1.6)
chemical change   (1.4)
chemical property   (1.4)
chemistry   (1.1)
compound   (1.5)
conversion factor   (1.13)
cubic centimeter, cm^3 (1.11)
cubic meter, m^3 (1.11)
degree Celsius, °C   (1.10)
density   (1.11)
element   (1.5)
energy   (1.3)
extensive property   (1.4)
gas   (1.2)
group   (1.7)
heterogeneous mixture   (1.5)
homogeneous mixture   (1.5)
homogeneous substance   (1.5)
hypothesis   (1.1)

intensive property   (1.4)
kelvin, K   (1.10)
kilogram, kg   (1.10)
kinetic energy   (1.3)
law   (1.1)
law of conservation of energy   (1.3)
law of conservation of matter   (1.2)
length   (1.10)
liquid   (1.2)
liter, L   (1.11)
macroscopic domain   (1.8)
mass   (1.2, and 1.10)
matter   (1.2)
metal   (1.7)
meter, m   (1.10)
microscopic domain   (1.8)
milliliter, mL   (1.11)
mixture   (1.5)
molecule   (1.6)
nonmetals   (1.7)
period   (1.7)
periodic table   (1.7)
physical change   (1.4)

physical property   (1.4)
plasma   (1.2)
potential energy   (1.3)
pure substance   (1.5)
ratio   (1.13)
rounding numbers   (1.12)
scientific method   (1.1)
second   (1.10)
semi-metal   (1.7)
series   (1.7)
SI units   (1.9)
significant figures   (1.12)
solid   (1.2)
solution   (1.5)
symbolic domain   (1.8)
temperature   (1.14)
theory   (1.1)
uncertainty   (1.12)
unit conversion factor   (1.13)
units   (1.9)
volume   (1.11)
weight   (1.2)
work   (1.3

Exercises

Note: Answers to odd-numbered exercises appear at the end of the book. The solutions to all odd-numbered exercises are worked out in the manual, *Solutions Guide for General Chemistry.*

The authors have paid special attention to significant figures in the answers to problems involving calculations. Our answers were obtained by carrying all figures in our calculator and rounding the final answer. If you choose to do an exercise in steps, rounding after each step as illustrated in some examples in the text, you may get answers that differ from ours by one or two units in the least significant figure. These differences are not errors; they simply reflect the two different, but equally acceptable, ways of rounding.

Questions

1. With what is the science of chemistry concerned?

2. Why do we use an object's mass, rather than its weight, to indicate the amount of matter it contains?

3. What properties distinguish solids from liquids? Liquids from gases? Solids from gases?

4. How does a heterogeneous mixture differ from a homogeneous mixture? How are they similar?

5. How does a homogeneous mixture differ from a pure substance? How are they similar?

6. How does an element differ from a compound? How are they similar?

7. How do molecules of elements and molecules of compounds differ? In what ways are they similar?

8. How does an atom differ from a molecule? In what ways are they similar?

9. Explain how you could determine whether the outside temperature is higher or lower than 0°C without using a thermometer.

10. A burning match and a bonfire may have the same temperature, yet you would not sit around a burning match on a fall evening in order to stay warm. Why not?

11. Matter is everywhere around us. Make a list by name of fifteen different kinds of matter that you encounter every day. Your list should include at least one example of each of the following: a solid, a liquid, a gas, an element, a compound, a heterogeneous mixture, a homogeneous mixture, and a pure substance.

12. We refer to astronauts in space as weightless, but not as massless. Why?

13. As we drive an automobile, we don't think about the chemicals involved. Prepare a list of the principal chemicals consumed and produced during the operation of an automobile.

14. Prepare a table listing the initial forms of energy and the energy transitions that take place during the operation of an automobile. Where has all the energy gone when an automobile comes to a stop and is shut off?

Properties and Classification

15. Classify each of the six underlined properties in the following paragraph as chemical or physical:

    Fluorine is a pale yellow <u>gas</u> that <u>reacts with most substances</u>. The free element <u>melts at −220°C</u> and <u>boils at −188°C</u>. Finely divided <u>metals burn in fluorine</u> with a bright flame. <u>Nineteen grams of fluorine will react with 1.0 gram of hydrogen</u>.

16. Many of the items you purchase are mixtures of pure compounds. Select several of these commercial products and prepare a list of the ingredients that were pure compounds.

17. Classify each of the following homogeneous materials as an element or a compound.

    (a) copper
    (b) water
    (c) nitrogen
    (d) sulfur
    (e) sucrose
    (f) a substance composed of molecules each of which contains two iodine atoms

18. Classify each of the following homogeneous materials as an element or a compound.

    (a) iron
    (b) oxygen
    (c) mercury oxide
    (d) gold
    (e) carbon dioxide
    (f) a substance composed of molecules each of which contains one hydrogen atom and one chlorine atom

19. Classify each of the following changes as physical or chemical.

    (a) condensation of steam
    (b) burning of gasoline
    (c) souring of milk
    (d) dissolving of sugar in water
    (e) melting of gold

20. Classify each of the following changes as physical or chemical.

    (a) burning of coal
    (b) melting of ice
    (c) mixing of chocolate syrup with milk
    (d) explosion of a firecracker
    (e) magnetizing of a screwdriver

21. The volume of a sample of oxygen gas changed from 10 mL to 11 mL as the temperature changed. Is the change a chemical or a physical change?

22. A 2.0-L volume of oxygen gas combined with 1.0 L of hydrogen gas to produce 2.0 L of water vapor. Does oxygen undergo a chemical or a physical change?

The Domains of Chemistry

23. A sulfur atom and a sulfur molecule are not identical. What is the difference between them?

24. How are the molecules in oxygen gas, the molecules in hydrogen gas, and water molecules similar? How do they differ?

25. Identify each of the underlined items as a part of the macroscopic domain, the microscopic domain, or the symbolic domain of chemistry. For those in the symbolic domain, indicate whether they are symbols for a macroscopic or a microscopic feature.

    (a) The mass of a piece of <u>lead</u> is 2.4 g.

(b) The mass of a certain <u>chlorine atom</u> is 35 amu.

(c) A bottle with a label that reads <u>Al</u> contains aluminum metal.

(d) <u>Al</u> is the symbol for an aluminum atom.

26. Identify each of the underlined items as a part of the macroscopic domain, the microscopic domain, or the symbolic domain of chemistry. For those in the symbolic domain, indicate whether they are symbols for a macroscopic or a microscopic feature.

   (a) A certain molecule contains one <u>H</u> atom and one Cl atom.

   (b) <u>Copper</u> has a density of about 8 g/cm^3.

   (c) The bottle contains 15 grams of <u>Ni</u>.

   (d) A <u>sulfur</u> molecule is composed of eight sulfur atoms.

27. Is the following a macroscopic or a microscopic description of chemical behavior? Explain your answer.

   According to one theory, the pressure of a gas increases as its volume decreases because the molecules in the gas have to move a shorter distance to hit the walls of the container.

28. Is the following a macroscopic or a microscopic description of chemical behavior? Explain your answer.

   According to one theory, if it requires 0.33 kJ of heat to melt 1.0 g of ice, 0.99 kJ of heat would be required to melt 3.0 g of ice.

The Periodic Table

29. Which groups in the periodic table contain both metals and nonmetals?

30. Which groups in the periodic table contain semi-metals?

31. Which of the following elements is likely to be the best conductor of electricity? magnesium (Mg), sulfur (S), or silicon (Si)

32. Which of the following elements is likely to be the poorest conductor of electricity? bromine (Br), germanium (Ge), or zinc (Zn)

33. Which of the following elements conduct electricity? bromine (Br), calcium (Ca), fluorine (F), iron (Fe), phosphorus (P), potassium (K), sulfur (S)

34. Which of the following elements conduct electricity? cesium (Cs), chlorine (Cl), iodine (I), krypton (Kr), lead (Pb), magnesium (Mg), strontium (Sr)

35. The element chlorine (Cl) forms a compound composed of molecules that contain one chlorine atom, one carbon atom, and three hydrogen atoms. What other elements are likely to form molecules with one atom of the element, one carbon atom, and three hydrogen atoms?

36. The element oxygen (O) forms a compound that contains two oxygen atoms for each carbon atom. What other elements are likely to form molecules with two atoms of the element and one carbon atom?

37. Which elements are members of Group 2A?

38. Which elements are members of Group 6A?

39. Which elements are members of Group 3?

40. Which elements are members of Group 15?

SI Units

41. Is one meter about an inch, a foot, a yard, or a mile?

42. Is one liter about an ounce, a pint, a quart, or a gallon?

43. Indicate what SI base units would appropriately be used to express each of the following measurements.

   (a) the length of a marathon race (26 miles 385 yards)
   (b) the mass of an automobile
   (c) the volume of a swimming pool
   (d) the speed of an airplane
   (e) the density of gold
   (f) the area of a football field
   (g) the maximum temperature at the South Pole on April 1, 1913

44. Indicate what SI base units would appropriately be used to express each of the following measurements.

   (a) the mass of the moon
   (b) the distance from Dallas to Oklahoma City
   (c) the speed of sound
   (d) the density of air
   (e) the temperature at which alcohol boils
   (f) the area of the state of Delaware
   (g) the volume of a flu shot or a measles vaccination

45. Use exponential notation to express the following quantities in terms of the SI base units in Table 1.2.

   (a) 0.13 g
   (b) 232 Gg
   (c) 5.23 pm
   (d) 86.3 mg
   (e) 37.6 cm
   (f) 54 μm
   (g) 1 Ts
   (h) 27 ps
   (i) 0.15 mK

46. Complete the following conversions between SI units.

   (a) 612 g = ___ mg
   (b) 8.160 m = ___ cm
   (c) 3779 μg = ___ g
   (d) 781 mL = ___ L
   (e) 4.18 kg = ___ g
   (f) 27.8 m = ___ km
   (g) 0.13 mL = ___ L
   (h) 1738 km = ___ m
   (i) 1.9 Gg = ___ g

47. Give the name and symbol of the prefixes used with SI units to indicate multiplication by the following exact quantities: 10^3; 10^{-2}; 0.1; 10^{-3}; 1,000,000; 0.000001.

48. Give the name of the prefix and the quantity indicated by the following symbols that are used as prefixes with SI base units: c, d, G, k, m, M, n, p, T.

49. Many medical laboratory tests are run using 5.0 μL of blood serum. What is this volume in milliliters?

50. If an aspirin tablet contains 325 mg of aspirin, how many grams of aspirin does it contain?

Uncertainty and Significant Figures

51. Express each of the following numbers in exponential notation.
    (a) 711.0
    (b) 0.239
    (c) 90743
    (d) 134.2
    (e) 0.05499
    (f) 10000.0
    (g) 0.000000738592

52. Express each of the following numbers in exponential notation.
    (a) 704
    (b) 0.03344
    (c) 547.9
    (d) 22086
    (e) 100.000
    (f) 0.0000000651
    (g) 0.007157

53. Indicate whether each of the following can be determined exactly or must be measured with some degree of uncertainty.
    (a) the number of eggs in a basket
    (b) the mass of a dozen eggs
    (c) the number of gallons of gasoline necessary to fill an automobile gas tank
    (d) the number of centimeters in exactly 2 m
    (e) the mass of this textbook
    (f) the time required to drive from San Francisco to Kansas City at an average speed of 53 mi/h

54. Indicate whether each of the following can be determined exactly or must be measured with some degree of uncertainty.
    (a) the number of seconds in an hour
    (b) the number of pages in this book
    (c) the number of grams in your weight
    (d) the number of grams in 3 kilograms
    (e) the volume of water you drink in one day
    (f) the distance from San Francisco to Kansas City

55. How many significant figures are contained in each of the following measurements?
    (a) 38.7 g
    (b) 2×10^{18} m
    (c) 3,486,002 kg

(d) 9.74150×10^{-4} J
(e) 0.0613 cm^3
(f) 17.0 kg
(g) 0.01400 g/mL

56. How many significant figures are contained in each of the following measurements?
    (a) 53 cm
    (b) 2.05×10^8 m
    (c) 86.002 J
    (d) 9.740×10^4 m/s
    (e) 10.0613 m^3
    (f) 0.17 g/mL
    (g) 0.88400 s

57. Round off each of the following numbers to two significant figures.
    (a) 0.436
    (b) 9.000
    (c) 27.2
    (d) 135
    (e) 1.497×10^{-3}
    (f) 0.445

58. Round off each of the following numbers to two significant figures.
    (a) 517
    (b) 86.3
    (c) 6.382×10^3
    (d) 5.0008
    (e) 22.497
    (f) 0.885

59. Perform the following calculations, and report each answer with the correct number of significant figures.
    (a) 628×342
    (b) $(5.63 \times 10^2) \times (7.4 \times 10^3)$
    (c) 28.0/13.483
    (d) 8119×0.000023
    (e) $14.98 + 27{,}340 + 84.7593$
    (f) $42.7 + 0.259$

60. Perform the following calculations, and report each answer with the correct number of significant figures.
    (a) 62.8×34
    (b) $0.147 + 0.0066 + 0.012$
    (c) $38 \times 95 \times 1.792$
    (d) $15 - 0.15 - 0.6155$
    (e) $8.78 \times (0.0500/0.478)$
    (f) $140 + 7.68 + 0.014$

Conversion of Units

A table of useful conversion factors can be found in Appendix B and inside the back cover of this book.

61. What ratio expresses the number of yards in 1 meter? The number of liters in 1 liquid quart? The number of pounds in 1 kilogram?

62. What ratio expresses the number of kilometers in 1 mile? The number of liters in 1 cubic foot? The number of grams in 1 ounce?

63. Soccer is played with a round ball that has a circumference between 27 and 28 in. and a weight between 14 and 16 oz. What are these specifications in units of centimeters and grams?

64. A woman's basketball has a circumference between 28.5 and 29.0 in. and a maximum weight of 20 oz. What are these specifications in units of centimeters and grams?

65. How many milliliters of a soft drink are contained in a 12.0-oz can?

66. A barrel of oil is exactly 42 gal. How many liters of oil are there in a barrel?

67. The diameter of a red blood cell is about 3×10^{-4} in. What is its diameter in centimeters?

68. The distance between the centers of the two oxygen atoms in an oxygen molecule is 1.21×10^{-8} cm. What is this distance in inches?

69. Is a 197-lb weight lifter light enough to compete in a class limited to those weighing 90 kg or less?

70. A very good 197-lb weight lifter lifted 192 kg in the clean and jerk. What was the mass of the weight lifted in pounds?

71. Gasoline is sold by the liter in many countries. How many liters are required to fill a 12.0-gal gas tank?

72. Milk is sold by the liter in many countries. What is the volume of exactly 1/2 gal of milk in liters?

73. Recently, a quantity of 490,000,000 lb (2 significant figures) of rayon was manufactured in the United States for use in carpets, automobile tires, and fabrics. What is this mass in kilograms?

74. A long ton is 2240 lb. What is this mass in kilograms?

75. Make each conversion indicated.

    (a) the record long jump, 29 ft 2.5 in., to meters
    (b) the greatest depth of the ocean, about 6.5 mi, to kilometers
    (c) the area of the state of Oregon, 96,981 mi^2, to square kilometers
    (d) the volume of 1 gill, exactly 4 oz, to milliliters
    (e) the estimated volume of the oceans, 3.3×10^8 mi^3, to cubic kilometers
    (f) the mass of a 3525-lb car to kilograms
    (g) the mass of a 2.3-oz egg to grams

76. Make each conversion indicated.

    (a) the length of a soccer field, 120 m, to feet
    (b) the height of Mt. Kilimanjaro, at 19,565 ft the highest mountain in Africa, to kilometers
    (c) the area of an 8.5×11-inch sheet of paper in cm^2
    (d) the displacement volume of an automobile engine, 161 in^3, to liters

    (e) the estimated mass of the atmosphere, 5.6×10^{15} short tons, to kilograms
    (f) the mass of a bushel of rye, 32.0 lb, to kilograms
    (g) the mass of a 5.00-grain aspirin to milligrams (1 grain = 0.00229 oz)

77. Many chemists at a recent conference in Atlanta, Georgia, participated in the 50-Trillion Angstrom Run. How long is this run in kilometers and in miles? (1 Å = 1×10^{-10} m)

78. A chemist's 50-Trillion Angstrom Run (see Exercise 77) would be an archeologist's 10,900 cubit run. How long is 1 cubit in meters and in feet? (1 Å = 1×10^{-8} cm)

79. Calculate the density of aluminum if 27.6 cm^3 has a mass of 74.6 g.

80. Osmium is one of the densest elements known. What is its density if 2.72 g has a volume of 0.121 cm^3?

81. (a) What is the mass of 6.00 cm^3 of mercury (density = 13.5939 g/cm^3)?
    (b) What is the mass of 25.0 mL of octane (density = 0.702 g/cm^3)?

82. (a) What is the mass of 4.00 cm^3 of sodium (density = 0.97 g/cm^3)?
    (b) What is the mass of 125 mL of gaseous chlorine (density = 3.16 g/L)?

83. (a) What is the volume of 25 g of iodine (density = 4.93 g/cm^3)?
    (b) What is the volume of 3.28 g of gaseous hydrogen (density = 0.089 g/L)?

84. (a) What is the volume of 11.3 g of graphite (density = 2.25 g/cm^3)?
    (b) What is the volume of 39.657 g of bromine (density = 2.928 g/cm^3)?

Heat and Temperature

85. Convert the boiling temperature of gold, 2966°C, to degrees Fahrenheit and to kelvins.

86. Convert the temperature of scalding water, 54°C, to degrees Fahrenheit and to kelvins.

87. Convert the temperature of the coldest area in a freezer, −10°F, to degrees Celsius and to kelvins.

88. Convert the temperature of dry ice, −77°C, to degrees Fahrenheit and to kelvins.

89. Convert the boiling temperature of liquid ammonia, −28.1°F, to degrees Celsius and to kelvins.

90. The label on a pressurized can of spray disinfectant warns against heating the can above 130°F. What are the corresponding temperatures on the Celsius and Kelvin temperature scales?

Applications and Additional Exercises

91. The following measurements were taken from the labels of commercial products. Determine the number of significant figures in each.
    (a) 0.0055 g active ingredients
    (b) 12 tablets
    (c) 3% hydrogen peroxide
    (d) 5.5 oz
    (e) 473 mL
    (f) 1.75% bismuth
    (g) 0.001% phosphoric acid
    (h) 99.80% inert ingredients

92. The label on a soft-drink bottle gives the volume in two units: 2.0 L and 67.6 fl oz. Use this information to find a conversion factor between the English and metric units. How many significant figures can you justify including in your conversion factor?

93. The label on a box of cereal gives the mass of cereal in two units: 978 g and 34.5 oz. Use this information to find a conversion factor between the English and metric units. How many significant figures can you justify including in your conversion factor?

94. According to the owners manual, the gas tank of a certain luxury automobile holds 22.3 gal. If the density of gasoline is 0.8206 g/mL, determine the mass in kilograms and pounds of the fuel in a full tank.

95. An instructor who was preparing for an experiment needed 225 g of phosphoric acid. The only container readily available was a 150-mL Erlenmeyer flask. Was it large enough to contain the acid, whose density is 1.83 g/mL?

96. In order to prepare for a laboratory period, a student lab assistant needed 125 g of a compound. A bottle containing 1/4 lb was available. Did the student have enough of the compound?

97. The weather in Europe was unusually warm during the summer of 1995. The TV news reported temperatures as high as 45°C. What was the temperature on the Fahrenheit scale?

98. A track-and-field athlete is 159 cm tall and weighs 45.8 kg. Is she more likely to be a distance runner or a shot-putter?

99. In a recent Grand Prix, the leader turned a lap with an average speed of 182.83 km/h. What was his speed in miles per hour, in meters per second, and in feet per second?

100. (a) In order to describe to a European how houses are constructed in the United States, we must convert the dimensions of a two-by-four to the metric system. The dimensions are 1.50 in. by 3.50 in. by 8.00 ft in the United States. What are the dimensions in centimeters and meters?
    (b) The two-by-fours are used as vertical studs, which must be placed 16.0 in. apart "on center." What is that distance in centimeters?

101. The mercury content of a stream was believed to be above the minimum considered safe (1 part per billion by weight). An analysis indicated that the concentration was 0.68 parts per billion. What quantity of mercury in grams was present in 15.0 L of the water, the density of which is 0.998 g/mL?

CHAPTER OUTLINE

2.1 Dalton's Atomic Theory
2.2 The Composition of Atoms
2.3 Symbols and Formulas
2.4 Isotopes
2.5 The Periodic Table
2.6 Formation of Ions
2.7 Ionic and Covalent Compounds
2.8 Chemical Equations
2.9 Equations for Ionic Reactions
2.10 Classification of Chemical Reactions
2.11 Naming Inorganic Compounds
         (Nomenclature)

2

The Language of Chemistry

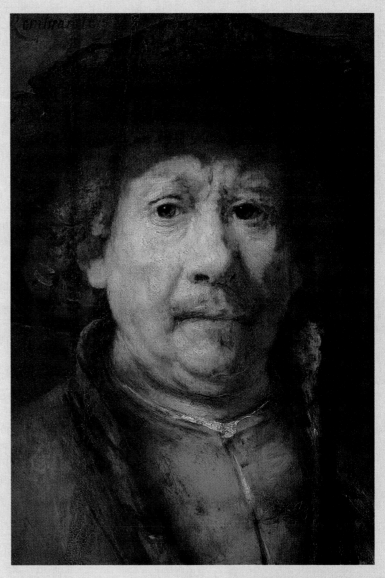

Chemical analysis of the pigments in paint can help in restoring paintings and in detecting forgeries.

The language used in chemistry is seen and heard in many disciplines ranging from agriculture to engineering to medicine to art. An art museum might be the last place we would expect to find it, but chemistry plays an important, though not an obvious, role in art. The pigments used in paintings are chemicals that have been selected for the tint, intensity, and longevity of their color. Art forgeries have been detected because the forger used pigments that contained the wrong substances—substances that were not available at the time the object was claimed to have been painted. Thus even in an art museum, we may hear the language of chemistry.

The language of chemistry includes its own vocabulary as well as its own form of shorthand. Symbols are used to indicate atoms. Formulas indicate elements and compounds as well as the composition of compounds. Chemical equations indicate how mixtures of elements and/or compounds change during chemical reactions and the relative amounts of all species involved in the change.

This chapter will lay the foundation for our study of the language of chemistry. The concepts included in this foundation include the atomic theory, the composition and mass of an atom, the variability of the composition of isotopes, ion formation, chemical bonding in ionic and covalent compounds, the kinds of chemical reactions, and the naming of compounds. We will also continue our introduction to one of the most powerful tools for organizing chemical knowledge, the periodic table.

2.1 Dalton's Atomic Theory

Figure 2.1

A crystal of the element copper and a drawing of its microscopic structure. There is only one type of atom in an element.

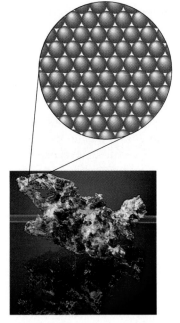

We are not sure when people first began asking questions about the microscopic nature of matter. We do know that some ancient Greek philosophers, Aristotle and Plato among them, came to the conclusion that matter could be infinitely divided. However, the Greek philosopher Democritus disagreed and argued that all matter was composed of small, finite particles that he called *atomos,* a term derived from the Greek word for "indivisible." These philosophers *thought* about atoms as a philosophical concept, but apparently they never considered performing experiments to prove their arguments. Over two thousand years later the English school teacher John Dalton studied the composition of matter. Using his own experimental data as well as those of others, he helped to revolutionize chemistry with his hypothesis that the behavior of matter could be explained using an atomic theory. Although first published in 1807, many of Dalton's hypotheses about the microscopic features of matter remain essentially unchanged in modern atomic theory.

1. Matter is composed of exceedingly small particles called atoms. An **atom** is the smallest unit of an element that can participate in a chemical change.

2. An **element** consists of only one type of atom (Fig. 2.1). In any sample of an element that is large enough to see, there are an incredibly large number of atoms, but all of the atoms of a particular element have identical chemical properties.

3. Atoms of one element differ in properties from atoms of all other elements.

4. A **compound** consists of a combination of atoms of two or more elements. In a given compound, the numbers of atoms of all types are always present in the same ratio (Fig. 2.2).

5. Atoms are neither created nor destroyed during a chemical change. They merely redistribute themselves to yield substances that are different from those present before the change (Fig. 2.3).

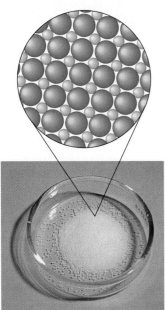

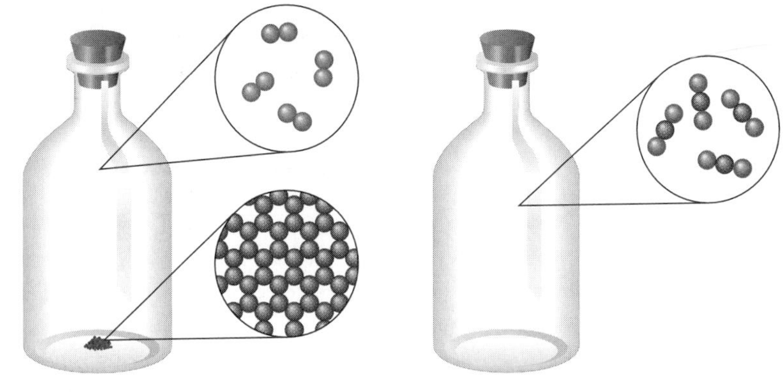

Figure 2.3 When the elements oxygen (a colorless gas) and carbon (a black solid) react, their atoms rearrange to form molecules of a colorless compound that contain a carbon atom (drawn as a black sphere) and two oxygen atoms (each drawn as a red sphere) as the drawings of the microscopic structures show.

The elements carbon and oxygen

The compound carbon dioxide

Figure 2.2

Crystals of the compound sodium chloride and a drawing of their microscopic structure. This compound results from the combination of two types of atoms.

Dalton's atomic theory provides a microscopic explanation for many macroscopic properties of matter, including several properties discussed in Chapter 1. For example, if an element such as copper (Fig. 2.1) consists of only one kind of atom, then we should not expect that it could be decomposed into simpler substances—that is, into substances composed of fewer types of atoms. The postulate that atoms are neither created nor destroyed during a chemical change explains why there is no detectable change in the total mass of matter present when matter changes from one type to another (the law of conservation of matter, Section 1.2 and Fig. 1.3).

Dalton knew of the experiments of the French chemist Joseph Proust, who demonstrated that *all samples of a pure compound contain the same elements in the same proportion by mass*. This statement is known as the **law of definite proportion** or the **law of constant composition.** The suggestion that the numbers of atoms of all types in a given compound always exist in the same ratio is consistent with these observations.

Dalton formulated the **law of multiple proportions** using his own data and data from Proust. The law states that *when two elements react to form more than one compound, a fixed mass of one element will react with masses of the other element that are in a ratio of small whole numbers*. For example, 1.0 g of carbon combines either with 1.3 g of oxygen to form carbon monoxide or with 2.6 g of oxygen to form carbon dioxide: The ratio of the masses of oxygen is 1 to 2. This can be explained by the atomic theory if the carbon-to-oxygen ratio in carbon monoxide is 1 carbon atom to 1 oxygen atom and the ratio in carbon dioxide is 1 carbon atom to 2 oxygen atoms. The ratio of oxygen atoms (and thus the ratio of their masses) is 1 to 2.

2.2 The Composition of Atoms

It is amazing that Dalton had such insight and that much of his atomic theory remains today. However, a number of elegant and sophisticated experiments performed since his theory was published have established that atoms themselves are composed of

Table 2.1

Fundamental Particles in an Atom

Particle	Mass, amu	Charge
Electron	0.00055	−1
Proton	1.0073	+1
Neutron	1.0087	0

smaller particles. Some of these experiments will be discussed in Chapter 5, but their results can be summarized here.

As noted in Section 1.6, an atom is extremely small. For example, a carbon atom weighs less than 2×10^{-23} gram. Rather than describe a quantity as small as this **atomic mass** in grams, we will use an arbitrary unit called the **atomic mass unit (amu).** The atomic mass unit is defined by assigning a particular type of carbon atom, a carbon-12 atom, a mass of exactly 12 amu. Thus an amu is exactly $\frac{1}{12}$ of the mass of a carbon-12 atom. Experiments have shown one atomic mass unit to be equal to 1.6605×10^{-24} gram.

We now know that atoms are assemblies of smaller particles called protons, neutrons, and electrons. A **proton** is a particle with a mass of 1.0073 amu and a single unit of positive charge. We say that a proton has a charge of +1. A **neutron** is a slightly heavier particle with a mass of 1.0087 amu and a charge of zero (as its name implies, it is neutral). The **electron** is a much lighter particle with a mass of about 0.00055 amu (it would take 1836 electrons to equal the mass of 1 proton) and a single unit of negative charge, a charge of −1. The properties of these fundamental particles are summarized in Table 2.1.

An atom consists of a very small, positively charged **nucleus** composed of protons and neutrons and surrounded by a large volume of space that contains the electrons necessary to balance the positive charge of the protons. The majority of the mass of an atom is located in the nucleus because protons and neutrons are much heavier than electrons. The diameter of the nucleus has been determined to be at least 100,000 times smaller than the diameter of the atom: The diameter of an atom is about 10^{-8} cm, whereas the diameter of the nucleus is only about 10^{-13} cm. To gain some perspective about their relative sizes, consider that if the nucleus were as large as the period at the end of this sentence, the diameter of the atom would be about 40 m (almost half the length of a football field). Because it is so small, the density of a nucleus is amazingly large—around 1×10^{14} g/cm^3.

The number of protons in the nucleus of an atom, its **atomic number,** determines the identity of the atom. A neutral atom must contain the same number of positive and negative charges, so the number of protons must equal the number of electrons in a neutral atom. Hence the atomic number of an atom also indicates the number of electrons in the neutral atom. The total number of protons and neutrons in the nucleus of an atom is approximately equal to the mass of the atom in atomic mass units, because the mass of a proton and that of a neutron are both about 1 amu and the mass of the electrons in an atom is negligible. The number of protons and neutrons in an atom is called its **mass number.**

Figure 2.4

Iodized salt.

Example 2.1   **Composition of an Atom**

Iodine is an essential trace element in our diet (Fig 2.4). Determine the numbers of protons, neutrons, and electrons in an iodine atom with a mass number of 127 and an atomic number of 53.

Solution: The atomic number of 53 tells us that a neutral iodine atom contains 53 protons in its nucleus and 53 electrons outside its nucleus. Because the sum of the numbers of protons and neutrons equals the mass number, 127, the number of neutrons is $127 - 53 = 74$.

Check your learning: A neutral atom of aluminum has a mass number of 27 and contains 13 electrons. What is the atomic number of this atom and how many protons and neutrons does it contain?

Answer: Atomic number is 13; 13 protons and 14 neutrons.

2.3   Symbols and Formulas

A chemical **symbol** is an abbreviation that we use to indicate an element or an atom of that element. For example, the symbol for silicon is Si. As shown in Fig. 2.5, we use the same symbol to label a bottle of the element silicon (macroscopic domain) or to indicate one silicon atom in a representation of a molecule (microscopic domain).

The symbols for several common elements and their atoms are listed in Table 2.2. Some symbols are derived from the common name of the element; others are abbreviations of the name in another language. Most symbols have one or two letters, but three-letter symbols have been used to describe elements that have atomic numbers greater than 104. To avoid confusion with other notations, only the first letter of a symbol is capitalized—Co is the symbol for the element cobalt; CO is the notation for the compound carbon monoxide, which contains atoms of the elements carbon (C) and oxygen (O). All known elements and their symbols are listed on the inside front cover of this book.

Traditionally, the discoverer (or discoverers) of a new element names the element. However, until the name is recognized by the International Union of Pure and Applied

Figure 2.5

Two very different uses of the symbol Si. (A) A label bears the symbol Si to indicate that the solid in the bottle is the element silicon. (B) In this formula, the symbols for silicon, Si, and chlorine, Cl, indicate that one Si atom is combined with four Cl atoms in a molecule.

(A)

$$Cl - Si - Cl$$

with Cl above and below Si

(B)

Table 2.2

Some Common Elements and
Their Symbols

Aluminum	Al
Bromine	Br
Calcium	Ca
Carbon	C
Chlorine	Cl
Chromium	Cr
Cobalt	Co
Copper	Cu (from *cuprum*)
Fluorine	F
Gold	Au (from *aurum*)
Helium	He
Hydrogen	H
Iodine	I
Iron	Fe (from *ferrum*)
Lead	Pb (from *plumbum*)
Magnesium	Mg
Mercury	Hg (from *hydrargyrum*)
Nitrogen	N
Oxygen	O
Potassium	K (from *kalium*)
Silicon	Si
Silver	Ag (from *argentum*)
Sodium	Na (from *natrium*)
Sulfur	S
Tin	Sn (from *stannum*)
Zinc	Zn

Chemistry (IUPAC), the recommended name of the new element is based on the Latin words for its atomic number. For example, for several years element 106 was called unnilhexium (Unh), element 107 unnilseptium (Uns), and element 108 unniloctium (Uno). These names are based on the following Latin words: *nil* = 0, *un* = 1, *hex* = 6, *sept* = 7, *oct* = 8, and so on.

The naming of an element can take several years and can be controversial. The original discoverers usually wait until their experiments have been independently confirmed by other scientists before proposing a name. For example, element 106 was first created and characterized in 1974 by a group of scientists in California. However, it was not until mid-1993 that experiments confirmed the existence of this element. As this text was written, the name *seaborgium* had been proposed by the discoverers in honor of Glenn Seaborg, a Nobel prize winner who has been active in the discovery of several heavy elements. This name and names for several other elements have been recommended by the nomenclature committee of the American Chemical Society (ACS), and these are the names we will use in this text. For several reasons,

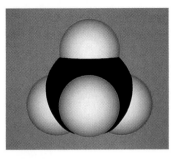

Figure 2.6

A space-filling model of a methane molecule, CH_4.

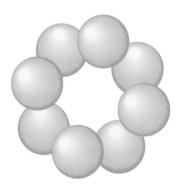

Figure 2.7

A sulfur molecule is composed of eight sulfur atoms (each drawn as a yellow sphere); thus it has a molecular formula of S_8.

including the fact that Seaborg is still alive and the fact that no element has yet been named for a living scientist, the IUPAC has proposed different names. The naming of elements 104 through 109 is under discussion, and different names appear in different places. In the period 1994–1996, scientists in Germany announced the discovery of elements 110, 111, and 112. Naming of these elements will await verification of the experiments.

A **molecular formula** is a symbolic representation of a molecule that uses chemical symbols to indicate the types of atoms and uses subscripts to show the numbers of atoms of each type in the molecule. (A subscript is used only when more than one atom of a given type is present.) Molecular formulas are also used as abbreviations for the names of compounds. We can use the same formula to represent a microscopic and a macroscopic item. For example, the molecular formula CH_4 represents a molecule (a microscopic entity) that contains one carbon atom (C is the symbol for a carbon atom) and four hydrogen atoms (H is the symbol for a hydrogen atom). Figure 2.6 illustrates a CH_4 molecule. These molecules compose the gas methane, so CH_4 is also an abbreviation for methane (a macroscopic entity), a component of natural gas and one of the gases involved in the greenhouse effect.

Some elements are composed of molecules that contain two or more atoms of the element. For example, samples of the elements hydrogen, oxygen, and nitrogen are composed of molecules that contain two atoms each and thus have the molecular formulas H_2, O_2, and N_2, respectively. Other elements composed of molecules containing two atoms are fluorine, F_2; chlorine, Cl_2; bromine, Br_2; and iodine, I_2. The most common form of the element sulfur is composed of molecules that consist of eight atoms of sulfur (Fig. 2.7); its molecular formula is S_8.

It is important to note that a subscript following a symbol and a number in front of a symbol do not represent the same thing; for example, H_2 and 2H indicate different things (Fig. 2.8). H_2 is a molecular formula; it represents a molecule of hydrogen consisting of two atoms of the element, chemically combined. The expression 2H, on the other hand, indicates two separate hydrogen atoms that are not combined as a unit. The expression $2H_2$ represents two molecules of hydrogen gas.

Many of the compounds that form when a metallic element combines with a non-metallic element (Section 1.7) do not contain molecules; they are composed of ions, which will be discussed in Section 2.6. We describe the composition of these compounds with an **empirical formula** (sometimes called a *simplest formula*), which indicates the types of atoms present and the simplest whole-number ratio of atoms in the compound. Titanium dioxide, the pigment in the white paint we use to paint our houses, is an ionic compound (Section 2.7); thus it does not contain molecules. Its formula, TiO_2, identifies the elements titanium (Ti) and oxygen (O) as the constituents of titanium dioxide. This formula also indicates the presence of twice as many atoms

Figure 2.8

The symbols H, 2H, and H_2 represent very different entities.

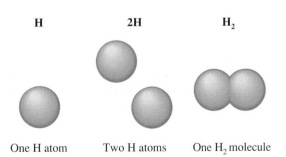

H   2H H_2

One H atom   Two H atoms   One H_2 molecule

Figure 2.9

(Left) The gaseous element chlorine, Cl_2, and the solid element sodium, Na. (Right) These elements react vigorously to form sodium chloride, NaCl, which contains equal numbers of the atoms of each element.

of the element oxygen as atoms of the element titanium: the ratio of oxygen atoms to titanium atoms is 2 to 1. The empirical formula of sodium chloride, table salt, is NaCl. This formula indicates that equal numbers of atoms of the elements sodium (Na) and chlorine (Cl) constitute salt (Fig 2.9).

We sometimes encounter an empirical formula written for a compound that is composed of molecules. In many cases, a molecular formula is based on an experimental determination of both the empirical formula and the molecular mass (which we will discuss in Chapter 3). For example, it can be determined experimentally that benzene contains only two elements, carbon (C) and hydrogen (H), and that for every carbon atom there is one hydrogen atom. Thus the empirical formula is CH. An experimental determination of the molecular mass reveals that the molecular formula of the benzene molecule is C_6H_6 (Fig. 2.10).

If we know a molecular formula, we can easily determine the empirical formula. For example, the molecular formula for ethyl acetate, the component that gives some nail polish removers its odor, is $C_4H_8O_2$. This formula indicates that a molecule of ethyl acetate (Fig. 2.11) contains four carbon atoms, eight hydrogen atoms, and two oxygen atoms. The ratio of atoms is 4 to 8 to 2. Dividing by the lowest common denominator 2 gives the simplest whole-number ratio of atoms, 2 to 4 to 1, so the empirical formula is C_2H_4O.

Parentheses in a formula indicate a group of atoms that behave as a unit. The formula for calcium phosphate, one of the minerals in our bones, is $Ca_3(PO_4)_2$. This formula indicates that there are three atoms of calcium (Ca) for every two phosphate

Figure 2.10

Two models of the benzene molecule, C_6H_6. (Left) A ball-and-stick model uses a different-colored ball for each kind of atom (C, black; H, white); sticks represent chemical bonds. Ball-and-stick models show geometrical relationships of the atoms and bonds, but the atomic sizes are not to scale. (Right) A space-filling model shows the relative sizes of the atoms as well as their geometric arrangement.

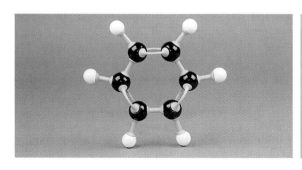
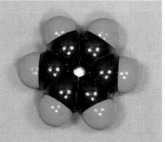

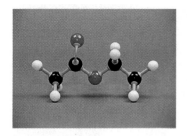

Figure 2.11

A ball-and-stick model of ethyl acetate.

(PO₄) groups. The PO_4 groups are discrete units, each of which consists of one P (phosphorus) atom and four O (oxygen) atoms. The formula shows a total count of three atoms of calcium, two atoms of phosphorus, and eight atoms of oxygen.

Example 2.2   **Empirical and Molecular Formulas**

Molecules of acetic acid, which are responsible for the smell and taste of vinegar, contain two carbon atoms, four hydrogen atoms, and two oxygen atoms. What are the empirical and molecular formulas of acetic acid?

Solution: The molecular formula is $C_2H_4O_2$, because one molecule actually contains two C, four H, and two O atoms. The simplest whole-number ratio of C to H to O atoms in acetic acid is 1 to 2 to 1, so the empirical formula is CH_2O.

Check your learning: A molecule of blood sugar, glucose, contains 6 carbon atoms, 12 hydrogen atoms, and 6 oxygen atoms. What are the empirical and molecular formulas of glucose?

Answer: Molecular formula, $C_6H_{12}O_6$; empirical formula, CH_2O.

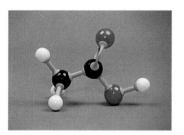

Figure 2.12

A ball-and-stick model of acetic acid.

A third type of formula that chemists often use, the structural formula, is even more informative than the molecular formula. The **structural formula** of a compound gives the same information as its molecular formula (the types and numbers of atoms in the molecule), but the structural formula also shows how the atoms are connected in the molecule. For example, we have seen that the molecular formula for acetic acid is $C_2H_4O_2$ (Example 2.2). Figure 2.12 shows an illustration of an acetic acid molecule. The structural formula that gives us that information is

$$H-\overset{\overset{\displaystyle H}{|}}{\underset{\underset{\displaystyle H}{|}}{C}}-C\overset{\displaystyle O}{\underset{\displaystyle O-H}{\diagup}}$$

This formula contains symbols for four H atoms, two C atoms, and two O atoms, which indicate the number of atoms in the molecule. The lines represent bonds that hold the atoms together. We will discuss chemical bonds and see how to predict the arrangement of atoms in a molecule in Chapter 6. For now, we will simply take the lines as an indication of how the atoms are connected in a molecule. Here we see that three of the four hydrogen atoms are bonded to one of the carbon atoms and that the fourth hydrogen is bonded to one of the oxygen atoms. As we consider chemical reactions in subsequent chapters, we will discover how structural formulas can give us excellent insights into how molecules behave chemically.

2.4   Isotopes

All atoms of a given element exhibit similar chemical properties; however, they can differ in mass. Because the numbers of protons (and electrons) in these atoms are the same (Section 2.2), the difference in mass must be due to differences in the number of neutrons present in the nuclei of these elements. Atoms of a particular element that differ only in the number of neutrons in the nucleus are called **isotopes.** Dalton had

Table 2.3

Nuclear Compositions of Atoms of the
Very Light Elements

	Symbol	Atomic Number	Number of Protons	Number of Neutrons	Mass, amu	% Natural Abundance
Hydrogen	$^{1}_{1}H$	1	1	0	1.0078	99.985
	$^{2}_{1}D$	1	1	1	2.0141	0.015
	$^{3}_{1}T$	1	1	2	3.01605	—
Helium	$^{3}_{2}He$	2	2	1	3.01603	0.00013
	$^{4}_{2}He$	2	2	2	4.0026	100
Lithium	$^{6}_{3}Li$	3	3	3	6.0151	7.42
	$^{7}_{3}Li$	3	3	4	7.0160	92.58
Beryllium	$^{9}_{4}Be$	4	4	5	9.0122	100
Boron	$^{10}_{5}B$	5	5	5	10.0129	19.6
	$^{11}_{5}B$	5	5	6	11.0093	80.4
Carbon	$^{12}_{6}C$	6	6	6	12.0000[a]	98.89
	$^{13}_{6}C$	6	6	7	13.0033	1.11
	$^{14}_{6}C$	6	6	8	14.0032	—
Nitrogen	$^{14}_{7}N$	7	7	7	14.0031	99.63
	$^{15}_{7}N$	7	7	8	15.0001	0.37
Oxygen	$^{16}_{8}O$	8	8	8	15.9949	99.759
	$^{17}_{8}O$	8	8	9	16.9991	0.037
	$^{18}_{8}O$	8	8	10	17.9992	0.204
Fluorine	$^{19}_{9}F$	9	9	10	18.9984	100
Neon	$^{20}_{10}Ne$	10	10	10	19.9924	90.92
	$^{21}_{10}Ne$	10	10	11	20.9940	0.257
	$^{22}_{10}Ne$	10	10	12	21.9914	8.82

[a]Mass assigned as exactly 12 by international agreement.

no way to know that isotopes exist, so they are not part of the original atomic theory.

We symbolize particular isotopes of an element with the atomic number as a subscript and the mass number as a superscript, both preceding the symbol for the element. For example, magnesium exists as a mixture of three isotopes, each with an atomic number of 12, and with mass numbers of 24, 25, and 26, respectively. These isotopes can be identified as $^{24}_{12}Mg$, $^{25}_{12}Mg$, and $^{26}_{12}Mg$. All magnesium atoms have 12 protons in their nucleus. They differ only because a $^{24}_{12}Mg$ atom has 12 neutrons in its nucleus, a $^{25}_{12}Mg$ atom has 13 neutrons, and a $^{26}_{12}Mg$ atom has 14 neutrons.

The numbers of protons and neutrons in the nuclei of the naturally occurring isotopes of the elements with atomic numbers 1 through 10 are given in Table 2.3. Note that the two heavier isotopes of hydrogen have common names and special symbols. Deuterium is $^{2}_{1}H$ and tritium is $^{3}_{1}H$.

Because each proton and each neutron contributes approximately one mass unit to the atomic mass of an atom and each electron contributes far less, the mass of a single atom is approximately equal to its mass number (a whole number). However, the

Figure 2.13

Diagram of a mass spectrometer. Ions are expelled from the ionization chamber by an electric field and move through a magnetic field, which causes their paths to curve. Only those ions with a specific mass can pass through the magnetic field into the detector; others collide with the walls of the tube. The mass of an ion can be measured by changing the strength of the magnetic field and seeing at what field strength the ion passes through to the detector.

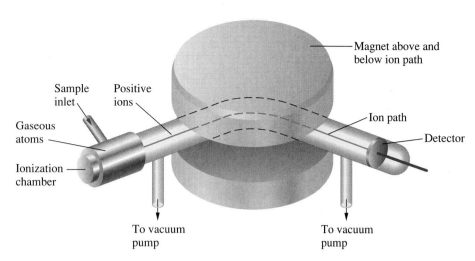

average masses of atoms of most elements are not whole numbers, because most elements exist naturally as mixtures of two or more isotopes. For example, the element boron, which can be used as a strengthening agent in graphite golf clubs and fishing rods, is composed of two isotopes; approximately 20% of these (1 atom in 5) are $^{10}_5$B, and the remaining 80% (4 atoms in 5) are $^{11}_5$B. The average mass of a boron atom is found as follows:

$$\frac{(1 \times 10 \text{ amu} + 4 \times 11 \text{ amu})}{5}$$

$$= \left(\frac{1}{5} \times 10 \text{ amu}\right) + \left(\frac{4}{5} \times 11 \text{ amu}\right)$$

$$= (0.20 \times 10 \text{ amu}) + (0.80 \times 11 \text{ amu})$$

$$= 2.0 \text{ amu} + 8.8 \text{ amu} = 10.8 \text{ amu}$$

It is important to understand that no single boron atom weighs exactly 10.8 amu; 10.8 amu is the average mass of all boron atoms.

The average mass of an atom of an element is equal to the sum of the masses of all isotopes in the element, each mass multiplied by the fraction that the isotope represents in the element. The values listed in tables of atomic masses, such as those given on the inside front cover of this book, are average masses of all isotopes present in a natural sample of an element.

The occurrence and natural abundances of isotopes can be experimentally determined using an instrument called a mass spectrometer. A diagram of a mass spectrometer is shown in Fig. 2.13, and a mass spectrum illustrating the three isotopes of magnesium is shown in Fig. 2.14.

Figure 2.14

The mass spectrum of magnesium has three peaks, which indicates that magnesium has three isotopes.

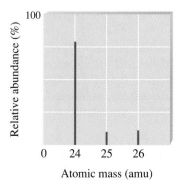

Example 2.3   Calculation of Average Atomic Mass

A meteorite found in central Indiana contains traces of the noble gas neon that appear to have been picked up from the solar wind as the meteorite traversed the solar system before it fell. A mass spectrum of a sample of the gas showed that it consisted of 91.84% $^{20}_{10}$Ne (mass 19.9924 amu), 0.47% $^{21}_{10}$Ne (mass 20.9940 amu), and 7.69% $^{22}_{10}$Ne (mass 21.9914 amu). What is the average mass of the neon in the solar wind?

Solution: Average mass = (0.9184 × 19.9924 amu) + (0.0047 × 20.9940 amu) +

(0.0769 × 21.9914 amu)

= (18.36 + 0.099 + 1.69) amu

= 20.15 amu

The average mass of a neon atom in the solar wind is 20.15 amu. (The average mass of a terrestrial neon atom is 20.1796 amu. See the inside front cover of the book. This result demonstrates that we may find slight differences in the natural abundance of isotopes depending on their origin.)

Check your learning: A mass spectrum of magnesium like that shown in Fig. 2.14 reveals that magnesium contains 78.70% $^{24}_{12}$Mg atoms (mass 23.98 amu), 10.13% $^{25}_{12}$Mg atoms (mass 24.99 amu), and 11.17% $^{26}_{12}$Mg atoms (mass 25.98 amu). Calculate the average mass of a Mg atom.

Answer: 24.31 amu

2.5   The Periodic Table

As early chemists worked with ores and purified elements, they realized that various elements could be grouped together by reason of their similar chemical behavior. The members of one such grouping are lithium (Li), sodium (Na), and potassium (K). These elements all look like metals, conduct electricity well, and have similar chemical properties. A second grouping includes calcium (Ca), strontium (Sr), and barium (Ba), which also look like metals, conduct electricity well, and have similar chemical properties. However, these properties are somewhat different, both chemically and physically, from those of the elements lithium, sodium, and potassium. Fluorine (F), chlorine (Cl), bromine (Br), and iodine (I) also exhibit similar properties, but these properties are very different from those of the elements above. They do not conduct electricity and are considered nonmetallic.

By the time 62 elements were known, their relationships could be observed. Dimitri Mendeleev in Russia (1872) and Lothar Meyer in Germany (1869) independently suggested a periodic relationship between the properties of the elements and their atomic masses. Later, it became apparent that the relationship actually involved atomic numbers rather than atomic masses. The modern statement of this relationship, **the periodic law,** is as follows: *The properties of the elements are periodic functions of their atomic numbers.* A periodic table (Fig. 2.15) is an arrangement of the elements in increasing order of their atomic numbers that collects atoms with similar properties in a group in the same vertical column.

Each block in the periodic table in Fig. 2.15 (and the one inside the front cover) contains three pieces of information. Starting at the top, we find the atomic number of the element (remember that this is the number of protons in the nucleus of an atom of the element), then the symbol of the element, and at the bottom, its average atomic mass. For example, for the element sodium in Group 1A, the information is

Atomic number	11
Symbol	Na
Average atomic mass	22.990

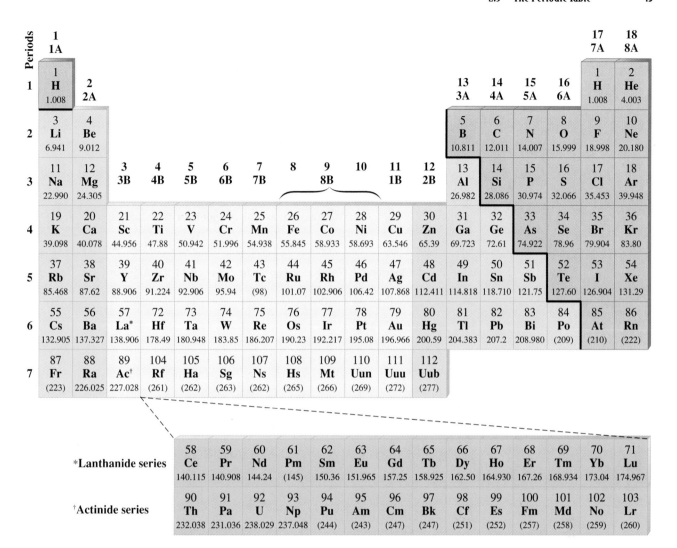

Periods	1 1A	2 2A	3 3B	4 4B	5 5B	6 6B	7 7B	8	9 8B	10	11 1B	12 2B	13 3A	14 4A	15 5A	16 6A	17 7A	18 8A
1	1 H 1.008																1 H 1.008	2 He 4.003
2	3 Li 6.941	4 Be 9.012											5 B 10.811	6 C 12.011	7 N 14.007	8 O 15.999	9 F 18.998	10 Ne 20.180
3	11 Na 22.990	12 Mg 24.305											13 Al 26.982	14 Si 28.086	15 P 30.974	16 S 32.066	17 Cl 35.453	18 Ar 39.948
4	19 K 39.098	20 Ca 40.078	21 Sc 44.956	22 Ti 47.88	23 V 50.942	24 Cr 51.996	25 Mn 54.938	26 Fe 55.845	27 Co 58.933	28 Ni 58.693	29 Cu 63.546	30 Zn 65.39	31 Ga 69.723	32 Ge 72.61	33 As 74.922	34 Se 78.96	35 Br 79.904	36 Kr 83.80
5	37 Rb 85.468	38 Sr 87.62	39 Y 88.906	40 Zr 91.224	41 Nb 92.906	42 Mo 95.94	43 Tc (98)	44 Ru 101.07	45 Rh 102.906	46 Pd 106.42	47 Ag 107.868	48 Cd 112.411	49 In 114.818	50 Sn 118.710	51 Sb 121.75	52 Te 127.60	53 I 126.904	54 Xe 131.29
6	55 Cs 132.905	56 Ba 137.327	57 La* 138.906	72 Hf 178.49	73 Ta 180.948	74 W 183.85	75 Re 186.207	76 Os 190.23	77 Ir 192.217	78 Pt 195.08	79 Au 196.966	80 Hg 200.59	81 Tl 204.383	82 Pb 207.2	83 Bi 208.980	84 Po (209)	85 At (210)	86 Rn (222)
7	87 Fr (223)	88 Ra 226.025	89 Ac† 227.028	104 Rf (261)	105 Ha (262)	106 Sg (263)	107 Ns (262)	108 Hs (265)	109 Mt (266)	110 Uun (269)	111 Uuu (272)	112 Uub (277)						

*Lanthanide series	58 Ce 140.115	59 Pr 140.908	60 Nd 144.24	61 Pm (145)	62 Sm 150.36	63 Eu 151.965	64 Gd 157.25	65 Tb 158.925	66 Dy 162.50	67 Ho 164.930	68 Er 167.26	69 Tm 168.934	70 Yb 173.04	71 Lu 174.967
†Actinide series	90 Th 232.038	91 Pa 231.036	92 U 238.029	93 Np 237.048	94 Pu (244)	95 Am (243)	96 Cm (247)	97 Bk (247)	98 Cf (251)	99 Es (252)	100 Fm (257)	101 Md (258)	102 No (259)	103 Lr (260)

Figure 2.15

This periodic table shows the nonmetals in green. The other elements are metals.

In Chapter 1 we saw how the elements in the periodic table are divided into metals, nonmetals, and semi-metals. Elements to the left of the heavy "stair-step" line in Fig. 2.15 are metals, elements to the right are nonmetals, and elements on the line are semi-metals. It is clear that the majority of the elements are metals. We can further classify the elements in the various groups (vertical columns) into **main-group elements** (or **representative elements**), the elements in the columns labeled 1A–8A (or 1, 2, and 12–18); **transition metals,** the elements in the columns labeled 1B and 2B–8B (3–11); and **inner transition metals,** the two rows at the bottom of the table called the lanthanides and actinides. This classification of the elements is based on their chemical behavior.

Several of the groups in the periodic table have special names that refer to the elements in a vertical group. All of the elements in Group 1A except hydrogen are known as **alkali metals,** and they all have similar chemical properties. Likewise, the elements in Group 2A are referred to as **alkaline earth metals.** Other group names are **chalcogens** for Group 6A and **halogens** for Group 7A. The Group 8A elements have several names. We will call them **noble gases,** although they are also called inert gases and rare gases. The groups can also be referred to by the first element of the group. For example, the chalcogens can be called the oxygen group or oxygen family. Hydrogen (H) is a unique, nonmetallic element that has properties similar to the elements of both Group 1A and Group 7A. For that reason, many periodic tables show it at the top of both groups.

Example 2.4   Naming Groups of Elements

Atoms of each of the following elements are essential for life. Give the group name for each. **(a)** chlorine, **(b)** calcium, **(c)** sodium, **(d)** sulfur.

Solution: The family names are **(a)** halogen, **(b)** alkaline earth metal, **(c)** alkali metal, **(d)** chalcogen.

Check your learning: Give the group name for each of the following elements. **(a)** krypton, **(b)** selenium, **(c)** barium, **(d)** lithium.

Answer: **(a)** noble gas **(b)** chalcogen **(c)** alkaline earth metal **(d)** alkali metal

---

Something about the atomic masses in the periodic table may have attracted your attention. Element 43 (technetium), element 61 (promethium), and all of the elements from element 94 (plutonium) "up" have their atomic mass in parentheses. These elements are synthetic and radioactive (we will consider radioactivity in Chapter 20). The average mass of the atoms in a sample of a synthetic element may vary depending on the sample's source. The number in parentheses is the atomic mass number (and approximate atomic mass) of the most stable isotope of that element.

2.6   Formation of Ions

Protons and neutrons in the nuclei of atoms are not involved in ordinary chemical reactions. However, the electrons of atoms are very much involved in the chemistry that occurs. During the formation of some compounds, atoms gain or lose electrons and form charged particles called **ions.** When a neutral atom *gains* one or more electrons, it will have a negative charge equal to the number of electrons gained because it has more negatively charged electrons than positively charged protons. The resulting ion is called an **anion.** When an atom *loses* one or more electrons, it will have a positive charge equal to the number of electrons lost because it has more protons than electrons (Fig. 2.16). The resulting ion is called a **cation.**

We can use the periodic table to predict whether an atom will form an anion or a cation, and we can even predict the charge on many of the resulting ions. We find that atoms of many main-group metals (Section 2.5) lose enough electrons to leave them with the same number of electrons as an atom of the *preceding* noble gas (Fig. 2.17). For example, an atom of an alkali metal (Group 1A) loses one electron and forms a cation with a $+1$ charge, whereas an alkaline earth atom (Group 2A) loses two electrons and forms a cation with a $+2$ charge. Thus, for example, a neutral calcium atom, with 20 protons and 20 electrons, readily loses 2 electrons. This gives a calcium ion with 20 protons, 18 electrons, a $+2$ charge, and the same number of electrons as atoms of the preceding noble gas, argon (Ar).

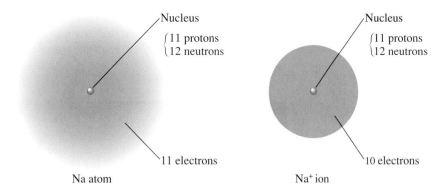

Figure 2.16

The composition of a sodium atom and that of a sodium ion.

When atoms of nonmetal elements form ions, they generally gain enough electrons to give them the same number of electrons as atoms of the *next* noble gas in the periodic table (Fig. 2.17). Atoms of Group 7A gain one electron and form anions with a -1 charge. Atoms of Group 6A gain two electrons and form ions with a -2 charge. For example, the neutral bromine atom, with 35 protons and 35 electrons, can gain 1 electron to provide it with 36 electrons. This gives a bromide ion with 35 protons, 36 electrons, a -1 charge, and the same number of electrons as atoms of the next noble gas, krypton (Kr).

The symbol for an ion consists of the symbol of the atom with the charge on the ion indicated as a superscript to the right of the symbol. The symbol for a potassium cation with a charge of $+1$ is K^+; that for a sulfur anion with a charge of -2 is S^{2-}.

Figure 2.17

Some elements exhibit a regular pattern of ionic charge when they form ions.

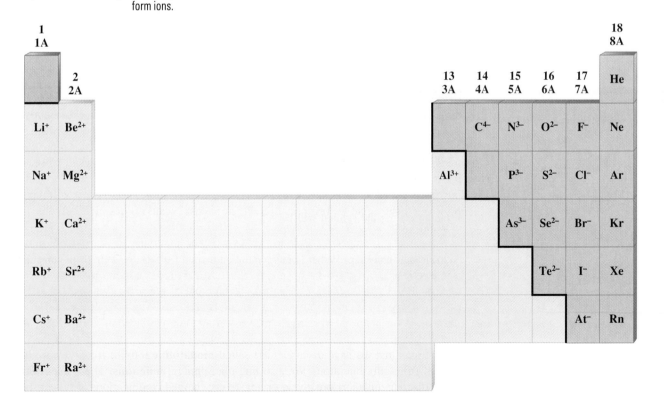

The name of a metal ion is the same as the name of the metal atom from which it forms; K^+ is a potassium ion, for example. An anion formed from a nonmetal atom is named by replacing the ending of the name of the element by the suffix *-ide*; S^{2-} is a sulfide ion.

Example 2.5   Formation of Ions

The table salt we eat every day is composed of ions. One type of ion in this salt is formed from chlorine atoms, the other from sodium atoms. Predict which atoms form cations, which form anions, and the charges on these ions.

Solution: Chlorine's position in the periodic table (in Group 7A and to the right of the heavy line) tells us that it is a nonmetal. Nonmetals form negative ions (anions). A chlorine atom must gain one electron to have the same number of electrons as an atom of the next noble gas, argon. Thus a chlorine atom will form an anion with one more electron than protons and a charge of -1. The ion is the Cl^- (chloride) ion.

Sodium's position in the periodic table (Group 1A) reveals that it is a metal. Metals form positive ions (cations). A sodium atom must lose one electron to have the same number of electrons as an atom of the preceding noble gas, neon. Thus a sodium atom will form a cation with one less electron than protons and a charge of $+1$, the Na^+ (sodium) ion.

Check your learning: Predict whether the magnesium and nitrogen atoms that react to form magnesium nitride will form an anion or a cation and the charges on each ion. Write the symbol for each ion and name it.

Answer: Mg will form a cation with a charge of $+2$: Mg^{2+}, the magnesium ion. Nitrogen will form an anion with a charge of -3: N^{3-}, the nitride ion.

Example 2.6   Composition of Ions

An ion found in some compounds used as antiperspirants contains 13 protons and 10 electrons. What is its symbol?

Solution: Because the number of protons remains unchanged when an atom forms an ion, the atomic number of the element must be 13. Knowing this lets us go to the periodic table and identify the element as Al (aluminum). The Al atom has lost 3 electrons and thus has three more positive charges (13) than it has electrons (10). This is the aluminum cation; the symbol will be Al^{3+}.

Check your learning: What is the symbol and name for the ion with 34 protons and 36 electrons.

Answer: Se^{2-}, the selenide ion.

The ions that we have discussed are called **monatomic ions;** that is, they are ions formed from only one atom. We also find many **polyatomic ions.** These ions, which act as discrete units, are positively or negatively charged combinations of two or more

Table 2.4

Common Polyatomic Ions

Charge	Name	Formula
+1	Ammonium	NH_4^+
−1	Acetate	$C_2H_3O_2^-$
−1	Cyanide	CN^-
−1	Dihydrogen phosphate	$H_2PO_4^-$
−1	Hydrogen carbonate	HCO_3^-
−1	Hydrogen sulfate	HSO_4^-
−1	Hydroxide	OH^-
−1	Nitrate	NO_3^-
−1	Nitrite	NO_2^-
−1	Perchlorate	ClO_4^-
−1	Permanganate	MnO_4^-
−2	Carbonate	CO_3^{2-}
−2	Hydrogen phosphate	HPO_4^{2-}
−2	Peroxide	O_2^{2-}
−2	Sulfate	SO_4^{2-}
−2	Sulfite	SO_3^{2-}
−3	Phosphate	PO_4^{3-}

atoms. At this point in our study, you will need to memorize the names, formulas, and charges of these polyatomic ions. Because we will use them repeatedly, they will soon become familiar. Some of the more important polyatomic ions are listed in Table 2.4.

2.7 Ionic and Covalent Compounds

We have seen that in ordinary chemical reactions, the nucleus of each atom remains unchanged. Electrons, however, can be added to atoms by transfer from other atoms, lost by transfer to other atoms, or shared with other atoms. The transfer and sharing of electrons among atoms govern the chemistry of the elements.

We can classify the forces that hold the ions or atoms together in compounds as ionic or covalent on the basis of the behavior of the electrons. When electrons are *transferred* and ions form, ionic bonds result. **Ionic bonds** are electrostatic forces of attraction between ions of opposite charge. The attractive forces that result from the *sharing* of electrons among atoms are called **covalent bonds.** Compounds are classified as ionic or covalent on the basis of the bonds present in them.

Ionic Compounds. When an element composed of atoms that readily lose electrons (a metal) reacts with an element composed of atoms that readily gain electrons (a nonmetal), a transfer of electrons usually occurs, producing ions. The compound formed by this transfer is stabilized by the electrostatic attractions (ionic bonds) between the ions of opposite charge present in the compound. For example, when each sodium atom in a sample of sodium metal (Group 1A) gives up one electron to

Figure 2.18
Sodium chloride melts at 801°C.

form a sodium cation, Na^+, and each chlorine atom in a sample of chlorine gas (Group 7A) accepts one electron to form a chloride anion, Cl^-, the resulting compound, NaCl, is composed of sodium ions and chloride ions in the ratio of one Na^+ ion for each Cl^- ion. Similarly, each calcium atom (Group 2A) can give up two electrons and can transfer one to each of two chlorine atoms to form $CaCl_2$, which is composed of Ca^{2+} and Cl^- ions in the ratio of one Ca^{2+} ion to two Cl^- ions.

A compound that contains ions and is held together by ionic bonds is called an **ionic compound.** The periodic table can help us recognize many of the compounds that are ionic: **When a metal is combined with one or more nonmetals, the compound is usually ionic.** This is not always true, but it is a good guideline for the compounds generally discussed in an introductory chemistry course.

We can often recognize ionic compounds because they are solids that often melt at high temperatures (Fig. 2.18) and boil at even higher temperatures. For example, sodium chloride melts at 801°C and boils at 1413°C. (For comparison, water melts at 0°C and boils at 100°C.) Just like ionic compounds dissolved in water, molten ionic compounds conduct electricity because the ions are free to move through the liquid.

In every ionic compound, the total number of positive charges on the cations equals the total number of negative charges on the anions. Thus ionic compounds are electrically neutral overall, even though they contain positive and negative ions. We can use this observation to help us write the formula of an ionic compound. *The formula of an ionic compound must have a ratio of ions such that the numbers of positive and negative charges are equal.*

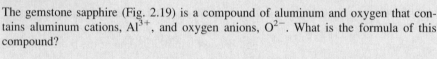

Example 2.7   Prediction of the Formula of an Ionic Compound

Figure 2.19
A sapphire.

The gemstone sapphire (Fig. 2.19) is a compound of aluminum and oxygen that contains aluminum cations, Al^{3+}, and oxygen anions, O^{2-}. What is the formula of this compound?

Solution: Because the ionic compound must be electrically neutral, it must have the same number of positive and negative charges. Two aluminum ions, each with a charge of +3, would give us six positive charges; and three oxide ions, each with a charge of −2, would give us six negative charges. The formula would be Al_2O_3.

Check your learning: Predict the formula of the ionic compound formed between the sodium cation, Na^+, and the sulfide anion, S^{2-}.

Answer: Na_2S

Many ionic compounds contain polyatomic ions (Table 2.4) as the cation or the anion or both. As with simple ionic compounds, these compounds must also be electrically neutral, so their formulas can be predicted by treating the polyatomic ions as discrete units.

Example 2.8   Prediction of the Formula of a Compound with a Polyatomic Anion

Baking powder contains calcium dihydrogen phosphate, an ionic compound composed of the ions Ca^{2+} and $H_2PO_4^-$. What is the formula of this compound?

Solution: The positive and negative charges must balance, and this ionic compound must be electrically neutral. Thus we must have two negative charges to balance the $+2$ charge of the calcium ion. This requires a ratio of one Ca^{2+} ion to two $H_2PO_4^-$ ions. We designate this by enclosing the formula for the dihydrogen phosphate ion in parentheses and adding a subscript 2. Thus the formula is $Ca(H_2PO_4)_2$. If we wanted to show the ionic charges, we would write $Ca^{2+}(H_2PO_4^-)_2$.

Check your learning: Predict the formula of the ionic compound formed between the magnesium ion and the phosphate ion, PO_4^{3-}. (*Hint:* Use the periodic table to predict the sign and the charge on the magnesium ion.)

Answer: $Mg_3(PO_4)_2$

Ionic compounds do not contain molecules. In almost all cases, we use empirical formulas for the formulas of ionic compounds.

Covalent Compounds.

Many compounds do not contain ions but instead consist of atoms bonded tightly together in molecules that result when atoms *share* electrons instead of transferring them from one atom to another. (Recall that molecules are uncharged groups of atoms that can behave as single discrete units.) Compounds made up of molecules are called **covalent compounds.** We can identify many covalent compounds from simple physical properties. They exist as gases, low-boiling liquids, and low-melting solids. The bonds that hold the atoms together in such molecules are called **covalent bonds.** We will consider covalent bonding and covalent compounds in considerable detail in Chapters 6 and 7.

Whereas ionic compounds are usually formed when a metal and a nonmetal combine, **covalent compounds are usually formed by a combination of nonmetals.** Thus the periodic table can help us recognize many of the compounds that are covalent. At this point in our study, we can use the positions of elements in the periodic table to determine whether compounds are ionic or covalent.

Example 2.9   Prediction of the Type of Bonding in Compounds

Predict whether the following compounds are ionic or covalent.
(a) KI, the compound used as a source of "iodine" in table salt
(b) H_2O_2, the bleach and disinfectant hydrogen peroxide
(c) $CHCl_3$, the anesthetic chloroform
(d) Li_2CO_3, a source of lithium in antidepressants

Solution:
(a) Potassium (Group 1A) is a metal, and iodine (Group 7A) is a nonmetal; KI is predicted to be ionic.
(b) Hydrogen (Group 1A and Group 7A) is a nonmetal, and oxygen (Group 6A) is a nonmetal; H_2O_2 is predicted to be covalent.
(c) Carbon (Group 4A) is a nonmetal, hydrogen (Group 1A and Group 7A) is a nonmetal, and chlorine (Group 7A) is a nonmetal; $CHCl_3$ is predicted to be covalent.
(d) Lithium (Group 1A) is a metal, and carbon (Group 4A) and oxygen (Group 6A) are nonmetals, Li_2CO_3 is predicted to be ionic. In each case, the compounds are ionic or covalent as predicted.

Check your learning: Using the periodic table, predict whether the following compounds are ionic or covalent. **(a)** SO_2, **(b)** CaF_2, **(c)** N_2H_4, **(d)** $Al_2(SO_4)_3$.

Answer: **(a)** covalent, **(b)** ionic, **(c)** covalent, **(d)** ionic.

2.8   Chemical Equations

When the atoms, molecules, or ions in one set of substances regroup to form other substances, we use a shorthand type of expression called a **chemical equation** to describe the chemical change. Symbols and formulas (Section 2.3) are used in equations to describe all the substances involved and their compositions. Thus chemical equations can describe behavior in both the macroscopic and the microscopic domains. Consider the reaction of methane with oxygen to form carbon dioxide and water, shown graphically in Fig. 2.20. The equation for this reaction is

$$CH_4 + 2O_2 \longrightarrow CO_2 + 2H_2O \tag{1}$$

Figure 2.20

Molecules of methane, CH_4, and oxygen, O_2, react to produce molecules of carbon dioxide, CO_2, and water, H_2O. The chemical equation for the reaction indicates the lowest whole-number ratio of atoms and molecules involved in the reaction. (A) This drawing of the microscopic domain illustrates how the atoms are redistributed during the reaction, beginning with the minimum number of molecules that can react. (B) Even when a larger number of molecules is involved, the ratio of reactants and products is still that shown by the chemical equation.

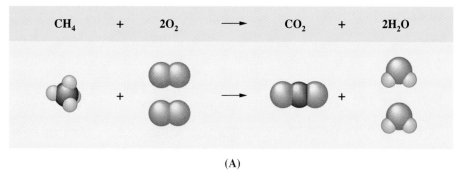

(A)

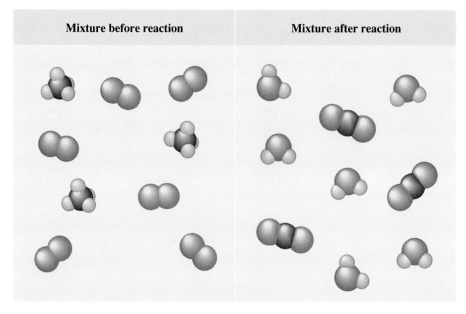

(B)

This equation tells us that methane (natural gas) and oxygen combine to produce carbon dioxide and water (macroscopic domain). The equation also indicates that for each one molecule of methane that reacts, two molecules of oxygen react and one molecule of carbon dioxide and two molecules of water are produced (microscopic domain).

Several features are common to most chemical equations:

1.   Chemical equations usually give the simplest whole-number ratio of molecules, atoms, or ions involved in a chemical change using numbers, called **coefficients,** that precede each symbol or formula in the equation. A coefficient of 1 is not written. In Equation 1, the coefficient before the formulas for the oxygen and water molecules is 2. The coefficient before CH_4 and CO_2 is 1, so it is not written. The subscripts in the formulas indicate the number of atoms of a specific type in the compound; they were explained in Section 2.3.

2.   An arrow separates substances to the left, which are called **reactants,** the substances present before any change occurs, and substances on the right, which are called **products,** the substances produced by the change. The arrow is read "to give," "to produce," "to yield," or "to form." A plus sign on the left side of an equation means "reacts with"; a plus sign on the right side is read "and." Thus Equation 1 can be read "One molecule of methane and two molecules of oxygen are reactants that form one molecule of the product carbon dioxide and two molecules of the product water."

3.   Because the law of conservation of matter states that matter can be neither created nor destroyed in a chemical reaction (Section 1.2), a chemical equation must be **balanced.** That is, the same number of atoms must be in the products as in the reactants. (Notice that in Equation 1, and Fig. 2.20(A), both sides of the arrow show one carbon atom, four hydrogen atoms, and four oxygen atoms.)

In actual reactions, vast numbers of each kind of molecule are involved, but the equation indicates the simplest ratio. When methane reacts with oxygen (burns in air) the reactants always combine in the ratio of one CH_4 molecule to two O_2 molecules to yield products in the ratio of one CO_2 molecule to two H_2O molecules. Figure 2.20(B) shows a reaction involving three CH_4 molecules. The equation for this reaction is still the one that gives the simplest ratios:

$$CH_4 + 2O_2 \longrightarrow CO_2 + 2H_2O$$

Balancing Equations.

Every equation must be balanced to show that the reactants and the products have (1) the same number of atoms and (2) the same number of net charges.

Consider the following statement: When water is decomposed by an electric current, hydrogen and oxygen are formed (Fig. 1.12). The macroscopic change could be expressed by the chemical equation

$$H_2O \longrightarrow H_2 + O_2 \qquad \text{(Equation is unbalanced.)}$$

However, this equation does not describe the microscopic domain correctly, because it does not show the same number of atoms in both the reactant and the products. Although it shows two atoms of hydrogen in both reactant and products, it shows one atom of oxygen in the reactant and two atoms of oxygen in the products. One molecule of water, which contains only one atom of oxygen, cannot be rearranged into a product that contains two atoms of oxygen. Two molecules of water are needed to provide two atoms of oxygen, but providing them also supplies a total of four atoms of hydrogen, which would appear as two hydrogen molecules. Thus we need to adjust the coefficients on each side of the arrow to write an equation that correctly describes the chemical change:

$$2H_2O \longrightarrow 2H_2 + O_2 \qquad \text{(Equation is balanced.)}$$

We should remember two important points when we balance an equation.

1. We balance an equation by changing the coefficients, not by changing the subscripts. The subscripts in a formula, such as the 2 in H_2O, cannot be changed to make an equation balance. The subscripts indicate the composition of each substance. If we change a subscript, we have a formula for a different substance.
2. Balancing equations by inspection is a trial-and-error process. We may have to try several sets of coefficients before we find the correct ones. Generally we start with the most complex formula and try giving it a coefficient of 1. Then we look at the equation and adjust the coefficients so that equal numbers of atoms of each type are present on both sides of the arrow.

Let's write an equation for the reaction of aluminum, Al, with hydrogen chloride, HCl, which produces hydrogen gas, H_2, and aluminum trichloride, $AlCl_3$, an important catalyst used to enhance the reactivity of hydrocarbons in the petrochemical industry. First we write an equation showing reactants and products:

$$Al + HCl \longrightarrow AlCl_3 + H_2 \qquad \text{(Equation is unbalanced.)}$$

Then we check the number of atoms of each type on each side to see whether the equation is balanced. The preceding equation shows one Al, one H, and one Cl atom on the left, but the right has one Al, two H, and three Cl atoms; the equation is not balanced. We start balancing with the substance that has the most atoms, $AlCl_3$. To get one unit of $AlCl_3$, we must have three Cl atoms from HCl. Therefore, we place the coefficient 3 in front of HCl:

$$Al + 3HCl \longrightarrow AlCl_3 + H_2 \qquad \text{(Equation is still unbalanced.)}$$

Still, the hydrogen atoms are not balanced. Three H atoms in three HCl molecules would yield one and one-half H_2 molecules:

$$Al + 3HCl \longrightarrow AlCl_3 + 1\tfrac{1}{2}H_2 \qquad \text{(Equation is balanced but not ideally.)}$$

Because we cannot have fractions of a molecule, we generally use whole numbers as coefficients in chemical equations. Therefore, we multiply each coefficient by 2 to obtain a balanced equation with the simplest whole-number coefficients:

$$2Al + 6HCl \longrightarrow 2AlCl_3 + 3H_2 \qquad \text{(Equation is balanced with smallest whole-number coefficients.)}$$

There are two Al, six H, and six Cl atoms on each side of the balanced question.

Other Information in Equations.
As part of its description of the macroscopic domain, an equation may indicate the *state* of reactants and products. Sodium metal is not found free in nature because solid sodium, Na, reacts with liquid water, H_2O, to give hydrogen gas, H_2, plus an aqueous (water) solution of sodium hydroxide, NaOH. The equation for this reaction is

$$2Na(s) + 2H_2O\ (l) \longrightarrow H_2(g) + 2NaOH(aq)$$

The letters in parentheses in this equation tell us something about the state of a reactant or product. A solid substance is indicated by adding (*s*) to the symbol or formula for that substance; a liquid substance is indicated by adding (l); a gas, by adding (g); and a substance that is dissolved in water, by adding (aq).

Special conditions that characterize the reaction may be written above or below the arrow. The electrolysis of water considered in this section is sometimes written as

$$2H_2O(l) \xrightarrow{\text{Elect.}} 2H_2(g) + O_2(g)$$

The abbreviation Elect. over the arrow stands for "electricity" or "electrolysis" and indicates that the reaction occurs when the energy of an electric current is added. A reaction carried out by heating may be indicated by a triangle (Δ) over the arrow:

$$CaCO_3(s) \xrightarrow{\Delta} CaO(s) + CO_2(g)$$

2.9   Equations for Ionic Reactions

Different methods of writing chemical equations are useful in different ways. Ionic equations are used to show which substances dissolve or react and release ions and which compounds do not form ions in solution. Net ionic equations are used to focus on the ions that actually react during a chemical change.

When ionic species dissolve in water, their ions move about independently in the solution. For example, a solution of the ionic compound sodium chloride consists of a mixture of freely moving Na^+ ions and Cl^- ions. In a complete **ionic equation,** ionic species in solution are written as ions; an ionic substance that does not dissolve (an insoluble ionic substance) is written as its empirical formula; and a molecular covalent compound that retains its form as a molecule in solution is written as its molecular formula. That is, in an ionic equation, each substance is written in its predominant form in the reaction solution. The simple equation for the reaction that occurs when solutions of the ionic compounds NaCl and $AgNO_3$ are mixed, giving a solution of the ionic compound $NaNO_3$ and solid AgCl, is

$$NaCl(aq) + AgNO_3(aq) \longrightarrow NaNO_3(aq) + AgCl(s)$$

The ionic equation is

$$\underbrace{Na^+(aq) + Cl^-(aq)}_{\substack{\text{Solution} \\ \text{of NaCl}}} + \underbrace{Ag^+(aq) + NO_3^-(aq)}_{\substack{\text{Solution} \\ \text{of AgNO}_3}} \longrightarrow \underbrace{Na^+(aq) + NO_3^-(aq)}_{\substack{\text{Solution} \\ \text{of NaNO}_3}} + \underbrace{AgCl(s)}_{\substack{\text{Solid} \\ \text{AgCl}}}$$

Sometimes we only write the species that change during a reaction; doing so gives a **net ionic equation.** In the previous equation, $Na^+(aq)$ and $NO_3^-(aq)$ are both present in the solutions of reactants and of products, but neither takes part in the reaction. Both $Na^+(aq)$ and $NO_3^-(aq)$ are present in the same form in the solutions of reactants and in the products.

Ions that appear in exactly the same form on both sides of an equation are called **spectator ions.** Spectator ions are omitted when we write a net ionic equation. The ions $Ag^+(aq)$ and $Cl^-(aq)$ take part in the reaction, combining to form solid silver chloride, AgCl(s). The net ionic equation, therefore, is

$$Ag^+(aq) + Cl^-(aq) \longrightarrow AgCl(s) \qquad (2)$$

This net ionic equation tells us that silver ions from *any* soluble ionic silver compound will combine with chloride ions from *any* soluble ionic chloride compound to form solid silver chloride. The full molecular or ionic equation tells us what *specific* substances furnish the reacting ions in a particular reaction.

One final rule is important. **For an ionic equation, the sum of total ion charges on each side, as well as the number of each kind of atom on each side, must balance.** Review Equation 2; the sum of the charges is the same on both sides. Here the sum is zero, but this is not always the case. The sums of the charges on both sides of an ionic equation must be equal.

Example 2.10   Net Ionic Equations

Baking powders contain a mixture of baking soda (sodium hydrogen carbonate, commonly called bicarbonate of soda, $NaHCO_3$) and an acid such as calcium dihydrogen phosphate, $Ca(H_2PO_4)_2$. When the powder is dry, no reaction occurs. As soon as water is added, the dihydrogen phosphate anion reacts with the hydrogen carbonate anion to form the monohydrogen phosphate anion $HPO_4{}^{2-}$, carbon dioxide, and water. The gaseous CO_2 is trapped in bread dough and causes the dough to rise. Write the net ionic equation for this reaction.

Solution: We recognize the sodium ion as Na^+ and the calcium ion as Ca^{2+}. This leads to the conclusion that $NaHCO_3$ contains the HCO_3 group with a -1 charge and $Ca(H_2PO_4)_2$ contains the H_2PO_4 group with a -1 charge (Section 2.6). Alternatively, we could remember the charges from the information in Table 2.4. The net ionic equation is

$$HCO_3{}^-(aq) + H_2PO_4{}^-(aq) \longrightarrow HPO_4{}^{2-}(aq) + CO_2(g) + H_2O(l)$$

An examination of the equation shows us that the atoms balance and the charges balance, with a -2 total charge on each side of the equation. The sodium and calcium ions are spectator ions.

Check your learning: Calcium hydrogen carbonate, $Ca(HCO_3)_2$, is soluble in water. If an aqueous solution of this ionic compound liberates carbon dioxide, solid calcium carbonate, $(CaCO_3)$ is formed as well as water. This reaction accounts for the formation of stalactites and stalagmites in caves. Write the net ionic equation.

Answer: $Ca^{2+}(aq) + 2HCO_3{}^-(aq) \longrightarrow CaCO_3(s) + CO_2(g) + H_2O(l)$

2.10   Classification of Chemical Reactions

Just as it is convenient to classify the elements as metals and nonmetals, it is convenient to classify chemical reactions. Several common types of reactions are discussed below. However, note that the classifications overlap and that some reactions fall into more than one class.

1. **Addition Reactions.** An **addition reaction,** or *combination reaction,* occurs when two or more substances combine to form another substance.

   $S(s) + O_2(g) \longrightarrow SO_2(g)$ (an atmospheric pollutant)

   $2Mg(s) + O_2(g) \longrightarrow 2MgO(s)$ (a solid used as a high-temperature electrical insulator)

   The reactions described by these two equations form a covalent compound and an ionic compound, respectively.

2. **Decomposition Reactions.** A **decomposition reaction** occurs when one compound breaks down (decomposes) into two or more substances.

$$2HgO(s) \longrightarrow 2Hg(l) + O_2(g) \qquad \text{(Hg is commonly used in thermometers.)}$$

$$CaCO_3(s) \longrightarrow CaO(s) + CO_2(g) \qquad \text{(CaO is sometimes called quicklime.)}$$

3. **Metathesis Reactions.** A **metathesis reaction,** or *double displacement reaction,* is a reaction in which two compounds exchange parts—usually ions. The most common type of metathesis reaction occurs because an insoluble product, a **precipitate,** is formed. For example, mixing aqueous solutions of the ionic solids calcium chloride and silver nitrate forms the white, insoluble solid silver chloride, a light-sensitive component of photographic film. This metathesis reaction can be represented as

$$CaCl_2(aq) + 2AgNO_3(aq) \longrightarrow 2AgCl(s) + Ca(NO_3)_2(aq)$$

Many metathesis reactions are written using ionic equations or net ionic equations. The ionic equation and the net ionic equation for the reaction of $CaCl_2$ with $AgNO_3$ are

Ionic: $\quad Ca^{2+}(aq) + 2Cl^-(aq) + 2Ag^+(aq) + 2NO_3^-(aq) \longrightarrow$
$$2AgCl(s) + Ca^{2+}(aq) + 2NO_3^-(aq)$$

Net Ionic: $\quad 2Cl^-(aq) + 2Ag^+(aq) \longrightarrow 2AgCl(s)$

Dividing each coefficient by 2 gives the lowest whole-number coefficients and yields

$$Cl^-(aq) + Ag^+(aq) \longrightarrow AgCl(s)$$

4. **Combustion Reactions.** A **combustion reaction** is a reaction of an element or a compound with oxygen that produces heat and, often, light. (This type of reaction is actually a special case of what is known as an oxidation–reduction reaction, which we will consider in Chapter 19.) We have already seen one combustion reaction, the reaction of methane and oxygen, in Fig. 2.20 and Equation 1 (Section 2.8). Other familiar combustion reactions involve carbon, compounds of carbon and hydrogen, or compounds of carbon, hydrogen, and oxygen. The products of these typical combustion reactions are carbon dioxide and water (if the reactants contain hydrogen).

   Combustion reactions of gasoline, heating oil, natural gas, and coal power our cars, heat our buildings, and produce much of our electricity. Each year we employ them to provide more and more useful energy. However, this increasing use also has a negative effect, because these reactions release carbon dioxide, CO_2, into the atmosphere. The increasing amounts of CO_2 in the atmosphere have a greenhouse effect, retaining heat close to the earth and slowly increasing the warmth of the earth to abnormal levels.

5. **Neutralization Reactions.** A **neutralization reaction** is a reaction between a solution of an acid and a solution of a base; it produces a salt and water. There are a number of ways to identify acids and bases. In this chapter, we will use definitions that involve the behavior that acids and bases exhibit when they dissolve in water.

Acids are compounds that react with water and increase the amount of **hydronium ion, H_3O^+,** present in the solution. When pure, most common acids (Table 2.5) are covalent compounds of hydrogen combined with one or more nonmetals—for example, HCl and H_2SO_4. Pure acids do not contain hydronium ions; they produce these

Table 2.5

Some Common Laboratory Acids and Bases, with the Products That Result When They Dissolve in Water

Formula of Acid or Base	Name of Compound	Ions Produced in Water
Acids		
HCl	Hydrogen chloride	$H_3O^+(aq) + Cl^-(aq)$
HNO_3	Nitric acid	$H_3O^+(aq) + NO_3^-(aq)$
H_2SO_4	Sulfuric acid	$H_3O^+(aq) + HSO_4^-(aq)$
CH_3CO_2H	Acetic acid	$H_3O^+(aq) + CH_3CO_2^-(aq)$
Bases		
NaOH	Sodium hydroxide	$Na^+(aq) + OH^-(aq)$
KOH	Potassium hydroxide	$K^+(aq) + OH^-(aq)$
$Ca(OH)_2$	Calcium hydroxide	$Ca^{2+}(aq) + 2OH^-(aq)$
NH_3	Ammonia	$NH_4^+(aq) + OH^-(aq)$

Figure 2.21

When gaseous hydrogen chloride dissolves in water, covalent HCl molecules (A), each composed of one H atom (blue) and one chlorine atom (green), react with water molecules to give a solution (B) of hydronium ions, H_3O^+, and chloride ions, Cl^-.

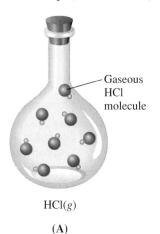

Gaseous HCl molecule

HCl(g)

(A)

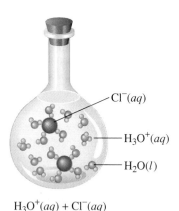

$Cl^-(aq)$

$H_3O^+(aq)$

$H_2O(l)$

$H_3O^+(aq) + Cl^-(aq)$

(B)

ions when they dissolve in, and react with, water. Pure HCl is a gas. When it dissolves in water, it reacts and produces hydronium ions and chloride ions (Fig. 2.21).

$$HCl(g) + H_2O(l) \longrightarrow H_3O^+(aq) + Cl^-(ag)$$

In some texts, the hydronium ion is written $H^+(aq)$.

Bases are compounds that dissolve in water or react with water and increase the amount of **hydroxide ion, OH^-,** in solution. The most common bases (Table 2.5) are ionic compounds that are composed of Group 1A or Group 2A metal cations combined with the hydroxide ion—for example, NaOH and $Ca(OH)_2$. When these compounds dissolve, hydroxide ions are released directly into the solution. For example, KOH dissolves and produces potassium ions, K^+, and hydroxide ions, OH^-.

$$KOH(s) \xrightarrow{H_2O} K^+(aq) + OH^-(aq)$$

Other bases produce small amounts of hydroxide ion by reacting with water. An example is ammonia, NH_3, which reacts with water to form the ammonium ion, NH_4^+, and the hydroxide ion. We will discuss both acids and bases in detail in Chapter 15.

A **salt** is an ionic compound that can be formed by the reaction of an acid with a base. When soluble salts dissolve in water, the ions separate and are free to move about independently (Fig. 2.22).

The general equation for a neutralization reaction in solution is

$$\text{Acid} + \text{base} \longrightarrow \text{salt} + \text{water}$$

An example is

$$HCl(aq) + KOH(aq) \longrightarrow KCl(aq) + H_2O(l)$$

We can understand this reaction by recognizing that hydronium ions produced by the HCl combine with hydroxide ions produced by the KOH, forming water and a solution of a salt composed of the ions K^+ and Cl^-. The complete ionic equation for the reaction is

$$H_3O^+(aq) + Cl^-(aq) + K^+(aq) + OH^-(aq) \longrightarrow 2H_2O(l) + K^+(aq) + Cl^-(aq)$$

The net ionic equation for the reaction is

$$H_3O^+(aq) + OH^-(aq) \longrightarrow 2H_2O(l)$$

NaCl(s)

(A)

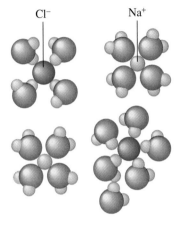

Cl^- Na^+

$Na^+(aq) + Cl^-(aq)$

(B)

Figure 2.22

The dissolution of the ionic salt sodium chloride in water. (A) The arrangement of the Na^+ (gray) and Cl^- ions (green) in the solid state. (B) The distribution of ions in an aqueous (water) solution of NaCl. Each ion is free to move independently because surrounding water molecules separate ions of opposite charge.

Example 2.11	Classifying Chemical Reactions

Classify each of the following chemical reactions according to type.
(a) $ZnCO_3(s) \longrightarrow ZnO(s) + CO_2(g)$
(b) $2Ga(l) + 3Br_2(l) \longrightarrow 2GaBr_3(s)$
(c) $Ca(OH)_2(aq) + 2HBr(aq) \longrightarrow CaBr_2(aq) + 2H_2O(l)$
(d) $BaCl_2(aq) + K_2SO_4(aq) \longrightarrow BaSO_4(s) + 2KCl(aq)$
(e) $C_2H_4(g) + 3O_2(g) \longrightarrow 2CO_2(g) + 2H_2O(l)$

Solution:
(a) Decomposition reaction
(b) Addition or combination reaction
(c) Neutralization reaction
(d) Metathesis reaction
(e) Combustion reaction

Check your learning: Write a balanced chemical equation demonstrating each of the following reaction types: (a) combustion of a compound containing only carbon and hydrogen, (b) addition, (c) decomposition, (d) metathesis, (e) combination.

Answer: See equations in Section 2.10.

2.11   Naming Inorganic Compounds (Nomenclature)

Nomenclature, a collection of rules for naming, is crucial to a science. In this section we will concentrate on naming **binary compounds,** compounds that contain two different kinds of atoms. Examples include NaCl, $CaBr_2$, and Li_3N. Of course, many compounds contain more than two kinds of atoms, and we will introduce rules that will give us a start at naming some of these compounds—compounds containing three different kinds of atoms, or **ternary compounds.** Compounds that contain oxygen as the third kind of atom will be named here. In this chapter we will name only *inorganic compounds,* compounds that are composed principally of elements other than carbon, using nomenclature proposed by the IUPAC. The rules for *organic compounds,* which contain bonded carbon atoms, will be described in Chapter 9.

To name an inorganic compound, we need to consider the answers to several questions: Is the compound ionic or covalent? If the compound is ionic, does the metal involved form ions of more than one type (variable charge) or of only one type (fixed charge)? Is the nonmetallic anion a simple one with a fixed charge? Finally, is the compound an acid, and if so, does the acid contain oxygen? From the answers we can assign an inorganic compound to one of the following four categories and name it accordingly.

1. **Binary Ionic Compounds Containing a Metal Ion (with a Fixed Charge) and a Nonmetal.** Group 1A and Group 2A metals form compounds that always contain metal ions with a charge of $+1$ and $+2$, respectively. There are a few other metal ions that also have a fixed ionic charge. Common examples of cations and anions are listed below (also see Fig. 2.17).

Name	Group	Charge
Alkali metals	1A	+1
Alkaline earth metals	2A	+2
Zinc (Zn)	2B	+2
Cadmium (Cd)	2B	+2
Aluminum (Al)	3A	+3
Chalcogens	6A	−2
Halogens	7A	−1

The name of a binary compound formed when one of these metals combine with a nonmetal consists of the name of the cation (the name of the metal) followed by the name of the anion (the name of the nonmetallic element with its ending replaced by the suffix -ide). Some examples are

NaCl, sodium chloride Na_2O, sodium oxide

KBr, potassium bromide   CdS, cadmium sulfide

CaI_2, calcium iodide Mg_3N_2, magnesium nitride

CsF, cesium fluoride Ca_3P_2, calcium phosphide

LiCl, lithium chloride Al_4C_3, aluminum carbide

2. **Binary Ionic Compounds Containing a Metal Ion (with a Variable Charge) and a Nonmetal.** With the exceptions noted in the category above, most of the metals in Groups 3B through 5A can form two or more cations with different charges. Compounds of these metals with nonmetals are named just like compounds in the first category, but the charge on the metal ion is specified by a Roman numeral in parentheses after the name of the metal. We determine the charge on the metal ion from the formula of the compound and the charge on the anion.

$FeCl_2$, iron (II) chloride

$FeCl_3$, iron (III) chloride

Hg_2O, mercury (I) oxide

HgO, mercury (II) oxide

CuS, copper (II) sulfide

In the past, we would have used the suffixes -ic and -ous to designate the metals with the higher and lower charges, respectively: $FeCl_3$ would have been ferric chloride, and $FeCl_2$ ferrous chloride. This is no longer the method of choice, but it is still used occasionally. For example, you may see the words *stannous fluoride* on a tube of toothpaste. This represents the formula SnF_2, which is more properly named tin(II) fluoride. The other fluoride of tin is SnF_4, which has been called stannic fluoride but should be named tin(IV) fluoride.

Example 2.12   Naming Ionic Compounds

Name the following ionic compounds, which contain a metal that can have more than one ionic charge: **(a)** Fe_2S_3, **(b)** CuSe, **(c)** GaN, **(d)** $CrCl_3$.

Solution: The anions in these compounds have a fixed negative charge (S^{2-}, Se^{2-}, N^{3-}, and Cl^{-}), and the compounds must be neutral. Because the total number of positive charges in each compound must equal the total number of negative charges, the positive ions must be Fe^{3+}, Cu^{2+}, Ga^{3+}, and Cr^{3+}. These charges are used in the names of the metal ions: **(a)** iron(III) sulfide, **(b)** copper(II) selenide, **(c)** gallium(III) nitride, **(d)** chromium(III) chloride.

Check your learning: Now write the formulas of the following ionic compounds. (a) chromium(III) phosphide, (b) mercury(II) sulfide, (c) manganese(II) fluoride, (d) copper(I) oxide.

Answer: **(a)** CrP, **(b)** HgS, **(c)** MnF_2; **(d)** Cu_2O.

3. **Binary Covalent Compounds Containing Two Nonmetals.** The name consists of the name of the less nonmetallic element (the element that is farther to the left or farther toward the bottom of the periodic table) followed by the name of the more nonmetallic element (farther to the right or farther toward the top of the periodic table). As in the foregoing two categories, the second-named nonmetal (the more nonmetallic element) is given the suffix *-ide*. The number of atoms of each nonmetal in the formula is designated by the following Greek prefixes:

mono-	one (sometimes omitted)
di-	two
tri-	three
tetra-	four
penta-	five
hexa-	six
hepta-	seven
octa-	eight
nona-	nine
deca-	ten

When only one atom of the first element is present, the prefix *mono-* is usually deleted. When two vowels are adjacent, the *a* in the Greek prefix is dropped. Some examples are

CO, carbon monoxide	SO_2, sulfur dioxide
CO_2, carbon dioxide	SO_3, sulfur trioxide
NO_2, nitrogen dioxide	BCl_3, boron trichloride
N_2O_4, dinitrogen tetroxide	PF_5, phosphorus pentafluoride
N_2O_5, dinitrogen pentoxide	IF_7, iodine heptafluoride

There are a few common names that you will encounter as you continue your study of chemistry. For example, although some chemists call NO nitric oxide, its proper name is nitrogen monoxide. Similarly, N_2O is known as nitrous oxide even though our rules would specify the name dinitrogen monoxide.

| Example 2.13 | Naming Covalent Compounds |

Name the following covalent compounds: **(a)** SF_6, **(b)** N_2O_3 **(c)** Cl_2O_7, **(d)** P_4O_6.

Solution: Because these compounds consist solely of nonmetals, we use Greek prefixes to designate the number of atoms of each element. **(a)** sulfur hexafluoride, **(b)** dinitrogen trioxide, **(c)** dichlorine heptoxide, **(d)** tetraphosphorus hexoxide.

Check your learning: Write the formulas for the following compounds: **(a)** phosphorus pentachloride, **(b)** dinitrogen monoxide, **(c)** iodine heptafluoride, **(d)** carbon tetrachloride.

Answer: **(a)** PCl_5, **(b)** N_2O, **(c)** IF_7, **(d)** CCl_4.

4. **Binary Acids and Ternary Oxyacids.** A pure binary acid is named like a binary compound that contains a metal and nonmetal (the first category above), with hydrogen named first. For example, the compound HCl is named hydrogen chloride. When a binary acid is dissolved in water, it is given a different name: *hydrogen* is changed to the prefix *hydro-*, and the *-ide* suffix is changed to *-ic*. The word *acid* is added. A solution of hydrogen chloride in water is called hydrochloric acid. The other common binary acids and their solutions are given below, where (*g*) is used to indicate the pure compound in its gaseous state, and (*aq*) an aqueous (water) solution.

HF(*g*), hydrogen fluoride	HF(*aq*), hydrofluoric acid
HCl(*g*), hydrogen chloride	HCl(*aq*), hydrochloric acid
HBr(*g*), hydrogen bromide	HBr(*aq*), hydrobromic acid
HI(*g*), hydrogen iodide	HI(*aq*), hydroiodic acid
H_2S(*g*), hydrogen sulfide	H_2S(*aq*), hydrosulfuric acid

Acids in which hydrogen is combined with oxygen in addition to other nonmetals are called **oxyacids.** These are ternary compounds. The pure compounds are named as acids, as are aqueous solutions of the compound. To name these acids, we use the root name of the anion. We replace the *-ate* suffix of the anion

Table 2.6

Names of Common Oxyacids

Formula	Anion Name	Acid Name
$HC_2H_3O_2$	Acetate	Acetic acid
HNO_3	Nitrate	Nitric acid
HNO_2	Nitrite	Nitrous acid
$HClO_4$	Perchlorate	Perchloric acid
H_2CO_3	Carbonate	Carbonic acid
H_2SO_4	Sulfate	Sulfuric acid
H_2SO_3	Sulfite	Sulfurous acid
H_3PO_4	Phosphate	Phosphoric acid

with an -ic suffix, or the -ite suffix with -ous, and add the word acid. Thus in H_2SO_4, the -ate of sulfate is replaced with -ic, and the name is sulfuric acid. In HNO_2, the anion is nitrite. The -ite is replaced with -ous, and the name is nitrous acid. The names of common oxyacids are given in Table 2.6.

For Review | Summary

Dalton's atomic theory states that all elements consist of extremely small particles called atoms. With the possible exception of their masses, all atoms of an element are identical, but they have different properties from atoms of other elements. Atoms of two or more elements combine in fixed proportions in a chemical reaction to form a compound. The **law of constant composition** states that all samples of a particular pure compound contain the same elements in the same proportion by mass. The **law of multiple proportions** states that when two elements react to form more than one compound, a fixed mass of one element will react with masses of the other element that are in a ratio of small whole numbers. The mass of one atom is usually expressed in **atomic mass units (amu)** and is referred to as the **atomic mass.** An amu is defined as exactly $\frac{1}{12}$ of the mass of a carbon-12 atom and is equal to 1.6605×10^{-24} g.

An atom contains a small, positively charged nucleus surrounded by electrons. The nucleus, which consists of protons and neutrons, is on the order of 100,000 times smaller than the atom. **Protons** are relatively heavy particles with a charge of +1 and a mass of 1.0073 amu. **Neutrons** are relatively heavy particles with no charge and a mass of 1.0087 amu, very close to that of protons. **Electrons** are light particles with a charge of −1 and a mass of 0.00055 amu. The number of protons in the nucleus is called the **atomic number.** The sum of the numbers of protons and neutrons in the nucleus is called the **mass number** and, expressed in atomic mass units, is approximately equal to the mass of the atom. Because an atom is neutral, the number of electrons in an atom must be equal to the number of protons.

A chemical formula identifies the atoms in a substance using **symbols,** which are one-, two-, or three-letter abbreviations for the atoms. Subscripts in formulas indicate the relative numbers of the different atoms. A **molecular formula** indicates the exact number of each type of atom in a molecule. An **empirical formula** gives the simplest whole-number ratio of atoms in a substance. A **structural formula** indicates the bonding arrangement of the atoms in the molecule.

Isotopes of an element are atoms with the same atomic number but different mass numbers; isotopes of an element, therefore, differ from each other only in the number of neutrons within the nucleus. When a naturally occurring element is composed of several isotopes, the atomic mass of the element represents the average of the masses of the isotopes involved.

The discovery of the periodic recurrence of similar properties among the elements led to formulation of the **periodic table,** in which the elements are arranged in order of increasing atomic number in rows known as **periods** and columns known as **groups.** Elements in the same group of the periodic table have similar chemical properties. Elements can be classified as metals, semi-metals and nonmetals, or as **main-group elements, transition metals,** and **inner transition metals.**

Groups are numbered, from left to right, either 1 through 18, or 1 through 8 along with the letter A or B. The elements in Group 1A are known as the **alkali metals;** those in Group 2A are the **alkaline earth metals;** those in 6A the **chalcogens;** those in 7A the **halogens;** and those in 8A the **noble gases.**

Metals (particularly those in Groups 1A and 2A) tend to lose the number of electrons that would leave them with the same number of electrons as occur in the preceding noble gas in the periodic table. By this means, a positively charged **ion** is formed. Similarly, nonmetals (especially those in Groups 6A and 7A, and to a lesser extent those in Group 5A) can gain the number of electrons needed to provide the atoms with the same number of electrons as occur in the *next* noble gas in the periodic table. Thus nonmetals tend to form negative ions. Positively charged ions are called **cations,** and negatively charged ions are called **anions.** Ions can be either **monatomic** (containing only one atom) or **polyatomic** (containing more than one atom).

Compounds that contain ions are called **ionic compounds.** Ionic compounds generally form from metals and nonmetals. Compounds that do not contain ions, but instead consist of atoms bonded tightly together in molecules (uncharged groups of atoms that behave as a single unit), are called **covalent compounds.** Covalent compounds usually form from two nonmetals.

When atoms, molecules, or ions regroup to form other substances, we use a shorthand type of expression called a **chemical equation** to describe the chemical reaction. Because matter can be neither created nor destroyed in a chemical reaction, the same number of atoms of each type must appear among both the reactants and the products of a reaction. The **coefficients** in a balanced chemical equation indicate the relative numbers of atoms, molecules, and ions involved in the reaction.

Most chemical reactions can be classified as **addition reactions, decomposition reactions, metathesis reactions, combustion reactions,** or **neutralization reactions.**

Key Terms and Concepts

acid (2.10)
addition reaction (2.10)
alkali metal (2.5)
alkaline earth metal (2.5)
anion (2.6)
atomic mass (2.2)
atomic mass unit, amu (2.2)
atomic number (2.2)
balanced equation (2.8)
base (2.10)
binary compound (2.11)
cation (2.6)
chalcogen (2.5)
chemical equation (2.8)
coefficients (2.8)
combustion reaction (2.10)
covalent bond (2.7)
covalent compound (2.7)
Dalton's atomic theory (2.1)
decomposition reaction (2.10)

electron (2.2)
empirical formula (2.3)
halogen (2.5)
hydronium ion (2.10)
hydroxide ion (2.10)
ion (2.6)
ionic bond (2.7)
ionic compound (2.7)
ionic equation (2.9)
inner transition metal (2.5)
isotope (2.4)
law of constant composition (2.1)
law of definite proportion (2.1)
law of multiple proportions (2.1)
main-group element (2.5)
mass number (2.2)
metathesis reaction (2.10)
molecular formula (2.3)
monatomic ion (2.6)
net ionic equation (2.9)

neutralization reaction (2.10)
neutron (2.2)
noble gas (2.5)
nomenclature (2.11)
nucleus (2.2)
oxyacid (2.11)
periodic law (2.5)
polyatomic ion (2.6)
precipitate (2.10)
products (2.8)
proton (2.2)
reactants (2.8)
representative element (2.5)
salt (2.10)
spectator ions (2.9)
structural formula (2.3)
symbol (2.3)
transition metal (2.5)

Exercises

Questions

1. Which of the following are properties of one atom of the metal lead? (a) It can combine with other atoms to form a molecule of a compound of lead. (b) It can be converted into foil or drawn into a wire. (c) It can be melted to make molten lead.

2. The $^{18}O/^{16}O$ ratio in some meteorites is greater than that used to calculate the average atomic mass of oxygen on earth. Is the average mass of an oxygen atom in these meteorites greater than, less than, or equal to that of a terrestrial oxygen atom?

3. Explain why the symbol for an atom of the element oxygen and the formula for a molecule of oxygen differ.

4. Explain why the symbol for an atom of the element sulfur and the formula for a molecule of sulfur differ.

5. Which of the following is the reason why we balance chemical equations? (a) To show that matter is composed of small indivisible particles called atoms. (b) To show that an element consists of one type of atom. (c) To show that compounds contain atoms of two or more different types. (d) To show that atoms are neither created nor destroyed in chemical reactions but are merely rearranged.

6. The existence of isotopes violates one of the original ideas of Dalton's atomic theory. Which one?

7. Which of the following statements is true of all the neutral atoms of an element? (a) They are identical in every way. (b) They have the same nuclear masses. (c) They have the same number of neutrons. (d) They have the same number of protons. (e) They have the same number of electrons.

8. How do isotopes of the same element always differ? (a) In number of neutrons. (b) In number of protons. (c) In number of electrons. (d) In the ionic charge. (e) In none of these.

9. Which of the following *always* differs between an Fe^{2+} ion and an Fe^{3+} ion? (a) The number of electrons. (b) The number of protons. (c) The number of neutrons. (d) All of these. (e) None of these.

10. Use the following equations to answer the next five questions.

    I. $H_2O(s) \longrightarrow H_2O(l)$
    II. $Na^+(aq) + Cl^-(aq) + Ag^+(aq) + NO_3^-(aq) \longrightarrow AgCl(s) + Na^+(aq) + NO_3^-(aq)$
    III. $CH_3OH(g) + O_2(g) \longrightarrow CO_2(g) + H_2O(g)$

IV. $2H_2O(l) \longrightarrow 2H_2(g) + O_2(g)$
V. $H^+(aq) + OH^-(aq) \longrightarrow H_2O(l)$

(a) Which equation describes a physical change?
(b) Which equation identifies the reactants and products of a combustion reaction?
(c) Which equation is not balanced?
(d) Which is a net ionic equation?
(e) Which equation describes a decomposition reaction?

Atomic Theory

11. In the following drawing, the symbol ○ represents one atom of a certain type. The symbol ● represents a single atom of another type. If the symbols for atoms touch, they are part of a single molecule. The following chemical change represented by these symbols may violate one of the ideas of Dalton's atomic theory. Which one?

OO + ●●   ⟶   ●○

(starting materials)   (products of the change)

12. The following drawing shows a chemical change. The symbol ○ represents one atom of a certain type, and the symbol ● represents a single atom of another type. If the symbols for atoms touch, they are part of a single molecule. The reaction represented by these symbols may violate one of the ideas of Dalton's atomic theory. Which one?

○ + ●●   ⟶   ●○ + ●○

(starting materials)   (products of the change)

13. Which postulate of Dalton's theory is consistent with (explains) the following observation about the weights of reactants and products: "When 100 grams of solid calcium carbonate is heated, 44 grams of carbon dioxide and 56 grams of calcium oxide are produced."

14. Identify the postulate of Dalton's theory that is violated by the following observations: "59.95% of one sample of titanium dioxide is titanium; 60.10% of a different sample of titanium dioxide is titanium."

15. Give the number of protons, electrons, and neutrons in each of the following isotopes: (a) $^{10}_{5}B$, (b) $^{199}_{80}Hg$, (c) $^{63}_{29}Cu$, (d) $^{13}_{6}C$, (e) $^{77}_{34}Se$.

16. Give the number of protons, electrons, and neutrons in each of the following isotopes: (a) $^{7}_{3}Li$, (b) $^{125}_{52}Te$, (c) $^{109}_{47}Ag$, (d) $^{15}_{7}N$, (e) $^{31}_{15}P$.

17. Atoms of a halogen have the following natural abundances and isotopic masses:

Abundance	Isotopic Mass
75.53%	34.9688 amu
24.47%	36.9659 amu

Calculate the average atomic mass of this halogen.

18. Atoms of an element have the following natural abundances and isotopic masses:

Abundance	Isotopic Mass
90.92%	19.99 amu
0.26%	20.99 amu
8.82%	21.99 amu

Calculate the average atomic mass of this element.

Symbols and Formulas

19. Name the following elements, whose symbols are based on Latin words.

Na, K, Fe, Hg, Ag, Cu, Pb, Sb, Sn, Au

20. Name the elements whose names are based on the names of scientists:

Cm, Es, Fm, Md, No

and those whose names are based on the names of planets, continents, countries or states.

Eu, Cf, Fr, U, Pu

21. Eating compounds that contain atoms of the following elements is considered essential for good health: calcium, chlorine, iodine, magnesium, phosphorus, sodium, zinc. What are the symbols of these elements?

22. Warnings of the toxicity of several elements have been reported in the media. Examples include cadmium, mercury, lead, barium, beryllium, and chromium. What are the symbols of these elements?

23. Write the molecular and empirical formulas of the following compounds.

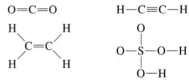

24. Write the molecular and empirical formulas of the following compounds.

25. Determine the empirical formula for the following compounds.

(a) caffeine, $C_8H_{10}N_4O_2$
(b) fructose, $C_{12}H_{22}O_{11}$
(c) hydrogen peroxide, H_2O_2
(d) glucose, $C_6H_{12}O_6$
(e) ascorbic acid (vitamin C), $C_6H_8O_6$

26. Determine the empirical formula for the following compounds.

    (a) acetic acid, $C_2H_4O_2$
    (b) citric acid, $C_6H_8O_7$
    (c) hydrazine, N_2H_4
    (d) nicotine, $C_{10}H_{14}N_2$
    (e) butane, C_4H_{10}

27. Write a symbol for each of the following neutral isotopes. Include the atomic number and mass number for each.

    (a) the alkali metal with 11 protons and a mass number of 23
    (b) the noble gas with 75 neutrons in its nucleus and 54 electrons in the neutral atom
    (c) the isotope with 33 protons and 40 neutrons in its nucleus
    (d) the alkaline earth metal with 88 electrons and 138 neutrons

28. Write a symbol for each of the following neutral isotopes. Include the atomic number and mass number for each.

    (a) the chalcogen with a mass number of 125
    (b) the halogen whose longest-lived isotope is radioactive
    (c) the noble gas, used in lighting, with 10 electrons and 10 neutrons
    (d) the lightest alkali metal with 3 neutrons

The Periodic Table

29. Using the periodic table, predict whether the following chlorides are ionic or covalent: KCl, NCl_3, ICl, $MgCl_2$, PCl_5, CCl_4.

30. Using the periodic table, predict whether the following chlorides are ionic or covalent, $SiCl_4$, PCl_3, $CaCl_2$, $CsCl$, $CuCl_2$, $CrCl_3$.

31. Using the periodic table, identify the lightest member of each of the following groups.

    (a) noble gases
    (b) alkaline earth metals
    (c) alkali metals
    (d) chalcogens

32. Using the periodic table, identify the heaviest member of each of the following groups.

    (a) alkali metals
    (b) chalcogens
    (c) noble gases
    (d) alkaline earth metals

33. Using the periodic table, classify each of the following elements as a metal or a nonmetal, and then further classify each as a main-group (representative) element, a transition metal, or an inner transition metal.

    (a) uranium   (b) bromine
    (c) strontium   (d) neon
    (e) gold   (f) americium

(g) rhodium   (h) sulfur
(i) carbon   (j) potassium

34. Using the periodic table, classify each of the following elements as a metal or a nonmetal, and then further classify each as a main-group (representative) element, a transition metal, or an inner transition metal.

    (a) cobalt   (b) europium
    (c) iodine   (d) indium
    (e) lithium   (f) oxygen
    (g) beryllium   (h) cadmium
    (i) terbium   (j) rhenium

35. Use the periodic table to give the name and symbol for each of the following elements.

    (a) the noble gas in the same period as germanium
    (b) the alkaline earth metal in the same period as selenium
    (c) the halogen in the same period as lithium
    (d) the chalcogen in the same period as cadmium

36. Use the periodic table to give the name and symbol for each of the following elements.

    (a) the halogen in the same period as the alkali metal with 11 protons
    (b) the alkaline earth metal in the same period with the neutral noble gas with 18 electrons
    (c) the noble gas in the same row as an isotope with 30 neutrons and 25 protons
    (d) the noble gas in the same period as gold

Molecules and Ions

37. For each of the following compounds, (a) state whether it is ionic or covalent and (b) if it is ionic, write the symbols for the ions involved: NF_3, BaO, $(NH_4)_2CO_3$, $Sr(H_2PO_4)_2$, IBr, Na_2O.

38. For each of the following compounds, (a) state whether it is ionic or covalent and (b) if it is ionic, write the symbols for the ions involved: $KClO_4$, $MgC_2H_3O_2$, H_2S, Ag_2S, N_2Cl_4, $Co(NO_3)_2$.

39. Write the formula of each of the following compounds.

    (a) rubidium bromide   (b) magnesium selenide
    (c) sodium oxide   (d) calcium chloride
    (e) hydrogen fluoride   (f) gallium phosphide
    (g) aluminum bromide   (h) ammonium sulfate

40. Write the formula of each of the following compounds.

    (a) lithium carbonate   (b) sodium perchlorate
    (c) barium hydroxide   (d) ammonium carbonate
    (e) sulfuric acid   (f) calcium acetate
    (g) magnesium phosphate   (h) sodium sulfite

41. Write the symbol for each of the following ions.

    (a) the alkali metal ion with a $+1$ charge and an atomic number of 55
    (b) the halide ion with 54 electrons
    (c) the anion with a mass number of 31 and a -3 charge

(d) the transition metal ion with 24 electrons and a +3 charge

42. Write the symbol for each of the following ions.

  (a) the Group 3A metal ion with a +3 charge, 28 electrons, and a mass number of 71

  (b) the anion with 36 electrons and 35 protons

  (c) the inner transition metal ion with 86 electrons and a +4 charge

  (d) the alkaline earth metal ion with a +2 charge and an atomic number of 38

Equations and Reactions

43. What is the difference between a chemical reaction and a chemical equation?

44. Describe the similarities and differences between an ionic equation and a net ionic equation.

45. Balance the following equations:

  (a) $PCl_5 + H_2O \longrightarrow POCl_3 + HCl$

  (b) $Cu + HNO_3 \longrightarrow Cu(NO_3)_2 + H_2O + NO$

  (c) $H_2 + I_2 \longrightarrow HI$

  (d) $Fe + O_2 \longrightarrow Fe_2O_3$

  (e) $Na + H_2O \longrightarrow NaOH + H_2$

  (f) $(NH_4)_2Cr_2O_7 \longrightarrow Cr_2O_3 + N_2 + H_2O$

  (g) $P_4 + Cl_2 \longrightarrow PCl_3$

  (h) $PtCl_4 \longrightarrow Pt + Cl_2$

46. Balance the following equations.

  (a) $Ag + H_2S + O_2 \longrightarrow Ag_2S + H_2O$

  (b) $P_4 + O_2 \longrightarrow P_4O_{10}$

  (c) $Pb + H_2O + O_2 \longrightarrow Pb(OH)_2$

  (d) $Fe + H_2O \longrightarrow Fe_3O_4 + H_2$

  (e) $Sc_2O_3 + SO_3 \longrightarrow Sc_2(SO_4)_3$

  (f) $Ca_3(PO_4)_2 + H_3PO_4 \longrightarrow Ca(H_2PO_4)_2$

  (g) $Al + H_2SO_4 \longrightarrow Al_2(SO_4)_3 + H_2$

  (h) $TiCl_4 + H_2O \longrightarrow TiO_2 + HCl$

47. Write a balanced equation that describes each of the following chemical reactions.

  (a) Solid calcium carbonate is heated and decomposes to white solid calcium oxide and carbon dioxide gas.

  (b) In a combustion reaction, butane, C_4H_{10}, burns in oxygen gas.

  (c) Aqueous solutions of magnesium chloride and sodium hydroxide react in a metathesis reaction to produce the solid antacid milk of magnesia and a solution of salt (sodium chloride).

  (d) Water vapor reacts with sodium metal to produce solid sodium hydroxide and hydrogen gas.

48. Write a balanced equation that describes each of the following chemical reactions.

  (a) Potassium chlorate, $KClO_3$, decomposes when heated to form potassium chloride and oxygen gas.

  (b) Aluminum metal reacts with elemental iodine in a combination reaction to form Al_2I_6.

  (c) When solid sodium chloride is added to liquid sulfuric acid, hydrogen chloride gas and solid sodium hydrogen sulfate are produced.

  (d) Solutions of phosphoric acid and potassium hydroxide react in a neutralization reaction with potassium dihydrogen phosphate as one of the products.

49. Indicate what type, or types, of reaction each of the following represents.

  (a) $Ca + Br_2 \longrightarrow CaBr_2$

  (b) $Ca(OH)_2 + 2HBr \longrightarrow CaBr_2 + 2H_2O$

  (c) $C_6H_{12} + 9O_2 \longrightarrow 6CO_2 + 6H_2O$

  (d) $CaCO_3 \longrightarrow CaO + CO_2$

  (e) $MgO + SiO_2 \longrightarrow MgSiO_3$

50. Indicate what type, of types, or reaction each of the following represents.

  (a) $H_2O + C \longrightarrow CO + H_2$

  (b) $SO_2 + H_2O \longrightarrow H_2SO_3$

  (c) $2KClO_3 \longrightarrow 2KCl + 3O_2$

  (d) $Al(OH)_3 + 3HBr \longrightarrow AlBr_3 + 3H_2O$

  (e) $Pb(NO_3)_2 + H_2SO_4 \longrightarrow PbSO_4 + 2HNO_3$

Naming Compounds

51. Name the following compounds: $CsCl$, BaO, K_2S, $BeCl_2$, HBr, AlF_3.

52. Name the following compounds: NaF, Rb_2O, BCl_3, H_2Se, P_4O_6, ICl_3.

53. Give the formula of the following compounds: chlorine dioxide, dinitrogen tetroxide, potassium phosphide, silver(I) sulfide, aluminum nitride, silicon dioxide.

54. Give the formula of the following compounds: barium chloride, magnesium nitride, sulfur dioxide, nitrogen trichloride, dinitrogen trioxide, tin(IV) chloride.

55. Each of the following compounds contains a metal that can exhibit more than one ionic charge. Name these compounds.

  (a) Cr_2O_3

  (b) $FeCl_2$

  (c) CrO_3

  (d) $TiCl_4$

  (e) CoO

  (f) MoS_2

56. Each of the following compounds contains a metal that can exhibit more than one ionic charge. Name these compounds.

  (a) $NiCO_3$

  (b) MoO_3

  (c) $Co(NO_3)_2$

  (d) V_2O_5

  (e) MnO_2

  (f) Fe_2O_3

57. The following ionic compounds are found in common household products. Write the formula for each compound.

    (a) potassium phosphate
    (b) copper(II) sulfate
    (c) calcium chloride
    (d) titanium dioxide
    (e) ammonium nitrate
    (f) sodium bisulfate (the common name for sodium hydrogen sulfate)

58. The following ionic compounds are found in common household products. Name each compound.

    (a) $Ca(H_2PO_4)_2$ (b) $FeSO_4$
    (c) $CaCO_3$ (d) MgO
    (e) $NaNO_2$ (f) KI

Applications and Additional Exercises

59. The following are properties of isotopes of two elements that are essential in our diet: (a) atomic number 26, mass number 58 and (b) atomic number 53, mass number 127. Determine the number of protons, neutrons, and electrons in each element, and name each element.

60. (a) Determine the number of protons, neutrons, and electrons in the following isotopes that are used in medical diagnoses: (i) atomic number 9, mass number 18; (ii) atomic number 43, mass number 99; (iii) atomic number 53, mass number 131; (iv) atomic number 81, mass number 201. (b) Name the elements in part (a).

61. Give the IUPAC names of the following compounds: manganese dioxide, mercurous chloride, Hg_2Cl_2; ferric nitrate, $Fe(NO_3)_3$; titanium tetrachloride; and cupric bromide, $CuBr_2$.

62. Calculate the average atomic mass of carbon, the isotopic composition of which is ^{12}C 98.89%, ^{13}C 1.11%. The atomic mass of ^{13}C has been determined to be 13.00335 amu.

63. Average atomic masses listed by IUPAC are based on a study of experimental results. Bromine has two isotopes, ^{79}Br and ^{81}Br, whose masses (78.9183 and 80.9163 amu) and abundances (50.537% and 49.46%) were determined in earlier experiments. Calculate the average atomic mass of bromine on the basis of these experiments and check it with the value inside the front cover.

64. The most recently determined values of isotopic masses and abundances of bromine are 78.9183 amu (50.69%) and 80.9163 amu (49.31%), respectively. Calculate the average atomic mass of bromine, and compare the result with the value inside the front cover.

65. Variations in atomic masses may occur because of isotope separation in industry. Lithium is an example. The isotopic composition of lithium from minerals is 7.5% ^6Li and 92.5% ^7Li. Their masses are 6.01512 amu and 7.01600 amu. A commercial source of lithium, recycled from a military

source, was 3.75% ^6Li. Calculate the difference in average atomic mass values from the two sources.

66. The average atomic masses of some elements may vary depending on the sources of their ores. The actual atomic mass of boron may vary from 10.807 to 10.819, depending on whether the mineral source is Turkey or the United States. The isotopic masses are known quite accurately (^{10}B, 10.0129 and ^{11}B, 11.0931). Calculate the percent abundances leading to the two values of the average atomic masses of boron from the two countries.

67. The list of ingredients on the box of a soluble fertilizer includes ammonium phosphate, potassium nitrate, copper(II) sulfate, and zinc sulfate. Write the formulas of these four compounds.

68. Colorful fireworks often involve the decomposition of barium nitrate and potassium chlorate and the reaction of the metals magnesium, aluminum, and iron with oxygen.

    (a) Write the formulas of barium nitrate and potassium chlorate.
    (b) The decomposition of potassium chlorate leads to the formation of potassium chloride and oxygen gas. Write an equation for this reaction.
    (c) The decomposition of barium nitrate leads to the formation of barium oxide, nitrogen gas, and oxygen gas. Write an equation for this reaction.
    (d) Write equations for the reactions of the metals magnesium, aluminum, and iron with oxygen.

69. What are the empirical formulas of the following compounds?

    (a)

    (b)

70. Fill in the blank.

71. (a) Aqueous hydrogen fluoride (hydrofluoric acid) is used to etch glass and to analyze minerals for their silicon content. Hydrogen fluoride will also react with sand (silicon dioxide). One of the products in each case is silicon tetrafluoride gas. Write an equation for the reaction of silicon dioxide with hydrofluoric acid.

(b) The insoluble mineral fluorite (calcium fluoride) occurs extensively in Illinois. Calcium fluoride can also be prepared by the reaction of a solution of calcium chloride with a solution of sodium fluoride. Write the complete ionic equation for the reaction.

(c) Write a net ionic equation for the reaction in part (b).

72. In the past, the principal source of one chemical fertilizer was a deposit of sodium nitrate in Chile. A series of discoveries has made it possible to manufacture an equivalent fertilizer from coal or natural gas, air, and water.

(a) Balance the equations for the following steps in the manufacture of the fertilizer ammonium nitrate.

   i. $H_2O(g) + C(s) \longrightarrow CO(g) + H_2(g)$

   ii. $CO(g) + H_2O(g) \longrightarrow CO_2(g) + H_2(g)$

   iii. $N_2(g) + H_2(g) \longrightarrow NH_3(g)$

   iv. $NH_3(g) + O_2(g) \longrightarrow NO(g) + H_2O(g)$

   v. $NO(g) + O_2(g) \longrightarrow NO_2(g)$

   vi. $NO_2(g) + H_2O(l) \longrightarrow HNO_3(aq) + NO(g)$

   vii. $NH_3(g) + HNO_3(aq) \longrightarrow NH_4NO_3(aq)$

(b) Name the compounds involved.

73. An ingenious process for obtaining magnesium from sea water involves several reactions. Write the chemical equation for each reaction.

(a) The first step is the decomposition of calcium carbonate from seashells to form calcium oxide.

(b) The second step is the formation of solid calcium hydroxide by the reaction of the oxide with water.

(c) Solid calcium hydroxide is added to the seawater, where magnesium is present as the chloride, precipitating magnesium hydroxide.

(d) The magnesium hydroxide is then dissolved in hydrochloric acid solution.

(e) Finally, magnesium metal and chlorine gas are produced by electrolysis of the molten chloride.

CHAPTER OUTLINE

3.1 Atomic and Molecular Masses
3.2 Moles of Atoms; Avogadro's Number
3.3 Moles of Molecules
3.4 Percent Composition from Formulas
3.5 Derivation of Empirical Formulas
3.6 Derivation of Molecular Formulas
3.7 Solutions
3.8 Dilution of Solutions

Stoichiometry and Chemical Change

3.9 Mole Relationships and Chemical Equations
3.10 Material Balances
3.11 Theoretical Yield, Actual Yield, and Percent Yield
3.12 Limiting Reactant

Quantitative Analysis

3.13 Titration
3.14 Gravimetric Analysis
3.15 Combustion Analysis

3

Chemical Stoichiometry

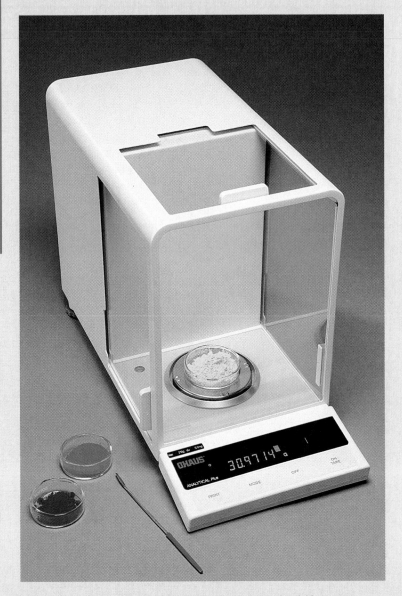

Weights and volumes are the experimental basis of chemical stoichiometry.

We turn to chemical stoichiometry for answers to many questions that ask "How much?" For example, we use stoichiometry to determine

- The fraction of the recommended daily allowance of copper contained in 1 milligram of copper(II) sulfate, $CuSO_4$
- The mass of platinum(II) chloride required to prepare a kilogram of the antitumor drug platinol, $Pt(NH_3)_2Cl_2$
- The amount of carbon dioxide, CO_2, a greenhouse gas, produced by combustion of 1 ton of coal
- The minimum amount of sodium azide, NaN_3, needed to inflate an air bag fully
- The most efficient scavenger to remove carbon dioxide from a space shuttle
- The concentration of acid in a sample of acid rain
- The fertilizer with the highest percent of nitrogen
- The protein content of a 4-ounce serving of fish
- The calorie content of a slice of pecan pie
- The purity of a sample of aspirin
- The fat content of a cereal

We can argue that modern chemistry started when scientists began asking how much. For example, Dalton's atomic theory (Section 2.1) was an attempt to explain the results of measurements that enabled him to calculate the relative masses of elements combined in various compounds.

In this chapter we shall consider how the masses of atoms, the chemical formulas of compounds, and quantities of matter are related and how chemical equations are used to compare quantities of substances involved in chemical changes. In future chapters, we will be concerned with the quantity of energy involved in chemical changes. These calculations fall into the category called **chemical stoichiometry**— the calculation of both material balances and energy balances in chemical systems.

Chemistry is a quantitative science based largely on experiments that involve calculation. However, the focus of these calculations is more than numbers: They provide quantitative predictions or descriptions of how matter behaves. Moreover, a correct calculation suggests that the theory underlying the calculation is correct. For example, our ability to calculate the volumes of gas samples by using the gas laws (Chapter 10) is one reason for believing that the theories that underlie these laws are correct. Calculations are tools that chemists use in understanding the behavior of matter, and chemical stoichiometry is one connection between the microscopic and macroscopic domains of chemistry.

3.1 Atomic and Molecular Masses

We have seen that the mass of a single atom in atomic mass units (amu) is approximately equal to the number of protons and neutrons in its nucleus (Section 2.2). However, atoms of the same element can differ in the number of neutrons in their nucleus (Section 2.4) and can have different masses. Thus we use the average mass of an atom in most situations.

For historical reasons, the term *atomic weight* is sometimes used instead of *atomic mass*. In that usage, an atomic weight is often expressed without units. For accuracy and currency, we will use **atomic mass** to refer to the average mass of an atom.

The average mass of a molecule is the sum of the average masses, in atomic mass units, of all the atoms in the molecule and is called the **molecular mass.** (The term *molecular weight* is also sometimes used, often without units.) The molecular formula (Section 2.3) of chloroform, the second anesthetic to be discovered and used in

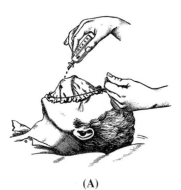

(A)

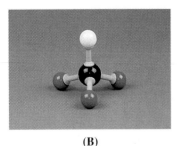

(B)

surgery (Fig. 3.1), is $CHCl_3$. This formula indicates that a chloroform molecule contains one carbon atom, one hydrogen atom, and three chlorine atoms. The (average) mass of a chloroform molecule is equal to the sum of the (average) atomic masses of these atoms. The molecular mass of an aspirin molecule, $C_9H_8O_4$ (Fig. 3.2), equals the sum of the atomic masses of nine carbon atoms, eight hydrogen atoms, and four oxygen atoms.

For $CHCl_3$:

$$1\ C\ = 1 \times 12.011\ amu =\ \ \ 12.011\ amu$$
$$1\ H\ = 1 \times\ \ 1.008\ amu =\ \ \ \ 1.008\ amu$$
$$3\ Cl = 3 \times 35.453\ amu = 106.359\ amu$$
$$\text{Molecular mass} = 119.378\ amu$$

For $C_9H_8O_4$:

$$9\ C = 9 \times 12.011\ amu = 108.099\ amu$$
$$8\ H = 8 \times\ \ 1.008\ amu =\ \ \ \ 8.064\ amu$$
$$4\ O = 4 \times 15.999\ amu =\ \ 63.996\ amu$$
$$\text{Molecular mass} = 180.159\ amu$$

Thus the average mass of a *single* $CHCl_3$ molecule is 119.378 amu, and that of a *single* molecule of $C_9H_8O_4$ molecule is 180.159 amu.

We cannot calculate the molecular mass of an ionic compound because an ionic compound does not contain molecules (Section 2.3). Instead, we calculate a **formula mass,** the sum of the atomic masses of the atoms found in one formula unit, as indicated by the empirical formula. Calculation of the formula masses of $CaCO_3$, an ionic compound commonly found in toothpaste as an abrasive, and of $Al_2(SO_4)_3$, an ionic compound used in water purification, follows.

For $CaCO_3$:

$$1\ Ca = 1 \times 40.078\ amu =\ \ 40.078\ amu$$
$$1\ C\ = 1 \times 12.011\ amu =\ \ 12.011\ amu$$
$$3\ O\ = 3 \times 15.999\ amu =\ \ 47.997\ amu$$
$$\text{Formula mass} = 100.086\ amu$$

For $Al_2(SO_4)_3$:

$$2\ Al =\ \ 2 \times 26.982\ amu =\ \ 53.964\ amu$$
$$3\ S\ =\ \ 3 \times 32.067\ amu =\ \ 96.201\ amu$$
$$12\ O = 12 \times 15.999\ amu = 191.988\ amu$$
$$\text{Formula mass} = 342.153\ amu$$

Note that atomic mass, molecular mass, and formula mass are quantities associated with the microscopic domain. The atomic mass unit is a microscopic unit that is used to measure the mass of single atoms, molecules, or formula units.

Figure 3.1

(A) This figure from an 1896 medical textbook shows one method used to administer the anesthetic chloroform during surgery. (B) A ball-and-stick model of a chloroform molecule, $CHCl_3$.

Figure 3.2

An aspirin tablet contains 325 mg of molecules of the type shown.

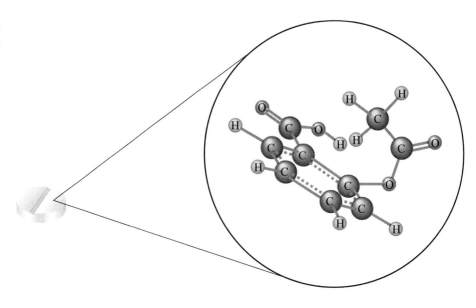

3.2 Moles of Atoms; Avogadro's Number

As we saw in Chapter 1, atoms are incredibly small and light. The mass of an average carbon atom, which is about 12 amu, corresponds to 2.0×10^{-23} g (0.000,000,000,000,000,000,000,020 g), and therefore it would take 5×10^{18} (5,000,000,000,000,000,000) atoms to equal the smallest mass that can be weighed on most laboratory balances, 10^{-4} g, or 0.0001 g. Thus when we work with enough of a substance to see, we are dealing with a tremendous number of atoms.

As noted in Chapter 2, a molecular formula tells us the number of atoms in a molecule. From this we can determine the relative numbers of atoms of each type that must be assembled to make a particular compound. However, we do not have to count out the actual number of atoms; we need only get the correct ratio of the numbers of atoms.

Carbon disulfide is a liquid that is used as one of the starting materials in the manufacture of rayon and cellophane. The formula of carbon disulfide, CS_2, tells us that one molecule contains one carbon atom and two sulfur atoms. Carbon disulfide can be prepared by the reaction of coke (essentially pure carbon) with the element sulfur at temperatures above 750°C. To make CS_2 we simply need to use enough carbon and sulfur so that we have twice as many sulfur atoms as carbon atoms.

We use the SI unit for amount of substance—the mole—to report numbers of atoms. One **mole** (1 mol) of atoms of any element contains the same number of atoms as there are in exactly 12 grams of the isotope $^{12}_{6}C$. Many experiments have shown that the number of atoms in exactly 1 mole of atoms of any element is 6.022×10^{23} atoms (to four significant figures). This number is called **Avogadro's number** in honor of the Italian professor of physics Amedeo Avogadro (1776–1856) whose studies of the properties of gases helped lay the foundations for clearing up much of the confusion that initially surrounded atomic masses.

Combining 1 mole of carbon atoms (1 mol of C atoms = 6.022×10^{23} C atoms) with 2 moles of sulfur atoms ($2 \times 6.022 \times 10^{23}$ S atoms) would produce 1 mole (6.022×10^{23} molecules) of carbon disulfide. We could make a smaller sample of CS_2 by using fewer moles of carbon and sulfur atoms (smaller numbers of C and S atoms), as long as there were two times as many moles of sulfur atoms as moles of carbon atoms.

A mole is a macroscopic unit. The number of atoms in a mole of atoms is so large that it is difficult to appreciate how big it really is. As a guide to the size of this number, consider that if the entire population of the United States (approximately 250,000,000 people) spent 12 hours a day, 365 days a year, counting atoms at a rate of 1 atom per second, they would need about 153 million years to count the atoms in a mole of atoms.

Fortunately, we do not have to count out the atoms necessary to make a sample of a compound; we can use a balance to get the number of atoms that we want. **The mass of 1 mole of atoms in grams is numerically equal to the atomic mass of the atom.** This mass is called the **molar mass.** Each of the following is a molar mass and contains 1 mole of atoms: 19 grams of fluorine, 31 grams of phosphorus, and the mass of each of the samples in Fig. 3.3. All of these samples contain the same number of atoms (6.022×10^{23} atoms) because a mole of atoms of any element contains the same number of atoms as a mole of atoms of any other element.

Now we can see that it is easy to count out atoms in the laboratory. Suppose, for example, that we have 32.1 grams of sulfur atoms and want an equal number of zinc atoms in order to make zinc sulfide, ZnS. From the atomic mass of sulfur, 32.1 amu,

Figure 3.3

Each sample contains 6.02×10^{23} atoms—1.00 mol of atoms (atomic masses are given in parentheses). Clockwise from sulfur, the yellow solid: 32.1 g of sulfur (32.1 amu), 65.4 g of zinc (65.4 amu), 12.0 g of carbon (12.0 amu), 24.3 g of magnesium (24.3 amu), 207 g of lead (207 amu), 28.1 g of silicon (28.1 amu), 63.5 g of copper (63.5 amu), and, in the center, 201 g of mercury (201 amu).

we can recognize that 32.1 grams of sulfur atoms is 1 mole of sulfur atoms. To get 1 mole of zinc atoms, we look up the atomic mass of zinc (65.4 amu) and then weigh out 1 mole of zinc atoms, 65.4 grams. Finding fractional amounts of moles just requires an extra arithmetical step.

We can indicate the amount of a substance using either moles or grams as units. The arithmetic involved with conversions among moles and grams is similar to that introduced in Section 1.13 for the conversion of other units. We can use either ratios or unit conversion factors to do this arithmetic.

Example 3.1   Conversion of Grams to Moles Using Ratios

The estimated daily requirement of 2.0 g of potassium may be obtained by eating potatoes and fruits—particularly bananas, watermelon, and peaches. What is the estimated daily requirement of potassium in moles of atoms?

Solution: This example requests the conversion of a quantity of potassium atoms in units of grams to units of moles.

The unknown quantity is the number of moles of potassium atoms; call it x mol K. The ratio of this quantity to 2.0 g of K atoms is x mol K/2.0 g K. From the atomic mass of potassium (39.098) we obtain the ratio between 1 mol of K atoms and the molar mass (1 mol K/39.098 g K). Equating these two ratios with the unit *mol K* in the numerators on both sides and with the unit *g K* in the denominators gives

$$\frac{x \text{ mol K}}{2.0 \text{ g K}} = \frac{1 \text{ mol K}}{39.098 \text{ g K}}$$

Now we solve the equation in order to find the number of moles of K atoms in 2.0 g.

$$x \text{ mol K} = 2.0 \text{ g K} \times \frac{1 \text{ mol K}}{39.098 \text{ g K}}$$

$$= 0.051 \text{ mol K}$$

A mass of 2.0 g of potassium atoms corresponds to 0.051 mol of potassium.

Check your learning: Use the ratio of 1 mol of chlorine atoms to the molar mass of chlorine and determine the moles of chlorine atoms in 0.6783 g of chlorine.

Answer: 0.01913 mol Cl

Figure 3.4

A ring with a 1-carat diamond.

Example 3.2   Conversion of Grams to Moles Using Unit Conversion Factors

A colorless diamond essentially is pure carbon. What quantity of carbon atoms in moles is contained in a 1.0-carat (0.20-g) diamond (Fig. 3.4)?

Solution: This example requires the conversion of a quantity of carbon atoms in units of grams to units of moles.

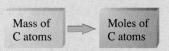

We can use the atomic mass of carbon (12.011) to obtain the unit conversion factors 1 mol C/12.011 g C and 12.011 g C/1 mol C. Then we determine the number of moles using the equation

$$x \text{ mol C} = 0.20 \text{ g C} \times \text{unit conversion factor}$$

The correct unit conversion factor is the ratio that cancels the units of grams and leaves moles.

$$x \text{ mol C} = 0.20 \text{ g C} \times \frac{1 \text{ mol C}}{12.011 \text{ g C}}$$

$$= 0.17 \text{ mol C}$$

Remember that applying a unit conversion factor is simply a shortcut to the second step used in converting units with ratios (Examples 1.5, 1.6, and 3.1).

Check your learning: Use a conversion factor and determine the moles of silicon atoms in 163.45 g of silicon.

Answer: 5.8196 mol Si

Example 3.3   **Conversion of Moles to Grams**

A liter of dry air contains about 9.2×10^{-4} mol of atoms of the element argon. What is the mass of argon in a liter of air?

Solution: This example requires the conversion of a quantity of argon atoms in units of moles to units of grams.

Moles of Ar atoms → Mass of Ar atoms

We can use the atomic mass of argon (39.948) to write the ratio 39.948 g Ar/1 mol Ar, which can be used to make the conversion by means of either a ratio or a unit conversion factor.

A. Using a Ratio

$$\frac{x \text{ g Ar}}{9.2 \times 10^{-4} \text{ mol Ar}} = \frac{39.948 \text{ g Ar}}{1 \text{ mol Ar}}$$

$$x \text{ g Ar} = 9.2 \times 10^{-4} \text{ mol Ar} \times \left(\frac{39.948 \text{ g Ar}}{1 \text{ mol Ar}} \right)$$

$$= 0.037 \text{ g Ar}$$

B. Using a Unit Conversion Factor

$$x \text{ g Ar} = 9.2 \times 10^{-4} \text{ mol Ar} \times \text{unit conversion factor}$$

$$= 9.2 \times 10^{-4} \text{ mol Ar} \times \left(\frac{39.948 \text{ g Ar}}{1 \text{ mol Ar}} \right)$$

$$= 0.037 \text{ g Ar}$$

Either method tells us that 9.2×10^{-4} mol of Ar atoms has a mass of 0.037 g.

Check your learning: Using either ratios or a unit conversion factor, determine the mass of 2.561 mol of gold.

Answer: 504.4 g Au

Example 3.4   Calculating Moles of Atoms and Number of Atoms

Copper wire is an example of the element copper. How many moles of copper atoms and how many copper atoms are contained in 5.00 g of copper wire?

Solution: This example requires the conversion of a quantity of copper metal in units of grams to units of moles, followed by conversion to the number of atoms.

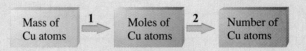

The numbers above the arrows refer to the numbered conversion steps below. Although we will illustrate the use of unit conversion factors, each step could be carried out using ratios, as in Example 3.1 or 3.3.

Step 1.   *Convert the mass of copper to moles of copper using the molar mass of copper (63.546 g/mol) as the unit conversion factor.*

$$\text{Moles of Cu} = 5.00 \text{ g Cu} \times \frac{1 \text{ mol Cu}}{63.546 \text{ g Cu}} = 0.0787 \text{ mol Cu}$$

Step 2.   *Convert the moles of copper to the number of atoms of copper using Avogadro's number (6.022×10^{23}) in the unit conversion factor.*

$$\text{Number Cu atoms} = 0.0787 \text{ mol Cu} \times \frac{6.022 \times 10^{23} \text{ Cu atoms}}{1 \text{ mol Cu}}$$

$$= 4.74 \times 10^{22} \text{ atoms}$$

Check your learning: A sample of sulfur contains 1.50×10^{24} S atoms. How many moles of sulfur atoms are in the sample? What is its mass?

Answer: 2.49 mol S; 79.9 g

3.3 Moles of Molecules

If we have a sample that contains 6.022×10^{23} molecules, we have a *mole of molecules*—the same number of molecules as there are atoms in exactly 12 grams of $^{12}_{6}C$. The **mass of one mole of molecules in grams is numerically equal to the molecular mass of the molecule.** This mass is sometimes called the **molar mass** of the substance. The molecular mass of chloroform, $CHCl_3$, is 119.377 amu; the molar mass is 119.377 grams per mole. Because a mole of molecules contains Avogadro's number of molecules, 119.377 grams of $CHCl_3$ contains 6.022×10^{23} molecules.

The molar mass of an ionic compound is the mass in grams numerically equal to the formula mass. The formula masses of the ionic compounds calcium carbonate and aluminum sulfate are 100.086 amu and 342.153 amu, respectively (Section 3.1); therefore, their molar masses are 100.086 grams per mole and 342.153 grams per mole, respectively.

A mole of an ionic compound contains 6.022×10^{23} of the units described by the formula. A mole of $CaCO_3$ therefore contains Avogadro's number of $CaCO_3$ units, each unit being composed of one Ca^{2+} ion and one CO_3^{2-} ion. A mole of $Al_2(SO_4)_3$ contains Avogadro's number of $Al_2(SO_4)_3$ units, each unit being composed of two Al^{3+} ions and three SO_4^{2-} ions. Each of the samples in Fig. 3.5 contains 1 mole.

We said earlier that a chemical formula indicates the number of atoms of each element in one molecular or formula unit of that compound (Section 2.3). For example, ethanol, C_2H_5OH, contains two carbon atoms in one molecule. *A chemical formula also indicates the number of moles of atoms in 1 mole of the compound.* Thus we also know that 2 moles of carbon atoms are contained in 1 mole of ethanol molecules ($2 \times 6.022 \times 10^{23}$ C atoms in 6.022×10^{23} C_2H_5OH molecules).

Figure 3.5

Each sample contains 6.02×10^{23} molecules or formula units—1.00 mol of the compound or element. Clockwise from the upper left: 130.2 g of $C_8H_{17}OH$ (1-octanol, formula mass 130.2 amu), 454.9 g of HgI_2 (mercury(II) iodide, formula mass 459.9 amu), 32.0 g of CH_3OH (methanol, formula mass 32.0 amu) and 256.5 g of S_8 (sulfur, formula mass 256.5 amu).

Example 3.5   Converting Grams of Molecules to Moles

Our bodies synthesize protein from amino acids. One of these amino acids is glycine, which has the molecular formula $C_2H_5O_2N$. How many moles of glycine molecules are contained in 28.35 g (1 oz) of glycine?

Solution: This example requires the conversion of a quantity of glycine molecules in units of grams to units of moles.

$$\boxed{\text{Mass of } C_2H_5O_2N} \longrightarrow \boxed{\text{Moles of } C_2H_5O_2N}$$

In order to make this conversion, we must determine the molecular mass of glycine as follows:

$$
\begin{aligned}
2\,C &= 2 \times 12.011 = 24.022 \text{ amu}\\
5\,H &= 5 \times \ 1.008 = \ 5.040 \text{ amu}\\
2\,O &= 2 \times 15.999 = 31.998 \text{ amu}\\
1\,N &= 1 \times 14.007 = 14.007 \text{ amu}\\
\hline
\text{Molecular mass} &= 75.067 \text{ amu}
\end{aligned}
$$

The mass of one $C_2H_5O_2N$ molecule is 75.067 amu, so 1 mol of $C_2H_5O_2N$ has a mass of 75.067 g. Now we use this to determine a ratio or a unit conversion factor to convert from grams to moles. We will illustrate the use of a unit conversion factor.

$$\text{Moles of } C_2H_5O_2N = \text{mass of } C_2H_5O_2N \times \text{unit conversion factor}$$

$$= 28.35 \ \cancel{\text{g}} \ C_2H_5O_2N \times \frac{1 \text{ mol } C_2H_5O_2N}{75.067 \ \cancel{\text{g}} \ C_2H_5O_2N}$$

$$= 0.3777 \text{ mol } C_2H_5O_2N$$

28.35 of glycine is the mass of 0.3777 mol of glycine.

Check your learning: How many moles of sucrose, $C_{12}H_{22}O_{11}$, are contained in 25 g of this sugar?

Answer: 0.073 mol $C_{12}H_{22}O_{11}$

Example 3.6   Converting Moles of Molecules to Grams

Vitamin C is a covalent compound with the molecular formula $C_6H_8O_6$. The minimum daily dietary allowance of vitamin C for a young woman of average weight is 4.6×10^{-4} mol. What is this allowance in grams?

Solution: This example requires the conversion of a quantity of vitamin C molecules in units of moles to units of grams.

Calculating the molecular mass of $C_6H_8O_6$ gives 176.124 amu, so 1 mol $C_6H_8O_6$ = 176.124 g $C_6H_8O_6$. Use of a unit conversion factor gives

$$\text{Mass of } C_6H_8O_6 = \text{moles of } C_6H_8O_6 \times \text{unit conversion factor}$$

$$= 4.6 \times 10^{-4} \ \overline{\text{mol } C_6H_8O_6} \times \frac{176.124 \text{ g } C_6H_8O_6}{1 \ \overline{\text{mol } C_6H_8O_6}}$$

$$= 0.081 \text{ g } C_6H_8O_6$$

Check your learning: What is the mass of 0.443 mol of hydrazine, N_2H_4?

Answer: 14.2 g

Example 3.7   Calculating Number of Moles and Number of Molecules

A packet of an artificial sweetener contains 40 mg of saccharin, which has the structural formula

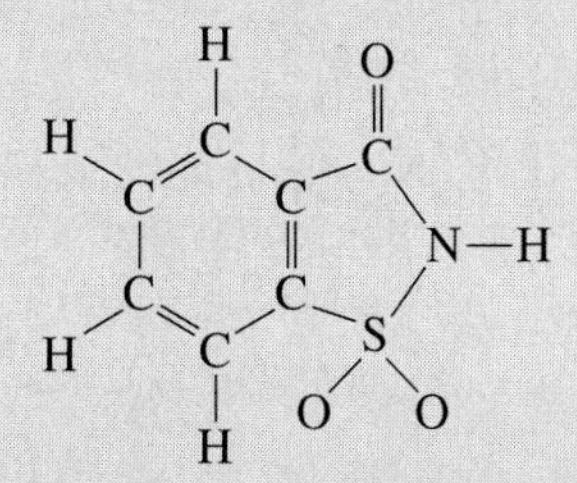

How many moles of saccharin molecules are contained in 40 mg (0.040 g) of saccharin? How many molecules? How many carbon atoms?

Solution: We need to convert the quantity of saccharin in units of grams to units of moles and then determine the number of molecules and the number of carbon atoms in the sample.

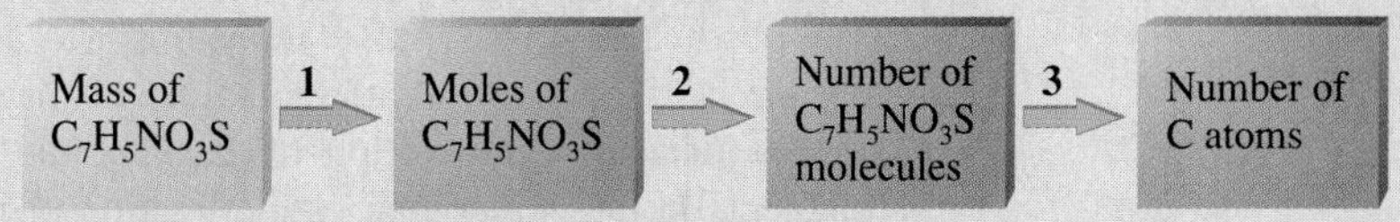

The numbers above the arrows refer to the conversion steps below.

Step 1.   *Convert the grams of saccharin to moles of saccharin using the molar mass (183.187 g/mol) in the unit conversion factor.* The molar mass is determined from the molecular formula $C_7H_5NO_3S$, which we find by counting the atoms in the structural formula.

$$\text{Moles of } C_7H_5NO_3S = 0.040 \ \overline{\text{g } C_7H_5NO_3S} \times \frac{1 \text{ mol } C_7H_5NO_3S}{183.187 \ \overline{\text{g } C_7H_5NO_3S}}$$

$$= 2.2 \times 10^{-4} \text{ mol } C_7H_5NO_3S$$

Step 2. *Convert the moles of $C_7H_5NO_3S$ to the number of molecules of $C_7H_5NO_3S$ using Avogadro's number (6.022×10^{23}) in the unit conversion factor.*

$$\text{Number of molecules} = 2.2 \times 10^{-4} \overline{\text{mol } C_7H_5NO_3S} \times \frac{6.022 \times 10^{23} \text{ molecules}}{1 \overline{\text{mol } C_7H_5NO_3S}}$$

$$= 1.3 \times 10^{20} \text{ molecules}$$

Step 3. *Convert the number of molecules of $C_7H_5NO_3S$ to the number of atoms of C using a ratio determined from the molecular formula, 7 C atoms/1 $C_7H_5NO_3S$ molecule, as the unit conversion factor.*

$$\text{Number of C atoms} = \text{number of } C_7H_5NO_3S \text{ molecules} \times \text{unit conversion factor}$$

$$= 1.3 \times 10^{20} \overline{C_7H_5NO_3S \text{ molecules}} \times \frac{7 \text{ C atoms}}{1 \overline{C_7H_5NO_3S \text{ molecule}}}$$

$$= 9.1 \times 10^{20} \text{ C atoms}$$

Alternatively, the number of carbon atoms could have been determined with the following sequence of conversions:

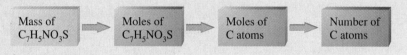

Mass of $C_7H_5NO_3S$ → Moles of $C_7H_5NO_3S$ → Moles of C atoms → Number of C atoms

Check your learning: How many moles of C_4H_{10} molecules are contained in 9.213 g of this compound? How many molecules? How many hydrogen atoms?

Answer: 0.1585 mol C_4H_{10}; 9.545×10^{22} molecules; 9.545×10^{23} H atoms

3.4   Percent Composition from Formulas

Years of experimentation have shown that all samples of a pure compound contain the same elements in the same proportion by mass. For example, 11% of the mass of any sample of pure water is always hydrogen; 89% of the mass is always oxygen. The observation of definite composition was a part of the evidence that led Dalton to outline his theory of the atomic nature of matter (Section 2.1).

The **percent composition** of a compound is the percent by mass of the various elements in the compound, or the fraction of the total mass of a sample of the compound that is due to a component element, multiplied by 100.

$$\text{Percent by mass of X} = \frac{\text{mass of X}}{\text{mass of sample}} \times 100$$

Example 3.8   Calculation of Percent Composition

Aspirin is a compound with the molecular formula $C_9H_8O_4$. What is its percent composition?

Solution: To calculate the percent composition (the percents by mass), we need to know the masses of C, H, and O in a known mass of $C_9H_8O_4$. It is convenient to work with 1 mol of $C_9H_8O_4$ because the formula tells us that 1 mol contains 9 mol C, 8 mol H, and 4 mol O atoms. We use the molar mass (180.159 g/mol, determined from the formula mass calculated in Section 3.1) to calculate the percentages.

$$\%C = \frac{9 \times \text{molar mass of C}}{\text{molar mass of } C_9H_8O_4} \times 100$$

$$= \frac{9 \times 12.011 \text{ g}}{180.159 \text{ g}} \times 100 = 60.002\%$$

$$\%H = \frac{8 \times \text{molar mass of H}}{\text{molar mass of } C_9H_8O_4} \times 100$$

$$= \frac{8 \times 1.008 \text{ g}}{180.159 \text{ g}} \times 100 = 4.476\%$$

$$\%O = \frac{4 \times \text{molar mass of O}}{\text{molar mass of } C_9H_8O_4} \times 100$$

$$= \frac{4 \times 15.999 \text{ g}}{180.159 \text{ g}} \times 100 = 35.522\%$$

Note that these percentages total 100%. One easy way to check a calculation of percent composition is to determine whether the percentages of the elements add up to 100%.

Check your learning: Calculate the percent composition of water to two significant figures, and check the values given in this section.

The percent composition of a compound is sometimes used to verify its identity or to determine its empirical formula, as will be described in the next section.

Figure 3.6

Gallium arsenide is used in fabrication of red light-emitting diodes in some calculators.

3.5 Derivation of Empirical Formulas

Chemical formulas are part of the symbolic domain of chemistry (Section 1.8) and serve as one of our links between the macroscopic domain and the microscopic domain. Gallium arsenide is a component in the production of the red light-emitting diodes found in many calculator displays and indicator lights (Fig. 3.6). The empirical formula (Section 2.3), GaAs, tells us that the simplest ratio of the number of gallium atoms to the number of arsenic atoms in a formula unit (microscopic domain) is 1 to 1. The empirical formula also tells us that the ratio of the numbers of moles of atoms in a gallium arsenide crystal (macroscopic domain) is also 1 to 1.

We can determine the empirical formula of a compound if we can determine the number of moles of atoms of each element in a sample. The empirical formula is the simplest mole ratio.

Example 3.9   Determining an Empirical Formula from Moles of Atoms

A sample of rock crystal contains 0.107 mol of Si atoms and 0.214 mol of O atoms. What is its empirical formula?

Solution: For every 0.107 mol of Si atoms in this sample of the compound there is 0.214 mol of O atoms; thus the simplest whole-number ratio of moles of Si atoms to moles of O atoms is 1 to 2. The empirical formula of the compound is SiO_2. Rock crystal, Fig. 3.7, is a form of quartz.

Note: To reduce a ratio of two nonintegers to integers, divide each by the smaller noninteger.

$$\frac{0.107}{0.107} = 1 \qquad \frac{0.214}{0.107} = 2$$

(Sometimes another step will be necessary, as will be illustrated in Example 3.10.)

Check your learning: A sample of methane contains 0.090 mol of C and 0.36 mol of H. What is the empirical formula of methane?

Answer: CH_4

As we will see in Sections 3.13–3.15, we can experimentally determine the mass of the atoms of each element present in a sample of a substance or its percent composition. We can determine an empirical formula from such information by calculating the number of moles of atoms in the same sample of the substance, determining the mole ratios, and reducing these ratios to simplest whole-number ratios. The string of conversions for a compound composed of any two elements, called A and X for reference, is

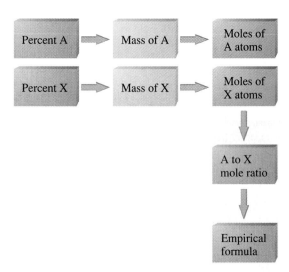

The first set of conversions, percent composition to mass, is not needed if the mass of the elements in a sample is known.

Example 3.10   Determining an Empirical Formula from Masses of Components

A sample of the black mineral hematite (Fig. 3.7), an oxide of iron found in many iron ores, contains 34.97 g of iron and 15.03 g of oxygen. What is the empirical formula of hematite?

Solution: The steps in the solution of this problem are shown in the following table.

Element	Mass of Element	Number of Moles	Divide by the Smaller Number	Smallest Integral Number of Moles
Iron	34.97 g	$34.97 \text{ g Fe} \times \dfrac{1 \text{ mol Fe}}{55.847 \text{ g Fe}}$ $= 0.6261 \text{ mol Fe}$	$\dfrac{0.6262}{0.6262} = 1.000$	$2 \times 1.0 = 2$
Oxygen	15.03 g	$15.03 \text{ g O} \times \dfrac{1 \text{ mol O}}{15.999 \text{ g O}}$ $= 0.9394 \text{ mol O}$	$\dfrac{0.9394}{0.6262} = 1.500$	$2 \times 1.5 = 3$

Here division by the smaller number of moles gives 1 to 1.5, not two integers, so an additional step is necessary. We multiply by the smallest whole number that will give whole numbers for the relative numbers of moles of each atom. The simplest whole-number ratio is 2 to 3, and the empirical formula is Fe_2O_3.

Check your learning: What is the empirical formula of a compound if a sample contains 0.130 g of nitrogen and 0.370 g of oxygen?

Answer: N_2O_5

Figure 3.8

Carbon dioxide is removed from these fermentation tanks through the large copper pipes at the top.

Example 3.11   Determining an Empirical Formula from Percent Composition

The percent composition of the gas formed during the bacterial fermentation of grain to produce ethanol (Fig. 3.8) is 27.29% C and 72.71% O. What is the empirical formula of this gas?

Solution: The percentage of an element in a compound is equal to the mass of that element in a 100-g sample of that compound. In order to determine an empirical formula from percent composition, we simply find the mass of each element present in some specific mass of a sample (100 g is a convenient mass) and then follow the procedure illustrated in the previous example.

$$100.0 \text{ g sample} \times \frac{27.29 \text{ g C}}{100.0 \text{ g sample}} = 27.29 \text{ g C}$$

$$100.0 \text{ g sample} \times \frac{72.71 \text{ g O}}{100.0 \text{ g sample}} = 72.71 \text{ g O}$$

Element	Mass of in 100.0 g of Sample	Number of Moles	Divide by the Smaller Number	Smallest Integral Number of Moles
Carbon	27.29 g	$27.29 \text{ g C} \times \dfrac{1 \text{ mol C}}{12.011 \text{ g C}}$ $= 2.272 \text{ mol C}$	$\dfrac{2.272}{2.272} = 1.000$	1
Oxygen	72.71 g	$72.71 \text{ g O} \times \dfrac{1 \text{ mol O}}{15.999 \text{ g O}}$ $= 4.545 \text{ mol O}$	$\dfrac{4.545}{2.272} = 2.000$	2

The simplest whole-number ratio is 1 to 2, and the empirical formula is CO2.

Check your learning: What is the empirical formula of a compound containing 10.1% Al and 89.9% Br?

Answer: $AlBr_3$

3.6   Derivation of Molecular Formulas

If we know the empirical formula of a compound along with the mass of a molecule of the compound (the molecular mass), we can determine the molecular formula of the compound. There are several ways to measure molecular masses—mass spectrometry (Section 2.4) is one. Often the heaviest peak observed in the mass spectrum of a molecular compound has a mass equal to the molecular mass of a molecule of the compound. Peaks at smaller masses result from decomposition of molecules into lighter fragments. Other techniques for determining molecular mass will be described in later chapters.

The molecular mass of a molecule is an integer multiple of the formula mass determined from the empirical formula. For example, the molecular mass of benzene, C_6H_6, is 78 amu. The empirical formula of benzene is CH, and the formula mass based on this empirical formula is 13 amu. The molecular mass of benzene is 6 times the formula mass (78 amu/13 amu = 6) because each benzene molecule is composed of 6 empirical formula units.

In general, we can say that a molecular formula is a whole-number multiple of the empirical formula. For benzene, the molecular formula, C_6H_6, is six times the empirical formula, CH, or

$$C_6H_6 = (CH)_n$$

where

$$n = \frac{\text{molecular mass}}{\text{empirical formula mass}}$$

Example 3.12   Determination of a Molecular Formula

Cyclopropane, a gas used as a general anesthetic, has an empirical formula of CH_2 and a formula mass of 14 amu. A mass spectrum of cyclopropane shows its molecular mass to be 42 amu. What is the molecular formula of the compound?

Solution: The compound has the molecular formula $(CH_2)_n$, where

$$n = \frac{\text{molecular mass}}{\text{empirical formula mass}}$$

$$= \frac{42}{14} = 3$$

The molecular formula is $(CH_2)_3$, or C_3H_6.

Check your learning: What is the molecular formula of a compound with a percent composition of 88.8% C and 11.2% H and a molecular mass of 54?

Answer: C_4H_6

3.7   Solutions

When we stir a little sugar in water, the sugar dissolves and a clear mixture, or solution, of sugar in water is formed. This solution consists of a solute (the dissolved sugar) and a solvent (the water). A **solution** is a homogeneous mixture (Section 1.5) of a **solute,** the substance that dissolves, and a **solvent,** the substance in which the solute dissolves. The solute or the solvent can be a gas, a liquid, or a solid, but in this chapter we will limit our attention to solutes dissolved in liquids.

If we make a solution that contains only a small amount of solute compared to the amount of solvent, we say the solution is **dilute;** the addition of more solute makes the solution more **concentrated.** The maximum amount of solute that can be dissolved in a particular amount of solvent depends on the nature of the solute and of the solvent.

The relative amount of solute dissolved in a solvent is called the **concentration** of the solution. A common unit used to express concentration is **molarity, M,** the number of moles of solute that would be contained in exactly one liter (1 L) of the solution. Molarity is calculated by dividing the amount of solute, in units of moles, in a given volume of solution by that volume, in units of liters.

$$\text{Molarity} = \frac{\text{amount of solute in moles}}{\text{volume of solution in liters}}$$

Example 3.13   Calculation of Concentration

White vinegar is a solution of acetic acid, CH_3CO_2H, in water. Vinegar, with an acidity of 5.00%, contains 50.4 g of acetic acid in 1.00 L of vinegar. What is this concentration in moles per liter?

Solution: The concentration of acetic acid is the number of moles of CH_3CO_2H dissolved in 1 L of solution. We use two steps to find the molarity of the solute.

Mass of CH_3CO_2H → **1** → Moles of CH_3CO_2H → **2** → Concentration of CH_3CO_2H

The numbers above the arrows refer to the conversion steps below.

Step 1. *Convert the grams of CH_3CO_2H to moles of CH_3CO_2H using the molar mass of 60.052 g/mol as the unit conversion factor.*

$$\text{Moles of } CH_3CO_2H = 50.4 \text{ g } CH_3CO_2H \times \frac{1 \text{ mol } CH_3CO_2H}{60.052 \text{ g } CH_3CO_2H}$$

$$= 0.839 \text{ mol } CH_3CO_2H$$

Step 2. *Convert the moles of CH_3CO_2H to concentration of CH_3CO_2H using the moles of CH_3CO_2H and the volume of solution.*

$$\text{Molarity} = \frac{0.839 \text{ mol } CH_3CO_2H}{1.00 \text{ L solution}} = 0.839 \text{ M}$$

Check your learning: $PbCrO_4$ has been used as a yellow pigment because it is not very soluble. What is the concentration of $PbCrO_4$ in a solution that contains 8.6×10^{-5} g of this solute in 2.0 L of solution?

Answer: 1.3×10^{-7} M

Molarity is a convenient concentration unit because the number of moles of solute in a known volume of solution can easily be calculated. This is important because many stoichiometry calculations require knowledge of the number of moles of one or more reactants or products. The molarity of a solution (moles of solute/1 L solution) is a ratio that can be used as a conversion factor to determine the number of moles of solute in a given volume of solution

$$\text{Number of moles} = \text{volume of solution} \times \frac{\text{moles of solute}}{1 \text{ L solution}}$$

or to determine the volume of a solution that contains a given number of moles of solute.

$$\text{Volume of solution} = \text{moles solute} \times \frac{1 \text{ L solution}}{\text{moles of solute}}$$

Example 3.14   Determining Moles of Solute in a Given Volume

At room temperature, the most concentrated solution of NaCl possible is about 5.3 M, because no more salt will dissolve in such a solution. How many moles of NaCl are contained in 0.250 L (approximately 1 cup) of a 5.3 M solution?

Solution: We can use the molarity of the solution as the conversion factor, 5.3 mol NaCl/1 L, in the following conversion.

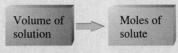

Moles of NaCl = volume of solution × conversion factor

$$= 0.250 \, \cancel{L} \times \frac{5.3 \, mol \, NaCl}{1 \, \cancel{L}}$$

$$= 1.3 \, mol \, NaCl$$

Check your learning: How many grams of sulfuric acid, H_2SO_4, are contained in 0.80 L of a 0.050 M solution of sulfuric acid?

Answer: 3.9 g H_2SO_4

Example 3.15   *Determining the Volume Containing a Given Number of Moles*

In Example 3.13 we saw that the concentration of 5.00% vinegar is 0.839 M. What volume of 5.00% vinegar contains 1.25 mol of acetic acid?

Solution: We can use the molarity of the solution, 0.839 M, to determine the conversion factor, 1.00 L/0.839 mol CH_3CO_2H, for the following conversion.

Volume of solution = mol CH_3CO_2H × conversion factor

$$= 1.25 \, \cancel{mol \, CH_3CO_2H} \times \frac{1.00 \, L}{0.839 \, \cancel{mol \, CH_3CO_2H}}$$

$$= 1.49 \, L$$

Check your learning: What volume of 1.50 M KBr solution contains 0.555 mol KBr?

Answer: 0.370 L

3.8   Dilution of Solutions

Figure 3.9 shows us one way to prepare 1 liter of a 0.0300 M solution of $KMnO_4$: Weigh out the $KMnO_4$ to get the 0.0300 mole necessary to make the solution, and then dissolve it. We could also prepare the solution from a more concentrated solution: Take the volume of the more concentrated solution that contained 0.0300 mole

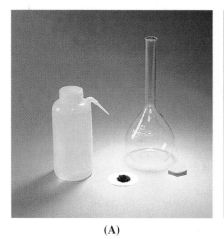
(A)

(B)

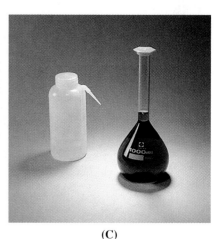
(C)

Figure 3.9

Preparation of a 0.0300 M solution of potassium permanganate, $KMnO_4$. (A) A carefully weighed 4.74 g of $KMnO_4$ (0.0300 mol) on a watch glass, an empty 1.000-L volumetric flask, and a wash bottle containing distilled water. (B) After being transferred into the volumetric flask, the solid $KMnO_4$ is dissolved in about 700 mL of distilled water and mixed well. (C) Enough water is added to fill the volumetric flask exactly to the 1.000-L mark. The flask is stoppered, and the solution is thoroughly mixed. The final solution contains 0.0300 mol of $KMnO_4$ in 1.000 L of solution and is thus 0.0300 M.

of $KMnO_4$, and then add enough water to get 1 liter of solution. For example, if we had a 2.00 M solution of $KMnO_4$, we could take 0.0150 liter of this solution (which would contain 0.0300 mole of $KMnO_4$) and add enough water to give 1 liter of solution (which would have a concentration of 0.0300 M).

We call preparation of a less concentrated solution from a more concentrated one **dilution** (Fig. 3.10). In dealing with dilution, it is helpful to remember that *the concentration of solute changes when a solution is diluted, but the total mass and number of moles of solute in the solution remain the same.*

Figure 3.10

Both solutions contain the same mass of copper nitrate. The solution on the right is more dilute because the copper nitrate in it is dissolved in a larger volume.

Example 3.16   Concentration of a Diluted Solution

If 0.040 L of a 5.00 M solution of copper nitrate, $Cu(NO_3)_2$, is diluted to a volume of 0.325 L by addition of water (Fig. 3.10), what is the molarity of the resulting diluted solution?

Solution: Because the number of moles of copper nitrate is the same in both the concentrated and the dilute solutions, we can use the following conversions to solve this problem.

The numbers above the arrows refer to the conversion steps below.

Step 1.   *Convert the volume of the concentrated solution of $Cu(NO_3)_2$ to moles of $Cu(NO_3)_2$ in that volume of the solution.* This conversion is like that in Example 3.14.

$$\text{Moles of } Cu(NO_3)_2 = \text{volume of concentrated solution} \times \frac{\text{moles of } Cu(NO_3)_2}{1 \text{ L}}$$

$$= 0.040 \text{ L} \times \frac{5.00 \text{ mol } Cu(NO_3)_2}{1.00 \text{ L}}$$

$$= 0.20 \text{ mol } Cu(NO_3)_2$$

Step 2. *Convert the moles of Cu(NO₃)₂ to concentration of Cu(NO₃)₂ in the dilute solution using the moles of solute and the volume of the dilute solution. This conversion is like that in Example 3.13.*

$$\text{Molarity} = \frac{\text{moles of Cu(NO}_3)_2}{\text{volume dilute solution}}$$

$$= \frac{0.20 \text{ mol Cu(NO}_3)_2}{0.325 \text{ L}}$$

$$= 0.62 \text{ M}$$

The solution was diluted from 5.00 M to 0.62 M.

Check your learning: What is the concentration of the solution that results from diluting 25.0 mL of a 2.04 M solution of CH_3OH to 500.0 mL?

Answer: 0.102 M

Example 3.17   Volume of a Diluted Solution

What volume of 0.12 M HBr can be prepared from 11 mL (0.011 L) of 0.45 M HBr?

Solution: Again the number of moles in the solutions does not change. The following conversions will solve this problem.

The numbers above the arrows refer to the conversion steps below.

Step 1. *Convert the volume of a solution of HBr of known concentration to moles of HBr in that volume of the solution.*

$$\text{Moles of HBr} = \text{volume of conc. soln.} \times \frac{\text{moles of HBr}}{1 \text{ L}}$$

$$= 0.011 \text{ L} \times \frac{0.45 \text{ mol HBr}}{1.00 \text{ L}}$$

$$= 5.0 \times 10^{-3} \text{ mol HBr}$$

Step 2. *Convert the moles of HBr to volume of dilute solution using the moles of solute and the concentration of the solution.*

$$\text{Volume of dilute soln.} = \text{moles of HBr} \times \text{unit conversion factor}$$

$$= 5.0 \times 10^{-3} \text{ mol HBr} \times \frac{1.00 \text{ L}}{0.12 \text{ mol HBr}}$$

$$= 0.042 \text{ L, or } 42 \text{ mL}$$

Check your learning: A laboratory experiment calls for 0.125 M HNO_3. What volume of 0.125 M HNO_3 can be prepared from 0.250 L of 1.88 M HNO_3?

Answer: 3.76 L

Example 3.18   Volume of Concentrated Solution Needed for Dilution

What volume of 1.59 M KOH is required to prepare 5.00 L of 0.100 M KOH?

Solution: Both the concentrated and the less concentrated solutions contain the same amount of KOH. Thus we can solve the problem with the following conversions.

The numbers above the arrows refer to the conversion steps below.

Step 1.   *Convert the volume of the (dilute) solution of known concentration to moles of solute in that volume.*

$$\text{Moles of KOH} = 5.00 \text{ L} \times \frac{0.100 \text{ mol KOH}}{1 \text{ L}}$$

$$= 0.500 \text{ mol KOH}$$

Step 2.   *Convert the moles of solute needed to the volume of the concentrated solution containing that amount of solute.*

$$\text{Volume conc. solution} = 0.500 \text{ mol} \times \frac{1 \text{ L}}{1.59 \text{ mol}}$$

$$= 0.314 \text{ L}$$

Check your learning: What volume of a 0.575 M solution of glucose, $C_6H_{12}O_6$, can be prepared from 50.00 mL of 3.00 M glucose solution?

Answer: 0.261 L

Stoichiometry and Chemical Change

3.9   Mole Relationships and Chemical Equations

Balanced chemical equations serve as a link between the events we see in the macroscopic domain and the microscopic entities we use to explain these events. For example, the chemical equation

$$2H_2O(l) \longrightarrow 2H_2(g) + O_2(g)$$

indicates the chemical change of water to hydrogen gas and oxygen gas (Fig. 3.11). This equation also indicates the composition of the molecules involved and the simplest whole-number ratio of these molecules in the reactants and products. The equation tells us that when two molecules of H_2O decompose, two molecules of H_2 and one molecule of O_2 are formed. Because the ratios of atoms, ions, or molecules are identical to the ratios of moles of atoms, ions, or molecules, the equation also indicates ratios of moles. Several possible different ratios of reactant and products are shown below the following equation.

$2H_2O(l)$	$\longrightarrow$	$2H_2(g)$	$+$	$O_2(g)$
2 molecules		2 molecules		1 molecule
12 molecules		12 molecules		6 molecules
$2 \times 6.022 \times 10^{23}$ molecules		$2 \times 6.022 \times 10^{23}$ molecules		6.022×10^{23} molecules
2 moles		2 moles		1 mole
0.216 mole		0.216 mole		0.108 mole

If we can describe a chemical change in the microscopic domain, we can use this information to predict the moles of quantities involved when macroscopic amounts of substances undergo a chemical change. Conversely, if we can experimentally determine the mole ratios of reactants and products in the macroscopic domain, we can use the same ratios to describe a reaction in terms of molecules in the microscopic domain.

We can use the coefficients in a chemical equation to determine ratios or conversion factors. And we can use these to find the numbers of atoms and molecules, or the numbers of moles of atoms and molecules, involved in a chemical change. For example, the foregoing equation indicates that 2 moles of H_2 are formed for every 2 moles of H_2O that decompose (2 mol H_2/2 mol H_2O), that 1 mole of O_2 is formed for every 2 moles of H_2O that decompose (1 mol O_2/2 mol H_2O), and that 1 mole of O_2 forms for every 2 moles of H_2 formed (1 mol O_2/2 mol H_2). These conversion factors are exact to any number of significant figures.

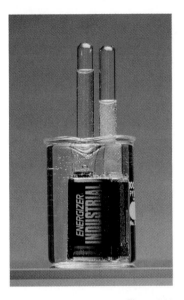

Figure 3.11

When an electric current from the battery passes through the water, the water decomposes and produces hydrogen gas (right tube) and oxygen gas (left tube) according to the equation
$2H_2O \longrightarrow 2H_2 + O_2$.

Example 3.19   Moles of Product Calculated from Moles of Reactant

The compound sodium aluminate, $NaAl(OH)_4$, is used in some antacids. How many moles of $NaAl(OH)_4$ can be prepared by the reaction of 3.0 mol of $Al_2(SO_4)_3$ according to the following equation?

$$Al_2(SO_4)_3 + 8NaOH \longrightarrow 2NaAl(OH)_4 + 3Na_2SO_4$$

Solution: This example requires that we convert from a quantity of $Al_2(SO_4)_3$ in moles to a quantity of $NaAl(OH)_4$ in moles. The chemical equation gives us the mole ratios.

$Al_2(SO_4)_3$	$+$	$8NaOH$	$\longrightarrow$	$2NaAl(OH)_4$	$+$	$3Na_2SO_4$
1 mole		8 moles		2 moles		3 moles

We can use it to determine the ratio 2 mol $NaAl(OH)_4$/1 mol $Al_2(SO_4)_3$ as a conversion factor for the conversion

$$\text{Moles of NaAl(OH)}_4 = \text{moles of Al}_2(\text{SO}_4)_3 \times \text{conversion factor}$$

$$= 3.0 \text{ mol Al}_2(\text{SO}_4)_3 \times \frac{2 \text{ mol NaAl(OH)}_4}{1 \text{ mol Al}_2(\text{SO}_4)_3}$$

$$= 6.0 \text{ mol NaAl(OH)}_4$$

Check your learning: How many moles of NH_3 are produced by the reaction of 4.0 mol of $Ca(OH)_2$ according to the following equation?

$$(\text{NH}_4)_2\text{SO}_4 + \text{Ca(OH)}_2 \longrightarrow 2\text{NH}_3 + \text{CaSO}_4 + 2\text{H}_2\text{O}$$

Answer: 8.0 mol NH_3

Example 3.20   Moles of Reactant Required in a Reaction

How many moles of I_2 are required to react with 0.429 mol of Al (Fig. 3.12) according to the following equation?

$$2\text{Al} + 3\text{I}_2 \longrightarrow 2\text{AlI}_3$$

Solution: We can use the chemical equation to determine the ratio 3 mol I_2/2 mol Al as a conversion factor for the conversion

$$\boxed{\begin{array}{c}\text{Moles}\\\text{of Al}\end{array}} \implies \boxed{\begin{array}{c}\text{Moles}\\\text{of I}_2\end{array}}$$

$$\text{mol I}_2 = 0.429 \text{ mol Al} \times \frac{3 \text{ mol I}_2}{2 \text{ mol Al}} = 0.644 \text{ mol I}_2$$

Check your learning: How many moles of $Ca(OH)_2$ are required to react with 1.36 mol of H_3PO_4 to produce $Ca_3(PO_4)_2$ according to the following equation?

$$3\text{Ca(OH)}_2 + 2\text{H}_3\text{PO}_4 \longrightarrow \text{Ca}_3(\text{PO}_4)_2 + 6\text{H}_2\text{O}$$

Answer: 2.04 mol $Ca(OH)_2$

Figure 3.12

Aluminum and iodine react to produce aluminum iodide. The heat of the reaction vaporizes some of the solid iodine as a purple vapor.

3.10   Material Balances

Most questions involving material balances are questions that ask "How much?" Answering these questions generally takes us through conversions in the order illustrated in Fig. 3.13, a stoichiometry flowchart, which summarizes many of the possible relationships in stoichiometry calculations.

Each box in the diagram in Fig. 3.13 represents one way to identify a quantity of a substance. The boxes are in different colors to emphasize that quantities can be expressed in different units. The words over the arrows indicate factors that are involved in the conversion from one quantity to the next. The double arrows remind

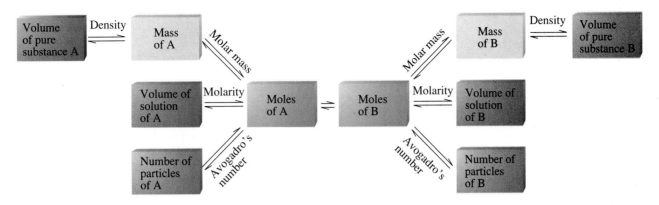

Figure 3.13

A stoichiometry flowchart.

us that the conversion can proceed in either direction. We have discussed each of the conversions in the diagram: Density was described in Section 1.11; the other conversions have been discussed in this chapter.

The overall conversion from a quantity of A to a quantity of B requires three steps (each of which may require one or more conversion factors).

1. Conversion of a quantity of substance A measured in units of mass or volume into a quantity measured in moles.

2. Determination, usually from a balanced chemical equation, of how many moles of substance B are equivalent to the number of moles of substance A.

3. Conversion of the number of moles of substance B into a quantity with units of mass or volume.

Example 3.21 **Masses of Reactants**

What mass of oxygen gas, O_2, from the air is consumed in the combustion of 702 g (1 L) of octane, C_8H_{18}, one of the principal components of gasoline?

$$2C_8H_{18} + 25O_2 \longrightarrow 16CO_2 + 18H_2O$$

Solution: We are asked to calculate the mass of O_2 that reacts with 702 g of octane. The overall conversion is

$$\boxed{\text{Mass of } C_8H_{18}} \longrightarrow \boxed{\text{Mass of } O_2}$$

As may be seen from the stoichiometry flowchart (Fig. 3.13) we cannot calculate the mass of O_2 directly from the mass of C_8H_{18}; two intermediate steps are required. The chain of calculations requires the following steps:

$$\boxed{\text{Mass of } C_8H_{18}} \xrightarrow{\;\;1\;\;} \boxed{\text{Moles of } C_8H_{18}} \xrightarrow{\;\;2\;\;} \boxed{\text{Moles of } O_2} \xrightarrow{\;\;3\;\;} \boxed{\text{Mass of } O_2}$$

The numbers above the arrows refer to the numbered conversion steps below.

Step 1.   *Convert the grams of C_8H_{18} to moles of C_8H_{18} using the molar mass of C_8H_{18} in the unit conversion factor.*

$$\text{Moles of } C_8H_{18} = 702 \text{ g } C_8H_{18} \times \frac{1 \text{ mol } C_8H_{18}}{114.232 \text{ g } C_8H_{18}} = 6.15 \text{ mol } C_8H_{18}$$

Step 2.   *Convert the moles of C_8H_{18} to moles of O_2 using the ratio determined from the balanced chemical equation as the conversion factor.*

$$\text{Moles of } O_2 = 6.15 \text{ mol } C_8H_{18} \times \frac{25 \text{ mol } O_2}{2 \text{ mol } C_8H_{18}} = 76.9 \text{ mol } O_2$$

Step 3.   *Convert the moles of O_2 to grams of O_2 using the molar mass of O_2 in the unit conversion factor.*

$$\text{Mass of } O_2 = 76.9 \text{ mol } O_2 \times \frac{31.998 \text{ g } O_2}{1 \text{ mol } O_2} = 2.46 \times 10^3 \text{ g } O_2$$

Although the logic of the problem involves three steps, we could do the calculations in one step, as follows:

$$\text{Mass of } O_2 = 702 \text{ g } C_8H_{18} \times \frac{1 \text{ mol } C_8H_{18}}{114.232 \text{ g } C_8H_{18}} \times \frac{25 \text{ mol } O_2}{2 \text{ mol } C_8H_{18}} \times \frac{31.998 \text{ g } O_2}{1 \text{ mol } O_2}$$

$$= 2.46 \times 10^3 \text{ g } O_2$$

Check your learning: What mass of CO is required to react with 25.13 g of Fe_2O_3 in the following reaction?

$$Fe_2O_3 + 3CO \longrightarrow 2Fe + 3CO_2$$

Answer: 13.22 g CO

Example 3.22   **Mass of Reactant from Mass of Product**

What mass of sodium hydroxide, NaOH, would be required to produce 16 g of the antacid milk of magnesia [magnesium hydroxide, $Mg(OH)_2$] by the reaction of magnesium chloride, $MgCl_2$, with NaOH?

$$MgCl_2(aq) + 2NaOH(aq) \longrightarrow Mg(OH)_2(s) + 2NaCl(aq)$$

Solution: This calculation is very similar to that in the previous example, except we are asked to calculate a mass of reactant needed to produce a certain amount of product. The overall conversion is

As may be seen from the stoichiometry flowchart (Fig. 3.13) we cannot calculate the required mass of NaOH directly from the mass of $Mg(OH)_2$; two intermediate steps are necessary.

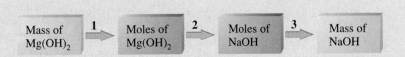

The numbers above the arrows refer to the conversion steps below. (In this example we will not work out the calculations, because they are very similar to previous conversions. However, you should perform each conversion as a self-check.)

Step 1.   *Convert the grams of Mg(OH)₂ to moles of Mg(OH)₂ using the molar mass of Mg(OH)₂ in the unit conversion factor.* Result: 0.27 mol Mg(OH)₂

Step 2.   *Convert the moles of Mg(OH)₂ to moles of NaOH using the ratio determined from the balanced chemical equation in the conversion factor.* Result: 0.54 mol NaOH

Step 3.   *Convert the moles of NaOH to grams of NaOH using the molar mass of NaOH in the unit conversion factor.* Result: 22 g NaOH

Check your learning: What mass of gallium oxide, Ga_2O_3, can be prepared from 29.0 g of gallium metal? The equation for the reaction is $4Ga + 3O_2 \longrightarrow 2Ga_2O_3$.

Answer: 39.0 g Ga_2O_3

Example 3.23   **Mass of Reactant from Volume of Product**

What volume of a 0.750 M solution of hydrochloric acid, a solution of HCl, can be prepared from the HCl produced by the reaction of 25.0 g of NaCl with an excess of sulfuric acid?

$$NaCl(s) + H_2SO_4(l) \longrightarrow HCl(g) + NaHSO_4(s)$$

Solution: Here we are asked to determine the volume of a solution of HCl that can be prepared from the HCl produced from a known amount of NaCl.

As may be seen from Fig. 3.13, two intermediate conversions are required.

The numbers above the arrows refer to the conversion steps below.

Step 1.   *Convert the grams of NaCl to moles of NaCl using the molar mass of NaCl in the unit conversion factor.*

$$\text{Moles of NaCl} = 25.0 \text{ g NaCl} \times \frac{1 \text{ mol NaCl}}{58.443 \text{ g NaCl}} = 0.428 \text{ mol NaCl}$$

Step 2. *Convert the moles of NaCl to moles of HC1 using the ratio determined from the balanced chemical equation as the conversion factor.*

$$\text{Moles of HCl} = 0.428 \; \overline{\text{mol NaCl}} \times \frac{1 \; \text{mol HCl}}{1 \; \overline{\text{mol NaCl}}} = 0.428 \; \text{mol HCl}$$

Step 3. *Convert the moles of HCl to volume of 0.750 M HCl solution using the moles of solute and the concentration of the solution.*

$$\text{Volume of solution} = 0.428 \; \overline{\text{mol HCl}} \times \frac{1 \; \text{L HCl solution}}{0.750 \; \overline{\text{mol HCl}}}$$

$$= 0.571 \; \text{L of HCl solution}$$

Check your learning: What volume of a 0.2089 M KI solution contains enough KI to react exactly with the $Cu(NO_3)_2$ in 43.88 mL of a 0.3842 M solution of $Cu(NO_3)_2$?

$$2Cu(NO_3)_2 + 4KI \longrightarrow 2CuI + I_2 + 4KNO_3$$

Solve this problem using the following conversions:

Answer: 0.1614 L, or 161.4 mL, of KI solution

The three preceding examples illustrate the basic flow of almost any stoichiometry problem involving material balances. However, some stoichiometry calculations involve additional steps that result from the nature of the material used in a reaction (a mixture rather than a pure compound, for example) or from the various ways of measuring it (measuring a volume rather than a mass, for example).

Example 3.24   Additional Steps in Stoichiometry Calculations

A mordant is a substance that combines with a dye to produce a stable fixed color in a dyed fabric. Calcium acetate is used as a mordant. It is prepared by the reaction of acetic acid with calcium hydroxide.

$$2CH_3CO_2H + Ca(OH)_2 \longrightarrow Ca(CH_3CO_2)_2 + 2H_2O$$

What mass of $Ca(OH)_2$ is required to react with the acetic acid in 25.0 mL of a solution that has a density of 1.065 g/mL and contains 58.0% acetic acid by mass?

Solution: Overall, this problem asks us to find the mass of a $Ca(OH)_2$ sample that will react with a given volume of an acetic acid solution. The conversion is

Because the solution contains CH_3CO_2H that reacts with $Ca(OH)_2$, we need a mole-to-mole conversion:

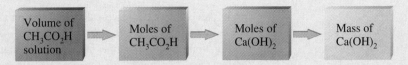

We are not given the molar concentration of the acetic acid solution. Instead we need to determine the mass of acetic acid from the density of the solution and the percent by mass of the acetic acid in the solution. Taking this into account gives the following overall string of conversions:

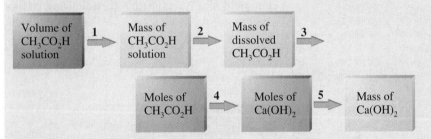

The calculations for these conversions are

Step 1. Mass of solution = $25.0 \; \overline{\text{mL solution}} \times \dfrac{1.065 \text{ g solution}}{1 \; \overline{\text{mL solution}}}$

$= 26.6$ g solution

Step 2. Mass of $CH_3CO_2H = 26.6 \; \overline{\text{g solution}} \times \dfrac{58.0 \text{ g } CH_3CO_2H}{100 \; \overline{\text{g solution}}}$

$= 15.4$ g CH_3CO_2H

Step 3. Moles of $CH_3CO_2H = 15.4 \; \overline{\text{g } CH_3CO_2H} \times \dfrac{1 \text{ mol } CH_3CO_2H}{60.052 \; \overline{\text{g } CH_3CO_2H}}$

$= 0.256$ mol CH_3CO_2H

Step 4. Moles of $Ca(OH)_2 = 0.256 \; \overline{\text{mol } CH_3CO_2H} \times \dfrac{1 \text{ mol } Ca(OH)_2}{2 \; \overline{\text{mol } CH_3CO_2H}}$

$= 0.128$ mol $Ca(OH)_2$

Step 5. Mass of $Ca(OH)_2 = 0.128 \; \overline{\text{mol } Ca(OH)_2} \times \dfrac{74.092 \text{ g } Ca(OH)_2}{1 \; \overline{\text{mol } Ca(OH)_2}}$

$= 9.48$ g $Ca(OH)_2$

Thus 9.48 g of $Ca(OH)_2$ is required.

Check your learning: The toxic pigment called white lead, $Pb_3(OH)_2(CO_3)_2$, has been replaced in white paints by rutile, TiO_2. How much rutile can be prepared from 379 g of an ore that contains 88.3% ilmenite ($FeTiO_3$) by mass?

$$2FeTiO_3 + 4HCl + Cl_2 \longrightarrow 2FeCl_3 + 2TiO_2 + 2H_2O$$

Answer: 176 g TiO_2

3.11 Theoretical Yield, Actual Yield, and Percent Yield

In the preceding section we calculated the maximum amount of product produced by a given chemical reaction. Such a calculation is based on three assumptions: (1) that the reaction is the only one involved; (2) that all of the reactant is converted into product; and (3) that all of the product can be collected. The calculated amount of product based on these assumptions is called the **theoretical yield.** We rarely find these conditions satisfied either in the laboratory or in industrial production; the **actual yield** or, more simply, the yield of product (the actual amount of product isolated), from a reaction is usually less than the theoretical yield. We can calculate the **percent yield** of a reaction from the following relationship:

$$\text{Percent yield} = \frac{\text{actual yield}}{\text{theoretical yield}} \times 100$$

Example 3.25   Calculation of Percent Yield

A general chemistry student, preparing copper metal by the reaction of 1.274 g of copper sulfate with zinc metal, obtained a yield of 0.392 g of copper. What was the percent yield?

$$CuSO_4(aq) + Zn(s) \longrightarrow Cu(s) + ZnSO_4(aq)$$

Solution: We can determine the percent yield if we know the theoretical yield and the actual yield. The actual yield is 0.392 g of copper metal. The theoretical yield is the amount that would have been obtained if all of the copper sulfate had been converted into copper and all of the copper had been recovered. The theoretical yield of copper can be calculated by the following string of conversions.

The numbers above the arrows refer to the conversion steps below. (In this example we will not work out the calculations, because they are very similar to previous conversions. However, you should check each conversion.)

Step 1.   *Convert the mass of $CuSO_4$ to moles of $CuSO_4$.* Result: 7.982×10^{-3} mol $CuSO_4$

Step 2.   *Convert the moles of $CuSO_4$ to moles of Cu.* Result: 7.982×10^{-3} mol Cu

Step 3.   *Convert the moles of Cu to mass of Cu.* Result: 0.5072 g Cu

The theoretical yield is 0.5072 g Cu. Now we can calculate the percent yield.

$$\text{Percent yield} = \frac{\text{actual yield}}{\text{theoretical yield}} \times 100$$

$$= \frac{0.392 \text{ g}}{0.5072 \text{ g}} \times 100 = 77.3\%$$

Check your learning: What is the percent yield of a reaction that produces 12.5 g of the Freon CF_2Cl_2 from 32.9 g of CCl_4?

$$CCl_4 + 2HF \longrightarrow CF_2Cl_2 + 2HCl$$

Answer: 48.3%

3.12 Limiting Reactant

Sometimes we find that not all of the reactants in a chemical reaction are completely consumed. Figure 3.14 shows one example, the reaction of silver nitrate with copper metal.

$$2AgNO_3 + Cu \longrightarrow 2Ag + Cu(NO_3)_2$$

The equation indicates that exactly 2 moles of $AgNO_3$ must react for each mole of Cu that reacts. If the ratio of reactants actually used in this reaction differs from 2 to 1, then one of the reactants will be present in excess and not all of it will be consumed. If, for example, we have 2 moles of $AgNO_3$ and 2 moles of Cu, then 1 mole of Cu must remain unreacted at the end of the reaction because 2 moles of $AgNO_3$ can react with only 1 mole of Cu. The reactant that is completely consumed ($AgNO_3$ in this case) is called the **limiting reactant.** This reactant limits the amount of product that can be formed and determines the theoretical yield of the reaction. Other reactants are said to be present in excess.

If we wish to determine which reactant is the limiting reactant, we calculate the amount of product that would be produced if each reactant were completely converted to product. The reactant that gives the smallest amount of product is the limiting reactant and will be completely consumed.

Figure 3.14

(A) A copper wire shortly after it was placed in a solution of silver nitrate. (B) The reaction is complete. Silver crystals cover the wire, and the solution has the blue color of copper nitrate. Some copper wire remains, so silver nitrate is the limiting reagent of this reaction.

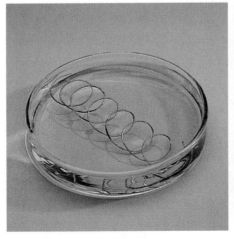

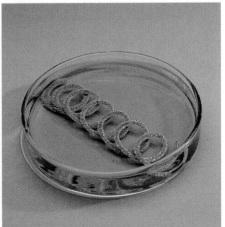

(A)   (B)

Example 3.26   Determining a Limiting Reactant

Silicon nitride is a very hard, high-temperature-resistant ceramic used as a component of turbine blades in jet engines. It is prepared according to the equation

$$3Si + 2N_2 \longrightarrow Si_3N_4$$

Which is the limiting reactant when 2.00 g of Si and 1.50 g of N_2 combine? How much silicon nitride will the reaction produce?

Solution: First, assume all of the Si will react and calculate the theoretical yield using the following steps:

| Mass of Si | →1 | Moles of Si | →2 | Moles of Si_3N_4 |

The numbers above the arrows refer to the conversion steps below. (In this example we will not work out the calculations, because they are very similar to previous conversions. However, you should check each conversion.)

Step 1.   *Convert the mass of Si to moles of Si.* Result: 7.12×10^{-2} mol Si

Step 2.   *Convert the moles of Si to moles of Si_3N_4.* Result: 2.37×10^{-2} mol Si_3N_4

Now assume all of the N_2 will react and calculate the theoretical yield using the following steps:

| Mass of N_2 | →1 | Moles of N_2 | →2 | Moles of Si_3N_4 |

Again, the numbers above the arrows refer to the conversion steps below.

Step 1.   *Convert the mass of N_2 to moles of N_2.* Result: 5.35×10^{-2} mol N_2

Step 2.   *Convert the moles of N_2 to moles of Si_3N_4.* Result: 2.68×10^{-2} mol Si_3N_4

At this point it should be clear that Si is the limiting reactant, because it produces the smaller amount of Si_3N_4 (2.37×10^{-2} mol). We can calculate the mass of Si_3N_4 using the following conversion:

| Moles of Si_3N_4 | → | Mass of Si_3N_4 |

$$\text{Mass of } Si_3N_4 = 2.37 \times 10^{-2} \text{ mol } Si_3N_4 \times \frac{140.286 \text{ g } Si_3N_4}{1 \text{ mol } Si_3N_4}$$

$$= 3.32 \text{ g } Si_3N_4$$

Check your learning: Which is the limiting reactant and how much water is formed when 5.00 g of H_2 and 10.0 g of O_2 react and form water?

Answer: O_2; 11.3 g H_2O

Quantitative Analysis

In the eighteenth century the "strength" (actually the concentration) of vinegar samples was determined by noting the amount of potassium carbonate, K_2CO_3, that had to be added, a little at a time, before bubbling ceased. The larger the weight of potassium carbonate added to reach the point where the bubbling ended, the more concentrated the vinegar.

We now know that the effervescence that occurred during this process was due to a reaction of acetic acid, CH_3CO_2H, the compound that is primarily responsible for the odor and taste of vinegar. Acetic acid reacts with potassium carbonate according to the equation

$$2CH_3CO_2H(aq) + K_2CO_3(s) \longrightarrow 2KCH_3CO_2(aq) + CO_2(g) + H_2O(l)$$

The bubbling was due to the evolution of gaseous CO_2.

The test of vinegar with potassium carbonate is one type of **quantitative analysis**—the determination of the amount or concentration of a substance in a sample. In the analysis of vinegar, the concentration of the solute (acetic acid) was determined from the amount of reactant that combined with the solute present in a known volume of the solution. In other types of chemical analyses, the amount of a substance present in a sample is determined by measuring the amount of product that results.

Figure 3.15

A pipet filled to a reference line and allowed to drain completely will deliver a precise volume. This is a 5-mL pipet, one of various sizes available.

Calibration mark

3.13 Titration

A **titration** is a determination of how much of one substance reacts with another. When we titrate with solutions, we add a solution of a reactant to a solution of a sample. When the exact amount of reactant necessary to react completely with the sample has been added, an indicator, or in some cases the color of the solution itself (as in Example 3.27), marks this **end point** of the titration. An **indicator** is a substance that is added to the sample to mark the end point of the titration—usually by a change of color. At the end point, we stop adding solution and determine the volume of solution added. This volume is used in determining the amount of, or the concentration of, the unknown solution.

A precise volume of a solution is usually measured with a pipet or a buret. A **pipet** (Fig. 3.15) is a tube that will deliver a known fixed volume of a liquid that is filled to a reference line. A **buret** is a cylinder that is graduated in fractions of a milliliter so that the volume of a liquid drained from it can be accurately determined.

A buret is used in a titration to add a solution of a reactant to a solution of a sample. At the end point, the titration is stopped, and the amount of solution delivered from the buret is determined by subtracting the initial reading from the final reading on the buret.

Figure 3.16

A colorless solution of $H_2C_2O_4$ is titrated with purple $KMnO_4$ solution of known strength. Colorless Mn^{2+} ion forms during the oxidation–reduction reaction. At the instant when all the $H_2C_2O_4$ has been reacted, adding the next drop of $KMnO_4$ turns the solution purple in the flask. The color change indicates the end point of the titration.

Example 3.27   Analysis of a Solution by Titration

A solution of oxalic acid, $H_2C_2O_4$, was added to a flask with a 20.00-mL pipet. After addition of sulfuric acid, H_2SO_4, the sample was titrated with a 0.09113 M solution of potassium permanganate, $KMnO_4$, (Fig. 3.16). A volume of 23.24 mL was required to reach the end point. What was the concentration of $H_2C_2O_4$?

$$5H_2C_2O_4 + 2KMnO_4 + 3H_2SO_4 \longrightarrow 2MnSO_4 + K_2SO_4 + 10CO_2 + 8H_2O$$

Solution: Titration calculations are fairly straightforward stoichiometry calculations. In this case the chain of conversions is

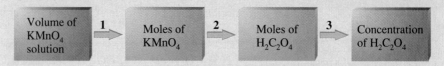

The numbers above the arrows refer to the conversion steps below.

Step 1.   *Convert the volume of a solution of $KMnO_4$ of known concentration to moles of $KMnO_4$ in that volume.*

$$\text{Moles of } KMnO_4 = 0.02324 \text{ L} \times \frac{0.09113 \text{ mol } KMnO_4}{1 \text{ L}}$$

$$= 2.118 \times 10^{-3} \text{ mol } KMnO_4$$

Step 2.   *Convert the moles of $KMnO_4$ to moles of $H_2C_2O_4$ using the ratio determined from the balanced chemical equation.*

$$\text{Moles of } H_2C_2O_4 = 2.118 \times 10^{-3} \text{ mol } KMnO_4 \times \frac{5 \text{ mol } H_2C_2O_4}{2 \text{ mol } KMnO_4}$$

$$= 5.295 \times 10^{-3} \text{ mol } H_2C_2O_4$$

Step 3.   *Convert the moles of $H_2C_2O_4$ to concentration of $H_2C_2O_4$ using the moles of solute and the volume of solution.*

$$\text{Molarity} = \frac{5.295 \times 10^{-3} \text{ mol } H_2C_2O_4}{0.02000 \text{ L}} = 0.2648 \text{ M}$$

Check your learning: What is the concentration of HCl in water if 48.47 mL of the HCl solution is required to titrate a solution of 0.5015 g of NaOH in water?

$$HCl + NaOH \longrightarrow NaCl + H_2O$$

Answer: 0.2587 M

3.14   Gravimetric Analysis

A **gravimetric analysis** is an analysis based on measurement of mass. A common gravimetric procedure involves dissolving a sample of unknown composition in water and allowing the solution to react with another substance to form a precipitate. The

Figure 3.17

When a solution containing a precipitate is filtered, the solid is retained on the filter paper in the funnel, and the solution passes through to the beaker below.

precipitate is filtered, dried, and weighed. We can determine the amount of one of the components of the unknown from the formula and mass of the precipitate.

A precipitation reaction occurs if a solid substance forms when solutions of two reactants are mixed. For example, mixing a solution of sodium sulfate with a solution of barium nitrate produces a precipitate of barium sulfate.

$$Na_2SO_4(aq) + Ba(NO_3)_2(aq) \longrightarrow BaSO_4(s) + 2NaNO_3(aq)$$

If an excess of sodium sulfate is used, essentially all of the Ba^{2+} ion in a solution can be precipitated as barium sulfate. Alternatively, if an excess of barium nitrate is used, essentially all of the SO_4^{2-} ion in a solution can be precipitated as barium sulfate.

If we recover all of the barium sulfate produced in a precipitation reaction by filtering the solution (Fig. 3.17), washing the precipitate until it is clean and then drying it, we can determine the mass of $BaSO_4$ produced. From this mass of $BaSO_4$ we could determine the mass of either barium ion or sulfate ion in a reactant solution.

Example 3.28   *Gravimetric Analysis*

Plaster of Paris is a white powder that is used to make casts for supporting broken bones. It consists of $CaSO_4$ and a little water. What is the percent of $CaSO_4$ in a 0.4550-g sample of plaster of Paris if reaction with excess $Ba(NO_3)_2$ produces 0.6168 g of dry $BaSO_4$?

$$CaSO_4(aq) + Ba(NO_3)_2(aq) \longrightarrow BaSO_4(s) + Ca(NO_3)_2(aq)$$

Solution: The string of conversions is

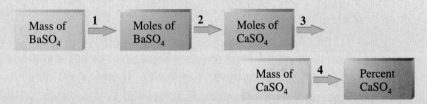

The numbers above the arrows refer to the conversion steps below. (In this example we will not work out the calculations, because they are very similar to previous conversions. However, you should check each calculation.)

Step 1. *Convert the mass of $BaSO_4$ to moles of $BaSO_4$.* Result: 2.643×10^{-3} mol $BaSO_4$

Step 2. *Convert the moles of $BaSO_4$ to moles of $CaSO_4$.* Result: 2.643×10^{-3} mol $CaSO_4$

Step 3. *Convert the moles of $CaSO_4$ to mass of $CaSO_4$.* Result: 0.3598 g $CaSO_4$

Step 4. Convert the mass of $CaSO_4$ to the percent by mass of $CaSO_4$ in the 0.4550-g sample.

$$\text{Percent } CaSO_4 = \frac{\text{mass } CaSO_4}{\text{mass sample}} \times 100$$

$$= \frac{0.3598 \text{ g}}{0.4550 \text{ g}} \times 100 = 79.08\%$$

Check your learning: What is the percent of chloride ion in a sample if 1.1324 g of the sample produces 1.088 g of AgCl in a gravimetric analysis? The net ionic equation for the reaction is $Ag^+(aq) + Cl^-(aq) \longrightarrow AgCl(s)$.

Answer: 23.76%

3.15 Combustion Analysis

A combustion reaction (Section 2.10) converts a compound that contains carbon and hydrogen (and sometimes oxygen) to carbon dioxide and water. For example, combustion of ethanol proceeds according to the equation

$$C_2H_5OH(l) + 3O_2(g) \longrightarrow 2CO_2(g) + 3H_2O(g)$$

At the temperature of the reaction, the carbon dioxide and water are gases.

Combustion reactions form the basis for the **combustion analysis** of the carbon and hydrogen content of many compounds. During the course of a combustion analysis, a carefully weighed sample of a compound is burned in an apparatus such as that shown in Fig. 3.18. An excess of oxygen is used to ensure complete conversion of all the carbon in the compound to CO_2 and of all of the hydrogen to H_2O. The H_2O produced is absorbed by a compound such as $Mg(ClO_4)_2$, and the CO_2 by a compound such as NaOH. The masses of the CO_2 and H_2O that form are determined by weighing the bulbs before and after the combustion. The moles of carbon and of hydrogen, or the masses of these elements, in the sample used in the combustion can be determined from the masses of CO_2 and H_2O produced.

The percent composition (Section 3.4) or the empirical formula (Section 3.5) of a compound composed of hydrogen and carbon may be determined as follows:

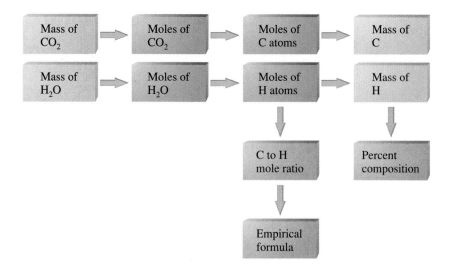

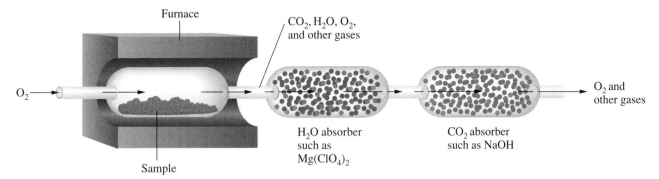

Figure 3.18   The combustion device consists of a furnace and two absorption bulbs, which contain different solids and which can be disconnected from the apparatus and weighed. A sample is heated and burns in a flowing stream of oxygen, and the resulting gases are carried into the bulbs, which are then reweighed. The difference in weight of the bulbs before and after an analysis gives the masses of carbon dioxide and water produced.

Example 3.29   Combustion Analysis

Polyethylene is a polymer that is used to produce food storage bags and many other flexible plastic items. A combustion analysis of a 0.001256-g sample of polyethylene yields 0.003940 g of CO_2 and 0.001612 g of H_2O. What is the empirical formula of polyethylene?

Solution: We can use the following two chains of conversions to calculate the moles of C atoms and the moles of H atoms in the sample and then determine the simplest C-to-H mole ratio.

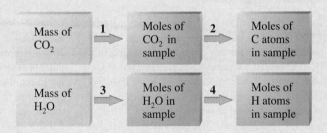

The numbers above the arrows refer to the conversion steps below. (In this example we will not work out the calculations, because they are very similar to previous conversions. However, you should check each conversion.)

Moles of C atoms:

Step 1.   *Convert the mass of CO_2 to moles of CO_2.* Result: 8.953×10^{-5} mol CO_2

Step 2.   *Convert the moles of CO_2 to moles of C atoms.* Result: 8.953×10^{-5} mol C atoms

Moles of H atoms:

Step 3. *Convert the mass of H_2O to moles of H_2O.* Result: 8.948×10^{-5} mol H_2O

Step 4. *Convert the moles of H_2O to moles of H atoms.* Result: 1.790×10^{-4} mol H atoms

The mole ratio:

Element	Number of Moles	Divide by the Smaller Number	Smallest Integral Number of Moles
C	8.953×10^{-5}	$\dfrac{8.953 \times 10^{-5}}{8.953 \times 10^{-5}} = 1.000$	1
H	1.790×10^{-4}	$\dfrac{1.790 \times 10^{-4}}{8.953 \times 10^{-5}} = 1.999$	2

The simplest whole-number ratio is 1 to 2. The empirical formula is CH_2.

Check your learning: A 0.002147-g sample of polystyrene, a polymer composed of carbon and hydrogen, produced 0.00726 g of CO_2 and 0.00148 g of H_2O in a combustion analysis. What is the experimental percent composition of polystyrene?

Answer: 7.71% H, 92.3% C

For Review   Summary

The calculations called **chemical stoichiometry** include material balances—that is, (1) finding quantitative relationships among atoms in compounds, (2) determining percent composition of a compound from its chemical formula, (3) finding empirical and molecular formulas from percent composition, (4) using the coefficients in a balanced equation to determine the quantities of reactants and products involved in the chemical reaction, (5) finding the concentrations of solutions, and (6) quantitative analysis. These concepts are the subject of this chapter. Chemical stoichiometry also includes the calculation of energy balances in a chemical system, which we will consider in Chapter 4.

The **atomic mass** of an atom is the average mass of a single atom in atomic mass units (amu). The **molecular mass** of a molecule is the average mass of a single molecule in atomic mass units and is equal to the sum of the atomic masses of its constituent atoms. Compounds that do not consist of individual molecules are characterized by a **formula mass,** which is equal to the sum of the atomic masses of the atoms shown in the empirical formula.

A **mole of atoms** of any substance contains the same number of atoms as are found in exactly 12 grams of the isotope $^{12}_{6}C$. A mole of any type of atoms contains **Avogadro's number** of atoms, 6.022×10^{23} atoms. The mass, in grams, of a mole of atoms is numerically equal to the atomic mass of the atom. A **mole of molecules** contains 6.022×10^{23} molecules, and the **molar mass,** the mass of a mole of molecules in grams, is numerically equal to the molecular mass. A **mole of an ionic compound,** which does not consist of individual molecules, is equal to 6.022×10^{23} formula units, as described by the **empirical formula.** Molecular formulas and empirical formulas provide information about both the microscopic and the macroscopic domains. The subscripts in a molecular formula indicate the number of atoms of each type in a molecule, as well as the number of moles of each type of atom in a mole of molecules. The subscripts in an empirical formula indicate the number of atoms of each type in a formula, as well as the number of moles of each type of atom in a mole of formula units.

The amount of a substance, in moles, can be determined from its mass by using its molar mass as a conversion factor. Conversely, its mass can be determined from the number of moles. A given quantity of a substance, in moles, can be obtained by weighing. The empirical formula of a substance can be determined from the numbers of moles of the various types of atoms that make up a sample of the substance. If the percent composition is known, then the numbers of moles of atoms—and hence the empirical formula—can be calculated. The molecular formula can be determined from the empirical formula when the molecular mass of a compound is known.

If a substance is dissolved, forming a solution, a given quantity of the substance, in moles, can be obtained by measuring out a calculated volume of the solution, provided that the molarity of the solution is known. The **molarity** of a solution, one common designation of the concentration of a solution, is equal to the number of moles of solute dissolved in exactly 1 liter of solution.

Stoichiometric calculations involving material balances reflect the fact that the balanced chemical equation that describes a reaction gives the relative numbers of moles of reactants and

products involved. Thus a balanced equation can be used to determine conversion factors relating moles of reactants and/or products. The amount of product produced by complete conversion of a reactant in a reaction is called the **theoretical yield** of the reaction. To calculate the theoretical yield of a reaction, we must know which reactant acts as the **limiting reactant.** The amount of product that is actually isolated from a reaction, either in the laboratory or in an industrial process, is called the **actual yield** of the reaction. The **percent yield** can be calculated from the actual yield and the theoretical yield.

Quantitative analysis is the determination of the amount or concentration of a substance in a sample. Techniques used in such analyses include **titration, gravimetric analysis,** and **combustion analysis.**

Key Terms and Concepts

actual yield   (3.11)
atomic mass   (3.1)
Avogadro's number   (3.2)
buret   (3.13)
combustion analysis   (3.15)
concentrated   (3.7)
concentration   (3.7)
dilute   (3.7)
dilution   (3.8)
empirical formula   (3.5)

end point   (3.13)
formula mass   (3.1)
gravimetric analysis   (3.14)
indicator   (3.13)
limiting reactant   (3.12)
molar mass   (3.2, 3.3)
molarity, M   (3.7)
mole   (3.2, 3.3)
molecular mass   (3.1)
molecular formula   (3.6)

percent composition   (3.4)
percent yield   (3.11)
pipet   (3.13)
solute   (3.7)
solution   (3.7)
solvent   (3.7)
theoretical yield   (3.11)
titration   (3.13)

Exercises

Questions

1. How are the molecular mass and the molar mass of a compound similar and how are they different?

2. Tell which of the following are microscopic quantities and which are macroscopic quantities (Section 1.8): atomic mass, 1 amu, formula mass, 1 gram, molar mass, mole, molecular mass.

3. What is the SI unit for amount of substance? How much substance corresponds to one of these units?

4. Write a sentence that describes how we can determine the number of moles of a compound in a known mass of the compound if we know its molecular formula.

5. Write a sentence that describes how we can determine the mass of a sample of a compound if we know its molecular formula and the number of moles of compound in the sample.

6. If we find a previously unknown mineral in a mine and wish to determine its percent composition, what information would we ask a lab technician to determine from a sample of the compound?

7. What information do we need in order to determine the molecular formula of a compound?

8. What additional information do we need to determine the mass of a compound if we know the number of moles in a sample of the compound?

9. What information do we need to calculate the concentration of sulfuric acid in a solution of sulfuric acid?

10. What do we mean when we say that a 200-mL sample and a 400-mL sample of a solution of salt have the same concentration? In what ways are these two samples identical? In what ways are these two samples different?

11. Explain what changes and what stays the same when the 0.850 L of copper nitrate solution in Fig. 3.10 is diluted to 1.80 L.

12. Why is molarity an intensive property (Section 1.4)?

13. (a) What information do we need to know in order to determine the mass of the burn cream silver sulfadiazine that can be prepared from sulfadiazine?
    (b) What information do we need to determine the percent yield of a reaction of silver oxide and sulfadiazine to produce the burn cream silver sulfadiazine? Which of this information must be determined experimentally for the particular reaction?
    (c) What information is required to determine the limiting reactant in the reaction of silver oxide and sulfadiazine described in part (b)?

14. In a reaction that produces sodium chloride from 2 g of sodium and 5 g of chlorine, sodium is the limiting reactant. Why isn't the chlorine completely consumed in the reaction?

15. What is a desirable yield from a reaction that is used in a gravimetric analysis?

16. Why is a titration stopped at the end point?

17. What is the limiting reactant in the reaction used for the analysis of calcium sulfate in Example 3.28?

18. What is the limiting reagent at the end point of the titration shown in Fig. 3.17?

19. Which of the postulates of Dalton's atomic theory (Section 2.1) explains why we can calculate a theoretical yield for a chemical reaction?

20. Which of the postulates of Dalton's atomic theory (Section 2.1) explains why we can determine the empirical formula of sulfur dioxide in one experiment and then use that formula to determine the yield of sulfur dioxide expected from a totally different experiment? Explain your answer.

Atomic and Molecular Masses; Molar Mass

21. Calculate the molecular mass of each of the following:
    (a) P_4
    (b) H_2O
    (c) $Ca(NO_3)_2$
    (d) CH_3CO_2H (acetic acid)
    (e) $C_{12}H_{22}O_{11}$ (sucrose, cane sugar)

22. Calculate the molar mass of each of the following:
    (a) S_8
    (b) C_5H_{12}
    (c) $Sc_2(SO_4)_3$
    (d) CH_3COCH_3 (acetone)
    (e) $C_6H_{12}O_6$ (glucose)

23. Calculate the formula mass and the molar mass of the following minerals.

    (a) limestone, $CaCO_3$
    (b) halite, $NaCl$
    (c) beryl, $Be_3Al_2Si_6O_{18}$
    (d) malachite, $Cu_2(OH)_2CO_3$
    (e) turquoise, $CuAl_6(PO_4)_4(OH)_8(H_2O)_4$

24. Calculate the formula mass and the molar mass of the following compounds.

    (a) the anesthetic halothane, $C_2HBrClF_3$
    (b) the herbicide paraquat, $C_{12}H_{14}N_2Cl_2$
    (c) caffeine, $C_8H_{10}N_4O_2$
    (d) urea, $CO(NH_2)_2$
    (e) a typical soap, $C_{17}H_{35}CO_2Na$

25. Determine the molar mass of the following compounds.

26. Determine the molar mass of the following compounds.

27. What is the average mass, in grams, of a bromine atom?

28. What is the average mass, in grams, of a water molecule?

29. Which of the following molecules has an average mass of 4.66×10^{-23} g?

30. Which element has atoms with an average mass of 5.14×10^{-23} g?

31. Which contains the greatest mass of oxygen: one molecule of ethanol (C_2H_5OH), one molecule of carbon dioxide, or 1 gram of water? Explain your answer.

32. The molecular mass of the DNA dimer in a human cell is 1.6×10^{12} amu. What mass of DNA, in grams, is present in an average human body, which contains 1×10^{13} cells?

Moles of Atoms and Molecules

33. (a) Which contains the greatest mass in 1 mol of atoms: hydrogen, oxygen, or fluorine? Explain why.

(b) Which contains the greatest mass in 1 mol of molecules: hydrogen gas (H_2), ozone gas (O_3), or fluorine gas (F_2)? Explain why.

34. (a) Which contains the greatest mass of oxygen: 0.75 mol of ethanol (C_2H_5OH), 0.60 mol of formic acid (HCO_2H), or 1.0 mol of water (H_2O)? Explain why.

    (b) Which contains the greatest number of moles of oxygen atoms: 1 mol of ethanol, 1 mol of formic acid, or 1 mol of water? Explain why.

35. (a) Calculate the mass, in grams, of 1.000 mol of each of the following compounds.
    i.   hydrogen fluoride, HF
    ii. ammonia, NH_3
    iii. nitric acid, HNO_3
    iv. silver sulfate, Ag_2SO_4
    v.   boric acid, $B(OH)_3$

    (b) Calculate the mass, in atomic mass units (amu), of 1.000 mol of each of the compounds in part (a).

36. (a) Calculate the mass, in grams, of 1.000 mol of each of the following compounds.
    i.   hydrogen bromide, HBr
    ii. methane, CH_4
    iii. sulfuric acid, H_2SO_4
    iv. hydrogen peroxide, H_2O_2
    v.   calcium hydroxide, $Ca(OH)_2$

    (b) Calculate the mass, in atomic mass units (amu), of 1.000 mol of each of the compounds in part (a).

37. Determine (a) the number of moles of compound and (b) the number of moles of each type of atom in each of the following:
    i.   25.0 g of propylene, C_3H_6
    ii. 3.06×10^{-3} g of the amino acid glycine, $C_2H_5NO_2$
    iii. 25 lb of the herbicide Treflan, $C_{13}H_{16}N_2O_4F$ (1 lb = 454 g)
    iv. 0.125 kg of the insecticide Paris Green, $Cu_4(AsO_3)_2(CH_3CO_2)_2$
    v.   325 mg of aspirin, $C_6H_4(CO_2H)(CO_2CH_3)$

38. Determine (a) the number of moles of compound and (b) the number of moles of each type of atom in each of the following:
    i.   2.12 g of potassium bromide, KBr
    ii. 0.1488 g of phosphoric acid, H_3PO_4
    iii. 23 kg of calcium carbonate, $CaCO_3$
    iv. 78.452 g of aluminum sulfate, $Al_2(SO_4)_3$
    v.   0.1250 mg of caffeine, $C_8H_{10}N_4O_2$

39. Determine the mass of each of the following:
    (a) 0.0146 mol KOH
    (b) 10.2 mol ethane, C_2H_6
    (c) 1.6×10^{-3} mol Na_2SO_4
    (d) 6.854×10^3 mol glucose, $C_6H_{12}O_6$
    (e) 2.86 mol $Co(NH_3)_6Cl_3$

40. Determine the mass of each of the following:
    (a) 2.345 mol LiCl
    (b) 0.0872 mol acetylene, C_2H_2
    (c) 3.3×10^{-2} mol Na_2CO_3
    (d) 1.23×10^3 mol fructose, $C_6H_{12}O_6$
    (e) 0.5758 mol $FeSO_4(H_2O)_7$

41. The approximate minimum daily dietary requirement of the amino acid leucine, $C_6H_{13}NO_2$, is 1.1 g. What is this requirement in moles?

42. Determine the mass in grams of each of the following:
    (a) 0.600 mol of oxygen atoms
    (b) 0.600 mol of oxygen molecules, O_2
    (c) 0.600 mol of ozone molecules, O_3

43. An average 55-kg woman has 7.5×10^{-3} mol of hemoglobin (molar mass = 64,456) in her blood. How many hemoglobin molecules is this? What is this quantity in grams? In pounds?

44. Determine the number of atoms and the mass of zirconium, silicon, and oxygen needed to make 0.3384 mol of zircon, $ZrSiO_4$, a semiprecious stone.

45. Determine which of the following contains the greatest mass of hydrogen: 1 mol of CH_4, 0.6 mol of C_6H_6, or 0.4 mol of C_3H_8.

46. Determine which of the following contains the greatest mass of aluminum: 122 g of $AlPO_4$, 266 g of Al_2Cl_6, or 225 g of Al_2S_3.

47. Determine which of the following contains the largest mass of carbon atoms: 0.10 mol of glucose, $C_6H_{12}O_6$; 3.0 g of ethane, C_2H_6; or a 1.0-g diamond (diamond is pure carbon).

48. (a) Determine which of the following has the greatest mass in 1.0 mol of atoms: phosphorus, sulfur, or bromine.

    (b) Determine which of the following has the greatest mass in 1.0 mol of molecules: P_4, S_8, or Br_2.

49. Weighing 3104 carats (1 carat = 200 mg), the Cullinan diamond was the largest natural diamond ever found (January 25, 1905). How many carbon atoms were present in the stone?

50. An engagement ring contains a diamond weighing 1.25 carats (1 carat = 200 mg). How many atoms are present in the diamond?

51. Up to 0.0100% by mass of copper(I) iodide may be added to table salt as a dietary source of iodine. How many moles of CuI would be contained in a 1.00 lb (454 g) of such table salt that contained 0.0100% CuI by mass?

52. The serving size of a popular raisin bran cereal is 55 g (1 cup). One serving supplies 270 mg of sodium, 11% of the Recommended Daily Value. How many moles and atoms of sodium are in the Recommended Daily Value?

53. A certain nut crunch cereal is listed as containing 11.0 g of sugar (sucrose, $C_{12}H_{22}O_{11}$) per serving size of 60.0 g. How many servings of this cereal must one eat to consume 0.0278 mol of sugar?

54. A cylinder of compressed gas contains nitrogen and oxygen in the ratio of 3 mol to 1 mol. If the cylinder is known to contain 2.50×10^4 g of oxygen, what is the total mass of the gas mixture?

Percent Composition; Empirical and Molecular Formulas

55. Calculate the following to four significant figures.

    (a) the percent composition of ammonia, NH_3
    (b) the percent composition of photographic hypo, $Na_2S_2O_3$
    (c) the percent of calcium ion in $Ca_3(PO_4)_2$

56. Determine the following to four significant figures.

    (a) the percent composition of hydrazoic acid, HN_3
    (b) the percent composition of TNT, $C_6H_2(CH_3)(NO_2)_3$
    (c) the percent of SO_4^{2-} in $Al_2(SO_4)_3$

57. Determine the percent ammonia, NH_3, in $Co(NH_3)_6Cl_3$, to three significant figures.

58. Determine the percent water in $CuSO_4(H_2O)_5$ to three significant figures.

59. Determine the empirical formulas for compounds with the following percent compositions.

    (a) 15.8% carbon and 84.2% sulfur
    (b) 40.0% carbon, 6.7% hydrogen, and 53.3% oxygen

60. Determine the empirical formulas for compounds with the following percent compositions:

    (c) 43.6% phosphorus and 56.4% oxygen
    (d) 28.7% K, 1.5% H, 22.8% P, and 47.0% O

61. A compound of carbon and hydrogen contains 92.3% C and has a molecular mass of 78.1 amu. What is its molecular formula?

62. Dichloroethane, a compound that is often used for dry cleaning, contains carbon, hydrogen, and chlorine. It has a molecular mass of 99 amu. Analysis of a sample shows that it contains 24.3% carbon and 4.1% hydrogen. What is its molecular formula?

63. Calculate the empirical formula of a crystalline salt that has the following percent composition: Na, 18.5%; S, 25.8%; O (except for that in the water), 19.3%; H_2O, 36.4%.

64. Most polymers are very large molecules composed of simple units repeated many times. Thus they often have relatively simple empirical formulas. Calculate the empirical formulas of the following polymers.

    (a) Lucite (Plexiglas): 59.9% C, 8.06% H, 32.0% O
    (b) Saran: 24.8% C, 2.0% H, 73.1% Cl

    (c) Polyethylene: 86% C, 14% H
    (d) Polystyrene: 92.3% C, 7.7% H
    (e) Orlon: 67.9% C, 5.70% H, 26.4% N

65. Calcium is necessary in the diet to promote strong bones. Tablets containing 500 mg (three significant figures) of calcium as calcium carbonate from sea shells are sold as a diet supplement. One of those tablets had a mass of 2.27 g. What was the percent calcium carbonate in the tablet?

66. A major textile dye manufacturer developed a new yellow azo-dye. The azo-dye has a composition of 75.95% C, 17.72% N, and 6.33% H by mass and a molar mass of about 240. Determine the molecular formula of the azo-dye.

Molarity and Solutions

67. Determine the concentration in moles per liter for each of the following solutions.

    (a) 0.444 mol of $CoCl_2$ in 0.654 L of solution
    (b) 98.0 g of phosphoric acid, H_3PO_4, in 1.00 L of solution
    (c) 0.2074 g of calcium hydroxide, $Ca(OH)_2$, in 40.00 mL of solution
    (d) 10.5 kg of $Na_2SO_4 \cdot 10H_2O$ in 18.60 L of solution
    (e) 7.0×10^{-3} mol of I_2 in 100.0 mL of solution
    (f) 1.8×10^4 mg of HCl in 0.075 L of solution

68. Determine the concentration in moles per liter for each of the following solutions.

    (a) 1.457 mol KCl in 1.500 L of solution
    (b) 0.515 g of H_2SO_4 in 1.00 L of solution
    (c) 20.54 g of $Al(NO_3)_3$ in 1575 mL of solution
    (d) 2.76 kg of $CuSO_4 \cdot 5H_2O$ in 1.45 L of solution
    (e) 0.005653 mol of Br_2 in 10.00 mL of solution
    (f) 0.000889 g of glycine, $C_2H_5NO_2$, in 1.05 mL of solution

69. Calculate the mass of solute present in each solution.

    (a) 2.0 L of 0.050 M sodium hydroxide, NaOH
    (b) 100.0 mL of 0.0020 M barium hydroxide, $Ba(OH)_2$
    (c) 0.080 L of 0.050 M sulfuric acid, H_2SO_4

70. Calculate the mass of solute present in each solution.

    (a) 1.50 L of 0.850 M phosphoric acid, H_3PO_4
    (b) 50.0 mL of 0.0300 M glucose, $C_{12}H_{22}O_{11}$
    (c) 0.02384 L of 0.1284 M HCl

71. Determine the moles of solute and the mass of solute required to make the indicated amount of solution.

    (a) 1.00 L of a 1.00 M $Ca(NO_3)_2$ solution
    (b) 465 mL of a 0.2543 M solution of C_3H_7OH, propyl alcohol
    (c) 0.075 L of a 0.1625 M $KClO_3$ solution
    (d) 0.075 L of a 0.1625 M $C_{12}H_{22}O_{11}$ solution

72. Determine the moles of solute and the mass of solute required to make the indicated amount of solution.

(a) 2.000 L of a 1.000 M Fe(NO₃)₃ solution

(a) 2.000 L of a 1.000 M $Fe(NO_3)_3$ solution

(b) 1233 mL of a 0.8842 M $CaSO_4$ solution

(c) 1.0750 L of a 1.1625 M H_3PO_4 solution

(d) 1.0750 L of a 1.1625 M C_3H_7OH solution

73. (a) What volume of a 1.00 M $Fe(NO_3)_3$ solution is required to make 1.00 L of a 0.250 M solution of $Fe(NO_3)_3$?

(b) What volume of a 0.3556 M C_3H_7OH solution is required to make 0.5000 L of a 0.1222 M solution of C_3H_7OH?

(c) What volume of a 0.850 M H_3PO_4 solution is required to make 10.00 L of a 0.350 M solution of H_3PO_4?

(d) What volume of a 0.33 M $C_{12}H_{22}O_{11}$ solution is required to make 25 mL of a 0.025 M solution of $C_{12}H_{22}O_{11}$?

(e) What concentration of NaCl results when 0.150 L of a 0.556 M solution is allowed to evaporate until the volume is reduced to 0.105 L?

74. (a) What concentration of $Fe(NO_3)_3$ results when 1.00 L of a 0.250 M solution of $Fe(NO_3)_3$ is diluted to 2.00 L?

(b) What concentration of C_3H_7OH results when 0.5000 L of a 0.1222 M solution of C_3H_7OH is diluted to 1.250 L?

(c) What concentration of H_3PO_4 results when 2.35 L of a 0.350 M solution of H_3PO_4 is diluted to 4.00 L?

(d) What concentration of $C_{12}H_{22}O_{11}$ results when 22.50 mL of a 0.025 M solution of $C_{12}H_{22}O_{11}$ is diluted to 100.0 mL?

(e) What is the resulting concentration when 225.5 mL of a 0.09988 M solution of Na_2CO_3 is allowed to evaporate until the volume is reduced to 45.00 mL?

75. A throat spray is 1.40% by mass phenol, C_6H_5OH, in water. The solution has a density of 0.9956 g/mL. Calculate the molarity of the solution.

76. A cough syrup contains 5.0% ethyl alcohol, C_2H_5OH, by mass. The density of the solution is 0.9928 g/mL. Determine the molarity of the alcohol in the cough syrup.

77. A 1.00-L bottle of a solution of concentrated HCl was purchased for the general chemistry laboratory. The solution contained 434.4 g of HCl. What is the molarity of the solution?

78. An experiment in a general chemistry laboratory called for a 2.00 M solution. How many mL of the solution in Exercise 77 would be required to make 250 mL of the dilute solution?

Chemical Calculations Involving Equations

79. (a) Write the balanced equation and then, using a procedure similar to that in Example 3.21, outline the steps necessary to determine:

i. The number of moles and the mass of chlorine, Cl_2, required to react with 10.0 g of sodium metal, Na, to produce sodium chloride, NaCl.

ii. The number of moles and the mass of oxygen formed by the decomposition of 1.252 g of mercury(II) oxide.

iii. The number of moles and the mass of sodium nitrate, $NaNO_3$, required to produce 128 g of oxygen. ($NaNO_2$ is the other product.)

iv. The number of moles and the mass of carbon dioxide formed by the combustion of 20.0 kg of carbon in an excess of oxygen.

v. The number of moles and the mass of copper(II) carbonate needed to produce 1.500 kg of copper(II) oxide. (CO_2 is the other product.)

vi. The number of moles and the mass of

12.85 g of with an excess of Br_2.

(b) Determine the number of moles and the mass requested for each reaction in part (a).

80. (a) Write the balanced equation and then, using a procedure similar to that in Example 3.21, outline the steps necessary to determine:

i. The number of moles and the mass of Mg required to react with 5.00 g of HCl and produce $MgCl_2$ and H_2.

ii. The number of moles and the mass of oxygen formed by the decomposition of 1.252 g of silver(I) oxide.

iii. The number of moles and the mass of magnesium carbonate, $MgCO_3$, required to produce 283 g of carbon dioxide. (MgO is the other product.)

iv. The number of moles and the mass of water formed by the combustion of 20.0 kg of acetylene, C_2H_2, in an excess of oxygen.

v. The number of moles and the mass of barium peroxide, BaO_2, needed to produce 2.500 kg of barium oxide, BaO. (O_2 is the other product.)

vi. The number of moles and the mass of

required to react with H_2O to

produce 9.55 g of .

(b) Determine the number of moles and the mass requested for each reaction in part (a).

81. (a) Using a procedure similar to that in Example 3.21, outline the steps necessary to determine the number of moles and the mass of H_2 produced by the reaction of 118.5 mL of a 0.8775 M solution of H_3PO_4 according to the equation

$$2Cr + 2H_3PO_4 \longrightarrow 3H_2 + 2CrPO_4$$

   (b) Do the calculations outlined.

82. (a) Using a procedure similar to that in Example 3.21, outline the steps necessary to determine the number of moles and the mass of gallium chloride formed by the reaction of 2.6 L of a 1.44 M solution of HCl according to the equation

$$2Ga + 6HCl \longrightarrow 2GaCl_3 + 3H_2$$

   (b) Do the calculations outlined.

83. (a) Determine how many molecules of I_2 are produced by the reaction of 0.4235 mol of $CuCl_2$ according to the equation

$$2CuCl_2 + 4KI \longrightarrow 2CuI + 4KCl + I_2$$

   (b) What mass of I_2 is produced?

84. Silver is often extracted from ores as $K[Ag(CN)_2]$ and then recovered by the reaction

$$2K[Ag(CN)_2](aq) + Zn(s) \longrightarrow$$
$$2Ag(s) + Zn(CN)_2(aq) + 2KCN(aq)$$

   (a) How many molecules of $Zn(CN)_2$ are produced by the reaction of 35.27 g of $K[Ag(CN)_2]$?
   (b) What mass of $Zn(CN)_2$ is produced?

85. Silver sulfadiazine burn cream creates a barrier against bacterial invasion and releases antimicrobial agents directly into a wound. What mass of silver oxide, Ag_2O, is required to produce 25.0 g of silver sulfadiazine, $AgC_{10}H_9N_4SO_2$, from the reaction of silver oxide and sulfadiazine?

$$2C_{10}H_{10}N_4SO_2 + Ag_2O \longrightarrow 2AgC_{10}H_9N_4SO_2 + H_2O$$

86. Carborundum is silicon carbide, SiC, a very hard material used as an abrasive on sandpaper and in other applications. It is prepared by the reaction of pure sand, SiO_2, with carbon at high temperature. Carbon monoxide, CO, is the other product of this reaction. Write the balanced equation for the reaction, and calculate how much SiO_2 is required to produce 3.00 kg of SiC.

87. Automotive air bags inflate when a sample of sodium azide, NaN_3, is very rapidly decomposed.

$$2NaN_3(s) \longrightarrow 2Na(s) + 3N_2(g)$$

What mass of sodium azide is required to produce 13.0 ft^3 (368 L) of nitrogen gas with a density of 1.25 g/L?

Percent Yield

88. A student spilled his ethanol preparation and consequently isolated only 25 g instead of the 81 g theoretically possible. What was his percent yield?

89. A sample of 0.53 g of carbon dioxide was obtained by heating 1.31 g of calcium carbonate. Outline the steps needed to determine the percent yield of the reaction, and then calculate the percent yield.

$$CaCO_3(s) \longrightarrow CaO(s) + CO_2(g)$$

90. Freon-12, CCl_2F_2, is prepared from CCl_4 by reaction with HF. The other product of this reaction is HCl. Outline the steps needed to determine the percent yield of a reaction that produces 12.5 g of CCl_2F_2 from 32.9 g of CCl_4. Determine the percent yield.

91. Citric acid, $C_6H_8O_7$, a component of jams, jellies, and fruity soft drinks, is prepared industrially via fermentation of sucrose by the mold *Aspergillus niger*. The overall reaction is

$$C_{12}H_{22}O_{11} + H_2O + 3O_2 \longrightarrow 2C_6H_8O_7 + 4H_2O$$

What mass of citric acid is produced from exactly 1 metric ton $(1.000 \times 10^3$ kg) of sucrose if the yield is 92.30%?

92. Toluene, $C_6H_5CH_3$, is oxidized by air under carefully controlled conditions to benzoic acid, $C_6H_5CO_2H$, which is used to prepare the food preservative sodium benzoate, $C_6H_5CO_2Na$. What is the percent yield of a reaction that converts 1.000 kg of toluene to 1.21 kg of benzoic acid?

$$2C_6H_5CH_3 + 3O_2 \longrightarrow 2C_6H_5CO_2H + 2H_2O$$

93. In a laboratory experiment, the reaction of 3.0 mol of H_2 with 2.0 mol of I_2 produced 1.0 mol of HI. Determine the theoretical yield in grams and the percent yield for this reaction.

94. Outline the steps needed to solve the following problem, and then do the calculations. Ether, $(C_2H_5)_2O$, for anesthetic use is prepared by the reaction of ethanol with sulfuric acid.

$$2C_2H_5OH + H_2SO_4 \longrightarrow (C_2H_5)_2O + H_2SO_4 \cdot H_2O$$

What is the percent yield of ether if 1.17 L $(d = 0.7134$ g/mL) is isolated from the reaction of 1.500 L of C_2H_5OH $(d = 0.7894$ g/mL)?

Limiting Reagents

95. Outline the steps needed to determine the limiting reactant when 30.0 g of propane, C_3H_8, is burned with 75.0 g of oxygen.

$$C_3H_8(g) + 5O_2(g) \longrightarrow 3CO_2(g) + 4H_2O(g)$$

Determine the limiting reactant.

96. Outline the steps needed to determine the limiting reagent when 0.50 g of Cr and 0.75 g of H_3PO_4 react according to the chemical equation

$$2Cr + 2H_3PO_4 \longrightarrow 2CrPO_4 + 3H_2$$

Determine the limiting reactant.

97. What is the limiting reagent when 1.50 g of lithium and 1.50 g of nitrogen combine to form lithium nitride, a com-

ponent of advanced batteries, according to the following unbalanced equation?

$$Li + N_2 \longrightarrow Li_3N \quad \text{(unbalanced)}$$

98. Uranium can be isolated from its ores by dissolving it as $UO_2(NO_3)_2$ and then separating it as solid $UO_2(C_2O_4) \cdot 3H_2O$. Addition of 0.4031 g of sodium oxalate, $Na_2C_2O_4$, to a solution containing 1.481 g of uranyl nitrate, $UO_2(NO_3)_2$, yields 1.073 g of solid $UO_2(C_2O_4) \cdot 3H_2O$.

$$Na_2C_2O_4 + UO_2(NO_3)_2 + 3H_2O \longrightarrow$$
$$UO_2(C_2O_4) \cdot 3H_2O + 2NaNO_3$$

Determine the limiting reagent and the percent yield of this reaction.

99. How many molecules of $C_2H_4Cl_2$ can be prepared from 15 C_2H_4 molecules and 8 Cl_2 molecules?

100. How many molecules of CH_2O can be prepared from 20 C atoms, 15 H_2 molecules, and 8 O_2 molecules?

101. How many molecules of the sweetener saccharin

can be prepared from 30 C atoms, 25 H atoms, 12 O atoms, 8 S atoms, and 14 N atoms?

102. How many moles of the anesthetic gas halothane

can be prepared from 30 mol of C atoms, 20 mol of H_2 molecules, 12 mol of F_2 molecules, 12 mol of Cl_2 molecules, and 10 mol of Br_2 molecules?

103. The phosphorus pentoxide used to produce phosphoric acid for cola soft drinks is prepared by burning phosphorus in oxygen.

(a) Determine the limiting reagent when 0.200 mol of P_4 and 0.200 mol of O_2 react according to the chemical equation

$$P_4 + 5O_2 \longrightarrow P_4O_{10}$$

(b) Calculate the percent yield if 10.0 g of P_4O_{10} is isolated from the reaction.

Quantitative Analysis

104. Outline the steps needed to determine the volume of 0.3446 M H_2SO_4 that would be required to titrate a solution that contains 2.474 g of KOH.

$$H_2SO_4 + 2KOH \longrightarrow K_2SO_4 + 2H_2O$$

Do the calculations and determine the volume.

105. Outline the steps needed to determine the volume of 0.0105 M HBr solution that would be required to titrate 125 mL of a 0.0100 M $Ca(OH)_2$ solution. Then determine the volume.

$$Ca(OH)_2 + 2HBr \longrightarrow CaBr_2 + 2H_2O$$

106. Titration of a 20.0-mL sample of a particularly acidic rain required 1.7 mL of 0.0811 M NaOH to reach the end point. If we assume that the acidity of the rain is due to the presence of sulfuric acid, what was the concentration of sulfuric acid in this sample of rain?

107. What is the concentration of NaCl in a solution if titration of 15.00 mL of the solution with 0.2503 M $AgNO_3$ requires 20.22 mL of the $AgNO_3$ solution to reach the end point?

$$AgNO_3(aq) + NaCl(aq) \longrightarrow AgCl(s) + NaNO_2(aq)$$

108. In a common medical laboratory determination of the concentration of free chloride ion in blood serum, a serum sample is titrated with a $Hg(NO_3)_2$ solution.

$$2Cl^- + Hg(NO_3)_2 \longrightarrow 2NO_3^- + HgCl_2$$

What is the Cl^- concentration in a 0.25-mL sample of normal serum that requires 1.46 mL of 8.25×10^{-4} M $Hg(NO_3)_2$ to reach the end point?

109. Potatoes can be peeled commercially by soaking them in a 3 M to 6 M solution of sodium hydroxide and then removing the loosened skins by spraying them with water. Does a sodium hydroxide solution have a suitable concentration if titration of 12.00 mL of the solution requires 30.6 mL of 1.65 M HCl to reach the end point?

110. Outline the steps needed to determine the percent gallium in gallium bromide from the following data: A sample of gallium bromide weighing 0.165 g reacted with silver nitrate, $AgNO_3$, giving gallium nitrate and 0.299 g of AgBr. Do the calculations.

111. When iodine was added to liquid chlorine cooled by a dry-ice–acetone mixture, orange crystals separated. The amount of chlorine in a sample of the crystals was analyzed by precipitation of AgCl. A 0.6548-g sample gave 1.2071 g of AgCl. Outline the steps needed to determine the empirical formula of the compound, and then do the calculations.

112. The principal component of mothballs is naphthalene, a compound that has a molecular mass of about 130 amu and contains only carbon and hydrogen. A 3.000-mg sample of naphthalene burns to give 10.3 mg of CO_2. Determine its empirical and molecular formulas.

113. A 0.025-g sample of a compound that is composed of boron and hydrogen and has a molecular mass of about 28 amu burns spontaneously when exposed to air, producing 0.063 g of B_2O_3. What are the empirical and molecular formulas of this compound?

Additional Exercises

114. Would you agree to buy a trillion (1,000,000,000,000) gold atoms for $5? Explain why or why not.

115. Assume that 45.0 g of Si reacts with N_2, giving a 100% yield, 0.533 mol of Si_3N_4, according to the equation

$$3Si + 2N_2 \longrightarrow Si_3N_4$$

Indicate which of the following can be determined by using only the information given in this exercise, and explain your reasoning: (a) the moles of Si reacting, (b) the moles of N_2 reacting, (c) the atomic mass of N, (d) the atomic mass of Si, (e) the mass of Si_3N_4 produced, (f) the limiting reactant.

116. A tough, low-density alloy of magnesium, ($d = 1.35$ g/cm^3) contains 14.0% Li, 1.0% Al, and 85.0% Mg. Calculate the number of moles of lithium in a sample of the alloy weighing 1.00 lb.

117. A tube of toothpaste contained 0.76 g of sodium monofluorophosphate, Na_2PO_3F, in 100 mL.

   (a) How much fluorine, in milligrams, was present?
   (b) How many fluorine atoms were present?

118. Nitric acid is prepared from ammonia in a three-step process.

   i. $4NH_3(g) + 5O_2(g) \longrightarrow 4NO(g) + 6H_2O(g)$
   ii. $2NO(g) + O_2(g) \longrightarrow 2NO_2(g)$
   iii. $3NO_2(g) + H_2O(l) \longrightarrow 2HNO_3(aq) + NO(g)$

   Calculate the number of tons (1 metric ton is defined as 1000 kg) of nitric acid that can be prepared from 100.0 tons of ammonia, assuming 100% efficiency in each of the reactions.

119. In one of the states of the United States, limits on the quantities of toxic substances that may be discharged into the sewer system have been established. One statute limits hexavalent chromium to 0.50 mg/L. If an industry in this state is discharging potasssium dichromate, $K_2Cr_2O_7$, what is the maximum concentration of that substance that may be discharged?

120. Urea, $CO(NH_2)_2$, is manufactured on a large scale for use in producing urea-formaldehyde plastics and as a fertilizer. Find the maximum mass of urea that can be manufactured from the CO_2 produced by combustion of 1.00×10^3 kg of carbon, followed by the reaction

$$CO_2(g) + 2NH_3(g) \longrightarrow CO(NH_2)_2(s) + H_2O(l)$$

121. In an accident, a solution containing 2.5 kg of nitric acid was spilled. Two kilograms (2.0 kg) of Na_2CO_3 was quickly spread on the area, and CO_2 was released by the reaction. Was sufficient Na_2CO_3 used to neutralize all of the acid?

122. D5W refers to one of the solutions used as an intravenous fluid. It is a 5.0%-by-mass solution of dextrose, $C_6H_{12}O_6$, in water. The density of D5W is 1.029 g/mL. Calculate the molarity of the solution.

123. Clorox is 5.0% sodium hypochlorite, NaOCl, by mass and has a solution density of 1.04 g/mL. The user is advised to add 1 cup (8.0 oz) per wash load of 8.0 gal. What is the molarity of the bleach in the wash load?

124. A compact car gets 37.5 miles per gallon on the highway. Gasoline contains 84.2% carbon by mass and has a density of 0.8205 g/mL. Determine the mass of carbon dioxide produced during a 500-mi trip.

125. The following quantities are placed in a container: 1.5×10^{24} atoms of hydrogen, 1.0 mol of sulfur, and 88.0 g of oxygen.

   (a) What is the total mass in grams for the collection of all three elements?
   (b) What is the total number of moles of atoms for the three elements?
   (c) If the mixture of the three elements formed a compound with molecules that contain two hydrogen atoms, one sulfur atom, and four oxygen atoms, which substance would be the limiting reactant?
   (d) How many grams of material would remain unreacted in the change described in part (c)?

126. Sodium bicarbonate (baking soda), $NaHCO_3$, can be purified by dissolving it in hot water (60°C), filtering to remove insoluble impurities, cooling to 0°C to precipitate solid $NaHCO_3$, and then filtering to remove the solid, leaving soluble impurities in solution. Any $NaHCO_3$ that remains in solution is not recovered. The solubility of $NaHCO_3$ in hot water at 60°C is 164 g/L. Its solubility in cold water at 0°C is 69 g/L. What is the percent yield of $NaHCO_3$ when it is purified by this method?

127. (a) The acid secreted by the cells of the stomach lining is a hydrochloric acid solution that typically contains 0.282 g of HCl per 50.00 mL of solution. What is the concentration of this acid?

   (b) The amount of active ingredient per tablet of several antacids and equations for the reactions of these ingredients with stomach acid (HCl) are given below. What volume of acid will react with the given mass of the active ingredient in each tablet?

   i.   Phillip's Tablets, 0.311 g $Mg(OH)_2$

$$Mg(OH)_2 + 2HCl \longrightarrow MgCl_2 + 2H_2O$$

   ii. Tums, 0.500 g $CaCO_3$

$$CaCO_3 + 2HCl \longrightarrow CaCl_2 + H_2O + CO_2$$

   iii. Rolaids, 0.334 g $NaAl(CO_3)(OH)_2$

$$NaAl(CO_3)(OH)_2 + 4HCl \longrightarrow$$
$$NaCl + AlCl_3 + 3H_2O + CO_2$$

   iv. Gelusil, 0.500 g $Mg_2Si_3O_8$ and 0.075 g $Mg(OH)_2$

$$Mg_2Si_3O_8 + 4HCl \longrightarrow 2MgCl_2 + 3SiO_2 + 2H_2O$$
$$Mg(OH)_2 + 2HCl \longrightarrow MgCl_2 + 2H_2O$$

128. Hach Chemical Company sells a test kit for determining the concentration of chloride ion in domestic water sup-

plies. The kit contains a silver nitrate, $AgNO_3$, solution that is added drop by drop to a 23.0-mL water sample to which an indicator has been added. When sufficient silver nitrate has been added to convert the chloride ion completely to silver chloride, AgCl, the solid produced turns orange. The concentration of the silver nitrate solution is such that each drop used to reach the color change corresponds to 12.5 mg of Cl^- per liter of water tested.

(a) What mass of chloride ion (mg/L) is contained in a water sample that requires 12 drops of test solution to reach the color change?

(b) What is the molar concentration of Cl^- in the sample tested in part (a)?

(c) If the sample size used is 5.75 mL instead of 23.0 mL, to what chloride ion concentration (mg/L) does 1 drop of the test solution correspond?

(d) This test kit can also be used for determination of bromide ion or iodide ion. If the reading obtained is multiplied by 2.25, the concentration of bromide ion (mg/L) is obtained. What factor should be used to obtain the concentration of iodide ion (mg/L)?

(e) If 20 drops of the silver nitrate test solution equals 1.00 mL, what is its molar concentration of $AgNO_3$?

CHAPTER OUTLINE

4.1 Thermal Energy, Heat, and Temperature
4.2 Measurement of Heat
4.3 Calorimetry
4.4 Thermochemistry and Thermodynamics
4.5 Enthalpy Changes
4.6 Hess's Law
4.7 Fuel and Food

4

Thermochemistry

Energy from the sun creates weather that drives these wind generators.

Our present society could not function without the heat produced by chemical reactions. We burn natural gas, petroleum products, coal, wood, and garbage. About 20% of the energy used in the United States is consumed in our homes, primarily for space heating, cooling, and producing hot water. Another 25% is used in transportation. Automobiles, airplanes, trains, and trucks make rapid travel possible and also transport our food, raw materials, and manufactured goods from production sites to consumption sites. In addition, generation of electricity uses another 15% in the combustion reactions of coal and oil used to provide steam to drive turbines to produce electricity.

About 40% of the energy consumed in the United States is used in manufacturing. Many industrial chemical reactions require enormous amounts of energy to proceed. These include reactions used in the manufacture of the fertilizer ammonia and the important industrial chemicals lime, sodium hydroxide, and chlorine. Each is produced in multi-billion-pound quantities. Production of iron requires high temperatures resulting from combustion of coke, a processed form of coal. The manufacture of aluminum consumes vast amounts of electrical energy.

Over 90% of the energy used by our society comes originally from the sun. Plants store solar energy as they convert carbon dioxide and water to glucose through photosynthetic reactions. These reactions are driven by energy from the sun. We release that stored energy when we burn wood, starch, cellulose, and other carbohydrate products, such as ethanol, produced from plants. Burning coal and petroleum also releases stored solar energy: These fuels are fossilized plant materials.

Useful forms of energy are also available from a variety of chemical reactions other than combustion reactions. For example, the energy produced by a flashlight battery results from chemical reactions.

Our world's increasing emphasis on recycling reflects both a desire to reduce the amount of energy invested in producing materials and a decreasing capacity for disposal of trash. For example, recovering aluminum by recycling requires only about 5% of the amount of energy necessary to produce the metal from the ore.

The sun is a storehouse of energy. Every day the sun provides the earth with about ten thousand times the amount of energy necessary to meet all of the world's energy needs for that day. The energy problem we read about is not a shortage of energy but rather a shortage of useful forms of energy. The challenge to us is to find ways to trap and store incoming solar energy so that it can be released in reactions or chemical processes that are both convenient and nonpolluting.

This chapter introduces many of the basic ideas that underlie the relationships between chemical changes and energy. Understanding these concepts will be necessary as humankind explores alternative ways to convert the sun's energy into useful forms. In this chapter we will find out how to measure the amount of heat absorbed or evolved during a chemical reaction or a physical change. In addition, we will see how this information is used to find the heat changes associated with other changes. This study of the amount of heat absorbed or given off during chemical or physical change is called **thermochemistry.**

Finally, we should note that the same concepts used to determine the amount of energy produced by combustion of a fuel also apply to food. Much of the food we eat is simply fuel for our bodies. Food warms our bodies because many of the body's chemical reactions, such as the metabolic combustion of sugar to carbon dioxide and water, produce heat. Thus the concepts introduced in this chapter also serve as an introduction to the study of the energetics of living things.

Figure 4.1

An oxyacetylene torch produces heat by the combustion of acetylene in oxygen. The heat enters, heats the cooler metal, and eventually melts it. The sparks result when the molten metal is blown away by the gases in the flame.

4.1   Thermal Energy, Heat, and Temperature

Thermal energy is kinetic energy (Section 1.3) associated with the random motion of atoms and molecules. The greater the amount of thermal energy in a sample of matter, the more vigorous the motion of its atoms and molecules. If we increase the amount of thermal energy in a sample of matter, then either the temperature of the substance increases or the substance melts or evaporates with no change in temperature. If we decrease the amount of thermal energy, then either the temperature of the substance decreases or the substance condenses or freezes with no change in temperature. We can detect a change in the amount of thermal energy in a substance by observing a temperature change or a phase change (melting, freezing, evaporation, and so on).

Heat is energy that is transferred between two bodies that are at different temperatures. Transfer of heat increases the thermal energy of one body and decreases the thermal energy of the other. Heat will move spontaneously from a warmer substance to a colder substance until the temperatures of the two substances are equal. However, heat can be forced to move from a cooler to a warmer substance, as it does in a refrigerator.

Chemical reactions and physical changes can release or absorb heat. A change that releases heat is called an **exothermic process.** The burning of charcoal, for example, is an exothermic process. A reaction or change that absorbs heat is an **endothermic process.** The boiling of water is an endothermic process, because heat is absorbed as water boils.

Most substances expand as their temperature increases and contract as their temperature decreases. This property is used to measure temperature changes. The level of a liquid in a thermometer rises and falls as the liquid expands and contracts in response to changes in temperature.

In the absence of a phase change, we can increase the temperature of an object by adding heat, or some other form of energy, to it (Fig. 4.1). Removing heat, or some other form of energy, reduces the temperature. For example, when hot coffee is poured into a mug, the coffee becomes cooler and the mug warmer as heat moves from the hot coffee into the cooler mug.

Figure 4.2

The periodic table is printed on thermochromic paper. An endothermic change converts the color from orange to yellow when the paper is warmed.

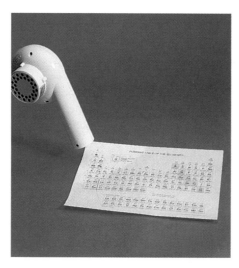

4.2 Measurement of Heat

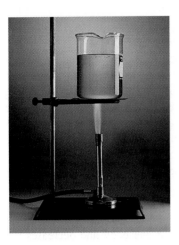

We use units of calories and joules to measure amounts of heat and other energy. A **calorie (cal)** is the amount of heat or energy necessary to raise the temperature of 1 gram of water by approximately 1 degree Celsius (Fig. 4.3). A **kilocalorie (kcal)** is 1000 calories. The SI unit of heat and energy is the **joule (J).*** One calorie is defined as 4.184 joules. A **kilojoule (kJ)** is 1000 joules. When we talk about heat produced or consumed by chemical reactions in the body, we use nutritional Calories. (Note that this unit of heat is capitalized.) One **nutritional Calorie** equals 1000 calories, or 1 kilocalorie.

The **specific heat** of a substance (an intensive property, Section 1.4) is the quantity of heat required to raise the temperature of 1 gram of the substance 1 degree Celsius (or 1 kelvin). The specific heat is also equal to the amount of heat lost by 1 gram of the substance when it cools by 1 degree Celsius. The specific heat of water, which is one of the largest, is 4.184 joules per gram per degree ($4.184 \text{ J g}^{-1}\,{}^\circ\text{C}^{-1}$) at 17°C. Metals have specific heats less than 1 joule per gram per degree; for example, the specific heat of aluminum is $0.88 \text{ J g}^{-1}\,{}^\circ\text{C}^{-1}$ at 25°C. The specific heat of a substance varies with temperature. However, we will usually make the simplifying assumption that this variation can be neglected over any range of temperatures of interest to us.

We can use the specific heat of a substance to measure heat. For example, if a 1-gram sample of aluminum picks up enough heat to raise its temperature by 1 degree, it must have absorbed 0.88 joule. If it cools by 2 degrees, it has lost 1.76 joule.

The **heat capacity** of a body of matter is the quantity of heat involved as it increases (or decreases) its temperature by 1 degree Celsius (1 kelvin). Heat capacity is an extensive property: It depends on the mass of the body. For example, 10 grams of aluminum (specific heat, $0.88 \text{ J g}^{-1}\,{}^\circ\text{C}^{-1}$) has a heat capacity of $8.8 \text{ J }{}^\circ\text{C}^{-1}$ ($10 \text{ g} \times 0.88 \text{ J g}^{-1}\,{}^\circ\text{C}^{-1}$), and 25 grams of aluminum has a heat capacity of $22 \text{ J }{}^\circ\text{C}^{-1}$.

If we know the mass of a substance and its specific heat, we can determine the amount of heat, q, entering or leaving the substance by measuring the temperature change before and after the heat is gained or lost.

$$q = \text{specific heat} \times \text{mass of substance} \times \text{temperature change}$$

or

$$q = cm\Delta T$$

where c is the specific heat of the substance, m is its mass, and ΔT (which is read "delta T") is the temperature change, $T_{\text{final}} - T_{\text{initial}}$. If heat enters a substance, then the temperature of that substance increases and its final temperature is higher than its initial temperature. $T_{\text{final}} - T_{\text{initial}}$ has a positive value, and the value of q is positive. If heat flows out of a substance, q is negative.

Figure 4.3

The temperature of this 800-g sample of water will increase by 1°C for each 800 cal, or 3350 J, absorbed.

| **Example 4.1** | **Measuring an Amount of Heat** |

A flask containing 800 g of water is heated as shown in Fig. 4.3. If the temperature of the water increases from 20°C to 85°C, how much heat (in joules) did the water absorb?

---

*A joule is defined as the amount of heat or other energy equal to the kinetic energy ($\frac{1}{2}mv^2$) of an object with a mass m of exactly 2 kilograms moving with a velocity v of exactly 1 meter per second. One joule is equivalent to $1 \text{ kg m}^2 \text{ s}^{-2}$, which is called 1 newton-meter.

Solution: One way we can answer this question involves three steps.

Step 1.   *Determine the amount of heat involved as 1 g of water changes temperature by 1°C (the specific heat of water).* Generally we look this up in a table of specific heats, but the preceding text gives the specific heat of water as 4.184 J g^{-1} °C^{-1}. To heat 1 g of water 1°C requires 4.184 J.

Step 2.   *Determine the amount of heat necessary to heat 800 g of water by 1°C.* To heat 800 g of water by 1 °C requires

$$800 \times 4.184 \text{ J} = 3350 \text{ J}$$

Step 3.   *Determine the amount of heat necessary to heat 800 g of water by 65°C (from 20°C to 85°C).* To increase the temperature of 800 g of water requires

$$65 \times 3350 \text{ J} = 220{,}000 \text{ J}$$

A second way to solve the problem is to use the equation

$$q = cm\Delta T = cm(T_{final} - T_{initial})$$
$$= 4.184 \text{ J g}^{-1} \text{°C}^{-1} \times 800 \text{ g} \times (85 - 20) \text{°C}$$
$$= 220{,}000 \text{ J}$$

Because the temperature increased, heat moved into the water and q is positive.

Check your learning: How much heat, in joules and calories, must be added to an iron skillet that weights 5.00×10^2 g to increase its temperature from 25°C to 250°C? The specific heat of iron is 0.451 J g^{-1} °C^{-1}.

Answer: 5.07×10^4 J; 1.21×10^4 cal

Figure 4.4

A calorimeter constructed from two polystyrene cups. A thermometer and stirrer extend through the cover into the reaction mixture.

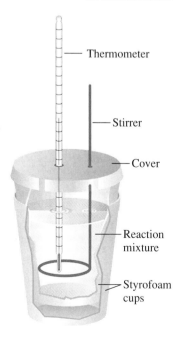

Thermometer

Stirrer

Cover

Reaction mixture

Styrofoam cups

4.3   Calorimetry

One technique we can use to measure the amount of heat involved in a chemical or physical change is called **calorimetry.** The simplest form of calorimetry is based on measuring the temperature change as heat is transferred into or out of a known quantity of water. We can determine the amount of heat from the temperature change of the water, as shown in Example 4.1 Of course, we must be sure that no heat other than that to be measured enters the water and that no heat is lost before the measurement is complete.

A **calorimeter** is a device used to measure the amount of heat involved in a chemical change. General chemistry students often use calorimeters constructed from two polystyrene cups like those used for coffee or other hot beverages (Fig. 4.4). Because the cups are good insulators, they prevent the loss (or gain) of heat to (or from) the surroundings. When an exothermic reaction occurs in solution in such a calorimeter, the heat produced by the reaction is trapped in the solution and increases its temperature. When an endothermic reaction occurs, the heat required is absorbed from the thermal energy of the solution and decreases its temperature. The change in temperature can be used to calculate the amount of heat involved in either case.

When we use polystyrene cups as a calorimeter, the amount of heat absorbed by the calorimeter is usually small enough that we can neglect it, and the amount of heat produced in the reaction, $q_{reaction}$, equals the amount of heat absorbed by the solution, q_{soln}. Because energy can be neither created nor destroyed in a chemical reaction,

$$q_{reaction} + q_{soln} = 0 \quad \text{and} \quad q_{reaction} = -q_{soln}$$

Example 4.2   **Heat Produced by an Exothermic Reaction**

What is the approximate amount of heat produced by the acid–base reaction when 50.0 mL (50 g) of 1.00 M HCl and 50.0 mL (50 g) of 1.00 M NaOH, both at 22.0°C, are added to a calorimeter like that shown in Fig. 4.4. The temperature increases to 28.9°C.

$$HCl(aq) + NaOH(aq) \longrightarrow NaCl(aq) + H_2O(l)$$

Neglect the heat capacity of the calorimeter. The specific heat of the solution produced is 4.18 J g^{-1} °C^{-1}.

Solution: We want to find the amount of heat produced by the reaction that results when two solutions are mixed. At the instant of mixing, we have 100 g of a mixture of HCl and NaOH at 22.0°C. The mixture undergoes an immediate change as the HCl and NaOH react and produce heat, which is trapped in the solution and which raises its temperature to 28.9°C. The amount of heat produced by the reaction, $q_{reaction}$, is equal to the amount trapped. And as we have noted,

$$q_{reaction} + q_{soln} = 0$$

so
$$q_{reaction} = -q_{soln}$$

The heat trapped in the solution can be determined by three steps similar to those used in Example 4.1.

Step 1. *Determine the amount of heat involved in changing the temperature of 1 g of the solution by 1°C (the specific heat of the solution). This example gives the specific heat as 4.18 J g^{-1} °C^{-1}.*

Step 2. *Determine the amount of heat necessary to heat 100 g of solution by 1°C. Result: 418 J*

Step 3. *Determine the amount of heat necessary to heat 100 g of solution from 22.0°C to 28.9°C. Result: 2.9×10^3 J*

The solution traps 2.9×10^3 J, thus the reaction produces 2.9×10^3 J, so $q_{reaction} = -2.9 \times 10^3$ J.

A second way to solve the problem is to use the equation

$$q_{reaction} = -q_{soln} = -[cm\Delta T] = -[cm(T_{final} - T_{initial})]$$
$$= -[4.18 \text{ J g}^{-1} \text{°C}^{-1} \times 100 \text{ g} \times (28.9 - 22.0) \text{ °C}]$$
$$= -2.9 \times 10^3 \text{ J}$$

The negative sign indicates that the reaction is exothermic. It produces 2.9 kJ of heat.

Check your learning: When 100 g of 0.200 M NaCl(aq) and 100 g of 0.200 M AgNO$_3$(aq), both at 21.91°C, are mixed in a polystyrene cup calorimeter, the temperature increases to 23.48°C as solid AgCl forms. How much heat is produced by the precipitation of AgCl(s) if the specific heat of the products is 4.20 J g^{-1} °C^{-1}?

Answer: 1.32×10^3 J is produced.

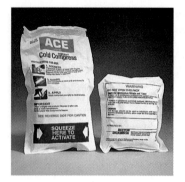

Figure 4.5

These instant cold packs consists of a bag containing solid ammonium nitrate and a second bag of water. When the bag of water is broken, the pack becomes cold because the dissolution of ammonium nitrate is an endothermic process that removes thermal energy from the water.

Example 4.3   Heat Absorbed by an Endothermic Reaction

When solid ammonium nitrate dissolves in water, the solution becomes cold. This is the basis for an "instant ice pack" (Fig. 4.5). When 2.0 g of solid NH_4NO_3 dissolves in 50.0 g of water at 25.00°C in a calorimeter, the temperature decreases to 22.00°C. Calculate the approximate heat involved in this change if the specific heat of the resulting solution is 4.2 J^{-1} °C^{-1}. We can neglect the heat capacity of the calorimeter.

Solution: When solid NH_4NO_3 dissolves in water in a calorimeter, no heat can enter from the surroundings, so the energy necessary to dissolve the NH_4NO_3 must be extracted from the solution. This lowers the temperature of the solution. We can calculate the amount of heat lost by the solution, q_{soln}, from the decrease in temperature of the solution. The amount of heat absorbed by the dissolution of NH_4NO_3 is equal to the amount removed from the solution but opposite in sign.

$$q_{reaction} + q_{soln} = 0$$

so

$$q_{reaction} = -q_{soln}$$

The heat removed from the solution can be evaluated by three steps similar to those used in Example 4.1.

Step 1. *Determine the amount of heat involved in changing the temperature of 1 g of the solution by 1°C (the specific heat of the solution). The example gives this as 4.2 J g^{-1} °C^{-1}.*

Step 2. *Determine the amount of heat necessary to cool 52.0 g of solution by 1°C. (The mass of the solution is due to 50.0 g of water and 2.0 g of NH_4NO_3.) Result: 220 J*

Step 3. *Determine the amount of heat absorbed during the dissolution as 52.0 g of solution cools from 25.00°C to 22.00°C. Result: 660 J*

A second way to solve the problem is to use the equation

$$q_{reaction} = -q_{soln} = -[cm\Delta T] = -[cm(T_{final} - T_{initial})]$$
$$= -[4.2 \text{ J g}^{-1} °C^{-1} \times 52.0 \text{ g} \times (22.00 - 25.00) °C]$$
$$= -(-660 \text{ J}) = 660 \text{ J}$$

The dissolution is an endothermic process. (q is positive.)

Check your learning: When a 3.00-g sample of KCl was added to 3.00×10^2 g of water in a polystyrene cup calorimeter, the temperature decreased by 1.05°C. If the specific heat of the resulting solution is 4.18 J g^{-1} °C^{-1}, how much heat is involved in the dissolution of the KCl?

Answer: 1.33 kJ is absorbed.

If the amount of heat absorbed by a calorimeter is too large to neglect or if we want more accurate results, then we must take into account the heat absorbed both by the solution and by the calorimeter.

Not all calorimeters trap the heat of a reaction in a solution of the products. If a reaction were run in the steel container of the calorimeter shown in Fig. 4.6, the heat of the reaction would be trapped in the container and in the water surrounding the container. Such calorimeters are used to measure the heats of reactions that involve

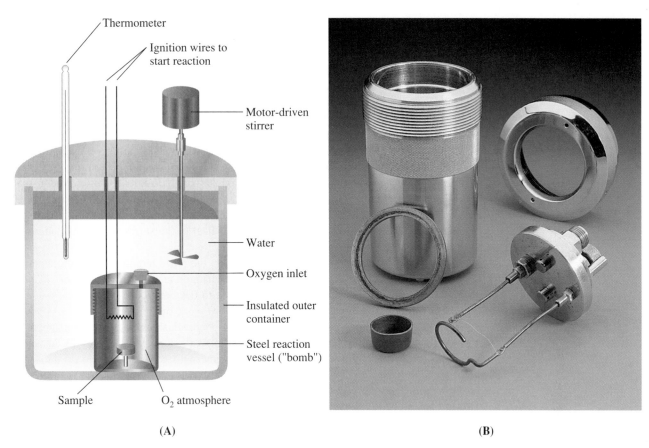

Thermometer

Ignition wires to
start reaction

Motor-driven
stirrer

Water

Oxygen inlet

Insulated outer
container

Steel reaction
vessel ("bomb")

Sample O_2 atmosphere

(A)   **(B)**

Figure 4.6

(A) A type of calorimeter used to measure heat produced or absorbed by reactions involving gases. The gases are contained in the pressure-proof "bomb." Brief electric heating by the ignition wire starts the reaction. The heat produced by the reaction is absorbed by the surroundings: the bomb, stirrer, thermometer, and water. The amount of heat can be determined if the heat capacity of the surroundings has been measured. (B) The steel container used in the calorimeter is about 10 cm (4 in.) tall.

gases. A modification of such a calorimeter was used to measure the heat produced by a living person. A man lived for four days inside a small room surrounded by water. From the increase in water temperature, it was found that the heat produced by his metabolism was approximately 2400 nutritional Calories per day.

4.4   Thermochemistry and Thermodynamics

Thermochemistry is a branch of **chemical thermodynamics,** the science that deals with the relationships between heat and forms of energy known as work. We will concentrate on thermochemistry in this chapter and will postpone our broader consideration of thermodynamics until Chapter 18. However, we do need to consider some widely used thermodynamic ideas.

Substances act as reservoirs of energy; energy can be added to them or removed from them. Substances generally store thermal energy by increasing the amount of diffusion, vibration, or rotation of their molecules; this increases the kinetic energy of the molecules. When thermal energy is lost, these motions decrease and the kinetic energy falls. The total of all possible kinds of energy present in a substance is called the **internal energy.** The substance (or substances) we choose to study is called a **system.** Everything else is called the **surroundings.**

As a system undergoes a change, its internal energy can change, and energy can be transferred from the system to the surroundings or from the surroundings to the

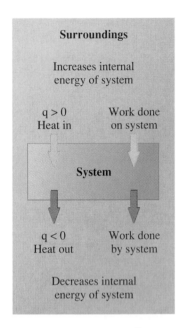

Figure 4.7

The effect of heat and work on the internal energy of a system.

system. Energy is transferred *into* a system when it absorbs heat from the surroundings or when the surroundings do **work** on the system (Fig. 4.7). As an example, consider what happens when you do work on a straightened paper clip or a hair pin by rapidly bending it back and forth several times. It becomes hot, not because you have added heat, but because you have done work on it and the work increased its thermal energy. The increased amount of thermal energy is reflected in an increase in the wire's temperature. Energy is transferred *out of* a system when heat is lost from the system or when the system does work on the surroundings.

A type of work called **expansion work** occurs when a system pushes back the surroundings against a restraining pressure, or when the surroundings compress the system against a pressure. As an example of such work, let us consider a system that consists of 1 gram of liquid water at 0°C in a container open to the atmosphere (Fig. 4.8). The water is the system. The container and the atmosphere are part of the surroundings. When the water loses 334 joules of heat to the surroundings, it changes to ice at 0°C.

Because this system loses energy as heat to the surroundings during the change, the total energy in the system is less after the change than it was before. Loss of heat is one way for the internal energy of a system to decrease.

The system also loses energy because it does work on the surroundings. As water freezes, it expands and pushes back the atmosphere. Pushing back the atmosphere is work, just as pumping air into a bicycle tire is work. It requires energy to do work, and as the system freezes, it uses some of its internal (stored) energy to do the work of pushing back the atmosphere.

We will consider how to determine the amount of work involved in a change when we discuss chemical thermodynamics more fully in Chapter 18.

4.5   Enthalpy Changes

An **enthalpy change, ΔH,** is the amount of heat gained or lost by a system that undergoes a change at a constant pressure (Fig. 4.9), the only work done being due to a volume change of the system. The heat lost when 1 gram of water freezes as described in Section 4.4 is directly proportional to the enthalpy change, because the water freezes at the essentially constant pressure of the atmosphere. On the other hand, the heat produced by a reaction measured in a calorimeter such as that shown in Fig. 4.6 is not directly proportional to ΔH, because the closed metal container prevents any work from being done as a result of volume changes.

The following conventions apply when we use ΔH:

1. The enthalpy change of a reaction is shown as a ΔH value following the equation for the reaction. This ΔH value indicates the amount of heat associated with the reaction of the number of moles of reactants shown in the chemical equation. For example, the equation

$$H_2(g) + \tfrac{1}{2}O_2(g) \longrightarrow H_2O(l) \qquad \Delta H = -286 \text{ kJ}$$

    indicates that when 1 mole of hydrogen gas and $\frac{1}{2}$ mole of oxygen gas at some temperature and pressure change to 1 mole of liquid water at the same temperature and pressure, 286 kJ of heat are given up to the surroundings.

2. The enthalpy change of a reaction depends on the state of the reactants and products of the reaction (whether we have gases, liquids, solids, or aqueous solutions), so these must be shown. For example, when 1 mole of hydrogen gas and $\frac{1}{2}$ mole

Figure 4.8

When a system consisting of 1 gram of liquid water at 0°C (A) changes to 1 gram of ice 0°C, (B) 334 J are given up to the surroundings, and the system does work on the surroundings as it expands.

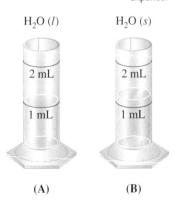

H₂O (*l*)   H₂O (*s*)

(A)   (B)

of oxygen gas change to 1 mole of liquid water at the same temperature and pressure, 286 kJ of heat are evolved; if water vapor (a gas) forms, only 242 kJ of heat are evolved.

3. A negative value of ΔH indicates an exothermic reaction; a positive value of ΔH indicates an endothermic reaction.

4. Sometimes the amount of heat is written as though it were a reactant or a product. For a reaction with a negative ΔH, as in item 1, we would write the heat given off as a product.

$$H_2(g) + \tfrac{1}{2}O_2(g) \longrightarrow H_2O(l) + 286 \text{ kJ}$$

For a reaction with a positive ΔH, we would write the heat that is absorbed as a reactant. For example, the reaction of nitrogen with oxygen could be written

$$N_2(g) + O_2(g) + 180.5 \text{ kJ} \longrightarrow 2NO(g)$$

or

$$N_2(g) + O_2(g) \longrightarrow 2NO(g) \qquad \Delta H = 180.5 \text{ kJ}$$

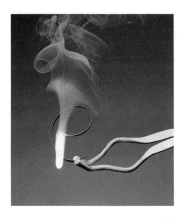

Figure 4.9

A burning magnesium ribbon. The heat released when 1 mole of Mg burns in air is the ΔH value for the reaction.

Example 4.4   Measurement of an Enthalpy Change

When 0.0500 mol of HCl(aq) reacts with 0.0500 mol of NaOH(aq) to form 0.0500 mol of NaCl(aq), 2.9 kJ of heat is produced (Example 4.2). What is ΔH, the enthalpy change, per mole of acid reacting for the acid–base reaction

$$HCl(aq) + NaOH(aq) \longrightarrow NaCl(aq) + H_2O(l)$$

run under the conditions described in Example 4.2?

Solution: For the reaction of 0.0500-mol amounts, $q = -2.9$ kJ. This gives us the ratio -2.9 kJ/0.0500 mol, which we can use as a conversion factor to find the heat produced when 1 mole of acid reacts.

$$x \text{ kJ} = 1 \text{ mol} \times \frac{-2.9 \text{ kJ}}{0.0500 \text{ mol}}$$

$$= -58 \text{ kJ}$$

The enthalpy change for the reaction is -58 kJ, which we could write

$$HCl(aq) + NaOH(aq) \longrightarrow NaCl(aq) + H_2O(l) \qquad \Delta H = -58 \text{ kJ}$$

Check your learning: When 1.42 g of iron reacts with 1.80 g of chlorine, 3.22 g of FeCl$_2$(s) and 8.60 kJ of heat are produced. What is the enthalpy change for the reaction when 1 mole of FeCl$_2$(s) is produced?

Answer: $\Delta H = -339$ kJ

The National Institute of Standards and Technology has been particularly active in measuring and tabulating enthalpy changes. Data are reported for reactions in which both the reactants and the products are at the same conditions. A commonly accepted set of conditions is called a **standard state.** Since 1981 the Institute has used a standard state of 298.15 K (25°C) and a pressure of 100 kilopascals (0.987 atmosphere). A common standard state used by the Institute before 1981 is 298.15 K and

Substance	Combustion Reaction	Enthalpy of Combustion, ΔH°_{298} (kJ mol^{-1})
Carbon	$C(s) + \frac{1}{2}O_2(g) \longrightarrow CO(g)$	-111
	$C(s) + O_2(g) \longrightarrow CO_2(g)$	-394
Hydrogen	$H_2(g) + \frac{1}{2}O_2(g) \longrightarrow H_2O(g)$	-242
	$H_2(g) + \frac{1}{2}O_2(g) \longrightarrow H_2O(l)$	-286
Magnesium	$Mg(s) + \frac{1}{2}O_2(g) \longrightarrow MgO(s)$	-602
Sulfur	$S(s) + O_2(g) \longrightarrow SO_2(g)$	-297
Carbon monoxide	$CO(g) + \frac{1}{2}O_2(g) \longrightarrow CO_2(g)$	-283
Methane	$CH_4(g) + 2O_2(g) \longrightarrow CO_2(g) + 2H_2O(g)$	-802
Acetylene	$C_2H_2(g) + \frac{5}{2}O_2(g) \longrightarrow 2CO_2(g) + H_2O(g)$	-1256
Methanol	$CH_3OH(l) + \frac{3}{2}O_2(g) \longrightarrow CO_2(g) + 2H_2O(g)$	-638
Isooctane	$C_8H_{18}(l) + \frac{25}{2}O_2(g) \longrightarrow 8CO_2(g) + 9H_2O(g)$	-5460

Table 4.1

Enthalpy of Combustion for 1 Mole of a Substance Under Standard State Conditions

1 atmosphere of pressure. Because ΔH of a reaction changes very little with such small changes in pressure, ΔH values (except for the most precisely measured values) are the same under both sets of standard conditions. We will use the standard state of 298.15 K and 1 atmosphere and will employ the symbol ΔH°_{298} to indicate an enthalpy change for these conditions. (The symbol ΔH is used to indicate an enthalpy change for a reaction where the conditions are not specified.)

The values of many different kinds of enthalpy changes have been determined. Four of the most common kinds are enthalpy of combustion, enthalpy of fusion, enthalpy of vaporization, and standard molar enthalpy of formation.

Enthalpy of Combustion.

Sometimes called heat of combustion, the **enthalpy of combustion** is the enthalpy change when exactly 1 mole of a substance burns (combines with oxygen) under standard state conditions. For example, the enthalpy of combustion of carbon to carbon dioxide, -394 kJ mol^{-1}, is the amount of heat produced when 1 mole of carbon and 1 mole of oxygen gas, both at 25°C and 1 atmosphere of pressure, react and produce 1 mole of carbon dioxide gas at 25°C and 1 atmosphere of pressure.

$$C(s) + O_2(g) \longrightarrow CO_2(g) \qquad \Delta H = -394 \text{ kJ}$$

Figure 4.10

The uncontrolled combustion of gasoline.

Enthalpies of combustion for several substances have been measured and are listed in Table 4.1. Several readily available substances with large enthalpies of combustion—hydrogen, carbon (as coal or charcoal), methane, acetylene, methanol, and isooctane (a component of gasoline)—are used as fuels.

Example 4.5	Using Enthalpy of Combustion

As Fig. 4.10 shows, the combustion of gasoline is an exothermic process. Let us determine the approximate amount of heat produced in burning 1.00 L of gasoline by assuming the enthalpy of combustion of gasoline is the same as that of isooctane, a common component of gasoline. The density of isooctane is 0.692 g/mL.

Solution: We need to determine how much heat can be produced by burning 1.00 L of isooctane under standard state conditions. We can look up the amount of heat produced by burning 1 mol of isooctane under standard conditions; this is the enthalpy of combustion of isooctane. Table 4.1 gives -5460 kJ mol^{-1} as the value. If we determine how many moles of isooctane are contained in 1.00 L of isooctane and multiply that by the amount of heat produced by combustion of 1 mol of isooctane, we have answered our question. We use the following steps to solve this problem:

Step 1. *Determine the mass of isooctane in 1.00 L from its volume and density (Section 1.11). Result: 692 g*

Step 2. *Determine the number of moles of isooctane from its molar mass (114 g mol^{-1}) and the mass of isooctane in 1.00 L. Result: 6.07 mol*

Step 3. *Determine the heat produced by multiplying the amount of heat produced by 1 mol by the number of moles.*

$$q = 6.07 \ \overline{\text{mol}} \ \text{C}_8\text{H}_{18} \times \frac{-5460 \text{ kJ}}{1 \ \overline{\text{mol}} \ \text{C}_8\text{H}_{18}} = -3.31 \times 10^4 \text{ kJ}$$

Combustion of 1.00 L of isooctane produces 33,100 kJ of heat.

Check your learning: How much heat is produced by combustion of 125 g of acetylene under standard state conditions?

Answer: 6.03×10^4 kJ

Figure 4.11

(A) This beaker of ice, just removed from a freezer, has a temperature of $-18.3°C$. (B) After 10 minutes the ice has absorbed enough heat from the air to warm to 0°C. A small amount has melted. (C) Thirty minutes later the ice has absorbed more heat, but its temperature is still 0°C. The ice melts without changing its temperature. (D) Only after all the ice has melted does the heat absorbed cause the temperature to increase to 20.0°C

Enthalpy of Fusion. When we add heat to a crystalline solid, its temperature increases until the solid reaches the temperature at which it will melt. At this point any additional heat causes the solid to begin to melt (fuse). As we add more heat, the temperature remains constant until all of the solid has melted (Fig. 4.11). After that the temperature rises again.

The amount of heat required to change exactly 1 mole of a substance from the solid state to the liquid state is called the **enthalpy of fusion, ΔH_{fus},** of the substance. The enthalpy of fusion of ice is 6.0 kJ mol^{-1} at 0°C.

$$\text{H}_2\text{O}(s) \longrightarrow \text{H}_2\text{O}(l) \qquad \Delta H = -\Delta H_{\text{fus}} = 6.0 \text{ kJ mol}^{-1}$$

Fusion (melting) is an endothermic process.

(A)

(B)

(C)

(D)

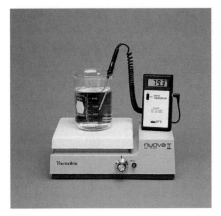

(A)

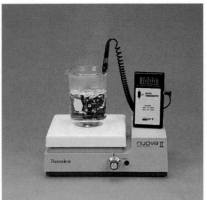

(B)

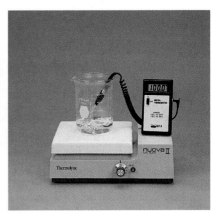

(C)

Figure 4.12

(A) The water heated by the hot plate has absorbed enough heat to increase its temperature from room temperature to 79.3°C. (B) After 5 more minutes the water has absorbed enough heat to warm to 100°C and then begin to boil. (C) After 30 more minutes enough heat has been added to boil away most of the water, but the temperature of the water is still 100°C. The heat converted the water from a liquid to a gas without changing its temperature.

Freezing is an exothermic process. In order for a substance to freeze, it must lose a quantity of heat equal to the amount it must gain to melt. Thus the enthalpy change for freezing of water is -6.0 kJ mol^{-1} at 0°C.

$$H_2O(l) \longrightarrow H_2O(s) \qquad \Delta H = -\Delta H_{fus} = -6.0 \text{ kJ mol}^{-1}$$

As heat is removed from water at 0°C and the water changes to ice, its temperature does not change until all of the liquid has frozen. Only at this point does the temperature fall as additional heat is removed.

Enthalpy of Vaporization.

Heat is required for a liquid to evaporate. The cooling effect can be evident when you leave a swimming pool or a shower. When the water on your skin evaporates, it removes heat from your skin and causes you to feel cold.

For the temperature of a liquid to remain constant as it evaporates, the liquid must absorb heat to offset the cooling due to the evaporation. The heat that must be supplied to keep the temperature constant as a mole of liquid evaporates is called the **enthalpy of vaporization, ΔH_{vap},** of the substance. The enthalpy of vaporization of water is 44.01 kJ mol^{-1} at 25°C.

$$H_2O(l) \longrightarrow H_2O(g) \qquad \Delta H_{vap} \text{ (at 25°C)} = 44.01 \text{ kJ mol}^{-1}$$

The quantity of heat lost as a gas condenses to a liquid equals that absorbed during evaporation.

At higher temperatures, less heat is required per mole of liquid evaporated. At its boiling point (100°C), ΔH_{vap} of water is 40.67 kJ mol^{-1}, compared to 44.01 kJ mol^{-1} at 25°C. The added heat converts the liquid water to the vapor, but it does not increase the temperature of the liquid water that remains (Fig. 4.12).

Figure 4.13

Evaporation of sweat helps cool the body.

Example 4.6	Using Enthalpy of Vaporization

One way our body is cooled is by evaporation of the water in sweat (Fig. 4.13). In very hot climates we can lose as much as 1.5 L of sweat per day. Although sweat is not pure water, we can get an approximate value of the amount of heat removed by evaporation by assuming that it is. Determine how much heat is required to evaporate 1.5 L of water (1.5 kg) at $T = 37$°C (body temperature); $\Delta H_{vap} = 43.46$ kJ mol^{-1} at 37°C.

Solution: We want to calculate the amount of heat necessary to evaporate 1.5 kg of water at 37°C. We are given the amount of heat (43.46 kJ) that 1 mol of water at 37°C absorbs as it evaporates. If we determine how many moles of water are contained in 1.5 kg of water and multiply that by the amount of heat absorbed by evaporation of 1 mol of water, we have answered our question. We can use the following steps to solve this problem:

Step 1.   *Determine the number of moles of water in 1.5 kg.* Result: 83 mol H_2O

Step 2.   *Determine the amount of heat necessary to evaporate this much water.*

$$83 \text{ mol } H_2O \times \frac{43.46 \text{ kJ}}{1 \text{ mol } H_2O} = 3.6 \times 10^3 \text{ kJ}$$

Thus 3600 kJ of heat is removed by the evaporation of 1.5 L of water.

Check your learning: How much heat is required to evaporate 100.0 g of liquid ammonia, NH_3, if its enthalpy of vaporization is 4.8 kJ mol^{-1}?

Answer: 28 kJ

The enthalpy changes associated with evaporation and condensation of a refrigerant are used to cool a refrigerator. In older machines the refrigerant is a Freon (a chlorofluorocarbon, CFC) such as CF_2Cl_2; in newer refrigerators, hydrofluorocarbons (HFCs), such as CH_2F_2, replace CFCs. Heat is removed from the refrigerator when the liquid CFC or HFC absorbs heat as it evaporates. The gas is then circulated through a compressor outside the refrigerator and condensed back to a liquid by compression. The heat given up during condensation is lost to the room through the cooling coils on the back of the refrigerator, and the liquid is recycled through the refrigerated compartment where it is allowed to evaporate, repeating the cycle. The cooling unit of a refrigerated air conditioner works the same way.

Standard Enthalpy of Formation.   A **standard enthalpy of formation, ΔH_f°,** is an enthalpy change for a reaction in which exactly 1 mole of a pure substance is formed from free elements in their most stable states under standard state conditions. The standard enthalpy of formation of $CO_2(g)$ is -394 kJ mol^{-1}, which is the enthalpy change for the exothermic reaction

$$C(s) + O_2(g) \longrightarrow CO_2(g) \qquad \Delta H_f^\circ = \Delta H_{298}^\circ = -394 \text{ kJ mol}^{-1}$$

starting with the reactants at a pressure of 1 atmosphere and 25°C with the carbon present as graphite (the most stable form of carbon under this set of standard conditions) and ending with 1 mole of gaseous product, CO_2, also at 1 atmosphere and a temperature of 25°C. For nitrogen dioxide, $NO_2(g)$, ΔH_f° is 33.2 kJ mol^{-1}, which is ΔH for the endothermic reaction

$$\tfrac{1}{2}N_2(g) + O_2(g) \longrightarrow NO_2(g) \qquad \Delta H_f^\circ = \Delta H_{298}^\circ = 33.2 \text{ kJ mol}^{-1}$$

A reaction with $\frac{1}{2}$ mole of N_2 and 1 mole of O_2 is correct in this case, because the standard enthalpy of formation refers to 1 mole of product.

You will find a table of standard enthalpies of formation of many common substances in Appendix I. The enthalpy values indicate that formation reactions range from very exothermic (-2984 kJ mol^{-1} for formation of P_4O_{10}, for example) to very endothermic (226.7 kJ mol^{-1} for formation of acetylene, C_2H_2, for example). By convention, the standard enthalpy of formation of an element in its most stable form is equal to zero.

Example 4.7   Evaluating an Enthalpy of Formation

Ozone, $O_3(g)$, forms from oxygen, $O_2(g)$, by an endothermic process. In the upper atmosphere, ultraviolet radiation is the source of the energy that drives the reaction. Assuming that both the reactants and the products of the reaction are in their standard states, determine the standard enthalpy of formation, ΔH_f°, of ozone from the following information.

$$3O_2(g) \longrightarrow 2O_3(g) \qquad \Delta H_{298}^\circ = 286 \text{ kJ}$$

Solution: ΔH_f° is the enthalpy change for the formation of 1 mol of a substance in its standard state from the elements in their standard states. Thus ΔH_f° for $O_3(g)$ is the enthalpy change for the reaction

$$\tfrac{3}{2}O_2(g) \longrightarrow O_3(g)$$

For the formation of 2 mol of $O_3(g)$, $\Delta H_{298}^\circ = 286$ kJ. This gives 286 kJ/2 mol, a ratio that we can use as a conversion factor to find the heat produced when 1 mol is formed.

$$x \text{ kJ} = 1 \text{ mol} \times \frac{286 \text{ kJ}}{2 \text{ mol}}$$

$$= 143 \text{ kJ}$$

The enthalpy change accompanying the formation of 1 mol of $O_3(g)$ from $O_2(g)$ is 143 KJ, so

$$\Delta H_f^\circ = 143 \text{ kJ mol}^{-1}$$

Check your learning: Hydrogen gas, H_2, reacts explosively with gaseous chlorine, Cl_2, to form hydrogen chloride, $HCl(g)$. What is the enthalpy change for the reaction of 1 mol of $H_2(g)$ with 1 mol of $Cl_2(g)$ if both the reactants and the products are at standard state conditions? The standard enthalpy of formation of $HCl(g)$ is -92.3 KJ mol^{-1} (Appendix I).

Answer: For the reaction $H_2(g) + Cl_2(g) \longrightarrow 2HCl(g)$, $\Delta H_{298}^\circ = -184.6$ kJ.

Note the difference in units between ΔH (as well as ΔH_{298}°) and ΔH_f°. The enthalpy change for a reaction, ΔH (or ΔH_{298}°), gives the amount of heat produced by the reaction as it is written and has units of kilojoules. The standard enthalpy of formation of a compound, ΔH_f°, identifies the enthalpy change associated with formation of exactly 1 mole of the compound and has units of kilojoules per mole. The magnitudes of ΔH and ΔH_{298}° change as the amounts of reactant change. For example, combustion of 1 mole of $CO(g)$ produces 283 kilojoules ($\Delta H = -283$ kJ), whereas combustion of 2 moles of $CO(g)$ produces 566 kilojoules ($\Delta H = -566$ kJ). However, the standard enthalpy of formation of a particular substance has a fixed value because it always refers to 1 mole of the substance.

4.6   Hess's Law

There are three ways to determine the amount of heat involved in a chemical change: Measure it experimentally, look it up in a table of experimental values, or calculate it from other experimentally determined enthalpy changes. Calculations usually involve the use of **Hess's law:** *If a process can be written as the sum of several step-wise processes, the enthalpy change of the total process equals the sum of the enthalpy changes of the various steps.* For example, carbon can react with oxygen to form carbon dioxide in a two-step process. First, carbon monoxide, CO, is formed.

$$C(s) + \tfrac{1}{2}O_2(g) \longrightarrow CO(g)$$

Then carbon dioxide, CO_2, is formed from the carbon monoxide.

$$CO(g) + \tfrac{1}{2}O_2(g) \longrightarrow CO_2(g)$$

The equation describing the overall change of C to CO_2 is the sum of these two chemical changes.

Step 1 $\qquad\qquad\qquad\qquad C(s) + \tfrac{1}{2}O_2(g) \longrightarrow CO(g)$

Step 2 $\qquad\qquad\qquad\qquad CO(g) + \tfrac{1}{2}O_2(g) \longrightarrow CO_2(g)$

Sum $\qquad C(s) + \tfrac{1}{2}O_2(g) + \cancel{CO(g)} + \tfrac{1}{2}O_2(g) \longrightarrow \cancel{CO(g)} + CO_2(g)$

Because the CO produced in Step 1 is consumed in Step 2, the net change is

$$C(s) + O_2(g) \longrightarrow CO_2(g)$$

According to Hess's law, the enthalpy change of the reaction is equal to the sum of the enthalpy changes of the steps. From the experimental enthalpies of combustion (Table 4.1), we have

$$
\begin{aligned}
C(s) + \tfrac{1}{2}O_2(g) &\longrightarrow CO(g) & \Delta H^\circ_{298} &= -111 \text{ kJ} \\
CO(g) + \tfrac{1}{2}O_2(g) &\longrightarrow CO_2(g) & \Delta H^\circ_{298} &= -283 \text{ kJ} \\
\hline
C(s) + O_2(g) &\longrightarrow CO_2(g) & \Delta H^\circ_{298} &= -394 \text{ kJ}
\end{aligned}
$$

We can see that ΔH for the bottom reaction is the sum of the ΔH values for the two reactions above it.

Before we use Hess's law, let us recall two important features of ΔH.

1. ΔH for a reaction in one direction is equal in magnitude and opposite in sign to ΔH for the reaction in the reverse direction. For example, ΔH for water melting is 6.0 kJ for 1 mole of ice.

$$H_2O(s) \longrightarrow H_2O(l) \qquad \Delta H = 6.0 \text{ kJ}$$

ΔH is -6.0 kJ for the reverse reaction, 1 mole of water freezing.

$$H_2O(l) \longrightarrow H_2O(s) \qquad \Delta H = -6.0 \text{ kJ}$$

2. ΔH is directly proportional to the quantities of reactants or products. This statement indicates that if the enthalpy of fusion of 1 mole of water is 6.0 kJ,

$$H_2O(s) \longrightarrow H_2O(l) \qquad \Delta H = 6 \text{ kJ}$$

then the enthalpy of fusion of 2 moles of water is twice as great:

$$2H_2O(s) \longrightarrow 2H_2O(l) \qquad \Delta H = 12 \text{ kJ}$$

Example 4.8   Stepwise Calculation of ΔH_f° Using Hess's Law

Determine the enthalpy of formation, ΔH_f°, of $FeCl_3(s)$ from the enthalpy changes of the following two-step process that occurs under standard state conditions.

$$Fe(s) + Cl_2(g) \longrightarrow FeCl_2(s)$$
$$FeCl_2(s) + \tfrac{1}{2}Cl_2(g) \longrightarrow FeCl_3(s)$$

The standard enthalpy of formation of $FeCl_2(s)$ is -341.8 kJ mol^{-1}. The enthalpy change for the second reaction is -57.7 kJ for each mole of solid $FeCl_3$ formed.

Solution: We are trying to find the standard enthalpy of formation of $FeCl_3(s)$, which is equal to ΔH_{298}° for the following combination reaction:

$$Fe(s) + \tfrac{3}{2}Cl_2(g) \longrightarrow FeCl_3(s) \qquad \Delta H_f^\circ = \Delta H_{298}^\circ$$

We can write this equation as the sum of the two steps with known ΔH values. The sum of the enthalpy changes of these two steps is equal to ΔH_f° for $FeCl_3(s)$.

$$
\begin{array}{ll}
Fe(s) + Cl_2(g) \longrightarrow FeCl_2(s) & \Delta H_{298}^\circ = -341.8 \text{ kJ} \\
\underline{FeCl_2(s) + \tfrac{1}{2}Cl_2(g) \longrightarrow FeCl_3(s)} & \underline{\Delta H_{298}^\circ = \ \ -57.7 \text{ kJ}} \\
Fe(s) + \tfrac{3}{2}Cl_2(g) \longrightarrow FeCl_3(s) & \Delta H_f^\circ = \Delta H_{298}^\circ = (-341.8 \text{ kJ}) + (-57.7 \text{ kJ}) \\
& \qquad \qquad \quad = -399.5 \text{ kJ}
\end{array}
$$

The enthalpy change for the formation of 1 mol of $FeCl_3(s)$ is -399.5 kJ, so the enthalpy of formation, ΔH_f°, is -399.5 kJ mol^{-1}.

Check your learning: Calculate ΔH for the process

$$N_2(g) + 2O_2(g) \longrightarrow 2NO_2(g)$$

from the following information:

$$N_2(g) + O_2(g) \longrightarrow 2NO(g) \qquad \Delta H = 180.5 \text{ kJ}$$
$$NO(g) + \tfrac{1}{2}O_2(g) \longrightarrow NO_2(g) \qquad \Delta H = -57.06 \text{ kJ}$$

Answer: 66.4 kJ

We also can use Hess's law to determine the enthalpy change of any reaction if the corresponding enthalpies of formation of the reactants and products are available. The stepwise reactions we consider are (1) decompositions of the reactants into their component elements (for which the enthalpy changes are proportional to the negative of the enthalpies of formation of the reactants), followed by (2) recombinations of the elements to give the products (for which the enthalpy changes are proportional to the enthalpies of formation of the products). The standard enthalpy change of the overall reaction is equal to the sum of the standard enthalpies of formation of all products minus the corresponding sum of the reactants. We may see Hess's law written as

$$\Delta H_{298}^\circ = \Sigma \Delta H_{f_{products}}^\circ - \Sigma \Delta H_{f_{reactants}}^\circ$$

when used for the special case of determining a standard enthalpy change from standard enthalpies of formation. In this equation, Σ means "the sum of."

Figure 4.14

The combustion of natural gas, CH_4, is an exothermic reaction.

Example 4.9   **Using Hess's Law**

What is the enthalpy of combustion when natural gas (methane, CH_4) burns (Fig. 4.14) and produces gaseous carbon dioxide and liquid water?

Solution: The enthalpy of combustion is equal to the standard enthalpy change for the reaction of methane with oxygen to produce carbon dioxide gas and liquid water.

$$CH_4(g) + 2O_2(g) \longrightarrow CO_2(g) + 2H_2O(l) \qquad \Delta H^\circ_{combustion} = \Delta H^\circ_{298}$$

We can write this equation as the sum of three steps: (1) decomposition of CH_4 to $C(s)$ and $H_2(g)$, (2) reaction of H_2 and O_2 to form H_2O, and (3) reaction of C and O_2 to form CO_2. The enthalpy changes of these steps are proportional to the standard enthalpies of formation found in Appendix I. The enthalpy of combustion is equal to the sum of the enthalpy changes of these steps.

$$
\begin{array}{lll}
CH_4(g) \longrightarrow C(s) + 2H_2(g) & \Delta H^\circ_1 = -\Delta H^\circ_{f_{CH_4(g)}} = & 74.81 \text{ kJ} \\
2H_2(g) + O_2(g) \longrightarrow 2H_2O(l) & \Delta H^\circ_2 = 2 \times \Delta H^\circ_{f_{H_2O(l)}} = 2 \times (-285.83 \text{ kJ}) \\
C(s) + O_2(g) \longrightarrow CO_2(g) & \Delta H^\circ_3 = \Delta H^\circ_{f_{CO_2(g)}} = & -393.51 \text{ kJ} \\
\hline
CH_4(g) + 2O_2(g) \longrightarrow CO_2(g) + 2H_2O(l) & \Delta H^\circ_{comb} = \Delta H^\circ_{298} = \\
& \Delta H^\circ_1 + \Delta H^\circ_2 + \Delta H^\circ_3 = & -890.36 \text{ kJ}
\end{array}
$$

Alternatively, we could use the special form of Hess's law given in the paragraph preceding this example to find ΔH°_{298}.

$$\Delta H^\circ_{298} = \Sigma \Delta H^\circ_{f_{products}} - \Sigma \Delta H^\circ_{f_{reactants}}$$

$$\Delta H^\circ_{comb} = \Delta H^\circ_{298} = + \left(2 \text{ mol } H_2O(l) \times \frac{-285.83 \text{ kJ}}{1 \text{ mol } H_2O(l)} \right)$$

$$+ \left(1 \text{ mol } CO_2(g) \times \frac{-393.51 \text{ kJ}}{1 \text{ mol } CO_2(g)} \right)$$

$$- \left(1 \text{ mol } CH_4(g) \times \frac{-74.81 \text{ kJ}}{1 \text{ mol } CH_4(g)} \right)$$

$$- \left(2 \text{ mol } O_2(g) \times \frac{0 \text{ kJ}}{1 \text{ mol } O_2(g)} \right)$$

$$= -571.66 \text{ kJ} - 393.51 \text{ kJ} + 74.81 \text{ kJ} - 0 \text{ kJ}$$

$$= -890.36 \text{ kJ}$$

The combustion of 1 mol of methane produces 890.36 kJ of heat when the water produced is a liquid. (The value of -802 kJ given in Table 4.1 for the enthalpy of combustion of methane is for the formation of gaseous water.)

Check your learning: Calculate the heat of combustion of 1 mol of ethanol, $C_2H_5OH(l)$, when $H_2O(l)$ and $CO_2(g)$ are formed. Use the following enthalpies of formation: $C_2H_5OH(l)$, -278 kJ mol^{-1}; $H_2O(l)$, -286 kJ mol^{-1}; and $CO_2(g)$, -394 kJ mol^{-1}.

Answer: -1368 kJ mol^{-1}

We can also use Hess's law to calculate enthalpy changes associated with phase changes. For example, evaporation is a phase change that can be described as

$$\text{Liquid} \longrightarrow \text{Gas}$$

The enthalpy of vaporization of the liquid is equal to the enthalpy change of the process and can be determined by applying Hess's law.

Example 4.10   Calculating Heat of Vaporization

Ethyl chloride has been used as a local anesthetic. When sprayed on the skin, it evaporates quickly. The heat needed to bring about the evaporation is removed from the skin, causing it to become numb. What is the enthalpy of vaporization of ethyl chloride, C_2H_5Cl, under standard state conditions?

Solution: The enthalpy of vaporization is the enthalpy change of the process described by the equation

$$C_2H_5Cl(l) \longrightarrow C_2H_5Cl(g) \qquad \Delta H^\circ_{vap} = \Delta H^\circ_{298}$$

We can determine the enthalpy change of this process by using the standard enthalpies of formation of liquid C_2H_5Cl and of gaseous C_2H_5Cl with Hess's law. In Appendix I we find that ΔH°_f for $C_2H_5Cl(l)$ is -136.5 kJ mol^{-1} and that ΔH°_f for $C_2H_5Cl(g)$ is -112.2 kJ mol^{-1}.

$$\Delta H^\circ_{vap} = \Delta H^\circ_{298} = \Sigma \Delta H^\circ_{f\,products} - \Sigma \Delta H^\circ_{f\,reactants}$$

$$= \left(1 \text{ mol } C_2H_5Cl(g) \times \frac{-112.2 \text{ kJ}}{1 \text{ mol } C_2H_5Cl(g)} \right)$$

$$- \left(1 \text{ mol } C_2H_5Cl(l) \times \frac{-136.5 \text{ kJ}}{1 \text{ mol } C_2H_5Cl(l)} \right)$$

$$= 24.3 \text{ kJ}$$

24.3 kJ of heat is required to evaporate 1 mol of liquid C_2H_5Cl. (The positive value of ΔH is consistent with evaporation being an endothermic process.)

Check your learning: Calculate the enthalpy of vaporization of $I_2(s)$ under standard state conditions, using the data in Appendix I.

Answer: 62.438 kJ

4.7   Fuel and Food

The major fuels in our society are fossil fuels, which form very slowly from the remains of plants and animals. All fossil fuels are being depleted much more rapidly than they are being formed. They consist primarily of **hydrocarbons,** compounds composed only of hydrogen and carbon.

   Natural gas is composed of low-molecular-mass hydrocarbons, principally methane (CH_4), with small amounts of ethane (C_2H_6), propane (C_3H_8), and butane (C_4H_{10}). These hydrocarbons are almost odorless, so small amounts of foul-smelling compounds called mercaptans are added to the gas as an aid in detecting leaks. (Some natural gas contains hydrogen sulfide, a contaminant that is removed before the gas is sold for commercial or residential use.) **Petroleum,** from which oil and gasoline are produced, is a liquid that varies in its composition, depending on the location of the wells from which it is pumped. It is a mixture that contains hundreds of different compounds, primarily hydrocarbons, ranging from methane to compounds containing 50 carbon atoms per molecule. **Coal** is a solid fuel composed mainly of hydrocarbons with high molecular masses. Compounds containing oxygen, nitrogen, and sulfur are also found in petroleum and coal.

There are problems with relying on fossil fuels as a principal source of energy. We are dependent on foreign reserves to meet our petroleum needs. As those supplies eventually dwindle, liquid fuels will become increasingly expensive. Combustion of fossil fuels produces carbon dioxide in quantities that are overwhelming the earth's ability to remove it from the atmosphere. The amount of carbon dioxide in the atmosphere has increased steadily over the past 40 years. This increase contributes to warming of the earth because carbon dioxide absorbs heat radiated from the earth's surface, thus reducing the amount of heat lost into space and warming the atmosphere. (This phenomenon is called the greenhouse effect.)

Incomplete combustion reactions and combustion reactions of impurities such as those that contain sulfur both lead to increased air pollution. There is growing recognition that clean fuels are of great importance and that we need to find substitutes to eliminate or reduce our dependence on fossil fuels.

A synthetic fuel of much interest is hydrogen, H_2, one of the fuels used in the space shuttle (Fig. 4.15). When hydrogen burns it forms water, a compound with no negative environmental impact. Free hydrogen does not occur naturally. It is a by-product of refining petroleum, and it can also be prepared from water by an endothermic reaction that requires energy.

$$2H_2O(l) \longrightarrow 2H_2(g) + O_2(g) \qquad \Delta H^\circ_{298} = 572 \text{ kJ}$$

As long as this energy must be obtained by combustion of fossil fuels, or as long as hydrogen is obtained from fossil fuels, it is unlikely to become a common fuel. However, if it becomes practical to decompose water into hydrogen and oxygen via nuclear or solar energy technology, then hydrogen could serve as a convenient source of energy.

The amount of heat that a fuel can produce can be determined from its enthalpy of combustion (Section 4.4) or by the use of Hess's law (Section 4.6, Example 4.9).

Food is the body's fuel. In addition, food provides the nutrients necessary to maintain the physiological functions of the body. Each day, a normally active, healthy adult requires about 130 kJ of energy from food for each kilogram of body weight. This corresponds to about 14 kilocalories, or 14 Calories, per pound of body weight. (Remember that the Calorie used in nutrition is what chemists call a kilocalorie; 1 Calorie = 1 kcal = 4.184 kJ.)

Most of the energy used by our bodies comes from carbohydrates. These are broken down in the mouth and intestine to glucose, which is transported to cells by the blood. In the cells glucose reacts with O_2 in a series of steps, eventually producing carbon dioxide, liquid water, and energy.

$$C_6H_{12}O_6(s) + 6O_2(g) \longrightarrow 6CO_2(g) + 6H_2O(l) \qquad \Delta H^\circ_{298} = -2870 \text{ kJ}$$

The amount of heat that can be produced in the body is the same as the amount that would be produced in a calorimeter burning the same amount of glucose under constant pressure. The average enthalpy of combustion per gram of carbohydrate is -17 kJ (-4.1 kcal g^{-1}), assuming liquid water is the product of the oxidation.

Fats also produce carbon dioxide and water when metabolized in the body. Palmitic acid, a typical fatty acid, reacts as follows:

$$C_{15}H_{21}CO_2H(s) + 23O_2(g) \longrightarrow 16CO_2(g) + 16H_2O(l) \qquad \Delta H^\circ_{298} = -9977 \text{ kJ}$$

Figure 4.15

The large central fuel tank of the space shuttle contains liquid hydrogen and liquid oxygen, which are burned in the engines at the rear of the space craft. A mixture of aluminum powder and ammonium perchlorate is the principal ingredient in the boosters. The large clouds of smoke contain aluminum oxide and aluminum chloride.

Figure 4.16

The Calorie content on this label is determined from the masses of carbohydrate, fat, and protein in a serving.

Fats produce significantly more energy per gram than either carbohydrates or proteins; their enthalpies of combustion average -38 kilojoules per gram (-9.1 kcal g^{-1}). This high enthalpy of combustion per gram makes fats ideal for storing energy. Excess energy from overeating is stored by the production of fat.

Proteins are not generally oxidized in the body; instead, their components are used to build body proteins. However, in cases where proteins are used to provide energy for the body, they produce about -17 kilojoules per gram (-4.1 kcal g^{-1}).

The Calorie content in a serving of a food (Fig. 4.16) is the sum of the Calories contained in the carbohydrates, fats, and proteins present in the serving.

Example 4.11   Heat Produced by Fuels and Food

Which will produce more heat, combustion of 1.0 g of hydrogen, H_2, or combustion of 1.0 g of methane, CH_4, assuming that the water produced is in the gas phase? Enthalpies of combustion can be found Table 4.1.

Solution: At first glance it might appear that methane is the right choice; the enthalpy of combustion of methane is -802 kJ mol^{-1}, and that of H_2 is -242 kJ mol^{-1}. However, we need the heat produced by 1.0 g of each, so we must convert these values to find the enthalpy of combustion per gram.

For H_2:
$$\frac{-242 \text{ kJ}}{1 \text{ mol}} \times \frac{1 \text{ mol}}{2.016 \text{ g}} = \frac{-120 \text{ kJ}}{1 \text{ g}}$$

For CH_4:
$$\frac{-802 \text{ kJ}}{1 \text{ mol}} \times \frac{1 \text{ mol}}{16.043 \text{ g}} = \frac{-50.0 \text{ kJ}}{1 \text{ g}}$$

Hydrogen provides 120 kJ of heat per gram of fuel; methane provides 50.0 kJ per gram.

Check your learning: What is the maximum mass of sugar (a carbohydrate) in a 12-oz glass of diet soda that contains less than 1 (nutritional) Calorie?

Answer: 0.25 g

For Review   Summary

Thermochemistry is the determination and study of the heat absorbed or given off during chemical and physical changes. **Thermal energy** is the energy associated with the random motion of atoms and molecules. Increasing the amount of thermal energy in a sample of matter causes the sample either to rise in temperature or to melt or evaporate with no change in temperature. Decreasing the amount of thermal energy in a sample of matter either decreases the temperature of the substance or causes it to condense or freeze with no change in temperature. **Heat,** q, is energy that is transferred between two bodies that are at different temperatures. Heat moves spontaneously from a hotter object to a cooler object. Whether measured in units of **calories** or **joules,** heat is measured in a **calorimeter,** usually by determining the temperature change of water or of a solution of known **specific heat.**

As a system loses heat by an **exothermic process,** its internal energy is decreased. Heat is added to a system by an **endothermic process,** which increases the internal energy of the system. When a system does **work,** the internal energy of the system is decreased. When work is done on a system, its internal energy increases.

If a chemical change is carried out at constant pressure and the only work done is due to expansion or contraction, q for the change is called the **enthalpy change** and is represented by the symbol ΔH (ΔH°_{298} for reactions that occur under standard state conditions). Examples of enthalpy changes include **enthalpy of combustion, enthalpy of fusion, enthalpy of vaporization,** and **standard enthalpy of formation.** The standard enthalpy of formation, ΔH°_f, is the enthalpy change accompanying the formation of exactly 1 mole of a substance from the elements in their most stable states at 298.15 K and 1 atmosphere (a **standard state**). If the enthalpies of formation are available for the reactants and products of a reaction, its enthalpy change can be calculated by using **Hess's law**: If a process can be written as the sum of several stepwise processes, the enthalpy change of the total process equals the sum of the enthalpy changes of the various steps.

$$\Delta H^\circ_{298} = \Sigma \Delta H^\circ_{f_{products}} - \Sigma \Delta H^\circ_{f_{reactants}}$$

ΔH for a reaction in one direction is equal in magnitude, but opposite in sign, to ΔH for the reaction in the opposite direction, and ΔH is directly proportional to the quantity of reactants.

Heat is produced from fuels by combustion reactions. Common fuels include the fossil fuels—natural gas, petroleum, and coal. Food is the fuel used by the body.

Key Terms and Concepts

calorie (4.2)	exothermic process (4.1)	nutritional Calorie (4.2)
calorimeter (4.3)	expansion work (4.4)	petroleum (4.7)
calorimetry (4.3)	heat (4.1)	specific heat (4.2)
chemical thermodynamics (4.4)	heat capacity (4.2)	standard enthalpy of formation, ΔH°_f
coal (4.7)	Hess's law (4.6)	(4.5)
endothermic process (4.1)	hydrocarbons (4.7)	standard state (4.5)
enthalpy change, ΔH (4.5)	internal energy (4.4)	surroundings (4.4)
enthalpy of combustion (4.5)	joule (4.2)	system (4.4)
enthalpy of fusion (4.5)	measurement of heat (4.2)	thermal energy (4.1)
enthalpy of vaporization (4.5)	natural gas (4.7)	work (4.4)

Exercises

Questions

1. A 500-mL bottle of water at room temperature and a 2-L bottle of water at the same temperature were placed in a refrigerator. After half an hour the 500 mL of water had cooled to the temperature of the refrigerator. An hour later the 2 L of water had cooled to the same temperature. When asked which sample of water lost the most heat, one student replied that both bottles lost the same amount of heat because they started at the same temperature and finished at the same temperature. A second student thought that the 2 L of water lost more heat because there was more water. A third student believed that the 500 mL of water lost more heat because it cooled more quickly. A fourth student thought that it was not possible to tell because we do not know the initial temperature and the final temperature of the water. Indicate which of these answers is correct, and explain the error in each of the other answers.

2. Explain the difference between heat capacity and specific heat of a substance.

3. Would the amount of heat produced by the reaction in Example 4.2 appear greater, appear smaller, or appear the same if we used a calorimeter that was a poorer insulator than a polystyrene cup calorimeter? Explain your answer.

4. Would the amount of heat absorbed by the dissolution in Example 4.3 appear greater, appear smaller, or appear the same if we used a calorimeter that was a poorer insulator than a polystyrene cup calorimeter? Explain your answer.

5. Explain how the heat measured in Example 4.2 differs from the enthalpy change for the exothermic reaction described by the equation

    $HCl(aq)$ and $NaOH(aq) \longrightarrow NaCl(aq) + H_2O(l)$

6. Which of the enthalpies of combustion in Table 4.1 are also standard enthalpies of formation?

7. Heat is added to a mixture of solid and liquid water. Explain why the temperature of the mixture does not change.

8. Does the standard enthalpy of formation of $H_2O(g)$ differ from ΔH°_{298} for the following reaction? If so, how?

    $$2H_2(g) + O_2(g) \longrightarrow 2H_2O(g)$$

9. What can be done to change the nitrogen in a sample of air to standard state conditions?

10. The gas burned in an oxyacetylene torch (Fig. 4.2) is essentially pure acetylene. Explain why the heat produced by combustion of exactly 1 mol of acetylene in such a torch is not equal to the enthalpy of combustion of acetylene listed in Table 4.1.

11. How could the accuracy of the enthalpy change calculated in Example 4.4 from the data of Example 4.2 be improved by changing the experiment described in Example 4.2?

12. Would the amount of heat absorbed by the dissolution in Example 4.3 appear greater, appear smaller, or appear the same if the specific heat of the calorimeter were taken into account? Explain your answer.

Heat

13. How much would the temperature of the 8.00×10^2 g of water shown in Fig. 4.3 increase if 36.5 kJ of heat were added?

14. If 14.5 kJ of heat were added to the 8.00×10^2 g of water in Fig. 4.3, how much would its temperature increase?

15. How much heat, in joules and in calories, must be added to a 75.0-g iron block with a specific heat of $0.451 \ J \ g^{-1} \ °C^{-1}$ to increase its temperature from 25°C to its melting temperature of 1535°C?

16. The specific heat of ice is $1.95 \ J \ g^{-1} \ °C^{-1}$. How much heat, in joules and in calories, is required to heat a 28.4-g (1-oz) ice cube from $-23.0°C$ to $-1.0°C$?

17. Calculate the heat capacity, in joules and in calories, of
    (a) 28.4 g (1 oz) of water (specific heat, $4.184 \ J \ g^{-1} \ °C^{-1}$)
    (b) 28.4 g (1 oz) of lead (specific heat, $0.129 \ J \ g^{-1} \ °C^{-1}$)

18. Calculate the heat capacity, in joules and in calories, of
    (a) 45.8 g of nitrogen gas (specific heat, $1.04 \ J \ g^{-1} \ °C^{-1}$)
    (b) 1.00 lb (454 g) of aluminum metal (specific heat, $0.88 \ J \ g^{-1} \ °C^{-1}$)

19. How many milliliters of water at 23°C with a density of $1.00 \ g \ mL^{-1}$ must be mixed with 180 mL (about 6 oz) of coffee at 95°C so that the resulting combination will have a temperature of 60°C? Assume that coffee and water have the same density and the same specific heat.

20. How much will the temperature of a cup (180 g) of coffee at 95°C be reduced when a 45-g silver spoon (specific heat, $0.24 \ J \ g^{-1} \ °C^{-1}$) at 25°C is placed in the coffee and the two are allowed to reach the same temperature? Assume that the coffee has the same density and specific heat as water.

21. The temperature of the cooling water as it leaves the hot engine of an automobile is 240°F. After it passes through the radiator it has a temperature of 175°F. Calculate the amount of heat transferred from the engine to the surroundings by 1.0 gal of water with a specific heat of $4.184 \ J \ g^{-1} \ °C^{-1}$.

Calorimetry

22. If a reaction produces 1.506 kJ of heat, which is trapped in 30.0 g of water initially at 26.5°C in a calorimeter like that in Fig. 4.4, what is the resulting temperature of the water?

23. Dissolving 3.0 g of $CaCl_2(s)$ in 150.0 g of water in a polystyrene cup calorimeter (Fig. 4.4) at 22.4°C causes the temperature to rise to 25.8°C. What is the approximate amount of heat involved in the dissolution, assuming the heat capacity of the resulting solution is $4.2 \ J \ g^{-1} \ °C^{-1}$? Is the reaction exothermic or endothermic?

24. A 0.500-g sample of KCl is added to 50.0 g of water in a polystyrene cup calorimeter (Fig. 4.4). If the temperature decreases by 1.05°C, what is the approximate amount of heat involved in the dissolution of the KCl, assuming the heat capacity of the resulting solution is $4.18 \ J \ g^{-1} \ °C^{-1}$? Is the reaction exothermic or endothermic?

25. The addition of 3.15 g of $Ba(OH)_2 \cdot 8H_2O$ to a solution of 1.52 g of NH_4SCN in 100 g of water in a polystyrene cup calorimeter caused the temperature to fall by 3.1°C. Assuming the specific heat of the solution and products is $4.20 \ J \ g^{-1} \ °C^{-1}$, calculate the approximate amount of heat absorbed by the reaction, which can be represented by the following equation:

$$Ba(OH)_2 \cdot 8H_2O(s) + 2NH_4SCN(aq) \longrightarrow Ba(SCN)_2(aq) + 2NH_3(aq) + 10H_2O(l)$$

26. When 50.0 g of 0.200 M $NaCl(aq)$ at 24.10°C is added to 100.0 g of 0.100 M $AgNO_3(aq)$ at 24.10°C in a polystyrene cup calorimeter, the temperature increases to 25.15°C as $AgCl(s)$ forms. Assuming the specific heat of the solution and products is $4.20 \ J \ g^{-1} \ °C^{-1}$, calculate the amount of heat absorbed by the solution.

27. The reaction of 50 mL of acid and 50 mL of base described in Example 4.2 increased the temperature of the solution by 6.9°C. How much would the temperature have increased if 100 mL of the acid and 100 mL of the base had been used in the same calorimeter starting at the same temperature of 22.0°C? Explain your answer.

28. If the 2.0 g of NH_4NO_3 in Example 4.3 were dissolved in 102.0 g of water under the same conditions, how much would the temperature change? Explain your answer.

29. A 70.0-g piece of metal at 80.0°C is placed in 100 g of water at 22.0°C contained in a calorimeter like that shown in Fig. 4.4. The metal and water come to the same temperature at 24.6°C. How much heat did the metal give up to the water? What is the specific heat of the metal?

30. One method of generating electricity is by a steam generating plant that burns coal. To determine the rate at which coal is to be fed into the burner, the heat of combustion per ton of coal must be determined using a calorimeter like the one in Fig. 4.6. When 1.00 g of coal is burned in a calorimeter, the temperature increases by 1.48°C. If the heat capacity of the calorimeter is $21.6 \ kJ \ °C^{-1}$, determine the heat produced by combustion of a ton of coal (2.000×10^3 pounds).

Enthalpy Changes

31. How much heat is produced by burning 4.00 mol of acetylene?

32. How much heat is produced by combustion of 125 g of methanol?

33. How many moles of isooctane must be burned to produce 100 kJ of heat?

34. What mass of carbon monoxide must be burned to produce 175 kJ of heat?

35. When 2.50 g of methane burns in oxygen, 124 kJ of heat is produced. What is the enthalpy of combustion per mole of methane under these conditions?

36. Joseph Priestly prepared oxygen in 1774 by heating red mercury(II) oxide with sunlight focused through a lens. How much heat is required to decompose exactly 1 mol of red $HgO(s)$ to $Hg(l)$ and $O_2(g)$ under standard conditions? See Appendix I for useful information.

37. Calculate the enthalpy of solution (ΔH for the dissolution) per mole of NH_4NO_3 under the conditions described in Example 4.3.

38. Calculate ΔH for the reaction described by the equation

$$NaCl(aq) + AgNO_3(aq) \longrightarrow AgCl(s) + NaNO_3(aq)$$

using the data in Example 4.2.

39. Calculate ΔH for the reaction described by the equation

$$Ba(OH)_2 \cdot 8H_2O(s) + 2NH_4SCN(aq) \longrightarrow Ba(SCN)_2(aq)$$
$$+ 2NH_3(aq) + 10H_2O(l)$$

under the conditions described in Exercise 25.

40. Calculate the enthalpy of solution (ΔH for the dissolution) per mole of $CaCl_2$ under the conditions described in Exercise 23.

41. How many kilojoules of heat will be liberated when exactly 1 mol of manganese, Mn, is burned to form $Mn_3O_4(s)$ at standard state conditions? ΔH_f° of Mn_3O_4 is equal to -1388 kJ mol^{-1}.

42. How many kilojoules of heat will be liberated when exactly 1 mol of iron, Fe, is burned to form $Fe_2O_3(s)$ at standard state conditions? ΔH_f° of Fe_2O_3 is equal to -824.2 kJ mol^{-1}.

43. How much heat is required to convert a 1-oz (28.4-g) ice cube at 0°C to liquid water at 0°C?

44. The volume of liquid water used to make an average ice cube is 25.0 mL. Water (density, 0.997 g mL^{-1}) is transferred to the ice maker of a refrigerator at 25.0 °C. The final temperature of the ice cube is -10.0°C. Determine the amount of heat lost by the water in the process.

45. Before the use of CFC, sulfur dioxide (enthalpy of vaporization, 6.00 kcal mol^{-1}) was used in refrigerators. What mass of SO_2 must be evaporated to remove as much heat as evaporation of 1.00 kg of CCl_2F_2 (enthalpy of vaporization, 17.4 kJ mol^{-1})? *Hint:* Note the different units.

Hess's Law

46. Calculate the standard molar enthalpy of formation of $NO(g)$ from the following data:

$$N_2(g) + 2O_2(g) \longrightarrow 2NO_2(g) \qquad \Delta H_{298}^\circ = 66.4 \text{ kJ}$$

$$2NO(g) + O_2(g) \longrightarrow 2NO_2(g) \qquad \Delta H_{298}^\circ = 114.1 \text{ kJ}$$

47. Which produces more heat, formation of $OsO_4(s)$ or formation of $OsO_4(g)$?

$$Os(s) + 2O_2(g) \longrightarrow OsO_4(s)$$

or $\qquad Os(s) + 2O_2(g) \longrightarrow OsO_4(g)$

for the phase change $OsO_4(s) \longrightarrow OsO_4(g)$, $\Delta H = 56.4$ kJ.

48. Both graphite and diamond burn.

$$C(s,\text{graphite}) + O_2(g) \longrightarrow CO_2(g)$$
$$C(s,\text{diamond}) + O_2(g) \longrightarrow CO_2(g)$$

For the conversion of graphite to diamond,

$$C(s,\text{graphite}) \longrightarrow C(s,\text{diamond}) \qquad \Delta H_{298}^\circ = 1.90 \text{ kJ}$$

Which produces more heat, the combustion of graphite or the combustion of diamond?

49. Calculate ΔH_{298}° for the process

$$Zn(s) + S(s) + 2O_2(g) \longrightarrow ZnSO_4(s)$$

from the following information:

$$Zn(s) + S(s) \longrightarrow ZnS(s) \qquad \Delta H_{298}^\circ = -206.0 \text{ kJ}$$
$$ZnS(s) + 2O_2(g) \longrightarrow ZnSO_4(s) \qquad \Delta H_{298}^\circ = -776.8 \text{ kJ}$$

50. Calculate ΔH_{298}° for the process

$$Sb(s) + \tfrac{5}{2}Cl_2(g) \longrightarrow SbCl_5(g)$$

from the following information:

$$Sb(s) + \tfrac{3}{2}Cl_2(g) \longrightarrow SbCl_3(g) \qquad \Delta H_{298}^\circ = -314 \text{ kJ}$$
$$SbCl_3(s) + Cl_2(g) \longrightarrow SbCl_5(g) \qquad \Delta H_{298}^\circ = -80 \text{ kJ}$$

51. Calculate ΔH_{298}° for the process

$$Co_3O_4(s) \longrightarrow 3Co(s) + 2O_2(g)$$

from the following information:

$$Co(s) + \tfrac{1}{2}O_2(g) \longrightarrow CoO(s) \qquad \Delta H_{298}^\circ = -237.9 \text{ kJ}$$
$$3CoO(s) + \tfrac{1}{2}O_2(g) \longrightarrow Co_3O_4(s) \qquad \Delta H_{298}^\circ = -177.5 \text{ kJ}$$

52. Calculate ΔH for the process

$$Hg_2Cl_2(s) \longrightarrow 2Hg(l) + Cl_2(g)$$

from the following information:

$$Hg(l) + Cl_2(g) \longrightarrow HgCl_2(s) \qquad \Delta H = -224 \text{ kJ}$$
$$Hg(l) + HgCl_2(s) \longrightarrow Hg_2Cl_2(s) \qquad \Delta H = -41.2 \text{ kJ}$$

53. From the molar heats of formation in Appendix I, determine how much heat is required to evaporate exactly 1 mol of water.

$$H_2O(l) \longrightarrow H_2O(g)$$

54. The white pigment TiO_2 is prepared by the reaction of titanium tetrachloride, $TiCl_4$, with water vapor in the gas phase.

$$TiCl_4(g) + 2H_2O(g) \longrightarrow TiO_2(s) + 4HCl(g)$$

How much heat is evolved in the production of exactly 1 mol of $TiO_2(s)$ under standard state conditions?

55. Using the data in Appendix I, calculate the standard enthalpy change for each of the following reactions.

(a) $Si(s) + 2F_2(g) \longrightarrow SiF_4(g)$

(b) $2C(s) + 2H_2(g) + O_2(g) \longrightarrow CH_3CO_2H(l)$
(c) $CH_4(g) + N_2(g) \longrightarrow HCN(g) + NH_3(g)$
(d) $CS_2(g) + 3Cl_2(g) \longrightarrow CCl_4(g) + S_2Cl_2(g)$

56. Using the data in Appendix I, calculate the standard enthalpy change for each of the following reactions.

(a) $N_2(g) + O_2(g) \longrightarrow 2NO(g)$
(b) $Si(s) + 2Cl_2(g) \longrightarrow SiCl_4(g)$
(c) $Fe_2O_3(s) + 3H_2(g) \longrightarrow 2Fe(s) + 3H_2O(l)$
(d) $2LiOH(s) + CO_2(g) \longrightarrow Li_2CO_3(s) + H_2O(g)$

57. (a) Calculate the enthalpy of combustion of propane, $C_3H_8(g)$, for the formation of $H_2O(g)$ and $CO_2(g)$. The enthalpy of formation of propane is -104 kJ mol^{-1}.
(b) Calculate the enthalpy of combustion of butane, $C_4H_{10}(g)$, for the formation of $H_2O(g)$ and $CO_2(g)$. The enthalpy of formation of butane is -126 kJ mol^{-1}.
(c) Both propane and butane are used as gaseous fuels. Which compound produces more heat per gram when burned?

58. The decomposition of hydrogen peroxide, H_2O_2, has been used to provide thrust in the control jets of various space vehicles. Using the data in Appendix I, determine how much heat is produced by the decomposition of exactly 1 mol of H_2O_2 under standard conditions.

$$2H_2O_2(l) \longrightarrow 2H_2O(g) + O_2(g)$$

59. In the early of days of automobiles, illumination at night was provided by burning acetylene, C_2H_2; some cavers still use acetylene as a source of light. The acetylene is prepared in the lamp by the reaction of water with calcium carbide, CaC_2.

$$CaC_2(s) + H_2O(l) \longrightarrow Ca(OH)_2(s) + C_2H_2(g)$$

Calculate the standard enthalpy of the reaction. The ΔH_f° of CaC_2 is -15.14 kcal mol^{-1}. (Note the units.)

Fuel and Food

60. From the data in Table 4.1, determine which of the following fuels produces the greatest amount of heat per gram when burned under standard conditions: $CO(g)$, $CH_4(g)$, or $C_2H_2(g)$.

61. The enthalpy of combustion of hard coal averages -35 kJ g^{-1}; that of gasoline averages -1.28×10^5 kJ gal^{-1}. How many kilograms of hard coal provide the same amount of heat as is available from 1.0 gal of gasoline?

62. Ethanol, C_2H_5OH, is used as a fuel for motor vehicles, particularly in Brazil.

(a) Write the balanced equation for the combustion of ethanol to $CO_2(g)$ and $H_2O(g)$, and using the data in Appendix I, calculate the enthalpy of combustion of exactly 1 mol of ethanol.
(b) The density of ethanol is 0.7893 g mL^{-1}. Calculate the enthalpy of combustion of exactly 1 L of ethanol.
(c) Assuming that the mileage an automobile gets is directly proportional to the heat of combustion of the fuel,

calculate how many times farther an automobile could be expected to go on 1 L of gasoline than on 1 L of ethanol. Assume that gasoline has the heat of combustion and the density of n-octane, C_8H_{18} ($\Delta H_f^\circ = -208.4$ kJ mol^{-1}; density, 0.7025 g mL^{-1}).

63. A pint of premium ice cream can contain 1100 Calories. What mass of fat, in grams and pounds, must be produced in the body to store an extra 1.1×10^3 Calories?

64. What is the maximum mass of carbohydrate in a can of Diet Coke that contains less than 1 Calorie per can?

65. The amount of fat recommended for someone with a daily diet of 2000 Calories is 65 g. What percent of the Calories in this diet would be supplied by this amount of fat?

66. A serving of a breakfast cereal contains 3 g of protein, 18 g of carbohydrates, and 6 g of fat. What is the Calorie content of a serving of this cereal?

67. A teaspoon of the carbohydrate sucrose (common sugar) contains 16 nutritional Calories (16 kcal). What is the mass of 1 teaspoon of sucrose?

68. The oxidation of the sugar glucose, $C_6H_{12}O_6$, is described by the equation

$$C_6H_{12}O_6(s) + 6O_2(g) \longrightarrow 6CO_2(g) + 6H_2O(l)$$
$$\Delta H = -2816 \text{ kJ}$$

Metabolism of glucose gives the same products, although the glucose reacts with oxygen in a series of steps in the body.

(a) How much heat, in kilojoules, can be produced by the metabolism of 1.0 g of glucose?
(b) How many nutritional Calories can be produced by the metabolism of 1.0 g of glucose?

Applications and Additional Exercises

69. (a) What is the final temperature when a 45-g aluminum spoon (specific heat, 0.88 J g^{-1} °C^{-1}) at 24°C is placed in 180 mL (180 g) of coffee at 85°C and the temperatures of the two become equal? Assume that coffee has the same specific heat as water.
(b) The first time one of the authors solved this problem, he got an answer of 88°C. Besides reworking the problem, how else could he recognize that this is an incorrect answer?

70. (a) What is the heat capacity of a 1.05-kg aluminum kettle? The specific heat of aluminum is 0.88 J g^{-1} °C^{-1}.
(b) How much heat is required to increase the temperature of this kettle from 23.0°C to 100.0°C?
(c) How much heat is required to heat a filled kettle and its contents from 23.0°C to 99.0°C if it contains 1.25 L of water with a density of 0.997 g mL^{-1} and a specific heat of 4.184 J g^{-1} °C^{-1}?

71. Most people find waterbeds uncomfortable unless the water temperature is maintained at about 85°F. Unless it is heated, a waterbed that measures 72 in. by 84 in. by 9 in. cools from

85°F to 72°F in 24 h. Estimate the amount of electrical energy that is required to keep the bed from cooling. What assumptions did you make in your estimate? (1 kWh = 3.6 × 10^6 J)

72. Homes may be heated by passing steam or pumping hot water through radiators.

(a) Calculate the quantity of heat released when 100 g of steam at 110°C cools and condenses to water at 80°C. (The specific heat of steam is 1.987 J g^{-1} $°C^{-1}$. Other values can be found in the chapter.)

(b) Calculate the quantity of water that will provide the same amount of heat if cooled from 95.0°C to 35.0°C.

73. When 1.0 g of fructose, $C_6H_{12}O_6(s)$, a sugar commonly found in fruits, is burned in oxygen in a calorimeter (Fig. 4.6), the temperature of the calorimeter increases by 1.58°C. If the heat capacity of the calorimeter and its contents is 9.90 kJ $°C^{-1}$, what is q for this combustion?

74. A sample of 0.562 g of carbon is burned in oxygen in a calorimeter (Fig. 4.6), producing carbon dioxide. Assume that both the reactants and the products are under standard state conditions and that the heat released is directly proportional to the enthalpy of combustion of graphite. The temperature of the calorimeter increases from 26.74°C to 27.93°C. What is the specific heat of the calorimeter and its contents?

75. Among the substances that react with oxygen and that have been considered as potential rocket fuels are diborane, B_2H_6, which produces $B_2O_3(s)$ and $H_2O(g)$; methane, CH_4, which produces $CO_2(g)$ and $H_2O(g)$; and hydrazine, N_2H_4, which produces $N_2(g)$ and $H_2O(g)$. On the basis of the heat evolved by 1.00 g of each substance in its reaction with oxygen, determine which of these compounds has the best prospects as a rocket fuel. ΔH_f° values for $B_2H_6(g)$, $CH_4(g)$, and $N_2H_4(l)$ can be found in Appendix I.

76. Water gas, a mixture of H_2 and CO, is an important industrial fuel produced by the reaction of steam with red-hot coke, essentially pure carbon.

$$C(s) + H_2O(g) \longrightarrow CO(g) + H_2(g)$$

(a) Assuming that coke has the same enthalpy of formation as graphite, calculate ΔH_{298}° for this reaction.

(b) Methanol, a liquid fuel that could possibly replace gasoline, can be prepared from water gas and additional hydrogen at high temperature and pressure in the presence of a suitable catalyst.

$$2H_2(g) + CO(g) \longrightarrow CH_3OH(g)$$

Under the conditions of the reaction, methanol forms as a gas. Calculate ΔH_{298}° for this reaction and for the condensation of gaseous methanol to liquid methanol.

(c) Calculate the heat of combustion of 1 mol of liquid methanol to $H_2O(g)$ and $CO_2(g)$.

77. Which is the least expensive source of energy in kilojoules per dollar, a box of breakfast cereal that weighs 32 oz and costs $4.23 or a liter of isooctane (density, 0.6919 g mL^{-1}) that costs $0.45? Compare the nutritional value of the cereal with the heat produced by combustion of the isooctane under standard conditions. A 1.0-oz serving of the cereal provides 130 (nutritional) Calories.

78. The following reactions can be used to prepare samples of metals. Determine the enthalpy change under standard state conditions for each.

(a) $2Ag_2O(s) \longrightarrow 4Ag(s) + O_2(g)$
(b) $SnO(s) + CO(g) \longrightarrow Sn(s) + CO_2(g)$
(c) $Cr_2O_3(s) + 3H_2(g) \longrightarrow 2Cr(s) + 3H_2O(l)$
(d) $2Al(s) + Fe_2O_3(s) \longrightarrow Al_2O_3(s) + 2Fe(s)$

79. How much heat is produced when 100 mL of 0.250 M HCl (density, 1.00 g mL^{-1}) and 200 mL of 0.150 M NaOH (density, 1.00 g mL^{-1}) are mixed?

$$HCl(aq) + NaOH(aq) \longrightarrow NaCl(aq) + H_2O(l)$$
$$\Delta H_{298} = -58 \text{ kJ}$$

If both solutions are at the same temperature and the heat capacity of the products is 4.19 J g^{-1} $°C^{-1}$, how much will the temperature increase? What assumption did you make in your calculation?

80. How much heat is produced when 1.25 g of chromium metal and oxygen gas, both at standard conditions, react to produce 1.83 g of product at standard conditions?

81. Ethylene, a by-product from the fractional distillation of petroleum, is fourth among the 50 chemical compounds produced commercially in the largest quantities. About 80% of synthetic ethanol is manufactured from ethylene by its reaction with water in the presence of a suitable catalyst.

$$C_2H_4(g) + H_2O(g) \longrightarrow C_2H_5OH(l)$$

Using the data in the table in Appendix I, calculate ΔH° for the reaction.

82. The following sequence of reactions occurs in the commercial production of aqueous nitric acid.

$$4NH_3(g) + 5O_2(g) \longrightarrow 4NO(g) + 6H_2O(l)$$
$$\Delta H = -907 \text{ kJ}$$

$$2NO(g) + O_2(g) \longrightarrow 2NO_2(g) \qquad \Delta H = -113 \text{ kJ}$$

$$3NO_2(g) + H_2O(l) \longrightarrow 2HNO_3(aq) + NO(g)$$
$$\Delta H = -139 \text{ kJ}$$

Determine the total energy change for the production of exactly 1 mol of aqueous nitric acid by this process.

83. (This problem is taken from the 1995 Chemistry Advanced Placement Examination and is used with the permission of the Educational Testing Service.) Propane, C_3H_8, is a hydrocarbon that is commonly used as a fuel for cooking.

(a) Write a balanced equation for the complete combustion of propane gas, which yields $CO_2(g)$ and $H_2O(l)$.

(b) Calculate the volume of air at 30°C and 1.00 atmosphere that is needed to burn completely 10.0 grams of propane. Assume that air is 21.0 percent O_2 by volume.

(*Hint:* In Chapter 10, we will see how to do this calculation using gas laws. For now, use the information that 1.00 L of air at 30°C and 1 atmosphere pressure contains 0.268 g of O_2 per liter.)

(c) The heat of combustion of propane is $-2,220.1$ kJ/mol. Calculate the heat of formation, ΔH_f°, of propane, given that ΔH_f° of $H_2O(l) = -285.3$ kJ/mol and ΔH_f° of $CO_2(g) = -393.5$ kJ/mol.

(d) Assuming that all of the heat evolved in burning 30.0 grams of propane is transferred to 8.00 kilograms of water (specific heat = 4.18 J/g·k), calculate the increase in temperature of the water.

84. Some gas companies calculate their charges on the basis of the number of therms used by a customer. A therm is defined as 100,000 B.T.U. One British thermal unit (B.T.U.) is the amount of heat that will heat exactly 1 lb of water by exactly 1°F. During a recent winter month in Indiana, it required 134 therms of gas to heat a small house using natural gas as the source of heat in a gas furnace that was 89% efficient. (The efficiency of a gas furnace is the percent of the heat produced by combustion that is transferred into the house; the remaining heat goes up the vent pipe to the outside.) The products of the combustion are carbon dioxide and water vapor.

(a) Assume that natural gas is pure methane, and determine the volume of natural gas, in cubic feet, that was required to heat the house. The average temperature of the natural gas was 56°F. At this temperature, natural gas has a density of 0.681 g L^{-1}.

(b) How many gallons of LPG (liquefied petroleum gas) would be required to replace the natural gas used? Assume that LPG is liquid propane [C_3H_8: density, 0.5318 g mL^{-1}; enthalpy of combustion, -2044 kJ mol^{-1} for the formation of $CO_2(g)$ and $H_2O(g)$] and that the furnace used to burn LPG has the same efficiency as the gas furnace.

(c) What mass of carbon dioxide is produced by combustion of the methane used to heat the house?

(d) What mass of water is produced by combustion of the methane used to heat the house?

(e) What volume of air is required to provide the oxygen for the combustion of the methane used to heat the house? Air contains 23% oxygen by mass. The average density of air during the month was 1.22 g L^{-1}.

(f) How many kilowatt-hours (1 kWh = 3.6×10^6 J) of electricity would be required to provide the heat necessary to heat the house? Note that electricity is 100% efficient in producing heat inside a house.

(g) Although electricity is 100% efficient in producing heat inside a house, the production and distribution of electricity are not 100% efficient. The efficiency of production and distribution of electricity produced in a coal-fired power plant is about 40%. A certain type of coal provides 12,000 B.T.U. per pound upon combustion. What mass of this coal, in kilograms, will be required to produce the electrical energy necessary to heat the house if the efficiency of generation and distribution is 40%?

(h) Why is the enthalpy of combustion used in part (b) of this problem different from that used in Exercise 83?

CHAPTER OUTLINE

**Historical Development of
the Atomic Theory**

5.1 Atomic Architecture: The Nuclear Atom
5.2 Light, Atomic Spectra, and Atomic
    Structure
5.3 The Bohr Model of the Atom

Quantum Mechanics

5.4 Behavior in the Microscopic World
5.5 The Quantum-Mechanical Model of
    the Atom
5.6 Results of the Quantum-Mechanical Model
    of the Atom
5.7 Orbital Energies and Atomic Structure

**The Periodic Table: Electron
Configuration and Atomic Properties**

5.8 The Aufbau Process
5.9 Electron Configurations and the
    Periodic Table
5.10 Variation of Atomic Properties Within
    Periods and Groups

5

Atoms: Their Structure, Spectra, and Properties

Rings of glowing gas encircling the site of a 1987 supernova explosion were photographed in 1994 by the Hubble Space Telescope, using the light of one wavelength from the spectrum of hydrogen.

What do the following items have in common: the supernova shown in the opening photograph of this chapter, the laser in a scalpel for cataract surgery or in a compact disc player, a color television screen, the colors of fireworks, a dentist's X-ray machine, and a neon sign? The answer is that all involve light, or some other form of electromagnetic radiation, produced when electrons in an atom or ion move from one energy level to another.

At first glance, we may see little relationship between light or other forms of radiation and the electronic structure of atoms, which is the focus of this chapter. However, toward the middle of the chapter, we will see how that all comes together. Finally, we will see how atomic structure explains the arrangement of elements in the periodic table, and we will take our first steps toward seeing how the structure of their atoms is responsible for the behavior of the elements.

In Chapter 2 we introduced protons, neutrons, and electrons—the particles that make up an atom. In this chapter we will examine some of the elegant experiments with which investigators identified and characterized these components of the atom and their arrangement as a nucleus surrounded by electrons. We will consider how the colors emitted by atoms or ions when they are heated (their atomic spectra) provide experimental support for a theory of the behavior of the electrons in such atoms.

Table 5.1

Highlights in the Development of Contemporary Atomic Theory

1803	John Dalton (English) proposed an atomic theory to explain the laws of definite composition and multiple proportions (Section 2.1).
1879	William Crookes (English) repeated the observations of earlier workers and concluded that rays produced by an electric current in an evacuated tube consist of a stream of charged particles produced at the cathode.
1885	Johann J. Balmer (Swiss) developed a simple empirical equation that could be used to calculate the wavelengths of lines in the spectrum of hydrogen atoms.
1886	Eugen Goldstein (German) characterized "canal rays" (positive ions) produced by an electric current in an evacuated tube.
1895	Wilhelm Roentgen (German) discovered X rays produced by an electric current in an evacuated tube.
1896	Antoine Becquerel (French) discovered natural radioactivity.
1890	Johannes Rydberg (German) modified Balmer's equation to describe the frequencies of the lines in the hydrogen spectrum.
1897	J. J. Thomson (British) showed that Crookes's cathode rays are beams of negatively charged particles (now known to be electrons) and calculated their charge-to-mass ratio.
1898	Marie Curie (Polish) and her husband, Pierre (French), isolated polonium and radium from pitchblende (a lead ore) by chemical processes.
1900	Max Planck (German) proposed a quantum theory that described the light emitted from a hot object as composed of discrete units called *quanta* or *photons*.
1903	Hantaro Nagaoka (Japanese) postulated a saturn-like atom, a positively charged sphere surrounded by a halo of electrons.
1904	J. J. Thomson proposed a model of the atom with electrons embedded in a sea of positive charges, the "plum-pudding model."
1905	Albert Einstein (German) described the relationship of mass and energy.

The chemical behavior of atoms is primarily determined by the arrangement of their electrons outside the atomic nucleus. Our goal is to introduce a model of the atom that helps us understand this arrangement, to understand why elements exhibit their characteristic kinds of chemical behavior, and to recall and predict this behavior in a systematic way. We will see that the periodic table is a powerful tool for understanding the electronic structure of the elements and the regularity of their atomic properties.

Historical Development of the Atomic Theory

Many scientists in different countries played a role in discovering the ideas that led to our present conception of the structure of the atom. Highlights of those discoveries are outlined in Table 5.1. In the following sections we will see how these ideas evolved into the contemporary model of the atom.

Table 5.1
(continued)

1909	Robert Millikan (American) determined the charge on the electron in his oil drop experiment.
1911	Ernest Rutherford (New Zealander) attributed the scattering of α particles by a thin gold foil to an atomic structure in which an atom's mass is concentrated in a tiny, positively charged nucleus.
1912	J. J. Thomson detected the isotopes neon-20 and neon-22.
1913	Niels Bohr (Danish) showed that the hydrogen spectrum could be explained by a previously unrecognized property of matter—the energy of electrons in atoms is limited to certain values (quantized).
1913	Henry Moseley (English) observed that an element emits characteristic X rays that depend on its nuclear charge (atomic number).
1920	Orme Masson (Australian), William Harkins (American), and Rutherford independently postulated the existence of an uncharged particle with the mass of a proton (the neutron).
1924	Louis de Broglie (French) combined equations from Einstein and Planck to suggest that electrons have wave-like properties.
1925	Wolfgang Pauli (German) stated that no two electrons in the same atom can have the same set of quantum numbers (Pauli's exclusion principle).
1926	Erwin Schrödinger (Austrian) combined the particle nature of an electron, its wave properties, and quantum restrictions and developed an equation that described the energy and likely location of electrons in a probability relationship.
1927	Werner Heisenberg (German) showed that we cannot determine simultaneously both the exact position and the momentum of an electron (Heisenberg's uncertainty principle).
1927	Frederick Hund (German) determined that electrons in subshells have maximum unpairing and that unpaired electrons have the same spin (Hund's rule).
1932	James Chadwick (British) characterized neutrons.

5.1   Atomic Architecture: The Nuclear Atom

Although Dalton postulated the existence of atoms in 1803, almost 80 years passed before information that led to an understanding of their architecture began to appear. Then, over the period from 1879 to 1932, the components of the atom were identified and characterized. These are the electron, a very light particle with a mass of 5.5×10^{-4} atomic mass unit (9.1×10^{-28} gram) and a single negative charge; the proton, a heavier particle with a mass of 1.0073 atomic mass units (1.6726×10^{-24} gram) and a positive charge of the same magnitude as the negative charge on the electron; and the neutron, a particle with a mass of 1.0087 atomic mass units (1.6749×10^{-24} gram, about equal to the mass of a proton) and no charge (Section 2.2).

The distribution and stability of both heavy positive particles and light negative particles in an atom have been clarified by a number of interrelated discoveries and experiments. Most notable of these were the proposal of a *nuclear atom* by Ernest Rutherford and the *quantum model* of the hydrogen atom by Neils Bohr.

Rutherford, who was born in New Zealand but spent his scientific career in England, used positively charged particles as high-speed energetic probes to travel within atoms. He observed the scattering of the particles. The particles used were **alpha particles** (α particles), or helium nuclei, which have a +2 charge and a mass of approximately 4 atomic mass units. They were produced by the radioactive decay of radium. (We will consider radioactivity and radioactive decay in Chapter 20.)

After making his initial observation that α particles were scattered by air, Rutherford asked his co-workers Hans Geiger and Ernest Marsden to examine the scattering of a beam of α particles by a very thin piece of gold foil. They used a luminescent screen to detect the particles and found that most particles passed right through the foil without deflection (Fig. 5.1). However, a few were diverted from their straight paths, and a very small number were deflected back toward their source. Analyzing a series of such experiments, Rutherford concluded that (1) the volume occupied by an atom must consist of a large amount of empty space, but (2) a heavy, positively charged body must be at the center of each atom. Because most of the fast-moving α particles passed through the gold atoms undeflected, they must have traveled through essentially empty space inside the atom (Fig. 5.2). Like charges repel one another, so the few positively charged α particles that changed paths abruptly must have hit, or closely approached, another body that also had a highly concentrated, positive charge.

Therefore, Rutherford proposed that an atom consists of a very small, positively charged nucleus, in which most of the mass of the atom is concentrated. Surround-

Figure 5.1

The experimental setup used to detect the scattering of α particles by a gold foil. The tedium of counting the flashes produced as individual α particles struck the luminescent screen inspired Geiger to invent his famous counter.

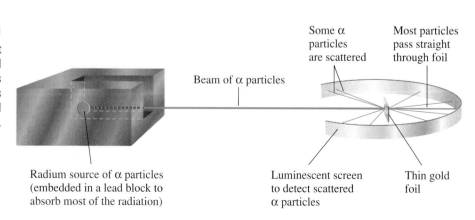

Beam of α particles

Some α particles are scattered

Most particles pass straight through foil

Radium source of α particles (embedded in a lead block to absorb most of the radiation)

Luminescent screen to detect scattered α particles

Thin gold foil

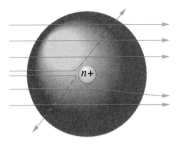

Figure 5.2

Scattering of a α particles by a much heavier, positively charged gold nucleus. Because the nucleus is very small compared to the size of the atom (if drawn to scale on this figure, the nucleus would be invisible), only a few α particles are deflected. Particles that collide with the nucleus are sharply deflected; the path of a particle that passes close by it is bent slightly. Most α particles pass through the relatively large region occupied by electrons, which are too light to deflect the rapidly moving particles.

ing the nucleus must be the negative electrons so that the atom is electrically neutral. This nuclear model of the atom, proposed in 1911, is one we still use today.

Rutherford extended the experiments to other, lighter elements. From the data obtained, he noticed that for many lighter elements, the number of positive charges in the nucleus was approximately equal to one-half of the atomic mass of the element. Henry Moseley's X-ray studies confirmed these conclusions regarding the charge of the nucleus. We saw in Section 2.2 that this nuclear charge is the atomic number of the element. Because a neutral atom contains the same number of protons and electrons, its atomic number also represents the number of electrons in the neutral atom.

At about the same time, J. J. Thomson and others showed that many elements consist of atoms with different masses. We saw in Chapter 2 that these are called isotopes and that isotopes differ from each other only by the number of neutrons within the nucleus. Interestingly, although their existence had been postulated as early as 1920 to explain the difference between the number of protons in an atom and its mass, neutrons were not detected until 1932. The sum of the protons and neutrons in an atom of an isotope is the mass number of that particular isotope (Section 2.2).

5.2 Light, Atomic Spectra, and Atomic Structure

Visible light and other forms of **electromagnetic radiation** are used to probe the energies of electrons within atoms. Radio waves from a cordless telephone, the X rays used by dentists, the energy used to cook food in a microwave oven, the radiant heat from a red-hot object, and the light from your color television screen—all are examples of electromagnetic radiation. These forms of energy may seem quite different, but they all exhibit wave-like behavior, and in a vacuum, they all travel with the speed of light. They all can be characterized by a **wavelength,** λ (the lower-case Greek letter lambda), and a **frequency,** ν (the lower-case Greek letter nu). The wavelength is the distance between any two consecutive peaks or troughs in a wave (Fig. 5.3). The

Figure 5.3

A comparison of wavelength and frequency. The wave with the shortest wavelength has the highest frequency.

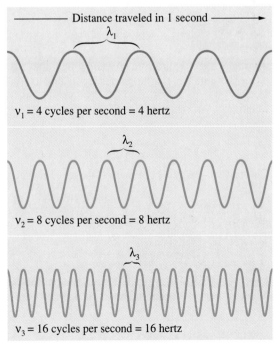

Distance traveled in 1 second

λ_1

$\nu_1 = 4$ cycles per second = 4 hertz

λ_2

$\nu_2 = 8$ cycles per second = 8 hertz

λ_3

$\nu_3 = 16$ cycles per second = 16 hertz

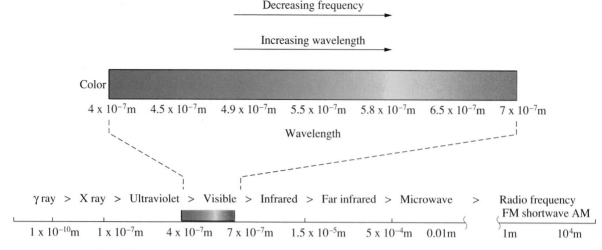

Figure 5.4

Portions of the electromagnetic spectrum in order of decreasing frequency and increasing wavelength. Wavelength and frequency are inversely proportional. Only wavelengths in the visible region can be seen by the human eye.

frequency (Fig. 5.3) is the number of waves (peaks or troughs) that pass a specified point in space in 1 second. We generally use units of cycles per second ($1/s$ or s^{-1}) for frequency. The unit s^{-1} is called the **hertz,** abbreviated Hz. A common multiple of this unit is the megahertz, MHz (1 MHz = 1×10^6 Hz).

The product of the wavelength times the frequency of a wave, or $\lambda\nu$, is equal to the speed of the wave. For electromagnetic radiation this speed is the speed of light, c, which is 2.998×10^8 m s^{-1} in a vacuum. Thus, for electromagnetic radiation,

$$c = 2.998 \times 10^8 \text{ m s}^{-1} = \lambda\nu$$

Because c is a constant, the wavelength and frequency are inversely proportional. As the wavelength increases, the frequency decreases. This inverse proportionality is illustrated in Fig. 5.4, which also illustrates a large portion of the electromagnetic spectrum.

Example 5.1   Determining the Frequency and Wavelength of Radiation

A sodium street light gives off yellow light that has a wavelength of 589 nm (1 nm = 1×10^{-9} m). What is the frequency of this light?

Solution: We can rearrange the equation $c = \lambda\nu$ to solve for the frequency:

$$\nu = \frac{c}{\lambda}$$

Because c is expressed in meters, we must also convert 589 nm to meters.

$$\nu = \left(\frac{2.998 \times 10^8 \text{ m s}^{-1}}{589 \text{ nm}}\right)\left(\frac{1 \times 10^9 \text{ nm}}{1 \text{ m}}\right)$$
$$= 5.09 \times 10^{14} \text{ s}^{-1}$$

Check your learning: An FM radio station broadcasts at a frequency of 101.3 MHz. What is the wavelength in meters of these radio waves?

Answer: 2.960 m

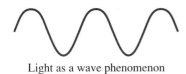

Light as a wave phenomenon

Light as a stream of photons

Figure 5.5

Electromagnetic radiation exhibits wave-like and particle-like properties.

Electromagnetic radiation also has properties associated with particles (Fig. 5.5). At the beginning of the twentieth century, matter was thought to be able to absorb or emit any quantity of energy. However, experiments by Max Planck could be explained only if energy could be gained or lost in individual units, or packets, with energies that are whole-number multiples of a constant, h, which is now called Planck's constant and has a value of 6.626×10^{-34} joule seconds (J s). These packets of radiation are called **quanta.** Albert Einstein subsequently proposed that quanta of electromagnetic radiation can be considered as a stream of particles called **photons.** The energy of a photon in joules can be determined from the expression

$$E = h\nu = \frac{hc}{\lambda}$$

where h is Planck's constant, c is the velocity of the radiation, λ is its wavelength, and ν is its frequency.

The utility of this relationship for our present purposes is that we can now calculate the energy of a photon of light from its wavelength, which can be measured in the laboratory.

Example 5.2   Calculating the Energy of Radiation

When we see light from a neon sign, we are observing radiation from excited neon atoms. If this radiation has a wavelength of 640 nm, what is the energy of the photon being emitted?

Solution: We use the part of Einstein's equation that includes the wavelength, λ, and convert units of nanometers to meters so that the units of λ and c are the same.

$$E = \frac{hc}{\lambda}$$

$$= \frac{(6.626 \times 10^{-34}\ \text{J s})(2.998 \times 10^{8}\ \text{m s}^{-1})}{(640\ \text{nm})\left(\dfrac{1\ \text{m}}{1 \times 10^{9}\ \text{nm}}\right)}$$

$$= 3.10 \times 10^{-19}\ \text{J}$$

Check your learning: The microwaves in an oven are of a specific frequency that will heat the water molecules contained in food. (This is why most plastics and glass do not become hot in a microwave oven—they do not contain water molecules.) This frequency is about 3×10^{11} Hz. What is the energy of one photon of these microwaves?

Answer: 2×10^{-22} J

In the mid-1600s, Sir Isaac Newton separated sunlight into its component colors by passing it through a glass prism (Fig. 5.6). He observed a **continuous spectrum**—a spectrum that contains all wavelengths, and thus all energies, of visible light. (Strictly speaking, sunlight is composed of energies that are integer multiples of Planck's constant. However, the difference between two adjacent energies of light is so small that it cannot be detected by eye.)

We see a continuous spectrum of sunlight in a rainbow when the light is diffracted through raindrops, when the angles of a cut-glass object in a sunny window "catch

Figure 5.6

Sunlight passing through a prism is separated into its component colors, resulting a continuous spectrum.

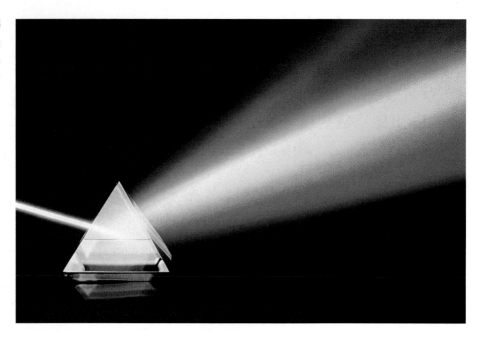

the light," and when sunlight or some other white light reflects off a compact disc (Fig. 5.7). Sunlight also contains ultraviolet light (very short wavelengths) and infrared light (long wavelengths), which can be detected with instruments but are invisible to the human eye. Incandescent (glowing) solids such as the filament inside a light bulb also give off light that contains all wavelengths of visible light.

When an electric current is passed through a gas at low pressure, the gas gives off, or emits, light. Passed through a prism and to a detector, the light is seen to be made up of a number of lines of discrete wavelength. These are called **emission spectra,** or *line spectra*. Each element displays its own characteristic set of lines, as demonstrated by the spectra in Fig. 5.8. Each line consists of a single wavelength of light, so the light emitted by a gas consists of light of discrete energies. For example, when an electric discharge passes through a tube containing hydrogen gas at low pressure, we see a blue-pink color. Passing the light through a prism produces a line spectrum, indicating that it is composed of light of several energies.

Figure 5.7

Diffraction by a grating, such as the closely spaced grooves on this compact disc, produces a spectrum.

Rydberg's equation (Table 5.1) gives the energies, in joules, of the lines in the visible spectrum of hydrogen.

$$E = h\nu = 2.179 \times 10^{-18}\,\text{J}\left(\frac{1}{n_1^2} - \frac{1}{n_2^2}\right)$$

where n_1 and n_2 are integers and $n_1 < n_2$. This equation is empirical; that is, it was derived from observation rather than theory.

At this point (the early part of the twentieth century), there was a great deal of uncertainty about the source of the line spectra of the elements. There was regularity, as evidenced by the spectrum of hydrogen and the Balmer and Rydberg equations that described that spectrum, but there was no consensus about the origin of these spectra. The resolution of the issue required Bohr's postulating a new type of behavior that was previously unknown in matter.

Figure 5.8

A comparison of the continuous spectrum of white light (the top spectrum) and the line spectra of the light from excited sodium, hydrogen, calcium, and mercury atoms.

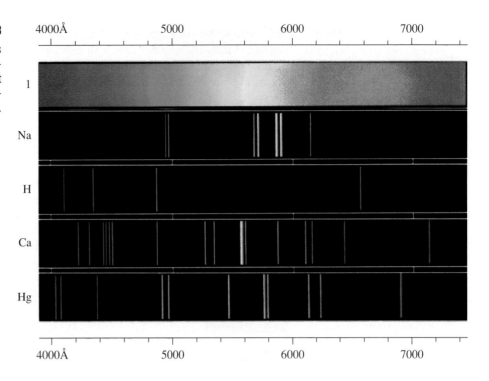

5.3 The Bohr Model of the Atom

After the work of Rutherford and his colleagues, the arrangement of the components of the atom seemed clear. However, one serious problem remained: According to the science known at the time, Rutherford's nuclear atom should not be stable. It was thought that the attractive force between the negatively charged electrons and the positively charged nucleus should cause the electrons to fall into the nucleus. If it was assumed that the electrons moved around the nucleus in circular orbits, then the inertia of the electrons could counterbalance the force of attraction, and the electrons would stay in their orbits. However, according to classical physics, an electron moving in such a circular orbit would radiate energy continuously. If this were to occur, the electron should move in smaller and smaller orbits as it lost energy and should finally fall into the nucleus.

In 1913 Neils Bohr proposed a new theory for the behavior of matter that appeared to solve the problem. His important contribution was to suggest that the energy of an electron in an atom is restricted to discrete, or distinct, values; this energy, like the energy of electromagnetic radiation, cannot vary continuously and take on any value. Another way we express this idea is to say that the energy is **quantized.** The success of this idea in explaining the line spectrum of the hydrogen atom ultimately led to its general acceptance.

Bohr's model of the hydrogen atom includes the following assumptions: (1) A single electron moves around the nucleus in a circular orbit. (2) The inertia of the electron would tend to move it away from the nucleus, but the inertia is counterbalanced by the electrostatic attraction which exists between the positive nucleus and the negative electron. (3) The energy of the electron is restricted to certain values. (This is the new assumption in the model.) Bohr's model is valid for the hydrogen atom

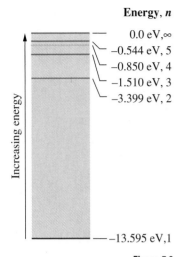

Energy, n

— 0.0 eV, ∞
— −0.544 eV, 5
— −0.850 eV, 4
— −1.510 eV, 3
— −3.399 eV, 2

— −13.595 eV, 1

Figure 5.9

Quantum numbers and energy levels in a hydrogen atom. The more negative the calculated value, the lower the energy.

and for hydrogen-like ions—that is, ions such as He^+ or Li^{2+} that have only one electron. Though the model fails with multi-electron atoms, it is still instructive.

Bohr derived an equation that gave the energy, E, of the electron when it was in any of the possible orbits in a hydrogen atom or hydrogen-like ion. The equation is

$$E = \frac{-kZ^2}{n^2}$$

where k is a constant (2.179×10^{-18} J or 13.595 eV), Z is the atomic number (or the number of protons in the nucleus, 1 for the hydrogen atom), and n is any integer from 1 to infinity (∞). Thus we can see that the energy of an electron in an orbit is characterized by an integer, n, which is called a **quantum number.** Do not let the negative sign in the previous equation confuse you. A negative energy simply indicates that the calculated energy is lower than that of the energy of an electron at an infinite distance from the nucleus (when $n = \infty$), where, by convention, the electron's energy is taken to be zero (Fig. 5.9). The units of energy in the Bohr equation depend on the value used for k. We will use values of k with units of joules or electron-volts (1 eV $= 1.602 \times 10^{-19}$ J).

Example 5.3   Calculating the Energy of an Electron in a Bohr Orbit

Early researchers were very excited when they were able to predict the energy of an electron at a particular distance from the nucleus in a hydrogen atom. If a spark promotes the electron in a hydrogen atom into an orbit with $n = 3$, what is the calculated energy, in joules, of the electron?

Solution: The energy of the electron is given by the equation

$$E = \frac{-kZ^2}{n^2}$$

The atomic number, Z, of hydrogen is 1, $k = 2.179 \times 10^{-18}$ J, and the electron is characterized by an n value of 3. Thus

$$E = \frac{(-2.179 \times 10^{-18} \text{ J}) \times (1)^2}{(3)^2}$$

$$= -2.421 \times 10^{-19} \text{ J}$$

Check your learning: The electron in this example is promoted even further to an orbit with $n = 6$. What is its new energy in electron-volts?

Answer: −0.3776 eV

Bohr also derived an equation that relates the distance of the electron from the nucleus, r, to the value of n

$$r = \frac{n^2 a_0}{Z}$$

where a_0 is a constant, the radius of the orbit in the hydrogen atom for which $n = 1$. This value of a_0 is 0.529 Å (1 Å $= 1 \times 10^{-10}$ m or 100 pm). According to this equa-

$n = 6$ $r = 36a_O$

$n = 5$ $r = 25a_O$

$n = 4$ $r = 16a_O$

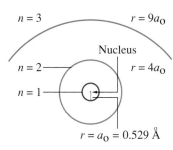

$n = 3$ $r = 9a_O$

Nucleus

$n = 2$ $r = 4a_O$

$n = 1$

$r = a_O = 0.529$ Å

Figure 5.10

A sketch of the hypothetical circular orbits of the Bohr model of the hydrogen atom, drawn to scale. If the nucleus were drawn to scale, it would be invisible.

tion, an electron in a hydrogen atom can get no closer to the nucleus than 0.529 Å, in an orbit with $n = 1$; thus the electron cannot fall into the nucleus. The equation also shows us that as an electron's energy increases (as n increases), the electron is found farther and farther from the nucleus (Fig. 5.10). This is consistent with the idea that as the electron moves away from the nucleus, the electrostatic attraction between it and the nucleus decreases, and it is held less tightly in the atom.

One of the fundamental laws of physics is that matter is most stable with the lowest possible energy. Thus the electron in a hydrogen atom usually moves in the $n = 1$ orbit, the orbit in which it has the lowest energy. When the electron is in this lowest-energy orbit, the atom is said to be in its **ground electronic state** (or simply **ground state**). If the atom receives energy from an outside source, it is possible for the electron to move to an orbit with a higher n value, in which case the atom is in an **excited electronic state** (or simply an **excited state**) with a higher energy.

We can relate the energy of electrons in atoms to what we learned previously about energy (Section 1.3). The law of conservation of energy says that we cannot create or destroy energy. Thus, if a certain amount of external energy is required to excite an electron from one energy level to another, then that same amount of energy will be liberated when the electron returns to its initial state (Fig. 5.11). In effect, an atom can "store" energy by using it to promote an electron to a state with a higher energy and can release energy when the electron returns to a lower state. The energy can be released as one larger quantum of energy as the electron returns to its ground state (say, from $n = 5$ to $n = 1$), or it can be released as two or more smaller quanta as the electron falls to an intermediate state and then to the ground state (say, from $n = 5$ to $n = 4$, emitting one quantum, and then to $n = 1$, emitting a second quantum).

The energy of the photon released as an electron falls from a higher-energy orbit (with energy E_2) to one of lower energy (with energy E_1) equals the absolute value of the difference in energy between those orbits.

$$E = \left| E_1 - E_2 \right|$$

If we call the quantum number of the outer, higher-energy orbit n_2 and that of the lower energy orbit n_1, then we can write

Figure 5.11

Relative energies of orbits in the Bohr model of the hydrogen atom and some of the electronic transitions that give rise to its atomic spectrum.

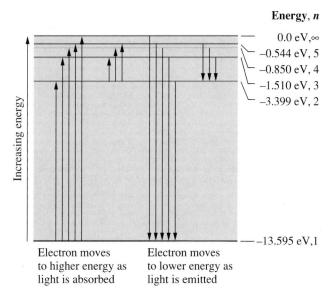

Energy, n

0.0 eV, ∞
−0.544 eV, 5
−0.850 eV, 4
−1.510 eV, 3
−3.399 eV, 2

Increasing energy

−13.595 eV, 1

Electron moves to higher energy as light is absorbed

Electron moves to lower energy as light is emitted

$$E = |E_1 - E_2|$$
$$= \left|\left(\frac{-kZ^2}{n_1^{\,2}}\right) - \left(\frac{-kZ^2}{n_2^{\,2}}\right)\right|$$

Because $Z = 1$ for a hydrogen atom, we have

$$E = 2.179 \times 10^{-18}\,\text{J}\left(\frac{1}{n_1^{\,2}} - \frac{1}{n_2^{\,2}}\right)$$

This theoretical expression is identical to Rydberg's empirical equation (Section 5.2). This provides strong evidence for the validity of the Bohr concept of discrete energy levels for electrons in the hydrogen atom.

Example 5.4   **Calculating the Energy and Wavelength of Electron Transitions**

What are the energy (in joules) and the wavelength (in meters) of the line in the spectrum of hydrogen that represents the movement of an electron from a Bohr orbit with n = 6 to the orbit with n = 4? In what part of the electromagnetic spectrum do we find this radiation?

Solution: In this case the electron starts out with $n = 6$, so $n_2 = 6$. It comes to rest in the $n = 4$ orbit, so $n_1 = 4$. The difference in energy between the two states is given by the expression

$$E = |E_1 - E_2| = 2.179 \times 10^{-18}\,\text{J}\left(\frac{1}{n_1^{\,2}} - \frac{1}{n_2^{\,2}}\right)$$
$$= (2.179 \times 10^{-18})\left(\frac{1}{4^2} - \frac{1}{6^2}\right)\text{J}$$
$$= (2.179 \times 10^{-18})\left(\frac{1}{16} - \frac{1}{36}\right)\text{J}$$
$$= 7.566 \times 10^{-20}\,\text{J}$$

The wavelength of a photon with this energy is found by the expression $E = \frac{hc}{\lambda}$. Rearrangement gives

$$\lambda = \frac{hc}{E}$$
$$= (6.626 \times 10^{-34}\,\text{J s}) \times \frac{(2.998 \times 10^8\,\text{m s}^{-1})}{7.566 \times 10^{-20}\,\text{J}}$$
$$= 2.626 \times 10^{-6}\,\text{m}$$

From Fig. 5.4, we can see that this wavelength is found in the infrared portion of the electromagnetic spectrum.

Check your learning: What are the energy in joules and the wavelength in meters of the photon produced when an electron falls from the $n = 5$ level to the $n = 3$ level in a He^+ ion?

Answer: 1.550×10^{-19} J; 1.282×10^{-6} m

Bohr's model of the hydrogen atom provides a great deal of insight into the behavior of matter at the microscopic level, but it does not provide a satisfactory explanation of many atomic properties. However, it does introduce several important features of all models used to describe the distribution of electrons in an atom.

1. The energies of electrons (energy levels) in an atom are quantized.
2. Quantum numbers are necessary to describe certain properties of electrons in an atom, such as their energy and location.
3. An electron's energy increases with increasing distance from the nucleus.
4. The discrete energies (lines) in the spectra of the elements result from quantized electronic energies.

Of these features, the most important is the postulate of quantized energy levels for an electron in an atom. As a consequence, the model laid the foundation for the quantum-mechanical model of the atom (Section 5.5). Bohr won a Nobel prize for his contributions, as did Roentgen, Bequerel, Marie and Pierre Curie, Rutherford, Thomson, Planck, Einstein, Millikan, de Broglie, Heisenberg, Schrödinger, Chadwick, and Pauli.

Quantum Mechanics

5.4   Behavior in the Microscopic World

The behavior of electrons and other microscopic components of atoms is different from the behavior we see in the macroscopic world. A microscopic object can behave both like a particle and like a wave. Because of this behavior, there are limits on how precisely we can measure or describe the properties of a microscopic particle.

Louis de Broglie asked the following question: If electromagnetic radiation can have particle-like character (as we noted in Section 5.2), can electrons and other microscopic particles exhibit wave-like character? In his 1925 Ph.D. dissertation, de Broglie predicted that a particle with mass m and velocity v should exhibit a wavelength λ given by

$$\lambda = \frac{h}{mv}$$

where h = Planck's constant. Shortly thereafter, two scientists at Bell Laboratories, C. J. Davisson and L. H. Germer, demonstrated experimentally that electrons could be diffracted like light. Diffraction is a property of waves; thus the Davisson–Germer experiments demonstrated that electrons do have wave-like properties. This phenomenon of matter exhibiting both particle-like and wave-like properties is known as the **dual nature of matter.**

Example 5.5   Calculating the Wavelength of a Particle

If an electron travels at a velocity of 1.000×10^7 m s^{-1} and has a mass of 9.109×10^{-28} g, what is its wavelength?

Solution: We can use de Broglie's equation to solve this problem, but we first must do a unit conversion of Planck's constant. Remember 1 J = 1 kg m^2 s^{-2} (Section 4.1); thus we can write h = 6.626 × 10^{-34} J s as 6.626 × 10^{-34} kg m^2 s^{-1}.

$$\lambda = \frac{h}{mv}$$

$$= \frac{(6.626 \times 10^{-34} \text{ kg m}^2 \text{ s}^{-1})}{(9.109 \times 10^{-31} \text{ kg})(1.000 \times 10^7 \text{ m s}^{-1})}$$

$$= 7.274 \times 10^{-11} \text{ m}$$

Check your learning: Calculate the wavelength of a softball with a mass of 100 g traveling at a velocity of 35 m s^{-1}.

Answer: 1.9 × 10^{-34} m. (We never think of a thrown softball as having a wavelength, because this wavelength is impossible for our senses or any instrument to detect.)

Werner Heisenberg considered the limits of how precisely we can measure properties of an electron or other microscopic particle. He determined that there is a fundamental limit to how closely we can measure both position and momentum. (The momentum of a particle is given by the product of its mass and velocity, *mv*.) The more accurately we measure the momentum of a particle, the less accurately we can determine its position. The converse is also true. Any experiment designed to measure the exact position of a particle will alter its momentum. If position and momentum are measured at the same time, the values are inexact for one or the other or both. This is summed up in what we now call the **Heisenberg uncertainty principle:** *It is impossible to determine simultaneously and precisely both the momentum and the position of a particle.* The product of the uncertainty in the position, Δx, and the uncertainty in the momentum, $\Delta(mv)$, must be greater than or equal to $h/2\pi$, the value of Planck's constant divided by 2π. The equation is

$$\Delta x \cdot \Delta(mv) \geq \frac{h}{2\pi}$$

This equation tells us that there is a limit to how well we can measure both the position of an object and its momentum. For example, if we improve our measurement of an electron's position so that the uncertainty in the position (Δx) has a value of, say, 1 pm (10^{-12} m, about 1% of the diameter of a hydrogen atom), then we find the uncertainty in its momentum (Δmv) as follows.

$$\Delta(mv) = \frac{h}{2\pi} \times \frac{1}{\Delta x}$$

$$= \frac{6.626 \times 10^{-34} \text{ kg m}^2 \text{ s}^{-1}}{2 \times 3.1416} \times \frac{1}{1 \times 10^{-12} \text{ m}}$$

$$= 1 \times 10^{-22} \text{ kg m s}^{-1}$$

Because the product $\Delta x \cdot \Delta(mv)$ must be greater than or equal to $h/2\pi$ and the value of $h/2\pi$ is quite small, the uncertainty in the position or in the momentum of a macroscopic object like a baseball is too small to observe. However, the mass of a microscopic object such as an electron is small enough for the uncertainty to be relatively large and significant.

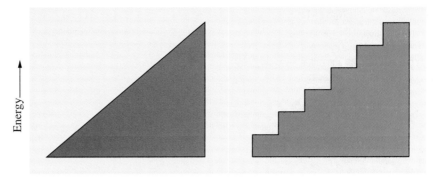

The dual nature of matter and Heisenberg's uncertainty principle are reflected in the model that must be used to describe the behavior of an atom's electrons and of other components of the microscopic world. The approach that is used is called **quantum mechanics** or **wave mechanics.**

A pictorial representation of the difference between the older classical description of matter and energy that preceded the work of Planck and Bohr, and the modern description that followed, may help us understand what we can determine about the electronic structure of the atom by using quantum mechanics. Figure 5.12 demonstrates this difference with a ramp and a set of stairs. Let us envision an electron, for a moment, as a particle. The classical description leads us to believe that the electron can be moved to *any place* farther up the ramp by adding energy. The electron can move to any position down the ramp by losing energy. Quantum mechanics, on the other hand, says that we can move the electron only to *certain steps,* or energy levels. Moreover, we must add the amount of energy exactly necessary to move to a higher-energy level or the electron must emit the exact amount of energy necessary to move to a lower-energy level. The electron cannot have an intermediate energy, nor can it absorb or emit intermediate energies.

In the classical view, we could determine exactly where the electron is on the ramp. In the quantum-mechanical view, we can determine which step the electron occupies, but we cannot determine exactly where the electron is on that step. Quantum mechanics lets us identify only the *probability* of finding the electron in a particular region on a step.

5.5   The Quantum-Mechanical Model of an Atom

Although we cannot pinpoint exactly an electron's position and momentum, we can calculate the probability of finding it at a given location within an atom. By using the mathematical tools of quantum mechanics, we can determine both the energy of an electron in an atom and the region of space it is most likely to occupy. The Austrian physicist Erwin Schrödinger showed that each of an atom's electrons can be visualized as occupying one of several volumes of space located around the nucleus. Each of these volumes is called an **orbital,** or an **atomic orbital.** Quantum-mechanical *orbitals* are very different from a Bohr *orbit.* A quantum-mechanical orbital is a three-dimensional region of space, somewhat like the space inside a very small balloon: An electron occupies a quantum-mechanical orbital much as a fly occupies a balloon— it could be moving anywhere inside. A Bohr orbit is a two-dimensional circular path: If an electron occupied a Bohr orbit, it would behave like a train on a circular track— it would simply go round and round.

Figure 5.13

(A) Cross section of the electron
density for an electron in a hydrogen
atom in its ground state. (B) The radial
probability distribution for the electron
in part (A).

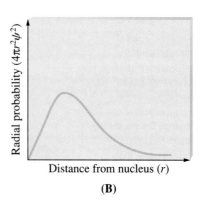

(A)   **(B)**

We can describe the distribution of an electron in an orbital in terms of the **electron density** in various regions of the orbital. The electron density is high in those regions of the orbital where the probability of finding an electron is relatively high, and it is low in those regions where that probability is low. Although the electron might be located anywhere within an orbital at any instant in time, it spends more of its time in certain high-probability regions. For example, in an isolated hydrogen atom in its ground state (Section 5.3), the electron is almost always found within a sphere that has a radius of about 1 angstrom and is centered about the nucleus (Fig. 5.13). Within this spherical orbital, the electron has the greatest probability of being approximately 0.529 angstrom from the nucleus.

Each electron in an atom can be described by a mathematical expression called a **wave function,** which is given the symbol ψ. The wave function describes properties of the orbital and the electron that occupies the orbital. From ψ we can determine the type of orbital that the electron occupies, the energy of the electron in that orbital (this energy is sometimes called the energy of the orbital), the shape of the orbital, and the probability of finding the electron in any particular region within the orbital. For instance, we can determine the probability of finding the electron at a given point a distance r from the nucleus if we calculate the value of ψ^2 at that point. The radial probability density $4\pi r^2 \psi^2$ indicates the probability of finding the electron within the volume of a very thin spherical shell at a distance r from the nucleus. This spherical shell is analogous to the skin of an onion: The skin is at a distance r from the center of the onion, and the thickness of the skin is very small compared to r.

Figure 5.13 illustrates radial probability density in the occupied orbital of a hydrogen atom in the ground state. The probability of finding the electron in a thin shell very close to the nucleus is practically zero; the probability increases rapidly just beyond the nucleus and becomes highest in a thin shell at a distance of 0.529 angstrom. The probability then decreases rapidly and becomes exceedingly small at any distance greater than about 1 angstrom. Thus most of the time, but not always, the electron is located within a sphere with a radius of 1 angstrom.

5.6   Results of the Quantum-Mechanical Model of the Atom

The quantum-mechanical model is a powerful tool for describing the behavior of electrons both in atoms and in molecules. Unlike the Bohr model, the quantum-mechanical model can be applied to all atoms. However, it does have some similarities to the

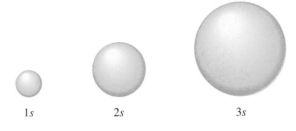

$1s$ $2s$ $3s$

Bohr model. Application of quantum mechanics to an atom indicates that the energy levels in the atom must be quantized and that certain properties of electrons in atoms are associated with quantum numbers. Quantum mechanics also explains the source of discrete energies (lines) in the spectra of the elements (see Fig. 5.8).

The general nature of the quantum-mechanical model of the atom is conveyed by the following six points:

1. The location of an electron cannot be determined exactly; all we can determine is the probability of finding an electron in a given region of space.

2. In an atom, electrons occupy orbitals that are characterized by three quantum numbers called the principal, angular momentum, and magnetic quantum numbers. Each of these quantum numbers is limited to integer values. This means that the mathematical expression, ψ, that describes an orbital contains integer values for these quantum numbers, much as Bohr's equations for the energy and radius of a hydrogen orbit contain a quantum number (Section 5.3).

   - The **principal quantum number,** n, can take on any integral value: $n = 1$, 2, 3, 4, 5, . . . to infinity. A larger n value indicates that an orbit is larger and generally has a higher energy than orbital with a lower n value. We can consider the principal quantum number to be a *distance-and-energy* quantum number. Its value gives us an indication of how far an electron is from the nucleus. As we have seen in the Bohr model (Section 5.3), an electron is less strongly bound in the atom, and the energy of an electron increases, as its distance from the nucleus increases. Figure 5.14 shows three orbitals of a hydrogen atom labeled 1s, 2s, and 3s. (Don't worry yet about the labels, we will discuss them in the next section.) These orbitals have n values of 1, 2, and 3, respectively. Note how their size increases as the value of n increases.

     In an atom, a set of orbitals with the same value of n is called a **shell.** In a hydrogen atom or a hydrogen-like ion, all orbitals in the same shell have the same energy. However, in an atom with two or more electrons, orbitals in the same shell can have different energies. These energies depend on the value of n and on the value of a second quantum number, l.

   - The **angular momentum quantum number** or **azimuthal quantum number,** l, has values that depend on the value of n, the principal quantum number. The possible values of l range from 0 to $n - 1$ in integral steps. We can consider this quantum number to be an *orbital-shape* quantum number because atomic orbitals with different l values have different shapes. We call the set of orbitals with the same l value in a shell a **subshell.**

     An electron with $l = 0$ occupies a spherical orbital. Such an orbital is called an *s* **orbital** (Fig. 5.14), and the electron is called an *s* **electron.** An electron with $l = 1$ occupies a dumbbell-shaped orbital (Fig. 5.15) called a *p* **orbital** and is called a *p* **electron.** A *d* **electron,** with $l = 2$, occupies an orbital that is usually drawn with four lobes, a *d* **orbital** (Fig. 5.16). An *f*

Figure 5.15

Boundary surface representations of the three 2p orbitals.

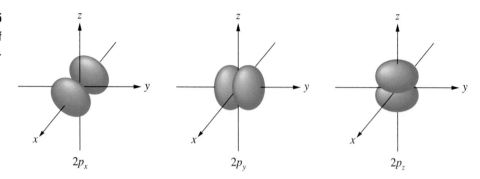

$2p_x$ $2p_y$ $2p_z$

electron, with $l = 3$, occupies a rather complex-looking volume of space with four or eight lobes, an **f orbital** (Fig. 5.17). Orbitals with l values greater than 3 are possible, but these are not occupied unless an electron has been promoted. For l values greater than 3, the letters representing the orbital type follow alphabetically.

l	0	1	2	3	4	5	6
Letter designation	s	p	d	f	g	h	i

In an atom with two or more electrons, the energies of electrons in orbitals *within a given shell* (orbitals with the same value of n) increase in the following order:

$$s \text{ electrons} < p \text{ electrons} < d \text{ electrons} < f \text{ electrons}$$

Increasing energy

We often use a shorthand notation to identify the principal quantum number and the azimuthal quantum number for an orbital or a subshell. We combine the value of n with the letter designation for l. For example, an orbital with $n = 2$ and $l = 1$ is labeled a $2p$ orbital, and an orbital with $n = 4$ and $l = 2$ is labeled a $4d$ orbital.

• The third quantum number is the **magnetic quantum number, m_l.** This quantum number, which may have integral values from $-l$ to $+l$ including zero, can be thought of as an *orientational* quantum number. That is, the number

Figure 5.16

Boundary surface representations of the five 3d orbitals.

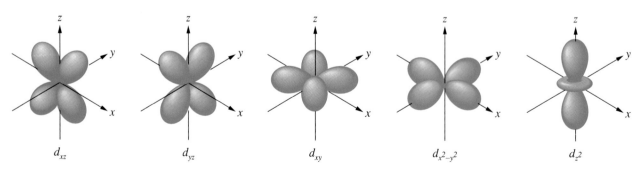

d_{xz} d_{yz} d_{xy} $d_{x^2-y^2}$ d_{z^2}

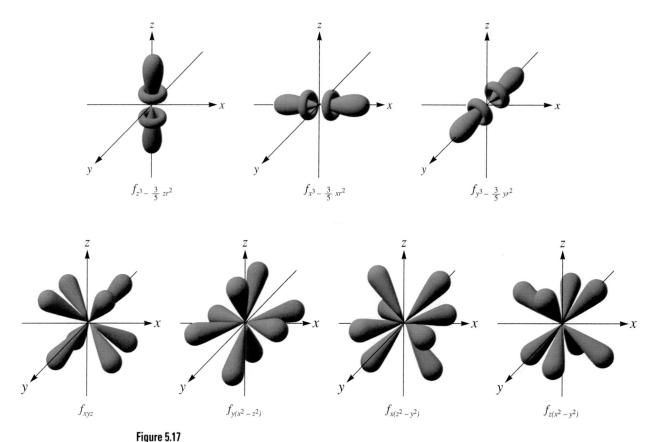

$f_{z^3 - \frac{3}{5} zr^2}$ $f_{x^3 - \frac{3}{5} xr^2}$ $f_{y^3 - \frac{3}{5} yr^2}$

f_{xyz} $f_{y(x^2 - z^2)}$ $f_{x(z^2 - y^2)}$ $f_{z(x^2 - y^2)}$

Figure 5.17

Boundary surface representations of the seven 4f orbitals.

Figure 5.18

A representation of electrons spinning in the two allowed directions.

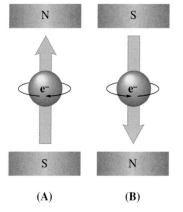

(A)   (B)

of possible m_l values tells us the number of possible orientations in space for each type of orbital (which is specified by the l quantum number). For $l = 0$, m_l can have only the value of zero. This tells us that there can be only one s-type orbital, which is spherical. When $l = 1$, the possible values of m_l are -1, 0, and $+1$. These three values tell us that a p-type orbital may have three possible orientations in space; these are illustrated in Fig. 5.15. Similarly, when $l = 2$ (a d-type orbital), there are five orientations in space represented by the five m_l values -2, -1, 0, $+1$, $+2$, as shown in Fig. 5.16. For $l = 3$ there are seven m_l values (-3, -2, -1, 0, $+1$, $+2$, $+3$) and thus seven orientations in space for f-type orbitals (Fig. 5.17). Table 5.2 summarizes the possible values of l and m_l allowed for each value of the principal quantum number, n, for the values $n = 1$ through 4.

Although orbitals in the same subshell (same l value) in an isolated atom have different orientations about the nucleus, they have the same energy. Orbitals with the same energy are called **degenerate orbitals.** The three p orbitals in any given subshell of an atom that is not bonded to another atom are degenerate. Likewise, the five d orbitals in a given subshell are degenerate, as are the seven f orbitals.

3. Electrons have a fourth quantum number called the **spin quantum number, s.** An electron behaves as though it spins about its own axis (Fig. 5.18). The spin quantum number is used to distinguish the two possible directions of spin, counterclockwise or clockwise, and is designated arbitrarily by either $s = +\frac{1}{2}$ or $s = -\frac{1}{2}$. As we depict electrons pictorially in this book, we will use an arrow pointing up ($\uparrow$) to symbolize an electron with $s = +\frac{1}{2}$ and will use an arrow

Table 5.2

Quantum Numbers for the First Four Levels of Orbitals in the Hydrogen Atom.

n	l	Orbital Designation	m_l	Number of Orbitals
1	0	$1s$	0	1
2	0	$2s$	0	1
	1	$2p$	$-1, 0, +1$	3
3	0	$3s$	0	1
	1	$3p$	$-1, 0, +1$	3
	2	$3d$	$-2, -1, 0, +1, +2$	5
4	0	$4s$	0	1
	1	$4p$	$-1, 0, +1$	3
	2	$4d$	$-2, -1, 0, +1, +2$	5
	3	$4f$	$-3, -2, -1, 0, +1, +2, +3$	7

pointing down ($\downarrow$) to symbolize an electron with $s = -\frac{1}{2}$. Electrons with the same spin quantum number are said to have **parallel spins.**

4. Atoms have an orbital for each allowed combination of n, l, and m_l. Any one of these orbitals can contain at most only two electrons, which must differ in their spin. Thus each orbital can contain a maximum of two electrons that have identical n, l, and m_l values but differ in their s values (a consequence of Pauli's exclusion principle, Table 5.1). Two such electrons are called **paired electrons** and are depicted in their orbital as $\uparrow\downarrow$.

5. The energy of an electron in an atom is limited to discrete values. We could calculate the energy of an electron from ψ, its wave function, but the mathematics would be much more complicated than in the Bohr model.

6. The maximum number of electrons that may be found in a shell with a principal quantum number of n is $2n^2$. The number of orbitals in a shell is n^2, and each orbital can hold a maximum of two electrons.

Example 5.6   **Working with Quantum Numbers**

Indicate the number of subshells, the number of orbitals in each subshell, and the values of l and m for the orbitals in the $n = 4$ shell of an atom.

Solution: For $n = 4$, l can have values of 0, 1, 2, and 3. Thus s, p, d, and f orbitals are found in the $n = 4$ shell of an atom. For $l = 0$ (the s subshell), m_l can be only 0, so there is only one $4s$ orbital. For $l = 1$ (p-type orbitals), m can have values of -1, 0, and $+1$, so we find three $4p$ orbitals. For $l = 2$ (d-type orbitals), m_l can have values of -2, -1, 0, $+1$, and $+2$, so we have five $4d$ orbitals. When $l = 3$ (f-type orbitals), m_l can have values of -3, -2, -1, 0, $+1$, $+2$, and $+3$, so we have seven $4f$ orbitals. Thus there are a total of 16 orbitals in the $n = 4$ shell of an atom.

Check your learning: Identify the subshell in which electrons with the following quantum numbers are found: **(a)** $n = 3$, $l = 1$; **(b)** $n = 5$, $l = 3$; **(c)** $n = 2$, $l = 0$.

Answer: **(a)** $3p$; **(b)** $5f$; **(c)** $2s$.

Figure 5.19

Generalized energy-level diagram for
atomic orbitals in an atom with two or
more electrons (not to scale).

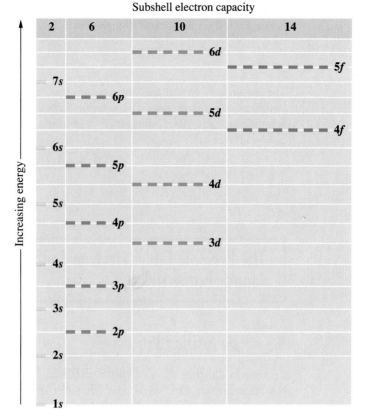

Figure 5.19

Generalized energy-level diagram for
atomic orbitals in an atom with two or
more electrons (not to scale).

5.7 Orbital Energies and Atomic Structure

The energies of atomic orbitals increase as the principal quantum number, n, increases.
In any atom with two or more electrons, the energies of the orbitals also increase
within a shell, in the order $s < p < d < f$. That is, the energy of the orbital type
increases as the l quantum number increases. Although the relative energies of certain orbitals vary from atom to atom, the increasing order of energy is roughly that
shown in Fig. 5.19. The energy of an electron in an orbital is indicated by the vertical coordinate in the figure. Thus the $1s$ orbital at the bottom of the diagram is the
orbital with electrons of lowest energy. Orbitals that have about the same energy are
indicated by the brackets at the right of the figure.

The arrangement of electrons in the orbitals of an atom is called the **electron configuration** of the atom. We describe an electron configuration with a symbol that contains three pieces of information:

1. The number of the principal quantum shell, n
2. The letter that designates the orbital type (the subshell, l)
3. A superscript that designates the number of electrons in that particular subshell

For example, the notation $2p^4$ (which is read "two-p-four") indicates four electrons
in a p subshell ($l = 1$) with a principal quantum number (n) of 2. Because there are

three p orbitals ($m = -1$, 0, and $+1$), the pictorial representation of these electrons in what is called an **orbital diagram** is

$$2p^4 \qquad \boxed{\uparrow\downarrow}\,\boxed{\uparrow}\,\boxed{\uparrow}$$
$$2p$$

The reason why two of the electrons are paired and the other two are not will be discussed in the next section. The notation $3d^8$ (which is read "three-d-eight") indicates eight electrons in the d subshell ($l = 2$) of the principal shell for which $n = 3$. This can be shown as

$$3d^8 \qquad \boxed{\uparrow\downarrow}\,\boxed{\uparrow\downarrow}\,\boxed{\uparrow\downarrow}\,\boxed{\uparrow}\,\boxed{\uparrow}$$
$$3d$$

The Periodic Table: Electron Configuration and Atomic Properties

5.8   The Aufbau Process

In order to illustrate the systematic variations in the electronic structures of the various elements, we can "build" the structures in atomic order. Beginning with hydrogen, and continuing across the periods of the periodic table, we mentally add one proton at a time to the nucleus and one electron to the proper subshell, until they have described the electron configurations of all the elements. This process is called the **aufbau process,** from the German word *aufbau* ("to build up"). Each added electron occupies the lowest-energy subshell available, subject to the limitations imposed by the allowed quantum numbers (Section 5.6). Electrons enter higher-energy subshells only after lower-energy subshells have been filled to capacity. Figure 5.20 illustrates a way to remember the order of filling of atomic orbitals.

Figure 5.20

Order of occupancy of atomic orbitals. The orbitals fill in order, starting with the 1s orbital, in the direction of each arrow, going down the arrows from top to bottom.

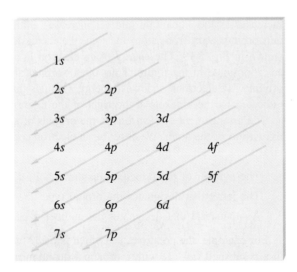

Look at the first and second periods of the periodic table as we carry out this mental building. A single hydrogen atom (atomic number 1) consists of one proton and one electron. Referring to Fig. 5.19 or 5.20, we would expect to find the electron in the $1s$ orbital. The electron configuration and the orbital diagram are given as

H $1s^1$ ↑
                          $1s$

Following hydrogen is the noble gas helium, which has an atomic number of 2. The helium atom contains two protons and two electrons. The first electron has the same four quantum numbers as the hydrogen atom electron ($n = 1$, $l = 0$, $m = 0$, and $s = +\frac{1}{2}$). The second electron also goes into the $1s$ orbital and fills that orbital. The second electron has the same n, l, and m quantum numbers but must have the opposite spin quantum number, $-\frac{1}{2}$. This is in accordance with the **Pauli exclusion principle** (Table 5.1): *No two electrons in the same atom can have the same set of four quantum numbers.* The electron configuration and the orbital diagram of helium are

He $1s^2$ ↑↓
                          $1s$

The $n = 1$ shell is completely filled in a helium atom.

The next atom is the alkali metal lithium with an atomic number of 3. The first two electrons in lithium fill the $1s$ orbital and have the same sets of four quantum numbers as the first two electrons in helium. The remaining electron must occupy the orbital of next lowest energy, the $2s$ orbital (Fig. 5.19 or 5.20). Thus the electron configuration and the orbital diagram of lithium are

Li $1s^2 2s^1$ ↑↓ ↑
                              $1s$ $2s$

An atom of the alkaline earth metal beryllium, with an atomic number of 4, contains four protons in the nucleus and four electrons surrounding the nucleus. The fourth electron fills the $2s$ orbital.

Be $1s^2 2s^2$ ↑↓ ↑↓
                              $1s$ $2s$

An atom of boron (atomic number 5) contains five electrons. The $n = 1$ shell is filled with two electrons, and three electrons will occupy the $n = 2$ shell. Because any s subshell can contain only two electrons, the fifth electron must occupy the next energy level, which will be a $2p$ orbital. There are three $2p$ orbitals that are degenerate, and the electron can occupy any one of these p orbitals.

B $1s^2 2s^2 2p^1$ ↑↓ ↑↓ ↑
                                  $1s$ $2s$ $2p$

Carbon (atomic number 6) has six electrons. Four of them fill the $1s$ and $2s$ orbitals. The remaining two electrons occupy the $2p$ subshell. We now have a choice of filling one of the $2p$ orbitals and pairing the electrons or leaving the electrons unpaired in two different, but degenerate, p orbitals. The orbitals are filled as described by **Hund's rule** (Table 5.1): *The lowest-energy configuration for an atom with electrons in a set of degenerate orbitals is that having the maximum number of unpaired electrons with the same spin.* The spin quantum numbers of the unpaired electrons are arbitrary; here, we will use $+\frac{1}{2}$. Thus the two electrons in the carbon $2p$ orbitals have identical n, l, and s quantum numbers and differ in their m_l quantum number (in accordance with the Pauli exclusion principle). The electron configuration and orbital diagram for carbon are

C $1s^2 2s^2 2p^2$ 1s 2s 2p

Nitrogen (atomic number 7) fills the $1s$ and $2s$ subshells and has one electron in each of the three $2p$ orbitals, in accordance with Hund's rule. These three electrons have parallel spins. Oxygen (atomic number 8) has a pair of electrons in any one of the $2p$ orbitals (the electrons have opposite spins) and a single electron in each of the other two. Fluorine (atomic number 9) has only one $2p$ orbital containing an unpaired electron. All of the electrons in the noble gas neon (atomic number 10) are paired, and all of the orbitals in the $n = 1$ and $n = 2$ shells are filled. The electron configurations and orbital diagrams of these four elements are shown below.

N $1s^2 2s^2 2p^3$ 1s 2s 2p

O $1s^2 2s^2 2p^4$ 1s 2s 2p

F $1s^2 2s^2 2p^5$ 1s 2s 2p

Ne $1s^2 2s^2 2p^6$ 1s 2s 2p

The alkali metal sodium (atomic number 11) has one more electron than the neon atom. This electron must go into the lowest-energy subshell available, the $3s$ orbital, giving a $1s^2 2s^2 2p^6 3s^1$ configuration. We can abbreviate this as [Ne]$3s^1$. The symbol [Ne] represents the configuration of the two filled shells in neon ($1s^2 2s^2 2p^6$). Similarly, the configuration of lithium can be represented as [He]$2s^1$, where [He] represents the configuration of the helium atom, which is identical to that of the filled inner shell of lithium. Writing the configurations in this way emphasizes the similarity of the configurations of lithium and sodium. Both atoms, which are in the alkali metal family, have only one electron in an s subshell outside a filled set of inner shells.

Li   [He]$2s^1$
Na   [Ne]$3s^1$

The alkaline earth metal magnesium (atomic number 12), with its 12 electrons in a [Ne]$3s^2$ configuration, is analogous to its family member beryllium, [He]$2s^2$. Both atoms have a filled s subshell outside their filled inner shells. Aluminum (atomic number 13), with 13 electrons and the electron configuration [Ne]$3s^2 3p^1$, is analogous to its family member boron, [He]$2s^2 2p^1$. The electron configurations of silicon (14 electrons), phosphorus (15 electrons), sulfur (16 electrons), chlorine (17 electrons), and argon (18 electrons) are analogous in the electron configurations of their outer shells to their corresponding family members carbon, nitrogen, oxygen, fluorine, and neon, except that the principal quantum number of the outer shell of the heavier elements has increased by 1 to $n = 3$. Table 5.3 lists the lowest-energy, or ground-state, elec-

Atomic Number	Symbol	Electron Configuration	Atomic Number	Symbol	Electron Configuration	Atomic Number	Symbol	Electron Configuration
1	H	$1s^1$	38	Sr	$[Kr]5s^2$	78	Pt	$[Xe]6s^14f^{14}5d^9$
2	He	$1s^2$	39	Y	$[Kr]5s^24d^1$	79	Au	$[Xe]6s^14f^{14}5d^{10}$
			40	Zr	$[Kr]5s^24d^2$	80	Hg	$[Xe]6s^24f^{14}5d^{10}$
3	Li	$1s^22s^1 = [He]2s^1$	41	Nb	$[Kr]5s^14d^4$	81	Tl	$[Xe]6s^24f^{14}5d^{10}6p^1$
4	Be	$[He]2s^2$	42	Mo	$[Kr]5s^14d^5$	82	Pb	$[Xe]6s^24f^{14}5d^{10}6p^2$
5	B	$[He]2s^22p^1$	43	Tc	$[Kr]5s^14d^6$	83	Bi	$[Xe]6s^24f^{14}5d^{10}6p^3$
6	C	$[He]2s^22p^2$	44	Ru	$[Kr]5s^14d^7$	84	Po	$[Xe]6s^24f^{14}5d^{10}6p^4$
7	N	$[He]2s^22p^3$	45	Rh	$[Kr]5s^14d^8$	85	At	$[Xe]6s^24f^{14}5d^{10}6p^5$
8	O	$[He]2s^22p^4$	46	Pd	$[Kr]4d^{10}$	86	Rn	$[Xe]6s^24f^{14}5d^{10}6p^6$
9	F	$[He]2s^22p^5$	47	Ag	$[Kr]5s^14d^{10}$			
10	Ne	$[He]2s^22p^6$	48	Cd	$[Kr]5s^24d^{10}$	87	Fr	$[Rn]7s^1$
			49	In	$[Kr]5s^24d^{10}5p^1$	88	Ra	$[Rn]7s^2$
11	Na	$[Ne]3s^1$	50	Sn	$[Kr]5s^24d^{10}5p^2$	89	Ac	$[Rn]7s^26d^1$
12	Mg	$[Ne]3s^2$	51	Sb	$[Kr]5s^24d^{10}5p^3$	90	Th	$[Rn]7s^26d^2$
13	Al	$[Ne]3s^23p^1$	52	Te	$[Kr]5s^24d^{10}5p^4$	91	Pa	$[Rn]7s^25f^26d^1$
14	Si	$[Ne]3s^23p^2$	53	I	$[Kr]5s^24d^{10}5p^5$	92	U	$[Rn]7s^25f^36d^1$
15	P	$[Ne]3s^23p^3$	54	Xe	$[Kr]5s^24d^{10}5p^6$	93	Np	$[Rn]7s^25f^46d^1$
16	S	$[Ne]3s^23p^4$				94	Pu	$[Rn]7s^25f^6$
17	Cl	$[Ne]3s^23p^5$	55	Cs	$[Xe]6s^1$	95	Am	$[Rn]7s^25f^7$
18	Ar	$[Ne]3s^23p^6$	56	Ba	$[Xe]6s^2$	96	Cm	$[Rn]7s^25f^76d^1$
			57	La	$[Xe]6s^25d^1$	97	Bk	$[Rn]7s^25f^86d^1$
19	K	$[Ar]4s^1$	58	Ce	$[Xe]6s^24f^2$	98	Cf	$[Rn]7s^25f^{10}$
20	Ca	$[Ar]4s^2$	59	Pr	$[Xe]6s^24f^3$	99	Es	$[Rn]7s^25f^{11}$
21	Sc	$[Ar]4s^23d^1$	60	Nd	$[Xe]6s^24f^4$	100	Fm	$[Rn]7s^25f^{12}$
22	Ti	$[Ar]4s^23d^2$	61	Pm	$[Xe]6s^24f^5$	101	Md	$[Rn]7s^25f^{13}$
23	V	$[Ar]4s^23d^3$	62	Sm	$[Xe]6s^24f^6$	102	No	$[Rn]7s^25f^{14}$
24	Cr	$[Ar]4s^13d^5$	63	Eu	$[Xe]6s^24f^7$	103	Lr	$[Rn]7s^25f^{14}6d^1$
25	Mn	$[Ar]4s^23d^5$	64	Gd	$[Xe]6s^24f^75d^1$	104	Rf	$[Rn]7s^25f^{14}6d^2$
26	Fe	$[Ar]4s^23d^6$	65	Tb	$[Xe]6s^24f^9$	105	Ha	$[Rn]7s^25f^{14}6d^3$
27	Co	$[Ar]4s^23d^7$	66	Dy	$[Xe]6s^24f^{10}$	106	Sg	$[Rn]7s^25f^{14}6d^4$
28	Ni	$[Ar]4s^23d^8$	67	Ho	$[Xe]6s^24f^{11}$	107	Ns	$[Rn]7s^25f^{14}6d^5$
29	Cu	$[Ar]4s^13d^{10}$	68	Er	$[Xe]6s^24f^{12}$	108	Hs	$[Rn]7s^25f^{14}6d^6$
30	Zn	$[Ar]4s^23d^{10}$	69	Tm	$[Xe]6s^24f^{13}$	109	Mt	$[Rn]7s^25f^{14}6d^7$
31	Ga	$[Ar]4s^23d^{10}4p^1$	70	Yb	$[Xe]6s^24f^{14}$	110	Uun	$[Rn]7s^25f^{14}6d^8$
32	Ge	$[Ar]4s^23d^{10}4p^2$	71	Lu	$[Xe]6s^24f^{14}5d^1$	111	Uuu	$[Rn]7s^25f^{14}6d^9$
33	As	$[Ar]4s^23d^{10}4p^3$	72	Hf	$[Xe]6s^24f^{14}5d^2$	112	Uub	$[Rn]7s^25f^{14}6d^{10}$
34	Se	$[Ar]4s^23d^{10}4p^4$	73	Ta	$[Xe]6s^24f^{14}5d^3$			
35	Br	$[Ar]4s^23d^{10}4p^5$	74	W	$[Xe]6s^24f^{14}5d^4$			
36	Kr	$[Ar]4s^23d^{10}4p^6$	75	Re	$[Xe]6s^24f^{14}5d^5$			
37	Rb	$[Kr]5s^1$	76	Os	$[Xe]6s^24f^{14}5d^6$			
			77	Ir	$[Xe]6s^24f^{14}5d^7$			

Table 5.3
Electron Configurations of the Elements

tron configuration for these atoms, as well as that for atoms of each of the other elements.

When we come to the next element in the periodic table, the alkali metal potassium (atomic number 19), we might expect that we would begin to add electrons to the $3d$ subshell. However, all available chemical and physical evidence indicates that potassium is like lithium and sodium and that the next electron is not added to the

$3d$ level but is added to the $4s$ level instead. Thus potassium has an electron configuration of $[Ar]4s^1$. Hence, potassium corresponds to Li and Na in its outer-shell configuration. The next electron is added to complete the $4s$ subshell, and calcium has an electron configuration of $[Ar]4s^2$. This gives calcium an outer-shell electron configuration corresponding to that of beryllium and magnesium.

Beginning with the transition metal scandium (atomic number 21), additional electrons are added successively to the $3d$ subshell. This subshell is filled to its capacity with 10 electrons (remember that there are five d orbitals with a total capacity of 10 electrons). The $4p$ subshell fills next. Note that for three series of elements, scandium (Sc) through copper (Cu), yttrium (Y) through silver (Ag), and lutetium (Lu) through gold (Au), a total of 10 d electrons are successively added to the $(n-1)$ shell next to the n shell to bring that $(n-1)$ shell from 8 to 18 electrons. For two series, lanthanum (La) through lutetium (Lu) and actinium (Ac) through lawrencium (Lr), 14 f electrons (in seven orbitals with a total capacity of 14 electrons) are successively added to the third shell from the outside, the $(n-2)$ shell, to bring that shell from 18 electrons to a total of 32 electrons.

Example 5.7   Quantum Numbers and Electron Configurations

Determine the electron configuration and orbital diagram for a phosphorus atom. What are four quantum numbers for the last electron added?

Solution: The atomic number of phosphorus is 15. Thus a phosphorus atom contains 15 electrons. The order of filling of the energy levels is $1s$, $2s$, $2p$, $3s$, $3p$, $4s$, ... The 15 electrons of the phosphorus atom will fill up to the $3p$ orbital, which will contain three electrons:

P $1s^2 2s^2 2p^6 3s^2 3p^3$ $\boxed{\uparrow\downarrow}$ $\boxed{\uparrow\downarrow}$ $\boxed{\uparrow\downarrow}\boxed{\uparrow\downarrow}\boxed{\uparrow\downarrow}$ $\boxed{\uparrow\downarrow}$ $\boxed{\uparrow}\boxed{\uparrow}\boxed{\uparrow}$

$1s$ $2s$ $2p$ $3s$ $3p$

The last electron added is a $3p$ electron. Therefore $n = 3$, and for a p-type orbital, $l = 1$. The m_l value could be -1, 0, or $+1$. The three p orbitals are degenerate, so any of these values is correct. However, your instructor may ask that you fill degenerate orbitals in a certain order. If the orbitals were filled, for example, from most negative m_l value to most positive m value, the last electron in phosphorus would be placed in the $m_l = +1$ orbital. For unpaired electrons, we take the value of the spin quantum number as $+\frac{1}{2}$; thus $s = +\frac{1}{2}$.

Check your learning: Identify the atoms from the electron configurations given:
(a) $[Ar]4s^2 3d^5$, (b) $[Kr]5s^2 4d^{10} 5p^6$.

Answer: (a) Mn, (b) Xe.

The periodic table can be a powerful tool in helping us predict the electron configuration of an element. However, we do find exceptions to the order of filling of orbitals that is shown in Figs. 5.19 and 5.20. For instance, the electron configurations of the transition metals chromium (Cr, atomic number 24) and copper (Cu, atomic number 29) and the inner transition metal lanthanum (La, atomic number 57), among others, are not those we would expect. In general, such exceptions involve subshells with very similar energies, and small effects can lead to changes in the order of filling.

In the case of Cr and Cu, we find that half-filled and completely filled subshells apparently represent conditions of preferred stability. This stability is such that an electron shifts from the $4s$ into the $3d$ orbital in order to gain the extra stability of a half-filled $3d$ subshell (in Cr) or a filled $3d$ subshell (in Cu). In other atoms, certain combinations of repulsions between electrons lead to minor exceptions in the expected order of filling, which are explained by the magnitude of the repulsions being greater than the small differences in energy between subshells.

5.9 Electron Configurations and the Periodic Table

In Section 2.5 we saw that if we arrange the elements in order of increasing atomic number and group those elements with similar chemical properties, then elements with the same properties reoccur periodically. In other words, elements can be arranged into a periodic table. When their electron configurations are added to the table (Fig. 5.21), we also see a periodic recurrence of similar electron configurations in the outer shells of these elements. These electrons in the outermost shells are called **valence electrons,** and the outermost shell is called the **valence shell.** Because they are on the surface of an atom, valence electrons play the most important role in chemical reactions. They are also the determining factor in most physical properties of the elements.

Elements in any one group (or family) have the same number of valence electrons; the alkali metals lithium and sodium have only one, the alkaline earth metals beryllium and magnesium have two, and the halogens fluorine and chlorine have seven. The similarity in chemical properties among elements of the same group occurs because they have the same number of valence electrons. **It is the loss, gain, or sharing of valence electrons that determines how elements react.**

It is important to remember that the periodic table was developed on the basis of the chemical behavior of the elements, well before any idea of their atomic structure was entertained. Now we can understand why the periodic table has the arrangement it has: The arrangement puts elements whose atoms have the same number of valence electrons in the same group. This arrangement is emphasized in Fig. 5.22, which

Figure 5.21

The electron configurations of the first 18 elements, arranged in the format of the periodic table.

1A								8A
H $1s^1$								**He** $1s^2$
	2A		3A	4A	5A	6A	7A	
Li [He}$2s^1$	**Be** [He}$2s^2$		**B** [He}$2s^22p^1$	**C** [He}$2s^22p^2$	**N** [He}$2s^22p^3$	**O** [He}$2s^22p^4$	**F** [He}$2s^22p^5$	**Ne** [He]$2s^22p^6$
Na [Ne}$3s^1$	**Mg** [Ne}$3s^2$		**Al** [Ne}$3s^23p^1$	**Si** [Ne}$3s^23p^2$	**P** [Ne}$3s^23p^3$	**S** [Ne}$3s^23p^4$	**Cl** [Ne}$3s^23p^5$	**Ar** [Ne}$3s^23p^6$

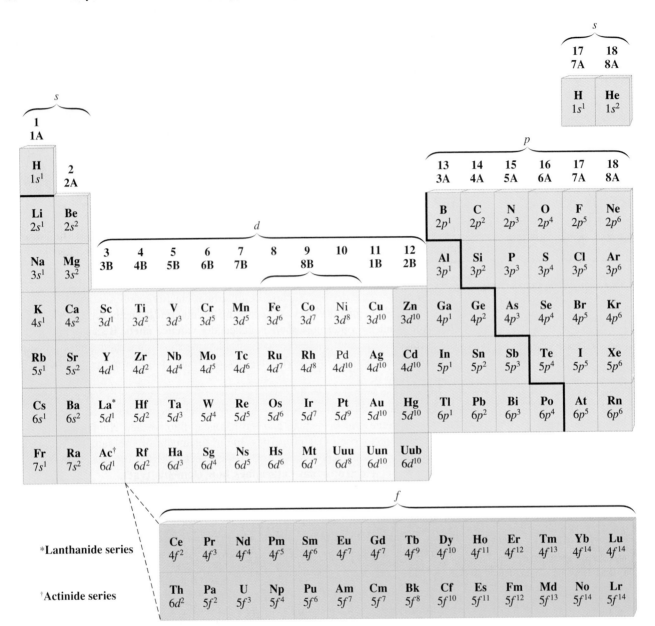

Figure 5.22

The order of occupancy of atomic orbitals in the periodic table.

shows in periodic table form the electron configuration of the last subshell to be filled by the aufbau process (also see Table 5.3). The colored sections of Fig. 5.22 show the three categories of elements classified in terms of the orbitals being filled by the valence electrons. We introduced the names of these elements in Section 2.5.

Representative Elements.   The **representative elements** are metallic and non-metallic elements in which the last electron added enters an *s* or a *p* orbital in the outermost shell. These elements are sometimes called **main-group elements** and are those shown in yellow in Fig. 5.22.

Transition Metals or Transition Elements.

The **transition metals** are metallic elements in which the last electron added enters a d orbital. The valence electrons in these elements are usually the ns and $(n-1)d$ electrons. These elements are shown in pink in Fig. 5.22. There are four transition metal series.

1. First transition series: scandium (Sc) through copper (Cu); $3d$ subshell filling.
2. Second transition series: yttrium (Y) through a silver (Ag); $4d$ subshell filling.
3. Third transition series: lanthanum (La) through gold (Au); $5d$ subshell filling.
4. Fourth transition series (an incomplete series): actinium (Ac) through element 111 (Uuu); $6d$ subshell filling.

Inner Transition Elements.

The **inner transition elements** are metallic elements that are filling their f orbitals; they are shown in dark pink in Fig. 5.22. The valence shells of the inner transition elements consist of the $(n-2)f$, $(n-1)d$, and ns subshells. There are two inner transition series.

1. The lanthanide series: cerium (Ce) through lutetium (Lu); $4f$ subshell filling.
2. The actinide series: thorium (Th) through lawrencium (Lr); $5f$ subshell filling.

Because of their similarities to the other members of the series, lanthanum and actinum are sometimes included as the first elements of the first and second inner transition series, respectively.

5.10   Variation of Atomic Properties Within Periods and Groups

The elements in many groups (vertical columns) of the periodic table exhibit similar chemical behavior. This similarity occurs because the members of a group have the same numbers and distributions of electrons in their valence shells. However, as we go to heavier elements in other groups, there is a gradual change from nonmetallic to metallic behavior that reflects changes in atomic size.

Across a period, we find a smoothly varying change in electron configuration. As we go across a period from left to right, we add a proton to the nucleus and an electron to the valence shell with each successive element. As we go down the elements in a group, the number of electrons in the valence shell remains constant, but the principal quantum number increases by 1 each time. Certain physical properties of the elements vary *periodically* as their electron configurations change. They are sizes (radii) of atoms and ions, ionization energies, and electron affinities. In subsequent chapters we will see that these periodic changes are responsible for the periodic changes in the chemical behavior of the elements.

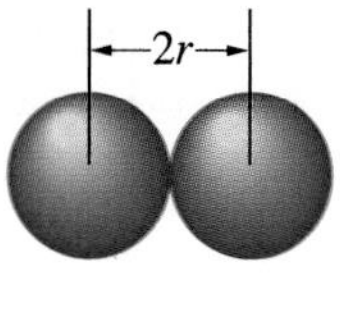

Variation in Covalent Radii.

There are several ways to define the radii of atoms and thus to determine their relative sizes. We will use the **covalent radius** (Fig. 5.23), which is defined as one-half the distance between the nuclei of two identical atoms when they are joined by a single covalent bond (Section 2.7). We know that as we scan down a group or family, the principal quantum number, n, increases by 1 for each element. Thus the electrons are being added to a region of space that is increasingly farther from the nucleus. Consequently, the size of the atom (and its covalent

Figure 5.23

The radius of an atom (r) is defined as one-half the distance between the nuclei in a molecule consisting of two identical atoms joined by a single covalent bond.

Table 5.4

Covalent Radii of the Halogen Group Elements

Atom	Covalent Radius, Å	Nuclear Charge	Number of Electrons in Each Shell
F	0.64	+ 9	2, 7
Cl	0.99	+17	2, 8, 7
Br	1.14	+35	2, 8, 18, 7
I	1.33	+53	2, 8, 18, 18, 7
At	1.4	+85	2, 8, 18, 32, 18, 7

radius) must increase as we increase the distance of the outermost electrons from the nucleus. However, the change is not as large as we might expect because the attraction of the nucleus for an atom's electrons increases as the nuclear charge increases. The changes are illustrated for the covalent radii of the halogens in Table 5.4 and can be observed for the entire periodic table in Fig. 5.24.

In general, as we move across a period from left to right, we find that each element has a smaller covalent radius than the one preceding it (Table 5.5 shows the data for period 3; also see Fig. 5.24). This decrease in size is due to the increasing nuclear charge across the period, which results in a larger force of electrostatic attraction between the nucleus and the electrons. Within a period the number of shells is constant, and the decreased size is due to the electrons being pulled closer to the nucleus.

Figure 5.24

Covalent and ionic radii of the elements. Covalent radii (red circles) are shown at the top of each block, ionic radii (blue circles) at the bottom. Numerical values of radii are in angstrom units (Å).

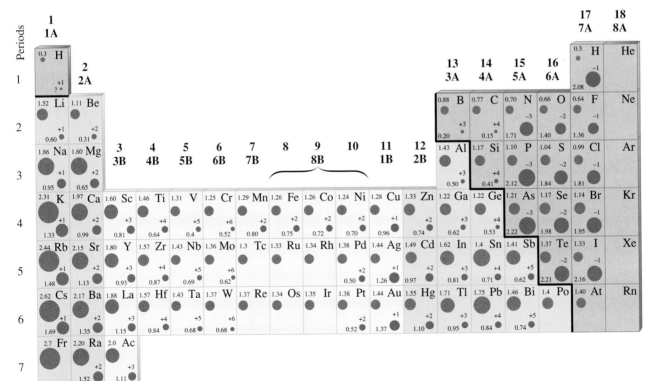

Table 5.5

Covalent Radii of Third Period Elements

Atom	Covalent Radius, Å	Nuclear Charge	Electron Configuration
Na	1.86	+11	$[Ne]3s^1$
Mg	1.60	+12	$[Ne]3s^2$
Al	1.43	+13	$[Ne]3s^23p^1$
Si	1.17	+14	$[Ne]3s^23p^2$
P	1.10	+15	$[Ne]3s^23p^3$
S	1.04	+16	$[Ne]3s^23p^4$
Cl	0.99	+17	$[Ne]3s^23p^5$

Variation in Ionic Radii.   The sizes of ions are defined by their ionic radii. We have seen that ions are formed when atoms gain or lose electrons (Chapter 2). A cation (positive ion) forms when one or more electrons are removed from the parent atom. As we would expect, the outermost, or valence, electrons are easiest to remove because they have the highest energies and are farthest from the nucleus. As a general rule, when the representative elements form cations, they do so by losing the ns or np electrons that are added last. The transition elements, on the other hand, lose the ns electrons before they begin to lose the $(n-1)d$ electrons, even though the ns electrons are added first in the aufbau process (Section 5.8).

A cation, because it always has fewer electrons than the parent atom and the same number of protons, is smaller than the atom from which it is derived (Fig. 5.24). The loss of all electrons from the outermost shell results in a smaller radius, because the remaining electrons occupy regions of space with smaller principal quantum numbers. For example, the covalent radius of a sodium atom ($1s^22s^22p^63s^1$) is 1.86 angstroms, whereas the ionic radius of a sodium cation ($1s^22s^22p^6$) is 0.95 angstrom. In fact, even the radii of these remaining filled electron shells decrease relative to their size in the neutral atom because of the decrease in the total number of electron–electron repulsions within the atom. The decreasing repulsions give rise to a greater average attraction of the nucleus per remaining electron, an effect referred to as an increase in the *effective nuclear charge*. Proceeding down the groups of the periodic table, we find that cations of succeeding elements generally have larger radii, corresponding to an increase in the principal quantum number, n.

An anion (negative ion) is formed by the addition of one or more electrons to the valence shell of an atom. This results in a greater amount of repulsion among the electrons and a decrease in the effective nuclear charge per electron. Both effects (the increased number of electrons and the decreased effective nuclear charge) cause the radius of an anion to be larger than that of the parent atom (Fig. 5.24). For example, a chlorine atom ($[Ne]3s^23p^5$) has a covalent radius of 0.99 angstrom, whereas the ionic radius of the chloride anion ($[Ne]3s^23p^6$) is 1.81 angstroms. For succeeding elements proceeding down any group, anions have larger principal quantum numbers and thus larger radii.

Atoms and ions that have the same electron configuration are called **isoelectronic species**. One series of isoelectronic species is N^{3-}, O^{2-}, F^-, Ne, Na^+, Mg^{2+}, and Al^{3+} ($1s^22s^22p^6$). Another isoelectronic series is P^{3-}, S^{2-}, Cl^-, Ar, K^+, Ca^{2+}, and Sc^{3+} ($[Ne]3s^23p^6$). The only difference in these species that would determine their size is different numbers of protons in the nucleus. The greater the nuclear charge,

Table 5.6

Radii of Two Sets of Isoelectronic Ions and Atoms

Electron configuration	$1s^2 2s^2 2p^6$						
Species	N^{3-}	O^{2-}	F^-	Ne	Na^+	Mg^{2+}	Al^{3+}
Nuclear charge	+7	+8	+9	+10	+11	+12	+13
Radius, Å	1.71	1.40	1.36	1.12	0.95	0.65	0.50
Electron configuration	$1s^2 2s^2 2p^6 3s^2 3p^6$						
Species	P^{3-}	S^{2-}	Cl^-	Ar	K^+	Ca^{2+}	Sc^{3+}
Nuclear charge	+15	+16	+17	+18	+19	+20	+21
Radius, Å	2.12	1.84	1.81	1.54	1.33	0.99	0.81

the smaller the radius in a series of isoelectronic ions and atoms. This trend is illustrated in Table 5.6.

Variation in Ionization Energies.

The amount of energy required to remove *the most loosely bound* electron from a gaseous atom is called its **first ionization energy.** The first ionization energy for an element, X, is the enthalpy change (ΔH) in the equation

$$X(g) + \text{energy} \longrightarrow X^+(g) + e^-$$

The energy required to remove the second most loosely bound electron is called the **second ionization energy** and is the ΔH for the reaction

$$X^+(g) \longrightarrow X^{2+}(g) + e^-$$

The energy required to remove the third electron is the **third ionization energy,** and so on. Energy is always required to remove electrons from atoms, and the foregoing ionizations are always endothermic—that is, ΔH is always positive. Looking at the relative sizes of atoms helps us understand overall periodic variations in ionization energies. The larger the atom, the further from the nucleus is the most loosely bound electron. Thus as size (atomic radius) increases, the ionization energy should decrease. Relating this logic to what we have just learned about radii, we would expect first ionization energies to decrease down a group and to increase across a period.

Figure 5.25 is a graph of the relationship between the first ionization energies and the atomic numbers of several elements. The values of first ionization energies for the elements are given in Table 5.7. The periodicity of the graph is clear, but note a curious feature: The ionization energy of boron (atomic number 5) is less than that of beryllium (atomic number 4) even though the nuclear charge of boron is greater by one proton. This can be explained by differences in the attraction of the positive nucleus for electrons in different subshells. On average, an *s* electron is attracted to the nucleus more than is a *p* electron in the same principal shell. This means that an *s* electron is harder to remove from an atom than a *p* electron in the same shell. The electron removed during the ionization of beryllium ([He]$2s^2$) is an *s* electron, whereas the electron removed during the ionization of boron ([He]$2s^2 2p^1$) is a *p* electron; this results in a smaller first ionization energy for boron, even though its nuclear charge is greater by one proton. The two additional nuclear charges in carbon are sufficient to make its first ionization energy larger than that of beryllium. The first ionization energy for oxygen is slightly less than that for nitrogen because of the repulsion between the two electrons that occupy the same 2p orbital in the oxygen atom. Because these two electrons occupy the same region of space, their repulsion will

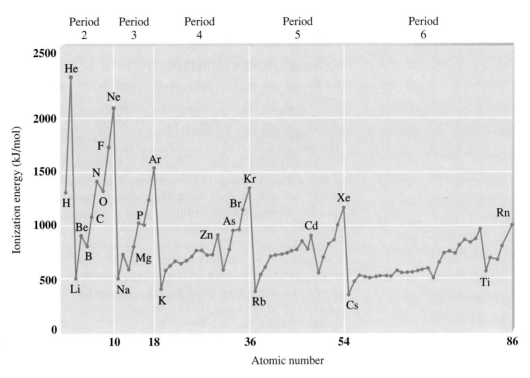

Figure 5.25 The first ionization energies of the elements in the first six periods plotted against their atomic numbers.

Table 5.7

First Ionization Energies of Some Elements, in Kilojoules per Mole

H 1310																	He 2370
Li 520	Be 900											B 800	C 1090	N 1400	O 1310	F 1680	Ne 2080
Na 490	Mg 730											Al 580	Si 780	P 1060	S 1000	Cl 1250	Ar 1520
K 420	Ca 590	Sc 630	Ti 660	V 650	Cr 660	Mn 710	Fe 760	Co 760	Ni 730	Cu 740	Zn 910	Ga 580	Ge 780	As 960	Se 950	Br 1140	Kr 1350
Rb 400	Sr 550	Y 620	Zr 660	Nb 670	Mo 680	Tc 700	Ru 710	Rh 720	Pd 800	Ag 730	Cd 870	In 560	Sn 700	Sb 830	Te 870	I 1010	Xe 1170
Cs 380	Ba 500	La 540	Hf 700	Ta 760	W 770	Re 760	Os 840	Ir 890	Pt 870	Au 890	Hg 1000	Tl 590	Pb 710	Bi 800	Po 810	At ...	Rn 1030
Fr ...	Ra 510																

overcome the additional nuclear charge of the oxygen nucleus. Analogous changes occur in succeeding periods (note aluminum and sulfur in Fig. 5.25).

Variation in Electron Affinities.

The **electron affinity** is the enthalpy change (ΔH) for the process of adding an electron to a gaseous atom to form an anion (negative ion).

$$X(g) + e^- \longrightarrow X^-(g) \qquad \text{E.A.} = \Delta H$$

This process can be either endothermic or exothermic, depending on the element. The electron affinities of some of the elements are given in Table 5.8. You can see that many of these elements have negative values of electron affinities, which means that energy is released when the gaseous atom accepts an electron. However, for some elements energy is required for the atom to become negatively charged, and the values of their electron affinities are positive.

We might expect it to be easier to add an electron to a series of atoms as the effective nuclear charge of the atoms increases. We find, as we go from left to right across a period, that electron affinities tend to become more negative (Table 5.8). The exceptions that occur among the elements of Group 2A, Group 5A, and Group 8A can be understood on the basis of electron configuration. Group 2A has a filled ns subshell, Group 5A has a half-filled np subshell, and Group 8A has all subshells filled. These electron configurations are relatively stable.

We also might expect the atom at the top of each group to have the largest electron affinity; their first ionization potentials suggest that these atoms have the largest effective nuclear charges. As we move down a group, however, we see that the second element in the group most often has the greatest electron affinity. The reduction of the electron affinity of the first member can be attributed to the small size of the $n = 2$ shell and the resulting large electron–electron repulsions. For example, chlorine has the highest electron affinity value in the periodic table: -348 kJ mol^{-1}. The electron affinity value of fluorine is -322 kJ mol^{-1}. When we add an electron to a fluorine atom to form a fluoride anion (F^-), we add an electron to the $n = 2$ shell. The electron is attracted to the nucleus, but there is also significant repulsion from

Table 5.8

Electron Affinities of Some Elements, in Kilojoules per Mole

1A							8A
H							**He**
-72	2A	3A	4A	5A	6A	7A	$+20^a$
Li	**Be**	**B**	**C**	**N**	**O**	**F**	**Ne**
-60	$+240^a$	-23	-123	0	-141	-322	$+30$
Na	**Mg**	**Al**	**Si**	**P**	**S**	**Cl**	**Ar**
-53	$+230^a$	-44	-120	-74	-201	-348	$+35^a$
K	**Ca**	**Ga**	**Ge**	**As**	**Se**	**Br**	**Kr**
-48	$+150^a$	-40^a	-116	-77	-195	-324	$+40^a$
Rb	**Sr**	**In**	**Sn**	**Sb**	**Te**	**I**	**Xe**
-46	$+160^a$	-40^a	-121	-101	-190	-295	$+40^a$
Cs	**Ba**	**Tl**	**Pb**	**Bi**	**Po**	**At**	**Rn**
-45	$+50^a$	-50	-101	-101	-170^a	-270^a	$+40^a$

[a]Calculated value.

the other electrons already present in the valence shell. The chlorine atom has the same electron configuration in the valence shell, but because the entering electron is going into the $n = 3$ shell, it occupies a considerably larger region of space, and the electron–electron repulsions are reduced. The entering electron does not experience so much repulsion, and the chlorine atom accepts an additional electron more readily.

For Review   Summary

A number of experiments, including Rutherford's α particle scattering experiments, have shown the architecture of the atom to be a very small, high-mass nucleus made up of protons and neutrons surrounded by largely empty space that contains electrons. Bohr described the hydrogen atom in terms of an electron moving in a circular orbit about a nucleus. In order to account for the stability of the hydrogen atom, he postulated that the electron was restricted to certain orbits characterized by discrete energies. Thus the energy of the electron was assumed to be **quantized.**

Microscopic particles exhibit both particle-like and wave-like behavior. Light and other forms of **electromagnetic radiation** move through a vacuum with a speed, c, of 2.998×10^8 m s^{-1}. This radiation shows wave-like behavior, which can be characterized by a **frequency,** ν, and a **wavelength,** λ, such that $c = \lambda\nu$. Electromagnetic radiation also has the properties of particles called **photons.** The energy of a photon is related to the frequency of the radiation: $E = h\nu$, where h is Planck's constant. The atomic spectrum of hydrogen is obtained by passing the light from a discharge tube through a prism. From a measurement of the wavelength of each line, the energy of each can be determined.

The Bohr model of the hydrogen atom explains the hydrogen spectrum. Using the Bohr model, we can calculate the energy of an electron and the radius of its orbit in any one-electron system. Bohr suggested that an atom emits energy, in the form of a photon, only when an electron falls from a higher-energy orbit to a lower-energy orbit. The predicted energy differences correspond to the values found in the spectrum of hydrogen. Because the Bohr model accounted for the **line spectrum** of hydrogen gas, his postulation of quantized energies for the electrons in atoms was widely accepted.

Atoms with two or more electrons cannot be described satisfactorily by the Bohr model, and it has been replaced by the **quantum-mechanical model.** This model describes the behavior of an electron in terms of a **wave function,** ψ, which identifies the **orbital,** or region of space, occupied by the electron. It is possible to calculate the energy of the electron from ψ or to determine the probability of finding the electron in a given location in the orbital. The distribution of an electron within an orbital may be described in terms of **electron density.**

An orbital is characterized by three quantum numbers. The **principal quantum number,** n, may be any positive integer. The energy of the orbital and its average distance from the nucleus are related to n. The **angular momentum quantum number,** l, can have any integral value from 0 to $(n - 1)$. The shape or type of the orbital is indicated by l, and within a multielectron atom, the energies of orbitals with the same value of n increase as l increases. The **magnetic quantum number,** m_l, with values ranging from $-l$ to $+l$, describes the orientation of the orbital in space. A fourth quantum number, the **spin quantum number,** s ($s = \pm\frac{1}{2}$), describes the spin of an electron about its own axis. Orbitals with the same value of n occupy the same shell. Orbitals with the same values of n and l occupy the same **subshell.** There are $2n^2$ electrons in any shell and $(2l + 1)$ orbitals in the l subshell of any shell.

By adding electrons one by one to atomic orbitals in the order 1s, 2s, 2p, 3s, 3p, 4s, 3d, 4p, and so on, and by following the **Pauli exclusion principle** (no two electrons can have the same set of four quantum numbers) and **Hund's rule** (whenever possible, electrons remain unpaired in degenerate orbitals), we can predict the **electron configuration** of most atoms. The electron configuration can be described in pictorial form by an **orbital diagram.**

In the periodic table, the elements are arranged in order of increasing atomic number. Elements with the same or similar electron configurations (and, consequently, similar chemical behavior) are found in the same **group. Valence electrons** in the outermost ns and np orbitals are responsible for the chemical behavior of the **representative,** or **main-group, elements. Transition metals** have their valence electrons in the ns and $(n - 1)d$ orbitals. The **inner transition metals** have their valence electrons in three shells: the ns, $(n - 1)d$, and $(n - 2)f$ orbitals.

The chemical properties of elements and many of their physical properties, such as **covalent radii, ionic radii, ionization energies,** and **electron affinities,** vary in a periodic way with the nature of the valence electrons involved. Radii increase down a group and decrease across a period. Cations are smaller than the corresponding neutral atom; anions are larger. Ionization energies decrease down a group and increase across a period, as size considerations would suggest. In general, electron affinities increase across a period and decrease down a group.

Key Terms and Concepts

alpha particles (5.1)

angular momentum quantum number (5.6)

aufbau process (5.8)

Bohr's model of the atom (5.3)

covalent radius (5.10)

degenerate orbitals (5.6)

dual nature of matter (5.4)

electromagnetic radiation (5.2)

electron affinity (5.10)

electron configuration (5.7)

electron density (5.5)

emission spectra (5.2)

excited state (5.3)

frequency (5.2)

ground state (5.3)

Heisenberg uncertainty principle (5.4)

Hund's rule (5.8)

inner transition metal (5.9)

ionization energy (5.10)

isoelectronic species (5.10)

magnetic quantum number (5.6)

main-group element (5.9)

orbital (5.5)

orbital diagram (5.7)

paired electrons (5.6)

parallel spins (5.6)

Pauli exclusion principle (5.8)

photon (5.2)

principal quantum number (5.6)

quanta (5.2)

quantized (5.3)

quantum-mechanical model (5.4)

quantum mechanics (5.4)

quantum number (5.3)

representative element (5.9)

shell (5.6)

spin quantum number (5.6)

subshell (5.6)

transition metal (5.9)

valence electrons (5.9)

valence shell (5.9)

wave function (5.5)

wavelength (5.2)

wave mechanics (5.4)

Exercises

Questions

1. How are electrons and protons similar? How are they different?

2. How are protons and neutrons similar? How are they different?

3. The light produced by a red neon sign is due to the emission of light by excited neon atoms. Qualitatively describe the spectrum produced by passing light from a neon lamp through a prism.

4. The ionization energy of hydrogen is reported as 13.595 eV and as 1310 kJ mol^{-1}. Why do these values differ? Which is the microscopic value and which the macroscopic value?

5. Why is the electron in a hydrogen atom bound less tightly when it has a quantum number of 3 than when it has a quantum number of 1?

6. Which of the following equations describe microscopic properties of atoms, elements, and/or light? Which describe macroscopic properties? Can any describe both?

   (a) Ionization energy = 800 kJ mol^{-1}

   (b) $c = \lambda \nu$

   (c) $E = h\nu$

   (d) $E = \dfrac{-kZ^2}{n^2}$

   (e) $r = \dfrac{n^2 a_0}{Z}$

   (f) $E = |E_1 - E_2|$

   (g) $\lambda = \dfrac{h}{mv}$

   (h) Electron affinity = -322 kJ mol^{-1}

7. Which of the following equations describe particle-like behavior? Which describe wave-like behavior? Do any involve both types of behavior? Give the reasons for your choices.

   (a) $c = \lambda \nu$

   (b) $E = mv$

   (c) $r = \dfrac{n^2 a_0}{Z}$

   (d) $E = h\nu$

   (e) $\lambda = \dfrac{h}{mv}$

8. What does it mean to say that the energies of the electrons in an atom are quantized?

9. Is $1s^2 2s^2 2p^6$ the symbol for a macroscopic property or a microscopic property of an element? Explain your answer.

10. The spectrum of hydrogen and that of calcium are shown in Fig. 5.8. What causes the lines in these spectra? Why are the colors of the lines different? Suggest a reason for the observation that the spectrum of calcium is more complicated than the spectrum of hydrogen.

11. What additional information do we need to answer the following question: "Which ion has the electron configuration $1s^2 2s^2 2p^6 3s^2 3p^6$?"

Atomic Structure

12. What part did the isolation of radium play in the development of our understanding of atomic structure?

13. What did the study of the scattering of α particles by metal foils play in the development of our knowledge of atomic structure?

14. What is an α particle? What is its approximate molar mass?

15. How are the Bohr model and the Rutherford model of the atom similar? How are they different?

16. Before the discovery of neutrons, some scientists believed that the nuclei consisted of a mixture of protons and electrons with more protons than electrons. What properties of atoms did this theory explain? Discuss how this theory explained these properties.

Light and Atomic Spectra

17. An FM radio station found at 103.1 on the FM dial broadcasts at a frequency of 1.031×10^8 s^{-1} (103.1 MHz). What is the wavelength of these radio waves in meters?

18. FM-95, an FM radio station, broadcasts at a frequency of 9.51×10^7 s^{-1} (95.1 MHz). What is the wavelength of these radio waves in meters?

19. A bright violet line occurs at 435.8 nm in the emission spectrum of mercury vapor. What amount of energy, in joules and in electron-volts, must be released by an electron in a mercury atom to produce a photon of this light?

20. Light with a wavelength of 614.5 nm looks orange. What is the energy, in joules and in electron-volts, of a photon of this orange light?

21. Heated lithium atoms emit photons of light with an energy of 2.961×10^{-19} J. Calculate the frequency and wavelength of the photon. What is the total energy in 1.00 mol of these photons? What is the color of the emitted light?

22. A photon of light produced by a surgical laser has an energy of 3.027×10^{-19} J. Calculate the frequency and wavelength of the photon. What is the total energy in 1.00 mol of photons? What is the color of the emitted light?

23. When rubidium ions are heated to a high temperature, two lines are observed in its line spectrum at wavelengths (a) 7.9×10^{-7} m and (b) 4.2×10^{-7} m. What are the frequencies of the two lines? What color do we see when we heat a rubidium compound?

24. The emission spectrum of cesium contains two lines whose wave numbers are (a) 3.45×10^{14} Hz and (b) 6.53×10^{14} Hz. What are the wavelengths and energies of the two lines?

The Bohr Model of the Atom

25. (a) From the data given in Fig. 5.9, determine the energy, in electron-volts, necessary to ionize a hydrogen atom. Explain how you used the data.

(b) Using the Bohr model, determine the energy, in electron-volts, necessary to ionize a hydrogen atom. Show your calculations.

26. (a) From the data in Fig. 5.9, determine the energy, in electron-volts, of the photon produced when an electron in a hydrogen atom moves from the orbit with $n = 5$ to the orbit with $n = 2$. Explain how you used the data.

(b) Using the Bohr model, determine the energy, in electron-volts, of the photon produced when an electron in a hydrogen atom moves from the orbit with $n = 5$ to the orbit with $n = 2$. Show your calculations.

27. Using the Bohr model, determine the lowest possible energy, in electron-volts, for the electron in the Li^{2+} ion.

28. Using the Bohr model, determine the lowest possible energy, in electron-volts, for the electron in the He$^+$ ion.

29. Using the Bohr model, determine the energy of an electron with $n = 6$ in a hydrogen atom.

30. Using the Bohr model, determine the energy of an electron with $n = 8$ in a hydrogen atom.

31. How far from the nucleus is the electron in a hydrogen atom if it has an energy of -0.850 eV?

32. What is the radius of the orbital of an electron with $n = 8$ in a hydrogen atom?

33. Using the Bohr model, determine the energy, in electron-volts and in joules, of the photon produced when an electron in a He$^+$ ion moves from the orbit with $n = 5$ to the orbit with $n = 2$.

34. Using the Bohr model, determine the energy, in electron-volts and in joules, of the photon produced when an electron in a Li^{2+} ion moves from the orbit with $n = 2$ to the orbit with $n = 1$.

35. (a) Consider a large number of hydrogen atoms with electrons randomly distributed in the $n = 1, 2, 3$, and 4 orbits. How many different wavelengths of light are emitted by these atoms as the electrons fall into lower-energy orbitals?

(b) Calculate the lowest and highest energies, in electron-volts, of light produced by the transitions described in part (a).

(c) Calculate the frequencies and wavelengths of the light produced by the transitions described in part (b).

The Quantum-Mechanical Model of the Atom

36. How are the Bohr model and the quantum-mechanical model of the atom similar? How are they different?

37. Without using quantum numbers, describe the differences among the shells, subshells, and orbitals of an atom.

38. How do the quantum numbers of the shells, subshells, and orbitals of an atom differ?

39. Tell what values are allowed for each of the four quantum numbers n, l, m_l and s.

40. Describe the properties of an electron that are associated with each of the four quantum numbers n, l, m_l, and s.

41. Identify the subshell in which electrons with the following quantum numbers are found.
    (a) $n = 2, l = 1$
    (b) $n = 4, l = 2$
    (c) $n = 6, l = 0$

42. Identify the subshell in which electrons with the following quantum numbers are found.
    (a) $n = 3, l = 2$
    (b) $n = 1, l = 0$
    (c) $n = 4, l = 3$

43. Which of the subshells in Exercise 41 contain degenerate orbitals? How many degenerate orbitals are found in each?

44. Which of the subshells in Exercise 42 contain degenerate orbitals? How many degenerate orbitals are found in each?

45. Write a set of quantum numbers for each of the electrons with an n of 3 in a Sc atom.

46. Write a set of quantum numbers for each of the electrons with an n of 4 in a Se atom.

47. Sketch the boundary surface (the shape) of a $d_{x^2-y^2}$ and a p_y orbital. Be sure to show and label the axes.

48. Sketch the boundary surface of a p_x and a d_{xz} orbital. Be sure to show and label the coordinates.

49. Consider the orbitals with the shapes shown here in outline.

(i)

(ii)

(iii)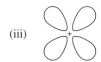

(a) What is the maximum number of electrons contained in an orbital of type (i)? Of type (ii)? Of type (iii)?
(b) How many orbitals of type (i) are found in a shell with $n = 2$? How many of type (ii)? How many of type (iii)?
(c) Write a set of quantum numbers for an electron in an orbital of type (i) in a shell with $n = 4$. In an orbital of type (ii) in a shell with $n = 2$. In an orbital of type (iii) in a shell with $n = 3$.
(d) What is the smallest possible n value for an orbital of type (i)? Of type (ii)? Of type (iii)?

(e) What are the possible l and m_l values for an orbital of type (i)? Of type (ii)? Of type (iii)?

50. (a) State the Heisenberg uncertainty principle, and define each of the terms that appears in it.
    (b) Describe briefly in words what the principle implies.

Electron Configurations

51. Using complete subshell notation ($1s^2 2s^2 2p^6$ and so on), predict the electron configuration of each of the following atoms: (a) C,   (b) P,   (c) V,   (d) Sb,   (e) Sm

52. Using complete subshell notation ($1s^2 2s^2 2p^6$ and so on), predict the electron configuration of each of the following atoms: (a) N,   (b) Si,   (c) Fe,   (d) Te,   (e) Tb

53. An orbital diagram for the valence shell of the N atom is

Use a similar diagram to describe the electron configuration of the valence shell of each of the following atoms: (a) C, (b) P, (c) V, (d) Sb, (e) Ru.

54. Use an orbital diagram (see Exercise 53) to describe the electron configuration of the valence shell of each of the following atoms: (a) N, (b) Si, (c) Fe, (d) Te, (e) Mo.

55. Using complete subshell notation ($1s^2 2s^2 2p^6$ and so on), predict the electron configuration of each of these ions: (a) Na^+, (b) P^{3-}, (c) Al^{2+}, (d) C^{2-}, (e) Fe^{3+}, (f) Sm^{3+}.

56. Using complete subshell notation ($1s^2 2s^2 2p^6$ and so on), predict the electron configuration of each of the following ions: (a) N^{3-}, (b) Ca^{2+}, (c). S^-, (d) Cs^{2+}, (e) Cr^{2+}, (f) Gd^{3+}.

57. Which atom has the following electron configuration?
$$1s^2 2s^2 2p^6 3s^2 3p^6 4s^2 3d^{10} 4p^6 5s^2 4d^2$$

58. Which atom has the following electron configuration?
$$1s^2 2s^2 2p^6 3s^2 3p^6 3d^7 4s^2$$

59. Which ion with a $+2$ charge has the following electron configuration?
$$1s^2 2s^2 2p^6 3s^2 3p^6 3d^{10} 4s^2 4p^6 4d^5$$
Which ion with a $+3$ charge has this configuration?

60. Which ion with a $+1$ charge has the following electron configuration?
$$1s^2 2s^2 2p^6 3s^2 3p^6 3d^{10} 4s^2 4p^6$$

61. Which of the following atoms contains only three valence electrons: Li, B, N, F, or Ne?

62. Which of the following has two unpaired electrons? (a) Mg; (b) Si; (c) S; (d) both Mg and S; (e) both Si and S.

63. Which atom would be expected to have a half-filled $6p$ subshell?

64. Which atom would be expected to have a half-filled $4s$ subshell?

65. In one area of Australia, the cattle did not thrive despite the presence of suitable forage. An investigation showed that the problem was the absence of sufficient cobalt in the soil. Cobalt forms cations in two oxidation states, +2 and +3. Write the electron configuration of the two cations.

66. Thallium was used as a poison in the Agatha Christie mystery *The Pale Horse*. Thallium has two oxidation states, +1 and +3. The +1 compounds are the more stable. Write the electron configuration of the +1 cation of thallium.

Periodic Properties

67. On the basis of their positions in the periodic table, predict which has the smallest atomic radius: Mg, Sr, Si, Cl, or I.

68. On the basis of their positions in the periodic table, predict which has the largest atomic radius: Li, Rb, N, F, or I.

69. On the basis of their positions in the periodic table, predict which has the largest first ionization energy: Mg, Ba, B, O, or Te.

70. On the basis of their positions in the periodic table, predict which has the smallest first ionization energy: Li, Cs, N, F, or I.

71. On the basis of their positions in the periodic table, rank the following atoms in order of increasing first ionization energy: F, Li, N, Rb.

72. On the basis of their positions in the periodic table, rank the following atoms in order of increasing first ionization energy: Mg, O, S, Si.

73. Atoms of which group in the periodic table have a valence shell electron configuration of ns^2np^3?

74. Atoms of which group in the periodic table have a valence shell electron configuration of ns^2?

75. On the basis of their positions in the periodic table, list the following atoms in order of increasing radius: Mg, Ca, Rb, Cs.

76. On the basis of their positions in the periodic table, list the following atoms in order of increasing radius: Sr, Ca, Si, Cl.

77. On the basis of their positions in the periodic table, list the following ions in order of increasing radius: K^+, Ca^{2+}, Al^{3+}, Si^{4+}.

78. List the following ions in order of increasing radius: Li^+, Mg^{2+}, Br^-, Te^{2-}.

79. Tell which of the following atoms and ions is (are) isoelectronic with Br^+: Se^{2+}, Se, As^-, Kr, Ga^{3+}, Cl^-.

80. Tell which of the following atoms and ions is (are) isoelectronic with S^{2+}: Si^{4+}, Cl^{3+}, Ar, As^{3+}, Si, Al^{3+}.

81. List the following ions in order of increasing radius: As^{3-}, Br^-, K^+, Mg^{2+}.

Applications and Additional Exercises

82. Read the labels of several commercial products and identify monatomic ions of at least four transition elements contained in the products. Write the complete electron configurations of these cations.

83. Read the labels of several commercial products and identify monatomic ions of at least six representative elements contained in the products. Write the complete electron configurations of these cations and anions.

84. Photons of infrared radiation are responsible for much of the warmth we feel when holding our hands before a fire. These photons will also warm other objects. How many infrared photons with a wavelength of 1.5×10^{-6} m must be absorbed by the water in order to warm a cup of water (175 g) from 25.0°C to 40°C?

85. One of the X rays used in a dentist's office has a wavelength of 0.2090 Å (2.090×10^{-11} m). What are the energy, in joules and in electron-volts, and the frequency of this X ray?

86. The eyes of certain members of the reptile family pass a single visual signal to the brain when the visual receptors are struck by photons of wavelength 850 nm. If a total energy of 3.15×10^{-14} J is required to trip the signal, what is the minimum number of photons that must strike the receptor?

87. RGB color television and computer displays use cathode ray tubes that produce colors by mixing red, green, and blue light. If we look at the screen with a magnifying glass, we can see individual dots turn on and off as the colors change. Using a spectrum of visible light, determine the approximate wavelength of each of these colors. What are the frequency and the energy of a photon of each of these colors?

88. Using the relative dimensions of the nucleus and atom (Section 2.2), estimate the diameter of the hydrogen atom, in kilometers and miles, if the nucleus were the size of a ping pong ball (35 mm in diameter).

89. Excited hydrogen atoms with very large radii have been detected. How large is an H atom with an electron characterized by a quantum number of 106? How many times larger is that than the radius of an H atom in its ground state?

90. Sketch the radial probability function $4\pi^2r^2\psi^2$ against r for the 1s orbital of a hydrogen atom. On the same graph, sketch the radial probability function for the 1s orbital of a lithium atom. (We cannot determine the exact features of the Li plot, but it should be qualitatively correct.)

91. Cobalt-60 and iodine-131 are radioactive isotopes commonly used in nuclear medicine. Write the complete electron configuration for an atom of each isotope.

92. Which is larger, an He^+ ion with an electron in an orbit with $n = 3$ or an Li^{2+} ion with an electron in an orbit with $n = 5$?

93. How many elements would be in the second period of the periodic table if the spin quantum number s could have the value $-\frac{1}{2}$, 0, or $+\frac{1}{2}$?

94. Of the five elements Al, Cl, I, Na, and Rb, which has the most exothermic reaction? (E represents an atom.)

$$E^+(g) + e^- \longrightarrow E(g)$$

95. Of the five elements Sn, Si, Sb, O, and Te, which has the most endothermic reaction? (E represents an atom.)

$$E(g) \longrightarrow E^+(g) + e^-$$

96. The metallic radii of the elements Na, Mg, and Al are 1.86, 1.60, and 1.43 Å, respectively. Explain why Al is not the one with the largest metallic radius, even though it contains the most electrons.

97. The ionic radii of the ions S^{2-}, Cl^-, and K^+ are 1.84, 1.81, and 1.33 Å, respectively. Explain why these ions have different sizes, even though they contain the same number of electrons.

98. Write the electron configuration for each of the following atoms and ions: a. B^{3+}, b. O^-, c. Cl^{3+}, d. Ca^{2+}, e. Ti.

99. Which main-group atom would be expected to have the lowest second ionization energy?

100. Explain why Al is a member of Group 13 rather than Group 3.

101. (a) The laser on a CD player or CD-ROM disk reader has a wavelength of 780 nm. In what region of the electromagnetic spectrum is this radiation? What is its frequency?

(b) A CD-ROM laser has a power of 5 milliwatts (1 watt = 1 J s^{-1}). How many photons of light are produced by the laser in 1 h?

(c) The ideal resolution of a CD-ROM player (which determines the how close together data can be stored on a compact disc) is determined using the formula

$$\text{Resolution} = 0.60\left(\frac{\lambda}{\text{NA}}\right)$$

where λ is the wavelength of the laser and NA is the numerical aperture. Numerical aperture is a measure of the size of the spot of light on the disc; the larger NA, the smaller the spot. In a typical CD-ROM system, NA = 0.45. If the 780-nm laser is used in a CD-ROM reader, what is the closest together that data can be stored in a CD?

(d) The data density of a CD-ROM disc using a 780-nm laser is 1.0×10^6 bits mm^{-2}. CDs have an outside diameter of 120 mm and a hole of diameter 15 mm. How many data bits can be contained on the disk? If a CD can hold 250,000 pages of text, how many data bits are needed for a typed page? *Hint:* Determine the area of the disc that is available to hold data. The area inside a circle is given by the formula $A = \pi\rho^2$.

CHAPTER OUTLINE

6.1 Ionic Bonds: Electrostatic Attractions Between Ions
6.2 The Electronic Structures of Ions
6.3 The Lattice Energies of Ionic Crystals
6.4 Covalent Bonds: Chemical Bonding by Electron Sharing
6.5 Polar Covalent Bonds: Electronegativity
6.6 Lewis Structures
6.7 Writing Lewis Structures: Covalently Bonded Atoms with a Noble Gas Configuration
6.8 Writing Lewis Structures: Covalently Bonded Atoms without a Noble Gas Configuration
6.9 Formal Charge
6.10 Oxidation State
6.11 Resonance
6.12 The Strengths of Covalent Bonds

6

Chemical Bonding: General Concepts

Computer models of the C_{60} molecule.

A computer drawing of a C_{60} molecule opens this chapter. These molecules, called *buckyballs*, are the components of buckminsterfullerene, an elemental form of carbon first detected in 1985. Buckminsterfullerene can be isolated from the soot produced by high-temperature evaporation of graphite in an atmosphere of helium.

The structure of the C_{60} molecule resembles that of the geodesic domes designed by Buckminster Fuller, from whom the substance gets its name. The molecule is spherical, with carbon atoms at the vertices of interconnecting pentagons and hexagons. This spherical structure is unusual: Molecules composed of atoms arranged in a closed cluster with an empty interior are known, but they are uncommon. Thus it is all the more remarkable that the structure of this molecule was predicted from its composition. At that time it was known only that this form of carbon consists of C_{60} molecules. This structure has since been verified by a variety of experimental determinations.

The initial speculations about the structure of the C_{60} molecule were guided by bonding theory, the theory of the forces that hold atoms together in compounds. Our ability to predict the structure of the C_{60} molecule is one of many examples that bonding theory provides us with a powerful tool for understanding and predicting the behavior of matter.

Bonding theory plays an important role in chemistry, because we rarely deal with isolated atoms but rather study groups of two or more atoms and the attractive forces, called **chemical bonds,** that hold atoms together. Chemical bonds change when a chemical change occurs. When elements react to form a compound, chemical bonds between the atoms of the elements are broken, and new chemical bonds form between the different types of atoms that make up the compound. When compounds undergo chemical reactions, the bonds between atoms are rearranged. When the atoms separate, the bonds are destroyed and the original compounds no longer exist.

Although we will generally work with groups of atoms rather than with single atoms, the properties of single atoms are important. The properties of the individual atoms in a compound determine the type of chemical bonds present. In this chapter we will use the electronic structure of atoms to predict which of two extreme types of chemical bonds (ionic bonds and covalent bonds) is likely to form. We will introduce Lewis electron-dot formulas, which are a useful way to describe how electrons are involved in these two kinds of chemical bonds. We will also examine the energy changes that accompany the formation or rearrangement of chemical bonds in chemical reactions.

6.1 Ionic Bonds: Electrostatic Attractions Between Ions

In Section 2.6 we discussed **ions** and identified compounds that are stabilized by ionic bonds. In Section 5.10 we discussed the electronic structure of ions and saw that a monatomic **cation** (a positive ion) forms when a neutral atom loses one or more electrons, usually from its valence shell. We also saw that a monatomic **anion** (a negative ion) forms when a neutral atom gains one or more electrons, usually in its valence shell.

Electrostatic attractions between ions are called **ionic bonds. Ionic compounds** are compounds that are stabilized by ionic bonds. Ionic bonds in binary compounds are usually found in those binary compounds that are formed from a metal (which forms

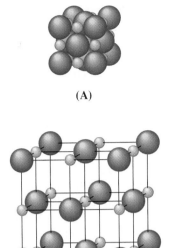

(A)

(B)

Figure 6.1

(A) The arrangement of sodium and chloride ions in a crystal of sodium chloride (common table salt). The smaller spheres represent sodium ions, the larger ones chloride ions. (B) An expanded view that shows the geometry more clearly.

the cations) and a nonmetal (which forms the anions). Thus we can generally identify an ionic compound by the location of its constituent elements in the periodic table.

We can think about the formation of ionic compounds in terms of the periodic properties we discussed in Chapter 5. The formation of ions that results from the transfer of electrons between atoms occurs most readily when a metal reacts with a nonmetal. The atoms of many metallic elements have relatively low ionization energies (Section 5.10) and lose electrons without too much difficulty. These elements lie to the left in a period or toward the bottom of a group in the periodic table. Nonmetal atoms have relatively high electron affinities and thus readily pick up electrons lost by metal atoms, thereby filling their outer s and p orbitals. Nonmetallic elements lie in the upper right-hand portion of the periodic table.

Ionic compounds are electrically neutral because the total number of positive charges on the cations and the total number of negative charges on the anions are equal. The formula of an ionic compound (Section 2.7) represents the simplest ratio of the numbers of ions necessary to give identical numbers of positive and negative charges. For example, the formula of aluminum oxide, Al_2O_3, indicates that this ionic compound contains two aluminum cations, Al^{3+}, for every three oxide anions, O^{2-}; thus $(2 \times +3) + (3 \times -2) = 0$.

Sodium chloride, NaCl, an ionic compound, consists of a regular arrangement of equal numbers of Na^+ cations and Cl^- anions (Fig. 6.1). This arrangement arises because each ion attracts the maximum number of oppositely charged ions about itself. In the sodium chloride crystal structure, we find that each Na^+ cation is surrounded by six Cl^- anions and that each Cl^- anion is surrounded by six Na^+ cations. The force that holds these ions together in the solid is the strong electrostatic attraction between ions of opposite charge. It requires 769 kilojoules of energy to convert 1 mole of NaCl to separated gaseous Na^+ and Cl^- ions.

$$NaCl(s) \longrightarrow Na^+(g) + Cl^-(g) \qquad \Delta H = 769 \text{ kJ}$$

This is about 20 times the energy needed to evaporate 1 mole of water.

6.2 The Electronic Structures of Ions

We began to determine the electronic structures of ions in Section 5.10; now we can expand that discussion. The atoms of the metallic elements in Groups 1A through 5A and 2B, *when they form cations,* tend to lose all valence electrons. This means we can quickly learn or recall the charges on the positive ions of these representative elements. The charge is equal to the group number of the element forming the ion, when we use the 1A to 8A numbering system. This group number is the same as the number of electrons in the outermost shell of a representative element or a member of Group 2B. Members of Group 1A have ionic charges of +1; members of Groups 2A and 2B, +2; and so on (see Fig. 2.17). We should also remember that, generally, metals form cations and nonmetals form anions. For example, we would not expect to see C^{4+} or N^{5+} cations, because carbon and nitrogen are nonmetals.

The exceptions to the expected behavior involve elements toward the bottom of the groups: Tl^+, Sn^{2+}, Pb^{2+}, and Bi^{3+} form in addition to the expected ions Tl^{3+}, Sn^{4+}, Pb^{4+}, and Bi^{5+}. We call the formation of these +1, +2, and +3 cations of the heavier elements an **inert-pair effect,** which reflects the relative nonreactivity of the pair of s electrons in the valence shell of atoms of these heavier elements of Groups

3A, 4A, and 5A. Mercury (Group 2B) also exhibits unexpected behavior: It forms a diatomic ion, Hg_2^{2+} (an ion formed from two mercury atoms), in addition to the expected monatomic ion Hg^{2+} (formed from only one mercury atom).

The positive ion produced by the loss of all valence electrons from a representative metal or a member of Group 2B will have either a **noble gas electron configuration,** in which the outermost shell has the configuration ns^2np^6 ($1s^2$ for cations of the second period), or a **pseudo–noble gas electron configuration,** in which the outermost shell has the configuration $ns^2np^6nd^{10}$. An example of the latter is the zinc ion, Zn^{2+}, from Group 2B that has the electron configuration $1s^22s^22p^63s^23p^63d^{10}$.

Transition and inner transition metal elements behave differently from representative elements. Most transition metal cations have +2 or +3 charges resulting from the loss of their outermost s electrons (or electron) first, sometimes followed by the loss of one or two d electrons from the next-to-outermost shell. For example, copper ($1s^22s^22p^63s^23p^63d^{10}4s^1$) forms the ion Cu^+ ($1s^22s^22p^63s^23p^63d^{10}$) by losing the $4s$ electron and forms the ion Cu^{2+} ($1s^22s^22p^63s^23p^63d^9$) by losing the $4s$ electron and one of the $3d$ electrons. Although the d orbitals of the transition elements are the last to fill when electron configurations are built up by the aufbau process (Section 5.8), the outermost s electrons are the first to be lost when these atoms ionize. When the inner transition metals form ions, they usually have a +3 charge resulting from the loss of their outermost s electrons and a d or f electron.

Example 6.1   Determining the Electronic Structures of Cations

There are at least 14 elements that are called essential trace elements for the human body: "essential trace" because they are required in small amounts daily in our diet, "elements" in spite of the fact they are usually ions. Thus two of these essential trace elements, chromium and zinc, are required as Cr^{3+} and Zn^{2+}. Write the electron configurations of these cations.

Solution: Zinc is a member of Group 2B, so it should have a charge of +2 and a pseudo–noble gas electron configuration. Chromium is a transition element and should lose its s electrons and then its d electrons when forming a cation. Thus we find the following electron configurations of the ions:

$$Zn^{2+} \quad (1s^22s^22p^63s^23p^63d^{10})$$
$$Cr^{3+} \quad (1s^22s^22p^63s^23p^63d^3)$$

Check your learning: Potassium and magnesium are required in our diet. Write the electron configurations of the ions expected from these elements.

Answer: K^+ $1s^22s^22p^63s^23p^6$; Mg^{2+} $1s^22s^22p^6$

Most monatomic anions form when a neutral nonmetal atom gains enough electrons to fill its outer s and p orbitals completely. Thus it is simple to determine the charge on such a negative ion; the charge is equal to the number of electrons that must be gained to fill the s and p orbitals of the parent atom. Oxygen, for example, has the electron configuration $1s^22s^22p^4$, whereas the oxygen anion has the noble gas electron configuration $1s^22s^22p^6$. The two additional electrons required to fill the valence orbitals give the oxide ion, O^{2-}, the charge of −2.

(A)

(B)

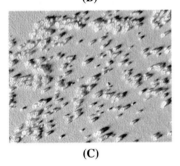

(C)

Figure 6.2

(A) Sodium is a soft metal. (B) Chlorine is a pale yellow gas. (C) Crystals of sodium chloride (table salt).

Example 6.2	Determining the Electronic Structure of Anions

Selenium and iodine are two essential trace elements that form anions. Write the electron configurations of the anions.

Solution:

Se^{2-} $1s^2 2s^2 2p^6 3s^2 3p^6 3d^{10} 4s^2 4p^6$

I^- $1s^2 2s^2 2p^6 3s^2 3p^6 3d^{10} 4s^2 4p^6 4d^{10} 5s^2 5p^6$

Check your learning: Write the electron configurations of a phosphorus atom and its negative ion. Give the charge on the anion.

Answer: P $1s^2 2s^2 2p^6 3s^2 3p^3$; P^{3-} $1s^2 2s^2 2p^6 3s^2 3p^6$

The electronic differences between an atom and its ion give them very different physical and chemical properties. Sodium *atoms* form sodium metal, a soft, silvery-white metal that burns vigorously in air and reacts rapidly with water. Chlorine *atoms* form chlorine gas, Cl_2, a yellow gas that is extremely corrosive to most metals and very poisonous to animals and plants. The vigorous reaction between the elements sodium and chlorine forms the white, crystalline compound sodium chloride, common table salt, which contains sodium ions and chloride ions (Fig. 6.2). The compound composed of these ions exhibits properties entirely different from those of the elements sodium and chlorine. Chlorine is poisonous, but sodium chloride is essential to life; sodium atoms react vigorously with water, but sodium chloride simply dissolves in water.

6.3 The Lattice Energies of Ionic Crystals

A solid ionic compound is stable because of the electrostatic attraction between its positive and negative ions. One measure of the strength of this attraction is the lattice energy of the compound. The **lattice energy,** U, of an ionic compound is defined as the energy required to separate exactly 1 mole of the solid into its component gaseous ions. For the ionic solid MX, the lattice energy is the enthalpy change of the *endothermic* process

$$\text{MX}(s) \longrightarrow \text{M}^{n+}(g) + \text{X}^{n-}(g) \qquad U = \Delta H^\circ_{298}$$

When a mole of a solid ionic compound is formed from gaseous ions, we would expect the same amount of energy to be released: The reaction is *exothermic* and ΔH is negative. For sodium chloride, $U = 769$ kJ. Thus it requires 769 kJ to separate 1 mole of solid NaCl into gaseous Na^+ and Cl^- ions. When 1 mole of gaseous Na^+ and Cl^- ions form solid NaCl, 769 kJ of heat is released. Lattice energies can be calculated from basic principles (which we will discuss in more detail in Chapter 11), or they can be measured experimentally.

The lattice energy, U, of an ionic crystal can be expressed by the equation

$$U = \frac{-C(Z^+)(Z^-)}{R_0}$$

where C is a constant that depends on the type of crystal structure and the electronic structures of the ions, Z^+ and Z^- are the charges on the ions, and R_0 is the interionic distance (the sum of the radii of the positive and negative ions). Thus the lattice energy of an ionic crystal increases rapidly as the charges on the ions increase and as the sizes of the ions decrease. When all other parameters are kept constant, doubling the charge on both cation and anion quadruples the lattice energy. For example, the lattice energy of LiF ($Z^+ = 1$ and $Z^- = -1$) is 1023 kJ mol^{-1}, whereas that of MgO ($Z^+ = 2$ and $Z^- = -2$) is 3900 kJ mol^{-1} (R_0 is nearly the same—2.00 Å for both compounds).

Different interionic distances give different lattice energies. For example, we can compare the lattice energies of MgF$_2$ (2957 kJ mol^{-1}) and MgI$_2$ (2327 kJ mol^{-1}) to observe the effect of the smaller ionic size of F$^-$. (See Fig. 5.24 for a comparison of ionic radii.)

Example 6.3   Lattice Energy Comparisons.

The precious gem ruby is aluminum oxide, Al$_2$O$_3$, containing traces of Cr^{3+}. The compound Al$_2$Se$_3$ is used in the fabrication of some semiconductor devices. Which has the larger lattice energy, Al$_2$O$_3$ or Al$_2$Se$_3$?

Solution: In these two ionic compounds, the charges Z^+ and Z^- are the same, so the difference in lattice energy will depend on R_0. The O^{2-} ion is smaller than the Se^{2-} ion (see Fig. 5.24). Thus Al$_2$O$_3$ would have a shorter interionic distance than Al$_2$Se$_3$, so Al$_2$O$_3$ would have the larger lattice energy.

Check your learning: Zinc oxide, ZnO, is a very effective sun screen. How would the lattice energy of ZnO compare to that of NaCl?

Answer: ZnO would have the larger lattice energy, because the Z values of both the cation and the anion in ZnO are greater, and the ZnO interionic distance is smaller than the NaCl interionic distance.

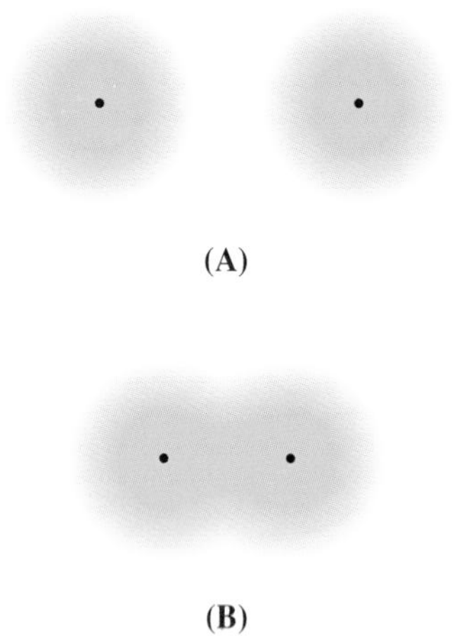

Figure 6.3

(A) The 1s orbitals of two separated hydrogen atoms. (B) The electron density in the overlapping 1s orbitals in a hydrogen molecule.

(A)

(B)

6.4   Covalent Bonds: Chemical Bonding by Electron Sharing

Covalent bonds form when electrons are shared between atoms—that is, when the same electrons occupy orbitals on each atom involved in a bond. Electrons are shared, and covalent bonds are formed between two atoms, when both atoms have about the same tendency to give up or to pick up electrons (when both elements have fairly similar ionization energies and electron affinities). Nonmetal atoms frequently form covalent bonds with other nonmetal atoms.

The hydrogen molecule, H$_2$, contains a covalent bond between two hydrogen atoms. Figure 6.3 illustrates how this bond is formed. When two hydrogen atoms approach each other, their 1s orbitals begin to overlap. The single electron on each hydrogen atom can then spill over into the other 1s orbital and can occupy the space around both atoms. The strong attraction of each shared electron to both nuclei provides the force that holds the molecule together. The resulting bond is very strong; a large amount of energy—436 kJ—must be added to break the bonds in 1 mole of hydrogen molecules, thus allowing the atoms to separate.

$$H_2(g) \longrightarrow 2H(g) \qquad \Delta H = 436 \text{ kJ}$$

Conversely, the same amount of energy is released when 1 mole of H_2 molecules forms from 2 moles of H atoms.

$$2H(g) \longrightarrow H_2(g) \qquad \Delta H = -436 \text{ kJ}$$

This is an important idea: Breaking chemical bonds requires energy, and forming chemical bonds releases energy.

In most cases, when atoms form covalent bonds they share enough electrons to assume noble gas electron configurations. Each electron in the bond in a hydrogen molecule has an equal probability of being near each nucleus. That is, the electron density is evenly distributed in the bond. Thus the $1s$ orbital of each hydrogen atom in the molecule is occupied by both electrons. In effect, each atom has the electronic structure of a helium atom ($1s^2$).

6.5   Polar Covalent Bonds: Electronegativity

If the atoms that form a covalent bond are identical, as in H_2, then the electrons in the bond are shared equally. On the other hand, if the atoms in a bond are different, then the bonding electrons need not be shared equally and a polar covalent bond can result. A **polar covalent bond** is a bond that has a partial positively charged end and a partial negatively charged end. In a polar covalent bond, the bonding electrons are unequally distributed, and the electron density in the bond is greater near one atom than near the other. The atom that has the electrons near it more of the time acquires a partial negative charge, and the other atom in the bond acquires a partial positive charge. For example, the electrons in the H—Cl bond in a hydrogen chloride molecule spend more time near the chlorine atom than near the hydrogen atom. Thus in a HCl molecule, the chlorine atom is somewhat negative and the hydrogen atom somewhat positive. Figure 6.4(A) shows the distribution of electrons in the H—Cl bond. Compare the uneven distribution of the shared electrons in Fig. 6.4 to Fig. 6.3, which shows the distribution of electrons in the nonpolar bond in H_2.

We sometimes designate the positive and negative atoms in a polar covalent bond by using a lowercase Greek letter delta, δ, with a plus sign or a minus sign to indicate whether the atom has a partial positive charge ($\delta+$) or a partial negative charge ($\delta-$). This symbolism is shown for the HCl molecule in Fig. 6.4(B).

Electronegativity is a measure of the attraction of an atom for the electrons in a chemical bond. The more strongly an atom attracts the electrons in its bonds, the larger its electronegativity. Electrons in a polar covalent bond spend more time in the vicinity of the more electronegative atom; thus the more electronegative atom is the one with the partial negative charge. The more electronegative this atom, the more time the bonding electrons spend near it, and the larger its partial negative charge.

Figure 6.5 on page 190 shows the electronegativity values of the elements as proposed by one of the most famous chemists of the twentieth century, Linus Pauling. In general, electronegativity increases from left to right across a period in the periodic table and decreases down a group. Thus the nonmetals, which lie in the upper right portion of the periodic table, tend to have the highest electronegativities, fluorine being the most electronegative. Metals tend to be the least electronegative elements, and the Group 1A metals have the lowest electronegativities.

We should be careful not to confuse electronegativity and electron affinity (Section 5.10). The electron affinity of an element is the energy released or absorbed when

Figure 6.4

(A) Electron density of the electrons in the H—Cl bond in the HCl molecule. The electron density in the bond is greater near the chlorine atom. The small black dots indicate the location of the hydrogen and chlorine nuclei in the molecule. (B) The symbols $\delta+$ and $\delta-$ indicate the polarity of the H—Cl bond.

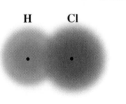

H   Cl

(A)

$\delta+$ $\delta-$
H — Cl

(B)

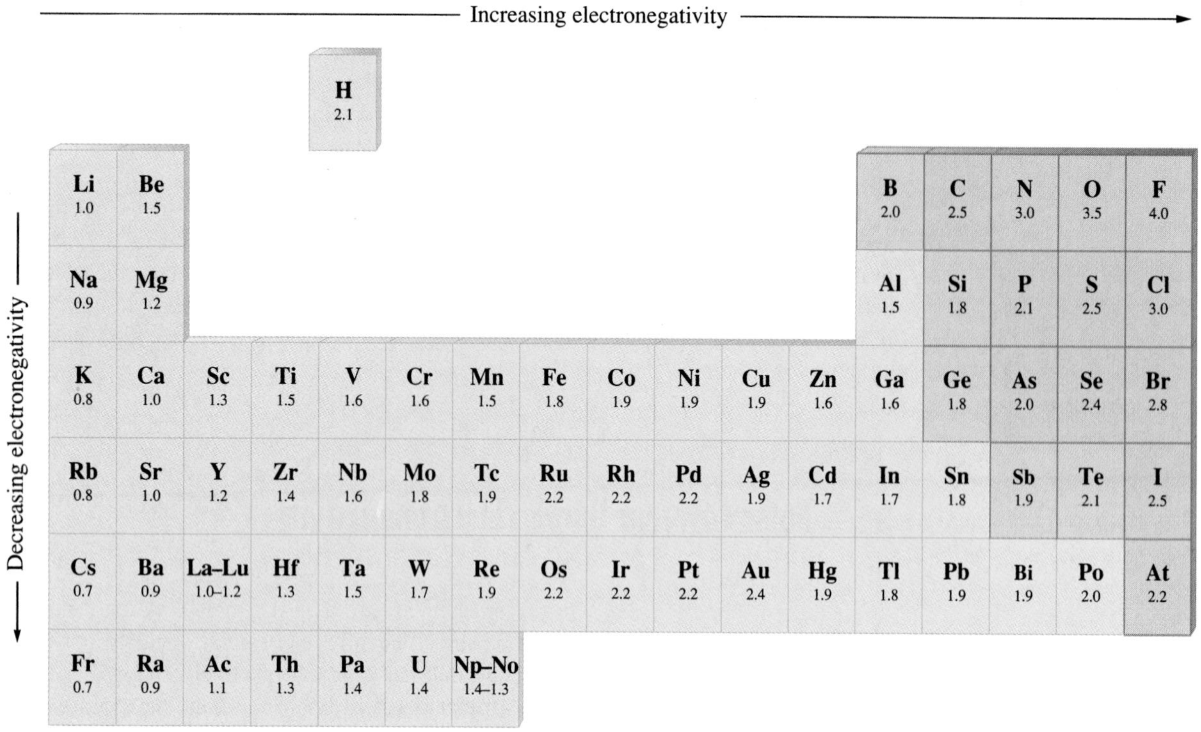

— Increasing electronegativity ⟶

Decreasing electronegativity ↓

						H 2.1										
Li 1.0	Be 1.5											B 2.0	C 2.5	N 3.0	O 3.5	F 4.0
Na 0.9	Mg 1.2											Al 1.5	Si 1.8	P 2.1	S 2.5	Cl 3.0
K 0.8	Ca 1.0	Sc 1.3	Ti 1.5	V 1.6	Cr 1.6	Mn 1.5	Fe 1.8	Co 1.9	Ni 1.9	Cu 1.9	Zn 1.6	Ga 1.6	Ge 1.8	As 2.0	Se 2.4	Br 2.8
Rb 0.8	Sr 1.0	Y 1.2	Zr 1.4	Nb 1.6	Mo 1.8	Tc 1.9	Ru 2.2	Rh 2.2	Pd 2.2	Ag 1.9	Cd 1.7	In 1.7	Sn 1.8	Sb 1.9	Te 2.1	I 2.5
Cs 0.7	Ba 0.9	La–Lu 1.0–1.2	Hf 1.3	Ta 1.5	W 1.7	Re 1.9	Os 2.2	Ir 2.2	Pt 2.2	Au 2.4	Hg 1.9	Tl 1.8	Pb 1.9	Bi 1.9	Po 2.0	At 2.2
Fr 0.7	Ra 0.9	Ac 1.1	Th 1.3	Pa 1.4	U 1.4	Np–No 1.4–1.3										

Figure 6.5

Pauling's values of the electronegativities of the elements.

an isolated gas-phase atom acquires an electron. Electronegativity describes a property of an atom in a bond.

The absolute value of the difference in electronegativity of two bonded atoms provides a rough measure of the polarity to be expected in the bond and, thus, the bond type. When the difference is very small or zero, the bond is nonpolar. When it is large, the bond is polar covalent or ionic. The absolute values of the electronegativity differences between the atoms in the bonds H—H, H—Cl, and Na^+Cl^- are 0, 0.9, and 2.1; the bonds are nonpolar, polar covalent, and ionic, respectively. Figure 6.6 shows the relationship between electronegativity and bond type.

Figure 6.6

As the electronegativity difference between two atoms increases, the bond becomes more ionic.

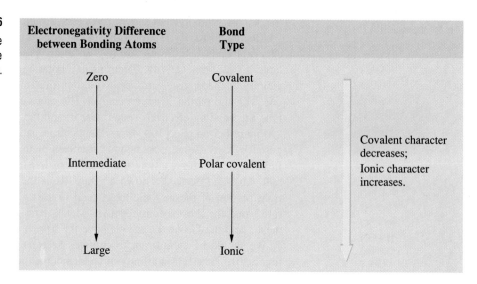

Electronegativity Difference between Bonding Atoms	Bond Type	
Zero	Covalent	
Intermediate	Polar covalent	Covalent character decreases; Ionic character increases.
Large	Ionic	

Is there a single value of electronegativity difference that marks the distinction between a polar covalent bond and an ionic bond? Unfortunately not. For example, the covalent bond in HF has an electronegativity difference of 1.9, the covalent bond in NH_3 a difference of 0.9. The ionic bond in NaCl has an electronegativity difference of 2.1, and that in MnI_2 a difference of 1.0. The best guide to the covalent or ionic character of a bond is the types of atoms involved and their position in the periodic table: Bonds between two nonmetals are generally covalent; bonds between a metal and a nonmetal are often ionic.

Example 6.4   Electronegativity and Bond Polarity

Nearly all proteins in our bodies are made from the same 20 amino acids. Using the electronegativity values in Fig. 6.5, arrange the following common covalent bonds that are found in these amino acids in order of increasing polarity. Then designate the positive and negative atoms by the symbols $\delta+$ and $\delta-$: C—H, C—N, C—O, N—H, O—H, S—H.

Solution: The polarity of these bonds increases as the absolute value of the electronegativity difference between the atoms increases. The atom with the $\delta-$ designation is the more electronegative of the two in the bond. The following list shows these bonds in order of increasing polarity.

Bond	Electronegativity Difference	Polarity
C—H	0.4	$^{\delta-}C—H^{\delta+}$
S—H	0.4	$^{\delta-}S—H^{\delta+}$
C—N	0.5	$^{\delta+}C—N^{\delta-}$
N—H	0.9	$^{\delta-}N—H^{\delta+}$
C—O	1.0	$^{\delta+}C—O^{\delta-}$
O—H	1.4	$^{\delta-}O—H^{\delta+}$

Check your learning: Silicones are polymeric compounds that contain, among others, the following types of covalent bonds: Si—O, Si—C, C—H, and C—C. Using the electronegativity values in Fig. 6.5, arrange these bonds in order of increasing polarity, and designate the positive and negative atoms by the symbols $\delta+$ and $\delta-$.

Answer:

Bond	Electronegativity Difference	Polarity
C—C	0.0	nonpolar
C—H	0.4	$^{\delta-}C—H^{\delta+}$
Si—C	0.7	$^{\delta+}Si—C^{\delta-}$
Si—O	1.7	$^{\delta+}Si—O^{\delta-}$

6.6   Lewis Structures

We use **Lewis symbols** to describe the valence electron configurations of atoms and monatomic ions. These symbols consist of the elemental symbol and one dot for each valence electron present. When two dots are written adjacent to each other, as in Ca:, they represent a pair of electrons in the same orbital. Table 6.1 shows the Lewis symbols for the elements of the third period of the periodic table.

Lewis symbols can be used to demonstrate the formation of cations from atoms, as shown here for sodium and calcium.

$$Na \cdot \longrightarrow Na^+ + e^- \qquad Ca: \longrightarrow Ca^{2+} + 2e^-$$

| Sodium atom | Sodium cation | Calcium atom | Calcium cation |

They can also be used to show the formation of anions from atoms, as shown here for chlorine and sulfur.

$$:\!\ddot{C}l\cdot + e^- \longrightarrow :\!\ddot{C}l\!:^- \qquad :\!\ddot{S}\cdot + 2e^- \longrightarrow :\!\ddot{S}\!:^{2-}$$

| Chlorine atom | Chloride anion | Sulfur atom | Sulfide anion |

When metals combine with nonmetals, electrons are transferred from the metals to the nonmetals, and ionic compounds are formed. The following examples illustrate the use of Lewis symbols to show the transfer of electrons during the formation of ionic compounds.

Metal	Nonmetal	Ionic Compound
Na · Sodium atom	+	:$\ddot{C}$l· Chlorine atom	$\longrightarrow$	Na$^+$[:$\ddot{C}$l:]$^-$ Sodium chloride (sodium ion and chloride ion)
Mg: Magnesium atom	+	:$\ddot{O}$· Oxygen atom	$\longrightarrow$	Mg^{2+}[:$\ddot{O}$:]$^{2-}$ Magnesium oxide (magnesium ion and oxide ion)
Ca: Calcium atom	+	2 :$\ddot{F}$· Fluorine atoms	$\longrightarrow$	Ca^{2+}[:$\ddot{F}$:]$_2^-$ Calcium fluoride (calcium ion and two fluoride ions)

We also use Lewis symbols to indicate the formation of covalent bonds. For example, when two chlorine atoms form a chlorine molecule, they share a pair of electrons (one from each atom) and fill their valence shell *s* and *p* orbitals, as indicated by the following Lewis structure:

$$:\!\ddot{C}l\cdot + :\!\ddot{C}l\cdot \longrightarrow :\!\ddot{C}l\!:\!\ddot{C}l\!:$$

| Chlorine atoms | Chlorine molecule |

Table 6.1

Lewis Symbol for the Atoms of the Third Period.

Atom	Electron Configuration	Lewis Symbol
Sodium	$[\text{Ne}]3s^1$	Na·
Magnesium	$[\text{Ne}]3s^2$	Mg :
Aluminum	$[\text{Ne}]3s^23p^1$	Al :
Silicon	$[\text{Ne}]3s^23p^2$	·Si·
Phosphorus	$[\text{Ne}]3s^23p^3$	·P·
Sulfur	$[\text{Ne}]3s^23p^4$	:S·
Chlorine	$[\text{Ne}]3s^23p^5$	:Cl·
Argon	$[\text{Ne}]3s^23p^6$	:Ar:

Lewis structures utilize Lewis symbols to show the distribution of electrons in covalently bonded molecules and polyatomic ions. The Lewis structure shown above for Cl_2 indicates that each Cl atom has three pairs of electrons that are not used in bonding (called **unshared pairs** or **lone pairs**) and one shared pair of electrons (written between the atoms). A dash is sometimes used to indicate a shared pair of electrons.

$$H-H \qquad :\overset{..}{\underset{..}{Cl}}-\overset{..}{\underset{..}{Cl}}:$$

A single shared pair of electrons is called a **single bond.**

As we might expect because the atoms involved are in the same group of the periodic table, the bonding in the other halogen molecules (F_2, Br_2, I_2, and At_2) is like that in the chlorine molecule: one single bond between atoms and three unshared pairs of electrons per atom. Like the chlorine atoms in Cl_2, each atom in these molecules has filled its s and p valence orbitals and thus has the electronic structure of a noble gas.

The number of single bonds that an atom can form can often be predicted from the number of electrons needed to fill its outer s and p orbitals. An atom often forms one single bond for each electron it needs. For example, each atom of a Group 4A element has four electrons in its outer shell and requires four more electrons to fill its outer s and p orbitals. These four electrons can be gained by forming four single covalent bonds, as illustrated for carbon in CCl_4 (carbon tetrachloride) and silicon in SiH_4 (silane).

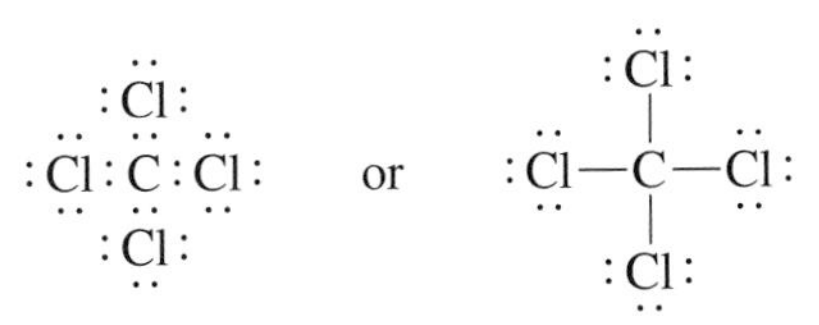

Carbon tetrachloride 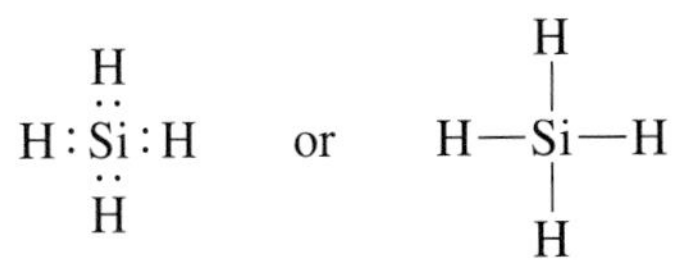   Silane

Note that the Lewis structure of a molecule in general does not indicate its three-dimensional shape. We will see in Chapter 7 that carbon tetrachloride and silane are not flat molecules.

In most covalent molecules, atoms are stable with filled s and p valence orbitals—the electron configuration is that of the nearest noble gas. Nitrogen and the other elements of Group 5A fill their outer s and p orbitals and attain a noble gas configuration with the addition of three electrons. These three electrons can be gained by the formation of three single covalent bonds, as in NH_3 (ammonia). Oxygen and other atoms in Group 6A fill their outer s and p orbitals with the addition of two electrons; thus they can form two single covalent bonds. The elements in Group 7A, such as fluorine, must form only one single covalent bond in order to fill their outer s and p orbitals.

$$H-\overset{\cdot\cdot}{\underset{\underset{H}{|}}{N}}-H \qquad H-\overset{\cdot\cdot}{\underset{\cdot\cdot}{O}}-H \qquad H-\overset{\cdot\cdot}{\underset{\cdot\cdot}{F}}:$$

Ammonia   Water   Hydrogen fluoride

One pair of atoms may share more than one pair of electrons. For example, a **double bond** forms when two pairs of electrons are shared between two atoms, as between the carbon and oxygen atoms in CH_2O (formaldehyde) and between the two carbon atoms in C_2H_4 (ethylene).

$$\overset{H}{\underset{H}{>}}C::\overset{\cdot\cdot}{O}: \quad \text{or} \quad \overset{H}{\underset{H}{>}}C=\overset{\cdot\cdot}{O}: \qquad\qquad \overset{H}{\underset{H}{>}}C::C\overset{H}{\underset{H}{<}} \quad \text{or} \quad \overset{H}{\underset{H}{>}}C=C\overset{H}{\underset{H}{<}}$$

Formaldehyde   Ethylene

A **triple bond** forms when three electron pairs are shared by two atoms, as in CO (carbon monoxide) and in the cyanide ion (CN^-).

$$:C:::O: \quad \text{or} \quad :C\equiv O: \qquad\qquad :C:::N:^- \quad \text{or} \quad :C\equiv N:^-$$

Carbon monoxide   Cyanide ion

Under normal conditions, the ability to form double or triple bonds is limited almost exclusively to bonds between carbon, nitrogen, and oxygen atoms. For example, the element nitrogen forms the N_2 molecule, which contains a triple bond (Fig. 6.7 A), whereas the element phosphorus (also in Group 5A) forms the P_4 molecule, which contains only single bonds (Fig. 6.7 B). Phosphorus, sulfur, and selenium sometimes form double bonds with carbon, nitrogen, and oxygen.

The atoms in polyatomic ions, such as OH^-, NO_3^-, and NH_4^+ (Section 2.5), are held together by covalent bonds. Thus compounds that contain polyatomic ions are stabilized by both covalent bonds and ionic bonds. For example, potassium nitrate, KNO_3, contains the K^+ cation and the polyatomic NO_3^- anion. Potassium nitrate has ionic bonds resulting from the electrostatic attraction between the ions K^+ and NO_3^- and covalent bonds between the nitrogen and oxygen atoms in NO_3^-.

Figure 6.7

(A) Elemental nitrogen, N_2, forms a triple bond. (B) Elemental phosphorus, P_4, which lies directly below nitrogen in Group 5A in the periodic table, forms single bonds.

$$:N \equiv N:$$

(A)

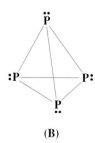

(B)

6.7 Writing Lewis Structures: Covalently Bonded Atoms with a Noble Gas Configuration

Sometimes we can write the Lewis structure for a molecule or ion by pairing up the unpaired electrons on the reactant atoms involved.

$$H \cdot + : \overset{\cdot\cdot}{\underset{\cdot\cdot}{Br}} \cdot \longrightarrow H : \overset{\cdot\cdot}{\underset{\cdot\cdot}{Br}} :$$

$$2H \cdot + : \overset{\cdot\cdot}{\underset{\cdot}{S}} \cdot \longrightarrow H : \overset{\cdot\cdot}{\underset{\cdot\cdot}{S}} :$$
$$ H$$

$$\cdot \overset{\cdot\cdot}{\underset{\cdot}{N}} + \cdot \overset{\cdot\cdot}{\underset{\cdot}{N}} \cdot \longrightarrow : N ::: N :$$

In other cases, when a molecule or ion is more complicated, it is helpful to follow the general procedure outlined below. Let us determine the Lewis structures of SiH_4, $PO_2F_2^-$, and NO^+ as examples in following this procedure.

Step 1.   *Determine the total number of valence (outer-shell) electrons in the molecule or ion.*

   a.   For a *molecule,* we add the number of valence electrons on each atom in the molecule.

   SiH_4:

1 Si atom =	4
4 H atoms =	4
Number of valence electrons =	8

   b.   For a *negative ion,* such as $PO_2F_2^-$, we add the number of valence electrons on the atoms in the ion and the number of negative charges on the ion (one electron is gained for each single negative charge).

   $PO_2F_2^-$:

1 P atom =	5
2 O atoms =	12
2 F atoms =	14
Add for negative charges =	1
Number of valence electrons =	32

   c.   For a *positive ion,* such as NO^+, we add the number of valence electrons on the atoms in the ion and then subtract the number of positive charges on the ion (one electron is lost for each single positive charge).

   NO^+:

1 N atom =	5
1 O atom =	6
Subtract for positive charges =	−1
Number of valence electrons =	10

Step 2.   *Draw a skeleton structure of the molecule or ion, showing the arrangement of atoms, and connect each atom to another with a single bond.*

$$\begin{array}{c} H \\ | \\ H - Si - H \\ | \\ H \end{array} \qquad \left[\begin{array}{c} F \\ | \\ F - P - O \\ | \\ O \end{array} \right]^- \qquad \left[N - O \right]^+$$

When several arrangements of atoms are possible, as for $PO_2F_2^-$, we must use experimental evidence to choose the correct one. As a rule, however, the less electronegative element is the central atom (for example, P in

$PO_2F_2^-$ and PCl_5, S in SF_6, and Cl in ClO_4^-), with the exception that hydrogen cannot be a central atom. As the most electronegative element, fluorine cannot be a central atom either.

Step 3.   *Deduct the 2 valence electrons for each bond written in Step 1.*

$$SiH_4 \quad 8 - 8 = 0$$
$$PO_2F_2^- \quad 32 - 8 = 24$$
$$NO^+ \quad 10 - 2 = 8$$

Step 4.   *Distribute the remaining electrons as unshared pairs so that each atom (except hydrogen) has 8 electrons if possible.* If there are too few electrons to give each atom 8 electrons, we convert single bonds to multiple bonds, where possible. However, we should remember that among the representative elements, the ability to form multiple bonds is limited almost exclusively to bonds between carbon, nitrogen, and oxygen, although phosphorus, sulfur, and selenium sometimes form double bonds with carbon, nitrogen, and oxygen.

SiH_4:   We have distributed the 8 valence electrons as 4 electron pairs in 4 single bonds; no electrons are left to form unshared pairs.

$PO_2F_2^-$:   We have distributed 8 of the 32 valence electrons as 4 electron pairs in 4 single bonds, leaving 24 electrons to be distributed as 3 unshared pairs around each of the fluorine and oxygen atoms.

NO^+:   In this ion, we used 2 of the 10 valence electrons in the single bond in the skeleton structure, leaving 8 to be distributed. We could distribute these 8 electrons as 4 unshared pairs, but this would leave at least 1 atom with fewer than 8 electrons. We can fill the valence shells about the nitrogen and oxygen atoms only if we distribute the 10 electrons as 3 electron pairs in a triple bond, 1 unshared pair on the nitrogen atom, and 1 unshared pair on the oxygen atom.

Figure 6.8

The atmosphere of Jupiter contains molecules not normally found in the earth's atmosphere.

$$H—Si—H$$ with H above and H below Si

$$\begin{bmatrix} :\ddot{F}: \\ | \\ :\ddot{F}—P—\ddot{O}: \\ | \\ :\ddot{O}: \end{bmatrix}^-$$

$$[:N≡O:]^+$$

As we will see in the next section, upon occasion a central atom may have more (or fewer) than 8 electrons in its valence shell; however, the outer atoms will contain 8 electrons (2 electrons for H atoms) in their valence shells.

Example 6.5   **Writing Lewis Structures: Central Atoms with Filled *s* and *p* Valence Orbitals**

The Galileo space craft detected small amounts of the compounds methane, CH_4, acetylene, C_2H_2, and phosphine, PH_3, in Jupiter's atmosphere (Fig. 6.8). What are the Lewis structures of these molecules?

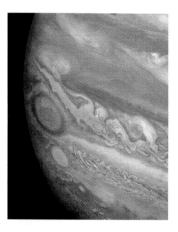

Solution:

Step 1.   *Calculate the number of valence electrons.*

$$CH_4 \quad 4 + (4 \times 1) = 8$$
$$C_2H_2 \quad (2 \times 4) + (2 \times 1) = 10$$
$$PH_3 \quad 5 + (3 \times 1) = 8$$

Step 2.   *Draw a skeleton and connect the atoms with single bonds.* Remember that H is never a central atom. This leads us to the HCCH skeleton.

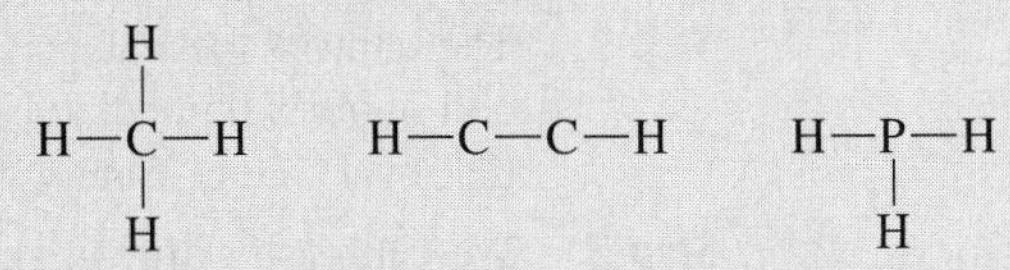

$$
\begin{array}{ccc}
\quad\;\; H & & \\
\quad\;\; | & & \\
H-C-H \quad & H-C-C-H \quad & H-P-H \\
\quad\;\; | & & \quad\; | \\
\quad\;\; H & & \quad\; H \\
\end{array}
$$

Step 3.   *Deduct two electrons per bond drawn in the skeleton.*

$$CH_4 \quad 8 - 8 = 0$$
$$C_2H_2 \quad 10 - 6 = 4$$
$$PH_3 \quad 8 - 6 = 2$$

Step 4.   *Distribute the remaining electrons in such a way as to give each atom a noble gas electron configuration.*
We find that the CH_4 molecule has no electrons to distribute and that both carbon and hydrogen have a noble gas configuration: The structure is complete. The C_2H_2 molecule has four remaining electrons, so a carbon–carbon triple bond must be formed in order for both carbon and hydrogen to have noble gas configurations. There are two valence electrons remaining in the PH_3 molecule. In order for P to have a noble gas configuration, the two electrons must appear on P as an unshared pair.

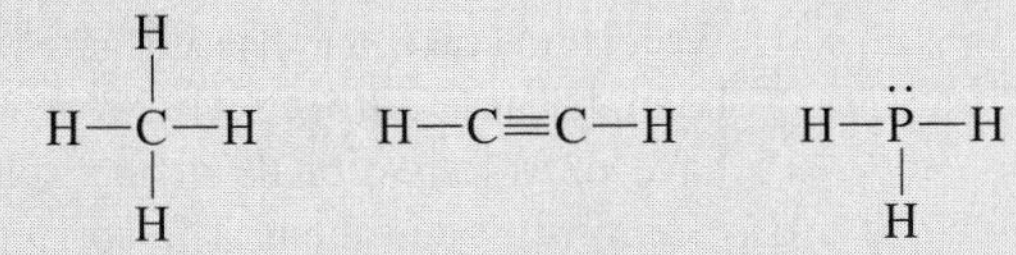

$$
\begin{array}{ccc}
\quad\;\; H & & \quad\;\; .. \\
\quad\;\; | & & \quad\;\; | \\
H-C-H \quad & H-C\equiv C-H \quad & H-P-H \\
\quad\;\; | & & \quad\; | \\
\quad\;\; H & & \quad\; H \\
\end{array}
$$

Check your learning: Both carbon monoxide, CO, and carbon dioxide, CO_2, are products of the combustion of fossil fuels. Both of these gases also cause problems—CO is toxic ad CO_2 has been implicated in the "greenhouse effect" and in global warming. What are the Lewis structures of these two molecules?

Answer: :C≡O: ; :Ö=C=Ö:

6.8   Writing Lewis Structures: Covalently Bonded Atoms without a Noble Gas Configuration

Many covalent molecules have central atoms that do not have a noble gas configuration in their Lewis structures. These molecules fall into three categories: (1) molecules that have an odd number of valence electrons and thus have an unpaired electron, (2) molecules whose central atom has fewer electrons than needed for a noble gas configuration, and (3) molecules whose central atom has more electrons than needed for a noble gas configuration.

Odd-Electron Molecules.

We call any molecule that contains one or more unpaired electrons **paramagnetic;** a molecule that has all electrons paired is called **diamagnetic.** We will discuss these terms and their meanings more in Chapters 22 and 23.

Nitric oxide, NO, is an example of an odd-electron molecule. This molecule is produced in internal combustion engines when oxygen and nitrogen react at high temperatures. To draw the Lewis structure of NO, we follow the steps outlined in Section 6.7.

Step 1.   The sum of the valence electrons is 5 (from N) + 6 (from O) = 11. The odd number immediately tells us that we have a paramagnetic molecule and that every atom cannot have eight electrons in its valence shell.

Step 2.   We can easily draw a skeleton with a nitrogen–oxygen single bond.

$$N—O$$

Step 3.   Deduct two electrons per bond (11 − 2 = 9).

Step 4.   Distribute the remaining electrons.

We give eight electrons to the more electronegative atom in these situations; thus oxygen has the filled valence shell.

$$:\dot{N}—\ddot{\underset{..}{O}}:$$

Nitrogen has only five electrons around it. To move closer to a filled shell for nitrogen, we use an unshared pair of electrons on oxygen to form a nitrogen–oxygen double bond.

$$:\dot{N}=\ddot{\underset{..}{O}}:$$

We cannot take another unshared pair of electrons from oxygen and form a triple bond, because nitrogen would then have nine electrons. Elements in the second period have only four orbitals in the valence shell and thus cannot have more than eight electrons around them. The correct Lewis structure for NO is that shown above.

Electron-Deficient Molecules.

We will encounter a few molecules that contain central atoms that do not have a filled valence shell. Generally, these will be molecules whose central atoms are Group 2A and Group 3A atoms and whose outer atoms are hydrogen or other atoms that do not readily form multiple bonds. For example, in the Lewis structures of beryllium dihydride, BeH$_2$, and boron trichloride, BCl$_3$, the beryllium and boron atoms have around them only four and six electrons, respectively, because they cannot pick up additional electrons by forming multiple bonds.

$$H—Be—H \qquad \overset{\displaystyle :\ddot{Cl}:}{\underset{\displaystyle :\ddot{Cl} \quad \ddot{Cl}:}{\overset{|}{B}}}$$

An atom like the boron atom in BCl$_3$, which does not have filled outer *s* and *p* orbitals, is very reactive. It readily combines with a molecule containing an atom that has an unshared pair of electrons. For example, NH$_3$ reacts with BCl$_3$ because the unshared pair on nitrogen can be shared with the boron atom.

$$:\overset{..}{\underset{..}{Cl}}: \qquad H$$

We call the bond formed when one of the atoms in the bond provides both bonding electrons (like the B—N bond in Cl_3BNH_3) a **coordinate covalent bond.** A coordinate covalent bond is formed, for example, when a water molecule combines with a hydrogen ion to form a hydronium ion and when an ammonia molecule combines with a hydrogen ion to form an ammonium ion. The equations are

Molecules with Extra Electrons.

We have seen that elements in the second period of the periodic table ($n = 2$) can accommodate only 8 electrons in their valence shell orbitals. However, elements in the third and higher periods ($n > 2$) have more than 4 valence orbitals and can share more than 4 pairs of electrons with other atoms. For example, in the Lewis structure of PCl_5, phosphorus shares 5 pairs of electrons, and in the Lewis structure of SF_6, sulfur shares 6 pairs of electrons.

In some molecules, such as IF_5 and XeF_4, some of the electrons in the outer shell of the central atom are unshared pairs.

When we write the Lewis structures for molecules such as XeF_4, we find that we have electrons left over after filling the valence shells of the outer atoms with eight electrons. These additional electrons are assigned to the central atom.

Example 6.6   Writing Lewis Structures: Central Atoms That Do Not Have Eight Electrons

Xenon is the only noble gas that forms a number of compounds (we examined XeF_4 above). What are the Lewis structures of XeF_2 and XeF_6?

Solution: We can draw the Lewis structure of any covalent molecule by following the four steps discussed above.

Step 1. *Calculate the number of valence electrons.*

$$XeF_2 \qquad 8 + (2 \times 7) = 22$$
$$XeF_6 \qquad 8 + (6 \times 7) = 50$$

Step 2. *Draw a skeleton joining the atoms by single bonds.* Xenon will be the central atom because fluorine cannot be a central atom.

$$F-Xe-F \qquad \begin{matrix} & F & F \\ & | & / \\ F- & Xe & -F \\ & / & | \\ & F & F \end{matrix}$$

Step 3. *Deduct 2 electrons for each bond drawn.*

$$XeF_2 \qquad 22 - 4 = 18$$
$$XeF_6 \qquad 50 - 12 = 38$$

Step 4. *Distribute the remaining electrons.*

XeF_2: We place 3 unshared pairs of electrons around each F atom, which accounts for 12 electrons and gives each F atom 8 electrons. Thus 6 electrons (3 unshared pairs) remain. These unshared pairs must be placed on the xenon atom. This is acceptable because Xe atoms have empty valence shell *d* orbitals and can accommodate more than 8 electrons. The Lewis structure of XeF_2 shows 2 bonding pairs and 3 unshared pairs of electrons around the Xe atom.

$$:\!\ddot{F}\!-\!\ddot{X}\!e\!-\!\ddot{F}\!:$$

XeF_6: We place 3 unshared pairs of electrons around each F atom, which accounts for 36 electrons. Two electrons remain and are placed on the Xe atom as an unshared pair.

$$\begin{matrix} :\ddot{F}: & :\ddot{F}: \\ | & / \\ :\ddot{F}-\ddot{X}e-\ddot{F}: \\ / & | \\ :\ddot{F}: & :\ddot{F}: \end{matrix}$$

Check your learning: The halogens form an interesting class of compounds called the interhalogens in which halogen atoms covalently bond to each other. Write the Lewis structures for $BrCl_3$ and $ICl_4{}^-$.

Answer:

$$:\!\ddot{C}\!l\!-\!\ddot{B}\!r\!-\!\ddot{C}\!l\!: \qquad \left[\begin{matrix} & :\ddot{C}l: \\ & | \\ :\ddot{C}l-I-\ddot{C}l: \\ & / \\ & :\ddot{C}l: \end{matrix}\right]^-$$
$$\qquad\quad :\ddot{C}l:$$

6.9   Formal Charge

Determination of the arrangement of atoms in a molecule requires experimental study. However, a concept called formal charge can help us predict the most reasonable arrangement of atoms and write the Lewis structure. The **formal charge** of an atom in a molecule is the *hypothetical* charge the atom would have if we could redistribute the electrons in the bonds evenly between the atoms in the bond. We can calculate formal charge using valence electrons as follows:

Step 1.   *Assign half of the electrons in a bond to one of the atoms in the bond and the other half of the electrons to the other atom.*

Step 2.   *Assign both electrons of an unshared pair to the atom to which the unshared pair belongs.*

Step 3.   *Determine the formal charge of each atom by subtracting the number of valence electrons assigned to the atom in Steps 1 and 2 from the number of valence electrons in the neutral atom.* As a check of our calculation, we can determine the sum of the formal charges: The sum of the formal charges of all atoms in a molecule should be zero; the sum of the formal charges in an ion should equal the charge on the ion.

It is very important to remember that the formal charge calculated for an atom is not the charge on the atom in the molecule. Formal charge is only a useful "bookkeeping" procedure that is helpful in predicting atomic arrangements and checking the validity of a Lewis structure.

Example 6.7   Calculating Formal Charge from Lewis Structures

Assign formal charges to the two interhalogens described in Example 6.6, Check your learning.

Solution: $BrCl_3$:

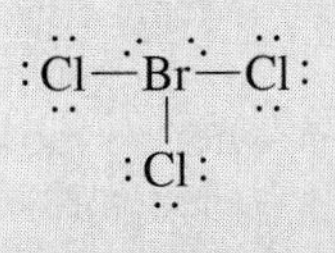

$$:\ddot{C}l-\ddot{B}r-\ddot{C}l:$$
$$|$$
$$:\ddot{C}l:$$

Step 1.   We assign one of the electrons in each Br—Cl bond to the Br atom and one to the Cl atom in that bond.

Step 2.   We assign the unshared pairs to their atom. Now each Cl atom has 7 valence electrons and the Br atom has 7 valence electrons.

Step 3.   We subtract the number of electrons assigned to each atom in Steps 1 and 2 from the number of valence electrons for the neutral atom (that is, from the number of electrons we assign to an atom when we construct the Lewis structure). This gives the formal charge.

$$\text{Br} \quad 7-7=0 \qquad \text{Cl} \quad 7-7=0$$

All atoms in $BrCl_3$ have a formal charge of zero, and the sum of the formal charges equals zero, as it must in a neutral molecule.

ICl_4^-:

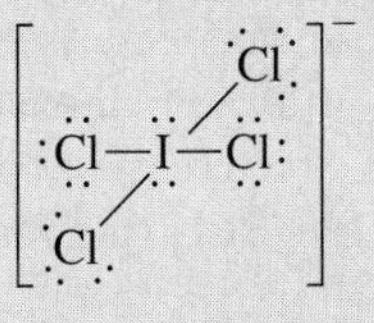

$$\left[\begin{array}{c} \ddot{C}l: \\ / \\ :\ddot{C}l-I-\ddot{C}l: \\ \ddot{C}l: \end{array} \right]^-$$

Step 1.   We divide the bonding electrons equally in all I—Cl bonds.

Step 2.   We assign unshared pairs of electrons to their atoms. Each Cl atom now has 7 electrons assigned to it, and the I atom has 8 valence electrons assigned to it.

Step 3.   Subtracting the assigned electrons from the number of valence electrons gives the following formal charges for these atoms:

$$\text{I} \quad 7 - 8 = -1 \qquad \text{Cl} \quad 7 - 7 = 0$$

The sum of the formal charges on all the atoms equals -1, which is identical to the charge on the ion (-1).

Check your learning: Ammonium nitrate is a common fertilizer. Calculate the formal charge on atoms in the ammonium cation, $NH_4{}^+$.

Answer: N = +1; H = 0

In many cases we can draw more than one structure for a molecule, with different arrangements of atoms. (The arrangement of atoms in a molecule or ion is called its **molecular structure**.) For example, the thiocyanate ion, an ion formed from a carbon atom, a nitrogen atom, and a sulfur atom, could have any of three different molecular structures: CNS^-, NCS^-, or CSN^-. The formal charges present in each of these molecular structures can help us pick the most likely arrangement of atoms.

We can use the following guidelines involving formal charge to decide between two or more possible structures for a molecule or ion.

1. A molecular structure that gives a Lewis structure in which all formal charges are zero is preferable to one in which some formal charges are not zero.

2. If the Lewis structure must have nonzero formal charges, the structure with the lowest number of nonzero formal charges is preferable, and a Lewis structure with no more than one large formal charge (-2, $+2$, and so on) is preferable to a structure with several large formal charges.

3. Lewis structures should have adjacent formal charges of zero or of opposite sign.

4. When we must choose among several Lewis structures with similar distributions of formal charges, the structure with the negative formal charges on the more electronegative atoms is preferable.

Possible Lewis structures and the corresponding formal charges for each of the three possible structures for the thiocyanite ion are

Structure	$[:\!N\!=\!C\!=\!S:]^-$	$[:\!C\!=\!N\!=\!S:]^-$	$[:\!C\!=\!S\!=\!N:]^-$
Formal Charge	$-1 \quad 0 \quad 0$	$-2 \quad +1 \quad 0$	$-2 \quad +2 \quad -1$

The first arrangement of atoms is preferred because it has the lowest number of atoms with nonzero formal charges (Guideline 2). Note that the sum of the formal charges in each case is equal to the charge on the ion (-1).

A second example shows how formal charge gives us insight into why one Lewis structure may be preferable to another. Consider the following two Lewis structures for BCl_3 (the formal charges are written beside the atoms).

In the Lewis structure with the double bond, the boron atom has a formal charge of -1 and the chlorine atom a formal charge of $+1$. This distribution of formal charge is unsatisfactory, because the negative formal charge is on the least electronegative atom (which violates Guideline 4) and the structure has a larger number of atoms with nonzero formal charge (which violates Guideline 2).

Example 6.8   Using Formal Charge to Determine Molecular Structure

Nitrous oxide (laughing gas), N_2O, is used as an anesthetic in minor surgery, such as the routine extraction of wisdom teeth. Which of the following is more likely to be the molecular structure for nitrous oxide, $:N=N=O:$ or $:N=O=N:$?

Solution: Determining formal charge yields

$$\overset{-1 \quad +1 \quad 0}{:N=N=O:} \qquad \overset{-1 \quad +2 \quad -1}{:N=O=N:}$$

The structure with a terminal oxygen atom, $:N=N=O:$, best satisfies the criteria for the most stable distribution of formal charge; the number of atoms with formal charges is minimized (Guideline 2). Thus we expect the structure of nitrous oxide to be

$$:N=N=O:$$

Check your learning: Which is more likely to be the structure for the NO_2^- ion?

$$\left[:N=O-O:\right]^- \quad \text{or} \quad \left[:O=N-O:\right]^-$$

Answer: ONO^-

6.10   Oxidation State

Like formal charge (Section 6.9), the **oxidation state** (sometimes called the **oxidation number**) of an atom is a bookkeeping concept. We will find oxidation states useful for naming compounds, classifying chemical reactions, writing chemical formulas, and keeping track of electrons that may be redistributed during some chemical reactions. We will discuss oxidation states of the elements in more detail in Chapter 8. At this point, let us consider how to determine oxidation states from Lewis structures.

The oxidation state of an atom in a molecule or in a polyatomic ion is a *hypothetical* charge—the charge the atom would have if the electrons in each bond were located on the more electronegative atom. The oxidation state of the atom that forms a monatomic ion is equal to the charge on the ion. Thus the oxidation states of calcium in the Ca^{2+} ion and of chlorine in the Cl^- ion are $+2$ and -1, respectively.

We can determine oxidation states in molecules or polyatomic ions using valence electrons as follows:

Step 1.   *Assign both of the electrons in a bond to the more electronegative atom in the bond.* (We distribute electrons equally, half to one atom and half to the other atom, when the two atoms are identical, as in a C—C bond.)

Step 2. *Assign both electrons of an unshared pair to the atom to which the unshared pair belongs.*

Step 3. *Determine the oxidation state of each atom by subtracting the number of electrons assigned to the atom in Steps 1 and 2 from the number of valence electrons in the neutral atom.* As a check of our calculation, we can determine the sum of the oxidation states. The sum should be zero for a molecule and should equal the charge of an ion.

These rules are similar to those given in Section 6.9 for calculating formal charge. In fact, the only real difference is in Step 1, in which we assign *both* electrons to the more electronegative atom when calculating oxidation state. For formal charge calculations, bonding electrons are *divided evenly* between the two atoms. We should remember that the oxidation state calculated for an atom is not the charge on that atom. Like a formal charge, oxidation state is a hypothetical charge; it is the charge an atom would have if valence electrons were distributed in accordance with an arbitrary set of rules.

Example 6.9   **Determining Oxidation States from Lewis Structures**

From the Lewis structure, determine the oxidation state of each of the atoms of $BrCl_3$ and ICl_4^-, which we considered in Example 6.7.

Solution: We determined that the best Lewis structures are

Now we determine the oxidation states from these Lewis structures.

Step 1. *Assign both of the electrons in each bond to the more electronegative atom in the bond, the chlorine atom.*

Step 2. *Assign unshared pairs to the atom on which they have been drawn.* We assign three pairs of electrons to each chlorine, two pairs to the bromine, and two pairs to the iodine. This gives the atoms the following numbers of electrons:

$BrCl_3$:	Br	4	ICl_4^-:	I	4
	Cl	8		Cl	8

Step 3. *Subtract the number of electrons on each atom after redistribution from the number of valence electrons in the free atom.*

$BrCl_3$:	Br	$7 - 4 = +3$	ICl_4^-:	I	$7 - 4 = +3$
	Cl	$7 - 8 = -1$		Cl	$7 - 8 = -1$

The sum of the oxidation states must be zero for a neutral molecule and must equal the charge of an ion. In this case, for $BrCl_3$, $3 + (3 \times -1) = 0$, and for ICl_4^-, $3 + (4 \times -1) = -1$. Note that these oxidation state values are quite different from the formal charges of the atoms in these species (Example 6.7).

Check your learning:.   Write the Lewis structure and determine the oxidation state of each of the atoms in the acid HONO and the anion NO_2^-.

Answer: $H-\ddot{O}-\ddot{N}=\ddot{O}:$ $:\ddot{O}=\ddot{N}-\ddot{O}:^-$

Oxidation state: +1 −2 +3 −2   −2 +3 −2

6.11   Resonance

You may have noticed that the nitrite anion, NO_2^-, can have two possible structures in which the atoms are in the same positions. However, the electrons involved in the nitrogen–oxygen double bond are in different positions.

$$\left[:\ddot{O}-\ddot{N}=\ddot{O}:\right]^- \qquad \left[:\ddot{O}=\ddot{N}-\ddot{O}:\right]^-$$

There is an experimental way to test whether one of the bonds in the NO_2^- ion is a double bond and the other a single bond. We can measure nitrogen–oxygen bond lengths by using X-ray diffraction as noted in Section 11.20. A double bond between two atoms is shorter (and stronger) than a single bond between the same two atoms. Such experiments show that both nitrogen–oxygen bonds in NO_2^- have the same strength and length and are identical in all other properties.

It is not possible to write a single Lewis structure for NO_2^- in which nitrogen has an octet and both bonds are equivalent. Instead we use the concept of **resonance:** *If two or more Lewis structures with the same arrangement of atoms can be written for a molecule or ion, then the actual distribution of electrons is an average of that shown by the various Lewis structures.* The actual distribution of electrons in each of the nitrogen–oxygen bonds in NO_2^- is an average of a double bond and a single bond. We call the individual Lewis structures **resonance forms.** The actual electronic structure of the molecule (the average of the resonance forms) is called a **resonance hybrid** of the individual resonance forms. A double-headed arrow is used between Lewis structures to indicate that they are resonance forms. The electronic structure of the NO_2^- ion is shown as

$$\left[:\ddot{O}=\ddot{N}-\ddot{O}:\right]^- \longleftrightarrow \left[:\ddot{O}-\ddot{N}=\ddot{O}:\right]^-$$

We should remember that a molecule or ion described as a resonance hybrid *never* possesses an electronic structure described by a single resonance form. Nor does it fluctuate between possible resonance forms. The actual electronic structure is *always* an average of that shown by all resonance forms. Think of a resonance hybrid as analogous to the mule, which is the hybrid offspring of a male donkey and a female horse. Just as the characteristics of a mule are fixed, so are the properties of a resonance hybrid. The mule is not a donkey part of the time and a horse part of the time. It is always a mule.

The carbonate anion, CO_3^{2-}, provides a second example of resonance.

We must use an unshared pair of electrons from an oxygen atom to fill the valence shell of the central carbon atom. However, all oxygens atoms are equivalent, and we can use an unshared pair from any one of the three atoms. This gives rise to three resonance forms of the carbonate ion. Because we can write three identical resonance structures, we know that the actual arrangement of electrons in the carbonate ion is the average of the three structures. Again, experiments show that all three carbon–oxygen bonds are exactly the same.

6.12   The Strengths of Covalent Bonds

Stable molecules exist because covalent bonds hold the atoms together; energy must be added to break the bonds and separate the atoms. We measure the strength of a covalent bond between two atoms by the energy required to break it—that is, the energy necessary to separate the bonded atoms. The stronger a bond, the more energy required to break it.

The energy required to break a specific covalent bond in exactly 1 mole of gaseous molecules is called the **bond energy** or the **bond dissociation energy** of that bond. The bond energy, D_{X-Y}, for a diatomic molecule is defined as the standard enthalpy change (Section 4.5) for the endothermic reaction.

$$XY(g) \longrightarrow X(g) + Y(g) \qquad D_{X-Y} = \Delta H^\circ_{298}$$

For example, the bond energy of the nonpolar covalent H—H bond, D_{H-H}, is 436 kilojoules per mole of H—H bonds broken:

$$H_2(g) \longrightarrow 2H(g) \qquad D_{H-H} = \Delta H^\circ_{298} = 436 \text{ kJ}$$

We can determine bond energies from enthalpies of formation (ΔH°_f, Section 4.5 and Appendix I). For example, breaking 1 mole of H—H bonds produces 2 moles of H atoms, so the enthalpy change of the reaction (the H—H bond energy) equals twice the heat of formation of 1 mole of H atoms (2×217.97 kJ, or 436 kJ as a rounded value). Bond energies for common diatomic molecules range from 946 kilojoules per mole for N_2 (triple bond) to 150 kilojoules per mole for I_2 (single bond) and from 569 kilojoules per mole for HF to 295 kilojoules per mole for HI.

Molecules with three or more atoms have two or more bonds. The sum of all bond energies in such a molecule is equal to the standard enthalpy change for the endothermic reaction that breaks all the bonds in the molecule. For example, the sum of the four C—H bond energies in CH_4, 1660 kilojoules, is equal to the standard enthalpy change of the reaction

$$\begin{array}{c} \text{H} \\ | \\ \text{H}-\overset{\displaystyle |}{\underset{\displaystyle |}{\text{C}}}-\text{H}(g) \\ | \\ \text{H} \end{array} \longrightarrow \text{C}(g) + 4\text{H}(g) \qquad \Delta H^\circ_{298} = 1660 \text{ kJ}$$

The *average* C—H bond energy, D_{C-H}, is $1660/4 = 415$ kilojoules per mole, because there are 4 moles of C—H bonds broken in the reaction.

The strength of a bond between two atoms increases as the number of electron pairs in the bond increases. Generally, as the bond strength increases, the bond length decreases. Thus we find that triple bonds are stronger and shorter than double bonds between the same two atoms and that double bonds are stronger and shorter than single bonds between the same two atoms. Table 6.2 gives average bond energies for

Table 6.2

Some Average Bond Energies (kJ mol^{-1}).

					Single Bonds						
H	C	N	O	F	Si	P	S	Cl	Br	I	
436	415	390	464	569	395	320	340	432	370	295	H
	345	290	350	439	360	265	260	330	275	240	C
		160	200	270	—	210	—	200	245	—	N
			140	185	370	350	—	205	—	200	O
				160	540	489	285	255	235	—	F
					230	215	225	359	290	215	Si
						215	230	330	270	215	P
							215	250	215	—	S
								243	220	210	Cl
									190	180	Br
										150	I

					Multiple Bonds				
C=C,	611	C=N,	615	C=O,	741	N=N,	418	O=O,	498
C≡C,	837	C≡N,	891	C≡O,	1080	N≡N,	946		

To determine a single-bond energy, find one of the atoms in the bond in the right-hand column and the other atom in the top row. The value of the bond energy is found at the intersection of that column and that row. For example, the average P—O bond energy is 350 kJ mol^{-1}.

some common bonds. A comparison of bond lengths and bond strengths for some common bonds appears in Table 6.3.

We can use bond energies to calculate approximate enthalpy changes in situations where enthalpies of formation are not available. Calculations of this type will also tell us whether a reaction is exothermic or endothermic (Section 4.1). An exothermic reaction (ΔH negative, heat produced) results when the bonds in the products are stronger than the bonds in the reactants. An endothermic reaction (ΔH positive, heat absorbed) results when the bonds in the products are weaker than those in the reactants.

Table 6.3

Average Bond Lengths and Bond Energies for Some Common Bonds.

Bond	Bond Length, Å	Bond Energy, kJ mol^{-1}
C—C	1.54	345
C=C	1.34	611
C≡C	1.20	837
C—N	1.43	290
C=N	1.38	615
C≡N	1.16	891
C—O	1.43	350
C=O	1.23	741
C≡O	1.13	1080

The approximate enthalpy change, ΔH, for a chemical reaction is equal to the sum of the energy required to break all bonds in the reactants (energy "in," positive sign) and the energy released when all bonds are formed in the products (energy "out," negative sign). This can be expressed mathematically in the following way:

$$\Delta H = \Sigma D_{\text{bonds broken}} - \Sigma D_{\text{bonds formed}}$$

In this expression, the symbol Σ means "the sum of" and D represents the bond energy in kilojoules per mole, which is always a positive number. The bond energy is obtained from a table (such as Table 6.3) and depends on whether the particular bond is a single, double, or triple bond. Thus, in calculating enthalpies in this manner, it is important that we consider the bonding in all reactants and products.

Consider the following reaction:

$$H_2(g) \ + \ Cl_2(g) \ \longrightarrow \ 2HCl(g)$$

or

$$H{-}H(g) + Cl{-}Cl(g) \longrightarrow 2 \ H{-}Cl(g)$$

To form the 2 moles of HCl, 1 mole of H—H bonds and 1 mole of Cl—Cl bonds must be broken. This requires the input of 679 kJ, the sum of the bond energy of the H—H bond (436 kJ mol^{-1}) and that of the Cl—Cl bond (243 kJ mol^{-1}). During the reaction, 2 moles of H—Cl bonds are formed (bond energy = 432 kJ mol^{-1}), releasing 2×432 kilojoules, or 864 kilojoules. Because the bonds in the products are stronger than those in the reactants by 185 kilojoules, the reaction releases 185 kilojoules more energy than it consumes. This excess energy is released as heat, so the reaction is exothermic. Appendix I gives a value of -92.307 kilojoules per mole for the standard molar enthalpy of formation, ΔH_f°. Twice that value is -184.6 kilojoules, which agrees well with the answer we obtained above for the formation of 2 moles of HCl.

If we use Equation 1, we get the same results.

$$\Delta H = \Sigma D_{\text{bonds broken}} - \Sigma D_{\text{bonds formed}}$$
$$\Delta H = \Sigma[D_{\text{H}-\text{H}} + D_{\text{Cl}-\text{Cl}}] - \Sigma 2D_{\text{H}-\text{Cl}}$$
$$= [436 + 243] - 2(432) = -185 \text{ kJ}$$

Example 6.10   Using Bond Energies to Calculate Approximate Enthalpy Changes

Some people believe that methanol, CH_3OH, may be an excellent alternative fuel. The high-temperature reaction of steam and carbon produces a mixture of the gases carbon monoxide, CO, and hydrogen, H_2, from which methanol can be produced. Using the bond energies in Table 6.2, calculate the approximate enthalpy change, ΔH, for the reaction

$$CO(g) + 2H_2(g) \longrightarrow CH_3OH(g)$$

Solution: First, we need to write Lewis structures of the reactants and the product.

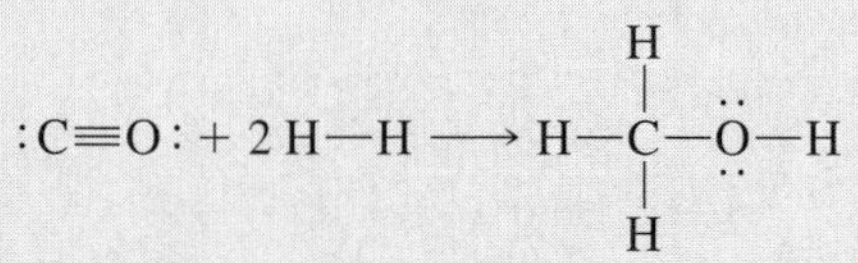

$$:C\equiv O: + 2\,H{-}H \longrightarrow \underset{\displaystyle H}{\overset{\displaystyle H}{H{-}\overset{|}{\underset{|}{C}}{-}\ddot{\underset{\cdot\cdot}{O}}{-}H}}$$

Thus we see that ΔH for this reaction involves the energy required to break a carbon–oxygen triple bond and two hydrogen–hydrogen single bonds and the energy produced by formation of three carbon–hydrogen single bonds, a carbon–oxygen single bond, and an oxygen–hydrogen single bond. We have the following expression:

$$\Delta H = \Sigma D_{\text{bonds broken}} - \Sigma D_{\text{bonds formed}}$$

$$\Delta H = [D_{C\equiv O} + 2(D_{H-H})] - [3(D_{C-H}) + D_{C-O} + D_{O-H}]$$

Using the bond energy values in Table 6.3, we obtain

$$\Delta H = [1080 \text{ kJ} + 2(436 \text{ kJ})] - [3(415 \text{ kJ}) + 350 \text{ kJ} + 464 \text{ kJ}]$$

$$= -107 \text{ kJ}$$

Check your learning: Ethyl alcohol, CH_3CH_2OH, is one of the oldest organic chemicals produced by humans. It has many uses in industry, and it is the alcohol of alcoholic beverages. It can be obtained by the fermentation of sugar or synthesized by the hydration of ethylene in the following reaction:

Using the bond energies in Table 6.2, calculate an approximate enthalpy change, ΔH, for this reaction.

Answer: -35 kJ

For Review   Summary

When atoms react, their electronic structures change and chemical bonds form. **Ionic bonds** result from electrostatic attractions between **ions** of opposite charge. **Ionic compounds** consist of regular three-dimensional arrangements of ions held together by these strong electrostatic attractions. The charges of cations formed by the representative metals can be determined readily, because with few exceptions, the electronic structures of these ions have a **noble gas configuration** or a **pseudo–noble gas configuration.** The charges of anions formed by the nonmetals can be determined readily, because these ions form when nonmetal atoms pick up enough electrons to fill their valence shells.

Covalent bonds result when electrons are shared as electron pairs between atoms, and the electrons are attracted by the nuclei of both atoms. Pairs of electrons need not be shared equally; they may spend more time near one atom than near the other in **polar covalent bonds.** The ability of an atom to attract a pair of electrons in a chemical bond is called its **electronegativity.** The

greater the difference in electronegativity between two atoms in a covalent bond, the more polar the bond.

When only one atom furnishes both electrons of an electron-pair bond, the bond is called a **coordinate covalent bond.** A **single bond** results when one pair of electrons is shared between two atoms. The sharing of two pairs of electrons between two atoms gives a **double bond;** the sharing of three pairs of electrons gives a **triple bond.** The distribution of valence electrons in bonds and **unshared pairs** in a molecule can be indicated with a **Lewis structure.**

Lewis structures can be written in a series of steps by starting with a skeleton structure consisting of single bonds and then distributing the remaining electrons as unshared pairs or in multiple bonds to give all atoms a noble gas configuration, if possible. In some instances, a central atom may have an empty valence orbital. Also, in other cases, a central atom may have more than eight electrons in its valence shell if it is an atom for which the principle quantum number of the valence shell, n, is greater than or equal to 3. The hypothetical **formal charges** of the atoms in a Lewis structure are a guide to determining the correct Lewis structure. The **oxidation states** of atoms can also be determined from Lewis structures.

If two or more Lewis structures with identical arrangements of atoms but different distributions of electrons can be written, the actual distribution of electrons is an average of the distributions indicated by the individual Lewis structures. The actual electron distribution is called a **resonance hybrid** of the individual Lewis structures (which are called **resonance forms**).

The strength of a covalent bond is measured by its **bond energy,** the enthalpy change of the reaction that breaks that particular bond in a mole of molecules.

Key Terms and Concepts

anion   (6.1)
bond energy   (6.12)
cation   (6.1)
coordinate covalent bond   (6.8)
covalent bond   (6.4)
diamagnetic   (6.8)
double bond   (6.6)
electronegativity   (6.5)
formal charge   (6.9)
inert-pair effect   (6.2)

ionic bond   (6.1)
ionic compound   (6.1)
lattice energy   (6.3)
Lewis structure   (6.6)
Lewis symbol   (6.6)
lone pairs   (6.6)
noble gas electron configuration   (6.2)
oxidation state   (6.10)
paramagnetic   (6.8)

polar covalent bond   (6.5)
pseudo–noble gas electron
   configuration   (6.2)
resonance   (6.11)
resonance forms   (6.11)
resonance hybrid   (6.11)
single bond   (6.6)
triple bond   (6.6)
unshared pairs   (6.6)

Exercises

Questions

1. Identify each of the following as part of chemistry's macroscopic domain, microscopic domain, or symbolic domain: (a) bond energy in $kJ\ mol^{-1}$, (b) cation, (c) covalent compound, (d) electron configuration, (e) formal charge, (f) ionic charge, (g) ionic compound, (h) Lewis structure, (i) lattice energy in $kJ\ mol^{-1}$. If an item is part of the symbolic domain, identify whether it describes a macroscopic property, a microscopic property, neither, or both.

2. Why does a positive ion have a positive charge?

3. Why does a nitrate ion have a negative charge?

4. Iron(III) sulfate is composed of Fe^{3+} and $SO_4{}^{2-}$ ions. Explain why a sample of iron(III) sulfate is uncharged.

5. Why is it incorrect to speak of the molecular weight of NaCl?

6. What information can you use to predict whether a bond between two atoms is covalent or ionic?

7. Predict which of the following compounds are ionic and which are covalent on the basis of the location of their constituent atoms in the periodic table? (a) Cl_2CO, (b) MnO, (c) NCl_3, (d) $CoBr_2$, (e) K_2S, (f) CO, (g) CaF_2, (h) HI, (i) CaO, (j) IBr, (k) CO_2.

8. Explain the difference between a covalent bond and an ionic bond.

9. Correct the following statement: "The bonds in solid $PbCl_2$ are ionic; the bond in a HCl molecule is covalent. Thus all of the valence electrons in $PbCl_2$ are located on the Cl^- ions, and all of the valence electrons in an HCl molecule are shared between the H and Cl atoms."

10. How are single, double, and triple bonds similar? How do they differ?

11. Why are the following two Lewis structures not resonance forms of the SCN^- ion?

$$\left[\,:\!\overset{..}{S}\!=\!C\!=\!\overset{..}{N}\!:\,\right]^{-}\qquad\left[\,:\!C\!\equiv\!N\!-\!\overset{..}{\underset{..}{S}}\!:\,\right]^{-}$$

12. What property of a covalent bond depends on the electronegativities of the atoms involved in the bond?

13. What information can you use to predict whether a bond is polar or nonpolar?

14. What is the difference between the hypothetical distributions of electrons used to calculate oxidation number and to calculate formal charge?

15. In what class of substances—elements or compounds—are the formal charges and the oxidation numbers of the atoms always equal?

16. Explain why fluorine always has an oxidation state of -1 in compounds.

17. Explain why oxygen has an oxidation state of -1 in hydrogen peroxide, H_2O_2.

18. Explain why hydrogen has an oxidation state of $+1$ in compounds with more electronegative elements.

Ions and Ionic Bonding

19. Which of the following atoms would be expected to form negative ions in binary ionic compounds and which would be expected to form positive ions? P, I, Mg, Cl, In, Cs, O, Pb, Co.

20. Which of the following atoms would be expected to form negative ions in binary ionic compounds and which would be expected to form positive ions? Br, Ca, Na, N, F, Al, Sn, S, Cd.

21. Predict the charge on the monatomic ions formed from the following atoms in binary ionic compounds: (a) P, (b) Mg, (c) Al, (d) O, (e) Cl, (f) Cs.

22. Predict the charge on the monatomic ions formed from the following atoms in binary ionic compounds: (a) I, (b) Sr, (c) K, (d) N, (e) S, (f) In.

23. Write the electron configuration of each of the following ions, and identify them as having (i) a noble gas electron configuration, (ii) a pseudo–noble gas electron configuration, or (iii) neither (i) nor (ii): (a) As^{3-}, (b) I^-, (c) Be^{2+}, (d) Cd^{2+}, (e) O^{2-}, (f) Ga^{3+}, (g) Li^+, (h) N^{3-}, (i) Sn^{2+}, (j) Co^{2+}.

24. Write the electron configuration for the monatomic ions formed from the following elements (the elements that form the greatest concentration of monatomic ions in seawater), and identify them as having (i) a noble gas electron configuration, (ii) a pseudo–noble gas electron configuration, or (iii) neither (i) nor (ii): (a) Cl, (b) Na, (c) Mg, (d) Ca, (e) K, (f) Br, (g) Sr, (h) F.

25. Write the Lewis symbols for each of the following ions: (a) As^{3-}, (b) I^-, (c) Be^{2+}, (d) O^{2-}, (e) Ga^{3+}, (f) Li^+, (g) N^{3-}, (h) Sn^{2+}.

26. Write the Lewis symbols for the monatomic ions formed from the following elements, the elements that form the greatest concentration of monatomic ions in seawater: (a) Cl, (b) Na, (c) Mg, (d) Ca, (e) K, (f) Br, (g) Sr, (h) F.

27. Write the electron configuration and the Lewis symbols for each of the following atoms and for the monatomic ion found in binary ionic compounds containing the element: (a) Al, (b) Br, (c) Sr, (d) Li, (e) As, (f) S.

28. Write the Lewis symbol of the ions in each of the following ionic compounds and the Lewis symbol of the atom from which it formed: (a) MgS, (b) Al_2O_3, (c) $GaCl_3$, (d) K_2O, (e) Li_3N, (f) KF.

29. M and X in the Lewis structures listed below represent various elements in the third period of the periodic table. Write the formula of each compound using the chemical symbols of each element.

   (a) $\left[M^{2+}\right]\left[:\overset{..}{\underset{..}{X}}:\right]^{2-}$ (b) $\left[M^{3+}\right]\left[:\overset{..}{\underset{..}{X}}:\right]^{-}_{3}$

   (c) $\left[M^{+}\right]_2\left[:\overset{..}{\underset{..}{X}}:\right]^{2-}$ (d) $\left[M^{3+}\right]_2\left[:\overset{..}{\underset{..}{X}}:\right]^{2-}_{3}$

30. The lattice energy of LiF is 1023 kJ mol^{-1}, and the interionic distance is 2.01 Å. MgO crystallizes in the same structure as LiF but with an interionic distance of 2.05 Å. Which of the following values most closely approximates the lattice energy of MgO: 255, 890, 1023, 2046, or 4008 kJ mol^{-1}? Explain your choice.

31. Which compound in each of the following pairs has the larger lattice energy? (*Note:* Mg^{2+} and Li^+ have similar radii, and O^{2-} and F^- have similar radii.) Explain your choice.

   (a) MgO or MgSe   (b) LiF or MgO
   (c) Li_2O or LiCl   (d) Li_2Se or MgO

32. Which compound in each of the following pairs has the larger lattice energy? (*Note:* Ba^{2+} and K^+ have similar radii, and S^{2-} and Cl^- have similar radii.) Explain your choice.

   (a) K_2O or Na_2O (b) K_2S or BaS
   (c) KCl or BaS   (d) BaS or $BaCl_2$

33. For which of the following compounds is the most energy required to convert 1 mole of the solid into separated ions: MgO, SrO, KF, CsF, or MgF_2?

34. For which of the following compounds is the most energy required to convert 1 mole of the solid into separated ions: K_2S, K_2O, CaS, Cs_2S, or CaO?

35. The lattice energy of KF is 794 kJ mol^{-1}, and the interionic distance is 2.69 Å. The interionic distance in NaF, which has the same structure as KF, is 2.31 Å. Is the lattice energy of NaF about 682, 794, 924, 1588, or 3175 kJ mol^{-1}? Explain your answer.

Covalent Bonding

36. Write the Lewis structure for the diatomic molecule P_2, an unstable form of phosphorus found in high-temperature phosphorus vapor.

37. Write Lewis structures for the following: (a) H_2, (b) HBr, (c) PCl_3, (d) SF_2, (e) H_2CCH_2, (f) HNNH, (g) H_2CNH, (h) NO^-, (i) N_2, (j) CO, (k) CN^-.

38. Write Lewis structures for the following: (a) O_2, (b) H_2CO, (c) AsF_3, (d) ClNO, (e) $SiCl_4$, (f) H_3O^+, (g) NH_4^+, (h) BF_4^-, (i) HCCH, (j) ClCN, (k) C_2^{2+}.

39. Write Lewis structures for the following: (a) ClF_3, (b) PCl_5, (c) BF_3, (d) PF_6^-.

40. Write Lewis structures for the following: (a) SeF_6, (b) XeF_4, (c) $SeCl_3^+$, (d) Cl_2BBCl_2 (contains a B—B bond).

41. Write Lewis structures for the following: (a) PO_4^{3-}, (b) ICl_4^-, (c) SO_3^{2-}, (d) HONO.

42. Write Lewis structures for the following molecules or ions: (a) SbH_3, (b) XeF_2, (c) FSO_2^-, (d) Se_8 (a cyclic molecule with a ring of eight Se atoms).

43. Which of the following can form a coordinate covalent bond? (a) CH_4, (b) BF_3, (c) CO, (d) H_2O, (e) O^{2-} (f), SiF_4.

44. Which of the following can form a coordinate covalent bond? (a) $AlCl_3$, (b) Cl^-, (c) NH_3, (d) C_2H_6, (e) H_2S, (f) PCl_5.

45. Write the following reactions using Lewis structures of the reactants and products. Indicate the coordinate covalent bond (or bonds) in the product.

    (a) $BF_3 + NH_3 \longrightarrow F_3BNH_3$
    (b) $Pb^{4+} + 6Cl^- \longrightarrow PbCl_6^{2-}$
    (c) $SnCl_2 + Br^- \longrightarrow SnCl_2Br^-$
    (d) $SO_3^{2-} + S \longrightarrow S_2O_3^{2-}$

46. Write the following reactions using Lewis structures of the reactants and products. Indicate the coordinate covalent bond (or bonds) in the product.

    (a) $Al^{3+} + 6H_2O \longrightarrow Al(OH_2)_6^{3+}$
    (b) $CO + BF_3 \longrightarrow OCBF_3$
    (c) $Sn^{2+} + 3Cl^- \longrightarrow SnCl_3^-$
    (d) $H^+ + Cl^- \longrightarrow HCl$

47. Although we cannot always tell which bond in a compound or ion is a coordinate covalent bond, sometimes we can identify the species that donates the electron pair and the species that accepts the electron pair. Identify the electron-pair donor and the electron-pair acceptor in each of the following reactions.

    (a) $H_2O + H^+ \longrightarrow H_3O^+$
    (b) $OH^- + H^+ \longrightarrow H_2O$
    (c) $Ag^+ + 2NH_3 \longrightarrow H_3NAgNH_3^+$
    (d) $PCl_5 + Cl^- \longrightarrow PCl_6^-$
    (e) $SO_3 + O^{2-} \longrightarrow SO_4^{2-}$
    (f) $Na_2O + CO_2 \longrightarrow Na_2CO_3$

48. Methanol, H_3COH, is used as the fuel in some race cars; ethanol,
    is used extensively as motor fuel in Brazil. Both produce CO_2 and H_2O when they burn. Write the chemical equations for these combustion reactions using Lewis structures instead of chemical formulas.

49. Sulfuric acid is the industrial chemical produced in greatest quantity worldwide: About 90 billion pounds is produced each year in the United States alone. Write the Lewis structure for sulfuric acid, H_2SO_4, which has two oxygen atoms and two OH groups bonded to the sulfur.

50. The atmosphere of Titan, the largest moon of Saturn, contains methane (CH_4) and traces of ethylene (C_2H_4), ethane (C_2H_6), hydrogen cyanide (HCN), propyne (H_3CCCH), and diacetylene (HCCCCH). Write the Lewis structures for these molecules.

51. Carbon tetrachloride was formerly used in fire extinguishers for electrical fires. It is no longer used for this purpose because of the formation of the toxic gas phosgene, Cl_2CO. Write the Lewis structures for carbon tetrachloride and phosgene.

52. Which of the following can provide an electron pair for the formation of a coordinate covalent bond: H^+, H^-, H_2?

Resonance

53. Formulas or skeleton structures of several molecules or ions are given. Write resonance forms that describe the distribution of electrons in each.

    (a) selenium dioxide, OSeO
    (b) nitrate ion, NO_3^-
    (c) nitric acid, HNO_3 (N is bonded to an OH group and two O atoms.)
    (d) benzene,
    (e) the formate ion,

54. Formulas or skeleton structures of several molecules or ions are given. Write resonance forms that describe the distribution of electrons in each.

    (a) sulfur dioxide, SO_2
    (b) carbonate ion, CO_3^{2-}
    (c) hydrogen carbonate ion, HCO_3^- (C is bonded to an OH group and two O atoms.)
    (d) pyridine,
    (e) the allyl ion,

55. Write the resonance forms of ozone, O_3, the component of the upper atmosphere that protects the earth from ultraviolet radiation.

56. Sodium nitrite, which has been used to preserve bacon and other meats, is an ionic compound. Write the resonance forms of the nitrite ion, NO_2^-, and show that this ion is isoelectronic with ozone, O_3.

57. Explain in terms of the bonds present, why acetic acid, CH_3CO_2H, contains two distinct types of carbon–oxygen bonds, whereas the acetate ion, formed by loss of a hydrogen ion from acetic acid, contains only one type of carbon–oxygen bond. The skeleton structures of these species are shown.

58. (a) Write the Lewis structures for CO_3^{2-}, CO_2, and CO, including resonance structures where appropriate.
    (b) Which of the three has the strongest carbon–oxygen bond?

59. Use Lewis structures and explain why the bond energy of the nitrogen–oxygen bond changes between NO_2^- and NO_3^-.

60. Toothpastes that contain sodium hydrogen carbonate (sodium bicarbonate) and hydrogen peroxide are widely used. Write Lewis structures for the hydrogen carbonate ion and the hydrogen peroxide molecule with resonance forms, where appropriate.

Formal Charge

61. Determine the formal charge of each element in the following: (a) HCl, (b) CF_4, (c) ClO_3^-, (d) PCl_3, (e) PF_5.

62. Determine the formal charge of each element in the following: (a) H_3O^+, (b) SO_4^{2-}, (c) NH_3, (d) O_2^{2-}, (e) H_2O_2.

63. Calculate the formal charge of chlorine in the molecules Cl_2, $BeCl_2$, and ClF_5.

64. Calculate the formal charge of each element in the following compounds and ions: (a) F_2CO, (b) NO^-, (c) BF_4^-, (d) $SnCl_3^-$, (e) H_2CCH_2, (f) ClF_3, (g) SeF_6, (h) PO_4^{3-}.

65. Recalling that it is necessary to average the properties of each resonance form to describe the properties of a molecule or ion that must be described with resonance forms, determine the formal charge on each atom in the following compounds or ions: (a) O_3, (b) SO_2, (c) NO_2^-, (d) NO_3^-, (e) $HONO_2$.

66. On the basis of formal charge considerations, which of the following would be expected to be the correct arrangement of atoms in nitrosyl chloride: ClNO or ClON?

67. On the basis of formal charge considerations, which of the following would be expected to be the correct arrangement of atoms in hypochlorous acid: HOCl or OClH?

68. On the basis of formal charge considerations, which of the following would be expected to be the correct arrangement of atoms in sulfur dioxide: OSO or SOO?

Electronegativity and Polar Bonds

69. From its position in the periodic table, determine which atom in each pair is more electronegative.
    (a) Br or Cl   (b) N or O
    (c) S or O   (d) P or S
    (e) Si or N   (f) Ba or P
    (g) N or K

70. From its position in the periodic table, determine which atom in each pair is more electronegative.
    (a) N or P   (b) N or Ge
    (c) S or F   (d) Cl or S
    (e) H or C   (f) Se or P
    (g) C or Si

71. From their positions in the periodic table, arrange the atoms in each of the following series in order of increasing electronegativity.
    (a) C, F, H, N, O   (b) Br, Cl, F, H, I
    (c) F, H, O, P, S   (d) Al, H, Na, O, P
    (e) Ba, H, N, O, As

72. From their positions in the periodic table, arrange the atoms in each of the following series in order of increasing electronegativity.
    (a) As, H, N, P, Sb   (b) Cl, H, P, S, Si
    (c) Br, Cl, Ge, H, Sr (d) Ca, H, K, N, Si
    (e) Cl, Cs, Ge, H, Sr

73. Which atoms attract electrons in the bond most strongly when they form a bond with a sulfur atom?

74. Which atoms attract electrons in the bond most strongly when they form a bond with a phosphorus atom?

75. Write the Lewis structure of a molecule that contains a polar single bond, of one that contains a polar double bond, and of one that contains a polar triple bond.

76. Write the Lewis structure of a molecule that contains a nonpolar single bond, of one that contains a nonpolar double bond, and of one that contains a nonpolar triple bond.

77. Which is the most polar bond: $C{=}C$, $C{-}H$, $N{-}H$, $O{-}H$, or $Se{-}H$?

78. Identify the more polar bond in each of the following pairs of bonds.
    (a) HF or HCl   (b) NO or CO
    (c) SH or OH   (d) PCl or SCl
    (e) CH or NH   (f) SO or PO
    (g) CN or NN

79. Which of the following molecules or ions contain polar bonds: O_3, S_8, O_2^{2-}, NO_3^-, CO_2, H_2S, BH_4^-?

Bond Energies

80. Which bond in each of the following pairs of bonds is the strongest?

(a) C—C or C=C　　　(b) C—N or C≡N
(c) C≡O or C=O　　　(d) H—F or H—Cl
(e) C—H or O—H　　　(f) C—N or C—O

81. Using the bond energies in Table 6.2, determine the approximate enthalpy change for each of the following reactions.

(a) $H_2(g) + Br_2(g) \longrightarrow 2HBr(g)$
(b) $CH_4(g) + I_2(g) \longrightarrow CH_3I(g) + HI(g)$
(c) $C_2H_4(g) + 3O_2(g) \longrightarrow 2CO_2(g) + 2H_2O(g)$

82. Using the bond energies in Table 6.2, determine the approximate enthalpy change for each of the following reactions.

(a) $Cl_2(g) + 3F_2(g) \longrightarrow 2ClF_3(g)$
(b) $H_2C=CH_2(g) + H_2(g) \longrightarrow H_3CCH_3(g)$
(c) $2C_2H_6(g) + 7O_2(g) \longrightarrow 4CO_2(g) + 6H_2O(g)$

83. When a molecule can form with two different structures, the structure with the stronger bonds is usually the stable form. Use bond energies to predict the correct structure of the hydroxylamine molecule. Is it

(a) H—N̈—Ö—H　　or　　(b) H—N̈—Ö̈:　?
　　　　|　　　　　　　　　　　|
　　　　H　　　　　　　　　　　H

84. The common forms of N and P are N_2 and P_4, respectively.

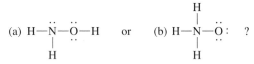

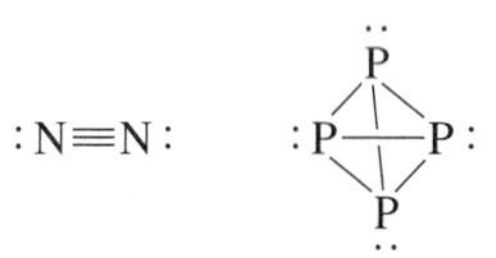

Using the following data, explain why the stable forms are N_2, not N_4, and P_4, not P_2.

Mean bond enthalpy, kJ mol⁻¹

N≡N	946	N—N	160
P≡P	481	P—P	215

85. Using the data in Appendix I, calculate the bond energies of the diatomic molecules F_2, O_2, and N_2.

86. Using the data in Appendix I, calculate the bond energy of the carbon–sulfur double bond in CS_2.

87. Using the data in Appendix I, determine which bond is stronger, the S—F bond in $SF_4(g)$ or in $SF_6(g)$.

88. Using the data in Appendix I, determine which bond is stronger, the P—Cl bond in $PCl_3(g)$ or in $PCl_5(g)$.

89. How does the bond energy of HCl(g) differ from the standard enthalpy of formation of HCl(g)? How can the standard enthalpy of formation of HCl(g) be used in determining the bond energy?

Oxidation State

90. With which elements will sulfur form binary compounds in which it has a positive oxidation state?

91. Bromine forms binary compounds with each of the following elements: K, Co, F, Cl, I, Zn, Ca, O, C, P. In which of the compounds will bromine have a positive oxidation state?

92. Phosphorus forms binary compounds with each of the following elements: K, Co, F, Cl, I, Zn, Ca, O, C, Br. In which of the compounds will phosphorus have a positive oxidation state?

93. Determine the oxidation state of each of the atoms in the following: (a) HCl, (b) CF_4, (c) NH_3, (d) PCl_3, (e) PCl_5, (f) H_3O^+, (g) ClO_3^-, (h) O_2^{2-}, (i) CO, (j) HONO, (k) Cl_2PPCl_2, (l) MgF_2, (m) Na_3P.

94. Determine the oxidation state of each of the atoms in the following: (a) F_2CO, (b) K_2O, (c) O_2, (d) $MgCl_2$, (e) HBr, (f) NO_3^-, (g) NH_4^+, (h) BF_4^-, (i) ClNO, (j) H_2CCH_2, (k) HNNH, (l) ClF_3, (m) Cl_2BBCl_2.

95. Which of the following elements will exhibit a positive oxidation state when combined with silicon? (a) Mg, (b) F, (c) Al, (d) O, (e) Na, (f) Cl, (g) Fe.

Applications and Additional Exercises

96. By reading the labels of several commercial products, prepare a list of six simple molecular compounds in these household products.

97. From the labels of several commercial products, prepare a list of six ionic compounds in the products. For each compound, write the formula. (You may need to look up some formulas in a suitable reference book.)

98. Write the symbol for the common ion and the Lewis structure of the ion formed from the atoms with the following electron configurations.

(a) $1s^2 2s^2 2p^1$
(b) $1s^2 2s^2 2p^5$
(c) $1s^2 2s^2 2p^6 3s^2$
(d) $1s^2 2s^2 2p^6 3s^2 3p^6 4s^2 3d^{10}$
(e) $1s^2 2s^2 2p^6 3s^2 3p^6 4s^2 3d^{10} 4p^4$
(f) $1s^2 2s^2 2p^6 3s^2 3p^6 4s^2 3d^{10} 4p^1$

99. The arrangements of atoms in several biologically important molecules are given below. Complete the Lewis structures of these molecules.

(a) The amino acid serine

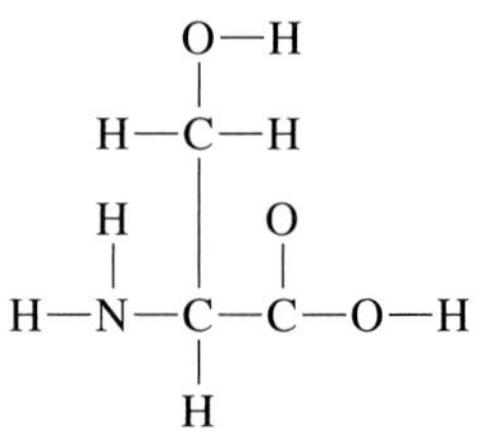

```
            O—H
             |
          H—C—H
             |
      H      O
      |      |
  H—N—C——C—O—H
      |
      H
```

(b) Urea

$$H-N-C-N-H$$

with H and O above the C—N—C chain (H, O, H above N, C, N)

(c) Pyruvic acid

$$H-C-C-C-O-H$$

with H, O, O above the first three carbons and H below the first carbon

(d) Uracil

a six-membered ring structure with O double bonded to top C, two N atoms with H, and C with H substituents

(e) Carbonic acid

$$H-O-C-O-H$$

with O above the central C

100. Compare the oxidation state and formal charge of bromine in the molecule Br_2, the molecule $BeBr_2$, and the ion BrO_3^-.

101. Write the Lewis structure (with resonance forms if appropriate) of each molecule or ion, and give the oxidation state (oxidation number) of the N, P, or S atoms in each of the following compounds. In each case show how you arrived at your answers.

(a) H_3PO_4, phosphoric acid, used in cola soft drinks
(b) NH_4Cl, ammonium chloride, used in instant ice packs
(c) S_2Cl_2, disulfur dichloride, used in vulcanizing rubber
(d) $K_4[O_3POPO_3]$, potassium pyrophosphate, an ingredient of some toothpastes

102. Should we expect the structure of nitrous acid to be

$$:\overset{..}{O}-N=\overset{..}{O}: \quad \text{or} \quad H-\overset{..}{O}-N=\overset{..}{O} \quad ?$$

with an H above the first O in the left structure

103. Iodine forms a series of fluorides: IF, IF_3, IF_5, and IF_7. Write Lewis structures for these four compounds, and determine the oxidation state and formal charge of the iodine atom in each molecule.

104. Complete the Lewis structure of ethanol and identify the most polar bond in the molecule. The skeleton structure is

$$H-C-C-O-H$$

with H H above and H H below the two carbons

105. Complete the Lewis structure and then indicate the longest bond in the following molecule.

$$H \quad C \quad C \quad C \quad C \quad C \quad C \quad H$$

with H H above and H H below

106. Use the bond energy to find an approximate value of ΔH for the reaction

$$:\overset{..}{O}=N-\overset{..}{O}: \longrightarrow :\overset{..}{O}=N-\overset{..}{O}-\overset{..}{F}:$$

with :F: above the N in both structures

Which is the more stable form of FNO_2?

107. Does a calculation of formal charge predict the correct structure for hydroxylamine (Exercise 83)?

108. A compound with a molar mass of about 28 contains 85.7% carbon and 14.3% hydrogen. Write the Lewis structure for a molecule of the compound.

109. A compound with a molar mass of about 42 contains 85.7% carbon and 14.3% hydrogen. Write the Lewis structure for a molecule of the compound.

110. Two arrangements of atoms are possible for a compound with a molar mass of about 45 that contains 52.2% C, 13.1% H, and 34.7% O. Write the Lewis structures for the two molecules.

111. Write the Lewis structure and chemical formula of the compound with a molar mass of about 70 that contains 19.7% nitrogen and 80.3% fluorine, and determine the formal charge and oxidation state of the atoms in this compound.

112. Determine the oxidation states of the elements in a compound that contains 32.9% Na, 12.9% Al, and 54.3% F.

113. Write the chemical formula of a compound of potassium, platinum(IV), and chlorine that contains a 2:1 ratio of potassium to platinum.

114. (This question is taken from the 1994 Chemistry Advanced Placement Examination and is used with the permission of the Educational Testing Service.) Use principles of atomic structure to answer each of the following:

(a) The radius of the Ca atom is 0.197 nanometer; the radius of the Ca^{2+} ion is 0.099 nanometer. Account for the difference.

(b) The lattice energy of $CaO(s)$ is $-3,460$ kilojoules per mole; the lattice energy of K_2O is $-2,240$ kilojoules per mole. Account for this difference.

(c)

	Ionization Energy, kJ mol^{-1}	
	First	Second
K	419	3,050
Ca	590	1,140

Explain the difference between Ca and K in regard to (i) their first ionization energies. (ii) their second ionization energies.

(d) The first ionization energy of Mg is 738 kilojoules per mole and that of Al is 578 kilojoules per mole. Account for this difference.

CHAPTER OUTLINE

Valence Shell Electron-Pair Repulsion Theory

7.1   Predicting Molecular Structures
7.2   Rules for Predicting Electron-Pair Geometry and Molecular Geometry
7.3   Molecular Polarity and Dipole Moment

Valence Bond Theory: Hybridization of Atomic Orbitals

7.4   Valence Bond Theory
7.5   Hybridization of Atomic Orbitals
7.6 sp Hybridization
7.7 sp^2 Hybridization
7.8 sp^3 Hybridization
7.9 sp^3d and sp^3d^2 Hybridization
7.10 Assigning Hybrid Orbitals to Central Atoms
7.11 Hybridization Involving Double and Triple Bonds

Molecular Orbital Theory

7.12 Molecular Orbitals
7.13 Molecular Orbital Energy Diagrams
7.14 Bond Order
7.15 H_2 and He_2 Molecules
7.16 Diatomic Molecules of the Second Period

7

Molecular Structures and Models of Covalent Bonds

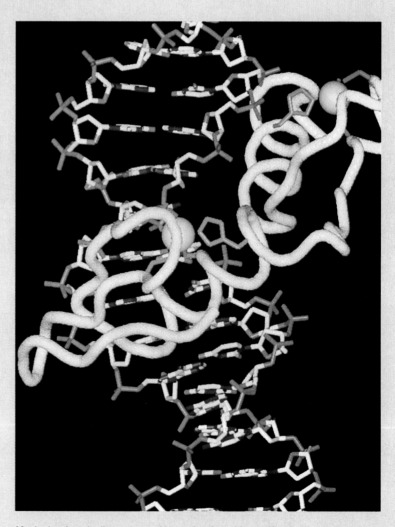

Much chemistry is directed at making molecules with specific shapes, such as this drug that interacts with the DNA helix.

This chapter is about theories of structure and bonding. It has been said that a theory is either too good to be true or too true to be good. Here "good" refers to ease of use, and "true" refers to the precision and accuracy of the theory. The first theory we will consider—a simple version of the valence shell electron-pair repulsion theory—is probably too good to be true. The second statement ("too true to be good") describes the molecular orbital theory when applied to molecules other than very small ones.

In Chapter 1 we defined a theory as an explanation put forward to account for laws or large bodies of experimental data. For a theory to be accepted, we want it to explain experimental data, and we want to be able to use it to predict behavior. The experimental data that are used to test one of the theories in this chapter are the actual three-dimensional structures of molecules. If this theory can accurately predict the structures of large numbers of molecules, we consider it useful. This is certainly the case for the valence shell electron-pair repulsion theory, despite questions of its accuracy.

At this point you may be wondering how we determine the three-dimensional structures of covalent molecules. In Section 11.20, we will discuss X-ray diffraction, an excellent method for determining the structures of crystalline materials. If you take more chemistry courses, you will encounter spectroscopic techniques such as infrared spectroscopy and nuclear magnetic resonance spectroscopy that also can be used for structure determination.

In Chapter 6, we examined Lewis structures of covalent molecules. These structures were presented in two dimensions, but we were reminded that molecules are three-dimensional. In this chapter we will learn how to predict the three-dimensional shapes of molecules as well as the geometrical arrangement of electron pairs around the central atom. We will also examine two different theories that are used to describe bonding in covalent molecules.

The orbitals in an atom that is part of a molecule behave differently from orbitals in the free atom, which we discussed in Chapter 5. For example, the atomic orbitals on a free carbon atom are different from those on a carbon atom in a CO_2 molecule or a CH_4 molecule. Once we are able to predict the geometry about an atom using a theory called the valence shell electron-pair repulsion theory, we will be able to describe the changes in its atomic orbitals using a second theory called valence bond theory. This theory utilizes the concept of hybridization to describe these changes. Alternatively, molecular orbital theory can be used to describe the bonding in a molecule in terms of orbitals extending over all of the atoms in that molecule.

Figure 7.1

Bond distances and angles in the formaldehyde molecule, H_2CO (1 Å = 100 pm = 1×10^{-10} m).

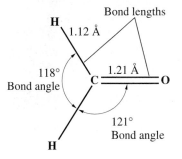

Valence Shell Electron-Paor Repulsion Theory

7.1 Predicting Molecular Structures

The three-dimensional arrangement of the atoms in a molecule is called its **molecular structure.** The structure of a molecule is described by its geometry—that is, its bond angles and bond distances (Fig. 7.1). A **bond angle** is the angle between any two bonds that include a common atom. It is usually measured in degrees. A **bond distance** is the distance between the nuclei of two bonded atoms along the straight line joining the nuclei. Bond distances are measured in angstroms (1 Å = 10^{-10} m) or in picometers (1 pm = 10^{-12} m, 100 pm = 1 Å).

We can predict approximate bond angles around a central atom in a molecule from an examination of the number of bonds and unshared electron pairs in its Lewis structure by using the **valence shell electron-pair repulsion theory (VSEPR theory).** This theory states that electron pairs in the valence shell of a central atom will adopt a geometry that minimizes repulsions between these electron pairs by maximizing the distance between them. The electrons in the valence shell of a central atom in a molecule form regions of high electron density either as bonding pairs of electrons, located primarily between bonded atoms, or as unshared pairs occupying regions of space shaped rather like those occupied by the bonding pairs. The electrostatic repulsion of these electrons is reduced to a minimum when the various regions of high electron density assume positions as far from each other as possible. If we use VSEPR theory, we can usually predict the correct arrangement of electron pairs around a central atom and, subsequently, the arrangement of atoms in a molecule. However, we must remember that the theory considers only electron-pair repulsions. Other interactions, such as nucleus–nucleus repulsions and nucleus–electron attractions are also involved in the final arrangement that atoms adopt in any molecular structure. Also VSEPR theory says nothing about the orbitals that are used for bonding between atoms.

As the simplest example of using VSEPR theory, let us predict the structure of a gaseous BeF_2 molecule. The Lewis structure of BeF_2 (Fig. 7.2) shows only two electron pairs around the central beryllium atom. With two bonds and no unshared pairs of electrons on the central atom, the bonds are as far apart as possible, and the electrostatic repulsion between these regions of high electron density is reduced to a minimum when they are on opposite sides of the central atom. The bond angle is 180° (Fig. 7.2). This and other geometries that minimize the repulsions among regions of high electron density (bonds and/or unshared pairs) are illustrated in Table 7.1. Two regions of electron density form a *linear* structure around a central atom in a molecule; three regions form a *trigonal planar* structure; four regions a *tetrahedral* structure; five regions a *trigonal bipyramidal* structure; and six regions an *octahedral* structure.

If the regions of high electron density are not identical, we may see bond angles that differ by several degrees from the ideal values given in Table 7.1. Nevertheless, the structures given are usually good approximations for the distribution of electron density around a central atom. Small distortions from the angles in Table 7.1 can result from differences in repulsion between various regions of electron density. VSEPR theory predicts these distortions by establishing an order of repulsions and an order of the amount of space occupied by different kinds of electron pairs. Electron-pair repulsions decrease in the order:

Unshared pair–unshared pair > unshared pair–bonding pair >
bonding pair–bonding pair

This order of repulsions determines the amount of space occupied by different regions of electrons. An unshared pair of electrons occupies a region of space that is larger than that occupied by the electrons in a triple bond, which is larger than that for a double bond, which, in turn, is larger than that for a single bond. The sizes decrease in the following order:

Unshared pair > triple bond > double bond > single bond

Formaldehyde, H_2CO, which is used as a preservative for biological and anatomical specimens, has regions of high electron density that consist of two single bonds and one double bond, and the bond angles differ by about 3° (Fig. 7.1). The larger dou-

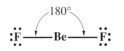

Figure 7.2

The linear structure of the BeF_2 molecule. The bonds are on opposite sides of the Be atom.

Table 7.1

Arrangement of Unshared Pairs and Bonds as a Result of Electron-Pair Repulsions

Number of Regions	Spatial Arrangement	Electron-Pair Geometry
Two regions of high electron density (bonds and/or unshared pairs)		**Linear.** 180° angle.
Three regions of high electron density (bonds and/or unshared pairs)		**Trigonal planar.** All angles 120°.
Four regions of high electron density (bonds and/or unshared pairs)		**Tetrahedral.** All angles 109.5°.
Five regions of high electron density (bonds and/or unshared pairs)		**Trigonal bipyramidal.** Angles of 90° or 120°. An attached atom may be equatorial (in the plane of the triangle) or axial (above or below the plane of the triangle).
Six regions of high electron density (bonds and/or unshared pairs)		**Octahedral.** All angles 90° or 180°.

ble bond occupies the larger space, so the smaller single bonds are pushed together, producing the smaller H—C—H bond angle.

Even though the arrangement of unshared pairs of electrons and bonds in a molecule corresponds to one of the arrangements in Table 7.1, its molecular structure may look different. The presence of an unshared pair affects the structure of the molecule, but the unshared pair cannot be detected by most experimental techniques used to

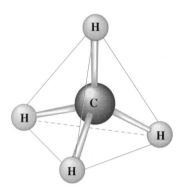

Figure 7.3

The molecular geometry of the methane molecule, CH_4. The tetrahedral arrangement of bonds produces a tetrahedral arrangement of hydrogen atoms.

determine structure. Consequently, the molecular structure is described as though the unshared pair were not there.

We differentiate between these two situations by naming the geometry that includes *all* electron pairs the **electron-pair geometry**. The structure that includes only the placement of the atoms in the molecule is called the **molecular geometry.** The two geometries are the same when there are no unshared electron pairs around the central atom; they differ when there are one or more unshared pairs. For example, the methane molecule, CH_4, which is the major component of natural gas, has four bonding pairs of electrons around the central carbon atom. The electron-pair geometry is tetrahedral, as is the molecular geometry (Fig. 7.3). On the other hand, the gaseous ammonia molecule, NH_3, which is dissolved in water and sold as household ammonia, also has a tetrahedral arrangement of regions of electron density and thus a tetrahedral electron-pair geometry (Fig. 7.4 A). However, one of these regions is an unshared pair and is not observed when the molecular structure is determined experimentally. The molecular geometry (the arrangement of atoms only) is a trigonal pyramid (Fig. 7.4 B) with the nitrogen atom at the apex and three hydrogen atoms forming the base. The H—N—H bond angle in NH_3 is smaller than the 109.5° angle in a regular tetrahedron (Table 7.1), because the unshared-pair–bonding-pair repulsion is greater than the bonding-pair–bonding-pair repulsion, leading to the larger size of the unshared pair of electrons relative to the single bonds (Fig. 7.4 C). Table 7.2 illustrates the ideal electron-pair geometries and molecular geometries that are predicted for various combinations of unshared pairs and bonding pairs.

We should note that different molecular geometries are possible for trigonal bipyramidal electron-pair geometries containing unshared pairs of electrons. This is because there are two distinct positions in a trigonal bipyramid: (1) a smaller **axial position** (if we hold a model of a trigonal bipyramid by the two axial positions, we have an *axis* around which we can rotate the model) and (2) a larger **equatorial position** (three positions form an equator around the middle of the molecule). Figure 7.5 A (p. 222) identifies the two positions. In a trigonal bipyramidal structure, *unshared pairs always occupy equatorial positions* because these larger positions more easily accommodate the larger unshared pairs. We can write three possible arrangements for the three bonds and two unshared pairs for the ClF_3 molecule (Fig. 7.5 B–D). The stable structure (Fig. 7.5 B) is the one that puts the unshared pairs in equatorial locations, giving a "T-shaped" molecular structure.

When a central atom has two unshared electron pairs and four bonding regions, we have an octahedral electron-pair geometry. The two unshared pairs are on opposite sides of the octahedron (180° apart), giving a square planar molecular geometry (Table 7.2) that minimizes unshared-pair–unshared-pair repulsions.

Figure 7.4

(A) Regions occupied by the unshared pair (shown in gray) and bonds (blue) in the ammonia molecule, NH_3. (B) The resulting trigonal pyramidal molecular geometry. (C) Unshared-pair–bonded-pair repulsion is stronger than bonded-pair–bonded-pair repulsion, causing the H—N—H angle to be slightly smaller than 109.5°.

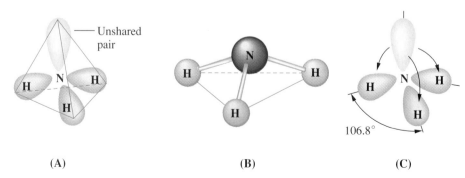

(A)   (B)   (C)

Table 7.2 Electron-Pair Geometries and Molecular Geometries Based on the Valence Shell Electron-Pair Repulsion Theory

Regions of High Electron Density (bonds and unshared pairs.)	Molecular Geometries and Examples (chemical bonds are indicated in black, unshared pairs in red)
Three: trigonal planar arrangement of bonds and/or unshared pairs	 3 Bonds / 0 Unshared pairs / Trigonal planar / CO_3^{2-}, BF_3, NO_3^-　　　2 Bonds / 1 Unshared pair / Bent (about 120°) / NO_2^-, $ClNO$
Four: tetrahedral arrangement of bonds and/or unshared pairs	 4 Bonds / 0 Unshared pairs / Tetrahedral / NH_4^+, CH_4　　　3 Bonds / 1 Unshared pair / Trigonal pyramidal / H_3O^+, PCl_3, NH_3　　　2 Bonds / 2 Unshared pairs / Bent (about 109.5°) / NH_2^-, H_2O
Five: trigonal bipyramidal arrangement of bonds and/or unshared pairs	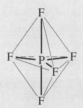  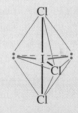  5 Bonds / 0 Unshared pairs / Trigonal bipyramidal / PF_5, $SnCl_5^-$　　4 Bonds / 1 Unshared pair / Seesaw / SF_4, ClF_4^+　　3 Bonds / 2 Unshared pairs / T-shaped / ICl_3, ClF_3　　2 Bonds / 3 Unshared pairs / Linear / I_3^-, ClF_2^-
Six: octahedral arrangement of bonds and/or unshared pairs	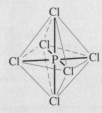 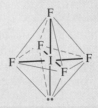 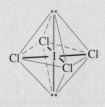 6 Bonds / 0 Unshared pairs / Octahedral / PCl_6^-, SF_6, IF_6^+　　5 Bonds / 1 Unshared pair / Square pyramidal / IF_5, XeF_5^+　　4 Bonds / 2 Unshared pairs / Square planar / ICl_4^-, XeF_4

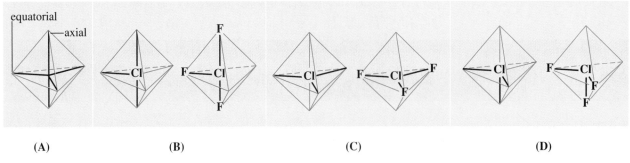

| (A) | (B) | (C) | (D) |

Figure 7.5

(A) Axial and equatorial positions in a trigonal bipyramid. (B)–(D) Possible arrangements of two unshared pairs (shown in red) and three bonds (black) in ClF_3 and the resulting molecular geometries, showing only the positions of the atoms without the unshared pairs. Part (B) shows the stable T-shaped arrangement.

7.2   Rules for Predicting Electron-Pair Geometry and Molecular Geometry

We can employ the following procedure to use the VSEPR theory to determine the electron-pair geometries and the molecular geometries of covalent molecules.

1.   Write the Lewis structure of the molecule (Sections 6.7 and 6.8).

2.   Count the number of regions of high electron density (unshared pairs and bonds) around the central atom in the Lewis structure. A single, double, or triple bond counts as one region of high electron density. A single unpaired electron counts as an unshared pair.

3.   Identify the most stable arrangement of the regions of high electron density as linear, trigonal planar, tetrahedral, trigonal bipyramidal, or octahedral (Table 7.1). This is the electron-pair geometry.

4.   If more than one arrangement of unshared pairs and chemical bonds is possible, choose the one that will minimize unshared pair repulsions. In trigonal bipyramidal arrangements, repulsion is minimized when every unshared pair is in equatorial position. In an octahedral arrangement with two unshared pairs, repulsion is minimized when the unshared pairs are on opposite sides of the central atom.

5.   Identify the molecular geometry (the arrangement of atoms) from the locations of the atoms at the ends of bonds (Table 7.2).

The following examples illustrate the use of VSEPR theory to predict the molecular geometry of molecules or ions that have no unshared pairs of electrons. In this case the electron-pair geometry is identical to the molecular geometry.

| **Example 7.1** | Predicting Electron-Pair and Molecular Geometries |

The element beryllium and its compounds are extremely toxic because beryllium replaces magnesium, its neighbor in the periodic table, in enzymes in the body. Beryllium also forms simple compounds with the halogens. Predict the electron-pair and molecular geometries of a gaseous $BeCl_2$ molecule.

Solution: We write the Lewis structure of $BeCl_2$ as

$$:\overset{..}{\underset{..}{Cl}}—Be—\overset{..}{\underset{..}{Cl}}:$$

This shows us two regions of high density around the beryllium atom: two bonds and no unshared pairs of electrons. Using VSEPR theory, we predict that the two regions of high electron density arrange themselves on opposite sides of the central atom with a bond angle of 180° (Table 7.1). This suggests that the electron-pair geometry and molecular geometry are identical and that gaseous $BeCl_2$ molecules are linear.

Check your learning: The increase of gaseous carbon dioxide, CO_2, in our atmosphere, which is due primarily to the combustion of carbon-containing fuels, should be of concern to everyone. What are the electron-pair and molecular geometries of this gaseous molecule?

Answer: Electron-pair geometry, linear; molecular geometry, linear.

Example 7.2   Predicting Electron-Pair and Molecular Geometries

The boron trihalides are very important industrial chemicals that act as catalysts for a wide variety of chemical reactions. (A catalyst is a substance that changes the speed of a reaction without itself undergoing permanent chemical change; Section 8.9.) Predict the electron-pair and molecular geometries of a molecule of boron trichloride, BCl_3.

Solution: We write the Lewis structure of BCl_3 as

$$:\!\overset{..}{\underset{..}{Cl}}\!:$$
$$\mid$$
$$:\!\overset{..}{\underset{..}{Cl}}\!-\!B\!-\!\overset{..}{\underset{..}{Cl}}\!:$$

Thus we see that BCl_3 contains three bonds and that there are no unshared pairs of electrons on boron. We expect the arrangement of three regions of high electron density to be trigonal planar (Table 7.1), giving a trigonal planar electron-pair geometry. The bonds in BCl_3 should lie in a plane with 120° angles between them, also a trigonal planar molecular geometry (Fig. 7.6).

Figure 7.6

(A) Trigonal planar arrangement of three bonds in BCl_3. (B) The resulting trigonal planar molecular geometry. (C) A model of BCl_3.

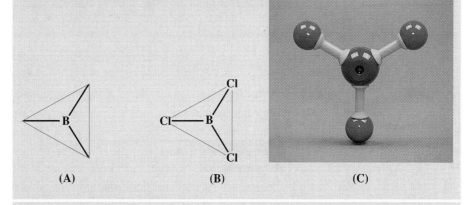

(A)   (B)   (C)

Check your learning: Aluminum trichloride, $AlCl_3$, is another important industrial catalyst. Predict the electron-pair and molecular geometries of a gaseous $AlCl_3$ molecule.

Answer: Electron-pair geometry, trigonal planar; molecular geometry, trigonal planar.

Example 7.3   Predicting Electron-Pair and Molecular Geometries

Two of the 50 chemicals produced in greatest quantity in the United States, ammonium nitrate and ammonium sulfate, contain the ammonium ion. Both are used for fertilizer. Predict the electron-pair and molecular geometries of the NH_4^+ cation.

Solution: We write the Lewis structure of NH_4^+ as

$$\left[\begin{array}{c} H \\ | \\ H-N-H \\ | \\ H \end{array} \right]^+$$

Now we can see that NH_4^+ contains four bonds from the nitrogen atom to hydrogen atoms and no unshared pairs. We expect the four regions of high electron density to arrange themselves such that they point to the corners of a tetrahedron with the central nitrogen atom in the middle (Table 7.1). The hydrogen atoms at the ends of the bonds are located at the corners of the tetrahedron. Therefore, the electron-pair geometry of NH_4^+ is tetrahedral and the molecular geometry of NH_4^+ (Fig. 7.7) is also tetrahedral.

Figure 7.7

(A) Tetrahedral arrangement of four bonds in NH_4^+. (B) The resulting tetrahedral molecular geometry. (C) A model of the NH_4^+ ion.

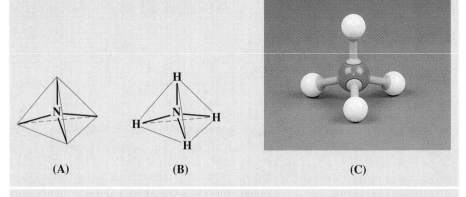

(A)   (B)   (C)

Check your learning: The starting material for the synthesis of many silicone polymers is silicon tetrachloride, $SiCl_4$. Predict the electron-pair and molecular geometries of a $SiCl_4$ molecule.

Answer: Electron-pair geometry, tetrahedral; molecular geometry, tetrahedral.

The effect of unshared pairs of electrons on molecular structure is illustrated in the following examples.

Example 7.4   Predicting Electron-Pair and Molecular Geometries

Predict the electron-pair and molecular geometries of a water molecule.

Solution: Our Lewis structure of H_2O,

$$\begin{array}{c} H \\ | \\ H-\overset{\displaystyle }{\underset{\displaystyle \cdot\cdot}{O}}: \end{array}$$

indicates that there are four regions of high electron density around the oxygen atom: two unshared pairs and two chemical bonds. We predict that these four regions are arranged in a tetrahedral fashion (Fig. 7.8 A), as indicated in Table 7.1. Therefore the electron-pair geometry is tetrahedral and the molecular structure is bent with an approximately 109.5° angle (Fig. 7.8 B). (In fact, the bond angle is 104.5°, which is attributed to unshared-pair–unshared-pair and unshared-pair–bond-pair repulsions.)

Figure 7.8

(A) Tetrahedral arrangement of two unshared pairs (shown in red) and two bonds (black) in H_2O. (B) The resulting bent (angular) molecular geometry.

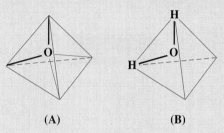

(A)   (B)

Check your learning: In Chapter 2, we defined an acid as any compund that, when dissolved in water, increased the concentration of the hydronium ion, H_3O^+. Predict the electron-pair and molecular geometries of this cation.

Answer: Electron-pair geometry, tetrahedral; molecular geometry, trigonal pyramidal.

Example 7.5	Predicting Electron-Pair and Molecular Geometries

Sulfur tetrafluoride, SF_4, is extremely valuable for the preparation of compounds containing fluorine such as herbicides (that is, SF_4 is used as a fluorinating agent). Predict the electron-pair and molecular geometries of an SF_4 molecule.

Solution: Our Lewis structure of SF_4,

indicates five regions of high electron density around the sulfur atom: one unshared pair and four chemical bonds. We predict that these five regions are directed toward the corners of a trigonal bipyramidal arrangement (Table 7.1), which is the electron-pair geometry. In order to minimize unshared-pair–bond-pair repulsions, the unshared pair occupies one of the equatorial positions. The molecular structure (Fig. 7.9) is that of a seesaw (Table 7.2).

Figure 7.9

(A) Trigonal bipyramidal arrangement of one unshared pair (shown in red) and four bonds (black) in SF_4. (B) The resulting seesaw-shaped molecular geometry.

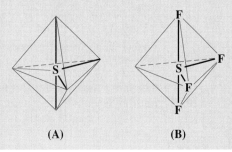

(A)   (B)

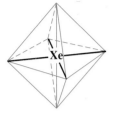

(A)

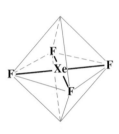

(B)

Figure 7.10

(A) Octahedral arrangement of two unshared pairs (shown in red) and four bonds (black) in XeF_4. (B) The stable square planar molecular geometry.

Check your learning: The interhalogen compound bromine trifluoride, BrF_3, can be used to fluorinate uranium and produce gaseous uranium hexafluoride, UF_6, for the processing of nuclear fuel. Predict the electron-pair and molecular geometries of a molecule of BrF_3.

Answer: Electron-pair geometry, trigonal bipyramidal; molecular geometry, T-shaped.

Example 7.6   Predicting Electron-Pair and Molecular Geometries

Of all the noble gases, xenon is the most reactive and its chemistry the best characterized. Oxygen and fluorine are the elements most commonly bonded to Xe. Predict the electron-pair and molecular geometries of the XeF_4 molecule.

Solution: The Lewis structure of XeF_4,

$$:\!\ddot{F} - Xe - \ddot{F}\!:$$

indicates six regions of high electron density around the xenon atom: two unshared pairs and four bonds. These six regions adopt an octahedral arrangement (Table 7.1), which is the electron-pair geometry. In order to minimize repulsions, the unshared pairs should be on opposite sides of the central atom (Fig. 7.10 A). The five atoms are all in the same plane and have a square planar molecular geometry (Fig. 7.10 B).

Check your learning: Another xenon–fluorine molecule is xenon difluoride, XeF_2. Predict the electron-pair and molecular geometries of this molecule.

Answer: Electron-pair geometry, trigonal bipyramidal; molecular geometry, linear.

Example 7.7   Predicting Geometries in Multicenter Molecules

Amino acids are called the building blocks of nature, because nearly all proteins are made from the same 20 amino acids. The simplest amino acid is glycine, $H_2NCH_2CO_2H$. The Lewis structure for glycine is

$$\begin{array}{cccc} & H & :\ddot{O} & \\ & | & \| & \\ H - N & - C - C & - \ddot{O} - H \\ & | & | & \\ & H & H & \end{array}$$

Predict the molecular geometry for the nitrogen atom, the two carbon atoms, and the oxygen atom with a hydrogen atom attached.

Solution: Just as with the simpler molecules in Examples 7.2 through 7.6, we use the rules in Section 7.2 to determine the electron-pair geometry and molecular geometry around each central atom.

Electron-pair geometries

Nitrogen	four regions of electron density; tetrahedral
Carbon ($\underline{C}H_2$)	four regions of electron density; tetrahedral
Carbon ($\underline{C}O_2$)	three regions of electron density; trigonal planar
Oxygen ($\underline{O}H$)	four regions of electron density; tetrahedral

Molecular geometries

Nitrogen	three bonds, one unshared pair; trigonal pyramidal
Carbon ($\underline{C}H_2$)	four bonds, no unshared pairs; tetrahedral
Carbon ($\underline{C}O_2$)	three bonds (double bond counts as one bond), no unshared pairs; trigonal planar
Oxygen ($\underline{O}H$)	two bonds, two unshared pairs; bent (109°)

Figure 7.11

(A) Polar molecules, such as hydrogen fluoride, are randomly oriented in the absence of an electric field. (B) The dipoles tend to align in an electric field, with their positive ends oriented toward the negative plate and their negative ends toward the positive plate.

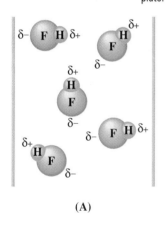

(A)

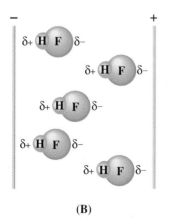

(B)

Check your learning: Another amino acid is alanine, which has the Lewis structure

$$
\begin{array}{ccccc}
 & & CH_3 & \overset{..}{\underset{..}{O}} & \\
 & & | & \| & \\
H - \overset{..}{N} - & C & - & C & - \overset{..}{\underset{..}{O}} - H \\
 & | & | & & \\
 & H & H & &
\end{array}
$$

Predict the electron-pair geometry and molecular geometry of the nitrogen atom, the three carbon atoms, and the oxygen atom with a hydrogen atom attached.

Answer: Electron-pair geometries: nitrogen, tetrahedral; carbon ($\underline{C}H$), tetrahedral; carbon ($\underline{C}H_3$), tetrahedral; carbon ($\underline{C}O_2$), trigonal planar; oxygen ($\underline{O}H$), tetrahedral. Molecular geometries: nitrogen, trigonal pyramidal; carbon ($\underline{C}H$), tetrahedral; carbon ($\underline{C}H_3$), tetrahedral; carbon ($\underline{C}O_2$), trigonal planar; oxygen ($\underline{O}H$), bent (109°).

7.3 Molecular Polarity and Dipole Moment

Polar bonds have a positive end ($\delta+$) and a negative end ($\delta-$). We say these bonds possess a *dipole,* a separation of charge described in Section 6.5. Some molecules also have a separation of charge. If such a charge separation exists, a molecule is said to be polar and the extent of the polarity is measured by its **dipole moment.**

Any diatomic molecule with a polar bond is a **polar molecule.** We have seen that a molecule of hydrogen chloride falls in this category. The hydrogen atom is at the positively charged end of the molecule, and the chlorine atom is at the negatively charged end (Section 6.5). Polar molecules tend to align when placed in an electric field, with the positive end of the molecule oriented toward the negative plate and the negative end toward the positive plate (Fig. 7.11). We can use an electrically charged object to attract polar molecules, but nonpolar molecules are not attracted (Fig. 7.12, p. 228).

Molecules that contain two or more polar bonds may or may not be polar. We must examine their molecular geometry to make this determination. If the bonds in a molecule are arranged such that their dipoles cancel, then the molecule is nonpolar. This is the situation in CO_2; each of the bonds is polar, but the molecule as a whole is nonpolar. From the Lewis structure, we determine that the CO_2 molecule is linear

Figure 7.12

(A) The stream of water is deflected toward the electrically charged rod because the polar water molecules are attracted by the charge. (B) The stream of nonpolar carbon tetrachloride molecules, CCl_4, is unaffected.

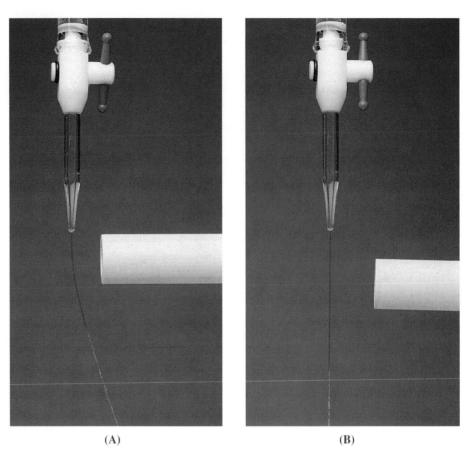

(A)   (B)

with polar $C{=}O$ bonds on opposite sides of the carbon atom. The bond polarities cancel because they are pointed in opposite directions.

$$\overset{\delta-\ \ \delta+\ \ \delta+\ \ \delta-}{O{=}C{=}O}$$

We can also symbolize the bond polarity in another way by using an arrow to show the direction of increasing electron density.

$$\underset{\longleftarrow\ \ \longrightarrow}{O{=}C{=}O}$$

If we apply the same criteria to the COS molecule, in which an S atom has replaced one of the O atoms, we again find that the Lewis structure predicts a linear molecule.

$$O{=}C{=}S$$

Although the $C{=}O$ and $C{=}S$ bonds are polar, the electronegativities of oxygen and sulfur are different (Fig. 6.5, Section 6.5). Thus the bonds have differing degrees of polarity and the dipoles do not cancel. The COS molecule is polar. Because oxygen is more electronegative than sulfur, the oxygen end of the molecule is the negative end. Chloromethane, CH_3Cl, is another example. Although the polar $C{-}Cl$ and $C{-}H$ bonds are arranged in a tetrahedral geometry, the $C{-}Cl$ bond has a dipole that is different from that of the $C{-}H$ bonds, and the dipoles do not completely cancel each other.

When we examine other molecules, such as BF_3 (trigonal planar), CH_4 (tetrahedral), PF_5 (trigonal bipyramidal), and SF_6 (octahedral), in which all the polar bonds are identical, we find that the molecules are nonpolar. The bonds in these molecules are arranged such that their dipoles cancel. However, if the dipoles do not cancel, then the molecule is polar; this is the case in many molecules that have both lone pairs and bonding pairs of electrons. Examples include H_2S and NH_3. A hydrogen atom is at the positive end, and the nitrogen or sulfur atom is at the negative end of the polar bonds in these molecules.

$$\delta+ \; H \diagdown \underset{\delta+ \; H \diagup}{\overset{..}{\underset{..}{S}}} \; \delta- \qquad\qquad \underset{\delta+ \; H}{\overset{\delta+ \; H \diagdown}{\delta+ \; H-N}} : \; \delta-$$

To summarize, to be polar a molecule must (1) contain at least one polar covalent bond and (2) not be totally symmetrical.

Example 7.8   **Explaining Polar Molecules**

Bringing a stream of liquid near an electrically charged rod (Fig. 7.12) demonstrates that water molecules are polar and carbon tetrachloride molecules are not. Explain the difference.

Solution: As shown in Table 6.1, the electronegativities of oxygen (3.5) and hydrogen (2.1) are different, so an O—H bond is polar. Similarly, the electronegativities of carbon (2.5) and chlorine (3.0) indicate that the C—Cl bond is polar. Both molecules contain polar bonds, so the difference in their properties must reflect a difference in their molecular geometries. The Lewis structures of these molecules are

$$\begin{array}{cc} \underset{H-O:}{\overset{\overset{\displaystyle H}{|}}{}} & \underset{:Cl:}{\overset{\overset{\displaystyle :Cl:}{|}}{:Cl-C-Cl:}} \end{array}$$

From the Lewis structures we can predict that H_2O has a bent molecular geometry (Fig. 7.8 B) and that CCl_4 has a tetrahedral molecular geometry. As in H_2S, the dipoles of the bonds in H_2O do not cancel, but the bond dipoles in the tetrahedral CCl_4 molecule do.

Check your learning: In an experiment similar to that shown in Fig. 7.12, the liquids $SnCl_4$, a starting material for the synthesis of many tin-containing pesticides, and CCl_3F, better known as CFC-11 or Freon-11, are caused to flow past a charged rod. Predict which liquid will be deflected by the rod and which will not.

Answer: CCl_3F will be deflected; $SnCl_4$ will not.

Valence Bond Theory: Hybridization of Atomic Orbitals

Lewis structures are usually satisfactory for identifying the types of bonds present in a molecule, and VSEPR theory is satisfactory for predicting electron-pair and molecular geometries. However, to achieve a more thorough understanding of electron distributions, we must use one of the models of bonding that is based on quantum mechanics. As we noted in Section 6.4, a covalent bond forms when one or more pairs of electrons are shared by two atoms and are simultaneously attracted by the nuclei of both. In the following sections, we will discuss how valence bond theory and hybridization are used to describe such bonds.

7.4   Valence Bond Theory

Valence bond theory attributes covalent bonding to overlapping atomic orbitals that share pairs of electrons. We say that orbitals **overlap** when a portion of one orbital and a portion of a second occupy the same region of space. According to valence bond theory, we expect a covalent bond to result when two conditions are met: (1) an orbital on one atom overlaps an orbital on a second atom and (2) an electron pair occupies both orbitals. Because of the overlap, the electrons are simultaneously attracted to both nuclei, and this attraction leads to bonding. The strength of a covalent bond depends on the amount of overlap of the orbitals involved. The strongest bond results when two bonded atoms are arranged such that the orbitals forming the bond have the maximum possible overlap.

Overlap increases as atoms approach each other (Fig. 7.13 A). However, there is an optimum distance between two bonded nuclei that leads to the strongest bond

Figure 7.13

(A) The interaction of two hydrogen atoms as a function of distance. (B) The sum of the energies of the electrons in two hydrogen atoms. At a distance of 0.74 Å, the H—H bond distance, the energy is a minimum of -7.24×10^{-19} J per bond (-436 kJ per mole of H_2 molecules).

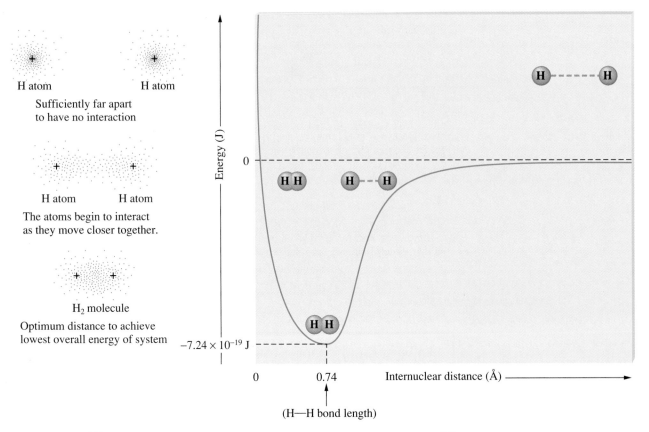

Figure 7.14

(A) The overlap of two *p* orbitals is greatest when the orbitals are directed end to end. (B) Any other arrangement results in less overlap. The plus signs indicate the locations of the nuclei.

(A)   (B)

between them. Figure 7.13 B illustrates how the sum of the energies of two hydrogen atoms (the colored line) changes as they approach each other. When the atoms are far apart there is no overlap, and we say the sum of the energies is zero. As the atoms move together their orbitals begin to overlap. Each electron begins to feel the attraction of the nucleus in the other atom. In addition, the electrons begin to repel each other, as do the nuclei. While the atoms are still widely separated, the attractions are a little stronger than the repulsions, and the energy of the system decreases (a bond begins to form). As the atoms move closer together the overlap increases, so the attraction of the nuclei for the electrons continues to increase (as do the repulsions among electrons and between the nuclei). At some specific distance between the atoms, which varies from bond to bond, the energy reaches its lowest (most stable) value. This specific distance corresponds to the bond distance between the two atoms (Section 7.1). The bond is stable because the attractive and repulsive forces are balanced. If the distance between the nuclei were to decrease further, the repulsions would become larger than the attractions, the energy of the system would increase rapidly, and the system would become less stable.

We call the difference between the energy of the two separated atoms and the energy when the atoms are at the bond distance the bond's energy. This is the quantity of energy released when the bond is formed. Conversely, the same amount of energy is required to break the bond. (We should remember that tables of bond energies, like that in Section 6.12, contain the energy required to break 1 mole of bonds, not to break just one bond. For example, it requires 7.24×10^{-19} joule to break one H—H bond, but 436,000 joules to break 1 *mole* of H—H bonds.)

In addition to the distance between two orbitals, their orientation also affects their overlap (except for two *s* orbitals, which are spherically symmetrical). The maximum overlap is obtained when orbitals are oriented such that they overlap end to end. Figure 7.14 illustrates this for two *p* orbitals; the overlap is greater when the orbitals overlap end to end rather than at an angle.

The overlap of two *s* orbitals (as in H_2), the overlap of an *s* orbital and a *p* orbital (as in HCl), and the end-to-end overlap of two *p* orbitals (as in Cl_2) all produce **sigma bonds (σ bonds),** as illustrated in Fig. 7.15. A sigma bond is a covalent bond in which the electron density is concentrated in the region along the internuclear axis; that is, a line between the nuclei would pass through the center of the overlap region. Single bonds in Lewis structures are described as sigma bonds in valence bond theory.

Figure 7.15

Sigma (σ) bonds form from the overlap of two *s* orbitals (A), of an *s* orbital and a *p* orbital (B) , and of two *p* orbitals (C). The plus signs indicate the locations of the nuclei.

(A)   (B)   (C)

Figure 7.16

Pi (π) bonds form from the side-by-side overlap of two *p* orbitals. The plus signs indicate the locations of the nuclei.

A **pi bond (π bond)** is a covalent bond that results from the side-by-side overlap of two *p* orbitals, as illustrated in Fig. 7.16. In a pi bond the regions of orbital overlap lie above and below the internuclear axis. A double bond in a Lewis structure, such as that found in O_2, consists of a sigma bond and a pi bond. A triple bond, such as that in N_2, consists of a sigma bond and two pi bonds. Only one bond in any multiple bond can be a sigma bond; the remaining bonds are pi bonds.

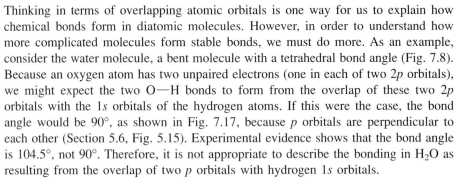

One sigma bond   One sigma bond   One sigma bond
No pi bonds   One pi bond   Two pi bonds

The overlap in a pi bond is less than that in a sigma bond, so pi bonds are generally weaker than sigma bonds. This is why a double bond between two atoms generally is not twice as strong as a single bond between the corresponding atoms (Section 6.12).

7.5   Hybridization of Atomic Orbitals

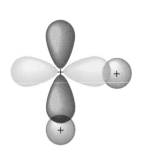

Figure 7.17

The hypothetical overlap of two of the 2*p* orbitals on an O atom (one shown in blue and the other in yellow with the 1*s* orbitals of two H atoms (shown in pink) produces a bond angle of 90°. The plus signs indicate the locations of the nuclei.

Thinking in terms of overlapping atomic orbitals is one way for us to explain how chemical bonds form in diatomic molecules. However, in order to understand how more complicated molecules form stable bonds, we must do more. As an example, consider the water molecule, a bent molecule with a tetrahedral bond angle (Fig. 7.8). Because an oxygen atom has two unpaired electrons (one in each of two 2*p* orbitals), we might expect the two O—H bonds to form from the overlap of these two 2*p* orbitals with the 1*s* orbitals of the hydrogen atoms. If this were the case, the bond angle would be 90°, as shown in Fig. 7.17, because *p* orbitals are perpendicular to each other (Section 5.6, Fig. 5.15). Experimental evidence shows that the bond angle is 104.5°, not 90°. Therefore, it is not appropriate to describe the bonding in H_2O as resulting from the overlap of two *p* orbitals with hydrogen 1*s* orbitals.

Quantum-mechanical calculations suggest why the observed bond angles in H_2O differ from those predicted by the overlap of the 1*s* orbital of the H atoms with the 2*p* orbitals of the O atom. These calculations lead to an important conclusion: The arrangement of the orbitals of an atom in a molecule is not the same as the arrangement of the atomic orbitals in the isolated (unbonded) atom.

The valence orbitals in an isolated oxygen atom are those described in Chapter 5; a 2*s* orbital and three 2*p* orbitals. The valence orbitals in an oxygen atom in a water molecule differ; they consist of four equivalent orbitals that point approximately toward the corners of a tetrahedron (Fig. 7.18). Consequently, the overlap of the O and H orbitals should result in a tetrahedral bond angle (109.5°). The observed angle of 104.5° is experimental evidence that quantum-mechanical calculations offer a useful explanation. (We shall see how the orbitals on the O atom can be described when we consider sp^3 hybridization in Section 7.8.)

We can describe the valence orbitals in a bonded atom by using the concept of **hybridization,** the mixing of the atomic orbitals of an isolated atom to generate a new set of atomic orbitals called **hybrid orbitals.** Hybrid orbitals are new types of atomic orbitals that result when two or more orbitals of an isolated atom mix. They are used to describe the orbitals in bonded atoms. The following six ideas are important in understanding hybridization.

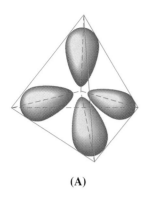

(A)

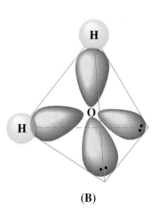

(B)

Figure 7.18

(A) The tetrahedral arrangement of the electron-pair geometry around the O atom in the water molecule. (B) Two of the hybrid orbitals on oxygen overlap with the 1*s* orbitals of hydrogen atoms to form the O—H bonds in H_2O.

Figure 7.19

Hybridization of an *s* orbital (yellow) and a *p* orbital of the same atom (blue) to produce two *sp* hybrid orbitals (red) oriented at 180° and on the same atom.

1. Hybrid orbitals do not exist in isolated atoms. They are found only in covalently bonded atoms.

2. Hybrid orbitals have shapes and orientations that are very different from those of the atomic orbitals in isolated atoms.

3. All orbitals in a set of hybrid orbitals are equivalent and form identical bonds (when the bonds are to a set of identical atoms).

4. The number of hybrid orbitals in a bonded atom is equal to the number of atomic orbitals that are used to form the hybrid orbitals.

5. The type of hybrid orbitals on a bonded atom depends on its electron-pair geometry as predicted by the VSEPR theory.

6. Sigma bonds in polyatomic molecules usually involve the overlap of hybrid orbitals. Pi bonds result from the overlap of unhybridized orbitals.

In the following sections, we shall discuss the common types of hybrid orbitals.

7.6 *sp* Hybridization

The beryllium atom in a gaseous $BeCl_2$ molecule (Example 7.1) is an example of a central atom with no unshared pairs of electrons in a linear arrangement of three atoms. The four valence orbitals of such a central atom consist of two hybrid orbitals and two remaining *p* orbitals that are not involved in the hybridization. The hybrid orbitals result from the mixing of one *s* orbital and one *p* orbital in the valence shell of the atom. This *sp* **hybridization** produces two equivalent *sp* **hybrid orbitals** that lie in a linear geometry (Fig. 7.19). In the linear $BeCl_2$ molecule, these hybrid orbitals overlap with orbitals of the chlorine atoms to form two identical sigma bonds. We illustrate the electronic differences between an isolated Be atom and the bonded Be atom in the orbital energy level diagram in Fig. 7.20. We represent each orbital by a line and each electron by an arrow; arrows pointing in opposite directions designate electrons of opposite spin. The colored arrows are electrons from the chlorine atoms in the Be—Cl bonds.

The electrons in atomic orbitals are redistributed when the orbitals hybridize. When a Be atom hybridizes, we redistribute the electrons such that there is one electron per hybrid orbital. Each of these electrons pairs up with the unpaired electron on a chlorine atom when a hybrid orbital and a chlorine orbital overlap during the formation of the Be—Cl bonds.

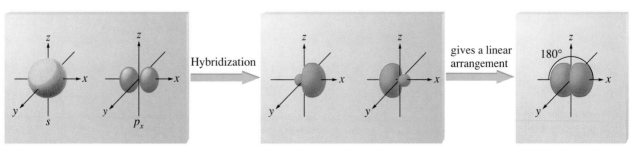

Figure 7.20

An orbital energy level diagram showing orbitals and electrons in an isolated Be atom and in the sp hybridized Be atom in $BeCl_2$.

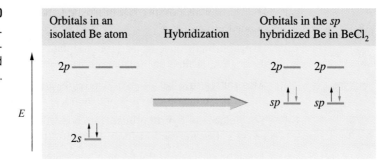

Other atoms that exhibit sp hybridization include the mercury atom in the linear $HgCl_2$ molecule, the zinc atom in $Zn(CH_3)_2$, which contains a linear C—Zn—C arrangement, and the carbon atoms in CO_2 and HCCH (acetylene), Section 7.11.

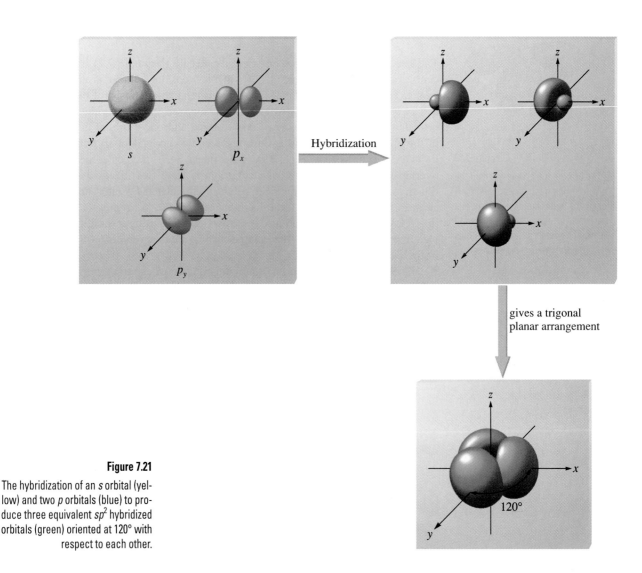

Figure 7.21

The hybridization of an s orbital (yellow) and two p orbitals (blue) to produce three equivalent sp^2 hybridized orbitals (green) oriented at 120° with respect to each other.

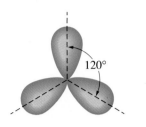

Figure 7.22

This alternative way of drawing the trigonal planar *sp²* hybrid orbitals is sometimes used in more crowded figures.

7.7   *sp²* Hybridization

The valence orbitals of a central atom surrounded by a trigonal planar arrangement of bonding pairs and unshared pairs consist of a set of three *sp²* **hybrid orbitals** and one unhybridized *p* orbital. This arrangement results from *sp²* **hybridization,** the mixing of one *s* orbital and two *p* orbitals that produces three identical hybrid orbitals (Fig. 7.21). Each of these three hybrid orbitals points toward a different corner of an equilateral triangle.

An alternative way to draw the lobes of the set of *sp²* hybrid orbitals is shown in Fig. 7.22. We sometimes use these thinner lobes when we want to draw a structure with several overlapping orbitals, because drawing thinner lobes gives a less crowded figure.

The structure of the boron trifluoride molecule, BF_3, suggests *sp²* hybridization for boron in this compound. The molecule is trigonal planar, and the boron atom is involved in three bonds to fluorine atoms (Fig. 7.23). Figure 7.24 shows the orbitals and electron distribution in an isolated boron atom and in the bonded atom in BF_3. We redistribute the three valence electrons of the boron atom in the three *sp²* hybrid orbitals, and each boron electron pairs with a fluorine electron when B—F bonds form.

Other atoms that exhibit *sp²* hybridization include the boron atom in BCl_3 (Fig. 7.6), the nitrogen atom in NO_3^- and ClNO, the sulfur atom in SO_2, and the carbon atoms in H_2CCH_2 and H_2CO (Fig. 7.1).

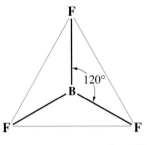

Figure 7.23

The trigonal planar structure of BF_3.

7.8   *sp³* Hybridization

The valence orbitals of an atom surrounded by a tetrahedral arrangement of bonding pairs and unshared pairs consist of a set of four *sp³* **hybrid orbitals.** The hybrids result from *sp³* **hybridization,** the mixing of one *s* orbital and all three *p* orbitals that produces four identical *sp³* hybrid orbitals (Fig. 7.25). Each of these hybrid orbitals points toward a different corner of a tetrahedron.

A molecule of methane, CH_4, consists of a carbon atom surrounded by four hydrogen atoms at the corners of a tetrahedron. This indicates that the carbon atom in methane exhibits *sp³* hybridization. Figure 7.26 shows the orbitals and electron distribution in an isolated carbon atom and in the bonded atom in CH_4. Note that we

Figure 7.24

An orbital energy level diagram showing orbitals and electrons in an isolated B atom and in the *sp²* hybridized B atom in BF_3.

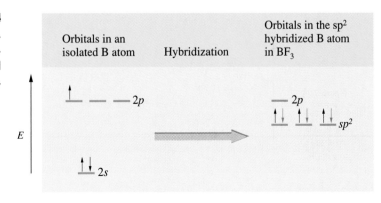

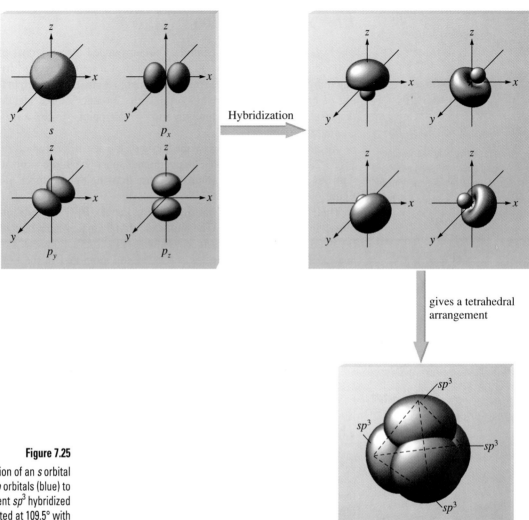

Figure 7.25

The hybridization of an *s* orbital (yellow) and three *p* orbitals (blue) to produce four equivalent sp^3 hybridized orbitals (purple) oriented at 109.5° with respect to each other.

Hybridization

gives a tetrahedral arrangement

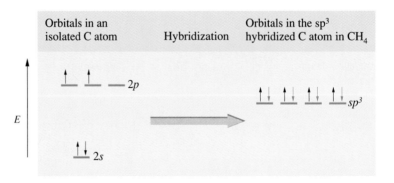

Figure 7.26

An orbital energy level diagram showing orbitals and electrons in an isolated C atom and in a sp^3 hybridized C atom in CH_4.

Figure 7.27

The ethane molecule. (A) The overlap diagram for C_2H_6. (B) The overall outline of the seven bonds in C_2H_6.

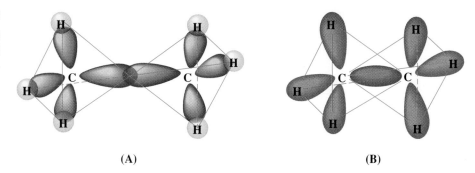

Figure 7.27

The ethane molecule. (A) The overlap diagram for C_2H_6. (B) The overall outline of the seven bonds in C_2H_6.

(A)   **(B)**

have redistributed the four valence electrons of the carbon atom in the hybrid orbitals and that each carbon electron pairs with a hydrogen electron when the C—H bonds form.

In a methane molecule, the $1s$ orbital of each of the four hydrogen atoms overlaps with one of the four sp^3 orbitals of the carbon atom to form an sp^3–s sigma (σ) bond. This results in the formation of four very strong, equivalent covalent bonds between the carbon atom and each of the hydrogen atoms to produce the methane molecule (Fig. 7.3).

The structure of ethane, C_2H_6, is similar to that of methane in that each carbon in ethane has four neighboring atoms (three H atoms and one C atom) arranged at the corners of a tetrahedron (Fig. 7.27 A). In ethane an sp^3 orbital of one carbon atom overlaps end to end with an sp^3 orbital of a second carbon atom to form a σ bond between the two atoms. Each of the remaining sp^3 hybrid orbitals overlaps with an s orbital of a hydrogen atom to form carbon–hydrogen σ bonds. The structure and over-all outline of the bonding orbitals of ethane are shown in Fig. 7.27 B.

A hybrid orbital can also hold an unshared pair of electrons. For example, the nitrogen atom in ammonia (Fig. 7.4 A) is surrounded by three bonding pairs of electrons and an unshared pair, all directed to the corners of a tetrahedron. The nitrogen atom is sp^3 hybridized with one hybrid orbital occupied by the unshared pair.

The molecular structure of water (Fig. 7.18) is consistent with a tetrahedral arrangement of two unshared pairs and two bonding pairs of electrons (Fig. 7.18 B). Thus we say that the oxygen atom is sp^3 hybridized, with two of the hybrid orbitals occupied by unshared pairs and two by bonding pairs.

Any atom with a tetrahedral arrangement of unshared pairs and bonding pairs may be regarded as being sp^3 hybridized. But hybridization occurs *only* when the atomic orbitals involved have very similar energies. It is possible to hybridize $2s$ and $2p$ orbitals or $3s$ and $3p$ orbitals, for example, but not $2s$ and $3p$ orbitals.

7.9   *sp³d* and *sp³d²* Hybridization

To describe the five bonding orbitals in a trigonal bipyramidal arrangement, we must use five of the valence shell atomic orbitals—the s orbital, the three p orbitals, and one of the d orbitals—giving five **sp^3d hybrid orbitals.** With an octahedral arrangement of six hybrid orbitals, we must use six valence shell atomic orbitals—the s orbital, the three p orbitals, and two of the d orbitals in its valence shell—giving six **sp^3d^2 hybrid orbitals.**

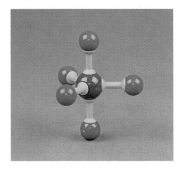

Figure 7.28

The trigonal bipyramidal structure of phosphorus pentachloride, PCl_5.

In a molecule of phosphorus pentachloride, PCl_5 (Fig. 7.28), there are five P—Cl bonds (and thus five pairs of valence electrons around the phosphorus atom) directed toward the corners of a trigonal bipyramid. We use the $3s$ orbital, the three $3p$ orbitals, and one of the $3d$ orbitals to form the set of five sp^3d hybrid orbitals (Fig. 7.29) involved in the P—Cl bonds. Other atoms that exhibit sp^3d hybridization include the sulfur atom in SF_4 (Fig. 7.9) and the chlorine atom in ClF_3 (Fig. 7.5 A) and ClF_4^+.

The sulfur atom in sulfur hexafluoride, SF_6, exhibits sp^3d^2 hybridization. A molecule of sulfur hexafluoride has six bonding pairs of electrons bonding six fluorine atoms surrounding a single sulfur atom (Fig. 7.30). There are no unshared pairs of electrons. To bond six fluorine atoms, the $3s$ orbital, the three $3p$ orbitals, and two of the $3d$ orbitals form six equivalent sp^3d^2 hybrid orbitals, each directed toward a different apex of an octahedron (Fig. 7.31). Other atoms that exhibit sp^3d^2 hybridization include the phosphorus atom in PCl_6^-, the iodine atom in the interhalogens IF_6^+, IF_5, ICl_4^- (Table 7.2), and IF_4^-, and the xenon atom in XeF_4 (Fig. 7.10).

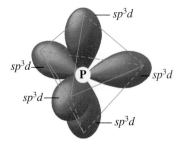

Figure 7.29

The five sp^3d orbitals on a phosphorus atom in PCl_5. These orbitals form a trigonal bipyramidal structure, and each has a small lobe that is not shown.

7.10   Assigning Hybrid Orbitals to Central Atoms

We recognize the hybridization of an atom from its molecular geometry. The geometrical arrangements characteristic of the various sets of hybrid orbitals are shown in Table 7.3. These arrangements are identical to those of the electron-pair geometries predicted by the VSEPR theory (Table 7.2). To find the hybridization of a central atom, we can use the following guidelines:

1. Determine the Lewis structure of the molecule.

2. Determine the electron-pair geometry from the number of bonding and unshared electron pairs.

3. Assign the set of hybridized orbitals that corresponds to this geometry. See Table 7.3.

Remember that *only* sigma bonds form from hybridized orbitals.

Figure 7.30

The octahedral structure of sulfur hexafluoride, SF_6; sp^3d^2 hybridization.

Figure 7.31

The six sp^3d^2 orbitals on a sulfur atom. These orbitals form an octahedral structure, and each has a small lobe that is not shown.

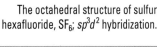

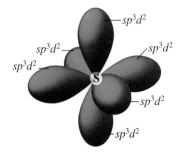

Table 7.3
Hybrid Orbitals

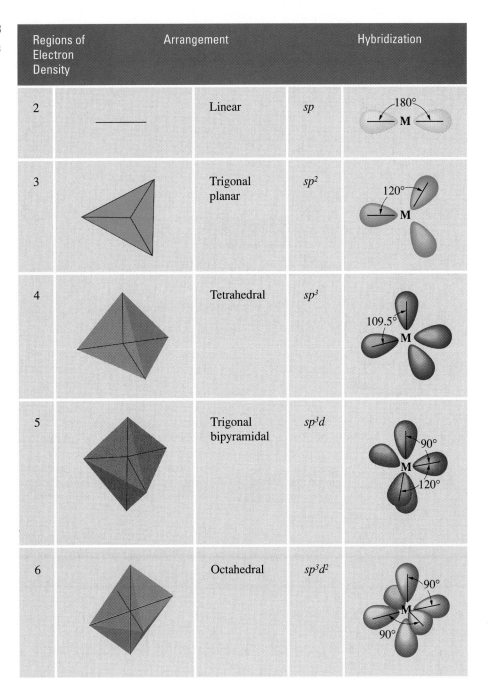

Regions of Electron Density	Arrangement		Hybridization
2		Linear	*sp*
3		Trigonal planar	*sp²*
4		Tetrahedral	*sp³*
5		Trigonal bipyramidal	*sp³d*
6		Octahedral	*sp³d²*

Example 7.9   **Assigning Hybridization**

Ammonium sulfate is important as a fertilizer (Example 7.3). What is the hybridization of the nitrogen atom in the ammonium ion, NH_4^+?

Solution: As described in Example 7.3, the nitrogen atom in NH_4^+ has a tetrahedral electron-pair geometry (arrangement of regions of high electron density; Fig. 7.4). This corresponds to sp^3 hybridization of nitrogen (Table 7.3).

Check your learning: The importance of boron trichloride, BCl_3, as an industrial catalyst was noted in Example 7.2. What is the hybridization of the boron atom in BCl_3?

Answer: The boron atom is sp^2 hybridized.

Example 7.10   Assigning Hybridization

Urea, $NH_2C(O)NH_2$, is sometimes used as a source of nitrogen in fertilizers. What is the hybridization of the nitrogen and carbon atoms in urea?

Solution: The Lewis structure of urea is

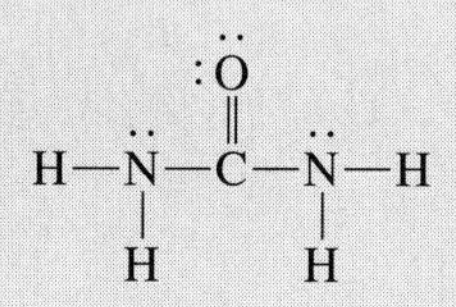

$$
\begin{array}{c}
\ddot{\text{O}}: \\
\|\\
\text{H}-\ddot{\text{N}}-\text{C}-\ddot{\text{N}}-\text{H} \\
\ \ \ | \qquad\quad | \\
\ \ \ \text{H} \qquad\ \ \text{H}
\end{array}
$$

The nitrogen atoms are surrounded by four regions of high electron density, which arrange themselves in a tetrahedral electron-pair geometry (Table 7.1). The hybridization in a tetrahedral arrangement is sp^3 (Table 7.3). This is the hybridization of the nitrogen atoms in urea.

 The carbon atom is surrounded by three regions of electron density in a trigonal planar arrangement (Table 7.1). The hybridization in a trigonal planar electron-pair geometry is sp^2 (Table 7.3), which is the hybridization of the carbon atom in urea.

Check your learning: Acetic acid, $H_3CC(O)OH$, is the molecule that gives vinegar its odor and sour taste. What is the hybridization of the two carbon atoms in acetic acid?

Answer: $H_3\underline{C}$, sp^3; $\underline{C}(O)OH$, sp^2

7.11   Hybridization Involving Double and Triple Bonds

The Lewis structure of ethylene, C_2H_4,

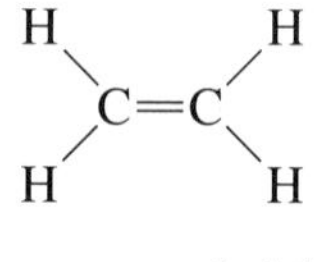

$$
\begin{array}{c}
\text{H} \qquad\qquad \text{H} \\
\ \diagdown \qquad\quad \diagup \\
\ \ \ \text{C}=\text{C} \\
\ \diagup \qquad\quad \diagdown \\
\text{H} \qquad\qquad \text{H}
\end{array}
$$

shows us that each carbon atom is surrounded by one other carbon atom and two hydrogen atoms. The three bonds form a trigonal planar electron-pair geometry. Thus

Figure 7.32

The ethylene molecule, C_2H_4. (A) The σ bonds result from overlap of sp^2 hybrid orbitals on the C atom with one sp^2 hybrid orbital on the other C atom and with s orbitals on the H atoms. There are four carbon–hydrogen bonds and one C—C σ bond. Note that the regions of orbital overlap are directly between the atoms. (B) Inclusion of the carbon–carbon π bond in C_2H_4. The π bond is formed by the side-to-side overlap of the unhybridized p orbitals in the two C atoms. The two portions of the π bond are above and below the plane of the σ system.

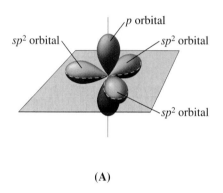

(A)

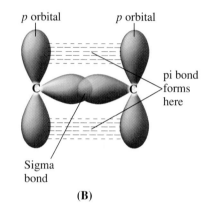

One π bond

(B)

Figure 7.33

(A) Diagram illustrating the three trigonal sp^2 hybrid orbitals of one carbon atom in C_2H_4. The hybrid orbitals lie in the same plane, and the one unhybridized p orbital is perpendicular to that plane. (B) The C=C double bond of C_2H_4 consists of a C—C σ bond resulting from end-to-end overlap of two sp^2 hybrids and a C—C π bond resulting from the side-by-side overlap of the p orbitals.

Figure 7.34

Diagram of the two linear sp hybrid orbitals of a carbon atom, which lie in a straight line, and the two unhybridized p orbitals.

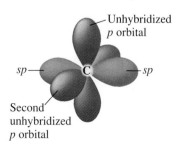

Unhybridized p orbital

sp —   — sp

Second unhybridized p orbital

we expect that the sigma (σ) bonds from each carbon atom are formed using a set of sp^2 hybrid orbitals. These orbitals form the carbon-hydrogen single bonds and the σ bond in the carbon-carbon double bond (Fig. 7.32). The pi (π) bond in the carbon-carbon double bond results from the overlap of the $2p$ orbital on each carbon atom that is not involved in the sp^2 hybridization.

Only two $2p$ orbitals hybridize with a $2s$ orbital, so we have one unhybridized $2p$ orbital left on each C atom. This unhybridized p orbital (shown in blue in Fig. 7.33 A) is perpendicular to the plane of the sp^2 hybrid orbitals. Thus when two sp^2 hybridized carbon atoms come together, one sp^2 orbital on each of them overlaps to form a σ bond, while the unhybridized $2p$ orbitals overlap in a side-by-side fashion (Fig. 7.33 B) and form a π bond (Section 7.4). The two carbon atoms of ethylene are thus bound together by two kinds of bonds—one σ and one π—giving the double bond.

In an ethylene molecule, the four hydrogen atoms and the two carbon atoms are all in the same plane. If the two planes of sp^2 hybrid orbitals were tilted, the π bond would be weakened because the p orbitals that form it cannot overlap effectively if they are not parallel. We find that a planar configuration for the ethylene molecule is the most stable form because the bonds are stronger in the planar arrangement.

As we saw in Section 7.6, formation of sp hybrid orbitals leaves two p orbitals unhybridized (Figs. 7.20 and 7.34). We find this situation in acetylene, H—C≡C—H, a linear molecule. Two sp hybrid orbitals, one on each of the two carbon atoms, overlap end to end to form a σ bond between the carbon atoms (Fig.

Figure 7.35

The acetylene molecule, C_2H_2. (A) The overlap diagram of two *sp* hybrid carbon atoms and two *s* orbitals from two hydrogen atoms. There are two carbon–hydrogen σ bonds and a carbon–carbon triple bond involving one carbon–carbon σ bond and two C—C π bonds. The dashed lines, each connecting two lobes, indicate the side-by-side overlap of the four unhybridized *p* orbitals. (B) The overall outline of the bonds in C_2H_2. The two portions of the π bonds are positioned with one above and below the line of the carbon–carbon σ bond and the other behind and in front of the line of the carbon–carbon σ bond.

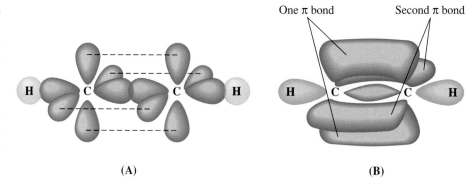

(A)   (B)

7.35). The remaining *sp* orbitals form σ bonds with hydrogen atoms. The two unhybridized *p* orbitals per carbon are positioned such that they overlap side by side and hence form two π bonds (Fig. 7.35). The two carbon atoms of acetylene are thus bound together by one σ bond and two π bonds, giving the **triple bond.**

The rules for determining hybridization also apply to molecules that are described with resonance forms (Section 6.11). When resonance occurs in a molecule or ion, hybridization of the central atom is unaffected. Resonance hybrids involve different placements of π bonds. Hybridization involves only sigma bonds and unshared pairs of electrons.

Example 7.11   Assignment of Hybridization Involving Resonance

Some acid rain results from the reaction of sulfur dioxide with atmospheric water vapor, followed by the formation of sulfuric acid. Sulfur dioxide, SO_2, is a major component of volcanic gases as well as a product of the combustion of sulfur-containing coal. What is the hybridization of the S atom in SO_2?

Solution: The resonance structures of SO_2 are

The sulfur atom is surrounded by two bonds and one unshared pair of electrons in both resonance structures. Therefore, the electron-pair geometry is trigonal planar and the hybridization of the sulfur atom is sp^2.

Check your learning: Another acid in acid rain is nitric acid, HNO_3, which is produced by the reaction of nitrogen dioxide, NO_2, with atmospheric water vapor. What is the hybridization of the N atom in NO_2? *Note:* The lone electron on N occupies a hybridized orbital just as an unshared pair would.

Answer: sp^2.

Molecular Orbital Theory

For almost every covalent molecule that exists, we can now draw the Lewis structure, predict the electron-pair geometry, predict the molecular geometry, and come very close to predicting approximate bond angles. However, one of the most important molecules we know, the oxygen molecule, O_2, presents a problem with respect to its Lewis structure. We would predict the following Lewis structure for O_2.

$$:\overset{..}{O}=\overset{..}{O}:$$

This electronic structure is very reasonable. There is an oxygen-oxygen double bond, and each oxygen atom has eight electrons around it. However, a variety of experiments (Fig. 7.36 shows one) reveal that oxygen is paramagnetic; that is, it has unpaired electrons. In fact, we find that an O_2 molecule has two unpaired electrons.

Molecular orbital theory (MO theory) provides a model that explains the experimental findings for the oxygen molecule. It also explains the bonding in a number of other molecules that are difficult to describe with Lewis structures. Additionally, it provides a model for describing the energies of electrons in a molecule as well as the probable location of these electrons. MO theory also helps us understand why some substances are electrical conductors, others semiconductors, and still others insulators (Chapter 24).

7.12 Molecular Orbitals

Molecular orbital theory describes the distribution of electrons in molecules in much the same way in which the distribution of electrons in atoms is described using atomic orbitals (Sections 5.4 and 5.5). Employing quantum mechanics, we describe the behavior of an electron in a molecule by a wave function, ψ, that can be used to determine the energy of the electron and the shape of the region of space within which it moves. As in an atom, we find that an electron in a molecule is limited to discrete (quantized) energies. The region of space in which a valence electron in a molecule is likely to be found often extends over all of the atoms in the molecule and is called a **molecular orbital.** Like an atomic orbital, a molecular orbital is full when it contains two electrons with opposite spin.

Figure 7.36

When liquid oxygen is poured between the poles of a strong magnet, some liquid remains there because of the attraction of its unpaired electrons for the magnetic field.

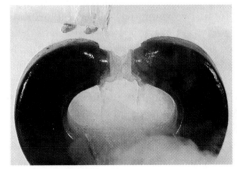

Figure 7.37

A representation of the formation of sigma (σ) molecular orbitals by the combination of two *s* atomic orbitals. The bonding molecular orbital is shown in purple, the antibonding molecular orbital in blue. The plus signs indicate the locations of nuclei.

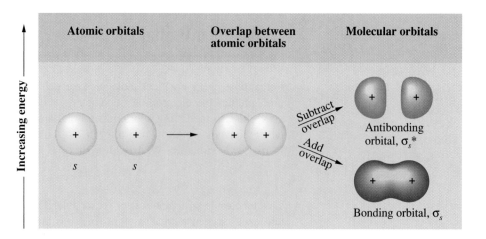

We will consider the orbitals in molecules composed of two identical atoms (H_2 or Cl_2, for example). Such molecules are called *homonuclear diatomic molecules.* Several different types of molecular orbitals occur in these diatomic molecules.

The exact wave function of a molecular orbital is difficult to determine. Approximate wave functions are generally used instead. Our approximation involves using the sum or the difference of the wave functions of two overlapping valence atomic orbitals of the constituent atoms to describe a molecular orbital.

Figure 7.37 illustrates the two types of molecular orbitals that can be formed from the overlap of two atomic *s* orbitals on adjacent atoms: a lower-energy σ_s (read as "sigma-*s*") molecular orbital is formed by addition of the *s* orbitals, and a higher-energy σ_s^* (read as "sigma-*s*-star") molecular orbital is formed by subtraction of the *s* orbitals. Electrons in a σ_s orbital are attracted by both nuclei at the same time and are more stable (of lower energy) than they would be in the isolated atoms. Adding electrons to these orbitals stabilizes a molecule, so we call these orbitals **bonding orbitals.** Electrons in the σ_s^* orbitals are located well away from the region between the two nuclei and destabilize the molecule; hence these orbitals are called **antibonding orbitals.** Electrons fill the lower-energy bonding orbital before the higher-energy antibonding orbital, just as they fill lower-energy atomic orbitals before they fill higher-energy atomic orbitals.

Figure 7.38 illustrates the four kinds of molecular orbitals that can be formed by the overlap of the *p* atomic orbitals on adjacent atoms. Each atom contains three *p* atomic orbitals in its valence shell: p_x, p_y, and p_z (Section 5.5). One of these (for example, p_x) overlaps end to end with a corresponding p_x atomic orbital of another atom and forms two **sigma (σ) molecular orbitals,** σ_{p_x} and $\sigma_{p_x}^*$ (read as "sigma-*p*-*x*" and "sigma-*p*-*x* star", respectively). Each of the other two *p* atomic orbitals, p_y and p_z, overlaps side by side with a corresponding *p* atomic orbital of another atom, giving rise to two **pi (π) bonding molecular orbitals** (only one of which is shown in Fig. 7.38) and two π^* antibonding molecular orbitals (only one of which is shown in Fig. 7.38). The two π bonding orbitals are oriented at right angles to each other, as are the two π^* antibonding orbitals. It is as though we have π and π^* molecular orbitals of one set oriented along the *y* axis of a set of coordinates and the π and π^* orbitals of the other set along the *z* axis. We use the notations π_{p_y} and π_{p_z} (read as "pi-*p*-*y*" and "pi-*p*-*z*") for the bonding π orbitals, and we use $\pi_{p_y}^*$ and $\pi_{p_z}^*$ ("pi-*p*-*y* star" and "pi-*p*-*z* star") for the antibonding π orbitals. Except for their orientation, the

Figure 7.38

A representation of the formation of sigma (σ) and pi (π) molecular orbitals by the combination of p atomic orbitals. The plus signs indicate the locations of nuclei. (A) The bonding and antibonding σ molecular orbitals formed by the end-to-end overlap of two p atomic orbitals. (B) The bonding and antibonding π molecular orbitals formed by the side-to-side overlap of two p atomic orbitals.

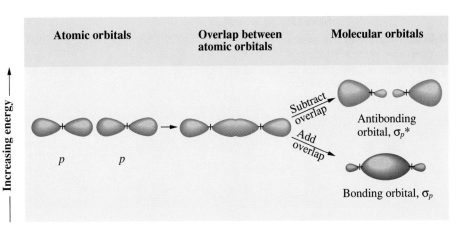

(A)

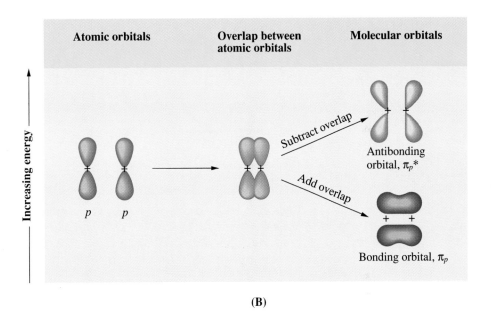

(B)

π_{p_y} and π_{p_z} orbitals are identical and have the same energy; they are **degenerate orbitals** (Section 5.5). The $\pi_{p_y}^*$ and $\pi_{p_z}^*$ antibonding orbitals are also degenerate and are identical except for their orientation. A total of six molecular orbitals result from the combination of the six atomic p orbitals in two atoms: σ_{p_x} and $\sigma_{p_x}^*$, π_{p_y} and $\pi_{p_y}^*$, and π_{p_z} and $\pi_{p_z}^*$.

7.13 Molecular Orbital Energy Diagrams

The relative energy levels of the lower-energy atomic and molecular orbitals of a homonuclear diatomic molecule are typically as shown in Fig. 7.39. In this figure each atomic or molecular orbital is represented by a line. Bonding and antibonding orbitals are joined by dashed lines to the atomic orbitals that combine to form them.

Figure 7.39

Molecular orbital energy diagram for a diatomic molecule containing identical atoms. Each solid line represents an atomic or molecular orbital that can hold one or two electrons.

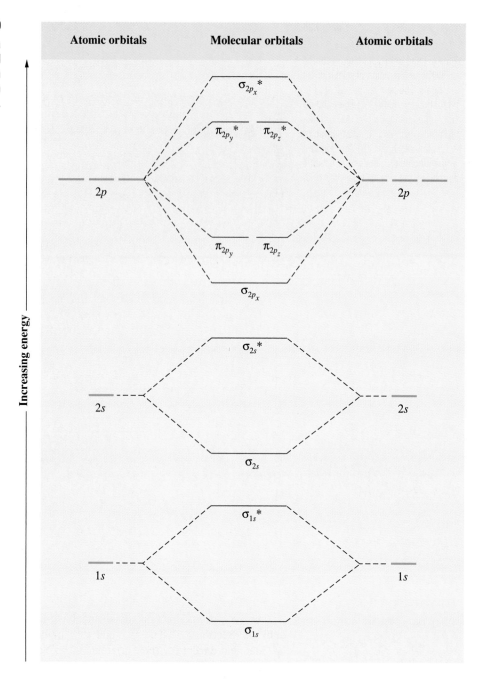

We predict the distribution of electrons in these molecular orbitals by filling the orbitals in the same way atomic orbitals are filled, by the aufbau process (Section 5.7). The number of electrons in each molecular orbital is indicated with a superscript. Thus we would expect a diatomic molecule or ion containing seven electrons to have the molecular electron configuration $(\sigma_{1s})^2(\sigma_{1s}*)^2(\sigma_{2s})^2(\sigma_{2s}*)^1$.

7.14 Bond Order

As electrons fill molecular orbitals in a diatomic molecule, some electrons enter bonding molecular orbitals, and others may enter antibonding molecular orbitals. Thus some electrons contribute to the stability of the molecule, whereas others may destabilize it. The net contribution of the electrons to the stability of a molecule can be identified by determining the net order of the bond that results from the filling of the molecular orbitals by electrons.

When using Lewis structures to describe the distribution of electrons in molecules, we define the order of a bond (bond order) as the number of bonding pairs of electrons between two atoms. Thus a single bond has a bond order of 1.0, a double bond has a bond order of 2.0, and a triple bond has a bond order of 3.0. Bond order is defined differently when the molecular orbital description of the distribution of electrons is used, but the resulting bond order is the same. In the molecular orbital model, the **bond order** for a given bond is the net number of pairs of bonding electrons and is equal to half the difference between the number of bonding electrons and the number of antibonding electrons. We can determine bond order with this equation:

$$\text{Bond order} = \frac{\left(\begin{array}{c}\text{number of} \\ \text{bonding electrons}\end{array}\right) - \left(\begin{array}{c}\text{number of} \\ \text{antibonding electrons}\end{array}\right)}{2}$$

The order of a covalent bond is a guide to its strength; a bond between two given atoms becomes stronger as the bond order increases. If the distribution of electrons in the molecular orbitals between two atoms is such that the resulting bond would have a bond order of zero, a stable bond does not form.

7.15 The H₂ and He₂ Molecules

A dihydrogen molecule (H_2) forms from two hydrogen atoms, each with one electron in a $1s$ atomic orbital. When the atomic orbitals of the two atoms combine, the electrons seek the molecular orbital of lowest energy, the σ_{1s} bonding orbital. A dihydrogen molecule, H_2, readily forms from two hydrogen atoms, because the energy of a H_2 molecule is lower than that of two H atoms. The σ_{1s} orbital that contains both electrons is lower in energy than either of the two $1s$ atomic orbitals.

A molecular orbital can hold two electrons, so both electrons in the H_2 molecule are in the σ_{1s} bonding orbital; the electron configuration is $(\sigma_{1s})^2$. We represent this configuration by a molecular orbital energy diagram (Fig. 7.40) in which we use an arrow, ↑, to indicate one electron in an orbital and two arrows, ↑↓, to indicate two electrons of opposite spin.

The bond order in a dihydrogen molecule is equal to half the difference between the number of bonding electrons and the number of antibonding electrons (Section 7.14). A dihydrogen molecule contains two bonding electrons and no antibonding electrons, so

$$\text{Bond order in } H_2 = \frac{2-0}{2} = 1$$

Because the bond order for the hydrogen–hydrogen bond is equal to 1, the bond is a single bond.

Figure 7.40

Molecular orbital energy diagram for the dihydrogen molecule, H_2.

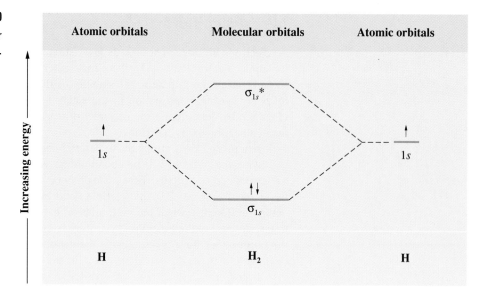

A helium atom has two electrons, both of which are in its $1s$ orbital (Table 5.4). Two helium atoms do not combine to form a dihelium molecule, He_2, with four electrons, because the stabilizing effect of the two electrons in the lower-energy bonding orbital would be offset by the destabilizing effect of the two electrons in the higher-energy molecular orbital. We would write the hypothetical electron configuration as $(\sigma_{1s})^2(\sigma_{1s}{}^*)^2$. The net energy change would be zero, so there is no driving force for helium atoms to form the diatomic molecule. In fact, helium exists as discrete atoms rather than as diatomic molecules. The bond order in a hypothetical He_2 molecule would be zero.

$$\text{Bond order in } He_2 = \frac{2 - 2}{2} = 0$$

A bond order of zero indicates that no bond is formed between two atoms.

7.16   Diatomic Molecules of the Second Period

Eight possible homonuclear diatomic molecules might be formed by the atoms of the second period of the periodic table: Li_2, Be_2, B_2, C_2, N_2, O_2, F_2, and Ne_2. However, we can predict that the Be_2 molecule and the Ne_2 molecule would not be stable. This can be explained by considering the molecular electron configurations (Table 7.4).

We predict valence molecular orbital electron configurations just as we predict electron configurations of atoms. Valence electrons are assigned to valence molecular orbitals with the lowest possible energies. Whenever there are two or more molecular orbitals of the same energy, electrons fill each orbital of that type singly before any pairing of electrons takes place within these orbitals.

The general order of increasing energy of the molecular orbitals in a homonuclear diatomic molecule is either

$$(\sigma_{1s})(\sigma_{1s}{}^*)(\sigma_{2s})(\sigma_{2s}{}^*)(\sigma_{2p_x})(\pi_{2p_y},\ \pi_{2p_z})(\pi_{2p_y}{}^*,\ \pi_{2p_z}{}^*)(\sigma_{2p_x}{}^*)$$

Table 7.4

Molecular Orbital Valence Electron Configurations for Diatomic Molecules of the Second Period

Molecule	Electron Configuration	Bond Order
Li_2	$(\sigma_{2s})^2$	1
Be_2 (unstable)	$(\sigma_{2s})^2(\sigma_{2s}*)^2$	0
B_2	$(\sigma_{2s})^2(\sigma_{2s}*)^2(\pi_{2p_y}, \pi_{2p_z})^2$	1
C_2	$(\sigma_{2s})^2(\sigma_{2s}*)^2(\pi_{2p_y}, \pi_{2p_z})^4$	2
N_2	$(\sigma_{2s})^2(\sigma_{2s}*)^2(\pi_{2p_y}, \pi_{2p_z})^4(\sigma_{2p_x})^2$	3
O_2	$(\sigma_{2s})^2(\sigma_{2s}*)^2(\sigma_{2p_x})^2(\pi_{2p_y}, \pi_{2p_z})^4(\pi_{2p_y}*, \pi_{2p_z}*)^2$	2
F_2	$(\sigma_{2s})^2(\sigma_{2s}*)^2(\sigma_{2p_x})^2(\pi_{2p_y}, \pi_{2p_z})^4(\pi_{2p_y}*, \pi_{2p_z}*)^4$	1
Ne_2 (unstable)	$(\sigma_{2s})^2(\sigma_{2s}*)^2(\sigma_{2p_x})^2(\pi_{2p_y}, \pi_{2p_z})^4(\pi_{2p_y}*, \pi_{2p_z}*)^4(\sigma_{2p_x}*)^2$	0

or

$$(\sigma_{1s})(\sigma_{1s}*)(\sigma_{2s})(\sigma_{2s}*)(\pi_{2p_y}, \pi_{2p_z})(\sigma_{2p_x})(\pi_{2p_y}*, \pi_{2p_z}*)(\sigma_{2p_x}*)$$

The order of the (π_{2p_y}, π_{2p_z}), and (σ_{2p_x}) levels changes between N_2 and O_2.

The atoms of the second period of the periodic table have electrons in the $1s$ shell ($n = 1$) and in the valence shell ($n = 2$). Because of the relatively high effective nuclear charge experienced by the electrons in inner shells, inner shells have small radii. Thus inner shells on adjacent atoms do not overlap and, consequently, do not form molecular orbitals. Electrons in inner shells do not enter into the bonding.

The combination of two lithium atoms to form a lithium molecule, Li_2, is analogous to the formation of H_2, but the atomic orbitals involved are the valence $2s$ orbitals. Each of the two lithium atoms, with an electron configuration $1s^2 2s^1$, has one valence electron. Hence we have two valence electrons available for the σ_{2s} bonding molecular orbital. Because both valence electrons would be in a bonding orbital, we would predict the Li_2 molecule to be stable. The molecule is, in fact, present in appreciable concentration in lithium vapor at temperatures near the boiling point of the element. All the other molecules in Table 7.4 with a bond order greater than zero are also known. Those molecules with a bond order of zero are not stable.

The O_2 molecule has enough electrons to half fill the $(\pi_{2p_y}*, \pi_{2p_z}*)$ level. We expect the two electrons that occupy these two degenerate orbitals to be unpaired, and this molecular electronic configuration for O_2 is consistent with the fact that the oxygen molecule has two unpaired electrons (Fig. 7.36). The presence of two unpaired electrons has proved difficult to explain using Lewis structures, but the molecular orbital theory explains it quite well. Furthermore, the unpaired electrons of the oxygen molecule provide one of the strong pieces of evidence in support of the molecular orbital theory.

Example 7.12 **Molecular Orbital Diagrams, Bond Order, and Number of Unpaired Electrons**

Draw the molecular orbital diagram for the O_2 molecule. From this diagram calculate the bond order for O_2. How does this diagram account for the paramagnetism of O_2?

Figure 7.41

Molecular orbital energy diagram for the valence orbitals of the dioxygen molecule, O_2.

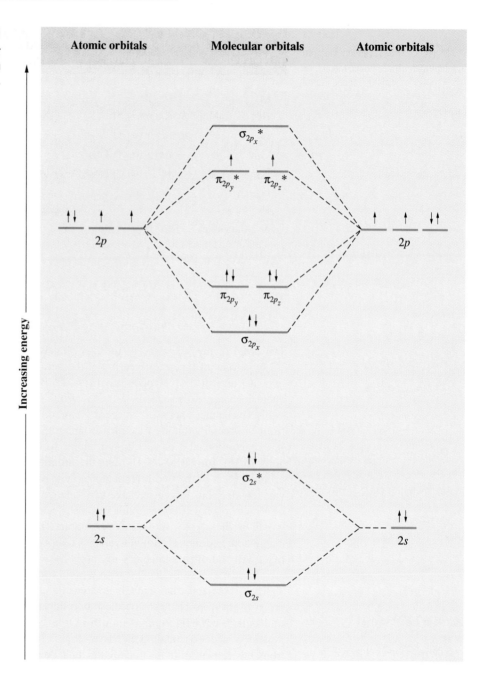

Solution: We draw a molecular orbital energy diagram similar to that shown in Fig. 7.39 in Section 7.13 except that, as noted in this section, we do not need to consider the inner $1s$ shells. Each oxygen atom contributes six valence electrons, so the diagram appears as shown in Fig. 7.41.

We calculate the bond order as follows:

$$\text{Bond order in } O_2 = \frac{8 - 4}{2} = 2$$

Oxygen's paramagnetism is explained by the presence of two unpaired electrons in the $(\pi_{2p_y}, \pi_{2p_z})^*$ molecular orbitals.

Check your learning: The accidental discovery of the formation of the dioxygenyl cation, O_2^+, and the role it played in the discovery of the entire field of noble gas chemistry, is an interesting part of the history of chemistry. From the molecular orbital diagram of O_2^+, predict its bond order and whether it is diamagnetic or paramagnetic.

Answer: O_2^+ has a bond order of 2.5 and is paramagnetic (with one unpaired electron).

For Review   Summary

The **molecular structure,** or three-dimensional arrangement of the atoms in a molecule or ion, is described in terms of **bond distances** and **bond angles.** We can predict approximate bond angles by using the **valence shell electron-pair repulsion (VSEPR) theory.** According to this theory, regions of high electron density (either bonding pairs or unshared pairs) located about a central atom repel each other, and the most stable arrangement is reached when they are as far away from each other as possible. For two regions of high electron density, the most stable arrangement is a **linear structure;** for three a **trigonal planar structure;** for four a **tetrahedral structure;** for five a **trigonal bipyramidal structure;** and for six an **octahedral structure.** Because an unshared pair of electrons is undetectable by the techniques used to determine molecular structure, the expected arrangement of the regions of high electron density and the actual molecular structure may appear different.

Valence bond theory describes bonding as resulting from the presence of a pair of electrons in two overlapping orbitals. Two orbitals on different atoms **overlap** when they occupy the same region of space. A bond results from the simultaneous attraction of the electrons to both of the bonded nuclei.

The orbitals of a central atom that participate in the formation of bonds with other atoms may be described as **hybrid orbitals,** combinations of some or all of its valence atomic orbitals. These hybrid orbitals either form **sigma (σ) bonds** directed toward other atoms of the molecule or contain unshared pairs. The type of **hybridization** around a central atom can be determined from

the geometry of the regions of high electron density about it. Two such regions imply sp hybridization; three, sp^2 hybridization; four, sp^3 hybridization; five, sp^3d hybridization; and six, sp^3d^2 hybridization. Atomic orbitals that are not used in hybridization are available to form **pi (π) bonds.**

Molecular orbital (MO) theory describes the behavior of an electron in a molecule by a wave function, which can be used to determine the energy of the electron and the region of space in the molecule where it is most likely to be found. The electron occupies an orbital that is called a **molecular orbital** because it may extend over all the atoms in the molecule. An electron in a **bonding molecular orbital** stabilizes a molecule. An electron in an **antibonding molecular orbital** makes a molecule less stable. A σ_s molecular orbital is a bonding orbital described by the addition of the overlap of two s orbitals of adjacent atoms. A σ_p molecular orbital is a bonding orbital described by the addition of the end-to-end overlap of p orbitals of adjacent atoms. A π_p molecular orbital is a bonding orbital described by the addition of the side-to-side overlap of p orbitals of adjacent atoms. The σ_s^*, σ_p^*, and π_p^* molecular orbitals are antibonding orbitals described by the subtraction of the overlap of atomic orbitals.

The filling of molecular orbitals with electrons is similar to the filling of atomic orbitals. Electrons fill orbitals of lowest energy first, and both π_{2p} orbitals (or π_{2p}^* orbitals) must be occupied by one electron before either will accept two electrons. According to their molecular orbital electron configurations, H_2 is stable but He_2 is not. Similarly, Li_2, B_2, C_2, N_2, O_2, and F_2 are expected to be stable, whereas Be_2 and Ne_2 are not. Experimental evidence confirms these expectations. In molecular orbital theory, the **bond order** of a bond between two atoms equals half the difference between the number of bonding electrons and the number of antibonding electrons the bond contains.

Key Terms and Concepts

antibonding orbital   (7.12)
axial position   (7.1)
bond angle   (7.1)
bond distance   (7.1)
bond order   (7.14)
bonding orbital   (7.12)
degenerate orbitals   (7.12)
dipole moment   (7.3)
double bond   (7.11)
electron-pair geometry   (7.1)

equatorial position   (7.1)
hybrid orbitals   (7.5–7.8)
hybridization   (7.5–7.10)
molecular geometry   (7.1)
molecular orbital   (7.12)
molecular orbital theory   (7.12)
molecular structure   (7.1)
overlap   (7.4)
pi (π) bond   (7.4, 7.11)
pi (π) molecular orbital   (7.12)

polar molecules   (7.3)
sigma (σ) bond   (7.4, 7.11)
sigma (σ) molecular orbital   (7.12)
triple bond   (7.11)
valence bond theory   (7.4)
valence shell electron-pair repulsion
    (VSEPR) theory   (7.1)

Exercises

Questions

1. Identify each of the following as part of chemistry's macroscopic domain or microscopic domain: (a) VSEPR theory, (b) hybridization, (c) a σ bond, (d) overlap of atomic orbitals, (e) an sp^3 hybrid orbital, (f) a trigonal planar molecular geometry, (g) a π_p molecular orbital, (h) bond order.

2. Explain why the HOH molecule is bent, whereas the HBeH molecule is linear.

3. What feature of its Lewis structure can be used to determine whether a molecule's (or an ion's) electron-pair geometry and molecular geometry will be identical?

4. Explain the difference between electron-pair geometry and molecular geometry.

5. Why is the H—N—H bond angle in NH_3 smaller than the H—C—H bond angle in CH_4? Why is the H—N—H bond angle in NH_4^+ identical to the H—C—H bond angle in CH_4?

6. Explain how a molecule that contains polar bonds can be nonpolar.

7. As a general rule, MX_n molecules (where M represents a central atom and X represents terminal atoms; $n = 2–5$) are polar if there is one or more unshared pairs of electrons on M. NH_3 is an example ($M = N$, $X = H$, $n = 3$). What two distributions of unshared pairs and bonding electrons are exceptions to this general rule?

8. Why is the concept of hybridization required in valence bond theory?

9. Explain how sigma bonds and pi bonds are similar and how they are different.

10. Sketch the distribution of electron density in one sp^3 hybrid orbital.

11. Explain why a carbon atom cannot form five bonds using sp^3d hybrid orbitals.

12. Sketch the distribution of electron density in the bond formed by the overlap of a sp^3 hybrid orbital with the $1s$ orbital of a hydrogen atom.

13. The bond energy of a C—C single bond averages 415 kJ mol^{-1}; that of a C≡C triple bond averages 837 kJ mol^{-1}. Explain why the triple bond is not three times as strong as a single bond.

14. Explain why an electron in the bonding molecular orbital in the H_2 molecule has a lower energy than an electron in the $1s$ atomic orbital of either of the separated H atoms.

15. Sketch the distribution of electron density in the bonding and antibonding molecular orbitals formed from two s orbitals and from two p orbitals.

16. How are the following similar and how do they differ?
    (a) σ molecular orbitals and π molecular orbitals.
    (b) ψ for an atomic orbital and ψ for a molecular orbital.
    (c) Bonding orbitals and antibonding orbitals.

Molecular Structure

17. Predict the electron-pair geometry and the molecular geometry of each of the following molecules and ions.

(a) SF_6 (b) PCl_5
(c) PO_4^{3-} (d) BeH_2
(e) CH_3^+

18. Identify the electron-pair geometry and the molecular geometry of each of the following molecules and ions.

(a) IF_6^+ (b) CF_4
(c) BF_3 (d) SiF_5^-
(e) $BeCl_2$

19. What are the electron-pair geometry and the molecular geometry of each of the following molecules and ions?

(a) ClF_5 (b) ClO_2^-
(c) $TeCl_4^{2-}$ (d) PCl_3
(e) SeF_4 (f) PH_2^-
(g) XeF_2

20. Predict the electron-pair geometry and the molecular geometry of each of the following molecules and ions.

(a) H_3O^+ (b) PCl_4^-
(c) $SnCl_3^-$ (d) $BrCl_4^-$
(e) ICl_3 (f) XeF_4
(g) SF_2

21. Identify the electron-pair geometry and the molecular geometry of each of the following molecules and ions.

(a) $ClNO$ (N is the central atom.)
(b) CS_2
(c) Cl_2CO (C is the central atom.)
(d) Cl_2SO (S is the central atom.)
(e) SO_2F_2 (S is the central atom.)
(f) XeO_2F_2 (Xe is the central atom.)
(g) $ClOF_2^+$ (Cl is the central atom.)

22. Predict the electron-pair geometry and the molecular geometry of each of the following molecules and ions.

(a) IOF_5 (I is the central atom.)
(b) $POCl_3$ (P is the central atom.)
(c) Cl_2SeO (Se is the central atom.)
(d) $ClSO^+$ (S is the central atom.)
(e) F_2SO (S is the central atom.)
(f) NO_2^-
(g) SiO_4^{4-}

23. Which of the molecules and ions in Exercises 17 and 19 contain polar bonds? Which of these molecules and ions have a dipole moment?

24. Which of the molecules and ions in Exercises 18 and 20 contain polar bonds? Which of these molecules and ions have a dipole moment?

25. Which of the following molecules have a dipole moment?

(a) CS_2
(b) SeS_2
(c) CCl_2F_2
(d) $OPCl_3$ (P is the central atom.)
(e) $ClNO$ (N is the central atom.)

26. Identify the molecules that have a dipole moment.

(a) SF_4 (b) CF_4
(c) Cl_2CCBr_2 (d) CH_3Cl
(e) H_2CO

27. The molecule XF_3 has a dipole moment. Is X boron or phosphorus?

28. The molecule XCl_2 has a dipole moment. Is X beryllium or sulfur?

29. Is the molecule Cl_2BBCl_2 polar or nonpolar?

30. There are three possible structures for PCl_2F_3. Draw them and discuss how measurements of dipole moments could help to distinguish among them.

31. Describe the molecular geometry around the indicated atom or atoms.

(a) The sulfur atom in sulfuric acid, H_2SO_4 [$(HO)_2SO_2$].
(b) The chlorine atom in chloric acid, $HClO_3$ [$HOClO_2$].
(c) The oxygen atom in hydrogen peroxide, HOOH.
(d) The nitrogen atom in nitric acid, HNO_3 [$HONO_2$].
(e) The oxygen atom in the OH group in nitric acid, HNO_3 [$HONO_2$].
(f) The central oxygen atom in the ozone molecule, O_3.
(g) Each of the carbon atoms in propyne, CH_3CCH.
(h) The carbon atom in the Freon CCl_2F_2.
(i) Each of the carbon atoms in allene, H_2CCCH_2.

Valence Bond Theory

32. Draw a curve that describes the change in the sum of the energies of H and Cl during the bond formation in HCl. What is the approximate energy at the lowest point on the curve?

33. Use valence bond theory to explain the bonding in H_2, HCl, and Cl_2. Sketch the overlap of the atomic orbitals involved in the bonds.

34. Use valence bond theory to explain the bonding in O_2. Sketch the overlap of the atomic orbitals involved in the bonds in O_2.

35. What is the hybridization of the central atom in each of the following?

(a) SF_6 (b) PCl_5
(c) PO_4^{3-} (d) BeH_2
(e) CH_3^+

36. Describe the hybridization of the central atom in each of the following molecules and ions.

(a) IF_6^+ (b) CF_4
(c) BF_3 (d) SiF_5^-
(e) $BeCl_2$

37. What is the hybridization of the central atom in each of the following molecules and ions?

(a) ClF_5 (b) ClO_2^-
(c) $TeCl_4^{2-}$ (d) PCl_3

(e) SeF_4
(f) PH_2^-
(g) XeF_2

38. Describe the hybridization of the central atom in each of the following molecules and ions.

(a) H_3O^+ (b) PCl_4^-
(c) $SnCl_3^-$ (d) $BrCl_4^-$
(e) ICl_3 (f) XeF_4
(g) SF_2

39. Identify the hybridization of the central atom in each of the following molecules and ions.

(a) $ClNO$ (N is the central atom.)
(b) CS_2
(c) Cl_2CO (C is the central atom.)
(d) Cl_2SO (S is the central atom.)
(e) SO_2F_2 (S is the central atom.)
(f) XeO_2F_2 (Xe is the central atom.)
(g) $ClOF_2^+$ (Cl is the central atom.)

40. What is the hybridization of the central atom in each of the following molecules and ions?

(a) IOF_5 (I is the central atom.)
(b) $POCl_3$ (P is the central atom.)
(c) Cl_2SeO (Se is the central atom.)
(d) $ClSO^+$ (S is the central atom.)
(e) F_2SO (S is the central atom.)
(f) NO_2^-
(g) SiO_4^{4-}

41. Describe the hybridization of the indicated atom in each of the following molecules and ions.

(a) The sulfur atom in sulfuric acid, H_2SO_4 [$(HO)_2SO_2$].
(b) The chlorine atom in chloric acid, $HClO_3$ [$HOClO_2$].
(c) The oxygen atom in hydrogen peroxide, $HOOH$.
(d) The nitrogen atom in nitric acid, HNO_3 [$HONO_2$].
(e) The oxygen atom in the OH group in nitric acid, HNO_3 [$HONO_2$].
(f) The central oxygen atom in the ozone molecule, O_3.
(g) Each of the carbon atoms in propyne, CH_3CCH.
(h) The carbon atom in the Freon CCl_2F_2.
(i) Each of the carbon atoms in allene, H_2CCCH_2.

Molecular Orbitals

42. Using molecular orbital energy diagrams and the electron occupancy of the molecular orbitals, compare the stability of H_2, HHe, and He_2. What is the bond order of each of these molecules?

43. Write the valence molecular orbital electron configurations for the diatomic ions X_2^{2+} where X is one of the elements of the third period (Na to Ar) of the periodic table. Assume the order of orbital energies is $(\sigma_{3s})(\sigma_{3s}^*)(\sigma_{3p_x})(\pi_{3p_y}, \pi_{3p_z})(\pi_{3p_y}^*, \pi_{3p_z}^*)(\sigma_{3p_x}^*)$ in each of these ions. On the basis of the bond order of these ions, tell which would not be expected to be stable.

44. The peroxide ion, O_2^{2-}, is found in the ionic compound sodium peroxide, Na_2O_2. Draw the molecular orbital energy diagram of this ion and write its molecular orbital electron configuration.

45. Potassium superoxide, KO_2, is used in respirators because it reacts with the moisture in the breath and releases oxygen gas. KO_2 contains the superoxide ion, O_2^-. Draw the molecular orbital energy diagram of this ion and write its valence molecular orbital electron configuration.

46. Determine the bond order of each member of the following groups. Which member of each group is predicted by the molecular orbital model to have the strongest bond?

(a) H_2, H_2^+, H_2^-
(b) O_2, O_2^{2+}, O_2^{2-}
(c) Li_2, Be_2^+, Be_2
(d) F_2, F_2^+, F_2^-
(e) N_2, N_2^+, N_2^-

47. Identify the member of each of the following pairs that has the higher first ionization energy (indicating the most tightly bound electron) in the gas phase.

(a) H and H_2 (b) N and N_2
(c) O and O_2 (d) C and C_2
(e) B and B_2

Applications and Additional Exercises

48. Draw Lewis structures for CO_2, NO_2^-, SO_3, and SO_3^{2-} and predict the shape of each.

49. A molecule with the formula AB_2, where A and B represent different atoms, could have one of three different shapes. Sketch and name the three different shapes that this molecule might have. Give an example of a molecule or ion that has each shape.

50. A molecule with the formula AB_3, where A and B represent different atoms, could have one of three different shapes. Sketch and name the three different shapes that this molecule might have. Give an example of a molecule or ion that has each shape.

51. (a) Draw the Lewis structures for CS_3^{2-}, CS_2, and CS, including resonance structures where appropriate. (b) Predict the molecular shapes for the first two species, and explain how you arrived at your predictions.

52. What is a compound or ion with a sp^3d hybridized phosphorus atom? With a sp^3d^2 hybridized phosphorus atom? With a sp^3 hybridized sulfur atom?

53. Write Lewis structures for NF_3 and PF_5. On the basis of bonding considerations, explain the fact that NF_3, PF_3, and PF_5 are stable molecules but NF_5 does not exist.

54. Methionine, $CH_3SCH_2CH_2CH(NH_2)CO_2H$, is an amino acid obtained from proteins. Draw a Lewis structure of this compound. What is the hybridization type of each carbon, of the nitrogen, and of the sulfur?

55. Sulfuric acid is manufactured by a series of reactions represented by the following equations:

$$S_8(s) + 8O_2(g) \longrightarrow 8SO_2(g)$$

$$2SO_2(g) + O_2(g) \longrightarrow 2SO_3(g)$$

$$SO_3(g) + H_2O(l) \longrightarrow H_2SO_4(l)$$

For each of the sulfur-containing molecules, (a) draw a Lewis structure, (b) predict the molecular geometry by VSEPR, and (c) determine the hybridization of sulfur. (The S_8 molecule consists of a ring containing eight sulfur atoms.)

56. Two important industrial chemicals, ethylene, C_2H_4, and propylene, C_3H_6, are produced by the steam (or thermal) cracking process.

$$2C_3H_8(g) \longrightarrow C_2H_4(g) + C_3H_6(g) + CH_4(g) + H_2(g)$$

For each of the four carbon compounds, (a) draw a Lewis structure, (b) predict the geometry about the carbons, and (c) determine the hybridization of each type of carbon atom.

57. For many years after they were discovered, it was believed that the noble gases could form no compounds. Now we know that belief was incorrect. A mixture of xenon and fluorine gases, confined in a quartz bulb and placed on a window sill, slowly produces a white solid. Analysis of the solid indicates that it contains 77.55% Xe and 22.58% F.

 (a) What is the formula of the compound?
 (b) Write a Lewis structure for the compound, assuming one Xe atom per molecule.
 (c) Predict the shape of the molecules of the compound.
 (d) What hybridization is consistent with the shape you predicted?

58. Hydrazine, N_2H_4, is used as a liquid rocket fuel. Draw a Lewis structure for the compound. Identify the hybridization of the nitrogen atoms. What geometry would be expected for each nitrogen? What bond angle would be expected for the H—N—H and H—N—N bonds?

59. In addition to NF_3, two other fluorine compounds of nitrogen are known: N_2F_4 and N_2F_2, each with a nitrogen–nitrogen bond. What shapes do you predict for these two molecules. What hybridization do you assign to the nitrogen in each molecule?

60. A useful solvent that will dissolve salts as well as organic compounds is the compound acetonitrile, H_3CCN. It is also present in paint strippers.

 (a) Write the Lewis structure for acetonitrile, and indicate the direction of the dipole moment in the molecule.
 (b) Identify the hybrid orbitals used by the carbon atoms in the molecule to form sigma bonds.
 (c) Describe the atomic orbitals that form the pi bonds in the molecule. Note that it is not necessary to hybridize the nitrogen atom.

61. (a) Write a Lewis structure for nitrous acid, HNO_2 (HONO).
 (b) What are the electron-pair and molecular geometries of the internal oxygen and nitrogen atoms in the HNO_2 molecule?
 (c) What hybridization may be assigned to the internal oxygen and nitrogen atoms in HNO_2?

62. Strike-anywhere matches contain a layer of $KClO_3$ and a layer of P_4S_3. The heat produced by the friction of striking the match causes these two compounds to react and to inflame, setting fire to the wooden stem of the match. $KClO_3$ contains the ClO_3^- ion. P_4S_3 is an unusual molecule with the skeleton structure

 (a) Write Lewis structures for P_4S_3 and the ClO_3^- ion
 (b) Describe the geometry about the P atoms, about the S atoms, and about the Cl atom in these species.
 (c) Assign a hybridization to the P atoms, to the S atoms, and to the Cl atom in these species.
 (d) Determine the oxidation states and formal charge of the atoms in P_4S_3 and in the ClO_3^- ion.

63. One of the resonance forms of the CO_3^{2-} ion is

Even though double bonds are bulkier than single bonds, the O—C—O bond angles in the carbonate ion are equal. Explain why.

64. Three molecules contain only two carbon atoms and several hydrogen atoms. Which contains the strongest carbon–carbon bond: the molecule with sp hybridized carbon atoms, the molecule with sp^2 hybridized carbon atoms, or the molecule with sp^3 hybridized carbon atoms?

65. A 0.10-mol sample of a compund contains 0.10 mol of carbon atoms, 0.10 mol of nitrogen atoms, and 0.10 mol of hydrogen atoms. Identify the molecular structure of the compound and the hybridization of the central atom. Describe the bonding in the compound in terms of valence bond theory, and show the overlap of the various orbitals that are used in forming the bonds.

66. One mole of S reacts with 2 mol of F_2 to give a compound containing one S atom and several F atoms per molecule. Identify the molecular structure of the compound and the hybridization of the sulfur. Describe the bonding in the compound in terms of valence bond theory, and show the overlap of the various orbitals that are used in forming the bonds.

67. Describe the molecular geometry and hybridization of the N, P, or S atoms in each of the following compounds.

(i) H_3PO_4, phosphoric acid, used in cola soft drinks.
(ii) NH_4NO_3, ammonium nitrate, a fertilizer and explosive.
(iii) ClSSCl, used in vulcanizing rubber.
(iv) $K_4[O_3POPO_3]$, potassium pyrophosphate, an ingredient of some toothpastes.

68. Describe the molecular geometry and hybridization of the C and O atoms in a molecule of ethanol. The skeleton structure is

69. Identify the hybridization of each carbon atom in the following molecule. (The arrangement of atoms is given.)

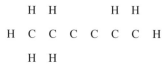

70. What is the geometry of the stable form of FNO_2? (N is the central atom.)

71. What are the molecular geometry and hybridization of the carbon atoms in a compound with a molar mass of about 28 that contains 85.7% carbon and 14.3% hydrogen.

72. A compound with a molar mass of about 42 contains 85.7% carbon and 14.3% hydrogen. Thus, what is its molecular geometry?

73. Identify the molecular geometry and hybridization of the N atom in a compound with a molar mass of about 70 that contains 19.7% nitrogen and 80.3% fluorine. Also determine the formal charge and oxidation state of the atoms in this compound.

74. (This question is taken from the 1994 Chemistry Advanced Placement Examination and is used with the permission of the Educational Testing Service.)

Consider the species

$$NO_2 \qquad NO_2^- \qquad NO_2^+$$

(a) Draw the Lewis electron-dot structure for each of the three species.
(b) List the species in order of increasing bond angle. Justify your answer.
(c) Select one of the species and give the hybridization of the nitrogen atom in it.
(d) Identify the only one of the species that dimerizes, and explain what causes it to do so. (*Note:* We will see in Chapter 14 that the molecule resulting from the dimerization has a bond formed by the pairing of unpaired electrons on the reactant molecules.)

CHAPTER OUTLINE

8.1 Chemical Behavior and the Periodic Table
8.2 Metals, Nonmetals, and Semi-Metals
8.3 Oxidation States
8.4 Periodic Variations in Oxidation State
8.5 Chemical Reactions
8.6 Variation in Metallic and Nonmetallic
    Behavior of the Representative Elements
8.7 The Activity Series
8.8 Prediction of Reaction Products
8.9 Chemical Properties of Some Important
    Industrial Chemicals

8

Chemical Reactions and the Periodic Table

The productivity of many farms results from the use of ammonia as a fertilizer.

For the past 10 years, an average of over 1.5×10^{10} kilograms (34,000,000,000 pounds) of ammonia, NH_3, has been produced in North America each year. This is about 100 pounds per year for every person on the continent. Some of the nitrogen from this ammonia ends up in our food—ammonia is an important nitrogen-rich fertilizer and is used extensively. Plants incorporate the nitrogen from ammonia into protein, vitamins, and other nitrogen-containing nutrients, which we eat when we enjoy grains, fruits, and vegetables. Other plant matter is eaten by animals, which produce meat that has a high protein content. Large quantities of ammonia are converted to nitric acid, another source of fertilizers and an important industrial chemical.

Ammonia is only a small part of the chemical industry. Each year, over 2.0×10^{11} kilograms (450 billion pounds) of various elements and inorganic chemicals and 1.3×10^{11} kilograms (280 billion pounds) of organic chemicals are produced in the United States alone. Large amounts are also produced in other industrialized countries. We will see that the chemistry and applications of these elements and inorganic compounds illustrate common and important types of chemical behavior.

In this chapter we will begin our study of the chemical reactions of many common elements and compounds. First, we will consider the differing chemical behaviors of metals, semi-metals, and nonmetals. This behavior, coupled with the common oxidation states that the elements exhibit, will serve as a valuable guide to the types of compounds that elements form. Next we will extend our knowledge of types of reactions beyond the introduction that appears in Chapter 2. We will then consider how we can predict the products of many common types of chemical reactions using the periodic table as a guide. Finally, we will conclude with a brief survey of the production and behavior of several industrially important inorganic compounds.

Figure 8.1

Mendeleev's early periodic table, published in 1872. Note the gaps left for missing elements with atomic masses 44, 68, 72, and 100 and the other gaps for elements with unpredicted masses.

8.1 Chemical Behavior and the Periodic Table

We can predict much of the behavior of an element from its position in the periodic table. As an example, consider the use that Mendeleev made of his periodic table in predicting the properties of unknown elements. Mendeleev's first table (Fig. 8.1)

TABELLE II

REIHEN	GRUPPE I. — R²O	GRUPPE II. — RO	GRUPPE III. — R²O³	GRUPPE IV. RH⁴ RO²	GRUPPE V. RH³ R²O⁵	GRUPPE VI. RH² RO³	GRUPPE VII. RH R²O⁷	GRUPPE VIII. — RO⁴
1	H = 1							
2	Li = 7	Be = 9,4	B = 11	C = 12	N = 14	O = 16	F = 19	
3	Na = 23	Mg = 24	Al = 27,3	Si = 28	P = 31	S = 32	Cl = 35,5	
4	K = 39	Ca = 40	— = 44	Ti = 48	V = 51	Cr = 52	Mn = 55	Fe = 56, Co = 59, Ni = 59, Cu = 63.
5	(Cu = 63)	Zn = 65	— = 68	— = 72	As = 75	Se = 78	Br = 80	
6	Rb = 85	Sr = 87	?Yt = 88	Zr = 90	Nb = 94	Mo = 96	— = 100	Ru = 104, Rh = 104, Pd = 106, Ag = 108.
7	(Ag = 108)	Cd = 112	In = 113	Sn = 118	Sb = 122	Te = 125	J = 127	
8	Cs = 133	Ba = 137	?Di = 138	?Ce = 140	—	—	—	— — —
9	(—)	—	—	—	—	—	—	
10	—	—	?Er = 178	?La = 180	Ta = 182	W = 184	—	Os = 195, Ir = 197, Pt = 198, Au = 199.
11	(Au = 199)	Hg = 200	Tl = 204	Pb = 207	Bi = 208	—	—	
12	—	—	—	Th = 231	—	U = 240	—	— — —

	Predicted for Eka-silicon	Silicon	Germanium	Tin
Atomic Weight	72	28	72.59	118
Specific Gravity	5.5	2.3	5.3	7.3
Color	Gray metal	Gray nonmetal	Gray metal	White metal
Oxidation State with Oxygen	+4	+4	+4	+4
Reaction with Acid	Very slow reaction	No reaction	Slow reaction with conc. acid	Slow reaction
Formula of Chloride	$EkCl_4$	$SiCl_4$	$GeCl_4$	$SnCl_4$

Table 8.1

Predicted Properties for Eka-silicon and Observed Properties of Silicon, Germanium, and Tin

included the elements known at that time. In this table, elements with similar chemical properties were placed in the same group, but gaps remained where other elements appeared to be missing. Mendeleev believed that undiscovered elements would fill the gaps, and assuming that chemical properties vary smoothly, he predicted the properties of the unknown elements from the known chemical behavior of their neighbors in the table. Time has proved Mendeleev right: The elements that now fill the gaps have properties similar to what he predicted.

Table 8.1 gives a comparison of the properties predicted by Mendeleev for germanium, which he called eka-silicon, and those determined for the element once it was isolated. You can see that the properties are intermediate between those of silicon and tin, the neighbors of germanium in Group 4A.

Although a modern periodic table looks a little different from Mendeleev's table, we can correlate and recall the behavior of the elements and their compounds by using the same principles Mendeleev used to predict their properties. The periodic table provides a powerful framework for organizing the chemical behavior of the elements.

8.2 Metals, Nonmetals, and Semi-Metals

We divide elements into three broad groups: metals, nonmetals, and semi-metals (Section 1.7 and Fig. 1.13). Metals lie to the left and below the broad line that zigzags from the left-hand side of the boron (B) block to the right-hand side of the polonium (Po) block in a periodic table (see the inside front cover of this text). Nonmetals lie to the right and above this broad line. Semi-metals lie along the line. A pure **metal** is generally a good conductor of heat and electricity. It has a metallic luster and is malleable and ductile (that is, we can bend it, hammer it into a sheet, or draw it into a wire without breaking it). A pure **nonmetal** is generally a poor conductor. It normally has no metallic luster and is brittle and nonductile in the solid state.

Metals and nonmetals also have different chemical properties. The general types of chemical behavior characteristic of metals and nonmetals are outlined in Table 8.2 on page 260. If we are familiar with the general behavior patterns of metals and nonmetals presented in this table, we can reasonably predict a great deal of the chemical behavior of an element simply from its position in the periodic table.

Some elements, called **semi-metals** or **metalloids,** cannot be clearly identified as being either metals or nonmetals; they possess some of the properties of each. The semi-metal silicon, for example, exhibits a bright metallic luster, but it is not a good conductor and it is brittle. Semi-metals may exhibit either metallic or nonmetallic

Table 8.2

Chemical Behavior of Metals and Nonmetals

Metals	Nonmetals
1. React with elemental nonmetals (except noble gases)	1. React with elemental metals and often with less electronegative nonmetals
2. Form oxides that, if soluble, react with water to give hydroxides	2. Form oxides that may react with water to give acids
3. Form basic hydroxides	3. Form acidic hydroxides (oxyacids)
4. React with O_2, F_2, H_2, and other nonmetals, usually giving ionic compounds	4. React with O_2, F_2, H_2, and other nonmetals, giving covalent compounds
5. Form binary hydrides that, if soluble, react with water to form hydrogen gas and the metal hydroxide	5. Form binary hydrides, which may be acidic
6. React with other metals, giving metallic compounds	6. React with metals, often giving ionic compounds
7. Exhibit lower electronegativity values	7. Exhibit higher electronegativity values
8. Readily form cations by loss of electrons	8. Readily form anions by accepting electrons to fill the outermost shell (except noble gases)

chemical properties, depending on the conditions under which they react. Figure 8.2 shows the metal aluminum, the nonmetal sulfur, and the semi-metal silicon.

As an example of metallic behavior, we consider the behavior of sodium and its compounds. We can recognize sodium as a metal because it lies in Group 1A at the left of the periodic table. It loses its one valence electron when it combines with nonmetals, giving ionic compounds that contain the Na^+ ion and the anion formed by the nonmetal.

Figure 8.2

Aluminum, a metal (left), silicon, a semi-metal (right), and sulfur, a nonmetal (top).

$$2Na(s) + H_2(g) \longrightarrow 2NaH(s) \qquad (Na^+ \text{ and } H^-)$$
$$2Na(s) + F_2(g) \longrightarrow 2NaF(s) \qquad (Na^+ \text{ and } F^-)$$
$$16Na(s) + S_8(s) \longrightarrow 8Na_2S(s) \qquad (Na^+ \text{ and } S^{2-})$$
$$12Na(s) + P_4(s) \longrightarrow 4Na_3P(s) \qquad (Na^+ \text{ and } P^{3-})$$

Sodium oxide, Na_2O, is a soluble ionic oxide, so we know it reacts with water giving sodium hydroxide, a base.

$$Na_2O(s) + H_2O(l) \longrightarrow 2Na^+(aq) + 2OH^-(aq)$$

The hydride of sodium, NaH, is a soluble ionic compound containing H^- ions. The hydride ion, H^-, reacts with water to form OH^- and H_2.

$$NaH(s) + H_2O(l) \longrightarrow NaOH(aq) + H_2(g)$$

As an example of nonmetallic behavior, we consider the behavior of chlorine and its compounds. We find chlorine at the upper right in the periodic table in Group 7A, and thus we can readily identify it as a nonmetal. We should expect this nonmetal to form ionic compounds that contain chloride ion when it combines with metals. The chloride ion forms when a chlorine atom picks up one electron and fills its valence shell.

$$2Li(s) + Cl_2(g) \longrightarrow 2LiCl(s) \qquad (Li^+ \text{ and } Cl^-)$$
$$Ca(s) + Cl_2(g) \longrightarrow CaCl_2(s) \qquad (Ca^{2+} \text{ and } Cl^-)$$
$$Mn(s) + Cl_2(g) \longrightarrow MnCl_2(s) \qquad (Mn^{2+} \text{ and } Cl^-)$$

We expect chlorine to combine with less electronegative nonmetals (hydrogen and the nonmetals lying to the left of and below chlorine in the periodic table) and form covalent molecules. The products in the following reactions are covalent.

$$H_2(g) + Cl_2(g) \longrightarrow 2HCl(g)$$
$$P_4(s) + 10Cl_2(g) \longrightarrow 4PCl_5(s)$$
$$Si(s) + 2Cl_2(g) \longrightarrow SiCl_4(l)$$

As is expected for a compound of hydrogen and a nonmetal, HCl can act as an acid.

8.3   Oxidation States

Many of the reactions that characterize the behavior of metals, nonmetals, and semi-metals are **oxidation–reduction reactions,** reactions that involve changes in **oxidation states** of atoms involved in the reactions. We have seen how to assign the oxidation state of an atom in a compound by using the Lewis structure of the compound (Section 6.10). In this section, we introduce a series of rules that we use to determine oxidation states whether we know the Lewis structure or not.

1. The oxidation state of an atom of any element in its elemental form is zero. For example, the oxidation state of an atom in $Na(s)$, $P_4(s)$, $Br_2(l)$ or $N_2(g)$ is zero.

2. The oxidation state of a monatomic ion is equal to the charge on the ion. For example, the oxidation state of the Ca^{2+} ion is $+2$, and that of O^{2-} is -2.

3. The oxidation state of a fluorine atom in a compound is always -1.

4. The elements of Group 1A (except hydrogen) have an oxidation state of $+1$ in compounds. For example, the oxidation state of Li is $+1$ in LiCl, LiOH, Li_4SiO_4, and other compounds of lithium.

5. The elements of Group 2A have an oxidation state of $+2$ in compounds. For example, the oxidation state of Ca is $+2$ in $CaCl_2$, $CaCO_3$, $Ca(H_2PO_4)_2$, and other compounds of Ca.

6. The elements of Group 7A have an oxidation state of -1 when combined with less electronegative elements. For example, Cl has an oxidation state of -1 in NaCl, CCl_4, PCl_3, and HCl.

7. Oxygen usually has an oxidation state of -2, but there are three exceptions:

   a. In compounds with fluorine, oxygen has a positive oxidation state (fluorine is more electronegative than oxygen). For example, oxygen has an oxidation state of $+2$ in OF_2.

   b. In peroxides (compounds that contain an O—O single bond), oxygen has an oxidation state of -1. An example is hydrogen peroxide, H_2O_2; its Lewis structure is

$$H-\overset{..}{\underset{..}{O}}-\overset{..}{\underset{..}{O}}-H$$

c. In compounds that contain O_2^- (superoxides), oxygen has an oxidation state of $-\frac{1}{2}$. An example is KO_2.

8. Hydrogen has an oxidation state of -1 in compounds with less electronegative elements and an oxidation state of $+1$ in compounds with more electronegative elements. For example, Na, Mg, and B are less electronegative than hydrogen (Fig. 6.5) and H has an oxidation state of -1 in NaH, MgH_2, and BH_3. The elements C, P, O, and Br are more electronegative than hydrogen; H has an oxidation state of $+1$ in CH_4, PH_3, H_2O, and HBr.

9. The sum of the oxidation states of all the atoms in a compound is zero. The sum of the oxidation states of all the atoms in an ion is equal to the charge on the ion. Thus if we know the oxidation states of all but one kind of atom in a compound or ion, we can calculate that unknown oxidation state.

Example 8.1   Calculation of Oxidation State

Sodium sulfate, Na_2SO_4, is important in the pulp and paper industry. It is reduced to sodium sulfide, Na_2S, which is used in making paper because a solution of Na_2S will dissolve the lignin in wood and release the cellulose. A by-product in the reaction is sodium sulfite, an ionic compound containing the SO_3^{2-} ion. Hydrogen sulfide can result if the paper process is allowed to become acidic. Calculate the oxidation state of sulfur in Na_2SO_4, in H_2S, and in the SO_3^{2-} ion.

Solution: Na_2SO_4: The oxidation state of sulfur can be calculated from the oxidation states of sodium and oxygen, because the oxidation states of all the atoms in the compound must add up to zero. The two sodium atoms, each with an oxidation state of $+1$ (Group 1A, Rule 4), total $+2$; the four oxygen atoms, each with an oxidation state of -2 (Rule 7), total -8. We can calculate the oxidation state of sulfur as follows:

Atom	Oxidation State	Number of Atoms × Oxidation State
Na	$+1$	$2 \times +1 = +2$
O	-2	$4 \times -2 = -8$
S	x	$1 \times \ x = \ x$
		Charge $= \ \ 0$

For the sum of the oxidation states of all seven atoms to be zero, sulfur must have an oxidation state of $+6$. In summary,

$$(2 \times +1) + (4 \times -2) + (1 \times x) = 0$$
$$x = +6$$

H_2S: Hydrogen is bonded to a more electronegative element, so it will have an oxidation state of $+1$ (Rule 8).

Atom	Oxidation State	Number of Atoms × Oxidation State
H	+1	$2 \times +1 = +2$
S	x	$\dfrac{1 \times \quad x = \quad x}{\text{Charge} = \quad 0}$

For the sum of the oxidation states of all three atoms to be zero, sulfur must have an oxidation state of -2. In summary,

$$(2 \times +1) + (1 \times x) = 0$$
$$x = -2$$

$SO_3{}^{2-}$: Each oxygen atom has an oxidation state of -2.

Atom	Oxidation State	Number of Atoms × Oxidation State
O	-2	$3 \times -2 = -6$
S	x	$\dfrac{1 \times \quad x = \quad x}{\text{Charge} = -2}$

For the sum of the oxidation states of all four atoms to equal -2, the charge on the ion, sulfur must have an oxidation state of $+4$. In summary,

$$(3 \times -2) + (1 \times x) = -2$$
$$x = +4$$

Check your learning: Determine the oxidation state of the underlined atom in each of the following compounds and ions: $K\underline{N}O_3$, $\underline{Al}H_3$, $\underline{N}H_4{}^+$, $H_2\underline{P}O_4{}^-$.

Answer: N, +5; Al, +3; N, −3; P, +5.

Figure 8.3

Lodestone, Fe_3O_4, is a black magnetic oxide of iron.

Some compounds may contain ions with different charges. If we determine the oxidation state of these ions using the rules in this section, we obtain the average oxidation state of the element. For example, the mineral lodestone (Fig. 8.3), a magnetic oxide with the formula Fe_3O_4, contains both Fe^{2+} and Fe^{3+} ions with the ratio of one Fe^{2+} ion to two Fe^{3+} ions. In order to calculate the average oxidation state of iron in lodestone, we assign oxygen an oxidation state of -2, and we have

$$(3 \times x) + (4 \times -2) = 0$$
$$3x = 8$$
or
$$x = +2\tfrac{2}{3}$$

The average oxidation state for iron in lodestone is $+\tfrac{8}{3}$, or $+2\tfrac{2}{3}$.

We can write the formulas of many compounds by using the oxidation states of the atoms involved, because the sum of the oxidation states of the atoms in the compound must equal zero (Rule 9).

Example 8.2   Writing Chemical Formulas

Calcium chloride is used to melt ice on roads, sidewalks, and bridges. Write the formula for calcium chloride, using the oxidation states of its constituent elements.

Solution: Calcium is a member of Group 2A and so has an oxidation state of +2 (Rule 5). Chlorine, a member of Group 7A, is combined with a less electronegative element and so has an oxidation state of −1 (Rule 6). The formula of calcium chloride cannot be CaCl, because +2 and −1 do not add up to 0. For the sum of the oxidation states to be zero for the compound, the atoms must be in a ratio of one calcium ion to two chloride ions, or $CaCl_2$.

Check your learning: What is the formula of potassium iodide, a food additive used in iodized salt?

Answer: KI

Example 8.3   Writing Chemical Formulas

The components of many detergents include sodium tripolyphosphate, an ionic compound that contains the $P_3O_{10}^{5-}$ ion. Use the oxidation state of sodium to determine the formula of this compound.

Solution: We use a combination of our knowledge of metals and nonmetals, oxidation states, and formulas of ionic compounds (Section 2.7) to answer this question.

Sodium is a metal (it is in Group 1A); consequently, it forms ionic compounds with nonmetals such as phosphorus and oxygen. As a member of Group 1A, sodium will exhibit an oxidation state of +1 (Rule 4). The oxidation state of a monatomic ion is equal to the charge on the ion (Rule 2), so sodium must be present as the Na^+ ion. For the sum of the ionic charges on the Na^+ and $P_3O_{10}^{5-}$ ions to be zero for the compound, the ions must be in a ratio of five sodium ions to one tripolyphosphate ion, or $Na_5P_3O_{10}$.

Check your learning: Determine the formula of strontium perchlorate, a component of fireworks that contains the perchlorate ion, ClO_4^-.

Answer: $Sr(ClO_4)_2$

In Section 2.11 we discussed naming compounds that contain two nonmetals by using Greek prefixes to describe the number of atoms in a molecule of the compound. Now we can use a second method of nomenclature for binary compounds—a method that uses oxidation states. If a compound contains an element that can have more than one oxidation state, then we place a Roman numeral in parentheses after the name of the element to indicate its oxidation state. For example,

NO	nitrogen(II) oxide	N_2O_3	nitrogen(III) oxide
SO_2	sulfur(IV) oxide	SO_3	sulfur(VI) oxide
SF_4	sulfur(IV) fluoride	SF_6	sulfur(VI) fluoride
ICl	iodine(I) chloride	ICl_3	iodine(III) chloride

8.4   Periodic Variations in Oxidation State

As shown in Section 8.3, if we know the oxidation states of the elements in a compound, we can write the formula of the compound. In addition, knowing the possible oxidation states for an element can help us select reasonable products for a reaction.

Many regularities in the common oxidation states of the representative elements are related to their positions in the periodic table, as shown in Table 8.3 on page 266. This array of numbers may look formidable at first, but there are many regularities that can simplify our recall of them. Refer to Table 8.3 as we outline those regularities.

1. The maximum positive oxidation state found in any group of representative elements is equal to the group number. Thus the maximum possible positive oxidation state increases from $+1$ for the alkali metals (Group 1A) to $+7$ for all of the halogens (Group 7A) except fluorine, which only has an oxidation state of -1 in its compounds.

2. Metallic elements usually exhibit only positive oxidation states.

3. With the exceptions of thallium at the bottom of Group 3A and mercury at the bottom of Group 2B, the maximum positive oxidation state is the only common oxidation state displayed by the metals of Groups 1A, 2A, 2B, and 3A.

4. The most negative oxidation state of a group of representative elements is equal to the group number minus 8. Thus, for the elements in Group 5A, the most negative oxidation state possible is $5 - 8 = -3$.

5. Negative oxidation states are commonly limited to nonmetals and semi-metals and are observed only when these elements are combined with less electronegative elements.

6. Elements commonly exhibit positive oxidation states only when combined with more electronegative elements. Oxygen exhibits a positive oxidation state only in the few compounds it forms with the more electronegative element fluorine. Because fluorine is the most electronegative element known, it never has a positive oxidation state.

7. With the exceptions of nonmetals in the second period (B, C, N, and O) and mercury, each representative element that exhibits multiple oxidation states in its compounds commonly has either all even or all odd oxidation states.

Table 8.3

Common Oxidation States of the Representative Elements in Compounds

1 1A	2 2A		12 2B	13 3A	14 4A	15 5A	16 6A	17 7A	18 8A
H +1 −1								H +1 −1	He
Li +1	Be +2			B +3	C +4 to −4	N +5 to −3	O −1 −2	F −1	Ne
Na +1	Mg +2			Al +3	Si +4	P +5 +3 −3	S +6 +4 −2	Cl +7 +5 +3 +1 −1	Ar
K +1	Ca +2		Zn +2	Ga +3	Ge +4	As +5 +3 −3	Se +6 +4 −2	Br +7 +5 +3 +1 −1	Kr +4 +2
Rb +1	Sr +2		Cd +2	In +3	Sn +4 +2	Sb +5 +3 −3	Te +6 +4 −2	I +7 +5 +3 +1 −1	Xe +8 +6 +4 +2
Cs +1	Ba +2		Hg +2 +1	Tl +3 +1	Pb +4 +2	Bi +5 +3	Po +2	At	Rn
Fr +1	Ra +2								

Before we proceed, a word of explanation about the meaning of the phrase *common oxidation states* is necessary. The common oxidation states of an element are those observed in a majority of its compounds. In some cases, such as for the elements in Groups 1A and 2A, this majority may be all or nearly all of the known compounds of the element. Of all the thousands of known sodium compounds, for example, there are only a very few, very reactive compounds in which sodium has been shown to exhibit an oxidation state of −1. For other elements, the minority may be relatively large. Boron has an oxidation state of +3 in the majority of its compounds, but in many other compounds it has a +2, +1, or noninteger oxidation state. In most compounds discussed in this text, the elements exhibit their common oxidation states.

8.5 Chemical Reactions

In this section we consider five important types of reactions that are used to classify the behavior of the elements and their compounds.

Precipitation Reactions. The most common form of metathesis reaction (Section 2.10) is a **precipitation reaction,** a reaction in which a solid is formed. For example, if we mix aqueous solutions of sodium iodide and of silver nitrate, then an insoluble solid, silver iodide, immediately forms (Fig. 8.4).

$$NaI(aq) + AgNO_3(aq) \longrightarrow AgI(s) + NaNO_3(aq)$$

Figure 8.4

Silver iodide, an insoluble solid, forms when solutions of silver nitrate and potassium iodide are mixed.

A precipitation reaction occurs if one of the products of the reaction is insoluble. General rules for the solubilities of common ionic solids are listed in Table 8.4. (We should remember that these rules are for the simple compounds of the more common metals; there are exceptions for the less common metals and for more complex compounds.)

Acid–Base Reactions. An **acid–base reaction** is a reaction in which a hydrogen ion, H^+, is transferred from one molecule or ion to another. We saw one example of an acid–base reaction in Section 2.10: the neutralization reaction, a reaction between a solution of an acid and a solution of a base.

In Section 2.10 we also saw that acids are compounds that dissolve in water and increase the amount of hydronium ion, H_3O^+, present in the solution. This is one way to identify an acid when it is dissolved in water. We can also use a more general definition: An **acid** is a compound that donates a hydrogen ion (H^+), or proton, to another compound, whether dissolved in water or not. Sometimes we call such an acid a **Brønsted acid,** after Johannes Brønsted, a Danish chemist. A compound that accepts a hydrogen ion is called a **base** (or a **Brønsted base**). The following reactions illustrate the transfer of a hydrogen ion from a Brønsted acid to a Brønsted base.

$$HCl(g) + NH_3(g) \longrightarrow NH_4Cl(s)$$
$$H_2SO_4(l) + NaOH(s) \longrightarrow NaHSO_4(s) + H_2O(l)$$

Table 8.4

Solubilities of Common Ionic Compounds in Water

1. Most nitrates and acetates are soluble in water.
2. All chlorides are soluble except Hg_2Cl_2, $AgCl$, $PbCl_2$, and $CuCl$.
3. All sulfates except $SrSO_4$, $BaSO_4$, and $PbSO_4$ are soluble; $CaSO_4$ and Ag_2SO_4 are slightly soluble.
4. Carbonates, phosphates, borates, arsenates, and arsenites of the ammonium ion and the alkali metals are soluble; others are insoluble.
5. The hydroxides of the alkali metals, barium, and strontium are soluble; other hydroxides are insoluble, but calcium hydroxide is slightly soluble.
6. Most sulfides are insoluble. However, the sulfides of the ammonium ion and the alkali metals are soluble, although they hydrolize (react with water) to give solutions of the hydroxides and hydrogen sulfide ion (HS^-). The sulfides of the alkaline earths and aluminum also hydrolize, giving the hydroxides and HS^- (or H_2S if the hydroxide is insoluble).

Figure 8.5

The beakers on the left contain the colorless gases HCl and NH_3. When these gases are mixed (right), they react and produce a cloud of fine particles of solid NH_4Cl.

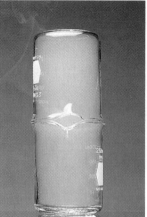

Figure 8.6

(A) Molecules of hydrogen chloride in the gas phase. Each chlorine atom (green sphere) is bonded to a hydrogen atom (blue sphere) by a covalent bond. (B) A solution of hydrogen chloride in water (hydrochloric acid). The hydrogen ions are bonded to water molecules by a coordinate covalent bond, giving a solution of H_3O^+ and Cl^- ions.

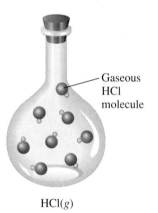

Gaseous HCl molecule

$HCl(g)$

(A)

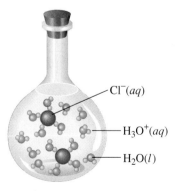

$Cl^-(aq)$

$H_3O^+(aq)$

$H_2O(l)$

$H_3O^+(aq) + Cl^-(aq)$

(B)

Ionic compounds called **salts** are produced in these reactions: NH_4Cl (Fig. 8.5) contains the ions NH_4^+ and Cl^-; $NaHSO_4$ contains Na^+ and HSO_4^-.

When dissolved in water, all common acids donate hydrogen ions to water molecules, forming **hydronium ions,** H_3O^+, plus whatever anion is produced when the acid loses a hydrogen ion (Fig. 8.6).

$$HCl(g) + H_2O(l) \longrightarrow H_3O^+(aq) + Cl^-(aq)$$
$$H_2SO_4(l) + H_2O(l) \longrightarrow H_3O^+(aq) + HSO_4^-(aq)$$
$$HSO_4^-(aq) + H_2O(l) \longrightarrow H_3O^+(aq) + SO_4^{2-}(aq)$$

This is why common acids form hydronium ions when dissolved in water, which leads to the definition of an acid used in Section 2.10.

For convenience, we sometimes write the hydronium ion as H^+ or $H^+(aq)$, but we should always remember that H^+ and $H^+(aq)$ are abbreviations; a hydrogen ion in water is always associated with at least one water molecule. In forming a hydronium ion, water accepts a hydrogen ion and therefore behaves as a base.

When we add a base to a solution of an acid, protons are donated from the hydronium ions to the base. For example, a solution of hydrogen chloride in water contains hydronium ions and chloride ions. If gaseous ammonia is bubbled through the solution, the ammonia molecules pick up protons from the hydronium ions and form ammonium ions, NH_4^+, which dissolve in the solution.

$$NH_3(g) + H_3O^+(aq) + Cl^-(aq) \longrightarrow NH_4^+(aq) + Cl^-(aq) + H_2O(l)$$

The products of this reaction are water and the salt ammonium chloride, NH_4Cl, which is soluble.

We can classify an acid as strong or weak on the basis of the extent to which it reacts with water to form hydronium ions. A **strong acid** dissolves in water giving a 100% yield (or very nearly so) of hydronium ion and the anion of the acid. Hydrochloric acid, stomach acid, is a strong acid.

$$HCl(g) + H_2O(l) \longrightarrow H_3O^+(aq) + Cl^-(aq)$$

100% yield of product

The six common strong acids and the principal anions they form in water are listed in Table 8.5.

Table 8.5

Common Strong Acids and Their Anions

Formula	Name	Anion	Name of Anion
HBr	Hydrogen bromide	Br^-	Bromide
HCl	Hydrogen chloride	Cl^-	Chloride
HI	Hydrogen iodide	I^-	Iodide
HNO_3	Nitric acid	NO_3^-	Nitrate
$HClO_4$	Perchloric acid	ClO_4^-	Perchlorate
H_2SO_4	Sulfuric acid	HSO_4^-	Hydrogen sulfate

A **weak acid** ionizes in dilute solution, giving a small percent yield of H_3O^+ ion (ordinarily 10% or less). Thus a solution of a weak acid consists primarily of covalent molecules (or ions) of the original acid, with lesser amounts of hydronium ion and the anion of the acid. The most familiar weak acid is found in vinegar, a solution of acetic acid, CH_3CO_2H, in water. Because acetic acid is a weak acid, most of the acid in vinegar is present as CH_3CO_2H molecules.

$$CH_3CO_2H(aq) + H_2O(l) \longrightarrow CH_3CO_2^-(aq) + H_3O^+(aq)$$
1% yield of products

Only about 1% of the acid molecules react with water to form hydronium ions and acetate ions, $CH_3CO_2^-$. Sodium hydrogen sulfate, $NaHSO_4$, is a weak acid that is used in some tub and tile cleansers. When we add $NaHSO_4$ to water, the weakly acidic hydrogen sulfate ions, HSO_4^-, react with water to give a small yield of H_3O^+ and SO_4^{2-} ions, but the majority do not react and remain as HSO_4^-. When weak acids react with bases, hydrogen ions, H^+, from both the H_3O^+ ion and the molecular acid are donated to the base.

Some common weak acids and the principal anions they form in water are listed in Table 8.6. A more extensive list of weak acids appears in Appendix F.

Table 8.6

Common Weak Acids and Their Anions

Formula	Name	Anion	Name of Anion
CH_3CO_2H	Acetic acid	$CH_3CO_2^-$	Acetate
H_2CO_3	Carbonic acid	HCO_3^-	Hydrogen carbonate
HCO_3^-	Hydrogen carbonate	CO_3^{2-}	Carbonate
HF	Hydrogen fluoride	F^-	Fluoride
H_2S	Hydrogen sulfide	HS^-	Hydrogen sulfide
HNO_2	Nitrous acid	NO_2^-	Nitrite
H_3PO_4	Phosphoric acid	$H_2PO_4^-$	Dihydrogen phosphate
$H_2PO_4^{2-}$	Dihydrogen phosphate	HPO_4^{2-}	Hydrogen phosphate
HPO_4^-	Hydrogen phosphate	PO_4^{3-}	Phosphate
HSO_4^-	Hydrogen sulfate ion	SO_4^{2-}	Sulfate

A **strong base** dissolves in water and gives a 100% yield (or very nearly so) of hydroxide ions and the cation of the base. Two common types of strong bases are soluble metal hydroxides and soluble ionic metal oxides.

Metal hydroxides are ionic; they are composed of cations and hydroxide ions. If such hydroxides dissolve, they give a 100% yield of hydroxide ions in solution. The soluble metal hydroxides include the hydroxides of the metals of Group 1A and of strontium and barium (Group 2A).

$$NaOH(s) \xrightarrow{H_2O} Na^+(aq) + OH^-(aq)$$

$$Ba(OH)_2(s) \xrightarrow{H_2O} Ba^{2+}(aq) + 2OH^-(aq)$$

Soluble ionic metal oxides contain oxide ions, O^{2-}. These are strong bases because the oxide ion reacts with water to give hydroxide ions:

$$O^{2-} + H_2O \longrightarrow 2OH^-$$

Potassium oxide, K_2O, for example, is a strong base; it contains the O^{2-} ion and dissolves in water giving hydroxide ion in 100% yield because of the reaction

$$K_2O(s) + H_2O(l) \longrightarrow 2K^+(aq) + 2OH^-(aq)$$

When we add an acid to a solution that contains hydroxide ions, the acid donates protons to the hydroxide ions. Adding nitric acid to a solution of barium hydroxide, for example, gives a solution of the salt barium nitrate, $Ba(NO_3)_2$, and water.

$$2HNO_3(l) + Ba^{2+}(aq) + 2OH^-(aq) \longrightarrow Ba^{2+}(aq) + 2NO_3^-(aq) + 2H_2O(l)$$

A **weak base** is a base that gives only a low concentration of hydroxide ions in water. Hydroxides such as $Mg(OH)_2$, $Ca(OH)_2$, and $Al(OH)_3$ are weak bases, even though they are ionic, because they are not very soluble. Only small amounts of hydroxide ion enter the solution when they dissolve. However, most weak bases are molecules or ions that may be quite soluble but give only a low yield (ordinarily 10% or less) of hydroxide ion when they react with water. Ammonia is a weak base of this kind; a solution of ammonia in water consists primarily of solvated ammonia molecules, $NH_3(aq)$. However, water acts as an acid with ammonia, and ammonia accepts protons from water to a very limited extent.

$$NH_3(aq) + H_2O(l) \longrightarrow NH_4^+(aq) + OH^-(aq)$$

A one molar solution of NH_3 contains 99.6% $NH_3(aq)$ and only 0.4% $NH_4^+(aq)$. When an acid is added to a solution of NH_3, protons are transferred both to the ammonia molecules and to the hydroxide ions.

Note that water can behave as either an acid (in a solution of a weak base, for example) or a base (in a solution of an acid, for example), depending on the nature of the substance dissolved in it.

Reactions of Oxides with Water.

Soluble ionic oxides of the metals are basic; that is, they react with water to form hydroxides (bases). These compounds are called **basic oxides.**

Many oxides of the nonmetals are acidic; they react with water and form acids *when the nonmetal is in a higher oxidation state*. Sulfur trioxide, SO_3, an oxide of sulfur with sulfur in its highest oxidation state (+6), reacts with water to give a strong acid, sulfuric acid, H_2SO_4.

$$SO_3(g) + H_2O(l) \longrightarrow H_2SO_4(l)$$

Sulfur dioxide, SO_2, with sulfur in an intermediate oxidation state of $+4$, reacts with water to give a weak acid, sulfurous acid, H_2SO_3. Sulfuric acid contains two oxygen atoms and two hydroxide groups (OH groups) covalently bonded to the sulfur. Sulfurous acid has one oxygen atom and two hydroxide groups bonded to the sulfur.

$$\ddot{:}\ddot{O}\ddot{:}$$

Sulfuric acid   Sulfurous acid

Thus H_2SO_4 and H_2SO_3 may be regarded as hydroxides of the nonmetal sulfur. Such molecules are more commonly called **oxyacids.**

Nitrogen(V) oxide, N_2O_5, reacts with water to form a strong acid, nitric acid, which can be represented by either the formula HNO_3 or the formula $HONO_2$. Nitrogen(III) oxide, N_2O_3, gives a solution of the weak acid HNO_2 (or HONO), nitrous acid, when it reacts with water. Carbon dioxide gives a weak acid, carbonic acid, H_2CO_3 [or $(HO)_2CO$]. Nonmetal oxides, such as SO_3 and CO_2, that react with water and produce solutions of acids are called **acidic oxides.**

Nitrogen(II) oxide, a lower-oxidation-state nitrogen oxide, and carbon monoxide, a lower-oxidation-state carbon oxide, do not react with water. These oxides are nonacidic.

Reactions of Basic Oxides and Acidic Oxides.

Basic oxides react with acidic nonmetal oxides and produce the same salt that would be produced by the reaction of solutions of these substances in water. For example, K_2O dissolves in water and produces a solution of potassium hydroxide, KOH. N_2O_5 dissolves in water and produces nitric acid, HNO_3. The reaction of a solution of KOH and HNO_3 produces the salt potassium nitrate, KNO_3, and water.

$$KOH(aq) + HNO_3(aq) \longrightarrow KNO_3(aq) + H_2O(l)$$

We can also prepare the salt KNO_3 by combining solid K_2O with gaseous N_2O_5.

$$K_2O(s) + N_2O_5(g) \longrightarrow 2KNO_3(s)$$

Oxidation–Reduction Reactions.

We can classify combustion reactions and many addition and decomposition reactions (Section 2.10) as **oxidation–reduction reactions:** reactions that involve changes in oxidation states (Section 8.3). When an atom, either free or in a molecule or ion, loses electrons, it is **oxidized;** *its oxidation state increases.* When an atom, either free or in a molecule or ion, gains electrons, it is **reduced;** *its oxidation state decreases.* Reactions involving oxidation and reduction are referred to as oxidation–reduction reactions, or **redox reactions.** An example is the combustion reaction of sulfur in oxygen (Fig. 8.7), in which sulfur is oxidized and oxygen is reduced.

$$\overset{0}{S} + \overset{0}{O_2} \longrightarrow \overset{+4\,-2}{SO_2}$$

As sulfur is oxidized in this reaction, its oxidation state increases from 0 to $+4$, as indicated by the small numbers above the equation. As oxygen is reduced, its oxidation state decreases from 0 to -2. Other examples of oxidation–reduction reactions are given on page 272. The first is a decomposition reaction, the second a displacement reaction.

Figure 8.7

Sulfur is oxidized by oxygen in a combustion reaction.

$$\overset{+2\ -2}{2HgO} \longrightarrow \overset{0}{2Hg} + \overset{0}{O_2}$$

$$\overset{-4+1}{CH_4} + \overset{0}{4Cl_2} \longrightarrow \overset{+4-1}{CCl_4} + \overset{+1-1}{4HCl}$$

Not all atoms in an oxidation–reduction reaction need change oxidation state; in the displacement reaction, the hydrogen atoms do not change oxidation state. However, *at least one atom must be oxidized and another reduced in any oxidation–reduction reaction,* because oxidation and reduction always occur simultaneously. If one atom gains electrons, a second atom must provide those electrons.

We call the species that gives up electrons to another reactant in an oxidation–reduction reaction the **reducing agent.** It causes the other reactant to be reduced. Because a reducing agent loses electrons, it is oxidized. In the three oxidation–reduction reactions in this section, S, the O^{2-} ion, and CH_4, respectively, are the reducing agents. The **oxidizing agent** in a redox reaction gains electrons and causes another reactant (the reducing agent) to be oxidized. The oxidizing agent picks up electrons during a redox reaction, so it is reduced. The oxidizing agents in the three equations in this section are O_2, the Hg^{2+} ion, and Cl_2, respectively.

8.6 Variation in Metallic and Nonmetallic Behavior of the Representative Elements

In Section 8.2 we saw that sodium, located at the left end of the third period in the periodic table, is an active metal. Likewise we saw that chlorine, at the right end of the third period, is a typical nonmetal. Proceeding across the third period from sodium to chlorine, we encounter five other elements whose chemical behavior becomes decreasingly metallic and increasingly nonmetallic as we go from left to right. The changeover from metallic to nonmetallic behavior is gradual, but most properties change from metallic to nonmetallic between aluminum and silicon. Both aluminum and silicon, however, may exhibit metallic or nonmetallic properties under the appropriate conditions.

To demonstrate the gradual changeover, let us examine the variation in the properties of the elements in the third period. Sodium, magnesium, and aluminum are shiny metals that conduct heat and electricity well. Silicon is a semi-metal. It has a luster characteristic of a metal, but it is a semiconductor (a poor conductor of electricity). Phosphorus and sulfur are dull in appearance and are nonconducting solids. Chlorine is a gas. Sodium oxide, Na_2O, reacts with water, giving the strong base sodium hydroxide, NaOH. Magnesium oxide, MgO, reacts slowly with water, giving the less strongly basic magnesium hydroxide, $Mg(OH)_2$. Aluminum oxide, Al_2O_3, does not react with water, but the reaction of a solution of an aluminum salt such as aluminum nitrate, $Al(NO_3)_3$, with a solution of a base produces aluminum hydroxide, $Al(OH)_3$, an **amphoteric compound.** Aluminum hydroxide can act either as a very weak base (reacting with strong acids) or as a weak acid (reacting with strong bases). Silicon dioxide, SiO_2, will not react with water, but many covalent silicon compounds (such as silicon tetrachloride, $SiCl_4$) will, giving gelatinous precipitates of silicic acid, $Si(OH)_4$, a very weak acid. The covalent oxides of phosphorus, sulfur, and chlorine in which these elements exhibit their highest oxidation states (P_4O_{10}, SO_3, and Cl_2O_7) react with water, giving the oxyacids phosphoric acid, H_3PO_4, sulfuric acid, H_2SO_4,

Figure 8.8

The solid elements of Group 5A. From left to right, phosphorus and arsenic (nonmetals), antimony (a semi-metal), and bismuth (a metal). Nitrogen, the lightest member of the group, is a colorless gas and is a nonmetal. (White phosphorus, which is shown, is so reactive with air that it must be stored under water.)

and perchloric acid, $HClO_4$, respectively. Of these acids, phosphoric acid is the weakest, perchloric acid the strongest. Thus we find that *the properties that are characteristic of metallic behavior become less pronounced and the properties that are characteristic of nonmetallic behavior become more pronounced from left to right across a period of representative elements.* For example, base strength, ionic character, and strength as a reducing agent decrease, whereas acid strength, covalent character, and strength as an oxidizing agent increase.

We also find that *the metallic character of the elements decreases and the nonmetallic character increases as we go up a group of representative elements.* For instance, bismuth, at the bottom of Group 5A, is a metallic element; antimony is a semi-metal; and nitrogen, phosphorus, and arsenic, at the top of the group, are nonmetals (Fig. 8.8). Elemental fluorine, at the top of Group 7A, is a stronger oxidizing agent than elemental iodine, near the bottom of the group.

The metallic behavior of an element decreases and its nonmetallic behavior increases as the positive oxidation state of that element in its compounds increases. For example, in perchloric acid, $HClO_4$, the oxidation state of chlorine is $+7$; in hypochlorous acid, HOCl, chlorine has an oxidation state of $+1$. Perchloric acid is a very strong acid (a characteristic of a compound of an element with pronounced nonmetallic behavior), whereas hypochlorous acid is a weak acid (less pronounced nonmetallic behavior). Thallium(III) chloride, $TlCl_3$, is a covalent compound (a characteristic of a nonmetal chloride) whereas thallium(I) chloride, TlCl, is ionic (a characteristic of a metal chloride). The chlorides of tin(IV) and lead(IV), $SnCl_4$ and $PbCl_4$, are liquids that contain tetrahedral covalent molecules. The chlorides of tin(II) and lead(II), $SnCl_2$ and $PbCl_2$, are solids with much more ionic character in their bonds than the chlorides of tin(IV) and lead(IV). Thus, when predicting the behavior of an element's compounds, we must consider the element's oxidation state as well as its position in the periodic table.

8.7 The Activity Series

Certain metals, such as sodium and potassium (see Fig. 8.9, page 274), react readily with cold water, forming hydrogen and metal hydroxides. Other metals, such as magnesium and iron, react with water only when heated. Magnesium metal reacts with iron ions to produce metallic iron and magnesium ions, but magnesium does not react

Figure 8.9

The reaction of potassium with water is so exothermic that the metal inflames.

Figure 8.10

When an iron nail is immersed in a solution of copper sulfate (top), copper deposits as the iron reduces the Cu^{2+} ion to metallic copper and is itself oxidized to Fe^{2+} (bottom).

with sodium ions or potassium ions. We use experimental observations like these to arrange the metals in order of their chemical activities, in an activity series. A brief form of the series, containing only hydrogen and the common metals, is given in Table 8.7 on page 275.

Potassium, the most reactive of the common metals, heads the activity series. Each succeeding metal in the series is less reactive, and gold, the least reactive of all, is found at the bottom. In theory, any metal in the series, in its elemental form, will reduce the ion of any metal below it in the series when that ion is present in solution in water. We find, for example, that the copper ion in an aqueous solution of a copper(II) salt is reduced to copper metal by iron, which is above copper in the series (Fig. 8.10).

$$Fe(s) + Cu^{2+}(aq) \longrightarrow Cu(s) + Fe^{2+}(aq)$$

In a similar manner, aqueous silver ions are reduced by metallic iron or copper, and aqueous mercury ions are reduced by iron, copper, or silver. Any metal above hydrogen in the series liberates hydrogen from aqueous acids (solutions of acids in water); those from cadmium to the top liberate hydrogen from hot water or steam; and those from sodium to the top liberate hydrogen even from cold water. The metals below hydrogen do not displace hydrogen from water or from aqueous acids.

The reactivity of the metals toward oxygen decreases down the series, as does the stability of the oxides formed. It is evident, then, that the activity series is very useful; it indicates the possibility of a reaction of a given metal with water, with oxygen, and with aqueous solutions of acids or of salts of other metals.

8.8   Prediction of Reaction Products

Many factors determine whether a chemical reaction will occur and what the products will be. These include the kinds of reactants and the conditions under which the reaction is run. Some chemists spend years developing a "feel" for the types of reactants and conditions that are likely to give a desired product. However, even they must resort to the ultimate test of their ideas; they try a proposed reaction in the laboratory to see if it works. The design and testing of reactions in order to prepare specific types of compounds or to improve the ease, yield, or simplicity with which known compounds are prepared constitute the field known as **chemical synthesis.**

Table 8.7

Activity Series of Common Metals

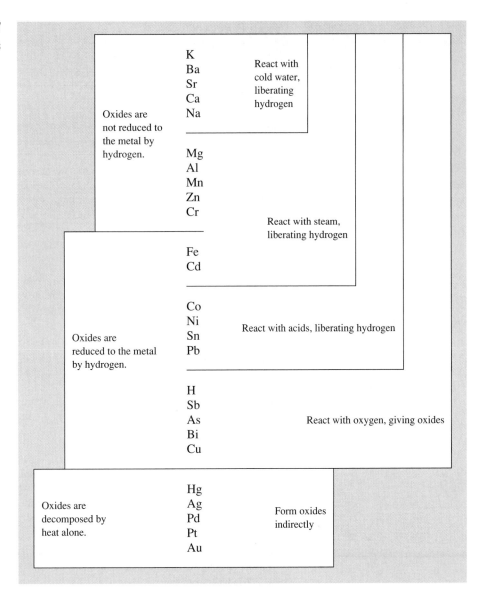

In this section we shall examine some general guidelines that can be helpful in answering the question "What products, if any, are likely to result from the reaction of two or more substances?" These guidelines are based on the metallic or nonmetallic behavior of the representative elements involved, the common oxidation states they are likely to exhibit, and the similarities of the compounds involved to those types already discussed. We shall consider only those reactions that occur at room temperature in water or that occur when the pure substances are heated. Although different conditions may lead to the formation of different products, these guidelines can serve as a foundation for our consideration of chemical reactions in subsequent chapters.

As a general approach to predicting the products of a reaction, we examine the reactants to see whether they are analogous to compounds or elements whose behavior we are familiar with. If, as in the next examples, the behavior is obvious, we can readily predict products.

Figure 8.11

The oxidation of magnesium in air.

Example 8.4   Predicting the Product of an Oxidation–Reduction Reaction

As shown in Fig. 8.11, the metal magnesium burns with a white flame in air, giving an ionic oxide. What is the principal product that forms when the metal is oxidized by the nonmetal oxygen?

Solution: Magnesium is a member of Group 2A and so will have an oxidation state of $+2$. Oxygen is combined with a less electronegative element and has an oxidation state of -2.

   We write the formula using the technique described in Section 8.3. For the sum of the oxidation states to be zero for the compound, the atoms must be in a ratio of one magnesium ion to one oxide ion: MgO.

Check your learning: A second substance produced in small amounts when the metal magnesium is heated in air results when magnesium combines with nitrogen, a nonmetal. What is this product?

Answer: Mg_3N_2

Example 8.5   Predicting the Product of an Acid–Base Reaction

Gaseous ammonia, NH_3, reacts with gaseous hydrogen iodide, HI, in a reaction similar to that of NH_3 and HCl (Fig. 8.5). What is the likely product of the reaction of NH_3 and HI?

Solution: From our discussion of acids and bases (Section 8.5), we recognize that ammonia is a base and that hydrogen iodide is one of the six strong acids. Thus we can expect an acid–base reaction in which a hydrogen ion is transferred from HI, the acid, to NH_3, the base, forming the salt ammonium iodide, NH_4I.

$$NH_3(g) + HI(g) \longrightarrow NH_4I(s)$$

This reaction is analogous to the reaction of ammonia with HCl ($NH_3 + HCl \longrightarrow NH_4Cl$). The similarity between the predicted reaction and known behavior supports our prediction.

Check your learning: What are the products of the reaction of a solution of the ionic hydroxide CsOH with perchloric acid, $HClO_4$?

Answer: $CsClO_4$ and H_2O

   If parallels with known behavior are not obvious, try the following method of analysis of the system.

1. Identify each element in the reactants as a metal or a nonmetal (from its position in the periodic table), and find its oxidation state as well as the oxidation states it commonly exhibits in its compounds (Table 8.3).

2. Consider the possibility of an oxidation–reduction reaction. The following general considerations are helpful.

    a. An elemental metal will be oxidized by an elemental nonmetal.

    $$Mg + Cl_2 \longrightarrow MgCl_2$$

    b. A more active metal in its elemental form will reduce the ion of a less active metal.

    $$Mg + SnCl_2 \longrightarrow MgCl_2 + Sn$$

    c. A more electronegative nonmetal in its elemental form will oxidize a less electronegative nonmetal, either as an element or when the less electronegative element is in an oxidation state that can increase.

    $$S + O_2 \longrightarrow SO_2$$
    $$2SO_2 + O_2 \longrightarrow 2SO_3$$
    $$2Na_2S + 3O_2 \longrightarrow 2Na_2O + 2SO_2$$

    d. The metals of Group 1A and calcium, strontium, and barium of Group 2A are active enough to reduce the hydrogen in water or acids to hydrogen gas.

    $$2Na + 2H_2O \longrightarrow 2NaOH + H_2$$
    $$Ba + 2HCl \longrightarrow BaCl_2 + H_2$$

    The other representative metals, except bismuth and lead (the least metallic representative metals), will only reduce the hydrogen in acids.

    $$2Al + 3H_2SO_4 \longrightarrow Al_2(SO_4)_3 + 3H_2$$

3. Consider the possibility of an acid–base reaction. Remember that the oxides and hydroxides of nonmetals are generally acidic and that the oxides and hydroxides of metals are generally basic. Thus we might find acid–base reactions involving

    a. Familiar types of acids and bases:

    $$2HI + Sr(OH)_2 \longrightarrow SrI_2 + 2H_2O$$

    b. Acidic oxides with bases:

    $$SO_3 + Ca(OH)_2 \longrightarrow CaSO_4 + H_2O$$

    c. Acids with basic oxides:

    $$6HCl + Al_2O_3 \longrightarrow 2AlCl_3 + 3H_2O$$

    d. Acidic oxides with basic oxides:

    $$SO_2 + MgO \longrightarrow MgSO_3$$

4. Consider the possibility of a metathetical reaction. Metathetical reactions between ionic compounds generally occur when one product is a solid (a precipitation reaction, Section 8.5), a gas, a weak acid, or a weak base.

5. Consider whether the products of the reaction may react with each other or with the solvent (if any).

    The following examples provide some specific illustrations of the application of these ideas.

Example 8.6   Predicting the Product of an Addition Reaction

Write the balanced chemical equation for the reaction of elemental calcium, Ca, with iodine, I_2.

Solution: Let us assume that we have no specific knowledge of the chemistry of this system and analyze it using the steps discussed above. Calcium is a member of Group 2A and is thus an active metal. Iodine is a member of Group 7A and is thus a non-metal. An active metal reacts with a nonmetal in an oxidation–reduction reaction. Iodine will be reduced to -1, its only available negative oxidation state as a member of Group 7A, and calcium will be oxidized to $+2$, the only oxidation state that a Group 2A element exhibits in compounds. With these oxidation states, the product must be CaI_2. The balanced equation is

$$Ca + I_2 \longrightarrow CaI_2$$

Check your learning: A compound used in some red calculator displays is the product of the reaction of gallium with sulfur. What is this compound?

Answer: Ga_2S_3

Example 8.7   Predicting the Product of an Acid–Base Reaction

Write a balanced chemical equation for the reaction that occurs when barium oxide, BaO, is added to an excess of a solution of perchloric acid, $HClO_4$.

Solution: Barium is a member of Group 2A. It is located near the bottom of the group and should exhibit metallic properties. Barium has the maximum possible oxidation state for a member of Group 2A, $+2$; thus it cannot be oxidized. It is located toward the top of the activity series, so it is unlikely that it will be reduced. Thus an oxidation–reduction is improbable. Perchloric acid is one of the strong acids (Table 8.5), and the oxides of metals are basic. Hence we can expect an acid–base reaction between BaO and the acid $HClO_4$ to produce a salt and water.

$$BaO(s) + 2HClO_4(aq) \longrightarrow Ba(ClO_4)_2(aq) + H_2O(l)$$

Check your learning: What are the two possible sets of products of the reaction of potassium hydroxide with sulfuric acid?

Answer: $KHSO_4$ and H_2O (from a 1-to-1 mole ratio of KOH to H_2SO_4) or K_2SO_4 and H_2O (from a 2-to-1 mole ratio of KOH to H_2SO_4).

Figure 8.12

Phosphorus burns in an atmosphere of chlorine.

Example 8.8   Reactions of Nonmetals

As shown in Fig. 8.12, phosphorus burns in a chlorine atmosphere. What are the two possible products of the reaction of phosphorus, P_4, with chlorine, Cl_2?

Solution: Both phosphorus (Group 5A) and chlorine (Group 7A) are nonmetals. Chlorine lies to the right of phosphorus in the third period, so it is more electronegative than phosphorus. Thus we recognize that it can oxidize phosphorus. Because phosphorus is oxidized, it will exhibit a positive oxidation state—probably $+3$ or $+5$, because these are the common positive oxidation states exhibited by members of Group 5A. Chlorine will exhibit a negative oxidation state because it is combined with a less electronegative element. The only negative oxidation state for chlorine is -1. Thus P_4 and Cl_2 could react to give two compounds. In one, phosphorus has an oxidation state of $+3$, and chlorine has an oxidation state of -1. In the other, phosphorus has an oxidation state of $+5$, and chlorine has an oxidation state of -1. The two compounds are PCl_3 and PCl_5.

Note that either phosphorus(III) chloride or phosphorus(V) chloride can in fact be formed, depending on the choice of reaction conditions. An excess of Cl_2 favors the formation of PCl_5; conversely, an excess of P_4 favors the formation of PCl_3. The reaction of 1 mol of P_4 with 6 mol of Cl_2 gives PCl_3.

$$P_4 + 6Cl_2 \longrightarrow 4PCl_3$$

If the amount of Cl_2 is increased to 10 mol, then PCl_5 is produced.

$$P_4 + 10Cl_2 \longrightarrow 4PCl_5$$

Check your learning: When elemental fluorine, F_2, reacts with a second element, the compounds that are produced generally contain the second element in its highest oxidation state. What compounds result when an excess of F_2 reacts with I_2? With S_8?

Answer: IF_7; SF_6.

Figure 8.13

Sulfuric acid. (A) A sketch of the molecular structure. (B) A ball-and-stick model. (C) A space-filling model.

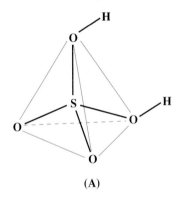

(A)

(B)

(C)

8.9   Chemical Properties of Some Important Industrial Chemicals

The compounds and elements listed in Table 8.8 on page 280 constituted about 97% of the elements and inorganic compounds produced in the United States during 1995. The chemical properties of these compounds and elements show the differences between the chemistry of metals and that of nonmetals. They also show the types of compounds and reactions, as well as the oxidation states, described in the preceding sections. The behavior of organic materials will be described in the next chapter.

Sulfuric Acid.   With 20% of the industrial production of inorganic chemicals in the United States devoted to its manufacture, sulfuric acid (Fig. 8.13) ranks as the single most significant industrial chemical. Per capita use of sulfuric acid has been taken as one index of the technical development of a nation. Perhaps surprisingly, sulfuric acid and the sulfate ion rarely appear in finished materials. Instead, sulfuric acid is used extensively as an acid (a source of hydrogen ions) because it is the cheapest strong acid. It is used to manufacture fertilizer, leather, tin plate, and other chemicals; to purify petroleum; and to make and dye fabrics.

Sulfuric acid is prepared from elemental sulfur or sulfides (Fig. 8.14). Sulfur is burned in air, and the nonmetal sulfur is oxidized by the more electronegative nonmetal oxygen. Covalent molecules of gaseous sulfur dioxide, with sulfur having one

Table 8.8

Important Industrial Elements and Inorganic Chemicals

Element or Compound	1995 U.S. Production (in billions of pounds)
H_2SO_4	95.36
N_2	68.04
O_2	54.38
CaO, $Ca(OH)_2$	41.23
NH_3	35.60
NaOH	26.19
H_3PO_4	26.19
Cl_2	25.09
Na_2CO_3	22.28
HNO_3	17.24
NH_4NO_3	15.99
CO_2	10.89
HCl	7.33
$(NH_4)_2SO_4$	5.24

Reprinted with permission from *Chemical and Engineering News,* April 8, 1996, p. 17. Copyright 1996 American Chemical Society.

Figure 8.14

(A) A sulfuric acid plant with mounds of sulfur, which is used as a starting material in the manufacture of sulfuric acid. (B) A close-up view of sulfur.

(A)

(B)

of its common oxidation states ($+4$), result. The sulfur is then oxidized to its highest oxidation state ($+6$) by additional oxygen, giving sulfur trioxide.

$$2SO_2 + O_2 \longrightarrow 2SO_3$$

Even though the yield from this reaction is highest at lower temperatures, sulfur trioxide forms slowly at these temperatures. At higher temperatures it forms more rapidly, but the yield is lower. In order to get as high a yield as possible, the reaction is run at lower temperatures with vanadium(V) oxide, V_2O_5, as a catalyst. A **catalyst** is a substance that changes the speed of a chemical reaction without affecting the yield and without undergoing a permanent chemical change itself.

Because sulfur trioxide is an oxide of a nonmetal in which the nonmetal has a high oxidation state, we expect it to react with water to form an acid—in this case, sulfuric acid.

$$H_2O + SO_3 \longrightarrow H_2SO_4$$

This reaction does occur, for example, when SO_3 from polluted air dissolves in raindrops—one source of acid rain. However, it is more efficient industrially to combine sulfur trioxide with sulfuric acid to produce pyrosulfuric acid, $H_2S_2O_7$. This acid then reacts with water to form sulfuric acid. The equations are

$$SO_3 + H_2SO_4 \longrightarrow H_2S_2O_7$$
$$H_2S_2O_7 + H_2O \longrightarrow 2H_2SO_4$$

Oxygen will also oxidize sulfur with an oxidation state of -2 in compounds. Thus sulfuric acid is prepared sometimes from the sulfur dioxide produced by burning hydrogen sulfide, an impurity separated from natural gas, and sometimes from the sulfur dioxide produced by heating metal sulfides such as Cu_2S in air.

Figure 8.15

The insoluble phosphate rock from phosphate mines is converted to soluble phosphates by treatment with sulfuric acid.

Figure 8.16

Phosphoric acid. (A) A sketch of the molecular structure. (B) A ball-and-stick model. (C) A space-filling model.

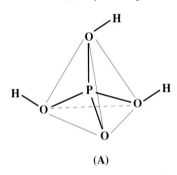

(A)

(B)

(C)

$$2H_2S + 3O_2 \longrightarrow 2SO_2 + 2H_2O$$

$$Cu_2S + 2O_2 \xrightarrow{\Delta} 2CuO + SO_2$$

Sulfuric acid is both a strong acid and an oxidizing agent. Its largest use is as a source of hydrogen ion. About 65% of the sulfuric acid produced in the United States is used to manufacture fertilizers by acid–base reactions with ammonia or with calcium phosphates [typically $Ca_5(PO_4)_3F$ and $Ca_5(PO_4)_3(OH)$, found principally in North Carolina and Florida as phosphate rock (Fig. 8.15)]. The reaction with calcium phosphates converts these extremely insoluble compounds into somewhat soluble dihydrogen phosphates that can enter a plant's roots and provide it with necessary phosphorus.

$$2Ca_5(PO_4)_3F(s) + 7H_2SO_4(l) + 3H_2O(l) \longrightarrow$$
$$3Ca(H_2PO_4)_2 \cdot H_2O(s) + 2HF(g) + 7CaSO_4(s)$$

This is an acid–base reaction in which a strong acid reacts with the anions (PO_4^{3-} and F^-) of two weak acids. The solid mixture of the salts $Ca(H_2PO_4)_2 \cdot H_2O$ and $CaSO_4$ is used directly as the fertilizer. The hydrogen fluoride is recovered and used in the preparation of fluorocarbons, such as Teflon, and other fluorides. If the amount of sulfuric acid is increased, phosphoric acid is produced in a related reaction.

$$Ca_5(PO_4)_3F(s) + 5H_2SO_4(l) \longrightarrow 5CaSO_4(s) + 3H_3PO_4(l) + HF(g)$$

Phosphoric acid (Fig. 8.16) is used primarily to make ammonium hydrogen phosphate and "triple superphosphate" fertilizers. The reactions are acid–base reactions.

$$2NH_3(g) + H_3PO_4(l) \longrightarrow (NH_4)_2HPO_4(s)$$
$$Ca_5(PO_4)_3F(s) + 7H_3PO_4(l) \longrightarrow 5Ca(H_2PO_4)_2(s) + HF(g)$$

Phosphoric acid and dihydrogen phosphate salts are also used in colas—read the label on a cola can.

The acid behavior of sulfuric acid is also evident in the acid–base reaction used to prepare ammonium sulfate, another soluble solid fertilizer providing nitrogen.

$$2NH_3(g) + H_2SO_4(l) \longrightarrow (NH_4)_2SO_4(s)$$

In 1995 about 5 billion pounds of $(NH_4)_2SO_4$ were prepared in the United States.

Many metal oxides behave as bases and react with sulfuric acid to form ionic sulfates. Some examples are copper sulfate (used as a fungicide and in electroplating),

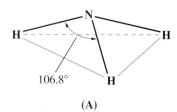

106.8°

(A)

(B)

(C)

Figure 8.17

Ammonia. (A) A sketch of the molecular structure. (B) A ball-and-stick model. (C) A space-filling model.

aluminum sulfate (used in water treatment and in paper making), and magnesium sulfate (Epsom salts) are prepared this way.

$$CuO(s) + H_2SO_4(aq) \longrightarrow Cu^{2+}(aq) + SO_4{}^{2-}(aq) + H_2O(l)$$
$$Al_2O_3(s) + 3H_2SO_4(aq) \longrightarrow 2Al^{3+}(aq) + 3SO_4{}^{2-}(aq) + 3H_2O(l)$$
$$MgO(s) + H_2SO_4(aq) \longrightarrow Mg^{2+}(aq) + SO_4{}^{2-}(aq) + H_2O(l)$$

Because these products are ionic, they exist in solution as ions. They can be recovered as solids by evaporating the water.

Sulfuric acid also undergoes an acid–base reaction with sodium chloride, giving hydrogen chloride and sodium hydrogen sulfate, a solid acid used in many household cleansers. The sodium hydrogen sulfate will then react with additional sodium chloride if the mixture is heated.

$$H_2SO_4(l) + NaCl(s) \longrightarrow NaHSO_4(s) + HCl(g)$$
$$NaHSO_4(s) + NaCl(s) \longrightarrow Na_2SO_4(s) + HCl(g)$$

Even though chloride ion is a very weak base, these metathetical reactions proceed because covalent gaseous HCl is formed and escapes from the reaction.

Sulfuric acid is an oxidizing agent, but it is not often used industrially in this way.

Ammonia.   Ammonia (Fig. 8.17) is one of the few compounds that can be prepared readily from elemental nitrogen. It is prepared by the reaction of a mixture of hydrogen and nitrogen at high pressure and temperature (400°C–600°C) in the presence of a catalyst that contains iron. The catalyst speeds up the reaction but does not increase the yield.

$$N_2(g) + 3H_2(g) \xrightarrow[\text{High pressure}]{\substack{400°C–600°C \\ \text{Fe catalyst}}} 2NH_3(g)$$

This reaction is simply the oxidation of a less electronegative nonmetal by a more electronegative nonmetal, giving a covalent compound.

Pure ammonia is a gas that is very soluble in water. Its uses can be attributed principally to two chemical properties: (1) ammonia is a base, and (2) the nitrogen atom in ammonia is reactive and can readily be oxidized to higher oxidation states in the production of other nitrogen-containing compounds.

Ammonia and its derivatives are an important source of nitrogen fertilizers (Fig. 8.18). Over 20 billion pounds of ammonium salts are prepared as fertilizers each year

Figure 8.18

Ammonia is used as a fertilizer. Pure ammonia is a gas at room temperature and pressure; thus it is transported in a pressurized container. The ammonia stays in the soil because it is very soluble in the water present and because it reacts with acids in the soil.

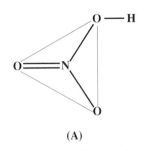

(A)

(B)

(C)

Figure 8.19

Nitric acid. (A) A sketch of the molecular structure. (B) A ball-and-stick model. (C) A space-filling model.

in the United States. The reactions are acid–base reactions that give salts containing the NH_4^+ ion. The principal reactions involve nitric acid, HNO_3, sulfuric acid, H_2SO_4, and phosphoric acid, H_3PO_4.

$$NH_3(g) + HNO_3(l) \longrightarrow NH_4NO_3(s)$$
$$2NH_3(g) + H_2SO_4(l) \longrightarrow (NH_4)_2SO_4(s)$$
$$NH_3(g) + H_3PO_4(l) \longrightarrow NH_4H_2PO_4(s)$$

The nitrogen in ammonia can be oxidized by oxygen. Ammonia burns in oxygen, forming nitrogen(II) oxide, NO, which can then react with additional oxygen to give nitrogen(IV) oxide, NO_2. The latter oxide reacts with water to give the oxyacid nitric acid. This reaction is more complex than most reactions of water with nonmetal oxides, because oxidation–reduction is also involved. The series of reactions is

$$4NH_3 + 5O_2 \longrightarrow 4NO + 6H_2O$$
$$2NO + O_2 \longrightarrow 2NO_2$$
$$3NO_2 + H_2O \longrightarrow 2HNO_3 + NO$$

Nitric acid (Fig. 8.19) is important industrially as a strong acid and oxidizing agent.

Calcium Oxide and Calcium Hydroxide.

Calcium oxide and calcium hydroxide are inexpensive bases used extensively in chemical processing, although most of the useful products prepared from them do not contain calcium.

Calcium oxide, CaO, is made by heating calcium carbonate, $CaCO_3$, which is widely and inexpensively available as limestone or oyster shells.

$$CaCO_3(s) \longrightarrow CaO(s) + CO_2(g)$$

Although this decomposition reaction is reversible, and CaO and CO_2 react to give $CaCO_3$, a 100% yield of CaO is obtained if the CO_2 is allowed to escape. Calcium hydroxide is prepared by the familiar acid–base reaction of a metal oxide with water.

$$CaO(s) + H_2O(l) \longrightarrow Ca(OH)_2(s)$$

Because Ca^{2+} is the ion of an active metal and is therefore very difficult to reduce, and because oxygen (with its oxidation state of -2) is difficult to oxidize, the oxidation–reduction chemistry of both CaO and $Ca(OH)_2$ is limited. These compounds are useful for their basic behavior; they accept protons and neutralize acids.

Oxygen and Nitrogen.

Both pure oxygen and pure nitrogen can be isolated from air. Nitrogen is surprisingly unreactive for so electronegative a nonmetal. The principal uses of nitrogen gas are actually based on this nonreactivity. It is used as an inert atmosphere blanket in the food industry to prevent spoilage due to oxidation. Liquid nitrogen boils at 77 K ($-196°C$). It is used to store biological materials such as blood and tissue samples at very low temperatures and as a coolant to produce low temperatures (see Fig. 8.20 on page 284).

Oxygen behaves as one would expect of a very electronegative nonmetal: It is a strong oxidizing agent. Its principal uses are in oxidations in the steel and petrochemical industries. During steel production, oxygen is blown through hot liquid iron to oxidize impurities. Two common reactions are

$$C + O_2 \longrightarrow CO_2$$
$$Si + O_2 \longrightarrow SiO_2$$

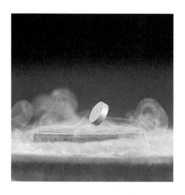

Figure 8.20

Liquid nitrogen boiling at −196°C cools a superconductor to a temperature at which it can exhibit its superconductivity and float a magnet. At higher temperatures, the magnet no longer floats because the material below it is no longer superconducting.

Silicon dioxide, SiO_2, is less dense than liquid iron and floats to the surface where it is removed; CO_2 escapes as a gas. Note that the two Group 4A elements carbon and silicon are both oxidized to their maximum oxidation states. During these reactions oxygen is reduced to its most negative oxidation state (-2).

Chlorine and Sodium Hydroxide.

Although they are very different chemically, chlorine and sodium hydroxide are paired here because they are prepared simultaneously in a very important electrochemical industrial process. The process utilizes sodium chloride, which occurs in large deposits in many parts of the world. Sodium chloride is composed of sodium ions and chloride ions, both of which are very resistant to chemical reduction or oxidation. An electrochemical process must be used to oxidize chloride ion to chlorine (Fig. 8.21).

When a direct current of electricity is passed through a solution of NaCl, the chloride ions migrate to the positive electrode and are oxidized to gaseous chlorine by giving up an electron to the electrode.

$$2Cl^-(aq) \longrightarrow Cl_2(g) + 2e^- \quad \text{(at the positive electrode)}$$

The electrons are transferred through the outside electrical circuit to the negative electrode. Although the positive sodium ions migrate toward this negative electrode, metallic sodium is not produced because sodium ions are too difficult to reduce under the conditions used. (Recall that metallic sodium is active enough to react with water and hence, even if produced, would immediately react with water to produce sodium ions again.) Instead, water molecules pick up electrons from the electrode and are reduced, giving hydrogen gas and hydroxide ions.

$$2H_2O(l) + 2e^- \text{ (from the negative electrode)} \longrightarrow H_2(g) + 2OH^-(aq)$$

These changes convert the aqueous solution of NaCl into an aqueous solution of NaOH, gaseous Cl_2, and gaseous H_2.

Figure 8.21

A cell for the industrial production of NaOH. A solution of NaCl enters the cell at the right and flows through the diaphragm. An electric current produces chlorine gas at the positive electrode and hydrogen gas and hydroxide ions at the negative electrode, resulting in a solution that contains sodium hydroxide.

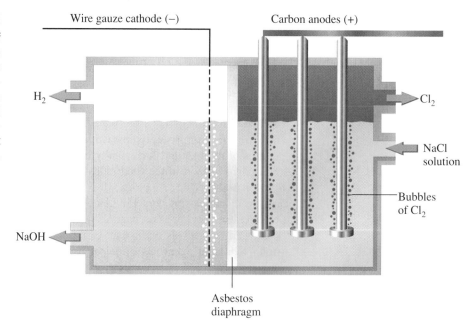

$$2Na^+(aq) + 2Cl^-(aq) + 2H_2O(l) \xrightarrow{\text{Electrolysis}} 2Na^+(aq) + 2OH^-(aq) + Cl_2(g) + H_2(g)$$

Sodium chloride
solution

Sodium hydroxide
solution

The nonmetal chlorine is more electronegative than any other element except fluorine, oxygen, and nitrogen (Fig. 6.5). Thus we would expect elemental chlorine to oxidize all the other elements except these three (and the noble gases, which are quite nonreactive). Its oxidizing property, in fact, is responsible for its principal use, as an oxidizing agent. For example, phosphorus(V) chloride, an important intermediate in the preparation of insecticides, is manufactured by oxidizing the less electronegative nonmetal phosphorus with chlorine.

$$P_4 + 10Cl_2 \longrightarrow 4PCl_5$$

A great deal of chlorine is also used to oxidize, and thus destroy, organic or biological materials in water purification and in bleaching.

Sodium hydroxide is a strong, very water-soluble (420 grams per liter) base used to make concentrated solutions of base.

$$NaOH(s) \xrightarrow{H_2O} Na^+(aq) + OH^-(aq)$$

It is used in acid–base reactions during chemical processing and in the preparation of soaps, rayon, and paper.

Nitric Acid.

The United States produced about 17 billion pounds of nitric acid (Fig. 8.19) in 1995, primarily by oxidation of ammonia as discussed earlier in this section.

The nitrate ion is readily metabolized by most plants. Thus nitrates, particularly ammonium nitrate, NH_4NO_3, are good fertilizers. In addition, nitric acid is both a strong acid and a strong oxidizing agent and is therefore used in both acid–base and oxidation–reduction reactions to prepare soluble metal salts. Its acid–base reactions include those with hydroxides, oxides, and carbonates.

$$KOH + HNO_3 \longrightarrow KNO_3 + H_2O$$
$$CuO + 2HNO_3 \longrightarrow Cu(NO_3)_2 + H_2O$$
$$CaCO_3 + 2HNO_3 \longrightarrow Ca(NO_3)_2 + H_2CO_3$$
$$ \hookrightarrow H_2O + CO_2$$

Carbonic acid, H_2CO_3, is not very stable in a concentrated solution and decomposes into gaseous carbon dioxide and water.

Nitrogen exhibits a wide range of oxidation states in its compounds. Oxidation–reduction reactions of nitric acid generally involve reduction of the nitrogen (oxidation state of +5) either to an oxidation state of +4 with formation of NO_2 or to an oxidation state of +2 with formation of NO. Reactions involving concentrated HNO_3 usually give NO_2.

$$Cu + 4HNO_3(\text{conc.}) \longrightarrow Cu(NO_3)_2 + 2NO_2 + 2H_2O$$

Dilute HNO_3 often gives NO.

$$3Ag + 4HNO_3(\text{dilute}) \longrightarrow 3AgNO_3 + NO + 2H_2O$$

Because gold is not oxidized by nitric acid, the latter reaction is used to separate silver from gold. Note that whereas most acids react with active metals such as zinc with evolution of hydrogen, nitric acid reacts by reduction of the nitrate group.

For Review   Summary

The chemical behavior of an element reflects its position in the periodic table. Those elements that lie to the left or toward the bottom of the table are **metals,** and all have similar characteristic behavior. **Nonmetals,** which lie in the upper right portion of the table, have characteristic nonmetallic behavior. The chemical behavior characteristic of metals becomes less pronounced from left to right across a period and from bottom to top in a group. The chemical behavior characteristic of nonmetals becomes more pronounced from left to right and from bottom to top. An element exhibits behavior that is increasingly characteristic of a nonmetal as its oxidation state in its compounds increases. The elements that lie near the dividing line between metals and nonmetals, the **semi-metals,** cannot be identified clearly as metals or nonmetals, and they may exhibit metallic or nonmetallic behavior depending on the conditions under which they react.

Oxidation states are assigned to atoms in compounds in accordance with rules cited in this chapter. Oxidation states are used in naming compounds, classifying chemical reactions, writing chemical formulas, and predicting chemical behavior.

Metals generally react with nonmetals, resulting in **oxidation** of the metal and **reduction** of the nonmetal. The oxidation states of the elements in the products can often be predicted from their positions in the periodic table. Most representative metals have a single common positive oxidation state in their compounds, and nonmetals have a single common negative oxidation state when they form monatomic anions.

Generally, representative metals and nonmetals form ionic **salts.** The hydroxides of metals act as **bases** unless the metals exhibit unusually high oxidation states. Basic hydroxides will accept protons from **acids.** Soluble ionic metal oxides are also bases because the oxide ion will pick up a hydrogen ion from water to form a hydroxide ion.

In addition to oxidizing metals, a nonmetal will often oxidize less electronegative nonmetals, giving covalent compounds. Fluorine, chlorine, and oxygen are particularly good oxidizing agents. Oxidation of hydrogen by nonmetals gives binary hydrides, which can act as acids, although HCl, HBr, and HI are the only binary **strong acids.** The other nonmetal hydrides range from **weak acids** to essentially nonacidic compounds as the nonmetallic character of the element involved decreases. Water and ammonia will also act as bases. Some oxides of the nonmetals react with water to produce **oxyacids** (nonmetal hydroxides), of which $HClO_4$, H_2SO_4, and HNO_3 are common strong acids. Most other oxyacids are weak.

The reactivity of unfamiliar species can be predicted by comparison with the chemistry of analogous compounds whose behavior is known or by consideration of the metallic or nonmetallic character of the elements involved, using the analysis discussed in this chapter.

The inorganic chemicals used industrially illustrate the differences between metals and nonmetals. Sulfuric acid and nitric acid are oxyacids formed from nonmetal oxides. Hydrochloric acid is a nonmetal hydride. Calcium oxide is a metal oxide that is used as a base and that reacts with water to give calcium hydroxide, another base. The hydroxide of sodium is also a base. In the reactions described that involve the metal ions Na^+ and Ca^{2+}, the oxidation states of the metal ions do not change. The oxidation states of the nonmetals, however, vary widely. The nitrogen atom, for example, varies from an oxidation state of -3 in NH_3 to $+5$ in HNO_3. The stability of the electronegative nonmetals in their reduced states is illustrated by the difficulty of preparing elemental chlorine.

Key Terms and Concepts

acid   (8.5)
acid–base reaction   (8.5)
acidic oxides   (8.5)
activity series   (8.7)
amphoteric compound   (8.6)
base   (8.5)
basic oxides   (8.5)
Brønsted acid   (8.5)
Brønsted base   (8.5)
catalyst   (8.9)
hydronium ions   (8.5)
industrial chemicals   (8.9)

metals   (8.2)
nonmetals   (8.2)
oxidation state   (8.3)
oxidation–reduction reactions   (8.3, 8.5)
oxidized   (8.5)
oxidizing agent   (8.5)
oxyacids   (8.5)
precipitation reaction   (8.5)
prediction of reaction products   (8.8)
redox reaction   (8.5)
reduced   (8.5)
reducing agent   (8.5)

salts   (8.5)
semi-metals   (8.2)
strong acid   (8.5)
strong base   (8.5)
variation in metallic character   (8.6)
variation in nonmetallic character   (8.6)
variation in oxidation states   (8.4)
weak acid   (8.5)
weak base   (8.5)
writing formulas   (8.3)

Exercises

Questions

1. How do the bonding and chemical behavior of a compound containing a metal and a hydroxyl group (OH group) differ from those of a compound containing a nonmetal and a hydroxyl group?

2. Explain why sulfuric acid, H_2SO_4, which is a covalent molecule, dissolves in water and produces a solution that contains ions.

3. What information can we use to determine which oxide is most likely to be amphoteric: CaO, SiO_2, or SO_3?

4. What information can we use to determine which element is most likely to form an amphoteric oxide: K, Ge, or Br?

5. Table 8.3 contains no oxidation states for radon because the chemistry of this radioactive noble gas is unexplored. What oxidation states would be expected for radon?

6. What information can we use to distinguish addition reactions that are also oxidation–reduction reactions from addition reactions that are acid–base reactions?

7. How can we use the periodic table to determine whether a compound that contains a hydroxyl group (an OH group) dissolves in water to give a solution of an acid or of a base?

8. How can we use the periodic table to determine whether an oxide dissolves in water to give a solution of an acid or of a base?

9. A certain gas reacts with oxygen when heated. The gas also dissolves in water, reacts, and forms a solution of an acid. Could the substance be N_2? Could it be SO_2? Could it be SO_3? Explain your answers.

10. A sample of an element that conducts electricity is broken into small pieces by cracking it with a hammer. These pieces react with chlorine to form a covalent chloride. Could the element be aluminum? Silicon? Iodine? Explain your answers.

11. An element is oxidized by fluorine but not by chlorine. Could the element be sodium? Aluminum? Silicon? Sulfur? Oxygen? Explain your answers.

12. Sodium hydrogen carbonate, also called sodium bicarbonate or baking soda, $NaHCO_3$, is sometimes kept in chemical laboratories because it can neutralize either acid or base spills. Explain how it can act either as a base or as an acid.

13. Explain why the maximum oxidation state exhibited by an atom of a Group 2A element is equal to the number of electrons in the valence shell of the neutral atom.

14. Explain why the most negative oxidation state of an atom of a Group 5A element is equal to the number of electrons required to fill the valence shell of the neutral atom.

15. Silver can be separated from gold because silver dissolves in nitric acid and gold does not (Section 8.9). Is the dissolution of silver in nitric acid an acid–base reaction or an oxidation–reduction reaction? Explain your answer.

Metals, Nonmetals, and Semi-Metals

16. Would you expect arsenic to behave as a semi-metal? Barium? Iodine? Give convincing reasons to support each of your answers.

17. From the positions of the elements in the periodic table, predict which member of each of the following pairs will

    (a) conduct electricity: Ca or S?
    (b) form a negative ion: Co or Cl?
    (c) form a solution of an acid in water: N_2O_3 or CaO?
    (d) form an ionic compound with fluorine: Al or P?
    (e) form a covalent compound with O: K or N?
    (f) have a higher electronegativity: Ba or Br?

18. From the positions of the elements in the periodic table, predict which member of each of the following pairs will

    (a) conduct electricity: P or Tl?
    (b) form a positive ion: Fe or F?
    (c) form a solution of a base in water: BaO or CO_2?
    (d) form a covalent compound with F: S or Mg?
    (e) form an ionic compound with O: Cl or Cr?
    (f) have a lower electronegativity: Mg or O?

19. From the positions of the elements in the periodic table, predict which member of each of the following pairs will be

    (a) a base: $CsOH$ or $BrOH$?
    (b) an acid: $In(OH)_3$ or $B(OH)_3$?
    (c) an oxidizing agent: S or Na?
    (d) an ionic compound: $RbCl$ or $BrCl$?
    (e) more easily reduced: Ca or O?

20. From the positions of the elements in the periodic table, predict which member of each of the following pairs will be

    (a) an acid: KOH or $ClOH$?
    (b) more easily reduced: Cl_2 or Cs?
    (c) an ionic compound: CaO or NO?
    (d) an oxidizing agent: I_2 or Ga?
    (e) a base: $Ca(OH)_2$ or $SO_2(OH)_2$?

21. From the positions of their components in the periodic table, predict which member in each of the following pairs will

    (a) be ionic: KF or HF?
    (b) react as a base: $P(OH)_3$ or $Al(OH)_3$?
    (c) oxidize Si: O_2 or Al?
    (d) reduce Sn^{2+} to Sn: Mg or Cl_2?
    (e) oxidize S: Na or F_2?
    (f) contain ions: KNO_2 or $ClNO_2$?
    (g) react with water to give a solution of an acid: In_2O_3 or P_4O_6?

22. From the positions of the elements in the periodic table, predict which member of each of the following pairs will

   (a) reduce S: Cl_2 or Mg?
   (b) neutralize HCl: KOH or BrOH?
   (c) oxidize Si: Ca or Cl_2?
   (d) react as an acid: ClOH or KOH?
   (e) reduce SeO_2: Ca or Cl_2?
   (f) contain ions: NaH or HF?
   (g) react with water to give a solution of a base: Li_2O or N_2O?

Oxidation States

23. Which forms compounds in which it has a maximum positive oxidation state of

   (a) +1: Na or Cl?   (b) +2: O or Sr?
   (c) +3: In or P?   (d) +4: Si or Ne?
   (e) +5: B or N?

24. Which forms compounds in which it has a most negative oxidation state of

   (a) −1: Br or K?
   (b) −2: Ca or S?
   (c) −3: Al or P?

25. Which of the following elements will exhibit a negative oxidation state when combined with phosphorus?

   (a) Mg   (b) F   (c) Al
   (d) O   (e) Na   (f) Cl
   (g) Fe   (h) S   (i) N

26. List the elements that will react with sulfur and form binary compounds (compounds containing only two elements) in which sulfur has a positive oxidation state.

27. Write the formula of a binary compound (a compound containing only two elements) for each of the following elements so that the element has the indicated oxidation state.

   (a) Ca, +2   (b) N, −3
   (c) N, +3   (d) S, −2
   (e) Cl, −1   (f) Br, +3
   (g) P, +3   (h) Pb, +2

28. Write the formula of a binary compound (a compound containing only two elements) for each of the following elements so that the element has the indicated oxidation state.

   (a) Se, −2   (b) As, +5
   (c) P, +5   (d) C, +2
   (e) Cl, +1   (f) Br, −1
   (g) P, −3   (h) Sn, +2

29. Determine the oxidation states of the elements in the following compounds.

   (a) NaI
   (b) $GdCl_3$
   (c) $LiNO_3$
   (d) H_2Se
   (e) Mg_2Si

   (f) RbO_2, rubidium superoxide
   (g) HF

30. Determine the oxidation states of the elements in the following compounds.

   (a) $CaSO_4$
   (b) Nd_2O_3
   (c) MgSe
   (d) H_3As
   (e) Li_3N
   (f) CaO_2, calcium peroxide
   (g) HI

31. Determine the oxidation states of the elements in the compounds listed. None of the oxygen atoms are part of peroxides or superoxides.

   (a) H_3PO_4 (b) $Al(OH)_3$
   (c) SeO_2 (d) KNO_2
   (e) In_2S_3 (f) P_4O_6

32. Determine the oxidation states of the elements in the compounds listed. None of the oxygen atoms are part of peroxides or superoxides.

   (a) H_2SO_4 (b) $Ca(OH)_2$
   (c) BrOH   (d) $BrNO_2$
   (e) $TiCl_4$ (f) NaH

33. From their positions in the periodic table and without reference to Table 8.3, predict the common oxidation states of the following elements in their compounds.

   (a) Al   (b) Cl
   (c) Ca   (d) Cs
   (e) As   (f) S
   (g) Ge   (h) H

34. Without reference to Table 8.3, identify those representative metals that may display two or more common oxidation states in their compounds.

35. Name the following compounds, using oxidation states as appropriate:

   (a) $FeCl_3$ (b) NO_2
   (c) SF_6 (d) As_2O_3
   (e) $Co(NO_3)_2$ (f) $V_2(SO_4)_3$

36. Name the following compounds, using oxidation states as appropriate:

   (a) $TiBr_4$ (b) SO_3
   (c) $SeCl_4$ (d) P_2O_5
   (e) $Fe(ClO_4)_3$ (f) $Tl_2(SO_4)_3$

37. Rename the following compounds, using oxidation states as appropriate:

   (a) Sulfur trioxide
   (b) Dinitrogen pentoxide
   (c) Tetraphosphorus hexoxide
   (d) Indium monoiodide
   (e) Carbon dioxide
   (f) Tin tetrachloride

38. Rename the following compounds, using oxidation states as appropriate:
    (a) Sulfur dioxide
    (b) Nitrogen monoxide
    (c) Diboron tetrachloride
    (d) Dioxygen difluoride
    (e) Carbon disulfide
    (f) Phosphorus pentachloride

39. The commercial production of phosphorus involves the use of carbon and silicon dioxide:

    $$2Ca_3(PO_4)_2(s) + 10C(s) + 6SiO_2(s) \longrightarrow$$
    $$6CaSiO_3(s) + 10CO(g) + P_4(s)$$

    Determine the oxidation state of each atom in the reactants and products. Which element has been oxidized and which reduced?

40. The first laboratory preparation of chlorine gas involved the reaction

    $$4NaCl(aq) + 2H_2SO_4(aq) + MnO_2(s) \longrightarrow$$
    $$2Na_2SO_4(aq) + MnCl_2(aq) + 2H_2O(l) + Cl_2(g)$$

    Determine the oxidation states for the elements that have been oxidized and reduced.

41. Elemental fluorine, F_2, was not isolated until 1886, when it was prepared by the electrolysis of HF in a molten mixture of hydrogen fluoride and potassium fluoride. The other product was H_2. Efforts to prepare fluorine by a chemical reaction of compounds, which did not require elemental fluorine in their preparation, failed until 1986 when K_2MnF_6 was shown to react with antimony(V) fluoride to form fluorine. The other products were manganese(III) fluoride and the potassium salt of the SbF_6^- ion.
    (a) Write an equation for the two reactions which yield fluorine.
    (b) Indicate the oxidation states of the atoms in the reactants and in the products of these reactions.
    (c) Which elements are reduced and which oxidized?
    (d) Which elements are the oxidizing agents and which the reducing agents?

Chemical Reactions

42. Classify each of the following as an acid–base reaction or an oxidation–reduction reaction.
    (a) $Na_2S + 2HCl \longrightarrow 2NaCl + H_2S$
    (b) $2Na + 2HCl \longrightarrow 2NaCl + H_2$
    (c) $Mg + Cl_2 \longrightarrow MgCl_2$
    (d) $MgO + 2HCl \longrightarrow MgCl_2 + H_2O$
    (e) $K_3P + 2O_2 \longrightarrow K_3PO_4$
    (f) $3KOH + H_3PO_4 \longrightarrow K_3PO_4 + 3H_2O$

43. When each of the following compounds dissolves in water, does it give a solution of an acid or a base?
    (a) CaO   (b) P_4O_{10}
    (c) H_3PO_4 (d) HCl

    (e) Na_2O (f) RbOH
    (g) SO_3 (h) $HClO_2$
    (i) $Ba(OH)_2$

44. Identify the atoms that are oxidized and reduced, the change in oxidation state for each, and the oxidizing and reducing agents in each of the following equations.
    (a) $Mg + NiCl_2 \longrightarrow MgCl_2 + Ni$
    (b) $PCl_3 + Cl_2 \longrightarrow PCl_5$
    (c) $C_2H_4 + 3O_2 \longrightarrow 2CO_2 + 2H_2O$
    (d) $Zn + H_2SO_4 \longrightarrow ZnSO_4 + H_2$
    (e) $2K_2S_2O_3 + I_2 \longrightarrow K_2S_4O_6 + 2KI$
    (f) $3Cu + 8HNO_3 \longrightarrow 3Cu(NO_3)_2 + 2NO + 4H_2O$

45. Complete and balance the following acid–base equations.
    (a) HCl gas reacts with solid $Ca(OH)_2(s)$.
    (b) A solution of $Sr(OH)_2$ is added to a solution of HNO_3.
    (c) SO_2 gas reacts with solid KOH.
    (d) HBr gas reacts with solid CoO.
    (e) Na_2O reacts with CO_2.

46. Complete and balance the following acid–base equations.
    (a) A solution of $HClO_4$ is added to a solution of LiOH.
    (b) Pure liquid H_2SO_4 reacts with MnO.
    (c) $Ba(OH)_2$ reacts with HF gas.
    (d) N_2O_5 reacts with K_2O.
    (e) SO_3 gas reacts with CaO.

47. Complete and balance the following oxidation–reduction reactions, which give the highest possible oxidation state for the oxidized atoms.
    (a) $Al + F_2 \longrightarrow$ (b) $Al + CuBr_2 \longrightarrow$
    (c) $P_4 + O_2 \longrightarrow$ (d) $Ca + H_2O \longrightarrow$
    (e) $In + HCl \longrightarrow$

48. Complete and balance the following oxidation–reduction reactions, which give the highest possible oxidation state for the oxidized atoms.
    (a) $K + H_2O \longrightarrow$ (b) $Ba + HBr \longrightarrow$
    (c) $Mg + Pb(NO_3)_2 \longrightarrow$ (d) $CO + O_2 \longrightarrow$
    (e) $Sn + I_2 \longrightarrow$

49. Which of the following compounds could be formed in a metathetical reaction because they are not soluble in water?
    (a) $LiNO_3$ (b) $CaCl_2$ (c) $AlPO_4$
    (d) Na_2SO_4 (e) $CaCO_3$ (f) MnS
    (g) KOH   (h) RbBr   (i) $Al(OH)_3$
    (j) $Al(NO_3)_3$ (k) $PbSO_4$ (l) Li_2S

50. Which of the following compounds could be formed in a metathetical reaction because they are not soluble in water?
    (a) $Fe(NO_3)_3$ (b) $BaSO_4$ (c) Cs_2S
    (d) $NaNO_3$ (e) $CaBr_2$ (f) $Ca_3(PO_4)_2$
    (g) NaOH   (h) RbI   (i) $Co(OH)_2$
    (j) Li_2SO_4 (k) $SrCO_3$ (l) CuS

51. Compslete and balance the equations for the following acid–base reactions. If water is used as a solvent, write the

reactants and products as solvated ions. In some cases there may be more than one correct answer, depending on the amounts of reactants used.

(a) $HBr(g) + In_2O_3(s) \longrightarrow$ (Assume the product does not dissolve in the water produced.)

(b) $Mg(OH)_2(s) + HClO_4(aq) \longrightarrow$

(c) $SO_3(g) + H_2O(l) \longrightarrow$ (Assume an excess of water and that the product dissolves.)

(d) $Na_2O(s) + H_2O(l) \longrightarrow$

(e) $Li_2O(s) + CH_3CO_2H(l) \longrightarrow$

(f) $SrO(s) + H_2SO_4(l) \longrightarrow$

52. Complete and balance the equations for the following acid–base reactions. If water is used as a solvent, write the reactants and products as solvated ions. In some cases there may be more than one correct answer, depending on the amounts of reactants used.

(a) $NH_3(g) + HI(g) \longrightarrow$

(b) $NH_3(aq) + HBr(aq) \longrightarrow$

(c) $CaO(s) + SO_3(g) \longrightarrow$

(d) $Li_2O(s) + N_2O_5(l) \longrightarrow$

(e) $Al(OH)_3(s) + HNO_3(aq) \longrightarrow$

(f) $KCl(s) + H_2SO_4(l)$

53. The following salts contain anions of weak acids. These anions react as weak bases. Write a balanced equation for the reaction of a solution of each with an excess of the indicated acid.

(a) $Li[CH_3CO_2] + H_3PO_4 \longrightarrow$

(b) $SrF_2 + H_2SO_4 \longrightarrow$

(c) $NaCN + HCl \longrightarrow$

(d) $CaCO_3 + HCl \longrightarrow$

54. The following salts contain anions of weak acids. These anions react as weak bases. Write a balanced equation for the reaction of a solution of each with an excess of the indicated acid.

(a) $CaHPO_4 + HClO_4 \longrightarrow$

(b) $KHCO_3 + HCl \longrightarrow$

(c) $KNO_2 + HBr \longrightarrow$

(d) $Li_2SO_3 + HCl \longrightarrow$

55. When heated to 700°C–800°C, diamonds, which are pure carbon, are oxidized by atmospheric oxygen. (They burn!) Write the balanced equation for this reaction.

56. The following reactions are all similar to those of the industrial chemicals described in Section 8.9. Complete and balance the equations for these reactions.

(a) Reaction of a weak base and a strong acid

$$NH_3 + HClO_4 \longrightarrow$$

(b) Pickling of steel in hydrochloric acid

$$Fe_2O_3 + HCl \longrightarrow$$

(c) Preparation of a soluble source of calcium and phosphorus used in animal feeds

$$Ca_3(PO_4)_2 + H_3PO_4 \longrightarrow$$

(d) Formation of an air pollutant produced by burning coal that contains iron sulfide

$$FeS + O_2 \longrightarrow$$

(e) Preparation of a soluble silver salt for silver plating

$$Ag_2CO_3 + HNO_3 \longrightarrow$$

(f) Neutralization of a basic solution of nylon in order to precipitate the nylon

$$Na^+(aq) + OH^-(aq) + H_2SO_4 \longrightarrow$$

(g) Hardening of plaster containing slaked lime

$$Ca(OH)_2 + CO_2 \longrightarrow$$

(h) Removal of sulfur dioxide from the flue gas of power plants

$$CaO + SO_2 \longrightarrow$$

(i) Neutralization of acid drainage from coal mines

$$CaCO_3 + H_2SO_3 \longrightarrow$$

(j) The reaction of baking powder that produces carbon dioxide gas and causes bread to rise

$$NaHCO_3 + NaH_2PO_4 \longrightarrow$$

(k) Separation of silver from gold with concentrated nitric acid

$$Ag + HNO_3 \longrightarrow$$

(l) Preparation of strontium hydroxide by electrolysis of a solution of strontium chloride

$$SrCl_2(aq) + H_2O(l) \xrightarrow{\text{Electrolysis}}$$

57. The military has experimented with lasers that produce very intense light when fluorine combines explosively with hydrogen. What is the balanced equation for this reaction?

58. Small amounts of hydrogen gas and oxygen gas are often produced by electrolysis of water containing a little sulfuric acid as a catalyst. Write the balanced equation for this reaction.

59. Great Lakes Chemical Company produces bromine, Br_2, from bromide salts, such as NaBr, in Arkansas brine by treating the brine with chlorine gas. Write a balanced equation for the reaction of NaBr with Cl_2.

60. Gaseous hydrogen chloride may be produced by reacting sodium chloride with concentrated sulfuric acid. Complete and balance two equations describing the possible reactions between NaCl and H_2SO_4.

61. In a common experiment in the general chemistry laboratory, magnesium metal is heated in air to produce MgO. MgO is a white solid, but in these experiments it often looks yellow because of small amounts of Mg_3N_2, a compound formed as some of the magnesium reacts with nitrogen. Write a balanced equation for each reaction.

62. Lithium hydroxide may be used to absorb carbon dioxide in enclosed environments, such as the space shuttle. Write an

equation for the reaction that involves 2 mol of LiOH per 1 mol of CO_2.

The Activity Series

63. One of the elements in the activity series (Table 8.7) will not produce hydrogen when it reacts with steam, but it will form hydrogen when it reacts with hydrochloric acid. Reaction of this element with a solution of $Ni(NO_3)_2$ produces $Ni(s)$. What is this element?

64. Zinc and mercury are both members of Group 2B, but ZnS and HgS behave differently when roasted in air. Write the equations for the reactions of ZnS and HgS with O_2 when they are roasted.

65. With the aid of the activity series, predict whether each of the following reactions will take place.
    (a) $Mg + Co^{2+} \longrightarrow Mg^{2+} + Co$
    (b) $Al_2O_3 + 3H_2 \longrightarrow 2Al + 3H_2O$
    (c) $2Au + Fe^{2+} \longrightarrow 2Au^+ + Fe$
    (d) $Ni + 2H^+ \longrightarrow Ni^{2+} + H_2$
    (e) $2HgO \longrightarrow 2Hg + O_2$

66. With the aid of the activity series, predict whether each of the following reactions will take place.
    (a) $Pb + Ni^{2+} \longrightarrow Pb^{2+} + Ni$
    (b) $PtO_2 \xrightarrow{\Delta} Pt + O_2$
    (c) $K_2O + H_2 \longrightarrow 2K + H_2O$
    (d) $3Cd + 2Bi^{3+} \longrightarrow 2Bi + 3Cd^{2+}$
    (e) $3Pd + 2Au^{3+} \longrightarrow 2Au + 3Pd^{2+}$

67. Lithium and beryllium do not appear in the activity series given in Table 8.7. From the positions of these elements in the periodic table, locate their approximate positions in the activity series and write equations for their reactions, if any, with water, steam, oxygen, and acids. Will the oxides of Li and Be be reduced by hydrogen? Are the oxides reduced by heat alone?

68. What mass of copper can be recovered from a solution of copper(II) sulfate by addition of 2.11 kg of aluminum?

Applications and Additional Exercises

69. The label on a bottle of vitamin tablets lists elements essential for human nutrition. Prepare a list of these elements, classifying them as metals, nonmetals, or semi-metals. From information given on the label, try to determine the form in which each element is present in the vitamin.

70. Identify the cations and anions in each of the following compounds.
    (a) LiCl   (b) CaI_2 (c) NaOH
    (d) Ga_2O_3 (e) K_2SO_4 (f) NH_4Cl
    (g) $Ca(H_2PO_4)_2$ (h) $SrHPO_4$ (i) $[PCl_4][PCl_6]$

71. Calcium propionate is sometimes added to bread to retard spoilage. This compound can be prepared by the reaction of calcium carbonate, $CaCO_3$, with propionic acid, $C_2H_5CO_2H$, which has properties similar to those of acetic acid. Write the balanced equation for the formation of calcium propionate.

72. Complete and balance the equations of the following reactions, each of which could be used to remove hydrogen sulfide from natural gas.
    (a) $Ca(OH)_2\ (s) + H_2S\ (g) \longrightarrow$
    (b) $NaOH\ (s) + H_2S\ (g) \longrightarrow$
    (c) $Na_2CO_3\ (aq) + H_2S\ (g) \longrightarrow$
    (d) $K_3PO_4\ (aq) + H_2S\ (g) \longrightarrow$

73. Write balanced chemical equations for the preparation of each of the following compounds (a) by a metathetical reaction, (b) by an oxidation–reduction reaction, and (c) by an acid–base reaction: $CaCl_2$, K_2SO_4, and $Mg(H_2PO_4)_2$. Write different reactions for (a), (b), and (c) in each case.

74. Write a balanced chemical equation for the reaction used to prepare each of the following compounds from the given starting material(s). In some cases, several steps and additional reactants may be required.
    (a) H_2SO_4 from CuS
    (b) $(NH_4)NO_3$ from N_2
    (c) HBr from Br_2
    (d) H_2S from Zn and S
    (e) Na_2CO_3 from NaCl and CO_2
    (f) NaH from H_3PO_4 and Na

75. A medical laboratory test for cyanide ion, CN^-, involves separation of the cyanide ion from a blood, urine, or tissue sample by the addition of sulfuric acid. Cyanide ion acts as a weak base in this reaction. The gaseous product is absorbed in a sodium hydroxide solution and then analyzed. Write balanced equations that describe the separation and absorption.

76. (a) Hydrogen sulfide, H_2S, is removed from natural gas by passing the raw gas through a solution of ethanolamine, $HOCH_2CH_2NH_2$, whose behavior is similar to that of NH_3 (which can be viewed as HNH_2) in its acid–base reactions. After the solution is saturated, the H_2S can be recovered by heating the solution to reverse the reaction. Write the reaction of ethanolamine with hydrogen sulfide, showing the Lewis structures of the reactants and products.
    (b) Part of the recovered H_2S is burned with air, and the product is allowed to react with the remaining H_2S to produce sulfur, S_8—a cyclic molecule composed of eight sulfur atoms which is easier to ship than H_2S. Write the balanced equations for the reactions that lead to the formation of sulfur from hydrogen sulfide, showing the Lewis structures of the reactants and products.

77. Sodium dithionate, $Na_2S_2O_4$, sometimes used in the paper industry to bleach paper pulp, is produced by addition of $NaBH_4$ to a solution of SO_2 and NaOH.

$$NaBH_4 + 8NaHSO_3 \longrightarrow 4Na_2S_2O_4 + NaBO_2 + 6H_2O$$

(a) Write an equation for the reaction of SO_2 and NaOH that produces $NaHSO_3$.

(b) Which atoms are oxidized and which are reduced in the reaction of $NaBH_4$ and $NaHSO_3$? What are the initial and final oxidation states of these atoms?

(c) Write the Lewis structures of the BH_4^- ion, the HSO_3^- ion (which contains an OH group and two O atoms bonded to the S atom), and the $S_2O_4^{2-}$ ion (which contains two SO_2 groups joined by a sulfur–sulfur bond).

(d) What are the electron-pair geometry and the molecular geometry about the B atom in BH_4^- and about the S atoms in the two sulfur-containing species? (We have not asked about $NaBO_2$ because this species can contain several different anions with the empirical formula BO_2^-.)

78. Calcium cyclamate $Ca(C_6H_{11}NHSO_3)_2$, an artificial sweetener, can be purified industrially by converting it to the barium salt through reaction of the acid $C_6H_{11}NHSO_3H$ with barium carbonate, treatment with sulfuric acid (barium sulfate is very insoluble), and then neutralization with calcium hydroxide. Write the balanced equations for these reactions.

79. What is the percent by mass of phosphorus in the compound formed by the reaction of phosphorus with magnesium? (Use four significant figures.)

80. What is the percent by mass of sulfur in the compound formed when 0.050 mol of Tl reacts with H_2SO_4, producing a thallium(I) compound and 0.025 mol of H_2?

81. What is the hydroxide ion concentration in a solution formed by dissolution of 1.00 g of barium metal in enough water to form 250 mL of the solution?

82. What is the perchloric acid concentration in a solution formed by dissolution of 2.35 g of Cl_2O_7 in enough water to form 1.00 L of the solution?

83. Describe the molecular structure of the nonmetal oxide that reacts with water to form a solution of H_2CO_3.

84. What is the molecular structure of the molecules that form when Na_3P is added to water?

85. How does the hybridization of the sulfur atom change when SO_2 is dissolved in water?

86. Chlorine is obtained by the electrolysis of molten sodium chloride or of aqueous sodium chloride.

(a) Calculate the mass of chlorine produced from 3.00 kg of sodium chloride in each of the reactions.

(b) Calculate the mass of each of the other products produced.

87. Phosphoric acid, one of the acids used in some cola drinks, is produced by the reaction of phosphorus(V) oxide, an acidic oxide, with water. Phosphorus(V) oxide is prepared by the combustion of phosphorus.

(a) Write the empirical formula of phosphorus(V) oxide.

(b) What is the molecular formula of phosphorus(V) oxide, which has a molar mass of about 280?

(c) Write balanced equations for the production of phosphorus(V) oxide and phosphoric acid.

(d) Determine the mass of phosphorus required to make 10.0 tons (1.00×10^4 kg) of phosphoric acid, assuming a yield of 98.85%.

88. The thermite reaction is often used to weld metals. The equation for the reaction is

Iron(III) oxide + aluminum $\longrightarrow$ iron + aluminum oxide

This reaction produces so much heat that the iron produced is molten.

(a) Write the chemical equation that represents the thermite reaction.

(b) Calculate the amount of iron(III) oxide and aluminum required to prepare enough reactant mixture to produce 1 pound (454 g) of iron.

(c) Using Hess's law, calculate ΔH°_{298} for the reaction.

89. One of the older commercial processes used in the production of H_2SO_4 is represented by the following reactions:

  i. $S(g) + O_2(g) \longrightarrow SO_2(g)$

 ii. $2NO(g) + O_2(g) \longrightarrow 2NO_2(g)$

iii. $NO_2(g) + SO_2(g) + H_2O(l) \longrightarrow H_2SO_4(aq) + NO(g)$

The NO produced in the third reaction is recycled through the second reaction to produce more NO_2.

(a) Determine the oxidation state of nitrogen and that of sulfur in each of their compounds.

(b) Name NO_2, SO_2, and NO two ways: with Greek prefixes and with oxidation states.

(c) Write the Lewis structures of each of the N and S compounds in the reactions, describe their molecular geometry, and describe the hybridization of the internal N, S, and O atoms in NO_2, SO_2, and H_2SO_4.

(d) What mass of sulfur is required to prepare 1.000×10^3 L of 66.0% sulfuric acid, by mass, with a density of 1.567 g cm^{-3} if the percent yield of SO_2 is 96.44% and the percent yield of H_2SO_4 is 99.00%?

(e) What is the molar concentration of the acid described in part (d)?

(f) A producer of test kits sells a kit for determining alkalinity. The kit contains a 0.035 M solution of H_2SO_4. What volume of the acid described in part (d) would this company need in order to produce 1.000×10^4 L of 0.035 M solution of H_2SO_4?

(g) What is ΔH°_{298} for each of the reactions described for the production of H_2SO_4? Which ΔH°_{298} is equal to a heat of formation? Is each reaction exothermic or endothermic? The heat of formation of $H_2SO_4(aq)$ may be taken as -909 kJ mol^{-1}.

CHAPTER OUTLINE

Elemental Carbon and Its Inorganic Compounds

9.1   Elemental Carbon
9.2   Inorganic Compounds of Carbon

Organic Compounds of Carbon

9.3   Alkanes
9.4   The Basics of Organic Nomenclature
9.5   Alkenes
9.6   Alkynes
9.7   Aromatic Hydrocarbons

Derivatives of Hydrocarbons

9.8   The Behavior of Hydrocarbon Derivatives
9.9   Alcohols
9.10 Ethers
9.11 Amines
9.12 Aldehydes and Ketones
9.13 Carboxylic Acids and Esters
9.14 Natural Products
9.15 Polymers

9

Structure and Bonding at Carbon

Urea crystals photographed through a microscope using polarized light.

Life on earth would not exist without carbon. With the exception of water and a few ionic compounds such as salt and the phosphate components of bone, all the compounds that make up living matter contain carbon atoms. In fact, early chemists regarded substances isolated from plants and animals as a type of matter that could not be synthesized artificially. A widespread belief called vitalism held that these organic compounds were produced by a vital force that was present only in living organisms. It was believed that organic compounds could be interconverted, but not synthesized from elements or minerals. The German chemist Friedrich Wohler was one of the early chemists to demonstrate that the idea of vitalism was incorrect when in 1828 he reported the synthesis of urea, a component of many body fluids including urine, by the reaction of $Pb(OCN)_2$ with an NH_3 solution. Since that time, it has been recognized that organic molecules derived from plants and animals obey the same natural laws as other substances.

Today carbon atoms are key components of synthetic plastics, soaps, perfumes, sweeteners, fabrics, and most of the other synthetic substances we use every day. Most fuels are compounds of carbon or are essentially pure carbon itself. Carbon is also an important component of many rocks and building materials. In this chapter we will consider the inorganic and organic behavior of carbon compounds, discuss how carbon gives rise to a vast number and variety of different compounds, and look at the role of carbon in representative biological and industrial settings.

Elemental Carbon and Its Inorganic Compounds

Carbon is the first member of Group 4A of the periodic table. It is a nonmetal, and almost all compounds of carbon are covalent. The valence shell electron configuration of a carbon atom is $2s^2 2p^2$. A carbon atom has no d orbitals in its valence shell, so it cannot contain more than eight bonding electrons (Section 6.4). Carbon atoms bond to two, three, or four other atoms in most covalent compounds, usually employing all four of the carbon atom's valence electrons. Thus the molecular geometries in which most carbon atoms take part are linear, trigonal planar, or tetrahedral (Sections 7.1 and 7.2). The hybridization of the carbon atom is generally sp, sp^2, or sp^3 (Sections 7.6–7.8).

$$O=C=O \qquad H-C\equiv C-H \qquad \begin{array}{c} H \\ \diagdown \\ C=C \\ \diagup \\ H \end{array} \begin{array}{c} H \\ \diagup \\ \diagdown \\ H \end{array} \qquad H-\overset{\displaystyle H}{\underset{\displaystyle H}{C}}-H$$

A few carbon compounds are ionic. Such exceptions include combinations of carbon and metals with very low electronegativities (Groups 1A and 2A). With these metals, carbon forms carbides, which are essentially ionic compounds with carbon anions. Examples include CaC_2, which contains the $C\equiv C^{2-}$ ion and Al_4C_3, which contains the C^{4-} ion.

9.1 Elemental Carbon

Carbon occurs in the uncombined (elemental) state in familiar forms: diamond, graphite, and charcoal. Less familiar forms are coke, carbon black, and fullerenes.

Figure 9.1

Samples of diamond and graphite, two forms of carbon.

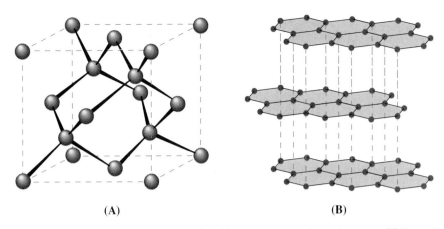

(A)   (B)

Figure 9.2 The structures of diamond and graphite with spheres representing carbon atoms. (A) The crystal structure of diamond. The covalent bonds are drawn wedge-shaped to help show the perspective. (B) The crystal structure of graphite. Each planar layer is composed of six-membered rings.

Figure 9.3

After these industrial diamonds are arranged in the mold, molten metal will be added to cast a diamond-studded bit used to drill through very hard rock formations.

Diamond (Fig. 9.1) is a very hard, crystalline material that is colorless and transparent when pure. The carbon atoms in a crystal of diamond are bonded into a giant molecule by covalent bonds (Fig. 9.2 A). Each atom forms four single bonds to four other atoms at the corners of a tetrahedron (sp^3 hybridization). Because the carbon–carbon single bonds are very strong (Section 6.12) and because they extend throughout the crystal to form a three-dimensional network, the crystals are very hard (Fig. 9.3) and have high melting points ($\sim 4400°C$).

Graphite (Fig. 9.1) is a soft, slippery, grayish-black solid that conducts electricity. Its properties are related to its structure (Fig. 9.2 B), which consists of layers of carbon atoms, each atom surrounded by three other carbon atoms in a trigonal planar arrangement. Each carbon atom in graphite forms three σ bonds, one to each of its nearest neighbors, by means of sp^2 hybrid orbitals. The unhybridized p orbital on each carbon atom projects above and below the layer (Fig. 9.4) and overlaps with unhybridized orbitals on adjacent carbon atoms in the same layer to form π bonds. Many resonance forms are necessary to describe the electronic structure of a graphite layer; Fig. 9.5 on page 296 illustrates two of these forms.

Atoms within a graphite layer are bonded together tightly by the σ and π bonds; however, the forces between layers are weak. The layers are held together by forces called London dispersion forces. We will discuss the nature of these forces more fully

Figure 9.4

The orientation of unhybridized p orbitals of the carbon atoms in graphite. Each p orbital is perpendicular to the plane of carbon atoms.

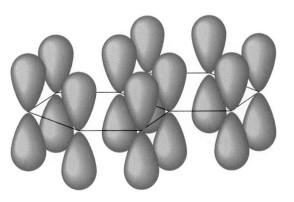

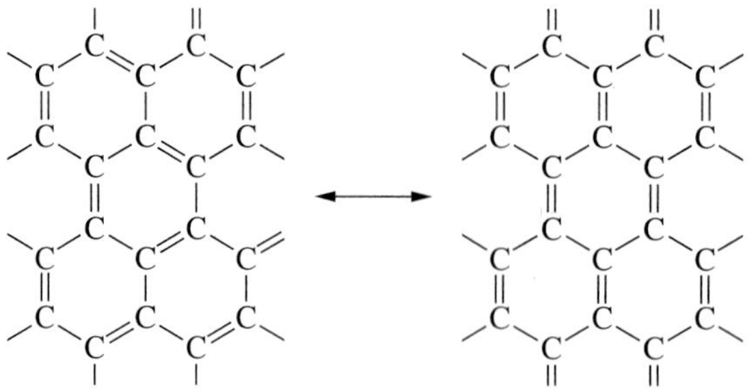

Figure 9.5

Two of the many resonance forms of graphite necessary to describe its electronic structure, that of a resonance hybrid.

in Chapter 11; for now we need only note that London forces are much weaker than covalent bonds and are easily broken. The weak forces between layers give graphite the soft, flaky character that makes it useful as the "lead" in pencils and the slippery character that makes it useful as a lubricant. The loosely held electrons in the π bonds can move throughout the solid; they are responsible for the electrical conductivity of graphite.

Other forms of elemental carbon include **carbon black, charcoal,** and **coke.** Carbon black is an amorphous form of carbon that is prepared commercially by burning natural gas, CH_4, in a very limited amount of air. We sometimes find carbon black as soot on crucibles that have been heated in a yellow flame. Charcoal and coke are produced by heating wood and coal, respectively, at high temperatures in the absence of air.

Recently, new forms of elemental carbon molecules have been identified in the soot generated by a smoky flame and in the vapor that results when graphite is heated to very high temperatures in a vacuum or in helium. The simplest of these new forms consists of icosahedral (soccer-ball-shaped) molecules that contain 60 carbon atoms, C_{60} (Fig. 9.6 and the opening photograph for Chapter 6). This form has been named **buckminsterfullerene** (the molecules are often called buckyballs) after the architect Buckminster Fuller, who designed domed structures that also contain five- and six-membered rings.

Figure 9.6

The molecular structure of C_{60}, buckminsterfullerene.

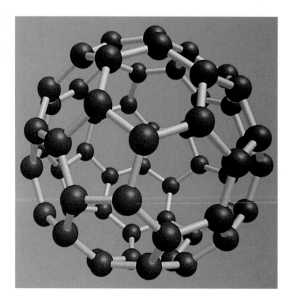

Figure 9.7

A carbon monoxide detector can indicate the presence of this dangerous invisible, odorless gas.

9.2   Inorganic Compounds of Carbon

We generally classify compounds that contain only one or two carbon atoms per formula unit and/or compounds that contain no carbon–hydrogen bonds as inorganic substances. Examples of such compounds include calcium carbide, CaC_2; carbon monoxide, CO; carbon dioxide, CO_2; hydrogen cyanide, HCN; salts containing the carbonate ion, CO_3^{2-}, or the cyanide ion, CN^-; and related ions and molecules.

Carbon monoxide, CO, forms when carbon burns in a limited amount of oxygen. It is produced commercially by the reaction of steam with hot coke in a reaction that also produces hydrogen.

$$H_2O(g) + C(s) \longrightarrow CO(g) + H_2(g)$$

This is an important industrial reaction because carbon monoxide is used in a variety of industrial synthesis, including the production of methanol, commonly known as wood alcohol.

$$CO(g) + 2H_2(g) \longrightarrow CH_3OH(l)$$

Carbon monoxide is a very dangerous poison. It bonds to the hemoglobin in blood more tightly than does oxygen; hence it reduces the blood's ability to carry oxygen to the brain and other body tissues. Carbon monoxide is an odorless and tasteless gas and therefore gives no warning of its presence. Household devices that detect carbon monoxide produced by faulty heating systems have recently become available (Fig. 9.7).

Carbon dioxide, CO₂ (Fig. 9.8), forms when carbon or organic compounds burn in an excess of oxygen.

$$C(s) + O_2(g) \longrightarrow CO_2(g)$$
$$C_7H_{16}(l) + 11O_2(g) \longrightarrow 7CO_2(g) + 8H_2O(g)$$

$$\ddot{O}=C=\ddot{O}$$

Figure 9.8

The molecular and Lewis structures of CO_2.

It also forms when the cells of a plant or animal convert a sugar such as glucose to carbon dioxide and water.

$$C_6H_{12}O_6(aq) + 6O_2(aq) \longrightarrow 6CO_2(g) + 6H_2O(l)$$
$$\text{Glucose}$$

Carbon dioxide is also a by-product of fermentation; in the absence of oxygen, sugars are converted to alcohol, acetic acid, or other products with shorter carbon chains, plus carbon dioxide. The reaction for the fermentation of the sugar glucose is

$$C_6H_{12}O_6(aq) \xrightarrow{\text{Yeast}} 2C_2H_5OH(aq) + 2CO_2(g)$$
$$\text{Glucose} \qquad\qquad \text{Ethanol}$$

Commercial amounts of carbon dioxide are produced as a by-product of ammonia synthesis. The reactions of carbon monoxide or methane that produce hydrogen for ammonia manufacture (Section 8.9) also produce CO_2.

$$CO(g) + H_2O(g) \longrightarrow CO_2(g) + H_2(g)$$
$$CH_4(g) + 2H_2O(g) \longrightarrow CO_2(g) + 4H_2(g)$$

The largest single use of carbon dioxide is as the refrigerant dry ice, solid CO_2. It is also used in the carbonation of beer and soft drinks, as a propellant in aerosol cans, as an inert atmosphere, and as a fire extinguisher. Liquid carbon dioxide is used to extract fats and oils from snack foods.

Figure 9.9
Carbon dioxide concentrations in the air have increased rapidly during this century. Concentrations prior to 1957 are estimated from air trapped in Antarctic ice. Since 1957 concentrations have been measured at a mountain site in Hawaii.

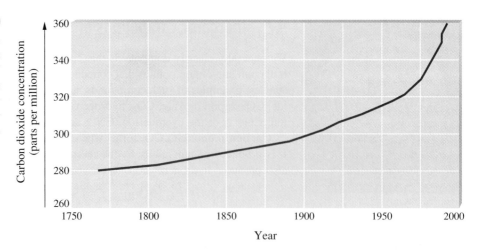

We can prepare carbon dioxide in the laboratory by the reaction of carbonates (salts containing the CO_3^{2-} ion) with acids.

$$CO_3^{2-}(s) + 2H_3O^+\,(aq) \longrightarrow 3H_2O(l) + CO_2(g)$$

Carbon dioxide is not a poison. However, a large concentration of carbon dioxide can cause suffocation (lack of oxygen).

The atmosphere contains about 0.04% carbon dioxide by volume, and is a huge reservoir of this compound. This content has increased significantly over the past century (Fig. 9.9) because of deforestation and the burning of fossil fuels. In the absence of any mitigating effects, a rise in the CO_2 concentration in the atmosphere causes a greenhouse effect in which radiation from the sun is absorbed by the earth's atmosphere rather than escaping back into space. This results in heating of the earth, much as a greenhouse or a closed automobile is heated by the sun's rays that enter and cannot escape. Some researchers believe that the resulting increased heat in the atmosphere could cause the earth's average temperature to increase 2°C–3°C during the next century. Such a change would have serious effects on climate, ocean levels, and agriculture. With the help of sunlight and chlorophyll (as a catalyst), green plants carry out photosynthesis. Although this reaction converts carbon dioxide and water into sugar and oxygen, the current rate at which carbon dioxide is produced is so great that the plant life on the earth cannot keep the CO_2 concentration sufficiently low.

The carbon–oxygen bonds in carbon dioxide are polar covalent bonds (Section 6.5). The high electronegativity of oxygen makes it negative with respect to carbon, which becomes the positive end of the bond.

$$\overset{\delta-}{O}=\overset{\delta+\ \delta+}{C}=\overset{\delta-}{O}$$

Thus carbon is susceptible to attack by negative species, and it reacts with bases such as the oxide ion and the hydroxide ion to form carbonate ions and hydrogen carbonate ions, respectively.

$$CO_2 + O^{2-} \longrightarrow CO_3^{2-}$$
$$CO_2 + OH^- \longrightarrow HCO_3^-$$

Figure 9.10

(A) The Lewis structure of carbonic acid. (B) One resonance form of the hydrogen carbonate ion. (C) One resonance form of the carbonate ion.

When carbon dioxide dissolves in water, some of the molecules react and form carbonic acid, H_2CO_3 (Fig. 9.10 A). Carbonic acid is a weak acid that produces the mildly sour taste of many carbonated beverages. The reaction of carbonic acid with a strong base produces one of two ions, depending on the stoichiometry. When we add 1 mole of hydroxide ion per mole of carbonic acid, the **hydrogen carbonate ion,** HCO_3^- (Fig. 9.10 B), and water are produced. Adding 2 moles of hydroxide ion per mole of acid gives the **carbonate ion,** CO_3^{2-} (Fig. 9.10 C), and water.

$$H_2CO_3 + OH^- \longrightarrow HCO_3^- + H_2O$$
$$H_2CO_3 + 2OH^- \longrightarrow CO_3^{2-} + 2H_2O$$

Approximately half of the carbon in the earth's crust occurs in the form of salts of the carbonate anion. The most common carbonate salts are calcium carbonate, $CaCO_3$, found in limestone, chalk, coral, and marble; magnesium carbonate, $MgCO_3$, found in magnesite; and $MgCa(CO_3)_2$, found in dolomite. Sodium hydrogen carbonate, $NaHCO_3$, sees household use as baking soda.

The oxygen atoms in carbonic acid, the hydrogen carbonate ion, and the carbonate ion surround the carbon atom in a trigonal planar array. Each carbon atom forms three σ bonds, one to each oxygen atom, by means of sp^2 hybrid orbitals. The unhybridized p orbital on each carbon atom projects above and below the plane of the molecule or ion and overlaps with orbitals on the oxygen atoms to form π bonds. The bonding in the two ions is best described as a resonance hybrid (Section 6.11).

The reaction of hot carbon with calcium oxide forms **calcium carbide,** CaC_2, an ionic compound containing the Ca^{2+} ion and the carbide ion, $:C\equiv C:^{2-}$. Calcium carbide reacts with water (Fig. 9.11) producing acetylene, $HC\equiv CH$.

$$CaC_2(s) + 2H_2O(l) \longrightarrow Ca(OH)_2(s) + C_2H_2(g)$$

The reaction of calcium carbide with nitrogen at about 1100°C gives **calcium cyanamide,** $CaCN_2$, which contains the linear cyanamide ion, NCN^{2-} (Fig. 9.12). Calcium cyanamide can be used as a fertilizer that is rich in nitrogen, and it is one of the components used to manufacture the plastic melamine. Melting calcium cyanamide with carbon and sodium carbonate produces **sodium cyanide,** $NaCN$.

$$CaCN_2 + C + Na_2CO_3 \longrightarrow CaCO_3 + 2NaCN$$

Cyanide ions react with acids to produce the very weak acid **hydrogen cyanide,** HCN, which is a deadly poisonous gas. Although NaCN and HCN are poisonous substances, they play important roles in the chemical industry. Hydrogen cyanide is used as a starting material in the manufacture of Lucite and Plexiglas. Sodium cyanide is used in the extraction of gold and silver from their ores (Section 23.3).

Figure 9.11

The reaction of solid calcium carbide with water produces acetylene, a flammable gas.

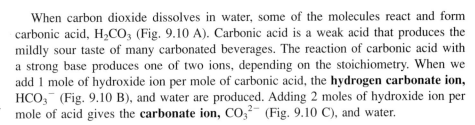

$\ddot{N}=C=\ddot{N}^{2-}$

Figure 9.12

The molecular and Lewis structures of the cyanamide ion.

Organic Compounds of Carbon

When we sum the number of organic compounds either found in living organisms or derived from them and add those synthesized by chemists, we get a total of about ten million. The existence of so many organic compounds is due primarily to the ability of carbon atoms to form strong bonds to other carbon atoms, giving chains and rings of many different sizes and complexities.

9.3   Alkanes

The simplest organic compounds contain only carbon and hydrogen and are called **hydrocarbons.** Even though they are composed of only two types of atoms, there are a wide variety of hydrocarbons because they may consist of chains, branched chains, and rings of carbon atoms, or combinations of these structures. In addition, hydrocarbons may differ in the types of carbon–carbon bonds present in their molecules. Many hydrocarbons are found in plants, animals, and their fossils; other hydrocarbons have been prepared in the laboratory. We use hydrocarbons every day, primarily as fuels: natural gas, acetylene, propane, butane, and the principal components of gasoline, diesel fuel, and heating oil. The familiar polymers polyethylene, polypropylene, and polystyrene (Section 9.15) are also hydrocarbons.

We can distinguish several different types of hydrocarbons by differences in the covalent bonds between their carbon atoms. **Alkanes,** or **saturated hydrocarbons,** contain only single bonds. Thus all of the carbon atoms in an alkane have sp^3 hybridization and are bonded to four other carbon and/or hydrogen atoms. The bond angles are close to 109.5° (the tetrahedral angle, Section 7.1). Thus chains of carbon atoms in an alkane have a staggered or zigzag, configuration. The Lewis structures and models of methane, ethane, and pentane are illustrated in Fig. 9.13. Note that the carbon atoms in the model of the pentane molecule do not lie in a straight line. Carbon atoms are usually written in straight lines in Lewis structures, but remember that Lewis structures are not intended to indicate the geometry of molecules.

Figure 9.13

Lewis structures and ball-and-stick models of molecules of methane, ethane, and pentane.

Methane
CH_4

Ethane
CH_3CH_3

Pentane
$CH_3CH_2CH_2CH_2CH_3$

All alkanes are composed of carbon and hydrogen atoms and have similar bonds and structures. Thus their properties, such as melting point and boiling point (Table 9.1), usually change smoothly as the number of carbon and hydrogen atoms in the molecules change.

Hydrocarbons with the same formula can have different structures. For example, two alkanes have the formula C_4H_{10}: they are called *n*-butane and 2-methylpropane and have the Lewis structures

Table 9.1

Properties of Some Alkanes[a]

	Molecular Formula	Melting Point, °C	Boiling Point, °C	Phase at STP[b]	Number of Structural Isomers
Methane	CH_4	-182.5	-161.5	Gas	1
Ethane	C_2H_6	-183.2	-88.6	Gas	1
Propane	C_3H_8	-187.7	-42.1	Gas	1
Butane	C_4H_{10}	-138.3	-0.5	Gas	2
Pentane	C_5H_{12}	-129.7	36.1	Liquid	3
Hexane	C_6H_{14}	-95.3	68.7	Liquid	5
Heptane	C_7H_{16}	-90.6	98.4	Liquid	9
Octane	C_8H_{18}	-56.8	125.7	Liquid	18
Nonane	C_9H_{20}	-53.6	150.8	Liquid	35
Decane	$C_{10}H_{22}$	-29.7	174.0	Liquid	75
Tetradecane	$C_{14}H_{30}$	5.9	253.5	Solid	1858
Octadecane	$C_{18}H_{38}$	28.2	316.1	Solid	60,523

[a]Physical properties for C_4H_{10} and the heavier molecules are those of the *normal* isomer, *n*-butane, *n*-pentane, etc.
[b]STP indicates a temperature of 0°C and a pressure of 1 atm.

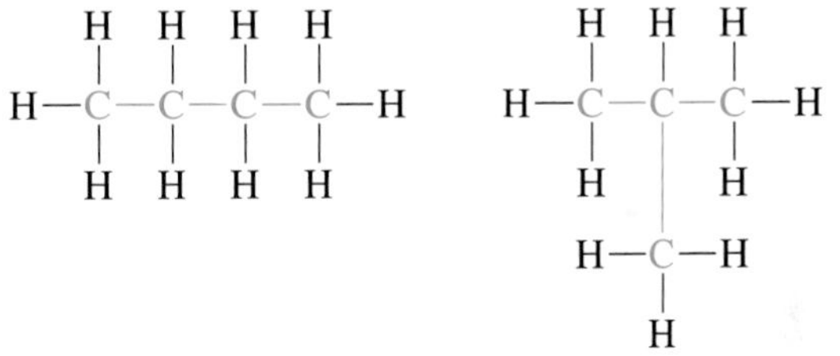

n-Butane   2-Methylpropane

n-Butane and 2-methylpropane are **structural isomers.** Structural isomers have the same molecular formula but different arrangements of the atoms in their molecules. The *n*-butane molecule contains an unbranched chain. We use the term *normal* or the prefix *n* to refer to a chain of carbon atoms with no branching. 2-Methylpropane is a branched-chain molecule.

Be careful when looking for isomers. Lewis structures that look different may actually represent the same isomers. For example, the following three structures all represent the same molecule, *n*-butane, and hence are not separate isomers. They are identical because each contains an unbranched chain of four carbon atoms. When you are unsure whether a hydrocarbon is branched, check each carbon atom in the Lewis structure. If any carbon atom has bonds with more than two other carbons, the molecule is branched.

Alkanes are not very reactive, but heat or light will activate reactions that involve the breaking of carbon–carbon or carbon–hydrogen single bonds. Combustion (Section 2.10) is one such reaction.

$$CH_4 + 2O_2 \longrightarrow CO_2 + 2H_2O$$

In a **substitution reaction,** another reaction typical of alkanes, a second type of atom is substituted for a hydrogen atom. The carbon atoms in a hydrocarbon do not change hybridization during a substitution reaction.

Ethane   Ethyl chloride

A reactive atom or group of atoms that replaces a hydrogen atom is called a **functional group.** The functional group, a halogen in this case, makes it possible for the ethyl chloride molecule to take part in many different kinds of reactions.

Alkanes burn in air; the reaction is a highly exothermic oxidation–reduction reaction that produces carbon dioxide and water. Alkanes are excellent fuels. For example, methane, CH_4, is the principal component of natural gas. The butane, C_4H_{10}, used in camping stoves and lighters is an alkane. Gasoline is a mixture of continuous- and branched-chain alkanes containing from five to nine carbon atoms plus various additives. Kerosene, diesel oil, and fuel oil are primarily mixtures of alkanes with higher molecular masses.

9.4   The Basics of Organic Nomenclature

The International Union of Pure and Applied Chemistry (IUPAC) has devised a system of nomenclature that is based on the names of the alkanes. The nomenclature for alkanes is based on two rules.

1. To name an alkane, we first locate the longest *continuous* chain of carbon atoms. A two-carbon chain is called ethane; a three-carbon chain, propane; and a four-carbon chain, butane. Longer chains are indicated as follows: five carbons, pentane; six carbons, hexane; seven carbons, heptane; eight carbons, octane; nine carbons, nonane; and ten carbons, decane.

2. We add prefixes to the name of the longest chain to indicate the positions and names of substituents. **Substituents** are branches or functional groups that replace hydrogen atoms on a chain. We number the carbon atoms in the chain by count-

ing from the end of the chain nearest the substituents. The position of attachment of a substituent or branch is identified by the number of the carbon atom in the chain.

① ② ③   ① ② ③   ③ ② ①   ⑥ ⑤   ④ ③   ② ①

$CH_3CH_2CH_3$ CH_3CHCH_3 CH_3CHCH_3 $CH_3CH_2CHCH_2CHCH_3$
                     | | F   F
                     Cl CH_3

Propane   2-Chloropropane   2-Methylpropane   2,4-Difluorohexane

When more than one substituent is present, either on the same carbon atom or on different carbon atoms, the substituents are usually listed alphabetically. In general, the longest continuous chain of carbon atoms is numbered in such a way as to produce the lowest number for the substituted atom(s) and/or group(s). Note that *-o* replaces *-ide* at the end of the name of an electronegative substituent and that the number of substituents of the same type is indicated by the prefixes *di-* (two), *tri-* (three), *tetra-* (four), and so on (for example, *difluoro-* indicates two fluoride substituents).

We call a substituent that contains one less hydrogen than the corresponding alkane an **alkyl group.** The name of an alkyl group is obtained by dropping the suffix *-ane* of the alkane name and adding *-yl*.

Methane   The methyl group   Ethane   The ethyl group

The open bonds in the methyl and ethyl groups indicate that these alkyl groups are bonded to another atom.

The four hydrogen atoms in a methane molecule are equivalent; they all have the same environment. (It may be easier to see the equivalency in the ball and stick model in Fig. 9.13.) Removal of any one of the four hydrogen atoms from methane gives a methyl group. Likewise, the six hydrogen atoms in ethane are equivalent (Fig. 9.13), and removing any one of these hydrogen atoms gives an ethyl group. However, in both propane and 2-methylpropane there are hydrogen atoms in two different environments, distinguished by the adjacent atoms or groups of atoms.

Propane   2-Methylpropane

Each of the six equivalent hydrogen atoms of the first type in propane and each of the nine equivalent hydrogen atoms of that type in 2-methylpropane (all shown in black) is bonded to a carbon atom bonded to only one other carbon atom. The two red hydrogen atoms in propane are of a second type. They differ from the six hydrogen atoms of the first type in that they are bonded to a carbon atom bonded to two other carbon atoms. The blue hydrogen atom in 2-methylpropane differs from the other nine hydrogen atoms in that molecule and from the red hydrogen atoms in

Table 9.2

Some Alkyl Groups

Alkyl Group	Structure
Methyl	CH_3-
Ethyl	CH_3CH_2-
n-Propyl	$CH_3CH_2CH_2-$
Isopropyl	$CH_3\overset{\mid}{C}HCH_3$
n-Butyl	$CH_3CH_2CH_2CH_2-$
sec-Butyl	$CH_3CH_2\overset{\mid}{C}HCH_3$
Isobutyl	CH_3CHCH_2- $\overset{\mid}{C}H_3$
tert-Butyl	$CH_3\overset{\mid}{\underset{\mid}{C}}CH_3$ CH_3

propane; it is bonded to a carbon atom bonded to three other carbon atoms. Two different alkyl groups can be formed from each of these molecules, depending on which hydrogen atom is removed. The names and structures of these and several other alkyl groups are listed in Table 9.2. Note that alkyl groups do not exist as stable independent entities. They are always part of some larger molecule. The location of an alkyl group on a hydrocarbon chain is indicated in the same way as that of any other substituent.

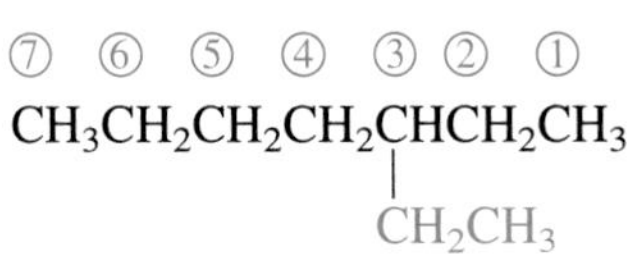

⑦ ⑥ ⑤ ④ ③ ② ①
$CH_3CH_2CH_2CH_2CHCH_2CH_3$
$\mid$
CH_2CH_3
3-Ethylheptane

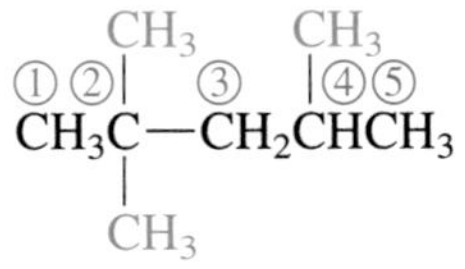

CH_3 CH_3
①② ③ ④⑤
$CH_3C-CH_2CHCH_3$
$\mid$
CH_3
2,2,4-Trimethylpentane

9.5   Alkenes

Organic compounds that contain one or more double or triple bonds between carbon atoms are called **unsaturated.** For example, unsaturated fats are composed of a variety of complex organic molecules, each of which contains at least one double bond between carbon atoms. Unsaturated hydrocarbon molecules that contain one or more double bonds are called **alkenes.** Unsaturated hydrocarbons that contain one or more triple bonds are called **alkynes** (Section 9.6). Carbon atoms linked by a double bond are bound together by two bonds, one σ bond and one π bond (Section 7.11).

Ethene, C_2H_4, is the simplest alkene. Each carbon atom in ethene, commonly called ethylene, has a trigonal planar structure (Fig. 7.32, Section 7.11). The second member of the series is propene (propylene; Fig. 9.14); the butene isomers follow in the series.

Figure 9.14

A ball-and-stick model of propene.

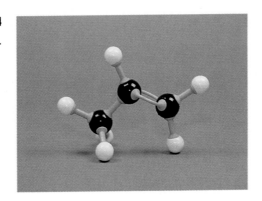

Ethylene (the common industrial name for ethene) is a basic raw material used in making polyethylene and in other industrial reactions. Over 23 million tons of ethylene were produced in the United States in 1995 for use in the polymer, petrochemical, and plastics industries.

Alkenes are much more reactive than alkanes, because the carbon–carbon double bond is a reactive functional group (Section 9.3). A π bond, being a weaker bond, is disrupted much more easily than a σ bond. Thus the reaction that is characteristic of alkenes is a type in which the π bond is broken and replaced by two σ bonds. Such a reaction is called an **addition reaction.** The hybridization of the carbon atoms in the double bond in an alkene changes from sp^2 to sp^3 during an addition reaction.

Halogens add to the double bond in an alkene instead of replacing a hydrogen, as occurs in an alkane. For example,

Ethene   1,2-Dichloroethane

Many other reagents react with the double bond in alkenes. An example is the acid-catalyzed addition of water to an alkene to yield an alcohol (Section 9.9).

Naming Alkenes. The name of an alkene is derived from that of the alkane with the same number of carbon atoms. The presence of the double bond is indicated by replacing the suffix *-ane* with the suffix *-ene,* and its location in the chain is indicated by the lowest number that specifies the position of the first carbon atom in the double bond.

Ethene
(ethylene)

Propene
(propylene)

1-Butene

2-Butene

Figure 9.15

Molecular models of (A) the *cis* and (B) the *trans* isomers of 2-butene.

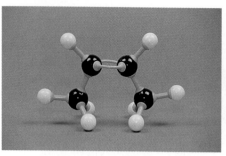
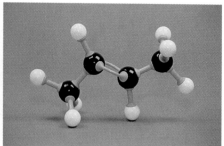

Isomers of Alkenes. Molecules of 1-butene and 2-butene are structural isomers (Section 9.3); they differ in the arrangement of the atoms. For example, the first carbon atom in 1-butene is bonded to two hydrogen atoms, that in 2-butene to three hydrogen atoms.

2-Butene and some other alkenes also form a second type of isomer called a **geometrical isomer.** In a set of geometrical isomers, the same types of atoms are attached to each other in the same order, but the geometries of the two molecules differ. Geometric isomers of alkenes differ in the orientation of the groups on either side of a C=C bond.

Carbon atoms are free to rotate around a single bond but not around a double bond; a double bond is quite rigid. This makes it possible to have two separate isomers of 2-butene, one with both methyl groups on the same side of the double bond and one with the methyl groups on opposite sides (Fig. 9.15). When formulas of butene are written with 120° bond angles around the doubly bonded carbon atoms, the isomers are apparent.

$$
\underset{\text{1-Butene}}{\overset{\displaystyle H}{\underset{\displaystyle H}{}}\!\!C\!\!=\!\!C\!\!\overset{\displaystyle H}{\underset{\displaystyle CH_2CH_3}{}}}
\qquad
\underset{\substack{cis \text{ isomer}}}{\overset{\displaystyle H}{\underset{\displaystyle CH_3}{}}\!\!C\!\!=\!\!C\!\!\overset{\displaystyle H}{\underset{\displaystyle CH_3}{}}}
\qquad
\underset{\substack{trans \text{ isomer}}}{\overset{\displaystyle CH_3}{\underset{\displaystyle H}{}}\!\!C\!\!=\!\!C\!\!\overset{\displaystyle H}{\underset{\displaystyle CH_3}{}}}
$$

2-Butene

The 2-butene isomer in which the two methyl groups are on the same side is called a ***cis* isomer;** the isomer in which the two methyl groups are on opposite sides is called a ***trans* isomer.** The different geometries produce different properties that make separation of the isomers possible.

9.6   Alkynes

Unsaturated hydrocarbon molecules with one or more triple bonds are called **alkynes;** they make up another series of unsaturated hydrocarbons. Two carbon atoms joined by a triple bond are bound together by one σ bond and two π bonds (Section 7.11).

The simplest member of the alkyne series is ethyne, C_2H_2, commonly called acetylene. The Lewis structure for ethyne, a linear molecule, is

$$H—C≡C—H$$

Ethyne (acetylene)

Chemically, the alkynes are similar to the alkenes except that, having two π bonds, they react even more readily, adding twice as much reagent in addition reactions. The reaction of acetylene with bromine is a typical example.

$$H-C\equiv C-H + 2Br_2 \longrightarrow \quad H-\underset{\underset{\displaystyle :\overset{..}{Br}\!::\!\overset{..}{Br}:}{|}}{\overset{\overset{\displaystyle :\overset{..}{Br}\!::\!\overset{..}{Br}:}{|}}{C}}-\underset{}{\overset{}{C}}-H$$

1,1,2,2-Tetrabromoethane

Acetylene and the other alkynes also burn readily. An acetylene torch (Fig. 4.1) takes advantage of the high enthalpy of combustion of acetylene (Section 4.5).

Naming Alkynes. The IUPAC nomenclature for alkynes is similar to that for alkenes except that the suffix *-yne* is used to indicate a triple bond in the chain. For example, $CH_3CH_2C\equiv CH$ is called 1-butyne.

9.7   Aromatic Hydrocarbons

Benzene, C_6H_6, is the simplest member of a large family of hydrocarbons called **aromatic hydrocarbons.** These compounds contain ring structures and exhibit electronic structures that must be described as resonance hybrids (Section 6.11). The resonance forms of benzene, C_6H_6, are

The benzene molecule and other planar aromatic hydrocarbon molecules contain hexagonal rings of sp^2 hybridized carbon atoms with the unhybridized p orbital of each carbon atom perpendicular to the ring. Three valence electrons in the sp^2 hybrid orbitals of each carbon atom and the valence electron of each hydrogen atom form the framework of σ bonds in the benzene molecule. The unhybridized carbon p orbitals, each with one electron, are perpendicular to the ring, like the unhybridized p orbitals in a graphite layer (see Fig. 9.4). These p orbitals combine to form π bonds. Benzene does not have alkene character. Each bond between two carbon atoms is neither a single nor a double bond but is intermediate in character (a hybrid). Although each bond is equivalent to the others, for convenience benzene is often written as one of its resonance forms.

There are many derivatives of benzene. The hydrogen atoms can be replaced by many different substituents. The following are typical examples.

Toluene   Xylene   Styrene

Toluene and xylene serve as important solvents and raw materials in the chemical industry. Styrene is used to produce the polymer polystyrene (Section 9.15).

Derivatives of Hydrocarbons

Derivatives of hydrocarbons are compounds that have the same carbon chain as the hydrocarbon but have one or more hydrogen atoms replaced by other atoms, or groups of atoms, called functional groups (Section 9.3). For example, an alcohol is a compound in which an —OH group replaces a hydrogen atom of a hydrocarbon. The alcohol methanol, CH_3OH, is related to methane, CH_4 (Fig. 9.16 A); the alcohol ethanol, C_2H_5OH, is related to ethane, C_2H_6 (Fig. 9.16 B).

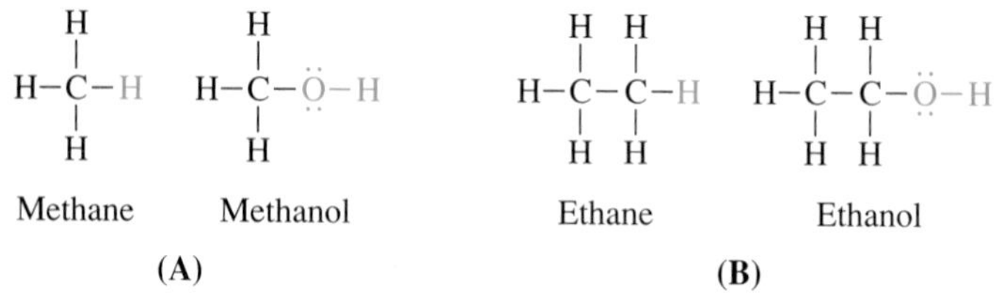

Figure 9.16

(A) The Lewis structures of methane and methanol. (B) The Lewis structures of ethane and ethanol.

Methane   Methanol   Ethane   Ethanol
(A)   (B)

9.8   The Behavior of Hydrocarbon Derivatives

Functional groups are the sites where reactions occur in organic molecules. Most chemical changes in organic molecules occur at the functional groups, because C—H bonds are not very reactive and require activation by heat or light to react. We can expect that organic molecules that contain the same functional group will undergo the same kinds of reactions. This is observed; for example, CH_3OH, C_2H_5OH, C_3H_7OH, and many other alcohols all react with acetic acid and form a class of compounds called esters (Section 9.13), because each alcohol contains the —OH group as its functional group. It is the —OH group that reacts with acetic acid.

We usually write the formula of an organic molecule such that the formula emphasizes the identity of any functional groups in the molecule. For example, we rarely write the molecular formula of ethanol (Fig. 9.16) as C_2H_6O; instead we write C_2H_5OH or CH_3CH_2OH. We often write the formula of an organic molecule to show the arrangement of the atoms in the molecule. The location of a hydrogen atom or of a functional group bonded to a chain of carbon atoms is indicated by writing the formula of the hydrogen atom or the substituent after the carbon atom to which it is bound. This is why we see the formula of ethanol written as CH_3CH_2OH and that of 2-chloropropane as $CH_3CHClCH_3$. Substituents with more than one atom are sometimes placed in parentheses. If the Cl atom in 2-chloropropane were replaced by an OH group, we would write the formula with parentheses: $CH_3CH(OH)CH_3$. We also commonly indicate the locations of double or triple bonds in chains of carbon atoms. For example, we write the formula of a 1-butyne molecule as $CH_3CH_2C{\equiv}CH$.

If we wish to identify a functional group and only identify the composition of the remaining hydrocarbon component of a molecule, a shorter formula can be used; for example, we sometimes see ethanol as C_2H_5OH instead of CH_3CH_2OH, and we may see 1-fluoropropane as C_3H_7F instead of $CH_3CH_2CH_2F$. Examples of three different ways to represent molecules with the functional groups Br, SH, and CO_2H follow.

$$H-\underset{\underset{H}{|}}{\overset{\overset{H}{|}}{C}}-\underset{\underset{H}{|}}{\overset{\overset{H}{|}}{C}}-\underset{\underset{H}{|}}{\overset{\overset{H}{|}}{C}}-Br$$

CH₃CH₂CH₂Br
C₃H₇Br

$$H-\underset{\underset{H}{|}}{\overset{\overset{H}{|}}{C}}-\underset{\underset{:S:}{|}}{\overset{\overset{H}{|}}{C}}-\underset{\underset{H}{|}}{\overset{\overset{H}{|}}{C}}-H$$
$$|$$
$$H$$

CH₃CH(SH)CH₃

$$H-\underset{\underset{H}{|}}{\overset{\overset{H}{|}}{C}}-\underset{\underset{H}{|}}{\overset{\overset{H}{|}}{C}}-\underset{\underset{H}{|}}{\overset{\overset{H}{|}}{C}}-CO_2H$$

CH₃CH₂CH₂CO₂H
C₃H₇CO₂H

Sometimes only the functional group in a molecule is of interest. In these cases we may indicate only the composition of the functional group and represent the remainder of the molecule by the letter R. A general formula for any alcohol can be written R—OH, where OH indicates the functional group characteristic of an alcohol and R represents the remainder of the molecule.

9.9 Alcohols

Alcohols are derivatives of hydrocarbons in which an OH group has replaced a hydrogen atom. Though all alcohols have one or more hydroxyl (—OH) functional groups, they do not behave like bases such as NaOH and KOH. Those strong bases are ionic compounds that contain OH^- ions. Alcohols are covalent molecules; the OH group is attached to a carbon atom in an alcohol molecule by a covalent bond.

Ethanol, CH_3CH_2OH, also called ethyl alcohol, is the most important of the alcohols. It has long been prepared by fermentation of starch, cellulose, and sugars.

$$C_6H_{12}O_6(aq) \xrightarrow{\text{Yeast}} 2C_2H_5OH(aq) + 2CO_2(g)$$

Glucose   Ethanol

Large quantities of ethanol are synthesized from the addition reaction of water with ethylene with an acid being used as a catalyst.

$$H-\underset{\underset{H}{|}}{\overset{\overset{H}{|}}{C}}=\underset{}{\overset{\overset{H}{|}}{C}}-H + HOH \xrightarrow{H_3O^+} H-\underset{\underset{H}{|}}{\overset{\overset{H}{|}}{C}}-\underset{\underset{H}{|}}{\overset{\overset{H}{|}}{C}}-\ddot{O}H$$

Alcohols containing two or more hydroxyl groups are known. Examples include 1,2-ethanediol (ethylene glycol, used in antifreeze), and 1,2,3-propanetriol (glycerine, used as a solvent for cosmetics and medicines).

$$H-\underset{\underset{H-\underset{}{\overset{}{C}}-\ddot{O}H}{|}}{\overset{\overset{H}{|}}{C}}-\ddot{O}H$$

1,2-Ethanediol

$$H-\underset{\underset{H-\underset{}{\overset{}{C}}-\ddot{O}H}{|}}{\overset{\overset{H}{|}}{C}}-\ddot{O}H$$

1,2,3-Propanetriol

Naming Alcohols. An alcohol's name comes from the hydrocarbon from which it is derived. The hydrocarbon's final *-e* is replaced by *-ol,* and the carbon atom to which the —OH group is bonded is indicated by a number placed before the name.

9.10   Ethers

Ethers are compounds that contain the functional group —O—. They are obtained from alcohols by the elimination of a molecule of water from two molecules of the alcohol. For example, when ethanol is treated with a limited amount of sulfuric acid and heated to 140°C, diethyl ether and water are formed.

Diethyl ether

In the general formula for ethers, R—O—R, the hydrocarbon groups (R) may be the same or different. Diethyl ether, the most important compound of this class, is a colorless, volatile liquid that is highly flammable. It has been used since 1846 as an anesthetic, but better anesthetics have largely taken its place. Now diethyl ether and other ethers are valuable solvents for gums, fats, waxes, and resins. Methyl *t*-butyl ether, $CH_3OC_4H_9$, is used as an additive in oxygenated gasolines.

9.11   Amines

Amines are molecules that contain carbon–nitrogen bonds. The nitrogen atom in an amine can bond to one, two, or three carbon atoms. Any additional bonds are to hydrogen atoms.

$CH_3—N—H$ $CH_3—N—CH_3$ $CH_3—N—CH_3$

Methyl amine   Dimethyl amine   Trimethyl amine

Figure 9.17

One of the resonance structures of pyridine.

In some amines the nitrogen atom replaces a carbon atom in an aromatic hydrocarbon (Section 9.7). Pyridine (Fig. 9.17) is one such amine.

Amines are bases. An amine contains a nitrogen atom with an unshared pair of electrons that will bind a hydrogen ion and form an ammonium ion. Like ammonia, amines are weak bases (Section 8.5).

Ammonia   Ammonium ion

Methyl amine   Methyl ammonium ion

Amines are responsible for the odor of spoiled fish. However, the amine functional group also appears in many useful molecules. For example, nylon (Section 9.15), many dyes, medications such as penicillin and codeine (Section 9.14), vitamins, and many of the essential molecules such as DNA and amino acids (Chapter 25) found in the body are amines.

9.12   Aldehydes and Ketones

Both **aldehydes** and **ketones** contain a **carbonyl group,** $\diagdown C\!=\!O$, a functional group with a carbon–oxygen double bond. In an aldehyde the carbonyl group is bonded to at least one hydrogen atom. In a ketone the carbonyl group is bonded to two carbon atoms.

$$
\begin{array}{ccc}
& \text{H} & \text{:O:} \\
& | & || \\
\text{H} & \!-\!\text{C}\!-\!\text{C}\!-\!\text{H} \\
& | \\
& \text{H}
\end{array}
\qquad
\begin{array}{ccccc}
\text{H} & \text{:O:} & \text{H} & \text{H} \\
| & || & | & | \\
\text{H}\!-\!\text{C}\!-\!\text{C}\!-\!\text{C}\!-\!\text{C}\!-\!\text{H} \\
| & & | & | \\
\text{H} & & \text{H} & \text{H}
\end{array}
$$

CH_3CHO $CH_3COCH_2CH_3$
An aldehyde   A ketone
Ethanal (acetaldehyde)   Methyl ethyl ketone

Thus the $-\!\overset{\overset{\displaystyle O}{||}}{C}\!-\!H$ functional group is the hallmark of an aldehyde, and the $\diagdown C\!=\!O$ group is the hallmark of a ketone. A simpler representation of an aldehyde group is $-CHO$; of a ketone, $-C(O)-$ or $-CO-$.

In both aldehydes and ketones, the geometry about the carbon atom in the carbonyl group is trigonal planar; the carbon atom exhibits sp^2 hybridization. Two of the sp^2 orbitals on the carbon atom in the carbonyl group are used to form σ bonds to the other carbon or hydrogen atoms in a molecule. The remaining sp^2 hybrid orbital forms a σ bond to the oxygen atom. The unhybridized p orbital on the carbon atom in the carbonyl group overlaps a p orbital on the oxygen atom to form the π bond in the double bond.

Like the $C\!=\!O$ bond in carbon dioxide (Section 9.2), the $C\!=\!O$ bond of a carbonyl group is polar. Many of the reactions of aldehydes and ketones start with the reaction between a Lewis base and the carbon atom at the positive end of the polar $C\!=\!O$ bond. This initial reaction is usually followed by one or more rearrangements to give the final product.

The importance of molecular structure in the reactivity of organic compounds is illustrated by the reactions that produce aldehydes and ketones. We can prepare a carbonyl group by oxidation of an alcohol. The reagents, other products, and reaction conditions are beyond the scope of this discussion, so we will represent the reaction as simply

$$
\begin{array}{c}
\text{H} \\
| \\
-\!\text{C}\!-\!\ddot{\text{O}}\!-\!\text{H} \\
|
\end{array}
\xrightarrow{\text{Oxidation}}
\begin{array}{c}
:\!\ddot{\text{O}} \\
|| \\
-\!\text{C}\!-
\end{array}
$$

Alcohol   Carbonyl group

We can prepare aldehydes by the oxidation of alcohols whose $-OH$ functional group is located on the carbon at the end of the chain of carbon atoms in the alcohol.

$$CH_3CH_2CH_2OH \longrightarrow CH_3CH_2CHO$$

Alcohol   Aldehyde

We use alcohols that have their —OH groups in the middle of the chain in the preparation of ketones, which require the carbonyl group to be bonded to two other carbon atoms.

$$CH_3CH(OH)CH_3 \longrightarrow CH_3COCH_3$$

Alcohol   Ketone

An alcohol with its OH group bonded to a carbon atom that is bonded to three other carbon atoms generally forms neither an aldehyde nor a ketone upon oxidation.

Formaldehyde, an aldehyde with the formula HCHO, is a colorless gas with a pungent and irritating odor. It is sold in an aqueous solution called formalin, which contains about 37% formaldehyde by weight. Formaldehyde causes coagulation of proteins, so it kills bacteria and stops many of the biological processes that cause tissue to decay. Thus formaldehyde is useful for preserving tissue specimens and embalming bodies. It is also used to sterilize soil. Formaldehyde is used in the manufacture of Bakelite, a hard plastic that has high chemical and electrical resistance.

Dimethyl ketone, CH_3COCH_3, a colorless liquid commonly called **acetone,** is the simplest and most important ketone. It is made commercially by fermenting corn or molasses or by oxidation of 2-propanol. Among other applications, it is used as a solvent for cellulose acetate, cellulose nitrate, acetylene, plastics, and varnishes; as a paint, varnish, and fingernail polish remover; and as a solvent in the manufacture of pharmaceuticals and chemicals.

9.13   Carboxylic Acids and Esters

The smell of vinegar is due to the presence of acetic acid, a carboxylic acid. The smell of ripe bananas and many other fruits is due to the presence of esters, compounds that can be prepared by the reaction of a carboxylic acid with an alcohol.

Both **carboxylic acids** and **esters** contain a carbonyl group with a second oxygen atom bonded to the carbon atom in the carbonyl group by a single bond. In a carboxylic acid, the second oxygen atom also bonds to a hydrogen atom. In an ester, the second oxygen atom bonds to another carbon atom.

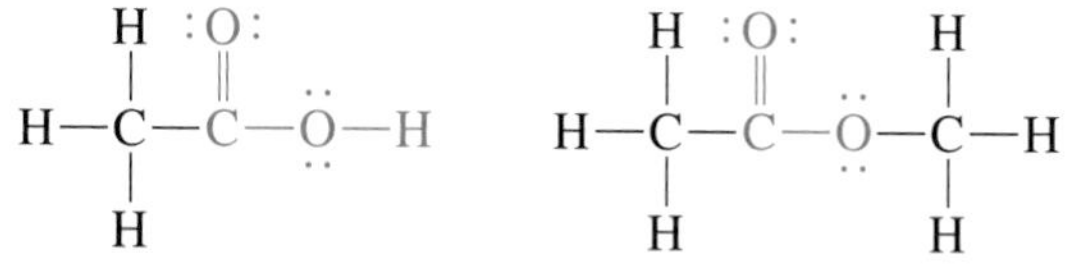

A carboxylic acid
Acetic acid

An ester
Methyl acetate

The functional group for an acid and that for an ester are highlighted in color in these formulas.

The hydrogen atom in the functional group of a carboxylic acid will react with a base to form an ionic salt.

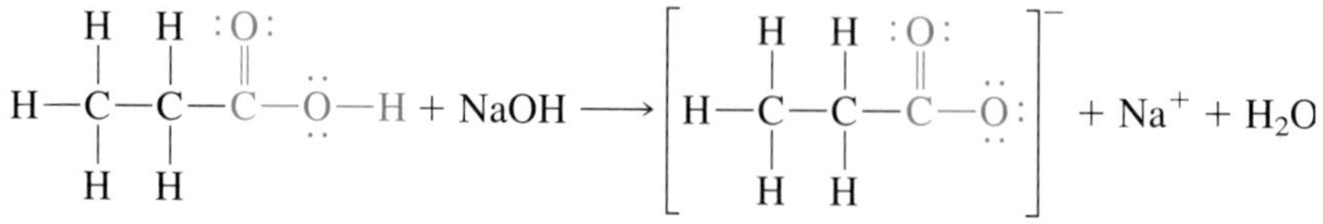

Propionic acid

Propionate ion

$+ Na^+ + H_2O$

Carboxylic acids are weak acids (Section 8.5); they are not 100% ionized in water. Generally, only about 1% of the molecules of a carboxylic acid dissolved in water are ionized at any given time. The remaining acid is present as neutral molecules.

We prepare carboxylic acids by the oxidation of aldehydes or alcohols whose —OH functional group is located on the carbon at the end of the carbon atoms chain in the alcohol.

$$R-CH_2-\ddot{O}-H \longrightarrow R-C\underset{\ddot{O}}{\overset{H}{\Big\langle}} \longrightarrow R-C\underset{\ddot{O}}{\overset{O-H}{\Big\langle}}$$

Alcohol   Aldehyde   Carboxylic acid

Esters are produced by the reaction of acids with alcohols. For example, the ester ethyl acetate, $CH_3CO_2CH_2CH_3$, is formed when acetic acid reacts with ethanol.

$$CH_3-C\underset{\ddot{O}-H}{\overset{\ddot{O}}{\Big\langle}} + HOCH_2CH_3 \longrightarrow CH_3-C\underset{\ddot{O}-CH_2CH_3}{\overset{\ddot{O}}{\Big\langle}} + H_2O$$

The simplest carboxylic acid is **formic acid,** HCO_2H, which has been known since 1670. Its name comes from the Latin word *formicus,* which means "ant"; it was first isolated via the distillation of red ants. It is partially responsible for the pain and irritation of ant and wasp stings.

Acetic acid, CH_3CO_2H, constitutes 3%–6% of vinegar. Cider vinegar is produced by allowing apple juice to ferment without oxygen present. This changes the sugar present in the juice to ethanol and then to acetic acid. Pure acetic acid has a penetrating odor and produces painful burns. It is an excellent solvent for many organic and some inorganic compounds, and it is essential in the production of cellulose acetate, a component of synthetic fibers such as rayon.

The distinctive and attractive odors and flavors of many flowers, perfumes, and ripe fruits are due to the presence of one or more esters (Fig. 9.18). Among the most

Figure 9.18

Chemists carry out sniff tests on perfumes produced by reactions of organic compounds.

important of the natural esters are fats (such as lard, tallow, and butter) and oils (such as linseed, cottonseed, and olive oils), which are esters of the trihydroxyl alcohol glycerine, $C_3H_5(OH)_3$ (Section 9.9), with large carboxylic acids such as palmitic acid, $CH_3(CH_2)_{14}CO_2H$; stearic acid, $CH_3(CH_2)_{16}CO_2H$; as well as oleic acid, $CH_3(CH_2)_7CH{=}CH(CH_2)_7CO_2H$. (Oleic acid is an unsaturated acid; it contains a $C{=}C$ double bond.)

9.14   Natural Products

Molecules that are found in plants and animals are called **natural products.** Many of these molecules have potent biological properties, and some (such as penicillin and aspirin) are important medically. A branch of organic chemistry called natural products chemistry is directed at the synthesis of these substances. There are many different classes of natural products; we will consider two: pheromones and alkaloids.

Human beings communicate orally, by using sign language, by writing, by drawing pictures, and to a limited extent by chemicals produced by our bodies. Other organisms with a more limited ability to communicate use chemicals more extensively. Some plants, for example, attract insects by producing a scent that announces they have flowered. Other organisms use **pheromones** (derived from the Greek *phero,* which means "carrier"), a chemical or mixture of chemicals that is secreted by an individual of a species and elicits a response in another individual of the same species.

We can classify insect pheromones by their functions. Alarm pheromones signify danger; sex-attractant pheromones help the different sexes of the same species to locate one another; and recruiting pheromones are used to alert other members of the species to the existence of a food source. A tiny amount of a pheromone will elicit the desired response. It has been reported that a typical female insect may carry only 10^{-8} grams of sex attractant, but that is enough to spread out and draw over a billion males from an area of several square miles and from as far away as 7 miles!

Pheromone chemistry is an exciting and vigorous field of research, and pheromones are of immediate practical value. For example, sex-attractant pheromones can be used for insect control. Male insects may be drawn by a pheromone to a trap where they can be either killed or sterilized and released to mate unproductively with wild females (Fig. 9.19). Two sex-attractant pheromones are

Figure 9.19

A trap for boll weevils that utilizes a sex-attractant pheromone.

$$(CH_3)_2CH(CH_2)_4\overset{\overset{\displaystyle O}{\diagup\!\!\diagup\!\!\backslash}}{C}H{-}CH(CH_2)_9CH_3 \qquad\qquad \textit{cis-}CH_3(CH_2)_{12}CH{=}CH(CH_2)CH_3$$

Gypsy moth pheromone   House fly pheromone

Since ancient times, plants have been used for medicinal purposes. One class of substances, called **alkaloids,** found in many of these plants has been isolated and found to contain cyclic molecules with an amine functional group. These amines are bases. They can react with H_3O^+ in dilute acid to form an ammonium salt, and this is how they are extracted from the plant.

$$R_3N\colon + H_3O^+ + Cl^- \longrightarrow [R_3NH^+]Cl^- + H_2O$$

The name *alkaloid* means "like an alkali"—that is, an alkaloid reacts with acid. The free compound can be obtained by reaction with a base.

$$[R_3NH^+]Cl^- + OH^- \longrightarrow R_3N\colon + H_2O + Cl^-$$

The structures of several naturally occurring alkaloids that have profound physiological effects in humans follow.

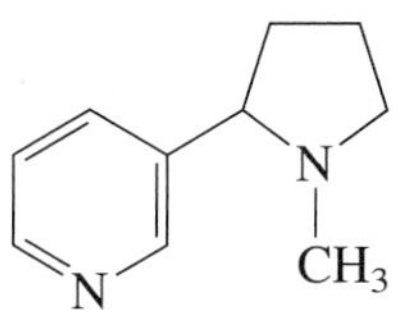

Nicotine

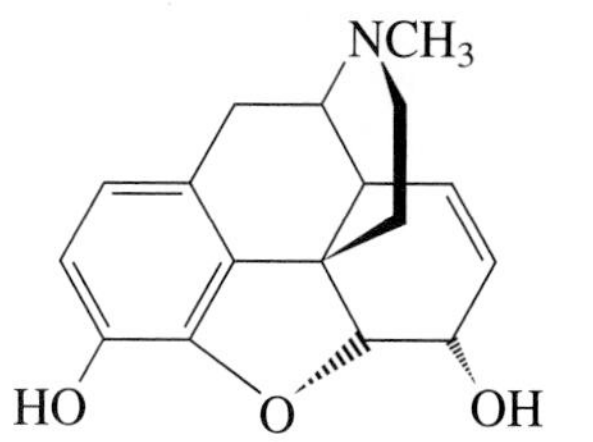

Morphine

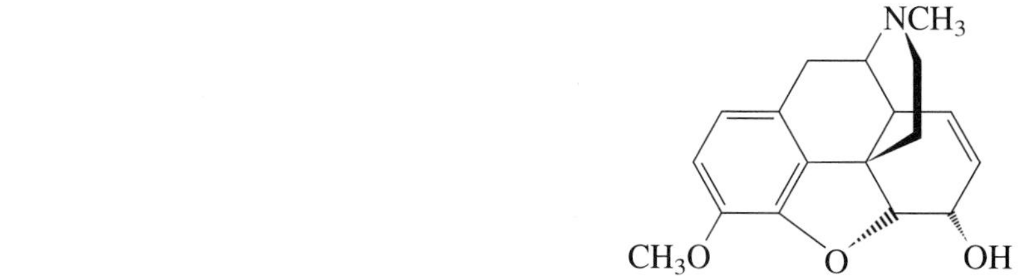

Codeine

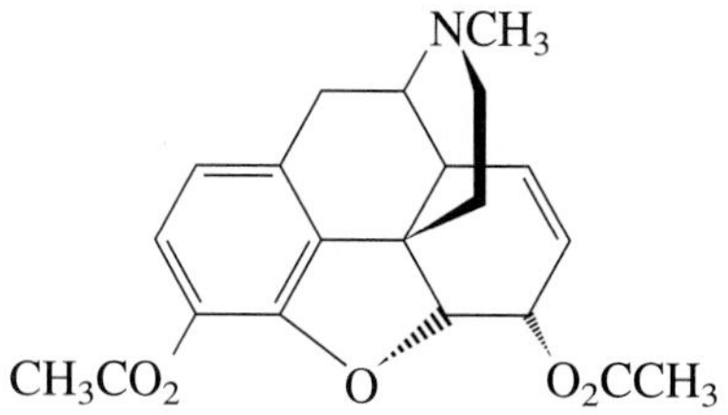

Heroin

Carbon atoms in the rings and the hydrogen atoms bonded to them have been omitted for clarity. The solid wedges indicate bonds that extend above the plane of the paper. Note that small changes in just one part of the molecule change the properties of morphine, codeine, and heroin. Morphine, a strong narcotic used to relieve pain, contains two alcohol functional groups, located at the bottom of the molecule in this structural formula. Changing one of these alcohol groups to a methyl ether gives codeine, a less potent drug used as a local anesthetic. If both alcohol groups are converted to esters of acetic acid, then the powerfully addictive drug heroin results.

9.15 Polymers

Polymers and **plastics,** compounds of very high molecular masses, are built up of a large number of smaller molecules, or **monomers,** that have reacted with one another. Cellulose, starch, proteins, and natural rubber are natural polymers (although most rubber in use is a synthetic polymer). Nylon, rayon, polyethylene, and Dacron are synthetic polymers.

Natural **rubber** comes mainly from latex, the sap of the rubber tree. Rubber consists of very long molecules, which are polymers formed by the union of isoprene units, C_5H_8.

$$\underset{\text{Isoprene}}{CH_2{=}\overset{\displaystyle CH_3}{\overset{|}{C}}{-}CH{=}CH_2} + \underset{\text{Isoprene}}{CH_2{=}\overset{\displaystyle CH_3}{\overset{|}{C}}{-}CH{=}CH_2} \longrightarrow$$

$$\underset{\text{Polyisoprene (rubber)}}{{-}CH_2{-}\overset{\displaystyle CH_3}{\overset{|}{C}}{=}CH{-}CH_2{-}CH_2{-}\overset{\displaystyle CH_3}{\overset{|}{C}}{=}CH{-}CH_2{-}}$$

The number of isoprene units in a rubber molecule is about 2000, so it has a molecular mass of approximately 136,000.

Rubber has the undesirable property of becoming sticky when warmed, but this can be eliminated by **vulcanization.** Rubber is vulcanized by heating it with sulfur to about 140°C. During the process, sulfur atoms are added at some of the double

Figure 9.20

A strand of raw nylon can be drawn from the film of nylon that forms at the interface between solutions of the two monomers involved in this synthesis.

bonds in the linear polymer and form bridges that bind one rubber molecule to another. In this way, a linear polymer is converted into a three-dimensional polymer.

Synthetic rubbers resemble natural rubber and are often superior to it in certain respects. For example, neoprene is a synthetic polymer with rubber-like properties.

$$n CH_2\!=\!\overset{\overset{\displaystyle :\ddot{C}l:}{|}}{C}\!-\!CH\!=\!CH_2 \xrightarrow{\text{Polymerization}} \left[\!-CH_2\!-\!\overset{\overset{\displaystyle :\ddot{C}l:}{|}}{C}\!=\!CH\!-\!CH_2\!-\!\right]_n$$

Chloroprene   Neoprene

The monomer chloroprene is similar to isoprene, except that a chlorine atom replaces the methyl group. Neoprene is used for making gasoline and oil hoses, automobile and refrigerator parts, electrical insulation, and sports clothing (wet suits).

Both natural rubber and neoprene are examples of **addition polymers,** polymers that form by addition reactions (Section 9.5).

Nylon is a **condensation polymer.** In a condensation polymerization reaction, some of the atoms of the monomers are lost as water, ammonia, carbon dioxide, and so on. Nylon (Fig. 9.20) is made from diamines that contain an amine group ($-NH_2$) at both ends, and acids that contain a carboxylic acid group ($-CO_2H$) at both ends. During condensation, linkages of the type $R-NH-CO-R$ are formed and water is eliminated. The part shown in color is an **amide linkage.**

$$(n+1)H_2\ddot{N}(CH_2)_6\ddot{N}H_2 + (n+1)HO_2C(CH_2)_4CO_2H \longrightarrow$$

Hexamethylenediamine   Adipic acid

$$H_2\ddot{N}(CH_2)_6\ddot{N}H\!\left[\!\overset{\overset{\displaystyle :O:}{\|}}{C}(CH_2)_4\overset{\overset{\displaystyle :O:}{\|}}{C}\ddot{N}H(CH_2)_6\ddot{N}H\!\right]_n\!\overset{\overset{\displaystyle :O:}{\|}}{C}(CH_2)_4CO_2H + 2nH_2O$$

Nylon

Fine nylon threads are produced by forcing melted nylon through small holes in a spinneret.

Dacron is made by a condensation process that forms an ester linkage. Dacron is one of the family of polyesters.

$$(n+1)HO\overset{\overset{\displaystyle ..H}{}}{\underset{\underset{\displaystyle ..H}{}}{C}}\!-\!\overset{\overset{\displaystyle H..}{}}{\underset{\underset{\displaystyle H..}{}}{C}}OH + (n+1)HO_2C\!-\!C\!\!\!\!\diagdown\!\!\!\!\diagup\!\!\!\!C\!-\!CO_2H \longrightarrow$$

Ethylene glycol   Terephthalic acid

Dacron

Polyethylene is widely used and results from the polymerization of ethylene. It is used in plastic bottles, bags for fruits and vegetables, and many other items (Fig. 9.21). If the ethylene molecules contain substituents, the polymer will also contain those substituents.

Figure 9.21

Plastic bags and strips of polymer films such as Saran Wrap are formed from large continuous tubes blown from the molten polymer.

$$nCH_2{=}CHR \xrightarrow{\text{Catalyst}} \left[CH_2{-}\underset{\underset{R}{|}}{CH} \right]_n$$

Polypropylene (R = CH$_3$) is the polymer of propylene, CH$_3$CH=CH$_2$. Polyvinyl chloride, or PVC (R = Cl), is the polymer of vinyl chloride, CH$_2$=CHCl. Teflon results from the polymerization of tetrafluoroethylene, CF$_2$=CF$_2$; polystyrene (R = C$_6$H$_5$) is the polymer of styrene, C$_6$H$_5$CH=CH$_2$. Saran is the polymer of 1,1-dichloroethene, CH$_2$=CCl$_2$.

For Review   Summary

Carbon is the first member of Group 4A of the periodic table; its valence shell electron configuration is $2s^2 2p^2$. Carbon generally completes its valence shell by sharing electrons in covalent bonds. A carbon atom has only s and p orbitals in its valence shell, so it cannot contain more than eight bonding electrons or form more than four bonds. Thus carbon atoms bond to two, three, or four other atoms in most covalent compounds. The molecular geometries about most carbon atoms are linear, trigonal planar, or tetrahedral. The hybridization of the carbon atom is limited to sp, sp^2, and sp^3.

Compounds that contain only one or two carbon atoms per formula unit or that contain no carbon–hydrogen bonds are called inorganic compounds. Here carbon combines with the more electronegative elements such as oxygen, nitrogen, sulfur, and the halogens. Examples include carbon monoxide, CO; carbon dioxide, CO$_2$; salts containing the carbonate ion, CO$_3^{2-}$, or the cyanide ion, CN$^-$; and related ions and molecules.

Strong, stable bonds between carbon atoms produce complex molecules containing chains and rings. The chemistry of these compounds is called **organic chemistry. Hydrocarbons** are organic compounds composed of only carbon and hydrogen. The **alkanes** are **saturated hydrocarbons**—that is, hydrocarbons that contain only single bonds. **Alkenes** contain one or more carbon–carbon double bonds. **Alkynes** contain one or more carbon–carbon triple bonds. **Aromatic hydrocarbons** contain ring structures with delocalized π-electron systems.

Most hydrocarbons have **structural isomers,** compounds with the same chemical formula but different arrangements of atoms. In addition, molecules of the alkenes and alkynes exhibit structural isomerism based on the position of the multiple bond in the molecule. Many alkenes also exhibit *cis–trans* **isomerism,** a type of geometric isomerism that results from the lack of rotation about a carbon–carbon double bond.

Organic compounds that are not hydrocarbons can be considered derivatives of hydrocarbons. A hydrocarbon derivative can be formed by replacing one or more hydrogen atoms of a hydrocarbon by a **functional group,** which contains at least one atom of an element other than carbon or hydrogen. The properties of hydrocarbon derivatives are determined largely by the functional group. The —OH group is the functional group of an **alcohol.** The —O— group is the functional group of an **ether.** Other functional groups include the —CHO group of an **aldehyde,** the —CO— group of a **ketone,** and the —CO₂H group of a **carboxylic acid.** A halogen atom can also act as a functional group.

Many naturally occurring organic compounds have potent biological properties and are important commercially and medically. Insect **pheromones** are chemicals used for communication among members of a species. **Alkaloids** are natural products, often medicinal, that are derived from plants and have a nitrogen atom contained in a cyclic system.

Polymers, compounds of very high molecular masses, form when a large number of smaller molecules (**monomers**) react with one another. **Addition reactions** form polymers such as rubber and polyethylene, whereas **condensation reactions** form polymers such as nylon and the polyesters.

Key Terms and Concepts

addition reaction   (9.5)
alcohol   (9.9)
aldehyde   (9.12)
alkane   (9.3)
alkene   (9.5)
alkyl group   (9.4)
alkyne   (9.6)
amide linkage   (9.15)
amine   (9.11)
aromatic hydrocarbon   (9.7)

buckminsterfullerene   (9.1)
carbonyl group   (9.12)
carboxylic acid   (9.13)
derivative   (9.8)
diamond   (9.1)
ester   (9.13)
ether   (9.10)
functional group   (9.3)
geometrical isomers   (9.5)
graphite   (9.1)

hydrocarbon   (9.3)
ketone   (9.12)
monomer   (9.15)
natural products   (9.14)
polymer   (9.15)
saturated hydrocarbon   (9.3)
structural isomer   (9.3)
substituent   (9.4)
substitution reaction   (9.3)

Exercises

Questions

1. Identify the microscopic properties, macroscopic properties, and/or symbols in each of the following statements.

    (a) Buckminsterfullerene, C_{60}, is a soluble form of the element carbon that is composed of soccer-ball-shaped molecules.
    (b) Graphite is slippery because the planes of carbon atoms in the crystal slide past each other very easily.
    (c) Even though diamond crystals are very hard, they will burn.
    (d) CO_2 and H_2O are produced when hexane burns in air.
    (e) When a Br_2 molecule adds to a $H_2C{=}CH_2$ molecule, the two C–Br σ bonds formed are stronger than the π bond between the carbon atoms.
    (f) Nylon is stable at higher temperatures than polyethylene because nylon contains polar functional groups and polyethylene does not.
    (g) Vinegar tastes sour because molecules of acetic acid in vinegar react with water to give hydronium ions.

2. Which contains more polar bonds: CO_2 or CS_2? Explain your answer.

3. Write the chemical formula and Lewis structure of an alkane, an alkene, and an alkyne, each of which contains five carbon atoms.

4. What is the difference between the electronic structures of saturated and unsaturated hydrocarbons?

5. On the microscopic level, how does the reaction of bromine with a saturated hydrocarbon differ from its reaction with an unsaturated hydrocarbon? How are they similar?

6. On the microscopic level, how does the reaction of bromine with an alkene differ from its reaction with an alkyne? How are they similar?

7. Why do the compounds hexane, hexanol, and hexene have such similar names?

8. Explain why unbranched alkenes exist as isomers whereas unbranched alkanes do not. Discuss whether this explanation involves the macroscopic domain or the microscopic domain.

9. Explain why the two molecules drawn on page 319 are not isomers.

$$
\begin{array}{c}
\text{H} \quad \text{H} \quad \text{H} \quad \text{H} \quad \text{H} \quad \text{H} \\
| \quad\; | \quad\; | \quad\; | \quad\; | \quad\; | \\
\text{H} - \text{C} - \text{C} - \text{C} - \text{C} = \text{C} - \text{C} - \text{H} \\
| \quad\; | \quad\; | \qquad\qquad\; | \\
\text{H} \quad \text{H} \quad \text{H} \qquad\qquad \text{H}
\end{array}
$$

$$
\begin{array}{c}
\text{H} \quad \text{H} \quad \text{H} \quad \text{H} \quad \text{H} \\
| \quad\; | \quad\; | \quad\; | \quad\; | \\
\text{H} - \text{C} - \text{C} - \text{C} - \text{C} - \text{C} - \text{H} \\
| \quad\; | \quad\; | \quad\; | \quad\; | \\
\text{H} \quad \text{H} \quad \text{H} \quad\; | \quad\; \text{H} \\
\qquad\qquad\quad \text{H} - \text{C} - \text{H} \\
\qquad\qquad\qquad\quad | \\
\qquad\qquad\qquad\quad \text{H}
\end{array}
$$

10. Explain why the two molecules drawn below are not isomers.

$$
\begin{array}{c}
\text{H} \quad \text{H} \quad \text{H} \quad \text{H} \quad \text{H} \quad \text{H} \\
| \quad\; | \quad\; | \quad\; | \quad\; | \quad\; | \\
\text{H} - \text{C} - \text{C} - \text{C} - \text{C} - \text{C} - \text{C} - \text{H} \\
| \quad\; | \quad\; | \quad\; | \quad\; | \quad\; | \\
\text{H} \quad \text{H} \quad \text{H} \quad \text{H} \quad \text{H} \quad \text{H}
\end{array}
$$

$$
\begin{array}{c}
\text{H} \quad \text{H} \quad \text{H} \quad \text{H} \quad \text{H} \\
| \quad\; | \quad\; | \quad\; | \quad\; | \\
\text{H} - \text{C} - \text{C} - \text{C} - \text{C} - \text{C} - \text{H} \\
| \quad\; | \quad\; | \quad\; | \\
\text{H} \quad \text{H} \quad \text{H} \quad \text{H} \quad\; | \\
\qquad\qquad\qquad\quad \text{H} - \text{C} - \text{H} \\
\qquad\qquad\qquad\qquad\quad | \\
\qquad\qquad\qquad\qquad\quad \text{H}
\end{array}
$$

11. Explain why it is not possible to prepare a ketone that contains only two carbon atoms.

12. How does hybridization of the substituted carbon atom change when an alcohol is converted to an aldehyde? An aldehyde to a carboxylic acid?

13. Fatty acids are carboxylic acids that have long hydrocarbon chains attached to a carboxylate group. How does a saturated fatty acid differ from an unsaturated fatty acid? How are they similar?

14. How does the carbon atom hybridization change when polyethylene is prepared from ethylene?

Carbon and Its Inorganic Compounds

15. Predict whether each of the following molecules is polar or nonpolar. (*Hint:* You may want to review Section 7.3.)

    (a) HCN   (b) CH_2F_2 (c) H_2CO
    (d) C_2H_2 (e) CH_3OH (f) CCl_4

16. Write balanced chemical equations for the following reactions.

    (a) The preparation of potassium cyanide from carbon, calcium cyanamide, and potassium carbonate.
    (b) The preparation of calcium carbide from carbon and calcium oxide.

    (c) The combustion of ethanol in a limited supply of oxygen to produce carbon monoxide and water.
    (d) The preparation of calcium cyanamide from calcium carbide and nitrogen.

17. What is the hybridization of carbon in CN_2^{2-}? In CO_3^{2-}? In HCN?

18. Write the resonance structures for the carbonate ion.

19. Which contains the strongest carbon–oxygen bond: CO, CO_2, or CO_3^{2-}? Explain your answer.

20. Write the formulas of the following compounds.

    (a) Lead(II) cyanide
    (b) Potassium cyanamide
    (c) Lithium carbonate
    (d) Carbon tetrachloride
    (e) Calcium hydrogen carbonate
    (f) Carbon disulfide

21. Using Lewis structures, show that the C_2^{2-} ion is isoelectronic with carbon monoxide.

22. Using molecular orbitals, explain why the C_2^{2-} ion is isoelectronic with nitrogen.

23. Give the oxidation number of carbon in each of the following: CO, CO_2, CN_2^{2-}, CO_3^{2-}, CaC_2, and HCN.

24. Which is a polar molecule, CO or CO_2? Explain your answer.

25. For loud events, a small cannon called a carbide cannon is available. Solid calcium carbide is placed in the cannon, a small amount of water is added, and the mixture of the resulting gas and air is ignited with a spark. A loud explosion results. Write equations for the two reactions that occur.

26. Bunsen burners in a general chemistry laboratory produce a yellow, sooty flame when insufficient air is mixed with the entering natural gas. Such a flame can deposit soot (carbon black) on a crucible or flask. A large industry that produces carbon black, used in the tire industry and as a pigment for black paints and inks, is based on a similar phenomenon. In such a plant, hundreds of gas jets burn with a minimum of oxygen present, and a belt of steel plates passes over them to collect the carbon black. Write an equation for the reaction that produces carbon black from methane.

Hydrocarbons

27. Write the Lewis structure and molecular formula for each of the following hydrocarbons.

    (a) Hexane
    (b) 3-Methylpentane
    (c) *cis*-3-Hexene
    (d) 4-Methyl-1-pentene
    (e) 3-Hexyne
    (f) 4-Methyl-2-pentyne

28. Write the Lewis structure and molecular formula for each of the following hydrocarbons.

 (a) Heptane
 (b) 3-Methylhexane
 (c) *trans*-3-Heptene
 (d) 4-Methyl-1-hexene
 (e) 2-Heptyne
 (f) 3,4-Dimethyl-1-pentyne

29. Give the complete IUPAC name for each of the following compounds.

 (a) $CH_3CH_2CBr_2CH_3$
 (b) $(CH_3)_3CCl$
 (c) $CH_3CHCH_2CH_3$
        |
        CH_3
 (d) $CH_3CH_2C{\equiv}CH$
 (e) $CH_3CFCH_2CH_2CH_2CH_3$
        |
        $CH_2CH{\equiv}CH$
 (f)
 (g) $(CH_3)_2CHCH_2CH{=}CH_2$

30. Give the complete IUPAC name for each of the following compounds.

 (a) $(CH_3)_2CHF$
 (b) $CH_3CHClCHClCH_3$
 (c) CH_3CHCH_3
       |
       CH_2CH_3
 (d) $CH_3CH_2CH{=}CHCH_3$
 (e) $CH_3CH_2CH_2CHBrCH_2CH_3$
       |
       $CH_2CH{=}CH_2$
 (f) F, F   C=C   H, CH_3
 (g) $(CH_3)_3CCH_2C{\equiv}CH$

31. Give the complete IUPAC name for each of the following compounds.

 (a) A Freon, CFC-12: CF_2Cl_2
 (b) A Freon, CFC-113: $CF_2ClCFCl_2$
 (c) A Freon replacement, HFC-134a: CF_3CH_2F
 (d) A Freon replacement, HFC-141b: CH_3CFCl_2

32. Butane is used as a lighter fuel. Write the Lewis structure for each isomer of butane.

33. Write Lewis structures for and name the five structural isomers of hexane.

34. Write Lewis structures for the *cis–trans* isomers of $CH_3CH{=}CHCl$.

35. Write structures for the three isomers of the aromatic hydrocarbon xylene, $C_6H_4(CH_3)_2$.

36. Isooctane is the common name of the isomer of C_8H_{18} used as the standard of 100 for the gasoline octane rating.

 $$CH_3CHCH_2CCH_3$$
 with CH_3 groups

 (a) What is the IUPAC name for the compound?
 (b) Name the other isomers of octane that contain a five-carbon chain.

37. Write Lewis structures and IUPAC names for the alkyne isomers of C_4H_6.

38. Write Lewis structures and IUPAC names for all isomers of C_4H_9Cl.

39. Name and write the structures for all isomers of the propyl and butyl alkyl groups.

40. Write the structures for all the isomers of the $-C_5H_{11}$ alkyl group.

41. Write Lewis structures and describe the molecular geometry at each carbon atom in the following compounds.

 (a) *cis*-3-Hexene
 (b) *cis*-1-Chloro-2-bromoethene
 (c) 2-Pentyne
 (d) 6-Ethyl-7-methyl-*trans*-2-octene

42. Benzene is one of the compounds used as octane enhancers in unleaded gasoline. It is manufactured by the catalytic conversion of acetylene to benzene:

 $$3C_2H_2 \longrightarrow C_6H_6$$

 Draw Lewis structures for these compounds, with resonance structures as appropriate, and determine the hybridization of the carbon atoms in each.

Derivatives of Hydrocarbons

43. Write a condensed structural formula, such as CH_3CH_3, and describe the molecular geometry at each carbon atom.

 (a) Propene
 (b) 1-Butanol
 (c) Ethyl propyl ether
 (d) *cis*-4-Bromo-2-heptene
 (e) 2,2,3-Trimethylhexane
 (f) Formaldehyde

44. Write a condensed structural formula, such as CH_3CH_3, and describe the molecular geometry at each carbon atom.

 (a) 2-Propanol   (b) Acetone
 (c) Methyl amine   (d) Dimethyl ether
 (e) Acetic acid   (f) 3-Methyl-1-hexyne

45. Identify the functional groups in the molecules listed below. *Note:* A functional group can be written front to back (for example, ROH for an alcohol) or back to front (HOR for an alcohol).

(a) The amino acid serine

$$
\begin{array}{c}
\text{O—H} \\
| \\
\text{H—C—H} \\
\\
\begin{array}{cc}
\text{H} & \text{O} \\
| & \| \\
\text{H—N—C—C—O—H} \\
| \\
\text{H}
\end{array}
\end{array}
$$

(b) Tartaric acid

$$
\begin{array}{c}
\text{OH} \\
| \\
\text{HCCO}_2\text{H} \\
| \\
\text{HCCO}_2\text{H} \\
| \\
\text{OH}
\end{array}
$$

(c) Tristearin, the common animal fat

$$
\begin{array}{c}
\text{CH}_3(\text{CH}_2)_{16}\text{CO}_2\text{CH}_2 \\
| \\
\text{CH}_3(\text{CH}_2)_{16}\text{CO}_2\text{CH} \\
| \\
\text{CH}_3(\text{CH}_2)_{16}\text{CO}_2\text{CH}_2
\end{array}
$$

(d) Aspirin

(e) Aspartame

(f) Ibuprofin

(g) MTBE, an additive in "oxygenated" automobile fuel

$$\text{CH}_3\text{OC}(\text{CH}_3)_3$$

(h) Glucose

(i) Biacetyl, a butter flavoring

$$
\begin{array}{c}
\text{O O} \\
\| \ \| \\
\text{CH}_3\text{C—CCH}_3
\end{array}
$$

46. Identify each of the following classes of organic compounds, in which R is an alkyl group.

(a) RH (b) ROH
(c) RCOR (d) RCO$_2$H
(e) RNH$_2$ (f) ROR
(g) RCHO

47. Draw Lewis structures for the following molecules.

(a) CH$_3$OH (b) CH$_3$CH(CO$_2$H)CH$_3$
(c) C$_2$H$_5$OCH$_3$ (d) CH$_3$CO$_2$H
(e) CH$_3$CO$_2$C$_2$H$_5$ (f) (CH$_3$)$_2$CHNH$_2$

48. Draw Lewis structures for the following molecules.

(a) CH$_3$NH$_2$ (b) CH$_2$CH(OH)CH$_3$
(c) C$_2$H$_5$NHCH$_3$ (d) C$_3$H$_7$OC$_3$H$_7$
(e) CH$_3$COC$_2$H$_5$ (f) (CH$_3$)$_2$CCl$_2$

49. (a) The foul odor of rancid butter is due to butyric acid, CH$_3$CH$_2$CH$_2$CO$_2$H. Draw the Lewis structure and determine the oxidation number and hybridization for each carbon atom in the molecule.

(b) The esters formed from buteric acid are pleasant-smelling compounds found in fruits and used in perfumes. Draw the Lewis structure for the ester formed from the reaction of buteric acid with 2-propanol.

50. Ethyl alcohol (in beverages), methyl alcohol (as a solvent, for example, in shellac), ethylene glycol (antifreeze), 2-propanol [isopropyl alcohol, CH$_3$CH(OH)CH$_3$, used in rubbing alcohol], and glycerine are apt to be found in the home. What are the structural formulas and IUPAC names of these compounds?

51. Write the condensed structures of both isomers with the formula C$_2$H$_6$O. Label the functional group of each isomer.

52. Write the condensed structures of all isomers with the formula C$_2$H$_6$O$_2$. Label the functional group (or groups) of each isomer.

53. Write the Lewis structures of both isomers with the formula C$_2$H$_7$N.

54. Write the Lewis structures of all isomers with the formula C$_3$H$_7$ON that contain an amide linkage.

55. Write two complete balanced equations for each of the following reactions, one using condensed formulas and one using Lewis structures.

(a) Ethanol reacts with propionic acid.
(b) One mole of 1-butyne reacts with 2 moles of iodine.
(c) Pentane is burned in air.
(d) Propanol is treated with concentrated H$_2$SO$_4$.
(e) Propene is treated with water in dilute acid.
(f) Methyl amine is added to a solution of HCl.
(g) Benzoic acid is added to a sodium hydroxide solution.

56. Write two complete balanced equations for each of the following reactions, one using condensed formulas and one using Lewis structures.

(a) 2-Butene is treated with water in dilute acid.
(b) 1-Butanol reacts with acetic acid.
(c) Propionic acid is poured onto solid calcium carbonate.
(d) Ethylammonium chloride is added to a solution of sodium hydroxide.
(e) Ethanol is treated with concentrated H_2SO_4.
(f) 2-Butene reacts with chlorine.
(g) Benzene burns in air.

57. Identify any carbon atoms that change hybridization, and show the change in hybridization, during the reactions in Exercise 55.

58. Identify any carbon atoms that change hybridization, and show the change in hybridization, during the reactions in Exercise 56.

59. What is the molecular structure about the nitrogen atom in trimethyl amine and in the trimethyl ammonium ion, $(CH_3)_3NH^+$? What is the hybridization of the nitrogen atom in trimethyl amine and in the trimethyl ammonium ion?

60. What is the molecular structure about the nitrogen atom in pyridine and in the pyridinium ion, $C_5H_5H^+$? What is the hybridization of the nitrogen atom in pyridine and in the pyridinium ion?

61. Write the two resonance structures for pyridine.

62. Write the two resonance structures for the pyridinium ion, $C_5H_5NH^+$.

63. Write the two resonance structures for the acetate ion.

64. Show by means of chemical reactions how nicotine could be extracted from tobacco using HCl and then regenerated as a free base.

65. Three compounds with the composition C_3H_8O are known. All three of the carbon atoms in compound A have different substituents, and there is an O—H bond. Compound B has two carbon atoms each bonded to three hydrogen atoms, and the third is bonded to only two hydrogen atoms; in addition, there is no O—H bond in the molecule. Compound C has two carbon atoms each bonded to three hydrogen atoms, and the third carbon atom is bonded to only one hydrogen atom. Molecules of compound C also contain an O—H bond. Write the Lewis structure of each of these three compounds, and classify them according to the functional groups present.

66. Alcohols A, B, and C all have the composition $C_4H_{10}O$. Molecules of alcohol A contain a branched carbon chain and can be oxidized to an aldehyde; molecules of alcohol B contain a linear carbon chain and can be oxidized to a ketone; and molecules of alcohol C can be oxidized to neither an aldehyde nor a ketone. Write the Lewis structures of these molecules.

67. What functional groups are present in morphine? In codeine? In heroin?

68. Morphine, codeine, and heroin differ only in the nature of two functional groups. Which are they?

69. Teflon is prepared by the addition polymerization of tetrafluoroethylene. Write the equation that describes the polymerization using Lewis symbols.

70. Kevlar, the polymer used in bulletproof vests, is made by the condensation polymerization of the monomers

Draw the structure of a portion of the Kevlar chain.

Applications and Additional Exercises

71. Many common substances are organic compounds. Using a handbook, encyclopedia, or other reference book, look up the formulas and/or structures of the principal organic component of the substances listed below and write the Lewis structure for each.

(a) Aspirin   (b) Antifreeze
(c) Grain alcohol   (d) Fingernail polish remover
(e) Rubbing alcohol   (f) Wood alcohol
(g) Vanilla   (h) Vinegar

72. The propellants used in aerosols are more environmentally "friendly" than the chlorofluorocarbons (Freons) used previously. Read the label of several aerosol products to determine the propellant. Using a handbook, encyclopedia, or other reference, look up the formulas and/or structures of these compounds. Then write their molecular formulas and draw their Lewis structures.

73. During 1994 the organic chemicals listed below ranked among the top 50 industrial chemicals in production in the United States. The numbers in parentheses are the amounts, in billion-pound units, produced that year. [Reference: *Chem. and Eng. News*, June 26, 1995, p. 39.] Using a handbook, encyclopedia, or other reference, look up the formulas and/or structures of these compounds, write the Lewis structure of each, and identify any functional groups present.

(a) Ethylene (48.52)
(b) Propylene (28.84)
(c) Ethylene dichloride (18.70)
(d) Vinyl chloride (14.81)
(e) Benzene (14.66)
(f) Methyl *tert*-butyl ether (13.67)
(g) Ethylbenzene (11.87)
(h) Styrene (11.27)
(i) Methanol (10.81)
(j) Xylene (9.06)
(k) Terephthalic acid (8.64)
(l) Formaldehyde (7.94)
(m) Ethylene oxide (6.78)

(n) Toluene (6.75)
(o) p-Xylene (6.23)
(p) Ethylene glycol (5.55)
(q) Cumene (5.16)
(r) Phenol (4.05)
(s) Acetic acid (3.82)
(t) Propylene oxide (3.70)
(u) Butadiene (3.40)
(v) Acrylonitrile (3.08)
(w) Vinyl acetate (3.02)
(x) Acetone (2.77)
(y) Adipic acid (1.80)
(z) n-Butyl alcohol (1.45)

74. There are many organic compounds that contain sulfur, phosphorus, and other nonmetal atoms. Some contain ions. Write the Lewis structures and determine the electron-pair geometry, molecular geometry, and hybridization of any main-group atoms other than C, O, and H in the compounds listed. Skeleton structures are given for some of these compounds.

(a) Methanethiol, a compound added to natural gas to give it an odor

$$CH_3SH$$

(b) A fungicide used to control potato blight and related fungi

$$(CH_3CH_2CH_2CH_2)_3SnOH$$

(c) Dimethyl sulfoxide, a solvent for some medicines

$$(CH_3)_2SO$$

(d) A portion of a silicone polymer used in implants

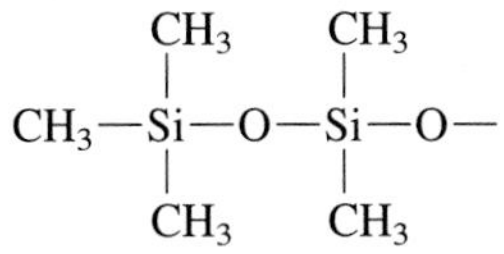

(e) A linear sodium alkylbenzene sulfonate detergent, an ionic compound

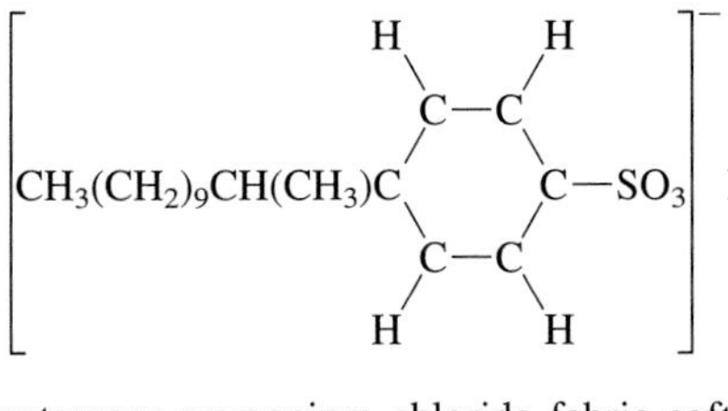

(f) A quaternary ammonium chloride fabric softener, an ionic compound

$$[CH_3(CH_2)_{10}N(CH_3)_3]^+Cl^-$$

(g) The artificial sweetener saccharin

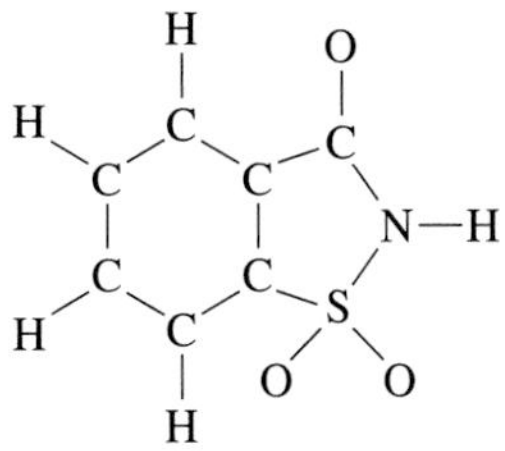

(h) trans-2-Butene-1-thiol, one component of a skunk's defensive odor

$$CH_3CH=CHCH_2SH$$

(i) Methyl parathion, an insecticide

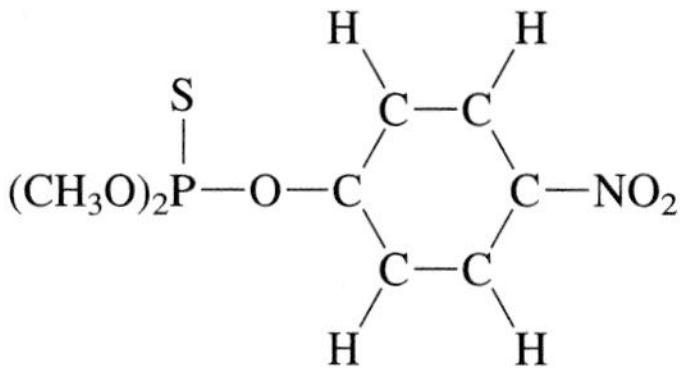

(j) 2,4-D, a herbicide

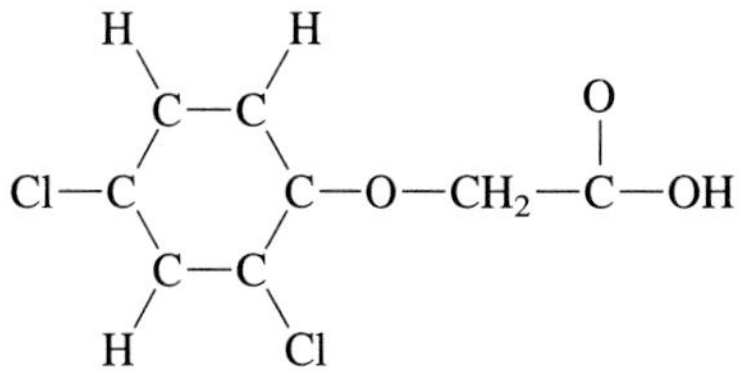

(k) Malathion, an insecticide

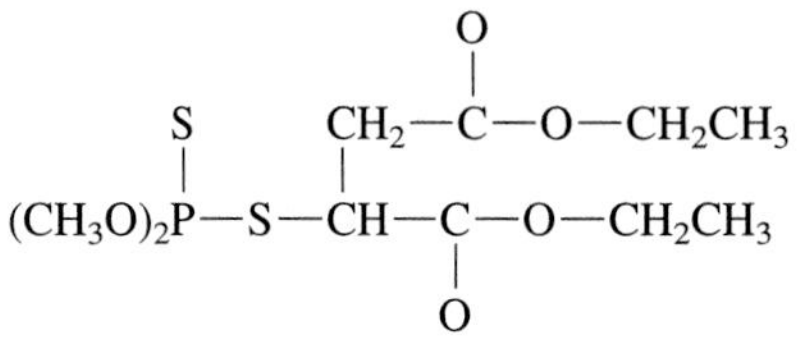

(l) Caffeine

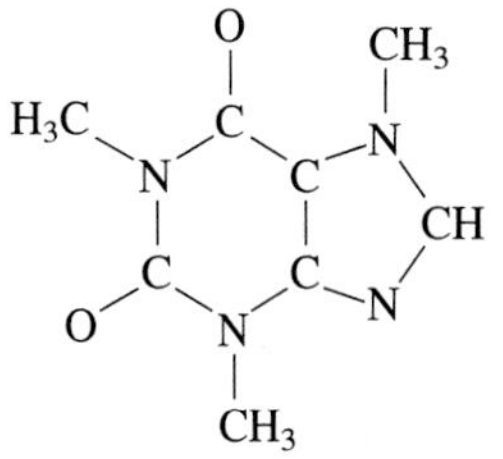

(m) Penicillin G

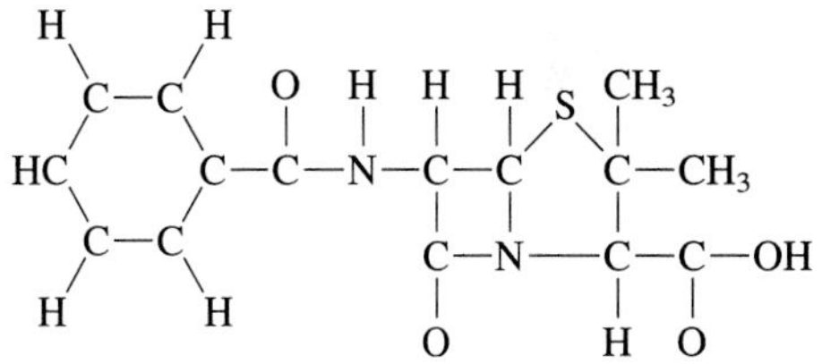

75. Indicate the types of hybridized orbitals used and the molecular geometry about each carbon atom in each of the following molecules.

(a) $CH_3CH=CH_2$
(b) $H_2C=C=O$
(c) C_2H_2
(d) H_2NCONH_2
(e) CH_3CO_2H
(f) CH_3COCH_3

76. The following represents a schematic for the oxidation of methane with the final product being carbon dioxide:

$$CH_4 \longrightarrow CH_3OH \longrightarrow CH_2O \longrightarrow HCO_2H \longrightarrow CO_2$$

For each compound in the sequence, (a) draw the Lewis structure, (b) describe the molecular geometry, and (c) determine the hybridization of the carbon atom.

77. Carbon reacts with sulfur to form carbon disulfide. Write the Lewis structure for carbon disulfide, predict its geometry, and determine the hybridization of the carbon atom.

78. The carbon–oxygen double bond in acetone, CH_3COCH_3, is composed of a σ bond and a π bond. Describe the hybridization of the central carbon atom in an acetone molecule and how the bonds to this atom are formed. Sketch the distribution of electron density in the carbon–oxygen σ bond and the carbon–oxygen π bond.

79. Laboratory wash bottles are made of polyethylene. How many moles of ethylene were required to make the polyethylene in a wash bottle that weighs 67.33 g?

80. What mass of 2-bromopropane could be prepared from 25.5 g of propene? Assume a 100% yield of product.

81. Yields in organic reactions are sometimes low. What is the percent yield of a process that produces 13.0 g of ethyl acetate from 10.0 g of CH_3CO_2H?

82. Acetylene is a very weak acid; however, it will react with moist silver(I) hydroxide, forming water and a compound composed of silver and carbon. Addition of a solution of HCl to a 0.2352-g sample of the compound of silver and carbon produced acetylene and 0.2822 g of AgCl.

    (a) What is the empirical formula of the compound of silver and carbon?

    (b) The production of acetylene upon addition of HCl to the compound of silver and carbon suggests that the carbon is present as the acetylide ion. Write the formula of the compound, showing the acetylide ion.

83. What volume of 0.505 M HCl is required to convert 3.58 g of NaCN to HCN?

$$NaCN + HCl \longrightarrow NaCl + HCN$$

84. How much acetic acid, in grams, is in exactly 1 qt of vinegar if the vinegar contains 3.00% acetic acid by volume? The density of acetic acid is 1.049 g/mL. (Conversion factors that may be helpful are given in Appendix B.)

85. Ethylene can be produced by the pyrolysis of ethane.

$$C_2H_6 \xrightarrow{\Delta} C_2H_4 + H_2$$

Assuming a 100.0% yield, determine how many kilograms of ethylene are produced by the pyrolysis of 1.000×10^3 kg of ethane.

86. Assuming a value of $n = 125$ in the formula for nylon, calculate how many pounds of hexamethylenediamine would be needed to produce 0.5000 ton lb (1.000×10^3 lb) of nylon.

87. Calculate ΔH°_{298} for the following reaction.

$$C(s, diamond) \longrightarrow C(s, graphite)$$

Is the reaction exothermic or endothermic?

88. Which gas produces the greatest amount of heat per gram when burned to produce $CO_2(g)$ and $H_2O(g)$: ethane, ethylene, or acetylene?

89. Calcium carbide reacts with water to form acetylene.

$$CaC_2(s) + 2H_2O(l) \longrightarrow Ca(OH)_2(s) + C_2H_2(g)$$

In early automobiles, the acetylene produced in this manner was burned in the headlights at night. How much heat is produced by combustion of 1.00 kg of acetylene under standard conditions if $CO_2(g)$ and $H_2O(g)$ are produced.

90. Carbon dioxide can be obtained on a commercial scale as a by-product of the fermentation of glucose or other sugars. Calculate the quantity of carbon dioxide that can be obtained from 175 pounds of glucose, given an 89.7% yield.

91. MTBE, Methyl *tert*-butyl ether, $CH_3OC(CH_3)_3$, is used as an oxygen source in oxygenated gasolines. MTBE is manufactured by reacting 1,1-dimethylethene with methanol.

    (a) Using Lewis structures, write the chemical equation representing the reaction.

    (b) What volume of methanol (density, 0.7915 g ml^{-1}) is required to produce exactly 1000 kg of MTBE, assuming a 100.0% yield?

92. The sugar substitute aspartame is the methyl ester of a dipeptide, a combination of the two amino acids aspartic acid and phenylalanine. When aspartame enters the body, it is dissociated into methanol and the amino acids, which are then used in the formation of proteins.

$$C_{12}H_{15}N_2O_3CO_2CH_3 + 2H_2O \longrightarrow$$
$$C_9H_{11}NO_2 + C_4H_7NO_4 + CH_3OH$$

Concern has been expressed that the ingestion of too much aspartame in diet sodas, for example, may produce a toxic reaction from the methanol.

    (a) What mass of methanol may be produced by 0.250 g of aspartame?

    (b) If significant amounts of methanol enter the bloodstream, blindness and even death may result. The toxic limit is 500 to 5000 mg per kilogram of body weight. Taking the minimum value (500 mg kg^{-1}), determine what mass of methanol is the lethal dose for a person weighing 140 lb.

    (c) The density of methanol is 0.7915 g mL^{-1}. What volume of methanol is a minimum lethal dose for a 140-lb person?

    (d) Methanol in the blood is absorbed by the liver where it is oxidized to formaldehyde, an aldehyde that is the toxic substance responsible for the liver damage caused by drinking methanol. What quantity of formaldehyde would be produced from the methanol in part (b)?

CHAPTER OUTLINE

The Macroscopic Behavior of Gases

10.1 The Pressure of a Gas
10.2 Volume and Pressure: Boyle's Law
10.3 Volume and Temperature: Charles's Law
10.4 The Kelvin Temperature Scale
10.5 Moles of Gas and Volume: Avogadro's Law
10.6 The Ideal Gas Equation
10.7 Standard Conditions of Temperature
and Pressure
10.8 Densities and Molar Masses of Gases
10.9 The Pressure of a Mixture of Gases:
Dalton's Law
10.10 Diffusion and Effusion of Gases:
Graham's Law
10.11 Stoichiometry Involving Gases
10.12 The Atmosphere

The Microscopic Behavior of Gases

10.13 The Kinetic-Molecular Theory
10.14 Molecular Velocities and Kinetic Energy
10.15 Relationship of the Behavior of Gases to
the Kinetic-Molecular Theory
10.16 Derivation of the Ideal Gas Equation from
Kinetic-Molecular Theory
10.17 Deviations from Ideal Gas Behavior:
Descriptions of Real Gases

10

Gases

The behavior of a hot air balloon illustrates many properties of gases.

We are surrounded by an ocean of gas—the atmosphere—and many of the properties of gases are familiar to us from our daily activities. The atmosphere and other gases exert a pressure. Although we normally do not notice the pressure of the atmosphere, sometimes it becomes apparent, such as when we try to pull a suction cup off a window. As we stick a suction cup to a window, we press the air out of the space between the window and the cup, leaving a vacuum. Pulling the cup away from the window is difficult, because the pressure of the air on the outside of the suction cup is not balanced by the pressure inside the cup. If the seal is tight so that no air leaks in, we must pull hard to overcome the pressure of the atmosphere on the suction cup.

Gases can be compressed. Scuba divers take advantage of the fact that enough air to last for about an hour under water can be compressed into a small cylinder. We compress air when we fill a bicycle tire, a basketball, or a volleyball with a hand pump. In fact, the volume of any gas (unlike that of a solid or a liquid) can be decreased greatly by increasing the pressure.

The pressure of a gas increases as it is heated. A can of hair spray contains the warning "Do not heat or incinerate the empty can," because heating could increase the pressure of the gas remaining in the can to the point where it ruptures or explodes. Sometimes, leaving a bicycle in the sun on a hot day will heat the air in the tires so much that the increased pressure can cause a blowout.

Gases expand. An airbag inflates when the nitrogen gas produced by the decomposition of sodium azide expands and fills the bag.

Gases diffuse. We can smell perfume because it evaporates and the resulting gas molecules enter our nose and trigger the sense of smell. If we release perfume in the corner of a room, in time we can smell it all over the room even if the air is still. Diffusion of the perfume occurs because the molecules of all gases are in constant high-speed motion. Two gases introduced into the same container mix by diffusion.

Gases have played an important part in the development of chemistry. Boyle's description of the change in volume of a sample of gas as its pressure varies was one of the first mathematical descriptions of the behavior of matter. The densities of gases played an important role in the determination of atomic masses. Conclusions regarding chemical stoichiometry and the molecular nature of matter followed from observations of the volumes of gases that combine in chemical reactions. The variations in the pressure and volume of a gas with temperature helped scientists develop the concept of absolute zero and the kinetic-molecular theory of gas behavior.

In this chapter we will examine how changes in temperature, pressure, volume, or mass of a sample of a gas affect the others of these properties. We will consider several simple equations that describe the behavior, bearing in mind that they are only approximations of the actual behavior of a gas. We will also describe a simple theoretical model of gases and compare the experimental behavior of gases with the predictions of this model. Comparing experiment and theory will help us assess the validity of the assumptions applied in the theory.

The Macroscopic Behavior of Gases

10.1   The Pressure of a Gas

An astronaut can survive in space because his or her space suit contains oxygen gas (Fig. 10.1). Although shielding and insulation disguise it somewhat, the flexible por-

Figure 10.1

The pressure of oxygen in astronaut Michael Gernhardt's space suit is about 5 psi.

tions of the suit bulge rather like a balloon. The gas in the suit presses against the inside with a force of about 5 pounds per square inch (5 psi), about one-third the pressure of the atmosphere at sea level. In space, there is no air or other gas outside the space suit. Thus there is no force on the outside to counterbalance the force on the inside of the suit.

We use the force a gas exerts on a surface (such as on the inside of the space suit) to measure its **pressure**—the force a gas exerts on a square unit of area (Fig. 10.2). In the United States, pressure is often measured in pounds of force on an area equal to one square inch (pounds per square inch, or psi). Pressure is also commonly measured with the unit of **atmosphere (atm).** One atmosphere of pressure equals the average pressure of the air at sea level at a latitude of 45°. The pressure of the atmosphere varies with the distance above sea level and with climatic changes. At higher elevations the pressure is less. The pressure at 16,000 feet is only about 0.5 atm (half of that at sea level) because about half of the atmosphere is below this elevation.

The pressure unit in the International System of Units (SI) is the **pascal (Pa).** One pascal is a very small pressure; in many cases, it is more convenient to use units of **kilopascals** (1 kPa = 1000 Pa). One atmosphere is equal to 101.325 kilopascals.

Figure 10.2

When the air was pumped out of this can, the pressure of the atmosphere, about 15 psi at sea level, crushed it.

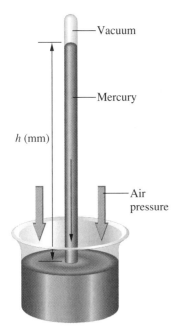

Figure 10.3

A mercury barometer. The height, h, of the mercury column is a measurement of the air pressure.

We can measure the atmospheric pressure, the pressure exerted by the column of air above us, with a **barometer** (Fig. 10.3). We can make a barometer by filling a glass tube 80 centimeters long, closed at one end, with mercury and inverting it in a container of mercury. The mercury in the tube falls until the pressure exerted by the atmosphere on the surface of the mercury in the container is just sufficient to support the weight of the mercury in the tube. Because the height of the mercury column is proportional to the pressure of the atmosphere, pressure is sometimes expressed in terms of **millimeters of mercury (mm Hg).** This pressure unit is generally referred to as a **torr** (the plural is also *torr*) after Evangelista Torricelli, who invented the barometer in 1643. A column of mercury exactly 1 millimeter high exerts a pressure of exactly 1 torr. At sea level, on average, the atmosphere will support a column of mercury 760 millimeters in height. Thus 1 atmosphere of pressure is exactly 760 millimeters of mercury, or 760 torr.

A **manometer** (Fig. 10.4) is a type of barometer that can be used to measure the pressure of a gas trapped in a container. It is a U-shaped tube that contains mercury or some other nonvolatile liquid. One of the arms is closed and contains a vacuum above the mercury. The other arm can be connected to a container filled with gas. Just as the height of a barometer measures the pressure of the atmosphere, the height of the mercury in a manometer measures the pressure of the gas in the container. The pressure is proportional to the distance between the mercury levels in the two arms of the tube (h in the diagram).

The conversions among units of pressure are as follows:

$$1 \text{ atm} = 760 \text{ mm Hg} = 760 \text{ torr} = 101.325 \text{ kPa}$$

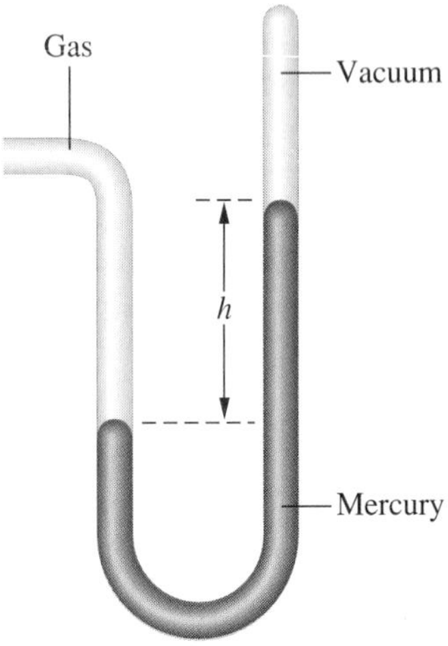

Figure 10.4

A mercury manometer.

Figure 10.5

A weather map shows pressure fronts measured in inches of mercury or millibars.

Example 10.1   **Conversion of Pressure Units**

Weather reports in the United States (Fig. 10.5) use pressure units of inches of mercury (in. Hg) rather than atm or torr (29.92 in. Hg = 760 torr). Convert a pressure of 29.2 in. Hg to torr, atmospheres, and kilopascals.

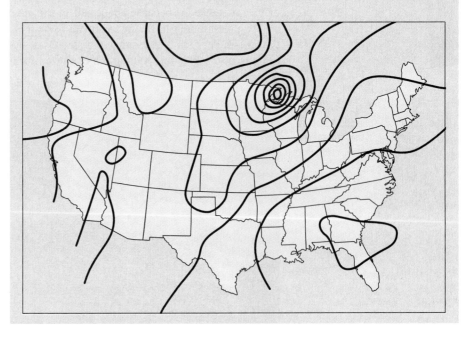

$$CH_3CH_2CH_2OH \longrightarrow CH_3CH_2CHO$$

Alcohol   Aldehyde

We use alcohols that have their —OH groups in the middle of the chain in the preparation of ketones, which require the carbonyl group to be bonded to two other carbon atoms.

$$CH_3CH(OH)CH_3 \longrightarrow CH_3COCH_3$$

Alcohol   Ketone

An alcohol with its OH group bonded to a carbon atom that is bonded to three other carbon atoms generally forms neither an aldehyde nor a ketone upon oxidation.

Formaldehyde, an aldehyde with the formula HCHO, is a colorless gas with a pungent and irritating odor. It is sold in an aqueous solution called formalin, which contains about 37% formaldehyde by weight. Formaldehyde causes coagulation of proteins, so it kills bacteria and stops many of the biological processes that cause tissue to decay. Thus formaldehyde is useful for preserving tissue specimens and embalming bodies. It is also used to sterilize soil. Formaldehyde is used in the manufacture of Bakelite, a hard plastic that has high chemical and electrical resistance.

Dimethyl ketone, CH_3COCH_3, a colorless liquid commonly called **acetone,** is the simplest and most important ketone. It is made commercially by fermenting corn or molasses or by oxidation of 2-propanol. Among other applications, it is used as a solvent for cellulose acetate, cellulose nitrate, acetylene, plastics, and varnishes; as a paint, varnish, and fingernail polish remover; and as a solvent in the manufacture of pharmaceuticals and chemicals.

9.13   Carboxylic Acids and Esters

The smell of vinegar is due to the presence of acetic acid, a carboxylic acid. The smell of ripe bananas and many other fruits is due to the presence of esters, compounds that can be prepared by the reaction of a carboxylic acid with an alcohol.

Both **carboxylic acids** and **esters** contain a carbonyl group with a second oxygen atom bonded to the carbon atom in the carbonyl group by a single bond. In a carboxylic acid, the second oxygen atom also bonds to a hydrogen atom. In an ester, the second oxygen atom bonds to another carbon atom.

A carboxylic acid   An ester
Acetic acid   Methyl acetate

The functional group for an acid and that for an ester are highlighted in color in these formulas.

The hydrogen atom in the functional group of a carboxylic acid will react with a base to form an ionic salt.

Propionic acid   Propionate ion

Amines are responsible for the odor of spoiled fish. However, the amine functional group also appears in many useful molecules. For example, nylon (Section 9.15), many dyes, medications such as penicillin and codeine (Section 9.14), vitamins, and many of the essential molecules such as DNA and amino acids (Chapter 25) found in the body are amines.

9.12   Aldehydes and Ketones

Both **aldehydes** and **ketones** contain a **carbonyl group,** $\diagdown\!\!C\!\!=\!\!O$, a functional group with a carbon–oxygen double bond. In an aldehyde the carbonyl group is bonded to at least one hydrogen atom. In a ketone the carbonyl group is bonded to two carbon atoms.

$$
\begin{array}{cc}
\underset{\substack{| \\ \text{H}}}{\overset{\substack{\text{H} \quad :\text{O}: \\ | \quad\quad || \\ }}{\text{H}\!-\!\text{C}\!-\!\text{C}\!-\!\text{H}}} &
\underset{\substack{\quad| \quad\quad | \quad | \\ \quad\text{H} \quad\; \text{H} \; \text{H}}}{\overset{\substack{\text{H} \;\; :\text{O}: \; \text{H} \; \text{H} \\ | \quad\quad || \quad | \quad | \\ }}{\text{H}\!-\!\text{C}\!-\!\text{C}\!-\!\text{C}\!-\!\text{C}\!-\!\text{H}}}
\end{array}
$$

CH_3CHO	$CH_3COCH_2CH_3$
An aldehyde	A ketone
Ethanal (acetaldehyde)	Methyl ethyl ketone

Thus the $-\overset{\overset{\textstyle O}{\textstyle ||}}{C}-H$ functional group is the hallmark of an aldehyde, and the $\diagdown\!\!C\!\!=\!\!O$ group is the hallmark of a ketone. A simpler representation of an aldehyde group is $-CHO$; of a ketone, $-C(O)-$ or $-CO-$.

In both aldehydes and ketones, the geometry about the carbon atom in the carbonyl group is trigonal planar; the carbon atom exhibits sp^2 hybridization. Two of the sp^2 orbitals on the carbon atom in the carbonyl group are used to form σ bonds to the other carbon or hydrogen atoms in a molecule. The remaining sp^2 hybrid orbital forms a σ bond to the oxygen atom. The unhybridized p orbital on the carbon atom in the carbonyl group overlaps a p orbital on the oxygen atom to form the π bond in the double bond.

Like the $C\!\!=\!\!O$ bond in carbon dioxide (Section 9.2), the $C\!\!=\!\!O$ bond of a carbonyl group is polar. Many of the reactions of aldehydes and ketones start with the reaction between a Lewis base and the carbon atom at the positive end of the polar $C\!\!=\!\!O$ bond. This initial reaction is usually followed by one or more rearrangements to give the final product.

The importance of molecular structure in the reactivity of organic compounds is illustrated by the reactions that produce aldehydes and ketones. We can prepare a carbonyl group by oxidation of an alcohol. The reagents, other products, and reaction conditions are beyond the scope of this discussion, so we will represent the reaction as simply

$$
-\underset{\substack{| \\ }}{\overset{\substack{\text{H} \\ |}}{\text{C}}}-\ddot{\text{O}}-\text{H} \xrightarrow{\text{Oxidation}} -\overset{\overset{\textstyle :\ddot{\text{O}}}{\textstyle ||}}{\text{C}}-
$$

Alcohol	Carbonyl group

We can prepare aldehydes by the oxidation of alcohols whose $-OH$ functional group is located on the carbon at the end of the chain of carbon atoms in the alcohol.

Solution: This problem involves conversion of a measurement in one set of units to other sets of units. It is similar to the conversions described in Section 1.13. The problem gives the relationship between inches of mercury and torr (760 torr/29.92 in. Hg), which we can use as a conversion factor to convert the pressure in inches of mercury to torr.

$$29.2 \text{ in. Hg} \times \frac{760 \text{ torr}}{29.92 \text{ in. Hg}} = 742 \text{ torr}$$

Now we convert torr into atmospheres and into kilopascals, using the conversion factors given in this section.

$$742 \text{ torr} \times \frac{1 \text{ atm}}{760 \text{ torr}} = 0.976 \text{ atm}$$

$$742 \text{ torr} \times \frac{101.325 \text{ kPa}}{760 \text{ torr}} = 98.9 \text{ kPa}$$

Check your learning: A typical barometric pressure in Kansas City is 740 torr. What is this pressure in atmospheres, in millimeters of mercury, and in kilopascals?

Answer: 0.974 atm, 740 mm Hg, 98.7 kPa

10.2   Volume and Pressure: Boyle's Law

Since the English natural philosopher Robert Boyle published the results of his observations of gases over 300 years ago, many scientists and thousands of general chemistry students have verified his observations: At a constant temperature, the volume of a sample of gas decreases when the pressure on the gas increases, and the volume increases when the pressure decreases. Many of these experiments used a J-shaped tube similar to that shown on page 330 in Fig. 10.6 to measure the change in volume of a gas as the height of a mercury column (and thus the pressure on the gas) changed. Table 10.1 reports data derived when air is used as the gas.

There are at least three ways to describe how the volume of a gas changes as the pressure changes: We can use a table (similar to Table 10.1), a graph, or a mathematical equation.

Figure 10.7 on page 330 contains two graphs constructed from the data in Table 10.1. The curve in Fig. 10.7 A is a hyperbola. Many of us find such a curved line difficult to read accurately at points representing low pressures or low volumes. Generally, it is easier to work with graphs that contain straight lines, and there is a way of graphing these data that does yield a straight line—plot the inverse of the volume, $\frac{1}{V}$, as the pressure changes (Fig. 10.7 B). A graph of the inverse of the pressure, $\frac{1}{P}$, as the volume changes also gives a straight line.

We can also describe the change in gas volume with a change in pressure by using a mathematical equation. If two variables, such as the volume and the pressure of our sample of gas, give a straight line when one is plotted against the inverse of the other, we call the two variables inversely proportional or say they exhibit **inverse proportionality:** One variable is proportional to the inverse of the other. We can write the mathematical relationship as

Table 10.1

Volume of a Sample of Air as the Pressure Varies: Data from Figure 10.6 and Two Additional Measurements

Volume, mL	Pressure, atm
2.40	0.993
2.26	1.06
2.10	1.14
1.83	1.30
1.57	1.53
1.24	1.90

Figure 10.6

Each of these identical J-tubes started with 2.40 mL of air at a pressure of 0.993 atm. Increasing the pressure by adding mercury reduced the volume of the air in their left arms.

Pressure of surrounding atmosphere

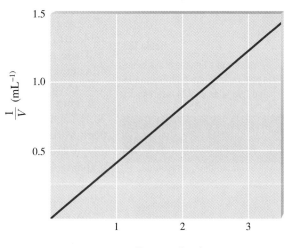

0.993 atm (755 torr)   0.993 atm (755 torr)   0.993 atm (755 torr)   0.993 atm (755 torr)

2.40 mL   2.26 mL   2.10 mL   1.83 mL

47 mm   108 mm   235 mm

0.993 atm (755 torr)   1.06 atm (802 torr)   1.14 atm (863 torr)   1.30 atm (990 torr)

Figure 10.7

Graphs of volume versus pressure for the sample of air illustrated in Fig. 10.6 and Table 10.1. (A) We can see that the volume decreases as the pressure increases and that it increases as the pressure decreases: an inversely proportional relationship. (B) A graph of the change in the inverse of the volume, $\frac{1}{V}$, as the pressure changes at constant temperature.

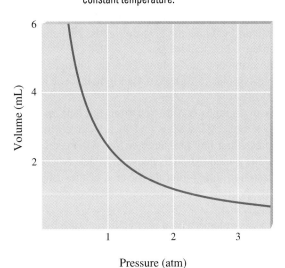

(A)

(B)

$$V \sim \frac{1}{P}$$

where V is the volume of the gas, P is its pressure, and $\sim$ means "is proportional to." An equation may be written by including a constant, k, referred to as a proportionality constant.

$$V = \text{constant} \times \frac{1}{P} \qquad \text{or} \qquad V = k \times \frac{1}{P}$$

This equation can be rearranged to give

$$PV = \text{constant} \qquad \text{or} \qquad PV = k$$

The meaning of this equation is that the pressure of a given amount of a particular gas times its volume is always equal to the same number (is constant) if the temperature does not change. When the amount of the gas or the temperature changes, a different value of the constant must be used. The equation was first used to describe the behavior of gases in 1660 by Robert Boyle. It is summarized in the statement now known as **Boyle's law:** *The volume of a given amount of gas held at constant temperature is inversely proportional to the pressure under which it is measured.* A gas described exactly by Boyle's law is called an **ideal gas.**

Example 10.2 Volume of a Gas Sample

The sample of gas in Fig. 10.6 has a volume of 2.40 mL at a pressure of 0.993 atm. Determine the volume of the gas at a pressure of 0.500 atm, using **(a)** Fig. 10.7 (A), **(b)** Fig. 10.7 (B), and **(c)** Boyle's law.

Solution:
(a) The curved line in Fig. 10.7 (A) shows the relationship between V and P for the sample of gas. The line indicates that at a pressure of 0.5 atm, V has a value of about 4.8 mL.

(b) The straight line in Fig. 10.7 (B) shows the relationship between $\frac{1}{V}$ and P for the sample of gas. The line indicates that at a pressure of 0.500 atm, $\frac{1}{V}$ has a value of about 0.21 mL^{-1}. Because $\frac{1}{V} = 0.21$ mL^{-1},

$$V = \frac{1}{0.21 \text{ mL}^{-1}} = 4.8 \text{ mL}$$

(c) From Boyle's law we know that the product of pressure and volume (PV) for a given sample of gas at a constant temperature is always equal to the same value. For two different conditions of pressure and volume, we have

$$P_1V_1 = k \qquad \text{and} \qquad P_2V_2 = k$$

Thus
$$P_1V_1 = P_2V_2$$

where the subscripts 1 and 2 indicate the pressure and volume at the two different conditions (with the same temperature). If we take P_1 and V_1 as the known values 0.993 atm and 2.40 mL, P_2 as the pressure where the volume is unknown, and V_2 as the unknown volume, we have the following information:

1st Conditions	2nd Conditions
$P_1 = 0.993$ atm	$P_2 = 0.500$ atm
$V_1 = 2.40$ mL	$V_2 = \,?$ mL

We can substitute these values into the equation and solve for V_2 as follows:

$$P_1V_1 = P_2V_2$$
$$0.993 \text{ atm} \times 2.40 \text{ mL} = 0.500 \text{ atm} \times V_2$$

Rearrangement gives

$$V_2 = \frac{0.993 \text{ atm} \times 2.40 \text{ mL}}{0.500 \text{ atm}}$$
$$= 4.77 \text{ mL}$$

Check your learning: The sample of gas in Fig. 10.6 has a volume of 2.40 mL at a pressure of 0.993 atm. Determine the volume of the gas at 2.50 atm using **(a)** Fig. 10.7 (A), **(b)** Fig. 10.7 (B), and **(c)** Boyle's law.

Answer: about 0.9 mL, 0.92 mL, 0.953 mL.

Figure 10.8

Sport divers use compressed air.

Example 10.3   Volume of a Compressed Gas

When filled with air, a typical scuba tank (Fig. 10.8) with a volume of 13.2 L has a pressure of 153 atm. How many liters of air will such a tank provide at a depth of 30 feet in the ocean where the pressure is 1.91 atm, assuming no change in temperature?

Solution: From Boyle's law at two different conditions of P and V, we have

$$P_1V_1 = P_2V_2$$

The conditions are

1st Conditions	2nd Conditions
$P_1 = 153$ atm	$P_2 = 1.91$ atm
$V_1 = 13.2$ L	$V_2 = ?$ L

Substitution and rearrangement give

$$153 \text{ atm} \times 13.2 \text{ L} = 1.91 \text{ atm} \times V_2$$
$$V_2 = \frac{153 \text{ atm} \times 13.2 \text{ L}}{1.91 \text{ atm}}$$
$$= 1.06 \times 10^3 \text{ L}$$

Check your learning: The volume of air contained in a bicycle pump is 655 cm^3. To what volume must this air be compressed to increase the pressure to 4.08 atm (60 psi) if its initial pressure is 0.950 atm?

Answer: 153 mL

In the three centuries since Boyle carried out his studies, measuring techniques have improved tremendously. The results of these improved measurements show that Boyle's law is only an approximation of the behavior of **real gases.** It holds precisely only at very low pressures or high temperatures. Painstaking measurements at higher

Table 10.2

Pressure–Volume Data for a Sample of Methane That Occupies 22.414 L at 1.0000 atm and 273.15 K

Measured Pressure, atm	Measured Volume, L	$P \times V$ Calculated from the Experimental Data, L atm
1.0000×10^{-3}	2.2415×10^4	22.415
1.0000×10^{-1}	2.2410×10^2	22.410
5.0000×10^{-1}	44.804	22.402
1.5000	14.966	22.449
2.0000	11.253	22.506
3.0000	7.5570	22.671
4.0000	5.7237	22.895

Figure 10.9

A plot of *PV* versus *V* shows that real gases exhibit deviations from the behavior predicted by Boyle's law.

pressures reveal that the product *PV* is not constant but rather changes as the pressure changes. Table 10.2 shows how the product *PV* changes for measured values of pressure and volume as the pressure increases in a sample of methane, the gas many of us use to heat our homes. Figure 10.9 contains graphs of *PV*, which Boyle's law suggests should be constant, versus *P* for methane, carbon dioxide, and helium. We can see that as the pressure increases, the difference between the behavior predicted by Boyle's law and the actual behavior increases. But we should note that the difference is small, and at low pressures we can use Boyle's law as a good approximation of the behavior of a real gas.

There are equations that can be used to describe pressure-versus-temperature behavior more precisely than Boyle's law, but they are complicated and differ for each gas. Boyle's law gives values that are good enough for most purposes.

10.3 Volume and Temperature: Charles's Law

Figure 10.10

The effect of temperature on a gas. When a balloon at room temperature, *T* = 25°C, is placed in contact with liquid nitrogen, *T* = −196°C, the volume of the gas in the balloon decreases as it cools.

When we put an air-filled balloon in a refrigerator and let the gas inside get cold, the balloon shrinks. It expands again when it warms up. When we make the balloon very cold, it shrinks a great deal (Fig. 10.10). These are examples of the effect of temperature on the volume of a given amount of a confined gas: The volume increases

Temperature, K	Volume, L	$\frac{V}{T}$, L K^{-1}
173	14.10	0.0815
223	18.26	0.0819
273	22.40	0.0821
373	30.65	0.0822
473	38.88	0.0822

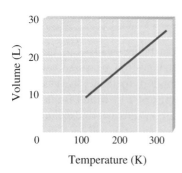

Figure 10.11

A graph of the change in the volume of 1 mol of methane gas as the temperature changes at a constant pressure of 1 atm. We can see that the volume decreases as the temperature decreases. The line stops at 111 K because methane liquefies at this temperature.

as the temperature increases, and it decreases as the temperature decreases. Measurements of the volume of a sample of methane at different temperatures, but at a constant pressure, are listed in Table 10.3. Figure 10.11 shows the same data in a graph. Because the graph of the volume and temperature is a straight line, the volume and temperature of the gas are said to exhibit **direct proportionality.**

The mathematical equation that we most commonly use to describe the relationship between the volume and temperature of a given mass of gas is known as Charles's law in recognition of the French physicist S.A.C. Charles, who first reported his studies of changes in volume with temperature in 1787. **Charles's law** states that *the volume of a given mass of gas is directly proportional to its temperature on the Kelvin scale when the pressure is held constant;* that is,

$$V \sim T$$

If we use a proportionality constant that depends on the mass of gas and its pressure, we get the equation

$$V = \text{constant} \times T \qquad \text{or} \qquad V = k \times T$$

Hence

$$\frac{V}{T} = \text{constant} \qquad \text{or} \qquad \frac{V}{T} = k$$

This constant has a different value and units from the constant in Boyle's law. It does not vary unless the amount of the gas or the pressure changes. The Charles's law equation means that as the Kelvin temperature of a given sample of gas changes, with no change in pressure, its volume also changes such that the ratio $\frac{V}{T}$ remains the same. A decrease in T results in a decrease in V, and an increase in T results in an increase in V.

Charles's law applies to the volume of a gas and its Kelvin temperature. In Section 1.14 we saw that the relationship between the Kelvin and Celsius temperature scales is K = °C + 273.15. (Remember that temperatures on the Kelvin scale are by convention reported without a degree sign.)

Example 10.4   Predicting Change in Volume with Temperature

A sample of carbon dioxide, CO_2, occupies 0.300 L at 10°C and 750 torr. What volume will the gas have at 30°C and 750 torr?

Solution: The mathematical form of Charles's law indicates that the ratio $\frac{V}{T}$ is constant for any sample of gas, provided that the temperature does not change. Thus for two different conditions of pressure and volume, we would have

$$\frac{V_1}{T_1} = k \qquad \text{and} \qquad \frac{V_2}{T_2} = k$$

Thus

$$\frac{V_1}{T_1} = \frac{V_2}{T_2}$$

where the subscripts 1 and 2 indicate the volume and temperature at the two different conditions (with the same pressure). If we convert the temperatures to the Kelvin scale and take V_1 and T_1 as the initial values, T_2 as the temperature where the volume is unknown, and V_2 as the unknown volume, we have

1st Conditions
$V_1 = 0.300 \text{ L}$
$T_1 = 10°C + 273.15 = 283 \text{ K}$

2nd Conditions
$V_2 = ? \text{ L}$
$T_2 = 30°C + 273.15 = 303 \text{ K}$

We can use these values and solve for V_2 as follows:

$$\frac{V_1}{T_1} = \frac{V_2}{T_2}$$

$$\frac{0.300 \text{ L}}{283 \text{ K}} = \frac{V_2}{303 \text{ K}}$$

Rearrangement gives

$$V_2 = \frac{0.300 \text{ L} \times 303 \text{ K}}{283 \text{ K}}$$

$$= 0.321 \text{ L}$$

Check your learning: A sample of oxygen, O_2, occupies 32.2 mL at 30°C and 452 torr. What volume will it occupy at $-70°C$ and the same pressure?

Answer: 21.6 mL

Example 10.5   Measuring Temperature with a Volume Change

Temperature is sometimes measured with a gas thermometer by observing the change in the volume of the gas as the temperature changes at constant pressure. The hydrogen in a particular hydrogen gas thermometer has a volume of 150.0 cm^3 when immersed in a mixture of ice and water (0.00°C). When immersed in boiling liquid ammonia, the volume of the hydrogen, at the same pressure, is 131.7 cm^3. Find the temperature of boiling ammonia on the Kelvin and Celsius scales.

Solution: This problem asks us to find the temperature of a 131.7-cm^3 sample of hydrogen gas that has a volume of 150.0 cm^3 at a temperature of 0.00°C and the same pressure. We have

1st Conditions
$V_1 = 150.0 \text{ cm}^3$
$T_1 = 0.00°C + 273.15 = 273.15 \text{ K}$

2nd Conditions
$V_2 = 131.7 \text{ cm}^3$
$T_2 = ? \text{ K}$

We can use these values and solve for V_2 as follows:

$$\frac{V_1}{T_1} = \frac{V_2}{T_2}$$

$$\frac{150.0 \text{ cm}^3}{273.15 \text{ K}} = \frac{131.7 \text{ cm}^3}{T_2}$$

Rearrangement gives

$$T_2 = \frac{131.7 \text{ cm}^3 \times 273.15 \text{ K}}{150.0 \text{ cm}^3}$$

$$= 239.8 \text{ K}$$

Subtracting 273.15 from 239.8 K, we find that the temperature of the boiling ammonia on the Celsius scale is $-33.4°C$.

Check your learning: What is the volume of a sample of ethane at 467 K and 1.10 atm if it occupies 405 mL at 298 K and 1.10 atm?

Answer: 635 mL

As is the case with Boyle's law, more precise measurements have shown Charles's law to be an approximation of the behavior of real gases. The values of $\frac{V}{T}$ calculated from the data in Table 10.3 are constant to two significant figures, but they begin to deviate in the third significant figure. However, at low pressures and above 0°C, the deviations from ideal behavior are small and the approximation is very useful.

10.4   The Kelvin Temperature Scale

In Section 1.14 we discussed the Kelvin temperature scale. Now we can use the effect of temperature on gases to determine zero on the Kelvin scale.

As we have seen, when a sample of a gas is cooled at constant pressure, its volume decreases. Graphs of volume and temperature data for samples of several different gases at constant pressure are shown in Fig. 10.12. In this figure, the temperature is in degrees Celsius. The data were collected at low pressures where the gases exhibit behavior very close to that of an ideal gas. They stop at those temperatures at which the gases condense and form a liquid. However, if we extrapolate (extend) the straight lines to the temperatures at which the gases would appear to have a zero volume if they did not condense, they all intersect the temperature axis at $-273.15°C$, regardless of the gas. Below this temperature, gases would theoretically have a negative volume (which is impossible), so $-273.15°C$ must be the lowest temperature possible. This temperature, called **absolute zero,** is taken as the zero point of the Kelvin temperature scale (0 K). The freezing point of water (0.00°C) is therefore 273.15 K, and its normal boiling point (100.00°C) is 373.15 K.

We can think about the impossibility of reaching a temperature below zero kelvins in another way. To lower the temperature of a substance, we reduce its thermal energy (Section 4.1). Absolute zero (0 K) is the temperature reached when all possible thermal energy has been removed from a substance. Obviously, a substance cannot be cooled any further after all thermal energy has been removed.

Figure 10.12

Charles's law of behavior for several gases, each at constant pressure. The volume of each gas extrapolates to 0 L at $-273.15°C$. The dashed lines represent the extrapolated behavior below the temperatures at which the gases condense and form liquids.

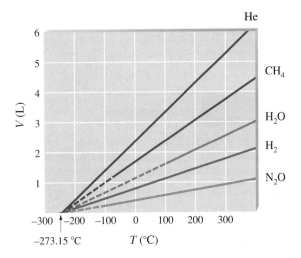

10.5   Moles of Gas and Volume: Avogadro's Law

The Italian physicist Amadeo Avogadro (1776–1856) advanced a hypothesis in 1811 to account for the behavior of gases. His hypothesis, which has since been experimentally confirmed, is known as **Avogadro's law:** *Equal volumes of all gases, measured under the same conditions of temperature and pressure, contain the same number of molecules.*

Avogadro's hypothesis was based on observations of the volumes of gases that react one with the other. He used the observations of Joseph Louis Gay-Lussac, who reported that the volumes of gases involved in a reaction can be expressed as a ratio of small whole numbers. These observations led to the general idea that gases combine, or react, in definite and simple proportions by volume, provided that all gas volumes are measured at the same temperature and pressure. For example, one volume of the gas ammonia reacts with the same volume of the gas hydrogen chloride to form ammonium chloride. One volume of oxygen combines with two volumes of carbon monoxide to give two volumes of carbon dioxide.

The explanation of Gay-Lussac's observations by Avogadro's law may be illustrated by the reaction of the gases carbon monoxide and oxygen (Fig. 10.13). According to Avogadro's law, equal volumes of gaseous CO, O_2, and CO_2 contain the same number of molecules. Because two molecules of carbon dioxide are formed from two molecules of carbon monoxide and one molecule of oxygen, the volume of carbon monoxide required and the volume of carbon dioxide produced are both twice as great as the volume of oxygen required. We now know that the small whole numbers in

Figure 10.13

Two volumes of CO combine with one volume of O_2 and form two volumes of CO_2.

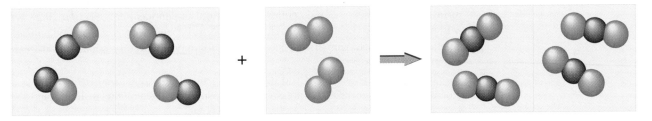

the ratios observed by Gay-Lussac are equal to the coefficients in the balanced equation that describes the reaction

$$2CO(g) + O_2(g) \longrightarrow 2CO_2$$

Example 10.6   Reaction of Gases

We burn methane, $CH_4(g)$, as a source of energy to heat and cook. What volume of $O_2(g)$ measured at 25°C and 760 torr is required to react with 1.0 L of methane measured under the same conditions of temperature and pressure? The products are $CO_2(g)$ and $H_2O(g)$.

Solution: The ratio of the volumes of CH_4 and O_2 will be equal to the ratio of their coefficients in the balanced equation for the reaction.

$$CH_4(g) + 2O_2(g) \longrightarrow CO_2(g) + 2H_2O(g)$$

1	2	1	2
volume	volumes	volume	volumes

From the equation we see that one volume of CH_4, in this case 1.0 L, reacts with two volumes (liters) of O_2.

$$1.0 \text{ L } CH_4 \times \frac{2 \text{ L } O_2}{1 \text{ L } CH_4} = 2.0 \text{ L } O_2$$

A volume of 2.0 L of O_2 will be required to react with 1.0 L of CH_4.

Check your learning: An acetylene tank for an oxyacetylene welding torch provides 9340 L of acetylene gas, C_2H_2, at 0°C and 1 atm. How many tanks of oxygen, each providing 7.00×10^3 L of O_2 at 0°C and 1 atm, will be required to burn the acetylene?

$$2C_2H_2 + 5O_2 \longrightarrow 4CO_2 + 2H_2O$$

Answer: 3.34 tanks (2.34×10^4 L)

10.6   The Ideal Gas Equation

We can combine Boyle's law, Charles's law, and Avogadro's law into one equation, the ideal gas equation, that relates pressure, volume, temperature, and number of moles of an ideal gas. The **ideal gas equation** is

$$PV = nRT$$

where P is the pressure of a gas, V is its volume, n is the number of moles of the gas, T is its temperature on the Kelvin scale, and R is a constant called the **gas constant.** Although this equation applies precisely only to an ideal gas (Section 10.2), we will use it to approximate the behavior of all gases, because it provides a good approximation for the behavior of real gases under common laboratory conditions of temperature and pressure.

The numerical value for R and its units depend on the units used for P, V, and T. For example, if we take the pressure in atmospheres, the volume in liters, and the

0.08206 L atm mol^{-1} K^{-1}
82.06 mL atm mol^{-1} K^{-1}
8.314 L kPa mol^{-1} K^{-1}

Table 10.4

Values of the Gas Constant, *R*, for Different Sets of Units

temperature in kelvins, *R* has the value 0.08206 L atm mol^{-1} K^{-1}. Other values of *R* are given in Table 10.4. The value and units of *R* used with the ideal gas equation must match the units of pressure, volume, temperature, and amount of gas, or these quantities must be converted to match the units for *R*.

Let's consider how the ideal gas equation reduces to Boyle's law, to Charles's law, or to Avogadro's law. Boyle's law (Section 10.2) states that for a given amount of an ideal gas, the product of its volume *V* and its pressure *P* is a constant at constant temperature. Because *n*, *R*, and *T* do not change under these conditions, their product is a constant. Thus the ideal gas equation, $PV = nRT$, becomes

$$PV = \text{constant}$$

which is the mathematical expression for Boyle's law.

Charles's law (Section 10.3) states that the volume *V* of a given amount of an ideal gas divided by its temperature *T* on the Kelvin scale is equal to a constant at constant pressure. If we rearrange the ideal gas equation so that the terms that do not vary under these conditions (*n*, *R*, and *P*) are on the right and *V* and *T* are on the left, we obtain the equation expressing Charles's law:

$$\frac{V}{T} = \frac{nR}{P} = \text{constant}$$

Avogadro's law (Section 10.4) states that equal volumes of gases, under the same conditions of temperature and pressure, contain the same number of molecules. Thus they contain the same number of moles, *n*, of gas. This law may be written in the form

$$n = \text{constant} \times V$$

We can rearrange the ideal gas equation to yield this expression by placing the constant quantities *R*, *T*, and *P* together:

$$n = \frac{P}{RT} \times V = \text{constant} \times V$$

The ideal gas equation contains five terms (*P*, *V*, *n*, *R*, and *T*). If we know any four of these, we can rearrange the equation and find the fifth. The following examples illustrate some of the ways in which the ideal gas equation can be used to relate *P*, *V*, *T*, and *n*.

Example 10.7   Volume of an Ideal Gas

Calculate the volume occupied by 0.54 mol of any ideal gas at 15°C and 0.967 atm.

Solution: In this problem we are given *n* = 0.54 mol, *T* = 15°C, and *P* = 0.967 atm and are asked to find *V*. Thus we must rearrange $PV = nRT$ to solve for *V*.

$$V = \frac{nRT}{P}$$

The equation requires that the temperature be in kelvins, so we convert the temperature from °C to K.

$$T = 15°C + 273.15 = 288 \text{ K}$$

To determine a volume in liters from data with units of atmospheres, moles, and kelvins, we need to use the value of R that goes with these units. It is 0.08206 L atm mol^{-1} K^{-1}. Substitution gives

$$V = \frac{nRT}{P} = \frac{(0.54 \text{ mol})(0.08206 \text{ L atm mol}^{-1} \text{ K}^{-1})(288 \text{ K})}{0.967 \text{ atm}}$$
$$= 13 \text{ L} \quad (\text{2 significant figures are justified by the data})$$

Note that the units cancel to liters, the unit for volume. Cancellation of units is a very simple way of checking to see that the units for R are consistent with those for P, V, n, and T.

Check your learning: Rearrange the ideal gas equation to solve for P, and then calculate the pressure of 10.5 mol of an ideal gas that occupies a volume of 565 L at 452°C. Use 0.08206 L atm mol^{-1} K^{-1} as the value of R.

Answer: $P = \frac{nRT}{V}$, 1.11 atm

Example 10.8   **Volume of Methane**

Methane, CH_4, can be used as fuel for an automobile; however, it is a gas at normal temperatures and pressures, which causes some problems with storage. One gallon of gasoline could be replaced by 655 g of CH_4. What is the volume of this much methane at 25°C and 745 torr? Use the value of R with the units of L atm mol^{-1} K^{-1}.

Solution: As shown in Example 10.7, the volume of an ideal gas may be obtained from the equation

$$V = \frac{nRT}{P}$$

The units of R (liters, atmospheres, moles, and kelvins) require that the pressure be expressed in atmospheres rather than in torr as given in the problem, the amount of gas in moles rather than in grams, and the temperature in K. Once these conversions are made, the values can be substituted into the equation and the volume calculated.

Step 1.   *Convert pressure to units of atmospheres.*

$$P = 745 \text{ torr} \times \frac{1 \text{ atm}}{760 \text{ torr}} = 0.980 \text{ atm}$$

Step 2.   *Convert quantity of gas to units of moles.*

$$n = 655 \text{ g } CH_4 \times \frac{1 \text{ mol}}{16.043 \text{ g } CH_4} = 40.8 \text{ mol}$$

Step 3.   *Convert temperature to kelvins.*

$$T = 25°C + 273.15 = 298 \text{ K}$$

Step 4.   *Substitute these values into the rearranged equation and find the volume.*

$$V = \frac{nRT}{P} = \frac{(40.8 \text{ mol})(0.08206 \text{ L atm mol}^{-1} \text{ K}^{-1})(298 \text{ K})}{0.980 \text{ atm}} = 1.02 \times 10^3 \text{ L}$$

Figure 10.14

During the Second World War, some British drivers used fuel gas for their automobiles because gasoline was tightly rationed. The bag on top of this car was required to hold the large volume of gas necessary.

It would require 1020 L (269 gal) of gaseous methane at about 1 atm of pressure to replace 1 gal of gasoline. It requires a large container to hold enough methane at 1 atm to replace several gallons of gasoline (Fig 10.14).

Check your learning: Calculate the volume in liters occupied by 0.044 mol of helium gas at $-77°C$ and 0.907 atm.

Answer: 0.78 L

Example 10.9 **Moles of Gas**

While resting, the average 70-kg human male consumes 14 L of O_2 per hour at 25°C and 100 kPa. How many moles of O_2 are consumed by a 70-kg man while resting for 1.0 h?

Solution: To solve for moles, we must rearrange the ideal gas equation, $n = \frac{PV}{RT}$. The pressure and volume are given in kilopascals and liters, so we use $R = 8.314$ L kPa mol^{-1} K^{-1}. The temperature is 25°C + 273.15 = 298 K. These values can now be used to find n.

$$n = \frac{PV}{RT} = \frac{(100 \text{ kPa})(14 \text{ L})}{(8.314 \text{ L kPa mol}^{-1} \text{ K}^{-1})(298 \text{ K})} = 0.57 \text{ mol } O_2$$

Check your learning: How many moles of H_2 gas are required to fill a 16.80-L balloon with a pressure of 1.050 atm and a temperature of 38°C?

Answer: 0.691 mol

10.7 Standard Conditions of Temperature and Pressure

We have seen that the volume of a given quantity of gas and the number of molecules (moles) in a given volume of gas vary with changes in pressure and temperature. Thus we use a set of **standard conditions (STP)** of temperature and pressure for reporting properties of gases: STP is 273.15 K (0.00°C) and exactly 1 atm (760 torr, or 101.325 kilopascals).

Any gas that approximates ideal behavior has a volume of 22.4 L at 0°C and 1 atm. This volume is referred to as the **standard molar volume** (Fig. 10.15).

Figure 10.15

A mole of any gas occupies a volume of approximately 22.4 L at 0°C and 1 atm. A cube with an edge length of 28.2 cm, or 11.1 in., contains 22.4 L.

32 g of O_2
(1 mole)

2 g of H_2
(1 mole)

17 g of NH_3
(1 mole)

We can convert the volume of a gas at any temperature and pressure to its volume at standard conditions or at any other temperature and pressure by using the ideal gas equation.

Example 10.10   Conversion to Standard Conditions

A sample of ammonia is found to occupy 0.250 L under laboratory conditions of 27°C and 0.850 atm. Find the volume at standard conditions of 0°C and 1.00 atm.

Solution: The pressure, volume, and temperature all change in this example, so we rearrange the ideal gas law so that all the variable terms are on the left.

$$\frac{P_1 V_1}{T_1} = nR \text{ (a constant)} \quad \text{and} \quad \frac{P_2 V_2}{T_2} = nR$$

Thus

$$\frac{P_1 V_1}{T_1} = \frac{P_2 V_2}{T_2}$$

We have

1st Conditions	2nd Conditions
$P_1 = 0.850$ atm	$P_2 = 1.00$ atm
$V_1 = 0.250$ L	$V_2 = ?$ L
$T_1 = 27°C + 273.15 = 300$ K	$T_2 = 0°C + 273.15 = 273$ K

Rearranging the equation gives

$$V_2 = \frac{P_1 V_1}{T_1} \times \frac{T_2}{P_2}$$

Substitution gives

$$V_2 = \frac{0.850 \text{ atm} \times 0.250 \text{ L} \times 273 \text{ K}}{300 \text{ K} \times 1.00 \text{ atm}} = 0.193 \text{ L}$$

Note that the units cancel to leave L, a correct unit for volume.

Check your learning: A weather balloon (Fig. 10.16) contained 217 L of helium at a pressure of 0.992 atm and a temperature of 25°C when it was released at ground level.

Figure 10.16

Weather balloons are not filled completely before they are released, because they expand at the lower pressures they encounter when they reach high altitudes.

Find the volume of the balloon after it reached an altitude of 10 km, where the temperature was $-43°C$ and the pressure 0.276 atm.

Answer: 602 L

10.8   Densities and Molar Masses of Gases

We can use the ideal gas equation to relate the pressure, volume, temperature, and number of moles of a gas. Now we will see that we can combine the ideal gas equation with other equations and find the density and the molar mass of a gas. We do not intend to provide more equations to be memorized but to show how we can obtain additional information about a gas by combining relationships that we have seen up to this point.

Density of a Gas.   The density d of a gas is the mass of exactly 1 liter of the gas. If we can determine the mass of 1 liter of a gas, then we can determine its density. From the ideal gas equation we can determine the number of moles of a gas in 1 liter. We can then determine the mass of the gas by using the relationship $m = n \times M$, which relates m (the mass of a sample), n (the number of moles in the sample), and M (its molar mass).

Example 10.11   Calculation of Density of Ethane

Ethane is an important raw material in the petrochemical industry. What is the density of ethane gas, C_2H_6, at a pressure of 183.4 kPa and a temperature of 25°C?

Solution:

Step 1.   Calculate the number of moles of ethane in exactly 1 liter of the gas.

$$n = \frac{PV}{RT}$$

$$= \frac{183.4 \text{ kPa} \times 1\text{L}}{8.314 \text{ L kPa mol}^{-1} \text{ K}^{-1} \times 298 \text{ K}}$$

$$= 0.0740 \text{ mol C}_2H_6$$

Step 2.   Determine the mass of the gas.

$$m = n \times M = 0.0740 \text{ mol C}_2H_6 \times \frac{30.070 \text{ g C}_2H_6}{1 \text{ mol C}_2H_6} = 2.23 \text{ g C}_2H_6$$

Step 3.   Calculate the density.

$$d = \frac{m}{V} = \frac{2.23 \text{ g}}{1 \text{ L}} = 2.23 \text{ g L}^{-1}$$

Check your learning: Calculate the density of fluorine gas, F_2, at 30.0°C and 725 torr.

Answer: 1.46 g L^{-1}

We must specify both the temperature and the pressure of a gas when reporting its density, because the number of moles of a gas (and thus the mass of the gas) in a liter changes with temperature or pressure. Gas densities are often reported at STP.

Molar Mass of a Gas.

If we know the mass m of a sample of a compound and the number of moles n in the sample, we have two of the three quantities in the equation that relates mass, number of moles, and molar mass: $m = n \times M$. Thus we can rearrange the equation to find M, the molar mass of the compound. If we weigh a sample of a gas at a known volume, temperature, and pressure, we can determine the number of moles of gas by using the ideal gas equation and then can determine the molar mass of the gas.

Example 10.12   The Molar Mass of Butane

Figure 10.17 shows how to set up a plastic syringe so it will hold a vacuum. The opening of the syringe is covered, and a nail keeps the pressure of the atmosphere from pushing the barrel into the vacuum.

Figure 10.17

A syringe that contains a vacuum.

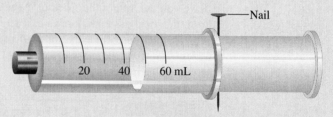

A syringe (with its cap and nail) containing 50 mL of vacuum weighs 75.212 g. The same syringe (with its cap and nail) containing 50 mL of gaseous butane at a pressure of 0.923 atm and a temperature of 24°C weighs 75.322 g. What is the molar mass of butane?

Solution:

Step 1.   Determine the mass of butane, m, *in the syringe.*

$$m = 75.322 \text{ g} - 75.212 \text{ g} = 0.110 \text{ g}$$

Step 2.   Determine the number of moles, n, *of butane in the syringe by using the ideal gas equation.*

$$n = \frac{PV}{RT} = \frac{0.923 \text{ atm} \times 0.050 \text{ L}}{0.08206 \text{ L atm mol}^{-1} \text{ K}^{-1} \times 297 \text{ K}}$$
$$= 1.9 \times 10^{-3} \text{ mol}$$

Step 3.   Determine the molar mass from the mass of the gas and the number of moles.

$$M = \frac{m}{n} = \frac{0.110 \text{ g}}{1.9 \times 10^{-3} \text{ mol}} = 58 \text{ g mol}^{-1}$$

Check your learning: A sample of phosphorus that weighs 3.243×10^{-2} g exerts a pressure of 31.89 kPa in a 56.0-mL bulb at 550°C. What are the molar mass and the molecular formula of phosphorus vapor?

Answer: 124 g mol^{-1}, P$_4$

10.9   The Pressure of a Mixture of Gases: Dalton's Law

The individual gases in a mixture of gases do not affect each other's pressure, and each individual gas exerts the same pressure that it would exert if it were not mixed with other gases. Thus if we know the pressure of each of the individual gases in a mixture, we can determine the total pressure of the mixture by adding the pressures exerted by the individual gases. This is **Dalton's law of partial pressures:** *The total pressure of a mixture of ideal gases is equal to the sum of the partial pressures of the component gases.*

$$P_T = P_A + P_B + P_C + \cdots$$

The pressure exerted by each individual gas in a mixture is called the **partial pressure** of that gas; this is the pressure the individual gas would exert if it were alone in the same container at the same temperature. In the equation P_T is the total pressure of a mixture of gases, P_A is the partial pressure of a gas A; P_B is the partial pressure of gas B; etc. Thus if nitrogen gas at a pressure of 1 atmosphere in a 1-liter flask is transferred into a second 1-liter flask containing oxygen gas at a pressure of 1 atmosphere, the mixture will have a pressure of 2 atmospheres (provided the temperature does not change).

Example 10.13   The Pressure of a Mixture of Gases

What is the total pressure in atmospheres in a 10.0-L vessel that contains 2.50×10^{-3} mol of H_2, 1.00×10^{-3} mol of He, and 3.00×10^{-4} mol of Ne at 35°C (308 K)?

Solution: The total pressure in the 10.0-L vessel is the sum of the partial pressures of the gases, because they do not react with each other.

$$P_T = P_{H_2} + P_{He} + P_{Ne}$$

The partial pressure of each gas can be determined from the ideal gas equation, using $P = \frac{nRT}{V}$.

$$P_{H_2} = \frac{(2.50 \times 10^{-3} \text{ mol})(0.08206 \text{ L atm mol}^{-1} \text{ K}^{-1})(308 \text{ K})}{10.0 \text{ L}} = 6.32 \times 10^{-3} \text{ atm}$$

$$P_{He} = \frac{(1.00 \times 10^{-3} \text{ mol})(0.08206 \text{ L atm mol}^{-1} \text{ K}^{-1})(308 \text{ K})}{10.0 \text{ L}} = 2.53 \times 10^{-3} \text{ atm}$$

$$P_{Ne} = \frac{(3.00 \times 10^{-4} \text{ mol})(0.08206 \text{ L atm mol}^{-1} \text{ K}^{-1})(308 \text{ K})}{10.0 \text{ L}} = 7.58 \times 10^{-4} \text{ atm}$$

$$P_T = (0.00632 + 0.00253 + 0.00076) \text{ atm} = 9.61 \times 10^{-3} \text{ atm}$$

Check your learning: Find the pressure of a mixture of 0.200 g of H_2, 1.00 g of N_2, and 0.820 g of Ar in a container with a volume of 2.00 L at 20°C.

Answer: 1.87 atm

Figure 10.18

If the level of water inside and that outside the vessel are the same, the pressure inside the vessel is equal to the atmospheric pressure outside the vessel.

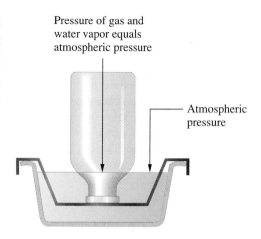

Pressure of gas and water vapor equals atmospheric pressure

Atmospheric pressure

A simple way to collect gases that do not react with water is to bubble them into a bottle that has been filled with water and inverted into a dish filled with water. The pressure of the gas inside the bottle can be made equal to the air pressure outside by raising or lowering the bottle. When the water level is the same both inside and outside the bottle (Fig. 10.18), the pressure of the gas is equal to the atmospheric pressure, which can be measured with a barometer.

However, there is another factor we must consider when we measure the pressure of the gas by this method. Water evaporates, and there is always gaseous water (water vapor) above a sample of liquid water. As a gas is collected over water, it becomes saturated with water vapor, and the total pressure of the mixture equals the sum of the partial pressures of the gas and the water vapor. The pressure of the pure gas is therefore equal to the total pressure minus the pressure of the water vapor.

The pressure of the water vapor above a sample of water in a closed container like that in Fig. 10.18 depends on the temperature, as shown in Fig. 10.19 and Table 10.5.

Figure 10.19

A graph of the vapor pressure of water as a function of temperature.

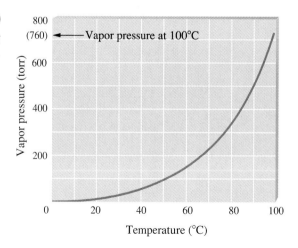

Table 10.5

Vapor Pressure of Ice and Water at Various Temperatures

Temperature, °C	Pressure, torr	Temperature, °C	Pressure, torr	Temperature, °C	Pressure, torr
−10	1.95	18	15.5	30	31.8
−5	3.0	19	16.5	35	42.2
−2	3.9	20	17.5	40	55.3
0	4.6	21	18.7	50	92.5
2	5.3	22	19.8	60	149.4
4	6.1	23	21.1	70	233.7
6	7.0	24	22.4	80	355.1
8	8.0	25	23.8	90	525.8
10	9.2	26	25.2	95	633.9
12	10.5	27	26.7	99	733.2
14	12.0	28	28.3	100.0	760.0
16	13.6	29	30.0	101.0	787.6

Example 10.14   Pressure of a Gas Collected Over Water

If 0.200 L of argon is collected over water at a temperature of 26°C and a pressure of 750 torr in a system like that shown in Fig. 10.18, what is the partial pressure of argon?

Solution: According to Dalton's law, the total pressure in the bottle (750 torr) is the sum of the partial pressure of argon gas and the partial pressure of gaseous water.

$$P_T = P_{Ar} + P_{H_2O}$$

Rearranging this equation to solve for the pressure of argon gives

$$P_{Ar} = P_T - P_{H_2O}$$

The pressure of water vapor above a sample of liquid water at 26°C is 25.2 torr (Table 10.5), so

$$P_{Ar} = 750 \text{ torr} - 25.2 \text{ torr}$$

$$= 725 \text{ torr}$$

Check your learning: A sample of oxygen collected over water at a temperature of 29.0°C and a pressure of 764 torr has a volume of 0.560 L. What is the partial pressure of oxygen?

Answer: 734 torr

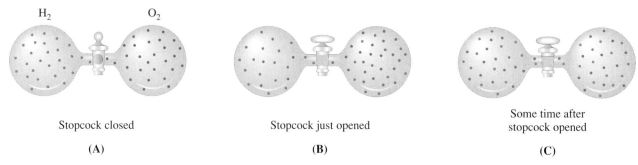

Stopcock closed

(A)

Stopcock just opened

(B)

Some time after
stopcock opened

(C)

Figure 10.20

(A) Two gases, H_2 and O_2, separated by a stopcock intermingle when the stopcock is opened; they mix by diffusing together. The rapid motion of the molecules and the relatively large spaces between them explain why diffusion occurs. Note that H_2, the lighter of the two gases, effuses through the opening faster than O_2. Thus just after the stopcock is opened (B), more H_2 molecules move to the O_2 side than O_2 molecules move to the H_2 side. After a short time (C), both the slower-moving O_2 molecules and the faster-moving H_2 molecules have distributed themselves evenly on both sides of the vessel.

10.10   Diffusion and Effusion of Gases: Graham's Law

We know that when a sample of gas is set free in one part of a closed container, its molecules very quickly diffuse throughout the container (Fig. 10.20). In some cases we can smell a gas as it diffuses throughout a closed room.

Diffusion, the spread of one substance throughout a space or through a second substance, is one of the properties of a gas. A second property is called **effusion,** which is the escape of gas molecules through a tiny hole such as a pinhole in a balloon (Fig. 10.21). Both the **rate of diffusion** of a gas and its **rate of effusion** depend on its molar mass.

When a mixture of gases is placed in a container with porous walls, effusion of the gases through the small openings in the walls occurs. The ligher gases effuse through the small openings more rapidly than the heavier ones. In 1832 Thomas Graham studied the rates of effusion of different gases and formulated **Graham's law:** *The rates of effusion of gases are inversely proportional to the square roots of their densities (or their molar masses).*

$$\frac{\text{Rate of effusion of gas A}}{\text{Rate of effusion of gas B}} = \frac{\sqrt{\text{density B}}}{\sqrt{\text{density A}}} = \frac{\sqrt{\text{molar mass B}}}{\sqrt{\text{molar mass A}}}$$

Effusion through a porous barrier from a region of higher pressure to one of lower pressure was used for the large-scale separation of gaseous $^{235}UF_6$ from $^{238}UF_6$ during the Second World War at the atomic energy installation in Oak Ridge, Tennessee.

Figure 10.21

A balloon filled with air remains full overnight. A balloon filled with helium deflates, because the lighter helium atoms effuse through small holes in the rubber much more rapidly than the heavier molecules of nitrogen and oxygen found in air.

For separation to be complete, a given volume of UF_6 had to effuse through the barriers some two million times.

Example 10.15 **Effusion Ratios**

Calculate the ratio of the rate of effusion of hydrogen to the rate of effusion of oxygen, using two different methods.

Solution: *Using densities:* The density of hydrogen is 0.0899 g L^{-1}, and that of oxygen is 1.43 g L^{-1}.

$$\frac{\text{Rate of effusion of hydrogen}}{\text{Rate of effusion of oxygen}} = \frac{\sqrt{1.43 \text{ g } L^{-1}}}{\sqrt{0.0899 \text{ g } L^{-1}}} = \frac{1.20}{0.300} = \frac{4}{1}$$

Using molar masses:

$$\frac{\text{Rate of effusion of hydrogen}}{\text{Rate of effusion of oxygen}} = \frac{\sqrt{32 \text{ g mol}^{-1}}}{\sqrt{2 \text{ g mol}^{-1}}} = \frac{\sqrt{16}}{\sqrt{1}} = \frac{4}{1}$$

Hydrogen effuses four times as rapidly as oxygen.

Check your learning: A gas of unknown identity effuses at the rate of 169 mL s^{-1} in an effusion apparatus in which carbon dioxide effuses at the rate of 102 mL s^{-1}. Calculate the molar mass of the unknown gas.

Answer: 16.0 g mol^{-1}

10.11   Stoichiometry Involving Gases

In Chapter 3 we saw that we can turn to chemical stoichiometry for answers to many of the questions that ask "how much?" We saw how to answer the question with masses or with volumes of solutions in Section 3.10. Now we will see that we can answer this question another way: with volumes of gases.

Chemical stoichiometry generally takes us through conversions in the order illustrated in Fig. 3.13, which summarizes many of the possible relationships in stoichiometry calculations. In brief, the figure may be summarized as

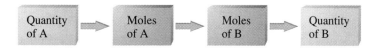

We have previously measured quantities by using the mass of a substance or the volume of a solution; now we can also use gas volume to indicate quantities. If we know the volume, pressure, and temperature of a gas, we can use the ideal gas equation to calculate how many moles of the gas are present. If we know how many moles of a gas are involved, we can calculate the volume of a gas at any temperature and pressure.

Example 10.16   Volume of a Gaseous Reactant

In Example 3.21 we explored the question "What mass of oxygen gas, O_2, is consumed in the combustion of 702 g (1 L) of octane, C_8H_{18}, one of the principal components of gasoline?" Now let us ask about the *volume* of oxygen: What volume of oxygen gas at 27°C and 0.899 atm is consumed in the combustion of 702 g (1 L) of octane, C_8H_{18}?

$$2C_8H_{18}(l) + 25O_2(g) \longrightarrow 16CO_2(g) + 18H_2O(g)$$

Solution: We are asked to calculate the volume of oxygen gas that reacts with 702 g of octane. The overall conversion is

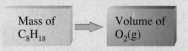

where the $O_2(g)$ is at 27°C and 0.899 atm. The chain of calculations requires the following conversions:

| Mass of C_8H_{18} | **1** → | Moles of C_8H_{18} | **2** → | Moles of $O_2(g)$ | **3** → | Volume of $O_2(g)$ |

We saw Steps 1 and 2 in this problem in Example 3.21. The new step is calculation of the volume of oxygen in Step 3. (The numbers above the arrows refer to the following conversion steps.)

Step 1. Convert the grams of C_8H_{18} to moles of C_8H_{18}, using the molar mass of C_8H_{18} in the unit conversion factor.

$$\text{mol } C_8H_{18} = 702 \text{ g } C_8H_{18} \times \frac{1 \text{ mol } C_8H_{18}}{114.232 \text{ g } C_8H_{18}} = 6.15 \text{ mol } C_8H_{18}$$

Step 2. Convert the moles of C_8H_{18} to moles of O_2, using the ratio determined from the balanced chemical equation as the conversion factor.

$$\text{mol } O_2 = 6.15 \text{ mol } C_8H_{18} \times \frac{25 \text{ mol } O_2}{2 \text{ mol } C_8H_{18}} = 76.9 \text{ mol } O_2$$

Step 3. Convert the moles of O_2 to volume of O_2 gas at 27°C and 0.899 atm, using the ideal gas equation.

$$PV = nRT$$

$$V = \frac{nRT}{P} = \frac{76.9 \text{ mol} \times 0.08206 \text{ L atm mol}^{-1} \text{ K}^{-1} \times 300 \text{ K}}{0.899 \text{ atm}}$$

$$= 2.11 \times 10^3 \text{ L}$$

Check your learning: What is the pressure in kilopascals in a 35.0-L balloon at 25°C filled with pure hydrogen gas produced by the reaction of 34.11 g of CaH_2 with water?

$$CaH_2 + 2H_2O \longrightarrow Ca(OH)_2 + 2H_2$$

Answer: 115 kPa

Example 10.17   **Volume of a Gaseous Product**

What volume of hydrogen at 27°C and 723 torr may be prepared by the reaction of 8.88 g of gallium with an excess of hydrochloric acid?

$$2Ga(s) + 6HCl(aq) \longrightarrow 2GaCl_3(aq) + 3H_2(g)$$

Solution: We are asked to calculate the volume of hydrogen gas that is produced by the reaction of 8.88 g of gallium with hydrochloric acid. The overall conversion is

where the $H_2(g)$ is at 27°C and 723 torr. The chain of calculations is very similar to that in the previous exercise.

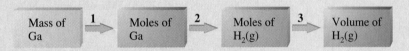

This problem requires the following steps:

Step 1.   *Convert the grams of Ga to moles of Ga, using the molar mass of Ga in the unit conversion factor.* Result: 0.127 mol Ga

Step 2.   *Convert the moles of Ga to moles of H_2, using the ratio determined from the balanced chemical equation as the conversion factor.* Result: 0.191 mol H_2

Step 3.   *Use the ideal gas equation to get the volume of H_2 after converting the temperature to kelvins and the pressure to atmospheres.*

$$V = \frac{nRT}{P}$$

$$= \frac{0.191 \text{ mol} \times 0.08206 \text{ L atm mol}^{-1} \text{ K}^{-1} \times 300 \text{ K}}{0.951 \text{ atm}}$$

$$= 4.94 \text{ L}$$

Check your learning: Sulfur dioxide is an intermediate in the preparation of sulfuric acid. What volume of SO_2 at 343°C and 1.21 atm is produced by burning 1.00 kg of sulfur in oxygen?

Answer: 1.30×10^3 L

Sometimes we can take advantage of a simplifying feature of the stoichiometry of gases that solids and solutions do not exhibit: All gases that show ideal behavior contain the same number of molecules in the same volume (at the same temperature and pressure). Thus the ratios of volumes of gases involved in a chemical reaction are given by the coefficients in the equation for the reaction, provided that the gas volumes are measured at the same temperature and pressure (Section 10.5).

Example 10.18   Volumes of Reacting Gases

Ammonia is an important fertilizer and industrial chemical. In a recent year a volume of 683 billion ft^3 of gaseous ammonia, measured at 25°C and 1 atm, was manufactured in the United States. What volume of $H_2(g)$, measured under the same conditions, was required to prepare this amount of ammonia by reaction with N_2?

$$N_2(g) + 3H_2(g) \longrightarrow 2NH_3(g)$$

Solution: Because equal volumes of H_2 and NH_3 contain equal numbers of molecules, and because every three molecules of H_2 that react produce two molecules of NH_3, the ratio of the volumes of H_2 and NH_3 will be equal to 3:2. Two volumes of NH_3, in units of billion ft^3, will be formed from three volumes of H_2. Thus

$$683 \text{ billion ft}^3 \text{ NH}_3 \times \frac{3 \text{ billion ft}^3 \text{ H}_2}{2 \text{ billion ft}^3 \text{ NH}_3} = 1.02 \times 10^3 \text{ billion ft}^3 \text{ H}_2$$

The manufacture of 683 billion ft^3 of NH_3 required 1020 billion ft^3 of H_2. (At 25°C and 1 atm, this is the volume of a cube with an edge length of approximately 1.9 miles!)

Check your learning: What volume of $O_2(g)$ measured at 25°C and 760 torr is required to react with 17.0 L of ethylene, $C_2H_4(g)$, measured under the same conditions of temperature and pressure? The products are CO_2 and water vapor.

Answer: 51.0 L

10.12   The Atmosphere

The **atmosphere,** or air, is the mixture of gases that surrounds the earth (Fig. 10.22). We cannot say exactly how deep the atmosphere is, but we know that it extends over 1000 kilometers (600 miles) above the earth's surface. The mass of the atmosphere is approximately 6×10^{15} tons. Fifty percent of that lies within 5 kilometers (3 miles)

Figure 10.22

The thin shell of the atmosphere is visible in this picture of the earth.

Component	Percent by Volume
Nitrogen	78.03
Oxygen	20.99
Argon	0.93
Carbon dioxide	0.035
Neon	0.0015
Hydrogen	0.0010
Helium	0.0005
Krypton	0.0001
Xenon	0.000008

Table 10.6

Principal Components of Dry Air at Sea Level

of the surface, 99.99% within 80 kilometers. For the most part the atmosphere consists of uncombined nitrogen, oxygen, and argon (Table 10.6). There are small amounts of water vapor and carbon dioxide, as well as traces of other gases. The percentage of carbon dioxide in the air varies somewhat. Water vapor content varies widely, from about 4% in tropical jungles to very low values in cold and desert climates. Other small but significant variations in the composition of the air result from volcanic and human activities.

The atmosphere is divided into layers on the basis of the way the temperature changes with altitude. The two lowest layers are the troposphere and the stratosphere. The layer nearest the earth, the *troposphere,* ranges in thickness from 10 kilometers in the polar regions to 16 kilometers near the equator. The pressure falls to about 0.1 atmosphere and the temperature to about 220 K at the top of the troposphere. The next region, the *stratosphere,* is where we find the ozone layer that shields the earth from ultraviolet radiation. The stratosphere extends to an altitude of about 60 kilometers, where the temperature is about 260 K and the pressure about 0.001 atmosphere.

The atmosphere is the source of six industrial gases: N_2, O_2, Ar, Ne, Kr, and Xe. These gases are separated by cooling air until it liquefies and then carefully distilling the liquid to separate the components at their different boiling temperatures. The process is similar to the separation of petroleum by fractional distillation (Section 12.15). Nitrogen is used as an inert atmosphere in the iron and steel industry and in food processing, where air would produce unacceptable oxidation of products. Nitrogen is also used in the production of ammonia (Example 10.18), and liquid nitrogen is used as a refrigerant (for temperatures as low as $-196°C$). Pure oxygen has important medical uses, but the largest quantities of oxygen are consumed in steel making (Fig. 10.23). The noble gases Ne, Ar, Kr, and Xe are used in "neon" lights and in applications where a totally nonreactive gas is required.

In addition to useful components, air also contains pollutants, which may be finely divided solids (dusts) or gases. The gaseous pollutants include carbon monoxide, sulfur oxides, nitrogen oxides, ozone, carbon dioxide, hydrocarbons, and chlorofluorocarbons. The chlorofluorocarbons have been manufactured for use as propellants in

Figure 10.23

Much oxygen is used for the manufacture of steel.

spray cans, as cleaning solvents, and as refrigerants, among other uses. They are present in the air because they have been discarded into the atmosphere after being used. The principal sources of the other polluting gases are transportation and the production of electricity.

The combustion of petroleum in vehicles produces CO_2, and incomplete combustion in a poorly tuned engine can also produce CO and unburned hydrocarbon molecules from the petroleum. In addition, the nitrogen oxides NO and NO_2 are produced by the reaction of the nitrogen and oxygen of the air at the high temperatures produced by combustion. When this mixture is trapped in stagnant air, photochemical smog can arise by a variety of complex processes.

Photochemical smog (Fig. 10.24) is a mixture of ozone, O_3, and oxidized hydrocarbons that results from the interaction of sunlight with a mixture of nitrogen oxides and hydrocarbon molecules. During a typical morning rush hour, the sun cleaves NO_2 molecules to produce oxygen atoms.

$$NO_2(g) + \text{light} \longrightarrow NO(g) + O(g)$$

Oxygen atoms are extremely unstable and can react in several ways. The most important reactions that produce smog are

$$O(g) + O_2(g) \longrightarrow O_3(g)$$
$$O(g) + H_2O(g) \longrightarrow 2OH(g)$$

Figure 10.24
Photochemical smog.

The OH group (the hydroxyl radical) is an unstable, very reactive species. Both ozone and the OH radical react with unburned hydrocarbons to form oxidized hydrocarbons, such as acetaldehyde and peroxyacetyl nitrate (PAN), that are responsible for the eye and lung irritation caused by photochemical smog.

The interplay of sunlight, exhaust gases, and travel times is reflected in the creation of photochemical smog. Such smog does not reach serious proportions on cloudy days or during the winter, when sunlight is less intense. On sunny days, however, smog builds during the day, feeding on the gases produced during the morning rush hour. It usually begins to decrease in the evening. Figure 10.25 illustrates how the concentrations of the various components of smog vary during the day. The most serious build-up of smog occurs on those days when smog-filled air is trapped below a

Figure 10.25
Daily variation in the levels of several compounds involved in smog formation.

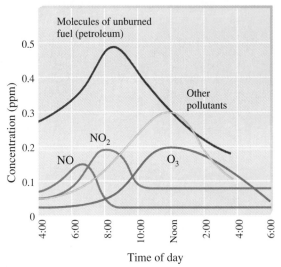

Figure 10.26

Trees damaged by acid rain.

layer of cooler air (an inversion) and the pollutants are unable to rise and disperse into the atmosphere.

We should note that ozone plays two very different roles in the atmosphere. Near the ground, in combination with nitrogen oxides and hydrocarbons, it can produce smog. In the stratosphere, ozone acts as a screen to prevent ultraviolet radiation, an excess of which can cause cancer and interfere with plant growth, from reaching the earth.

The second serious source of pollution is the combustion of coal to produce electricity. Much of the coal found in the Midwest contains sulfides, such as FeS, which produce sulfur dioxide when burned. The equation for the reaction of FeS is

$$4FeS(s) + 7O_2(g) \longrightarrow 4SO_2(g) + 2Fe_2O_3(s)$$

The sulfur dioxide is slowly oxidized to sulfur trioxide,

$$2SO_2(g) + O_2(g) \longrightarrow 2SO_3(g)$$

which combines rapidly with small drops of water in the air to form small drops of a solution of sulfuric acid.

$$SO_3(g) + H_2O(l) \longrightarrow H_2SO_4(aq)$$

These droplets of sulfuric acid may be incorporated into rain, producing **acid rain.**

Rainfall in clean air is slightly acidic as a result of the carbon dioxide in the air. Carbon dioxide dissolves in raindrops and produces carbonic acid, which gives rise to a low concentration of hydronium ion in the water. Depending on the amount of carbon dioxide dissolved, rain in clean air can have a concentration of hydronium ion, H_3O^+, up to 3×10^{-6} M. Acidic rain is rain with a hydronium ion concentration greater than 3×10^{-6} M. Incorporation of sulfuric acid droplets can readily produce rain with higher concentrations of hydronium ion.

Acid rain can increase the acidity of lakes and streams, and hydronium ion concentrations above 1×10^{-5} M can be fatal to fish and other aquatic life. Acid rain also damages trees (Fig. 10.26) and other nonaquatic plant life, as well as building materials (Fig. 10.27).

Figure 10.27

The damage caused by acid rain can be seen by comparing these photographs of the fountain figure "The Duck Girl." The first (left) was taken in 1911, the second (right) after the statue had been outdoors for 75 years.

The Microscopic Behavior of Gases

10.13   The Kinetic-Molecular Theory

The gas laws we have worked with up to this point, as well as the ideal gas equation, are empirical; that is, they have been derived from observations of experimental behavior. We have seen that the mathematical forms of these laws closely describe the macroscopic behavior of most gases at pressures less than about 1 or 2 atm. Although the gas laws describe relationships that have been verified by many experiments, they do not tell us *why* the behavior of gases exhibits these relationships.

Now let us consider a simple microscopic model that has been developed to explain the behavior of an ideal gas. This theory also provides an approximate picture of the behavior of a real gas. Although other, more complicated models are available, we use the ideal gas model for two reasons: It explains the most important aspects of the behavior of gases, and the more complicated models are based on the simple theory.

The **kinetic-molecular theory** is an explanation of the properties of an ideal gas in terms of the behavior of continuously moving molecules that are so small that they can be regarded as having no volume (Fig. 10.28). This theory can be summed up with the following five postulates about the molecules of an ideal gas.

1. Gases are composed of molecules that are in continuous motion. The molecules of an ideal gas move in straight lines and change direction only when they collide with other molecules or with the walls of a container.

2. The molecules of a gas are small compared to the distances between them; molecules of an ideal gas are considered to have no volume. Thus the average distance between the molecules of a gas is large compared to the size of the molecules.

3. The pressure of a gas in a container results from the bombardment of the walls of the container by the molecules of the gas.

Figure 10.28

The kinetic-molecular theory. Very small gas particles move rapidly in a container. Collisions of the particles against the walls of the container produce the pressure.

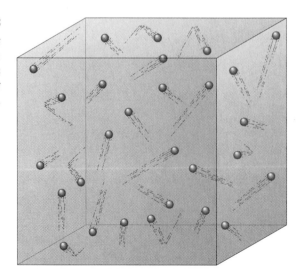

4.   Molecules of an ideal gas are assumed to exert no forces other than collision forces on each other. Thus the collisions among molecules and between molecules and walls must be elastic; that is, the collisions involve no loss of energy due to friction.

5.   The average kinetic energy of the molecules is proportional to the Kelvin temperature of the gas and is the same for all gases at the same temperature.

The test of the kinetic-molecular theory and its postulates is its ability to explain and describe the behavior of a gas. In Sections 10.15 and 10.16 we shall see that the various gas laws can be derived from the assumptions of the theory. This has led chemists to believe that the assumptions of the theory accurately represent the properties of gas molecules. However, before we can proceed, we need to consider the relationship between the average kinetic energy of a collection of gas molecules and the temperature.

10.14   Molecular Velocities and Kinetic Energy

The kinetic energy (Section 1.3) of a moving molecule or other moving particle is a function of its mass and velocity. In effect, kinetic energy is a measure of how hard a particle will hit you—the faster it moves or the heavier it is, the harder it will hit. For example, it hurts a lot more to be hit by a golf ball than by a ping pong ball moving at the same speed, because the golf ball is heavier and has the greater kinetic energy.

We can calculate the kinetic energy (KE) of a particle from its mass (m) and speed (u) by using the equation

$$KE = \frac{1}{2}mu^2$$

If we use units of kilograms for mass and units of meters per second for speed, we get the kinetic energy in units of joules.

The speed and kinetic energy of an individual gas molecule can change when it collides with another molecule. This is identical to an effect you can see on a pool table when two moving pool balls collide; after the collision, one may move faster (more kinetic energy) and the other slower (less kinetic energy). However, the total kinetic energies of the two balls are the same before and after the collision. That is why we call the collision elastic.

An enormous number of collisions per second occur between the molecules in a sample of a gas. Each of the molecules in a sample of hydrogen gas at standard conditions has about 11 billion collisions per second. Each collision can result in an exchange of energy between the molecules involved and can change the speed of these molecules over a wide range. Thus individual molecules in a sample of a gas travel at speeds ranging from practically zero to thousands of meters per second. However, because so many molecules and collisions are involved, the distribution of the molecular speeds of all molecules and their average speed do not vary.

We can see how the molecular speeds in a gas are distributed in any of the curves in Fig. 10.29. The vertical axis gives us the number of molecules that have any of the values of molecular speed, u, plotted along the horizontal axis. The graph shows us that very few molecules move at very low or very high speeds. The number of molecules with intermediate speeds increases rapidly up to a maximum and then drops

Figure 10.29

Distribution of molecular velocities at three different temperatures. The average velocity increases as the temperature increases.

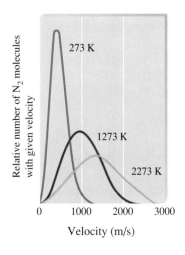

off rapidly. The average of the square of the speeds is u_{avg}^2, and the square root of this average, called the *root-mean-square speed,* is u_{rms} (it is equal to $\sqrt{u_{avg}^2}$).

As we have seen, the kinetic energy of a single molecule is given by the equation $KE = \frac{1}{2}mu^2$. We can find the average kinetic energy, KE_{avg}, of a collection of gas molecules by replacing the square of the speed of a single molecule, u^2, with the average of the square of the speeds of all the molecules, u_{avg}^2. Thus the average kinetic energy of a collection of molecules is given by

$$KE_{avg} = \frac{1}{2}mu_{avg}^2$$

At the temperature changes, the distribution of the molecular speeds in a gas changes. An increase in temperature shifts the distribution to higher speeds. Thus at a higher temperature, more molecules have higher speeds and fewer molecules have lower speeds (Fig. 10.29). Consequently, as the temperature of a gas increases, its average kinetic energy increases. The average kinetic energy of a collection of gas molecules is directly proportional to the temperature of the gas and can be described by the equation

$$KE_{avg} = kT$$

where k is a constant and T is the Kelvin temperature.

Experimental evidence shows that all gases have the same average kinetic energy at the same temperature. Measurements have shown that, on average, lighter molecules, such as hydrogen, move at higher speeds than heavier molecules, such as oxygen (Fig. 10.30). The faster average speeds of the lighter molecules compensate for their smaller masses; thus the kinetic energies are equal.

The Wake Shield Facility that NASA has designed to fly on a space shuttle takes advantage of the distribution of molecular velocities in its design. The performance of advanced thin-film semiconductor devices is greatly enhanced if they can be manufactured in near-perfect vacuum. The Wake Shield satellite (Fig. 10.31) is designed to provide the vacuum required. It is a stainless steel dish 4 meters (12 feet) in diameter) that carries the equipment necessary to fabricate the semiconductor materials. The shield will be deployed into low earth orbit at an altitude of about 300 kilometers. As the Wake Shield orbits the earth, its speed will exceed by a factor of 10 the speed of most of the gaseous atoms and molecules at that altitude. Consequently, it will sweep away nearly all of the residual gas in its path. Most gas particles outside its path will not move fast enough to catch up with the experimental package located at the center behind the shield. At an altitude of 300 kilometers, space is a fairly good

Figure 10.30

Distribution of molecular velocities of different gases. On average, the lighter gases move faster.

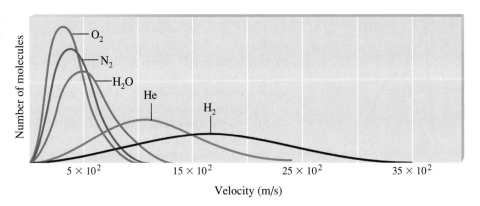

Figure 10.31

The Wake Shield sweeps out residual gases and enhances the vacuum immediately behind it as it orbits the earth.

vacuum with a pressure of about 10^{-7} torr (1×10^{-10} atm). In the vacuum behind the shield, the pressure will be reduced to below 10^{-14} torr. Thus it will be possible to evaporate gallium arsenide or other semiconductor materials from a thermal source and deposit them as thin films on a substrate without a reaction between these extremely active gases and gaseous contaminants.

10.15 Relationship of the Behavior of Gases to the Kinetic-Molecular Theory

We can use the kinetic-molecular theory and its postulates to explain the behavior of a gas as follows:

1. **Boyle's Law.** We have seen that the pressure exerted by a gas in a container is caused by bombardment of the walls of the container by rapidly moving molecules of the gas. The pressure varies directly with the number of molecules that hit a unit area of the wall per unit of time. Reducing the volume of a given mass of gas increases the number of molecules in a given volume. Reducing the volume also decreases the area of the walls of the container. Hence the number of impacts per unit of time on a unit area of wall surface increases, which increases the pressure. These results are in accordance with Boyle's law.

2. **Charles's Law.** We have seen that at constant volume, an increase in temperature increases the pressure of a gas. This increase in pressure reflects the increase in average speed and kinetic energy of the molecules as the temperature is raised. An increase in the average speed results in more frequent and harder impacts on the walls of the container—that is, it results in greater pressure.

    If we wish to keep the force per unit area (the pressure) constant as the temperature increases, we can increase the volume so that each molecule travels farther, on average, before hitting a wall and so that the total area of the walls is larger. The greater number of unit areas combined with the smaller number of molecules striking each unit area of wall at a given time offsets the greater force with which each molecule hits, making it possible for the pressure to remain constant.

3. **Dalton's Law.** Because of the relatively large distance between the molecules of a gas, the molecules of one component of a mixture of gases bombard the walls of the container with the same frequency in the presence of other kinds of molecules as in their absence. Thus the total pressure of a mixture of gases equals the sum of the partial pressures of the individual gases.

4. **Graham's Law.** The molecules of a gas are in rapid motion, and the molecules themselves are small. The average distance between the molecules of a gas is large compared to the size of the molecules. As a consequence, gas molecules can move past each other easily and they readily diffuse.

    The rate of effusion of a gas depends on the mass of its molecules, because at the same temperature, molecules of all gases have the same average kinetic energy. For two different gases at the same temperature,

$$\text{Average kinetic energy for first gas} = \frac{1}{2}m_1(u_1^2)_{\text{avg}}$$

$$\text{Average kinetic energy for second gas} = \frac{1}{2}m_2(u_2^2)_{\text{avg}}$$

where m_1 and m_2 are the masses of individual molecules of the two gases and $(u_1^2)_{\text{avg}}$ and $(u_2^2)_{\text{avg}}$ are the averages of the squares of their speeds.

According to the kinetic-molecular theory, the kinetic energies of the two gases are equal. Hence

$$\frac{1}{2}m_1(u_1^2)_{\text{avg}} = \frac{1}{2}m_2(u_2^2)_{\text{avg}}$$

Multiplying both sides of the equation by 2 gives

$$m_1(u_1^2)_{\text{avg}} = m_2(u_2^2)_{\text{avg}}$$

Rearranging and taking square roots of both sides gives

$$\frac{(u_1)_{\text{rms}}}{(u_2)_{\text{rms}}} = \frac{\sqrt{m_2}}{\sqrt{m_1}}$$

where $(u_1)_{\text{rms}}$ and $(u_2)_{\text{rms}}$ are the root-mean-square speeds of the respective molecules (Section 10.14). The rate at which a gas effuses, R, is proportional to the root-mean-square speed of its molecules. The molar mass, M, of a gas is proportional to the mass of the individual molecules. Thus

$$\frac{R_1}{R_2} = \frac{\sqrt{M_2}}{\sqrt{M_1}}$$

The ratio of the rates of effusion is thus proportional to the ratio of the root-mean-square speeds of the molecules or to the inverse ratio of the square roots of their masses. This is Graham's law.

Hydrogen is the lightest gas and therefore effuses the most rapidly. The root-mean-square speed of its molecules at room temperature is about 1 mile per second; that of oxygen molecules is about $\frac{1}{4}$ mile per second. Collisions between molecules make diffusion rates much lower than would be expected from such speeds. A molecule in hydrogen gas at standard conditions has about 11 billion collisions per second. Thus hydrogen molecules travel about 1.7×10^{-5} centimeter between collisions and, on average, collide about 60,000 times in traveling between two points 1 centimeter apart.

10.16   Derivation of the Ideal Gas Equation from Kinetic-Molecular Theory

In this section we will see how the ideal gas equation, $PV = nRT$, can be derived from the kinetic-molecular theory of gases. Let us assume that we have N molecules, each with a mass m, in a cubical container, as shown in Fig. 10.32, with an edge length of l centimeters. Although the molecules are moving in all possible directions, we may assume that, on average, $\frac{1}{3}$ of the molecules ($\frac{1}{3}N$) are moving in the direction of the x axis, $\frac{1}{3}$ in the direction of the y axis, and $\frac{1}{3}$ in the direction of the z axis. This assumption is valid, because the motion of the molecules is entirely random and no particular direction is preferred.

Initially, let us assume that all molecules are moving with the same speed, and let us consider the collisions of molecules with wall A (Fig. 10.32). Each molecule, on average, travels $2l$ centimeters between two consecutive collisions with wall A as it moves back and forth across the container. If the speed of the molecule is u centimeters per second, it will collide $\frac{u}{2l}$ times per second with wall A. The collisions are perfectly elastic (Section 10.13), so the molecule will rebound with a speed of $-u$, having lost no kinetic energy as a result of the collision. Because momentum is defined as the product of mass and speed, the average momentum before a collision is mu and after a collision is $-mu$. Thus the average change in momentum per molecule per collision is $2mu$. There are $\frac{u}{l}$ collisions per second, making the average change in momentum per molecule per second

$$2mu \times \frac{u}{2l} = \frac{mu^2}{l}$$

The total change in momentum per second for the $\frac{N}{3}$ molecules that can collide with wall A is

$$\frac{N}{3} \times \frac{mu^2}{l}$$

Figure 10.32

The container used in the derivation of the ideal gas equation from kinetic-molecular theory.

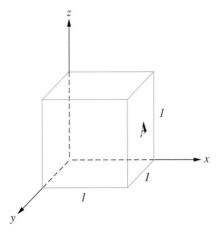

This is the average force on wall A, because force is equal to change of momentum per second. Pressure is force per unit area. Because the area of wall A is l^2, the pressure P on wall A is

$$P = \frac{force}{area} = \frac{l}{l^2} \times \frac{N}{3} \times \frac{mu^2}{l} = \frac{N}{3} \times \frac{mu^2}{l^3}$$

Because l^3 is equal to the volume, V, of the cubical container, we have

$$P = \frac{N}{3} \times \frac{mu^2}{V}$$

or

$$PV = \tfrac{1}{3}Nmu^2$$

Actually, molecules move with different speeds, so we must use the average of the squares of their speeds. Thus

$$PV = \tfrac{1}{3}Nm(u^2)_{avg}$$

This is the fundamental equation of the kinetic-molecular theory of gases. It describes the behavior of an ideal gas exactly, and it is a good approximation for the behavior of real gases at ordinary pressures.

The equation above may be written as follows:

$$PV = \tfrac{2}{3}N[\tfrac{1}{2}m(u^2)_{avg}]$$

The term in the set of brackets, $\tfrac{1}{2}m(u^2)_{avg}$, is the expression for the average kinetic energy of the gas, KE_{avg} (Section 10.14). Thus

$$PV = \tfrac{2}{3}N \, (KE_{avg})$$

The average kinetic energy of the molecules, KE_{avg}, is directly proportional to the Kelvin temperature, T.

$$KE_{avg} = kT$$

The number of molecules, N, is proportional to the number of moles of molecules, n.

$$N = k'n$$

Substituting the expressions for KE_{avg} and N in the equation for PV gives

$$PV = \tfrac{2}{3}(k'n)(kt) = n(\tfrac{2}{3}k'k)T$$

The expression $\tfrac{2}{3}k'k$ is the constant (being made up entirely of constants) that we write as R and call the gas constant. Therefore,

$$PV = nRT$$

Thus we can derive the ideal gas equation from the assumptions of the kinetic-molecular theory (Section 10.14). Because this equation describes the behavior of gases very well, the assumptions used in its derivation are strongly supported.

10.17   Deviations from Ideal Gas Behavior: Descriptions of Real Gases

Molecules of an ideal gas have no significant volume and do not attract each other (Section 10.13). Real gases approximate this behavior at low pressures and elevated temperatures.

When we compress a real gas at ordinary temperature and pressure, the volume is reduced as the molecules are crowded closer together. This reduction in volume is really a reduction in the amount of empty space between the molecules. If we do the compression at high pressures, the molecules are crowded so close together that the actual volume of the molecules—the molecular volume—is a relatively large fraction of the total volume, including the empty space between molecules. Because we cannot compress the molecules themselves, only a fraction of the entire volume is affected by a further increase in pressure. Thus at higher pressures the whole volume is not inversely proportional to the pressure, as predicted by Boyle's law.

The molecules in a real gas at relatively low pressures have practically no attraction for one another, because they are far apart. Thus they behave almost like molecules of ideal gases. However, if we crowd the molecules closer together by increasing the pressure, then the effect on the force of attraction between the molecules increases. This has the same effect as an increase in external pressure. Consequently, when we apply an external pressure to a volume of gas, especially at low temperatures, we find a slightly greater decrease in volume than should be achieved by the pressure alone. This slightly greater decrease in volume caused by intermolecular attraction is more pronounced at low temperatures because the molecules move more slowly, their kinetic energy is smaller relative to the attractive forces, and they fly apart less easily after collisions with one another.

There are several different equations that are better approximations of the behavior of a gas than the ideal gas equation. These equations include the effects of the volume of the molecules and the attractions between molecules in describing the relationship among pressure, volume, and temperature. The simplest of these approximations was reported by the Dutch scientist Johannes van der Waals in 1879.

The ideal gas equation predicts that the pressure of an ideal gas is

$$PV = nRT$$

The van der Waals equation corrects for the volume of the gas molecules and the attractive forces between them by adding two terms to this equation:

$$\left(P + \frac{n^2 a}{V^2}\right)(V - nb) = nRT$$

Correction for molecular attraction

Correction for volume of molecules

where a and b are constants that depend on the gas, and the other terms have their usual meaning in a gas equation. Tables of experimental values of the van der Waals constants a and b for many different gases appear in most chemistry handbooks.

We use the constant a to introduce the effect of the attraction between molecules, which van der Waals assumed varied inversely as the square of the total volume of the gas. (See Sections 11.2 and 11.3 for a discussion of the nature of the van der Waals force.) Because this force augments the pressure and thus tends to make the volume smaller, we add it to the term P.

We use the constant b to represent the total volume of the molecules themselves; it is subtracted from V, the total volume of the gas. When V is large, both nb and $n^2 a/V^2$ are negligible, and the van der Waals equation reduces to the simple gas equation, $PV = nRT$.

At low pressures the correction for intermolecular attraction, a, is more important than the correction for molecular volume, b. At high pressures the correction for the volume of the molecules becomes important, because the molecules themselves are incompressible and constitute an appreciable fraction of the total volume. At some intermediate pressure, the two corrections cancel each other, and the gas appears to follow the relationship given by $PV = nRT$ over a small range of pressures.

Strictly speaking, then, we should expect the ideal gas equation to apply exactly only to gases whose molecules do not attract one another and do not occupy an appreciable part of the whole volume. Because no real gases have these properties, we speak of such hypothetical gases as ideal gases. Under ordinary conditions, however, the deviations from the gas laws are small enough that they may be disregarded.

For Review   Summary

A gas takes the shape and volume of its container. Gases are readily compressible and capable of infinite expansion. The pressure of a gas may be expressed (in terms of force per unit area) in the SI units of pascals or kilopascals. Other units used for pressure include torr and atmospheres (1 atm = 760 torr = 101.325 kPa).

The behavior of real gases can be described by a number of approximate laws based on experimental observations of their properties. A gas that follows these laws exactly is called an **ideal gas.** The volume of a given quantity of a gas is inversely proportional to the pressure of the gas, provided that the temperature does not change (**Boyle's law**). The volume of a given quantity of a gas is directly proportional to its temperature on the Kelvin scale, provided that the pressure does not change (**Charles's law**). Under the same conditions of temperature and pressure, equal volumes of all gases contain the same number of molecules (**Avogadro's law**). The rates of diffusion of gases are inversely proportional to the square roots of their densities or to the square roots of their molar masses (**Graham's law**). In a mixture of gases, the total pressure is equal to the sum of the partial pressures of the individual gases present. The partial pressure of a gas is the pressure that the gas would exert if it were the only gas present. The volumes and densities of gases are often reported under **standard conditions** of temperature and pressure (STP): 0°C and 1 atmosphere.

The equations describing Boyle's law, Charles's law, and Avogadro's law are special cases of the **ideal gas equation,** $PV = nRT$, where P is the pressure of the gas, V is its volume, n is the number of moles of the gas, and T is its Kelvin temperature. R is the gas constant. The value and units we use for the gas constant in the equation depend on the units used for P and V.

The ideal gas equation can be derived by applying the assumptions of the **kinetic-molecular theory** of gases. This theory assumes that gases consist of molecules of negligible volume, which are so widely separated that there is no attraction between them. The molecules move randomly and change direction only when they collide with one another or with the walls of the container. These collisions are elastic. The pressure of a gas results from bombardment of the walls of the container by the gas molecules. The average kinetic energy of the molecules is proportional to the temperature of the gas and is the same for all gases at a given temperature.

The molecules of a gas do not all move at the same speed at the same instant of time. Some move relatively slowly, others relatively rapidly. The speeds are distributed over a wide range. The average of the squares of the molecular speeds, $(u^2)_{avg}$, is directly proportional to the temperature of the gas on the Kelvin scale and inversely proportional to the masses of the individual molecules. Heating a gas increases both the value of $(u^2)_{avg}$ and the number of molecules moving at higher speeds. As shown by Graham's law, heavier molecules effuse more slowly than lighter ones. Heavier molecules move more slowly, on average.

The molecules in a real gas (in contrast to those in an ideal gas) possess a finite volume and attract each other slightly. At relatively low pressures the molecular volume can be neglected. And at temperatures well above the temperature at which the gas liquefies, the attractions between molecules can be neglected. Under these conditions the ideal gas equation is a good approximation of the behavior of a real gas. However, at lower temperatures, higher pressures, or both, corrections for molecular volume and molecular attractions are required. The **van der Waals equation** can be used to describe the behavior of real gases under these conditions.

Key Terms and Concepts

absolute zero (10.4)	Boyle's law (10.2, 10.15)	direct proportionality (10.3)
acid rain (10.12)	Charles's law (10.3, 10.15)	effusion (10.10, 10.15)
atmosphere (10.12)	Dalton's law of partial pressures (10.9, 10.15)	gas constant (10.6)
Avogadro's law (10.5)	diffusion (10.10, 10.15)	Graham's law (10.10, 10.15)
barometer (10.1)		Kelvin temperature (10.4)

kinetic energy   (10.14)
kinetic-molecular theory   (10.13)
manometer   (10.1)
ideal gas   (10.2, 10.13)
ideal gas equation   (10.6, 10.16)
inverse proportionality   (10.2)

partial pressure   (10.9)
photochemical smog   (10.12)
pressure   (10.1)
rate of diffusion   (10.10, 10.15)
rate of effusion   (10.10, 10.15)
real gas   (10.2, 10.3, 10.17)

standard conditions (STP)   (10.7)
standard molar volume   (10.7)
van der Waals equation   (10.17)
volume   (10.2, 10.3, 10.5)

Exercises

Assume the gases in the following exercises exhibit ideal behavior, unless the exercise indicates otherwise.

Questions

1. Explain why cooling a balloon as shown in Fig. 10.10 causes it to shrink.

2. Explain why and how the volumes of the bubbles exhausted by a scuba diver (Fig. 10.8) change as they rise to the surface, assuming that they do not break up.

3. An alternative way to state Boyle's law is "All other things being equal, the pressure of a gas is inversely proportional to its volume."

    (a) What is the meaning of the phrase *inversely proportional*?

    (b) What are the "other things" that must be equal?

4. An alternative way to state Avogadro's law is "All other things being equal, the number of molecules in a gas is directly proportional to the volume of the gas."

    (a) What is the meaning of the phrase *directly proportional*?

    (b) What are the "other things" that must be equal?

5. Which will sink to the table first, a floating balloon filled with helium gas, like the one in Fig. 10.21, or an identical balloon filled with the same volume of hydrogen gas instead of helium? Explain your answer.

6. Explain why the number of molecules are not identical in the left-hand and right-hand bulbs shown in Fig. 10.20 (B).

7. The distribution of molecular velocities in a sample of helium is shown in Fig. 10.30. If the sample is cooled, will the distribution of velocities look more like that of H_2 or that of H_2O? Explain your answer.

8. In addition to the data found in Fig. 10.12, what other information do we need to find the mass of the sample of He used to determine the line for He?

9. How would Fig. 10.7 (B) change if the number of moles of gas in the sample used to determine the curve were doubled?

10. Using the postulates of the kinetic-molecular theory, explain why a gas uniformly fills a container of any shape.

11. Can the speed of a given molecule in a gas double at constant temperature? Explain your answer.

12. Describe what happens to the average kinetic energy of ideal gas molecules when the conditions are changed as follows:

    (a) The pressure of the gas is increased by reducing the volume at constant temperature.

    (b) The pressure of the gas is increased by increasing the temperature at constant volume.

    (c) The average velocity of the molecules is increased by a factor of 2.

13. Explain why the plot of PV for CO_2 (Fig. 10.9) differs from that of an ideal gas.

14. Explain why polluting chlorofluorocarbon molecules do not simply fall out of the atmosphere even though the average molecular mass of an air molecule is about 29 amu and the molecular mass of a common chlorofluorocarbon molecule is 121 amu.

15. Which of the following equations involve(s) microscopic properties? Which involve(s) macroscopic properties? Do any involve both?

    (a) $PV = nRT$

    (b) $P_T = P_A + P_B + P_C$

    (c) $PV = k$

    (d) $u_{rms} = \sqrt{(u^2)_{avg}}$

    (e) $\dfrac{R_1}{R_2} = \sqrt{\dfrac{M_2}{M_1}}$

    (f) $KE = \frac{1}{2}mv^2$

16. Which of the following figures illustrate a macroscopic feature of gases, and which illustrate a microscopic feature? Figs. 10.2, 10.6, 10.9, 10.13, 10.19, 10.25, 10.28

Pressure

17. A typical barometric pressure in Denver, Colorado, is 615 mm Hg. What is this pressure in atmospheres and in kilopascals?

18. A typical barometric pressure in Kansas City is 740 torr. What is this pressure in atmospheres, in millimeters of mercury, and in kilopascals?

19. Canadian tire gauges are marked in units of kilopascals. What reading on such a gauge corresponds to 32 lb / in^2 (psi)? (1 atm = 14.7 psi)

20. A medical laboratory catalog describes the pressure in a cylinder of a gas as 14.82 MPa (1 MPa = 10^6 Pa). What is the pressure of this gas in atmospheres and in torr?

21. The barometric pressure at sea level is sometimes about 750 mm Hg. Calculate the pressure in atmospheres and kilopascals.

22. (a) On a mid-August day in the northeastern United States, the following information appeared in the local newspaper: atmospheric pressure at sea level, 29.97 in. What was the pressure in kilopascals?
    (b) The pressure near the seacoast in the northeastern United States is usually about 30.0 in. During a hurricane, however, the pressure may fall to near 28.0 in. Calculate this lower pressure in torr.

23. During the Viking landings on Mars, the atmospheric pressure was determined to be on average about 6.50 millibars (1 bar = 0.987 atm). What is that pressure in torr and in kilopascals?

24. The pressure of the atmosphere on the surface of the planet Venus is about 88.8 atm. Compare that pressure in pounds per square inch to the normal pressure on earth at sea level in pounds per square inch.

25. Arrange the following gases in order of increasing pressure: He at 375 torr, N_2O at 0.25 atm, CO_2 at 198 Pa.

26. Arrange the following gases in order of increasing pressure: H_2 at 250 torr, N_2 at 1.4 atm, O_2 at 98 kPa.

The Gas Laws

27. What is the height of the mercury column in Fig. 10.6 if the volume of the gas is reduced to 1.20 mL?

28. Determine the pressure of the gas in the container shown in Fig. 10.6 when its volume is 0.66 mL (a) using Boyle's law and (b) using a graph of $\frac{1}{P}$ against V prepared from the data in the figure.

29. Determine the volume of 1 mol of CH_4 gas at 200 K and 1 atm, using Fig. 10.11.

30. Say the gas in Fig. 10.6 is pure nitrogen instead of air. What is the mass of gas in the container if the temperature is 23°C?

31. The volume of a sample of carbon monoxide, CO, is 405 mL at 10.0 atm and 467 K. What volume will it occupy at 4.29 atm and 467 K?

32. A sample of oxygen, O_2, occupies 32.2 mL at 0°C and 917 torr. What volume will it occupy at 0°C and 760 torr?

33. What is the volume of a sample of carbon monoxide, CO, at 3°C and 744 torr if it occupies 13.3 L at 55°C and 744 torr?

34. A 2.50-L volume of hydrogen measured at the normal boiling point of nitrogen, −196°C, is warmed to the normal boiling point of water, 100°C. Calculate the new volume of the gas, assuming ideal behavior and no change in pressure.

35. A high-altitude balloon is filled with 1.41×10^4 L of hydrogen at a temperature of 21°C and a pressure of 745 torr. What is the volume of the balloon at a height of 20 km, where the temperature is −48°C and the pressure is 63.1 torr?

36. A cylinder of oxygen for medical use contains 35.4 L of oxygen at a pressure of 149.6 atm at 25°C. What is the pressure of this oxygen if the cylinder is stored in a warehouse with no air conditioning at a temperature of 49.5°C?

37. A spray can is used until it is empty except for the propellant gas, which has a pressure of 1344 torr at 23°C. If the can is thrown into a fire (T = 475°C), what will be the pressure in the hot can?

38. Assume the pressure remains the same and determine the temperature of the gas in the balloon in Fig. 10.10 when the balloon shrinks to one-half its original volume.

The Ideal Gas Equation and Its Applications

39. Assume the temperature is 25°C, and determine the approximate number of moles of gas contained in the sample of gas used in Fig. 10.6.

40. Assume the pressure is 0.917 atm, and determine the approximate number of moles of H_2O contained in the sample used to gather data for Fig. 10.12.

41. How many moles of gaseous boron trifluoride, BF_3, are contained in a 4.3410-L bulb at 788.0 K if the pressure is 1.220 atm? How many grams of BF_3?

42. How many moles of chlorine gas, Cl_2, are contained in a 10.3-L tank at 21.2°C if the pressure is 633 torr? How many grams of Cl_2?

43. What is the volume of a bulb that contains 8.17 g of helium, He, at 13°C and a pressure of 8.73 atm?

44. What is the volume of a bulb that contains 8.17 g of neon at 13°C and a pressure of 8.73 atm?

45. What is the temperature of a 0.274-g sample of methane, CH_4, confined in a 300.0-mL bulb at a pressure of 198.7 kPa?

46. Iodine, I_2, is a solid at room temperature but sublimes (converts to a gas) when warmed. What is the temperature in a 73.3-mL bulb that contains 0.292 g of I_2 vapor at a pressure of 0.462 atm?

47. What is the pressure, in atmospheres, in a 3.785-L tank (a 1-gal tank) that contains 453.6 g (1 lb) of CO_2 at a temperature of 33°C?

48. The volume of an automobile air bag was 66.8 L when it was inflated at 25°C with 77.8 g of nitrogen gas. What was the pressure in the bag in kilopascals?

49. What is the density of laughing gas, dinitrogen monoxide, N_2O, at a temperature of 325 K and a pressure of 113.0 kPa?

50. Calculate the density of Freon 12, CF_2Cl_2, at 30.0°C and 0.954 atm.

51. How many moles of oxygen gas, O_2, are contained in a cylinder of medical oxygen with a volume of 35.4 L, a pressure of 151 atm, and a temperature of 30.5°C?

52. A small cylinder of SF_6 for use in chemistry lectures has a volume of 0.334 L. How many moles of SF_6 are contained in such a cylinder at a pressure of 154 atm and a temperature of 22°C?

53. How may grams of gas are present in each of the following cases?
    (a) 0.100 L of CO_2 at 307 torr and 26°C
    (b) 8.75 L of C_2H_4 at 378.3 kPa and 483 K
    (c) 221 mL of Ar at 0.23 torr and −54°C

54. (a) What is the concentration of the atmosphere in molecules per milliliter at STP?
    (b) At a height of 150 km (about 94 mi), the atmospheric pressure is 3.0×10^{-6} torr and the temperature is 420 K. What is the concentration of the atmosphere in molecules per milliliter at 150 km?

55. A 20.0-L cylinder contained 11.34 kg of butane. Calculate the mass of the gas remaining in the cylinder if it was opened and the gas escaped until the pressure in the cylinder was equal to the atmospheric pressure, 0.983 atm, at a temperature of 27°C.

56. Calculate the volume occupied by 1.00 g of water vapor at 100°C and a pressure of 1.00 atm. Estimate the percentage of the gas volume occupied by molecules. Water molecules are as closely packed as possible at 1.00 atm in liquid water with a density of 1.000 g mL^{-1}.

Dalton's Law of Partial Pressures

57. If all three of the gases in Fig. 10.15 were placed together in a 11.2-L container at 0°C, what would the pressure be if no reaction occurred?

58. What is the temperature of the water if the partial pressure of the gas contained in the bottle shown in Fig. 10.18 is 720 torr on a day when the atmospheric pressure is 745 torr?

59. A 5.73-L flask at 25°C contains 0.0388 mol of N_2, 0.147 mol of CO, and 0.0803 mol of H_2. What is the pressure in the flask in atmospheres, in torr, and in kilopascals?

60. A 36.0-L cylinder of a gas used for calibration of blood gas analyzers in medical laboratories contains 350 g of CO_2, 805 g of O_2, and 4880 g of N_2. What is the pressure in the cylinder in atmospheres, in torr, and in kilopascals?

61. A cylinder of a gas mixture used for calibration of blood gas analyzers in medical laboratories contains 5.0% CO_2, 12.0% O_2, and the remainder N_2 at a total pressure of 146 atm. What is the partial pressure of each component of this gas? (The percentages given indicate the percent of the total pressure that is due to each component.)

62. A sample of gas isolated from unrefined petroleum contains 90.0% CH_4, 8.9% C_2H_6, and 1.1% C_3H_8 at a total pressure of 307.2 kPa. What is the partial pressure of each component of this gas? (The percentages given indicate the percent of the total pressure that is due to each component.)

63. A sample of carbon monoxide was collected over water at a total pressure of 756 torr and a temperature of 18°C. What is the pressure of the carbon monoxide?

64. A sample of hydrogen collected over water at 27°C has a pressure of 1.087 atm. What is the pressure of the hydrogen?

65. The volume of a sample of a gas collected over water at 30.0°C and 0.932 atm is 627 mL. What will the volume of the gas be when it is dried and measured at STP?

66. Most mixtures of hydrogen gas with oxygen gas are explosive. However, a mixture that contains less than 3.0% O_2 is not. If enough O_2 is added to a cylinder of H_2 at 33.2 atm to bring the total pressure to 34.5 atm, is the mixture explosive?

67. Which is denser at the same temperature and pressure, dry air or air saturated with water vapor? Explain.

68. A mixture of 0.200 g of H_2, 1.00 g of N_2, and 0.820 g of Ar is stored in a closed container at STP. Find the volume of the container, assuming ideal gas behavior.

Stoichiometry Involving Gases

69. Joseph Priestley first prepared pure oxygen by heating mercuric oxide, HgO.

$$2HgO(s) \xrightarrow{\Delta} 2Hg(l) + O_2(g)$$

    (a) Outline the steps necessary to answer the following question: What volume of O_2 at 23°C and 0.975 atm is produced by the decomposition of 5.36 g of HgO?
    (b) Answer the question.

70. Lime, CaO, is produced by heating calcium carbonate, $CaCO_3$; carbon dioxide is the other product.

    (a) Outline the steps necessary to answer the following question: What volume of carbon dioxide at 875°C and 0.966 atm is produced by the decomposition of 1 ton (1.000×10^3 kg) of calcium carbonate?
    (b) Answer the question.

71. Cavendish prepared hydrogen in 1766 by the novel method of passing steam through a red-hot gun barrel.

$$4H_2O(g) + 3Fe(s) \xrightarrow{\Delta} Fe_3O_4(s) + 4H_2(g)$$

    (a) Outline the steps necessary to answer the following question: What volume of H_2 at a pressure of 745 torr and a temperature of 20°C can be prepared from the reaction of 15.0 g of H_2O?
    (b) Answer the question.

72. Before small batteries were available, carbide lamps were used for bicycle lights. Acetylene gas, C_2H_2, and solid calcium hydroxide were formed by the reaction of calcium carbide CaC_2 with water. The ignition of the acetylene gas provided the light. Currently, the same lamps are used by some cavers, and calcium carbide is used to produce acetylene for carbide cannons.

    (a) Outline the steps necessary to answer the following question: What volume of C_2H_2 at 1.005 atm and 12.2°C is formed by the reaction of 15.48 g of CaC_2 with water?
    (b) Answer the question.

73. What volume of O_2 at STP is required to oxidize 8.0 L of NO at STP to NO_2? What volume of NO_2 is produced at STP?

74. Calculate the volume of oxygen required to burn 12.00 L of ethane gas, C_2H_6, to produce carbon dioxide and water, if the volumes of C_2H_6 and O_2 are measured under the same conditions of temperature and pressure.

75. (a) Outline the steps necessary to answer the following question: If the oxygen consumed by a resting human male (Section 10.6, Example 10.9) is used to produce energy via the oxidation of glucose,

$$C_6H_{12}O_6 + 6O_2 \longrightarrow 6CO_2 + 6H_2O$$

    what is the mass of glucose required per hour?
    (b) Answer the question.

76. The chlorofluorocarbon CCl_2F_2 can be recycled into a different compound by reaction with hydrogen to produce $CH_2F_2(g)$, a compound useful in chemical manufacturing.

$$CCl_2F_2(g) + 4H_2(g) \longrightarrow CH_2F_2(g) + 2HCl(g)$$

    (a) Outline the steps necessary to answer the following question: What volume of hydrogen at 225 atm and 35.5°C would be required to react with 1 ton (1.000 × 10^3 kg) of CCl_2F_2?
    (b) Answer the question.

77. What is the molar mass of a gas if 0.0494 g of the gas occupies a volume of 0.100 L at a temperature 26°C and a pressure of 307 torr?

78. What is the molar mass of a gas if 0.281 g of the gas occupies a volume of 125 mL at a temperature of 126°C and a pressure of 777 torr?

79. Methanol, CH_3OH, is produced industrially by the following reaction

$$CO(g) + 2H_2(g) \xrightarrow[300°C, 300\ atm]{\text{Copper catalyst}} CH_3OH(g)$$

Assuming that the gases behave as ideal gases, find the ratio of the total volume of the reactants to the final volume.

80. What volume of oxygen at 423.0 K and a pressure of 127.4 kPa is produced by the decomposition of 129.7 g of BaO_2 to BaO and O_2?

81. A 2.50-L sample of a colorless gas at STP decomposed to give 2.50 L of N_2 and 1.25 L of O_2 at STP. What is the colorless gas?

82. A cylinder of $O_2(g)$ used in breathing by emphysema patients has a volume of 3.00 L at a pressure of 10.0 atm. If the temperature of the cylinder is 28.0°C, what mass of oxygen is in the cylinder?

83. Automobile air bags are inflated with nitrogen gas, which is formed by the decomposition of solid sodium azide (NaN_3). The other product is sodium metal. Calculate the volume of nitrogen gas at 27°C and 756 torr formed by the decomposition of 125 g of sodium azide.

84. How could you show experimentally that the molecular formula of propene is C_3H_6, not CH_2?

85. The density of a certain gaseous fluoride of phosphorus is 3.93 g L^{-1} at STP. Calculate the molar mass of this fluoride, and determine its molecular formula.

86. In an experiment in a general chemistry laboratory, a student collected a sample of a gas over water. The volume of the sample was 265 mL at a pressure of 753 torr and a temperature of 27°C. The mass of the gas in the sample was 0.472 g. What was the molecular mass of the gas?

87. In a laboratory determination, a 0.1009-g sample of a compound containing boron and chlorine gave 0.3544 g of silver chloride upon reaction with silver nitrate. A 0.06237-g sample of the compound exerted a pressure of 6.52 kPa at 27°C in a volume of 147 mL. What is the molecular formula of the compound?

Kinetic-Molecular Theory; Graham's Law

88. Show how Boyle's, Charles's, and Dalton's laws follow from the assumptions of the kinetic-molecular theory.

89. What is the ratio of the average kinetic energy of a SO_2 molecule to that of an O_2 molecule in a mixture of the two gases? What is the ratio of the root-mean-square speeds, u_{rms}, of the two gases?

90. A 1-L sample of CO initially at STP is heated to 546°C, and its volume is increased to 2 L.

    (a) What effect do these changes have on the number of collisions of the molecules of the gas per unit area of the container wall?
    (b) What is the effect on the average kinetic energy of the molecules?
    (c) What is the effect on the root-mean-square speed of the molecules?

91. The root-mean-square speed of H_2 molecules at 25°C is about 1.6 km s^{-1}. What is the root-mean-square speed of a N_2 molecule at 25°C?

92. Show that the ratio of the rate of diffusion of gas 1 to the rate of diffusion of gas 2, $\frac{R_1}{R_2}$, is the same at 0°C and 100°C.

93. Heavy water, D_2O (molar mass, 20.03), can be separated from ordinary water, H_2O (molar mass 18.01), as a result of the difference in the relative rates of diffusion of the molecules in the gas phase. Calculate the relative rates of diffusion of H_2O and D_2O.

94. Which of the following gases diffuse more slowly than oxygen? F_2, Ne, N_2O, C_2H_2, NO, Cl_2, H_2S

95. Calculate the relative rate of diffusion of H_2 compared to that of D_2 (molar mass, 4.0) and the relative rate of diffusion of O_2 (molar mass, 32) compared to that of O_3 (molar mass, 48).

96. A gas of unknown identity effuses at the rate of 83.3 mL s^{-1} in an effusion apparatus in which carbon dioxide effuses at the rate of 102 mL s^{-1}. Calculate the molecular mass of the unknown gas.

97. When two cotton plugs, one moistened with ammonia and the other with hydrochloric acid, are simultaneously inserted into opposite ends of a glass tube 87.0 cm long, a white ring of NH_4Cl forms where gaseous NH_3 and gaseous HCl first come into contact.

$$NH_3(g) + HCl(g) \longrightarrow NH_4Cl(s)$$

At what distance from the ammonia-moistened plug does this occur?

Nonideal Behavior of Gases

98. Graphs showing the behavior of several different gases follow. Which of these gases exhibit(s) behavior significantly different from that expected for ideal gases?

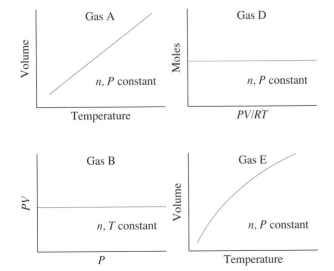

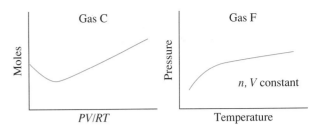

99. Under which of the following sets of conditions does a real gas behave most like an ideal gas, and for which conditions is a real gas expected to deviate from ideal behavior? Explain.

   (a) High pressure, small volume
   (b) High temperature, low pressure
   (c) Low temperature, high pressure

100. Describe the factors responsible for the deviation of the behavior of real gases from that of an ideal gas.

101. For which of the following gases should the correction for the molecular volume be largest: CO, CO_2, H_2, He, NH_3, or SF_6?

Applications and Additional Exercises

102. For 1 mol of H_2 showing ideal gas behavior, draw labeled graphs of

   (a) The variation of P with V at T = 273 K
   (b) The variation of V with T at P = 1.00 atm
   (c) The variation of P with T at V = 22.4 L
   (d) The variation of the average velocity of the gas molecules with T
   (e) The variation of $\frac{1}{P}$ with V at T = 273 K

103. A liter of methane gas, CH_4, at STP contains more atoms of hydrogen than does a liter of pure hydrogen gas, H_2, at STP. Using Avogadro's law as a starting point, explain why.

104. The effect of chlorofluorocarbons (such as CCl_2F_2) on the depletion of the ozone layer is well known. One of the solutions to the problem is to use substitutes, such as $CH_3CH_2F(g)$, for the chlorofluorocarbons. Calculate the volume occupied by 10.0 g of each of the compounds $CCl_2F_2(g)$ and $CH_3CH_2F(g)$ at STP.

105. (a) Outline the steps necessary to answer the following question: What is the molecular formula of a compound that contains 39% C, 45% N and 16% H if 0.157 g of the compound occupies 125 mL with a pressure of 99.5 kPa at 22°C?
    (b) Answer the question.

106. (a) Outline the steps necessary to answer the following question: What are the molecular geometry and the hybridization of the carbon atoms in a compound with a molar mass of about 28 that contains 85.7% carbon and 14.3% hydrogen?
    (b) Answer the question.

107. (a) Outline the steps necessary to answer the following question: What is the geometry about the carbon atoms in a compound with a molar mass of about 44 that contains 81.7% carbon and 18.3% hydrogen?
(b) Answer the question.

108. (a) Outline the steps necessary to answer the following questions: What are the molecular geometry and the hybridization of the N atom in a compound with a molar mass of about 70 that contains 19.7% nitrogen and 80.3% fluorine? What are the formal charge and the oxidation number of the atoms in this compound?
(b) Answer the questions.

109. A commercial mercury vapor analyzer can detect, in air, concentrations of gaseous Hg atoms (which are poisonous) as low as 2×10^{-6} mg L^{-1} of air. What is the partial pressure of gaseous mercury if the atmospheric pressure is 733 torr at 26°C?

110. As 1 g of the radioactive element radium decays over 1 year, it produces 1.16×10^{18} alpha particles (helium nuclei). Each alpha particle becomes an atom of helium gas. What is the pressure in pascals of the helium gas produced if it occupies a volume of 125 mL at a temperature of 25°C?

111. (a) What is the total volume of the $CO_2(g)$ and $H_2O(g)$ at 600°C and 0.888 atm produced by the combustion of 1.00 L of $C_2H_6(g)$ measured at STP?
(b) What is the partial pressure of H_2O in the product gases?

112. Ethanol, C_2H_5OH, is produced industrially from ethylene, C_2H_4, by the following sequence of reactions.

$$3C_2H_4 + 2H_2SO_4 \longrightarrow C_2H_5HSO_4 + (C_2H_5)_2SO_4$$
$$C_2H_5HSO_4 + (C_2H_5)_2SO_4 + 3H_2O \longrightarrow$$
$$3C_2H_5OH + 2H_2SO_4$$

What volume of ethylene at STP is required to produce 1.000 metric ton (1000 kg) of ethanol if the overall yield of ethanol is 90.1%?

113. One molecule of hemoglobin will combine with four molecules of oxygen. If 1.0 g of hemoglobin combines with 1.53 mL of oxygen at body temperature (37°C) and a pressure of 743 torr, what is the molar mass of hemoglobin?

114. A sample of a compound of xenon and fluorine was confined in a bulb with a pressure of 18 torr. Hydrogen was added to the bulb until the pressure was 72 torr. Passage of an electric spark through the mixture produced Xe and HF.

After the HF was removed by reaction with solid KOH, the final pressure of xenon and unreacted hydrogen in the bulb was 36 torr. What is the empirical formula of the xenon fluoride in the original sample? (*Note:* Xenon fluorides contain only one xenon atom per molecule.)

115. One method of analyzing amino acids is the van Slyke method. The characteristic amino groups ($-NH_2$) in protein material are allowed to react with nitrous acid, HNO_2, to form N_2 gas. From the volume of the gas, the amount of amino acid can be determined. A 0.0604-g sample of a biological sample containing glycine, $CH_2(NH_2)CO_2H$, was analyzed by the van Slyke method and yielded 3.70 mL of N_2 collected over water at a pressure of 735 torr and 29°C. What was the percentage of glycine in the sample?

$$CH_2(NH_2)CO_2H + HNO_2 \longrightarrow$$
$$CH_2(OH)CO_2H + H_2O + N_2$$

116. (a) Is the pressure of the gas in the hot air balloon shown at the opening of this chapter greater than, less than, or equal to that of the atmosphere outside the balloon?
(b) Is the density of the gas in the hot air balloon shown at the opening of this chapter greater than, less than, or equal to that of the atmosphere outside the balloon?
(c) At a pressure of 1 atm and a temperature of 20°C, dry air has a density of 1.2256 g L^{-1}. What is the (average) molar mass of dry air?
(d) The average temperature of the gas in a hot air balloon is (1.30×10^2)°F. Calculate its density, assuming the molar mass equals that of dry air.
(e) The lifting capacity of a hot air balloon is equal to the difference between the mass of the cool air displaced by the balloon and the mass of the gas in the balloon. What is the difference between the mass of 1.00 L of the cool air in part (c) and the mass of 1.00 L of the hot air in part (d)?
(f) An average balloon has a diameter of 60 feet and a volume of 1.1×10^5 ft^3. What is the lifting power of such a balloon? If the weight of the balloon and its rigging is 500 lb, what is its capacity for carrying passengers and cargo?
(g) A balloon carries 40.0 gal of liquid propane (density, 0.5005 g L^{-1}). What volume of CO_2 and H_2O gas is produced by the combustion of this propane?
(h) A balloon flight can last about 90 min. If all of the fuel is burned during this time, what is the approximate rate of heat loss (in kilojoules per minute) from the hot air in the bag during the flight?

CHAPTER OUTLINE

11.1   Kinetic-Molecular Theory: Liquids and Solids

Forces Between Molecules

11.2   Dipole–Dipole Attractions
11.3   Dispersion Forces
11.4   Hydrogen Bonding

Properties of Liquids and Solids

11.5   Evaporation of Liquids and Solids
11.6   Boiling of Liquids
11.7   Distillation
11.8   Melting of Solids
11.9   Critical Temperature and Pressure
11.10   Phase Diagrams
11.11   Cohesive Forces and Adhesive Forces

The Structures of Crystalline Solids

11.12   Types of Solids
11.13   Crystal Defects
11.14   The Structures of Metals
11.15   The Structures of Ionic Crystals
11.16   The Radius Ratio Rule
11.17   Unit Cells
11.18   Calculation of Ionic Radii
11.19   The Born–Haber Cycle
11.20   X-Ray Diffraction

11

Intermolecular Forces, Liquids, and Solids

Molten lava solidifies when it cools.

We are familiar with many macroscopic differences between solids and liquids. We would not try to build a house out of water or try to swim in granite. We know that solids are rigid and that liquids flow and assume the shape of any container into which they are poured. We know that many liquids evaporate, but that most solids do not. However, in spite of the apparent differences in their behavior, many of the properties of both liquids and solids are similar. For example, solids and liquids are both essentially incompressible; we cannot change the volume of a sample of ice or of liquid water by pressing on it. Both solids and liquids have much larger densities than gases. For example, 44 grams of solid CO_2 (dry ice) or of liquid CO_2 has a volume of about 30 milliliters, a volume about the size of a golf ball. By contrast, the volume of 44 grams of CO_2 gas is about 25,000 milliliters, the volume of a beach ball with a diameter of about 14 inches. This is an increase of almost 1000 times from the volume of the solid to the volume of the gas. Both solids and liquids can be used for refrigeration: When either liquid or solid CO_2 is converted to a gas, heat is absorbed and the surroundings are cooled.

The similarities and differences in the behavior of liquids and solids can be explained by an extension of the same kinetic-molecular theory that we used to describe the properties of gases (Section 10.13). In this chapter we will use the kinetic-molecular theory to explore the behavior of liquids and solids. The forces that cause molecules to form liquids or solids will be examined, as will the regularities in structure that result when atoms, molecules, or ions arrange themselves into crystalline materials.

11.1 Kinetic-Molecular Theory: Liquids and Solids

We can use the kinetic-molecular theory to describe solids, liquids, and gases (Fig. 11.1). In the following description we use the term *particle* to mean an atom, a molecule, or an ion.

1. Solids, liquids, and gases are all made up of very small particles (atoms, molecules, and/or ions).

2. The arrangement of particles differs.

   • Particles in a solid are tightly packed, usually in a regular pattern.

Figure 11.1

The arrangement of molecules in a gas, in a liquid, and in a crystalline solid. The density of the gaseous molecules is exaggerated.

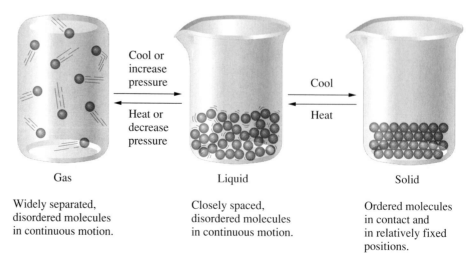

Gas

Widely separated, disordered molecules in continuous motion.

Liquid

Closely spaced, disordered molecules in continuous motion.

Solid

Ordered molecules in contact and in relatively fixed positions.

- Particles in a liquid are close together with no regular arrangement.
- Particles in a gas are well separated with no regular arrangement.

3. The amount of motion of the particles differs.
   - Particles in a solid vibrate (jiggle) but generally do not translate (move from place to place).
   - Particles in a liquid vibrate, move about, and slide past each other.
   - Particles in a gas vibrate and move freely at high speeds.

4. As the temperature of a solid, a liquid, or a gas increases, the particles move more rapidly. As the temperature falls, the particles slow down.

5. There is very little empty space between particles of a solid or a liquid, whereas much larger amounts of empty space separate the particles of a gas (although this can vary with temperature and pressure).

The differences in the properties that solids, liquids, and gases exhibit reflect the strength of the **intermolecular attractions** between the atoms, molecules, or ions that make up each phase.

In Chapter 10 we saw that molecules in a real gas have enough kinetic energy to overcome the forces of attraction between them, so they move apart if they collide. But we also found that when a gas is cooled, the average speed and kinetic energy of the molecules decrease (Section 10.14). Thus we should not be surprised that when a gas is cooled sufficiently, the intermolecular attractions prevent the molecules from moving apart, the gas condenses, and the molecules come into contact. This condensation produces either a liquid or a solid. We have seen liquid water form on the outside of a cold glass or soft drink can. This happens as the water vapor in the air is cooled by the cold glass or can. The frost that forms in a freezer or outdoors during some winter nights (Fig. 11.2) results when water vapor comes in contact with a very cold surface and is converted directly to solid water without going through the liquid phase.

If the temperature is not too high, we can also liquefy many gases by compressing them. The increased pressure brings the molecules of a gas closer together such that the attractions between the molecules become strong relative to their kinetic energy. Consequently, they form liquids. Carbon dioxide is a gas at room temperature and 1 atmosphere, but it generally is a liquid in CO_2 fire extinguishers, because the pressure is greater than about 65 atmospheres. At this pressure, gaseous carbon dioxide condenses to a liquid at room temperature.

Although the molecules in a liquid are in contact, they still move about. Thus a liquid can change its shape, take the shape of its container, evaporate, and diffuse. However, because of the much shorter distances that molecules in a liquid can move before they collide with other molecules, they diffuse much more slowly than in gases. Liquids are relatively incompressible, because any increase in pressure can only slightly reduce the distance between the closely packed molecules. Thus the volume of a liquid decreases very little with increased pressure.

It is difficult to describe the arrangement of molecules in a liquid because they move continuously. We can only say that the molecules that make up a liquid are relatively closely packed in a random arrangement.

When the temperature of a liquid becomes sufficiently low, or the pressure on the liquid becomes sufficiently high, the molecules of the liquid no longer have enough kinetic energy to move past each other, and a solid forms. Solids are rigid. They cannot expand like gases or be poured like liquids, because the molecules cannot change position easily.

In spite of the fact that solids are rigid, their molecules are not still. Although most molecules in a solid do not move about, they do vibrate. Diffusion takes place to a

Figure 11.2

Frost forms when water vapor in the air is cooled and is converted directly to ice.

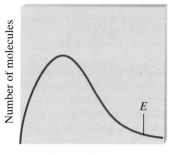

Molecular speed

Figure 11.3

Distribution of molecular speeds in a liquid. Molecules with speeds greater than E possess sufficient kinetic energy to escape from the surface of the liquid into the gas phase.

very limited extent in the solid state, and some solids even evaporate. You may have seen dry ice (solid carbon dioxide) evaporate without melting. We can smell moth balls and solid deodorants, because these solids slowly evaporate and the gaseous molecules move into our noses. The motion of the molecules in a solid gradually decreases to a minimum at 0 kelvins (absolute zero) as the temperature of the solid is lowered.

The evaporation of liquids and solids can be explained by the kinetic-molecular theory. In a liquid, just as in a gas, the molecules move at various rates—some slowly, many at intermediate rates, and some very rapidly (Fig. 11.3). A few rapidly moving molecules near the surface of a liquid may have enough kinetic energy to overcome the attraction of their neighbors and escape—that is, evaporate—into the gas above the liquid, becoming molecules in the gas phase. Although the molecules in a solid do not move through the solid, they do vibrate—some gently, many with intermediate energy, and some wildly. A few wildly vibrating molecules on the surface of some solids may possess enough energy to overcome the attraction of their neighbors and escape into the gas phase.

Forces Between Molecules

Under appropriate conditions, the attractions between all gas molecules will cause them to form liquids or solids. The strengths of these attractive forces vary widely, although the intermolecular forces between small molecules are usually weak compared to the intramolecular forces that bond atoms together within a molecule. For example, to overcome the intermolecular forces in 1 mole of liquid HCl and convert it to gaseous HCl requires only about 17 kilojoules. However, to break the covalent bonds between the hydrogen and chlorine atoms in 1 mole of HCl requires about 25 times more energy, or 430 kilojoules.

We will consider the attractive forces between molecules, which are collectively called van der Waals forces, in the next three sections.

11.2   Dipole–Dipole Attractions

One type of **van der Waals force** is the **dipole–dipole attraction,** an attractive force that results from the electrostatic attraction of the positive end of one polar molecule (Section 7.3) for the negative end of another (Fig. 11.4). The iodine monochloride molecule, ICl, is a molecule with a dipole. The more electronegative chlorine atom bears the partial negative charge, whereas the less electronegative iodine atom bears

Figure 11.4

Two arrangements of polar molecules that allow an attraction between the negative end of one molecule and the positive end of another.

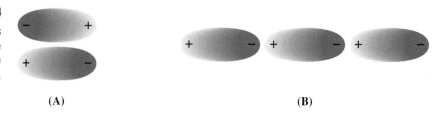

(A)   (B)

the partial positive charge. An attractive force between ICl molecules results from the attraction of the positive end of one ICl molecule for the negative end of another.

The effect of a dipole–dipole attraction is apparent when we compare the properties of ICl molecules to those of nonpolar Br_2 molecules. ICl is a solid at $0°C$; Br_2 is a liquid. Both ICl and Br_2 have the same number of atoms and approximately the same molecular mass. At $0°C$, in the gas phase, both molecules would have approximately the same kinetic energy. However, the dipole–dipole attractions between ICl molecules are sufficient to cause them to form a solid, whereas the intermolecular attractions between nonpolar Br_2 molecules are not.

11.3   Dispersion Forces

A second component of the van der Waals attractions is called the **London dispersion force** (after Fritz London, who in 1928 first explained it). Dispersion forces cause nonpolar substances such as Br_2, the other halogens, and the noble gases to condense to liquids and to freeze into solids when the temperature is lowered sufficiently.

Because of the constant motion of its electrons, an atom can develop a temporary dipole moment when its electrons are distributed unsymmetrically about the nucleus. A second atom, in turn, can be distorted by the appearance of the dipole in the first atom. The electrons of the second atom are attracted toward the positive end of the first atom (Fig. 11.5). Thus a dipole appears on the second atom, and the two rapidly fluctuating, temporary dipoles attract each other. Dispersion forces are the result of these attractions.

When two atoms are located in different molecules, dispersion forces that develop between the atoms can attract the two molecules to each other. The forces are weak, however, and become significant only when the molecules are almost touching. Dispersion forces between molecules are present when any two molecules are close together. Although the principal forces that stabilize a liquid or solid may be due to dipole–dipole or other attractive forces, dispersion forces always play some role in their stability.

Larger and heavier atoms and molecules exhibit stronger dispersion forces than smaller and lighter atoms and molecules. F_2 and Cl_2 are gases at room temperature (weaker attractive forces), Br_2 is a liquid, and I_2 is a solid (stronger attractive forces). The increase in attractive forces as the atoms and molecules involved become larger is readily explained by London's theory. In a larger atom the valence electrons are, on average, farther from the nuclei than in a smaller atom. Thus they are less tightly held and can more easily form the temporary dipoles that produce the attraction.

Figure 11.5

Dispersion forces result from the formation of temporary dipoles. (A) The temporary dipole in the atom on the right produces a dipole in the atom on the left. (B) The attraction of the two resulting dipoles produces the attractive force.

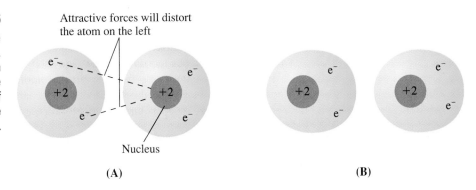

(A)   (B)

Figure 11.6

Ball-and-stick molecular models and Lewis structures of (A) *n*-pentane and (B) neopentane.

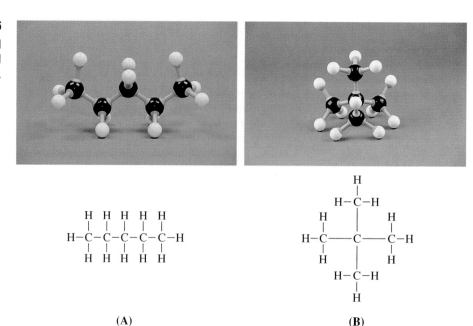

(A)

(B)

The shapes of molecules also affect the magnitudes of the dispersion forces between them. For example, *n*-pentane (Fig. 11.6 A) is a liquid at room temperature, whereas neopentane (Fig. 11.6 B) is a gas. Thus we see that the dispersion forces in *n*-pentane are larger than those in neopentane, even though both compounds are composed of molecules with the same chemical formula, C_5H_{12}. The neopentane molecule is shaped in such a way that the valence electrons in the carbon–carbon bonds are well inside the molecule. The *n*-pentane molecule is shaped in such a way that the valence electrons in the carbon–carbon bonds are closer to the surface of the molecule. Thus any dispersion forces that involve electrons participating in carbon–carbon bonds must act over a longer distance with neopentane, and this longer distance results in weaker overall attractive forces.

11.4   Hydrogen Bonding

Nitrosyl fluoride (ONF, molecular mass 49 amu) is a gas at room temperature. Water (molecular mass 18 amu) is a liquid, even though it has a lower molecular mass. We cannot attribute this difference between the two compounds to dispersion forces; both molecules have about the same shape, and ONF is the heavier and larger molecule. Nor can we attribute it to differences in the dipole moments of the molecules; both molecules have the same dipole moment. The large difference between the boiling points is due to a special kind of dipole–dipole attraction that is called **hydrogen bonding.**

Hydrogen bonds can form whenever a hydrogen atom is bonded to one of the more electronegative atoms: a fluorine, oxygen, or, nitrogen atom (Section 6.5). The large

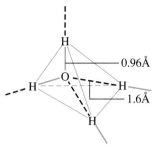

— Covalent bonds

-- Hydrogen bonds

Figure 11.7

Covalent bonds and hydrogen bonds about an oxygen atom in liquid or solid water form a distorted tetrahedron. Two hydrogen atoms participate in covalent bonds, and two participate in hydrogen bonds to the oxygen. Each bond extending from a hydrogen atom away from the tetrahedron connects to an oxygen atom of another water molecule.

difference in electronegativity between hydrogen atoms (2.1) and the second atom (4.1 for an F atom, 3.5 for an O atom, 3.0 for a N atom) leads to a highly polar covalent bond in which the hydrogen bears a large partial positive charge and the second element bears a large partial negative charge. The electrostatic attraction between the partially positive hydrogen atom in one molecule and the partially negative atom in another molecule gives rise to the strong dipole–dipole attraction called a **hydrogen bond.** Hydrogen bonds are weaker than ordinary covalent bonds (only about 5%–10% as strong) but stronger than other dipole–dipole attractions and London forces. Examples of hydrogen bonds include $HF \cdots HF$, $H_2O \cdots HOH$, $H_3N \cdots HNH_2$, $H_3N \cdots HOH$, and $H_2O \cdots HOCH_3$. Figure 11.7 illustrates the two hydrogen bonds about a water molecule in liquid water.

Hydrogen bonds have a pronounced effect on the properties of condensed phases. For example, we generally expect the amount of energy necessary to overcome intramolecular attractive forces to decrease with decreasing molecular mass (Section 11.3). With the exception of H_2O, HF, and NH_3, the enthalpies of vaporization (Section 4.5) of the hydrides and the noble gases shown in Fig. 11.8 exhibit this trend. The amount of heat required to evaporate 1 mole of each substance in a given class generally decreases as the molar mass decreases. The compounds H_2O, HF, and NH_3 deviate because they possess appreciable hydrogen bonding. Hence they have abnormally large enthalpies of vaporization.

Figure 11.8 The enthalpies of vaporization of several binary hydrides and noble gases. Each line connects compounds of elements in one group of the periodic table.

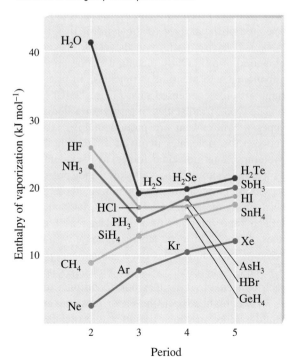

Figure 11.9

Sublimation of iodine produces a purple gas that condenses to give crystals farther up the tube.

Properties of Liquids and Solids

11.5   Evaporation of Liquids and Solids

All of us have seen wet clothes dry because of **evaporation,** the change of condensed matter to a gas. We know that some liquids, such as gasoline, alcohol, and acetone, evaporate more rapidly than water under the same conditions. Others evaporate more slowly. Motor oil and ethylene glycol (antifreeze) evaporate so slowly that they seem not to evaporate at all.

Some solids evaporate too. Over a period of time, an ice cube in a freezer becomes smaller as it evaporates. Chemists refer to this direct conversion from the solid phase to the gas phase (without passing through the liquid phase) as **sublimation.** Snow sublimes at temperatures below its melting point. When solid iodine is warmed, the solid sublimes and a beautiful purple vapor forms (Fig. 11.9).

When a solid or a liquid evaporates in a closed container, molecules cannot escape. Eventually, some molecules strike the condensed phase and move back into it. This change from the vapor back to a condensed state is called **condensation.** When the rate of condensation becomes equal to the rate of evaporation, neither the amount of the solid or liquid nor the amount of the vapor in the container changes. The vapor in the container is said to be in **equilibrium** with the solid or liquid. At equilibrium, however, some molecules from the vapor condense at the same time that an equal number of molecules from the solid or liquid evaporate; the equilibrium is a **dynamic equilibrium.**

The pressure exerted by the vapor in equilibrium with a solid or a liquid in a closed container at a given temperature is called the **vapor pressure** of the condensed phase. The area of the surface of the solid or liquid in contact with a vapor and the size of the vessel have no effect on the vapor pressure. We can measure the vapor pressure of a solid or a liquid by placing a sample in an evacuated, closed container and using a manometer to measure the increase in pressure that is due to the vapor in equilibrium with the condensed phase (Fig. 11.10).

The type of molecules in a solid or a liquid determine its vapor pressure. The four vapor pressures that are plotted in Fig. 11.11 differ because the attractive forces between the molecules involved differ. Ethyl ether has the smallest (and most easily overcome) intermolecular forces, as is evident from its higher vapor pressure at every temperature shown. The very low vapor pressure of ethylene glycol (too low even to see on the graph at room temperature, about 25°C) is a reflection of the strong forces between its molecules.

The vapor pressure of a liquid increases as its temperature increases because the rates of motion of its molecules increase with increasing temperature. At a higher temperature, more molecules have enough energy to escape from the liquid, as shown in Fig. 11.12. The escape of more molecules per unit of time and the greater speed of each molecule that escapes both contribute to the higher vapor pressure.

As we might expect, the vapor pressures of solids behave analogously to those of liquids. Solids in which the intermolecular forces are weak exhibit measurable vapor pressures at room temperature, and the vapor pressures of solids increase with increasing temperature. For example, the vapor pressure of solid iodine is 0.2 torr at 20°C and 90 torr at 114°C.

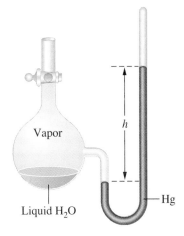

Figure 11.10

A closed vessel containing only liquid water and water vapor. The vapor pressure is equal to the difference in height, h, of the mercury in the two columns of the manometer.

Figure 11.11

The vapor pressures of four common liquids at various temperatures. The normal boiling points of these substances are the temperatures at which their vapor pressures reach 760 torr (1 atm): 34.6°C for ethyl ether, 78.4°C for ethyl alcohol, 100°C for water, and 198°C for ethylene glycol (which is off this scale).

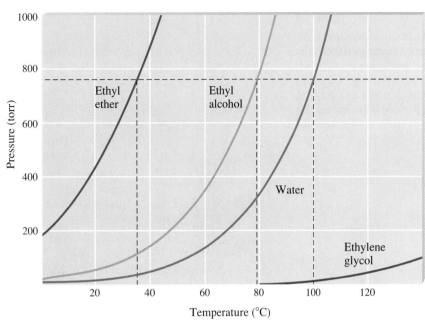

Figure 11.12

Distribution of molecular speeds in a liquid at two temperatures. At the higher temperature, more molecules have the necessary speed, E, to escape from the liquid into the gas phase.

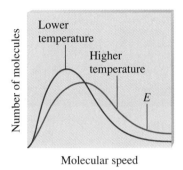

At the beginning of this section we discussed one process that is called sublimation: the direct conversion from the solid to the gas phase. There is another process that we also call sublimation: the conversion of a solid directly into gas followed by condensation of the gas back to the solid. Many substances, such as iodine, can be purified by this second type of sublimation (Fig. 11.9) if the impurities have low vapor pressures.

11.6 Boiling of Liquids

As a liquid is heated, its vapor pressure increases until the vapor pressure equals the pressure of the gas above it. At this point the liquid begins to boil. When a liquid boils, bubbles of vaporized liquid form within it and then rise to the surface, where they burst and release the vapor. If we stop heating during boiling and insulate the mixture of gas and liquid so that no heat can enter or escape, the gas and liquid phases remain in equilibrium. The results may be unexpected if this experiment is carried out in a gravity-free environment (Fig. 11.13).

If we put a thermometer in a boiling liquid and then add more heat, we see that the temperature of the remaining liquid stays constant (Fig. 4.12). The additional heat converts some of the liquid to the gas phase rather than raising its temperature. The amount of heat needed to convert a mole of a boiling liquid to vapor is the enthalpy of vaporization (Section 4.5) of the liquid at its boiling temperature.

The **normal boiling point** of a liquid is the temperature at which its vapor pressure equals 1 atmosphere (760 torr, Fig. 11.11). A liquid boils at temperatures higher than its normal boiling point when the external pressure is greater than 1 atmosphere; conversely, the boiling point is lowered by reduced pressure (see Fig. 11.14 on page 380). For example, at high altitudes where the atmospheric pressure is less than 760 torr, water boils at temperatures below its normal boiling point of 100°C. Food in

Figure 11.13

Boiling of liquid freon during a space shuttle flight. After the heat is turned off, in microgravity the vapor forms a spherical bubble that remains suspended in the liquid.

Figure 11.14

A liquid such as ethyl alcohol that does not boil at room temperature under 1 atmosphere of pressure (A) may boil at that same temperature when we reduce the pressure by placing the liquid in a jar and pumping out the air (B).

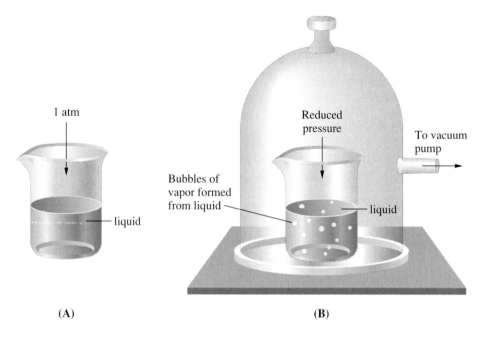

(A)   (B)

boiling water cooks more slowly at high altitudes than at sea level, because the temperature of boiling water is lower at the higher altitude.

Example 11.1   A Boiling Point at Reduced Pressure

Figure 11.15

Liquid air boils when exposed to room temperature, because the intermolecular forces are very weak.

A common atmospheric pressure in Leadville, Colorado (elevation 10,200 ft), is 510 torr. Use the graph in Fig. 11.11 to determine the approximate boiling point of water at this temperature.

Solution: The graph of the vapor pressure of water versus temperature in Fig. 11.11 indicates that the vapor pressure of water is 510 torr at about 90°C. Thus at about 90°C, the vapor pressure of water will equal the atmospheric pressure in Leadville and water will boil.

Check your learning: Use Fig. 11.11 to determine the approximate temperature at which the vapor pressure of ethyl ether is 1000 torr.

Answer: Approximately 44°C

Boiling points increase as the forces of attraction between molecules increase, and they decrease with decreasing intermolecular attractions (Fig. 11.15). The trend of increases in intermolecular attraction with increasing size (Section 11.3) is reflected in the boiling points of the substances shown in Table 11.1.

11.7   Distillation

Dissolved materials in a liquid may make it unsuitable for some purposes. We know that hard water (water that contains dissolved minerals) should not be used in auto-

Table 11.1

Molecular Masses and Boiling Points

Substance	He	Ne	Ar	Kr	Xe	Rn
Molecular mass, amu	4.0	20.2	39.9	83.8	131.3	222
Boiling point, °C	−268.9	−245.9	−185.7	−152.9	−107.1	−61.8
Substance	H_2	F_2	Cl_2	Br_2	I_2	
Molecular mass, amu	2.0	38.0	70.9	159.8	253.8	
Boiling point, °C	−252.7	−187	−34.6	58.8	184.4	

mobile batteries or steam irons because, over time, it interferes with their effective operation. Some hard water has an unpleasant taste because of dissolved minerals.

We can separate water and other liquids from nonvolatile impurities (impurities with very low vapor pressures) by a process known as **distillation** (Fig. 11.16). The impure liquid is boiled, and the vapor is condensed in another part of the apparatus. Dissolved matter that does not evaporate stays behind. Distillation takes advantage of the facts that (1) heating a liquid speeds up its rate of evaporation and (2) cooling a vapor favors its condensation.

One way to obtain water for emergency survival on the desert involves distillation (see Fig. 11.17 on page 382). This method uses solar energy to evaporate water from crushed vegetation placed in a pit; the evaporated water collects on a transparent plastic sheet placed over the pit and drips into a collection cup.

Figure 11.16

Laboratory distillation apparatus. When an impure liquid is distilled, nonvolatile substances remain in the distilling flask (the flask that is heated). Pure liquid evaporates, condenses in the water-cooled condenser, and collects in the receiving flask.

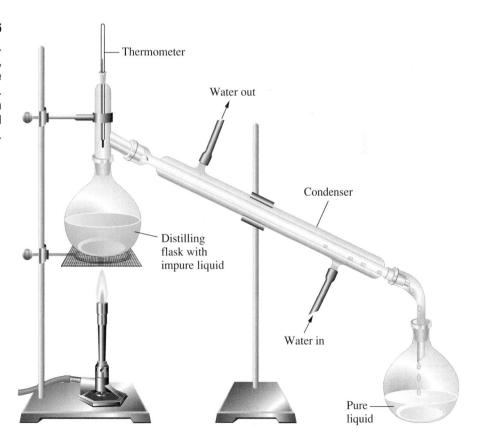

Thermometer

Water out

Condenser

Distilling flask with impure liquid

Water in

Pure liquid

Figure 11.17

Solar distillation of water. The heat produced by the sun shining into a hole covered with a clear plastic sheet is sufficient to evaporate water from crushed vegetation. The water vapor condenses on the slightly cooler sheet and drips into the cup.

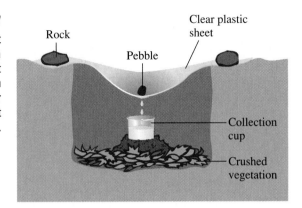

Figure 11.17

Solar distillation of water. The heat produced by the sun shining into a hole covered with a clear plastic sheet is sufficient to evaporate water from crushed vegetation. The water vapor condenses on the slightly cooler sheet and drips into the cup.

11.8   Melting of Solids

When we heat a crystalline solid, we increase the average energy of its molecules or ions, and the solid gets hotter. At some point, the added energy becomes large enough to overcome some of the forces holding the molecules or ions of the solid in their fixed positions, and the solid begins to melt. At this point the temperature of the solid stops changing, and if we continue to apply heat, the temperature of the mixture of solid and liquid does not change until all of the solid is melted. Only after all of the solid has melted does our heating increase the temperature of the liquid (Figs. 11.18 and 4.11).

If we stop heating during melting and place the mixture of solid and liquid in a perfectly insulated container so that no heat can enter or escape, the solid and liquid phases remain in equilibrium. This is almost the situation with a mixture of ice and water in a very good thermos bottle; almost no heat gets in or out and the mixture of solid ice and liquid water remains for hours. In a mixture of solid and liquid at equilibrium, changes from solid to liquid and from liquid to solid continue, but the rate of melting is equal to the rate of freezing, and the quantities of solid and liquid remain constant. The temperature at which the solid and liquid phases of a given substance are in equilibrium is called the **melting point** of the solid and the **freezing point** of the liquid.

Figure 11.18

Heating curve for water. When ice is heated, the temperature increases until the melting point is reached. The temperature remains constant until all the ice has melted; then it increases again. When the water begins to boil, the temperature again remains constant until all of the water has vaporized. The temperature then increases again.

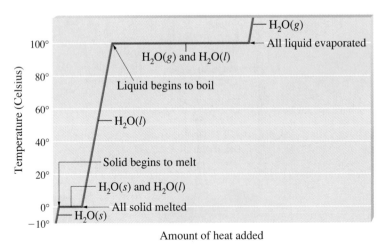

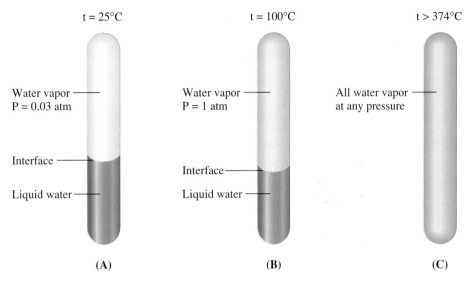

Figure 11.19

A sample of liquid water and water vapor at 25°C (A), at 100°C (B), and at 375°C (C). Just above the critical temperature, 374°C, the boundary between liquid and vapor disappears.

The amount of heat needed to melt a mole of a solid is the enthalpy of fusion of the solid (Section 4.5). The enthalpy of fusion and the melting point of a crystalline solid are determined by the strength of the attractive forces between the units present in the crystal. Molecules with weak attractive forces form crystals with low melting points. Crystals that consist of particles with stronger attractive forces melt at higher temperatures.

Water and a few metals that are used for casting expand when they freeze and contract when they melt. However, most substances contract when they freeze and expand when they melt.

11.9   Critical Temperature and Pressure

If we place a sample of water in a tube at 25°C and remove the air, we are left with a mixture of liquid water and water vapor (with a pressure of 23.8 torr, Table 10.5). There is a clear boundary between the two phases (Fig. 11.19 A). As we increase the temperature, the pressure of the water vapor increases (for example, to 760 torr at 100°C), but liquid water and the vapor are present and the boundary between them is still distinct (Fig. 11.19 B). However, if we heat to just above 374°C (in a *very* strong tube), the boundary disappears (Fig. 11.19 C) and it is no longer possible to distinguish liquid and vapor. All of the water in the tube is physically identical and exhibits the properties of a gas. This temperature, 374°C, is the critical temperature of water. Above its **critical temperature,** a gas cannot be liquefied, no matter how much pressure is applied. The pressure required to liquefy a gas at its critical temperature is called the **critical pressure.** The critical temperatures and critical pressures of some common substances are given in Table 11.2 on page 384.

A substance at a temperature and pressure above its critical temperature and pressure is called a **supercritical fluid.** Supercritical fluids exhibit properties that are intermediate between those of a liquid and those of a gas. Like a gas, a supercritical fluid will expand and fill a container, but its density is approximately that of a liquid. Consequently, these fluids have the solvent properties of a liquid combined with the low viscosity of a gas. They can penetrate very small openings in a solid mixture and

Table 11.2

Critical Temperatures and Pressures of
Some Common Substances

	Critical Temperature, K	Critical Pressure, atm
Hydrogen	33.2	12.8
Nitrogen	126.0	33.5
Oxygen	154.3	49.7
Carbon dioxide	304.2	73.0
Ammonia	405.5	111.5
Sulfur dioxide	430.3	77.7
Water	647.1	217.7

remove soluble components. Supercritical carbon dioxide has proved to be an extremely useful solvent in the food industry. It is nontoxic, noncarcinogenic, and relatively inexpensive, and it is not considered to be a pollutant. The CO_2 can be easily recovered simply by reducing the pressure and separating the resulting gas. Supercritical carbon dioxide is used to decaffeinate coffee and tea, to remove fats from potato chips, and to extract the compounds that produce flavors and fragrances from citrus oils.

Example 11.2   The Critical Temperature of Carbon Dioxide

If we shake a carbon dioxide fire extinguisher on a cool day (say 18°C), we can hear liquid CO_2 sloshing around inside the cylinder. However, the same cylinder appears to contain no liquid on a hot summer day (say 35°C). Explain these observations.

Solution: On the cool day, the temperature of the CO_2 is below the critical temperature of CO_2, 304 K or 31°C (Table 11.2), so liquid CO_2 is present in the cylinder. On the hot day, the temperature of the CO_2 is greater than its critical temperature of 31°C. Above this temperature no amount of pressure can liquefy CO_2, so no liquid CO_2 exists in the fire extinguisher.

Check your learning: Ammonia can be liquefied by compression at room temperature; oxygen cannot be liquefied under these conditions. Why do the two gases exhibit different behavior?

Answer: The critical temperature of ammonia is 405.5 K, which is higher than room temperature. The critical temperature of oxygen is below room temperature; thus oxygen cannot be liquefied at room temperature.

11.10   Phase Diagrams

If we fill a syringe with cold water and warm it, then as long as the temperature is lower than 100°C and the pressure is 760 torr (1 atmosphere), the syringe will contain only liquid water (Fig. 11.20 A on page 385). However, when the temperature reaches 100°C, adding a little more heat produces water vapor with a temperature of 100°C and a pressure of 760 torr (Fig. 11.20 B). In this way we can show that at tem-

Figure 11.20

Syringes containing water at 760 torr. (A) At 80°C and 760 torr, only liquid water is contained in the syringe. (B) When the water has warmed to 100°C and a little additional heat is added, some water evaporates, and then the syringe contains both liquid and vapor at 100°C and 760 torr.

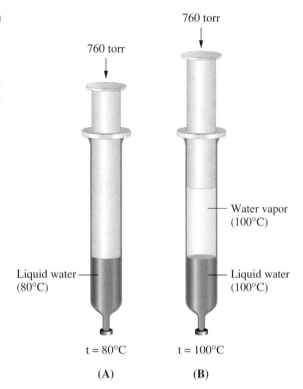

760 torr

760 torr

Water vapor (100°C)

Liquid water (80°C)

Liquid water (100°C)

t = 80°C

t = 100°C

(A)

(B)

peratures between 0°C and 100°C and at a pressure of 760 torr, only liquid water exists. At 100°C and a pressure of 760 torr, liquid and gaseous water (both at a pressure of 760 torr) are in equilibrium. If we begin heating cold water and reduce the pressure on the barrel of the syringe to 355 torr, water vapor appears at 80°C. At a pressure of 355 torr, liquid water exists between 0°C and 80°C; at 80°C, water vapor begins to form and the liquid and vapor exist at equilibrium (with both at a pressure of 355 torr).

When we cool water vapor at 110°C and keep the pressure at 760 torr, it remains a gas until it cools to 100°C, at which point liquid water begins to form. If more heat is removed, the temperature does not change until all the vapor condenses. Thus 100°C and 760 torr are one "point" at which liquid water and water vapor are at equilibrium. At a lower pressure of 355 torr, the equilibrium between liquid and vapor is found at 80°C.

Water at 10°C and a pressure of 380 torr cools as heat is removed. When the temperature reaches 0.005°C, further loss of heat results in the formation of ice. Ice and liquid water are in equilibrium at 0.005°C at a pressure of 380 torr. At 760 torr, the equilibrium between ice and water occurs at 0°C.

We can summarize in a **phase diagram** all of the information about the temperatures and pressures at which the various phases of a substance are in equilibrium. A phase diagram shows the pressures and temperatures at which gaseous, liquid, and solid phases of a substance can exist. We construct a phase diagram by plotting points that represent the pressures and temperatures at which two different phases of a substance are in equilibrium. A part of the phase diagram for water is given in Fig. 11.21 A. An expanded portion is given in Fig. 11.21 B on page 386.

We can use a point on the phase diagram to identify the phase of a sample of water under the conditions of pressure and temperature that the point represents. A point at

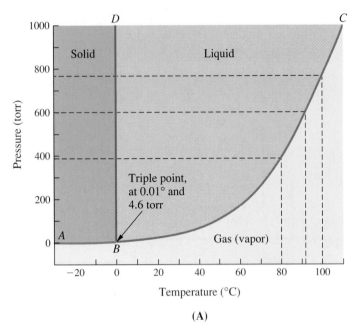

(A)

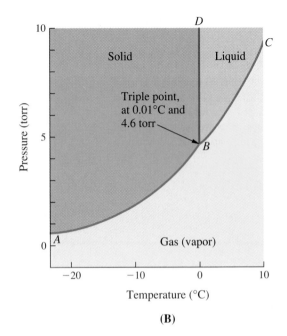

(B)

Figure 11.21

(A) Phase diagram for water. (B) An expanded portion between −20°C and 10°C, showing that the equilibrium between the solid and gas phases varies with temperature.

600 torr and −10°C, for example, is found in the portion of the phase diagram (Fig. 11.21 A) labeled "Solid." Under these conditions water exists only as a solid (ice). At 600 torr and 50°C (within the portion labeled "Liquid"), water exists only as a liquid. For any combination of pressure and temperature in the region labeled "Gas" (600 torr and 110°C, for example), water exists only in the gaseous state.

The solid lines in a phase diagram identify temperatures and pressures at which an equilibrium exists between phases. The red line BC in Fig. 11.21 A shows the combinations of temperature and pressure at which liquid water and water vapor are in equilibrium. It is a plot of the vapor pressure of water versus temperature and it also shows how the boiling point of water changes with pressure. This diagram shows that when the pressure is 760 torr, the liquid and vapor are at equilibrium only when the temperature is 100°C, as we saw in the experiment with the syringes. The figure also indicates that the equilibrium occurs at 94°C when the pressure is 600 torr and at 80°C when the pressure is 355 torr.

If the figure were large enough, line BC would be seen to terminate at a temperature of 374°C and a pressure of 165,500 torr (217.7 atmospheres), the critical temperature and pressure of water. Beyond the critical temperature, liquid and gaseous water cannot be in equilibrium because liquid water cannot exist above the critical temperature.

The blue line AB in Fig. 11.21 A indicates the temperatures and pressures at which ice and water vapor are in equilibrium. The purple line BD shows the temperature at which ice and liquid water are in equilibrium as the pressure changes. If we look at the expanded diagram in Fig. 11.21 B, we see that ice has a vapor pressure of about 1.5 torr at −10°C. Thus if we place a frozen sample in a vacuum with a pressure less than 1.5 torr, ice will evaporate. This is the basis for freeze-drying (Fig. 11.22).

The almost vertical purple line BD in Fig. 11.21 shows us that as the pressure increases, the melting temperature remains almost constant (it actually decreases very slightly). On the other hand, the diagram indicates clearly that the boiling tempera-

Figure 11.22

Campers prepare dinner. To produce their freeze-dried meal, a water-containing food was cooled, and then the frozen water was removed as a vapor by sublimation at a decreased pressure.

ture increases markedly with an increase in pressure. Representative vapor pressures for water are given in Table 10.5.

The lines that separate the various regions in Fig. 11.21 represent points at which an equilibrium exists between two phases. One point exists where all three phases are in equilibrium. This point, referred to as the **triple point,** occurs at point *B* in the figure, where the three lines intersect. At the pressure and temperature of the triple point (4.6 torr and 0.01°C), all three states are in equilibrium with each other.

11.11   Cohesive Forces and Adhesive Forces

Figure 11.23

Honey is a viscous liquid.

The **viscosity** of a liquid is a measure of its resistance to flow. Water, gasoline, and other liquids that flow freely have a low viscosity. Honey (Fig. 11.23), syrup, and other liquids that do not flow freely have higher viscosities. We can measure viscosity by measuring the rate at which a metal ball falls through a liquid (the ball falls more slowly through a more viscous liquid) or by measuring the rate at which a liquid flows through a narrow tube (more viscous liquids flow more slowly).

The intermolecular forces between the molecules of a liquid, the size and shape of the molecules, and the temperature determine how easily a liquid flows. The more complex the molecules in a liquid and the stronger the intermolecular forces between them, the more difficult it is for them to move past each other and the greater the viscosity of the liquid. As the temperature increases, the molecules move more rapidly and their kinetic energies are better able to overcome the forces that hold them together. Thus the viscosity of the liquid decreases. We call all of the various intermolecular forces holding a liquid together **cohesive forces.**

The molecules within a liquid are attracted equally in all directions by the cohesive forces within the liquid. However, the molecules on the surface of a liquid are

Surface

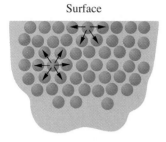

Figure 11.24

Attractive forces experienced by a molecule at the surface and by a molecule in the bulk of a liquid.

Figure 11.25   Cohesive forces produce the surface tension that gives the spherical shape to these water drops on a spider web.

Figure 11.26   Surface tension keeps this insect afloat.

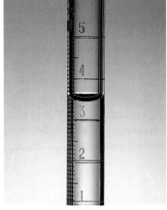

Figure 11.27

Note the difference between the shape of the meniscus of water (above) and that of the meniscus of mercury (below) in these glass burets.

attracted only into the liquid and to either side (Fig. 11.24). This unbalanced molecular attraction pulls the surface molecules back into the liquid, leaving the minimum number of molecules possible on the surface. Thus the surface area is reduced to a minimum. A small drop of liquid tends to assume a spherical shape because in a sphere, the ratio of surface area to volume is at a minimum (Fig. 11.25).

Surface tension is a phenomenon explained by the cohesive forces that cause the surface of a liquid to contract. The surface of a liquid acts as though it were a stretched membrane. A steel needle carefully placed on water will float. Some insects, even though they are denser than water, move on its surface because they are supported by the surface tension (Fig. 11.26).

The forces of attraction between a liquid and a surface are called **adhesive forces.** When the adhesive forces between water and a surface are weak, water does not wet the surface. For example, water does not wet waxed surfaces and polyethylene. Water forms drops on these surfaces, because the cohesive forces within the drops are greater than the adhesive forces between the water and the plastic. Water wets glass and spreads out on it, because the adhesive force between water and glass is greater than the cohesive forces within the water. **Capillary action** occurs when one end of a small-diameter tube is immersed in a liquid that wets the tube. The liquid creeps up the inside of the tube until the weight of the liquid and the adhesive forces are in balance. The smaller the diameter of the tube, the higher the liquid climbs. It is partly by capillary action that water and dissolved nutrients are brought from the soil up through the roots and into a plant.

When water is confined in a glass tube, its meniscus (surface) has a concave shape, because the water wets the glass and creeps up the side of the tube. On the other hand, the meniscus of mercury confined in a tube is convex, because mercury does not wet glass and the cohesive forces in the mercury tend to draw it into a drop (Fig. 11.27).

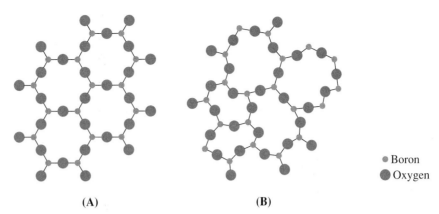

Figure 11.28

(A) A two-dimensional illustration of the ordered arrangement of atoms in crystalline boric oxide. (B) An illustration showing the disorder in amorphous boric oxide.

(A)   (B)

• Boron
• Oxygen

The Structures of Crystalline Solids

11.12   Types of Solids

When most liquids are cooled, they eventually freeze and form **crystalline solids**—solids in which the atoms, ions, or molecules are arranged in a definite repeating pattern (Fig. 11.28 A). Although some solids (such as diamonds and the individual grains in sugar, sand, and table salt) are single crystals, most common crystalline solids are aggregates of many small crystals. Examples of the latter include sandstone, chunks of ice, granite, and metal objects.

Liquids such as tar, molten glass, molten plastics, and molten butter, which consist of large molecules or a mixture of molecules that cannot move readily, do not form crystalline solids when cooled. Instead, as the temperature is lowered, their molecules move more and more slowly and finally stop in random positions before they can move into an ordered arrangement. The resulting materials are called **amorphous solids** or *glasses* (Fig. 11.28 B). Such solids lack an ordered internal structure. Common examples of amorphous solids include candle wax, butter, glass, and plastics such as those in polyethylene bags, compact discs, and recording tape.

There are several different types of crystalline solids: ionic solids, metallic solids, covalent network solids, and molecular solids. **Ionic solids,** such as sodium chloride (Fig. 11.29) and nickel oxide, are composed of positive and negative ions that are held together by electrostatic attractions, which can be quite strong. Many ionic crystals also have high melting points (Fig. 11.30). Although they are hard, they also tend to be brittle, and they shatter rather than bend. Ionic solids do not conduct electricity; but they do conduct when molten or dissolved because their ions are free to move. Many simple compounds formed by the reaction of a metallic element with a nonmetallic element are ionic.

Metallic solids such as crystals of copper, aluminum, and iron are formed by metal atoms. These crystals have many useful and varied properties. All exhibit high thermal and electrical conductivity, metallic luster, and malleability. Many are very hard and quite strong. Because of their malleability (the ability to deform under pressure or hammering), they do not shatter and make useful construction materials. The melting points of the metals vary widely. Mercury is a liquid at room temperature, and

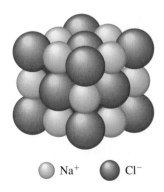

○ Na⁺   ● Cl⁻

Figure 11.29

The structure of sodium chloride, an ionic solid, consists of sodium ions and chloride ions.

Figure 11.30

Ionic metal oxides are used to line this crucible. They do not melt when in contact with molten iron because they have very high melting temperatures.

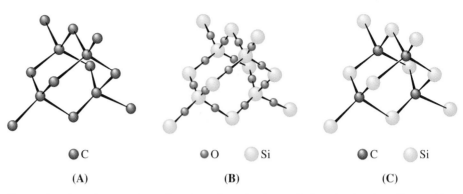

● C
(A)

● O   ○ Si
(B)

● C   ○ Si
(C)

Figure 11.31   A covalent crystal contains a three-dimensional network of covalent bonds, as illustrated by the structures of diamond (A), silicon dioxide (B), and silicon carbide (C). Lines between atoms indicate covalent bonds.

the alkali metals melt below 200°C. Several post-transition metals also have low melting points, whereas most transition metals melt at temperatures above 1000°C. These differences reflect differences in strength of metallic bonding (Chapter 24) among the metals.

Covalent network solids include crystals of diamond, silicon, some other non-metals, and some covalent compounds such as silicon dioxide (sand) and silicon carbide (carborundum, the abrasive on sandpaper). Many minerals have networks of covalent bonds. The atoms in these solids are held together by a three-dimensional network of covalent bonds, but it is not possible to identify individual molecules in them (Fig. 11.31). For a covalent network solid to break or to melt, covalent bonds must be broken. Because covalent bonds are relatively strong, covalent network solids are characterized by hardness, strength, and high melting points. Diamond, one of the hardest substances known, melts above 3500°C.

Molecular solids, such as ice (Fig. 11.32), sucrose (table sugar), and iodine, are composed of neutral molecules. The strengths of the attractive forces between the units present in different crystals vary widely, as indicated by the melting points of the crystals. Small, symmetrical molecules (nonpolar molecules), such as H_2, N_2, O_2, and F_2, have weak intermolecular attractive forces and form molecular solids with very low melting points (below −200°C). Molecular solids that consist of larger, non-

Figure 11.32

The arrangement of water molecules in ice, a molecular solid. Here the oxygen atoms are red, and the hydrogen atoms are blue. Covalent bonds are shown in red and hydrogen bonds in purple. The oxygen atoms at the corners of a prism are also at the corners of adjoining prisms; this is only a portion of a continuous array.

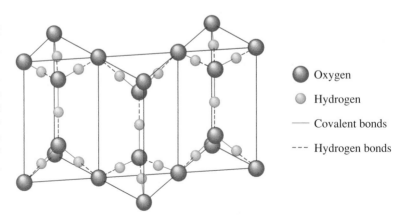

● Oxygen
○ Hydrogen
— Covalent bonds
--- Hydrogen bonds

Figure 11.33
Two types of crystal defects.

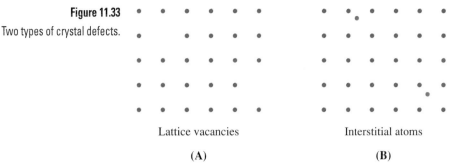

Lattice vacancies   Interstitial atoms

(A)   **(B)**

polar molecules have more attractive forces and melt at higher temperatures, whereas molecular solids composed of asymmetrical molecules with permanent dipole moments (polar molecules) melt at still higher temperatures. Examples include ice (melting point, 0°C) and table sugar (melting point, 185°C).

A crystalline solid has a sharp melting temperature because the forces holding its atoms, ions, or molecules together are all of the same strength. Thus the attractions between the units that make up the crystal all have the same strength, and all require the same amount of energy to be broken. The gradual softening of an amorphous material differs dramatically from the sharp melting of a crystalline solid. This results from the structural nonequivalence of the molecules in the amorphous solid. Some forces are weaker than others, and when an amorphous material is heated, the weakest intermolecular attractions break first. As the temperature is increased further, the stronger attractions are broken. Thus amorphous materials soften over a range of temperatures.

11.13   Crystal Defects

In a crystalline solid the atoms, ions, or molecules are arranged in a definite repeating pattern, but occasional defects may occur in the pattern. Several types of **crystal defects** are known. **Vacancies** are defects that occur when positions that should contain atoms or ions are vacant (Fig. 11.33 A). Less commonly, some atoms or ions in a crystal may occupy positions, called **interstitial sites,** that are located between the regular positions for atoms (Fig. 11.33 B). Other distortions are found in impure crystals, as for example when the cations, anions, or molecules of the impurity are too large to fit into the regular positions without distorting the structure. Minute amounts of impurities are sometimes added to a crystal to cause imperfections in the structure so that the electrical conductivity (see Chapter 24) or some other physical properties of the crystal will change. The controlled formation of defects in silicon crystals has practical application in the manufacture of semiconductors and computer chips.

Figure 11.34
Spheres (in this case, spherical pills) arrange themselves in a closest packed array.

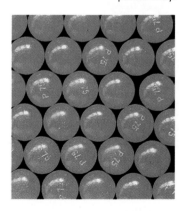

11.14   The Structures of Metals

A pure metal is a crystalline solid with metal atoms packed closely together in a repeating pattern. In most metals the atoms behave as though they were spheres of equal size and arrange themselves as shown in Fig. 11.34. Each sphere touches six others in a plane. This arrangement, called **closest packing,** can extend indefinitely

Figure 11.35

A closest packed structure. (A) In one layer of closest packed spheres, each sphere touches six others. (B) In two closest packed layers, each sphere in layer B (the blue layer) touches three spheres in layer A (the red layer).

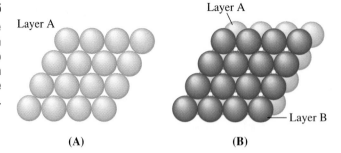

(A)   (B)

in a single layer. Crystals of many metals can be described as stacks of such closest packed layers.

We find two types of stacking of closest packed layers in simple metallic crystalline structures. In both types a second layer (B) is placed on the first layer (A) so that each sphere in the second layer is in contact with three spheres in the first layer, as shown in Fig. 11.35. The third layer is positioned in one of two ways.

In one structure we find each sphere in the third layer directly above a sphere in the first layer (Fig. 11.36 A and C). The third layer is also type A. The stacking continues with type B and type A close packed layers alternating (ABABAB · · ·). This arrangement is called **hexagonal closest packing,** and metals that crystallize this way are said to have a **hexagonal closest packed structure.** Examples include Cd, Co,

Figure 11.36

In both of these different structures (A and B), spheres are packed as compactly as possible. The lower diagrams (C and D) show expanded structures. Note that the first and third layers have identical orientations in structure C and different orientations in structure D. When either structure is infinitely extended, each sphere is surrounded by 12 others.

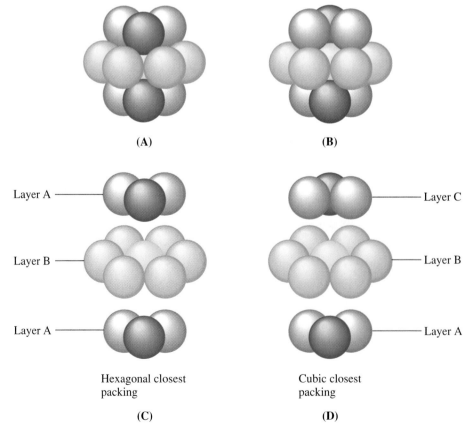

(A)   (B)

Layer A   Layer C

Layer B   Layer B

Layer A   Layer A

Hexagonal closest   Cubic closest
packing   packing

(C)   (D)

Figure 11.37

A portion of a body-centered cubic structure. (A) Note that in one layer of spheres, the spheres do not touch. (B) In two layers, each sphere in one layer touches four spheres in the adjacent layer but none in its own layer.

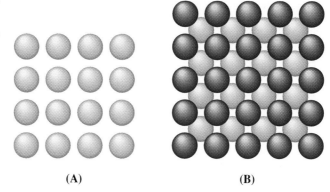

(A)   (B)

Li, Mg, Na, and Zn. (Those elements or compounds that crystallize with the same structure have **isomorphous structures.**)

In a second structure, the third layer is located such that its spheres are not directly above those in either layer A or layer B (Fig. 11.36 B and D). This is a layer of type C. The stacking continues with alternating layers of type A, type B, and type C (ABCABC···), an arrangement we call **cubic closest packing.** Metals that crystallize this way are called **cubic closest packed,** or **face-centered cubic.** Examples include Ag, Al, Ca, Cu, Ni, Pb, and Pt.

In both hexagonal closest packed and cubic closest packed metals, each atom touches 12 near neighbors: 6 in its own layer and 3 in each adjacent layer. This gives each atom a coordination number of 12. The **coordination number** of an atom or ion is the number of neighbors it touches. About two-thirds of all metals crystallize in closest packed arrays with coordination numbers of 12.

Most of the remaining metals crystallize in a **body-centered cubic structure,** which contains planes of spheres that are not closest packed. Each sphere in a plane has four neighbors that do not touch (Fig. 11.37 A). The structure consists of repeating layers of these planes. The second layer is stacked on top of the first such that a sphere in the second layer touches four spheres in the first layer (Fig. 11.37 B). The spheres of the third layer are positioned directly above the spheres of the first layer (Fig. 11.38); those of the fourth, above the second; and so on. Any atom in this structure touches four atoms in the layer above it and four atoms in the layer below it. Thus an atom in a body-centered cubic structure has a coordination number of 8. Isomorphous metals with a body-centered cubic structure include Ba, Cr, Mo, W, and Fe at room temperature.

Polonium (Po) crystallizes in the **simple cubic structure,** which is rare for metals. It contains planes in which each sphere touches its four nearest neighbors (Fig. 11.39 A). The structure is not closest packed. The planes are stacked directly above

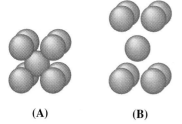

(A)   (B)

Figure 11.38

A portion of a body-centered cubic structure, showing parts of three layers (A) and an expanded view (B). In an infinite extension of this structure, each sphere touches four spheres in the layer above it and four spheres in the layer below.

Figure 11.39

(A) A portion of a plane of spheres found in a simple cubic structure. Note that the spheres are in contact. (B) A portion showing two planes.

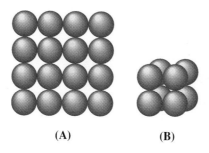

(A)   (B)

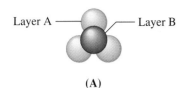

(A)

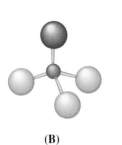

(B)

Figure 11.40

(A) Four spheres form a tetrahedral hole between two adjacent closest packed layers. (B) A smaller cation located in a tetrahedral hole surrounded by four larger anions, with the structure expanded to show the geometrical relationships.

each other such that an atom in the second layer touches only one atom in the first layer (Fig. 11.39 B). The coordination number of a polonium atom in a simple cubic array is 6: An atom touches four other atoms in its own layer, one atom in the layer above, and one atom in the layer below.

11.15   The Structures of Ionic Crystals

Ionic crystals consist of two or more different kinds of ions that usually have different sizes. The packing of these ions into a crystal structure is more complex than the packing of metal atoms that are the same size.

Most monatomic ions behave as charged spheres, and their attraction for ions of opposite charge is the same in every direction. Consequently, stable structures for ionic compounds result (1) when ions of one charge are surrounded by as many ions as possible of the opposite charge and (2) when the cations and anions are in contact with each other. Two principal factors determine structures: the relative sizes of the ions and the ratio of the numbers of positive and negative ions in the compound.

In simple ionic structures we often find the anions, which are normally larger than the cations, arranged in a closest packed array. The cations commonly occupy one of two types of holes (or interstices) remaining between the anions. The smaller of the holes is found between three anions in one plane and one anion in an adjacent plane (Fig. 11.40 A). The four anions that form this hole are arranged at the corners of a tetrahedron (Fig. 11.40 B); the hole is called a **tetrahedral hole.** The larger type of hole is found at the center of six anions (three in one layer and three in an adjacent layer) located at the corners of an octahedron (Fig. 11.41). Such a hole is called an **octahedral hole.**

Depending on the relative sizes of the cations and anions, the cations of an ionic compound may occupy tetrahedral or octahedral holes. As we will see in Section 11.16, relatively small cations occupy tetrahedral holes, and larger cations occupy octahedral holes. If the cations are too large to fit into the octahedral holes, the anions may adopt a more open structure, such as a simple cubic array (Fig. 11.39). The larger cations can then occupy the larger **cubic holes** made possible by the more open spacing (Fig. 11.42).

There are two tetrahedral holes for each anion in either a hexagonal closest packed or a cubic closest packed array of anions. A compound that crystallizes in a closest packed array of anions with cations in the tetrahedral holes can have a maximum cation-to-anion ratio of 2 to 1; all of the tetrahedral holes are filled at this ratio. Exam-

Figure 11.41

(A) Six spheres form an octahedral hole between two adjacent closest packed layers. (B) A smaller cation located in an octahedral hole surrounded by six larger anions, with the structure expanded and rotated to show the geometrical relationships.

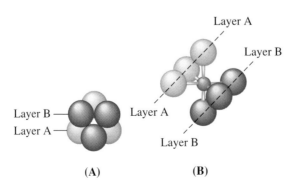

(A)   (B)

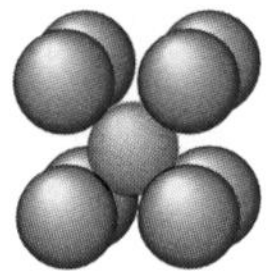

Figure 11.42
A cation in the cubic hole in a simple cubic array of anions.

ples include Li_2O, Na_2O, Li_2S, and Na_2S. Compounds with a cation-to-anion ratio of less than 2 to 1 may also crystallize in a closest packed array of anions with cations in the tetrahedral holes, if the ionic sizes fit. In these compounds, however, some of the tetrahedral holes remain vacant.

Example 11.3   Occupancy of Tetrahedral Holes

Zinc sulfide is an important industrial source of zinc and is also used as a white pigment in paint. Zinc sulfide crystallizes with zinc ions occupying one-half of the tetrahedral holes in a closest packed array of sulfide ions. What is the formula of zinc sulfide?

Solution: Because there are two tetrahedral holes per anion (sulfide ion) and one-half of these holes are occupied by zinc ions, there must be $\frac{1}{2} \times 2$, or 1, zinc ion per sulfide ion. Thus the formula is ZnS.

Check your learning: Lithium selenide can be described as a closest-packed array of selenide ions with lithium ions in all of the tetrahedral holes. What is the formula of lithium selenide?

Answer: Li_2Se

The ratio of octahedral holes to anions in either a hexagonal or a cubic closest packed structure is 1 to 1. Thus compounds with cations in octahedral holes in a closest packed array of anions can have a maximum cation-to-anion ratio of 1 to 1. In NiO, MnS, NaCl, and KH, for example, all the octahedral holes are filled. Ratios of less than 1 to 1 are observed when some of the octahedral holes remain empty.

Example 11.4   Stoichiometry of Ionic Compounds

Sapphire is aluminum oxide. Aluminum oxide crystallizes with aluminum ions in two-thirds of the octahedral holes in a closest packed array of oxide ions. What is the formula of aluminum oxide?

Solution: Because there is one octahedral hole per anion (oxide ion) and only two-thirds of these holes are occupied, the ratio of aluminum to oxygen must be $\frac{2}{3}$ to 1, which would give $Al_{2/3}O$. The simplest whole-number ratio is 2 to 3, so the formula is Al_2O_3.

Check your learning: The white pigment titanium oxide crystallizes with titanium ions in one-half of the octahedral holes in a closest packed array of oxide ions. What is the formula of titanium oxide?

Answer: TiO_2

In a simple cubic array of anions, there is one cubic hole that can be occupied by a cation for each anion in the array. In CsCl, and in other compounds with the same structure, all of the cubic holes are occupied. Half of the cubic holes are occupied in SrH_2, UO_2, $SrCl_2$, and CaF_2.

Figure 11.43

Packing of anions (green spheres) around cations (red spheres) of decreasing size.

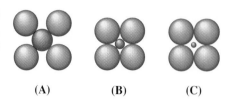

(A)   (B)   (C)

11.16   The Radius Ratio Rule

The structure of an ionic compound is largely the result of stoichiometry and of simple geometric and electrostatic relationships that depend on the relative sizes of the cation and anion. Different types of ionic compounds often crystallize in the same structure when the relative sizes of their ions and their stoichiometries are similar.

A relatively large cation can touch a large number of anions and so occupies a cubic or an octahedral hole, whereas a relatively small cation can touch only a few anions and so occupies a tetrahedral hole. A cation, M^+, with a coordination number of 6 touches four X^- ions in a plane (Fig. 11.43 A). In addition, an X^- ion touches the M^+ ion from above, and another touches it from below (these are not shown in the figure). The M^+ ion shown is large enough to expand the array of X^- ions so they do not touch one another. As long as the expansion is not great enough to allow still another anion to touch the cation, this is a stable situation; the cation–anion contacts are maintained.

Figure 11.43 B illustrates what happens when the size of the M^+ ion is somewhat smaller; the X^- ions touch each other, as do the M^+ and X^- ions. If the size of the M^+ ion is even smaller, it becomes impossible to get a structure with a coordination number of 6. The anions touch (Fig. 11.43 C), but there is no contact between the M^+ and X^- ions—this is an unstable structure. In this case, a more stable structure would be formed with fewer anions about the cation. The limiting condition for the formation of a structure with a coordination number of 6 is illustrated in Fig. 11.43 B: The X^- ions touch one another, and the M^+ and X^- ions touch. This occurs when the sizes of the ions are such that the radius ratio (the radius of the positive ion, r^+, divided by the radius of the negative ion, r^-) is equal to 0.414.

There is a minimum radius ratio (r^+/r^-) for each coordination number below which an ionic structure with that coordination number is generally not stable. This principle is known as the **radius ratio rule.** Table 11.3 lists the approximate limiting values of the radius ratios for several ionic structures.

Table 11.3

Limiting Values of the Radius Ratio for Ionic Compounds

Coordination Number	Type of Hole Occupied	Limiting Values of r^+/r^-
8	Cubic	Above 0.732
6	Octahedral	0.414 to 0.732
4	Tetrahedral	0.225 to 0.414

Example 11.5   Using the Radius Ratio Rule to Predict Coordination Number

Predict the coordination number of Cs^+ (r^+ = 1.69 Å) in CsCl. The radius of a chloride ion is 1.81 Å.

Solution: For CsCl,

$$\frac{r^+}{r^-} = \frac{1.69\ \text{Å}}{1.81\ \text{Å}} = 0.934$$

The radius ratio is greater than 0.732 (Table 11.3), which indicates that a coordination number of 8 is likely for Cs^+ in CsCl.

Check your learning: Predict the coordination number of Na^+ (r^+ = 0.95 Å) in NaCl. The radius of a chlorine ion is 1.81 Å.

Answer: 6

The radius ratio rule is only a guide to the type of structure that may form. It applies strictly only to ionic crystals and in some cases fails with them. Moreover, the rule may not hold for crystals in which the bonds are covalent. In spite of its limitations, however, the radius ratio rule is a useful guide for predicting many structures. It also underlines one of the most significant features responsible for the structures of ionic solids: the relative sizes of the cations and anions.

11.17   Unit Cells

The components of a crystal are arranged in a definite repeating pattern, so we can describe the structure of a crystal by describing one of the repeating units. Our description starts with a **space lattice**, the set of all of the points within the crystal that have identical environments. These points are also arranged in a definite repeating pattern.

A simple three-dimensional cubic space lattice is shown in Fig. 11.44. (The atoms around the points have been omitted so that we can see the lattice.) There are an infinite number of ways in which we can construct a space lattice in any given crystal.

Figure 11.44

A portion of a simple cubic lattice, with one unit cell shaded.

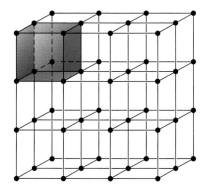

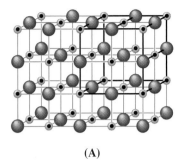

(A)

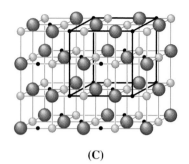

(B)

(C)

Figure 11.45

A portion of the structure of NaCl, showing three positions for a space lattice. Small gray spheres represent Na$^+$, large green spheres Cl$^-$. Black lines connect points that define a unit cell. (A) Lattice points in the center of the Na$^+$ ions. (B) Lattice points in the center of the Cl$^-$ ions. (C) Lattice points between Na$^+$ and Cl$^-$ ions. Each structure has been expanded for clarity; in the actual structure, the ions touch.

For example, we could put the points of a space lattice in the sodium chloride structure at the centers of the sodium ions (Fig. 11.45 A), at the centers of the chloride ions (Fig. 11.45 B), or between the sodium and chloride ions (Fig. 11.45 C). In each case our space lattice would be the same; only the locations of the points of the lattice would differ. We call that part of a space lattice that will generate the entire lattice if repeated in three dimensions a **unit cell** (Fig. 11.44). The cubes outlined in Fig. 11.45 illustrate unit cells in the lattices we selected for the sodium chloride structure. We can now describe the structure of a sodium chloride crystal by describing the size and shape of one of the unit cells in Fig. 11.45 and indicating the arrangement of its contents.

Thus far we have considered only unit cells shaped like a cube, but there are others. In general, a unit cell is defined by the lengths of three axes (a, b, and c) and the angles (α, β, and γ) between them (Fig. 11.46). The axes are defined as the lengths between points in the space lattice. Consequently, unit cell axes join points with identical environments. Unit cells have one of the seven shapes indicated in Fig. 11.47 and Table 11.4.

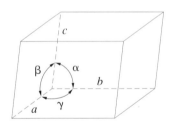

Figure 11.46
A unit cell.

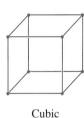

Cubic

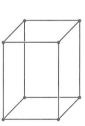

Tetragonal

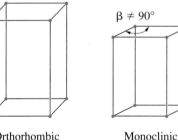

Orthorhombic   Monoclinic

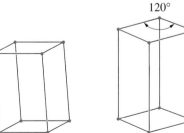

Figure 11.47
The seven unit cell shapes.

Triclinic   Hexagonal   Rhombohedral

Table 11.4

Unit Cells of the Seven Crystal
Systems

System	Axes	Angles
Cubic	$a = b = c$	$\alpha = \beta = \gamma = 90°$
Tetragonal	$a = b \neq c$	$\alpha = \beta = \gamma = 90°$
Orthorhombic	$a \neq b \neq c$	$\alpha = \beta = \gamma = 90°$
Monoclinic	$a \neq b \neq c$	$\alpha = \gamma = 90°; \beta \neq 90°$
Triclinic	$a \neq b \neq c$	$\alpha \neq \beta \neq \gamma \neq 90°$
Hexagonal	$a = b \neq c$	$\alpha = \beta = 90°; \gamma = 120°$
Rhombohedral	$a = b = c$	$\alpha = \beta = \gamma \neq 90°$

We can count the number of lattice points or atoms in a unit cell by using the following rules.

1.   A point or atom lying completely within a unit cell belongs to that unit cell only; count it as 1 when totaling the number of points or atoms in the cell.

2.   A point or atom lying on a face of a unit cell is shared equally by two unit cells; count it at $\frac{1}{2}$ when totaling the number of points or atoms in a cell (Fig. 11.48 A).

3.   A point or atom lying on an edge is shared by four unit cells and is counted as $\frac{1}{4}$ (Fig. 11.48 B).

4.   A point or atom at a corner is shared by eight unit cells and is counted as $\frac{1}{8}$ (Fig. 11.48 C).

Unit Cells of Metals.

Most metals crystallize in one of four unit cells (see Fig. 11.49 on page 400). Polonium, the simple cubic metal, crystallizes with polonium atoms on the lattice points at the corners of a **simple cubic unit cell** (Fig. 11.49 A on page 400). There is $8 \times \frac{1}{8} = 1$ atom (and lattice point) in the simple cubic unit cell of polonium. Because a unit cell containing one lattice point is called a primitive cell, the simple cubic lattice is sometimes called a primitive cubic lattice.

Body-centered cubic metals crystallize with a cubic unit cell that has metal atoms on the lattice points at the corners of the unit cell and a second atom on an identical lattice point in the center of the cell (Fig. 11.49 B on page 400). This is called a **body-centered cubic unit cell.** These cells contain 2 atoms (and lattice points): $8 \times \frac{1}{8} =$

Figure 11.48

(A) Two unit cells share an atom on a face. (B) Four unit cells share an atom on an edge. (C) Eight unit cells share an atom on a corner.

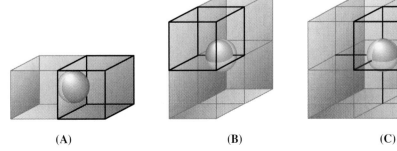

(A)   (B)   (C)

Figure 11.49

Unit cells of metals, showing (in the upper figures) the locations of lattice points and (in the lower figures) metal atoms located in the unit cell.

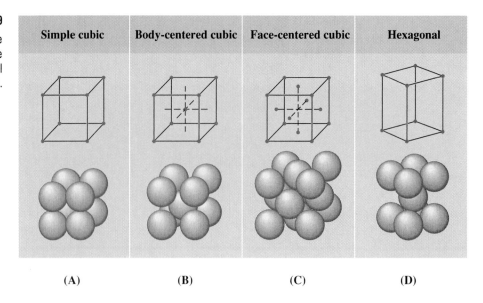

Simple cubic	Body-centered cubic	Face-centered cubic	Hexagonal
(A)	(B)	(C)	(D)

Figure 11.49

Unit cells of metals, showing (in the upper figures) the locations of lattice points and (in the lower figures) metal atoms located in the unit cell.

1 at the corners and 1 in the center of the cube. The atoms at the corners touch the atom in the center along the body diagonals of the cubic unit cell. Because the atoms are on identical lattice points, they have identical environments.

Face-centered cubic metals crystallize with a cubic unit cell that has metal atoms on lattice points at the corners of the unit cell and metal atoms on identical lattice points at the centers of each face of the unit cell (Fig. 11.49 C). This is called a **face-centered cubic unit cell.** These cells contain 4 atoms (and lattice points): $8 \times \frac{1}{8} = 1$ at the corners and $6 \times \frac{1}{2} = 3$ in the centers of the faces. The atoms at the corners touch the atoms in the centers of adjacent faces along the face diagonals of the cube. The metal atoms are on lattice points, they have identical environments.

The common hexagonal metals crystallize with a hexagonal unit cell that has metal atoms on the lattice points at the corners of the unit cell and a second atom inside the cell (Fig. 11.49 D). These cells contain 2 atoms (and lattice points): 1 at the corners and 1 inside the cell. Because the atom in the center of the cell is not on a lattice point, it can have a different environment from the atoms on the corners. In fact, the orientations of the nearest neighbors differ for the two atoms in a hexagonal cell.

Unit Cells of Ionic Compounds.

Many ionic compounds crystallize with cubic unit cells, and we will use these compounds to describe the general features of ionic structures.

The structure of CsCl is simple cubic. Chloride ions are located on the lattice points at the corners of a simple cubic unit cell, and the cesium ion is located at the center of the cell (Fig. 11.50). The cesium ion and the chloride ion touch along the body diagonals of the cubic cell. There are one cesium ion and one chloride ion per unit cell, giving the 1-to-1 stoichiometry required by the formula for cesium chloride. There is no lattice point in the center of the cell, and CsCl is not a body-centered structure because a cesium ion is not identical to a chloride ion.

We have said that the location of lattice points is arbitrary. This is illustrated by an alternative description of the CsCl structure in which the lattice points are located

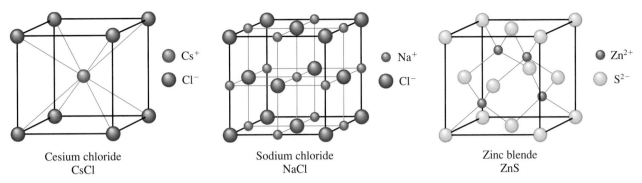

Cesium chloride
CsCl

Na⁺
Cl⁻

Sodium chloride
NaCl

Zinc blende
ZnS

Figure 11.50

The unit cells of some ionic compounds of the general formula MX. The red spheres represent positive ions (cations), and the green or yellow spheres represent negative ions (anions). These structures have been expanded to show the geometrical relationships. In the crystal, the cations and anions touch.

in the centers of the cesium ions. In this description, the cesium ions are located on the lattice points at the corners of the cell, and the chloride ion is located at the center of the cell. The two unit cells are different, but they describe identical structures.

Sodium chloride crystallizes with a face-centered cubic unit cell (Fig. 11.50). Chloride ions are located on the lattice points of the cell. Sodium ions are located in the octahedral holes in the middle of the cell edges and in the center of the cell. The sodium and chloride ions touch each other along the cell edges. The unit cell contains four sodium ions and four chloride ions, giving the 1-to-1 stoichiometry required by the formula, NaCl.

The cubic form of zinc sulfide, zinc blende, also crystallizes in a face-centered cubic unit cell (Fig. 11.50). This structure contains sulfide ions on the lattice points of a face-centered cubic lattice. (The arrangement of sulfide ions is identical to the arrangement of chloride ions in sodium chloride.) Zinc ions are located in alternating tetrahedral holes—that is, in one-half of the tetrahedral holes. There are four zinc and four sulfide ions in the unit cell, giving the empirical formula ZnS.

A calcium fluoride unit cell is also a face-centered cubic unit cell (Fig. 11.51), but in this case the cations are located on the lattice points; equivalent calcium ions are located on the lattice points of a face-centered cubic lattice. All of the tetrahedral sites in the face-centered cubic array of calcium ions are occupied by fluoride ions. There are four calcium ions and eight fluoride ions in a unit cell, a calcium-to-fluorine ratio of 1 to 2, as required by the chemical formula, CaF_2. Close examination of Fig. 11.51 will reveal a simple cubic array of fluoride ions with calcium ions in one-half of the cubic holes. The structure cannot be described in terms of a space lattice of points on the fluoride ions, because the fluoride ions do not all have identical environments. The orientations of the four calcium ions about the fluoride ions differ.

Figure 11.51

The unit cell of CaF_2. The red spheres represent calcium ions, Ca^{2+}, and the yellow-green spheres represent fluoride ions, F^-. This structure has been expanded to show the geometrical relationships. In the crystal, the cations and anions touch.

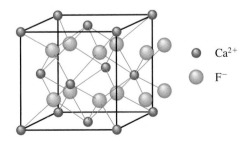

Ca^{2+}

F^-

11.18   Calculation of Ionic Radii

If we know the edge length of a unit cell of an ionic compound and the position of the ions in the cell, we can calculate **ionic radii** for the ions in the compound if we make assumptions about individual ionic shapes and contacts.

Example 11.6   Calculation of Ionic Radii

The edge length of the unit cell of LiCl (NaCl-like structure, face-centered cubic) is 5.14 Å. Assuming that the lithium ion is small enough that the chloride ions are in contact, as in Fig. 11.43 B, calculate the ionic radius for the chloride ion.

Solution: The NaCl structure (see Fig. 11.49) contains a right triangle involving two chloride ions and one sodium ion. In the isomorphous LiCl structure, the lithium ion is so small that all ions in the structure touch, as shown in Fig. 11.43 B.

Because a, the distance between the center of a chloride ion and the center of a lithium ion, is one-half of the edge length of the cubic unit cell

$$a = \frac{5.14 \text{ Å}}{2}$$

Similarly, b is the distance between the center of a chloride ion and the center of a lithium ion and hence is also one-half of the edge length of the cubic unit cell.

$$b = \frac{5.14 \text{ Å}}{2}$$

By the Pythagorean theorem, c, the distance between the centers of two chloride ions, can be calculated. (See Fig. 11.52.)

Figure 11.52

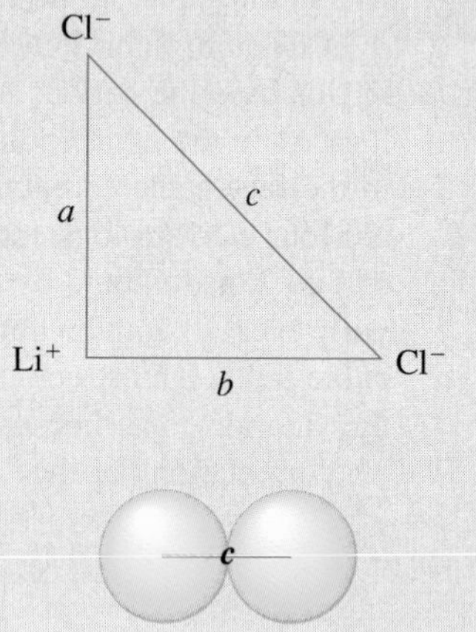

$$c^2 = a^2 + b^2$$

$$= \left(\frac{5.14}{2}\right)^2 + \left(\frac{5.14}{2}\right)^2 = 13.21$$

$$c = \sqrt{13.21} = 3.63 \text{ Å}$$

Because the anions are assumed to touch each other, c is twice the radius of one chloride ion. Hence the radius of the chloride ion is $\frac{1}{2}c$.

$$r_{Cl^-} = \tfrac{1}{2}c = \tfrac{1}{2} \times 3.63 \text{ Å} = 1.81 \text{ Å}$$

It is important that we realize that values for ionic radii calculated from the edge lengths of unit cells depend on numerous assumptions, such as a spherical shape for ions, which are approximations at best. Hence such calculated values are themselves approximate, and comparisons cannot be pushed too far. Nevertheless, this method has proved useful for calculating ionic radii from experimental measurements such as X-ray crystallographic determinations.

11.19 The Born–Haber Cycle

We define the lattice energy, U, of an ionic compound as the energy required to separate the ions in 1 mole of the compound by infinite distances (Section 6.3). For the ionic solid MX, the **lattice energy** is the enthalpy change of the endothermic process

$$MX(s) \longrightarrow M^+(g) + X^-(g) \qquad U = \Delta H^\circ_{298}$$

It is not possible to measure most lattice energies directly. However, a cyclic process can be used to calculate the lattice energy from other quantities. The **Born–Haber cycle** is such a calculation that involves ΔH°_f, the enthalpy of formation of the compound (Section 4.5); I, the ionization energy of the metal (Section 5.10); E.A., the electron affinity of the nonmetal (Section 5.10); ΔH°_s, the enthalpy of sublimation of the metal; D, the bond dissociation energy of the nonmetal (Section 6.12); and U, the lattice energy of the compound. The Born–Haber cycle for sodium chloride analyzes the formation of solid sodium chloride from 1 mole of metallic sodium and $\frac{1}{2}$ mole of chlorine gas as a step-by-step process that can be expressed diagrammatically as follows, the overall change being shown in color and the individual steps in black.

$$
\begin{array}{llll}
Na^+(g) & + & Cl^-(g) & \\
\uparrow I & & \uparrow -E.A. & \searrow -U \\
Na(g) & + & Cl(g) & \\
\uparrow \Delta H^\circ_s & & \uparrow \frac{1}{2}D & \\
Na(s) & + & \frac{1}{2}Cl_2(g) & \xrightarrow{\Delta H^\circ_f} NaCl(s)
\end{array}
$$

First we assume that the sodium metal is vaporized and the bonds in the diatomic chlorine molecules are broken. Then the gaseous sodium atoms are ionized, and the electrons from them are transferred to the gaseous chlorine atoms to form chloride ions. The gaseous sodium ions and chloride ions thus formed come together to give solid sodium chloride. The enthalpy change in this step is the negative of the lattice energy, which is the amount of energy required to produce 1 mole of gaseous sodium ions and 1 mole of gaseous chloride ions from 1 mole of solid sodium chloride. The total energy evolved in this hypothetical preparation of sodium chloride is equal to the experimentally determined enthalpy of formation, ΔH_f, of the compound from its elements. Hess's law (Section 4.6) can be used to show the relationship between the enthalpies of the individual steps and the enthalpy of formation as follows:

1. Enthalpy of sublimation of $Na(s)$

$$Na(s) \longrightarrow Na(g) \qquad\qquad \Delta H = \Delta H_s^\circ \quad = 109 \text{ kJ}$$

2. One-half of the bond energy of Cl_2

$$\tfrac{1}{2}Cl_2(g) \longrightarrow Cl(g) \qquad\qquad \Delta H = \tfrac{1}{2}D \quad = 122 \text{ kJ}$$

3. Ionization energy of $Na(g)$

$$Na(g) \longrightarrow Na^+(g) + e^- \qquad\qquad \Delta H = I \quad = 496 \text{ kJ}$$

4. Negative of the electron affinity of Cl

$$Cl(g) + e^- \longrightarrow Cl^-(g) \qquad\qquad \Delta H = -\text{E.A.} = -368 \text{ kJ}$$

5. Negative of the lattice energy of $NaCl(s)$

$$Na^+(g) + Cl^-(g) \longrightarrow NaCl(s) \qquad\qquad \Delta H = -U \quad = ?$$

Enthalpy of formation of $NaCl(s)$, add Steps 1–5

$$\Delta H = \Delta H_f^\circ = \Delta H_s^\circ + \tfrac{1}{2}D + I + (-\text{E.A.}) + (-U)$$

$$Na(s) + \tfrac{1}{2}Cl_2(g) \longrightarrow NaCl(s) \quad = -411 \text{ kJ (from Appendix I)}$$

Thus the enthalpy of formation can be calculated from other values.

Often we use the Born–Haber cycle to determine other enthalpy values because values of ΔH_f° can be determined by other methods. We can solve for the lattice energy, U, by rearranging the equation for ΔH_f° as follows:

$$U = -\Delta H_f^\circ + \Delta H_s^\circ + \tfrac{1}{2}D + I - \text{E.A.}$$

For sodium chloride, using the above data, we find that the lattice energy is

$$U = (411 + 109 + 122 + 496 - 368) \text{ kJ} = 770 \text{ kJ}$$

The Born–Haber cycle can also be used to calculate any one of the other quantities in the equation for lattice energy, provided that all of the rest are known. Usually ΔH_f°, ΔH_s°, I, and D are known. Chemists often use the cycle to calculate electron affinities that cannot be measured directly by using values of U calculated from the equation described in Section 6.3.

11.20   X-Ray Diffraction

We can determine the size of the unit cell and the arrangement of atoms in a crystal from measurements of the **diffraction of X rays** by the crystal. X rays are electromagnetic radiation with wavelengths (Section 5.2) about as long as the distance between neighboring atoms in crystals (about 2 Å).

When a beam of monochromatic X rays strikes two planes of atoms in a crystal at a certain angle θ, it is reflected (Fig. 11.53). There is a simple mathematical relationship among the wavelength λ of the X rays, the distance between the planes in the crystal, and the angle of diffraction (reflection).

$$n\lambda = 2d \sin \theta$$

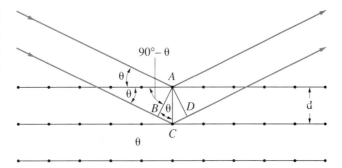

Figure 11.53

Diffraction of a monochromatic beam of X rays by planes of atoms in a crystal.

This equation is called the **Bragg equation** after W. H. Bragg, the English physicist who showed that the diffraction of X rays can be interpreted as though a crystal were a stack of planes. When the X rays strike the crystal at any angle other than θ, there is interference of the reflected rays. This destroys the intensity of the reflected rays.

The reflection corresponding to $n = 1$ is called the first-order reflection, that corresponding to $n = 2$ is the second-order reflection, and so on. Each successive order has a larger angle.

By rotating a crystal and varying the angle θ at which the X-ray beam strikes it, we can determine the values of θ at which diffraction occurs by looking for the maximum diffracted intensities. If the wavelength of the X rays is known, the spacing d of the planes within the crystal can be determined from these values.

We can use values of d to determine the size and shape of the unit cell. The location of atoms in a unit cell can be determined from the relative intensities of the diffraction (Fig. 11.54).

Figure 11.54

X-ray diffraction photograph of a grain of salt, a single crystal of sodium chloride. When the crystal was rotated in a beam of X rays, the X rays were diffracted from many different sets of planes in the crystal. The spacing of the spots (where the diffracted beams struck the film) indicates the size of the unit cell of the crystal. The relative intensity (brightness) of each spot depends on the location of the atoms in the unit cell. It is possible to determine the location of the atoms in the unit cell from the relative intensities of the spots.

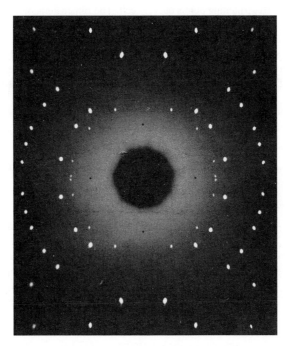

For Review   Summary

The physical properties of condensed matter (liquids and solids) can be explained in terms of the kinetic-molecular theory, which we used in Chapter 10 to explain the physical properties of gases. In a liquid, the **intermolecular attractive forces** hold the molecules in contact, although they still have sufficient kinetic energy to move past each other (diffuse). Because the molecules are in contact, liquids are essentially incompressible. The speeds of the molecules in a liquid vary, so some molecules have enough kinetic energy to escape from the liquid, leading to **evaporation.** The pressure of a vapor in **dynamic equilibrium** with a liquid is called the **vapor pressure** of the liquid. The vapor pressure of a liquid increases with increasing temperature, and the **boiling point** of the liquid is reached when its vapor pressure equals the external pressure on the liquid. The **normal boiling point** of a liquid is the temperature at which its vapor pressure is equal to 1 atmosphere.

When the temperature of a liquid becomes sufficiently low, or the pressure on the liquid sufficiently high, the molecules can no longer move past each other, and a rigid solid forms. Many solids are **crystalline solids;** they are composed of a repeating pattern of atoms, molecules, or ions. Others are **amorphous solids,** in which the disordered arrangement of molecules found in a liquid is retained. Crystalline solids melt at that specific temperature at which enough energy has been added for the kinetic energy of their molecules just to overcome the intermolecular attractive forces holding the molecules in their fixed positions in the crystal. The temperature at which a solid and its liquid are in equilibrium is called the **melting point** of the solid and the **freezing point** of the liquid. Amorphous solids do not have sharp melting points but rather soften over a range of temperatures. Some solids exhibit a measurable vapor pressure and can **sublime.**

The conditions under which a solid and a liquid, a liquid and a gas, or a solid and a vapor are in equilibrium are described by a **phase diagram.** At the **triple point,** all three phases are in equilibrium.

Intermolecular attractive forces, collectively referred to as **van der Waals forces,** are responsible for the behavior of liquids and solids and are electrostatic in nature. **Dipole–dipole attractions** result from the electrostatic attraction of the negative end of one dipolar molecule for the positive end of another. The temporary dipole that results from the motion of the electrons in an atom can induce a dipole in an adjacent atom and give rise to the **London dispersion force.** London forces increase with increasing molecular size. **Hydrogen bonds** are a special type of dipole–dipole attraction that results when hydrogen is bonded to one of the more electronegative elements such as F, O, or N.

The various intermolecular forces are responsible for the **critical temperature** and **critical pressure** of a substance. The critical temperature of a substance is that temperature above which the substance cannot be liquefied, no matter how much pressure is applied. The critical pressure is the pressure required to liquefy a gas at its critical temperature.

The structures of crystalline metals and simple ionic compounds can be described in terms of packing of spheres. Metal atoms can pack in **hexagonal closest packed structures, cubic closest packed structures, body-centered cubic structures,** and **simple cubic structures.** The anions in simple ionic structures commonly adopt one of these structures, and the cations occupy the spaces remaining between the anions. Small cations usually occupy **tetrahedral holes** in a closest packed array of anions. Larger cations usually occupy **octahedral holes.** Still larger cations can occupy **cubic holes** in a simple cubic array of anions. The **radius ratio rule** serves as a guide to the **coordination number** of the cation. The structure of a solid can be described by indicating the size and shape of a **unit cell** and the contents of the cell. The dimensions of a unit cell can be determined by **X-ray diffraction.**

The energy required to separate the ions in a mole of an ionic compound by an infinite distance is called the **lattice energy of the compound.** Lattice energies can be calculated, or they can be determined by using Hess's law in a **Born–Haber cycle.**

Key Terms and Concepts

adhesive force   (11.11)
amorphous solid   (11.12)
body-centered cubic structure   (11.14)
body-centered cubic unit cell   (11.17)
Born–Haber cycle   (11.19)
Bragg equation   (11.20)
capillary action   (11.11)
closest packing   (11.14)
cohesive force   (11.11)

condensation   (11.5)
coordination number   (11.14)
covalent network solid   (11.12)
cubic hole   (11.15)
critical pressure   (11.9)
critical temperature   (11.9)
crystal defects   (11.13)
crystalline solid   (11.12)
cubic closest packed structure   (11.14)

dipole–dipole attraction   (11.2)
distillation   (11.7)
dynamic equilibrium   (11.5)
equilibrium   (11.5)
evaporation   (11.5)
face-centered cubic structure   (11.14)
face-centered cubic unit cell   (11.17)
freezing point   (11.8)
glass   (11.12)

hexagonal closest packed structure (11.14)
hydrogen bond (11.4)
intermolecular attractions (11.1)
interstitial sites (11.13)
ionic radius (11.16, 11.18)
ionic solid (11.12)
isomorphous structures (11.14)
lattice energy (11.19)
London dispersion force (11.3)

melting point (11.8)
metallic solid (11.12)
molecular solid (11.12)
normal boiling point (11.6)
octahedral hole (11.15)
phase diagram (11.10)
radius ratio rule (11.16)
simple cubic structure (11.14)
space lattice (11.17)
sublimation (11.5)

supercritical fluid (11.9)
surface tension (11.11)
tetrahedral hole (11.15)
triple point (11.10)
unit cell (11.17)
vacancies (11.13)
van der Waals force (11.2)
vapor pressure (11.5)
viscosity (11.11)
X-ray diffraction (11.20)

Exercises

Questions

1. List the properties of solids and liquids identified in the introduction to this chapter.

2. In terms of the properties identified in the introduction to this chapter, how do liquids and solids differ? How are they similar?

3. Which of the following belong to the macroscopic domain of liquids and which to the microscopic domain? the kinetic-molecular description of a liquid (Section 11.1), a dipole–dipole attraction, critical temperature, boiling point, a hydrogen bond, phase diagram, surface tension, vapor pressure, variable shape.

4. Which of the following belong to the macroscopic domain of solids and which to the microscopic domain? the cubic shape of a sodium chloride crystal, an ionic bond, the kinetic-molecular description of a solid (Section 11.1), radius ratio, melting point, coordination number, a face-centered-cubic structure, unit cell, a molecular solid, sublimation.

5. Which of the following figures illustrate a macroscopic feature of liquids and which illustrate a microscopic feature? Figs. 11.1, 11.3, 11.5, 11.7, 11.8, 11.11, 11.15, 11.18, 11.21, 11.24.

6. Which of the following figures illustrate a macroscopic feature of solids and which illustrate a microscopic feature? Figs. 11.28, 11.31, 11.33, 11.41, 11.45, 11.53.

7. What are the macroscopic features and what are the microscopic features in the Bragg equation?

8. Water evaporates at room temperature, so why doesn't a bubble of water vapor form inside a completely full water balloon that contains no air bubbles?

9. Heat is added to boiling water. Explain why the temperature of the boiling water does not change. What does change?

10. Heat is added to ice at 0°C. Explain why the temperature of the ice does not change. What does change?

11. Sodium hydrogen carbonate, $NaHCO_3$, is found in the home as baking soda. Describe four types of "intramolecular" and "intermolecular" forces that are present in crystals of this compound.

12. Explain the difference beween intramolecular forces and intermolecular forces.

Kinetic-Molecular Theory and the Condensed State

13. In terms of the kinetic-molecular theory (Section 11.1), in what ways are liquids similar to solids? In what ways are liquids different from solids?

14. In terms of the kinetic-molecular theory (Section 11.1), in what ways are liquids similar to gases? In what ways are liquids different from gases?

15. Describe how the motions of the molecules change as a substance changes from a solid to a liquid.

16. Describe how the motions of the molecules change as a substance changes from a liquid to a gas.

17. How does an increase in pressure lead to liquefaction of a gas?

18. How does a decrease in temperature lead to liquefaction of a gas?

19. Explain why liquids assume the shape of any container into which they are poured, whereas solids are rigid and retain their shape.

Forces Between Molecules

20. What is the evidence that all neutral atoms and molecules exert attractive forces on each other?

21. Why do the boiling points of the noble gases increase in the order He < Ne < Ar < Kr < Xe?

22. Define the following and give an example of each: (a) dispersion force, (b) dipole–dipole attraction, (c) hydrogen bond.

23. On the basis of dipole moments and/or hydrogen bonding, explain in a qualitative way the differences in the boiling points of acetone (56.2°C) and 1-propanol (97.4°C), which have similar molar masses.

24. On the basis of dipole moments and/or hydrogen bonding, explain the differences in the boiling points of n-butane (−1°C) and chloroethane (12°C), which have similar molar masses.

25. Explain why the molar enthalpies of vaporization of the following liquids increase in the order $CH_4 < C_2H_6 < C_3H_8$, even though all three liquids are stabilized by the same dispersion forces (forces between the hydrogen atoms on the outside of the molecules).

26. Explain why the enthalpies of vaporization of the following liquids increase in the order $CH_4 < NH_3 < H_2O$, even though all three liquids have approximately the same molar mass.

27. The melting point of $H_2O(s)$ is 0°C. Would you expect the melting point of $H_2S(s)$ to be −85°C, 0°C, or 185°C? Explain your answer.

28. The melting point of $O_2(s)$ is −218°C. Would you expect the melting point of $N_2(s)$ to be −226°C, −218°C, or −210°C? Explain your answer.

29. Silane (SiH_4), phosphine (PH_3), and hydrogen sulfide (H_2S) melt at −185°C, −133°C, and −85°C, respectively. What does this suggest about the polar character and intermolecular attractions in the three compounds?

30. The enthalpy of vaporization of $CO_2(l)$ is 9.8 kJ mol^{-1}. Would you expect the enthalpy of vaporization of $CS_2(l)$ to be 28 kJ mol^{-1}, 9.8 kJ mol^{-1}, or −8.4 kJ mol^{-1}? Discuss the plausibility of each of these answers.

31. Identify and describe the most important interatomic or intermolecular force responsible for forming each of the following solids: Cu, O_2, NO, HF, Si, CaO.

32. Identify the two most important forces that cause the following to crystallize: (a) CH_3CH_2OH (b) NH_4F, an ionic solid, (c) CH_3CH_2Cl.

33. Which types of forces are responsible for the surface tension of liquid bromine, Br_2: (a) hydrogen bonds, (b) covalent bonds, (c) dipole–dipole attractions, (d) London forces, or (e) ionic bonds? Explain your answer.

34. Explain why a hydrogen bond between two water molecules is weaker than a hydrogen bond between two hydrogen fluoride molecules.

35. Which member of each of the following pairs has the higher vapor pressure at a given temperature?
    (a) $Br_2(s)$ or $I_2(s)$
    (b) $LiCl(l)$ or $MgS(l)$
    (c) $HBr(s)$ or Kr (s)
    (d) $CaCl_2(s)$ or $Cl_2(s)$

36. The hydrogen bonds in liquid hydrogen fluoride, HF, are stronger than those in liquid water, yet the molar enthalpy of vaporization of liquid hydrogen fluoride is less than that of water. Explain.

37. The density of liquid NH_3 is 0.64 g mL^{-1}; the density of gaseous NH_3 at STP is 0.0007 g mL^{-1}. Explain the difference between the densities of these two phases.

38. The density of liquid ethyl chloride, C_2H_5Cl, is 0.903 g mL^{-1} at its boiling temperature of 13°C. What is the density of gaseous ethyl chloride, in grams per milliliter, at the same temperature and a pressure of 1 atm? (*Hint:* Use the ideal gas equation.)

39. Ethyl chloride (boiling point, 13°C) is used as a local anesthetic. When the liquid is sprayed on the skin, it cools the skin enough to freeze and numb it. Explain the cooling effect of liquid ethyl chloride.

40. Why does iodine sublime more rapidly at 110°C than at 25°C?

41. What explanation can you offer for the differences in properties of diethyl ether and ethanol, which have the same molecular formula? Diethyl ether, boiling point 35°C; ethanol, boiling point 78.5°C.

Melting, Evaporation, and Boiling

42. What is the characteristic feature of the dynamic equilibrium between a liquid and its vapor in a closed container?

43. Identify two observations that indicate that some liquids have vapor pressures sufficient to evaporate under ordinary conditions.

44. Identify two observations that indicate that some solids (such as ice, naphthalene, moth balls, and solid air fresheners) have vapor pressures sufficient to evaporate under ordinary conditions.

45. What is the relationship between the intermolecular forces in a liquid and its vapor pressure?

46. What is the relationship between the intermolecular forces in a solid and its melting temperature?

47. Explain why the vapor pressure of a liquid increases as its temperature increases.

48. Explain why the vapor pressure of a solid increases as its temperature increases.

49. Why does spilled gasoline evaporate more rapidly on a hot day than on a cold day?

50. (a) Look up the boiling points of HF, HCl, HBr, and HI in a handbook and explain why they vary as they do.
    (b) Look up the boiling points of Ne, Ar, Kr, and Xe in a handbook and explain why they vary as they do.
    (c) Neon and HF have approximately the same molecular masses. Explain why their boiling points differ.
    (d) Compare the change in the boiling points of Ne, Ar, Kr, and Xe with the change in the boiling points of HF, HCl, HBr, and HI and explain the difference between the changes with increasing atomic or molecular mass.

51. When is the boiling point of a liquid equal to its normal boiling point?

52. How does the boiling of a liquid differ from its evaporation?

53. Arrange each of the following sets of compounds in order of increasing boiling temperature.
    (a) HCl, H_2S, SiH_4
    (b) F_2, Cl_2, Br_2
    (c) CH_4, C_2H_6, C_3H_8
    (d) O_2, NO, CaO

54. Use the information in Fig. 11.11 and estimate the boiling point of water in Denver when the atmospheric pressure is 625 torr.

55. A syringe like that shown in Fig. 11.20 A at a temperature of 20°C is filled with liquid ether in such a way that there is no space for any vapor.
    (a) In one experiment the temperature is kept constant and the plunger is withdrawn somewhat, forming a volume that can be occupied by vapor. What is the approximate pressure of the vapor produced?
    (b) If the system were insulated and no heat could enter or leave, would the temperature of the liquid increase, decrease, or remain constant as the plunger was withdrawn? Explain.

56. The molecular mass of butanol, C_4H_9OH, is 74, that of ethylene glycol, $CH_2(OH)CH_2OH$, is 62, yet their boiling points are 117.2°C and 174°C, respectively. Explain the reason for the difference.

57. If a severe storm results in the loss of electricity, it may be necessary to use an old-fashioned clothes line to dry laundry. In many parts of the country in the dead of winter, the clothes quickly freeze when they are hung on the line. If it doesn't snow, will they dry anyway? Explain your answer.

58. Explain the following observations.
    (a) It takes longer to cook an egg in Ft. Davis, Texas (altitude, 5000 ft above sea level) than in Boston (at sea level).
    (b) Perspiring is a mechanism for cooling the body.

Phase Diagrams, Critical Temperature, and Critical Pressure

59. From the phase diagram for water (Fig. 11.21), determine the state of water at
    (a) −10°C and 400 torr   (b) 25°C and 700 torr
    (c) 50°C and 300 torr   (d) 80°C and 50 torr
    (e) −10°C and 2 torr   (f) 50°C and 2 torr

60. From the phase diagram for water (Fig. 11.21), determine the state of water at
    (a) 35°C and 650 torr   (b) −15°C and 300 torr
    (c) −15°C and 1 torr   (d) 75°C and 25 torr
    (e) 40°C and 1 torr   (f) 60°C and 400 torr

61. What phase changes can water undergo as the temperature changes if the pressure is held at 2 torr? If the pressure is held at 400 torr?

62. What phase changes can water undergo as the pressure changes if the temperature is held at 0.005°C? If the temperature is held at 40°C? At −40°C?

63. From the accompanying phase diagram for carbon dioxide, determine the state of CO_2 at
    (a) −30°C and 20 atm   (b) −60°C and 10 atm
    (c) −60°C and 1 atm   (d) 20°C and 15 atm
    (e) 0°C and 1 atm   (f) 20°C and 1 atm

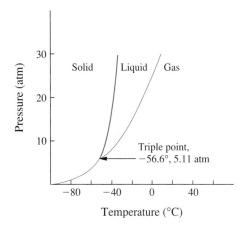

64. From the phase diagram for carbon dioxide in Exercise 63, determine the state of CO_2 at
    (a) 20°C and 10 atm   (b) 10°C and 20 atm
    (c) 10°C and 1 atm   (d) −40°C and 5 atm
    (e) −80°C and 15 atm   (f) −80°C and 0.1 atm

65. What phase changes can carbon dioxide undergo as the temperature changes if the pressure is held at 15 atm? If the pressure is held at 5 atm? What is the approximate temperature of these changes? (See the phase diagram in Exercise 63.)

66. What phase changes can carbon dioxide undergo as the pressure changes if the temperature is held at −50°C? If the temperature is held at −40°C? At 20°C? (See the phase diagram in Exercise 63.)

67. Draw a rough graph that shows how the pressure changes inside a cylinder of carbon dioxide at a pressure of 65 atm and a temperature of 20°C, as gaseous carbon dioxide is released from the cylinder at constant temperature.

68. Draw a rough graph that shows how the pressure changes inside a cylinder of carbon dioxide at a pressure of 65 atm and a temperature of 40°C, as gaseous carbon dioxide is released from the cylinder at constant temperature.

69. What is the triple point of a substance?

70. Is it possible to liquefy nitrogen at room temperature (about 25°C)? Is it possible to liquefy sulfur dioxide at room temperature? Explain your answers.

71. If one continues beyond point C on the line BC in Fig. 11.21 A, at what temperature and pressure does the line end?

72. Dry ice, $CO_2(s)$, does not melt at atmospheric pressure. It sublimes at a temperature of $-78°C$. What is the lowest pressure at which $CO_2(s)$ will melt to give $CO_2(l)$? At approximately what temperature will this occur? (See Exercise 63 for the phase diagram.)

Properties of Solids

73. Solids can be classified into the following four types: 1) ionic, 2) metallic, 3) covalent network, 4) molecular. For each of these types of solids, indicate what kinds of particles (atoms, molecules, or ions) occupy the unit cells, and identify forces among these particles.

74. What types of liquids form amorphous solids?

75. At very low temperatures, oxygen, O_2, freezes and forms a crystalline solid. Which term best describes these crystals: (a) ionic, (b) covalent network, (c) metallic, (d) amorphous, or (e) molecular crystals?

76. As it cools, olive oil slowly solidifies and forms a solid over a range of temperatures. Which best describes the solid: (a) ionic, (b) covalent network, (c) metallic, (d) amorphous, or (e) molecular crystals?

77. Explain why a sample of amorphous boric oxide, (B_2O_3, Fig. 11.28) could be considered a form of a liquid even though it looks like a solid, is rigid, and shatters when struck with a hammer.

78. Explain why ice, which is a crystalline solid, has a melting temperature of $0°C$, whereas butter, which is an amorphous solid, softens over a range of temperatures.

79. Identify the type of crystalline solid (metallic, network covalent, ionic, or molecular) formed by each of the following substances.

(a) SiO_2 (b) KCl
(c) Cu (d) CO
(e) C (diamond)   (f) $BaSO_4$
(g) NH_3 (h) NH_4F
(i) C_2H_5OH

80. Identify the type of crystalline solid (metallic, network covalent, ionic, or molecular) formed by each of the following substances.

(a) $CaCl_2$ (b) SiC
(c) N_2 (d) Fe
(e) C (graphite)   (f) $CH_3CH_2CH_2CH_3$
(g) HCl (h) NH_4NO_3
(i) K_3PO_4

81. What is the coordination number of a chromium atom in the body-centered cubic structure of chromium?

82. What is the coordination number of an aluminum atom in the face-centered cubic structure of aluminum?

83. Cobalt metal crystallizes in a hexagonal closest packed structure. What is the coordination number of a cobalt atom?

84. Nickel metal crystallizes in a cubic closest packed structure. What is the coordination number of a nickel atom?

85. Describe the crystal structure of Pt, which crystallizes with four equivalent metal atoms in a cubic unit cell.

86. Describe the crystal structure of iron, which crystallizes with two equivalent metal atoms in a cubic unit cell.

87. Barium crystallizes in a body-centered cubic unit cell with an edge length of 5.025 Å.
(a) What is the atomic radius of barium in this structure?
(b) Calculate the density of barium.

88. Tungsten crystallizes in a body-centered cubic unit cell with an edge length of 3.165 Å.
(a) What is the atomic radius of tungsten in this structure?
(b) Calculate the density of tungsten.

89. Platinum (atomic radius = 1.38 Å) crystallizes in a cubic closest packed structure. Calculate the edge length of the face-centered cubic unit cell and the density of platinum.

90. Aluminum (atomic radius = 1.43 Å) crystallizes in a cubic closest packed structure. Calculate the edge length of the face-centered cubic unit cell and the density of aluminum.

91. Cadmium sulfide, which artists sometimes use as a yellow pigment, crystallizes with cadmium occupying one-half of the tetrahedral holes in a closest packed array of sulfide ions. What is the formula of cadmium sulfide? Explain your answer.

92. A compound of cadmium, tin, and phosphorus is used in the fabrication of some semiconductors. It crystallizes with cadmium occupying one-quarter of the tetrahedral holes and tin occupying one-quarter of the tetrahedral holes in a closest packed array of phosphide ions. What is the formula of the compound? Explain your answer.

93. What is the formula of the magnetic oxide of cobalt, used in recording tapes, that crystallizes with cobalt atoms occupying one-eighth of the tetrahedral holes and one-half of the octahedral holes in a closest packed array of oxide ions?

94. A compound containing zinc, aluminum, and sulfur crystallizes with a closest packed array of sulfide ions. Zinc ions are found in one-eighth of the tetrahedral holes, and aluminum ions in one-half of the octahedral holes. What is the empirical formula of the compound?

95. A compound of thallium and iodine crystallizes in a simple cubic array of iodide ions with thallium ions in all of the cubic holes. What is the formula of this iodide? Explain your answer.

96. Explain why the chemically similar alkali metal chlorides NaCl and CsCl have different structures, whereas the chemically different NaCl and MnS have the same structure.

97. Each of the following compounds crystallizes in a structure matching that of NaCl, CsCl, ZnS, or CaF_2. From the radius ratio, predict which structure is formed by each. Show your work.

   (a) NaF   (b) AlAs   (c) CoO
   (d) CsBr   (e) BeO   (f) BaF_2

98. Each of the following compounds crystallizes in a structure matching that of NaCl, CsCl, ZnS, or CaF_2. From the radius ratio, predict which structure is formed by each. Show your work.

   (a) ZnSe   (b) NiO   (c) SrSe
   (d) SrF_2 (e) TlBr   (f) CaS

99. Rubidium iodide crystallizes with a cubic unit cell that contains iodide ions at the corners and a rubidium ion in the center. What is the formula of the compound?

100. One of the various manganese oxides crystallizes with a cubic unit cell that contains manganese ions at the corners and in the center. Oxide ions are located at the center of each edge of the unit cell. What is the formula of the compound?

101. NaH crystallizes with the same crystal structure as NaCl. The edge length of the cubic unit cell of NaH is 4.880 Å.

   (a) Calculate the ionic radius of H^-. (The ionic radius of Na^+ is 0.95 Å.)
   (b) Calculate the density of NaH.

102. Thallium(I) iodide crystallizes with the same structure as CsCl. The edge length of the unit cell of TlI is 4.20 Å.

   (a) Calculate the ionic radius of Tl^+. (The ionic radius of I^- is 1.81 Å.)
   (b) Calculate the density of TlI.

103. A cubic unit cell contains manganese ions at the corners and fluoride ions at the center of each edge.

   (a) What is the empirical formula of this compound? Explain your answer.
   (b) What is the coordination number of the Mn ion?
   (c) Calculate the edge length of the unit cell if the radius of a Mn ion is 0.65 Å.
   (d) Calculate the density of the compound.

104. As minerals were formed from the molten magma, different ions occupied the same sites in the crystals. Lithium often occurs along with magnesium in minerals, despite the difference in the charge on their ions. Suggest an explanation.

105. The lattice energy of LiF is 1023 kJ mol^{-1} and the distance between Li and F is 2.008 Å. NaF crystallizes in the same structure as LiF but with a distance between Na and F of 2.31 Å. Which of the following values most closely ap-

proximates the lattice energy of NaF: 510, 890, 1023, 1175, or 4090 kJ mol^{-1}? Explain your choice.

106. For which is the least energy required to convert 1 mol of the solid into separated ions? (a) MgO, (b) SrO, (c) KF, (d) CsF, (e) MgF_2.

107. The reaction of a metal, M, with a halogen, X_2, proceeds as indicated by the equation

$$M(s) + X_2(g) \longrightarrow MX_2(s)$$

If the reaction is exothermic, indicate the effect of the following factors on the enthalpy change for this reaction. Explain your answers.

   (a) A large radius versus a small radius for M^{+2}.
   (b) A high ionization energy versus a low ionization energy for M.
   (c) An increasing bond energy for the halogen.
   (d) A decreasing electron affinity for the halogen.
   (e) An increasing size of the anion formed by the halogen.

108. What is the spacing between crystal planes that diffract X rays with a wavelength of 1.541 Å at an angle θ of 15.55° (first-order reflection)?

109. Gold crystallizes in a face-centered cubic unit cell. The second-order reflection ($n = 2$) of X rays for the planes that make up the tops and bottoms of the unit cells is at $\theta = 22.20°$. The wavelength of the X rays is 1.54 Å. What is the density of metallic gold?

Applications and Additional Exercises

110. The types of intermolecular forces in a substance are identical whether it is a solid, a liquid, or a gas. Why, then, does a substance change phase from a gas to a liquid or to a solid?

111. In which of the following lists are the compounds arranged in order of increasing boiling points?

   (a) $N_2 < CO_2 < H_2O < KCl$
   (b) $H_2O < N_2 < CO_2 < KCl$
   (c) $N_2 < KCl < CO_2 < H_2O$
   (d) $CO_2 < N_2 < KCl < H_2O$
   (e) $KCl < H_2O < CO_2 < N_2$.

112. Explain why some molecules in a solid may have enough energy to sublime away from the solid, even though the solid does not contain enough energy to melt.

113. (a) The density of aluminum is 2.7 g cm^{-3}; that of silicon is 2.3 g cm^{-3}. Explain why Si has the lower density even though it has heavier atoms.
   (b) Which would be expected to have the higher melting point, Al or Si? Explain your answer.

114. How much energy is released when 250 g of steam at 135°C is converted to ice at −20°C? The specific heat of steam is 2.00 J g^{-1} K^{-1}; that of liquid water is 4.18 J g^{-1} K^{-1}; that of ice is 2.04 J g^{-1} K^{-1}.

115. The enthalpy of vaporization of water is larger than its enthalpy of fusion. Explain why.

116. Which of the following elements reacts with sulfur to form a solid in which the sulfur atoms form a closest packed array with all of the octahedral holes occupied: Li, Na, Be, Ca, or Al?

117. When an electron in an excited molybdenum atom falls from the L to the K shell, an X ray is emitted. These X rays are diffracted at an angle of 7.75° by planes with a separation of 2.64 Å. What is the difference in energy between the K shell and the L shell in molybdenum, assuming a first-order diffraction?

118. The free space in a metal can be found by subtracting the volume of the atoms in a unit cell from the volume of the cell. Calculate the percentage of free space in each of the three cubic lattices if all atoms in each are of equal size and touch their nearest neighbors. Which of these structures represents the most efficient packing? That is, which packs with the least amount of unused space?

119. What is the percent by mass of titanium in rutile, a mineral that contains titanium and oxygen, if structure can be described as a closest packed array of oxide ions with titanium ions in one-half of the octahedral holes? What is the oxidation number of titanium?

CHAPTER OUTLINE

12.1   The Nature of Solutions

The Process of Dissolution
12.2   The Formation of Solutions
12.3   Dissolution of Ionic Compounds
12.4   Dissolution of Molecular Electrolytes

Macroscopic Properties of Solutions
12.5   Solutions of Gases in Liquids
12.6   Solutions of Liquids in Liquids (Miscibility)
12.7   The Effect of Temperature on the Solubility
       of Solids in Water
12.8   Solid Solutions

Expressing Concentration
12.9   Percent Composition
12.10 Molarity
12.11 Molality
12.12 Mole Fraction

Colligative Properties of Solutions
12.13 Lowering of the Vapor Pressure of
       a Solvent
12.14 Elevation of the Boiling Point of a Solvent
12.15 Distillation of Solutions
12.16 Depression of the Freezing Point of
       a Solvent
12.17 Phase Diagram for an Aqueous Solution
       of a Nonelectrolyte
12.18 Osmosis and Osmotic Pressure of
       Solutions
12.19 Determination of Molecular Masses
12.20 Colligative Properties of Electrolytes

Colloid Chemistry
12.21 Colloids
12.22 Preparation of Colloidal Systems
12.23 Soaps and Detergents
12.24 Electrical Properties of Colloidal Particles
12.25 Gels

12

Solutions and Colloids

Many different properties of solutions are reflected in this photograph of cod beneath the ice in the Arctic Ocean.

Solutions are crucial to the processes that sustain our lives and to many other processes that involve chemical reactions. When we digest food, for example, the nutrients must go into solution before they can pass through the walls of our intestines into our blood. There, they are carried throughout our bodies in solution. The air we breathe is a solution of nitrogen, oxygen, and other gases. Acid rain is a solution of oxides of sulfur in rain water. The dissolution of substances from the air and the earth is important in converting rocks to soil, in altering the fertility of the soil, and in changing the form of the earth's surface. Many minerals are the result of reactions that have taken place in solution.

Most chemical reactions take place in solution. In a gaseous or liquid solution, molecules and ions can move freely, come into contact with each other, and react. On the other hand, molecules and ions cannot move freely in solids; chemical reactions between solids, if they occur at all, are generally very slow.

In this chapter we will consider the nature of solutions. We will examine factors that determine whether a solution will form and the properties of the solutions that may result. In addition, we will discuss colloids—systems that resemble solutions but consist of dispersions of particles that are somewhat larger than ordinary molecules or ions.

12.1   The Nature of Solutions

When we stir a little sugar with water, the sugar dissolves and forms a solution. As do all solutions, this solution consists of a **solute** (the substance that dissolves—in this case, sugar) and a **solvent** (the substance in which a solute dissolves—in this case, water). Sugar is a molecular solid. When it dissolves, sugar molecules become uniformly distributed among the molecules of water; thus the solution is a homogeneous mixture of sugar and water (solute and solvent) molecules. The molecules of sugar diffuse continuously through the water, and although sugar molecules are heavier than water molecules, the sugar does not settle out upon standing.

Potassium dichromate, $K_2Cr_2O_7$, is an ionic compound composed of colorless potassium ions, K^+, and orange dichromate ions, $Cr_2O_7^{2-}$. When we stir a little solid potassium dichromate in water (Fig. 12.1), the potassium ions and dichromate ions become uniformly distributed among the molecules of water. This orange solution is a homogeneous mixture of potassium ions, dichromate ions, and water molecules. The solute particles (ions in this case) diffuse through the water just as molecular solutes do, and they do not settle out upon standing.

Figure 12.1

Two solutions and a sample of solid potassium dichromate, $K_2Cr_2O_7$. The colors show that the orange $Cr_2O_7^{2-}$ ion is uniformly distributed; thus both solutions are homogeneous. The more intensely colored solution contains more $K_2Cr_2O_7$ than the less intensely colored solution and is more concentrated.

We call a solution that contains a small proportion of solute relative to solvent **dilute;** we call one with a large proportion of solute to solvent **concentrated** (Fig. 12.1). A solution is **unsaturated** if more solute will dissolve in it; it is **saturated** if no more solute will dissolve. The **solubility** of a solute is the quantity that will dissolve in a given amount of solvent to produce a saturated solution.

If we add solid sugar to a saturated solution of sugar, we see the solid fall to the bottom and no more seems to dissolve. Actually, molecules of sugar continue to leave the solid and dissolve, but at the same time, other molecules of sugar in solution collide with the solid and take up positions on the crystal. In some cases we can actually see the crystals of sugar slowly change shape as this process continues; however, the total mass of the crystals does not change. In a given amount of time (say, in 1 second), the number of molecules that return to the solid is equal to the number of molecules that dissolve; thus a state of **equilibrium** exists, much like the dynamic equilibrium that occurs when a liquid evaporates in a closed container (Section 11.5). That is, the amount of solid sugar and the amount of dissolved sugar do not change. An equilibrium between a solute and a solution is established any time excess solute and a saturated solution are in contact.

As we will see in Section 12.7, the solubilities of many solids increase as the temperature increases and decrease as the temperature decreases. Thus if we prepare a saturated solution at an elevated temperature (where the solute is more soluble) and then cool the solution, some solute usually crystallizes so the solution remains saturated as the solution cools. However, if we prepare a saturated solution at an elevated temperature and remove all traces of undissolved solute, we can sometimes cool the solution without crystallization of the solute. When the cool solution contains more solute than it would if the dissolved solute were in equilibrium with undissolved solute, it is called **supersaturated.** Some syrups prepared this way are supersaturated solutions of sugar in water.

Supersaturated solutions are unstable, and the excess dissolved solute may crystallize spontaneously. Alternatively, we can initiate crystallization of the excess solute by shaking or stirring the solution or by adding a small crystal of the solute (Fig. 12.2). After crystallization is complete, a saturated solution, in equilibrium with the crystals of solute, remains.

Water is used so often as a solvent that the word *solution* has come to imply an aqueous solution to many people (aqueous solutions have water as the solvent). However, almost any gas, liquid, or solid can act as a solvent for other gases, liquids, or

Figure 12.2

The excess sodium acetate in a saturated solution of sodium acetate crystallizes spontaneously when a "seed" crystal of sodium acetate is added and fills the container with solid sodium acetate.

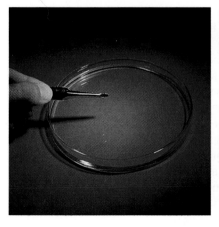
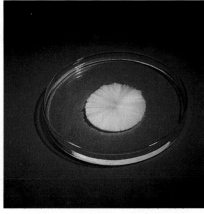
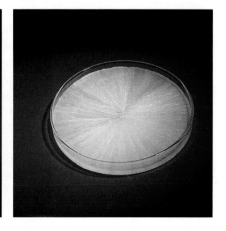

solids. Many alloys are solid solutions of one metal dissolved in another; for example, nickel coins contain nickel dissolved in copper. Air is a gaseous solution, a homogeneous mixture of gases. Oxygen (a gas), ethanol (a liquid), and sugar (a solid) all dissolve in water (a liquid) to form liquid solutions. All solutions have certain properties in common:

1. They are homogeneous; that is, after a solution is mixed, it has the same composition from point to point.

2. The components of a solution are mixed on a molecular scale; that is, they consist of a mixture of atoms, molecules, and/or ions.

3. The solute in a solution that is not supersaturated will not settle or separate from the solvent.

4. The composition of a solution can be varied continuously, within limits.

Although it is easy to identify the solute and solvent in most cases (as when 1 gram of sugar is dissolved in 100 milliliters of water), sometimes the identification is difficult. For example, we cannot distinguish solute from solvent in a solution of equal amounts of ethanol and water. In such cases, the choice is arbitrary.

The Process of Dissolution

12.2   The Formation of Solutions

Figure 12.3

Samples of helium and argon (A) spontaneously mix to give a solution in which the disorder of the molecules of the two gases is increased (B).

He   Ar

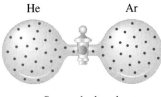

Stopcock closed

(A)

Stopcock open

(B)

When we pour ethanol into water, a solution forms by a **spontaneous process,** a process that occurs without the input of energy. Sometimes we stir to speed up the dissolution process or the mixing, but even without stirring a homogeneous solution would form if we waited long enough. Two factors favor a spontaneous process: (1) a decrease in the internal energy of the system (an exothermic change, Section 4.1) and (2) an increase in the disorder in the system. During dissolution, an internal energy change occurs as heat is absorbed or evolved when the solute and solvent mix. A change in the disorder results from the mixing of the solute and solvent.

Let us first consider the formation of a solution with no energy change, so we can concentrate on how changes in disorder contribute to the formation of solutions. Then we will consider the effect of energy changes.

When the strengths of the intermolecular forces of attraction between the solute and solvent molecules (or ions) are the same as the strengths of the forces between the molecules in the separate components, a solution is formed with no accompanying energy change. Such a solution is called an **ideal solution.** In Section 12.13 we will see that we can identify an ideal solution from its macroscopic behavior. Ideal gases (or gases such as helium and argon, which closely approach ideal behavior) contain molecules that have no significant intermolecular attractions, so a mixture of these gases is one example of an ideal solution.

When we connect bulbs that contain helium and argon (Fig. 12.3 A), the gases spontaneously diffuse together and form a solution (Fig. 12.3 B). This solution forms because the disorder of the helium and argon atoms increases when they mix. The atoms occupy a volume twice as large as that which each occupied before mixing, and they are randomly distributed among one another.

Figure 12.4

When we consider the steps in the formation of a solution, we see that the molecules of the solute and solvent must separate (Steps 1 and 2) before they can mix (Step 3). Thus intermolecular forces can play an important role in solution formation.

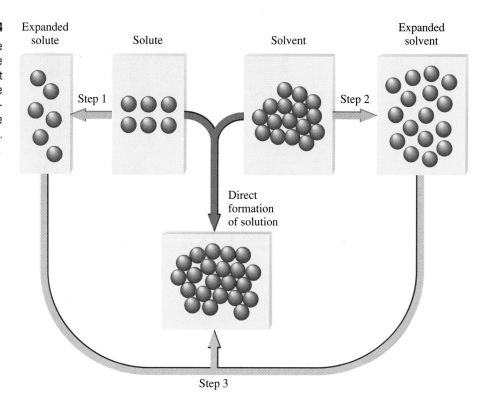

Figure 12.5

Oil and water are immiscible liquids.

We can also make ideal solutions from pairs of chemically similar liquids such as the alcohols methanol, CH_3OH, and ethanol, C_2H_5OH, or the liquid hydrocarbons pentane, C_5H_{12}, and hexane, C_6H_{14}. If we placed methanol and ethanol (or pentane and hexane) in the bulbs shown in Fig. 12.3 A, the molecules of the liquids would diffuse together spontaneously (although at a much slower rate than gases), giving solutions with a disorder greater than that of the pure liquids.

We can expect that moving molecules will become randomly distributed among one another and form a solution unless something holds them back. One microscopic effect that can prevent mixing is due to forces of attraction between molecules (intermolecular forces). These forces are negligible in gases, so gases dissolve in each other in any proportions. However, intermolecular forces play an important role in the dissolution of liquids and solids (Fig. 12.4). If solute molecules attract each other strongly but attract solvent molecules weakly, then the solute molecules will remain in contact with each other and not dissolve, even though formation of a solution would increase their disorder. If the solvent molecules attract each other strongly but do not attract the solute molecules, then the solvent molecules may not separate to let the solute molecules into the solvent. A solution forms only when the attractions between solute and solvent molecules are more or less equal to, or are greater than, the sum of the attractions between solute molecules and those between solvent molecules.

Oil and water don't mix (Fig. 12.5). We can see why hydrocarbons such as oil and gasoline don't mix with water if we consider what happens when octane, a typical hydrocarbon in gasoline, is added to water. Water molecules are held in contact by hydrogen bonds (Section 11.4); octane molecules are held in contact by London dispersion forces (Section 11.3). When the two liquids are mixed, the attraction between the octane molecules and the water molecules is not strong enough to overcome the

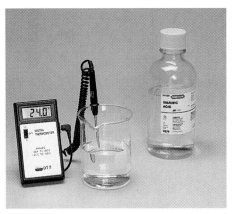 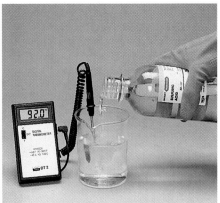

Figure 12.6   When concentrated sulfuric acid dissolves in water, the solution gets hot; here the temperature changes from 24°C to 92°C.

hydrogen bonds between the water molecules. The relatively strong hydrogen bonds keep the water molecules clustered together, and no mixing occurs. Eventually, the water and octane separate into two layers again.

On the other hand, when we add methanol to water, a solution forms readily. Water–water, methanol–methanol, and water–methanol hydrogen bonds are of about equal strength. There is no tendency for stronger hydrogen bonding to cause the water molecules or the methanol molecules to cluster together, and a solution forms because of the increase in disorder.

If stronger attractions replace weaker attractions as a solution forms, heat is released (the dissolution process is exothermic) as the stronger solute–solvent attractions replace weaker solute–solute and solvent–solvent attractions during dissolution. If weaker attractions replace stronger attractions as a solution forms, heat is absorbed (the dissolution process is endothermic).

Heat is evolved when sulfuric acid dissolves in water. When 1 mole of sulfuric acid is dissolved in 9 moles of water, the resulting acid–water interactions are stronger than the sum of the acid–acid attractive forces in pure sulfuric acid plus the water–water attractive forces in pure water. The solution becomes very hot (Fig. 12.6) because 63.2 kilojoules of heat is produced. The evolution of energy as heat indicates that the solution contains less energy than the separate components before mixing.

Both a loss of energy and an increase in disorder favor a spontaneous process such as the formation of a solution. However, it is not necessary for both the change in energy and the change in disorder to favor a spontaneous process for the process to occur. One effect can dominate the other.

Ammonium nitrate is used to make instant "ice packs" for the treatment of athletic injuries (Fig. 12.7). A thin-walled plastic bag of water is sealed inside a larger bag with crystalline NH_4NO_3. When we break the smaller bag, a cold solution of NH_4NO_3 forms, and the cold reduces swelling of the injured area. Ammonium nitrate dissolves in water even though the process is endothermic (the solution cools as thermal energy is absorbed from the water during the dissolution process). The solution forms because the increase in disorder is so large that it more than compensates for the increase in the energy content.

Figure 12.7

An instant ice pack gets cold when ammonium nitrate dissolves in water, an endothermic process.

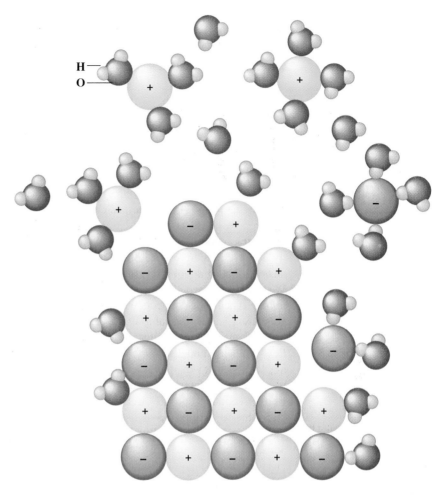

12.3 Dissolution of Ionic Compounds

Water and other polar molecules are attracted to ions (Fig. 12.8). The electrostatic attraction between an ion and a molecule with a dipole (Section 7.3) is called an **ion–dipole attraction.** These attractions play an important role in the dissolution of ionic compounds in water.

When ionic compounds dissolve in water, the ions in the solid separate because water molecules surround and insulate the ions, reducing the strong electrostatic forces between them. Let us consider what happens at the microscopic level when we add solid potassium chloride to water. Ion–dipole forces attract the positive (hydrogen) end of the polar water molecules to the negative chloride ions at the surface of the solid, and they attract the negative (oxygen) end to the positive potassium ions. The water molecules penetrate between individual K^+ and Cl^- ions and surround them, reducing the strong interionic forces that bind the ions together and letting them move off into solution as hydrated ions (Fig. 12.8). Several water molecules associate with (solvate) each ion in solution. The increase in the distance between ions that is due to the layers of water molecules around them reduces the electrostatic attraction between oppositely charged ions. In addition, the layers of water act as insulators, which further reduces the electrostatic attraction. The reduction of the electrostatic attraction permits the independent motion of each hydrated ion in a dilute solution, resulting in an increase in the disorder of the system as the ions change from their ordered positions in the crystal to much more disordered and mobile states in solution. This increased disorder is responsible for the dissolution of many ionic compounds, including potassium chloride, that dissolve with absorption of heat. In other cases, the electrostatic attractions between the ions in a crystal are so large, or the ion–dipole attractive forces between the ions and water molecules are so weak, that the increase in disorder cannot compensate for the energy required to separate the ions, and the crystal is insoluble. Such is the case with insoluble substances such as calcium carbonate (limestone), calcium phosphate (the inorganic component of bone), and iron oxide (rust).

Soluble ionic compounds are called **electrolytes;** their solutions conduct electricity because of the presence of the dissolved ions. Other substances, such as sugar and alcohol, that form solutions that do not conduct electricity are called **nonelectrolytes.** We can identify whether a solution contains an electrolyte or a nonelectrolyte by determining whether it conducts electricity. The apparatus consists of an electrical circuit containing an electric lamp and two electrodes in a beaker (Fig. 12.9). When we fill the beaker with pure water, the lamp does not light because essentially no current flows. When a nonelectrolyte such as sugar is added to the water, the lamp still does not light. However, if an electrolyte such as potassium chloride is dissolved in the water, then the lamp glows brightly. As will be seen in the next section, some electrolytes conduct, but only weakly.

Svante Arrhenius, a Swedish chemist, first successfully explained electrolytic conduction in his Ph.D. thesis published in 1883. The current theory of electrolytes embodies most of his postulates. According to the theory, when a solution conducts an electric current, positive ions move toward the negative electrode and negative ions move toward the positive electrode. The movement of the ions toward the electrodes of opposite charge accounts for electrolytic conduction. A solution of a nonelectrolyte contains molecules of the nonelectrolyte rather than ions, and thus it cannot conduct an electric current.

Pure water is an extremely poor conductor of electricity because it is only very slightly ionized—only about 2 out of every 1 billion molecules ionize at 25°C. Water

Figure 12.9

This apparatus is used to test the conductivity of solutions: The bulb lights if the solution conducts. The solution on the left contains sugar, a nonelectrolyte and has no conductivity. The solution in the middle contains potassium chloride (100% ionized—a strong electrolyte) in the water. On the right is a solution of acetic acid, a weak, partially ionized electrolyte.

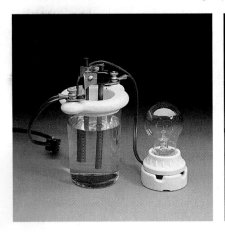

ionizes when one molecule of water gives up a proton to another molecule of water, yielding hydronium and hydroxide ions.

$$H_2O + H_2O \longrightarrow H_3O^+ + OH^-$$

12.4   Dissolution of Molecular Electrolytes

In some cases, we find that solutions prepared from compounds composed of covalent molecules conduct electricity because the molecules react with the solvent and produce ions. For example, pure hydrogen chloride is a gas consisting of covalent HCl molecules. This gas contains no ions. However, when we dissolve hydrogen chloride in water, we find that the solution is a very good conductor. The water molecules play an important part in forming ions; solutions of hydrogen chloride in many other solvents, such as benzene, do not conduct electricity and do not contain ions.

Hydrogen chloride molecules react with water to form hydronium ions (H_3O^+) and chloride ions (Cl^-). As shown by the Lewis formulas, a hydrogen ion (proton) shifts from a hydrogen chloride molecule to an unshared pair of electrons on the water molecule during the reaction.

$$H-\overset{\cdot\cdot}{\underset{H}{O}}: + H-\overset{\cdot\cdot}{\underset{\cdot\cdot}{Cl}}: \longrightarrow \left[H-\overset{\cdot\cdot}{\underset{H}{O}}-H\right]^+ + :\overset{\cdot\cdot}{\underset{\cdot\cdot}{Cl}}:^-$$

The hydronium ions and chloride ions conduct the current in the solution. All common strong acids react with water and give a 100% yield of hydronium ions when they dissolve, and all are **strong electrolytes.**

Many other compounds dissolve in water as **weak electrolytes** (only a small fraction ionizes and gives less than about a 10% yield of ions). Their solutions conduct electricity weakly (Fig. 12.9). For example, cyanic acid, HOCN, dissolves in water principally as hydrated molecules. Only a small percentage of these dissolved molecules ionize under ordinary conditions.

$$H-\overset{\cdot\cdot}{\underset{H}{O}}: + H-\overset{\cdot\cdot}{\underset{\cdot\cdot}{O}}-C\equiv N: \longrightarrow \left[H-\overset{\cdot\cdot}{\underset{H}{O}}-H\right]^+ + \left[:\overset{\cdot\cdot}{\underset{\cdot\cdot}{O}}-C\equiv N:\right]^-$$

The reaction proceeds only to the extent of 6% in a 0.1 M solution of HOCN. Other acids, such as acetic acid, CH_3CO_2H, nitrous acid, HNO_2, and hydrogen cyanide, HCN, are also weak electrolytes, and only a small fraction of their hydrated molecules ionize at any one time.

Some bases are also weak electrolytes; they dissolve to give solutions of hydrated molecules, some of which undergo ionization. For example, a solution of ammonia in water consists primarily of hydrated molecules, $NH_3(aq)$, with small amounts of ammonium ions, $NH_4^+(aq)$, and hydroxide ions, $OH^-(aq)$, which result from the reaction of ammonia with water.

$$H-\underset{\underset{H}{|}}{\overset{\overset{H}{|}}{N}}: + H-\overset{\cdot\cdot}{\underset{H}{O}}: \longrightarrow \left[H-\underset{\underset{H}{|}}{\overset{\overset{H}{|}}{N}}-H\right]^+ + \left[:\overset{\cdot\cdot}{\underset{\cdot\cdot}{O}}-H\right]^-$$

Macroscopic Properties of Solutions

12.5   Solutions of Gases in Liquids

We know that the solubility of a gas in a liquid can vary. When we open a warm Coca-Cola, we are much more likely to get foam than when we open a cold one. Carbon dioxide is less soluble in warm water than in cold water.

As we change the solvent but keep the gas the same, we find that the solubilities of the gas change. Only a little oxygen dissolves in water at room temperature—about 40 milliliters in 1 liter when the pressure of oxygen gas above the solution is 1 atmosphere. However, much more oxygen dissolves in a liter of perfluorododecane, $C_{12}F_{26}$, a substance that has been considered for use as an artificial substitute for blood plasma because it is inert and will dissolve a large amount of oxygen.

We also find that solubilities change as we change the gas but keep the solvent the same. For example, 1 liter of water will dissolve about 0.04 liter of oxygen, 1.7 liters of carbon dioxide, 80 liters of sulfur dioxide, or 1130 liters of ammonia, each at 0°C and 1 atmosphere.

The pressure of a gas and the temperature also affect its solubility. The solubility of a gas increases as the pressure of the gas increases. We see this effect in bottled carbonated beverages. Pressure is used to force carbon dioxide into solution in the beverage, and the bottle is capped to maintain this pressure. When we open the bottle, the pressure decreases and some of the gas escapes from the solution (Fig. 12.10).

If a solution of a gas exhibits ideal behavior (or approximately ideal behavior), then we can describe its solubility with the mathematical equation

$$C_g = kP_g$$

where C_g is the solubility of the gas in the solution, P_g is the pressure of the gas, and k is a proportionality constant that depends on the identity of the gas and that of the solvent. This is a mathematical statement of **Henry's law:** *The quantity of an ideal gas that dissolves in a definite volume of liquid is directly proportional to the pressure of the gas.* Here *quantity* refers to the grams or moles of the gas. This means, for example, that if 1 gram of a gas (or 1 mole of some other gas) at 1 atmosphere of pressure dissolves in 1 liter of water, 5 grams of the gas (or 5 moles of the other gas) will dissolve at 5 atmospheres.

Figure 12.10

Opening the bottle reduced the pressure of the *gaseous* carbon dioxide above this beverage, whereupon *dissolved* carbon dioxide escaped from the solution and formed the bubbles.

Example 12.1   Application of Henry's Law

The solubility of gaseous O_2 is 1.38×10^{-3} mol per liter of water at 20°C and when the gas has a pressure of 760 torr. Assuming an ideal solution, determine the solubility when the pressure is 155 torr, the approximate pressure of oxygen in the atmosphere.

Solution: According to Henry's law, for an ideal solution the solubility C_g of a gas (1.38×10^{-3} mol L^{-1} in this case) is directly proportional to the pressure P_g of the undissolved gas above the solution (760 torr in this case). Because we know both C_g and P_g, we can rearrange this expression to solve for k.

$$C_g = kP_g$$

$$k = \frac{C_g}{P_g}$$

$$= \frac{1.38 \times 10^{-3}\ \text{mol L}^{-1}}{760\ \text{torr}}$$

$$= 1.82 \times 10^{-6}\ \text{mol L}^{-1}\ \text{torr}^{-1}$$

Now we can use k to find the solubility at the lower pressure.

$$C_g = kP_g$$

$$= 1.82 \times 10^{-6}\ \text{mol L}^{-1}\ \text{torr}^{-1} \times 155\ \text{torr}$$

$$= 2.82 \times 10^{-4}\ \text{mol L}^{-1}$$

Pay careful attention to the units of k. Many different quantities are used to express the solubilities of gases.

Check your learning: 1.45×10^{-3} g of a gaseous N_2 will dissolve in 100 mL of water at 0°C and when the gas has a pressure of 0.200 atm. Assume the solution is ideal, and determine the solubility when the pressure is 1.00 atm.

Answer: 7.25×10^{-3} g

Figure 12.11

The small bubbles of air in this glass of water formed when the water warmed and the solubility of its dissolved air decreased.

Henry's law applies to gases that form ideal solutions. Ideal behavior is not observed when a chemical reaction takes place between the gas and the solvent. Thus, for example, the solubility of ammonia in water does not increase as rapidly with increasing pressure as predicted by the law, because ammonia, being a base, reacts to some extent with water to form ammonium ions and hydroxide ions.

$$
\underset{\substack{|\\H}}{\overset{\substack{H\\|}}{H-N}}: + \ H-\overset{..}{\underset{\substack{|\\H}}{O}}: \longrightarrow \left[\underset{\substack{|\\H}}{\overset{\substack{H\\|}}{H-N-H}}\right]^{+} + \left[:\overset{..}{\underset{..}{O}}-H\right]^{-}
$$

The solubilities of most gases in water decrease with an increase in temperature, if the gas does not react with water (Fig. 12.11 and Fig. 12.12 on page 424). For example, 48.9 milliliters of oxygen dissolve in 1 liter of water at 1 atmosphere and 0°C, but only 31.6 milliliters dissolve at 25°C, 24.6 milliliters at 50°C, and 23.0 milliliters at 100°C. This relationship is not one of inverse proportion, the variation of the solubility of a gas with temperature must be determined experimentally.

The decreasing solubility of gases such as oxygen with increasing temperature is a very important factor in thermal pollution of natural waters. A heat discharge that increases the temperature by a few degrees can reduce the solubility of oxygen in the water to a level too low for many forms of aquatic life to survive.

Gases can form supersaturated solutions. If a solution of a gas in a liquid is prepared either at low temperature or under pressure (or both), then as the solution warms or as the gas pressure is reduced, the solution may become supersaturated. For example, it is believed that the 1986 disaster that killed more than 1700 people near Lake

Figure 12.12

The solubilities of these gases in water decrease as the temperature increases. All solubilities were measured with a constant pressure of 1 atmosphere of gas above the solutions.

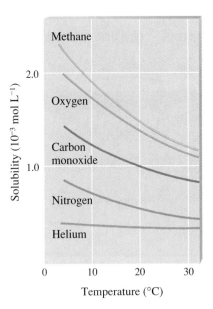

Nyos in West Africa (Fig. 12.13) resulted when a large volume of carbon dioxide gas was released from the depths of the lake.

12.6   Solutions of Liquids in Liquids (Miscibility)

We know that some liquids mix with each other in all proportions. These liquids are called **miscible**. Liquid soap, ethanol, sulfuric acid, and ethylene glycol (antifreeze) are examples of liquids that are completely miscible with water (Fig. 12.14). Two-cycle motor oil is miscible with gasoline. Nail polish is miscible with acetone.

Figure 12.13

In 1986 a cloud of gas bubbled from Lake Nyos in the African country of Cameroon and killed more than 1700 people. Lake Nyos is a deep lake in a volcanic crater, and the water at its bottom is saturated with carbon dioxide via a warm mineral-laden spring. It is believed that during the lake's seasonal turnover, water saturated with carbon dioxide reached the surface. The consequent reduction in pressure released tremendous quantities of dissolved CO_2. The colorless gas, which is denser than air, flowed down the valley below the lake and suffocated humans and other animals living there.

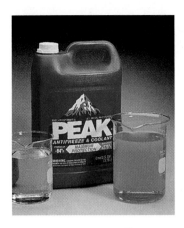

Figure 12.15 Water and oil are immiscible. Eventually the oil floats to the surface and two separate layers form.

Figure 12.14

Water and antifreeze are miscible; the mixture of the two is homogeneous.

Liquids that mix with water in all proportions are usually polar substances (Section 7.3) or substances that form hydrogen bonds (Section 11.4). For such liquids, the dipole–dipole attractions (or hydrogen bonds) of the solute molecules with the solvent molecules are at least as strong as those between molecules in the pure solute or in the pure solvent. Hence the two kinds of molecules mix easily. Nonpolar liquids are miscible with each other because there is no appreciable tendency for solvent molecules to attract other solvent molecules and squeeze out solute molecules. The solubility of polar molecules in polar solvents and of nonpolar molecules in nonpolar solvents is an illustration of the old chemical axiom "Like dissolves like."

Two liquids that do not mix are called **immiscible.** Layers are formed when we pour immiscible liquids into the same container. Gasoline, oil, benzene, carbon tetrachloride, some paints, and many other nonpolar liquids are immiscible with water (Fig. 12.15). There is no effective attraction between the molecules of such nonpolar liquids and polar water molecules. The only strong attractions in such a mixture are between the water molecules, so they effectively squeeze out the molecules of the nonpolar liquid.

Two liquids, such as bromine and water, that are slightly soluble in each other are said to be **partially miscible.** Two partially miscible liquids usually form two layers when mixed. Each layer is a saturated solution of one liquid in the other (Fig. 12.16).

12.7 The Effect of Temperature on the Solubility of Solids in Water

Figure 12.16

Bromine (a deep orange liquid) and water are partially miscible. The top layer in this mixture is a saturated solution of bromine in water; the bottom layer is a saturated solution of water in bromine.

The dependence of solubility on temperature for a number of inorganic solids in water is shown by the solubility curves in Fig. 12.17 on page 426. Generally, the solubility of a solid increases with increasing temperature, although there are exceptions. A sharp break in a solubility curve indicates the formation of a new compound with a different solubility. For example, when we heat a mixture of a saturated solution of $Na_2SO_4 \cdot 10H_2O$ in contact with the solid $Na_2SO_4 \cdot 10H_2O$ to above 32°C, the solid forms the anhydrous salt, Na_2SO_4, in equilibrium with a saturated solution. The red curve in Fig. 12.17 shows the effect of increasing temperature on the solubility of $Na_2SO_4 \cdot 10H_2O$ up to 32°C (it increases with temperature), and above this point the

Figure 12.17

This graph shows how the solubility of several solids changes with temperature. The break in the red curve occurs at 32.4°C, the temperature at which solid $Na_2SO_4 \cdot 10H_2O$ decomposes to Na_2SO_4 and water.

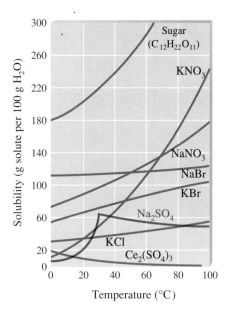

curve shows the effect on the solubility of anhydrous Na_2SO_4 (it decreases with increasing temperature).

Changes in solubility with temperature can be used to prepare supersaturated solutions. If we prepare a solution at an elevated temperature (where the solute is more soluble) and remove all traces of undissolved solute, we can sometimes cool the solution without crystallization of solute. Some hand warmers take advantage of this behavior (Fig. 12.18).

12.8 Solid Solutions

When we think of solutions we usually think of liquids. However, we can also prepare **solid solutions,** solids with the properties of solutions. If we melt a mixture of lithium chloride and sodium chloride, mix it well, and allow it to cool, a solid solution of the two ionic compounds forms. The resulting crystalline solid contains an array of chloride ions with a random distribution of lithium ions and sodium ions in holes in the array (Fig. 12.19). The crystal is a solid solution of LiCl and NaCl. It exhibits the properties of a solution: It is homogeneous, its composition can be varied (from pure LiCl to pure NaCl), and neither NaCl nor LiCl separates out on standing. Solid solutions of ionic compounds result when ions of one type randomly replace other ions of about the same size in a crystal.

Some ionic substances appear to be **nonstoichiometric compounds;** that is, their chemical formulas deviate from ideal ratios or are variable. In other respects, however, these substances resemble compounds; they are not inhomogeneous mixtures but are instead homogeneous throughout. Many of these so-called nonstoichiometric compounds are, in fact, solid solutions of two or more compounds. A sample of ruby, for example, with the formula $Cr_{0.02}Al_{1.98}O_3$, is a solid solution of Cr_2O_3 in Al_2O_3 with a 1-to-99 ratio of Cr to Al. Other nonstoichiometric compounds are solid solutions that contain one ion in two different oxidation states. For example, $TiO_{1.8}$ contains both Ti^{3+} and Ti^{4+} ions and can be considered a solid solution of Ti_2O_3 in TiO_2.

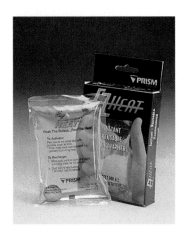

Figure 12.18

This hand warmer produces heat when the sodium acetate in a supersaturated solution crystallizes. (See Fig 12.2.) The hand warmer can be used repeatedly by placing it in hot water so that the solid dissolves and reforms the supersaturated solution.

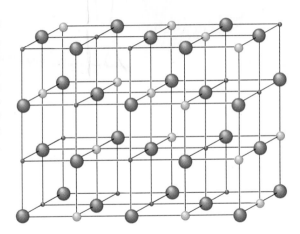

Figure 12.19

A portion of the structure of a solid solution of LiCl and NaCl. The chloride ions (large green spheres) form a face-centered cubic array, with the octahedral holes occupied by a random arrangement of lithium ions (smaller purple spheres) and sodium ions (larger gray spheres).

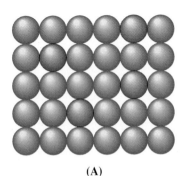

(A)

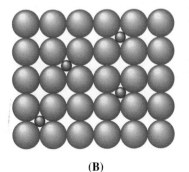

(B)

Figure 12.20

Two types of solid solutions. (A) In a substitutional solid solution, solute atoms (blue spheres) replace atoms of the solvent crystal (red spheres). (B) In an interstitial solid solution, small solute atoms (purple spheres) occupy holes in the lattice of the solvent crystal (red spheres).

Some **alloys** are solid solutions composed of two or more metals. In such an alloy, atoms of one of the component metals take up positions in the crystal lattice of the other. The solute atoms may randomly replace the atoms of the solvent crystal, forming a class of solid solutions called **substitutional solid solutions** (Fig. 12.20 A). For example, chromium dissolves in nickel and forms a solid solution in which the chromium atoms replace nickel atoms in the face-centered cubic structure of nickel. The solubility may be limited (zinc and copper; chromium and nickel) or the metals may be miscible in all proportions (nickel and copper).

Small atoms (hydrogen, carbon, boron, and nitrogen) may occupy the holes in the lattice of a metal, forming another class of solid solutions called **interstitial solid solutions** (Fig. 12.20 B). A solid solution of carbon in iron (austenite) is an example; the iron atoms are on the lattice points of a face-centered cubic lattice, and the carbon atoms occupy the interstitial positions.

Not all alloys are solid solutions. Some are heterogeneous mixtures in which the component metals are mutually insoluble and the solid alloy is composed of a mixture of small crystals of each metal. For example, tin and lead (plumber's solder) are insoluble in each other in the solid state. Other alloys, such as Cu_5Zn_8 and Ag_3Al, are actually compounds that form with only one specific stoichiometric composition. In general, the formulas of such intermetallic compounds are not those that would be predicted on the basis of the usual valence rules.

Expressing Concentration

The concentration of a solution is an intensive, macroscopic property—the amount of solute dissolved in a unit amount of solvent or of solution. We can report a concentration in many ways, depending on what we want to emphasize or how we want to use it. For example, earlier in this chapter we described the solubility of oxygen in terms of the number of milliliters of oxygen gas that would dissolve in 1 liter of water. In the following sections we will consider four other useful ways to report concentrations.

12.9   Percent Composition

If we read the labels we will find that the concentration of sodium hypochlorite, $NaOCl$, in household bleach is 5.25%; that of glyphosate, the active ingredient in a herbicide is 0.96%; and that of hydrochloric acid in a basin and tile cleaner is 15%. Each of these concentrations is a **percent by mass** (sometimes called **mass percent** or **weight percent**)—the percent of the solute in the solution. A 5.25% $NaOCl$ solution by mass contains 5.25 grams of $NaOCl$ in 100 grams of solution (5.25 grams of $NaOCl$ and 94.75 grams of water), 0.787 gram of $NaOCl$ in 15 grams of solution (0.787 gram of $NaOCl$ and 14.223 grams of water), or any other ratio of $NaOCl$ to solution for which the mass of $NaOCl$ is 5.25% of the total mass.

$$\text{Percent by mass} = \frac{\text{mass of solute}}{\text{mass of solution}} \times 100$$

When we use percent by mass in calculations, it is often convenient to use the ratio grams of solute to 100 grams of solution (as a fraction, g solute/100 g solution) as a conversion factor. The number of grams of solute in 100 grams of solution is equal to the percent by mass of solute.

Example 12.2   Calculation of Percent by Mass

Normal spinal fluid contains 3.75 mg (0.00375 g) of glucose in 5.0 g of spinal fluid. What is the percent by mass of glucose in spinal fluid?

Solution: We calculate the percent by mass from the mass of the solute and the mass of the solution.

$$\text{Percent by mass} = \frac{\text{mass of solute}}{\text{mass of solution}} \times 100$$

$$= \frac{0.00375 \text{ g}}{5.0 \text{ g}} \times 100$$

$$= 0.075\%$$

Check your learning: A bottle of a tile cleanser contains 130 g of HCl and 750 g of water. What is the percent by mass of HCl in this cleanser?

Answer: 14.8%

Example 12.3   Use of Percent by Mass

Concentrated hydrochloric acid, a saturated solution of HCl in water, is sometimes used in the general chemistry laboratory. Determine what mass of HCl is contained in 0.500 L of this solution, which has a density of 1.19 g mL^{-1} and contains 37.2% HCl by mass.

Solution: We can use the following steps to solve this problem.

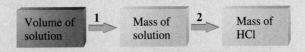

The numbers above the arrows refer to the conversion steps below.

Step 1.   *Determine the mass of the 500 mL of concentrated hydrochloric acid solution, using the density of this solution in the conversion factor.*

$$500 \text{ mL} \times \frac{1.19 \text{ g}}{1 \text{ mL}} = 595 \text{ g}$$

Step 2.   *Determine the mass of HCl dissolved in the solution, using the grams of HCl in 100 grams of solution (the percent composition) as the conversion factor.*

$$595 \text{ g solution} \times \frac{37.2 \text{ g HCl}}{100 \text{ g solution}} = 221 \text{ g HCl}$$

Check your learning: What volume of the concentrated HCl described in Example 12.3 must you use if you need 125 g of HCl to prepare a metal chloride?

Answer: 282 mL

12.10   Molarity

We saw in Section 3.7 that the **molarity, M,** of a solution is equal to the amount of solute (in moles) contained in exactly 1 liter of the solution. The use of molarity to express the concentration of a solution makes it easy to select a desired number of moles, molecules, or ions of the solute by measuring out the appropriate volume of solution. For example, if 1 mole of sodium hydroxide is needed for a given reaction, we can use 40 grams of solid NaOH (1 mole), or 1 liter of a 1 M solution of NaOH, or 2 liters of a 0.5 M NaOH solution.

We need two pieces of information to calculate the molarity, M, of a solution: the volume of a sample of the solution and the number of moles of solute dissolved in that volume. The molarity is determined by dividing the number of moles of solute by the volume.

$$\text{Molarity} = \frac{\text{moles of solute}}{\text{liters of solution}}$$

Examples of the use of molar concentrations in stoichiometry calculations were presented in Chapter 3. However, here is one more as a reminder.

Example 12.4   **Calculation of Molarity**

What is the molarity of HNO_3 in concentrated nitric acid, a solution with a density of 1.42 g/mL that contains 68.0% HNO_3 by mass?

Solution: We can calculate the molarity (the moles of HNO_3 in exactly 1 L of concentrated nitric acid) by using the following steps:

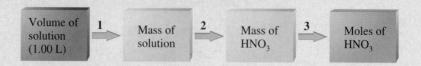

The numbers above the arrows refer to the conversion steps below.

Step 1.   *Determine the mass of exactly 1 L (exactly 1000 mL) of the solution, using the density of the solution in the conversion factor. Result: 1.42×10^3 g solution*

Step 2.   *Determine the mass of HNO_3 in 1.42×10^3 g of the solution, using the percent by mass as the conversion factor. Result: 966 g HNO_3*

Step 3.   *Convert from mass to moles of HNO_3 in 1.00 L of the solution to obtain the molarity using the mass of 1 mol of HNO_3 (63.012 g) in the unit conversion factor. Result: 15.3 M*

Check your learning: What is the molarity of water in the concentrated solution of nitric acid described in Exercise 12.4?

Answer: 25.2 M

A concentration unit called normality is sometimes used to describe the concentrations of acids and bases or of oxidizing and reducing agents. Normality will be described in detail in the chapters that deal with acid–base reactions (Section 15.12) and with electrochemical reactions (Section 19.15).

12.11   Molality

Later in this chapter we will see that the freezing and boiling points of a solution vary in direct proportion to the number of moles of solute dissolved in exactly 1 kilogram of the solvent used to prepare the solution. This unit of concentration, the number of moles of solute dissolved in exactly 1 kilogram of solvent, is called **molality, m.** Molality can be calculated by dividing the moles of solute in a solution by the mass of the solvent, in kilograms, used to prepare the solution.

$$\text{Molality} = \frac{\text{moles of solute}}{\text{kilograms of solvent}}$$

Note that we use kilograms of solvent rather than liters of solution to calculate molality. This is the important difference between the calculation of molality and that of molarity.

Example 12.5   Calculation of Molality

The antifreeze in most automobile radiators is a mixture of equal volumes of ethylene glycol and water with minor amounts of other additives that prevent corrosion. What is the molality of ethylene glycol, $C_2H_4(OH)_2$, in a solution prepared from 2218 g of ethylene glycol and 1996 g of water (approximately 2 L of glycol and 2 L of water)?

Solution: We can use the following steps to solve this problem.

$$\boxed{\begin{array}{c}\text{Mass of}\\C_2H_4(OH)_2\end{array}} \xrightarrow{\ \ \mathbf{1}\ \ } \boxed{\begin{array}{c}\text{Moles of}\\C_2H_4(OH)_2\end{array}} \xrightarrow{\ \ \mathbf{2}\ \ } \boxed{\begin{array}{c}\text{Molality of}\\\text{solution}\end{array}}$$

The numbers above the arrows refer to the conversion steps below.

Step 1.   *Convert the grams of $C_2H_4(OH)_2$ to moles of $C_2H_4(OH)_2$, using the molar mass of $C_2H_2(OH)_2$ in the unit conversion factor.*

$$\text{mol } C_2H_4(OH)_2 = 2218 \text{ g } C_2H_4(OH)_2 \times \frac{1 \text{ mol } C_2H_4(OH)_2}{62.068 \text{ g } C_2H_4(OH)_2}$$

$$= 35.74 \text{ mol } C_2H_4(OH)_2$$

Step 2.   *Determine the molality of the solution from the number of moles of solute and the mass of solvent, in kilograms:* 1996 g = 1.996 kg.

$$\text{Molality} = \frac{35.74 \text{ mol } C_2H_4(OH)_2}{1.996 \text{ kg solvent}}$$

$$= 17.91 \text{ m}$$

Check your learning: **(a)** What is the molality of a solution that contains 0.850 g of ammonia, NH_3, dissolved in 125 g of water? **(b)** Determine the molality of 0.250 kg of a solution that contains 40.0 g of NaCl.

Answer: (a) 0.399 m (b) 3.26 m

12.12   Mole Fraction

We will see in the next section that the vapor pressure of a solution often is directly proportional to the mole fraction of the solute and solvent present in the solution. The **mole fraction, X,** of a component in a solution is the number of moles of that individual component divided by the total number of moles of all components present.

The mole fraction of substance A in a solution of substances A, B, C, etc., in a solvent S is expressed as follows:

Mole fraction of A = X_A

$$= \frac{\text{moles of A}}{\text{moles of A} + \text{moles of B} + \text{moles of C} + \cdots + \text{moles of S}}$$

The sum of the mole fractions of all components of a mixture always equals 1.

Example 12.6   Calculation of Mole Fraction

At 25°C, 25.0 g of glycine, $C_2H_5O_2N$, the simplest amino acid found in proteins, will dissolve in exactly 100 g of water. Calculate the mole fraction of glycine and water in the solution.

Solution: We can use the following steps to solve this problem.

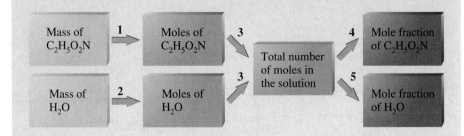

Step 1.   *Convert from grams to moles of $C_2H_5O_2N$.*

$$\text{Moles of } C_2H_5O_2N = 25.0 \text{ g } C_2H_5O_2N \times \frac{1 \text{ mol } C_2H_5O_2N}{75.067 \text{ g } C_2H_5O_2N}$$

$$= 0.333 \text{ mol } C_2H_5O_2N$$

Step 2.   *Convert from exactly 100 g of H_2O to moles of H_2O.*

$$\text{Moles of } H_2O = 100 \text{ g } H_2O \times \frac{1 \text{ mol } H_2O}{18.015 \text{ g } H_2O} = 5.5509 \text{ mol } H_2O$$

Step 3.   *Calculate the total number of moles in solution.*

Moles of $C_2H_5O_2N$ + moles of H_2O = 0.333 mol + 5.5509 mol = 5.884 mol

Step 4.   *Calculate $X_{C_2H_5O_2N}$, the mole fraction of $C_2H_5O_2N$.*

$$X_{C_2H_5O_2N} = \frac{\text{moles of } C_2H_5O_2N}{(\text{moles of } C_2H_5O_2N + \text{moles of } H_2O)}$$

$$= \frac{0.333 \text{ mol}}{5.884 \text{ mol}} = 0.0566$$

Step 5.   *Calculate X_{H_2O}, the mole fraction of H_2O.*

$$X_{H_2O} = \frac{\text{moles of } H_2O}{(\text{moles of } C_2H_5O_2N + \text{moles of } H_2O)}$$

$$= \frac{5.5509 \text{ mol}}{5.884 \text{ mol}} = 0.9434$$

Note, as a check on the work, that the sum of the mole fractions, 0.0566 + 0.9434, is 1.0000.

Check your learning: What is the mole fraction of each component in a solution that contains 42.0 g CH_3OH, 35.0 g C_2H_5OH, and 50.0 g C_3H_7OH?

Answer: X_{CH_3OH}, 0.452; $X_{C_2H_5OH}$, 0.262; $X_{C_3H_7OH}$, 0.287

Example 12.7   Conversion of Molality to Mole Fraction

Calculate the mole fraction of solute and solvent in a 3.000 m solution of sodium chloride.

Solution: A 3.000 m solution of sodium chloride contains 3.000 mol of NaCl dissolved in exactly 1 kg, or 1000 g, of water. With this information, we can use the following steps to solve this problem.

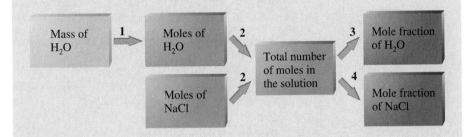

Step 1. *Convert from grams to moles of H_2O.* Result: 55.51 mol H_2O

Step 2. *Calculate the total number of moles in solution from the number of moles of H_2O and of NaCl.* Result: 58.51 mol

Step 3. *Calculate the mole fraction of H_2O.* Result: 0.9487

Step 4. *Calculate the mole fraction of NaCl.* Result: 0.0513

Note that the sum of the two mole fractions, 0.9487 + 0.0513, is 1.000. Hence we could have calculated one of these mole fractions by simply taking the difference. For example,

$$X_{NaCl} = 1.000 - X_{H_2O}$$
$$= 1.000 - 0.9487 = 0.0513$$

Check your learning: Calculate the mole fraction of solute and solvent in a 1.50 m solution of iodine, I_2, dissolved in dichloromethane, CH_2Cl_2.

Answer: X_{I_2} 0.115; $X_{CH_2Cl_2}$ 0.885

Example 12.8   Conversions among Percent by Mass, Molarity, Molality, and Mole Fraction

A sulfuric acid solution containing 200.1 g of H_2SO_4 in 0.3500 L of solution at 20°C has a density of 1.3294 g mL^{-1}. Calculate **(a)** the molarity, **(b)** the molality, **(c)** the percent by mass of H_2SO_4, and **(d)** the mole fraction of H_2SO_4 in the solution.

Solution: **(a)** We can use the following two steps to determine the molarity.

Step a1.   *Determine the moles of H_2SO_4 from the mass of H_2SO_4. Result: 2.040 mol H_2SO_4*

Step a2.   *Determine the molarity of H_2SO_4 from the number of moles of H_2SO_4 and the volume of the solution. Result: 5.829 M*

(b) We can use the following three steps to determine the molality.

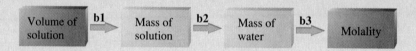

Step b1.   *Determine the mass of the solution. Result: 465.3 g*

Step b2.   *Determine the mass of water by subtracting the mass of H_2SO_4 from the mass of the solution. Result: 0.2652 kg (265.2 g) H_2O*

Step b3.   *Determine the molality of H_2SO_4 from the number of moles of H_2SO_4 (from Step a1) and the mass of the solvent. Result: 7.692 m*

(c) *Calculate the percent H_2SO_4 by mass from the mass of H_2SO_4 and the mass of the solution (from Step b1). Result: 43.00% H_2SO_4*

(d) We can calculate the mole fractions using the following steps:

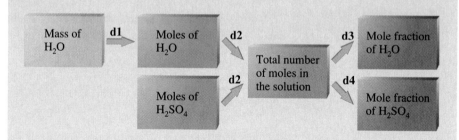

Step d1.   *Convert the mass of H_2O (from Step b2) to moles of H_2O. Result: 14.72 mol H_2O*

Step d2.   *Calculate the total number of moles in solution from the number of moles of H_2O and of H_2SO_4 (from Step a1). Result: 16.76 mol*

Step d3.   *Calculate the mole fraction of H_2O, using the number of moles of H_2O and the total number of moles in the solution. Result: 0.8783*

Step d4.   *Calculate the mole fraction of H_2SO_4, using the number of moles of H_2SO_4 and the total number of moles in the solution. Result: 0.1217*

Check your learning: Calculate **(a)** the percent composition and **(b)** the molality of an aqueous solution of $NaNO_3$ if the mole fraction of $NaNO_3$ is 0.200.

Answer: **(a)** 54.1% $NaNO_3$, 45.9% H_2O **(b)** 13.9 m

Colligative Properties of Solutions

12.13   Lowering of the Vapor Pressure of a Solvent

Solutes can affect the properties of a solvent. For example, the water in syrup evaporates much more slowly than pure water, because the vapor pressure of water in the syrup is lower than the vapor pressure of pure water. The presence of the solute sugar in the syrup solution is responsible for the difference.

The vapor pressure of the solvent above a solution, the freezing temperature of a solution, and the boiling temperature of a solution change as the concentration of solute particles changes. Interestingly, however, these changes are independent of the nature (kind, size, or charge) of the solute particles, provided that the solution approximates ideal behavior (Section 12.2). The changes depend only on the concentration of solute particles. For example, solutions of 1 mole of solid sugar, $C_{12}H_{22}O_{11}$, 1 mole of liquid ethylene glycol, $C_2H_4(OH)_2$, and 1 mole of gaseous nitrous oxide, N_2O, each dissolved in 1 kilogram of water, begin to freeze at $-1.86°C$. A solution prepared from 1 kilogram of water and 1 mole of ions (for instance, 0.5 mole of ammonium ions, NH_4^+, and 0.5 mole of chloride ions, Cl^-, resulting from the dissolution of ammonium chloride) also freezes at $-1.86°C$, even though the dissolved particles are ions. Changes in the properties of solutions that depend only on the concentration of solute species are called **colligative properties.**

If we were to measure the vapor pressures of a number of solutions with different concentrations of nonvolatile solutes, to a fairly good approximation we would find that the vapor pressures are directly proportional to the mole fraction of the solvent in the solution. The direct proportionality holds exactly for solutions that exhibit ideal behavior (solutions that form with no accompanying enthalpy change, Section 12.2). The approximation of direct proportionality improves with nonideal solutions as they become more dilute because a nonideal solution approaches more nearly ideal behavior as it becomes more dilute.

The proportionality of vapor pressure and mole fraction is summed up in **Raoult's law:** *The vapor pressure of any component of an ideal solution is equal to the vapor pressure of the pure component multiplied by its mole fraction in the solution.* For the vapor pressure of the solvent in a solution of a nonvolatile solute, the mathematical form of this law is

$$P_{solv} = X_{solv}P°_{solv} \tag{1}$$

where P_{solv} is the vapor pressure of the solvent in the solution, X_{solv} is the mole fraction of solvent in the solution, and $P°_{solv}$ is the vapor pressure of the pure solvent.

The decrease in vapor pressure, ΔP_{solv}, of the solvent in a solution that exhibits ideal behavior, compared to the vapor pressure of the pure solvent, is directly proportional to the mole fraction of solute. The equation for the change in vapor pressure is

$$\Delta P_{solv} = X_{solute}P°_{solv} \tag{2}$$

where X_{solute} is the mole fraction of solute and $P°_{solv}$ is the vapor pressure of the pure solvent. This **vapor-pressure lowering** is a colligative property; it depends only on the mole fraction of solute particles in an ideal solution, not on their identity.

Example 12.9   Calculation of a Vapor Pressure

Assume the solution is ideal and determine the vapor pressure of a solution of 92.1 g of glycerin, $C_3H_5(OH)_3$, in 184.4 g of ethanol, C_2H_5OH, at 40°C. The vapor pressure of pure ethanol is 0.178 atm at 40°C. Glycerin is essentially nonvolatile at this temperature.

Solution: We can find the vapor pressure of ethanol in the solution in two ways: (1) We can use the direct proportionality between the vapor pressure of the solvent (ethanol) and the mole fraction of solvent. (2) Or we can use the direct proportionality between the change in vapor pressure and the mole fraction of solute (glycerin). In either case, we need to determine the number of moles of glycerin and ethanol in the solution and then determine a mole fraction.

$$92.1 \text{ g } C_3H_5(OH)_3 \times \frac{1 \text{ mol } C_3H_5(OH)_3}{92.094 \text{ g } C_3H_5(OH)_3} = 1.00 \text{ mol } C_3H_5(OH)_3$$

$$184.4 \text{ g } C_2H_5OH \times \frac{1 \text{ mol } C_2H_5OH}{46.069 \text{ g } C_2H_5OH} = 4.000 \text{ mol } C_2H_5OH$$

Now we can calculate the mole fraction of the solvent (ethanol) and use Equation (1) to determine its vapor pressure

$$X_{C_2H_5OH} = \frac{4.000 \text{ mol}}{1.00 \text{ mol} + 4.000 \text{ mol}} = 0.800$$

$$P_{solv} = X_{solv}P^{\circ}_{solv} = 0.800 \times 0.178 \text{ atm} = 0.142 \text{ atm}$$

Alternatively, we could calculate the mole fraction of the nonvolatile solute (glycerin), use Equation 2 to determine the change in vapor pressure, and then determine the new vapor pressure from that.

$$X_{C_3H_5(OH)_3} = \frac{1.00 \text{ mol}}{1.00 \text{ mol} + 4.000 \text{ mol}} = 0.200$$

$$\Delta P = X_{solute}P^{\circ}_{solv} = 0.200 \times 0.178 \text{ atm} = 0.0356 \text{ atm}$$

$$\Delta P = P^{\circ}_{solv} - P_{solv}$$

$$P_{solv} = P^{\circ}_{solv} - \Delta P = 0.178 \text{ atm} - 0.0356 \text{ atm} = 0.142 \text{ atm}$$

Check your learning: A solution contains 5.00 g of urea, $CO(NH_2)_2$, a nonvolatile solute, per 0.100 kg of water. If the vapor pressure of pure water at 25°C is 23.7 torr, what is the vapor pressure of the solution?

Answer: 23.4 torr

If a solute has a vapor pressure of its own, then it contributes to the vapor pressure of a solution. The vapor pressure of an ideal solution of a volatile solute is the sum of the vapor pressures of the solvent and the volatile solute. We can determine the vapor pressure of an ideal solution that contains a volatile solute by using the expression

$$P_{solution} = X_{solute}P^{\circ}_{solute} + X_{solv}P^{\circ}_{solv}$$

Table 12.1

Boiling Points, Molal Boiling-Point Elevation Constants, Freezing Points, and Molal Freezing-Point Depression Constants for Several Solvents

Solvent	Boiling Point, °C (at 1 atm)	K_b, °C m^{-1}	Freezing Point, °C	K_f, °C m^{-1}
Water	100.00	0.512	0.00	1.86
Acetic acid	118.1	3.07	16.6	3.9
Benzene	80.1	2.53	5.5	5.12
Chloroform	61.26	3.63	−63.5	4.68
Nitrobenzene	210.9	5.24	5.67	8.1

where X_{solute} is the mole fraction of the solute, P°_{solute} is the vapor pressure of the pure solute, X_{solv} is the mole fraction of the solvent, and P°_{solv} is the vapor pressure of the pure solvent.

12.14 Elevation of the Boiling Point of a Solvent

As we heat a liquid, its vapor pressure increases. It boils when its vapor pressure equals the external pressure on its surface (Section 11.6). When we add a nonvolatile solute to the liquid, we lower its vapor pressure, so a higher temperature is needed to increase the vapor pressure to the point where the solution boils. This **boiling-point elevation** of the solvent is the same for all ideal solutions of the same concentration; for example, 1 mole of any nonvolatile nonelectrolyte dissolved in 1 kilogram of water raises the boiling point by 0.512°C.

The change in boiling point of a dilute solution, ΔT, from that of the pure solvent is directly proportional to the molal concentration of the solute:

$$\Delta T = K_b m$$

where m is the molal concentration of the solute and K_b is called the **molal boiling-point elevation constant.** This constant is the increase in boiling point for a 1 molal solution of a nonvolatile nonelectrolyte. The values of K_b for several solvents are listed in Table 12.1: We see that the value of K_b varies from solvent to solvent.

The extent to which the vapor pressure of a solvent is lowered and the boiling point is elevated depends on the number of solute particles present in a given amount of solvent, not on the mass or size of the particles. A mole of sodium chloride forms 2 moles of ions in solution and causes nearly twice as great a rise in boiling point as does 1 mole of nonelectrolyte. One mole of sugar contains 6.022×10^{23} particles (as molecules), whereas 1 mole of sodium chloride contains $2 \times 6.022 \times 10^{23}$ particles (as ions). Calcium chloride, $CaCl_2$, which consists of three ions, causes nearly three times as great a rise in boiling point as does sugar. In Section 12.20 we will consider why the elevation is not exactly twice (for NaCl) or exactly three times (for $CaCl_2$) that of the boiling-point elevation for a nonelectrolyte.

Example 12.10 Calculating the Boiling Point of a Solution

What is the boiling point of a 0.33 m solution of a nonvolatile solute in benzene?

Solution: If we assume that the solution is ideal, we can solve this problem in two steps.

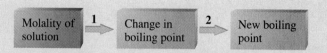

Step 1.   *Use the direct proportionality between the change in boiling point and molal concentration to determine how much the boiling point changes.*

$$\Delta T = K_b m = 2.53°C\ m^{-1} \times 0.33\ m = 0.83°C$$

Step 2.   *Determine the new boiling point from the boiling point of the pure solvent and the change.*

$$\text{Boiling temperature} = 80.1°C + 0.83°C = 80.9°C$$

Check your learning: What is the boiling point of a 1.93 m solution of a nonvolatile solute in nitrobenzene?

Answer: 221.0°C

Example 12.11   The Boiling Point of a Solution of Iodine

Find the boiling point of a solution of 92.1 g of iodine, I_2, in 800.0 g of chloroform, $CHCl_3$, assuming that the iodine is nonvolatile and that the solution is ideal.

Solution: We can solve this problem using four steps.

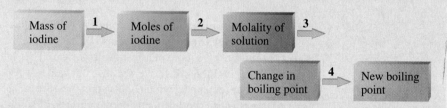

Step 1.   *Convert from mass to moles of I_2.* Result: 0.363 mol

Step 2.   *Determine the molality of the solution from the number of moles of solute and the mass of solvent, in kilograms.* Result: 0.454 m

Step 3.   *Use the direct proportionality between the change in boiling point and molal concentration to determine how much the boiling point changes.* Result: 1.65°C

Step 4.   *Determine the new boiling point from the boiling point of the pure solvent and the change.* Result: 62.91°C

Check your learning: What is the boiling point of a solution of 1.0 g of glycerin, $C_3H_5(OH)_3$, in 47.8 g of water? Assume an ideal solution.

Answer: 100.12°C

12.15 Distillation of Solutions

Distillation (Section 11.7) plays a central role in many separations both in the laboratory and in industrial settings. It is used in the separation of petroleum, petrochemicals, fermentation products, and many other materials. Very pure water is prepared by distillation. We can use distillation in the laboratory to separate solvent and solutes and recover pure solvent from many solutions. If the solute is nonvolatile, an apparatus like the one shown in Fig. 11.16 may be used. However, distillation of a mixture of volatile liquids is more complicated.

Oil refineries and other chemical plants that separate complex solutions by distillation use fractionating columns to achieve separation of liquids that would require a great number of simple fractional distillations of the type described in this section. Each year, millions of gallons of crude oil, a complex mixture of hydrocarbons, is separated into its components by fractional distillation (Fig. 12.21).

The boiling point of a solution is the temperature at which the total vapor pressure of the mixture is equal to the atmospheric pressure. The total vapor pressure of a solution composed of two volatile substances depends on the concentration and the vapor pressure of each substance and will have one of three values:

Figure 12.21

In a column for the fractional distillation of crude oil, oil heated to about 425°C in the furnace vaporizes when it enters the base of the tower. The vapors rise through bubble caps in a series of trays in the tower. As the vapors gradually cool, fractions of higher, then of lower, boiling points condense to liquids and are drawn off.

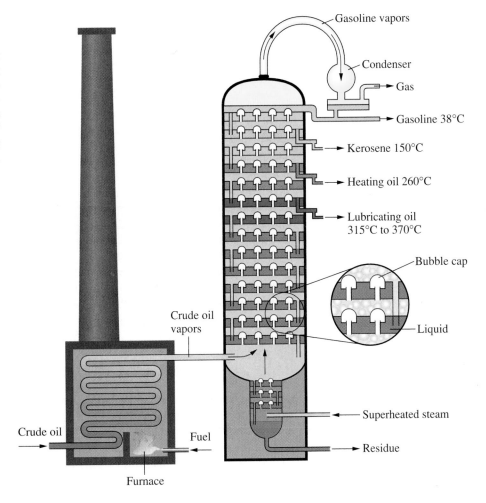

1. Between the vapor pressures of the pure components

2. Less than the vapor pressure of either pure component

3. Greater than the vapor pressure of either pure component

Usually the total vapor pressure—and thus the boiling point—of a mixture of two liquids lies between the vapor pressures of the two components.

If we distill a solution of type 1, the vapor produced at the boiling point is richer in the lower-boiling component (the component with the higher vapor pressure) than was the original solution. When this vapor is condensed, the resulting distillate contains more of the lower-boiling component than did the original mixture. This means that the composition of the boiling mixture constantly changes, the amount of the lower-boiling component of the mixture constantly decreases, and the boiling point rises as distillation continues. The vapor (and hence the distillate) contains more and more of the higher-boiling component and less and less of the lower-boiling component as distillation proceeds. Changing the receiver at intervals yields successive fractions, each one increasingly richer in the less volatile (higher-boiling) component. If this process of **fractional distillation** is repeated several times, relatively pure samples of the two liquids can be obtained.

The vapor pressure of a solution of nitric acid and water is less than that of either component (a solution of type 2). When a *dilute* solution is distilled, the first fraction of distillate consists mostly of water, because water has a higher vapor pressure than nitric acid under distillation conditions. As distillation is continued, the solution that remains in the distilling flask becomes richer in nitric acid. When a concentration of 68% HNO_3 by mass is reached, the solution boils at a constant temperature of 120.5°C (at 1 atmosphere). At this temperature, the liquid and the vapor have the same composition, and the liquid distills without any further change in composition.

When a nitric acid solution more concentrated than 68% is distilled, the vapor that is first formed contains a large amount of HNO_3. The solution that remains in the distilling flask contains a greater percentage of water than at first, and the concentration of the nitric acid in the distilling flask decreases as the distillation is continued. Finally, a concentration of 68% HNO_3 is reached, and the solution again boils at the constant temperature of 120.5°C (at 1 atmosphere). The 68% solution of nitric acid is referred to as a constant boiling solution.

Solutions of both type 2 and type 3 produce constant boiling solutions when distilled. A **constant boiling solution** forms a vapor with the same composition as the solution; thus it distills without a change in concentration. Constant boiling solutions are also called **azeotropic mixtures.** A 68% solution of HNO_3 in water is an azeotropic mixture. Other common and important substances that form azeotropic mixtures with water are HCl (20.24%, 110°C at 1 atmosphere), H_2SO_4 (98.3%, 338°C at 1 atmosphere), and ethanol (96% C_2H_5OH, 78.2°C at 1 atmosphere).

12.16   Depression of the Freezing Point of a Solvent

Solutions freeze at lower temperatures than pure liquids. We can use salt, calcium chloride, or urea to melt ice (Fig. 12.22) because solutions of these substances freeze at lower temperatures than pure water. Antifreezes such as ethylene glycol are used in automobile radiators to reduce the freezing point of the coolant. Sea water freezes at a lower temperature than fresh water, so the Arctic and Antarctic oceans have tem-

Figure 12.22

Rock salt (NaCl), calcium chloride, or a mixture of the two is used to melt ice.

peratures below 0°C. The body fluids of fish and other cold-blooded sea animals that live in these oceans freeze below 0°C.

The change in the freezing point of a solvent is directly proportional to the molal concentration of solute particles in the solution. The **freezing-point depression** of a solvent is the same for all ideal solutions of the same concentration; for example, 1 mole of any nonvolatile nonelectrolyte dissolved in 1 kilogram of water depresses the boiling point by 1.86°C.

The change in freezing point of a dilute solution, ΔT, from that of the pure solvent is directly proportional to the molal concentration of the solute:

$$\Delta T = K_f m$$

where m is the molal concentration of the solute in the solvent and K_f is called the **molal freezing-point depression constant.** This constant is the depression of the freezing point for a 1 molal solution of a nonvolatile nonelectrolyte. The values of K_f for several solvents are listed in Table 12.1: We can see that the value of K_f varies from solvent to solvent.

Example 12.12   Calculation of the Freezing Point of a Solution

What is the freezing point of the 0.33 m solution of a nonvolatile solute in benzene described in Example 12.10?

Solution: If we assume the solution is ideal, we can solve this problem in two steps.

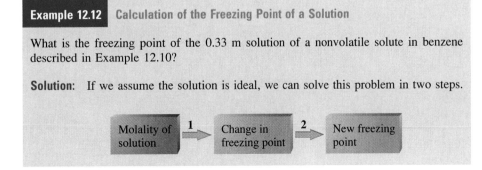

Step 1. *Use the direct proportionality between the change in freezing point and molal concentration to determine how much the freezing point changes.*

$$\Delta T = K_f m = 5.12°C\ m^{-1} \times 0.33\ m = 1.7C$$

Step 2. *Determine the new freezing point from the freezing point of the pure solvent and the change.*

$$\text{Freezing temperature} = 5.5°C - 1.7°C = 3.8°C$$

Check your learning: What is the freezing point of a 1.85 m solution of a nonvolatile solute in nitrobenzene?

Answer: −9°C

The extent to which the freezing point is depressed depends on the number of solute particles present in a given amount of solvent, not on the mass or size of the particles. A mole of sodium chloride forms 2 moles of ions in solution and causes nearly twice as great a freezing-point depression as does 1 mole of nonelectrolyte. Each individual ion has the same effect on the freezing point as a single molecule does. In Section 12.20 we will consider why the lowering produced by salts is not exactly an integer product of that produced by a similar amount of a nonelectrolyte.

Example 12.13   The Freezing Point of a Solution of an Electrolyte

The concentration of ions in sea water is approximately the same as that in a solution of 4.2 g of NaCl in 125 g of water. Assume that each of the ions in the NaCl solution has the same effect on the freezing point of water as a nonelectrolyte molecule, and determine the freezing temperature of the solution (which is approximately equal to the freezing temperature of sea water).

Solution: We can solve this problem by using the following steps:

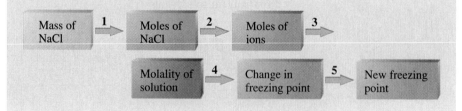

Step 1. *Convert from mass to moles of NaCl.* Result: 0.072 mol NaCl

Step 2. *Determine the number of moles of ions present in the solution, using the number of moles of ions in 1 mol of NaCl as the conversion factor (2 mol ions/1 mol NaCl).* Result: 0.14 mol ions

Step 3. *Determine the molality of the ions in the solution from the number of moles of ions and the mass of solvent, in kilograms.* Result: 1.1 m

Step 4. *Use the direct proportionality between the change in freezing point and molal concentration to determine how much the freezing point changes.* Result: 2.0°C

Step 5. *Determine the new freezing point from the freezing point of the pure solvent and the change.* Result: −2.0°C

Check your learning: Assume that each of the ions in calcium chloride, $CaCl_2$, has the same effect on the freezing point of water as a nonelectrolyte molecule. Calculate the freezing point of a solution of 0.724 g of $CaCl_2$ in 175 g of water.

Answer: −0.208°C

12.17   Phase Diagram for an Aqueous Solution of a Nonelectrolyte

If we compare the phase diagram of an aqueous solution of a nonelectrolyte (red lines in Fig. 12.23) with the phase diagram for pure water (black lines), we can see how the differences in their properties are reflected in the diagram. The freezing point for the solution is lower than that for pure water, so the line separating the solid and liquid states for the solution (the red line) is displaced to the left of the line for pure water (the black line). Correspondingly, the higher boiling point for the solution is shown by the displacement of the red line separating the liquid and gas states to the right of the black one. The decrease in vapor pressure of the solution, at any given temperature, is indicated by the vertical distance (shown in orange) between the black line and the red line. On the diagram, the freezing-point depression and the boiling-point elevation are the horizontal distances (shown in yellow) between the black line and the red line and labeled ΔT_f and ΔT_b, respectively.

Figure 12.23

Phase diagram for a 1 m aqueous solution of a nonelectrolyte (red solid lines) compared to that for pure water (black solid lines). In order to show more clearly the differences between the solution and pure water, the diagram is not to scale.

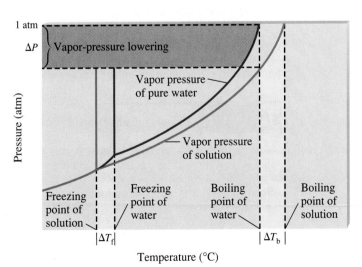

Figure 12.24

In this apparatus for demonstrating osmosis, the levels are equal at the start (A); but at equilibrium (B), the level of the solution is higher because of the net transfer into it of molecules of solvent through the semipermeable membrane.

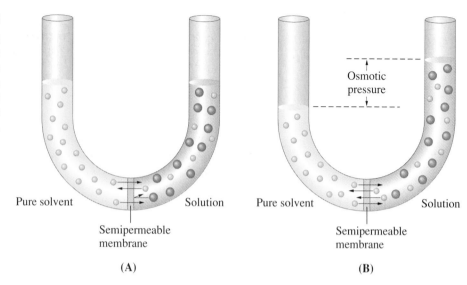

12.18   Osmosis and Osmotic Pressure of Solutions

Certain membranes (such as cellophane, animal bladders, cell membranes, and some polymer films) permit a solvent to pass through, but not a solute. Such membranes are called **semipermeable membranes.** When we separate a solution and a sample of the pure solvent by a semipermeable membrane, the pure solvent diffuses through the membrane and dilutes the solution (Fig. 12.24), a process known as **osmosis.** During osmosis, the volume of the solution increases because solvent molecules move from the pure solvent into the solution.

When osmosis is carried out in an apparatus like that shown in Fig. 12.24, the level of the solution rises until its hydrostatic pressure (which is due to the weight of the column of solution in the tube) is great enough to prevent the further osmosis of solvent molecules into the solution. The pressure required to stop the osmosis from a pure solvent into a solution is called the **osmotic pressure, π,** of the solution. This pressure is directly proportional to the molar (not the molal) concentration of the solution and to its temperature. The osmotic pressure of a dilute solution can be calculated from the expression

$$\pi = MRT$$

where M is the molar concentration of the solute, R is the gas constant, and T is the temperature of the solution on the Kelvin scale.

Example 12.14   Calculation of Osmotic Pressure

What is the osmotic pressure in atmospheres of a 0.30 M solution of glucose in water that is used for intravenous infusion at body temperature, 37°C?

Solution: We can find the osmotic pressure, π, in atmospheres using the formula $\pi = MRT$, where T is on the Kelvin scale (310 K) and the value of R includes the unit of atmospheres (0.08206 L atm mol^{-1} K^{-1}).

$$\pi = MRT = 0.30 \text{ mol } L^{-1} \times 0.08206 \text{ } L \text{ atm mol}^{-1} K^{-1} \times 310 \text{ K} = 7.6 \text{ atm}$$

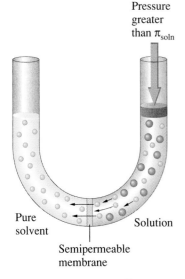

Pressure greater than π_{soln}

Pure solvent

Solution

Semipermeable membrane

Figure 12.25

Applying a pressure greater than the osmotic pressure of a solution will reverse osmosis.

Figure 12.26

Premium drinking water is produced by reverse osmosis in dispensers in stores.

Figure 12.27

(A) Red blood cells swell in a hypotonic solution. (B) Normal red blood cells in an isotonic solution. (C) Red blood cells shrivel in a hypertonic solution.

Check your learning: What is the osmotic pressure in atmospheres of a solution with a volume of 0.750 L that contains 5.0 g of methanol, CH_3OH, in water at 37°C?

Answer: 5.3 atm

When a solution is placed in an apparatus like the one shown in Fig. 12.25, applying pressure greater than the osmotic pressure of the solution reverses the osmosis and pushes solvent molecules from the solution into the pure solvent. This technique of reverse osmosis is used for desalting sea water and producing the water that is available from machines in many grocery stores (Fig. 12.26). Plants that use this principle produce water for the city of Key West and for other places in the world.

Biologists and others in the life sciences find the effects of osmosis particularly evident in biological systems, because cells are surrounded by semipermeable membranes. Carrots and celery that have become limp because they have lost water can be made crisp again by placing them in water. Water moves into the carrot or celery cells by osmosis. A cucumber placed in a concentrated salt solution loses water by osmosis and becomes a pickle. Osmosis can also affect animal cells. Solute concentrations are particularly important when solutions are injected into the body. Solutes in body cell fluids and blood serum give these solutions an osmotic pressure of approximately 7.7 atmospheres. Solutions injected into the body must have the same osmotic pressure as blood serum; that is, they should be *isotonic* with blood serum. When a less concentrated solution, a *hypotonic* solution, is injected in sufficient quantity to dilute the blood serum, water from the diluted serum passes into the blood cells by osmosis, causing the cells to expand and rupture. When a more concentrated solution, a *hypertonic* solution, is injected, the cells lose water to the more concentrated solution, shrivel, and may even die (Fig. 12.27).

12.19 Determination of Molecular Masses

Osmotic pressure and changes in freezing point, boiling point, and vapor pressure are directly proportional to the concentration of solute present. Consequently, we can use a measurement of one of these properties to determine the molar mass of a solute.

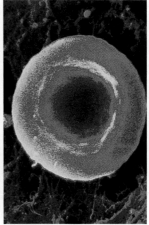

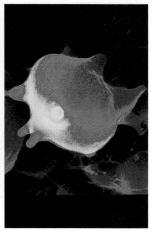

(A)

(B)

(C)

Example 12.15   Determination of a Molar Mass from a Freezing-Point Depression

A solution of 4.00 g of a nonelectrolyte dissolved in 55.0 g of benzene is found to freeze at 2.32°C. What is the molar mass of this compound?

Solution: We can solve this problem by using the following steps:

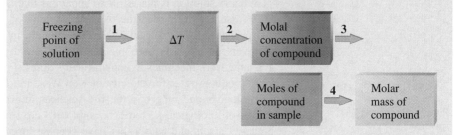

Step 1.   Determine the change in freezing point from the observed freezing point and the freezing point of pure benzene (Table 12.1 on page 437).

$$\Delta T = 5.5°C - 2.32°C = 3.2°C$$

Step 2.   Determine the molal concentration from ΔT and K_f, the molal freezing-point depression constant for benzene (Table 12.1).

$$\Delta T = K_f m$$

$$m = \frac{\Delta T}{K_f} = \frac{3.2°C}{5.12°C\ m^{-1}} = 0.62\ m$$

Step 3.   Determine the number of moles of compound in the solution from the molal concentration and the mass of solvent used to make the solution.

$$\text{Moles of solute} = \frac{0.62\ \text{mol solute}}{1.00\ \text{kg solvent}} \times 0.0550\ \text{kg solvent}$$

$$= 0.034\ \text{mol}$$

Step 4.   Determine the molar mass from the mass of the solute and the number of moles in that mass.

$$\text{Molar mass} = \frac{4.00\ \text{g}}{0.034\ \text{mol}} = 1.2 \times 10^2\ \text{g mol}^{-1}$$

Check your learning.   A solution of 35.7 g of a nonelectrolyte in 220.0 g of chloroform has a boiling point of 64.5°C. What is the molar mass of this compound?

Answer: 1.8×10^2 g mol^{-1}

Example 12.16   Determination of a Molar Mass from Osmotic Pressure

0.500 L of an aqueous solution that contains 10.0 g of hemoglobin has an osmotic pressure of 5.9 torr at 22°C. What is the molar mass of hemoglobin?

Solution: One set of steps that can be used to solve this problem follows.

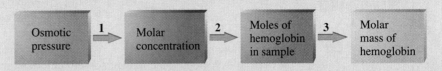

Step 1.   *Convert the osmotic pressure to atmospheres, and then determine the molar concentration from the osmotic pressure.*

$$\pi = 5.9 \text{ torr} \times \frac{1 \text{ atm}}{760 \text{ torr}} = 7.8 \times 10^{-3} \text{ atm}$$

$$\pi = MRT$$

$$M = \frac{\pi}{RT} = \frac{7.8 \times 10^{-3} \text{ atm}}{(0.08206 \text{ L atm mol}^{-1} \text{ K}^{-1})(295 \text{ K})}$$

$$= 3.2 \times 10^{-4} \text{ M}$$

Step 2.   *Determine the number of moles of hemoglobin in the solution from the concentration and the volume of the solution.*

$$\text{Moles of hemoglobin} = \frac{3.2 \times 10^{-4} \text{ mol}}{1 \text{ L solution}} \times 0.500 \text{ L solution}$$

$$= 1.6 \times 10^{-4} \text{ mol}$$

Step 3.   *Determine the molar mass from the mass of hemoglobin and the number of moles in that mass.*

$$\text{Molar mass} = \frac{10.0 \text{ g}}{1.6 \times 10^{-4} \text{ mol}} = 6.2 \times 10^4 \text{ g mol}^{-1}$$

Check your learning: What is the molar mass of a protein if a solution of 0.0200 g of the protein in 25.0 mL of solution has an osmotic pressure of 0.56 torr at 25°C?

Answer: 2.7×10^4 g mol^{-1}

12.20   Colligative Properties of Electrolytes

We have seen that the effect of nonelectrolytes on the colligative properties of a solution depends only on the number—not on the kind—of particles dissolved. For example, 1 mole of any nonelectrolyte dissolved in 1 kilogram of solvent produces the same lowering of the freezing point as does 1 mole of any other nonelectrolyte.

The lowering of the freezing point produced by 1 mole of a strong electrolyte is greater than that produced by 1 mole of a nonelectrolyte, because the electrolyte ionizes when it dissolves. When 1 mole of sodium chloride is dissolved in 1 kilogram of water, the solution freezes at −3.37°C. The freezing point is lowered 1.81 times as much as for a nonelectrolyte. This illustrates the fact that 1 mole of an electrolyte produces more than 1 mole of solute particles. Almost all ionic compounds and compounds that react with water to form ions behave this way when dissolved.

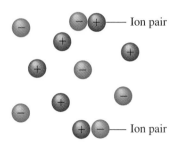

We should expect that if the ions of sodium chloride were completely separated in aqueous solution, they would lower the freezing point, and raise the boiling point, of the solvent twice as much as would an equal concentration of a nonelectrolyte, because sodium chloride gives 2 moles of ions per mole of compound. However, 1 mole of sodium chloride lowers the freezing point of water only 1.81 times as much as 1 mole of a nonelectrolyte. A similar discrepancy occurs for the boiling-point elevation. Apparently, the ions of sodium chloride (and other strong electrolytes) are not completely separated in solution.

In 1923, the chemists Peter Debye and Erich Huckel proposed a theory to explain the apparent incomplete ionization of strong electrolytes. They suggested that although interionic attraction in an aqueous solution is very greatly reduced by hydration of the ions and by the insulating action of the polar solvent (Section 12.3), it is not completely nullified. The residual attractions prevent the ions from behaving as totally independent particles (Fig. 12.28). In some cases a positive and a negative ion may actually touch, giving a solvated unit called an **ion pair.** Thus the **ion activity,** or the effective concentration, of any particular kind of ion is less than that indicated by the actual concentration. Ions become more and more widely separated the more dilute the solution, and the residual interionic attractions become less and less. Thus in extremely dilute solutions, the effective concentrations of the ions (their activities) are essentially equal to the actual concentrations.

Colloid Chemistry

12.21   Colloids

As children, we all made suspensions such as mixtures of mud and water or flour and water. Sometimes we used tempera paint, a suspension of a solid paint in water. These suspensions are heterogeneous mixtures composed of particles that are visible (or that can be seen with a magnifying glass). They are cloudy, and the suspended particles settle out after mixing. On the other hand, when we make a solution, we prepare a homogeneous mixture in which no settling occurs and in which the dissolved species are molecules or ions. Solutions exhibit completely different behavior from suspensions. A solution may be colored, but it is transparent, the molecules or ions are invisible, and they do not settle out on standing.

There is a group of substances called **colloids** that exhibit properties intermediate between those of suspensions and solutions. The particles in a colloid are larger than most simple molecules; however, colloidal particles are small enough that they do not settle out. The particles in a colloid are large enough to scatter light, a phenomenon called the **Tyndall effect.** This can make colloidal mixtures appear cloudy or opaque (Fig. 12.29). Clouds are colloidal mixtures. They are composed of air and water droplets that are much larger than molecules but that are small enough that they do not settle out. (Raindrops form when the colloidal particles of water coalesce into particles that are large enough to fall.)

Another colloidal mixture is produced by heating starch in water. This mixture is opaque and is composed of insoluble starch particles that are much larger than molecules, but the particles do not settle out. A colloidal mixture is called a **colloidal dis-**

Figure 12.29
A laser beam is scattered by colloidal-size water droplets when it passes through clouds, or fog, as this light show in Houston demonstrates.

persion; the finely divided component is called the *dispersed phase,* and the other component is called the *dispersion medium.*

The term *colloid*—from the Greek words *kolla,* meaning "glue," and *eidos,* meaning "like"—was first used in 1861 by Thomas Graham to classify substances such as starch and gelatin. We now know that colloidal properties are not limited to this class of substances; any substance can exist in colloidal form. Many colloidal particles are aggregates of hundreds, thousands, or even more molecules, but others (such as proteins and polymer molecules) consist of a single extremely large molecule. The protein and synthetic polymer molecules that form colloids may have molecular masses ranging from a few thousand to many million atomic mass units.

Colloids may be dispersed in a gas, a liquid, or a solid, and the dispersed phase may be a gas, a liquid, or a solid. However, a gas dispersed in another gas is not a colloidal system, because the particles are of molecular dimensions. Examples of colloidal systems are given in Table 12.2.

Table 12.2
Colloidal Systems

Dispersed Phase	Dispersion Medium	Common Examples	Name
Solid	Gas	Smoke, dust	
Solid	Liquid	Starch suspension, some inks, paints, milk of magnesia	Sol
Solid	Solid	Some colored gems, some alloys	
Liquid	Gas	Clouds, fogs, mists, sprays	Aerosol
Liquid	Liquid	Milk, mayonnaise, butter	Emulsion
Liquid	Solid	Jellies, gels, pearl, opal (H_2O in SiO_2)	Gel
Gas	Liquid	Foams, whipped cream, beaten egg whites	Foam
Gas	Solid	Pumice, floating soaps	

12.22   Preparation of Colloidal Systems

We can prepare a colloidal system by producing particles of colloidal dimensions and distributing these particles throughout a dispersion medium. Particles of colloidal size are formed by two methods:

1. Dispersion methods—that is, by breaking down larger particles. For example, paint pigments are produced by dispersing large particles by grinding in special mills.

2. Condensation methods—that is, growth from smaller units, such as molecules or ions. For example, clouds form when water molecules condense and produce very small droplets.

A few solid substances, when brought into contact with water, disperse spontaneously and form colloidal systems. Gelatin, glue, starch, and dehydrated milk powder behave in this manner. The particles are already of colloidal size; the water simply disperses them. Powdered milk particles of colloidal size are produced by dehydrating milk spray. Some atomizers produce colloidal dispersions of a liquid in air.

We can prepare an emulsion by shaking together or blending two immiscible liquids. This breaks one liquid into droplets of colloidal size, which then disperse throughout the other liquid. Oil spills in the ocean may be difficult to clean up, partly because wave action can cause the oil and water to form an emulsion. In many emulsions, however, the dispersed phase tends to coalesce, form large drops, and separate. Therefore, emulsions are usually stabilized by an **emulsifying agent,** a substance that inhibits the coalescence of the dispersed liquid. For example, a little soap will stabilize an emulsion of kerosene in water. Milk is an emulsion of butterfat in water with the protein casein as the emulsifying agent. Mayonnaise is an emulsion of oil in vinegar with egg yolk as the emulsifying agent.

Condensation methods form colloidal particles by aggregation of molecules or ions. If the particles grow beyond the colloidal size range, drops or precipitates form, and no colloidal system results. Clouds appear when water molecules aggregate and form colloid-sized drops. If the drops increase in size, rain results. Many condensation methods employ chemical reactions. We can prepare a red colloidal suspension of iron(III) hydroxide by mixing a concentrated solution of iron(III) chloride with hot water.

$$Fe^{3+} + 3Cl^- + 6H_2O \longrightarrow Fe(OH)_3 + 3H_3O^+ + 3Cl^-$$

A colloidal gold sol results from the reduction of a very dilute solution of gold(III) chloride by a reducing agent such as formaldehyde, tin(II) chloride, or iron(II) sulfate.

$$Au^{3+} + 3e^- \longrightarrow Au$$

Some gold sols prepared in 1857 are still intact.

12.23   Soaps and Detergents

Pioneers made soap by boiling fats with a strongly basic solution made by leaching potassium carbonate, K_2CO_3, from wood ashes with hot water. Animal fats contain esters of fatty acids (long-chain carboxylic acids, Section 9.13). When animal fats are treated with a base such as potassium carbonate or sodium hydroxide, glycerol and salts of fatty acids such as palmitic, oleic, and stearic acid are formed. The salts of fatty acids are called **soaps.** The sodium salt of stearic acid, sodium stearate, has the

Figure 12.30

Both soaps and detergents contain a nonpolar hydrocarbon end (blue) and an ionic end (red). The important difference between soaps and detergents is the formula of the ionic end. The length of the hydrocarbon end can vary from soap to soap and from detergent to detergent.

Hydrocarbon chain   Ionic end

CH$_2$ CH$_2$ CH$_2$ CH$_2$ CH$_2$ CH$_2$ CH$_2$ CH$_2$ CO$_2^-$Na$^+$

CH$_3$ CH$_2$ CH$_2$ CH$_2$ CH$_2$ CH$_2$ CH$_2$ CH$_2$ CH$_2$

Sodium stearate (soap)

CH$_3$ CH$_2$ CH$_2$ CH$_2$ CH$_2$ CH$_2$ OSO$_3^-$Na$^+$

CH$_2$ CH$_2$ CH$_2$ CH$_2$ CH$_2$ CH$_2$

Sodium lauryl sulfate (detergent)

formula $C_{17}H_{35}CO_2Na$ and contains an uncharged nonpolar hydrocarbon chain, the $C_{17}H_{35}$— unit, and an ionic carboxylate group, the —CO_2^- unit (Fig. 12.30).

Detergents (soap substitutes) also contain nonpolar hydrocarbon chains, such as $C_{12}H_{25}$—, and an ionic group, such as a sulfate, —OSO_3^-, or a sulfonate, —SO_3^- (Fig. 12.30). Soaps form insoluble calcium and magnesium compounds in hard water; detergents form water-soluble products—a definite advantage for detergents.

The cleansing action of soaps and detergents can be explained in terms of the structures of the molecules involved. The hydrocarbon end of a soap or detergent molecule dissolves in, or is attracted to, oil, grease, or dirt particles. The ionic end is attracted by water (Fig. 12.31). As a result, the soap or detergent molecules become

Figure 12.31

This diagrammatic cross section of an emulsified drop of oil in water shows how a soap or detergent acts as an emulsifier. The nonpolar hydrocarbon end of the soap or detergent dissolves in the oil; the ionic end dissolves in water. The positive ions are solvated and move independently through the water.

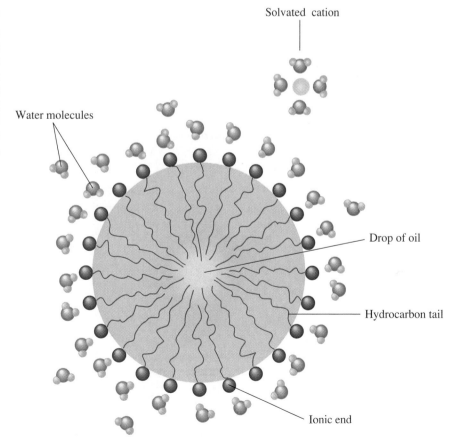

Solvated cation

Water molecules

Drop of oil

Hydrocarbon tail

Ionic end

oriented at the interface between the dirt particles and the water, so they act as a kind of bridge between two different kinds of matter, nonpolar and polar. As a consequence, dirt particles become suspended as colloidal particles and are readily washed away.

12.24 Electrical Properties of Colloidal Particles

Dispersed colloidal particles are usually electrically charged. A colloidal particle of iron(III) hydroxide, for example, does not contain enough hydroxide ions to compensate exactly for the positive charges on the iron(III) ions. Thus each individual colloidal particle bears a positive charge, and the colloidal dispersion consists of charged colloidal particles and some free hydroxide ions, which keep the dispersion electrically neutral. Most metal hydroxide colloids have positive charges, whereas most metals and metal sulfides form negatively charged dispersions. All colloidal particles in any one system have charges of the same sign. This helps to keep them dispersed, because particles that contain like charges repel each other.

We can take advantage of the charge on colloidal particles to remove them from a variety of mixtures. When we place a colloidal dispersion in a container with charged electrodes, positively charged particles, such as iron(III) hydroxide particles, move to the negative electrode. There the colloidal particles lose their charge and coagulate as a precipitate.

The carbon and dust particles in smoke are often colloidally dispersed and electrically charged. Frederick Cottrell, an American chemist, developed a process to

Figure 12.32

In a Cottrell precipitator, positively and negatively charged particles in smoke are precipitated as they pass the electrodes.

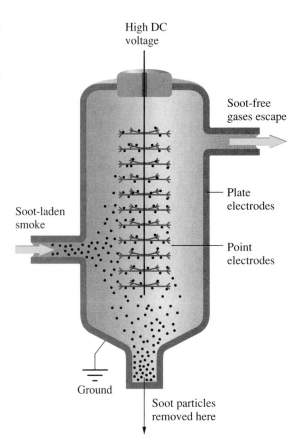

remove these particles. The charged particles are attracted to highly charged electrodes, where they are neutralized and deposited as dust (Fig. 12.32). This is one of the methods used to clean up the smoke from a variety of industrial processes. The process is also important in the recovery of valuable products from the smoke and flue dust of smelters, furnaces, and kilns.

12.25   Gels

When we make gelatin desserts, we are making a type of colloid. Jello sets on cooling because the hot aqueous "solution" of gelatin coagulates as it cools, and the whole mass, including the liquid, sets to an extremely viscous body known as a **gel.** It appears that the fibers of the dispersed substance form a complex three-dimensional network, the interstices being filled with the liquid medium or a dilute solution of the dispersed phase. Because the formation of a gel is accompanied by the taking up of water or some other solvent, the gel is said to be hydrated or solvated.

Pectin, a carbohydrate from fruit juices, is a gel-forming substance important in jelly making. Silica gel, a colloidal dispersion of hydrated silicon dioxide, is formed when dilute hydrochloric acid is added to a dilute solution of sodium silicate. Canned Heat is a gel made by mixing alcohol and a saturated aqueous solution of calcium acetate. The membrane of a living cell is colloidal in character, and within the cell is a gel. In fact, all living tissue is colloidal, and life processes depend on the chemistry of colloids.

For Review   Summary

A **solution** forms when one substance, a **solute,** dissolves in a second substance, a **solvent,** giving a homogeneous mixture of atoms, molecules, or ions. The solute may be a solid, a liquid, or a gas. The solvent is usually a liquid, although solutions of gases in other gases and solutions of gases, liquids, and solids in solids are possible. Substances that dissolve and give solutions that contain ions and conduct electricity are called **electrolytes.** Electrolytes may be molecular compounds that react with the solvent to give ions, or they may be ionic compounds. **Nonelectrolytes** are substances that dissolve to give solutions of molecules. These solutions do not conduct electricity.

The extent to which a gas dissolves in a liquid solvent is proportional to the pressure of the gas, provided that the gas does not react with the solvent. Generally, the solubility of a gas decreases with increasing temperature. Liquids that mix with each other in all proportions are called **miscible.** Polar liquids tend to be miscible with water, whereas nonpolar liquids tend to be **immiscible** with water. Generally, the solubility of solids in water increases with increasing temperature, but there are exceptions.

Solutions that obey **Raoult's law** exactly—ideal solutions—form when the average strength of the intermolecular forces of attraction between solute and solvent is equal to the average strength of the forces of attraction between pure solute molecules and between pure solvent molecules. Ideal solutions form by a spontaneous process, because the disorder of the molecules of solute and solvent increases when the solution forms. An increase in the average strength of the solute–solvent intermolecular attractive forces, compared to those in the pure solute and solvent, also favors spontaneous formation of a solution. The relative strengths of the solute–solute and solvent–solvent attractions compared to the solute–solvent attractions can be measured by the enthalpy of solution of the solute in the solvent. In many cases, the increase in disorder compensates for the decrease in the average strength of the solute–solvent attractions compared to those of the pure components of the solution.

The relative amounts of solute and solvent in a solution can be described quantitatively as the **concentration** of the solution. Units of concentration include **percent composition, molar concentration, molal concentration,** and **mole fraction.**

Properties of a solution that depend only on the concentration of solute particles are called **colligative properties.** They include changes in the vapor pressure, boiling point, and freezing point of the solvent in the solution. The magnitudes of these properties depend only on the total concentration of solute particles in solution, not on the type of particles. The total concentration of solute particles in a solution also affects its **osmotic pressure.** This is the pressure that must be applied to the solution in order to prevent diffusion of molecules of pure solvent through a semipermeable membrane into the solution.

A **colloid** is a suspension of small, insoluble particles, which are usually larger than molecules but small enough not to settle from the dispersion medium. Examples of colloids include smoke, clouds, paints, mayonnaise, foams, and living tissue.

Key Terms and Concepts

azeotropic mixture   (12.15)
boiling-point elevation   (12.14)
colligative properties   (12.13)
colloid   (12.21)
concentrated solution   (12.1)
constant boiling solution   (12.15)
detergent   (12.23)
dilute solution   (12.1)
dynamic equilibrium   (12.1)
electrolyte   (12.3)
emulsion   (12.22)
fractional distillation   (12.15)
freezing-point depression   (12.16)
gel   (12.25)
Henry's law   (12.5)

ideal solution   (12.2, 12.13)
immiscible   (12.6)
interstitial solid solution   (12.8)
ion activity   (12.20)
ion–dipole attraction   (12.3)
ion pair   (12.20)
miscible   (12.6)
molality, m   (12.11)
molarity, M   (12.10)
mole fraction   (12.12)
nonelectrolyte   (12.3)
nonstoichiometric compounds   (12.8)
osmosis   (12.18)
osmotic pressure   (12.18)
percent by mass   (12.9)

percent composition   (12.9)
Raoult's law   (12.13)
saturated solution   (12.1)
semipermeable membrane   (12.18)
soap   (12.23)
solid solution   (12.8)
solubility   (12.1)
solute   (12.1)
solvent   (12.1)
spontaneous process   (12.2)
substitutional solid solution   (12.8)
supersaturated solution   (12.1)
Tyndall effect   (12.21)
unsaturated solution   (12.1)
vapor-pressure lowering   (12.13)

Exercises

Questions

1. How do solutions differ from compounds? From other mixtures?

2. Generally, chemical reactions occur in solution. Provide an example of a chemical reaction that does not occur in solution. (Remember, solutions include more than liquids.)

3. Which of the principal characteristics of solutions can we see in the solutions of $K_2Cr_2O_7$ shown in Fig. 12.1?

4. How do the solutions of $K_2Cr_2O_7$ shown in Fig. 12.1 and the solution of Br_2 in water shown in Fig. 12.16 (the top solution) differ?

5. Which are a part of the macroscopic domain of solutions and which are a part of the microscopic domain? boiling-point elevation, Henry's law, hydrogen bond, ion–dipole attraction, molarity, nonelectrolyte, nonstoichiometric compound, osmosis, solvated ion.

6. Which are a part of the macroscopic domain of solids and which are a part of the microscopic domain? azeotropic mixture, electrolyte, emulsion, freezing-point depression, hydronium ion, London force, molality, Raoult's law, saturated solution, spontaneous process.

7. Which of the following figures illustrate a macroscopic feature of solutions or colloids and which illustrate a microscopic feature? Fig. 12.3, 12.6, 12.8, 12.12, 12.17, 12.24

8. Which of the following figures illustrate a macroscopic feature of solutions or colloids and which illustrate a microscopic feature? Fig. 12.1, 12.4, 12.19, 12.23, 12.28, 12.29, 12.31

9. What is the microscopic explanation for the macroscopic behavior illustrated in Fig. 12.9?

10. What is the microscopic explanation for the macroscopic behavior illustrated in Fig. 12.15?

11. In order to prepare supersaturated solutions of most solids in water, we cool saturated solutions. Supersaturated solutions of most gases in water are prepared by warming saturated solutions. Explain the reasons for the difference between the two procedures.

12. What is the difference between a 1 M solution and a 1 m solution?

13. How did the concentration of CO_2 in the beverage shown in Fig. 12.10 change when the bottle was opened?

14. Sketch a qualitative graph of the pressure of water vapor above a sample of pure water and of that above a sugar solution as the liquids evaporate to half their original volume.

15. (a) Which of the solutions shown in Fig. 12.1 has the higher vapor pressure? (b) The lower freezing temperature? (c) The higher boiling temperature?

16. A solution of potassium nitrate (an electrolyte) and a solution of glycerine (a nonelectrolyte) both boil at 100.3°C. What other physical properties of the two solutions are identical?

17. (a) What is a constant boiling solution? (b) Describe two ways in which to prepare a constant boiling solution of HNO_3.

18. Explain the following terms as applied to solutions: solute, solvent, electrolyte, nonelectrolyte, saturated, supersaturated, unsaturated, concentrated, dilute.

19. Suppose you are presented with a clear solution of sodium thiosulfate, $Na_2S_2O_3$. How can you determine whether the solution is unsaturated, saturated, or supersaturated?

The Process of Dissolution

20. Which of the following spontaneous processes occurs with an increase in disorder?

    (a) Evaporation of water
    (b) Precipitation of AgCl from solution
    (c) Condensation of water vapor to ice
    (d) Mixing of natural gas, $CH_4(g)$, and air
    (e) Dissolution of KBr in water

21. Indicate the most important type of solute–solvent attractions (Sections 11.2–11.4, 12.2) in each of the following solutions.

    (a) The solutions in Fig. 12.1
    (b) Methanol, CH_3OH, in ethanol, C_2H_5OH
    (c) CH_4 in benzene, C_6H_6
    (d) The Freon CF_2Cl_2 in the Freon $CF_2ClCFCl_2$
    (e) $O_2(l)$ in $N_2(l)$

22. Indicate the most important types of solute–solvent attractions (Sections 11.2–11.4, 12.2) in each of the following solutions.

    (a) The solutions in Fig. 12.6
    (b) NO(l) in CO(l)
    (c) $Cl_2(g)$ in $Br_2(l)$
    (d) HCl in benzene, C_6H_6
    (e) Methanol, CH_3OH, in water

23. Explain why HBr, a gas, is a nonelectrolyte when dissolved in benzene and an electrolyte when dissolved in water.

24. Explain why the ions Na^+ and Cl^- are strongly solvated in water but not in hexane, a solvent composed of nonpolar molecules.

25. Heat is released when some solutions form; heat is absorbed when other solutions form. Provide a molecular explanation for the difference between these two types of spontaneous processes.

26. Compare the processes that occur when methanol, CH_3OH, hydrogen chloride, HCl, and sodium hydroxide, NaOH, dissolve in water. Write equations and prepare sketches showing the form in which each of these compounds is present in its respective solution.

27. Suggest an explanation for the observations that ethanol, C_2H_5OH, is completely miscible with water and that ethanethiol, C_2H_5SH, is soluble only to the extent of 1.5 g per 100 mL of water.

28. Explain, in terms of the intermolecular attractions between solute and solvent, why methanol, CH_3OH, is miscible with water in all proportions but only 0.11 g of butanol, $CH_3CH_2CH_2CH_2OH$, dissolves in 100 g of water at 25°C.

29. Why are solid ionic compounds nonconductors, whereas ionic compounds that are fused (melted) are good conductors?

30. At 0°C, 0.70 g of O_2 at 1.00 atm dissolves in exactly 1 L of water. At 0°C and 4.00 atm, how many grams of O_2 dissolve in exactly 1 L of water?

Units of Concentration

31. (a) Outline the steps necessary to answer the following question: What mass of a 95% by mass solution of sulfuric acid is needed to prepare 200.0 g of a 20.0% solution by mass? (b) Answer the question.

32. (a) Outline the steps necessary to answer the following question: What mass of a concentrated solution of nitric acid (68.0% HNO_3 by mass) is needed to prepare 400.0 g of a 10.0% solution of HNO_3 by mass? (b) Answer the question.

33. What mass of a 4.00% NaOH solution by mass contains 15.0 g of NaOH?

34. What mass of solid NaOH (97.0% NaOH by mass) is required to prepare 1.00 L of a 10.0% solution of NaOH by mass? The density of the 10.0% solution is 1.109 g cm^{-3}.

35. What mass of HCl is contained in 45.0 mL of an HCl solution that has a density of 1.19 g cm^{-3} and contains 37.21% HCl by mass?

36. Calculate what volume of a sulfuric acid solution (density = 1.070 g cm^{-3} and containing 10.00% H_2SO_4 by mass) contains 18.50 g of pure H_2SO_4.

37. (a) Outline the steps necessary to answer the following question: What is the mass of the solute in 0.500 L of 0.30 M glucose, $C_6H_{12}O_6$, used for intravenous injection? (b) Answer the question.

38. (a) Outline the steps necessary to answer the following question: What is the mass of solute in 200.0 L of a 1.556 M solution of KBr? (b) Answer the question.

39. Calculate the number of moles and the mass of the solute in each of the following solutions.

   (a) 2.00 L of 18.5 M H_2SO_4, concentrated sulfuric acid
   (b) 100 mL of 3.8×10^{-5} M NaCN, the minimum lethal concentration of sodium cyanide in blood serum
   (c) 5.50 L of 13.3 M formaldehyde, H_2CO, used to "fix" tissue samples
   (d) 325 mL of 1.8×10^{-6} M $FeSO_4$, the minimum concentration of iron sulfate detectable by taste in drinking water

40. Calculate the number of moles and the mass of the solute in each of the following solutions.

   (a) 325 mL of 8.23×10^{-5} M KI, a source of iodine in the diet
   (b) 75.0 mL of 2.2×10^{-5} M H_2SO_4, a sample of acid rain
   (c) 0.2500 L of 0.1135 M K_2CrO_4, an analytical reagent used for determining iron
   (d) 10.5 L of 3.716 M $(NH_4)_2SO_4$, a liquid fertilizer

41. (a) Outline the steps necessary to answer the following question: What is the molarity of $KMnO_4$ in a solution of 0.0908 g of $KMnO_4$ in 0.500 L of solution? (b) Answer the question.

42. (a) Outline the steps necessary to answer the following question: What is the molarity of HCl if 35.23 mL of a solution of HCl contains 0.3366 g of HCl? (b) Answer the question.

43. Calculate the molarity of each of the following solutions.

   (a) 0.195 g of cholesterol, $C_{27}H_{46}O$, in 0.100 L of serum, the average concentration of cholesterol in human serum
   (b) 4.25 g of NH_3 in 0.500 L of solution, the concentration of NH_3 in household ammonia
   (c) 1.49 kg of isopropyl alcohol, C_3H_7OH, in 2.50 L of solution, the concentration of isopropyl alcohol in rubbing alcohol
   (d) 0.029 g of I_2 in 0.100 L of solution, the solubility of I_2 in water at 20°C

44. Calculate the molarity of each of the following solutions.

   (a) 293 g HCl in 666 mL of solution, a concentrated HCl solution
   (b) 2.026 g $FeCl_3$ in 0.1250 L of a solution used as an unknown in general chemistry laboratories
   (c) 0.001 mg Cd^{2+} in 0.100 L, the maximum permissible concentration of cadmium in drinking water
   (d) 0.0079 g $C_7H_5SNO_3$ in 1 oz (29.6 mL), the concentration of saccharin in a diet soft drink.

45. There is about 1.0 g of calcium, as Ca^{2+}, in 1.0 L of milk. What is the molarity of Ca^{2+} in milk?

46. What are the approximate molar concentrations of the solutions shown in Fig. 12.1? These solutions are contained in 500-mL beakers.

47. The hardness of water (hardness count) is usually expressed as parts per million (by mass) of $CaCO_3$, which is equivalent to milligrams of $CaCO_3$ per liter of water. What is the molar concentration of Ca^{2+} ions in a water sample with a hardness count of 175?

48. The level of mercury in a stream was believed to be above the minimum considered safe (1 part per billion by weight). However, an analysis indicated that the concentration was 0.68 part per billion. Assume a density of 1.0 g cm^{-3} and calculate the molarity of mercury in the stream.

49. (a) Outline the steps necessary to answer the following question: What is the molality of phosphoric acid, H_3PO_4, in a solution of 14.5 g of H_3PO_4 in 125 g of water? (b) Answer the question.

50. (a) Outline the steps necessary to answer the following question: What is the molality of nitric acid in a concentrated solution of nitric acid (68.0% HNO_3 by mass)? (b) Answer the question.

51. Calculate the molality of each of the following solutions.

   (a) 583 g of H_2SO_4 in 1.50 kg of water, the acid solution used in an automobile battery
   (b) 0.86 g of NaCl in 1.00×10^2 g of water, a solution of sodium chloride for intravenous injection
   (c) 46.85 g of codeine, $C_{18}H_{21}NO_3$, in 125.5 g of ethanol, C_2H_5OH
   (d) 25 g of I_2 in 125 g of ethanol, C_2H_5OH

52. Calculate the molality of each of the following solutions.

   (a) 0.710 kg of sodium carbonate (washing soda), Na_2CO_3, in 10.0 kg of water, a saturated solution at 0°C
   (b) 125 g of NH_4NO_3 in 275 g of water, a mixture used to make an instant ice pack
   (c) 25 g of Cl_2 in 125 g of dichloromethane, CH_2Cl_2
   (d) 0.372 g of histamine, C_5H_9N, in 125 g of chloroform, $CHCl_3$

53. The concentration of glucose, $C_6H_{12}O_6$, in normal spinal fluid is 75 mg per 100 g. What is the molal concentration?

54. A 13.0% solution of K_2CO_3 by mass has a density of 1.09 g cm^{-3}. Calculate the molality of the solution.

55. (a) Outline the steps necessary to answer the following question: What are the mole fractions of H_3PO_4 and water in a solution of 14.5 g of H_3PO_4 in 125 g of water? (b) Answer the question.

56. (a) Outline the steps necessary to answer the following question: What are the mole fractions of HNO_3 and water in a concentrated solution of nitric acid (68.0% HNO_3 by mass) (b) Answer the question.

57. Calculate the mole fraction of each solute and solvent in Exercise 51.

58. Calculate the mole fraction of each solute and solvent in Exercise 52.

59. Calculate the mole fractions of methanol, CH_3OH, ethanol, C_2H_5OH, and water in a solution that is 40% methanol, 40% ethanol, and 20% water by mass. (Assume the data are good to two significant figures.)

60. A sample of lead glass is prepared by melting together 20.0 g of silica, SiO_2, and 80.0 g of lead(II) oxide, PbO. What is the mole fraction of SiO_2 and that of PbO in the glass?

61. Many diabetics must determine their blood sugar content daily. The measurements provided by the devices used for such purpose are in milligrams per deciliter (0.1 L).

   (a) What is the concentration of glucose, $C_6H_{12}O_6$, in grams per liter when a reading of 95 is observed?
   (b) What is the concentration in moles per liter?

62. In Canada and the British Isles, the units used on blood sugar measuring devices are millimoles per liter. When a measurement of 5.3 is observed, what is the concentration of glucose in milligrams per deciliter?

63. What volume of a 0.20 M K_2SO_4 solution contains 57 g of K_2SO_4?

64. Equal volumes of 0.050 M $Ca(OH)_2$ and 0.400 M HCl are mixed. Calculate the molarity of each ion present in the final solution.

65. What volume of 0.600 M HCl is required to react completely with 2.50 g of sodium hydrogen carbonate?

$$NaHCO_3 + HCl \longrightarrow NaCl + CO_2 + H_2O$$

66. Calculate the volume of 0.050 M HCl necessary to precipitate the silver contained in 12.0 mL of 0.050 M $AgNO_3$.

$$Ag^+(aq) + Cl^-(aq) \longrightarrow AgCl(s)$$

67. What volume of a 0.33 M solution of hydrobromic acid would be required to neutralize completely 1.00 L of 0.15 M barium hydroxide?

68. To 10.0 mL of a 0.100 M $K_2Cr_2O_7$ solution, we add 7.0 mL of a 0.200 M $Pb(NO_3)_2$ solution. What mass of $PbCr_2O_7$ forms?

69. A gaseous solution is found to contain 15% H_2, 10% CO, and 75% CO_2 by mass. What is the mole fraction of each component?

70. Concentrated hydrochloric acid is 37.0% HCl by mass and has a density of 1.19 g mL^{-1}. Calculate (a) the molarity, (b) the molality, and (c) the mole fraction of HCl and of H_2O.

71. Calculate the percent composition and the molality of an aqueous solution of $NaNO_3$ if the mole fraction of $NaNO_3$ is 0.20.

Colligative Properties

72. When equal masses of two compounds, A and B, are dissolved in equal amounts of a solvent, the solution of A freezes at the lower temperature. Which of the following statements are true? (a) A has a larger molecular weight than B. (b) B has a larger molecular weight than A. (c) If A and B have the same molecular weight, A is molecular and B is ionic. (d) The solution of A has a lower boiling temperature than the solution of B.

73. Which evaporates faster under the same conditions, 50 mL of distilled water or 50 mL of sea water? Explain your answer.

74. Why does 1 mol of sodium chloride depress the freezing point of 1 kg of water almost twice as much as 1 mol of glycerin?

75. (a) Outline the steps necessary to answer the following question: What is the boiling point of a solution of 115.0 g of sucrose, $C_{12}H_{22}O_{11}$, in 350.0 g of water? (b) Answer the question.

76. (a) Outline the steps necessary to answer the following question: What is the boiling point of a solution of 9.04 g of I_2 in 75.5 g of benzene, assuming the I_2 is nonvolatile? (b) Answer the question.

77. (a) Outline the steps necessary to answer the following question: What is the freezing temperature of a solution of 115.0 g of sucrose, $C_{12}H_{22}O_{11}$, in 350.0 g of water, which freezes at 0.0°C when pure? (b) Answer the question.

78. (a) Outline the steps necessary to answer the following question: What is the boiling point of a solution of 9.04 g of I_2 in 75.5 g of benzene? (b) Answer the question.

79. (a) Outline the steps necessary to answer the following question: What is the osmotic pressure of an aqueous solution of 1.64 g of $Ca(NO_3)_2$ in water at 25°C? The volume of the solution is 275 mL. (b) Answer the question.

80. (a) Outline the steps necessary to answer the following question: What is the osmotic pressure of a solution of bovine insulin (molar mass, 5700 g mol^{-1}) at 18°C if 100.0 mL of the solution contains 0.103 g of the insulin? (b) Answer the question.

81. (a) Outline the steps necessary to solve the following problem. A solution of 5.00 g of a compound in 25.00 g of carbon tetrachloride (boiling point, 76.8°C; K_b = 5.02°C m^{-1}) boils at 81.5°C at 1 atm. What is the molar mass of the compound? (b) Solve the problem.

82. A sample of an organic compound (a nonelectrolyte) weighing 1.350 g lowered the freezing point of 10.0 g of benzene by 3.66°C. Calculate the molar mass of the compound.

83. A 1 m solution of HCl in benzene has a freezing point of 0.4°C. Is HCl an electrolyte in benzene? Explain.

84. A solution contains 5.00 g of urea, $CO(NH_2)_2$, per 0.100 kg of water. If the vapor pressure of pure water at 25°C is 23.7 torr, what is the vapor pressure of the solution?

85. A 12.0-g sample of a nonelectrolyte is dissolved in 80.0 g of water. The solution freezes at −1.94°C. Calculate the molar mass of the substance.

86. Arrange the following solutions in order of decreasing freezing points: 0.1 m Na_3PO_4, 0.1 m C_2H_5OH, 0.01 m CO_2, 0.15 m NaCl, 0.2 m $CaCl_2$.

87. How could you prepare a 3.08 m aqueous solution of glycerin, $C_3H_8O_3$? What is the freezing point of this solution?

88. Calculate the boiling-point elevation of 0.100 kg of water that contains 0.010 mol of NaCl, 0.020 mol of Na_2SO_4, and 0.030 mol of $MgCl_2$, assuming complete dissociation of these electrolytes.

89. A sample of sulfur weighing 0.210 g was dissolved in 17.8 g of carbon disulfide, CS_2 (K_b = 2.34°C m^{-1}). If the boiling-point elevation was 0.107°C, what is the formula of a sulfur molecule in carbon disulfide?

90. In a significant experiment performed many years ago, 5.6977 g of cadmium iodide in 44.69 g of water raised the boiling point 0.181°C. What does this suggest about the nature of a solution of CdI_2?

91. Lysozyme is an enzyme that cleaves cell walls. A 0.100-L sample of a solution of lysozyme that contains 0.0750 g of the enzyme exhibits an osmotic pressure of 1.32×10^{-3} atm at 25°C. What is the molecular mass of lysozyme?

92. The osmotic pressure of a solution that contains 7.0 g of insulin per liter is 23 torr at 25°C. What is the molecular mass of insulin?

93. The osmotic pressure of human blood is 7.6 atm at 37°C. What mass of glucose, $C_6H_{12}O_6$, is required to make 1.00 L of aqueous solution for intravenous feeding if the solution must have the same osmotic pressure as blood at body temperature, 37°C?

94. What is the freezing point of a solution of dibromobenzene, $C_6H_4Br_2$, in 0.250 kg of benzene, if the solution boils at 83.5°C?

95. What is the boiling point of a solution of NaCl in water if the solution freezes at −0.93°C?

Colloids

96. Distinguish between dispersion methods and condensation methods for preparing colloidal systems.

97. How do colloidal "solutions" differ from true solutions with regard to dispersed particle size and homogeneity?

98. Identify the dispersed phase and the dispersion medium in each of the following colloidal systems: starch dispersion, smoke, fog, pearl, whipped cream, floating soap, jelly, and milk.

99. Explain the cleansing action of soap.

100. How can it be demonstrated that colloidal particles are electrically charged?

101. Give an example of an opaque liquid that is a colloidal suspension and an example of an opaque liquid that is a true solution or a pure substance.

102. What is the structure of a gel?

Applications and Additional Exercises

103. The triple point of air-free water (Section 11.10) is defined as 273.15 K. Why is it important that the water be free of air?

104. Meat can be classified as fresh (not frozen) even though it is stored at −1°C. Why wouldn't meat freeze at this temperature?

105. Calculate the approximate percent by mass of KBr in a saturated solution of KBr in water at 40°C. See Fig. 12.17 for useful data.

106. How many liters of $NH_3(g)$ at 25°C and 1.46 atm are required to prepare 3.00 L of a 2.50 M solution of NH_3?

107. Hydrogen gas dissolves in the metal palladium with hydrogen atoms going into the holes between metal atoms. Determine the molarity, molality, and percent by mass of hydrogen atoms in a solution (density = 10.8 g cm^{-3}) of 0.94 g of hydrogen gas in 215 g of palladium metal.

108. A 0.200-L volume of gaseous ammonia measured at 30.0°C and 764 torr was absorbed in 0.100 L of water. How many milliliters of 0.0100 M hydrochloric acid are required in the neutralization of this aqueous ammonia? What is the molarity of the aqueous ammonia solution? (Assume there is no change in volume when the gaseous ammonia is added to the water.)

109. A 1.80-g sample of an acid, H_2X, required 14.00 mL of KOH solution for neutralization of all the hydrogen ions. Exactly 14.2 mL of this same KOH solution was found to neutralize 10.0 mL of 0.750 M H_2SO_4. Calculate the molecular mass of H_2X.

110. A solution of sodium carbonate that had a volume of 0.400 L was prepared from 4.032 g of $Na_2CO_3 \cdot 10H_2O$. Calculate the molarity of this solution.

111. Calculate the percent by mass and the molality in terms of $CuSO_4$ for a solution prepared by dissolving 11.5 g of $CuSO_4 \cdot 5H_2O$ in 0.100 kg of water. Remember to consider the water released from the hydrate.

112. The vapor pressure of methanol, CH_3OH, is 94 torr at 20°C. The vapor pressure of ethanol, C_2H_5OH, is 44 torr at the same temperature. Calculate the mole fraction of methanol and of ethanol in a solution of 50.0 g of methanol and 50.0 g of ethanol. Ethanol and methanol form a solution that behaves like an ideal solution. Calculate the vapor pressure of methanol and of ethanol above the solution at 20°C. Calculate the mole fraction of methanol and of ethanol in the vapor above the solution.

113. A solution of $Ba(OH)_2$ is 0.1055 M. What volume of 0.211 M nitric acid is required in the neutralization of 15.5 mL of the $Ba(OH)_2$ solution?

114. The sulfate in 50.0 mL of dilute sulfuric acid was precipitated using an excess of barium chloride. The mass of $BaSO_4$ formed was 0.482 g. Calculate the molarity of the sulfuric acid solution.

115. A sample of $HgCl_2$ weighing 9.41 g is dissolved in 32.75 g of ethanol, C_2H_5OH ($K_b = 1.20°C$ m^{-1}). The boiling-point elevation of the solution is 1.27°C. Is $HgCl_2$ an electrolyte in ethanol? Show your calculations.

116. A salt is known to be an alkali metal fluoride. A quick approximate determination of freezing point indicates that 4 g of the salt dissolved in 100 g of water produces a solution that freezes at about −1.4°C. What is the formula of the salt? Show your calculations.

117. An organic compound has a composition of 93.46% C and 6.54% H by mass. A solution of 0.090 g of this compound in 1.10 g of camphor melts at 158.4°C. The melting point of pure camphor is 178.4°C. K_f for camphor is 37.7°C m^{-1}. What is the molecular formula of the solute? Show your calculations.

118. How many liters of HCl gas, measured at 30.0°C and 745 torr, are required to prepare 1.25 L of a 3.20 M solution of hydrochloric acid?

119. The sugar fructose contains 40.0% C, 6.7% H, and 53.3% O by mass. A solution of 11.7 g of fructose in 325 g of ethanol has a boiling point of 78.59°C. The boiling point of ethanol is 78.35°C, and K_b for ethanol is 1.20°C m^{-1}. What is the molecular formula of fructose?

120. The radius of a water molecule, assuming a spherical shape, is approximately 1.40 Å. Assume that water molecules cluster around each metal ion in a solution such that the water molecules essentially touch both the metal ion and each other. On this basis, and assuming that 4, 6, 8, and 12 are the only possible coordination numbers, find the maximum number of water molecules that can hydrate each of the following ions.

   (a) Mg^{2+} (radius 0.65 Å)
   (b) Al^{3+} (0.50 Å)
   (c) Rb^+ (1.48 Å)
   (d) Sr^{2+} (1.13 Å)

121. What is the molarity of H_3PO_4 in a solution that is prepared by dissolving 10.0 g of P_4O_{10} in sufficient water to make 0.500 L of solution?

122. This question is taken from the 1994 Chemistry Advanced Placement Examination and is used with the permission of the Educational Testing Service.

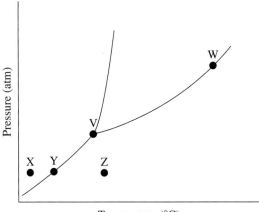

The phase diagram for a pure substance is shown above. Use this diagram and your knowledge about changes of phase to answer the following questions.

   (a) What does point V represent? What characteristics are specific to the system only at point V?
   (b) What does each point on the curve between V and W represent?
   (c) Describe the changes that the system undergoes as the temperature slowly increases from X to Y to Z at 1.0 atm.
   (d) In a solid–liquid mixture of the substance, will the solid float or sink? Explain.

CHAPTER OUTLINE

Rates of Chemical Reactions

13.1 Rate of Reaction
13.2 Relative Rates of Reaction and Reaction Velocity
13.3 Factors That Affect Reaction Rates
13.4 Rate Laws
13.5 Integrated Rate Laws
13.6 The Half-life of a Reaction

The Microscopic Explanation of Reaction Rates

13.7 Collision Theory of the Reaction Rate
13.8 Activation Energy and the Arrhenius Equation
13.9 Elementary Reactions
13.10 Reaction Mechanisms
13.11 Catalysts

13

Chemical Kinetics

A lizard warms itself in order to speed up the chemical reactions of its metabolism.

A lizard suns itself so it can survive. A warm lizard runs faster than a cold one because the chemical reactions that move its muscles occur more rapidly at higher temperatures than at lower ones. A lizard must be warm in order to eat and to avoid being eaten.

The speed of chemical reactions also play an important role in many other situations. The instructions for a test strip that medical laboratories use to test for the presence of glucose (sugar), blood, and other factors in the urine state, "Proper read time is critical for optimal results." The glucose test is read 30 seconds after dipping the test strip in a urine sample, the blood test at 60 seconds.

The test for urine glucose is based on two sequential reactions that are catalyzed by enzymes. In the first reaction, glucose is converted to gluconic acid (a carboxylic acid) and hydrogen peroxide. In the second reaction, the hydrogen peroxide reacts with potassium iodide to give colors ranging from green to brown, depending on the concentration of glucose present. The blood test is based on a reaction of a peroxide with a colored indicator, and the reaction is catalyzed by the hemoglobin in blood. The color that results may range from orange to green, depending on the concentration of blood present. The timing of these tests is critical because they are based on the speed of the reactions that produce the color.

Whenever we plan to run a chemical reaction, we should ask at least two questions. One is "Will the reaction produce the desired products in useful quantities?" We can answer this question by making equilibrium and thermodynamic calculations, as discussed in the next chapter and in Chapters 16–18, or we can use qualitative considerations, such as the fact that the reaction of an acid with a base yields a salt or the fact that an active metal generally reacts with a nonmetal. The second question is "How rapidly will the reaction occur?" A reaction that takes 50 years to produce a product is about as useless as one that never gives a product at all!

This chapter considers the second question—the rate at which a chemical reaction yields products (chemical kinetics). We will examine factors that influence the rates of chemical reactions, the mechanisms by which reactions proceed, and the quantitative techniques used to determine and to describe the rate at which reactions occur.

Rates of Chemical Reactions

13.1 Rate of Reaction

The rate of a chemical reaction, like many other rates, involves production or consumption during some unit of time. For example, we can determine the rate of production of a well in gallons *per minute* or the rate of an electric generating plant's use of coal in tons *per hour.*

We can determine the **rate of a reaction** by measuring the change in the concentration of a reactant or a product per unit of time. We could also measure a change that is proportional to the concentration, such as a change in the number of molecules per cubic centimeter or a change in the pressure of a gas. In some cases the color of a reaction mixture changes with time, and this too can be related to concentration changes. However, we must be sure the change we measure is due only to the reaction and does not reflect a change in volume of the system (which would also change the concentration) or the presence of one or more competing reactions.

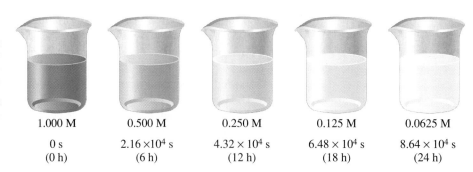

1.000 M	0.500 M	0.250 M	0.125 M	0.0625 M
0 s (0 h)	2.16×10^4 s (6 h)	4.32×10^4 s (12 h)	6.48×10^4 s (18 h)	8.64×10^4 s (24 h)

When we measure the concentration of H_2O_2 in a solution of hydrogen peroxide in water, we find that the concentration changes slowly as the H_2O_2 decomposes according to the equation

$$2H_2O_2(aq) \longrightarrow 2H_2O(l) + O_2(g) \qquad (1)$$

The change in concentration with time for one such solution is shown in Fig. 13.1 and in Table 13.1. From these experimental data, we can determine the rate at which the hydrogen peroxide decomposes.

$$\text{Rate of decomposition of } H_2O_2 = -\frac{\text{change in concentration of reactant}}{\text{time interval}}$$

$$= -\frac{[H_2O_2]_{t_2} - [H_2O_2]_{t_1}}{t_2 - t_1} = -\frac{\Delta[H_2O_2]}{\Delta t}$$

We use brackets here to indicate molar concentrations and the symbol delta (Δ) to indicate "the change in." Thus $[H_2O_2]_{t_1}$ represents the molar concentration of hydrogen peroxide at some time t_1, $[H_2O_2]_{t_2}$ represents the molar concentration of hydrogen peroxide at a later time t_2, and $\Delta[H_2O_2]$ represents the change in molar concentration of hydrogen peroxide during the time interval Δt (that is, $t_2 - t_1$). The minus sign preceding the expression is used to convert the change in concentration of hydrogen peroxide to a positive rate. Reaction rates are positive numbers.

A change in the number of moles of reactant or product in a liter of volume during some unit of time has units of moles per liter per second, which may be written mol L^{-1} s^{-1}. Because mol L^{-1} is identical to molarity, sometimes we see rates written as M s^{-1} (molar per second).

Table 13.1

Variation in the Rate of Decomposition of H_2O_2 in an Aqueous Solution at 40°C

Time, s	$[H_2O_2]$, mol L^{-1}	$\Delta[H_2O_2]$, mol L^{-1}	Δt, s	Rate of decomposition, mol L^{-1} s^{-1}
0	1.000			
		−0.500	2.16×10^4	2.31×10^{-5}
2.16×10^4	0.500			
		−0.250	2.16×10^4	1.16×10^{-5}
4.32×10^4	0.250			
		−0.125	2.16×10^4	5.79×10^{-6}
6.48×10^4	0.125			
		−0.062	2.16×10^4	2.9×10^{-6}
8.64×10^4	0.0625			

Example 13.1   Calculation of a Rate from Changes in Concentration

When we use hydrogen peroxide on a cut, it foams. The foaming occurs because the rate of the decomposition reaction is increased by the presence of an enzyme (a catalyst, Section 13.3) found in body tissue. The rapid evolution of oxygen gas produces the foam. In some cases it takes only about 15 seconds for the concentration of H_2O_2 to fall from 1.00 M to 0.12 M. What is the rate of decomposition of H_2O_2 under these conditions?

Solution: We can use the following steps to solve this problem.

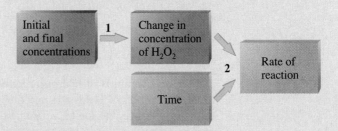

The numbers at the arrows refer to the steps below.

Step 1. *Find the change in concentration, $\Delta[H_2O_2]$, by subtracting the initial concentration, $[H_2O_2]_{t_1}$, from the final concentration, $[H_2O_2]_{t_2}$.*

$$\Delta[H_2O_2] = [H_2O_2]_{t_2} - [H_2O_2]_{t_1}$$
$$= 0.12 \text{ mol L}^{-1} - 1.00 \text{ mol L}^{-1}$$
$$= -0.88 \text{ mol L}^{-1}$$

Step 2. *Calculate the rate by dividing the negative of the change in concentration by the time required for the reaction, Δt.* The time required for the reaction is 15 s, so

$$\text{Rate} = -\frac{\Delta[H_2O_2]}{\Delta t} = -\frac{(-0.88 \text{ mol L}^{-1})}{15 \text{ s}}$$
$$= 5.9 \times 10^{-2} \text{ mol L}^{-1} \text{ s}^{-1}$$

We can see that the presence of the enzyme significantly increases the rate of decomposition compared to that reported in Table 13.1.

Check your learning: Calculate the rate of decomposition of H_2O_2 with no enzyme if it requires 6.48×10^4 s for the concentration of H_2O_2 to fall from 1.000 M to 0.125 M.

Answer: $1.35 \times 10^{-5} \text{ mol L}^{-1} \text{ s}^{-1}$

The reaction rates we find in Table 13.1 are averages of the rates during each 21,600-second interval. The average rate of decomposition between 0 and 21,600 seconds is $2.31 \times 10^{-5} \text{ mol L}^{-1} \text{ s}^{-1}$. If we calculate the rate for the time interval from 0 to 86,400 seconds, we find a lower average rate of $1.09 \times 10^{-5} \text{ mol L}^{-1} \text{ s}^{-1}$.

There are several ways to determine the **instantaneous rate** of a reaction, the rate of the reaction at any instant in time. One way is to use a graphical procedure. If we

Figure 13.2

A graph of concentration versus time for the solution of H_2O_2 shown in Fig. 13.1. The rate at any instant is equal to the negative of the slope of a line tangent to this curve at that time. A tangent (purple line) is shown at $t = 40{,}000$ s.

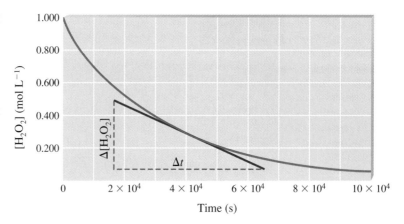

plot the concentration of hydrogen peroxide against time, then the instantaneous rate of decomposition of H_2O_2 at some time t is equal to the negative of the slope of a straight line tangent to the curve at that time. Such a graph is shown in Fig. 13.2 with a tangent drawn at $t = 40{,}000$ seconds. The slope of this line is $\Delta[H_2O_2]/\Delta t$. The negative of the slope, $-\Delta[H_2O_2]/\Delta t$, is the instantaneous rate of decomposition at 40,000 seconds (rate $= 8.90 \times 10^{-6}$ mol L^{-1} s^{-1}). We can determine the **initial rate** of the reaction—the rate when no products are present—from the slope of the tangent at zero time.

The rates at which the concentrations of reactants or products change during reactions are rarely constant. As the reactants are consumed, the rates at which the concentrations change usually decrease until they reach zero.

13.2 Relative Rates of Reaction and Reaction Velocity

We have been careful to refer to the rate of Reaction 1 in Section 13.1 as the rate of decomposition of hydrogen peroxide. It is very important that we identify the reactant or product used to measure the rate of a reaction, because the rates of change of the concentrations of all the reactants and products need not be the same. If reactants or products differ in phase, then there is no fixed relationship between the rates of change of the concentrations of reactants and products. Such is the case with the decomposition of H_2O_2. We could use two different types of measurements to determine the rate of the reaction

$$2H_2O_2(aq) \longrightarrow 2H_2O(l) + O_2(g)$$

We could measure the change in volume or concentration of oxygen gas in an apparatus similar to that shown in Fig. 13.3. Alternatively, we could measure the changes in the concentration of hydrogen peroxide by titrating samples from the solution. We would probably find different values for the rate of the reaction by the two methods, because the relationship between $\Delta[H_2O_2]/\Delta t$ and $\Delta[O_2]/\Delta t$ depends on the volume of the system used to collect the oxygen gas. For example, although $\Delta[H_2O_2]/\Delta t$ would not change, $\Delta[O_2]/\Delta t$ would decrease by a factor of 2 if the volume of the gas collection apparatus were twice as large.

Figure 13.3

An apparatus for measuring the rate of evolution of a gas from a reaction mixture held at constant temperature. The volume of gas produced is measured in the gas buret.

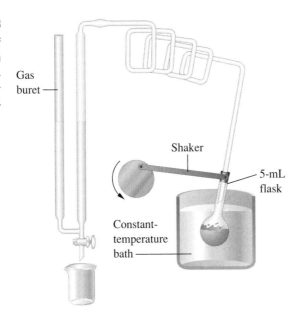

Gas buret

Shaker

5-mL flask

Constant-temperature bath

If all of the reactants and products are present in solution or in the gas phase, we can write equations that describe the relative rates of change of their concentrations or pressures. The following chemical equation describes the decomposition of ammonia upon heating; here, all the reactants and products are in the gas phase.

$$2NH_3(g) \longrightarrow N_2(g) + 3H_2(g)$$

This equation tells us that 1 mole of N_2 forms every time 2 moles of NH_3 decompose. Thus during the time required for 1 mole of N_2 to form, 2 moles of NH_3 will decompose, so the rate of decomposition of NH_3 is 2 times faster than the rate of formation of N_2.

$$-\frac{\Delta[NH_3]}{\Delta t} = \frac{2\Delta[N_2]}{\Delta t}$$

Similarly, the rate of formation of H_2 is 3 times the rate of formation of N_2 because 3 moles of H_2 form during the same amount of time that it takes 1 mole of N_2 to form.

$$\frac{\Delta[H_2]}{\Delta t} = \frac{3\Delta[N_2]}{\Delta t}$$

Figure 13.4 on page 466 illustrates the change in concentrations with time for the decomposition of ammonia into nitrogen and hydrogen at 1100°C. We can see from the slopes drawn at $t = 500$ seconds that the instantaneous rates of change in the concentrations of the reactants and products are different.

We can identify the rate of decomposition of ammonia by using a rate of change in concentration of NH_3, of N_2, or of H_2, but we must be very careful to identify which rate we are using. Alternatively, we can report the **reaction velocity, v,** which is defined as the rate of change of concentration of any reactant or product in a reaction divided by its stoichiometric coefficient. The reaction velocity is independent of

Figure 13.4

A graph of changes in concentrations during the reaction $2NH_3 \longrightarrow N_2 + 3H_2$. The rates of change of the three concentrations are not equal, as the different slopes of the tangents at $t = 500$ s show.

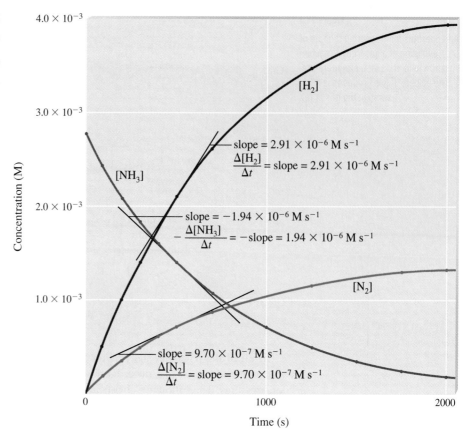

the reactant or product used to determine it. For the decomposition of ammonia in contact with hot quartz, we have

$$2NH_3(g) \longrightarrow 1N_2(g) + 3H_2(g)$$

and

$$\text{Reaction velocity} = v = -\frac{1}{2} \cdot \frac{\Delta[NH_3]}{\Delta t} = \frac{1}{1} \cdot \frac{\Delta[N_2]}{\Delta t} = \frac{1}{3} \cdot \frac{\Delta[H_2]}{\Delta t}$$

The negative sign in the term $-\frac{1}{2}\Delta[NH_3]/\Delta t$ is needed to convert a change in a concentration of a reactant to a positive number. Reaction velocities are positive numbers.

Example 13.2   **Expressions for a Reaction Velocity**

The first step in the production of nitric acid is the combustion of ammonia.

$$4NH_3(g) + 5O_2(g) \longrightarrow 4NO(g) + 6H_2O(g)$$

Write the equations that relate the reaction velocity of this reaction to both the rates of combination of the reactants and the rates of formation of the products.

Solution:

$$\text{Reaction velocity} = v = -\frac{1}{4} \cdot \frac{\Delta[NH_3]}{\Delta t}$$

$$= -\frac{1}{5} \cdot \frac{\Delta[O_2]}{\Delta t}$$

$$= \frac{1}{4} \cdot \frac{\Delta[NO]}{\Delta t}$$

$$= \frac{1}{6} \cdot \frac{\Delta[H_2O]}{\Delta t}$$

Check your learning: The rate of formation of Br_2 is 6.0×10^{-6} mol L^{-1} s^{-1} in a reaction in water described by the following net ionic equation:

$$5Br^- + BrO_3^- + 6H^+ \longrightarrow 3Br_2 + 3H_2O$$

What are the rates of reaction of the reactants and the reaction velocity?

Answer: $-\Delta[Br^-]/\Delta t = 1.0 \times 10^{-5}$ mol L^{-1} s^{-1}; $-\Delta[BrO_3^-]/\Delta t = 2.0 \times 10^{-6}$ mol L^{-1} s^{-1};
$-\Delta[H^+]/\Delta t = 1.2 \times 10^{-5}$ mol L^{-1} s^{-1}; $v = 2.0 \times 10^{-6}$ mol L^{-1} s^{-1}

Example 13.3   **Reaction Velocity for Decomposition of H_2O_2**

What is the instantaneous reaction velocity determined from the decomposition of H_2O_2 at $t = 40{,}000$ s for the reaction

$$2H_2O_2 \longrightarrow 2H_2O + O_2$$

which exhibits the behavior illustrated in Figure 13.2?

Solution: The instantaneous rate of decomposition of H_2O_2 at $t = 40{,}000$ s is 8.90×10^{-6} mol L^{-1} s^{-1}; that is,

$$-\frac{\Delta[H_2O_2]}{\Delta t} = 8.90 \times 10^{-6} \text{ mol } L^{-1} s^{-1}$$

The reaction velocity for this reaction is given by

$$v = -\frac{1}{2} \cdot \frac{\Delta[H_2O_2]}{\Delta t}$$

Thus we may determine that

$$v = -\frac{1}{2} \cdot \frac{\Delta[H_2O_2]}{\Delta t} = \frac{1}{2} \cdot (8.90 \times 10^{-6} \text{ mol } L^{-1} s^{-1})$$

$$= 4.45 \times 10^{-6} \text{ mol } L^{-1} s^{-1}$$

Check your learning: If the rate of decomposition of ammonia, NH_3, at 1150 K is 2.10×10^{-6} mol L^{-1} s^{-1}, what is the reaction velocity for the decomposition reaction of ammonia into nitrogen and hydrogen?

Answer: 1.05×10^{-6} mol L^{-1} s^{-1}

Figure 13.5

Twenty-four hours before this photograph was taken, a piece of iron (right) and a piece of sodium (left) were placed in air. The sodium has reacted completely with air during this time; the iron has not.

There are subtle complications that accompany the use of reaction velocities to describe the rates of more complex reactions than those we have considered in this section. We will not discuss these complications in this introductory chapter. Instead, the remainder of this chapter describes reaction rates in terms of reactants or products. We will be very specific about which reactant or product is being used to describe the rate. We will also see that in many situations, the outcome is independent of the species used to identify the rate.

13.3   Factors That Affect Reaction Rates

The rates at which reactants are consumed and products are formed during chemical reactions vary greatly. We can identify five factors that affect the speeds of chemical reactions: the nature of the reacting substances, the state of subdivision of the reactants, the temperature of the reactants, the concentration of the reactants, and the presence of a catalyst.

The Nature of the Reacting Substances.
The rate of a reaction depends on the nature of the participating substances. Any simple combination of two simple ions of opposite charge usually occurs very rapidly. The reaction of an acid with a base is an example.

$$H_3O^+ + OH^- \longrightarrow 2H_2O$$

Water forms as rapidly as we can mix solutions of hydrochloric acid and sodium hydroxide; the acid–base reaction is almost instantaneous. Water also forms when hydrogen peroxide decomposes:

$$2H_2O_2 \longrightarrow 2H_2O + O_2$$

Figure 13.6

Iron powder reacts rapidly with dilute hydrochloric acid and produces bubbles of hydrogen gas because the powder has a large total surface area. An iron horseshoe nail reacts more slowly.

However, this reaction is much slower than the acid–base reaction; in some cases it can take months to occur. Reactions that involve the breaking of chemical bonds or the rearrangement of molecules commonly are slower than reactions that involve the simple combination of two ions of opposite charge.

Similar reactions may have different rates under the same conditions if different reactants are involved. For example, when small pieces of the metals iron and sodium are left in air, the sodium reacts completely with air overnight, whereas the iron is barely affected (Fig. 13.5). The active metals calcium and sodium both react with water to form hydrogen gas and a base. Yet calcium reacts at a moderate rate, whereas sodium reacts so rapidly that the reaction is of almost explosive violence.

The State of Subdivision of the Reactants.
Except for substances in the gaseous state or in solution, reactions occur at the boundary, or interface, between two phases. Hence the rate of a reaction between two phases depends to a large extent on the surface contact between them. Finely divided liquids and solids, because of the greater surface area available, react more rapidly than the same amount of the substance in a large body. For example, large pieces of iron react slowly with acids; finely divided iron reacts much more rapidly (Fig. 13.6). Large pieces of wood smolder, smaller pieces burn rapidly, and grain dust (which, like wood, is composed of cellulose) may burn at an explosive rate, as seen in Fig. 13.7.

Temperature of the Reactants.
Chemical reactions proceed faster at higher temperatures. Wood will react with air to rot at ordinary temperatures, but at high tem-

Figure 13.7

Grain dust (very finely divided grain) can burn at an explosive rate. When the dust is suspended in a confined space, an explosion can result. A grain dust explosion in this storage elevator destroyed several silos.

(A)

(B)

Figure 13.8

(A) A statue as it appeared in 1913. In the 100 years before that time, the concentration of air pollutants was low and the rate of damage to the statue was slow. (B) The same statue in 1984. In the period between 1913 and 1984, the concentration of air pollutants increased and the rate of damage to the statue increased.

Figure 13.9

Phosphorus burns much more rapidly in an atmosphere of 100% oxygen (left) than in air (right), which is only about 20% oxygen.

peratures it reacts fast enough to burn. Foods cook faster at higher temperatures than at lower ones. We use a burner or a hot plate in the laboratory to increase the speed of reactions that proceed slowly at ordinary temperatures.

In many cases, the rate of a reaction in a homogeneous system is approximately doubled by an increase in temperature of only 10°C. This rule is a rough approximation, however, and applies only to reactions that last longer than a second or two.

Concentration of the Reactants. The rates of many reactions depend on the concentration of the reactants. Rates usually increase when the concentration of one or more of the reactants increases. For example, the rate at which limestone, $CaCO_3$, deteriorates as a result of reaction with the pollutant sulfur dioxide depends on the amount of sulfur dioxide in the air (Fig. 13.8). In a polluted atmosphere where the concentration of sulfur dioxide is high, it deteriorates more rapidly than in less polluted air. Similarly, Fig. 13.9 shows that phosphorus burns much more rapidly in an atmosphere of pure oxygen than in air, which is only about 20% oxygen.

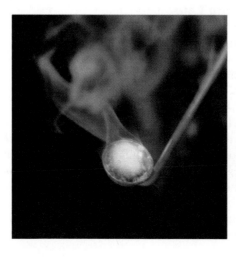
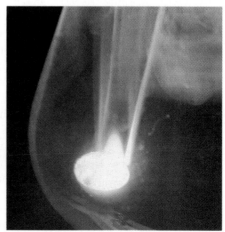

Figure 13.10

Hydrogen peroxide foams when chopped liver is added (right) because an enzyme in the liver catalyzes its decomposition to oxygen and water. Note that the H_2O_2 does not appear to be decomposing (left) before the catalyst is added.

Figure 13.11

Liquid-filled candy starts as a gritty paste of sucrose, water, flavoring agents, and a small amount of the enzyme invertase, all coated with chocolate. The candy is allowed to stand until the enzyme-catalyzed reaction of sucrose with water forms glucose and fructose, sugars that are much more soluble in water than is sucrose and that dissolve to give the liquid centers.

The Presence of a Catalyst. We have seen that hydrogen peroxide foams when it is applied to a cut, because an enzyme in the body catalyzes its decomposition (Fig. 13.10). However, in the absence of the catalyst (for example, in a bottle in the medicine cabinet), complete decomposition can take months. A **catalyst** is a substance that affects the rate of a chemical reaction but is not itself used up by the reaction. Many biological reactions are catalyzed by enzymes, which are complex substances produced by living organisms. In solution in water, the common table sugar sucrose does not react with the water, even over a period of years. However, an enzyme called invertase catalyzes the reaction between sucrose and water, which forms a mixture of the sugars glucose and fructose—a mixture called invert sugar. Candy manufacturers use this reaction to produce the liquid centers in liquid-filled candies (Fig. 13.11).

The catalytic converter (Fig. 13.12) in an automobile exhaust system contains a catalyst that increases the rates of oxidation of CO and unburned hydrocarbons (to CO_2 and H_2O) as well as the rates of decomposition of nitrogen oxides (to N_2 and O_2). Thus the polluting carbon monoxide, hydrocarbons, and nitrogen oxides are destroyed during the brief time they spend in the exhaust system.

13.4   Rate Laws

We use **rate laws** or **rate equations** to describe the relationship between the rate of a chemical change and the concentration of its reactants. In general, a rate law (or **differential rate law,** as it is sometimes called) has the form

$$\text{Rate} = k[A]^m[B]^n[C]^p\ldots$$

in which [A], [B], and [C] represent molar concentrations of reactants (or sometimes products or other substances); k is the **rate constant,** which is specific for a particular reaction at a particular temperature; and the exponents m, n, and p are usually positive integers (although fractions and negative numbers sometimes appear). *Both* k *and the exponents* m, n, *and* p *must be determined experimentally by observing how the rate of a reaction changes as the concentrations of the reactants are changed.* The rate constant k is independent of the concentration of A, B, or C, but it does vary with temperature.

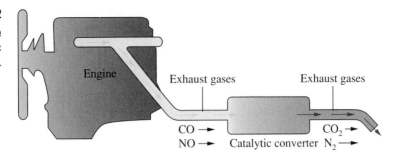

Figure 13.12

The exhaust from an automobile engine is passed through a catalytic converter to decrease pollutants.

If we were to determine the rate of decomposition of H_2O_2 at several different points along the curve shown in Fig. 13.2, we would find each rate to be directly proportional to the concentration of hydrogen peroxide at that time. These rates can be described by the rate law

$$Rate = k[H_2O_2] \qquad (2)$$

with $k = 3.2 \times 10^{-5}$ s^{-1} at 40°C.

The rate for the reaction of hydrochloric acid with sodium hydroxide is described by a similar expression:

$$Rate = k[H_3O^+][OH^-] \qquad (3)$$

In this equation, the value of k differs from that in Equation 2, because rate constants depend on the nature of the reacting substances. The value of k in Equation 3 is 1.4 $\times 10^{11}$ L mol^{-1} s^{-1}, a value that indicates an extremely fast reaction.

Example 13.4   **Calculation of a Rate from a Rate Law**

Butadiene is a petrochemical that is used in the manufacture of artificial rubber—in solution, molecules of butadiene react with each other to form very large molecules that are converted to artificial rubber. In the gas phase, two butadiene molecules can react (dimerize) to give larger molecules according to the reaction

$$2C_4H_6(g) \longrightarrow C_8H_{12}(g)$$

The rate law for this reaction is

$$Rate = k[C_4H_6]^2$$

with $k = 6.14 \times 10^{-2}$ L mol^{-1} s^{-1} at the temperature of the reaction. Find the rate of formation of C_8H_{12} when the concentration of C_4H_6 is 0.0200 M (this corresponds to a pressure of approximately 0.5 atm).

Solution: The rate is determined by substituting into the rate law.

$$\begin{aligned} Rate &= k[C_4H_6]^2 \\ &= 6.14 \times 10^{-2} \text{ L mol}^{-1} \text{ s}^{-1} (0.0200 \text{ mol L}^{-1})^2 \\ &= 2.46 \times 10^{-5} \text{ mol L}^{-1} \text{ s}^{-1} \end{aligned}$$

The concentration of C_8H_{12} increases by 2.46×10^{-5} moles per liter each second when the concentration of C_4H_6 is 0.020 M.

Check your learning: The rate of reaction of NH_4^+ according to the equation

$$NH_4^+(aq) + NO_2^-(aq) \longrightarrow N_2(g) + 2H_2O(l)$$

is given by the equation

$$Rate = k[NH_4^+][NO_2^-]$$

with $k = 2.7 \times 10^{-4}$ L mol^{-1} s^{-1}. What is the rate when $[NH_4^+] = 0.10$ M and $[NO_2^-] = 0.0040$ M?

Answer: 1.1×10^{-7} mol L^{-1} s^{-1}

When, during a decomposition reaction of H_2O_2, we add more H_2O_2 and double its concentration, the rate doubles because it is directly proportional to the concentration of H_2O_2. On the other hand, when we double the concentration of C_4H_6 in a dimerization reaction, the rate increases four times because it is proportional to the square of the concentration of C_4H_6.

We describe this effect of the concentrations of reactants on the rate of a reaction in terms of **reaction order.** Consider a reaction for which the rate law is

$$Rate = k[A]^m[B]^n$$

If the exponent m is 1, then the reaction is **first-order** in A. If m is 2, then the reaction is **second-order** in A. If n is 1, then the reaction is first order in B. If n is 2, then the reaction is second order in B. If m or n is zero, the reaction is **zero-order** in A or B, respectively, and the rate of the reaction does not change as the concentration of that reactant changes. The **overall order of a reaction** is the sum of the orders with respect to each reactant. If $m = 1$ and $n = 1$, the overall order of the reaction is second order ($m + n = 1 + 1 = 2$).

The rate of the decomposition of H_2O_2 with the rate law

$$Rate = k[H_2O_2]$$

is first order in hydrogen peroxide and first order overall. The rate of dimerization of C_4H_6,

$$Rate = k[C_4H_6]^2$$

is second order in C_4H_6 and second order overall. The reaction of H_3O^+ and OH^-,

$$Rate = k[H_3O^+][OH^-]$$

is first order in H_3O^+, first order in OH^-, and second order overall.

Example 13.5   Writing Rate Laws from Reaction Orders

Experiment shows that the reaction of nitrogen dioxide with carbon monoxide,

$$NO_2(g) + CO(g) \longrightarrow NO(g) + CO_2(g)$$

is second order in NO_2 and zero order in CO at 100°C. What is the rate law for the reaction?

Solution: The reaction will have the form

$$Rate = k[NO_2]^m[CO]^n$$

The reaction is second order in NO_2; thus $m = 2$. The reaction is zero order in CO; thus $n = 0$. The rate law is

$$\text{Rate} = k[NO_2]^2[CO]^0 = k[NO_2]^2$$

(Remember that a number raised to the zero power is equal to 1, so $[CO]^0 = 1$.)

Check your learning: The rate law for the reaction

$$H_2(g) + 2NO(g) \longrightarrow N_2O(g) + H_2O(g)$$

has been determined to be

$$\text{Rate} = k[NO]^2[H_2]$$

What are the orders with respect to each reactant, and what is the overall order of the reaction?

Answer: Order in NO = 2; order in H_2 = 1; overall order = 3

The following examples illustrate how experimental data can be used to determine rate laws.

Example 13.6   **Determining a Rate Law from Initial Rates**

The rates of the reactions of nitrogen oxides with ozone are important factors in deciding how significant these reactions are in ozone depletion in the upper atmosphere (Fig. 13.13). One such reaction is the combination of nitric oxide, NO, with ozone, O_3.

$$NO(g) + O_3(g) \longrightarrow NO_2(g) + O_2(g)$$

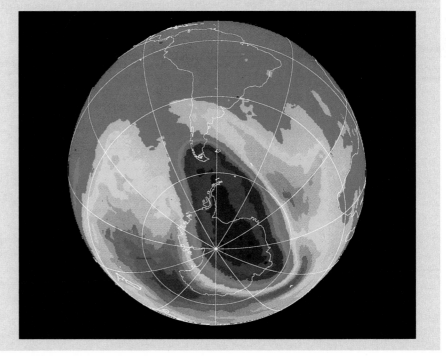

Figure 13.13

Over the past several years, the atmospheric ozone concentration over Antarctica has decreased during the winter. This map shows the decreased concentration as the dark area over Antarctica and the tip of South America.

This reaction has been studied in the laboratory, and the following rate data have been determined at 25°C.

[NO], mol L^{-1}	[O$_3$], mol L^{-1}	$\dfrac{\Delta[NO_2]}{\Delta t}$, mol L^{-1} s^{-1}
1.00×10^{-6}	3.00×10^{-6}	0.660×10^{-4}
1.00×10^{-6}	6.00×10^{-6}	1.32×10^{-4}
1.00×10^{-6}	9.00×10^{-6}	1.98×10^{-4}
2.00×10^{-6}	9.00×10^{-6}	3.96×10^{-4}
3.00×10^{-6}	9.00×10^{-6}	5.94×10^{-4}

Determine the rate law and the rate constant for the reaction at 25°C.

Solution: The rate law will have the form

$$\text{Rate} = k[NO]^m[O_3]^n$$

We can determine the values of m, n, and k from the experimental data by using the following three-step process.

Step 1.   *Determine the value of m from the data in which [NO] varies and [O$_3$] is constant.*

In the last three lines in the data table, [NO] varies and [O$_3$] remains constant. When [NO] doubles, the rate doubles, and when [NO] triples, the rate also triples. Thus the rate is also directly proportional to [NO], and m in the rate law is equal to 1.

Step 2.   *Determine the value of n from the data in which [O$_3$] varies and [NO] is constant.*

In the first three lines in the data table, [NO] is constant and [O$_3$] varies. The reaction rate changes in direct proportion to the change in [O$_3$]. When [O$_3$] doubles, the rate doubles; when [O$_3$] increases by a factor of 3, the rate increases by a factor of 3. Thus the rate is directly proportional to [O$_3$], and n is equal to 1.

The rate law is thus

$$\text{Rate} = k[NO]^1[O_3]^1 = k[NO][O_3]$$

Step 3.   *Determine the value of k from one set of concentrations and the corresponding rate.*

$$k = \frac{\text{rate}}{[NO][O_3]}$$

$$= \frac{0.660 \times 10^{-4}\ \text{mol L}^{-1}\ \text{s}^{-1}}{(1.00 \times 10^{-6}\ \text{mol L}^{-1})(3.00 \times 10^{-6}\ \text{mol L}^{-1})}$$

$$= 2.20 \times 10^7\ \text{L mol}^{-1}\ \text{s}^{-1}$$

This fast reaction could play an important role in ozone depletion if [NO] were large enough.

Check your learning: Acetaldehyde decomposes when heated to yield methane and carbon monoxide according to the equation

$$CH_3CHO(g) \xrightarrow{\Delta} CH_4(g) + CO(g)$$

Determine the rate law and the rate constant for the reaction from the following experimental data.

[CH$_3$CHO], mol L^{-1}	$-\dfrac{\Delta[\text{CH}_3\text{CHO}]}{\Delta t}$, mol L^{-1} s^{-1}
1.75×10^{-3}	2.06×10^{-11}
3.50×10^{-3}	8.24×10^{-11}
7.00×10^{-3}	3.30×10^{-10}

Answer: Rate = $k[\text{CH}_3\text{CHO}]^2$ with $k = 6.73 \times 10^{-6}$ L mol^{-1} s^{-1}

Some of us find an algebraic method helpful in determining the coefficients in rate laws. To use this method, we select two sets of rate data that differ in the concentration of only one reactant and set up a ratio of the two rates and the two rate laws. After canceling terms that are equal, we are left with an equation that contains only one unknown, the coefficient of the concentration that varies. We solve this equation for the coefficient.

Example 13.7   The Algebraic Method for Determining Rate Laws

Determine the rate law for the reaction

$$2\text{NO}(g) + \text{Cl}_2(g) \longrightarrow 2\text{NOCl}(g)$$

from the following experimental data using the algebraic method.

[NO], mol L^{-1}	[Cl$_2$], mol L^{-1}	$-\dfrac{\Delta[\text{NO}]}{\Delta t}$, mol L^{-1} s^{-1}
0.10	0.10	3.0×10^{-3}
0.10	0.20	6.0×10^{-3}
0.20	0.20	2.42×10^{-2}

Solution: The rate law for this reaction will have the form

$$\text{Rate} = k[\text{NO}]^m[\text{Cl}_2]^n$$

As in Example 13.6, we can approach this problem in a stepwise fashion and determine the values of m and n from the experimental data by using a two-step process. The difference between this example and Example 13.6 is in how we find m and n.

Step 1.   *Determine the value of m from the data in which [NO] varies and [Cl$_2$] is constant.*
     We can write the ratios

$$\frac{\text{Rate}_a}{\text{Rate}_b} = \frac{k[\text{NO}]_a^m[\text{Cl}_2]_a^n}{k[\text{NO}]_b^m[\text{Cl}_2]_b^n}$$

where the subscripts a and b indicate data from two different determinations. Using the last two sets of data, in which $[Cl_2]$ does not vary, gives

$$\frac{\text{Rate}_2}{\text{Rate}_3} = \frac{6.0 \times 10^{-3} \text{ mol L}^{-1}\text{ s}^{-1}}{2.4 \times 10^{-2} \text{ mol L}^{-1}\text{ s}^{-1}} = \frac{k(0.10 \text{ M})^m(0.20 \text{ M})^n}{k(0.20 \text{ M})^m(0.20 \text{ M})^n}$$

After we cancel equal terms in the top and bottom of this equation, we are left with

$$\frac{6.0 \times 10^{-3}}{2.4 \times 10^{-2}} = \frac{(0.10)^m}{(0.20)^m} = \left(\frac{0.10}{0.20}\right)^m$$

which simplifies to

$$0.25 = (0.50)^m$$

which means m must be 2.

Step 2. *Determine the value of* n *from the data in which [Cl$_2$] varies and [NO] is constant.*

$$\frac{\text{Rate}_1}{\text{Rate}_2} = \frac{3.0 \times 10^{-3} \text{ mol L}^{-1}\text{ s}^{-1}}{6.0 \times 10^{-3} \text{ mol L}^{-1}\text{ s}^{-1}} = \frac{k(0.10 \text{ M})^m(0.10 \text{ M})^n}{k(0.10 \text{ M})^m(0.20 \text{ M})^n}$$

Cancellation gives

$$\frac{3.0 \times 10^{-3}}{6.0 \times 10^{-3}} = \frac{(0.10)^n}{(0.20)^n} = \left(\frac{0.10}{0.20}\right)^n$$

which simplifies to

$$0.50 = (0.50)^n$$

Thus n must be 1, and the form of the rate law is

$$\text{Rate} = k[\text{NO}]^m[\text{Cl}_2]^n = k[\text{NO}]^2[\text{Cl}_2]$$

Check your learning: Use the algebraic method and check the rate law determined in Example 13.6 for the reaction

$$\text{NO}(g) + \text{O}_3(g) \longrightarrow \text{NO}_2(g) + \text{O}_2(g)$$

Answer: Rate $= k[\text{NO}][\text{O}_3]$

In some of our examples, the exponents m and n in the rate law happen to be the same as the coefficients in the chemical equation for the reaction. *This is an accident* and is not always the case, as the following gas-phase examples illustrate.

$$\text{NO}_2 + \text{CO} \longrightarrow \text{NO} + \text{CO}_2 \qquad \text{Rate} = k[\text{NO}_2]^2$$
$$\text{CH}_3\text{CHO} \longrightarrow \text{CH}_4 + \text{CO} \qquad \text{Rate} = k[\text{CH}_3\text{CHO}]^2$$
$$2\text{N}_2\text{O}_5 \longrightarrow 2\text{NO}_2 + \text{O}_2 \qquad \text{Rate} = k[\text{N}_2\text{O}_5]$$
$$2\text{NO}_2 + \text{F}_2 \longrightarrow 2\text{NO}_2\text{F} \qquad \text{Rate} = k[\text{NO}_2][\text{F}_2]$$
$$2\text{NO}_2\text{Cl} \longrightarrow 2\text{NO}_2 + \text{Cl}_2 \qquad \text{Rate} = k[\text{NO}_2\text{Cl}]$$

It is common for a rate law to differ from the overall stoichiometry of a chemical reaction. In some cases, one or more reactants may not even appear in the rate law.

13.5   Integrated Rate Laws

The rate laws we have seen thus far express the relationship between the rate and the concentrations of reactants. We can also determine a second form of each rate law that relates the concentrations of reactants and time. These are called **integrated rate laws.** We can use an integrated rate law to determine the amount of reactant or product present after a period of time or to determine the length of time required for a reaction to occur. For example, an integrated rate law is used to determine how long a radioactive material must be stored for its radioactivity to decay to a safe level.

First-order Reactions.

An equation relating the rate constant k to the initial concentration $[A_0]$ and the concentration $[A]$ present after any given time t can be derived for a first-order reaction. The derivation requires more advanced mathematics than we have been using, so we will not present it, but for a first-order reaction of the form

$$aA \longrightarrow products$$

with the differential rate law

$$Rate = -\frac{\Delta[A]}{\Delta t} = k[A]$$

where k is the rate constant based on the change in concentration of the reactant A, the integrated rate law is

$$\ln [A] = -kt + \ln [A_0] \qquad (4)$$

In this equation, ln is the notation for a natural logarithm, a logarithm to the base e. If $\ln x = y$, then $x = e^y$, where e is a constant equal to 2.7183 (to five significant figures). (See Appendix A.3 for more discussion on logarithms and exponents.) The integrated rate law is also expressed as a ratio of the concentrations $[A_0]$ and $[A]$.

$$\ln \frac{[A_0]}{[A]} = kt \qquad (5)$$

Example 13.8   **The Integrated Rate Law for a First-order Reaction**

On the basis of the decomposition of C_4H_8, the rate constant for the first-order decomposition of cyclobutane, C_4H_8, at 500°C is 9.2×10^{-3} s^{-1}.

$$C_4H_8 \longrightarrow 2C_2H_4$$

How long will it take for 80.0% of a sample of 0.100 M C_4H_8 to decompose—that is, for the concentration of C_4H_8 to decrease to 0.0200 M?

Solution: Use the integrated form of the rate law to answer questions regarding time.

$$\ln \frac{[A_0]}{[A]} = kt$$

There are four variables in the rate law; if we know three of them, we can determine the fourth. In this case we know $[A_0]$, $[A]$, and k, and we need to find t.

The initial concentration of C_4H_8, $[A_0]$, is 0.100 mol L^{-1}; at time t, $[A]$ is 0.0200 mol L^{-1}; and k is 9.2×10^{-3} s^{-1}.

$$t = \ln \frac{[A_0]}{[A]} \times \frac{1}{k}$$

$$= \ln \frac{0.100 \text{ mol L}^{-1}}{0.0200 \text{ mol L}^{-1}} \times \frac{1}{9.2 \times 10^{-3} \text{ s}^{-1}}$$

$$= 1.609 \times \frac{1}{9.2 \times 10^{-3} \text{ s}^{-1}}$$

$$= 1.7 \times 10^2 \text{ s}$$

Check your learning: Iodine-131 is a radioactive isotope that is used to diagnose and treat some forms of thyroid cancer. Iodine-131 decays to xenon-131 according to the equation

$$\text{I-131} \longrightarrow \text{Xe-131} + \text{electron}$$

The decay is first order with a rate constant of 0.138 d^{-1}. How long will it take for 90% of the iodine-131 in a 0.500 M solution of this substance to decay to Xe-131?

Answer: 16.7 days

We can use integrated rate laws with experimental data that consist of time and concentration information to determine the order and the rate constant of a reaction. The integrated rate law shown in Equation 4 has the form of an equation of a straight line. This is apparent when we compare the two equations

$$\ln [A] = (-k)(t) + \ln [A_0]$$
$$y \quad = \quad m\ x + \quad b$$

A plot of y versus x is a straight line with a slope m and an intercept b. Likewise, a plot of ln [A] versus t for a first-order reaction is a straight line with a slope of $-k$ and an intercept of ln [A$_0$]. If the plot is not a straight line, the reaction is not first order in A.

Example 13.9   Determination of Reaction Order by Graphing

Show that the data in Table 13.1 can be represented by a first-order rate law by graphing ln [H$_2$O$_2$] versus time. Determine the rate constant for the rate of decomposition of H$_2$O$_2$ from these data.

Solution: The data from Table 13.1, with the addition of values of ln [H$_2$O$_2$], are given below. The plot is presented in Fig. 13.14.

Time, s	[H$_2$O$_2$], M	ln [H$_2$O$_2$]
0	1.000	0.0
2.16 × 10^4	0.500	−0.631
4.32 × 10^4	0.250	−1.386
6.48 × 10^4	0.125	−2.079
8.64 × 10^4	0.0625	−2.772

Figure 13.14

A graph of the linear relationship between ln [H$_2$O$_2$] and time for the decomposition of hydrogen peroxide.

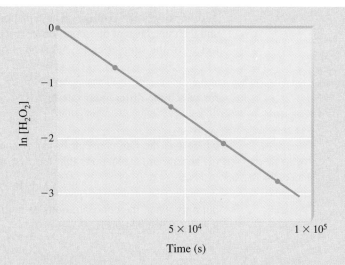

The plot of ln [H$_2$O$_2$] versus time is linear, so we have verified that the reaction may be described by a first-order rate law.

The rate constant for a first-order reaction is equal to the negative of the slope of the plot of ln [H$_2$O$_2$] versus time, where

$$\text{Slope} = \frac{\text{change in } y}{\text{change in } x} = \frac{\Delta y}{\Delta x} = \frac{\Delta \ln [\text{H}_2\text{O}_2]}{\Delta t}$$

In order to determine the slope of the line, we need two values of ln [H$_2$O$_2$], which are determined *from the line* at two values of t (choosing one near each end of the line is preferable). For example the value of ln [H$_2$O$_2$] determined from the line when t is 1.00×10^4 s is -0.320, and the value when $t = 8.00 \times 10^4$ s is -2.567.

$$\text{Slope} = \frac{-2.567 - (-0.320)}{(8.00 \times 10^4 - 1.00 \times 10^4)\,\text{s}}$$

$$= \frac{-2.150}{7.00 \times 10^4\,\text{s}}$$

$$= -3.21 \times 10^{-5}\,\text{s}^{-1}$$

$$k = -\text{slope} = -(-3.21 \times 10^{-5}\,\text{s}^{-1}) = 3.21 \times 10^{-5}\,\text{s}^{-1}$$

Check your learning: Graph the following data to determine whether the reaction A $\longrightarrow$ B + C is first order.

Time, s	[A], M
4.0	0.220
8.0	0.144
12.0	0.110
16.0	0.088
20.0	0.074

Answer: The plot of ln [A] versus t is not a straight line. The equation is not first order.

Second-Order Reactions.

The equations that relate the concentrations of reactants and the rate constant of second-order reactions are complicated. We will limit ourselves to two of the simpler cases: (1) second-order reactions that involve one reactant,

$$aA \longrightarrow products$$

for which the differential rate law is

$$Rate = k[A]^2$$

where k is the rate constant based on the change in concentration of reactant A, and (2) second-order reactions that are first order with respect to each of two reactants,

$$A + B \longrightarrow products$$

but only when the concentrations of both reactants are the same. The differential rate law for this equation is

$$Rate = k[A][B]$$

Again, k is the rate constant based on the change in concentration of the reactant A. In these two cases, the equation that relates the rate constant k to the initial concentration $[A_0]$ of the reactant A, and to the concentration $[A]$ present after any given time t, is

$$\frac{1}{[A]} = kt + \frac{1}{[A_0]} \tag{6}$$

Example 13.10   **The Integrated Rate Law for a Second-Order Reaction**

The reaction of ethyl acetate, an ester, with a strong base produces a solution of acetate ion and ethanol.

$$CH_3CO_2CH_2CH_3(aq) + OH^-(aq) \longrightarrow CH_3CO_2^-(aq) + CH_3CH_2OH(aq)$$

The reaction is second order with a rate constant equal to 4.50 L mol^{-1} min^{-1}. If the initial concentrations of ester and base are both 0.0200 M, what is the concentration remaining after 10.0 min?

Solution: We use the integrated form of the rate law to answer questions regarding time. For a second-order reaction with equal concentrations of two reactants, we have

$$\frac{1}{[A]} = kt + \frac{1}{[A_0]}$$

We know three variables in this equation: $[A_0]$ = 0.0200 mol L^{-1}, k = 4.50 L mol^{-1} min^{-1}, and t = 10.0 min. Therefore, we can solve for $[A]$, the fourth variable.

$$\frac{1}{[A]} = 4.50 \text{ L mol}^{-1} \text{ min}^{-1} \times 10.0 \text{ min} + \frac{1}{0.0200 \text{ mol L}^{-1}}$$

$$= 45.0 \text{ L mol}^{-1} + 50.0 \text{ L mol}^{-1}$$

$$= 95.0 \text{ L mol}^{-1}$$

Then $\qquad [\text{A}] = \dfrac{1}{95.0 \text{ L mol}^{-1}} = 0.0105 \text{ mol L}^{-1}$

Hence 0.0105 mol of both the ester and the base remains per liter at the end of 10.0 min, compared to the 0.0200 mol of each that was originally present per liter.

Check your learning: If the initial concentrations of ethyl acetate and base in Example 13.10 are both 0.0200 M, what is the concentration remaining after 20.0 min?

Answer: 0.00714 mol L^{-1}

The integrated rate law for our second-order reactions has the form of the equation of a straight line.

$$\frac{1}{[\text{A}]} = kt + \frac{1}{[\text{A}_0]}$$
$$y = mx + b$$

A plot of 1/[A] versus t for a first-order reaction is a straight line with a slope of k and an intercept of 1/[A$_0$]. If the plot is not a straight line, the reaction is not second order.

Example 13.11   Determination of Reaction Order by Graphing

Test the data given to show if the dimerization of C_4H_6 is a first-order or a second-order reaction.

Time, s	[C$_4$H$_6$], M
0	1.00×10^{-2}
1600	5.04×10^{-3}
3200	3.37×10^{-3}
4800	2.53×10^{-3}
6200	2.08×10^{-3}

Solution: In order to distinguish a first-order reaction from a second-order reaction, we plot ln [C$_4$H$_6$] versus t and compare that plot with a plot of 1/[C$_4$H$_6$] versus t. The values needed for these plots follow.

Time, s	$\dfrac{1}{[C_4H_6]}$, M^{-1}	ln [C$_4$H$_6$]
0	100	−4.605
1600	198	−5.289
3200	296	−5.692
4800	395	−5.978
6200	481	−6.175

Figure 13.15

First- and second-order plots for the
dimerization of C_4H_6.

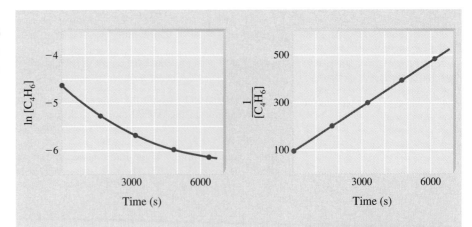

The plots are shown in Fig. 13.15. The plot of $1/[C_4H_6]$ versus t is linear, which indicates that the reaction is second order. If the reaction were first order, the plot of $\ln [C_4H_6]$ versus t would be linear; it is not.

Check your learning: Do the following data fit a second-order rate law?

Time, s	[A], M
5	0.952
10	0.625
15	0.465
20	0.370
25	0.308
35	0.230

Answer: Yes; the plot of $1/[A]$ versus t is linear.

Zero-Order Reactions.   The rate law for a zero-order reaction is

$$\text{Rate} = k[A]^0 = k$$

where k is the rate constant based on the change in concentration of the reactant A. The rate of a zero-order reaction is constant.

The integrated rate law for a zero-order reaction also has the form of the equation of a straight line.

$$[A] = -kt + [A_0] \tag{7}$$
$$y = mx + b$$

A plot of $[A]$ versus t for a zero-order reaction is a straight line with a slope of $-k$ and an intercept of $[A_0]$. Figure 13.16 shows a plot of $[NH_3]$ versus t for the decomposition of ammonia on a hot tungsten wire and for the decomposition of ammonia on hot quartz, SiO_2. The decomposition of NH_3 on hot tungsten is zero order; the plot is a straight line. From the slope of the line for the zero-order decomposition, we can determine the rate constant: $k = -\text{slope} = 1.31 \times 10^{-6}$ mol L^{-1} s^{-1}. The decomposition of NH_3 on hot quartz is not zero order (it is first order).

Figure 13.16

The decomposition of NH_3 on a tungsten (W) surface and that on a silica (SiO_2) surface show different behaviors. The decomposition on tungsten is zero order.

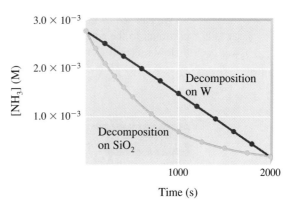

13.6   The Half-life of a Reaction

The **half-life of a reaction, $t_{1/2}$,** is the time required for half of the initial concentration of the limiting reactant to be consumed. In each succeeding half-life, half of the remaining concentration of the reactant is used up. Figure 13.1, which illustrates the decomposition of hydrogen peroxide, displays the concentration after each of several successive half-lives. During the first half-life (from 0 seconds to 21,600 seconds), the concentration decreases from 1.000 M to 0.500 M. During the second half-life (from 21,600 seconds to 43,200 seconds), it decreases from 0.500 M to 0.250 M; during the third half-life, it decreases from 0.250 M to 0.125 M. The concentration decreases by half during each successive period of 21,600 seconds. The decomposition of hydrogen peroxide is a first-order reaction, and as shown below, the half-life of a first-order reaction is independent of the concentration of the reactant. However, half-lives of reactions with other orders depend on the concentrations of the reactants.

First-Order Reactions.   We can derive an equation for determining the half-life of a first-order reaction from the alternative form of the integrated rate law, Equation 5 in Section 13.5, as follows.

$$\ln \frac{[A_0]}{[A]} = kt$$

$$t = \ln \frac{[A_0]}{[A]} \times \frac{1}{k}$$

If we set the time t equal to the half-life, $t_{1/2}$, we have a concentration of A at this time equal to $\frac{1}{2}$ of the initial concentration. Hence, at time $t_{1/2}$, $[A] = \frac{1}{2}[A_0]$.

Therefore,

$$t_{1/2} = \ln \frac{[A_0]}{\frac{1}{2}[A_0]} \times \frac{1}{k}$$

$$= \ln 2 \times \frac{1}{k} = 0.693 \times \frac{1}{k}$$

Thus

$$t_{1/2} = \frac{0.693}{k} \tag{8}$$

We can see in Equation 8 that the half-life of a first-order reaction is inversely proportional to the rate constant k. A fast reaction (shorter half-life) will have a larger k; a slow reaction (longer half-life) will have a smaller k.

Example 13.12   Calculation of a First-order Half-life

Calculate the rate constant for the first-order decomposition of hydrogen peroxide in water at 40°C, using the data given in Fig. 13.1.

Solution: The half-life for the decomposition of H_2O_2 is 2.16×10^4 s.

$$t_{1/2} = \frac{0.693}{k}$$

$$k = \frac{0.693}{t_{1/2}} = \frac{0.693}{2.16 \times 10^4 \text{ s}} = 3.21 \times 10^{-5} \text{ s}^{-1}$$

Check your learning: We have seen that the first-order radioactive decay of iodine-131 has a rate constant of 0.138 d^{-1} (Example 13.8, Check your learning). What is the half-life for this decay?

Answer: 5.02 d

Second-Order Reactions.

We can derive the equation for calculating the half-life of a second-order reaction in which the initial concentrations of both reactants are the same by using Equation 6 in Section 13.5.

$$\frac{1}{[A]} = kt + \frac{1}{[A_0]}$$

or

$$\frac{1}{[A]} - \frac{1}{[A_0]} = kt$$

If $t = t_{1/2}$, then

$$[A] = \tfrac{1}{2}[A_0]$$

and we can write

$$\frac{1}{\frac{1}{2}[A_0]} - \frac{1}{[A_0]} = kt_{1/2}$$

$$\frac{2}{[A_0]} - \frac{1}{[A_0]} = kt_{1/2}$$

$$\frac{1}{[A_0]} = kt_{1/2}$$

Thus

$$t_{1/2} = \frac{1}{k[A_0]} \tag{9}$$

For a second-order reaction, $t_{1/2}$ is inversely proportional to the concentration, and the half-life increases as the reaction proceeds because the concentrations of reactants decrease. Consequently, using the half-life concept is more complex for second-order reactions than for first-order reactions. For example, the rate constant of a second-order reaction cannot be calculated directly from the half-life unless the initial concentration is known; this is not the case for first-order reactions.

Example 13.13   Calculation of a Second-order Half-life

Calculate the half-life for the second-order reaction described in Example 13.9 ($k = 4.50$ L mol^{-1} min^{-1}) if the initial concentrations of both reactants are 0.0200 M.

Solution: The half-life of the second-order reaction in Example 13.9 can be calculated from the equation

$$t_{1/2} = \frac{1}{k[A_0]}$$

Thus
$$t_{1/2} = \frac{1}{(4.50 \text{ L mol}^{-1} \text{ min}^{-1})(0.0200 \text{ mol L}^{-1})}$$
$$= 11.1 \text{ min}$$

Check your learning: What is the half-life for the reaction in this example if the initial concentrations of both reactants are 0.0300 M?

Answer: 7.41 min

Zero-Order Reactions.   We can derive an equation for calculating the half-life of a zero-order reaction from Equation 5 in Section 13.5.

$$[A] = -kt + [A_0]$$

When one-half of the initial amount of reactant has been consumed, $t = t_{1/2}$ and $[A] = \frac{1}{2}[A_0]$. Thus

$$\frac{[A_0]}{2} = -kt + [A_0]$$

$$kt_{1/2} = \frac{[A_0]}{2}$$

and
$$t_{1/2} = \frac{[A_0]}{2k}$$

The half-life of a zero-order reaction increases as the initial concentration increases.

Equations for both differential and integrated rate laws and the corresponding half-lives for zero-, first-, and second-order reactions are summarized in Table 13.2.

	Zero Order	First Order	Second Order
Rate law	Rate = k	Rate = $k[A]$	Rate = $k[A]^2$
Integrated rate law	$[A] = -kt + [A_0]$	$\ln [A] = -kt + \ln [A_0]$	$\dfrac{1}{[A]} = kt + \dfrac{1}{[A_0]}$
Plot needed to fit rate data	$[A]$ vs. t	$\ln [A]$ vs. t	$\dfrac{1}{[A]}$ vs. t
Relationship between slope and rate constant	$k = -$slope	$k = -$slope	$k =$ slope
Half-life	$t_{1/2} = \dfrac{[A_0]}{2k}$	$t_{1/2} = \dfrac{0.693}{k}$	$t_{1/2} = \dfrac{1}{[A_0]k}$

*The rate constant, k, in these equations is the rate constant based on the change in concentration of a reactant, A. Changes in concentrations of other reactants or products may require use of the reaction stoichiometry to relate the concentration of A to other concentrations.

The Microscopic Explanation of Reaction Rates

13.7 Collision Theory of the Reaction Rate

That atoms, molecules, or ions must collide before they can react with each other should not surprise us; after all, atoms must be close together to form chemical bonds. In a few reactions, called **diffusion-controlled reactions,** every collision between reactants leads to products. The rates of such reactions are determined solely by how rapidly the reactants can diffuse together. Diffusion-controlled reactions are very fast; hence they have large rate constants. For a typical diffusion-controlled second-order gas-phase reaction at 25°C, such as the reaction of an oxygen atom with a nitrogen molecule,

$$O + N_2 \longrightarrow NO + N$$

the rate constant ranges from 10^{10} to 10^{12} L mol^{-1} s^{-1}. The diffusion-controlled reaction between hydronium ions and hydroxide ions in water at 25°C,

$$H_3O^+ + OH^- \longrightarrow 2H_2O$$

has a rate constant of 1.4×10^{11} L mol^{-1} s^{-1}. At these rates, more than 95% of the reactants would be consumed in 10^{-11} second after mixing.

Most reactions occur at a much slower rate, however, because only a very small fraction of the collisions in these slower reactions give products. In most collisions the reactants simply bounce away unchanged. For a collision to lead to a reaction, the following two things must occur.

1. The colliding species must be oriented in such a way that the atoms that are bonded together in the product come into contact during the collision.

2. The collision must occur with enough energy for the valence shells of the reacting species to penetrate into each other so that the electrons can rearrange and form new bonds (and new chemical species).

We can see the importance of these factors in the reaction of nitrogen dioxide with carbon monoxide at elevated temperatures.

$$NO_2(g) + CO(g) \longrightarrow NO(g) + CO_2(g)$$

During this reaction, oxygen atoms are transferred from NO_2 molecules to CO molecules. There are many orientations of the NO_2 and CO molecules that do not place an oxygen atom of an NO_2 molecule close to the carbon atom of a CO molecule during collision. Three collisions that are *not* effective in producing a chemical reaction are indicated in Fig. 13.17 (A, B, and C). Only a collision in which an oxygen atom strikes the carbon atom (Fig. 13.17 D) can produce a reaction.

Even when the orientation is correct, a collision may not lead to reaction. As the oxygen atom of an NO_2 molecule approaches the carbon atom of a CO molecule, the electrons in the two molecules begin to repel each other. Unless the molecules possess enough kinetic energy, the two molecules will bounce away from each other before they can get close enough to react. If the molecules are moving fast enough, then the repulsion between their electrons is not strong enough to keep them apart, and the molecules can get close enough for a C—O bond to begin to form as the N—O bond begins to break. Using dots to represent partially formed or broken bonds, we can write the resulting species as follows:

$$NO_2 + CO \longrightarrow \underset{O}{\overset{\diagup\diagup}{N}}\cdots O\cdots C{=\!=}O$$

The species $O{=\!=}N\cdots O\cdots C{=\!=}O$, which contains the partially formed $O\cdots C$ and the partially broken $N\cdots O$ bonds, is called the **activated complex,** or **transition state.** An activated complex is a combination of reacting atoms and/or molecules that is intermediate between reactants and products and in which some bonds have weakened and new bonds have begun to form. Ordinarily, an activated complex cannot be isolated. It breaks down to give either reactants or products, depending on the conditions under which the reaction takes place.

Figure 13.17

Some possible collisions between NO_2 and CO molecules. Only in part (D) are the molecules correctly oriented for transfer of an oxygen atom from NO_2 to CO to give NO and CO_2.

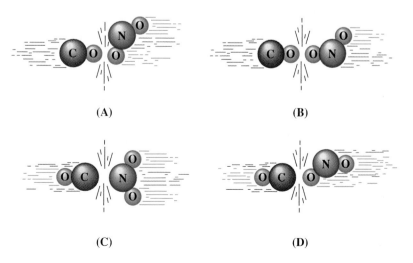

(A)   (B)

(C)   (D)

$$NO_2 + CO \longrightarrow \underset{O}{\overset{}{N}} \cdots O \cdots C \equiv O \begin{cases} \nearrow NO_2 + CO \text{ (reactants)} \\ \quad \text{or} \\ \searrow NO + CO_2 \text{ (products)} \end{cases}$$

The collision theory shows why reaction rates increase as concentrations increase. With increased concentration of any of the reacting substances, the chances for collisions between molecules are increased because more molecules are present per unit of volume. More collisions mean a faster reaction rate.

13.8   Activation Energy and the Arrhenius Equation

We call the minimum energy necessary to form an activated complex during a collision between reactants the **activation energy, E_a** (Fig. 13.18). Kinetic energy plays an important role in a reaction, because the activation energy is provided by a collision of a reactant molecule with another molecule, with the wall of the reaction vessel, or with an inert contaminant. If the activation energy is much larger than the average kinetic energy of the molecules, then the reaction will be slow. Only a few fast-moving molecules will have enough energy to react. If the activation energy is much smaller than the average kinetic energy of the molecules, then the fraction of molecules that have the necessary kinetic energy will be large and most collisions between molecules will result in reaction. The reaction will be fast.

Figure 13.19 shows us the energy relationships for the general reaction of a molecule of A with a molecule of B to form molecules of C and D. The figure shows that the energy of the transition state is higher than that of the reactants A and B by an amount equal to E_a, the activation energy. Thus the sum of the kinetic energies of A and B must be equal to or greater than E_a to reach the transition state. After the transition state has been reached, and as C and D begin to form, the system loses energy until its total energy is lower than that of the initial mixture. This lost energy is transferred to other molecules, giving them enough energy to reach the transition state. The forward reaction (that between molecules A and B) therefore tends to take place readily once the reaction has started. In Fig. 13.19, ΔE represents the difference in energy between the reactants (A and B) and the products (C and D). The sum of E_a and ΔE represents the activation energy for the *reverse* reaction,

$$C + D \longrightarrow A + B$$

We can use the **Arrhenius equation**,

$$k = A \times e^{-E_a/RT}$$

Figure 13.18

Model of activation energy. The boulder can release potential energy by falling the distance from height Y to height Z. However, activation energy must first be put into the system to lift the boulder over the barrier. The activation energy is released as the boulder falls through the distance X to Y, and additional energy or net energy is released as it falls from Y to Z.

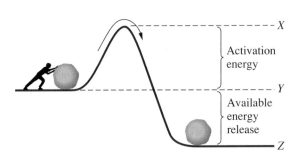

Figure 13.19

Potential energy relationships for the reaction A + B $\rightleftharpoons$ C + D. The energy represented by the broken portion of the curve is that for the system with a molecule of A and a molecule of B present. The energy represented by the solid portion of the curve is that for the system with a molecule of C and a molecule of D present. The activation energy for the forward reaction is represented by E_a; the activation energy for the reverse reaction is represented by $(E_a + \Delta E)$. The transition state is present at the peak.

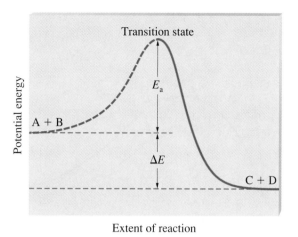

to express the relationship between the activation energy and the rate constant of a given reaction. In this equation, R is a constant with the value 8.314 J mol^{-1} K^{-1}, T is temperature on the Kelvin scale, E_a is the activation energy in joules per mole, e is the constant 2.7183, and A is a constant called the **frequency factor,** which is related to the frequency of collisions and the orientation of the reacting molecules.

Both aspects of the collision theory of reaction rates are reflected in the Arrhenius equation. A indicates how many collisions have the correct orientation to lead to products. The remainder of the equation, $e^{-E_a/RT}$, gives the fraction of the collisions in which the energy of the reacting species is greater than E_a, the activation energy for the reaction.

The Arrhenius equation describes, quantitatively, much of what we have already said about reaction rates. For two reactions at the same temperature, the reaction with the higher activation energy has the lower rate constant and the slower rate. The larger value of E_a results in a smaller value for $e^{-E_a/RT}$, reflecting the smaller fraction of molecules with sufficient energy to react. Alternatively, the reaction with the smaller E_a value has a larger fraction of molecules with enough energy to react (Fig. 13.20). This will be reflected as a larger value of $e^{-E_a/RT}$, a larger rate constant, and a faster rate for the reaction. An increase in temperature has the same effect as a decrease in activation energy. At a higher temperature a larger fraction of molecules has the necessary energy to react (Fig. 13.21 on page 490), as indicated by an increase in the value of $e^{-E_a/RT}$. The rate constant is also directly proportional to the frequency factor,

Figure 13.20

As the activation energy of a reaction decreases, the number of molecules with at least this much energy increases, as shown by the shaded areas.

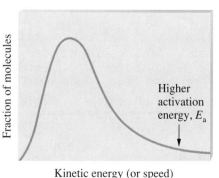
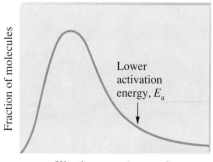

Figure 13.21

At a higher temperature, T_2, more molecules have an energy greater than E_a, as shown by the shaded area.

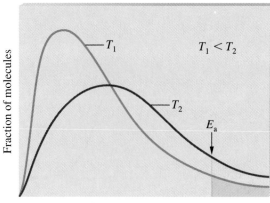

A. Hence a change in conditions or reactants that increases the number of collisions in which the orientation of the molecules is right for reaction results in an increase in A and, consequently, an increase in k.

In order to determine E_a for a reaction, we must measure k at different temperatures and evaluate E_a from an alternative version of the Arrhenius equation that has the form of an equation for a straight line.

$$\ln k = \left(-\frac{E_a}{R}\right)\left(\frac{1}{T}\right) + \ln A$$
$$y \;=\; m \quad x \;+\; b$$

Thus a plot of $\ln k$ versus $1/T$ gives a straight line with the slope $-E_a/R$, from which E_a can be determined. The intercept gives the value of $\ln A$.

Example 13.14   Determination of E_a

The variation of the rate constant with temperature for the decomposition of $HI(g)$ to $H_2(g)$ and $I_2(g)$ is given in the following table. What is the activation energy for the reaction?

T, K	k, L mol^{-1} s^{-1}
555	3.52×10^{-7}
575	1.22×10^{-6}
645	8.59×10^{-5}
700	1.16×10^{-3}
781	3.95×10^{-2}

Solution: Values of $1/T$ and $\ln k$ are

$1/T$, K^{-1}	$\ln k$
1.8×10^{-3}	-14.860
1.74×10^{-3}	-13.617
1.55×10^{-3}	-9.362
1.43×10^{-3}	-6.759
1.28×10^{-3}	-3.231

A graph of $\ln k$ against $1/T$ is given in Fig. 13.22. To determine the slope of the line, we need two values of $\ln k$, which are determined *from the line* at two values of $1/T$ (choosing one near each end of the line is preferable). For example, the value of $\ln k$ determined from the line when $1/T = 1.25 \times 10^{-3}$ is -2.593; the value when $1/T = 1.78 \times 10^{-3}$ is -14.447. The slope of this line is given by the following expression:

$$\text{Slope} = \frac{\Delta(\ln k)}{\Delta\left(\dfrac{1}{T}\right)}$$

$$= \frac{(-14.447) - (-2.593)}{(1.78 \times 10^{-3}\ K^{-1}) - (1.25 \times 10^{-3}\ K^{-1})}$$

$$= \frac{-11.854}{0.53 \times 10^{-3}\ K^{-1}} = -2.2 \times 10^{4}\ K$$

$$= -\frac{E_a}{R}$$

Thus
$$E_a = -\text{slope} \times R = -(-2.2 \times 10^{4}\ K \times 8.314\ J\ mol^{-1}\ K^{-1})$$
$$= 1.8 \times 10^{5}\ J\ mol^{-1} = 1.8 \times 10^{2}\ kJ\ mol^{-1}$$

Figure 13.22

A graph of the linear relationship between $\ln k$ and $1/T$ for the reaction $2HI \longrightarrow H_2 + I_2$ according to the Arrhenius equation.

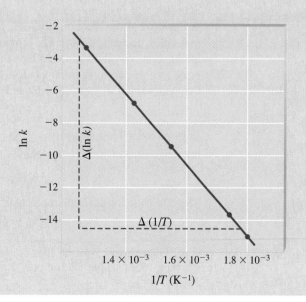

Check your learning: The rate constant for the rate of decomposition of N_2O_5 to NO and O_2 in the gas phase is 1.66 L mol^{-1} s^{-1} at 650 K and 7.39 L mol^{-1} s^{-1} at 700 K. What is the activation energy for this decomposition?

Answer: 113 kJ mol^{-1}

13.9   Elementary Reactions

When we write a balanced equation for a chemical reaction we indicate what is reacting and what is produced, but we say nothing abut how the reaction actually takes place. The process, or pathway, by which a reaction occurs is called the **reaction mechanism,** or the **reaction path.**

Although it may not be obvious to us as we watch, a reaction usually occurs in steps. The decomposition of ozone, for example, appears to follow a mechanism with two steps.

$$O_3 \longrightarrow O_2 + O \qquad (10)$$
$$O + O_3 \longrightarrow 2O_2 \qquad (11)$$

The two steps add up to the overall reaction for the decomposition,

$$2O_3 \longrightarrow 3O_2$$

The oxygen atom produced in the first step is used in the second and does not appear as a final product. Species that are produced in one step and consumed in another are called **intermediates.**

We call each of the steps in a reaction mechanism an **elementary reaction.** Elementary reactions occur exactly as they are written and cannot be broken down into simpler steps. Although the equation for the decomposition of ozone,

$$2O_3 \longrightarrow 3O_2$$

indicates that two molecules of ozone react to give three molecules of oxygen, the mechanism of the reaction does not involve the collision and reaction of two ozone molecules. Instead, a molecule of ozone decomposes to an oxygen molecule and an intermediate oxygen atom, and then the oxygen atom reacts with a second ozone molecule to give two oxygen molecules. These two elementary reactions occur exactly as they are written in Equations 10 and 11.

Unimolecular Elementary Reactions.   In a **unimolecular elementary reaction,** the rearrangement of a *single* molecule or ion gives one or more molecules of product.

$$A \longrightarrow products$$

The rate equation for a unimolecular reaction is

$$Rate = k[A]$$

A unimolecular reaction may be one of several elementary reactions in a complex mechanism, or it may be the only reaction in a mechanism. Reaction 10,

$$O_3 \longrightarrow O_2 + O$$

illustrates a unimolecular elementary reaction that occurs in a two-step reaction mechanism. The gas-phase decomposition of cyclobutane, C_4H_8, to ethylene, C_2H_4, occurs via a one-step unimolecular mechanism.

All that is required for these unimolecular reactions to occur is the separation of parts of single reactant molecules into products. The reaction of cyclobutane also shows that an overall reaction can be an elementary reaction as well.

Chemical bonds do not simply fall apart during chemical reactions. The activation energy for the decomposition of C_4H_8, for example, is 261 kilojoules of energy per mole. This means that it requires 261 kilojoules to distort 1 mole of these molecules into activated complexes that decompose into products.

In a sample of C_4H_8, a few of the rapidly moving C_4H_8 molecules collide with other rapidly moving molecules and pick up additional energy. If this energy is sufficient to transform the C_4H_8 molecule into an activated complex, ethylene molecules can form. In effect, a particularly energetic collision knocks a C_4H_8 molecule into the geometry of the activated complex. However, only a small fraction of gas molecules travel at sufficiently high speeds with large enough kinetic energies to accomplish this (see Fig. 13.20). Hence, at any one time, only a few molecules pick up enough energy from collisions to react.

The rate of decomposition of C_4H_8 is directly proportional to its concentration. Doubling the concentration of C_4H_8 in a sample gives twice as many molecules per liter. Although the fraction of molecules with enough energy to react remains the same, the total number of such molecules is twice as great. Consequently, the change in the amount of C_4H_8 per liter and the reaction rate are twice as great.

$$\text{Rate} = -\frac{\Delta[C_4H_8]}{\Delta t} = k[C_4H_8]$$

A similar relationship applies to any unimolecular elementary reaction; the reaction rate is directly proportional to the concentration of the reactant, and the reaction exhibits first-order behavior. The proportionality constant is the rate constant for the particular unimolecular reaction.

Bimolecular Elementary Reactions.   The collision *and combination* of two molecules or atoms to give an activated complex in an elementary reaction is called a **bimolecular elementary reaction.** There are two types of bimolecular elementary reactions:

$$A + B \longrightarrow \text{products} \quad \text{and} \quad 2A \longrightarrow \text{products}$$

For the bimolecular elementary reaction in which the two reactant molecules are different,

$$A + B \longrightarrow products$$

the rate law is first order in A and first order in B.

$$Rate = k[A][B]$$

For the bimolecular elementary reaction in which two identical molecules collide and react,

$$A + A \longrightarrow products$$

the rate law is second order in A.

$$Rate = k[A][A] = k[A]^2$$

Reaction 11,

$$O + O_3 \longrightarrow 2O_2$$

is an example of a bimolecular elementary reaction that is found in a two-step reaction mechanism. Other examples of reactions with mechanisms that consist of a single bimolecular elementary reaction include the reaction of nitrogen dioxide with carbon monoxide (Section 13.7),

$$NO_2 + CO \longrightarrow NO + CO_2$$

and the decomposition of two hydrogen iodide molecules to give hydrogen, H_2, and iodine, I_2 (Fig. 13.23),

$$2HI \longrightarrow H_2 + I_2$$

Termolecular Elementary Reactions.

A **termolecular elementary reaction** involves the simultaneous collision of three atoms, molecules, or ions. Termolecular elementary reactions are uncommon, because the probability of three particles colliding simultaneously is less than a thousandth of the probability of two particles colliding. There are, however, a few established termolecular elementary reactions. The reaction of nitric oxide with oxygen,

$$2NO + O_2 \longrightarrow 2NO_2 \qquad Rate = k[NO]^2[O_2]$$

Figure 13.23

Probable mechanism for the dissociation of two HI molecules to produce one molecule of H_2 and one molecule of I_2.

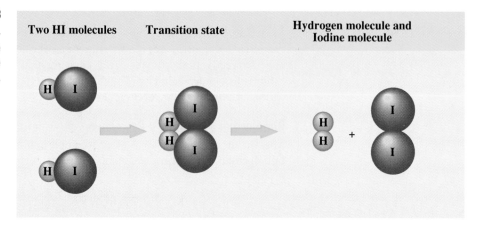

| Two HI molecules | Transition state | Hydrogen molecule and Iodine molecule |

and the reaction of nitric oxide with chlorine,

$$2NO + Cl_2 \longrightarrow 2NOCl \qquad Rate = k[NO]^2[Cl_2]$$

appear to involve termolecular steps.

13.10   Reaction Mechanisms

As noted in Section 13.9, we call the stepwise sequence of elementary reactions that converts reactants into products a **reaction mechanism,** or **reaction path.** The decomposition of C_4H_8 ($C_4H_8 \longrightarrow 2C_2H_4$), for example, has a one-step mechanism, and the decomposition of ozone ($2O_3 \longrightarrow 3O_2$) has a two-step mechanism.

Even an apparently simple reaction such as the formation of ethylene, C_2H_4, and hydrogen gas, H_2, from ethane, C_2H_6, may proceed by a complex mechanism. Ethylene may be prepared by the pyrolysis (decomposition by heating) of ethane, a component of natural gas, at temperatures of 500°–800°C.

$$C_2H_6(g) \xrightarrow{\Delta} C_2H_4(g) + H_2(g)$$

The actual mechanism is much more complex than is suggested by this equation, and it involves three stages:

1. *Initiation.* The reaction begins with a unimolecular reaction in which a C_2H_6 molecule splits at its weakest point, the C—C bond.

$$
\begin{array}{ccc}
\text{H} & \text{H} & \\
| & | & \\
\text{H}-\text{C}-\text{C}-\text{H} & \xrightarrow{\Delta} & \text{H}-\overset{\displaystyle |}{\underset{\displaystyle |}{\text{C}}}\cdot + \cdot\overset{\displaystyle |}{\underset{\displaystyle |}{\text{C}}}-\text{H} \\
| & | & \\
\text{H} & \text{H} &
\end{array}
\tag{12}
$$

   The CH_3 group extracts a hydrogen atom from another C_2H_6 molecule in a bimolecular reaction.

$$CH_3 + C_2H_6 \longrightarrow CH_4 + C_2H_5 \tag{13}$$

2. *Propagation.* The C_2H_5 group undergoes a unimolecular dissociation.

$$C_2H_5 \longrightarrow C_2H_4 + H \tag{14}$$

   The hydrogen atom produced in Equation 14 reacts with another C_2H_6 molecule in a bimolecular reaction.

$$H + C_2H_6 \longrightarrow C_2H_5 + H_2 \tag{15}$$

   The C_2H_5 group produced in Equation 15 reacts according to Equation 14, producing another hydrogen atom to undergo reaction as in Equation 15 and produce another C_2H_5. The reactions given by Equations 14 and 15 produce C_2H_4 and H_2, which are the principal products of the pyrolysis. Normally, each initiation (Equations 12 and 13) is followed by about 100 cycles of propagation (Equations 14 and 15) before termination.

3. *Termination.* The chain of propagation reactions may be stopped by one of the following bimolecular reactions, which remove C_2H_5 from the system.

$$2C_2H_5 \longrightarrow C_4H_{10}$$
$$2C_2H_5 \longrightarrow C_2H_4 + C_2H_6$$

The C_2H_5 produced in Equation 14 can react according to Equation 15 to produce another hydrogen atom for Equation 14. This reinitiates the series of elementary reactions represented by Equations 14 and 15, and they repeat themselves over and over. Such a mechanism, involving repeating reactions, is known as a **chain mechanism.**

Some of the elementary reactions in a reaction path are relatively slow. The slowest reaction step determines the maximum rate, because a reaction can proceed no faster than its slowest step. The slowest step, therefore, is the **rate-determining step** of the reaction.

We can write the rate law for each elementary reaction in a reaction mechanism (Section 13.9), but *we cannot write a correct rate law or establish the order for an overall reaction simply by inspection of the overall balanced equation.* We must determine the overall rate law from experimental data and deduce the mechanism from the rate law (and sometimes from other data). The reaction of NO_2 and CO provides us with an excellent example.

$$NO_2 + CO \longrightarrow CO_2 + NO$$

For temperatures above 225°C, the rate law has been found to be

$$Rate = k[NO_2][CO]$$

For temperatures above 225°C, therefore, the reaction is first order with respect to NO_2 and first order with respect to CO. This is consistent with a single-step bimolecular mechanism, and it is possible that this is the mechanism above 225°C.

At temperatures below 225°C, the reaction is described by a rate law that is second order in NO_2:

$$Rate = k[NO_2]^2$$

This is consistent with a mechanism that involves the following two elementary reactions, the first of which is slow and is therefore the rate-determining step:

$$NO_2 + NO_2 \longrightarrow NO_3 + NO \quad \text{(slow)}$$
$$NO_3 + CO \longrightarrow NO_2 + CO_2 \quad \text{(fast)}$$

(The sum of the two equations gives the net overall reaction.)

In general, when the slow step (the rate-determining step) is the first step in a mechanism, the rate law for the overall reaction is the same as the rate law for this step. However, when the rate-determining step occurs later in the mechanism, the rate law for the overall reaction may be complex. The oxidation of iodide ion by hydrogen peroxide illustrates this point.

$$H_2O_2 + 3I^- + 2H_3O^+ \longrightarrow 4H_2O + I_3^-$$

In a solution with a high concentration of acid, one reaction pathway has the following rate law:

$$Rate = k[H_2O_2][I^-][H_3O^+]$$

This rate law is consistent with several mechanisms, three of which follow.

Mechanism A

$$H_2O_2 + H_3O^+ + I^- \longrightarrow 2H_2O + HOI \quad \text{(slow)}$$
$$HOI + H_3O^+ + I^- \longrightarrow 2H_2O + I_2 \quad \text{(fast)}$$
$$I_2 + I^- \longrightarrow I_3^- \quad \text{(fast)}$$

Mechanism B

$$H_3O^+ + I^- \longrightarrow HI + H_2O \quad \text{(fast)}$$
$$H_2O_2 + HI \longrightarrow H_2O + HOI \quad \text{(slow)}$$
$$HOI + H_3O^+ + I^- \longrightarrow 2H_2O + I_2 \quad \text{(fast)}$$
$$I_2 + I^- \longrightarrow I_3^- \quad \text{(fast)}$$

Mechanism C

$$H_3O^+ + H_2O_2 \longrightarrow H_3O_2^+ + H_2O \quad \text{(fast)}$$
$$H_3O_2^+ + I^- \longrightarrow H_2O + HOI \quad \text{(slow)}$$
$$HOI + H_3O^+ + I^- \longrightarrow 2H_2O + I_2 \quad \text{(fast)}$$
$$I_2 + I^- \longrightarrow I_3^- \quad \text{(fast)}$$

In mechanism A the slow step is the first elementary reaction, so the rate law for the overall reaction is identical to the rate law for this step. In mechanisms B and C, the rate-determining step is the second elementary reaction, and the overall rate law differs from the rate law for this step. To derive the rate laws for mechanisms B and C requires some familiarity with equilibrium constants, which will be introduced in the next chapter. However, because mechanisms A, B, and C all have the same overall rate law, we cannot distinguish among them from the rate law alone. We need additional experimental information to determine which mechanism actually leads to product. Moreover, the rate law provides no information about what happens after the rate-determining step. The subsequent steps must be worked out from other chemical knowledge or from other measurements.

The determination of the mechanism of a reaction is important in selecting conditions that provide a good yield of the desired product. Knowing a reaction's mechanism sometimes helps us to prepare a previously unknown compound. The study of reaction mechanisms, or the **kinetics of reaction,** is a very active research area.

13.11 Catalysts

We have seen that the rate of many reactions can be accelerated by catalysts (Section 13.3). These substances may be divided into two general classes: homogeneous catalysts and heterogeneous catalysts.

Homogeneous Catalysts.
A **homogeneous catalyst** is present in the same phase as the reactants. It interacts with a reactant to form an intermediate substance, which then decomposes or reacts with another reactant in one or more steps to regenerate the original catalyst and give product.

The ozone in the upper atmosphere that protects the earth from ultraviolet radiation is formed when ultraviolet light, $h\nu$, interacts with oxygen molecules.

$$3O_2 \xrightarrow{h\nu} 2O_3$$

As shown in Section 13.9, ozone decomposes by the following mechanism:

$$O_3 \longrightarrow O_2 + O \quad (10)$$
$$O + O_3 \longrightarrow 2O_2 \quad (11)$$

The rate of decomposition of ozone is influenced by the presence of nitric oxide, NO, because it acts as a catalyst in the following mechanism:

$$NO(g) + O_3(g) \longrightarrow NO_2(g) + O_2(g) \tag{16}$$

$$O_3(g) \longrightarrow O_2(g) + O(g) \tag{17}$$

$$NO_2(g) + O(g) \longrightarrow NO(g) + O_2(g) \tag{18}$$

The overall chemical change for Equations 16 through 18 is the same as for Equations 10 and 11. It is

$$2O_3(g) \longrightarrow 3O_2(g)$$

The nitric oxide reacts and is regenerated in these reactions. It is not permanently used up; thus it acts as a catalyst. The rate of decomposition of ozone is greater in the presence of nitric oxide because of the catalytic activity of NO. Certain compounds that contain chlorine also catalyze the decomposition of ozone.

As we noted in Section 13.3, the rates of many biological reactions are increased by enzymes, organic molecules that act as catalysts (see Chapter 25).

Heterogeneous Catalysts.

Heterogeneous catalysts act by furnishing a surface at which a reaction can occur. Typically, gas-phase and liquid-phase reactions catalyzed by heterogeneous catalysts occur on the surface of the catalyst rather than within the gas or liquid phase.

Heterogeneous catalysis has at least four steps: (1) adsorption of the reactant onto the surface of the catalyst, (2) activation of the adsorbed reactant, (3) reaction of the adsorbed reactant, and (4) diffusion of the product from the surface into the gas or liquid phase (desorption). Any one of these steps may be slow and thus may be the rate-determining step. In general, however, the overall rate of the reaction is faster than it would be if the reactants were in the gas or liquid phase. The steps that are believed to occur in the reaction of compounds that contain a carbon–carbon double bond with hydrogen on a nickel catalyst are illustrated in Fig. 13.24. This is the catalyst used in the hydrogenation of polyunsaturated fats and oils (which contain several carbon–carbon double bonds) to produce saturated fats and oils (which contain only carbon–carbon single bonds).

Other significant industrial processes that involve the use of heterogeneous catalysts include the preparation of sulfuric acid (Section 8.10), the preparation of ammo-

Figure 13.24

Steps in the catalysis of the reaction $C_2H_4 + H_2 \longrightarrow C_2H_6$ by nickel. (A) Hydrogen is adsorbed on the surface, breaking the H—H bonds and forming Ni—H bonds. (B) Ethylene is adsorbed on the surface, breaking the π bond and forming Ni—C bonds. (C) Atoms diffuse across the surface and form new C—H bonds when they collide. (D) C_2H_6 molecules are not strongly attracted to the nickel surface, so they escape from the surface.

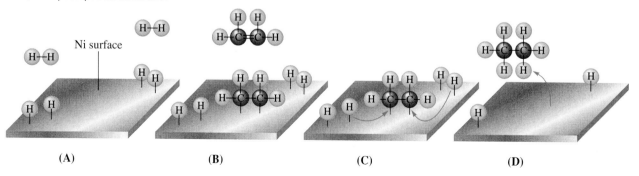

(A)   (B)   (C)   (D)

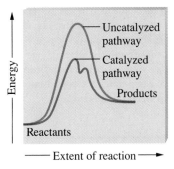

Figure 13.25

Potential energy diagram showing the effect of a catalyst on the activation energy. The catalyst provides a different reaction path that has a lower activation energy.

nia (Section 8.10), the oxidation of ammonia to nitric acid (Section 8.10), and the synthesis of methanol, CH_3OH (Section 9.2). Heterogeneous catalysts are also used in the catalytic converters found on most gasoline-powered automobiles (Fig. 13.12).

Both homogeneous and heterogeneous catalysts function by providing a reaction path that has a lower activation energy than would be found in the absence of the catalyst (Fig. 13.25). This lower activation energy results in an increase in rate (Section 13.7). Note that a catalyst decreases the activation energy for both the forward and the reverse reactions and hence *accelerates both the forward and the reverse reactions.*

Some substances, called **inhibitors,** decrease the rate of a chemical reaction. In many cases, these are substances that react with and "poison" some catalyst in the system and thereby prevent its action. Many biological poisons are inhibitors that reduce the catalytic activity of an organism's enzymes. Catalytic converters are poisoned by lead, so lead-free fuels are required for automobiles equipped with such converters.

For Review   Summary

The **rate** of a reaction can be identified either by the decrease in concentration of a reactant, or by the increase in concentration of a product, per unit of time. For the reaction $xA \longrightarrow yB$,

$$\text{Rate} = -\frac{\Delta[A]}{\Delta t} \quad \text{or} \quad \text{Rate} = \frac{\Delta[B]}{\Delta t}$$

where the values of the two rates differ if x and y, or the phases of A and B, differ. The **reaction velocity,** v, of a reaction is the rate of change of concentration of any reactant or product in a reaction divided by its stoichiometric coefficient.

$$v = -\frac{1}{x} \cdot \frac{\Delta[A]}{\Delta t} = \frac{1}{y} \cdot \frac{\Delta[B]}{\Delta t}$$

In general, the rate of a given reaction increases as the temperature or the concentration of a reactant increases. The rate of a reaction can also be increased by a **catalyst,** a substance that increases the rate of a reaction but is not permanently changed by the reaction. Reactions that involve two phases proceed more rapidly the more finely divided the condensed phase.

The rate of a given reaction can be described by a **rate law** of the form

$$\text{Rate} = k[A]^m[B]^n[C]^p \dots$$

in which [A], [B], and [C] represent molar concentrations of reactants (or sometimes products or other substances); m, n, and p are usually positive integers; and k is the rate constant. The exponents, m, n, p, and so on, which must be determined experimentally for each different reaction, describe the **order in each reactant.** The **overall order** of the reaction is the sum of the exponents. Two types of rate laws are used to represent changes

that accompany a chemical reaction. A **rate law,** or **differential rate law** (discussed above), describes how the rate of a reaction depends on the concentrations of reactants. An **integrated rate law** describes how concentration varies with time. Equations for both types of rate laws are shown in Table 13.2.

Experimental measurements are used to determine the appropriate rate law for a macroscopic chemical reaction. Differential rate laws are determined by the **method of initial rates.** We measure values for the initial rates of a reaction at different concentrations of the reactants and, from these measurements, determine the order of the reaction in each reactant. Integrated rate laws are determined from measurements of concentration at various times during a reaction. The appropriate integrated rate law for the reaction is the one that gives a straight line in the plots described in Table 13.2.

The **half-life of a reaction** is the time required for one-half of the limiting reagent to react. Equations for half-lives are given in Table 13.2. The half-life of a zero-order reaction decreases as the initial concentration of the reactant in the reaction decreases. The half-life of a first-order reaction is independent of concentration, and the half-life of a second-order reaction decreases as the concentration increases.

Before atoms, molecules, or ions can react, they must collide. When every collision leads to reaction, the rate is controlled by how rapidly the reactants diffuse together. These **diffusion-controlled reactions** are very fast. Most reactions occur more slowly, because the reacting species must be oriented correctly when they collide and because they must possess a certain minimum energy, the **activation energy,** in order to form an activated complex or transition state. The rate constant is related to the orientation of the colliding reactants and to the activation energy by the **Arrhenius equation,**

$$k = A \times e^{-E_a/RT}$$

The collection of individual steps, or **elementary reactions,** by which reactants are converted to products during the course of

a reaction is called the **reaction mechanism.** The overall rate of a reaction is determined by the rate of the slowest step, the **rate-determining step.** Although it is not possible to write a rate law for an overall reaction without using experimental data, a rate law can be written by inspection for an elementary reaction. **Unimolecular elementary reactions** have first-order rate laws; **bimolecular elementary reactions** have second-order rate laws.

Key Terms and Concepts

activated complex   (13.7)
activation energy, E_a (13.8)
bimolecular elementary reaction   (13.9)
Arrhenius equation   (13.8)
catalyst   (13.3, 13.11)
chain mechanism   (13.10)
collision theory   (13.7)
differential rate law   (13.3)
diffusion-controlled reaction   (13.7)
elementary reaction   (13.9)
first-order reaction   (13.4, 13.5)
frequency factor   (13.8)

half-life of a reaction, $t_{1/2}$ (13.6)
heterogeneous catalyst   (13.11)
homogeneous catalyst   (13.11)
initial rate   (13.1)
instantaneous rate   (13.1)
integrated rate laws   (13.5)
intermediate   (13.9)
overall order of a reaction   (13.4)
rate-determining step   (13.10)
rate constant   (13.4)
rate equation   (13.4)
rate law   (13.4)

rate of reaction   (13.1)
reaction mechanism   (13.9, 13.10)
reaction order   (13.4)
reaction path   (13.9, 13.10)
reaction velocity, v (13.2)
second-order reaction   (13.4, 13.5)
termolecular elementary reaction   (13.9)
transition state   (13.7)
unimolecular elementary reaction   (13.9)
zero-order reaction   (13.4, 13.5)

Exercises

Questions

1. How do the rate of a reaction and its rate constant differ?

2. Explain how each of the factors that determine the rate of a reaction is responsible for changing the rate.

3. Which of the following are part of the macroscopic domain of chemical kinetics, and which are part of the microscopic domain? activated complex, elementary reaction, first-order reaction, half-life of a reaction, homogeneous catalysts, integrated rate laws, rate of reaction, reaction mechanism, unimolecular elementary reaction.

4. Which of the following are part of the macroscopic domain of chemical kinetics, and which are part of the microscopic domain? bimolecular elementary reaction, chain mechanism, heterogeneous catalysts, instantaneous rate, intermediate, rate-determining step, reaction path, second-order reaction, transition state.

5. Which of the following figures illustrate a macroscopic feature of chemical kinetics, and which illustrate a microscopic feature? Figs. 13.1, 13.4, 13.9, 13.16, 13.19, 13.20, 13.23

6. Which of the following figures illustrate a macroscopic feature of chemical kinetics, and which illustrate a microscopic feature? Fig. 13.2, 13.7, 13.14, 13.17, 13.21, 13.24

7. Describe how graphical methods can be used to determine the order of a reaction and its rate constant from a series of data that include the concentration of a reactant at varying times.

8. Describe how graphical methods can be used to determine the activation energy of a reaction from a series of data that include the rate of reaction at varying temperatures.

9. Chemical reactions occur when reactants collide. For what reasons may a collision fail to produce a chemical reaction?

10. When every collision between reactants leads to a reaction, what determines the rate at which the reaction occurs?

11. What is the rate law (rate equation) for the elementary termolecular reaction A + 2B $\longrightarrow$ products? For 3A $\longrightarrow$ products?

12. Why are elementary reactions that involve three or more reactants very uncommon?

Reaction Rates and Reaction Velocity

13. Explain the differences among average rate, initial rate, and instantaneous rate.

14. Explain why $\Delta[H_2O_2]/\Delta t$ would not change, but $\Delta[O_2]/\Delta t$ would decrease by a factor of 2, if the volume of the gas collection apparatus in Fig. 13.3 were twice as large.

15. Explain why an egg cooks more slowly in boiling water in Denver than in New York City. *Hint:* Consider the effect of temperature on reaction rate and the effect of pressure on boiling point.

16. Describe the effect of each of the following on the rate of the reaction of magnesium metal with a solution of hydrochloric acid: the molarity of the hydrochloric acid, the temperature of the solution, the size of the pieces of magnesium.

17. Ozone decomposes to oxygen according to the equation $2O_3(g) \longrightarrow 3O_2(g)$. Write the equations that relate the reaction velocity of this reaction to the rate of disappearance of O_3 and the rate of formation of oxygen.

18. Chlorine trifluoride is used to prepare uranium hexafluoride, a volatile compound of uranium used in the separation of uranium isotopes in the nuclear industry. Chlorine trifluoride is prepared by the reaction $Cl_2(g) + 3F_2(g) \longrightarrow 2ClF_3(g)$. Write the equations that relate the reaction velocity of this reaction to the rate of disappearance of Cl_2 and F_2 and the rate of formation of ClF_3.

19. A study of the rate of dimerization of C_4H_6 ($2C_4H_6 \longrightarrow C_8H_{12}$) gave the following data:

Time, s	$[C_4H_6]$, M
0	1.00×10^{-2}
1600	5.04×10^{-3}
3200	3.37×10^{-3}
4800	2.53×10^{-3}
6200	2.08×10^{-3}

(a) Determine the average rate of dimerization between 0 s and 1600 s. Between 1600 s and 3200 s.

(b) Estimate the instantaneous rate of dimerization at 3200 s from a graph of time versus $[C_4H_6]$. What are the units of this rate?

(c) Determine the average rate of formation of C_8H_{12} between 0 s and 1600 s and the instantaneous rate of formation at 3200 s from the rates found in parts (a) and (b).

20. A study of the rate of a reaction that can be represented as $2A \longrightarrow B$ gave the following data:

Time, s	[A], M
5	0.952
10	0.625
15	0.465
20	0.370
25	0.308
35	0.230

(a) Determine the average rate of disappearance of A between 0 s and 10 s. Between 10 s and 20 s.

(b) Estimate the instantaneous rate of dimerization at 15 s from a graph of time versus [A]. What are the units of this rate?

(c) Determine the average rate of formation of B between 0 s and 10 s and the instantaneous rate of formation at 20 s from the rates found in parts (a) and (b).

21. Consider the following reaction:

$$5Br^-(aq) + BrO_3^-(aq) + 6H^+(aq) \longrightarrow 3Br_2(aq) + 3H_2O(l)$$

If the rate of disappearance of $Br^-(aq)$ at a particular moment during the reaction is 3.5×10^{-4} M s^{-1}, what is the rate of appearance of $Br_2(aq)$ at that moment?

22. What are the rate of decomposition of ammonia, the rates of formation of N_2 and H_2, and the reaction velocity for the decomposition of ammonia on a tungsten surface under the conditions reflected in Fig. 13.16?

Reaction Rates and Rate Equations

23. (a) Doubling the concentration of a reactant increases the rate of a reaction four times. What is the order of the reaction with respect to that reactant?

(b) Tripling the concentration of a reactant increases the rate of a reaction three times. What is the order of the reaction with respect to that reactant?

24. (a) Tripling the concentration of a reactant increases the rate of a reaction nine times. What is the order of the reaction with respect to that reactant?

(b) Increasing the concentration of a reactant by a factor of four increases the rate of a reaction four times. What is the order of the reaction with respect to that reactant?

25. Determine how each of the following will affect the rate of the reaction

$$CO(g) + NO_2(g) \longrightarrow CO_2(g) + NO(g)$$

if the rate law for the reaction is

$$\text{Rate} = k[NO_2]^2$$

(a) Decreasing the pressure of NO_2 from 0.50 atm to 0.250 atm

(b) Increasing the concentration of CO three times from 0.01 M to 0.03 M

(c) Increasing the temperature

26. Determine how each of the following will affect the rate of the reaction

$$CO(g) + NO_2(g) \longrightarrow CO_2(g) + NO(g)$$

if the rate law for the reaction is

$$\text{Rate} = k[NO_2][CO]$$

(a) Increasing the pressure of NO_2 from 0.1 atm to 0.3 atm

(b) Increasing [CO] from 0.02 M to 0.06 M

(c) Decreasing the temperature

27. Radioactive phosphorus is used in the study of biochemical reaction mechanisms, because phosphorus atoms are components of many biochemical molecules. The location of the phosphorus (and the location of the molecule it is bound in) can be detected from the electrons (beta particles) it produces.

$$^{32}_{15}P \longrightarrow {}^{32}_{16}S + e^-$$

$$\text{Rate} = 4.85 \times 10^{-2} \text{ day}^{-1}[^{32}P]$$

What is the instantaneous rate of production of electrons in a sample with a phosphorus concentration of 0.0033 M?

28. The rate constant for the radioactive decay of $^{14}_6C$ is 1.21×10^{-4} year^{-1}. The products of the decay are nitrogen atoms and electrons (beta particles)

$$^{14}_6C \longrightarrow {}^{14}_7N + e^-$$

$$\text{Rate} = k[^{14}_6C]$$

What is the instantaneous rate of production of N atoms in a sample with a carbon-14 content of 6.5×10^{-9} M?

29. Regular flights of supersonic aircraft in the stratosphere are a major concern because such aircraft produce NO as a by-product in the exhaust of their engines. Nitrogen(II) oxide reacts with ozone, and it has been suggested that this could contribute to depletion of the ozone layer.

$$NO + O_3 \longrightarrow NO_2 + O_2$$

The reaction is first order both in NO and in O_3 with a rate constant of 2.20×10^7 L mol^{-1} s^{-1} at 25°C. What is the instantaneous rate of disappearance of NO when [NO] = 3.3×10^{-6} M and $[O_3] = 5.9 \times 10^{-7}$ M?

30. The decomposition of acetaldehyde is a second-order reaction with a rate constant of 4.71×10^{-8} L mol^{-1} s^{-1}. What is the instantaneous rate of decomposition of acetaldehyde in a solution with a concentration of 5.55×10^{-4} M?

31. Alcohol is removed from the blood stream by a series of metabolic reactions. The first reaction produces acetaldehyde, and then other products are formed. The following data have been determined for the rate at which alcohol is removed from the blood of an average male, although individual rates can vary by 25%–30%. Women metabolize alcohol a little more slowly.

[C$_2$H$_5$OH], M	4.4×10^{-2}	3.3×10^{-2}	2.2×10^{-2}
Rate, mol L^{-1} h^{-1}	2.0×10^{-2}	2.0×10^{-2}	2.0×10^{-2}

Determine the rate equation, the rate constant, and the overall order for this reaction.

32. Under certain conditions, the decomposition of ammonia on a metal surface gives the following data:

[NH$_3$], M	1.0×10^{-3}	2.0×10^{-3}	3.0×10^{-3}
Rate, mol L^{-1} s^{-1}	1.5×10^{-6}	1.5×10^{-6}	1.5×10^{-6}

Determine the rate equation, the rate constant, and the overall order for this reaction.

33. Nitrosyl chloride, NOCl, decomposes to NO and Cl$_2$.

$$2NOCl(g) \longrightarrow 2NO(g) + Cl_2(g)$$

From the following data, determine the rate equation, the rate constant, and the overall order for this reaction.

[NOCl], M	0.10	0.20	0.30
Rate, mol L^{-1} s^{-1}	8.0×10^{-10}	3.2×10^{-9}	7.2×10^{-9}

34. From the following data, determine the rate equation, the rate constant, and the order with respect to A for the reaction A $\longrightarrow$ 2C.

[A], M	1.33×10^{-2}	2.66×10^{-2}	3.99×10^{-2}
Rate, mol L^{-1} s^{-1}	3.80×10^{-7}	1.52×10^{-6}	3.42×10^{-6}

35. Nitrogen(II) oxide reacts with chlorine according to the equation

$$2NO(g) + Cl_2(g) \longrightarrow 2NOCl(g)$$

The following initial rates of reaction have been observed for certain reactant concentrations.

[NO], mol L^{-1}	[Cl$_2$], mol L^{-1}	Rate, mol L^{-1} h^{-1}
0.50	0.50	1.14
1.00	0.50	4.56
1.00	1.00	9.12

What is the rate equation that describes the rate's dependence on the concentrations of NO and Cl$_2$? What is the rate constant? What are the orders with respect to each reactant?

36. Hydrogen reacts with nitrogen(II) oxide to form nitrogen(I) oxide (laughing gas) according to the equation

$$H_2(g) + 2NO(g) \longrightarrow N_2O(g) + H_2O(g)$$

From the following data, determine the rate equation, the rate constant, and the orders with respect to each reactant.

[NO], M	0.30	0.60	0.60
[H$_2$], M	0.35	0.35	0.70
Rate, mol L^{-1} s^{-1}	2.835×10^{-3}	1.134×10^{-2}	2.268×10^{-2}

37. For the reaction A $\longrightarrow$ B + C, the following data were obtained at 30°C.

[A], M	0.230	0.356	0.557
Rate, mol L^{-1} s^{-1}	4.17×10^{-4}	9.99×10^{-4}	2.44×10^{-3}

(a) What is the order of the reaction with respect to [A], and what is the rate equation?

(b) What is the rate constant?

38. For the reaction $Q \longrightarrow W + X$, the following data were obtained at 30°C.

[Q], M	0.170	0.212	0.357
Rate, mol $L^{-1}s^{-1}$	6.68×10^{-3}	1.04×10^{-2}	2.94×10^{-2}

(a) What is the order of the reaction with respect to [Q], and what is the rate equation?

(b) What is the rate constant?

39. Use the data provided to apply a graphical method and determine the order and rate constant of the reaction

$$SO_2Cl_2 \longrightarrow SO_2 + Cl_2$$

Time, s	$[SO_2Cl_2]$, M
0	0.100
5.00×10^3	0.0896
1.00×10^4	0.0802
1.50×10^4	0.0719
2.50×10^4	0.0577
3.00×10^4	0.0517
4.00×10^4	0.0415

40. Use the data provided to apply a graphical method and determine the order and rate constant of the reaction

$$2P \longrightarrow Q + W$$

Time, s	[P], M
4.0	1.09×10^{-3}
9.0	1.077×10^{-3}
13.0	1.068×10^{-3}
18.0	1.055×10^{-3}
22.0	1.046×10^{-3}
25.0	1.039×10^{-3}

41. When pure, ozone decomposes slowly to oxygen ($2O_3 \longrightarrow 3O_2$). Use the data provided to apply a graphical method and determine the order and rate constant of the reaction.

Time, h	$[O_3]$, M
0	1.00×10^{-5}
2.0×10^3	4.98×10^{-6}
7.6×10^3	2.07×10^{-6}
1.00×10^4	1.66×10^{-6}
1.23×10^4	1.39×10^{-6}
1.43×10^4	1.22×10^{-6}
1.70×10^4	1.05×10^{-6}

42. Use the data provided to apply a graphical method and determine the order and rate constant of the reaction

$$2X \longrightarrow Y + Z$$

Time, s	[X], M
5.0	0.0990
10.0	0.0497
15.0	0.0332
20.0	0.0249
25.0	0.0200
30.0	0.0166
35.0	0.0143
40.0	0.0125

43. What is the half-life for the first-order decay of phosphorus-32? (See Exercise 27 for the equation.) The rate constant for the decay is 4.85×10^{-2} day^{-1}.

44. What is the half-life for the first-order decay of carbon-14? (See Exercise 28 for the equation.) The rate constant for the decay is 1.21×10^{-4} year^{-1}.

45. What is the half-life for the decomposition of NOCl when the concentration of NOCl is 0.15 M? The rate constant for this second-order reaction is 8.0×10^{-8} L mol^{-1} s^{-1}.

46. What is the half-life for the decomposition of O_3 when the concentration of O_3 is 2.35×10^{-6} M? The rate constant for this second-order reaction is 50.4 L mol^{-1} h^{-1}.

47. The rate constant for the first-order decomposition at 45°C of dinitrogen pentoxide, N_2O_5, dissolved in chloroform, $CHCl_3$, is 6.2×10^{-4} min^{-1}.

$$2N_2O_5 \longrightarrow 4NO_2 + O_2$$

(a) What is the rate of decomposition when $[N_2O_5] = 0.40$ M?

(b) What are the rates of formation of NO_2 and of O_2 when $[N_2O_5] = 0.40$ M?

48. Most of the nearly 16 billion pounds of HNO_3 produced in the United States during 1989 was prepared by the following sequence of reactions, each run in a separate reaction vessel.

$$4NH_3(g) + 5O_2(g) \longrightarrow 4NO(g) + 6H_2O(g) \qquad (1)$$
$$2NO(g) + O_2(g) \longrightarrow 2NO_2(g) \qquad (2)$$
$$3NO_2(g) + H_2O(l) \longrightarrow 2HNO_3(aq) + NO(g) \qquad (3)$$

The first reaction is run by burning ammonia in air over a platinum catalyst. This reaction is fast. The reaction in Equation (3) is also fast. The second reaction limits the rate at which nitric acid can be prepared from ammonia. If Equation (2) is second order in NO and first order in O_2, what is the rate of formation of NO_2 when the oxygen concentration is 0.50 M and the nitric oxide concentration is 0.75 M? The rate constant for the reaction is $5.8 \times 10^{-6} \, L^2 \, mol^{-2} \, s^{-1}$.

49. The following data have been determined for the reaction

$$I^- + OCl^- \longrightarrow IO^- + Cl^-$$

[I$^-$], M	0.10	0.20	0.30
[OCl$^-$], M	0.050	0.050	0.010
Rate, mol L^{-1} s^{-1}	3.05×10^{-4}	6.10×10^{-4}	1.83×10^{-4}

Determine the rate equation and the rate constant for this reaction.

50. Determine the rate constant for the decomposition of H_2O_2 shown in Fig. 13.1 from the data given in the figure.

51. The reaction of compound A to give compounds C and D was found to be second order in A. The rate constant for the reaction was determined to be $2.42 \, L \, mol^{-1} \, s^{-1}$. If the initial concentration is $0.500 \, mol \, L^{-1}$, what is the value of $t_{1/2}$?

52. The half-life of a reaction of compound A to give compounds D and E is 8.50 min when the initial concentration of A is $0.150 \, mol \, L^{-1}$. How long will it take for the concentration to drop to $0.0300 \, mol \, L^{-1}$ if the reaction is first order with respect to A? Second order with respect to A?

Collision Theory of Reaction Rates

53. What is the activation energy of a reaction, and how is the energy related to the activated complex of the reaction?

54. Account for the relationship between the rate of a reaction and its activation energy.

55. How does an increase in temperature of 10°C affect the rate of many reactions? Explain this effect in terms of the collision theory of the reaction rate.

56. If the rate of a reaction doubles for every 10°C rise in temperature, how much faster does the reaction proceed at 45°C than at 25°C? At 95°C than at 25°C?

57. In an experiment, a sample of $NaClO_3$ was 90% decomposed in 48 min. Approximately how long would this decomposition have taken if the sample had been heated 20°C higher?

58. The rate constant at 325°C for the decomposition reaction $C_4H_8 \longrightarrow 2C_2H_4$ (Section 13.9) is $6.1 \times 10^{-8} \, s^{-1}$, and the activation energy is 261 kJ per mole of C_4H_8. Determine the frequency factor for the reaction.

59. The rate constant for the decomposition of acetaldehyde, CH_3CHO, to methane, CH_4, and carbon monoxide, CO, in the gas phase is $1.1 \times 10^{-2} \, L \, mol^{-1} \, s^{-1}$ at 703 K and 4.95 $L \, mol^{-1} \, s^{-1}$ at 865 K. Determine the activation energy for this decomposition.

60. An elevated level of the enzyme alkaline phosphatase (ALP) in the serum is an indication of possible liver or bone disorder. The level of serum ALP is so low that it is very difficult to measure directly. However, ALP catalyzes a number of reactions, and its relative concentration can be determined by measuring the rate of one of these reactions under controlled conditions. One such reaction is the conversion of p-nitrophenyl phosphate (PNPP) to p-nitrophenoxide ion (PNP) and phosphate ion. Control of temperature during the test is very important; the rate of the reaction increases 1.47 times if the temperature changes from 30°C to 37°C. What is the activation energy for the ALP-catalyzed conversion of PNPP to PNP and phosphate?

61. In terms of collision theory, to which one of the following is the rate of a chemical reaction proportional: (a) the change in free energy per second, (b) the change in temperature per second, (c) the number of collisions per second, or (d) the number of product molecules?

62. Hydrogen iodide, HI, decomposes in the gas phase to produce hydrogen, H_2, and iodine, I_2. Measuring the value of the rate constant, k, for the reaction at several different temperatures yielded the following data:

Temperature, K	k, M^{-1} s^{-1}
555	6.23×10^{-7}
575	2.42×10^{-6}
645	1.44×10^{-4}
700	2.01×10^{-3}

What is the value of the activation energy, in kilojoules per mole, for this reaction?

Elementary Reactions, Reaction Mechanisms, Catalysts

63. In general, can we predict the effect of doubling the concentration of A on the rate of the overall reaction A + B $\longrightarrow$ C? Can we predict the effect if the reaction is known to be an elementary reaction?

64. Define (a) unimolecular reaction, (b) bimolecular reaction, (c) elementary reaction, and (d) overall reaction.

65. Which of the following equations, as written, could describe elementary reactions?

(a) $Cl_2 + CO \longrightarrow Cl_2CO$
   Rate $= k[Cl_2]^{3/2}[CO]$

(b) $PCl_3 + Cl_2 \longrightarrow PCl_5$
Rate $= k[PCl_3][Cl_2]$
(c) $2NO + H_2 \longrightarrow N_2O + H_2O$
Rate $= k[NO][H_2]$
(d) $2NO + O_2 \longrightarrow 2NO_2$
Rate $= k[NO]^2[O_2]$
(e) $NO + O_3 \longrightarrow NO_2 + O_2$
Rate $= k[NO][O_3]$

66. Account for the increase in reaction rate brought about by a catalyst.

67. Describe how a homogeneous catalyst and a heterogeneous catalyst function.

68. (a) Chlorine atoms resulting from the decomposition of chlorofluoromethanes, such as CCl_2F_2, catalyze the decomposition of ozone in the ozone layer of the earth's atmosphere. One simplified mechanism for the decomposition is

$$O_3 \xrightarrow{\text{Sunlight}} O_2 + O$$
$$O_3 + Cl \longrightarrow O_2 + ClO$$
$$ClO + O \longrightarrow Cl + O_2$$

Explain why chlorine atoms are catalysts in the gas-phase transformation

$$2O_3 \longrightarrow 3O_2$$

(b) Nitric oxide is also involved in the decomposition of ozone by the mechanism

$$O_3 \xrightarrow{\text{Sunlight}} O_2 + O$$
$$O_3 + NO \longrightarrow NO_2 + O_2$$
$$NO_2 + O \longrightarrow NO + O_2$$

Is NO a catalyst for the decomposition? Explain your answer.

69. Write the rate equation for each of the elementary reactions given in both parts of Exercise 68.

70. Nitrogen(II) oxide, NO, reacts with hydrogen, H_2, according to the equation

$$2NO + 2H_2 \longrightarrow N_2 + 2H_2O$$

What would be the rate law if the mechanism for this reaction were

$$2NO + H_2 \longrightarrow N_2 + H_2O_2 \quad \text{(slow)}$$
$$H_2O_2 + H_2 \longrightarrow 2H_2O \quad \text{(fast)}$$

Applications and Additional Exercises

71. Some bacteria are resistant to the antibiotic penicillin because they produce penicillinase, an enzyme with a molecular mass of 30,000, which converts penicillin into inactive molecules. Although the kinetics of enzyme-catalyzed reactions can be complex, at low concentrations this reaction

can be described by a rate equation that is first order in the catalyst (penicillinase) and that also involves the concentration of penicillin. From the following data, collected using a solution containing 0.15 μg (0.15×10^{-6}g) of penicillinase per liter, determine the order of the reaction with respect to penicillin and the value of the rate constant.

[Penicillin], M	Rate, mol L^{-1} min^{-1}
2.0×10^{-6}	1.0×10^{-10}
3.0×10^{-6}	1.5×10^{-10}
4.0×10^{-6}	2.0×10^{-10}

72. The hydrolysis of the sugar sucrose to the sugars glucose and fructose,

$$C_{12}H_{22}O_{11} + H_2O \longrightarrow C_6H_{12}O_6 + C_6H_{12}O_6$$

follows a first-order rate equation for the disappearance of sucrose:

$$\text{Rate} = k[C_{12}H_{22}O_{11}]$$

(The products of the reaction, glucose and fructose, have the same molecular formulas but differ in the arrangement of the atoms in their molecules.)

(a) In neutral solution, $k = 2.1 \times 10^{-11}$ s^{-1} at 27°C and $k = 8.5 \times 10^{-11}$ s^{-1} at 37°C. Determine the activation energy, the frequency factor, and the rate constant for this equation at 47°C.

(b) When a solution of sucrose with an initial concentration of 0.150 M reaches equilibrium, the concentration of sucrose is 1.65×10^{-7} M. How long will it take the solution to reach equilibrium at 27°C in the absence of a catalyst? Because the concentration of sucrose at equilibrium is so low, assume that the reaction is irreversible.

(c) Why does assuming that the reaction is irreversible simplify the calculation in part (b)?

73. Either of two reactions could initiate the pyrolysis of ethane to ethylene:

$$C_2H_6 \longrightarrow 2CH_3$$

or

$$C_2H_6 \longrightarrow C_2H_5 + H$$

In these reactions, the activation energy is essentially equal to the bond energy (Section 6.12) of the bond broken in the reaction. Explain why the first of these reactions should be more effective than the second in initiating the pyrolysis of ethane to ethylene.

74. The element Co exists in two oxidation states, Co(II) and Co(III), and the ions form many complexes. The rate at which one of the complexes of Co(III) was reduced by Fe(II) in water was measured. Determine the activation energy of the reaction from the following data.

T, K	k, s^{-1}
293	0.054
298	0.100

75. Both technetium-99 and thallium-201 are used to image heart muscle in patients who may have heart problems. The half-lives are 6 h and 73 h, respectively. What percent of the radioactivity would remain for each of the isotopes after 2 days (48 h)?

76. There are two molecules with the formula C_3H_6. Propene, $CH_3CH=CH_2$, is the monomer of the polymer polypropylene, which is used for indoor–outdoor carpets. Cyclopropane,

$$CH_2-CH_2$$
$$\diagdown CH_2 \diagup$$

is used as an anesthetic. When heated to 499°C, cyclopropane rearranges (isomerizes) and forms propene, with a rate constant of 5.95×10^{-4} s^{-1}. What is the half-life of this reaction? What fraction of the cyclopropane remains after 0.75 h at 499°C?

77. For the past 10 years the unsaturated hydrocarbon 1,3-butadiene ($CH_2=CH-CH=CH_2$) has ranked number 38 among the top 50 industrial chemicals. It is used primarily for the manufacture of synthetic rubber. An isomer exists also as cyclobutene.

$$CH_2-CH_2$$
$$| \qquad |$$
$$CH=CH$$

The isomerization of cyclobutene to butadiene is first order, and the rate constant was measured as 2.0×10^{-4} s^{-1} at 150°C in a 0.53-L flask. Determine the partial pressure of cyclobutene and its concentration after 30.0 min if an isomerization reaction was carried out at 150°C with an initial pressure of 55 torr.

78. Nitroglycerine is an extremely sensitive explosive. In a series of carefully controlled experiments, samples of the explosive were heated to 160°C, and their first-order decomposition was studied. Determine the average rate constants using the following data from the experiments.

Initial [$C_3H_5N_3O_9$], M	t, s	% Decomposed
4.88	300	52.0
3.52	300	52.9
2.29	300	53.2
1.81	300	53.9
5.33	180	34.6
4.05	180	35.9
2.95	180	36.0
1.72	180	35.4

79. Fluorine-18 is a radioactive isotope that decays by positron emission to form oxygen-18 with a half-life of 109.7 min. (A positron is a particle that has the mass of an electron and a single unit of positive charge; the equation is $^{18}_9F \longrightarrow {}^{18}_8O + e^+$.) Physicians use ^{18}F to study the brain by injecting a quantity of fluoro-substituted glucose into the blood of a patient. The glucose accumulates in the regions where the brain is active and needs nourishment.

(a) What is the rate constant for the decomposition of fluorine-18?

(b) If a sample of glucose that contains fluorine-18 is injected into the blood, what percent will remain after 5.59 h?

(c) How long does it take for 99.99% of the ^{18}F to decay?

80. This question is taken from the 1994 Chemistry Advanced Placement Examination and is used with the permission of the Educational Testing Service.

$$2NO(g) + 2H_2(g) \longrightarrow N_2(g) + 2H_2O(g)$$

Experiments were conducted to study the rate of the reaction represented by the equation above. Initial concentrations and rates of reaction are given in the table below.

	Initial Concentration, mol/L		Initial Rate of Formation of N$_2$, mol/L min
Experiment	[NO]	[H$_2$]	
1	0.0060	0.0010	1.8×10^{-4}
2	0.0060	0.0020	3.6×10^{-4}
3	0.0010	0.0060	0.30×10^{-4}
4	0.0020	0.0060	1.2×10^{-4}

(a) (i) Determine the order for each of the reactants, NO and H$_2$, from the data given and show your reasoning.

(ii) Write the overall rate law for the reaction.

(b) Calculate the value of the rate constant, k, for the reaction. Include units.

(c) For experiment 2, calculate the concentration of NO remaining when exactly one-half of the original amount of H$_2$ has been consumed.

(d) The following sequence of elementary steps is a proposed mechanism for the reaction.

$$\text{I.} \quad NO + NO \rightleftharpoons N_2O_2$$
$$\text{II.} \quad N_2O_2 + H_2 \longrightarrow H_2O + N_2O$$
$$\text{III.} \quad N_2O + H_2 \longrightarrow N_2 + H_2O$$

Based on the data presented, which of the above is the rate-determining step? Show that the mechanism is consistent with (i) the observed rate law for the reaction, and (ii) the overall stoichiometry of the reaction.

CHAPTER OUTLINE

An Introduction to Equilibrium

14.1 The State of Equilibrium
14.2 Reaction Quotients and Equilibrium Constants
14.3 Homogeneous and Heterogeneous Equilibria
14.4 Changes in Concentrations at Equilibrium: Le Châtelier's Principle
14.5 Predicting the Direction of a Reversible Reaction

Kinetics and Equilibria

14.6 The Relationship of Reaction Rates and Equilibria
14.7 Reaction Mechanisms Involving Equilibria

Equilibrium Calculations

14.8 Concentration and Pressure Changes
14.9 Calculations Involving Equilibrium Concentrations
14.10 Calculation of Changes in Concentration
14.11 Techniques for Solving Equilibrium Problems

14

An Introduction to Chemical Equilibrium

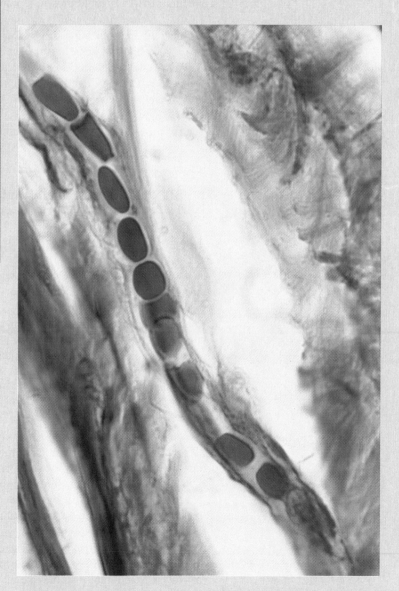

Red blood cells in a capillary in a human scalp. Multiple equilibria occur in such a biological system.

Some chemical reactions result in complete conversion of the reactants, giving a 100% yield of product. However, many other chemical reactions, particularly those that occur in our bodies and in other living systems, do not produce a complete conversion of reactants to products. These reactions appear to stop after only part of the reactants has been converted.

As an example, let us consider the behavior of the carbon dioxide found in our blood. The cells in our bodies produce carbon dioxide as a waste product of metabolism. The CO_2 dissolves in our blood and is transported to our lungs, where it comes out of solution and is exhaled. Carbon dioxide will react with the water in our blood serum and form carbonic acid, H_2CO_3, but this reaction does not give a 100% yield of H_2CO_3. Most of the CO_2 in our blood is present as dissolved molecules, of which only a very small percentage reacts and is converted to the acid. Even this small amount of H_2CO_3 could present a problem if it were to ionize and produce H_3O^+ and HCO_3^-; a concentration of H_3O^+ much above 10^{-7} M in our blood can prove fatal. Fortunately, the ionization of H_2CO_3 is another reaction that gives a very low yield of product, and the body has mechanisms that reduce the yield of H_3O^+ even more.

We can use the concepts of chemical equilibrium described in this chapter to understand how our body controls the concentration of H_3O^+ in our blood. We will see how to predict the yield of a reaction under specific conditions, how to change a reaction's conditions in order to increase or reduce that yield, and how living systems control the amounts of products present.

An Introduction to Equilibrium

14.1 The State of Equilibrium

Figure 14.1

Equilibrium in a mixture of NO_2 and N_2O_4. Brown NO_2 reacts with itself to form colorless N_2O_4. As the color indicates, at equilibrium, not all of the NO_2 has been converted to N_2O_4.

If we run a reaction in a closed system so that the products cannot escape, we often find that the reaction does not give a 100% yield of products. Instead, some reactants remain after the concentrations stop changing. At this point, when there is no further change in concentrations of reactant and product, we say the reaction is at **equilibrium.**

A mixture of reactants and products is present at equilibrium. For example, when we place a sample of nitrogen dioxide (NO_2, a brown gas) in a flask, it forms dinitrogen tetroxide (N_2O_4, a colorless gas) by the reaction

$$2NO_2(g) \longrightarrow N_2O_4(g)$$

The color becomes lighter as NO_2 is converted to colorless N_2O_4. However, even after the concentrations of reactant and product stop changing, the brown color of the mixture of gases (Fig. 14.1) shows that some NO_2 remains.

The formation of N_2O_4 from NO_2 is a **reversible reaction,** a reaction in which the reactants can combine to form products and the products can combine to form the reactants. Not only can NO_2 react and form N_2O_4, but the N_2O_4 produced can also decompose to form NO_2. As soon as the forward reaction produces any N_2O_4, the reverse reaction begins and N_2O_4 starts to decompose back to NO_2. At equilibrium, the concentrations of N_2O_4 and NO_2 no longer change, because the rate of formation of NO_2 is exactly equal to the rate of formation of N_2O_4. Careful investigations have shown that *equilibrium is a dynamic process; equilibrium is established in a system*

Figure 14.2

These jugglers provide an illustration of equilibrium. Each throws clubs to the other at the same rate at which he receives clubs from that person. Because clubs are thrown continuously in both directions, the number of clubs moving in each direction is constant, and the number of clubs each juggler has at a given time remains constant.

when reactants combine to form products at the same rate at which products combine to form reactants (Fig. 14.2).

We can detect a state of equilibrium because after it is reached, the concentrations of reactants and products do not appear to change. However, it is important that we verify that the absence of change is due to equilibrium and not to a reaction rate that is so slow that changes in concentration are difficult to detect.

We often use a double arrow when writing an equation for a reversible reaction. We also use a double arrow with reactions that are at equilibrium. For example, Fig. 14.1 shows the reaction

Figure 14.3

An equilibrium between liquid bromine and bromine vapor.

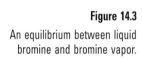

$$2NO_2(g) \rightleftharpoons N_2O_4(g)$$

When we wish to speak about one particular component of a reversible reaction, we use a single arrow. For example, at the equilibrium shown in Fig. 14.1, the rate of the reaction

$$2NO_2 \longrightarrow N_2O_4$$

is equal to the rate of the reaction

$$N_2O_4 \longrightarrow 2NO_2$$

Let us consider the evaporation of bromine as a second example of an equilibrium

$$Br_2(l) \rightleftharpoons Br_2(g)$$

Figure 14.3 shows a sample of liquid bromine in equilibrium with bromine vapor. When we pour liquid bromine into an empty bottle where there is no bromine vapor, some liquid evaporates, the amount of liquid decreases, and the amount of vapor increases. When we cap the bottle so that no vapor escapes, the amounts of liquid and vapor stop changing and an equilibrium between the liquid and the vapor is established. If the bottle were not capped, the bromine vapor would escape and no equilibrium would be reached.

14.2   Reaction Quotients and Equilibrium Constants

A general equation for a reversible reaction may be written

$$mA + nB + \cdots \rightleftharpoons xC + yD + \cdots$$

We can write the **reaction quotient,** Q, for this equation as

$$Q = \frac{[C]^x[D]^y \cdots}{[A]^m[B]^n \cdots}$$

where we use brackets to indicate "molar concentration of." The reaction quotient is a ratio of the molar concentrations of the products of the chemical equation (multiplied together) and of the reactants (also multiplied together), each raised to a power equal to the coefficient preceding that substance in the balanced chemical equation. The reaction quotient for the reversible reaction

$$2NO_2(g) \rightleftharpoons N_2O_4(g)$$

is given by the expression

$$Q = \frac{[N_2O_4]}{[NO_2]^2}$$

Example 14.1   **Writing Reaction Quotients**

Write the reaction quotient for each of the following reactions.
(a) $3O_2(g) \rightleftharpoons 2O_3(g)$
(b) $N_2(g) + 3H_2(g) \rightleftharpoons 2NH_3(g)$
(c) $4NH_3(g) + 7O_2(g) \rightleftharpoons 4NO_2(g) + 6H_2O(g)$

Solution:

(a) $Q = \dfrac{[O_3]^2}{[O_2]^3}$

(b) $Q = \dfrac{[NH_3]^2}{[N_2][H_2]^3}$

(c) $Q = \dfrac{[NO_2]^4[H_2O]^6}{[NH_3]^4[O_2]^7}$

Check your learning: Write the reaction quotient for each of the reactions.
(a) $2SO_2(g) + O_2(g) \rightleftharpoons 2SO_3(g)$
(b) $C_4H_8(g) \rightleftharpoons 2C_2H_4(g)$
(c) $2C_4H_{10}(g) + 13O_2(g) \rightleftharpoons 8CO_2(g) + 10H_2O(g)$

Answer: **(a)** $Q = [SO_3]^2/[SO_2]^2[O_2]$; **(b)** $Q = [C_2H_4]^2/[C_4H_8]$;
(c) $Q = [CO_2]^8[H_2O]^{10}/[C_4H_{10}]^2[O_2]^{13}$

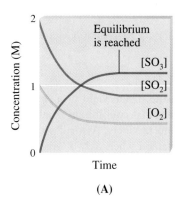

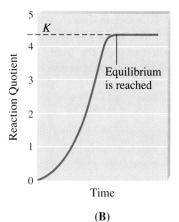

Figure 14.4

(A) The change in the concentrations of reactants and products as the reaction $2SO_2(g) + O_2(g) \rightleftharpoons 2SO_3(g)$ approaches equilibrium. (B) The change in the value of the reaction quotient as the reaction approaches equilibrium.

The numerical value of Q for a given reaction varies; it depends on the concentrations of products and reactants *present at the time when Q is determined.* At the instant pure reactants are mixed, Q equals zero because there are no products present at that point. As the reaction proceeds, the value of Q increases as the concentrations of the products increase and the concentrations of the reactants simultaneously decrease (Fig. 14.4). When the reaction reaches equilibrium, the value of the reaction quotient no longer changes because the concentrations no longer change.

When a mixture of reactants and products of a reaction reaches equilibrium at a given temperature, its reaction quotient always has the same value. This value is called the **equilibrium constant,** K, of the reaction at that temperature. *When a reaction is at equilibrium at a given temperature, the concentration of reactants and products is such that the value of reaction quotient, Q is always equal to the equilibrium constant, K, for that reaction at that temperature.*

The mathematical expression that indicates that a reaction quotient always assumes the same value at equilibrium,

$$Q = K = \frac{[C]^x[D]^{y}\cdots}{[A]^m[B]^{n}\cdots}$$

is a mathematical statement of the **law of mass action:** When a reaction has attained equilibrium at a given temperature, the reaction quotient for the reaction always has the same value.

Example 14.2   Evaluating a Reaction Quotient

Gaseous nitrogen dioxide forms dinitrogen tetroxide according to the equation

$$2NO_2(g) \rightleftharpoons N_2O_4(g)$$

When 0.10 mol of NO_2 is added to a 1.0-L flask at 25°C, the concentration changes so that at equilibrium $[NO_2] = 0.016$ M and $[N_2O_4] = 0.042$ M.
(a) What is the value of the reaction quotient before any reaction occurs?
(b) What is the value of the equilibrium constant for the reaction?

Solution:
(a) Before any product forms, $[NO_2] = 0.10$ mol $L^{-1} = 0.10$ M, and $[N_2O_4] = 0$ M. Thus

$$Q = \frac{[N_2O_4]}{[NO_2]^2} = \frac{(0)}{(0.10)^2} = 0$$

(b) At equilibrium, the value of the equilibrium constant is equal to the value of the reaction quotient. At equilibrium, then,

$$K = Q = \frac{[N_2O_4]}{[NO_2]^2} = \frac{(0.042)}{(0.016)^2}$$
$$= 1.6 \times 10^2$$

The equilibrium constant is 1.6×10^2.
Chemists cannot agree on whether to use units with equilibrium constants, so we see different styles in different places. Where units are used, they may differ depending on the expression for Q. In some cases the units cancel out. It is common practice to omit the units of Q or K; we will do so in most examples.

Check your learning: Show that the three sets of equilibrium concentrations in Table 14.1 give the value of the reaction quotient indicated.

The magnitude of an equilibrium constant is a measure of the yield of a reaction when it reaches equilibrium. A large value for K indicates that equilibrium is attained only after the reactants have been largely converted into products. A small value of K—much less than 1—indicates that equilibrium is attained when only a small proportion of the reactants have been converted into products.

Regardless of the initial mixture of reactants and products in a reversible reaction, the composition of a system will always adjust itself to a condition of equilibrium for which the value of the reaction quotient is equal to the equilibrium constant for the system, provided that the temperature does not change. The data in Table 14.1 illustrate this. Under the same conditions, different starting mixtures of CO, H_2O, CO_2, and H_2 react and give the same value of Q at equilibrium (within the limits of the significant figures of the data). This value is 0.640, the equilibrium constant for the reaction under these conditions.

$$CO(g) + H_2O(g) \rightleftharpoons CO_2(g) + H_2(g) \qquad K = 0.640$$

It is important to remember that an equilibrium can be established either starting from reactants or starting from products. For example, equilibrium was established from mixture 2 in Table 14.1 when the *products* of the reaction were heated in a closed container. In fact, one technique that is used to determine whether a reaction is truly at equilibrium is to approach equilibrium starting with reactants in one experiment and starting with products in another. If the same value of the reaction quotient is observed when the concentrations stop changing in both experiments, then we may be certain that the system has reached equilibrium.

As we finish this section, we should note that we make simplifying assumptions, as in many other cases in chemistry, when evaluating a reaction quotient or an equilibrium constant. Strictly speaking, we should calculate the value of Q or K from the

Table 14.1

Concentrations of Three Mixtures Before and After Reaching Equilibrium and the Value of Q at Equilibrium at 800°C

$CO(g) + H_2O(g) \rightleftharpoons CO_2(g) + H_2(g)$

	Mixture 1	Mixture 2	Mixture 3
Before Reaction*			
$[CO]_i$	0.0243 M	0 M	0.0094 M
$[H_2O]_i$	0.0243 M	0 M	0.0055 M
$[CO_2]_i$	0 M	0.0468 M	0.0005 M
$[H_2]_i$	0 M	0.0468 M	0.0046 M
At Equilibrium			
$[CO]$	0.0135 M	0.0260 M	0.0074 M
$[H_2O]$	0.0135 M	0.0260 M	0.0035 M
$[CO_2]$	0.0108 M	0.0208 M	0.0025 M
$[H_2]$	0.0108 M	0.0208 M	0.0066 M
$Q (= K)$	0.640	0.640	0.64

*The subscript i indicates initial concentrations, the concentrations before any reaction occurred.

activities of the reactants and products rather than from their concentrations. (The activity of a dissolved species is discussed in Section 12.20.) However, the activity of a dilute solute is usefully approximated by its molar concentration, so we will use concentrations as approximations of the activity of a dissolved species. The activity of a gas is approximated by its pressure (in atmospheres), so we use pressures for gases. However, we also can use molar concentrations of gases in our equilibrium calculations, because the molar concentration of a gas is directly proportional to its pressure. The activity of a pure solid or pure liquid is 1, and the activity of a solvent in a dilute solution is close to 1. Thus these species (solids, liquids, and solvents) are omitted from reaction quotients and equilibrium calculations.

Using concentrations and pressures instead of activities means that we calculate approximate values for reaction quotients and equilibrium constants. However, these approximations hold well for dilute solutions and for gases with pressures less than about 2 atmospheres.

14.3 Homogeneous and Heterogeneous Equilibria

We can divide equilibria into two classes: homogeneous equilibria and heterogeneous equilibria. Moreover, each of these classes can be further classified by the types of reactions involved. However, we should not lose track of the fact that all of these equilibria have common features. In particular,

1. The reaction quotient for all reversible reactions

$$m\text{A} + n\text{B} + \cdots \rightleftharpoons x\text{C} + y\text{D} + \cdots$$

    is written the same way:

$$Q = \frac{[\text{C}]^x[\text{D}]^y \cdots}{[\text{A}]^m[\text{B}]^n \cdots}$$

    Even if a reaction is unfamiliar, we can write the reaction quotient by using the approximations for activities discussed in the previous section. However, we should remember that a pure solid, a pure liquid, or a solvent in a dilute solution is not included in Q.

2. No matter how a reaction gets to equilibrium, at equilibrium at a given temperature, its value of Q is always equal to the equilibrium constant for the reaction at that temperature.

Homogeneous Equilibria. A **homogeneous equilibrium** is an equilibrium within a single phase. In this book we will concentrate on homogeneous equilibria that involve changes in solution or in mixtures of gases. The reaction shown in Fig. 14.1 is a homogeneous equilibrium.

Homogeneous equilibria in solution. Reactions between dissolved reactants that produce dissolved products give one type of homogeneous equilibrium. The chemical species involved can be molecules, ions, or a mixture of both. Several examples and the expressions for their equilibrium constants follow.

$$\text{C}_2\text{H}_2 + 2\text{Br}_2 \rightleftharpoons \text{C}_2\text{H}_2\text{Br}_4 \qquad\qquad K = \frac{[\text{C}_2\text{H}_2\text{Br}_4]}{[\text{C}_2\text{H}_2][\text{Br}_2]^2}$$

$$I_2(aq) + I^-(aq) \rightleftharpoons I_3^-(aq) \qquad K = \frac{[I_3^-]}{[I_2][I^-]}$$

$$Hg_2^{2+}(aq) + NO_3^-(aq) + 3H_3O^+(aq) \rightleftharpoons 2Hg^{2+}(aq) + HNO_2(aq) + 4H_2O(l)$$

$$K = \frac{[Hg^{2+}]^2[HNO_2]}{[Hg_2^{2+}][NO_3^-][H_3O^+]^3}$$

$$HF(aq) + H_2O(l) \rightleftharpoons H_3O^+(aq) + F^-(aq) \qquad K = \frac{[H_3O^+][F^-]}{[HF]}$$

$$NH_3(aq) + H_2O(l) \rightleftharpoons NH_4^+(aq) + OH^-(aq) \qquad K = \frac{[NH_4^+][OH^-]}{[NH_3]}$$

In the last three examples, water is the solvent, so we do not write it in the expression for the reaction quotient, even though it is a product or a reactant.

Some types of homogeneous equilibria are so common that they have individual names and their equilibrium constants have special symbols. However, these equilibria exhibit the same behavior as all other equilibria. We will discuss some of these equilibria later in the text (Chapters 16 and 17).

Homogeneous equilibria among gases. Gaseous reactants that produce gaseous products also give homogeneous equilibria. We use molar concentrations in the expressions for the equilibrium constants in the following examples, but we will see shortly that partial pressures of the gases could be used as well.

$$C_2H_6(g) \rightleftharpoons C_2H_4(g) + H_2(g) \qquad K = \frac{[C_2H_4][H_2]}{[C_2H_6]}$$

$$3O_2(g) \rightleftharpoons 2O_3(g) \qquad K = \frac{[O_3]^2}{[O_2]^3}$$

$$N_2(g) + 3H_2(g) \rightleftharpoons 2NH_3(g) \qquad K = \frac{[NH_3]^2}{[N_2][H_2]^3}$$

$$C_3H_8(g) + 5O_2(g) \rightleftharpoons 3CO_2(g) + 4H_2O(g) \qquad K = \frac{[CO_2]^3[H_2O]^4}{[C_3H_8][O_2]^5}$$

Note that the concentration of water has been included in the last example because water is not a solvent in a gas-phase reaction.

Whenever gases are involved in a reaction, the pressure of each gas can be used (instead of concentration) in the equation for the reaction quotient or equilibrium constant because the pressure of a gas is directly proportional to its concentration at constant temperature. This relationship can be derived from the ideal gas equation (Section 10.6).

$$PV = nRT$$

The molar concentration of the gas, designated C, is the number of moles, n, in exactly 1 liter of the gas. Substituting 1 L for V, and rearranging the equation, we get

$$P(1\ L) = nRT$$
$$P = \left(\frac{n}{1\ L}\right)RT$$
$$= CRT$$

Because R is a constant, the quantity RT is a constant when the temperature does not change. Thus at constant temperature, the pressure of a gas is directly proportional to its concentration, or

$$P = \text{constant} \times C$$

For the system

$$C_2H_6 \rightleftharpoons C_2H_4 + H_2$$

we can write the reaction quotient, using the gases's partial pressures (Section 10.9).

$$Q_p = \frac{(P_{C_2H_4})(P_{H_2})}{P_{C_2H_6}}$$

In this equation we use Q_p to indicate a reaction quotient written with partial pressures: $P_{C_2H_6}$ is the partial pressure of C_2H_6, P_{H_2} is the partial pressure of H_2, and $P_{C_2H_4}$ is the partial pressure of C_2H_4. At equilibrium,

$$K_p = Q_p = \frac{(P_{C_2H_4})(P_{H_2})}{P_{C_2H_6}}$$

The subscript p in the symbol K_p designates an equilibrium constant found by using the partial pressures instead of concentrations. The equilibrium constant K_p is still a constant, but its numerical value and units may differ from those of the equilibrium constant found for the same reaction by using concentrations.

Conversion between a value for K, an equilibrium constant expressed in terms of concentrations, and a value for K_p, an equilibrium constant expressed in terms of pressures, is straightforward. The relationship between K and K_p is

$$K_p = K(RT)^{\Delta g}$$

where Δg is the difference between the sum of the coefficients of the *gaseous* products and the sum of the coefficients of the *gaseous* reactants in the reaction. For the gas-phase reaction

$$mA + nB \rightleftharpoons xC + yD$$

we have

$$\Delta g = (x + y) - (m + n)$$

The equation relating K and K_p is not hard to derive.

$$\begin{aligned}
K_p &= \frac{(P_C)^x(P_D)^y}{(P_A)^m(P_B)^n} \\
&= \frac{(C_C \times RT)^x(C_D \times RT)^y}{(C_A \times RT)^m(C_B \times RT)^n} \\
&= \frac{(C_C)^x(C_D)^y}{(C_A)^m(C_B)^n} \times \frac{(RT)^{x+y}}{(RT)^{m+n}} \\
&= K(RT)^{(x+y)-(m+n)} \\
&= K(RT)^{\Delta g}
\end{aligned}$$

Example 14.3   Calculation from K to K_p

Write the equations for the conversion of K to K_p for each of the following reactions.
(a) $C_4H_8(g) \rightleftharpoons 2C_2H_4(g)$
(b) $CO(g) + H_2O(g) \rightleftharpoons CO_2(g) + H_2(g)$
(c) $N_2(g) + 3H_2(g) \rightleftharpoons 2NH_3(g)$

Solution:
(a) $\Delta g = (2) - (1) = 1$
$\quad K_p = K(RT)^{\Delta g} = K(RT)^1 = K(RT)$
(b) $\Delta g = (1 + 1) - (1 + 1) = 0$
$\quad K_p = K(RT)^{\Delta g} = K(RT)^0 = K$
(c) $\Delta g = (2) - (1 + 3) = -2$
$\quad K_p = K(RT)^{\Delta g} = K(RT)^{-2} = K(RT)^{-2}$

Check your learning: Write the equations for the conversion of K to K_p for each of the following reactions, which occur in the gas phase.
(a) $2SO_2 + O_2 \rightleftharpoons 2SO_3$
(b) $N_2O_4 \rightleftharpoons 2NO_2$
(c) $C_3H_8 + 5O_2 \rightleftharpoons 3CO_2 + 4H_2O$

Answer: (a) $K_p = K(RT)^{-1}$; (b) $K_p = K(RT)$; (c) $K_p = K(RT)$

Heterogeneous Equilibria.

A **heterogeneous equilibrium** is an equilibrium in which reactants and products are found in two or more phases. The phases maybe any combination of solid phases, liquid phases, gas phases, and solutions. When dealing with these equilibria, remember that the activities of pure solids, pure liquids, and solvents are 1 (Section 14.2), so these species do not appear in reaction quotients or equilibrium constants.

Some heterogeneous equilibria involve chemical changes. Examples include

$$PbCl_2(s) \rightleftharpoons Pb^{2+}(aq) + 2Cl^-(aq) \qquad K = [Pb^{2+}][Cl^-]^2$$

$$CaO(s) + CO_2(g) \rightleftharpoons CaCO_3(s) \qquad K = \frac{1}{[CO_2]}$$

$$C(s) + 2S(g) \rightleftharpoons CS_2(g) \qquad K = \frac{[CS_2]}{[S]^2}$$

Other heterogeneous equilibria involve phase changes; an example is the evaporation of liquid bromine (Fig. 14.3).

$$Br_2(l) \rightleftharpoons Br_2(g) \qquad K = [Br_2(g)]$$

We can write equations for reaction quotients of heterogeneous equilibria that involve gases by using pressures instead of concentrations. Two examples are

$$CaO(s) + CO_2(g) \rightleftharpoons CaCO_3(s) \qquad K_p = \frac{1}{P_{CO_2}}$$

$$C(s) + 2S(g) \rightleftharpoons CS_2(g) \qquad K_p = \frac{P_{CS_2}}{(P_S)^2}$$

Some heterogeneous equilibria are so common that they have individual names and their equilibrium constants have special symbols. However, these equilibria exhibit the same behavior as all other equilibria. We will discuss some of these equilibria in Chapter 17.

14.4   Changes in Concentrations at Equilibrium: Le Châtelier's Principle

When a system is at equilibrium, we can change the **position of equilibrium** (that is, the amounts of reactants and products present) by changing the concentration of a reactant or product, the pressure in certain gaseous systems, or the temperature of the system. The result of each change can be predicted from **Le Châtelier's principle:** *When a stress* (such as a change in concentration, pressure, or temperature) *is applied to a system at equilibrium, the equilibrium shifts in a way that tends to minimize the stress.*

Effect of Change in Concentration on Equilibrium.

A chemical system at equilibrium can be shifted out of equilibrium by adding or removing one or more of the reactants or products. The concentrations of both reactants and products then undergo additional changes to return the system to equilibrium. Figure 14.5 shows a shift in equilibrium in a mixture of Fe^{3+}, SCN^-, and $Fe(SCN)^{2+}$,

$$Fe^{3+}(aq) + SCN^-(aq) \rightleftharpoons Fe(SCN)^{2+}(aq)$$

The stress on the system in the figure is the reduction of the equilibrium concentration of SCN^-. As a consequence, Le Châtelier's principle leads us to predict that the concentration of $Fe(SCN)^{2+}$ should decrease, increasing the concentration of SCN^- partway back to its original concentration, and increasing the concentration of Fe^{3+} above its initial equilibrium concentration.

The effect of a change in concentration on a system at equilibrium is illustrated further by the equilibrium

$$H_2(g) + I_2(g) \rightleftharpoons 2HI(g) \qquad K = 50.0 \text{ (at 400°C)}$$

Figure 14.5

(A) The test tube contains 0.1 M Fe^{3+}. (B) Thiocyanate ion has been added to the solution in part (A), forming the red $Fe(SCN)^{2+}$ ion.
$Fe^{3+}(aq) + SCN^-(aq) \rightleftharpoons$
$Fe(SCN)^{2+}(aq)$
(C) Silver nitrate has been added to the solution in part (B), precipitating some of the SCN^- as the white $AgSCN(s)$ at the bottom of the tube.
$Ag^+(aq) + SCN^-(aq) \rightleftharpoons AgSCN(s)$
The decrease in the SCN^- concentration shifts the equilibrium in the solution to the left, decreasing the concentration (and color) of the red $Fe(SCN)^{2+}$.

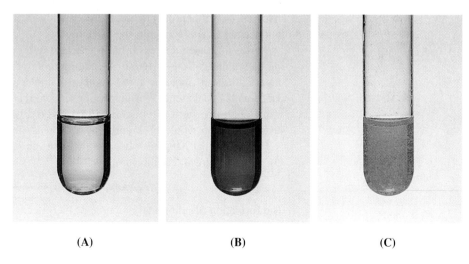

(A)   (B)   (C)

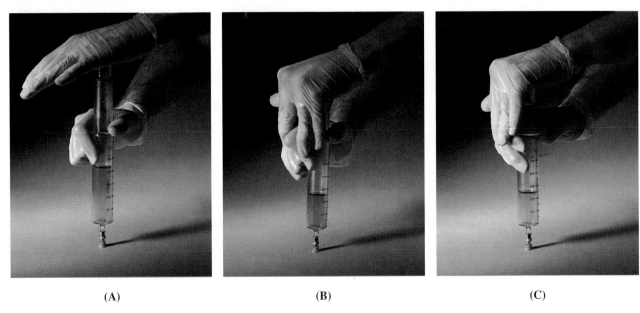

(A)   (B)   (C)

Figure 14.6

The effect of pressure on the equilibrium between the gas NO_2 and the gas N_2O_4: $2NO_2 \rightleftharpoons N_2O_4$. (A) Brown NO_2 and colorless N_2O_4 at equilibrium in a closed syringe. (B) A sudden decrease in volume increases the concentration of both NO_2 and N_2O_4. The color is darker before equilibrium is reestablished because the NO_2 concentration is higher. (C) After equilibrium is reestablished, the concentration of NO_2 has decreased (so the color is lighter) and the amount of N_2O_4 increased from that in part (B) as predicted by Le Châtelier's principle.

A mixture of gases at 400°C with $[H_2] = [I_2] = 0.221$ M and $[HI] = 1.563$ M is at equilibrium; for this mixture, $Q = K = 50.0$. If H_2 is introduced into the system so quickly that its concentration doubles before it begins to react (new $[H_2] = 0.442$ M), the reaction will shift so that a new equilibrium is reached, at which $[H_2] = 0.374$ M, $[I_2] = 0.153$ M, and $[HI] = 1.692$ M. This gives

$$Q = \frac{[HI]^2}{[H_2][I_2]} = \frac{(1.692)^2}{(0.374)(0.153)} = 50.0 = K$$

We have stressed this system by introducing additional H_2. The stress is relieved when the reaction shifts to the right, using up some (but not all) of the excess H_2, reducing the amount of uncombined I_2, and forming additional HI.

Effect of Change in Pressure on Equilibrium.

Sometimes we can change the position of equilibrium by changing the pressure on a system (Fig. 14.6). However, changes in pressure have a measurable effect only in systems where gases are involved—and then only when the chemical reaction produces a change in the total number of gas molecules in the system.

As we increase the pressure of a gaseous system at equilibrium, either by decreasing the volume of the system or by adding more of the equilibrium mixture, we introduce a stress by increasing the number of molecules per unit of volume. In accordance with Le Châtelier's principle, a chemical reaction that reduces the total number of molecules per unit of volume will be favored because this relieves the stress. The reverse reaction would be favored by a decrease in pressure.

Consider what happens when we increase the pressure on a system in which NO, O_2, and NO_2 are in equilibrium.

$$2NO(g) + O_2(g) \rightleftharpoons 2NO_2(g)$$

The formation of additional amounts of NO_2 decreases the total number of molecules in the system, because each time two molecules of NO_2 form, a total of three molecules of NO and O_2 react. This reduces the total pressure exerted by the system and reduces, but does not completely relieve, the stress of the increased pressure. On the other hand, a decrease in the pressure on the system favors decomposition of NO_2 into NO and O_2, which tends to restore the pressure.

Let us now consider the reaction

$$N_2(g) + O_2(g) \rightleftharpoons 2NO(g)$$

Because there is no change in the total number of molecules in the system during reaction, a change in pressure does not favor either formation or decomposition of gaseous nitric oxide.

Effect of Change in Temperature on Equilibrium.   Changing concentration or pressure upsets an equilibrium because the reaction quotient is shifted away from the equilibrium value. Changing the temperature of a system at equilibrium has a different effect: *A change in temperature changes the value of the equilibrium constant.* However, we can predict the effect of the temperature change by treating it as a stress on the system and applying Le Châtelier's principle.

When hydrogen reacts with gaseous iodine, energy is released as heat is evolved.

$$H_2(g) + I_2(g) \rightleftharpoons 2HI(g) \qquad \Delta H = -9.4 \text{ kJ (exothermic)}$$

Because this reaction is exothermic, we can write it with heat as a product.

$$H_2(g) + I_2(g) \rightleftharpoons 2HI(g) + 9.4 \text{ kJ}$$

Increasing the temperature of the reaction increases the amount of energy present. Thus increasing the temperature has the effect of increasing the amount of one of the products of this reaction. The reaction shifts to the left to relieve the stress, and there is an increase in the concentration of H_2 and I_2 and a reduction in the concentration of HI. Lowering the temperature of this system reduces the amount of energy present, favors the production of heat, and favors the formation of hydrogen iodide.

When we change the temperature of a system at equilibrium, the equilibrium constant for the reaction changes. Lowering the temperature in the HI system increases the equilibrium constant from 50.0 at 400°C to 67.5 at 357°C. At equilibrium at the lower temperature, the concentration of HI has increased and the concentrations of H_2 and I_2 have decreased. Raising the temperature decreases the value of the equilibrium constant from 67.5 at 357°C to 50.0 at 400°C.

The effect of temperature on the equilibrium between NO_2 and N_2O_4 may be seen in Fig. 14.7.

$$2NO_2(g) \rightleftharpoons N_2O_4(g) \qquad \Delta H = -57.20 \text{ kJ}$$

Figure 14.7

The effect of temperature on the equilibrium between the gases NO_2 and N_2O_4.
$2NO_2 \rightleftharpoons N_2O_4 \quad \Delta H = -57.20 \text{ kJ}$
(an exothermic reaction)
The brown color of the NO_2 is much fainter at the lower temperature (right) than at the higher temperature (left). This is because the equilibrium shifts from NO_2 to colorless N_2O_4 at the lower temperature.

The negative ΔH value tells us that the reaction is exothermic and could be written

$$2NO_2(g) \rightleftharpoons N_2O_4(g) + 57.20 \text{ kJ}$$

At higher temperatures, the gas mixture has a deep brown color, which indicates a significant amount of brown NO_2 molecules. If, however, we put a stress on the system by cooling the mixture (withdrawing energy), the equilibrium shifts to the right to supply some of the energy lost by cooling. The concentration of colorless N_2O_4 increases, and the concentration of brown NO_2 decreases, causing the brown color to fade.

Effect of a Catalyst on Equilibrium.

A catalyst has no effect on the value of an equilibrium constant or on equilibrium concentrations. The catalyst merely increases the rates of both the forward and the reverse reactions to the same extent so that equilibrium is reached more rapidly.

All of these effects—change in concentration or pressure, change in temperature, and the effect of a catalyst on a chemical equilibrium—play a role in the industrial synthesis of ammonia from nitrogen and hydrogen according to the equation

$$N_2 + 3H_2 \rightleftharpoons 2NH_3$$

A great deal of ammonia is manufactured by this reaction (Section 8.9). Each year ammonia is among the top ten chemicals, by mass, manufactured in the world. About 3.4 billion pounds are manufactured in the United States each year.

Ammonia plays a vital role in our global economy. It is used in the production of fertilizers and is, itself, a fertilizer for the growth of corn, cotton, and other crops (Fig. 14.8). Large quantities of ammonia are converted to nitric acid, which plays an important role in the production of fertilizers, explosives, plastics, dyes, and fibers and is also used in the steel industry.

It has been known for many years that nitrogen and hydrogen react to form ammonia. However, it became possible to manufacture ammonia in useful quantities by this reaction only in about 1915, after the factors that influence its equilibrium were understood. Fritz Haber, a German chemist, received the 1918 Nobel prize in chemistry for work in which he used equilibrium concepts to develop a means of synthesizing ammonia on a commercial scale.

To be practical, an industrial process must give a large yield of product relatively quickly, particularly when billions of pounds are needed. One way to increase the yield of ammonia is to increase the pressure on the system in which N_2, H_2, and NH_3 are in equilibrium or are coming to equilibrium.

$$N_2(g) + 3H_2(g) \rightleftharpoons 2NH_3(g)$$

Figure 14.8
Gaseous ammonia is injected into the soil to act as a fertilizer.

The formation of additional amounts of ammonia reduces the total pressure exerted by the system and somewhat reduces the stress of the increased pressure.

Although increasing the pressure of a mixture of N_2, H_2, and NH_3 increases the yield of ammonia, at low temperatures the rate of formation of ammonia is slow. At room temperature, for example, the reaction is so slow that if we prepared a mixture of N_2 and H_2, no detectable amount of ammonia would form during our lifetime. Attempts to increase the rate of the reaction by increasing the temperature are counterproductive. The formation of ammonia from hydrogen and nitrogen is an exothermic process:

$$N_2(g) + 3H_2(g) \longrightarrow 2NH_3(g) \qquad \Delta H = -92.2 \text{ kJ}$$

Figure 14.9

Commercial production of ammonia
requires heavy equipment to handle
the high temperatures and pressures
needed.

Thus increasing the temperature to increase the rate lowers the yield. If we lower the temperature to shift the equilibrium to the right to favor the formation of more ammonia, equilibrium is reached more slowly because of the large decrease of reaction rate with decreasing temperature. If we increase the temperature to increase the rate, equilibrium is reached more rapidly, but the yield is reduced because the equilibrium shifts to the left.

Part of the rate of formation lost by operating at lower temperatures can be recovered by using a catalyst to increase the reaction rate. Iron powder is one catalyst used. However, as we have seen, a catalyst serves equally well to increase the rate of a reverse reaction—in this case, the decomposition of ammonia into its constituent elements. Thus the net effect of the iron catalyst on the reaction is to cause equilibrium to be reached more rapidly.

In the commercial production of ammonia, it is not feasible to use temperatures much lower than 500°C. At lower temperatures the reaction proceeds too slowly to be practical, even in the presence of a catalyst. Conditions of about 500°C and 150–900 atmospheres are selected to give the best compromise among rate, yield, and the cost of the equipment necessary to produce and contain gases at high pressure and high temperatures (Fig. 14.9).

14.5   Predicting the Direction of a Reversible Reaction

Le Châtelier's principle can be used to predict changes in equilibrium concentrations when a system that is at equilibrium is subjected to a stress. However, if we have a mixture of reactants and products that have not yet reached equilibrium, the changes necessary to reach equilibrium may not be so obvious. In such a case we can compare the values of Q and K for the system to predict the changes.

We have seen that if the reversible reaction

$$mA + nB + \cdots \rightleftharpoons xC + yD + \cdots \qquad Q = \frac{[C]^x[D]^y \cdots}{[A]^m[B]^n \cdots}$$

is not at equilibrium, then the reaction quotient is not equal to the equilibrium constant. As the reaction comes to equilibrium, the value of Q changes so that it becomes equal to K. Thus we can predict the direction of the changes in concentrations as a reaction comes to equilibrium by comparing Q with K. If Q is less than K ($Q < K$), the amounts of reactants (A and B) will decrease and the amounts of products (C and D) will increase until Q is equal to K. We say that the reaction *shifts to the right* or *proceeds toward products*. If Q is greater than K ($Q > K$), the amounts of reactants will increase and the amounts of products will decrease until Q is equal to K. We say that the reaction *shifts to the left* or *proceeds toward reactants*. When the reaction reaches equilibrium, Q is equal to K and no change in concentrations occurs.

Example 14.4   **Predicting the Direction of Shift**

Given below are the starting concentrations of reactants and products for three experiments involving the reaction

$$CO(g) + H_2O(g) \rightleftharpoons CO_2(g) + H_2(g) \qquad K = 0.64$$

Determine in which direction the reaction shifts as it goes to equilibrium in each.

	Experiment 1	Experiment 2	Experiment 3
$[CO]_i$	0.0203 M	0.011 M	0.0094 M
$[H_2O]_i$	0.0203 M	0.0011 M	0.0025 M
$[CO_2]_i$	0.0040 M	0.037 M	0.0015 M
$[H_2]_i$	0.0040 M	0.046 M	0.0076 M

Solution:
Experiment 1

$$Q = \frac{[CO_2][H_2]}{[CO][H_2O]} = \frac{(0.0040)(0.0040)}{(0.0203)(0.0203)} = 0.039$$

Because $Q < K$ (0.039 < 0.64); the reaction will shift to the right.

Experiment 2

$$Q = \frac{[CO_2][H_2]}{[CO][H_2O]} = \frac{(0.037)(0.046)}{(0.011)(0.0011)} = 1.4 \times 10^2$$

Because $Q > K$ (140 > 0.64), the reaction will shift to the left.

Experiment 3

$$Q = \frac{[CO_2][H_2]}{[CO][H_2O]} = \frac{(0.0015)(0.0076)}{(0.0094)(0.0025)} = 0.48$$

Because $Q < K$ (0.48 < 0.64), the reaction will shift to the right.

Check your learning: Calculate the reaction quotient and determine the direction in which each of the following systems will shift to reach equilibrium.

(a) A 1.00-L flask containing 0.0500 mol of $NO(g)$, 0.0155 mol of $Cl_2(g)$, and 0.500 mol of $NOCl(g)$.

$$2NO(g) + Cl_2(g) \rightleftharpoons 2NOCl(g) \qquad K = 4.6 \times 10^4 \text{ mol}^{-1}$$

(b) A 5.0-L flask containing 17 g of NH_3, 14 g of N_2, and 12 g of H_2.

$$N_2(g) + 3H_2(g) \rightleftharpoons 2NH_3(g) \qquad K = 0.060 \text{ mol}^{-2}$$

(c) A 2.00-L flask containing 230 g of $SO_3(g)$.

$$2SO_3(g) \rightleftharpoons 2SO_2(g) + O_2(g) \qquad K = 0.230 \text{ mol}$$

Answer: **(a)** $Q = 6.45 \times 10^3$, shifts right; **(b)** $Q = 0.23$, shifts left; **(c)** $Q = 0$, shifts right

We can now see why reactions reach equilibrium only if the reactants and products cannot escape from the reaction mixture. For example, in an open container, calcium carbonate (limestone, $CaCO_3$) may be completely converted to calcium oxide, CaO, and carbon dioxide, CO_2, when heated at 900°C (Fig. 14.10).

$$CaCO_3(s) \rightleftharpoons CaO(s) + CO_2(g) \qquad Q_p = P_{CO_2}$$

When calcium carbonate is first exposed to heat, there is essentially no carbon dioxide present, and the equilibrium quotient is less than the equilibrium constant. Thus the reaction shifts to the right, producing calcium oxide and carbon dioxide. In an open container, the gaseous carbon dioxide escapes from the reaction mixture, so the value of the reaction quotient is *always* less than the value of the equilibrium constant. The reaction keeps producing products as long as the carbon dioxide escapes from the mixture (and as long as any $CaCO_3$ remains), because Q_p never reaches the value of K_p. In a closed container, on the other hand, the value of Q_p increases until the reaction reaches equilibrium because the carbon dioxide cannot escape.

Figure 14.10

Heating calcium carbonate, $CaCO_3$, in a container open to air produces lime (calcium oxide), CaO, and carbon dioxide, CO_2. This photograph shows a kiln used for commercial production of lime.

Kinetics and Equilibria

Now that we have introduced equilibria, we are in a position to revisit chemical kinetics and consider the relationships between rates and equilibria.

14.6   The Relationship of Reaction Rates and Equilibria

We have seen that equilibrium is established when the reactants in a system combine to form products at the same rate at which the products combine to form reactants. In the simplest case, and only for a reaction with a mechanism that consists of a single elementary step, we can readily derive the equilibrium constant from the rates of

Figure 14.11

Rates of reaction of forward and reverse reactions for $NO_2(g) + CO(g) \rightleftharpoons NO(g) + CO_2(g)$, with only NO_2 and CO present initially.

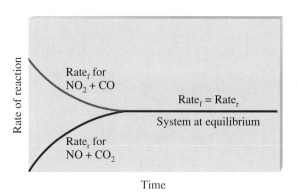

its forward and reverse reactions at equilibrium. We will use the oxidation of carbon monoxide by nitrogen dioxide to illustrate the relationship.

In a closed flask at 300°C, at the beginning of the reaction

$$NO_2(g) + CO(g) \rightleftharpoons NO(g) + CO_2(g)$$

the flask contains only NO_2 and CO. At this temperature the rate of the forward reaction, $rate_f$, (Section 13.9) is given by the expression

$$Rate_f = k_f[NO_2][CO]$$

At first no molecules of NO and CO_2 are present, so there can be no reverse reaction to re-form NO_2 and CO. The rate of the reverse reaction, $rate_r$, is zero (Fig. 14.11). However, as soon as some of the products are formed, they begin to react to re-form the reactants. Initially the rate of re-forming reactants is slow because the concentrations of products are low. However, as the concentrations of products increase, the rate of the reverse reaction, $rate_r$, increases. The rate of this reaction is given by

$$Rate_r = k_r[NO][CO_2]$$

Meanwhile, the concentrations of NO_2 and CO decrease, so the rate of the forward reaction, $rate_f$, decreases. Consequently, the two reaction rates approach each other and finally become equal. When $rate_f = rate_r$, equilibrium is established, and we can write

$$rate_f = rate_r$$
$$k_f[NO_2][CO] = k_r[NO][CO_2]$$

or, after rearrangement,

$$\frac{k_f}{k_r} = \frac{[NO][CO_2]}{[NO_2][CO]}$$

Because k_f and k_r are constants, the ratio k_f/k_r is also constant, and the expression may be written

$$K = \frac{k_f}{k_r} = \frac{[NO][CO_2]}{[NO_2][CO]}$$

K is the equilibrium constant for the reaction. Just as the rate constants k_f and k_r are specific for each reaction at a definite temperature, K is likewise a constant specific to this system in equilibrium at a given temperature.

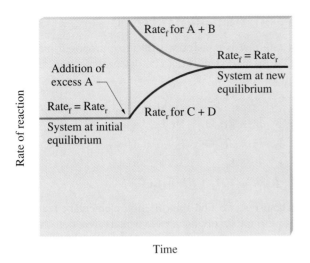

(A)

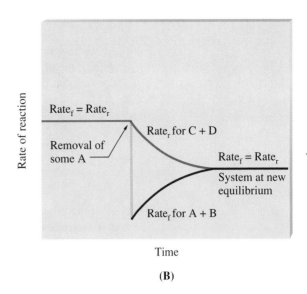

(B)

Figure 14.12

(A) The effect of adding reactant A on the rates of the forward and reverse reactions for A + B $\rightleftharpoons$ C + D when the system is initially at equilibrium. (B) The effect of removing some reactant A on the rates of the forward and reverse reactions for A + B $\rightleftharpoons$ C + D when the system is initially at equilibrium.

Addition of a reactant or product to a reaction at equilibrium causes a shift in the reaction, because the equality of the rates of the forward and reverse reactions is upset. For example, when the system

$$A + B \rightleftharpoons C + D$$

is in equilibrium and an additional quantity of either A or B (or both) is added, the rate of the forward reaction increases because the concentration of the reacting molecules increases (Fig. 14.12 A). The rate of the forward reaction becomes faster than that of the reverse reaction, so the system is out of equilibrium. However, as the concentrations of C and D increase, the rate of the reverse reaction increases, while the decrease in the concentrations of A and B causes the rate of the forward reaction to decrease. The rates of the two reactions become equal again, and a second state of equilibrium is reached. Although the concentrations of A, B, C, and D have changed, the reaction quotient [C][D]/[A][B] is again equal to its original value of K.

A system may be shifted out of equilibrium by decreasing the rate of the forward or the reverse reaction (Fig. 14.12 B).

14.7   Reaction Mechanisms Involving Equilibria

We have seen that if the slow step is the first step in a reaction mechanism, then the rate law for the overall reaction is the same as the rate law for this step (Section 13.10). However, when the rate-determining step occurs later in the mechanism, the rate law for the overall reaction may be complex. In many cases, there may be a fast step that comes to equilibrium before the slow step is reached, and we must use that equilibrium in deriving the rate law.

The oxidation of iodide ion by hydrogen peroxide illustrates this point.

$$H_2O_2 + 3I^- + 2H_3O^+ \longrightarrow 4H_2O + I_3^-$$

In a solution with a high concentration of acid, one reaction pathway has the following experimental rate law:

$$\text{Rate} = k[H_2O_2][I^-][H_3O^+]$$

As we saw in Section 13.10, this rate law is consistent with three proposed mechanisms. One of these mechanisms has the first step as the rate-determining step, so the rate law can be written by inspection.

Mechanism A

$$H_2O_2 + H_3O^+ + I^- \longrightarrow 2H_2O + HOI \qquad \text{(slow)}$$
$$HOI + H_3O^+ + I^- \longrightarrow 2H_2O + I_2 \qquad \text{(fast)}$$
$$I_2 + I^- \longrightarrow I_3^- \qquad \text{(fast)}$$

Proposed rate law: $\text{Rate} = k[H_2O_2][I^-][H_3O^+]$

In the other two proposed mechanisms, we find that the first elementary reaction is a fast **pre-equilibrium,** an equilibrium involving an elementary reaction that is established before the rate-determining step. The rate-determining step is the second elementary reaction. If the pre-equilibrium is established at a rate that is fast relative to that of the slow step, then we can derive the rate laws for these mechanisms. We calculate the concentrations of the products in the pre-equilibrium by using the equation for the reaction quotient and then use these concentrations in the rate law for the rate-determining step.

Example 14.5   Rate Law for a Mechanism with a Fast Pre-equilibrium

Determine the rate for Mechanism B (Section 13.10) for the oxidation of iodide ion by hydrogen peroxide in a strongly acidic solution.

$$H_3O^+ + I^- \rightleftharpoons HI + H_2O \qquad \text{(fast equilibrium)}$$
$$H_2O_2 + HI \longrightarrow H_2O + HOI \qquad \text{(slow)}$$
$$HOI + H_3O^+ + I^- \longrightarrow 2H_2O + I_2 \qquad \text{(fast)}$$
$$I_2 + I^- \longrightarrow I_3^- \qquad \text{(fast)}$$

Solution: If this is the mechanism for the reaction of I^- and H_2O_2 in acid solution, then the rate of disappearance of H_2O_2 will be equal to the rate of the slow step.

$$H_2O_2 + HI \longrightarrow H_2O + HOI \qquad \text{(slow)}$$

The rate law for this step may be written by inspection.

$$\text{Rate} = k[H_2O_2][HI]$$

HI is a strong acid and is present in very low concentration. We cannot measure [HI], but we can determine [HI] from $[H_3O^+]$ and $[I^-]$ through the pre-equilibrium that is the first step of the mechanism. For the equilibrium

$$H_3O^+ + I^- \rightleftharpoons HI + H_2O \qquad \text{(fast equilibrium)}$$

we have the equilibrium constant

$$K = \frac{[HI]}{[H_3O^+][I^-]}$$

This can be solved for [HI]

$$[HI] = K[H_3O^+][I^-]$$

Substituting into this rate law gives

$$\text{Rate} = k[\text{H}_2\text{O}_2][\text{HI}]$$
$$= kK[\text{H}_2\text{O}_2][\text{H}_3\text{O}^+][\text{I}^-]$$
$$= k'[\text{H}_2\text{O}_2][\text{I}^-][\text{H}_3\text{O}^+]$$

Thus Mechanism B leads to a proposed rate law that is first-order in H_2O_2, I^-, and H_3O^+, as observed for the reaction.

Check your learning: A third mechanism proposed for the oxidation of iodide ion by hydrogen peroxide in a strongly acidic solution is Mechanism C (Section 13.10)

$$\text{H}_3\text{O}^+ + \text{H}_2\text{O}_2 \rightleftharpoons \text{H}_3\text{O}_2^+ + \text{H}_2\text{O} \quad \text{(fast equilibrium)}$$
$$\text{H}_3\text{O}_2^+ + \text{I}^- \longrightarrow \text{H}_2\text{O} + \text{HOI} \quad \text{(slow)}$$
$$\text{HOI} + \text{H}_3\text{O}^+ + \text{I}^- \longrightarrow 2\text{H}_2\text{O} + \text{I}_2 \quad \text{(fast)}$$
$$\text{I}_2 + \text{I}^- \longrightarrow \text{I}_3^- \quad \text{(fast)}$$

Write the rate law for the slow step of this mechanism, then determine an equation for the concentration of H_3O_2^+ from the pre-equilibrium, and finally determine the rate law for the slow step in terms of the concentrations of H_2O_2, I^-, and H_3O^+.

Answer: Rate $= k[\text{H}_3\text{O}_2^+][\text{HI}]$; $[\text{H}_3\text{O}_2^+] = K[\text{H}_3\text{O}^+][\text{H}_2\text{O}_2]$; Rate $= k'[\text{H}_2\text{O}_2][\text{I}^-][\text{H}_3\text{O}^+]$

Equilibrium Calculations

We know that at equilibrium, the value of the reaction quotient of any reaction is equal to its equilibrium constant. Thus we can use the mathematical expression for Q to determine a number of quantities associated with a reaction at equilibrium or approaching equilibrium. However, before we consider how to do these calculations, we need to consider how changes in the concentration or pressure of one species is related to changes in the concentration or pressure of another.

14.8 Concentration and Pressure Changes

Changes in concentrations or pressures of reactants and products occur as a system approaches equilibrium. In this section, we will see that we can relate these changes to each other by using the coefficients in the balanced chemical equation that describes the system. We use the decomposition of ammonia as an example.

Upon heating, ammonia reversibly decomposes into nitrogen and hydrogen according to the equation

$$2\text{NH}_3(g) \rightleftharpoons \text{N}_2(g) + 3\text{H}_2(g)$$

If a sample of ammonia decomposes and the concentration of N_2 increases by 0.11 M, then the change in the N_2 concentration, $\Delta[\text{N}_2]$, is 0.11 M. The change is positive because the concentration of N_2 increases.

The change in the H_2 concentration, $\Delta[H_2]$, is also positive—the concentration of H_2 increases as ammonia decomposes. The chemical equation tells us that the change in the concentration of H_2 is three times the change in the concentration of N_2, because for each mole of N_2 that is produced, three moles of H_2 are produced.

$$\Delta[H_2] = 3 \times \Delta[N_2]$$
$$= 3 \times (0.11 \text{ M}) = 0.33 \text{ M}$$

The change in concentration of NH_3, $\Delta[NH_3]$, is twice that of $\Delta[N_2]$; the equation indicates that two moles of NH_3 must decompose for each mole of N_2 formed. However, the change in the NH_3 concentration is negative because the concentration of ammonia *decreases* as it decomposes.

$$\Delta[NH_3] = -2 \times \Delta[N_2] = -2 \times (0.11 \text{ M}) = -0.22 \text{ M}$$

We can relate these factors directly to the coefficients in the equation

$$2NH_3(g) \rightleftharpoons N_2(g) + 3H_2(g)$$
$$\Delta[NH_3] = -2 \times \Delta[N_2] \qquad \Delta[N_2] = 0.11 \text{ M} \qquad \Delta[H_2] = 3 \times \Delta[N_2]$$

Note that all the changes on one side of the arrows are of the same sign and that all the changes on the other side of the arrows are of the opposite sign.

If we did not know the magnitude of the change in the concentration of N_2, we could represent it by the symbol Δ.

$$\Delta[N_2] = \Delta$$

The changes in the other concentrations would then be represented

$$\Delta[H_2] = 3 \times \Delta[N_2] = 3\Delta$$
$$\Delta[NH_2] = -2 \times \Delta[N_2] = -2\Delta$$

The coefficients in the Δ terms are identical to those in the balanced equation for the reaction.

$$2NH_3(g) \rightleftharpoons N_2(g) + 3H_2(g)$$
$$\text{Changes: } -2\Delta \qquad \Delta \qquad 3\Delta$$

The simplest way for us to find the coefficients for the Δ values in any reaction is to use the coefficients in the balanced chemical equation. The sign of the coefficient is positive when the concentration increases; it is negative when the concentration decreases.

Example 14.6   Determining Relative Changes in Concentration

Complete the changes in concentrations for the following reactions on the lines below.

(a) $C_2H_2 + 2Br_2 \rightleftharpoons C_2H_2Br_4$
   $\quad \Delta \qquad \underline{\quad} \qquad \underline{\quad}$

(b) $I_2(aq) + I^-(aq) \rightleftharpoons I_3^-(aq)$
   $\quad \underline{\quad} \qquad \underline{\quad} \qquad \Delta$

(c) $C_3H_8(g) + 5O_2(g) \rightleftharpoons 3CO_2(g) + 4H_2O(g)$
   $\quad \Delta \qquad \underline{\quad} \qquad \underline{\quad} \qquad \underline{\quad}$

Solution:

(a) $C_2H_2 + 2Br_2 \rightleftharpoons C_2H_2Br_4$

$\quad\quad \Delta \quad\quad 2\Delta \quad\quad\quad -\Delta$

(b) $I_2(aq) + I^-(aq) \rightleftharpoons I_3^-(aq)$

$\quad\quad -\Delta \quad\quad -\Delta \quad\quad\quad \Delta$

(c) $C_3H_8(g) + 5O_2(g) \rightleftharpoons 3CO_2(g) + 4H_2O(g)$

$\quad\quad \Delta \quad\quad\quad 5\Delta \quad\quad\quad -3\Delta \quad\quad -4\Delta$

Check your learning: Complete the changes in concentrations for each of the following reactions.

(a) $2SO_2 + O_2 \rightleftharpoons 2SO_3$

$\quad\quad \underline{\quad} \quad\quad \Delta \quad\quad \underline{\quad}$

(b) $C_4H_8 \rightleftharpoons 2C_2H_4$

$\quad\quad \underline{\quad} \quad\quad -2\Delta$

(c) $4NH_3(g) + 7O_2(g) \rightleftharpoons 4NO_2(g) + 6H_2O(g)$

$\quad\quad \underline{\quad} \quad\quad 7\Delta \quad\quad\quad \underline{\quad} \quad\quad \underline{\quad}$

Answer: **(a)** 2Δ, Δ, -2Δ; **(b)** Δ, -2Δ; **(c)** 4Δ, 7Δ, -4Δ, -6Δ

14.9   Calculations Involving Equilibrium Concentrations

Because the value of the reaction quotient of any reaction at equilibrium is equal to its equilibrium constant, we can use the mathematical expression for Q to determine a number of quantities associated with a reaction at equilibrium. It may help if we keep in mind that $Q = K$ (at equilibrium) in all of these situations and that there are only three basic types of calculations:

1. *Calculation of an Equilibrium Constant.* When we know the concentrations of all reactants and products at equilibrium, we can determine the equilibrium constant for the reaction.

2. *Calculation of Missing Equilibrium Concentrations.* When we know the equilibrium constant, and either we know all of the equilibrium concentrations except one, or we know the concentrations of the reactants and products before equilibrium and at least one of the equilibrium concentrations, we can calculate the other concentrations at equilibrium.

3. *Calculation of Equilibrium Concentrations from Initial Concentrations.* When all we know is the equilibrium constant and a set of concentrations of reactants and products that are not at equilibrium, we can calculate the changes in concentrations as the system comes to equilibrium and, from those, the new concentrations at equilibrium.

We will consider the first two types of calculations in this section and save the third type for later. Calculations of the third type can be somewhat more involved.

Calculation of an Equilibrium Constant.

Generally, we can recognize that we must do this type of calculation when we need an equilibrium constant and are not given one. However, we will be given equilibrium concentrations or enough information to determine the equilibrium concentrations from the initial concentrations.

Example 14.2 showed how we determine the equilibrium constant of a reaction if we know the concentrations of reactants and products at equilibrium. The following example is a bit more complicated: It shows us how to use the stoichiometry of the reaction and a combination of initial concentrations and equilibrium concentrations to determine an equilibrium constant.

Example 14.7   Calculation of an Equilibrium Constant

Iodine molecules react reversibly with iodide ions to produce triiodide ions.

$$I_2(aq) + I^-(aq) \rightleftharpoons I_3^-(aq)$$

If a solution with the concentrations of I_2 and I^- both equal to 1.000×10^{-3} M before reaction gives an equilibrium concentration of I_2 of 6.61×10^{-4} M, what is the equilibrium constant for the reaction?

Solution: We begin by calculating the changes in concentrations as the system comes to equilibrium. Then we calculate the equilibrium concentrations and, finally, the equilibrium constant. First we set up a table with the initial concentrations, the changes in concentrations, and the final concentrations using $-\Delta$ as the change in concentration of I_2.

	I_2	+	I^-	$\rightleftharpoons$	I_3^-
Initial **Concentration, M**	1.000×10^{-3}		1.000×10^{-3}		0
Change, M	$-\Delta$		$-\Delta$		$+\Delta$
Equilibrium **Concentration, M**	$[I_2]_i + (-\Delta)$		$[I^-]_i + (-\Delta)$		$[I_2]_i + \Delta$

We find Δ from $[I_2]_i$, the initial concentration of I_2, and $[I_2]$, its final concentration.

$$-\Delta = \Delta[I_2] = [I_2] - [I_2]_i$$

$$= 6.61 \times 10^{-4} \text{ M} - 1.000 \times 10^{-3} \text{ M}$$

$$= -3.39 \times 10^{-4} \text{ M}$$

Now we can fill in the table with the concentrations at equilibrium.

	I_2	+	I^-	$\rightleftharpoons$	I_3^-
Initial **Concentration, M**	1.000×10^{-3}		1.000×10^{-3}		0
Change, M	$-\Delta = -3.39 \times 10^{-4}$		$-\Delta$		$+\Delta$
Equilibrium **Concentration, M**	6.61×10^{-4}		6.61×10^{-4}		3.39×10^{-4}

We now calculate the value of the equilibrium constant.

$$K = Q = \frac{[I_3^-]}{[I_2][I^-]}$$

$$= \frac{3.39 \times 10^{-4}\,M}{(6.61 \times 10^{-4}\,M)(6.61 \times 10^{-4}\,M)}$$

$$= 776\,M^{-1} = 776\,(mol\,L^{-1})^{-1} = 776\,mol^{-1}\,L$$

Check your learning: Ethanol and acetic acid react and form water and ethyl acetate, the solvent responsible for the odor of some nail polish removers.

$$C_2H_5OH + CH_3CO_2H \rightleftharpoons CH_3CO_2C_2H_5 + H_2O$$

When 1 mol each of C_2H_5OH and CH_3CO_2H are allowed to react in 1 L of the solvent dioxane, equilibrium is established when one-third of a mole of each of the reactants remains. Calculate the equilibrium constant for the reaction. Note: Water is not a solvent in this reaction.

Answer: K = 4

Calculation of a Missing Equilibrium Concentration.

When we know the equilibrium constant for a reaction and know the concentrations at *equilibrium* of all reactants and products except one, we can calculate the missing concentration.

Example 14.8 Calculation of a Missing Equilibrium Concentration

Nitrogen oxides are air pollutants that are produced by the reaction of nitrogen and oxygen at high temperatures. At 2000°C the equilibrium constant for the reaction

$$N_2(g) + O_2(g) \rightleftharpoons 2NO(g)$$

is 4.1×10^{-4}. Find the concentration of $NO(g)$ in an equilibrium mixture with air at 1 atm pressure at this temperature. In air, $[N_2] = 0.036$ mol L^{-1} and $[O_2] = 0.0089$ mol L^{-1}.

Solution: We are given all the equilibrium concentrations except that of NO. Thus we can solve for the missing equilibrium concentration by rearranging the equation for the equilibrium constant.

$$K = Q = \frac{[NO]^2}{[N_2][O_2]}$$

$$[NO]^2 = K[N_2][O_2]$$

$$[NO] = \sqrt{K[N_2][O_2]}$$

$$= \sqrt{(4.1 \times 10^{-4})(0.036\,mol\,L^{-1})(0.0089\,mol\,L^{-1})}$$

$$= \sqrt{1.31 \times 10^{-7}\,(mol\,L^{-1})^2}$$

$$= 3.6 \times 10^{-4}\,mol\,L^{-1}$$

Thus [NO] is 3.6×10^{-4} mol L^{-1} at equilibrium under these conditions.

We can check our answer by substituting all equilibrium concentrations into the expression for the reaction quotient to see whether it is equal to the equilibrium constant.

$$K = Q = \frac{[NO]^2}{[N_2][O_2]}$$

$$= \frac{(3.6 \times 10^{-4} \text{ mol L}^{-1})^2}{(0.036 \text{ mol L}^{-1})(0.0089 \text{ mol L}^{-1})}$$

$$= 4.0 \times 10^{-4}$$

The answer checks. Our calculated value gives the equilibrium constant within the error associated with the significant figures in the problem.

Check your learning: The equilibrium constant for the reaction of nitrogen and hydrogen to produce ammonia at a certain temperature is $6.00 \times 10^{-2} \text{ mol}^{-2} \text{ L}^2$. Calculate the equilibrium concentration of ammonia if the equilibrium concentrations of nitrogen and hydrogen are 4.26 M and 2.09 M, respectively.

Answer: 1.53 mol L^{-1}

14.10   Calculation of Changes in Concentration

If we know the equilibrium constant for a reaction and a set of concentrations of reactants and products that are *not at equilibrium,* we can calculate the changes in concentrations as the system comes to equilibrium as well as the new concentrations at equilibrium. The typical procedure can be summarized in four steps.

Step 1.   *Determine in what direction the reaction shifts to come to equilibrium.*

    a.   Write a balanced chemical equation for the reaction.

    b.   If the direction in which the reaction must shift to reach equilibrium is not obvious, calculate Q from the initial concentrations in order to determine the direction of shift.

Step 2.   *Determine the relative changes needed to reach equilibrium, and then write the equilibrium concentrations in terms of these changes.*

    a.   Define the changes in the initial concentrations that are needed for the reaction to reach equilibrium. Generally, we represent the *smallest* change with the symbol Δ and express the other changes in terms of Δ.

    b.   Define missing equilibrium concentrations in terms of the initial concentrations and the changes in concentration determined in a.

Step 3.   *Solve for the change and the equilibrium concentrations.*

    a.   Substitute the equilibrium concentrations into the expression for the equilibrium constant, solve for Δ, and check any assumptions used to find Δ.

    b.   Calculate the equilibrium concentrations.

Step 4.   *Check the arithmetic.*

    Check the calculated equilibrium concentrations by substituting them into the equilibrium expression and determining whether they give the equilibrium constant.

Sometimes a particular step may differ from problem to problem—it may be more complex in some problems and less complex in others. However, every calculation of equilibrium concentrations from a set of initial concentrations will involve these steps.

In solving equilibrium problems that involve changes in concentration, sometimes it is convenient to set up a table of initial concentrations, the changes in these concentrations as the reaction reaches equilibrium, and the final concentrations at equilibrium. Some textbooks use the symbol x to represent the change. However, we use the symbol Δ to emphasize the fact that we solve for the changes in concentrations and then use the changes to find the concentrations at the final equilibrium.

Example 14.9 Calculating Changes in Concentration as a Reaction Goes to Equilibrium

Under certain conditions, the equilibrium constant for the decomposition of $PCl_5(g)$ into $PCl_3(g)$ and $Cl_2(g)$ is 0.0211 mol L^{-1}. What are the equilibrium concentrations of PCl_5, PCl_3, and Cl_2 if the initial concentration of PCl_5 was 1.00 M?

Solution: Two clues suggest that we need to calculate changes in concentration as the reaction goes to equilibrium:

Clue 1. We are given the equilibrium constant, K.

Clue 2. We are given concentrations of the system when it is not at equilibrium.

This suggests that we can use the stepwise process described above.

Step 1. *Determine in what direction the reaction shifts.*

The balanced equation for the decomposition of PCl_5 is

$$PCl_5(g) \rightleftharpoons PCl_3(g) + Cl_2(g)$$

Because we have no products initially, $Q = 0$ and the reaction will shift to the right.

Step 2. *Determine the relative changes needed to reach equilibrium, and then write the equilibrium concentrations in terms of these changes.*

Let us represent the increase in concentration of PCl_3 by the symbol Δ. Then we can express the other changes in terms of Δ by considering the coefficients in the chemical equation a discussed in Section 14.8.

$$PCl_5(g) \rightleftharpoons PCl_3(g) + Cl_2(g)$$
$$\quad -\Delta \qquad\quad \Delta \qquad\quad \Delta$$

The changes in concentration and the expressions for the equilibrium concentrations are summarized in the following table (the concentrations given in the problem are shown in color).

	PCl_5	$\rightleftharpoons$	PCl_3	+	Cl_2
Initial Concentration, M	1.00		0		0
Change, M	$-\Delta$		Δ		Δ
Equilibrium Concentration, M	$1.00 + (-\Delta)$ $= 1.00 - \Delta$		$0 + \Delta = \Delta$		$0 + \Delta = \Delta$

Step 3. *Solve for the change and the equilibrium concentrations.*

Substituting the equilibrium concentrations into the equilibrium constant equation gives

$$K = \frac{[PCl_3][Cl_2]}{[PCl_5]} = 0.0211$$

$$= \frac{(\Delta)(\Delta)}{(1.00 - \Delta)}$$

This equation contains only one variable, Δ, the change in concentration. We can write the equation as a quadratic equation and solve for Δ by using the *quadratic formula*.

$$0.0211 = \frac{(\Delta)(\Delta)}{(1.00 - \Delta)}$$

$$0.0211(1.00 - \Delta) = \Delta^2$$

$$\Delta^2 + 0.0211\Delta - 0.0211 = 0$$

Section A.4 of the Appendix shows that the solution of an equation of the form

$$ax^2 + bx + c = 0$$

is
$$x = \frac{-b \pm \sqrt{b^2 - 4ac}}{2a}$$

In this problem, $x = \Delta$, $a = 1$, $b = 0.0211$, and $c = -0.0211$. Substituting in the quadratic formula yields

$$\Delta = \frac{-0.0211 \pm \sqrt{(0.0211)^2 - 4(1)(-0.0211)}}{2(1)}$$

$$= \frac{-0.0211 \pm \sqrt{(4.45 \times 10^{-4}) + (8.44 \times 10^{-2})}}{2}$$

$$= \frac{-0.0211 \pm 0.291}{2}$$

Hence

$$\Delta = \frac{-0.0211 + 0.291}{2} \qquad \text{or} \qquad \Delta = \frac{-0.0211 - 0.291}{2}$$

$$= 0.135 \qquad\qquad\qquad\qquad = -0.156$$

Quadratic equations often have two different solutions: one that is physically possible and one that is physically impossible (an extraneous root). In this case, the second solution (-0.156) is physically impossible because we have defined the change Δ as a positive change. Thus

$$\Delta = 0.135 \text{ M}$$

The equilibrium concentrations are

$$[PCl_5] = 1.00 - 0.135 = 0.86 \text{ M}$$

$$[PCl_3] = \Delta = 0.135 \text{ M}$$

$$[Cl_2] = \Delta = 0.135 \text{ M}$$

Step 4. *Check the arithmetic.*

Substitution into the expression for K (to check the calculation) gives

$$K = \frac{[PCl_3][Cl_2]}{[PCl_5]} = \frac{(0.135)(0.135)}{(0.86)} = 0.021 \text{ M}$$

The equilibrium constant calculated from the equilibrium concentrations is equal to the value of K given in the problem, within the limits of the significant figures. Thus the calculated equilibrium concentrations check.

Check your learning: Acetic acid, CH_3CO_2H, reacts with ethanol, C_2H_5OH, to form water and ethyl acetate, $CH_3CO_2C_2H_5$.

$$CH_3CO_2H + C_2H_5OH \rightleftharpoons CH_3CO_2C_2H_5 + H_2O$$

The equilibrium constant for this reaction with dioxane as a solvent is 4.0. What are the equilibrium concentrations when a mixture that is 0.15 M in CH_3CO_2H, 0.15 M in C_2H_5OH, 0.40 M in $CH_3CO_2C_2H_5$, and 0.40 M in H_2O is mixed in enough dioxane to make 1.0 L of solution? *Hint:* The equation for Step 3 is

$$4.0 = \frac{(0.40 - \Delta)^2}{(0.15 + \Delta)^2}$$

To solve, we take the square root of both sides, giving

$$2.0 = \frac{0.40 - \Delta}{0.15 + \Delta}$$

and then find Δ.

Answer: $[CH_3CO_2H] = [C_2H_5OH] = 0.18$ M, $[CH_3CO_2C_2H_5] = [H_2O] = 0.37$ M

14.11   Techniques for Solving Equilibrium Problems

One thing that sometimes makes equilibrium problems difficult is solving the equilibrium constant equation for Δ. However, there are several different techniques we can use:

1.   Assume that the change is small compared with the initial concentrations, and neglect Δ when it is added to or subtracted from an initial concentration.

2.   Use successive approximations to converge on the solution.

3.   Assume the reaction goes to completion in one direction and then comes back to equilibrium.

4.   Solve the equilibrium equation with the quadratic formula, or take advantage of a special simplifying mathematical feature of the equation, as illustrated in Example 14.9, check your learning.

5.   Use a spreadsheet, a mathematics program, or an equilibrium program on a computer, or use the power of an advanced calculator.

There are at least three uncertainties that make the results of many equilibrium calculations somewhat inaccurate. (1) The concentrations and pressures used in the calculations are approximations of the activities of the reactants and products (Section 14.2). (2) Many equilibrium constants are known to only two significant figures, and this imposes uncertainty on the results of equilibrium calculations. (3) Equilibrium constants are determined experimentally and are subject to experimental error.

Most chemists accept the results of a simplified equilibrium calculation if it is within 5% of the value that a more thorough calculation using the same data would yield. We will use this "5% test" in the following examples, which illustrate how to use the techniques listed above for solving equilibrium constant equations.

The Assumption That Δ Is Small.

If Q and K are *both* much larger than 1 or *both* are much smaller than 1, then Q for the initial concentrations is close to K, and the concentration changes necessary for Q to become equal to K will be small. Under these conditions, it is possible to assume that Δ is small enough, relative to the initial concentrations, for us to neglect some Δ values in solving the equilibrium equation.

Example 14.10   Assuming Δ Is Small

Find the concentration of $NO(g)$ at equilibrium when a mixture of $O_2(g)$ with an initial concentration of 0.50 M and $N_2(g)$ with an initial concentration of 0.75 M is heated at 700°C.

$$N_2(g) + O_2(g) \rightleftharpoons 2NO(g) \qquad K = 4.1 \times 10^{-9}$$

Solution: Here we are given initial concentrations and the equilibrium constant. These are the clues we use to recognize that we need to calculate changes in concentration as the reaction goes to equilibrium, even though the problem asks for only one equilibrium concentration.

Step 1.   *Determine the direction that the reaction shifts.*

The equation is given. Before the reaction starts, Q is zero because no product is present. The reaction will shift to the right, increasing the concentration of product.

Step 2.   *Determine the relative changes needed to reach equilibrium.*

The table of initial concentrations, changes, and equilibrium concentrations for this problem follows. (The concentrations given in the problem are shown in color.)

	N_2	+	O_2	$\rightleftharpoons$	$2NO$
Initial Concentration, M	0.75		0.50		0
Change, M	$-\Delta$		$-\Delta$		2Δ
Equilibrium Concentration, M	$0.75 + (-\Delta)$		$0.50 + (-\Delta)$		$0 + 2\Delta = 2\Delta$

Step 3.   *Solve for the change and the equilibrium concentrations.*

At equilibrium,

$$K = 4.1 \times 10^{-9} = \frac{[NO]^2}{[N_2][O_2]} = \frac{(2\Delta)^2}{(0.75 - \Delta)(0.50 - \Delta)}$$

There are several ways to solve this equation. Here we note that both Q and K are much less than 1. Thus we assume that Δ is small—that is, that 0.75 is a close approximation to the value of $(0.75 - \Delta)$ and 0.50 is a close approximation to $(0.50 - \Delta)$. With these approximations,

$$K = 4.1 \times 10^{-9} = \frac{(2\Delta)^2}{(0.75)(0.50)}$$

so

$$\Delta^2 = \frac{(4.1 \times 10^{-9})(0.75)(0.50)}{4} = 3.8 \times 10^{-10}$$

$$\Delta = \sqrt{3.8 \times 10^{-10}}$$

$$= 1.9 \times 10^{-5}$$

Now check the assumption that Δ is small relative to 0.50 and 0.75.

$$\frac{\Delta}{0.50} = \frac{1.9 \times 10^{-5}}{0.50} = 3.8 \times 10^{-5} \qquad (3.9 \times 10^{-3}\%)$$

$$\frac{\Delta}{0.75} = \frac{1.9 \times 10^{-5}}{0.75} = 2.5 \times 10^{-5} \qquad (2.5 \times 10^{-3}\%)$$

Δ is less than 5% of either 0.50 or 0.75, so the assumption is valid.

The change in concentration of NO is 2Δ, or 3.8×10^{-5} M. The concentration of NO at equilibrium is equal to $0 + 2\Delta$, which is 3.8×10^{-5} M.

Step 4. *Check the arithmetic.*

$$K = 3.9 \times 10^{-9}$$

$$= \frac{[NO_2]}{[N_2][O_2]}$$

$$= \frac{(4.0 \times 10^{-5})^2}{(0.75)(0.50)}$$

$$= 4.3 \times 10^{-9}$$

Substituting the equilibrium concentrations into the expression for the equilibrium constant shows that the calculated values check, within the limits of the significant figures in the problem.

Check your learning: Calculate the concentrations of all species present at equilibrium when HF with an initial concentration of 0.50 M decomposes.

$$2HF(g) \rightleftharpoons H_2(g) + F_2(g) \qquad K = 1.0 \times 10^{-13}$$

Answer: [HF] = 0.50 M, [H$_2$] = [F$_2$] = 1.6 × 10^{-7} M

In Example 14.10, we did not assume that Δ was zero. Rather, we assumed that Δ was small enough for the term $(0.50 - \Delta)$ to be approximately equal to 0.50 and for the term $(0.75 - \Delta)$ to be approximately equal to 0.75. These two approximations simplified the equilibrium constant equation so that it could be solved much more easily. If the assumption that Δ is small does not prove true, then we can solve with successive approximations or with one of the other techniques.

The Use of Successive Approximations.

As an introduction to the use of successive approximations, let's consider the equilibrium for the decomposition of PCl_5 into PCl_3 and Cl_2.

Example 14.11 | Using Successive Approximations

In Example 14.9 we considered a set of conditions for the reaction

$$PCl_5(g) \rightleftharpoons PCl_3(g) + Cl_2(g) \qquad K = 0.0211$$

that led to the equilibrium equation

$$0.0211 = \frac{(\Delta)(\Delta)}{1.00 - \Delta}$$

Now solve this equation using successive approximations.

Solution: *First approximation.* As a first approximation, we assume that Δ is small relative to 1.00. This gives

$$0.0211 = \frac{(\Delta)(\Delta)}{1.00}$$

which is readily solved.

$$\Delta^2 = 0.0211$$
$$\Delta = \sqrt{0.0211} = 0.145$$

Now we check to see whether Δ is small enough to be neglected.

$$\frac{\Delta}{1.00} = \frac{0.145}{1.00} = 0.145 \qquad (14.5\%)$$

Δ is not small enough to neglect (it is 14.5% of 1.00), so this particular approximation is not satisfactory. However, it does give us useful information for a second approximation.

Second approximation. The first approximation shows us that the value of $(1.00 - \Delta)$ is not close to 1.00; that is, Δ is not approximately zero. Hence we use a second approximation, that Δ is about 0.145, as calculated in the previous approximation, so that $1.00 - \Delta = 0.86$.

$$0.0211 = \frac{(\Delta)(\Delta)}{1.00 - \Delta} = \frac{\Delta^2}{0.86}$$
$$\Delta^2 = 0.0211 \times 0.86 = 0.018$$
$$\Delta = \sqrt{0.018} = 0.13$$

The change in the values of Δ obtained in the first and second approximations (0.145 − 0.13 = 0.01) is 7% of the first value of Δ (0.145). Because these values differ by more than 5%, a third approximation is necessary.

Third approximation. A still better approximation of Δ, found using the second approximation, is 0.13. We use this value in our calculation.

$$0.0211 = \frac{(\Delta)(\Delta)}{1.00 - \Delta} = \frac{(\Delta)(\Delta)}{1.00 - 0.13} = \frac{\Delta^2}{0.87}$$
$$\Delta = \sqrt{0.0211 \times 0.87} = 0.13$$

This value for Δ differs by less than 5% from that obtained in the second approximation (it is the same value). Hence we may conclude that $\Delta = 0.13$. In Example 14.9 we obtained a value of 0.135 using the quadratic equation. The difference is due to the number of significant figures available for Δ.

Check your learning: What are the equilibrium concentrations in a mixture of PCl_5, PCl_3, and Cl_2 formed from PCl_5 with a concentration of 0.100 M?

$$PCl_5(g) \rightleftharpoons PCl_3(g) + Cl_2(g) \qquad K = 0.0211$$

Answer: $[PCl_5]$ = 0.063 M, $[PCl_3]$ = $[Cl_2]$ = 0.037 M

Assume the Reaction Goes to Completion and Then Comes to Equilibrium.

If Q for an initial reaction mixture is much smaller than 1 and K is much larger than 1, or vice versa, we can often simplify determination of the equilibrium concentrations by attacking the problem in two steps: (1) Assume that the shift in the reaction goes to completion, giving an intermediate mixture *that is not at equilibrium* and that contains a 100% yield of products as determined by any limiting reactant that may be present. (2) Calculate the equilibrium concentrations as the intermediate mixture of products (and reactants) comes to equilibirum. The reaction quotient for the intermediate mixture is closer to K than that for the initial mixture, and the approximations discussed in this section may be applicable for solving the equilibrium equation.

Example 14.12   Assuming a Reaction Goes to Completion

Determine the equilibrium concentrations that result from the reaction of a mixture of 0.100 mol of H_2 and 0.050 mol of F_2 in a 1.00-L flask according to the equation

$$H_2(g) + F_2(g) \rightleftharpoons 2HF(g) \qquad K = 115$$

Solution: Here Q is less than 1 (Q = 0) and K is greater than 1 (K = 115), so we will determine an intermediate set of concentrations that result from the formation of a 100% yield of product. We will then calculate the concentration of HF in the equilibrium mixture that results when the intermediate mixture comes to equilibrium.

Step 1.   *Determine in what direction the reaction shifts to come to equilibrium.*

The reaction shifts to the right because Q = 0.

Step 2.   *Determine the changes needed to reach equilibrium.*

Our 1.00-L sample initially contains 0.050 mol of F_2 and 0.100 mol of H_2. The 0.050 mol of F_2 will combine with 0.050 mol of H_2, producing 0.100 mol of HF and leaving 0050 mol of H_2 that did not react. Thus [HF] in the intermediate mixture would be 0.100 M, $[H_2]$ in the intermediate mixture would be 0.050 M, and $[F_2]$ in the initial mixture would be zero. The following table summarizes the concentrations in the initial mixture (shown in red) and in the intermediate mixture (shown in blue).

	H_2	+	F_2	$\rightleftharpoons$	2HF
Initial Concentration, M	0.100		0.050		0
Change, M	−0.050		−0.050		0.100
Intermediate Concentration, M	0.050		0		0.100

Step 3.   *Solve for the changes and the equilibrium concentrations.*

We use the concentrations in the intermediate mixture, which is not at equilibrium, to determine the concentrations in the equilibrium mixture, using the changes summarized in the following table.

	H_2	+	F_2	$\rightleftharpoons$	$2HF$
Intermediate Concentration, M	0.050		0		0.100
Change, M	Δ		Δ		-2Δ
Equilibrium Concentration, M	$0.050 + \Delta$		$0 + \Delta$		$0.100 + (-2\Delta)$

At equilibrium,
$$K = 115 = \frac{[HF]^2}{[H_2][F_2]}$$
$$= \frac{(0.100 - 2\Delta)^2}{(0.050 + \Delta)(\Delta)}$$

Assuming that Δ is small gives

$$115 = \frac{(0.100 - 2\Delta)^2}{(0.050 + \Delta)(\Delta)}$$
$$= \frac{(0.100)^2}{(0.050)(\Delta)}$$
$$\Delta = \frac{(0.100)^2}{(0.050)(115)}$$
$$= 0.0017$$

A check of the approximation shows that

$$\frac{2\Delta}{0.100} = \frac{0.0034}{0.100} = 0.034 \qquad (3.4\%)$$

$$\frac{\Delta}{0.050} = \frac{0.0017}{0.050} = 0.034 \qquad (3.4\%)$$

A second approximation is not necessary, because Δ is less than 5% of the intermediate concentrations.

Next we calculate the equilibrium concentrations.

$$[H_2] = 0.050 + \Delta = 0.050 + 0.0017 = 0.052 \text{ M}$$
$$[F_2] = 0 + \Delta = 0 + 0.0017 = 0.0017 \text{ M}$$
$$[HF] = 0.100 - 2\Delta = 0.100 - 2(0.0017) = 0.097 \text{ M}$$

Step 4.   *Check the arithmetic.*

$$K = \frac{[HF]^2}{[H_2][F_2]}$$
$$= \frac{(0.097)^2}{(0.052)(0.0017)}$$
$$= 1.1 \times 10^2$$

The values check within the limits of the significant figures.

Check your learning: Calculate the equilibrium concentration of NO_2 in 1.00 L of a solution prepared from 0.258 mol of NO_2 with chloroform as the solvent. For the reaction $2NO_2 \rightleftharpoons N_2O_4$ in chloroform, $K = 9.3 \times 10^4$.

Answer: $[NO_2] = 1.2 \times 10^{-3}$, $[N_2O_4] = 0.128$ M

We solve equilibrium problems involving pressures with the same procedures that we have applied to problems involving concentrations.

Example 14.13 Calculating Equilibrium Pressures

Nitrogen at a pressure of 2.00 atm, hydrogen at a pressure of 6.00 atm, and ammonia at a pressure of 5.00 atm are placed in a reaction vessel. What are the pressures of these gases when equilibrium is established?

$$N_2(g) + 3H_2(g) \rightleftharpoons 2NH_3(g) \qquad K_p = 1.645 \times 10^2 \text{ atm}^{-2}$$

Solution: We are given initial pressures and the equilibrium constant. These are the clues we use to recognize that we need to calculate pressure changes as the reaction goes to equilibrium.

Step 1. Determine in what direction the reaction shifts to come to equilibrium.

The reaction quotient for the initial mixture of gases is

$$Q_p = \frac{P_{NH_3}^2}{P_{N_2}P_{H_2}^3} = \frac{(5.00 \text{ atm})^2}{(2.00 \text{ atm})(6.00 \text{ atm})^3}$$
$$= 0.0579 \text{ atm}^{-2}$$

The reaction will shift to the right and increase the pressure of NH_3, because Q_p is less than K_p.

Step 2. Determine the changes needed to reach equilibrium.

Q_p and K_p are of different magnitudes ($Q_p < 1$; $K_p > 1$). Thus we determine an intermediate set of pressures that would result from the formation of a 100% yield of product giving $Q_p > 1$. Then we calculate the pressures in the equilibrium mixture that results when the intermediate mixture comes to equilibrium. This is similar to the approach used in Example 14.12.

The change in pressure of a gas, Δ_p, can be treated in the same way as the change in concentration, Δ. For example, in this system the concentration of NH_3 increases twice as much as the concentration of N_2 decreases, and the pressure of NH_3 also increases twice as much as the pressure of N_2 decreases.

When the mixture in this example gives a 100% yield of NH_3, it consumes all of the N_2 and all of the H_2. Thus Δ_p for N_2 is -2.00 atm, Δ_p for H_2 is -6.00 atm, and Δ_p for NH_3 is 4.00 atm. The following table summarizes the pressures in the initial mixture (in red) and the intermediate mixture (in blue).

	N_2	+	$3H_2$	$\rightleftharpoons$	$2NH_3$
Initial Pressure, atm	2.00		6.00		5.00
Change, atm	−2.00		−6.00		4.00
Intermediate Pressure, atm	0		0		9.00

Now we can set up a table of pressures and changes in pressures as the intermediate mixture comes to equilibrium.

	N_2	$+$	$3H_2$	$\rightleftharpoons$	$2NH_3$
Intermediate Pressure, atm	0		0		9.00
Change, atm	Δ_p		$3\Delta_p$		$-2\Delta_p$
Equilibrium Pressure, atm	Δ_p		$3\Delta_p$		$9.00 - 2\Delta_p$

Step 3. *Solve for the changes and the equilibrium concentrations.*

At equilibrium,
$$K_p = \frac{P_{NH_3}^2}{P_{N_2}P_{H_2}^3} = 1.645 \times 10^2$$

$$= \frac{(9.00 - 2\Delta_p)^2}{(\Delta_p)(3\Delta_p)^3}$$

We assume that Δ_p is small and that $(9.00 - 2\Delta_p) = 9.00$. This gives

$$1.645 \times 10^2 = \frac{(9.00)^2}{(\Delta_p)(3\Delta_p)^3}$$

$$27\Delta_p^4 = \frac{(9.00)^2}{1.645 \times 10^2}$$

$$\Delta_p^4 = 0.0182$$

$$\Delta_p^2 = \sqrt{\Delta_p^4} = 0.135$$

$$\Delta_p = \sqrt{\Delta_p^2} = 0.367$$

Now we check the assumption that Δ_p is small (whether $2\Delta_p$ is less than 5% of 9.00).

$$\frac{2\Delta_p}{9.00} = \frac{2 \times 0.367}{9.00} = 0.0816 \qquad (8.16\%)$$

$2\Delta_p$ is larger than 5% of the initial pressure, 9.00 atm, so we cannot neglect $2\Delta_p$ in $(9.00 - 2\Delta_p)$, and we need a second approximation.

As a second approximation, we assume that $\Delta_p = 0.367$ so that $(9.00 - 2\Delta_p) = 8.27$.

$$1.645 \times 10^2 = \frac{(9.00 - 2\Delta_p)^2}{(\Delta_p)(3\Delta_p)^3}$$

$$1.645 \times 10^2 = \frac{(8.27)^2}{(\Delta_p)(3\Delta_p)^3}$$

and
$$\Delta_p = 0.352$$

The change in Δ_p between the first and second approximations is

$$\frac{(0.367 - 0.352)}{0.352} = 0.045 \qquad (4.5\%)$$

which is less than a 5% change. Thus we can use $\Delta_p = 0.352$ atm to find the equilibrium pressures.

$$P_{N_2} = \Delta_p = 0.352 \text{ atm}$$
$$P_{H_2} = 3\Delta_p = 1.06 \text{ atm}$$
$$P_{NH_3} = 9.00 - 2\Delta_p = 9.00 - (2 \times 0.352) = 8.30 \text{ atm}$$

Step 4. *Check the arithmetic.*

$$K_p = \frac{P_{NH_3}^2}{P_{N_2}P_{H_2}^3}$$

$$= \frac{(8.30 \text{ atm})^2}{(0.352 \text{ atm})(1.06 \text{ atm})^3}$$

$$= 164 \text{ atm}^{-2}$$

The answer checks.

Check your learning: A vessel contains gaseous CO, CO_2, H_2, and H_2O at equilibrium at 980°C. The pressures of these gases are as follows: CO, 0.150 atm; CO_2, 0.200 atm; H_2, 0.090 atm; and H_2O, 0.200 atm. Hydrogen is pumped into the vessel, and the equilibrium pressure of CO changes to 0.230 atm. Calculate the partial pressures of the other substances at the new equilibrium, assuming no change in temperature and the following reaction:

$$CO_2(g) + H_2(g) \rightleftharpoons CO(g) + H_2O(g)$$

Hint: This is a two-step problem. First we calculate an equilibrium constant for a reaction at equilibrium under one set of conditions. Then we calculate the concentrations at a new equilibrium in a calculation of the second type noted in Section 14.9.

Answer: $K_p = 1.67$. The partial pressures for CO, H_2O, CO_2, and H_2 at the new equilibrium are 0.230 atm, 0.280 atm, 0.120 atm, and 0.321 atm, respectively

For Review   Summary

A reaction reaches **equilibrium** when there is no change in the concentrations of reactants or products. Equilibrium is dynamic, because the rate of formation of products via the forward reaction is equal to the rate at which the products re-form reactants via the reverse reaction. For any reaction

$$mA + nB + \cdots \rightleftharpoons xC + yD + \cdots$$

at *equilibrium*, the **reaction quotient** Q is equal to the **equilibrium constant** K for the reaction

$$K = Q = \frac{[C]^x[D]^y \cdots}{[A]^m[B]^n \cdots}$$

If a reactant or product is a pure solid, a pure liquid, or the solvent in a dilute solution, then the concentration of this component does not appear in the expression for the equilibrium constant. At equilibrium, the values of [A], [B], [C], and [D] may vary, but the value of the reaction quotient will always equal K. The addition or removal of a reactant or product may shift a reaction out of equilibrium, but the reaction will proceed—and the concentrations will change—such that the reaction quotient again becomes equal to the equilibrium constant.

A **homogeneous equilibrium** is an equilibrium within a single phase. A **heterogeneous equilibrium** is an equilibrium between two or more phases and involves a boundary surface between those phases. We can decide whether a reaction is at equilibrium by comparing the reaction quotient with the equilibrium constant for the reaction. If $Q = K$, then the reaction is at equilibrium. If $Q > K$, then the concentrations of products are greater than at equilibrium and the reaction will shift to the left, increasing the amounts of reactants and decreasing the amounts of products. If $Q < K$, then the concentrations of products are less than at equilibrium and the reaction will shift to the right, increasing the amounts of products and decreasing the amounts of reactants.

A change in temperature changes the value of the equilibrium constant. A change in pressure generally affects a system in equilibrium only when gases are involved. The effect of a change in conditions is described by **Le Châtelier's principle:** If a stress such as a change in temperature, pressure, or concentration is applied to a system at equilibrium, the equilibrium shifts in a way that relieves the effects of the stress.

A catalyst increases the rates of the forward and reverse reactions equally. Hence a catalyst in a reversible reaction causes the reaction to come to equilibrium more rapidly, but the catalyst has no effect on the value of the equilibrium constant or on the equilibrium concentrations.

For a reaction with a mechanism that consists of a single elementary reaction, the equation for the equilibrium constant is equal to the ratio of the equation for the rate of the forward reaction to the equation for the rate of the reverse reaction. If a reaction mechanism consists of a fast pre-equilibrium followed by a slow rate-determining step, the rate law for the mechanism is found from the rate law for the slow step with at least one concentration determined from the equation for the equilibrium in the preceding step.

The ratios of the changes in concentrations (the Δ values) of a reaction are equal to the coefficients in the balanced chemical equation. The sign of the coefficient of Δ is positive when the concentration increases, negative when it decreases. There are three basic types of equilibrium calculations: (1) calculation of an equilibrium constant from the concentrations of reactants and products at equilibrium, (2) calculation of a missing equilibrium concentration, given the equilibrium constant and all equilibrium concentrations except one, and (3) calculation of all equilibrium concentrations from initial concentrations and the equilibrium constant.

Key Terms and Concepts

activity   (14.2)
assumption that Δ is small   (14.11)
calculation of a missing equilibrium
    concentration   (14.10)
calculation of an equilibrium constant
    (14.9)
calculation of changes in concentration
    (14.10)
calculation of equilibrium concentrations
    (14.9)
calculation of equilibrium pressures
    (14.11)

change in concentration or pressure, Δ
    (14.8)
direction of shift   (14.5)
effect of a catalyst   (14.4)
effect of change in concentration   (14.4)
effect of change in pressure   (14.4)
effect of change in temperature   (14.4)
equilibrium   (14.1)
equilibrium constant   (14.2)
heterogeneous equilibria   (14.3)
homogeneous equilibria   (14.3)
K_p (14.3)

law of mass action   (14.2)
Le Châtelier's principle   (14.4)
Q_p (14.3)
quadratic formula   (14.11)
rate law   (14.7)
reaction mechanisms   (14.7)
reaction quotient   (14.2)
reaction rates and equilibrium   (14.6)
reversible reaction   (14.1)
successive approximation   (14.11)
uncertainties in equilibrium calculations
    (14.11)

Exercises

Questions

1. Is the study of the equilibrium

$$CO_2(g) + H_2(g) \rightleftharpoons CO(g) + H_2O(g)$$

   concerned with the macroscopic behavior or with the microscopic behavior of a mixture of CO_2, H_2, CO, and H_2O? Explain your answer.

2. Explain why there may be an infinite number of values for the reaction quotient of a reaction at a given temperature, but there can be only one value for the equilibrium constant.

3. Explain why an equilibrium between $Br_2(l)$ and $Br_2(g)$ would not be established if the top were removed from the bottle shown in Fig. 14.3.

4. Explain why it is not possible to tell whether the sample in Fig. 14.! was prepared from NO_2 or from N_2O_4.

5. Explain how to recognize the conditions under which changes in pressure would affect a system in equilibrium.

6. What property of a reaction can we use to predict the effect of a change in temperature on the magnitude of an equilibrium constant?

7. The following equation represents a reversible decomposition.

$$CaCO_3(s) \rightleftharpoons CaO(s) + CO_2(g)$$

   Under what conditions will decompositin proceed to completion so that no $CaCO_3$ remains in a closed container?

8. Explain why the equilibrium constants in Examples 14.8 and 14.10 have different values even though both examples use the reaction represented by the equation

$$N_2 + O_2 \rightleftharpoons 2NO$$

9. Among the solubility rules in Chapter 12 is the statement that all chlorides are soluble except Hg_2Cl_2, $AgCl$, $PbCl_2$, and $CuCl$.

   (a) Write the expression for the equilibrium constant for the reaction represented by the equation

$$AgCl(s) \rightleftharpoons Ag^+(aq) + Cl^-(aq)$$

Is K greater than 1, less than 1, or about equal to 1? Explain your answer.

(b) Write the expression for the equilibrium constant for the reaction represented by the equation

$$Pb^{2+}(aq) + 2Cl^-(aq) \rightleftharpoons PbCl_2(s)$$

Is K greater than 1, less than 1, or about equal to 1? Explain your answer.

10. Among the solubility rules in Chapter 12 is the statement that carbonates, phosphates, borates, arsenates, and arsenites—except those of the ammonium ion and the alkali metals—are insoluble.

(a) Write the expression for the equilibrium constant for the reaction represented by the equation

$$CaCO_3(s) \rightleftharpoons Ca^{2+}(aq) + CO_3^{2-}(aq)$$

Is K greater than 1, less than 1, or about equal to 1? Explain your answer.

(b) Write the expression for the equilibrium constant for the reaction represented by the equation

$$3Ba^{2+}(aq) + 2PO_4^{3-}(aq) \rightleftharpoons Ba_3(PO_4)_2(s)$$

Is K greater than 1, less than 1, or about equal to 1? Explain your answer.

11. Benzene is one of the compounds used as octane enhancers in unleaded gasoline. It is manufactured by the catalytic conversion of acetylene to benzene:

$$3C_2H_2 \longrightarrow C_6H_6$$

Would this reaction be most useful commercially if K were about 0.01, about 1, or about 10? Explain your answer.

12. Show that the complete chemical equation and the net ionic equation for the reaction represented by the equation

$$KI(aq) + I_2(aq) \rightleftharpoons KI_3(aq)$$

give the same expression for the reaction quotient. KI_3 is composed of the ions K^+ and I_3^-.

13. What would happen to the color of the solution in Fig. 14.5B if a small amount of NaOH were added and $Fe(OH)_3$ precipitated? Explain your answer.

14. Basing your answer on consideration of how reaction rates change with temperature, explain why equilibrium constants change with temperature.

15. When a burner on a gas stove is operating at one setting, the size of the flame does not change.

$$CH_4 + 2O_2 \rightleftharpoons CO_2 + 2H_2O$$

Is an equilibrium among CH_4, O_2, CO_2, and H_2O established under these conditions? Explain your answer.

16. (a) In the discussion of the manufacture of sulfuric acid (Section 8.9), it is noted that at higher temperatures SO_3 forms more rapidly from the reaction of SO_2 with O_2, but the yield is lower than at lower temperatures.

$$2SO_2 + O_2 \longrightarrow 2SO_3$$

Does the equilibrium constant for the reaction increase, decrease, or remain about the same as the temperature increases?

(b) Is the reaction endothermic or exothermic?

17. For a titration (Section 3.13) to be effective, the reaction must be rapid and the yield of the reaction must be 100%. Acid–base reactions are commonly used in titrations. Is K greater than 1, less than 1, or about equal to 1 for an acid–base reaction?

18. For a precipitation reaction to be useful in a gravimetric analysis (Section 3.14), the product of the reaction must be insoluble. Is K greater than 1, less than 1, or about equal to 1 for a useful precipitation reaction?

Reaction Quotients and Equilibrium Constants

19. Write the mathematical expression for the reaction quotient, Q, for each of the following reactions.

(a) $CH_4(g) + Cl_2(g) \rightleftharpoons CH_3Cl(g) + HCl(g)$
(b) $N_2(g) + O_2(g) \rightleftharpoons 2NO(g)$
(c) $2SO_2(g) + O_2(g) \rightleftharpoons 2SO_3(g)$
(d) $BaSO_3(s) \rightleftharpoons BaO(s) + SO_2(g)$
(e) $P_4(g) + 5O_2(g) \rightleftharpoons P_4O_{10}(s)$
(f) $Br_2(g) \rightleftharpoons 2Br(g)$
(g) $CH_4(g) + 2O_2(g) \rightleftharpoons CO_2(g) + 2H_2O(l)$
(h) $CuSO_4 \cdot 5H_2O(s) \rightleftharpoons CuSO_4(s) + 5H_2O(g)$

20. Write the mathematical expression for the reaction quotient, Q, for each of the following reactions.

(a) $N_2(g) + 3H_2(g) \rightleftharpoons 2NH_3(g)$
(b) $4NH_3(g) + 5O_2(g) \rightleftharpoons 4NO(g) + 6H_2O(g)$
(c) $N_2O_4(g) \rightleftharpoons 2NO_2(g)$
(d) $CO_2(g) + H_2(g) \rightleftharpoons CO(g) + H_2O(g)$
(e) $NH_4Cl(s) \rightleftharpoons NH_3(g) + HCl(g)$
(f) $2Pb(NO_3)_2(s) \rightleftharpoons 2PbO(s) + 4NO_2(g) + O_2(g)$
(g) $2H_2(g) + O_2(g) \rightleftharpoons 2H_2O(l)$
(h) $S_8(g) \rightleftharpoons 8S(g)$

21. The initial concentrations or pressures of reactants and products are given for each of the following systems. Calculate the reaction quotient and determine the direction in which each system will shift to reach equilibrium.

(a) $2NH_3(g) \rightleftharpoons N_2(g) + 3H_2(g)$ $K = 17$
   $[NH_3] = 0.20$ M; $[N_2] = 1.00$ M; $[H_2] = 1.00$ M
(b) $2NH_3(g) \rightleftharpoons N_2(g) + 3H_2(g)$ $K_p = 6.8 \times 10^4$ atm^2
   Initial pressures: $NH_3 = 3.0$ atm; $N_2 = 2.0$ atm; $H_2 = 1.0$ atm
(c) $2SO_3(g) \rightleftharpoons 2SO_2(g) + O_2(g)$ $K = 0.230$
   $[SO_3] = 0.00$ M; $[SO_2] = 1.00$ M; $[O_2] = 1.00$ M
(d) $2SO_3(g) \rightleftharpoons 2SO_2(g) + O_2(g)$ $K_p = 16.5$ atm
   Initial pressures: $SO_3 = 1.00$ atm; $SO_2 = 1.00$ atm; $O_2 = 1.00$ atm
(e) $2NO(g) + Cl_2(g) \rightleftharpoons 2NOCl(g)$ $K = 4.6 \times 10^4$
   $[NO] = 1.00$ M; $[Cl_2] = 1.00$ M; $[NOCl] = 0$ M

(f) $N_2(g) + O_2(g) \rightleftharpoons 2NO(g)$ $K_p = 0.050$
Initial pressures: NO = 10.0 atm; $N_2 = O_2 = 5$ atm

22. The initial concentrations or pressures of reactants and products are given for each of the following systems. Calculate the reaction quotient and determine the direction in which each system will shift to reach equilibrium.

(a) $2NH_3(g) \rightleftharpoons N_2(g) + 3H_2(g)$ $K = 17$
$[NH_3] = 0.50$ M; $[N_2] = 0.15$ M; $[H_2] = 0.12$ M

(b) $2NH_3(g) \rightleftharpoons N_2(g) + 3H_2(g)$ $K_p = 6.8 \times 10^4$ atm^2
Initial pressures: $NH_3 = 2.00$ atm; $N_2 = 10.00$ atm; $H_2 = 10.00$ atm

(c) $2SO_3(g) \rightleftharpoons 2SO_2(g) + O_2(g)$ $K = 0.230$
$[SO_3] = 2.00$ M; $[SO_2] = 2.00$ M; $[O_2] = 2.00$ M

(d) $2SO_3(g) \rightleftharpoons 2SO_2(g) + O_2(g)$ $K_p = 16.5$ atm
Initial pressures: $SO_3 = 0$ atm; $SO_2 = 1.00$ atm; $O_2 = 1.130$ atm

(e) $2NO(g) + Cl_2(g) \rightleftharpoons 2NOCl(g)$
$K_p = 2.5 \times 10^3$ atm^{-1}
Initial pressures: NOCl = 0 atm; NO = 1.00 atm; Cl_2 = 1.00 atm

(f) $N_2(g) + O_2(g) \rightleftharpoons 2NO(g)$ $K = 0.050$
$[N_2] = 0.100$ M; $[O_2] = 0.200$ M; $[NO] = 1.00$ M

23. Sketch graphs similar to Fig. 14.4 for Q and the concentrations of NO_2 and N_2O_4 as the 0.10 mol of NO_2 described in Example 14.2 comes to equilibrium. Plot points where $[NO_2] = 0.10$ M, 0.080 M, 0.060 M, 0.040 M, 0.020 M, and 0.016 M, the equilibrium concentration.

24. Sketch graphs similar to Fig. 14.4 for Q and the pressures of NH_3, N_2, and H_2 as the mixture of gases described in Example 14.13 comes to equilibrium. Start with $P_{NH_3} = 5.00$ atm, $P_{N_2} = 2.00$ atm, and $P_{H_2} = 6.00$ atm, and plot points as P_{NH_3} changes from 5 atm to 5.5 atm, 6 atm, 6.5 atm, 7 atm, 7.5 atm, and 8 atm and, finally, to the equilibrium pressure, 8.30 atm.

Homogeneous and Heterogeneous Equilibria

25. Which of the systems described in Exercise 19 give homogeneous equilibria? Which give heterogeneous equilibria?

26. Which of the systems described in Exercise 20 give homogeneous equilibria? Which give heterogeneous equilibria?

27. For which of the reactions in Exercise 19 does K (calculated using concentrations) equal K_p (calculated using pressures)?

28. For which of the reactions in Exercise 20 does K (calculated using concentrations) equal K_p (calculated using pressures)?

29. Convert the values of K to values of K_p and the values of K_p to values of K.

(a) $N_2(g) + 3H_2(g) \rightleftharpoons 2NH_3(g)$ $K = 0.50$ at 400°C

(b) $H_2 + I_2 \rightleftharpoons 2HI$ $K = 50.2$ at 448°C

(c) $Na_2SO_4 \cdot 10H_2O(s) \rightleftharpoons Na_2SO_4(s) + 10H_2O(g)$
$K_p = 4.08 \times 10^{-25}$ atm^{10} at 25°C

(d) $H_2O(l) \rightleftharpoons H_2O(g)$ $K_p = 0.122$ atm at 50°C

30. Convert the values of K to values of K_p and the values of K_p to values of K.

(a) $Cl_2(g) + Br_2(g) \rightleftharpoons 2BrCl(g)$
$K = 4.7 \times 10^{-2}$ at 25°C

(b) $2SO_2(g) + O_2(g) \rightleftharpoons 2SO_3(g)$
$K_p = 48.2$ atm at 500°C

(c) $CaCl_2 \cdot 6H_2O(s) \rightleftharpoons CaCl_2(s) + 6H_2O(g)$
$K_p = 5.09 \times 10^{-44}$ atm^6 at 25°C

(d) $H_2O(l) \rightleftharpoons H_2O(g)$ $K_p = 0.196$ atm at 60°C

Changes in Concentrations at Equilibrium: Le Châtelier's Principle

31. Suggest four ways in which the concentration of hydrazine, N_2H_4, could be increased in an equilibrium described by the equation
$$N_2(g) + 2H_2(g) \rightleftharpoons N_2H_4(g) \quad \Delta H = 95 \text{ kJ}$$

32. Suggest four ways in which the concentration of PH_3 could be increased in an equilibrium described by the equation
$$P_4(g) + 6H_2(g) \rightleftharpoons 4PH_3(g) \quad \Delta H = 110.5 \text{ kJ}$$

33. How will an increase in temperature affect each of the following equlibria? An increase in pressure?

(a) $2NH_3(g) \rightleftharpoons N_2(g) + 3H_2(g)$ $\Delta H = 92$ kJ
(b) $N_2(g) + O_2(g) \rightleftharpoons 2NO(g)$ $\Delta H = 181$ kJ
(c) $2O_3(g) \rightleftharpoons 3O_2(g)$ $\Delta H = -285$ kJ
(d) $CaO(s) + CO_2(g) \rightleftharpoons CaCO_3(s)$ $\Delta H = -176$ kJ

34. How will an increase in temperature affect each of the following equilibria? An increase in pressure?

(a) $2H_2O(g) \rightleftharpoons 2H_2(g) + O_2(g)$ $\Delta H = 484$ kJ
(b) $N_2(g) + 3H_2(g) \rightleftharpoons 2NH_3(g)$ $\Delta H = -92.2$ kJ
(c) $2Br(g) \rightleftharpoons Br_2(g)$ $\Delta H = -224$ kJ
(d) $H_2(g) + I_2(s) \rightleftharpoons 2HI(g)$ $\Delta H = 53$ kJ

35. (a) Methanol, a liquid fuel that could possibly replace gasoline, can be prepared from water gas and additional hydrogen at high temperature and pressure in the presence of a suitable catalyst. Write the expression for the equilibrium constant for the reversible reaction
$$2H_2(g) + CO(g) \rightleftharpoons CH_3OH(g) \quad \Delta H = -90.2 \text{ kJ}$$

(b) Assume that equilibrium has been established and predict how the concentrations of H_2, CO, and CH_3OH will differ at a new equilibrium if (1) more H_2 is added. (2) CO is removed. (3) CH_3OH is added. (4) the pressure on the system is increased. (5) the temperature of the system is increased. (6) more catalyst is added.

36. (a) Nitrogen and oxygen react at high temperatures. Write the expression for the equilibrium constant for the reversible reaction
$$N_2(g) + O_2(g) \rightleftharpoons 2NO(g) \quad \Delta H = 181 \text{ kJ}$$

(b) Assume that equilibrium has been established and predict how the concentrations of N_2, O_2, and NO will differ at a new equilibrium if (1) more O_2 is added. (2) N_2 is removed. (3) NO is added. (4) the pressure on the system is increased. (5) the temperature of the system is increased. (6) a catalyst is added.

37. (a) Water gas, a mixture of H_2 and CO, is an important industrial fuel produced by the reaction of steam with red-hot coke, essentially pure carbon. Write the expression for the equilibrium constant for the reversible reaction

$$C(s) + H_2O(g) \rightleftharpoons CO(g) + H_2(g) \quad \Delta H = 131.30 \text{ kJ}$$

(b) Assume that equilibrium has been established and predict how the concentration of each reactant and product will differ at a new equilibrium if (1) more C is added. (2) H_2O is removed. (3) CO is added. (4) the pressure on the system is increased. (5) the temperature of the system is increased.

38. (a) Pure iron metal can be produced by the reduction of iron(III) oxide with hydrogen gas. Write the expression for the equilibrium constant for the reversible reaction

$$Fe_2O_3(s) + 3H_2(g) \rightleftharpoons$$
$$2Fe(s) + 3H_2O(g) \quad \Delta H = 98.7 \text{ kJ}$$

(b) Assume that equilibrium has been established and predict how the concentration of each reactant and product will differ at a new equilibrium if (1) more Fe is added. (2) H_2O is removed. (3) H_2 is added. (4) the pressure on the system is increased. (5) the temperature of the system is increased.

39. Ammonia is a weak base that reacts with water according to the equation

$$NH_3(aq) + H_2O(l) \rightleftharpoons NH_4^+(aq) + OH^-(aq)$$

Will any of the following increase the percent of ammonia that is converted to the ammonium ion in water? (a) Addition of NaOH. (b) Addition of HCl. (c) Addition of NH_4Cl.

40. Acetic acid is a weak acid that reacts with water according to the equation

$$CH_3CO_2H(aq) + H_2O(aq) \rightleftharpoons$$
$$H_3O^+(aq) + CH_3CO_2^-(aq)$$

Will any of the following increase the percent of acetic acid in a solution of acetic acid that reacts and produces $CH_3CO_2^-$ ion? (a) Addition of HCl. (b) Addition of NaOH. (c) Addition of $NaCH_3CO_2$.

41. Suggest two ways in which the equilibrium concentration of Ag^+ can be reduced in a solution of Na^+, Cl^-, Ag^+, and NO_3^-, in contact with solid AgCl.

$$Na^+(aq) + Cl^-(aq) + Ag^+(aq) + NO_3^-(aq) \rightleftharpoons$$
$$AgCl(s) + Na^+(aq) + NO_3^-(aq) \quad \Delta H = -65.9 \text{ kJ}$$

42. How can the pressure of water vapor be increased in the following equilibrium?

$$H_2O(l) \rightleftharpoons H_2O(g) \quad \Delta H = 41 \text{ kJ}$$

43. Additional solid silver sulfate, a slightly soluble solid, is added to a solution of silver ion and sulfate ion in equilibrium with solid silver sulfate. Which of the following will occur? (a) The Ag^+ and SO_4^{2-} concentrations will not change. (b) The added silver sulfate will dissolve. (c) Additional silver sulfate will form and precipitate from solution as Ag^+ ions and SO_4^{2-} ions combine. (d) The Ag^+ ion concentration will increase and the SO_4^{2-} ion concentration will decrease.

44. Section 8.9 describes the preparation of CaO by heating calcium carbonate, $CaCO_3$, which is widely and inexpensively available as limestone or oyster shells.

$$CaCO_3(s) \longrightarrow CaO(s) + CO_2(g)$$

Although this decomposition reaction is reversible and CaO and CO_2 react to give $CaCO_3$, a 100% yield of CaO is obtained if the CO_2 is allowed to escape. Explain why and how a 100% yield is obtained from this reversible reaction.

Kinetics and Equilibria

45. The rate of the elementary reaction $H_2(g) + I_2(g) \longrightarrow 2HI(g)$ at 25°C is given by

$$\text{Rate} = 1.7 \times 10^{-18} [H_2][I_2]$$

The rate of decomposition of gaseous HI to $H_2(g)$ and $I_2(g)$ at 25°C is given by

$$\text{Rate} = 2.4 \times 10^{-21} [HI]^2$$

What is the equilibrium constant for the formation of gaseous HI from the gaseous elements at 25°C?

46. The rate of the elementary reaction $2NO(g) + O_2(g) \longrightarrow 2NO_2(g)$ at 380°C is given by

$$\text{Rate} = 2.6 \times 10^3 [NO]^2[O_2]$$

The rate of the reverse reaction at 380°C is given by

$$\text{Rate} = 4.1 [NO_2]^2$$

What is the equilibrium constant for the reaction at 380°C?

47. One possible mechanism for the reaction $CO(g) + Cl_2(g) \longrightarrow COCl_2(g)$ is

$$Cl_2 \rightleftharpoons 2Cl \quad \text{(fast)}$$
$$CO + Cl \longrightarrow COCl \quad \text{(slow)}$$
$$COCl + Cl \longrightarrow COCl_2 \quad \text{(fast)}$$

(a) Write the rate equation for the slow elementary reaction.
(b) Write the equilibrium constant expression for the first elementary reaction.
(c) Solve the equilibrium constant expression from part (b) for [Cl], and substitute this concentration into the rate equation from part (a) to obtain the overall rate equation for this mechanism.

48. A possible mechanism for the decomposition of ozone is

$$O_3 \rightleftharpoons O_2 + O \quad \text{(fast)}$$
$$O + O_3 \longrightarrow 2O_2 \quad \text{(slow)}$$

(a) Write the rate equation for the slow elementary reaction.

(b) Write the equilibrium constant expression for the first elementary reaction.

(c) Solve the equilibrium constant expression from part (b) for [O], and substitute this concentration into the rate equation from part (a) to obtain the overall rate equation for this mechanism.

49. For the reaction A $\longrightarrow$ B + C, the following data were obtained at 30°C.

Experiment	[A], mol L^{-1}	Rate, mol L^{-1} h^{-1}
1	0.170	0.0500
2	0.340	0.100
3	0.680	0.200

(a) What is the rate equation, and what is the order of the reaction?

(b) Calculate the rate constant for the reaction.

(c) The equilibrium constant for the reaction is 0.500. Assume that the reaction proceeds by a one-step mechanism and calculate the rate constant for the reverse reaction.

50. The formation of NOCl from NO and Cl$_2$ proceeds by the termolecular elementary reaction

$$2NO + Cl_2 \longrightarrow 2NOCl \quad Rate = k[NO]^2[Cl_2]$$

Use the mathematical equation for the equilibrium constant as a guide and write the rate law for the reverse reaction.

Calculation of Equilibrium Constants

51. A reaction is represented by the equation

$$A + 2B \rightleftharpoons 2C \quad K = 1 \times 10^3$$

(a) Write the mathematical expression for the equilibrium constant.

(b) Using concentrations of 1 M or less, make up values for two different sets of concentrations that describe a mixture of A, B, and C at equilibrium.

52. A reaction is represented by the equation

$$2W \rightleftharpoons X + 2Y \quad K = 5 \times 10^{-4}$$

(a) Write the mathematical expression for the equilibrium constant.

(b) Using concentrations of 1 M or less, make up values for two different sets of concentrations that describe a mixture of W, X, and Y at equilibrium.

53. For which of the reactions in Exercise 21 does $K = K_p$?

54. For which of the reactions in Exercise 22 does $K = K_p$?

55. What is the value of the equilibrium constant for the formation of NH$_3$ at 500°C?

$$N_2(g) + 3H_2(g) \rightleftharpoons 2NH_3(g)$$

An equilibrium mixture of NH$_3(g)$, H$_2(g)$, and N$_2(g)$ at 500°C was found to contain 1.35 mol H$_2$ per liter, 1.15 mol N$_2$ per liter, and 0.412 mol NH$_3$ per liter.

56. Hydrogen is prepared commercially by the reaction of methane and water vapor at elevated temperatures.

$$CH_4(g) + H_2O(g) \rightleftharpoons 3H_2(g) + CO(g)$$

What is the equilibrium constant for the reaction if a mixture at equilibrium contains gases with the following concentrations: CH$_4$, 0.126 M; H$_2$O, 0.242 M; CO, 0.126 M; and H$_2$, 1.15 M at a temperature of 760°C?

57. A 0.72-mol sample of PCl$_5$ is put into a 1.00-L vessel and heated. At equilibrium, the vessel contains 0.40 mol of PCl$_3(g)$ and 0.40 mol of Cl$_2(g)$. Calculate the value of the equilibrium constant for the decomposition of PCl$_5$ to PCl$_3$ and Cl$_2$ at this temperature.

58. At 1 atm and 25°C, NO$_2$ gas with an initial concentration of 1.00 M is 3.3×10^{-3} % decomposed into the gases NO and O$_2$. Calculate the equilibrium constant for the reaction.

$$2NO_2 \rightleftharpoons 2NO + O_2$$

59. Calculate the equilibrium constant K_p for the reaction

$$2NO(g) + Cl_2(g) \rightleftharpoons 2NOCl(g)$$

from the equilibrium pressures: NO, 0.050 atm; Cl$_2$, 0.30 atm; and NOCl, 1.2 atm.

60. When heated, iodine vapor dissociates according to the equation

$$I_2(g) \rightleftharpoons 2I(g)$$

At 1274 K a sample exhibits a partial pressure of I$_2$ of 0.1122 atm and a partial pressure due to I atoms of 0.1378 atm. Determine the equilibrium constant, K_p, for the decomposition at 1274 K.

61. A sample of ammonium chloride was heated in a closed container.

$$NH_4Cl(s) \rightleftharpoons NH_3(g) + HCl(g)$$

At equilibrium the pressure of NH$_3(g)$ was found to be 1.75 atm. What is the equilibrium constant K_p for the decomposition at this temperature?

62. At a temperature of 60°C the vapor pressure of water is 0.196 atm. What is the equilibrium constant K_p for the transformation at 60°C?

$$H_2O(l) \rightleftharpoons H_2O(g)$$

Concentration and Pressure Changes

63. Complete the changes in concentrations (or pressure if requested) under each of the following reactions.

(a) $2SO_3(g) \rightleftharpoons 2SO_2(g) + O_2(g)$

___	___	Δ
		0.125 M

(b) $4NH_3(g) + 3O_2(g) \rightleftharpoons 2N_2(g) + 6H_2O(g)$

	3Δ		
___	0.24 M	___	___

(c) Change in pressure:

$$2CH_4(g) \rightleftharpoons C_2H_2(g) + 3H_2(g)$$

| _____ | Δ | _____ |
| _____ | 25 torr | _____ |

(d) Change in pressure:

$$CH_4(g) + H_2O(g) \rightleftharpoons CO(g) + 3H_2(g)$$

| _____ | Δ | _____ | _____ |
| _____ | 5 atm | _____ | _____ |

(e) $NH_4Cl(s) \rightleftharpoons NH_3(g) + HCl(g)$

| | Δ | _____ |
| | 1.03×10^{-4} M | _____ |

(f) Change in pressure:

$$Ni(s) + 4CO(g) \rightleftharpoons Ni(CO)_4(g)$$

| | _____ | Δ |
| | _____ | 0.40 atm |

64. Complete the changes in concentrations (or pressure if requested) under each of the following reactions.

(a) $2H_2(g) + O_2(g) \rightleftharpoons 2H_2O(g)$

| _____ | _____ | 2Δ |
| _____ | _____ | 1.50 M |

(b) $CS_2(g) + 4H_2(g) \rightleftharpoons CH_4(g) + 2H_2S(g)$

| Δ | _____ | _____ | _____ |
| 0.020 M | _____ | _____ | _____ |

(c) Change in pressure:

$$H_2(g) + Cl_2(g) \rightleftharpoons 2HCl(g)$$

| Δ | _____ | _____ |
| 1.50 atm | _____ | _____ |

(d) Change in pressure:

$$2NH_3(g) + 2O_2(g) \rightleftharpoons N_2O(g) + 3H_2O(g)$$

| _____ | _____ | _____ | Δ |
| _____ | _____ | _____ | 60.6 torr |

(e) $NH_4HS(s) \rightleftharpoons NH_3(g) + H_2S(g)$

| | Δ | _____ |
| | 9.8×10^{-6} M | _____ |

(f) Change in pressure:

$$Fe(s) + 5CO(g) \rightleftharpoons Fe(CO)_5(g)$$

| | _____ | Δ |
| | _____ | 0.012 atm |

65. Why are there no changes specified for Ni in Exercise 63, part (f)? What property of Ni does change?

66. Why are there no changes specified for NH_4HS in Exercise 64, part (e)? What property of NH_4HS does change?

Calculations Involving Equilibrium Concentrations

67. Analysis of the gases in a sealed reaction vessel containing NH_3, N_2, and H_2 at equilibrium at 400°C established the concentration of N_2 to be 1.2 M and the concentration of H_2 to be 0.24 M.

$$N_2(g) + 3H_2(g) \rightleftharpoons 2NH_3(g) \quad K = 0.50 \text{ at } 400°C$$

Calculate the equilibrium molar concentration of NH_3.

68. Calculate the number of moles of HI that are in equilibrium with 1.25 mol of H_2 and 1.25 mol of I_2 in a 5.00-L flask at 448°C.

$$H_2 + I_2 \rightleftharpoons 2HI \quad K = 50.2 \text{ at } 448°C$$

69. What is the pressure of BrCl in an equilibrium mixture of Cl_2, Br_2, and BrCl if the Cl_2 pressure in the mixture is 0.115 atm and the Br_2 pressure is 0.450 atm?

$$Cl_2(g) + Br_2(g) \rightleftharpoons 2BrCl(g) \quad K_p = 4.7 \times 10^{-2}$$

70. What is the pressure of CO_2 in a mixture at equilibrium that contains H_2 at 0.50 atm, H_2O at 2.0 atm, and CO at 1.0 atm at 990°C?

$$H_2(g) + CO_2(g) \rightleftharpoons H_2O(g) + CO(g) \quad K_p = 1.6$$

71. Cobalt metal can be prepared by reducing cobalt(II) oxide with carbon monoxide at 550°C.

$$CoO(s) + CO(g) \rightleftharpoons Co(s) + CO_2(g) \quad K = 4.90 \times 10^2$$

What concentration of CO remains in an equilibrium mixture with $[CO_2] = 0.100$ M?

72. Carbon reacts with water vapor at elevated temperatures.

$$C(s) + H_2O(g) \rightleftharpoons CO(g) + H_2(g) \quad K = 0.2 \text{ at } 1000°C$$

What is the concentration of CO in a mixture prepared from C and H_2O at equilibrium with $[H_2O] = 0.500$ M?

73. Sodium sulfate 10-hydrate, $Na_2SO_4\cdot10H_2O$, dehydrates according to the equation

$$Na_2SO_4\cdot10H_2O(s) \rightleftharpoons Na_2SO_4(s) + 10H_2O(g)$$
$$K_p = 4.08 \times 10^{-25} \text{ at } 25°C$$

What is the pressure of water vapor in equilibrium with a mixture of $Na_2SO_4\cdot10H_2O$ and $NaSO_4$?

74. Calcium chloride 6-hydrate, $CaCl_2\cdot6H_2O$, dehydrates according to the equation

$$CaCl_2\cdot6H_2O(s) \rightleftharpoons CaCl_2(s) + 6H_2O(g)$$
$$K_p = 5.09 \times 10^{-44} \text{ at } 25°C$$

What is the pressure of water vapor in equilibrium with a mixture of $CaCl_2\cdot6H_2O$ and $CaCl_2$?

Calculation of Changes in Concentration or Pressure

75. A student solved the following problem and found the equilibrium concentrations to be $[SO_2] = 0.590$ M, $[O_2] = 0.0450$ M, and $[SO_3] = 0.260$ M. How could this student check the work without resolving the problem?

The problem: Consider the following reaction at 600°C.

$$2SO_2(g) + O_2(g) \rightleftharpoons 2SO_3(g) \quad K_c = 4.32$$

What are the equilibrium concentrations of all species in a mixture that was prepared with $[SO_3] = 0.500$ M, $[SO_2] = 0$ M, and $[O_2] = 0.350$ M?

76. A student solved the following problem and found $[N_2O_4] = 0.16$ M at equilibrium. How could this student recognize that the answer was wrong without reworking the problem?

The problem: What is the equilibrium concentration of N_2O_4 in a mixture formed from a sample of NO_2 with a concentration of 0.10 M?

$$2NO_2(g) \rightleftharpoons N_2O_4(g) \quad K = 160$$

77. (a) Assuming that the change in initial concentration is small enough to be neglected, calculate the equilibrium concentration of both species in 1.00 L of a solution prepared from 0.129 mol of N_2O_4 with chloroform as the solvent.

$$N_2O_4 \rightleftharpoons 2NO_2 \quad K = 1.07 \times 10^{-5} \text{ in chloroform,}$$

   (b) Show that the change is small enough to be neglected.

78. (a) Assuming that the change in initial concentration is small enough to be neglected, calculate the equilibrium concentration of all species in an equilibrium mixture that results from the decomposition of $COCl_2$ with an initial concentration of 0.3166 M.

$$COCl_2(g) \rightleftharpoons CO(g) + Cl_2(g) \quad K = 2.2 \times 10^{-10}$$

   (b) Show that the change is small enough to be neglected.

79. (a) Assuming that the change in initial pressure is small enough to be neglected, calculate the equilibrium pressures of all species in an equilibrium mixture that results from the decomposition of H_2S with an initial pressure of 0.824 atm.

$$2H_2S(g) \rightleftharpoons 2H_2(g) + S_2(g) \quad K_p = 2.2 \times 10^{-6}$$

   (b) Show that the change is small enough to be neglected.

80. (a) Assuming that the change in initial pressure is small enough to be neglected, calculate the equilibrium pressures of all species in an equilibrium mixture that results from the decomposition of $COCl_2$ with an initial pressure of 0.255 atm.

$$COCl_2(g) \rightleftharpoons CO(g) + Cl_2(g) \quad K_p = 5.4 \times 10^{-9}$$

   (b) Show that the change is small enough to be neglected.

81. What are all concentrations after a mixture that contains $[H_2O] = 1.00$ M and $[Cl_2O] = 1.00$ M comes to equilibrium at 25°C?

$$H_2O(g) + Cl_2O(g) \rightleftharpoons 2HOCl(g) \quad K = 0.0900$$

Hint: Changes in initial concentrations are too large to be neglected in this calculation.

82. What are the concentrations of PCl_5, PCl_3, and Cl_2 in an equilibrium mixture produced by the decomposition of a sample with $[PCl_5] = 2.00$ M?

$$PCl_5(g) \rightleftharpoons PCl_3(g) + Cl_2(g) \quad K = 0.0211$$

Hint: The change in the initial concentration is too large to be neglected in this calculation.

83. Calculate the pressures of all species at equilibrium in a mixture of NOCl, NO, and Cl_2 produced when a sample of NOCl with a pressure of 0.500 atm comes to equilibrium according to the reaction

$$2NOCl(g) \rightleftharpoons 2NO(g) + Cl_2(g) \quad K_p = 4.0 \times 10^{-4} \text{ atm}$$

Hint: The change in the initial pressure is too large to be neglected in this calculation.

84. The equilibrium constant K_p for the decomposition of nitrosyl bromide is 1.0×10^{-2} atm at 25°C.

$$2NOBr(g) \rightleftharpoons 2NO(g) + Br_2(g)$$

What percentage of NOBr with an initial pressure of 0.25 atm is decomposed at 25°C? *Hint:* The change in initial pressure is too large to be neglected in this calculation.

85. Calculate the equilibrium concentrations of NO, O_2, and NO_2 in a mixture at 250°C that results from the reaction of 0.20 M NO and 0.10 M O_2.

$$2NO(g) + O_2(g) + \rightleftharpoons 2NO_2(g)$$
$$K = 2.3 \times 10^5 \text{ at } 250°C$$

Hint: K is large. Assume the reaction goes to completion and then comes back to equilibrium.

86. Calculate the equilibrium concentrations that result when 0.25 M O_2 and 1.0 M HCl react and come to equilibrium.

$$4HCl(g) + O_2(g) \rightleftharpoons 2Cl_2(g) + 2H_2O(g)$$
$$K = 3.1 \times 10^{13}$$

Hint: K is large. Assume the reaction goes to completion and then comes back to equilibrium.

87. One of the important reactions in the formation of smog is represented by the equation

$$O_3(g) + NO(g) \rightleftharpoons NO_2(g) + O_2(g) \quad K_p = 6.0 \times 10^{34}$$

What is the pressure of O_3 remaining after a mixture of O_3 with a pressure of 1.2×10^{-8} atm and NO with a pressure of 1.2×10^{-8} atm comes to equilibrium? *Hint: K_p* is large. Assume the reaction goes to completion and then comes back to equilibrium.

88. Calculate the pressures of NO, Cl_2, and NOCl in an equilibrium mixture produced by the reaction of a starting mixture with 4.0 atm NO and 2.0 atm Cl_2.

$$2NO(g) + Cl_2(g) \rightleftharpoons 2NOCl(g)$$
$$K_p = 2.5 \times 10^3 \text{ atm}^{-1}$$

Hint: K_p is large. Assume the reaction goes to completion and then comes back to equilibrium.

89. Calculate the number of grams of HI that are in equilibrium with 1.25 mol of H_2 and 63.5 g of iodine in a 5.00-L flask at 448°C.

$$H_2 + I_2 \rightleftharpoons 2HI \quad K = 50.2 \text{ at } 448°C$$

90. Butane exists as two isomers, *n*-butane and *iso*-butane.

$$CH_3CH_2CH_2CH_3 \rightleftharpoons CH_3-\overset{\displaystyle CH_3}{\underset{\displaystyle CH_3}{\overset{|}{\underset{|}{C}}}}-H \quad K_p = 2.5 \text{ at } 25°C$$

What is the pressure of *iso*-butane in a container of the two isomers at equilibrium with a total pressure of 1.22 atm?

91. What is the minimum mass of $CaCO_3$ required to establish equilibrium at a certain temperature in a 6.50-L container if the equilibrium constant is 0.050 mol L^{-1} for the decomposition of $CaCO_3$ at that temperature?

$$CaCO_3(s) \rightleftharpoons CaO(s) + CO_2(g)$$

92. Calculate the equilibrium concentrations that result when 1.0 M Cl_2 and 1.0 M H_2O react and come to equilibrium according to the equation

$$2Cl_2(g) + 2H_2O(g) \rightleftharpoons 4HCl(g) + O_2(g) \quad K = 3.2 \times 10^{-14}$$

93. The equilibrium constant for the reaction

$$H_2(g) + CO_2(g) \rightleftharpoons H_2O(g) + CO(g)$$

is 1.6 at 990°C. Calculate the number of moles of each component in the final equilibrium mixture obtained from adding 1.00 mol of H_2, 2.00 mol of CO_2, 0.750 mol of H_2O, and 1.00 mol of CO to a 5.00-L container at 990°C.

94. At 25°C and at 1 atm, the partial pressures in an equilibrium mixture of N_2O_4 and NO_2 are $P_{N_2O_4} = 0.70$ atm and $P_{NO_2} = 0.30$ atm.
    (a) Predict how the pressures of NO_2 and N_2O_4 will change if the total pressure increases to 9.0 atm? Will they increase, decrease, or remain the same?
    (b) Calculate the partial pressures of NO_2 and N_2O_4 when they are in equilibrium at 9.0 atm and 25°C.

95. In a 3.0-L vessel, the following equilibrium partial pressures are measured: N_2, 190 torr; H_2, 317 torr; and NH_3, 1000 torr (3 significant figures).

$$N_2(g) + 3H_2(g) \rightleftharpoons 2NH_3(g)$$

    (a) How will the partial pressures of H_2, N_2, and NH_3 change if H_2 is removed from the system? Will they increase, decrease, or remain the same?
    (b) Hydrogen is removed from the vessel until the partial pressure of nitrogen, at equilibrium, is 250 torr. Calculate the partial pressures of the other substances under the new conditions.

Applications and Additional Exercises

96. The equilibrium constant for the reaction

$$CO + H_2O \rightleftharpoons CO_2 + H_2$$

is 5.0 at a given temperature.
    (a) Upon analysis, an equilibrium mixture of the substances present at the given temperature was found to contain 0.20 mol of CO, 0.30 mol of water vapor, and 0.90 mol of H_2 in a liter. How many moles of CO_2 were in the equilibrium mixture?
    (b) While the same temperature was maintained, additional H_2 was added to the system, and some water vapor was removed by drying. A new equilibrium mixture was thereby established containing 0.40 mol of CO, 0.30 mol of water vapor, and 1.2 mol of H_2 in a liter. How many moles of CO_2 were in the new equi-

librium mixture? Compare this with the quantity in part (a), and discuss whether the second value is reasonable. Explain how it is possible for the water vapor concentration to be the same in the two equilibrium solutions, even though some vapor was removed before the second equilibrium was established.

97. Antimony pentachloride decomposes according to the equation

$$SbCl_5(g) \rightleftharpoons SbCl_3(g) + Cl_2(g)$$

An equilibrium mixture in a 5.00-L flask at 448°C contains 3.85 g of $SbCl_5$, 9.14 g of $SbCl_3$, and 2.84 g of Cl_2. How many grams of each will be found if the mixture is transferred into a 2.00-L flask at the same temperature?

98. What are the concentrations of N_2, H_2, and NH_3 after a mixture that contains 1.00 M N_2 and 3.00 M H_2 comes to equilibrium at 650°C?

$$N_2(g) + 3H_2(g) \rightleftharpoons 2NH_3(g) \quad K = 0.040 \text{ at } 650°C$$

Hint: Changes in concentrations are too large to be neglected in this calculation. If you use successive approximations, many will be required.

99. An equilibrium is established according to the equation

$$Hg_2^{2+} + NO_3^- + 3H^+ \rightleftharpoons 2Hg^{2+} + HNO_2 + H_2O$$
$$K = 4.6$$

Which of the following statements best describes what will happen in a soluton that is 0.20 M each in Hg_2^{2+}, NO_3^-, H^+, Hg^{2+}, and HNO_2? (a) Hg_2^{2+} will be oxidized and NO_3^- reduced. (b) Hg_2^{2+} will be reduced and NO_3^- oxidized. (c) Hg^{2+} will be oxidized and HNO_2 reduced. (d) Hg^{2+} will be reduced and HNO_2 oxidized. (e) There will be no change because all reactants and products have an activity of 0.20.

100. What is the value of the equilibrium constant for the change $H_2O(l) \rightleftharpoons H_2O(g)$ at 30°C? (See Section 10.9 for useful information.)

101. Write the formula for the reaction quotient for the ionization of HOCN in water (see Section 12.4).

102. Write the formula of the reaction quotient for the ionization of NH_3 in water (see Section 12.4).

103. What is the approximate value of the equilibrium constant K_p for the change

$$C_2H_5OC_2H_5(l) \rightleftharpoons C_2H_5OC_2H_5(g)$$

at 25°C. (See Section 11.5 for useful information.)

104. Consider the equilibrium

$$4NO_2(g) + 6H_2O(g) \rightleftharpoons 4NH_3(g) + 7O_2(g)$$

    (a) What is the expression for the equilibrium constant of the reaction?
    (b) How must the concentration of NH_3 change to reach equilibrium if the reaction quotient is less than the equilibrium constant?

(c) If the reaction were at equilibrium, how would a decrease in pressure affect the pressure of NO_2?

(d) If the pressure of NO_2 changes by 28 torr as a mixture of the four gases reaches equilibrium, how much will the pressure of O_2 change?

105. The binding of oxygen by hemoglobin (Hb), giving oxyhemoglobin (HbO_2), is partially regulated by the concentration of H_3O^+ and dissolved CO_2 in the blood. Although the equilibrium is complicated, it can be summarized as

$$HbO_2 + H_3O^+ + CO_2 \rightleftharpoons CO_2\text{—}Hb\text{—}H^+ + O_2 + H_2O$$

(a) Write the equilibrium constant expression for this reaction.

(b) Explain why the production of lactic acid and CO_2 in a muscle during exertion stimulates release of O_2 from the oxyhemoglobin in the blood passing through the muscle.

106. (a) Acetic acid and other carboxylic acids form dimers in the vapor state.

$$2CH_3CO_2H(g) \rightleftharpoons CH_3C\begin{smallmatrix}O-H\cdots O\\ \\O\cdots H-O\end{smallmatrix}CCH_3(g)$$

$$K_p = 3.70 \times 10^{-3} \text{ torr}^{-1} \text{ at } 121.1°C$$

Calculate the equilibrium partial pressure of the dimer if the partial pressure of the monomer of $CH_3CO_2H(g)$ is 85.0 torr.

(b) Calculate K (calculated from concentrations) for the equilibrium.

107. The hydrolysis of the sugar sucrose to the sugars glucose and fructose,

$$C_{12}H_{22}O_{11} + H_2O \longrightarrow C_6H_{12}O_6 + C_6H_{12}O_6$$

follows a first-order rate equation for the disappearance of sucrose.

$$\text{Rate} = k[C_{12}H_{22}O_{11}]$$

In neutral solution, $k = 2.1 \times 10^{-11} \text{ s}^{-1}$ at 27°C. (The products of the reaction, glucose and fructose, have the same molecular formulas but differ in the arrangement of the atoms in their molecules.)

(a) The equilibrium constant for the reaction is 1.36×10^5 at 27°C. What are the concentrations of glucose, fructose, and sucrose after a 0.150 M aqueous solution of sucrose has reached equilibrium? Remember that the activity of a solvent is 1.

(b) How long will the reaction of a 0.150 M solution of sucrose require to reach equilibrium at 27°C in the absence of a catalyst? Because the concentration of sucrose at equilibrium is so low, assume that the reaction is irreversible.

108. (a) The density of trifluoroacetic acid vapor was determined at 118.1°C and 468.5 torr and found to be 2.784 $g L^{-1}$. Calculate K for the association of the acid.

$$2CF_3CO_2H(g) \rightleftharpoons CF_3C\begin{smallmatrix}O-H\cdots O\\ \\O\cdots H-O\end{smallmatrix}CCF_3(g)$$

(b) Calculate K_p for the association of the acid.

109. Chlorine molecules are 1.00% dissociated at 975 K at a pressure of 1.00 atm (1.00% of the pressure is due to Cl atoms).

$$Cl_2(g) \rightleftharpoons 2Cl(g)$$

Calculate K_p and K for the dissociation at that temperature.

110. Liquid N_2O_3 is dark blue at low temperatures, but the color fades and becomes greenish at higher temperatures as the compound decomposes to NO and NO_2. At 25°C a value of $K_p = 1.91$ atm has been established for this decomposition. If 0.236 mol of N_2O_3 is placed in a 1.52-L vessel at 25°C, calculate the equilibrium partial pressures of $N_2O_3(g)$, $NO_2(g)$, and $NO(g)$.

111. (a) A 0.010 M soluton of the weak acid HA has an osmotic pressure (see Section 12.18) of 0.293 atm at 25°C. A 0.010 M solution of the weak acid HB has an osmotic pressure of 0.345 atm under the same conditions. Which acid has the larger equilibrium constant for ionization ($HX \rightleftharpoons H^+ + X^-$)?

(b) What are the equilibrium constants for the ionization of these acids?

112. The amino acid alanine has two isomers, α-alanine and β-alanine. When equal masses of these two compounds are dissolved in equal amounts of a solvent, the solution of α-alanine freezes at the lower temperature (see Section 12.16). Which form, α-alanine or β-alanine, has the larger equilibrium constant for ionization ($HX \rightleftharpoons H^+ + X^-$)?

113. A 1.00-L vessel at 400°C contains the following equilibrium concentratoins: N_2, 1.00 M; H_2, 0.50 M; and NH_3, 0.25 M. How many moles of hydrogen must be removed from the vessel to increase the concentration of nitrogen to 1.1 M?

15

CHAPTER OUTLINE

The Brønsted–Lowry Concept of Acids and Bases

15.1 The Protonic Concept of Acids and Bases
15.2 Amphiprotic Species
15.3 The Strengths of Brønsted Acids and Bases
15.4 Acid–Base Neutralization
15.5 The Relative Strengths of Strong Acids and Bases
15.6 Properties of Brønsted Acids in Aqueous Solution
15.7 Preparation of Brønsted Acids
15.8 Monoprotic, Diprotic, and Triprotic Acids
15.9 Properties of Brønsted Bases in Aqueous Solution
15.10 Preparation of Hydroxide Bases
15.11 Quantitative Reactions of Acids and Bases
15.12 Equivalents of Acids and Bases

The Lewis Concept of Acids and Bases

15.13 Definitions and Examples

Acids and Bases

Cave formations result from acid–base reactions involving limestone, $CaCO_3$.

Our bodies, our homes, our industrial society—acids and bases play an important role in each. Proteins, enzymes, blood, genetic material, and other components of living matter contain both acids and bases. We seem to like the sour taste of acids; we add them to soft drinks, salad dressings, and spices. Many foods, including citrus fruits and some vegetables, contain acids. Cleaners in our homes contain acids or bases. Acids and bases play an important role in the chemical industry. For example, sulfuric acid is the chemical manufactured in the largest quantity, over 3 billion pounds per year in the United States alone. Of the 15 chemicals produced in the largest quantities, half are acids or bases.

Acids and bases have been known for a long time. When Robert Boyle characterized them in 1680, he noted that acids dissolve many substances, change the color of certain natural dyes (for example, they change litmus from blue to red), and lose these characteristic properties after coming in contact with alkalis (bases). In the eighteenth century it was recognized that acids have a sour taste, react with limestone to liberate a gaseous substance (now known to be CO_2), and interact with alkalis to form neutral substances. Humphry Davy contributed greatly to the development of the modern acid–base concept by demonstrating that hydrogen is the essential constituent of acids. In 1814 Gay-Lussac concluded that acids are substances that can neutralize bases and that these two classes of substances can be defined only in terms of each other. The significance of hydrogen was reemphasized in 1884 when Arrhenius defined an acid as a compound that dissolves in water to yield hydrogen ions (now recognized to be hydronium ions) and defined a base as a compound that dissolves in water to yield hydroxide ions.

From Chapter 2 up to this point we have defined acids and bases as Arrhenius did: We identified an acid as a compound that dissolves in water to yield hydronium ions and a base as a compound that dissolves in water to yield hydroxide ions (Section 2.10). This definition is not wrong; it is simply limited. Now we extend the definition of an acid or a base by using the more general definition proposed in 1923 by the Danish chemist Johannes Brønsted and the English chemist Thomas Lowry: Acids donate protons (hydrogen ions), and bases accept protons. Later in this chapter we will introduce a still more general model of acid–base behavior introduced by the American chemist G. N. Lewis.

The Brønsted–Lowry Concept of Acids and Bases

15.1   The Protonic Concept of Acids and Bases

A compound that donates a proton (a hydrogen ion, H^+) to another compound is called a **Brønsted acid,** and a compound that accepts a proton is called a **Brønsted base.** Stated simply, an acid is a proton donor, a base is a proton acceptor. An **acid–base reaction** is the transfer of a proton from a proton donor (acid) to a proton acceptor (base).

Acids may be molecules, such as HCl, H_2SO_4, organic acids (for example, acetic acid, CH_3CO_2H, Section 9.13), and H_2O. Anions, such as HSO_4^-, $H_2PO_4^-$, HS^-, and HCO_3^-, and cations, such as H_3O^+, NH_4^+, and $[Al(H_2O)_6]^{3+}$, also may act as acids.

Bases fall into the same three categories: molecules, such as H_2O, NH_3, and CH_3NH_2; anions, such as OH^-, HS^-, HCO_3^-, CO_3^{2-}, F^-, and PO_4^{3-}; and cations, such as $[Al(H_2O)_5OH]^{2+}$. The most familiar bases are ionic compounds such as NaOH and $Ca(OH)_2$ that contain the hydroxide ion, OH^-. The hydroxide ion in these compounds accepts protons to form water.

$$H_3O^+ + OH^- \longrightarrow 2H_2O$$

We call the product that remains after an acid donates a proton the acid's **conjugate base.** This species is a base because it can accept a proton (to re-form the acid).

Acid		Proton		Conjugate base
HCl	$\rightleftharpoons$	H^+	+	Cl^-
H_2SO_4	$\rightleftharpoons$	H^+	+	HSO_4^-
H_2O	$\rightleftharpoons$	H^+	+	OH^-
HSO_4^-	$\rightleftharpoons$	H^+	+	SO_4^{2-}
NH_4^+	$\rightleftharpoons$	H^+	+	NH_3

We call the product that results when a base accepts a proton the base's **conjugate acid.** This species is an acid because it can give up a proton (and thus re-form the base).

Base		Proton		Conjugate acid
OH^-	+	H^+	$\rightleftharpoons$	H_2O
H_2O	+	H^+	$\rightleftharpoons$	H_3O^+
NH_3	+	H^+	$\rightleftharpoons$	NH_4^+
S^{2-}	+	H^+	$\rightleftharpoons$	HS^-
CO_3^{2-}	+	H^+	$\rightleftharpoons$	HCO_3^-
F^-	+	H^+	$\rightleftharpoons$	HF

Figure 15.1

A white cloud of solid, finely divided NH_4Cl is produced by the acid–base reaction between the colorless gases HCl and NH_3. (The gases in this photograph escaped from concentrated solutions of HCl and NH_3.)

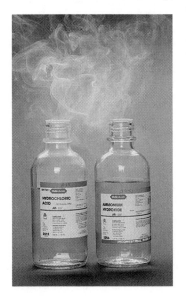

In order for an acid to act as a proton donor and form its conjugate base, a second base must be present to accept the proton from the acid. When the second base accepts the proton, it forms its conjugate acid, a second acid. For example, hydrogen chloride, HCl, reacts with anhydrous ammonia, NH_3, to form an ionic solid that contains ammonium ions, NH_4^+, and chloride ions, Cl^- (Fig. 15.1).

$$HCl + NH_3 \rightleftharpoons NH_4^+ + Cl^-$$

Acid₁   Base₂   Acid₂   Base₁

During this reaction, hydrogen chloride (acid₁) gives up a proton to form chloride ion, its conjugate base (base₁); ammonia acts as a proton acceptor and therefore is a base (base₂). The proton combines with ammonia to give its conjugate acid, the ammonium ion (acid₂). We use a double arrow ($\rightleftharpoons$) in this equation to indicate two things: (1) Hydrogen ion transfer is reversible, so the reaction is a reversible reaction. (2) Acid–base reactions come to equilibrium, although in some cases the equilibrium constants are so very large that the reaction gives essentially a 100% yield of products.

We can write equations for several other acid–base reactions as follows:

Acid₁	+	Base₂	⇌	Acid₂	+	Base₁
HNO_3	+	NH_3	⇌	NH_4^+	+	NO_3^-
HBr	+	F^-	⇌	HF	+	Br^-
HSO_4^-	+	CO_3^{2-}	⇌	HCO_3^-	+	SO_4^{2-}
NH_4^+	+	S^{2-}	⇌	HS^-	+	NH_3

In each of these acid–base reactions, the forward reaction is the transfer of a proton from acid₁ to base₂. The reverse reaction is the transfer of a proton from acid₂ to base₁. Base₁ and base₂ compete for the proton, and the base that has the stronger attraction for the proton determines which species are present in the greatest concentrations at equilibrium. In all the examples given in the foregoing list of reactions, base₂ is the stronger base. As a result, at equilibrium the system consists primarily of a mixture of acid₂ and base₁, and the equilibrium constant is greater than 1.

When we add a Brønsted acid to water, protons are transferred from the acid molecules to water molecules, giving hydronium ions, H_3O^+. The acid is said to **ionize.** The properties common to solutions of acids in water are those of the hydronium ions. For example, when hydrogen fluoride dissolves in water and ionizes, protons are transferred from hydrogen fluoride molecules to water molecules.

$$HF + H_2O \rightleftharpoons H_3O^+ + F^-$$

$$\text{Acid}_1 \quad \text{Base}_2 \quad\quad \text{Acid}_2 \quad \text{Base}_1$$

The fluoride ion is a stronger base than water, so this reaction gives only a small amount of H_3O^+ and F^-; at equilibrium, the majority of the HF molecules have not reacted. The yield of products in this reaction is about 3% in a 1 molar solution—the equilibrium constant is less than 1 ($K = 7.2 \times 10^{-4}$).

When we add a base such as ammonia to water, the base accepts protons from water molecules. Water functions as the proton donor and hence as an acid.

$$H_2O + NH_3 \rightleftharpoons NH_4^+ + OH^-$$

$$\text{Acid}_1 \quad \text{Base}_2 \quad\quad \text{Acid}_2 \quad\quad \text{Base}_1$$

The hydroxide ion is a stronger base than ammonia, so this reaction gives only a small amount of NH_4^+ and OH^-; at equilibrium, the majority of the ammonia molecules have not reacted. The yield of products in this reaction is about 0.5% in a 1 molar solution—the equilibrium constant is less than 1 ($K = 1.8 \times 10^{-5}$).

In the preceding paragraphs we saw that water can act either as an acid or as a base, depending on the nature of its solute. In fact, in pure water or in any solution, water acts both as an acid and a base. A very small fraction of water molecules donate protons to other water molecules to form hydronium ions and hydroxide ions.

$$\underset{\substack{H_2O \\ \text{Acid}_1}}{\overset{\underset{\displaystyle |}{\overset{\displaystyle H}{}}}{:\!\ddot{O}\!-\!H}} + \underset{\substack{H_2O \\ \text{Base}_2}}{\overset{\underset{\displaystyle |}{\overset{\displaystyle H}{}}}{:\!\ddot{O}\!-\!H}} \rightleftharpoons \underset{\substack{H_3O^+ \\ \text{Acid}_2}}{\left[\overset{\underset{\displaystyle |}{\overset{\displaystyle H}{}}}{H\!-\!\ddot{O}\!-\!H}\right]^+} + \underset{\substack{OH^- \\ \text{Base}_1}}{:\!\ddot{O}\!-\!H^-}$$

This type of reaction, in which a substance ionizes when one molecule of the substance reacts with another molecule of the same substance, is referred to as **autoionization,** or **self-ionization.** Pure water exhibits self-ionization to a very slight extent. Only about 2 out of every 10^9 molecules in a sample of pure water are ionized at 25°C.

The equilibrium constant for the ionization of water is called the **ion product for water,** and because of its importance, it is given the special symbol K_w.

$$H_2O(l) + H_2O(l) \rightleftharpoons H_3O^+(aq) + OH^-(aq) \qquad K_w = [H_3O^+][OH^-]$$

As in Chapter 14, we approximate activities with molar concentrations, so the brackets represent "molar concentration of." We should note that this is one of those equilibria that involve the solvent of a dilute solution as a reactant; water is both the reactant and the solvent for the hydronium and hydroxide ions. In such a case, the concentration of the solvent does not appear in the equilibrium equation (Section 14.2).

The very slight ionization of pure water is reflected in the small value of the equilibrium constant; at 25°C, K_w has the value 1.0×10^{-14}. The degree of ionization of water and the resulting concentrations of hydronium ion and hydroxide ion increase as the temperature increases. At 100°C, K_w is about 1×10^{-12}, 100 times larger than the value at 25°C.

Example 15.1   Ion Concentrations in Pure Water

What are the hydronium ion concentration and the hydroxide ion concentration in pure water at 25°C?

Solution: The self-ionization of water yields the same number of hydronium and hydroxide ions. Therefore, in pure water, $[H_3O^+] = [OH^-]$. At 25°C,

$$K_w = [H_3O^+][OH^-] = [H_3O^+]^2 = [OH^-]^2 = 1.0 \times 10^{-14}$$

so

$$[H_3O^+] = [OH^-] = \sqrt{1.0 \times 10^{-14}} = 1.0 \times 10^{-7}\ M$$

The hydronium ion concentration and the hydroxide ion concentration are equal, and we find that both are 1.0×10^{-7} M.

Check your learning: The ion product of water at 80°C is 2.44×10^{-13}. What are the concentrations of the hydronium ion and the hydroxide ion in pure water at 80°C?

Answer: $[H_3O^+] = [OH^-] = 4.94 \times 10^{-7}$ M

It would be difficult for us to prepare a sample of pure water with hydronium and hydroxide ion concentrations equal to those predicted by the equilibrium constant, because almost all water contains dissolved carbon dioxide from the air. Carbon dioxide reacts with water and forms carbonic acid, H_2CO_3, a weak acid which increases the hydronium ion concentration and decreases the hydroxide ion concentration.

All aqueous solutions contain both hydronium ions and hydroxide ions because water undergoes self-ionization. When an acid is added to pure water, the hydronium ion concentration, $[H_3O^+]$, becomes larger than 1.0×10^{-7} M, and the hydroxide ion concentration, $[OH^-]$, becomes less than 1.0×10^{-7} M but never zero. When a base is added to water, $[OH^-]$ becomes greater than 1.0×10^{-7} M, and $[H_3O^+]$ decreases but not to zero. In any aqueous solution (solution in water), the concentration of hydronium ion and that of hydroxide ion are inversely proportional.

Example 15.2   **The Inverse Proportionality of $[H_3O^+]$ and $[OH^-]$**

A solution of carbon dioxide in water has a hydronium ion concentration of 2.0×10^{-6} M. What is the concentration of hydroxide ion at 25°C?

Solution: We know the value of the ion product for water at 25°C.

$$2H_2O(l) \rightleftharpoons H_3O^+(aq) + OH^-(aq) \qquad K_w = [H_3O^+][OH^-] = 1.0 \times 10^{-14}$$

Thus we can calculate the missing equilibrium concentration.

$$K_w = [H_3O^+][OH^-] = 1.0 \times 10^{-14}$$
$$= (2.0 \times 10^{-6})[OH^-] = 1.0 \times 10^{-14}$$

Rearrangement yields

$$[OH^-] = \frac{1.0 \times 10^{-14}}{2.0 \times 10^{-6}} = 5.0 \times 10^{-9}$$

The hydroxide ion concentration in water is reduced to 5.0×10^{-9} M as the hydrogen ion concentration increases to 2.0×10^{-6} M. As expected from Le Châtelier's principle, the autoionization reaction shifts to the left to reduce the stress of the increased hydronium ion concentration, and $[OH^-]$ is reduced relative to that in pure water.

A check of these concentrations confirms that our arithmetic is correct.

$$K_w = [H_3O^+][OH^-] = (2.0 \times 10^{-6})(5.0 \times 10^{-9}) = 1.0 \times 10^{-14}$$

Check your learning: What is the hydronium ion concentration in an aqueous solution with a hydroxide ion concentration of 0.001 M at 25°C?

Answer: $[H_3O^+] = 1 \times 10^{-11}$ M.

15.2   Amphiprotic Species

Many molecules and ions can behave like water and may either gain or lose a proton under the appropriate conditions. Such species are said to be **amphiprotic.**

The proton-containing anions listed in Section 15.1 are amphiprotic, which can be seen from the following equations involving HS^- and HCO_3^-.

Acid$_1$	+	Base$_2$	$\rightleftharpoons$	Acid$_2$	+	Base$_1$
HS$^-$	+	OH$^-$	$\rightleftharpoons$	H$_2$O	+	S^{2-}
HBr	+	HS$^-$	$\rightleftharpoons$	H$_2$S	+	Br$^-$
HCO$_3^-$	+	CN$^-$	$\rightleftharpoons$	HCN	+	CO$_3^{2-}$
H$_3$O$^+$	+	HCO$_3^-$	$\rightleftharpoons$	H$_2$CO$_3$	+	H$_2$O

The hydroxides of metals near the boundary between metals and nonmetals in the periodic table, are amphiprotic and so react either as acids or as bases.

$$[Al(H_2O)_3(OH)_3] + OH^- \rightleftharpoons H_2O + [Al(H_2O)_2(OH)_4]^-$$
$$H_3O^+ + [Al(H_2O)_3(OH)_3] \rightleftharpoons [Al(H_2O)_4(OH)_2]^+ + H_2O$$

In the first reaction, one of the water molecules of the hydrated aluminum hydroxide loses a proton. In the second reaction, the hydrated aluminum hydroxide accepts a proton from the hydronium ion. In both cases the solid hydrated aluminum hydroxide dissolves.

15.3   The Strengths of Brønsted Acids and Bases

We can rank the strengths of acids by their tendency to form hydronium ions in aqueous solution. The reaction of an acid with water is given by the general expression

$$HA + H_2O \rightleftharpoons H_3O^+ + A^-$$

Water is the base that reacts with the acid HA, A$^-$ is the conjugate base of the acid HA, and the hydronium ion is the conjugate acid of water. A **strong acid** gives a 100% yield (or very nearly so) of H$_3$O$^+$ and A$^-$ when the acid ionizes in water. A **weak acid** gives small yields of H$_3$O$^+$ (ordinarily 10% or less).

We can measure the relative strengths of acids by measuring their percent ionization in aqueous solutions. In solutions of the same concentration, stronger acids ionize to a greater extent and so give higher yields of hydronium ions than do weaker acids. The following data on percent ionization in 0.1 molar solutions indicate the order of acid strength CH$_3$CO$_2$H < HNO$_2$ < HSO$_4^-$.

$$CH_3CO_2H + H_2O \rightleftharpoons H_3O^+ + CH_3CO_2^- \quad \text{(1.3\% ionized)}$$
$$HNO_2 + H_2O \rightleftharpoons H_3O^+ + NO_2^- \quad \text{(6.5\% ionized)}$$
$$HSO_4^- + H_2O \rightleftharpoons H_3O^+ + SO_4^{2-} \quad \text{(29\% ionized)}$$

Another measure of the strength of an acid is the magnitude of the equilibrium constant for its ionization. Although the ionization equilibrium behaves like any other homogeneous equilibrium (Section 14.3), the equilibrium constant is called the **ionization constant, K_a**. For the reaction of an acid HA,

$$HA + H_2O \rightleftharpoons H_3O^+ + A^-$$

we write the equation for the ionization constant as

$$K_a = \frac{[H_3O^+][A^-]}{[HA]}$$

where the concentrations are those at equilibrium. Although water is a reactant in the reaction, it is the solvent as well, so we do not include [H$_2$O] in the equation (Section 14.2). The larger the percent ionization of an acid, the larger the concentration of H_3O^+ and A^- relative to the concentration of the nonionized acid, HA. Thus a stronger acid has a larger ionization constant than a weaker acid. For the acids CH_3CO_2H, HNO_2, and HSO_4^-, the ionization constants are

$$CH_3CO_2H + H_2O \rightleftharpoons H_3O^+ + CH_3CO_2^- \qquad K_a = \frac{[H_3O^+][CH_3CO_2^-]}{[CH_3CO_2H]}$$
$$= 1.8 \times 10^{-5}$$

$$HNO_2 + H_2O \rightleftharpoons H_3O^+ + NO_2^- \qquad K_a = \frac{[H_3O^+][NO_2^-]}{[HNO_2]}$$
$$= 4.5 \times 10^{-4}$$

$$HSO_4^- + H_2O \rightleftharpoons H_3O^+ + SO_4^{2-} \qquad K_a = \frac{[H_3O^+][SO_4^{2-}]}{[HSO_4^-]}$$
$$= 1.2 \times 10^{-2}$$

The ionization constants increase as the strengths of the acids increase. (A table of ionization constants of weak acids appears in Appendix F.)

We can rank the strengths of bases by their tendency to form hydroxide ions in aqueous solution. The reaction of a Brønsted base with water is given by

$$B + H_2O \rightleftharpoons HB^+ + OH^-$$

Water is the acid that reacts with the base, HB^+ is the conjugate acid of the base B, and the hydroxide ion is the conjugate base of water. A **strong base** gives a 100% yield (or very nearly so) of OH^- and HB^+ when it reacts with water. A **weak base** gives a small yield of hydroxide ions (ordinarily 10% or less). We usually consider soluble ionic hydroxides such as NaOH to be strong bases, because they dissolve in water to give a 100% yield of hydroxide ions.

As we did with acids, we can measure the relative strengths of bases by measuring their percent ionization in aqueous solutions. In solutions of the same concentration, stronger bases ionize to a greater extent and so give higher yields of hydroxide ions than do weaker bases. The following data on percent ionization in 0.1 molar solutions indicate the order of base strength $NO_2^- < CH_3CO_2^- < NH_3$.

$$NO_2^- + H_2O \rightleftharpoons HNO_2 + OH^- \qquad (0.0015\% \text{ ionization})$$
$$CH_3CO_2^- + H_2O \rightleftharpoons CH_3CO_2H + OH^- \qquad (0.0075\% \text{ ionization})$$
$$NH_3 + H_2O \rightleftharpoons NH_4^+ + OH^- \qquad (1.3\% \text{ ionization})$$

Another measure of the strength of a base is the magnitude of its **ionization constant, K_b.** A stronger base has a larger ionization constant than a weaker base. For the reaction of a base B,

$$B + H_2O \rightleftharpoons HB^+ + OH^-$$

we write the equation for the ionization constant as

$$K_b = \frac{[HB^+][OH^-]}{[B]}$$

where the concentrations are those at equilibrium. Again, we do not include $[H_2O]$ in the equation because water is the solvent. The ionization constants of three bases with strengths that increase in the order $NO_2^- < CH_3CO_2^- < NH_3$ are

$$K_a \times K_b = K_w = 1.0 \times 10^{-14} \quad (\text{at } 25°C) \qquad (1)$$

The ionization constant, K_a, of acetic acid, CH_3CO_2H, is 1.8×10^{-5}. The ionization constant, K_b, of its conjugate base, $CH_3CO_2^-$, is 5.6×10^{-10}. If we multiply these two constants, we see that the product is equal to K_w.

$$K_a \times K_b = (1.8 \times 10^{-5}) \times (5.6 \times 10^{-10}) = 1.0 \times 10^{-14} = K_w$$

Example 15.3   **The Product $K_a \times K_b$**

Show that Equation 1 holds for the nitrite ion, NO_2^-, and its conjugate acid.

Solution: K_b for NO_2^- is given in this section as 2.22×10^{-11}. The conjugate acid of NO_2^- is HNO_2; K_a for HNO_2 is given as 4.5×10^{-4}.

$$K_a \times K_b = (4.5 \times 10^{-4}) \times (2.22 \times 10^{-11}) = 1.0 \times 10^{-14} = K_w$$

Check your learning: We can determine the relative acid strengths of NH_4^+ and HCN by comparing their ionization constants. The ionization constant of HCN is given in Appendix F as 4×10^{-10}. The ionization constant of NH_4^+ is not listed, but the ionization constant of its conjugate base, NH_3, is listed as 1.8×10^{-5}. Determine the ionization constant of NH_4^+, and decide which is the stronger acid, HCN or NH_4^+.

Answer: NH_4^+ is the slightly stronger acid. (K_a for $NH_4^+ = 5.6 \times 10^{-10}$)

An acid will donate a proton to the conjugate base of a weaker acid (any acid that has a smaller ionization constant and hence lies below it in Table 15.1). Nitrous acid, HNO_2, is a stronger acid than acetic acid, CH_3CO_2H, so nitrous acid reacts with acetate ion to form acetic acid and nitrite ion.

$$HNO_2 + CH_3CO_2^- \rightleftharpoons CH_3CO_2H + NO_2^-$$

The nitrite ion, NO_2^-, the conjugate base of nitrous acid, is a weaker base than the acetate ion, $CH_3CO_2^-$, the conjugate base of acetic acid. Thus the acetate ion competes more effectively for the proton than does the nitrite ion.

15.4 Acid–Base Neutralization

A **neutral solution** contains equal concentrations of hydronium and hydroxide ions. When we mix solutions of an acid and a base, an acid–base **neutralization reaction** occurs. However, even if we mix stoichiometrically equivalent quantities, we may find that the resulting solution is not neutral. It could contain either an excess of hydronium ions or an excess of hydroxide ions, because the nature of the salt that is formed determines whether the solution is acidic, neutral, or basic. The following three situations illustrate how acidic or basic solutions can arise.

1. *A strong acid and a weak base yield a weakly acidic solution.* A strong acid reacts with a weak base to form the conjugate base of the strong acid and the conjugate acid of the weak base. The conjugate base of the strong acid is a weaker base than water and has no effect on the acidity of the resulting solution. However, the conjugate acid of the weak base is a weak acid and ionizes slightly in

water. This increases the amount of hydrogen ion in the solution produced in the reaction and renders it slightly acidic. For example, the equation for the reaction between solutions of the strong acid HCl with the weak base ammonia is

$$H_3O^+(aq) + Cl^-(aq) + NH_3(aq) \rightleftharpoons NH_4^+(aq) + Cl^-(aq) + H_2O(l)$$

The conjugate acid of ammonia, the ammonium ion, is a weak acid. After the initial reaction of NH_3 with HCl, a small fraction of the ammonium ions formed give up protons to water molecules, thus producing a weakly acidic solution.

$$NH_4^+(aq) + H_2O(aq) \rightleftharpoons H_3O^+(aq) + NH_3(aq)$$

2. *A weak acid and a strong base yield a weakly basic solution.* A weak acid reacts with a strong base to form the conjugate base of the weak acid and the conjugate acid of the strong base. The conjugate acid of the strong base is a weaker acid than water and has no effect on the acidity of the resulting solution. However, the conjugate base of the weak acid is a weak base and ionizes slightly in water. This increases the amount of hydroxide ion in the solution produced in the reaction and renders it slightly basic. For example, the equation for the reaction of the weak acid acetic acid with sodium hydroxide is

$$CH_3CO_2H(aq) + Na(aq) + OH^-(aq) \rightleftharpoons Na^+(aq) + CH_3CO_2^-(aq) + H_2O(l)$$

The conjugate acid of the hydroxide ion is water, which does not affect the acidity of the solution. The conjugate base of acetic acid is the acetate ion, a weak base. A small fraction of the acetate ions react with water, produce OH^- ions, and give a weakly basic solution.

$$CH_3CO_2^-(aq) + H_2O(l) \rightleftharpoons CH_3CO_2H(aq) + OH^-(aq)$$

3. *A weak acid plus a weak base can yield an acidic, a basic, or a neutral solution.* This is the most complex of the three types of reactions. When the conjugate acid and the conjugate base are *unequal* in strength, the solution can be either acidic or basic, depending on the relative strengths of the two conjugates. Occasionally the weak acid and the weak base have the *same* strength. In this case their respective conjugate base and acid have the same strength, and the solution is neutral.

Figure 15.3

The beaker could contain a solution of HCl, HBr, or HI because each of these acids is a strong electrolyte and is 100% ionized in water.

15.5   The Relative Strengths of Strong Acids and Bases

The strongest acids, such as HCl, HBr, and HI, appear to have about the same strength in water (Fig. 15.3). The water molecule is such a strong base, compared to the conjugate bases Cl^-, Br^-, and I^-, that ionization of these strong acids is essentially complete in aqueous solutions and these acids appear to be of equal strength. In solvents less strongly basic than water, we find that HCl, HBr, and HI differ markedly in their tendency to give up a proton to the solvent. In ethanol, a weaker base than water, the extent of ionization increases in the order HCl < HBr < HI, and it is evident that HI is the strongest of these acids. This tendency of water to equalize any differences in strength among strong acids is known as the **leveling effect of water.**

Water also exerts a leveling effect on the strengths of very strong bases. For example, the oxide ion, O^{2-}, and the amide ion, NH_2^-, are such strong bases that they react completely with water.

$$O^{2-} + H_2O \longrightarrow OH^- + OH^-$$
$$NH_2^- + H_2O \longrightarrow NH_3 + OH^-$$

Thus O^{2-} and NH_2^- appear to have the same base strength in water; they both give a 100% yield of hydroxide ion.

In the absence of any leveling effect, the acid strength of binary compounds of hydrogen with nonmetals increases down a group in the periodic table. For Group 7A the order of increasing acidity is HF < HCl < HBr < HI. Likewise, for Group 6A the order of increasing acid strength is H_2O < H_2S < H_2Se < H_2Te.

Across a row in the periodic table, the acid strength of binary hydrogen compounds increases with increasing electronegativity of the nonmetal atom, because the polarity of the H—A bond increases. Thus the order of increasing acidity (for removal of one proton) across the second row is CH_4 < NH_3 < H_2O < HF; across the third row it is SiH_4 < PH_3 < H_2S < HCl.

Compounds that contain oxygen and one or more hydroxyl (OH) groups can be acidic, basic, or amphoteric, depending on the position in the periodic table (and thus the electronegativity, Section 6.5) of the central atom E, the atom bonded to the hydroxyl group. Such compounds have the general formula $O_nE(OH)_m$; they include sulfuric acid, $O_2S(OH)_2$, sulfurous acid, $OS(OH)_2$, nitric acid, O_2NOH, perchloric acid, O_3ClOH, aluminum hydroxide, $Al(OH)_3$, calcium hydroxide, $Ca(OH)_2$, and potassium hydroxide, KOH.

$$-E-O-H$$

Bond b

Bond a

If the central atom, E, has a low electronegativity, its attraction for electrons is low. Little tendency exists for the central atom to form a strong covalent bond with the oxygen atom. Hence bond a is ionic, hydroxide ions are released to the solution, and the material behaves as a base—this is the case with $Ca(OH)_2$ and KOH. Lower electronegativity is characteristic of the more metallic elements, and hence the metallic elements form ionic, basic hydroxides.

If, on the other hand, the atom E has a relatively high electronegativity, it strongly attracts the electrons it shares with the oxygen atom, making bond a relatively strongly covalent. The oxygen–hydrogen bond, bond b, is thereby weakened because electrons are displaced toward E. Bond b is polar and readily releases hydrogen ions to the solution, so the material behaves as an acid. High electronegativities are characteristic of the more nonmetallic elements. Thus nonmetallic elements form covalent, acidic hydroxides. These acids, which contain OH groups, are called **oxyacids.**

Increasing the oxidation state (Section 8.3) of the central atom E also increases the acidity of an oxyacid, because this increases the attraction of E for the electrons it shares with oxygen and thereby weakens the O—H bond. Sulfuric acid, $O_2S(OH)_2$ (with a sulfur oxidation state of $+6$), is more acidic than sulfurous acid, $OS(OH)_2$ (with a sulfur oxidation state of $+4$). Likewise nitric acid, O_2NOH (nitrogen oxidation state, $+5$), is more acidic than nitrous acid, $ONOH$ (nitrogen oxidation state, $+3$). In each of these pairs, the oxidation state of the central atom is higher for the stronger acid.

The hydroxides of elements with intermediate electronegativities and relatively high oxidation states (for example, elements near the diagonal line separating the metals from the nonmetals in the periodic table) are usually **amphoteric.** This means that the hydroxides act as acids when they react with strong bases and as bases when they

Figure 15.4

The blue-green solution contains the indicator bromcresol green in a neutral solution; the yellow solution contains the same indicator in an acidic solution.

Figure 15.5

Magnesium reacts with a solution of acetic acid, producing hydrogen gas and magnesium acetate.

Figure 15.6

When solid CuO is added to a colorless solution of HNO_3, a blue solution of $Cu(NO_3)_2$ is produced.

react with strong acids. The amphoterism of aluminum hydroxide is reflected in its solubility in both strong acids and strong bases. In strong bases, the relatively insoluble hydrated aluminum hydroxide, $[Al(H_2O)_3(OH)_3]$, is converted to the soluble ion, $[Al(H_2O)_2(OH)_4]^-$, by reaction with hydroxide ion (Section 15.2). In strong acids, it is converted to the soluble ion $[Al(H_2O)_6]^{3+}$ by reaction with hydronium ion. The net equation is

$$3H_3O^+ + [Al(H_2O)_3(OH)_3] \rightleftharpoons [Al(H_2O)_6]^{3+} + 3H_2O$$

15.6   Properties of Brønsted Acids in Aqueous Solution

Aqueous solutions of acids contain higher concentrations of hydronium ion than does pure water. Consequently, these solutions all exhibit the following properties, which are due to the presence of excess hydronium ion.

1. They have a sour taste.

2. They change the color of certain indicators (organic dyes). For example, they change litmus from blue to red and bromcresol green from blue-green to yellow (Fig. 15.4).

3. They react with most representative metals and with the first-row transition metals (except copper) to liberate hydrogen gas (Fig. 15.5). For example,

$$2H_3O^+(aq) + Mg(s) \longrightarrow Mg^{2+}(aq) + H_2(g) + 2H_2O(l)$$

4. They react with many basic metal oxides and hydroxides to form salts and water (Fig. 15.6).

$$2H_3O^+(aq) + 2Br(aq)^- + CuO(s) \longrightarrow Cu^{2+}(aq) + 2Br(aq)^- + 3H_2O(l)$$
$$2H_3O^+(aq) + 2NO_3^-(aq) + Cu(OH)_2(s) \longrightarrow Cu^{2+}(aq) + 2NO_3^-(aq) + 4H_2O(l)$$

5. They react with the salts of weaker acids, such as carbonates or sulfides, to give the weak acid and a new salt.

$$2H_3O^+(aq) + 2ClO_4^-(aq) + FeS(s) \longrightarrow$$
$$H_2S(g) + Fe^{2+}(aq) + 2ClO_4^-(aq) + 2H_2O(l)$$
$$2H_3O^+(aq) + 2Cl^-(aq) + CaCO_3(s) \longrightarrow$$
$$H_2CO_3(aq) + Ca^{2+}(aq) + 2Cl^-(aq) + 2H_2O(l)$$

Carbonic acid (H_2CO_3) is unstable and decomposes to carbon dioxide and water.

$$H_2CO_3(aq) \longrightarrow H_2O(l) + CO_2(g)$$

15.7   Preparation of Brønsted Acids

We can prepare Brønsted acids by any of the following methods.

1. *By direct union of the constituent elements.* Binary compounds of hydrogen with more electronegative nonmetals are acids, so the direct reaction of hydrogen with such nonmetals as F_2, Cl_2, Br_2, and S_8 yields acids.

$$H_2 + Br_2 \longrightarrow 2HBr$$
$$8H_2 + S_8 \longrightarrow 8H_2S$$

2. *By the reaction of an oxide of a nonmetal with water.* Most oxides of nonmetals with oxidation states of three or higher are acidic (Section 8.6). The action of water on such a nonmetal oxide forms an acid.

$$SO_3 + H_2O \longrightarrow H_2SO_4$$
$$P_4O_{10} + 6H_2O \longrightarrow 4H_3PO_4$$
$$CO_2 + H_2O \longrightarrow H_2CO_3$$

3. *By the metathesis reaction* (Section 2.10):
   a. of a salt of a volatile acid with a nonvolatile acid.

$$NaF(s) + H_2SO_4(l) \longrightarrow NaHSO_4(s) + HF(g)$$

   b. of a salt with an acid to produce a second acid and an insoluble solid.

$$Ca_3(PO_4)_2(s) + 3H_2SO_4(l) \longrightarrow 2H_3PO_4(l) + 3CaSO_4(s)$$

   c. of a salt of a weak acid with a strong acid.

$$Na^+(aq) + CH_3CO_2^-(aq) + H_3O^+(aq) + Cl^-(aq) \longrightarrow$$
$$CH_3CO_2H(aq) + Na^+(aq) + Cl^-(aq) + H_2O(l)$$

4. *By the reaction of water with certain nonmetal halides containing polar bonds.*

$$PBr_3 + 3H_2O \longrightarrow H_3PO_3 + 3HBr$$
$$PCl_5 + 4H_2O \longrightarrow H_3PO_4 + 5HCl$$
$$SiI_4 + 4H_2O \longrightarrow Si(OH)_4 + 4HI$$

15.8 Monoprotic, Diprotic, and Triprotic Acids

We can classify acids by the number of protons a molecule can give up in a reaction. Acids such as HCl, HNO_3, and HCN that contain one ionizable hydrogen atom in each molecule are called **monoprotic acids.** Their reactions with water are

$$HCl + H_2O \longrightarrow H_3O^+ + Cl^-$$
$$HNO_3 + H_2O \longrightarrow H_3O^+ + NO_3^-$$
$$HCN + H_2O \longrightarrow H_3O^+ + CN^-$$

Even though it contains four hydrogen atoms, acetic acid, CH_3CO_2H, is also monoprotic because only the hydrogen atom from the carboxyl group (Section 9.13) reacts with bases.

Diprotic acids contain two ionizable hydrogen atoms per molecule; ionization of such acids occurs in two stages. The first ionization always takes place to a greater extent than the second ionization. For example, sulfuric acid ionizes as follows:

First ionization: $H_2SO_4 + H_2O \rightleftharpoons H_3O^+ + HSO_4^-$ K_a = about 10^2

Second ionization: $HSO_4^- + H_2O \rightleftharpoons H_3O^+ + SO_4^{2-}$ $K_a = 1.2 \times 10^{-2}$

Triprotic acids, such as phosphoric acid, contain three ionizable hydrogen atoms per molecule. They ionize in three steps. The first ionization always takes place to a greater extent than the second ionization, and the second to a greater extent than the third.

First ionization: $H_3PO_4 + H_2O \rightleftharpoons H_3O^+ + H_2PO_4^-$ $K_a = 7.5 \times 10^{-3}$

Second ionization: $H_2PO_4^- + H_2O \rightleftharpoons H_3O^+ + HPO_4^{2-}$ $K_a = 6.3 \times 10^{-8}$

Third ionization: $HPO_4^{2-} + H_2O \rightleftharpoons H_3O^+ + PO_4^{3-}$ $K_a = 3.6 \times 10^{-13}$

15.9   Properties of Brønsted Bases in Aqueous Solution

A solution of a Brønsted base in water contains a higher concentration of hydroxide ion than that which is found in pure water, because bases either contain hydroxide ions or react with water to form hydroxide ions. Solutions of bases exhibit the following properties.

1. They have a bitter taste.
2. They change the colors of certain indicators. For example, they change litmus from red to blue, phenolphthalein from colorless to red, and alizarin from yellow to red (Fig. 15.7).
3. They neutralize acids. Solutions of metal hydroxides give a salt and water.

$$Na^+(aq) + OH^-(aq) + HCl(g) \longrightarrow Na^+(aq) + Cl^-(aq) + H_2O(l)$$
      Metal hydroxide

Solutions of other bases give a salt or give a salt and a weak electrolyte.

$$NH_3(aq) + HNO_3(l) \longrightarrow NH_4^+(aq) + NO_3^-(aq)$$
     Weak base

$$Na^+(aq) + CH_3CO_2^-(aq) + HBr(g) \longrightarrow Na^+(aq) + Br^-(aq) + CH_3CO_2H(aq)$$
   Salt of a weak acid

15.10   Preparation of Hydroxide Bases

We can prepare bases that contain hydroxide ions (hydroxide bases) by the following methods.

Figure 15.7

(A) The colorless liquid is a solution of the indicator phenolphthalein in a neutral solution. The pink solution contains the same indicator in a basic solution. (B) The red solution contains the indicator alizarin in base. The yellow liquid is a neutral solution of the same indicator.

(A)   (B)

Figure 15.8

Calcium reacts with water and produces hydrogen gas and the base $Ca(OH)_2$.

1. *By the reaction of oxides of active metals with water.* The soluble oxides of active metals react with water to give bases (Section 8.5).

$$Li_2O + H_2O \longrightarrow 2LiOH$$
$$SrO + 2H_2O \longrightarrow Sr(OH)_2$$

With an excess of water, aqueous solutions of the soluble bases will form.

2. *By the reaction of active metals with water.* The metals of Group 1A, and Ca, Sr, and Ba of Group 2A, react directly with a stoichiometric amount of water to give strong bases (Fig. 15.8).

$$2K + 2H_2O \longrightarrow 2KOH + H_2$$
$$Ca + 2H_2O \longrightarrow Ca(OH)_2 + H_2$$

With an excess of water, aqueous solutions of the soluble bases will form.

3. *By the electrolysis of certain salt solutions.*

$$2Na^+ + 2Cl^- + 2H_2O \xrightarrow{\text{Electrolysis}} 2Na^+ + 2OH^- + H_2(g) + Cl_2(g)$$

15.11   Quantitative Reactions of Acids and Bases

We often use a chemical reaction to **standardize** a solution—that is, to determine its concentration. We can use a **standard solution** (a solution of known concentration) in a titration and determine the amount of a substance present in a sample. Because the end point of a titration (Section 3.13) involving a neutralization reaction can be detected readily by the use of an indicator, solutions of acids and bases are often used in these analyses.

We discussed titration in Section 3.13. The following example indicates how to use a titration to standardize a solution.

Example 15.4   **Standardization of a Solution**

Sometimes we use solid potassium hydrogen phthalate, $KHC_8H_4O_4$, as a standard acid in the laboratory because it is easy to purify and to weigh. If we titrate 1.5024 g of this acid with 37.28 mL of NaOH solution, what is the concentration of the NaOH solution?

$$KHC_8H_4O_4 + NaOH \longrightarrow KNaC_8H_4O_4 + H_2O$$

Solution: Titration calculations are fairly straightforward stoichiometry calculations. In this case the chain of conversions is

| Mass of $KHC_8H_4O_4$ | →[1] | Moles of $KHC_8H_4O_4$ | →[2] | Moles of NaOH | →[3] | Concentration of NaOH |

The numbers above the arrows refer to the conversion steps below.

Step 1.   *Convert the mass of KHC$_8$H$_4$O$_4$ to moles of KHC$_8$H$_4$O$_4$.*

$$\text{Moles of KHC}_8\text{H}_4\text{O}_4 = 1.5024 \text{ g KHC}_8\text{H}_4\text{O}_4 \times \frac{1 \text{ mol KHC}_8\text{H}_4\text{O}_4}{204.223 \text{ g KHC}_8\text{H}_4\text{O}_4}$$

$$= 7.3567 \times 10^{-3} \text{ mol KHC}_8\text{H}_4\text{O}_4$$

Step 2.   *Convert the moles of KHC$_8$H$_4$O$_4$ to moles of NaOH.*

$$\text{Moles of NaOH} = 7.3567 \times 10^{-3} \text{ mol KHC}_8\text{H}_4\text{O}_4 \times \frac{1 \text{ mol NaOH}}{1 \text{ mol KHC}_8\text{H}_4\text{O}_4}$$

$$= 7.3567 \times 10^{-3} \text{ mol NaOH}$$

Step 3.   *Convert the moles of NaOH to concentration of NaOH.*

$$\text{Molarity of NaOH} = \frac{7.3567 \times 10^{-3} \text{ mol NaOH}}{0.03728 \text{ L}} = 0.1973 \text{ M}$$

Check your learning: Titration of a 40.00-mL sample of a solution of H_3PO_4 requires 35.00 mL (0.03500 L) of 0.1500 M KOH to reach the end point. Determine the molar concentration of H_3PO_4 if the reaction is $2KOH + H_3PO_4 \longrightarrow K_2HPO_4 + 2H_2O$.

Answer: 0.06562 M

We can solve titration problems using mole relationships, as described here and in Section 3.13. An alternative method in which we use equivalents is described in the next section.

15.12   Equivalents of Acids and Bases

An **equivalent of an acid** is the amount of the acid that provides 1 mole of protons (H^+) in an acid–base reaction. An **equivalent of a base** is the amount of the base that provides 1 mole of hydroxide ions in an acid–base reaction or that will react with 1 mole of protons in an acid–base reaction. One equivalent of an acid will react with 1 equivalent of a base.

We can see that 1 mole of a monoprotic acid, such as HCl, HNO_3, or CH_3CO_2H, is 1 equivalent of the acid because each molecule of the acid contains only one ionizable hydrogen ion. When hydrochloric acid reacts with sodium hydroxide, a hydrogen is transferred to the hydroxide ion.

$$HCl + NaOH \longrightarrow NaCl + H_2O$$

When 1 mole of hydrochloric acid reacts, 1 mole of hydrogen ion is transferred to 1 mole of hydroxide ions. It follows, then, that 1 equivalent (equiv) of hydrochloric acid is the same as 1 mol of HCl (1 equiv HCl/1 mol HCl). Similarly, 1 equivalent of sodium hydroxide is the same as 1 mole of NaOH (1 equiv NaOH/1 mol NaOH).

One mole of a diprotic acid such as H_2SO_4 can provide either 1 mole or 2 moles of protons, depending on the reaction. For example, 1 mole of sulfuric acid will react with 1 mole of sodium hydroxide, transferring 1 mole of protons to the base,

$$H_2SO_4 + NaOH \longrightarrow NaHSO_4 + H_2O$$

or will react with 2 moles of sodium hydroxide, transferring 2 moles of protons to the base.

$$H_2SO_4 + 2NaOH \longrightarrow Na_2SO_4 + 2H_2O$$

Note in the first reaction, 1 equivalent of sulfuric acid is equal to 1 mole of the acid (1 equiv H_2SO_4/1 mol). In the second reaction, 1 equivalent of sulfuric acid is equal to ½ mole of the acid (2 equiv H_2SO_4/1 mol).

The number of protons actually transferred in an acid–base reaction determines the mass of an equivalent of the acid or the base. Thus *the mass of an equivalent of an acid or a base must be deduced from the reaction, not merely from the formula of the substance.*

Example 15.5   **Calculation of the Mass of an Equivalent of an Acid**

Calculate the mass of an equivalent of H_3PO_4 when it combines with NaOH according to the equation

$$H_3PO_4 + 3NaOH \longrightarrow Na_3PO_4 + 3H_2O$$

Solution: The mass of 1 mol of H_3PO_4 is 97.994 g. A sample of 1 mol of H_3PO_4 provides 3 mol of H^+ (3 equiv H_3PO_4/1 mol H_3PO_4).

$$\frac{97.994 \text{ g } H_3PO_4}{1 \text{ mol } H_3PO_4} \times \frac{1 \text{ mol } H_3PO_4}{3 \text{ equiv } H_3PO_4} = \frac{32.665 \text{ g } H_3PO_4}{1 \text{ equiv } H_3PO_4}$$

The mass of an equivalent of H_3PO_4 in this reaction is 32.665 g.

Check your learning: Calculate the mass of an equivalent of H_3PO_4 when it combines with NaOH using the equation $H_3PO_4 + 2NaOH \longrightarrow Na_2HPO_4 + 2H_2O$.

Answer: 48.997 g

The **normality, N,** of a solution is the number of equivalents of solute contained in a liter of solution. (Compare this with the definition of molarity given in Section 3.7.)

$$\text{Normality} = \frac{\text{equivalents of solute}}{\text{liters of solution}} = \frac{\text{equiv}}{L}$$

Normality can also be determined from the molarity of the solution, provided that the reaction is known.

$$\text{Normality} = \frac{\text{moles of solute}}{\text{liters of solution}} \times \frac{\text{equivalents of solute}}{\text{moles of solute}} = \frac{\text{equiv}}{L}$$

A solution that contains 1 equivalent of solute in 1 liter of solution is a 1 N (one normal) solution. Provided that both hydrogens react, a 1 N solution of sulfuric acid contains 1 equivalent, or 98/2 grams = 49 grams, of H_2SO_4 per liter; a 2 N solution of H_2SO_4 contains 2 equivalents, or 98 grams, per liter; and a 0.01 N solution contains

0.49 gram per liter. A 1 N solution of sulfuric acid, if both hydrogens react, is identical to a 0.5 M solution of sulfuric acid; each contains 49 grams of solute per liter of solution.

Example 15.6   Normality of a Solution

Calculate the normality of a solution that contains 3.65 g of HCl per 0.500 L of solution. Use 36.5g as the mass of an equivalent of HCl.

Solution: The normality of HCl is the number of equivalents of HCl dissolved in 1 L of solution. We can find the normality with the following two steps.

$$\boxed{\text{Mass of HCl}} \xrightarrow{\ \ 1\ \ } \boxed{\text{Equivalents of HCl}} \xrightarrow{\ \ 2\ \ } \boxed{\text{Normality of HCl}}$$

The numbers above the arrows refer to the conversion steps below.

Step 1.   *Convert the mass of HCl to the number of equivalents of HCl.*

$$\text{Equivalents of HCl} = 3.65 \ \overline{\text{g HCl}} \times \frac{1 \ \text{equiv HCl}}{36.5 \ \overline{\text{g HCl}}} = 0.100 \ \text{equiv HCl}$$

Step 2.   *Convert the equivalents of HCl to normality of HCl.*

$$\text{Normality of HCl} = \frac{0.100 \ \text{equiv}}{0.500 \ \text{L}} = 0.200 \ \text{N HCl}$$

Check your learning: What is the normality of 0.600 L of a solution containing 2.35 g of H_3PO_4 for a reaction in which two of the hydrogens react?

Answer: 7.99×10^{-2} N H_3PO_4

The Lewis Concept of Acids and Bases

15.13   Definitions and Examples

In 1923, G. N. Lewis proposed a generalized model of acid–base behavior in which acids and bases are identified by their ability to accept or to donate a pair of electrons and form a coordinate covalent bond (Section 6.8). A **Lewis acid** is any species (molecule or ion) that can accept a pair of electrons, and a **Lewis base** is any species (molecule or ion) that can donate a pair of electrons.

An acid–base reaction occurs when a base donates a pair of electrons to an acid. A **Lewis acid–base adduct,** a compound that contains a coordinate covalent bond between the Lewis acid and the Lewis base, is formed. The following equations illustrate the general application of the Lewis model.

The boron atom in boron trifluoride, BF_3, has only six electrons in its valence shell. Consequently, BF_3 is a very good Lewis acid and reacts with many Lewis bases; fluoride ion is the Lewis base in this reaction:

$$:\ddot{F}:^- + B-\ddot{F}: \longrightarrow \left[:\ddot{F}-B-\ddot{F}:\right]^-$$

Lewis base   Lewis acid   Acid–base adduct

In the following reaction, each of two ammonia molecules, Lewis bases, donates a pair of electrons to a silver ion, the Lewis acid.

$$2\,H-\overset{H}{\underset{H}{N}}: + Ag^+ \longrightarrow \left[H-\overset{H}{\underset{H}{N}}-Ag-\overset{H}{\underset{H}{N}}-H\right]^+$$

Lewis base   Lewis acid   Acid–base adduct

Figure 15.9 shows an analogous reaction, in which four ammonia molecules serve as Lewis bases. Each ammonia molecule donates a pair of electrons to a copper ion that serves as the Lewis acid.

$$4NH_3 + Cu^{2+} \longrightarrow \left[\begin{matrix} H_3N & & NH_3 \\ & Cu & \\ H_3N & & NH_3 \end{matrix}\right]^{2+}$$

Lewis base   Lewis acid   Acid–base adduct

Nonmetal oxides act as Lewis acids and react with oxide ions, the Lewis base, to form oxyanions.

$$:\ddot{O}:^{2-} + S\!\!=\!\!\ddot{O}: \longrightarrow \left[:\ddot{O}-S-\ddot{O}:\right]^{2-}$$

Lewis base   Lewis acid   Acid–base adduct

Many Lewis acid–base reactions are displacement reactions in which one Lewis base displaces another Lewis base from an acid–base adduct or in which one Lewis acid displaces another Lewis acid.

$$\left[H-\overset{H}{\underset{H}{N}}-Ag-\overset{H}{\underset{H}{N}}-H\right]^+ + 2[:C\!\!\equiv\!\!N:]^- \longrightarrow [:N\!\!\equiv\!\!C-Ag-C\!\!\equiv\!\!N:]^- + 2\,:\overset{H}{\underset{H}{N}}-H$$

Acid–base adduct   Base   New adduct   New base

Figure 15.9

The tube on the left contains a 0.1 M solution of $CuSO_4$. The tube on the right contains a 0.1 M solution of $[Cu(NH_3)_4]^{2+}$, the Lewis acid–base adduct of Cu^{2+} and NH_3.

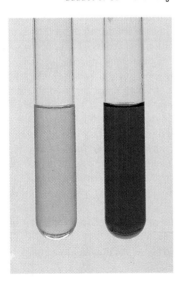

$$\left[\begin{array}{c} :\ddot{O}: \\ | \\ C=\ddot{O}: \\ | \\ :\ddot{O}: \end{array}\right]^{2-} + \begin{array}{c} :\ddot{O}: \\ || \\ S=\ddot{O}: \\ | \\ :\ddot{O}: \end{array} \longrightarrow \left[\begin{array}{c} :\ddot{O}: \\ | \\ :\ddot{O}-S-\ddot{O}: \\ | \\ :\ddot{O}: \end{array}\right]^{2-} + :\ddot{O}=C=\ddot{O}:$$

Acid–base adduct   Acid   New adduct   New acid

$$H-\ddot{Cl}: \ + :\ddot{O}-H \longrightarrow \left[H-\ddot{O}-H\right]^{+} + \ :\ddot{Cl}:^{-}$$
$$\qquad\qquad\quad | \qquad\qquad\qquad |$$
$$\qquad\qquad\quad H \qquad\qquad\qquad H$$

Acid–base adduct   Base   New adduct   New base

The last displacement reaction shows how the reaction of a Brønsted acid with a base fits into the Lewis model. A Brønsted acid such as HCl is an acid–base adduct according to the Lewis concept, and proton transfer occurs because a more stable acid–base adduct is formed. Thus, although the definitions of acids and bases in the two theories are quite different, the theories overlap considerably.

For Review   Summary

A compound that can donate a proton (a hydrogen ion) to another compound is called a **Brønsted acid.** The compound that accepts the proton is called a **Brønsted base.** The species remaining after a Brønsted acid has lost a proton is the **conjugate base** of the acid. The species formed when a Brønsted base gains a proton is the **conjugate acid** of the base. Thus an **acid–base reaction** occurs when a proton is transferred from an acid to a base, with formation of the conjugate base of the reactant acid and formation of the conjugate acid of the reactant base. **Amphiprotic species** can act as both proton donors and proton acceptors. Water is the most important amphiprotic species. It can form both the hydronium ion, H_3O^+, and the hydroxide ion, OH^-.

The strengths of Brønsted acids and bases in aqueous solutions can be determined by measurement of their percent ionization or of their ionization constants. Stronger acids form weaker conjugate bases, and weaker acids form stronger conjugate bases. Thus **strong acids** are completely ionized in aqueous solution because their conjugate bases are weaker bases than water ($K_a > 1$). **Weak acids** are only partially ionized ($K_a < 1$), because their conjugate bases are strong enough to compete successfully with water for possession of protons. **Strong bases** react with water to give a 100% yield of hydroxide ion ($K_b > 1$). **Weak bases** give only small yields of hydroxide ion ($K_b < 1$). The strengths of the binary acids increase from left to right across a period of the periodic table ($CH_4 < NH_3 < H_2O < HF$), and they increase down a

group ($HF < HCl < HBr < HI$). The strengths of oxyacids that contain the same central element increase as the oxidation number of the element increases ($H_2SO_3 < H_2SO_4$). The strengths of oxyacids also increase as the electronegativity of the central element increases ($H_2SeO_4 < H_2SO_4$).

The characteristic properties of aqueous solutions of Brønsted acids are due to the presence of hydronium ions; the properties of aqueous solutions of Brønsted bases are due to the presence of hydroxide ions. The **neutralization** that occurs when aqueous solutions of acids and bases are mixed results from the reaction of the hydronium and hydroxide ions to form water. Some salts formed in neutralization reactions may make the product solutions slightly acidic or slightly basic.

The concentration of an acid or a base in solution may be described in terms of the number of moles per liter (molarity) or in terms of the number of equivalents per liter (**normality**). An **equivalent of an acid** is the amount of the acid that will provide 1 mole of protons in a reaction. An **equivalent of a base** is the amount of the base that will provide 1 mole of hydroxide ions or will accept 1 mole of protons.

A **Lewis acid** is a molecule or ion that can accept a pair of electrons. A **Lewis base** is a molecule or ion that can donate a pair of electrons. Thus a Lewis acid–base reaction results in the formation of a coordinate covalent bond.

Key Terms and Concepts

acid–base reaction (15.1)
amphiprotic (15.2)
amphoteric (15.5)
autoionization (15.1)
Brønsted acid (15.1)
Brønsted base (15.1)
conjugate acid (15.1)
conjugate base (15.1)
diprotic acids (15.8)
equivalent of a base (15.12)

equivalent of an acid (15.12)
ion product for water, K_w (15.1)
ionization constant, K_a, K_b (15.3)
ionize (15.1)
leveling effect of water (15.5)
Lewis acid (15.13)
Lewis acid–base adduct (15.13)
Lewis base (15.13)
monoprotic acids (15.8)
neutral solution (15.4)

neutralization reaction (15.4)
normality, N (15.12)
oxyacid (15.5)
self-ionization (15.1)
standard solution (15.11)
strong acid (15.3)
strong base (15.3)
triprotic acids (15.8)
weak acid (15.3)
weak base (15.3)

Exercises

Questions

1. In Section 2.10 we saw an introductory definition of an acid: An acid is a compound that reacts with water and increases the amount of hydronium ion present. In this chapter we have seen two more definitions of acids: A compound that donates a proton (a hydrogen ion, H^+) to another compound is called a Brønsted acid, and a Lewis acid is any species that can accept a pair of electrons. Explain why the introductory definition is a macroscopic definition, whereas the Brønsted definition and the Lewis definition are microscopic definitions.

2. (a) Identify the strong Brønsted acids and the strong Brønsted bases in the list of important industrial inorganic compounds in Table 8.8.
   (b) List those compounds in Table 8.8 that can behave as Brønsted acids with strengths lying between those of H_3O^+ and H_2O.
   (c) List those compounds in Table 8.8 that can behave as Brønsted bases with strengths lying between those of H_2O and OH^-.

3. Describe the role played by water (acid, base, and/or solvent) when each of the following compounds dissolves, and explain why the resulting solution is acidic, basic, or neutral.

   (a) $HCl(g)$ (b) $NH_3(g)$
   (c) $BaO(s)$ (d) $LiNH_2(s)$
   (e) $(H_3O)ClO_4(s)$ (f) $KOH(s)$
   (g) $KCl(s)$

4. Describe the role played by water (acid, base, and/or solvent) when each of the following compounds dissolves, and explain why the resulting solution is acidic, basic, or neutral.

   (a) $Na_2O(s)$ (b) $NaI(s)$
   (c) $(H_3O)_2SO_4(s)$ (d) $LiOH(s)$
   (e) $NaH(s)$ (f) $HF(g)$
   (g) $NH_4NO_3(s)$

5. Write equations that show NH_3 as a conjugate acid and as a conjugate base.

6. Write equations that show $H_2PO_4^-$ acting both as an acid and as a base.

7. The odor of vinegar is due to the presence of acetic acid, CH_3CO_2H, a weak acid. List, in order of descending concentration, all of the ionic and molecular species present in a 1 M aqueous solution of this acid.

8. Household ammonia is a solution of the weak base NH_3 in water. List, in order of descending concentration, all of the ionic and molecular species present in a 1 M aqueous solution of this base.

9. Explain why the equilibrium constant for the ionization of an acid in aqueous solution can be used as a measure of the strength of the acid.

10. Explain why the equilibrium constant for the reaction of a base with the water in an aqueous solution can be used as a measure of the strength of the base.

11. (a) Explain why $LiNH_2$ and NaC_2H_5O appear to have the same base strength in water.
    (b) How can we experimentally determine which is the stronger base?

12. (a) Explain why $HClO_4$ and HI appear to have the same acid strength in water.
    (b) How can we determine which is the stronger acid?

13. Explain why the ionization constant, K_a, for H_2SO_4 is larger than the ionization constant for H_2SO_3.

14. Explain why the ionization constant, K_a, for HI is larger than the ionization constant for HF.

15. How do we best experimentally identify an amphoteric hydroxide?

16. Describe how the reaction of water with oxides of active metals differs from that of water with oxides of nonmetals.

17. Write the ionic equation for the reaction between aqueous solutions of a strong acid and a strong base.

18. One molecular compound in the list of important industrial compounds in Table 8.8 can act as a Lewis base, and one molecular compound can act as a Lewis acid. Identify these compounds.

Brønsted Acids and Bases

19. Show by suitable equations that each of the following species can act as a Brønsted acid.

    (a) H_3O^+ (b) HCl
    (c) NH_3 (d) CH_3CO_2H
    (e) NH_4^+ (f) HSO_4^-

20. Show by suitable equations that each of the following species can act as a Brønsted acid.

    (a) HNO_3 (b) PH_4^+
    (c) H_2S (d) $C_2H_5CO_2H$
    (e) $H_2PO_4^-$ (f) HS^-

21. Show by suitable equations that each of the following species can act as a Brønsted base.

    (a) H_2O (b) OH^-
    (c) NH_3 (d) CN^-
    (e) S^{2-} (f) $H_2PO_4^-$

22. Show by suitable equations that each of the following species can act as a Brønsted base.

    (a) HS^- (b) PO_4^{3-}
    (c) NH_2^- (d) C_2H_5OH
    (e) O^{2-} (f) HPO_4^{2-}

23. What is the conjugate acid of each of the following? What is the conjugate base of each?

    (a) OH^- (b) H_2O
    (c) HCO_3^- (d) NH_3
    (e) HSO_4^- (f) H_2O_2
    (g) HS^- (h) $H_5N_2^+$

24. What is the conjugate acid of each of the following? What is the conjugate base of each?

    (a) H_2S (b) $H_2PO_4^-$
    (c) PH_3 (d) HS^-
    (e) HSO_3^- (f) $H_3O_2^+$
    (g) H_4N_2 (h) CH_3OH

25. Identify and label the Brønsted acid, its conjugate base, the Brønsted base, and its conjugate acid in each of the following equations.

    (a) $HNO_3 + H_2O \longrightarrow H_3O^+ + NO_3^-$
    (b) $CN^- + H_2O \longrightarrow HCN + OH^-$
    (c) $H_2SO_4 + Cl^- \longrightarrow HCl + HSO_4^-$
    (d) $HSO_4^- + OH^- \longrightarrow SO_4^{2-} + H_2O$
    (e) $O^{2-} + H_2O \longrightarrow 2OH^-$
    (f) $[Cu(H_2O)_3(OH)]^+ + [Al(H_2O)_6]^{3+} \longrightarrow$
        $[Cu(H_2O)_4]^{2+} + [Al(H_2O)_5(OH)]^{2+}$
    (g) $H_2S + NH_2^- \longrightarrow HS^- + NH_3$

26. Identify and label the Brønsted acid, its conjugate base, the Brønsted base, and its conjugate acid in each of the following equations.

    (a) $NO_2^- + H_2O \longrightarrow HNO_2 + OH^-$
    (b) $HBr + H_2O \longrightarrow H_3O^+ + Br^-$
    (c) $HS^- + H_2O \longrightarrow H_2S + OH^-$
    (d) $H_2PO_4^- + OH^- \longrightarrow HPO_4^{2-} + H_2O$
    (e) $H_2PO_4^- + HCl \longrightarrow H_3PO_4 + Cl^-$
    (f) $[Fe(H_2O)_5(OH)]^{2+} + [Al(H_2O)_6]^{3+} \longrightarrow$
        $[Fe(H_2O)_6]^{3+} + [Al(H_2O)_5(OH)]^{2+}$
    (g) $CH_3OH + H^- \longrightarrow CH_3O^- + H_2$

27. Write equations for the reaction of each of the following with water.

    (a) HCl (b) HNO_3
    (c) NH_3 (d) NH_2^-
    (e) $HClO_4$ (f) F^-
    (g) NH_4^+

28. Gastric juice, the digestive fluid produced in the stomach, contains hydrochloric acid, HCl. Milk of magnesia, a suspension of solid $Mg(OH)_2$ in an aqueous medium, is sometimes used to neutralize excess stomach acid. Write a complete balanced equation for the neutralization reaction, and identify the conjugate acid–base pairs.

29. Nitric acid reacts with insoluble copper(II) oxide to form copper(II) nitrate, $Cu(NO_3)_2$, a soluble compound that has been used to prevent the growth of algae in swimming pools. Write the balanced chemical equation for the reaction of an aqueous solution of HNO_3 with CuO.

30. What are amphiprotic species? Illustrate with suitable equations.

31. State which of the following species are amphiprotic, and write chemical equations illustrating the amphiprotic character of these species.

    (a) H_2O (b) $H_2PO_4^-$
    (c) S^{2-} (d) CH_4
    (e) HSO_4^- (f) $Al(H_2O)_6^{3+}$

32. State which of the following species are amphiprotic, and write chemical equations illustrating the amphiprotic character of these species.

    (a) NH_3
    (b) $H_2PO_4^-$
    (c) Br^-
    (d) NH_4^+

(e) AsO_4^{3-}

(f) $Fe(H_2O)_5(OH)^{2+}$

Strengths of Acids and Bases

33. Determine the ionization constant at 25°C for the weak acid $CH_3NH_3^+$, the conjugate acid of the weak base CH_3NH_2, $K_b = 4.4 \times 10^{-4}$.

34. Determine the ionization constant at 25°C for the weak acid $(CH_3)_2NH_2^+$, the conjugate acid of the weak base $(CH_3)_2NH$, $K_b = 7.4 \times 10^{-4}$.

35. Determine the ionization constant at 25°C for the weak base HPO_4^{2-}, the conjugate base of the weak acid $H_2PO_4^-$, $K_a = 7 \times 10^{-7}$.

36. Determine the ionization constant at 25°C for the weak base HTe^-, the conjugate base of the weak acid H_2Te, $K_a = 2.3 \times 10^{-3}$.

37. Which of the bases in Exercises 33 and 34, CH_3NH_2 or $(CH_3)_2NH$, is the stronger base? Which of the conjugate acids in Exercises 33 and 34 is the stronger acid?

38. Which of the acids in Exercises 35 and 36, $H_2PO_4^-$ or H_2Te, is the stronger acid? Which of the conjugate bases in Exercises 35 and 36 is the stronger base?

39. Using ionization constants, determine which is the stronger acid, NH_4^+ or HBrO.

40. Using ionization constants, determine which is the stronger base, $(CH_3)_3N$ or $H_2BO_3^-$.

41. Predict which acid in each of the following pairs is the stronger, and explain your reasoning for each choice.

(a) H_2O or HF

(b) $B(OH)_3$ or $Al(OH)_3$

(c) HSO_3^- or HSO_4^-

(d) NH_3 or H_2S

(e) H_2O or H_2Te

42. Predict which acid in each of the following pairs is the stronger, and explain your reasoning for each choice.

(a) HSO_4^- or $HSeO_4^-$

(b) NH_3 or H_2O

(c) PH_3 or HI

(d) NH_3 or PH_3

(e) H_2S or HBr

43. Rank the compounds in each of the following groups in order of increasing acidity or basicity, as indicated, and explain the order you assign.

(a) Acidity: HCl, HBr, HI

(b) Basicity: H_2O, OH^-, H^-, Cl^-

(c) Basicity: $Mg(OH)_2$, $Si(OH)_4$, $ClO_3(OH)$

(d) Acidity: HF, H_2O, NH_3, CH_4

(e) Basicity: ClO^-, ClO_2^-, ClO_3^-, ClO_4^-

(f) Acidity: HOI, $HOIO$, $HOIO_2$, $HOIO_3$

44. Rank the compounds in each of the following groups in order of increasing acidity or basicity, as indicated, and explain the order you assign.

(a) Acidity: $NaHSO_3$, $NaHSeO_3$, $NaHSO_4$

(b) Basicity: BrO_2^-, ClO_2^-, IO_2^-

(c) Acidity: HOCl, HOBr, HOI

(d) Acidity: HOCl, HOClO, $HOClO_2$, $HOClO_3$

(e) Basicity: NH_2^-, HS^-, HTe^-, PH_2^-

(f) Basicity: BrO^-, BrO_2^-, BrO_3^-, BrO_4^-

45. List, in order of descending concentration, all of the ionic and molecular species, including solvent, present in the solution that results upon mixing equal volumes of 0.1 M solutions of each of the following.

(a) HNO_3 and NaOH

(b) HBr and NH_3

(c) NH_4NO_3 and NaOH

(d) HCl and KCH_3CO_2

46. List, in order of descending concentration, all of the ionic and molecular species, including solvent, present in the solution that results upon mixing equal volumes of 0.1 M solutions of each of the following.

(a) HF and KOH

(b) HBr and $NaNH_2$

(c) $NaHCO_3$ and NaOH

(d) H_2SO_4 and LiOH

47. Both HF and HCN ionize in water to a limited extent. Which of the conjugate bases, F^- or CN^-, is the stronger base? See Table 15.1.

48. Soaps are sodium and potassium salts of a family of acids called fatty acids, which are isolated from animal fats. These acids, which are related to acetic acid, CH_3CO_2H, all contain the carboxyl group, $—CO_2H$, and have about the same strength as acetic acid. Palmitic acid $C_{15}H_{31}CO_2H$, and stearic acid, $C_{17}H_{35}CO_2H$ are two examples.

(a) Write a balanced chemical equation indicating the formation of sodium palmitate, $C_{15}H_{31}CO_2Na$, from palmitic acid and sodium carbonate, and write a corresponding equation for the formation of sodium stearate.

(b) Is a soap solution acidic, basic, or neutral?

Solutions of Brønsted Acids and Bases

49. Write equations to illustrate three typical and characteristic reactions of aqueous acids.

50. Containers of $NaHCO_3$, sodium hydrogen carbonate (sometimes called sodium bicarbonate), are often kept in chemical laboratories for use in neutralizing any acids or bases spilled. Write balanced chemical equations for the neutralization by $NaHCO_3$ of (a) a solution of HCl and (b) a solution of KOH.

51. Write equations to show the stepwise ionization of the polyprotic acids H_2S and H_3PO_4.

52. Write equations to show the stepwise ionization of the polyprotic acids H_2CO_3 and H_3AsO_4.

53. Write the chemical equation for the reaction expected when a solution of $HClO_4$ is added to each of the following.

    (a) A potassium hydroxide solution
    (b) Ammonia gas
    (c) Aluminum
    (d) Solid aluminum sulfide, Al_2S_3
    (e) Solid sodium oxide
    (f) Solid calcium carbonate

54. Write the chemical equation for the reaction expected when a solution of H_2SO_4 is added to each of the following. There may be more than one equation possible, depending on the ratio of reactants you chose.

    (a) CH_3NH_2 vapor
    (b) A solution of sodium hydroxide
    (c) Magnesium
    (d) Solid calcium nitride, Ca_3N_2
    (e) Cobalt(II) carbonate
    (f) Cesium oxide

55. Is the solution that results when 1 mol of HCl is added to a solution of 1 mol of NH_3 neutral, acidic, or basic? Explain your answer.

56. Is the solution that results when 1 mol of HNO_3 is added to a solution of 1 mol of CH_3NH_2 neutral, acidic, or basic? Explain your answer.

57. Is the solution that results when 1 mol of NaOH is added to a solution of 1 mol of HF neutral, acidic, or basic? Explain your answer.

58. Is the solution that results when 1 mol of KOH is added to a solution of 1 mol of HNO_2 neutral, acidic, or basic? Explain your answer.

59. Is the solution that results when 1 mol of HBr is added to a solution of 1 mol of $NaClO_4$ neutral, strongly acidic, weakly acidic, strongly basic, or weakly basic? Explain your answer.

60. Is the solution that results when 1 mol of KOH is added to a solution of 1 mol of C_2H_5OH neutral, strongly acidic, weakly acidic, strongly basic, or weakly basic? Explain your answer.

61. List three methods of preparing hydroxide bases, and write a chemical equation that is an example of each method.

62. List four methods of preparing Brønsted acids, and write a chemical equation that is an example of each method.

Titration; Normality

63. Using a procedure similar to that in Example 15.4 or in the examples in Chapter 3, outline the steps necessary to solve the following problem.

    What volume of 0.600 M HCl is required to react completely with 2.50 g of sodium carbonate?

    $$Na_2CO_3 + 2HCl \longrightarrow 2NaCl + CO_2 + H_2O$$

64. Using a procedure similar to that in Example 15.4 or in the examples in Chapter 3, outline the steps necessary to solve the following problem.

    What volume of 0.08892 M HNO_3 is required to react completely with 0.2352 g of potassium hydrogen phosphate?

    $$2HNO_3 + KHPO_4 \longrightarrow H_3PO_4 + 2KNO_3$$

65. Using a procedure similar to that in Example 15.4 or in the examples in Chapter 3, outline the steps necessary to solve the following problem.

    What volume of a 0.3300 M solution of sodium hydroxide is required to titrate 15.00 mL of 0.1500 M oxalic acid?

    $$C_2O_4H_2 + 2NaOH \longrightarrow Na_2C_2O_4 + 2H_2O$$

66. Using a procedure similar to that in Example 15.4 or in the examples in Chapter 3, outline the steps necessary to solve the following problem.

    What volume of a 0.00945 M solution of potassium hydroxide is required to titrate 50.00 mL of a sample of acid rain with a H_2SO_4 concentration of 1.23×10^{-4} M?

    $$H_2SO_4 + 2KOH \longrightarrow K_2SO_4 + 2H_2O$$

67. Calculate, to one decimal place, the mass of an equivalent of each of the reactants in the following equations.

    (a) $H_2SO_4 + 2LiOH \longrightarrow Li_2SO_4 + 2H_2O$
    (b) $HBr + NH_3 \longrightarrow NH_4Br$
    (c) $KOH + H_2SO_3 \longrightarrow KHSO_3 + H_2O$
    (d) $Al(OH)_3 + 3HNO_3 \longrightarrow Al(NO_3)_3 + 3H_2O$

68. Calculate, to one decimal place, the mass of an equivalent of each of the reactants in the following equations.

    (a) $KOH + KHCO_3 \longrightarrow K_2CO_3 + H_2O$
    (b) $Mg(OH)_2 + 2HCl \longrightarrow MgCl_2 + H_2$
    (c) $2LiOH + LiH_2PO_4 \longrightarrow Li_3PO_4 + 2H_2O$
    (d) $3H_2S + 2Li_3N \longrightarrow 3Li_2S + 2NH_3$

69. Calculate the normality of each of the following solutions.

    (a) 4.0 equivalents of HCl in 2.0 L of solution
    (b) 2.5×10^{-3} equivalent of HBr in 50 mL of solution
    (c) 0.30 equivalent of H_2SO_4 in 400 mL of solution
    (d) 1.50×10^{-3} equivalent of NaOH in 1.00 mL of solution

70. Calculate the normality of each of the following solutions.

    (a) 5.0 equivalents of H_2SO_4 in 2.0 L of solution
    (b) 0.050 equivalent of HF in 0.67 L of solution
    (c) 1.5×10^{-3} equivalent of $H_2PO_4^-$ in 0.250 L of solution
    (d) 2.50×10^{-3} equivalent of NaOH in 100 mL of solution

71. (a) Calculate the molarity of a solution of hydrochloric acid that contains 3.65 g of HCl in 0.50 L of solution.
    (b) Calculate the normality of the solution in part (a).

72. (a) How many grams of H_2SO_4 are contained in 1.2 L of 0.50 M solution?
    (b) How many grams of H_2SO_4 are contained in 1.2 L of 0.50 N solution, assuming that both hydrogen ions react?

73. (a) What is the molarity of a solution of barium hydroxide, 92.62 mL of which contains 14.621 mg of $Ba(OH)_2$?
    (b) What is the normality of the solution in part (a) if all OH^- ions react?

74. Using a procedure similar to that in Example 15.4 or to the examples in Chapter 3, outline the steps necessary to solve the following problem, and then solve the problem.

    Titration of 15.60 mL of a 0.432 N HI solution with NH_3 requires 30.60 mL of the NH_3 solution to reach the end point. What is the normality of the NH_3 solution?

75. Using a procedure similar to that in Example 15.4 or to the examples in Chapter 3, outline the steps necessary to solve the following problem, and then solve the problem.

    A 0.366 N solution of KOH is titrated with H_2SO_4, producing K_2SO_4. If a 32.00-mL sample of the KOH solution is used, what volume of 0.366 M H_2SO_4 is required to reach the end point? What volume of 0.366 N H_2SO_4? (Assume both protons react.)

76. Using a procedure similar to that in Example 15.4 or to the examples in Chapter 3, outline the steps necessary to solve the following problem, and then solve the problem.

    A sample of solid calcium hydroxide, $Ca(OH)_2$, is allowed to stand in water until a saturated solution is formed. A titration of 75.00 mL of this solution with 5.00×10^{-2} N HCl requires 36.6 mL of the acid to reach the end point.

    $$Ca(OH)_2 + 2HCl \longrightarrow CaCl_2 + 2H_2O$$

    What is the normality of a saturated $Ca(OH)_2$ solution? The molarity? What is the solubility of $Ca(OH)_2$ in grams per liter of solution?

77. (a) A 0.1824-g sample of an acid is titrated with 0.3090 N NaOH, and 9.256 mL is required to reach the end point. What is the mass of an equivalent of the acid?
    (b) Which of the following acids was used: HCl, H_3PO_4, H_2SO_4, HNO_3, or CH_3CO_2H? Explain your answer.

78. What is the normality of a H_2SO_4 solution if 40.2 mL of the solution is required to titrate 6.50 g of $NaHCO_3$?

    $$H_2SO_4 + 2NaHCO_3 \longrightarrow Na_2SO_4 + 2H_2O + 2CO_2$$

79. Titration of 0.4500 g of an acid requires 282.00 mL of 6.00×10^{-3} N NaOH to reach the end point. What is the mass of an equivalent of the acid?

80. What is the normality of a 40.0% aqueous solution of sulfuric acid, for which the specific gravity is 1.3057?

Lewis Acids and Bases

81. Write the Lewis structures of the reactants and product of each of the following equations, and identify the Lewis acid and the Lewis base in each.
    (a) $CO_2 + OH^- \longrightarrow HCO_3^-$
    (b) $B(OH)_3 + OH^- \longrightarrow B(OH)_4^-$
    (c) $I^- + I_2 \longrightarrow I_3^-$
    (d) $AlCl_3 + Cl^- \longrightarrow AlCl_4^-$
        (Assume Al–Cl single bonds.)
    (e) $O^{2-} + SO_3 \longrightarrow SO_4^{2-}$

82. Write the Lewis structures of the reactants and product of each of the following equations, and identify the Lewis acid and the Lewis base in each.
    (a) $CS_2 + SH^- \longrightarrow HCS_3^-$
    (b) $BF_3 + F^- \longrightarrow BF_4^-$
    (c) $I^- + SnI_2 \longrightarrow SnI_3^-$ (Assume Sn–I single bonds.)
    (d) $Al(OH)_3 + OH^- \longrightarrow Al(OH)_4^-$
    (e) $F^- + SO_3 \longrightarrow SFO_3^-$

83. The dissolution of solid $Al(NO_3)_3$ in water is accompanied by both a Lewis and a Brønsted acid–base reaction. Write balanced chemical equations for the reactions.

84. (a) Write a balanced chemical equation for the Lewis acid–base reaction between $Al(OH)_4^-$ and CO_2, which causes the precipitation of $Al(OH)_3$ when CO_2 is added to an aqueous solution of $Al(OH)_4^-$.
    (b) Write the corresponding equation using the more accurate hydrated forms for the ion, $[Al(H_2O)_2(OH)_4]^-$, and for the precipitate, $[Al(H_2O)_3(OH)_3]$.

85. The reaction of SO_3 with H_2SO_4 to give pyrosulfuric acid, $H_2S_2O_7$ (Section 8.9), is a Lewis acid–base reaction that results in the formation of S—O—S bonds and one OH group per sulfur. Write the chemical equation for this reaction, showing the Lewis structures of the reactants and product, and identify the Lewis acid and the Lewis base.

86. Calcium oxide is prepared by the decomposition of calcium carbonate at elevated temperatures (Section 8.9). This is the reverse of the Lewis acid–base reaction between CaO and CO_2. Write the chemical equation for the reaction of CaO with CO, showing the Lewis structures of the reactants and product, and identify the Lewis acid and the Lewis base.

87. Each of the following species may be considered a Lewis acid–base adduct of a proton with a Lewis base. For each of the following pairs, indicate which adduct is the stronger acid.
    (a) HCl or HCN
    (b) HBr or H_3O^+
    (c) H_3O^+ or NH_4^+
    (d) H_2O or NH_3
    (e) HSO_4^- or HCO_3^-

88. Using Lewis structures, write balanced equations for each of the following reactions.

(a) $HCl(g) + PH_3(g) \longrightarrow$

(b) $H_3O^+ + CH_3^- \longrightarrow$

(c) $CaO + SO_3 \longrightarrow$

(d) $NH_4^+ + C_2H_5O^- \longrightarrow$

Applications and Additional Exercises

89. In dilute aqueous solution, HF acts as a weak acid. However, pure liquid HF (b.p. = 19.5°C) is a strong acid. In liquid HF, HNO_3 acts as a base and accepts protons. The acidity of liquid HF can be increased by adding one of several inorganic fluorides that are Lewis acids and accept F^- ion (for example, BF_3 or SbF_5). Write balanced chemical equations for the reaction of pure HNO_3 with pure HF and of pure HF with BF_3. Write the Lewis structures of the reactants and products.

90. In paper making, the use of wood pulp as a source of cellulose and of alum-rosin sizing (which keeps ink from soaking into the pages and blurring) produces a paper that deteriorates in 25–50 years. This is quite satisfactory for most purposes, but it is not suitable for the needs of libraries and archives. The deterioration is due to formation of acids. Alum-rosin slowly produces sulfuric acid under humid conditions. Also, wood pulp contains lignin, which forms carboxylic acids as it ages. These acids are similar to acetic acid in that they contain the acidic $-CO_2H$ group. In order to stop this acidification, books are sometimes soaked in magnesium hydrogen carbonate solution; treated with cyclohexylamine, $C_6H_{11}NH_2$, a base like ammonia but with one hydrogen atom replaced by a $-C_6H_{11}$ group; and treated with gaseous diethyl zinc, $(C_2H_5)_2Zn$, a covalent molecule containing $Zn-C$ bonds. Diethyl zinc is a source of $C_2H_5^-$, which is very much like CH_3^- in its properties. Write balanced equations for these reactions. Use the formula RCO_2H for the carboxylic acids formed from lignin; the exact nature of R is not important in these reactions.

91. Amino acids are biologically important molecules that are the building blocks of proteins. The simplest amino acid is glycine, $H_2NCH_2CO_2H$. The common feature of amino acids is that they contain the functional groups indicated in color: an amine group, $-NH_2$, and a carboxylic acid group, $-CO_2H$. An amino acid can function as either an acid or a base. For glycine, the acid strength of the carboxyl group is about the same as that of acetic acid, CH_3CO_2H, and the base strength of the amino group is slightly greater than that of ammonia, NH_3.

   (a) Write the Lewis structures of the ions that form when glycine is dissolved in 1 M HCl and in 1 M KOH.

   (b) Write the Lewis structure of glycine when this amino acid is dissolved in water. (*Hint:* Consider the relative base strengths of the $-NH_2$ and $-CO_2^-$ groups.)

92. Boric acid, H_3BO_3, reacts not as a Brønsted acid but as a Lewis acid.

   (a) Write an equation for its reaction with water to form H_3O^+ and an anion.

   (b) Predict the shape of the anion thus formed.

   (c) What hybridization of the boron is consistent with the shape you have predicted?

93. Are the concentrations of hydronium ion and hydroxide ion in solutions of an acid or a base in water directly proportional or inversely proportional? Explain your answer. (Inverse and direct proportionality are discussed in Sections 10.2 and 10.3.)

94. Determine what mass of $Ca(OH)_2$ will react with 25.0 g of propionic acid to form the preservative calcium propionate according to the equation

$$
\begin{array}{c}
\text{H H H O} \\
\text{| | | ||} \\
\text{H}-\text{C}-\text{C}-\text{C}-\text{C}-\text{O}-\text{H} + Ca(OH)_2 \longrightarrow \\
\text{| | |} \\
\text{H H H}
\end{array}
$$

$$
Ca\left(\begin{array}{c}
\text{H H H O} \\
\text{| | | ||} \\
\text{H}-\text{C}-\text{C}-\text{C}-\text{C}-\text{O} \\
\text{| | |} \\
\text{H H H}
\end{array}\right)_2 + 2H_2O
$$

95. How many milliliters of a 0.1500 M solution of KOH will be needed to titrate 40.00 mL of a 0.0656 M solution of H_3PO_4?

$$H_3PO_4 + 2KOH \longrightarrow KHPO_4 + 2H_2O$$

96. The active ingredient formed by aspirin in the body is salicylic acid, $C_6H_4OH(CO_2H)$. The carboxylate group (CO_2H, Section 9.13) acts as a weak acid. The phenol group (an OH group bonded to an aromatic ring) also acts as an acid but it is a much weaker acid. List, in order of descending concentration, all of the ionic and molecular species present in a 0.001 M aqueous solution of $C_6H_4OH(CO_2H)$.

97. The sodium salt of benzoic acid, $C_6H_5CO_2H$, is commonly used as a preservative in foods—in cider for example. List, in order of descending concentration, all of the ionic and molecular species present in a 0.001 M aqueous solution of this acid.

98. Trichloroacetic acid, CCl_3CO_2H, is amphiprotic.

$$CCl_3CO_2H + B \longrightarrow CCl_3CO_2H + BH$$

$$HA + CCl_3CO_2H \longrightarrow CCl_3CO_2H_2^+ + A^-$$

Write equations for the reaction of pure $CCl_3CO_2H(l)$ with $H_2O(l)$, with $HCl(g)$, with $NaNH_2(s)$, with $H_2SO_4(l)$, and with $CH_3CO_2H(l)$. Trichloroacetic acid has an acid strength between those of H_3O^+ and HSO_4^-.

99. Potassium acid phthalate, $KHC_6H_4O_4$ or KHP, is used in many laboratories, including general chemistry laboratories, to standardize solutions of base. KHP is one of only a few stable solid acids that can be dried by warming and weighed. A 0.3420-g sample of potassium acid phthalate, $KHC_6H_4O_4$, reacts with 35.73 mL of a NaOH solution in a

titration. What is the molar concentration of the NaOH?

$$KHC_6H_4O_4 + NaOH \longrightarrow KNaC_6H_4O_4 + H_2O$$

100. The reaction of WCl_6 with Al at about 400°C gives black crystals of a compound that contains only tungsten and chlorine. A sample of this compound, when reduced with hydrogen, gives 0.2232 g of tungsten metal and hydrogen chloride, which is absorbed in water. Titration of the hydrochloric acid thus produced requires 46.2 mL of 0.1051 M NaOH to reach the end point. What is the empirical formula of the black tungsten chloride?

101. The reaction of 0.871 g of sodium with an excess of liquid ammonia containing a trace of $FeCl_3$ as a catalyst produced 0.473 L of pure H_2 measured at 25°C and 745 torr. What is the equation for the reaction of sodium with liquid ammonia? Show how you determined the stoichiometry of the reaction.

102. Write equations for three different methods that employ acids to prepare the insoluble salt $AlPO_4$.

103. What mass of $CaCO_3$ is required to neutralize the acid produced when 0.766 L of gaseous SO_2 with a pressure of 0.944 atm and a temperature of 18°C dissolves in water?

$$CaCO_3 + H_2SO_3 \longrightarrow CaSO_3 + H_2O + CO_2$$

104. A compound that contains a metal, oxygen, and hydrogen crystallizes with hydroxide ions in an approximately closest packed structure and the metal ions in $\frac{1}{2}$ of the octahedral holes formed by the hydroxide ions. What is the empirical formula of this compound? Is it an acid or a base? Explain your answer.

105. Why does 1 mol of hydrogen chloride depress the freezing point of 1 kg of water almost twice as much as 1 mol of hydrogen fluoride? Explain your reasoning.

106. (a) Which has the higher vapor pressure, a 0.1 m solution of CH_3CO_2H or a 0.1 m solution of $C_6H_5NH_2$?

(b) Which has the lower freezing temperature?

(c) Which has the higher boiling temperature? Explain your reasoning in each case.

107. Sodium thiosulfate, hypo, decomposes in acidic solution.

$$S_2O_3^{2-}(aq) + 2H_3O^+(aq) \longrightarrow S(s) + H_2SO_3(aq)$$
$$Rate = k[S_2O_3^{2-}][H_3O^+]$$

Will this reaction proceed faster in a solution with 0.1 M HCl or with 0.1 M acetic acid? Explain the reason for your answer.

108. Adipic acid is used in the manufacture of nylon. The percent composition of this acid was determined to be 49.3% C and 6.8% H, the remainder being oxygen. Titration of 0.3752 g of the acid requires 37.89 mL of 0.1355 M NaOH. The freezing point lowering of a solution of 0.2461 g of the acid dissolved in 25.0 g of nitrobenzene ($K_f = 5.67°C\ m^{-1}$) is 0.38°C. The acid is a nonelectrolyte in nitrobenzene.

(a) Determine the empirical and molecular formulas of adipic acid.

(b) Write the chemical equation for the reaction of adipic acid with sodium hydroxide.

(b) Is all of the oxygen in adipic acid present in carboxylic acid groups ($-CO_2H$ groups), or could some oxygen be present in a nonacidic functional group? Explain your reasoning.

(c) Each carbon atom in a molecule of adipic acid either is located in a carboxylic acid group or is bonded to other carbon atoms and to two hydrogen atoms. Draw the Lewis structure of adipic acid.

(d) Describe the electron-pair geometry and molecular geometry around each of the carbon atoms in adipic acid.

(e) Is adipic acid likely to be a strong acid or a weak acid? Explain your reasoning.

16

Ionic Equilibria of Weak Electrolytes

CHAPTER OUTLINE

16.1 Ion Concentrations in Solutions of Strong Electrolytes
16.2 pH and pOH
16.3 The Ionization of Weak Acids and Weak Bases
16.4 Concentrations in Solutions of Weak Acids or Weak Bases
16.5 Acid–Base Properties of Solutions of Salts
16.6 Mixtures of Acids or Bases; Polyprotic Acids
16.7 The Ionization of Very Weak Acids
16.8 The Salt Effect and the Common Ion Effect
16.9 Buffer Solutions
16.10 The Ionization of Hydrated Metal Ions
16.11 Acid–Base Indicators
16.12 Titration Curves

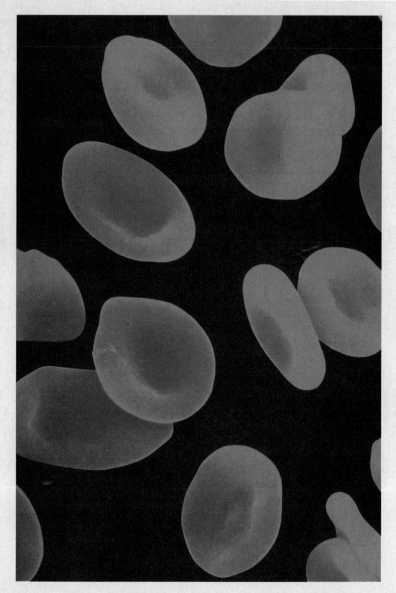

Red blood cells are immersed in plasma, a complex, weakly basic buffer system.

Water is the most important solvent on our planet. Our lives depend on reactions that occur in aqueous solution within our cells. Many of these solutions contain ions that result when electrolytes ionize. The concentrations of these ions are very important in determining the rates, mechanisms, and yields of reactions in such solutions.

The concentrations of ions in a solution of a strong electrolyte can be calculated from the concentration of the solute, because a strong electrolyte is 100% ionized in solution. However, the concentrations of ions in a solution of a weak electrolyte cannot be determined so easily. Both ions and nonionized molecules of the weak electrolyte are present, and the concentrations of all these species must be calculated from the concentration of the weak electrolyte and the extent to which it ionizes. These calculations involve the concepts of equilibria introduced in Chapter 14.

Weak acids and weak bases are important weak electrolytes. They are found in many purely chemical processes and in those of biological interest. Amino acids, for example, are both weak acids and weak bases. In this chapter we will consider some ways of expressing concentrations of hydronium ion and hydroxide ion in solutions of weak acids and weak bases. Then we will examine equilibria involving these weak electrolytes. We will also see that the indicators, such as phenolphthalein, used in titrations are weak acids or weak bases and will learn how to use these properties to select an appropriate indicator for a titration.

16.1 Ion Concentrations in Solutions of Strong Electrolytes

We usually speak of concentrations of solutions of electrolytes as though the electrolyte were not ionized. We use the number of moles of the electrolyte needed to make a liter of the solution, rather than giving the actual concentrations of the molecules and ions present in the solution (Fig. 16.1). For example, we call a solution prepared from 2.0 moles of HCl and enough water to give 1.0 liter of solution a 2.0 M solution of HCl.

We can determine the approximate concentrations of ions in a solution of a *strong* electrolyte directly from its molar concentration, because we assume that strong electrolytes are completely ionized in aqueous solution. If the HCl in our 2.0 M solution were 100% ionized, the concentration of HCl molecules would be zero. The concentrations of H_3O^+ and Cl^- would both equal 2.0 M. A solution prepared using 0.35 mole of the ionic compound K_2SO_4 in 0.70 liter of solution (0.35 mol K_2SO_4/0.70 L) is called a 0.50 M solution of potassium sulfate. However, the solution actually contains K^+ and SO_4^{2-} ions with $[K^+]$ essentially equal to 1.0 M and $[SO_4^{2-}]$ equal to 0.50 M, because each formula unit of potassium sulfate contains two potassium ions and one sulfate ion. We will use the approximation that strong electrolytes are 100% ionized in water.

If we were to make a *very* dilute solution of a strong acid (or a strong base), we might be surprised by the resulting concentration of hydronium ion (or hydroxide ion) in the solution. It could be greater than the concentration of acid (or base) used to make the solution. This can occur because the acid (or base) we add is not the only source of H_3O^+ (or OH^-) in water; the self-ionization of water also produces these ions.

$$H_2O + H_2O \rightleftharpoons H_3O^+ + OH^-$$

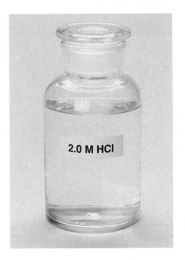

Figure 16.1

This solution contains no HCl molecules even though it is labeled 2.0 M HCl. In the solution, $[H_3O^+]$ = 2.0 M and $[Cl^-]$ = 2.0 M.

2.0 M HCl

Table 16.1

Concentrations in Aqueous Solutions of HCl

[HCl], M[a]		[H$_3$O$^+$], M[b]	pH
0.100	(1.00×10^{-1})	0.100	1.000
0.0100	(1.00×10^{-2})	0.0100	2.000
0.00100	(1.00×10^{-3})	0.00100	3.000
0.000100	(1.00×10^{-4})	0.000100	4.000
0.0000100	(1.00×10^{-5})	0.0000100	5.000
0.00000100	(1.00×10^{-6})	0.00000101	5.996
0.000000427	(4.27×10^{-7})	0.000000450	6.347
0.000000100	(1.00×10^{-7})	0.000000166	6.780
0.0000000100	(1.00×10^{-8})	0.000000101	6.996
0.00000000100	(1.00×10^{-9})	0.000000100	7.000
0.000000000100	(1.00×10^{-10})	0.000000100	7.000

[a]Concentration of HCl added to aqueous solution.
[b]Concentration of H$_3$O$^+$ in the resulting solution.

If the concentration or ionization of an acid or of a base is *very, very low,* the ionization of water can produce more H$_3$O$^+$ or OH$^-$ than the acid or the base.

As an example, let us consider the concentrations of hydronium ion in various aqueous solutions of hydrogen chloride, as shown in Table 16.1 and Figure 16.2. If we use 1.00×10^{-6} mole of HCl, or more, to make 1.00 liter of a solution at 25°C, the hydronium ion concentration essentially equals the concentration of the HCl used to make the solution. At lower concentrations of HCl, the concentration of hydronium ion in the solution is larger than the concentration of HCl used to make the solution. For concentrations of HCl less than 1.00×10^{-7} M, the concentration of hydronium ion stabilizes at 1.00×10^{-7} M. The additional hydronium ion (the hydronium ion that is not produced by the ionization of HCl) is produced by the self-ionization of water.

The situation with bases and the hydroxide ion is analogous to that with acids and the hydronium ion. Figure 16.3 on page 585 is a graph of the concentration of hydroxide ion in various solutions of sodium hydroxide. When the concentration of NaOH

Figure 16.2

A graph of the concentrations of hydronium ion in 1.00-L solutions prepared from various amounts of HCl.

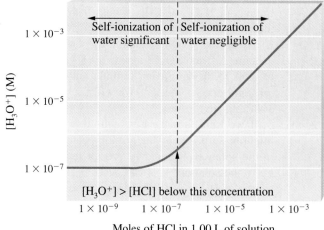

Figure 16.3

A graph of the concentrations of hydroxide ion in 1.00-L solutions prepared from various amounts of NaOH.

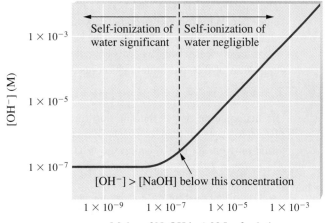

1×10^{-3}

Self-ionization of water significant | Self-ionization of water negligible

[OH⁻] (M)

1×10^{-5}

1×10^{-7}

[OH⁻] > [NaOH] below this concentration

1×10^{-9} 1×10^{-7} 1×10^{-5} 1×10^{-3}

Moles of NaOH in 1.00 L of solution

is greater than 1.00×10^{-6} M, the concentration of hydroxide ion essentially equals the concentration of NaOH used to make the solution. However, the self-ionization of water becomes important when the concentration of the hydroxide ion produced by the base is less than 1.00×10^{-6} M. When [NaOH] is less than 1.00×10^{-7} M, the concentration of OH⁻ levels off at 1.00×10^{-7} M. The self-ionization of water keeps the hydroxide ion concentration constant.

When the concentration of hydronium ion in a solution of an acid is greater than 4.5×10^{-7} M, the self-ionization of water produces less than 5% of the hydronium ion (Table 16.1). Therefore, when calculating [H₃O⁺] in these solutions, we can neglect the self-ionization of water. Similarly, when the concentration of hydroxide ion in a solution of a base is greater than 4.5×10^{-7} M, the self-ionization of water produces less than 5% of the hydroxide ion in solution, and the self-ionization of water can be neglected. Most solutions of acids or bases have hydronium ion or hydroxide ion concentrations greater than 4.5×10^{-7} M, so we usually neglect the self-ionization of water.

Figure 16.4

The concentration of hydronium ion in lime juice is much greater than 1×10^{-7} M; that of hydroxide ion is much less than 1×10^{-7} M. The excess of hydronium ion accounts for the sour taste of a lime.

16.2 pH and pOH

We have seen that water is both an acid and a base (Section 15.1). It undergoes self-ionization and forms very small *but equal* amounts of hydronium and hydroxide ions.

$$H_2O + H_2O \rightleftharpoons H_3O^+ + OH^- \qquad K_w = [H_3O^+][OH^-] = 1.0 \times 10^{-14} \text{ (at 25°C)}$$

The degree of ionization increases as the temperature increases. At 100°C, K_w is about 1×10^{-12}, 100 times larger than at 25°C.

Pure water is neutral—neither acidic nor basic—because the self-ionization of water yields the same numbers of hydronium and hydroxide ions at any temperature. At 25°C,

$$[H_3O^+] = [OH^-] = \sqrt{K_w} = 1.0 \times 10^{-7} \text{ M}$$

When we add an acid to water, the hydronium ion concentration, [H₃O⁺], becomes larger than 1.0×10^{-7} M, and the hydroxide ion concentration, [OH⁻], becomes less than 1.0×10^{-7} M, but not zero (Fig. 16.4). Similarly, when we add a base to water, [OH⁻] becomes greater than 1.0×10^{-7} M, and [H₃O⁺] decreases, but not to zero.

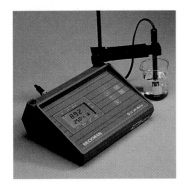

Figure 16.5

A pH meter displays the pH of a solution. The pH is determined electronically from the difference in electrical potential between two electrodes when they are dipped into the solution. This pair of pH electrodes is combined into a single unit for convenience.

The product of $[H_3O^+]$ and $[OH^-]$ is always a constant, so the value of one or the other concentration can become very small, but never zero.

One convenient way to represent very small concentrations in solution is to use a logarithmic scale, such as the pH scale. The **pH** of a solution is the negative of the logarithm of the hydronium ion concentration.

$$pH = -\log [H_3O^+]$$

In this equation, *log* is the notation for a common logarithm, a logarithm to the base 10. If $\log x = y$, then $x = 10^y$. We can find the hydronium ion concentration of a solution from its pH by using the relationship

$$[H_3O^+] = 10^{-pH}$$

The pH of a solution can be measured with a pH meter (Fig. 16.5), with a solution of an indicator (Fig. 16.6), or with pH paper (Fig. 16.7 on page 587). A neutral solution, one in which $[H_3O^+] = [OH^-]$, has a pH of 7.00 at 25°C. The pH of an acidic solution ($[H_3O^+] > [OH^-]$) is less than 7; the pH of a basic solution ($[OH^-] > [H_3O^+]$) is greater than 7.

We can use **pOH,** the negative of the logarithm of the hydroxide ion concentration, to represent $[OH^-]$.

$$pOH = -\log [OH^-]$$

and

$$[OH^-] = 10^{-pOH}$$

Normally we do not measure pOH values; they are calculated from pH values.

At 25°C, pH + pOH = 14.00, or pOH = 14.00 − pH. We can show this by starting with the equation

$$[H_3O^+][OH^-] = K_w$$

If we take the negative of the logarithm of both sides of this equation, we have

$$-\log ([H_3O^+][OH^-]) = -\log K_w$$

or

$$(-\log [H_3O^+]) + (-\log [OH^-]) = -\log K_w$$

Figure 16.6

Universal indicator assumes a different color in solutions of different pH. Thus it can be added to a solution to determine the pH of the solution. From left to right the test tubes contain universal indicator and 0.1 M solutions of the progressively weaker acids HCl (pH = 1), CH_3CO_2H (pH = 3), and NH_4Cl (pH = 5); pure water, a neutral substance (pH = 7); and 0.1 M solutions of the progressively stronger bases aniline, $C_6H_5NH_2$ (pH = 9), NH_3 (pH = 11), and NaOH (pH = 13).

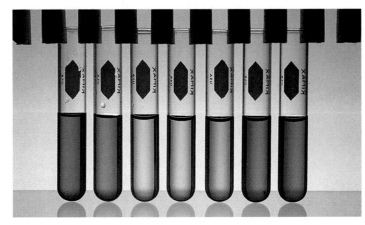

Figure 16.7

pH paper contains a mixture of indicators that give different colors in solutions of differing pH.

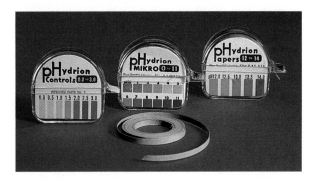

To be consistent with the definition of pH and pOH, we will define $-\log K_w$ as pK_w. Then at 25°C we have

$$pK_w = -\log (1.0 \times 10^{-14}) = -(-14.00) = 14.00$$

and

$$pH + pOH = pK_w = 14.00$$

Table 16.2 shows the relationships among $[H_3O^+]$, $[OH^-]$, pH, and pOH and gives the values for some common substances. The lower the pH (and the higher the pOH), the higher the acidity.

Table 16.2

Relationships of $[H_3O^+]$, $[OH^-]$, pH, and pOH

$[H_3O^+]$, M	$[OH^-]$, M	pH	pOH	Sample Solution
10^1	10^{-15}	−1	15	Strongly acidic
10^0 or 1	10^{-14}	0	14	← 1 M HCl
10^{-1}	10^{-13}	1	13	
10^{-2}	10^{-12}	2	12	← Gastric juice / ← Lime juice 1 M CH₃CO₂H
10^{-3}	10^{-11}	3	11	← Stomach acid
10^{-4}	10^{-10}	4	10	← Wine
10^{-5}	10^{-9}	5	9	← Coffee More acidic
10^{-6}	10^{-8}	6	8	
10^{-7}	10^{-7}	7	7	← Pure water / ← Blood Neutral
10^{-8}	10^{-6}	8	6	
10^{-9}	10^{-5}	9	5	More basic
10^{-10}	10^{-4}	10	4	
10^{-11}	10^{-3}	11	3	← Milk of magnesia
10^{-12}	10^{-2}	12	2	← Household ammonia, NH₃
10^{-13}	10^{-1}	13	1	
10^{-14}	10^0 or 1	14	0	← 1 M NaOH Strongly basic

Example 16.1   Calculation of pH from $[H_3O^+]$

Stomach acid is a solution of HCl with a concentration of 1.2×10^{-3} M. What is the pH of stomach acid?

Solution: Because HCl is a strong acid and is essentially 100% ionized, the hydronium ion concentration is the same as the concentration of HCl, $[H_3O^+] = 1.2 \times 10^{-3}$ M. Thus we have

$$pH = -\log [H_3O^+]$$
$$= -\log (1.2 \times 10^{-3})$$
$$= -(-2.92) = 2.92$$

(The use of logarithms is explained in Appendix A. On a calculator, take the logarithm of 1.2×10^{-3}; the pH is equal to the negative of this value.)

The use of significant figures in a pH is different from what we have seen up to this point. A pH of 2.92 represents a hydronium ion concentration with two significant figures rather than three. A pH is a logarithm, and the number to the left of the decimal point in a logarithm merely establishes the decimal place in the number that the logarithm represents. The number of figures to the right of the decimal in a logarithm should be rounded to the number of significant figures in the number from which the logarithm is obtained. There are two significant figures in the hydronium ion concentration (1.2×10^{-3}) and two to the right of the decimal point in the pH (2.92).

Check your learning: Calculate the pH of a 5.37×10^{-4} M solution of nitric acid, a strong acid.

Answer: 3.270

Example 16.2   Calculation of pH from Percent Ionization

Calculate the pH of a 0.125 M solution of nitrous acid (a weak acid, 6.5% ionized, Section 15.3).

Solution: The percent ionization for 0.125 M nitrous acid is 6.5%. Hence the number of moles of hydronium ion in exactly 1 L of solution is 6.5% of 0.125 mol, or $(6.5/100) (0.125) = 0.0081$ mol.

$$[H_3O^+] = 8.1 \times 10^{-3} \text{ mol/1 L} = 8.1 \times 10^{-3} \text{ M.}$$
$$pH = -\log [H_3O^+] = -\log (8.1 \times 10^{-3})$$
$$= -(-2.09) = 2.09$$

Remember, the logarithm 2.09 indicates a hydronium ion concentration with only two significant figures.

Check your learning: Calculate the pH of a 0.10 M solution of acetic acid, which is 1.3% ionized.

Answer: pH = 2.89

Example 16.3   Calculation of pH and pOH

We have seen that pure water is very difficult to obtain because water dissolves carbon dioxide (Section 15.1). Water in equilibrium with air contains $(4.4 \times 10^{-5})\%$ carbon dioxide. The resulting carbonic acid, H_2CO_3, gives the solution a hydronium ion concentration of 2.0×10^{-6} M, about 20 times greater than that of pure water. Calculate the pH and pOH of the solution at 25°C.

Solution:

$$pH = -\log [H_3O^+] = -\log (2.0 \times 10^{-6})$$
$$= -(-5.70) = 5.70$$
$$pH + pOH = 14.00$$
$$pOH = 14.00 - 5.70 = 8.30$$

Check your learning: Calculate the pH and pOH of a 7.88×10^{-4} M solution of HBr, a strong acid, at 25°C.

Answer: pH = 3.103 pOH = 10.90

Example 16.4   Calculation of Hydronium Ion Concentration from pH

Calculate the hydronium ion concentration of blood, the pH of which is 7.3 (slightly alkaline).

Solution:

$$pH = -\log [H_3O^+] = 7.3$$
$$\log [H_3O^+] = -7.3$$
$$[H_3O^+] = 10^{-7.3} \quad \text{or} \quad [H_3O^+] = \text{antilog of } -7.3$$
$$[H_3O^+] = 5 \times 10^{-8} \text{ M}$$

(On a calculator take the antilog, or the "inverse" log, of -7.3, or calculate $10^{-7.3}$.)

Check your learning: Calculate the hydronium ion concentration of a solution with a pH of 9.33.

Answer: 4.7×10^{-10}

Example 16.5   Calculation of pOH of a Basic Solution

What are the pOH and the pH of a 0.0125 M solution of potassium hydroxide, KOH?

Solution: Potassium hydroxide is a strong base and is completely ionized in dilute solution: $[OH^-] = 0.0125$ M.

$$pOH = -\log [OH^-] = -\log 0.0125 = -(-1.903) = 1.903$$

The pH can be found from the POH.

$$pH + pOH = 14.00$$

$$pH = 14.00 - pOH = 14.00 - 1.903 = 12.10$$

Check your learning: Calculate the pOH and the pH of a 0.10 M solution of acetate ion, $CH_3CO_2^-$, at 25°C. There is a 0.0075% yield of OH^- in the reaction of acetate ion with water.

$$CH_3CO_2^- + H_2O \rightleftharpoons CH_3CO_2H + OH^-$$

Answer: pOH = 5.12, pH = 8.88

16.3   The Ionization of Weak Acids and Weak Bases

We saw in Chapter 15 that many acids and bases are weak; that is, they do not ionize fully in aqueous solution. A solution of a weak acid in water is a mixture of the nonionized acid, hydronium ion, and the conjugate base of the acid, with the nonionized acid present in the greatest concentration. Thus a weak acid increases the hydronium ion concentration in an aqueous solution (but not as much as the same amount of a strong acid).

Acetic acid, CH_3CO_2H, is a weak acid. When we add acetic acid to water, it ionizes to a small extent according to the equation

$$\underset{\substack{\text{Nonionized} \\ \text{acid}}}{CH_3CO_2H(aq)} + H_2O(l) \rightleftharpoons \underset{\substack{\text{Hydronium} \\ \text{ion}}}{H_3O^+(aq)} + \underset{\substack{\text{Conjugate} \\ \text{base}}}{CH_3CO_2^-(aq)}$$

giving an equilibrium mixture with most of the acid present in the nonionized (molecular) form. Perhaps we should write the equation for the ionization of acetic acid as

$$\mathbf{CH_3CO_2H}(aq) + H_2O(l) \rightleftharpoons H_3O^+(aq) + CH_3CO_2^-(aq)$$

to remind ourselves that in a mixture of acetic acid molecules, hydronium ion, and acetate ion, nonionized acetic acid molecules are present in the greatest concentration. This equilibrium, like other equilibria, is dynamic (Section 14.1); acetic acid molecules donate hydrogen ions to water molecules and form hydronium ions and acetate ions at the same rate at which hydronium ions donate hydrogen ions to acetate ions to re-form acetic acid molecules and water molecules.

We can tell by measuring the pH of an aqueous solution of known concentration that only a fraction of the weak acid is ionized at any moment (Fig. 16.8). The remaining weak acid is present in the nonionized form. The equilibrium constant for the ionization of a weak acid, K_a, is called the **ionization constant** of the weak acid and is equal to the reaction quotient when the reaction is at equilibrium (Section 15.3). For acetic acid, at equilibrium

$$K_a = Q = \frac{[H_3O^+][CH_3CO_2^-]}{[CH_3CO_2H]} = 1.8 \times 10^{-5}$$

Figure 16.8

pH paper indicates that a 0.1 M solution of HCl (beaker on left) has a pH of 1 (also see Fig. 16.7). The acid is fully ionized and $[H_3O^+] = 0.1$ M. A 0.1 M solution of CH_3CO_2H (beaker on right) has a pH of 3 ($[H_3O^+] = 0.001$ M) because the weak acid CH_3CO_2H is only partially ionized.

Table 16.3

Ionization Constants of Some
Weak Acids

Ionization Reaction	K_a at 25°C
$HSO_4^- + H_2O \rightleftharpoons H_3O^+ + SO_4^{2-}$	1.2×10^{-2}
$HF + H_2O \rightleftharpoons H_3O^+ + F^-$	7.2×10^{-4}
$HNO_2 + H_2O \rightleftharpoons H_3O^+ + NO_2^-$	4.5×10^{-4}
$HNCO + H_2O \rightleftharpoons H_3O^+ + NCO^-$	3.46×10^{-4}
$HCO_2H + H_2O \rightleftharpoons H_3O^+ + HCO_2^-$	1.8×10^{-4}
$CH_3CO_2H + H_2O \rightleftharpoons H_3O^+ + CH_3CO_2^-$	1.8×10^{-5}
$HClO + H_2O \rightleftharpoons H_3O^+ + ClO^-$	3.5×10^{-8}
$HBrO + H_2O \rightleftharpoons H_3O^+ + BrO^-$	2×10^{-9}
$HCN + H_2O \rightleftharpoons H_3O^+ + CN^-$	4×10^{-10}

Although water is a reactant, it is the solvent as well, so $[H_2O]$ does not appear in the expression for K_a (Section 14.2). Table 16.3 gives the ionization constants for several weak acids; additional ionization constants can be found in Appendix F.

At equilibrium, a solution of a weak base in water is a mixture of the nonionized base, the conjugate acid of the weak base, and hydroxide ion, with the nonionized base present in the greatest concentration. Thus a weak base increases the hydroxide ion concentration in an aqueous solution (but not as much as the same amount of a strong base).

When we make a solution of the weak base trimethylamine, $(CH_3)_3N$, in water, it reacts according to the equation

$$\underset{\substack{\text{Nonionized} \\ \text{base}}}{(CH_3)_3N(aq)} + H_2O(l) \rightleftharpoons \underset{\substack{\text{Conjugate} \\ \text{acid}}}{(CH_3)_3NH^+(aq)} + \underset{\substack{\text{Hydroxide} \\ \text{ion}}}{OH^-(aq)}$$

Figure 16.9

pH paper indicates that a 0.1 M solution of NH_3 is basic. The solution has a pOH of 3 ($[OH^-] = 0.001$ M) because the weak base NH_3 only partially reacts with water. A 0.1 M solution of NaOH has a pOH of 1 because NaOH is a strong base.

giving an equilibrium mixture with most of the base present as the nonionized (molecular) form. Again, we should probably write the equation

$$\mathbf{(CH_3)_3N}(aq) + H_2O(l) \rightleftharpoons (CH_3)_3NH^+(aq) + OH^-(aq)$$

to emphasize that nonionized molecules of the weak base have the greatest concentration. This equilibrium, like other equilibria, is dynamic (Section 14.1); trimethylamine molecules accept hydrogen ions from water molecules and form trimethylammonium ions and hydroxide ions at the same rate at which trimethylammonium ions donate hydrogen ions to hydroxide ions to re-form trimethylamine molecules and water molecules.

We can tell by measuring the pH of an aqueous solution of a weak base of known concentration that only a fraction of the base reacts with water (Fig. 16.9). The remaining weak base is present as the unreacted form. The equilibrium constant for the ionization of a weak base, K_b, is called the ionization constant of the weak base and is equal to the reaction quotient when the reaction is at equilibrium. For trimethylamine, at equilibrium

$$K_b = Q = \frac{[(CH_3)_3NH^+][OH^-]}{[(CH_3)_3N]}$$

The concentration of water (the solvent for this reaction) does not appear in the equilibrium equation. The ionization constants of several weak bases are given in Table 16.4 on page 591 and in Appendix G.

Table 16.4

Ionization Constants of Some
Weak Bases

Base	Ionization	K_b at 25°C
$(CH_3)_2NH$ Dimethylamine	$+ H_2O \rightleftharpoons (CH_3)_2NH_2^+ + OH^-$	7.4×10^{-4}
CH_3NH_2 Methylamine	$+ H_2O \rightleftharpoons CH_3NH_3^+ + OH^-$	4.4×10^{-4}
$(CH_3)_3N$ Trimethylamine	$+ H_2O \rightleftharpoons (CH_3)_3NH^+ + OH^-$	7.4×10^{-5}
NH_3 Ammonia	$+ H_2O \rightleftharpoons NH_4^+ + OH^-$	1.8×10^{-5}
$C_6H_5NH_2$ Aniline	$+ H_2O \rightleftharpoons C_6H_5NH_3^+ + OH^-$	4.6×10^{-10}

Example 16.6   **Equilibria of Weak Acids and Weak Bases**

Determine the effect on the concentration of acetic acid, hydronium ion, and acetate ion when small amounts of each of the following is added to separate solutions of acetic acid: (a) HCl, (b) KCH_3CO_2, (c) NaCl, (d) KOH and (e) CH_3CO_2H.

Solution: The equilibrium among CH_3CO_2H, H_3O^+, and $CH_3CO_2^-$ is

$$CH_3CO_2H + H_2O \rightleftharpoons H_3O^+ + CH_3CO_2^-$$

We can use Le Châtelier's principle to predict the shift in equilibrium that would result from changing the concentration of any one of these species.

(a) HCl is a strong acid that forms H_3O^+ in water and increases the concentration of H_3O^+. To compensate, the equilibrium shifts to the left to return to equilibrium, decreasing the concentrations of $CH_3CO_2^-$ and H_3O^+ while increasing the concentration of CH_3CO_2H. However, the shift is not enough to reduce the concentration of H_3O^+ to its original value, so the final H_3O^+ concentration is higher than the initial concentration.

(b) KCH_3CO_2 is a salt of the acetate ion and a strong electrolyte. It increases the concentration of $CH_3CO_2^-$. To compensate, the equilibrium shifts to the left to return to equilibrium, the concentrations of H_3O^+ and $CH_3CO_2^-$ decrease, and that of CH_3CO_2H increases. However, the concentration of $CH_3CO_2^-$ is not reduced to its original value.

(c) NaCl contains none of the species involved in the equilibrium, so we should expect it to have no appreciable effect on the equilibrium concentrations. (In Section 16.8 we will see that salts can have an effect on the ionization of a weak electrolyte, but the effect is generally so small that we neglect it.)

(d) KOH is a base; it reacts with hydronium ion and reduces its concentration. As a consequence, the equilibrium shifts to the right to restore the hydronium ion concentration. The concentration of $CH_3CO_2^-$ also increases, while the concentration of CH_3CO_2H decreases. The hydronium ion concentration increases, but to to its original value.

(e) When CH_3CO_2H is added to the solution, the acetic acid concentration is increased, so the reaction shifts to the right to return to equilibrium. This shift increases the concentrations of H_3O^+ and $CH_3CO_2^-$. However, it does not reduce the concentration of acetic acid to its original value.

Check your learning: Determine the effect on the concentrations of NH_3, NH_4^+, and OH^- when small amounts of each of the following is added to a solution of NH_3 in water: **(a)** HCl, **(b)** NH_3, **(c)** NaOH, **(d)** NH_4Cl, **(e)** KNO_3. The equilibrium among NH_3, NH_4^+, and OH^- is

$$NH_3 + H_2O \rightleftharpoons NH_4^+ + OH^-$$

Answer: **(a)** $[NH_3]$ decreases, $[NH_4^+]$ increases, $[OH^-]$ decreases. **(b)** $[NH_3]$ increases, $[NH_4^+]$ increases, $[OH^-]$ increases. **(c)** $[NH_3]$ increases, $[NH_4^+]$ decreases, $[OH^-]$ increases. **(d)** $[NH_3]$ increases, $[NH_4^+]$ increases, $[OH^-]$ decreases. **(e)** no effect expected.

Although the reaction of a weak acid with water and the reaction of a weak base with water give different products, the equilibrium equations are similar. Both reactions have the form

$$\mathbf{W}(aq) + H_2O(l) \rightleftharpoons X(aq) + Y(aq)$$

with W, the nonionized weak acid or weak base, present in the greatest concentration and with water as a solvent. The equations for the reaction quotients have the same form in both cases:

$$Q = \frac{[X][Y]}{[W]}$$

and both K_a and K_b are less than 1. Of course, we must always remember that weak acids produce H_3O^+ ion whereas weak bases produce OH^- ion, but we can use the same type of arithmetic to calculate the equilibrium concentrations of H_3O^+ in a solution of a weak acid that we use to calculate the equilibrium concentrations of OH^- in a solution of a weak base.

As we finish this section, we should note that for many applications we should calculate the value of an ionization constant, as we should calculate many other equilibrium constants, by using the activities of the species involved (see Section 14.2) rather than their concentrations. However, the activity of a dilute solute is closely approximated by its molar concentration, so concentrations are commonly used.

Figure 16.10

Vinegar is a solution of acetic acid, a weak acid.

Example 16.7 Determination of K_a from Equilibrium Concentrations

Acetic acid is the principal ingredient in vinegar (Fig. 16.10); that's why vinegar tastes sour. At equilibrium, a solution contains $[CH_3CO_2H] = 0.0787$ M and $[H_3O^+] = [CH_3CO_2^-] = 0.00118$ M. What is the value of K_a for acetic acid?

Solution: We are asked to calculate an equilibrium constant from equilibrium concentrations—a type of calculation described in Section 14.9. At equilibrium the value of the equilibrium constant is equal to the reaction quotient for the reaction

$$CH_3CO_2H + H_2O \rightleftharpoons H_3O^+ + CH_3CO_2^-$$

$$K_a = Q = \frac{[H_3O^+][CH_3CO_2^-]}{[CH_3CO_2H]} = \frac{(0.00118)(0.00118)}{0.0787} = 1.77 \times 10^{-5}$$

Check your learning: What is the equilibrium constant for the ionization of the HSO_4^- ion, the weak acid used in some household cleansers?

$$HSO_4^- + H_2O \rightleftharpoons H_3O^+ + SO_4^{2-}$$

In one mixture of $NaHSO_4$ and Na_2SO_4 at equilibrium, $[H_3O^+] = 0.027$ M, $[HSO_4^-] = 0.29$ M, and $[SO_4^{2-}] = 0.13$ M.

Answer: K_a for $HSO_4^- = 1.2 \times 10^{-2}$

Example 16.8   Determination of K_b from Equilibrium Concentrations

Caffeine, $C_8H_{10}N_4O_2$, is a weak base (it tastes bitter). What is the value of K_b for caffeine if a solution at equilibrium has $[C_8H_{10}N_4O_2] = 0.050$ M, $[C_8H_{10}N_4O_2H^+] = 5.0 \times 10^{-3}$ M, and $[OH^-] = 2.5 \times 10^{-3}$ M?

Solution: At equilibrium the value of the equilibrium constant is equal to the reaction quotient for the reaction

$$C_8H_{10}N_4O_2 + H_2O \rightleftharpoons C_8H_{10}N_4O_2H^+ + OH^-$$

$$K_b = Q = \frac{[C_8H_{10}N_4O_2H^+][OH^-]}{[C_8H_{10}N_4O_2]} = \frac{(5.0 \times 10^{-3})(2.5 \times 10^{-3})}{(0.050)} = 2.5 \times 10^{-4}$$

Check your learning: What is the equilibrium constant for the ionization of the HPO_4^{2-} ion, a weak base?

$$HPO_4^{2-} + H_2O \rightleftharpoons H_2PO_4^- + OH^-$$

In a solution containing a mixture of NaH_2PO_4 and Na_2HPO_4 at equilibrium, $[OH^-] = 1.3 \times 10^{-6}$ M, $[H_2PO_4^-] = 0.042$ M, and $[HPO_4^{2-}] = 0.341$ M.

Answer: K_b for $HPO_4^{2-} = 1.6 \times 10^{-7}$

Example 16.9   Determination of K_a or K_b from pH

The pH of a 0.0516 M solution of nitrous acid, HNO_2, is 2.34. What is K_a for HNO_2?

$$HNO_2 + H_2O \rightleftharpoons H_3O^+ + NO_2^-$$

Solution: This calculation is similar to that of Example 14.7 in Section 14.9. We determine an equilibrium constant by starting with the initial concentrations of HNO_2, H_3O^+, and NO_2^- as well as one of the final concentrations, the concentration of hydronium ion at equilibrium. (Remember that pH is simply another way to express the concentration of hydronium ion.)

We can solve this problem with the following steps.

$$\boxed{\text{pH}} \longrightarrow \boxed{[H_3O^+]} \longrightarrow \boxed{\Delta[H_3O^+]} \longrightarrow \boxed{\begin{array}{c}\Delta[HNO_2]\\ \text{and}\\ \Delta[NO_2^-]\end{array}} \longrightarrow \boxed{\begin{array}{c}\text{Equilibrium}\\ \text{concentrations}\end{array}} \longrightarrow \boxed{K_a}$$

We can summarize the various concentrations and changes in the following table. (The concentration of water does not appear in the expression for the equilibrium constant, so we do not need to consider it.)

$$HNO_2 + H_2O \rightleftharpoons H_2O^+ + NO_2^-$$

	$HNO_2 + H_2O \rightleftharpoons$	H_2O^+ +	NO_2^-
Initial Concentration, M	0.0516	~0	0
Change, M	$-\Delta$	Δ	Δ
Equilibrium Concentration, M	$[HNO_2]_i + (-\Delta) =$ $0.0516 + (-\Delta)$	$[H_3O^+]_i + \Delta$ $\sim 0 + \Delta$	$[NO_2^-]_i + \Delta$ $0 + \Delta$

To get the various values in the table, we first calculate $[H_3O^+]$, the equilibrium concentration of H_3O^+, from the pH.

$$[H_3O^+] = 10^{-2.34} = 0.0046 \text{ M}$$

The change in concentration of H_3O^+, Δ, is the difference between the equilibrium concentration of H_3O^+, which we determined from the pH, and the initial concentration, $[H_3O^+]_i$. The initial concentration of H_3O^+ is its concentration in pure water, which is so much less than the final concentration that we approximate it as zero (~0).

The change in the concentration of NO_2^- is equal to the change in concentration of H_3O^+. For each 1 mole of H_3O^+ that forms, 1 mole of NO_2^- forms. The change in concentration of HNO_2 is $-\Delta$. (You may wish to review Section 14.8, a discussion of the relationships between the changes in concentrations of reactants and products.)

Now we can fill in the table with the concentrations at equilibrium.

	$HNO_2 + H_2O \rightleftharpoons$	H_2O^+ +	NO_2^-
Initial Concentration, M	0.0516	~0	0
Change, M	$-\Delta$	$\Delta = 0.0046$	Δ
Equilibrium Concentration, M	0.0470	0.0046	0.0046

We now calculate the value of the equilibrium constant using the concentrations at equilibrium from the table.

$$K_a = Q = \frac{[H_3O^-][NO_2^-]}{[HNO_2]} = \frac{(0.0046)(0.0046)}{(0.0470)}$$

$$= 4.5 \times 10^{-4}$$

Check your learning: The pH of a solution of household ammonia, a 0.950 M solution of NH_3, is 11.612. What is K_b for NH_3? *Hint:* Convert pH to pOH or $[H_3O^+]$ to $[OH^-]$ in the early stages of solving this problem.

Answer: $K_b = 1.8 \times 10^{-5}$

16.4   Concentrations in Solutions of Weak Acids or Weak Bases

If we prepare a solution of acetic acid of known concentration, we can calculate the concentration of nonionized acid, acetate ion, and hydronium ion present at equilibrium in the solution by using the techniques discussed in Sections 14.10 and 14.11. These techniques are very powerful because they can also be used to determine equilibrium concentrations in a solution of a weak base, in a solution that contains a mixture of acids, in a solution that contains a mixture of bases, or in a solution that contains a mixture of a weak acid (or weak base) and a salt of the acid (or base). As we saw in Section 14.10, these calculations can be divided into four steps:

Step 1.   *Determine in what direction the reaction shifts to come to equilibrium.*

Step 2.   *Determine the relative changes needed to reach equilibrium, and express the equilibrium concentrations in terms of these changes.*

Step 3.   *Solve for the changes and the equilibrium concentrations.*

Step 4.   *Check the arithmetic and assumptions.*

Sometimes a given step may differ from problem to problem—it may be more complex in some problems and less complex in others. But every calculation of equilibrium concentrations that begins with initial concentrations will involve these four steps.

Graphically, we can represent these steps as shown in the following blocks.

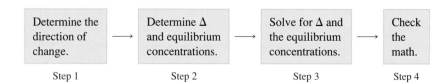

Determine the direction of change.		Determine Δ and equilibrium concentrations.		Solve for Δ and the equilibrium concentrations.		Check the math.
Step 1		Step 2		Step 3		Step 4

For now, we will consider solutions of acids in which the concentration of H_3O^+ is greater than 4.5×10^{-7} M or solutions of bases in which the concentration of OH^- is greater than 4.5×10^{-7} M so that we can neglect the contribution of water to these concentrations. The contribution of the self-ionization of water will be considered in Section 16.7.

Figure 16.11

The pain of an ant's sting is due to formic acid.

Example 16.10   Equilibrium Concentrations in a Solution of a Weak Acid

Formic acid, HCO_2H, is the irritant that causes the body's reaction to ant stings (Fig. 16.11). What is the concentration of hydronium ion in a 0.534 M solution of formic acid?

$$HCO_2H + H_2O \rightleftharpoons H_3O^+ + HCO_2^-   \qquad K_a = 1.8 \times 10^{-4}$$

Solution: The initial concentration of acid and its ionization constant are the clues we use to recognize that we need to calculate changes in concentration as the reaction goes to equilibrium. We can approach the solution through the following four steps.

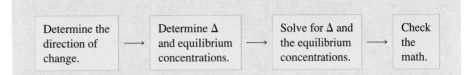

Step 1. *Determine the direction of change.*

The equation for the reaction is given in the statement of the problem. It may be obvious that the reaction will come to equilibrium by a shift to the right, because the initial concentration of one of the products is zero. However, it may be instructive for us to verify this shift by calculating Q before the reaction starts. The initial concentrations are $[HCO_2H] = 0.534$ M, $[H_3O^+] = {\sim}0$, and $[HCO_2^-] = 0$ M.

$$Q = \frac{[H_3O^+][HCO_2^-]}{[HCO_2H]} = \frac{({\sim}0)(0)}{(0.534)} = 0$$

Before any reaction occurs, Q is less than K, so the reaction will shift to the right, increasing the concentration of products and decreasing the concentration of reactants.

Step 2. *Determine Δ and equilibrium concentrations.*

The changes in concentration are

$$HCO_2H + H_2O \rightleftharpoons H_3O^+ + HCO_2^-$$
$$-\Delta \Delta \Delta$$

The concentration of water does not appear in the expression for the equilibrium constant, so we do not need to consider its change in concentration.

A table of initial concentrations (concentrations before the acid ionizes), changes in concentration, and equilibrium concentrations follows (the data given in the problem appear in color).

	HCO_2H	$+$	$H_2O \rightleftharpoons$	H_3O^+	$+$	HCO_2^-
Initial Concentration, M	0.534			${\sim}0$		0
Change	$-\Delta$			Δ		Δ
Equilibrium Concentration, M	$0.534 + (-\Delta)$			$0 + \Delta = \Delta$		$0 + \Delta = \Delta$

Step 3. *Solve for Δ and the equilibrium concentrations.*

At equilibrium,

$$K_a = 1.8 \times 10^{-4} = \frac{[H_3O^+][HCO_2^-]}{[HCO_2H]}$$

$$= \frac{(\Delta)(\Delta)}{0.534 - \Delta} = 1.8 \times 10^{-4}$$

Now we solve for Δ. Because both Q for the initial concentrations and K_a are much less than 1, we assume that the changes in concentrations needed to make $Q = K$ are so small that we can make the approximation that $(0.534 - \Delta) = 0.534$. This gives

$$K_a = 1.8 \times 10^{-4} = \frac{\Delta^2}{0.534}$$

We solve for Δ as follows:

$$\Delta^2 = 0.534 \times (1.8 \times 10^{-4}) = 9.6 \times 10^{-5}$$
$$\Delta = \sqrt{9.6 \times 10^{-5}}$$
$$= 9.8 \times 10^{-3}$$

To check the assumption that Δ is small compared to 0.534, we calculate

$$\frac{\Delta}{0.534} = \frac{9.8 \times 10^{-3}}{0.534} = 1.8 \times 10^{-2} \quad (1.8\% \text{ of } 0.534)$$

Δ is less than 5% of the initial concentration; as discussed in Section 14.11, the assumption is valid.

We find the equilibrium concentration of hydronium ion in this formic acid solution from its initial concentration and the change in that concentration as indicated in the last line of the foregoing table.

$$[H_3O^+] = {\sim}0 + \Delta = 0 + 9.8 \times 10^{-3} \text{ M}$$
$$= 9.8 \times 10^{-3} \text{ M}$$

Step 4.   *Check the work.*

If the arithmetic is correct, the value of Q at equilibrium will equal K_a. For the equilibrium concentrations, we have

$$[H_3O^+] = 9.8 \times 10^{-3} \text{ M}$$
$$[HCO_2^-] = \Delta = 9.8 \times 10^{-3} \text{ M}$$
$$[HCO_2H] = 0.534 - \Delta = 0.524 \text{ M}$$

The value of Q at equilibrium is

$$Q = \frac{[H_3O^+][HCO_2^-]}{[HCO_2H]} = \frac{(9.8 \times 10^{-3})^2}{0.524}$$
$$= 1.8 \times 10^{-4}$$
$$= K_a$$

This shows us that the calculated values check. [If we want to show that we can neglect the hydronium ion produced by water we use our calculated equilibrium concentration of hydronium ion (9.8×10^{-3} M) and verify that $[H_3O^+] > 4.5 \times 10^{-7}$ M.]

Check your learning: Only a small fraction of a weak acid ionizes in aqueous solution. What is the percent ionization of acetic acid in a 0.100 M solution of acetic acid, CH_3CO_2H?

$$CH_3CO_2H + H_2O \rightleftharpoons H_3O^+ + CH_3CO_2^- \qquad K_a = 1.8 \times 10^{-5}$$

Hint: Determine $[CH_3CO_2^-]$ at equilibrium. The percent ionization is the fraction of acetic acid that is ionized $\times$ 100, or $[CH_3CO_2^-]/[CH_3CO_2H]_i \times 100$.

Answer: Percent ionization = 1.3%

The following example shows that the concentration of products produced by the ionization of a weak base can be determined by the same series of steps used with a weak acid.

Example 16.11   Equilibrium Concentrations in a Solution of a Weak Base

Find the concentration of hydroxide ion in a 0.25 M solution of trimethylamine, a weak base.

$$(CH_3)_3N + H_2O \rightleftharpoons (CH_3)_3NH^+ + OH^- \qquad K_b = 7.4 \times 10^{-5}$$

Solution: This problem requires that we calculate an equilibrium concentration by determining concentration changes as the ionization of a base goes to equilibrium. The solution is approached in the same way as that for the ionization of formic acid in Example 16.10. The reactants and products will be different and the numbers will be different, but the logic is the same.

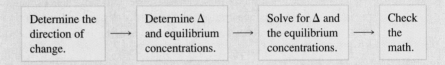

Step 1.   Determine the direction of change.

Before the reaction starts, Q is zero because no product is present. The reaction will shift to the right, increasing the concentration of products.

Step 2.   Determine Δ and equilibrium concentrations.

A table of changes and concentrations follows (the data given in the problem appear in color).

	$(CH_3)_3N$	+	H_2O $\rightleftharpoons$	$(CH_3)_3NH^+$	+	OH^-
Initial Concentration, M	0.25			0		~0
Change	$-\Delta$			Δ		Δ
Equilibrium Concentration, M	$0.25 + (-\Delta)$			$0 + \Delta$		$0 + \Delta$

Step 3.   Solve for Δ and the equilibrium concentrations.

At equilibrium,

$$K_b = \frac{[(CH_3)_3NH^+][OH^-]}{[(CH_3)_3N]} = \frac{(\Delta)(\Delta)}{0.25 - \Delta} = 7.4 \times 10^{-5}$$

If we assume that Δ is small relative to 0.25, then we can replace $(0.25 - \Delta)$ in the preceding equation with 0.25. Solving the simplified equation gives

$$\Delta = 4.3 \times 10^{-3}$$

This change is less than 5%, so the assumption is justified.
   Now we determine the concentration of hydroxide ion.

$$[OH^-] = \sim 0 + \Delta = 0 + 4.3 \times 10^{-3} \text{ M}$$
$$= 4.3 \times 10^{-3} \text{ M}$$

Step 4.   Check the work.

A check of our arithmetic shows that at equilibrium, $Q = 7.4 \times 10^{-5}$, which equals K_b.

Check your learning:
(a) Show that the calculation in Step 3 of this example gives a Δ value of 4.3×10^{-3} and that the calculation in Step 4 shows $Q = 7.4 \times 10^{-5}$.
(b) Find the concentration of hydroxide ion in a 0.0325 M solution of ammonia, a weak base with a K_b value of 1.76×10^{-5}. Calculate the percent ionization of ammonia, the fraction ionized $\times$ 100, or $[NH_4^+]/[NH_3]_i \times 100$.

Answer: **(a)** 7.56×10^{-4} M, **(b)** 2.33%

Some weak acids and weak bases ionize to such an extent that the simplifying assumption that Δ is small is inappropriate. As we solve for the equilibrium concentrations in such cases, we will see that we cannot neglect the change in the initial concentration of the acid or base, and we must solve the equilibrium equations by successive approximations or by using the quadratic equation as described in Section 14.11.

Figure 16.12

This disinfectant is acidic because it contains NaHSO₄.

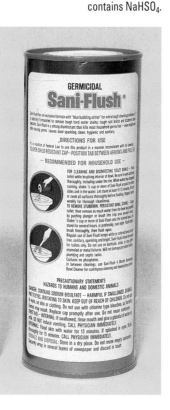

Example 16.12   Equilibrium Concentrations in a Solution of a Weak Acid

NaHSO₄ is used in some household cleansers because it contains the HSO₄⁻ ion, a weak acid (Fig. 16.12). What is the pH of a 0.50 M solution of HSO₄⁻?

$$HSO_4^- + H_2O \rightleftharpoons H_3O^+ + SO_4^{2-} \qquad K_a = 1.2 \times 10^{-2}$$

Solution: We need to determine the equilibrium concentration of the hydronium ion that results from the ionization of HSO₄⁻ so that we can use $[H_3O^+]$ to determine the pH. As in the previous examples, we can approach the solution by the following steps.

| Determine the direction of change. | → | Determine Δ and equilibrium concentrations. | → | Solve for Δ and the equilibrium concentrations. | → | Check the math. |

Step 1.   *Determine the direction of change.*

Before the reaction starts, Q is zero because no product is present. The reaction will shift to the right, increasing the concentration of product.

Step 2.   *Determine Δ and equilibrium concentrations.*

The table of changes and concentrations follows.

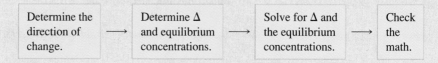

	HSO₄⁻	+	H₂O	⇌	H₃O⁺	+	SO₄²⁻
Initial Concentration, M	0.50				~0		0
Change	$-\Delta$				Δ		Δ
Equilibrium Concentration, M	$0.50 + (-\Delta)$ $= 0.50 - \Delta$				$0 + \Delta = \Delta$		$0 + \Delta = \Delta$

Step 3.   *Solve for Δ and the concentrations.*

As we begin solving for Δ, we find this calculation is more complicated than in previous examples. As we discuss these complications, we should not lose sight of the fact that it is still the purpose of this step to determine the value of Δ.

At equilibrium,

$$K_a = 1.2 \times 10^{-2} = \frac{[H_3O^+][SO_4^{2-}]}{[HSO_4^-]} = \frac{(\Delta)(\Delta)}{0.50 - \Delta}$$

If we assume that Δ is small and thus approximate $(0.50 - \Delta)$ as 0.50, we find

$$\Delta = 7.7 \times 10^{-2}$$

When we check the assumption, we find

$$\frac{\Delta}{0.50} = \frac{7.7 \times 10^{-2}}{0.50} = 0.15 \quad (15\%)$$

Δ is not less than 5% of 0.50, so the assumption is not valid. We need additional approximations or the quadratic formula to find Δ. We will use the quadratic formula.

The equation

$$K_a = 1.2 \times 10^{-2} = \frac{(\Delta)(\Delta)}{0.50 - \Delta}$$

gives

$$(6.0 \times 10^{-3}) - (1.2 \times 10^{-2})\Delta = \Delta^2$$

or

$$\Delta^2 + (1.2 \times 10^{-2})\Delta - 6.0 \times 10^{-3} = 0$$

This equation can be solved using the quadratic formula. For an equation of the form

$$ax^2 + bx + c = 0$$

x is given by the equation

$$x = \frac{-b \pm \sqrt{b^2 - 4ac}}{2a}$$

In this problem, $x = \Delta$, $a = 1$, $b = 1.2 \times 10^{-3}$, and $c = -6.0 \times 10^{-3}$. Solving for Δ gives a negative root, which cannot be correct (Section 14.11), and a positive root:

$$\Delta = 7.2 \times 10^{-2}$$

Now we determine the hydronium ion concentration and the pH.

$$[H_3O^+] = {\sim}0 + \Delta = 0 + 7.2 \times 10^{-2}\,M$$
$$= 7.2 \times 10^{-2}\,M$$

The pH of this solution is

$$pH = -\log [H_3O^+] = -\log 7.2 \times 10^{-2} = 1.14$$

Step 4.   *Check the work.*

Using the concentrations calculated with $\Delta = 7.2 \times 10^{-2}$, we find that $Q = 1.2 \times 10^{-2}$ at equilibrium. Because Q at equilibrium is equal to K_a, the arithmetic checks.

Check your learning: Calculate the pH in a 0.010 M solution of caffeine, a weak base.

$$C_8H_{10}N_4O_2 + H_2O \rightleftharpoons C_8H_{10}N_4O_2H^+ + OH^-   K_b = 2.5 \times 10^{-4}$$

Hint: It will be necessary to convert $[OH^-]$ to $[H_3O^+]$ or to convert pOH to pH toward the end of the calculation.

Answer: pH = 11.16

16.5   Acid–Base Properties of Solutions of Salts

Many salts exhibit behavior that we might not expect: They are acidic or basic. This type of behavior is generally observed with salts that are formed when we neutralize a weak acid or a weak base.

Salts of Weak Bases and Strong Acids   When we neutralize a weak base with a strong acid, the product is a salt that contains the conjugate acid of the weak base. This conjugate acid is a weak acid (Section 15.3). For example, ammonium chloride, NH_4Cl, is a salt formed by the reaction of the weak base ammonia with the strong acid HCl.

$$NH_3 + HCl \longrightarrow NH_4Cl$$

A solution of this salt contains ammonium ions and chloride ions. The chloride ion has no effect on the acidity of the solution. However, the ammonium ion, the conjugate acid of ammonia, reacts with water and increases the acidity of the solution by increasing the hydronium ion concentration (Fig. 16.13).

Figure 16.13

A solution of NH_4NO_3 is acidic because NH_4^+ is a weak acid.

$$NH_4^+ + H_2O \rightleftharpoons H_3O^+ + NH_3$$

The equilibrium equation for this reaction is simply the ionization constant, K_a, for the acid NH_4^+.

$$\frac{[H_3O^+][NH_3]}{[NH_4^+]} = K_a$$

We will not find a value of K_a for the ammonium ion in the appendix of this text. However, it is not difficult to determine K_a for NH_4^+. Its value can be calculated from the value of the ionization constant of water, K_w, and that of K_b, the ionization constant of its conjugate base, NH_3, by using the following relationship (Section 15.3):

$$K_w = K_a \times K_b$$

This relation holds for any base and its conjugate acid or for any acid and its conjugate base. Thus for ammonia and its conjugate base, we have

$$K_a(\text{for } NH_4^+) = \frac{K_w}{K_b(\text{for } NH_3)}$$
$$= \frac{1.0 \times 10^{-14}}{1.8 \times 10^{-5}}$$
$$= 5.6 \times 10^{-10}$$

Example 16.13   The pH of a Solution of a Salt of a Weak Base and a Strong Acid

Aniline is an amine that is used to manufacture dyes. It is isolated as aniline hydrochloride, $[C_6H_5NH_3]Cl$, a salt prepared by the reaction of the weak base aniline and hydrochloric acid. What is the pH of a 0.233 M solution of aniline hydrochloride?

$$C_6H_5NH_3^+ + H_2O \rightleftharpoons H_3O^+ + C_6H_5NH_2$$

Solution: The $C_6H_5NH_3^+$ ion is the conjugate acid of a weak base. The value of K_a for this acid is not listed in Appendix F, but we can determine it from the value of K_b for aniline, $C_6H_5NH_2$, which is given as 4.6×10^{-10} (Table 16.4 and Appendix G).

$$K_a(\text{for } C_6H_5NH_3^+) \times K_b(\text{for } C_6H_5NH_2) = K_w = 1.0 \times 10^{-14}$$

$$K_a(\text{for } C_6H_5NH_3^+) = \frac{K_w}{K_b(\text{for } C_6H_5NH_2)} = \frac{1.0 \times 10^{-14}}{4.6 \times 10^{-10}} = 2.2 \times 10^{-5}$$

Now we have the ionization constant and the initial concentration of the weak acid, which is the information we need to determine the equilibrium concentration of H_3O^+ and the pH using our customary approach.

Determine the direction of change.	→	Determine Δ and equilibrium concentrations.	→	Solve for Δ and the equilibrium concentrations.	→	Check the math.

With these steps, we find $[H_3O^+] = 2.3 \times 10^{-3}$ M and pH = 2.64

Check your learning:
(a) Do the calculations and show that the hydronium ion concentration for a 0.233 M solution of $C_6H_5NH_3^+$ is 2.3×10^{-3} and that the pH is 2.64.
(b) Determine the hydronium ion concentration in a 0.100 M solution of ammonium nitrate, NH_4NO_3, a salt composed of the ions NH_4^+ and NO_3^-. Use the data in Table 16.4 to determine K_a for the ammonium ion. Which is the stronger acid, $C_6H_5NH_3^+$ or NH_4^+?

Answer: **(b)** $K_a(\text{for } NH_4^+) = 5.6 \times 10^{-10}$, $[H_3O^+] = 7.5 \times 10^{-6}$ M; $C_6H_5NH_3^+$ is the stronger acid.

Figure 16.14

Addition of an indicator to a solution of $NaCH_3CO_2$ shows that the solution is slightly basic.

Salts of Weak Acids and Strong Bases   When we neutralize a weak acid with a strong base, we get a salt that contains the conjugate base of the weak acid. This conjugate base is usually a weak base (Section 15.3). For example, sodium acetate, $NaCH_3CO_2$, is a salt formed by the reaction of the weak acid acetic acid with the strong base sodium hydroxide.

$$CH_3CO_2H + NaOH \longrightarrow NaCH_3CO_2 + H_2O$$

A solution of this salt contains sodium ions and acetate ions. The sodium ion has no effect on the acidity of the solution. However, the acetate ion, the conjugate base of acetic acid, reacts with water and increases the concentration of hydroxide ion (Fig. 16.14).

$$CH_3CO_2^- + H_2O \rightleftharpoons CH_3CO_2H + OH^-$$

The equilibrium equation for this reaction is the ionization constant, K_b, for the base $CH_3CO_2^-$. The value of K_b can be calculated from the value of the ionization constant of water, K_w, and that of K_a, the ionization constant of the conjugate acid of the anion, by using the equation

$$K_w = K_a \times K_b$$

For the acetate and its conjugate acid, we have

$$K_b(\text{for } CH_3CO_2^-) = \frac{K_w}{K_a(\text{for } CH_3CO_2H)} = \frac{1.0 \times 10^{-14}}{1.8 \times 10^{-5}} = 5.6 \times 10^{-10}$$

Some handbooks do not report values of K_b. They only report ionization constants for acids. If we want to determine a K_b value using one of these handbooks, we must look up the value of K_a for the conjugate acid and convert it to a K_b value.

Example 16.14   Equilibrium in a Solution of a Salt of a Weak Acid and a Strong Base

Determine the acetic acid concentration in a solution with $[CH_3CO_2^-] = 0.050$ M and $[OH^-] = 2.5 \times 10^{-6}$ M at equilibrium. The reaction is

$$CH_3CO_2^- + H_2O \rightleftharpoons CH_3CO_2H + OH^-$$

Solution: In this problem, we are given two of three equilibrium concentrations and asked to find the missing concentration. If we can find the equilibrium constant for the reaction, the process is straightforward.

The acetate ion behaves as a base in this reaction; hydroxide ions are a product. We determine K_b as follows:

$$K_b(\text{for } CH_3CO_2^-) = \frac{K_w}{K_a(\text{for } CH_3CO_2H)} = \frac{1.0 \times 10^{-14}}{1.8 \times 10^{-5}} = 5.6 \times 10^{-10}$$

Now we find the missing concentration.

$$K_b = \frac{[CH_3CO_2H][OH^-]}{[CH_3CO_2^-]} = 5.6 \times 10^{-10}$$

$$= \frac{[CH_3CO_2H](2.5 \times 10^{-6})}{(0.050)} = 5.6 \times 10^{-10}$$

When we solve this equation, we get $[CH_3CO_2H] = 1.1 \times 10^{-5}$ M.

Check your learning: What is the pH of a 0.083 M solution of the basic ion CN^-? Use 4.0×10^{-10} as K_a for HCN. *Hint:* We will probably need to convert pOH to pH or find $[H_3O^+]$ using $[OH^-]$ in the final stages of this problem.

Answer: 11.16

16.6   Mixtures of Acids or Bases; Polyprotic Acids

When we make an aqueous solution that contains a mixture of acids, each acid reacts with water and forms hydronium ion. The total hydronium ion concentration is the sum of the contributions of all the acids; however, each acid need not contribute equally. Stronger acids contribute more hydronium ion than weaker acids. In fact, in a mixture of two weak acids, if K_a for the stronger acid is 20 or more times larger than K_a for the weaker acid, then the stronger acid is the dominant producer of hydronium ion. The weaker acid contributes hydronium ion to the solution, but the amount is 5% or less and is usually neglected. In such a solution the stronger acid determines the concentration of hydronium ion, and the ionization of the weaker acid is fixed by the concentration of hydronium ion produced by the stronger acid.

When we make a solution of a weak diprotic acid (Section 15.8), we get a solution that contains a mixture of acids. The ion that results from the first ionization is also an acid, and it ionizes in the second ionization. Carbonic acid, H_2CO_3, is an example of a weak diprotic acid. The first ionization of carbonic acid produces hydronium ions and hydrogen carbonate ions in low yield.

First ionization:

$$H_2CO_3 + H_2O \rightleftharpoons H_3O^+ + HCO_3^- \qquad K_{H_2CO_3} = \frac{[H_3O^+][HCO_3^-]}{[H_2CO_3]} = 4.3 \times 10^{-7}$$

The hydrogen carbonate ion is also an acid. It ionizes and forms hydronium ions and carbonate ions in even lower yields.

Second ionization:

$$HCO_3^- + H_2O \rightleftharpoons H_3O^+ + CO_3^{2-} \qquad K_{HCO_3^-} = \frac{[H_3O^+][CO_3^{2-}]}{[HCO_3^-]} = 7 \times 10^{-11}$$

$K_{H_2CO_3}$ is larger than $K_{HCO_3^-}$ by a factor of 10^4, so H_2CO_3 is the dominant producer of hydronium ion in the solution. This means that very little of the HCO_3^- formed by the ionization of H_2CO_3 ionizes to give hydronium ions (and carbonate ions), and the concentrations of H_3O^+ and HCO_3^- are practically equal in a pure aqueous solution of H_2CO_3.

If the first ionization constant of a weak diprotic acid is larger than the second by a factor of at least 20, then it is appropriate to treat the first ionization separately and calculate concentrations of species resulting from it before calculating concentrations of species resulting from subsequent ionization. This can simplify our work considerably, because we can determine the concentrations of H_3O^+ and the conjugate base from the first ionization and then determine the concentration of the conjugate base of the second ionization in a solution with concentrations determined by the first ionization.

Example 16.15   Ionization of a Diprotic Acid

When we buy soda water (carbonated water), we are buying a solution of carbon dioxide in water. The solution is acidic because CO_2 reacts with water to form carbonic acid, H_2CO_3. What are $[H_3O^+]$, $[HCO_3^-]$, and $[CO_3^{2-}]$ in a saturated solution of CO_2 with an initial $[H_2CO_3] = 0.033$ M?

$$H_2CO_3 + H_2O \rightleftharpoons H_3O^+ + HCO_3^- \qquad K_{H_2CO_3} = 4.3 \times 10^{-7}$$
$$HCO_3^- + H_2O \rightleftharpoons H_3O^+ + CO_3^{2-} \qquad K_{HCO_3} = 7 \times 10^{-11}$$

Solution: As indicated by the ionization constants, H_2CO_3 is a much stronger acid than HCO_3^-, so H_2CO_3 is the dominant producer of hydronium ion in solution. Thus there are two parts in the solution of this problem: (1) Using the customary four steps, we determine the concentrations of H_3O^+ and HCO_3^- produced by ionization of H_2CO_3. (2) Then we determine the concentration of CO_3^{2-} in a solution with the concentrations of H_3O^+ and HCO_3^- determined in part (1). To summarize:

$$[H_2CO_3] \longrightarrow \begin{array}{c} [H_3O^+] \text{ and } [HCO_3^-] \\ \text{from } H_2CO_3 \end{array} \longrightarrow \begin{array}{c} [CO_3^{2-}] \text{ from} \\ HCO_3^- \end{array}$$

Part (1)   We determine the concentration of H_3O^+ and HCO_3^-,

$$H_2CO_3 + H_2O \rightleftharpoons H_3O^+ + HCO_3^- \qquad K_{H_2CO_3} = 4.3 \times 10^{-7}$$

as we would for the ionization of any other weak acid.

$$\begin{array}{c}\text{Determine the} \\ \text{direction of} \\ \text{change.}\end{array} \longrightarrow \begin{array}{c}\text{Determine } \Delta \\ \text{and equilibrium} \\ \text{concentrations.}\end{array} \longrightarrow \begin{array}{c}\text{Solve for } \Delta \text{ and} \\ \text{the equilibrium} \\ \text{concentrations.}\end{array} \longrightarrow \begin{array}{c}\text{Check} \\ \text{the} \\ \text{math.}\end{array}$$

An abbreviated table of changes and concentrations follows.

	H_2CO_3	$+$	H_2O	$\rightleftharpoons$	H_3O^+	$+$	HCO_3^-
Initial Concentration, M	0.033				~0		0
Change	$-\Delta$				Δ		Δ
Equilibrium Concentration, M	$0.033 - \Delta$				Δ		Δ

Substituting the equilibrium concentrations into the equilibrium gives us

$$K_{H_2CO_3} = \frac{[H_3O^+][HCO_3^-]}{[H_2CO_3]} = \frac{(\Delta)(\Delta)}{0.033 - \Delta} = 4.3 \times 10^{-7}$$

Solving the preceding equation gives

$$\Delta = 1.2 \times 10^{-4}$$

Thus
$$[H_2CO_3] = 0.033 \text{ M}$$
$$[H_3O^+] = [HCO_3^-] = 1.2 \times 10^{-4} \text{ M}$$

Calculation of Q using these concentrations shows that the arithmetic is correct.

Part (2)   Now we determine the concentration of CO_3^{2-} in a solution at equilibrium with $[H_3O^+]$ and $[HCO_3^-]$ both equal to 1.2×10^{-4} M.

$$HCO_3^- + H_2O \rightleftharpoons H_3O^+ + CO_3^{2-} \qquad K_{HCO_3^-} = 7 \times 10^{-11}$$

$$K_{HCO_3^-} = \frac{[H_3O^+][CO_3^{2-}]}{[HCO_3^-]} = \frac{(1.2 \times 10^{-4})[CO_3^{2-}]}{1.2 \times 10^{-4}} = 7 \times 10^{-11}$$

or
$$[CO_3^{2-}] = \frac{(7 \times 10^{-11}) \times (1.2 \times 10^{-4})}{1.2 \times 10^{-4}}$$

$$= 7 \times 10^{-11} \text{ M}$$

To summarize: In part 1 of this example, we found that the H_2CO_3 in a 0.033 M solution ionizes slightly and that at equilibrium $[H_2CO_3] = 0.033$ M, $[H_3O^+] = 1.2 \times 10^{-4}$, and $[HCO_3^-] = 1.2 \times 10^{-4}$ M. In part 2 we determined that $[CO_3^{2-}] = 7 \times 10^{-11}$ M. The concentration of H_3O^+ produced by the ionization of HCO_3^- is also 7×10^{-11} M, a concentration far too small to make a significant contribution to $[H_3O^+]$ in this solution.

Check your learning: The concentration of H_2S in a saturated aqueous solution at room temperature is approximately 0.1 M. Calculate $[H_3O^+]$, $[HS^-]$, and $[S^{2-}]$ in a solution with a concentration of 0.1 M.

$$H_2S + H_2O \rightleftharpoons H_3O^+ + HS^- \qquad K_{H_2S} = 1.0 \times 10^{-7}$$
$$HS^- + H_2O \rightleftharpoons H_3O^+ + S^{2-} \qquad K_{HS^-} = 1.0 \times 10^{-19}$$

Answer: $[H_2S] = 0.1$ M, $[H_3O^+] = [HS^-] = 0.0001$ M, $[S^{2-}] = 1 \times 10^{-19}$ M

A triprotic acid is an acid that has three ionizable hydrogen ions that undergo stepwise ionization (Section 15.8). Phosphoric acid is a typical example.

First ionization: $H_3PO_4 + H_2O \rightleftharpoons H_3O^+ + H_2PO_4^- \qquad K_{H_3PO_4} = 7.5 \times 10^{-3}$

Second ionization: $H_2PO_4^- + H_2O \rightleftharpoons H_3O^+ + HPO_4^{2-} \qquad K_{H_2PO_4^-} = 6.3 \times 10^{-8}$

Third ionization: $HPO_4^{2-} + H_2O \rightleftharpoons H_3O^+ + PO_4^{3-} \qquad K_{HPO_4^{2-}} = 3.6 \times 10^{-13}$

Figure 16.15

Some soft drinks are mildly acidic because they contain a small amount of phosphoric acid.

The differences in the ionization constants of these reactions tell us that in each successive step, the degree of ionization is significantly less. This is a general characteristic of polyprotic acids, and successive ionization constants often differ by a factor of about 10^5 to 10^6.

This set of three dissociation reactions may appear to make calculations of equilibrium concentrations in a solution of H_3PO_4 very complicated. However, because the successive ionization constants differ by a factor of 10^5 to 10^6, the calculations can be broken down into a series of parts similar to those for diprotic acids.

Example 16.16   Ionization of a Triprotic Acid

Phosphoric acid is used in soft drinks to give them an acidic taste (Fig. 16.15). Calculate the concentrations of all species—H_3PO_4, $H_2PO_4^-$, HPO_4^{2-}, PO_4^{3-}, H_3O^+, and OH^-—present at equilibrium in a 0.100 M solution of phosphoric acid.

$$H_3PO_4 + H_2O \rightleftharpoons H_3O^+ + H_2PO_4^- \qquad K_{H_3PO_4} = 7.5 \times 10^{-3}$$
$$H_2PO_4^- + H_2O \rightleftharpoons H_3O^+ + HPO_4^{2-} \qquad K_{H_2PO_4^-} = 6.3 \times 10^{-8}$$
$$HPO_4^{2-} + H_2O \rightleftharpoons H_3O^+ + PO_4^{3-} \qquad K_{HPO_4^{2-}} = 3.6 \times 10^{-13}$$

Solution: As indicated by the ionization constants, H_3PO_4 is a much stronger acid than $H_2PO_4^-$ or HPO_4^{2-}, so it is the dominant producer of hydronium ion in solution. Consequently, we can solve the problem in the following four parts.

Part (1) We determine the concentration of H_3O^+, $H_2PO_4^-$, and H_3PO_4 in a 0.100 M solution of H_3PO_4.

$$H_3PO_4 + H_2O \rightleftharpoons H_3O^+ + H_2PO_4^- \qquad K_{H_3PO_4} = 7.5 \times 10^{-3}$$

We use the the same scheme we would with any other weak acid.

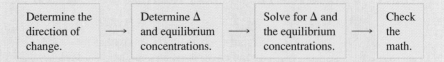

In this case we need to use successive approximations or the quadratic equation to find Δ. Result: $[H_3PO_4] = 0.076$ M, $[H_3O^+] = [H_2PO_4^-] = 0.024$ M

Part (2) Now we consider the second step of the ionization.

$$H_2PO_4^- + H_2O \rightleftharpoons H_3O^+ + HPO_4^{2-} \qquad K_{H_2PO_4^-} = 6.3 \times 10^{-8}$$

We know the concentration of $H_2PO_4^-$ and H_3O^+ produced by the first step of the ionization. We find the missing concentration in this equilibrium, the concentration of HPO_4^-, from the equilibrium constant and the equilibrium concentrations. Result: $[HPO_4^{2-}] = 6.3 \times 10^{-8}$

Part (3) Now we consider the third step of the ionization.

$$HPO_4^{2-} + H_2O \rightleftharpoons H_3O^+ + PO_4^{3-} \qquad K_{HPO_4^{2-}} = 3.6 \times 10^{-13}$$

We know the concentration of HPO_4^{2-} (formed by the second step in the ionization) and the H_3O^+ concentration (formed by the first step of the ionization). Again we find the missing concentration in this equilibrium from the equilibrium constant and the other equilibrium concentrations. Result: $[PO_4^{3-}] = 9.4 \times 10^{-19}$ M

Part (4) Calculation of $[OH^-]$ completes this example. We know $[H_3O^+]$, so we can readily find $[OH^-]$.

$$K_w = [H_3O^+][OH^-] = 1.0 \times 10^{-14}$$

$$[OH^-] = \frac{K_w}{[H_3O^+]} = \frac{1.0 \times 10^{-14}}{2.4 \times 10^{-2}} = 4.2 \times 10^{-13} \text{ M}$$

Check your learning: Carry out the calculations described in this exercise and verify the results.

The behavior of a solution that contains a mixture of bases is similar to that of a solution of a mixture of acids, except that hydroxide ions are formed instead of hydronium ions. In a mixture of bases, the total hydroxide ion concentration is the sum of the contributions from all the bases; however, each base need not contribute equally. Stronger bases contribute more hydroxide ion than weaker bases. In fact, in a mixture of two weak bases, if K_b for the stronger base is 20 or more times larger than K_b for the weaker base, than the stronger base is the dominant producer of hydroxide ion. The weaker base contributes hydroxide ion to the solution, but the amount is 5% or less. In such a solution, the stronger base determines the concentration of hydroxide ion, and the ionization of the weaker base is fixed by the concentration of hydroxide ion produced by the stronger base.

16.7   The Ionization of Very Weak Acids

In any aqueous solution of a monoprotic acid, we have two simultaneous equilibria that produce hydronium ions. These are the ionization of the acid:

$$HA + H_2O \rightleftharpoons H_3O^+ + A^-$$

and the self-ionization of water:

$$H_2O + H_2O \rightleftharpoons H_3O^+ + OH^-$$

The total concentration of hydronium ion, $[H_3O^+]_{total}$, in a solution of an acid is the sum of the hydronium ion concentrations from the acid, $[H_3O^+]_{acid}$, and from the water, $[H_3O^+]_{water}$.

$$[H_3O^+]_{total} = [H_3O^+]_{acid} + [H_3O^+]_{water}$$

To this point we have neglected $[H_3O^+]_{water}$ because the concentration of hydronium ion produced by an acid has been greater than 4.5×10^{-7} M. However, as we can see in Table 16.1, if an acid is present in low enough concentration (or is very weak), $[H_3O^+]_{acid}$ may not be greater than 4.5×10^{-7} M. Under these conditions, we must consider the ionization of water when calculating the concentrations of molecules and ions in solution.

Before we consider the effect of water on ion concentrations, let us consider why the ionization of water is significant when $[H_3O^+]_{total}$ is less than 4.5×10^{-7} M. In a solution of acid, the total hydronium ion concentration is given by the equation

$$[H_3O^+]_{total} = [H_3O^+]_{acid} + [H_3O^+]_{water}$$

Looking at the equation for the self-ionization of water,

$$H_2O + H_2O \rightleftharpoons H_3O^+ + OH^-$$

we can see that for each mole of H_3O^+ formed by the self-ionization of water, 1 mole of OH^- is produced. Thus the concentration of hydronium ion *produced by water* is equal to the concentration of hydroxide ion produced by water, because the only source of hydroxide ion in a solution of an acid is the ionization of water. Thus

$$[H_3O^+]_{water} = [OH^-]_{water}$$

In order for $[H_3O^+]_{water}$ (which equals $[OH^-]_{water}$) to be negligible, it must pass the 5% test; that is, $[H_3O^+]_{water}$ must be less than 5% of $[H_3O^+]_{total}$.

$$[H_3O^+]_{water} = [OH^-]_{water} = 0.05 \times [H_3O^+]_{total} \text{ (or less)}$$

The product of the *total* hydronium ion concentration in the solution and the hydroxide ion concentration equals K_w.

$$K_w = [H_3O^+]_{total}[OH^-]_{water}$$

Because

$$[H_3O^+]_{water} = [OH^-]_{water},$$

we can write

$$K_w = [H_3O^+]_{total}[H_3O^+]_{water}$$

and when

$$[H_3O^+]_{water} = 0.05[H_3O^+]_{total},$$

we have

$$K_w = [H_3O^+]_{total} \times (0.05 \times [H_3O^+]_{total})$$

We can now solve this equation for the minimum total hydronium ion concentration at which the contribution of hydronium ion by water can be neglected.

$$K_w = [H_3O^+]_{total} \times (0.05 \times [H_3O^+]_{total})$$

$$[H_3O^+]^2_{total} = \frac{K_w}{0.05} = \frac{1.0 \times 10^{-14}}{0.05}$$

$$[H_3O^+]_{total} = \sqrt{\frac{1.0 \times 10^{-14}}{0.05}} = 4.5 \times 10^{-7} \text{ M}$$

Example 16.17   Neglecting the Contribution of Water

Can the contribution of water to the hydronium ion concentration be neglected in a 1.0×10^{-5} M solution of HCN?

$$HCN + H_2O \rightleftharpoons H_3O^+ + CN^- \qquad K_a = 4 \times 10^{-10}$$

Solution: In order to answer this question, we find the concentration of hydronium ion produced by the ionization of HCN and determine whether it is large enough ($[H_3O^+]_{acid} > 4.5 \times 10^{-7}$ M) for us to neglect the self-ionization of water. We use the familiar scheme.

Determine the direction of change.	→	Determine Δ and equilibrium concentrations.	→	Solve for Δ and the equilibrium concentrations.	→	Check the math.

	HCN	**+ H₂O ⇌ H₃O⁺**	**+ CN⁻**
Initial Concentration, M	0.000010	~0	0
Change	$-\Delta$	Δ	Δ
Equilibrium Concentration, M	$0.000010 - \Delta$	Δ	Δ

At equilibrium,

$$K_a = \frac{[H_3O^+][CN^-]}{[HCN]} = \frac{(\Delta)(\Delta)}{0.000010 - \Delta} = 4 \times 10^{-10}$$

$$\Delta = 6 \times 10^{-8} = [H_3O^+] = [CN^-]$$

A check of the calculated equilibrium concentrations gives $Q = 4 \times 10^{-10}$, so the arithmetic is correct.

Clearly this calculation does not give a reasonable answer. The concentration of H_3O^+ calculated for this solution of acid is less than that found in pure water. The contribution of the self-ionization of water cannot be neglected in this solution, because the calculated concentration of hydronium ion derived from the ionization of HCN (6×10^{-8} M) is less than 4.5×10^{-7} M.

Check your learning: Can the contribution of water to [OH⁻] be neglected in a 0.12 M solution of SO_4^{2-}?

$$SO_4^{2-} + H_2O \rightleftharpoons HSO_4^{2-} + OH^- \quad K_b = 8.3 \times 10^{-13}$$

Answer: No

Although the arithmetic used in Example 16.17 checks, the calculation shown does not give the actual concentration of hydronium ion at equilibrium, because the self-ionization of water was neglected. The check we performed (by evaluating Q) simply showed that the arithmetic was carried out correctly, not that we chose the correct procedure. The following discussion shows how we derive the equation to use when the self-ionization of water cannot be neglected.

In any aqueous solution of an acid, we have two simultaneous equilibria:

$$HA + H_2O \rightleftharpoons H_3O^+ + A^-$$
$$H_2O + H_2O \rightleftharpoons H_3O^+ + OH^-$$

The total concentration of hydronium ion, regardless of its source, is the correct concentration to use in the equilibrium equation either for the ionization of the acid or for the self-ionization of water. When an acid anion, A^-, or a hydroxide ion combines with a hydronium ion, the source of the H_3O^+ (either from HA or from H_2O) is irrelevant. Thus

$$K_a = \frac{[H_3O^+]_{total}[A^-]}{[HA]}$$

where the concentrations are at equilibrium. We have seen that

$$[H_3O^+]_{total} = [H_3O^+]_{acid} + [H_3O^+]_{water}$$

The concentration of A^- in the solution is equal to $[H_3O^+]_{acid}$ because for each molecule of HA that ionizes, one H_3O^+ ion and one A^- ion result. Likewise, the concentration of OH^- is equal to $[H_3O^+]_{water}$. Thus

$$[H_3O^+]_{total} = [A^-] + [OH^-]$$

If we rearrange this equation to solve for $[A^-]$, we have a term that can be substituted into the equilibrium equation.

$$[A^-] = [H_3O^+]_{total} - [OH^-]$$
$$K_a = \frac{[H_3O^+]_{total}[A^-]}{[HA]}$$
$$K_a = \frac{[H_3O^+]_{total}\{[H_3O^+]_{total} - [OH^-]\}}{[HA]}$$

Now let us solve this equation for the total hydronium ion concentration in the solution. Multiplication gives

$$K_a = \frac{\{[H_3O^+]_{total} \times [H_3O^+]_{total}\} - \{[H_3O^+]_{total} \times [OH^-]\}}{[HA]}$$
$$= \frac{\{[H_3O^+]^2_{total}\} - \{[H_3O^+]_{total} \times [OH^-]\}}{[HA]}$$

Because $K_w = [H_3O^+]_{total} \times [OH^-]$, we can write

$$K_a = \frac{[H_3O^+]^2_{total} - K_w}{[HA]}$$

Rearranging gives

$$K_a[HA] = [H_3O^+]^2_{total} - K_w$$

and

$$[H_3O^+]^2_{total} = K_a[HA] + K_w$$

Taking the square root of both sides of the equation yields

$$[H_3O^+]_{total} = \sqrt{K_a[HA] + K_w}$$

We now introduce the only assumption in this derivation. We assume that HA is such a weak acid that the change in concentration of HA upon ionization is negligible. Under this assumption, the concentration of the acid at equilibrium, [HA], is equal to the initial concentration of the acid, $[HA]_i$. Thus

$$[H_3O^+]_{total} = \sqrt{K_a[HA]_i + K_w} \tag{1}$$

This equation enables us to evaluate the total concentration of hydronium ion in a solution of a weak acid when the ionization of water cannot be neglected.

For a weak base, B, we use

$$[OH^-]_{total} = \sqrt{K_a[B]_i + K_w}$$

Example 16.18   **The Contribution of the Self-ionization of Water**

What is the concentration of H_3O^+ in a 1.0×10^{-5} M solution of HCN ($K_a = 4 \times 10^{-10}$)?

Solution: As shown in Example 16.17, the contribution of the concentration of H_3O^+ derived from water cannot be neglected in this calculation. Thus we must use Equation 1 to determine the concentration of H_3O^+ in this solution when $[HCN]_i = 1.0 \times 10^{-5}$ M.

$$\begin{aligned}
[H_3O^+]_{total} &= \sqrt{K_a[HA]_i + K_w} \\
&= \sqrt{(4 \times 10^{-10}) \times (1.0 \times 10^{-5}) + (1.0 \times 10^{-14})} \\
&= \sqrt{1.4 \times 10^{-14}} \\
&= 1.2 \times 10^{-7} \text{ M} \\
&= 1 \times 10^{-7} \text{ M} \quad \text{(to the number of significant figures justified by } K_a)
\end{aligned}$$

This $[H_3O^+]$ is twice as large as that calculated by ignoring the self-ionization of water (Example 16.17).

Check your learning: What is $[OH^-]$ in a 0.12 M solution of SO_4^{2-}?

$$SO_4^{2-} + H_2O \rightleftharpoons HSO_4^- + OH^- \quad K_b = 8.3 \times 10^{-13}$$

Answer: 3.3×10^{-7} M

16.8   The Salt Effect and the Common Ion Effect

The extent of ionization of a weak electrolyte can be influenced by the presence of other ions. For example, when we add a salt to a solution of a weak acid, its degree of ionization increases slightly. For example, the ionization constant of acetic acid changes from 1.8×10^{-5} in dilute solution to 2.2×10^{-5} in a 0.1 M solution of NaCl. This effect, which is called the **salt effect,** occurs because the addition of the salt changes the activities of the ions of the weak electrolyte. The change is due to attraction between the weak electrolyte's ions and the ions of the added salt. The salt effect is generally small, and we shall neglect the slight errors that may result from it.

A much more significant effect can result if the added salt has an ion in common with the weak electrolyte. As we saw in Example 16.6, the hydronium ion concentration of an aqueous solution of acetic acid decreases when the strong electrolyte potassium acetate, KCH_3CO_2, is added. We explain this effect by using Le Châtelier's principle (Section 14.4) The addition of acetate ions causes the equilibrium to shift to the left, reducing the concentration of H_3O^+ to compensate for the increased acetate ion concentration. This increases the concentration of CH_3CO_2H.

$$CH_3CO_2H + H_2O \rightleftharpoons H_3O^+ + CH_3CO_2^-$$

Because sodium acetate and acetic acid have the acetate ion in common, the influence on the equilibrium is sometimes called the **common ion effect.**

We can use the expression for the ionization constant to determine the extent to which the concentration of the hydronium ion decreases.

Example 16.19   Concentration in the Presence of a Common Ion

What is $[H_3O^+]$ in a solution that is initially 0.10 M in CH_3CO_2H and is 0.50 M in KCH_3CO_2?

$$CH_3CO_2H + H_2O \rightleftharpoons H_3O^+ + CH_3CO_2^- \qquad K_a = 1.8 \times 10^{-5}$$

Solution: We can use the standard path for an equilibrium calculation to determine the concentration of hydronium ion.

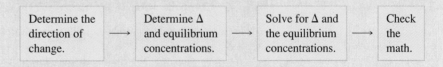

Step 1. *Determine the direction of change.*

Before any reaction, $[CH_3CO_2H]$ is 0.10 M, $[H_3O^+]$ is 1×10^{-7} M (the concentration of hydronium ion in pure water), and $[CH_3CO_2^-]$ is 0.50 M.

$$Q = \frac{[H_3O^+][CH_3CO_2^-]}{[CH_3CO_2H]} = \frac{(1 \times 10^{-7})(0.50)}{0.10} = 5 \times 10^{-7}$$

Because Q is less than K_a, the reaction will proceed to the right, increasing the concentration of H_3O^+ and $CH_3CO_2^-$.

Step 2.   *Determine* Δ *and equilibrium concentrations.*

We take the change in concentration of H_3O^+ as the mixture comes to equilibrium to be Δ. The table of initial concentrations, changes, and equilibrium concentrations for this problem follows.

$$CH_3CO_2H \; + \; H_2O \rightleftharpoons H_3O^+ \; + \; CH_3CO_2^-$$

Initial Concentration, M	0.10	~0	0.50
Change	$-\Delta$	Δ	Δ
Equilibrium Concentration, M	$0.10 + (-\Delta)$ $= 0.10 - \Delta$	$0 + \Delta = \Delta$	$0.50 + \Delta$

Step 3.   *Solve for* Δ *and the equilibrium concentrations.*

Substitution into the equilibrium equation gives

$$K_a = \frac{[H_3O^+][CH_3CO_2^-]}{[CH_3CO_2H]} = \frac{(\Delta)(0.50 + \Delta)}{(0.10 - \Delta)} = 1.8 \times 10^{-5}$$

Because both K_a and the initial value of Q are less than 1, we assume that the change necessary to reach equilibrium is small and that $(0.10 - \Delta) = 0.10$ and $(0.50 + \Delta) = 0.50$. This simplifies solving the equation. We have

$$1.8 \times 10^{-5} = \frac{(\Delta)(0.50)}{(0.10)}$$

which gives

$$\Delta = \frac{(1.8 \times 10^{-5})(0.10)}{0.50} = 3.6 \times 10^{-6}$$

The simplifying assumption that Δ is negligibly small compared to either 0.50 or 0.10 is well justified: $\Delta/0.10 = 3.6 \times 10^{-5}$ (0.0036%). Now we find

$$[H_3O^+] = 0 + \Delta = 3.6 \times 10^{-6}\,M$$
$$[CH_3CO_2H] = 0.10 - \Delta = 0.10\,M$$
$$[CH_3CO_2^-] = 0.50 + \Delta = 0.50\,M$$

Step 4.   *Check the work.*

When we calculate Q using the equilibrium concentrations we have calculated, we find $Q = K_a$.

   We have seen that the hydrogen ion concentration in a 0.10 M solution of acetic acid is 0.0013 M ("Check your learning," Example 16.10). This is reduced to 0.0000036 M by the presence of 0.50 M sodium acetate in the 0.10 M acetic acid solution—Le Châtelier's principle in action.

Check your learning: A 0.25 M solution of trimethylamine, $(CH_3)_3N$, has a hydroxide ion concentration of 4.3×10^{-3} M. Find the concentration of OH^- in a 0.25 M solution of trimethylamine that is also 0.10 M in trimethylamine hydrochloride, a strong electrolyte that provides the $(CH_3)_3NH^+$ ion.

$$(CH_3)_3N + H_2O \rightleftharpoons (CH_3)_3NH^+ + OH^- \qquad K_b = 7.4 \times 10^{-5}$$

Answer: 1.8×10^{-4} M

Addition of a base to a weak acid produces a salt of the acid. Hence, if the base is the limiting reactant (Section 3.12) and if there is not enough base to neutralize all of the weak acid, addition of a base to a solution of a weak acid can create a common ion effect. The solution that is produced contains both the weak acid and its salt. Similarly, addition of an acid to a solution of weak base can create a common ion effect if the acid is the limiting reactant.

Example 16.20   **Creation of a Common Ion Effect by Partial Neutralization**

Exactly 10.0 mL of a 0.100 M HCl solution is added to 25 mL of 0.100 M NH_3 solution. Calculate the hydroxide ion concentration of the resulting 35 mL of solution.

Solution: First we assume that the reaction of HCl with NH_3 gives a 100% yield of NH_4Cl and calculate the amount of NH_4Cl produced (a limiting reactant problem with HCl the limiting reactant; Section 3.12). Then we calculate the hydroxide ion concentration in this intermediate solution that contains a mixture of NH_3 and NH_4^+ (a common ion problem involving a weak base and its salt).

Part (1)   We can use the following steps to calculate the concentrations assuming a 100% yield of NH_4Cl.

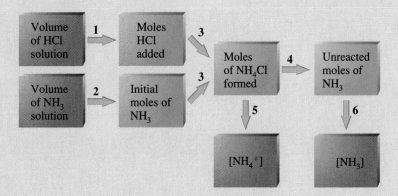

Step 1.   10.0 mL (0.0100 L) of 0.100 M HCl contains

$$0.0100 \text{ L} \times \left(\frac{0.100 \text{ mol HCl}}{1 \text{ L}} \right) = 1.0 \times 10^{-3} \text{ mol HCl}$$

Step 2.   25 mL (0.025 L) of 0.100 M NH_3 contains

$$0.025 \text{ L} \times \left(\frac{0.100 \text{ mol } NH_3}{1 \text{ L}} \right) = 2.5 \times 10^{-3} \text{ mol } NH_3$$

Steps 3 and 4.   The 1.0×10^{-3} mol of HCl neutralizes 1.0×10^{-3} mol of NH_3, according to the equation

$$HCl + NH_3 \longrightarrow NH_4^+ + Cl^-$$

producing 1.0×10^{-3} mol of NH_4Cl and leaving 1.5×10^{-3} mol of unreacted NH_3.

Steps 5 and 6.   We use the final volume of the solution, 35 mL, to determine the initial concentrations before equilibrium.

$$[NH_3] = \frac{1.5 \times 10^{-3} \text{ mol}}{0.035 \text{ L}} = 4.3 \times 10^{-2} \text{ M}$$

The concentration of NH_4^+ equals the concentration of NH_4Cl.

$$[NH_4^+] = [NH_4Cl] = \frac{1.0 \times 10^{-3} \text{ mol}}{0.035 \text{ L}} = 2.9 \times 10^{-2} \text{ M}$$

Part (2)   Calculation of equilibrium concentrations is a typical common-ion equilibrium calculation for a mixture of NH_3 and NH_4^+.

$$NH_3 + H_2O \rightleftharpoons NH_4^+ + OH^- \qquad K_b = 1.8 \times 10^{-5}$$

It is similar to the one illustrated in Example 16.19.

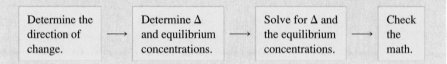

Step 1.   *Determine the direction of change.*

The reaction will shift to the right because Q for the mixture of NH_3 and NH_4^+ is less than K_b.

Step 2.   *Determine Δ and equilibrium concentrations.*

The table of changes and concentrations follows.

	$[NH_3]$	+	$H_2O \rightleftharpoons$	$[OH^-]$	+	$[NH_4^+]$
Concentration after HCl addition, M	4.3×10^{-2}			~0		2.9×10^{-2}
Change	$-\Delta$			Δ		Δ
Equilibrium Concentration, M	$4.3 \times 10^{-2} - \Delta$			$0 + \Delta = \Delta$		$2.9 \times 10^{-2} + \Delta$

Step 3.   *Solve for Δ and the equilibrium concentrations.*

$$K_b = \frac{[NH_4^+][OH^-]}{[NH_3]} = \frac{(2.9 \times 10^{-2} + \Delta)(\Delta)}{4.3 \times 10^{-2} - \Delta} = 1.8 \times 10^{-5}$$

If we assume that Δ is small, such that $(2.9 \times 10^{-2} + \Delta) = 2.9 \times 10^{-2}$ and $(4.3 \times 10^{-2} - \Delta) = 4.3 \times 10^{-2}$, we find

$$\Delta = 2.7 \times 10^{-5}$$

The value of $\Delta/2.9 \times 10^{-2}$ is small enough to justify our approximation that Δ can be neglected.
   Now we find $[OH^-]$.

$$[OH^-] = 0 + \Delta = 2.7 \times 10^{-5}$$

Step 4. *Check the work.*

When we calculate Q using the equilibrium concentrations, we find that $Q = K_a$ and the arithmetic checks.

Check your learning: Calculate the pH of a solution that results from mixing 0.125 L of 0.211 M NaOH and 0.125 L of 0.422 M HNCO.

Answer: 3.461

16.9   Buffer Solutions

Solutions of an acid and its common ion (that is, its conjugate base) or of a base and its common ion (its conjugate acid) as described in Section 16.8 are **buffered.** When we add a small amount of acid or base to any one of them, the pH of the solution changes very little.

A solution of a weak acid and its salt (or a solution of a weak base and its salt) is called a **buffer solution,** or a **buffer.** Buffer solutions resist a change in pH when small amounts of acid or base are added (Fig. 16.16). A solution of acetic acid and sodium acetate is an example of a buffer that consists of a weak acid and its salt. An example of a buffer that consists of a weak base and its salt is a solution of ammonia and ammonium chloride.

How Buffers Work.   A mixture of acetic acid and sodium acetate is acidic because it contains acetic acid. It is a buffer because it contains both the weak acid and its salt. Hence it acts to keep the hydronium ion concentration (and the pH) almost constant upon the addition of a small amount of either an acid or a base. If we add a base such as sodium hydroxide, then the hydroxide ions react with the few hydronium ions present. Then more of the acetic acid reacts with water, restoring the hydronium ion concentration almost to its original value.

$$CH_3CO_2H + H_2O \longrightarrow H_3O^+ + CH_3CO_2^-$$

Figure 16.16

(A) The buffered solution on the left and the unbuffered solution on the right have the same pH (pH 8); they are basic. (B) After the addition of 1 mL of a 0.01 M HCl solution, the pH of the buffered solution has not detectably changed, but the unbuffered solution has become acidic.

(A)

(B)

The pH changes very little. If we add an acid such as hydrochloric acid, then most of the hydronium ions from the hydrochloric acid combine with acetate ions, forming acetic acid molecules.

$$H_3O^+ + CH_3CO_2^- \longrightarrow CH_3CO_2H + H_2O$$

Thus there is very little increase in the concentration of the hydronium ion, and the pH remains practically unchanged.

A mixture of ammonia and ammonium chloride is basic because it contains the base ammonia. It is a buffer because it also contains the salt of the weak base. If we add a base (hydroxide ions), then ammonium ions in the buffer react with the hydroxide ions to form ammonia and water and reduce the hydroxide ion concentration almost to its original value.

$$NH_4^+ + OH^- \longrightarrow NH_3 + H_2O$$

If we add an acid (hydronium ions), then ammonia molecules in the buffer mixture react with the hydronium ions to form ammonium ions and reduce the hydronium ion concentration almost to its original value.

$$H_3O^+ + NH_3 \longrightarrow NH_4^+ + H_2O$$

The three parts of the following example illustrate the change in pH that accompanies the addition of base to a buffered solution of a weak acid and to an unbuffered solution of a strong acid.

Example 16.21   pH Changes in Buffered and Unbuffered Solutions

Acetate buffers are used in biochemical studies of enzymes and other chemical components of cells in order to prevent pH changes that might change the biochemical activity of these compounds.

(a) Calculate the pH of an acetate buffer that is a mixture with 0.100 M acetic acid and 0.100 M sodium acetate.

Solution: To determine the pH of the buffer solution, we use a typical common-ion equilibrium calculation, as illustrated in Example 16.19.

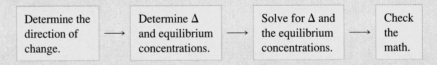

Step 1.   Determine the direction of change.

The equilibrium between a mixture of H_3O^+, $CH_3CO_2^-$, and CH_3CO_2H is

$$CH_3CO_2H + H_2O \rightleftharpoons H_3O^+ + CH_3CO_2^-$$

The equilibrium constant is not given, so we look it up in Appendix F: $K_a = 1.8 \times 10^{-5}$. With $[CH_3CO_2H] = [CH_3CO_2^-] = 0.10$ M and $[H_3O^+] = {\sim}0$ M, the initial value of Q is less than K_a, so the reaction shifts to the right.

Step 2.   Determine Δ and equilibrium concentrations.

A table of changes and concentrations follows.

Figure 16.17
A buffer that consists of equal concentrations of acetic acid and sodium acetate is acidic, as the indicator in this solution shows.

$$CH_3CO_2H \; + \; H_2O \rightleftharpoons H_3O^+ \; + \; CH_3CO_2^-$$

Initial Concentration, M	0.100	~0	0.100
Change	$-\Delta$	Δ	Δ
Equilibrium Concentration, M	$0.100 - \Delta$	Δ	$0.100 + \Delta$

Step 3. *Solve for Δ and the equilibrium concentrations.*

We find

$$\Delta = 1.8 \times 10^{-5} \text{ M}$$

and

$$[H_3O^+] = \Delta = 1.8 \times 10^{-5} \text{ M}$$

Thus

$$pH = -\log [H_3O^+] = -\log (1.8 \times 10^{-5})$$
$$= 4.74 \quad \text{(See Fig. 16.17.)}$$

Step 4. *Check the work.*

If we calculate all equilibrium concentrations, we find that the equilibrium value of Q is equal to K_a.

(b) Calculate the pH after 1.0 mL of 0.10 M NaOH is added to 100 mL of this buffer, giving a solution with a volume of 101 mL.

Solution: This calculation is similar to the one in Example 16.20. First we calculate the concentrations of an intermediate mixture resulting from the complete reaction between the acid in the buffer and the added base. Then we determine the concentrations as the intermediate mixture comes to equilibrium.

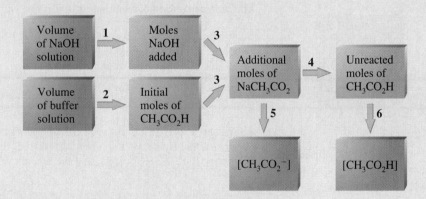

Step 1. 1.0 mL (0.0010 L) of 0.10 M NaOH contains

$$0.0010 \text{ L} \times \left(\frac{0.10 \text{ mol NaOH}}{1 \text{ L}}\right) = 1.0 \times 10^{-4} \text{ mol NaOH}$$

Step 2. Before reaction, 0.100 L of the buffer solution contains

$$0.100 \text{ L} \times \left(\frac{0.100 \text{ mol CH}_3\text{CO}_2\text{H}}{1 \text{ L}}\right) = 1.00 \times 10^{-2} \text{ mol CH}_3\text{CO}_2\text{H}$$

Steps 3 and 4. The 1.0×10^{-4} mol of NaOH neutralizes 1.0×10^{-4} mol of CH_3CO_2H, leaving

$$(1.00 \times 10^{-2}) - (0.01 \times 10^{-2}) = 0.99 \times 10^{-2} \text{ mol } CH_3CO_2H$$

and producing 1.0×10^{-4} mol of $NaCH_3CO_2$. This makes a total of

$$(1.00 \times 10^{-2}) + (0.01 \times 10^{-2}) = 1.01 \times 10^{-2} \text{ mol } NaCH_3CO_2$$

Steps 5 and 6. After reaction, CH_3CO_2H and $NaCH_3CO_2$ are contained in 101 mL of the intermediate solution, so

$$[CH_3CO_2H] = \frac{9.9 \times 10^{-3} \text{ mol}}{0.101 \text{ L}} = 0.098 \text{ M}$$

$$[NaCH_3CO_2] = \frac{1.01 \times 10^{-2} \text{ mol}}{0.101 \text{ L}} = 0.100 \text{ M}$$

Now we calculate the pH after the intermediate solution, which is 0.098 M in CH_3CO_2H and 0.100 M in $NaCH_3CO_2$, comes to equilibrium. The calculation is very similar to that in part (a) of this example.

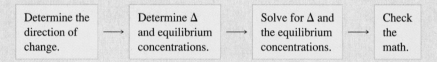

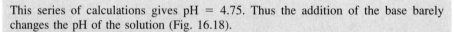

This series of calculations gives pH = 4.75. Thus the addition of the base barely changes the pH of the solution (Fig. 16.18).

(c) For comparison, calculate the pH after 1.0 mL of 0.10 M NaOH is added to 100 mL of an unbuffered solution with a pH of 4.74 (a 1.8×10^{-5} M solution of HCl). The volume of the final solution is 101 mL.

Solution: This 1.8×10^{-5} M solution of HCl has the same hydronium ion concentration as the 0.10 M solution of acetic-acid–sodium-acetate buffer described in part **(a)** of this example. The solution contains

$$0.100 \text{ L} \times \left(\frac{1.8 \times 10^{-5} \text{ mol HCl}}{1 \text{ L}} \right) = 1.8 \times 10^{-6} \text{ mol HCl}$$

As shown in part **(b)**, 1 mL of 0.10 M NaOH contains 1.0×10^{-4} mol of NaOH. When the NaOH and HCl solutions are mixed, the HCl is the limiting reagent in the reaction. All of the HCl reacts, and the amount of NaOH that remains is

$$(1.0 \times 10^{-4}) - (1.8 \times 10^{-6}) = 9.8 \times 10^{-5} \text{ M}$$

The concentration of NaOH is

$$\frac{9.8 \times 10^{-5} \text{ M NaOH}}{0.101 \text{ L}} = 9.7 \times 10^{-4} \text{ M}$$

The pOH of this solution is

$$pOH = -\log [OH^-] = -\log (9.7 \times 10^{-4}) = 3.01$$

The pH is

$$pH = 14.00 - pOH = 10.99$$

Figure 16.18

Here the buffer shown in Fig. 16.17 is still acidic, with very nearly the same pH, after the addition of 1 mL of 0.1 M NaOH.

The pH changes from 4.74 to 10.99 in this unbuffered solution. Compare this to the change of 4.74 to 4.75 that occurred when the same amount of NaOH was added to the buffered solution described in part (b).

Check your learning: (a) Check all the results in this example. (b) Show that 1.0 mL of 0.10 M HCl changes the pH of 100 mL of a 1.8×10^{-5} M HCl solution from 4.74 to 3.00.

If we add an acid or a base to a buffer that is a mixture of a weak base and its salt, the calculations of the changes in pH are analogous to those for a buffer that is a mixture of a weak acid and its salt.

Buffer Capacity.   Buffer solutions do not have an unlimited capacity to keep the pH relatively constant (Fig. 16.19). If we add so much base to a buffer that the weak acid is used up, no more buffering action toward base is possible (Fig. 16.20 on page 622). On the other hand, if we add an excess of acid, then the weak base is used up, and no more buffering action toward acid is possible. In fact, we do not even need to use up all the acid or base in a buffer to overwhelm it; its buffering action diminishes rapidly as a given component nears depletion.

The **buffer capacity** is the amount of acid or base that can be added to a given volume of a buffer solution before the pH changes significantly, usually by 1 unit. Buffer capacity depends on the amounts of the weak acid and its salt, or on the amounts of the weak base and its salt, that are in a buffer mixture. For example, 1 L of a solution that is 1.0 M in acetic acid and 1.0 M in sodium acetate has a greater buffer capacity than 1 L of a solution that is 0.10 M in acetic acid and 0.10 M in sodium acetate, even though both solutions have the same pH. The first solution has more buffer capacity because it contains more acetic acid and acetate ion.

Figure 16.19

The indicator color shows that adding a small amount of acid to a buffered solution of pH 10 (beaker on left) does not affect the pH of this buffered system (middle beaker). However, adding a large amount of acid exhausts the buffering capacity of the solution, and the pH changes dramatically (beaker on right).

Figure 16.20

An illustration of buffering action. The graph shows the change in pH as an increasing amount of 0.10 M NaOH solution is added to 100 mL of a buffer solution in which, initially $[CH_3CO_2H] = 0.10$ M and $[CH_3CO_2^-] = 0.10$ M.

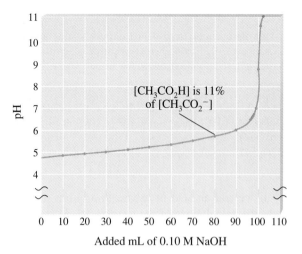

Selection of Suitable Buffer Mixtures.

There are two useful rules of thumb for selecting buffer mixtures.

1. A good buffer mixture should have about equal concentrations of both of its components. A buffer solution has generally lost its usefulness when one component of the buffer pair is less than about 10% of the other. Figure 16.20 shows an acetic-acid–acetate-ion buffer as base is added. The initial pH is 4.74. A change of 1 pH unit occurs when the acetic acid concentration is reduced to 11% of the acetate ion concentration.

2. Weak acids and their salts are better as buffers for pH < 7; weak bases and their salts are better as buffers for pH > 7.

For most effective buffering in the acid range, we should choose a weak acid and its salt that give a ratio $[H_3O^+]/K_a$ as close to 1 as possible. Similarly, for a weak base and its salt, the ratio $[OH^-]/K_b$ should be as close to 1 as possible.

The Henderson–Hasselbalch Equation.

Biological scientists use an expression called the **Henderson–Hasselbalch equation** to relate the pH (or pOH) of a buffered solution, the **pK** of the acid (or base) in the buffer, and the ratio of the concentrations of the acid (or base) and its salt. For an acid buffer, we have

$$pH = pK_a + \log \frac{[A^-]}{[HA]}$$

where pK_a is the negative of the common logarithm of the ionization constant of the weak acid ($pK_a = -\log K_a$). This equation relates the pH, the ionization constant of a weak acid, and the concentrations of the weak acid and its salt in a buffered solution. For a basic buffer,

$$pOH = pK_b + \log \frac{[HB^+]}{[B]}$$

where pK_b is the negative of the common logarithm of the ionization constant of the weak base ($pK_b = -\log K_b$).

Buffers play a very important role in chemical processes; they keep hydrogen ion and hydroxide ion concentrations approximately constant over a range of changes in condition, such as the addition of acid, base, or water. Common buffer pairs besides the acetic-acid–sodium-acetate and ammonia–ammonium-chloride buffers include H_2CO_3 and HCO_3^-, and $H_2PO_4^-$ and HPO_4^{2-}.

Blood is an important example of a buffered solution, the principal acid and ion responsible for the buffering action being carbonic acid, H_2CO_3, and the hydrogen carbonate ion, HCO_3^-. When an excess of hydrogen ion enters the blood stream, it is removed primarily by the reaction

$$H_3O^+ + HCO_3^- \longrightarrow H_2CO_3 + H_2O$$

When an excess of the hydroxide ion is present, it is removed by the reaction

$$OH^- + H_2CO_3 \longrightarrow HCO_3^- + H_2O$$

The pH of human blood thus remains very near 7.35—slightly alkaline. Variations are usually less than 0.1 of a pH unit. A change of 0.4 of a pH unit is likely to be fatal.

16.10 The Ionization of Hydrated Metal Ions

Figure 16.21
The pH paper shows that a 0.1 M solution of aluminum nitrate has a pH of 3.

When we measure the pH of the solutions of a variety of metal ions, we find that these ions act as weak acids when in solution. The aluminum ion is an example. When aluminum nitrate dissolves in water, the aluminum(III) ion reacts with water to give a hydrated aluminum ion, $Al(H_2O)_6^{3+}$.

$$Al(NO_3)_3(s) + 6H_2O(l) \longrightarrow Al(H_2O)_6^{3+}(aq) + 3NO_3^-(aq)$$

The hydrated aluminum ion is a weak acid (Fig. 16.21) and donates a proton to a water molecule.

$$Al(H_2O)_6^{3+} + H_2O \rightleftharpoons H_3O^+ + Al(H_2O)_5(OH)^{2+} \qquad K_a = 1.4 \times 10^{-5}$$

Like other polyprotic acids, the hydrated aluminum ion ionizes in stages, as shown by

$$Al(H_2O)_6^{3+} + H_2O \rightleftharpoons H_3O^+ + Al(H_2O)_5(OH)^{2+}$$
$$Al(H_2O)_5(OH)^{2+} + H_2O \rightleftharpoons H_3O^+ + Al(H_2O)_4(OH)_2^+$$
$$Al(H_2O)_4(OH)_2^+ + H_2O \rightleftharpoons H_3O^+ + Al(H_2O)_3(OH)_3$$

The ionization of a cation carrying more than one charge is not extensive beyond the first stage. Additional examples of the first stage in the ionization of hydrated metal ions follow.

$$Fe(H_2O)_6^{3+} + H_2O \rightleftharpoons H_3O^+ + Fe(H_2O)_5(OH)^{2+}$$
$$Cu(H_2O)_4^{2+} + H_2O \rightleftharpoons H_3O^+ + Cu(H_2O)_3(OH)^+$$
$$Zn(H_2O)_6^{2+} + H_2O \rightleftharpoons H_3O^+ + Zn(H_2O)_5(OH)^+$$

Ions known to be hydrated in solution are often indicated in abbreviated form—that is, without showing the hydration. For example, Al^{3+} is often written instead of $Al(H_2O)_6^{3+}$. However, when water participates in a reaction, the hydration becomes important and we use formulas that show the extent of hydration.

Example 16.22   Ionization of $Al(H_2O)_6^{3+}$

Calculate the pH of a 0.10 M solution of aluminum chloride that dissolves to give the hydrated aluminum ion $[Al(H_2O)_6]^{3+}$ in solution.

Solution: In spite of the unusual appearance of the acid, this is a typical acid ionization problem.

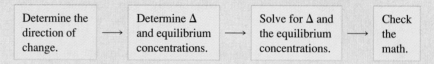

Step 1.   *Determine the direction of change.*

The equation for the reaction and K_a are

$$Al(H_2O)_6^{3+} + H_2O \rightleftharpoons H_3O^+ + Al(H_2O)_5(OH)^{2+} \qquad K_a = 1.4 \times 10^{-5}$$

The reaction shifts to the right to reach equilibrium.

Step 2.   *Determine Δ and equilibrium concentrations.*

The table of changes and concentrations follows:

$$Al(H_2O)_6^{3+} + H_2O \rightleftharpoons H_3O^+ + Al(H_2O)_5(OH)^{2+}$$

	$Al(H_2O)_6^{3+}$	H_3O^+	$Al(H_2O)_5(OH)^{2+}$
Initial Concentration, M	0.10	~0	0
Change	$-\Delta$	Δ	Δ
Equilibrium Concentration, M	$0.10 - \Delta$	Δ	Δ

Step 3.   *Solve for Δ and the equilibrium concentrations.*

Substituting the expressions for the equilibrium concentrations into the equation for the ionization constant yields

$$K_a = \frac{[H_3O^+][Al(H_2O)_5(OH)^{2+}]}{[Al(H_2O)_6^{3+}]}$$

$$= \frac{(\Delta)(\Delta)}{0.10 - \Delta} = 1.4 \times 10^{-5}$$

Solving this equation gives

$$\Delta = 1.2 \times 10^{-3} \text{ M}$$

From this we find

$$[H_3O^+] = 0 + \Delta = 1.2 \times 10^{-3} \text{ M}$$

$$pH = -\log [H_3O^+] = 2.92 \quad \text{(an acidic solution)}$$

Step 4.   *Check the work.*

The arithmetic checks; at equilibrium, $Q = K_a$.

Check your learning: What is $[Al(H_2O)_5(OH)^{2+}]$ in a 0.15 M solution of $Al(NO_3)_3$ that contains enough of the strong acid HNO_3 to bring $[H_3O^+]$ to 0.10 M?

Answer: 2.1×10^{-5} M

The constants for the different stages of ionization are not known for many metal ions, so we cannot calculate the extent of their ionization. However practically all metal ions other than those of the alkali metals ionize to give acidic solutions. Ionization increases as the charge of the metal ion increases or as the size of the metal ion decreases.

16.11   Acid–Base Indicators

Certain organic substances change color in dilute solution when the hydronium ion concentration reaches a particular value. For example, phenolphthalein is a colorless substance in any aqueous solution with a hydronium ion concentration greater than 5.0×10^{-9} M (pH less than 8.3). In more basic solutions where the hydronium ion concentration is less than 5.0×10^{-9} M (pH greater than 8.3), it is red or pink. Substances such as phenolphthalein, which can be used to determine the pH of a solution, are called **acid–base indicators.** Acid–base indicators are either weak organic acids or weak organic bases.

The equilibrium in a solution of the acid–base indicator methyl orange, a weak acid, can be represented by an equation in which we use HIn as a simple representation for the complex methyl orange molecule.

$$\underset{\text{Red}}{HIn} + H_2O \rightleftharpoons H_3O^+ + \underset{\text{Yellow}}{In^-} \qquad K_a = \frac{[H_3O^+][In^-]}{[HIn]} = 4.0 \times 10^{-4}$$

The anion of methyl orange, In^-, is yellow, and the nonionized form, HIn, is red. When we add acid to a solution of methyl orange, the increased hydronium ion concentration shifts the equilibrium toward the nonionized red form, in accordance with Le Châtelier's principle. If we add base, we shift the equilibrium toward the yellow form.

An indicator's color is the visible result of the ratio of the concentrations of the two species In^- and HIn. If most of the indicator (typically about 60%–90% or more) is present as In^-, then we see the color of the In^- ion, which is yellow for methyl orange. If most is present as HIn, then we see the color of the HIn molecule, red for methyl orange. For methyl orange, we can rearrange the equation for K_a and write

$$\frac{[In^-]}{[HIn]} = \frac{[\text{substance with yellow color}]}{[\text{substance with red color}]} = \frac{K_a}{[H_3O^+]}$$

This shows us how the ratio of $[In^-]$ to [HIn] varies with the concentration of hydronium ion. When $[H_3O^+]$ has the same numerical value as K_a, the ratio of $[In^-]$ to [HIn] is equal to 1, which means that 50% of the indicator is present in the red form (HIn) and 50% in the yellow ionic form (In^-), and the solution appears orange in color. When the hydronium ion concentration increases to 8×10^{-4} M (a pH of 3.1), the solution turns red. No change in color is visible for any further increase in the

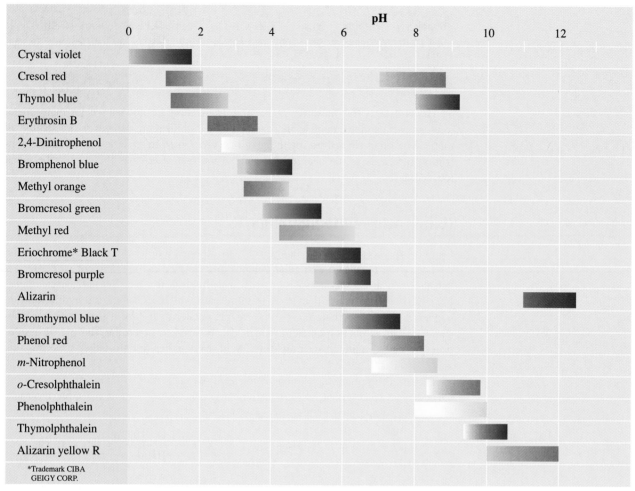

The pH ranges shown are approximate. Specific transition ranges depend on the solvent.

Figure 16.22

Ranges of color change for several acid–base indicators.

hydronium ion concentration (decrease in pH). At hydronium ion concentration of 4×10^{-5} M (a pH of 4.4), most of the indicator is in the yellow ionic form, and a further decrease in the hydronium ion concentration (increase in pH) does not produce a visible color change. The pH range between 3.1 (red) and 4.4 (yellow) is the *color-change interval* of methyl orange; the pronounced color change takes place between these pH values.

There are many different acid–base indicators that cover a wide range of pH values and can be used to determine the approximate pH of an unknown solution by a process of elimination. Universal indicator (Fig. 16.6) and pH paper (Fig. 16.7) contain a mixture of indicators and exhibit different colors at different pH values. Figure 16.22 shows several indicators, their colors, and their color-change intervals.

16.12   Titration Curves

We draw **titration curves,** plots of the pH against the volume of acid or base added in a titration, to show the point where equivalent quantities of acid and base are present (the equivalence point).

Table 16.5

pH Values in the Titrations of a Strong Acid with a Strong Base and of a Weak Acid with a Strong Base

Volume of 0.100 M NaOH Added, mL	Moles of NaOH Added	pH Values	
		0.100 M HCl[a]	0.100 M CH₃CO₂H[b]
0.0	0.0	1.00	2.87
5.0	0.00050	1.18	4.14
10.0	0.00100	1.37	4.57
15.0	0.00150	1.60	4.92
20.0	0.00200	1.95	5.35
22.0	0.00220	2.20	5.61
24.0	0.00240	2.69	6.13
24.5	0.00245	3.00	6.44
24.9	0.00249	3.70	7.14
25.0	0.00250	7.00	8.72
25.1	0.00251	10.30	10.30
25.5	0.00255	11.00	11.00
26.0	0.00260	11.29	11.29
28.0	0.00280	11.75	11.75
30.0	0.00300	11.96	11.96
35.0	0.00350	12.22	12.22
40.0	0.00400	12.36	12.36
45.0	0.00450	12.46	12.46
50.0	0.00500	12.52	12.52

[a]Titration of 25.00 mL of 0.100 M HCl (0.00250 mol of HCl) with 0.100 M NaOH.
[b]Titration of 25.00 mL of 0.100 M CH_3CO_2H (0.00250 mol of CH_3CO_2H) with 0.100 M NaOH.

The simplest acid–base reactions are those of a strong acid with a strong base. Table 16.5 shows us data for the titration of a 25.0-milliliter sample of 0.100 M hydrochloric acid with 0.100 M sodium hydroxide. The values of the pH measured after successive additions of small amounts of NaOH are listed in the first column of this table and are shown in Fig. 16.23. The pH increases very slowly at first, increases

Figure 16.23

Titration curve for the titration of 25.00 milliliters of 0.100 M HCl (strong acid) with 0.100 M NaOH (strong base). The pH ranges for the color change of phenolphthalein, litmus, and methyl orange are indicated by the shaded areas.

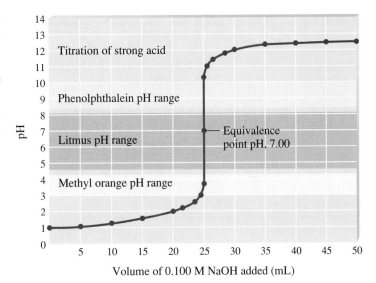

very rapidly in the middle portion of the curve, and then increases very slowly again. The point of inflection (the midpoint of the vertical part of the curve) is the equivalence point for the titration. It indicates when equivalent quantities of acid and base are present. For the titration of a strong acid and a strong base, the equivalence point occurs at a pH of 7 (Table 16.5 and Fig. 16.23).

The titration of a weak acid with a strong base (or of a weak base with a strong acid) is somewhat more complicated than that just discussed, but it follows the same basic principles. Let us consider the titration of 25.0 milliliters of 0.100 M acetic acid (a weak acid) with 0.100 M sodium hydroxide and compare the titration curve with that of the strong acid (Fig. 16.23). Table 16.5 gives the pH values during the titration, and Fig. 16.24 shows the titration curve.

There are important differences between the two titration curves. The titration curve for the weak acid begins at a higher pH value (less acidic) and maintains higher pH values up to the equivalence point. This is because acetic acid is a weak acid and is only fractionally ionized. The pH at the equivalence point is also higher (8.72 rather than 7.00) because the solution contains sodium acetate, a weak base that raises the pH as described in Section 16.5.

$$CH_3CO_2^- + H_2O \rightleftharpoons CH_3CO_2H + OH^-$$

After the equivalence point, the two curves are identical (Table 16.5 and Figs. 16.23 and 16.24) because the pH is dependent on the excess of hydroxide ion in both cases.

Titration curves help us pick an indicator that will provide a sharp color change at the equivalence point. The best selection would be an indicator that has a color-change interval that brackets the pH at the equivalence point of the titration. Alternatively, an indicator with a color-change interval in the steeply rising portion of the titration curve could be used.

The color-change intervals of three indicators are indicated in Figs. 16.23 and 16.24. The equivalence points of both the titration of the strong acid and that of the weak acid are located in the color-change interval of phenolphthalein. We can use it for titrations either of strong acid with strong base or of weak acid with strong base.

Figure 16.24

Titration curve for the titration of 25.00 milliliters of 0.100 M CH_3CO_2H (weak acid) with 0.100 M NaOH (strong base). The pH ranges for the color change of phenolphthalein, litmus, and methyl orange are indicated by the shaded areas.

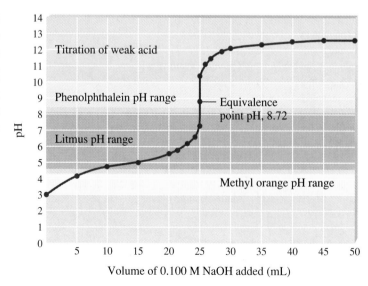

Litmus is a suitable indicator for the HCl titration because its color change brackets the equivalence point. However, we should not use litmus for the CH_3CO_2H titration, because the pH is within the color-change interval of litmus when only about 12 milliliters of NaOH have been added, and it does not leave the range until 25 millimeters have been added. The color change would be very gradual, taking place during the addition of 13 milliliters of NaOH, making litmus useless as an indicator of the equivalence point.

We could use methyl orange for the HCl titration, but it would not give very precise results: (1) It completes its color change slightly before the equivalence point is reached (but very close to it, so this is not too serious), and (2) it changes color, as Fig. 16.23 shows, during the addition of nearly 0.5 milliliter of NaOH, which is not so sharp a color change as that of litmus or phenolphthalein. Figure 16.24 shows us that methyl orange would be completely useless as an indicator for the CH_3CO_2H titration. Its color change begins after about 1 milliliter of NaOH has been added and ends when about 8 milliliters have been added. The color change is completed long before the equivalence point is reached and hence provides no indication of the equivalence point.

We can base our choice of indicator on a calculated pH, the pH at the equivalence point. At the equivalence point, equimolar amounts of acid and base have been mixed, and the calculation becomes that of the pH of a solution of the salt that results from the titration, a calculation described in Section 16.5.

For Review   Summary

Water is an extremely weak electrolyte that undergoes **self-ionization.**

$$2H_2O \rightleftharpoons H_3O^+ + OH^-$$

$$K_w = [H_2O^+][OH^-] = 1.0 \times 10^{-14} \text{ at } 25°C$$

Pure water contains equal concentrations of 1.0×10^{-7} M hydronium ion and hydroxide ion at 25°C. The **ion product of water,** K_w is the equilibrium constant for the self-ionization reaction. All aqueous solutions contain both H_3O^- and OH^-, and the product $[H_3O^+][OH^-]$ is equal to 1.0×10^{-14} at 25°C.

The concentration of hydronium ion in a solution of an acid in water is greater than 1.0×10^{-7} M and results from two reactions: the ionization of the acid and the self-ionization of water. However, the contribution of water to the total $[H_3O^+]$ in an acid solution can be neglected when the total $[H_3O^+]$ is greater than 4.5×10^{-7} M at 25°C.

The concentration of hydroxide ion in a solution of a base in water is greater than 1.0×10^{-7} M and also results from two reactions: the ionization of the base and the self-ionization of water. The contribution of water to the total $[OH^-]$ in a solution of a base is negligible when the total $[OH^-]$ is greater than 4.5×10^{-7} M at 25°C.

The concentration of H_3O^+ in a solution can be expressed as the pH of the solution; **pH $= -\log[H_3O^+]$**. The concentration of OH^- can be expressed as pOH: **pOH $= -\log[OH^-]$**. In pure water, pH $= 7$ and pOH $= 7$; however, most water has a pH less than 7 as a result of dissolved carbon dioxide from the air. In any aqueous solution at 25°C, pH $+$ pOH $= 14.00$.

Although strong electrolytes ionize completely in solution, a solution of a weak electrolyte contains a mixture of molecular and ionic species. The concentrations of the nonionized molecules of a weak acid (HA), of hydronium ion, and of the conjugate base (A^-) of the weak acid can be determined from K_a, the **ionization constant of the weak acid,** which is the equilibrium constant for the reaction.

$$HA + H_2O \rightleftharpoons H_3O^+ + A^- \qquad K_a = \frac{[H_3O^+][A^-]}{[HA]}$$

The large majority of the species in the equilibrium mixture are unreacted molecules of the nonionized acid. The concentrations of the nonionized molecules of a weak base (B), of hydroxide ion, and of the conjugate acid (HB^+) of the weak base can be determined from K_b, **the ionization constant for a weak base.** K_b is the equilibrium constant for the reaction.

$$B + H_2O \rightleftharpoons BH^+ + OH^- \qquad K_b = \frac{[BH^+][OH^-]}{[B]}$$

The large majority of the species in the equilibrium mixture are unreacted molecules of the nonionized base. Determination of

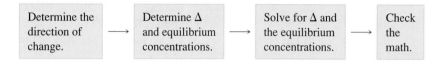

equilibrium concentrations of weak acids or weak bases involves the procedures described for other equilibria in Chapter 14 (see diagram above).

The conjugate base (A^-) of a weak acid (HA) exhibits basic behavior and reacts with water to give a basic solution. The conjugate acid (HB^+) of a weak base (B) gives an acid solution. The value of the product of the ionization constants of a weak acid and its conjugate base or of a weak base and its conjugate acid is equal to the value of K_w; that is, $K_a \times K_b = K_w$. Thus the ionization constant of a conjugate acid or conjugate base can be determined from that of its respective base or acid.

Polyprotic acids (or bases) and mixtures of weak acids (or bases) are treated as though the ionization proceeded in a stepwise fashion; we first calculate the concentrations produced by the stronger acid (or base) and then use those concentrations to determine the concentrations resulting from ionization of the weaker acid (or base).

The extent of ionization of a weak acid in solution can be reduced by adding a compound that contains the conjugate base of the weak acid (the **common ion effect**) or by adding an acid. Likewise, the ionization of a weak base can be reduced by adding the conjugate acid of the weak base or by adding hydroxide ion.

A solution containing a mixture of an acid and its conjugate base, or of a base and its conjugate acid, is called a **buffer solution.** Unlike that of the solution of an acid, a base, or most salts, the hydronium ion concentration of a buffer solution does not change greatly when a small amount of acid or base is added to the buffer solution. The base (or acid) in the buffer reacts with the added acid (or base).

The pH of a solution can be determined with a pH meter or with **acid–base indicators.** Acid–base indicators are also used to determine the equivalence point in an acid–base titration. The best indicator for titration of an acid and a base is selected by considering the **titration curve** for the acid–base pair.

Key Terms and Concepts

acid–base indicator   (16.11)
buffer   (16.9)
buffer capacity   (16.9)
buffer solution   (16.9)
common ion effect   (16.8)
diprotic acid   (16.6)
equivalence point   (16.12)
Henderson–Hasselbalch equation   (16.9)

ionization constant   (16.3)
ionization of hydrated metal ions   (16.10)
ionization of water   (16.2)
ionization of weak acids   (16.3–16.7)
ionization of weak bases   (16.3–16.6)
ion product for water   (16.2)
K_a (16.3)

K_b (16.3)
pH   (16.2)
pK   (16.9)
pOH   (16.2)
salt effect   (16.8)
stepwise ionization   (16.6)
titration curve   (16.12)
triprotic acid   (16.6)

Exercises

Questions

1. Which of the following symbols are used to represent the same thing? Which are used to represent different things?
   (a) The symbol $[H_3O^+]$ in Chapter 13 and the symbol $[H_3O^+]$ in this chapter.
   (b) The symbol k in Chapter 13 and the symbol K in this chapter.
   (c) The symbol K in Chapter 10 and the symbol K in this chapter.
   (d) The symbol K in Chapter 14 and the symbol K in this chapter.

2. Explain what we represent when we write
$$CH_3CO_2H(aq) + H_2O(l) \rightleftharpoons$$
$$H_3O^+(aq) + CH_3CO_2^-(aq)$$

3. Identify the microscopic species present in a solution labeled 0.1 M NaOH, and list these species in decreasing order of their concentrations.

4. Identify the microscopic species present in a solution labeled 0.1 M HCl, and list these species in decreasing order of their concentrations.

5. Identify the microscopic species present in a solution labeled 0.1 M NH_3, and list these species in decreasing order of their concentrations.

6. Identify the microscopic species present in a solution labled 0.1 HF, and list these species in decreasing order of their concentrations.

7. Identify the microscopic species present in a solution of $Ba(NO_2)_2$, and list these species in decreasing order of their concentrations. (*Hint:* Remember that the NO_2^- ion is a weak base.)

8. Identify the microscopic species present in a solution of $Al(NO_3)_3$, and list these species in decreasing order of their concentrations. (*Hint:* Remember that the $Al(H_2O)_6^{3+}$ ion is a polyprotic weak acid.)

9. Explain why equilibrium calculations are not necessary to determine ionic concentrations in solutions of certain strong electrolytes such as NaOH and HCl. When are equilibrium calculations necessary as part of the determination of the concentrations of all ions of some other strong electrolytes in solution?

10. Are the concentrations of hydronium ion and hydroxide ion in a solution of an acid or a base in water directly proportional or inversely proportional? Explain your answer. (Inverse and direct proportionality are discussed in Sections 10.2 and 10.3.)

11. Explain why a sample of pure water at 40°C is neutral even though $[H_3O^+] = 1.7 \times 10^{-7}$ M. K_w is 2.9×10^{-14} at 40°C.

12. Is the solution shown in Fig. 16.5 prepared from an acid or a base? Explain your answer.

13. What two common assumptions can simplify calculation of equilibrium concentrations in a solution of a weak acid?

14. What two assumptions can simplify calculation of equilibrium concentrations in a solution of a weak base?

15. Sketch graphs similar to Fig. 14.4 for Q and the concentrations of NH_3, NH_4^+, and OH^- at six conditions (including the initial $[NH_3]$ and the equilibrium concentration $[NH_3]$ = 0.099 M) when 0.100 mol of NH_3 is added to 1.00 L of water and comes to equilibrium.

16. Sketch graphs similar to Fig. 14.4 for Q and the concentrations of HSO_4^-, H_3O^+, and SO_4^{2-} at six conditions (including the initial concentration of HSO_4^- and the equilibrium concentration of HSO_4^-) when 0.15 mol of $NaHSO_4$ is added to 1.00 L of water and comes to equilibrium. At equilibrium, $[H_3O^+] = 0.037$ M.

17. Which of the following will increase the percent of NH_3 that is converted to the ammonium ion in water? (a) Addition of NaOH. (b) Addition of HCl. (c) Addition of NH_4Cl.

18. Which of the following will increase the percent of HF that is converted to the fluoride ion in water? (a) Addition of NaOH. (b) Addition of HCl. (c) Addition of NaF.

19. What is the effect on the concentrations of NO_2^-, HNO_2, and OH^- when each of the following is added to separate solutions of KNO_2 in water? (a) HCl, (b) HNO_2, (c) NaOH, (d) NaCl, (e) KNO_2. The equation for the equilibrium is

$$NO_2^- + H_2O \rightleftharpoons HNO_2 + OH^-$$

20. What is the effect on the concentration of hydrofluoric acid, hydronium ion, and fluoride ion when each of the following is added to separate solutions of hydrofluoric acid? (a) HCl, (b) KF, (c) NaCl, (d) KOH, and (e) HF.

The equation for the equilibrium is

$$HF + H_2O \rightleftharpoons H_3O^+ + F^-$$

21. Explain why the neutralization reaction of a strong acid and a weak base gives a weakly acidic solution.

22. Explain why the neutralization reaction of a weak acid and a strong base gives a weakly basic solution.

23. Why is the hydronium ion concentration in a solution that is 0.10 M in HCl and 0.10 M in HCO_2H determined by the concentration of HCl?

24. Explain why a buffer can be prepared from a mixture of NH_4Cl and NaOH but not from NH_3 and NaOH.

25. Explain why the pH does not change significantly when a small amount of an acid or a base is added to a solution that contains equal amounts of the acid H_3PO_4 and NaH_2PO_4, a salt of its conjugate base.

26. Explain why the pH does not change significantly when a small amount of an acid or a base is added to a solution that contains equal amounts of the base NH_3 and NH_4Cl, a salt of its conjugate acid.

27. Explain how to choose the appropriate acid–base indicator for the titration of a weak base with a strong acid.

28. Explain why an acid–base indicator changes color over a range of pH values rather than at a specific pH.

Ion Concentrations; pH and pOH

29. List the ions and their concentrations in the following solutions.

   (a) 0.10 M KNO_3
   (b) 0.050 M $Ba(OH)_2$
   (c) 1.3 M Na_2SO_4 (Neglect the reaction of the sulfate ion with water.)
   (d) 0.45 M HBr

30. List the ions and their concentrations in the following solutions.

   (a) 2.5 M $CaCl_2$ (b) 0.15 M $LiClO_4$
   (c) 0.250 M CsOH   (d) 0.10 M HNO_3

31. The ionization constant for water (K_w) is 2.9×10^{-14} at 40°C. Calculate $[H_3O^+]$, $[OH^-]$, pH, and POH for pure water at 40°C.

32. The ionization constant for water (K_w) is 9.614×10^{-14} at 60°C. Calculate $[H_3O^+]$, $[OH^-]$, pH, and pOH for pure water at 60°C.

33. Calculate the pH and pOH of each of the following solutions at 25°C.

    (a) 0.200 M HCl   (b) 0.0143 M NaOH
    (c) 3.0 M HNO_3 (d) 0.0031 M $Ca(OH)_2$

34. Calculate the pH and the pOH of each of the following solutions at 25°C.

    (a) 0.000259 M $HClO_4$ (b) 0.21 M NaOH
    (c) 0.000071 M $Ba(OH)_2$ (d) 2.5 M KOH

35. Calculate the pH and pOH of a 0.10 M solution of each of the following.

    (a) The weak acid CH_3CO_2H (1.3% ionized)
    (b) The weak base NO_2^- (0.0015% ionized)
    (c) The weak acid HSO_4^- (29% ionized)

36. Calculate the pH and pOH of a 0.10 M solution of each of the following.

    (a) The weak base NH_3 (1.3% ionization)
    (b) The weak base $CH_3CO_2^-$ (0.0075% ionization)
    (c) The weak acid HNO_2 (6.5% ionized)

37. What are the pH and pOH of the solution shown in Fig. 16.1?

38. What are the hydronium and hydroxide ion concentrations in the solution shown in Fig. 16.5?

39. Calculate the hydrogen ion concentration and the hydroxide ion concentration in wine from its pH. See Table 16.2 for useful information.

40. Calculate the hydronium ion concentration and the hydroxide ion concentration in lime juice from its pH. See Table 16.1 for useful information.

41. For which of the following solutions must we consider the ionization of water when calculating the pH or pOH?

    (a) 3×10^{-8} M HNO_3
    (b) 0.10 g of HCl in 1.0 L of solution
    (c) 0.00080 g of NaOH in 0.50 L of solution
    (d) 1×10^{-7} M $Ca(OH)_2$
    (e) 0.0245 M KNO_3

42. For which of the following solutions must we consider the ionization of water when calculating the pH or pOH?

    (a) 6×10^{-7} M KOH   (b) 3×10^{-5} M H_2SO_4
    (c) 1.3×10^{-3} M NaCl
    (d) 1×10^{-6} g of HBr in 1.0 L of solution
    (e) 0.034 g of $Sr(OH)_2$ in 0.234 L of solution

Determination of Equilibrium Constants

43. From the equilibrium concentrations given, calculate K_a for each of the weak acids and K_b for each of the weak bases.

    (a) CH_3CO_2H: $[H_3O^+] = 1.34 \times 10^{-3}$ M
        $[CH_3CO_2^-] = 1.34 \times 10^{-3}$ M
        $[CH_3CO_2H] = 9.866 \times 10^{-2}$ M
    (b) ClO^-: $[OH^-] = 4.0 \times 10^{-4}$ M
        $[HClO] = 2.38 \times 10^{-5}$ M
        $[ClO^-] = 0.273$ M
    (c) HCO_2H: $[HCO_2H] = 0.524$ M
        $[H_3O^+] = 9.8 \times 10^{-3}$ M
        $[HCO_2^-] = 9.8 \times 10^{-3}$ M
    (d) $C_6H_5NH_3^+$: $[C_6H_5NH_3^+] = 0.233$ M
        $[C_6H_5NH_2] = 2.3 \times 10^{-3}$ M
        $[H_3O^+] = 2.3 \times 10^{-3}$ M

44. From the equilibrium concentrations given, calculate K_a for each of the weak acids and K_b for each of the weak bases.

    (a) NH_3: $[OH^-] = 3.1 \times 10^{-3}$ M
        $[NH_4^+] = 3.1 \times 10^{-3}$ M
        $[NH_3] = 0.533$ M
    (b) HNO_2: $[H_3O^+] = 0.011$ M
        $[NO_2^-] = 0.0438$ M
        $[HNO_2] = 1.07$ M
    (c) $(CH_3)_3N$: $[(CH_3)_3N] = 0.25$ M
        $[(CH_3)_3NH^+] = 4.3 \times 10^{-3}$ M
        $[OH^-] = 4.3 \times 10^{-3}$ M
    (d) NH_4^+: $[NH_4^+] = 0.100$ M
        $[NH_3] = 7.5 \times 10^{-6}$ M
        $[H_3O^+] = 7.5 \times 10^{-6}$ M

45. Determine K_a for H_2CO_3 and for HCO_3^- and K_b for HCO_3^- and for CO_3^{2-} from the following concentrations present in an equilibrium mixture of H_2CO_3 in water. $[H_2CO_3] = 0.033$ M, $[HCO_3^-] = 1.2 \times 10^{-4}$ M, $[CO_3^{2-}] = 7 \times 10^{-11}$ M, $[H_3O^+] = 1.2 \times 10^{-4}$ M, $[OH^-] = 8.3 \times 10^{-11}$ M.

46. Determine K_a for H_2S and for HS^- and K_b for HS^- and for S^{2-} from the following concentrations present in an equilibrium mixture of H_2S in water. $[H_2S] = 0.1$ M, $[HS^-] = 0.0001$ M, $[S^{2-}] = 1 \times 10^{-19}$ M, $[OH^-] = 1 \times 10^{-10}$ M, $[H_3O^+] = 0.0001$ M.

47. Determine K_a for acetic acid, CH_3CO_2H. In a 0.10 M solution, the acid is 1.3% ionized.

48. Determine K_a for nitrous acid, HNO_2. In a 0.10 M solution, the acid is 6.5% ionized.

49. Determine K_b for ammonia, NH_3. In a 0.10 M solution, the base is 1.3% ionized.

50. Determine K_b for the nitrite ion, NO_2^-. In a 0.10 M solution, this base is 0.0015% ionized.

51. Determine K_a for the hydrogen sulfate ion, HSO_4^-. In a 0.10 M solution, the acid is 29% ionized.

52. Determine K_b for the acetate ion, $CH_3CO_2^-$. In a 0.10 M solution, the base is 0.0075% ionized.

53. The pH of a 0.20 M solution of HF is 1.92. Determine K_a for HF.

54. The pH of a 0.15 M solution of HSO_4^- is 1.43. Determine K_a for HSO_4^-.

55. The pH of a 0.10 M solution of caffeine is 11.16. Determine K_b for caffeine.

$$C_8H_{10}N_4O_2 + H_2O \rightleftharpoons C_8H_{10}N_4O_2H^+ + OH^-$$

56. The pH of a solution of household ammonia, a 0.950 M solution of NH_3, is 11.612. Determine K_b for NH_3.

57. Calculate the ionization constant for each of the following acids or bases from the ionization constant of its conjugate base or conjugate acid.

    (a) F^-
    (b) NH_4^+
    (c) AsO_4^{3-}
    (d) $(CH_3)_2NH_2^+$
    (e) NO_2^-
    (f) $HC_2O_4^-$ (as a base)

58. Calculate the ionization constant for each of the following acids or bases from the ionization constant of its conjugate base or conjugate acid.

    (a) HTe^- (as a base)
    (b) $(CH_3)_3NH^+$
    (c) $HAsO_4^{3-}$ (as a base)
    (d) HO_2^- (as a base)
    (e) $C_6H_5NH_3^+$
    (f) HSO_3^- (as a base)

Ionization of Weak Acids and Weak Bases

59. Calculate the concentrations of all solute species in each of the following solutions of acids or bases. Assume that the ionization of water can be neglected, and show that the change in the initial concentrations can be neglected. Ionization constants can be found in Appendices F and G.

    (a) 0.0092 M HClO, a weak acid
    (b) 0.0784 M $C_6H_5NH_2$, a weak base
    (c) 0.0810 M HCN, a weak acid
    (d) 0.11 M $(CH_3)_3N$, a weak base
    (e) 0.120 M $Fe(H_2O)_6^{2+}$, a weak acid, $K_a = 1.6 \times 10^{-7}$

60. Calculate the concentrations of all solute species in each of the following solutions of acids or bases. Assume that the ionization of water can be neglected, and show that the change in the initial concentration of the acid or base can be neglected. Ionization constants can be found in Appendixes F and G.

    (a) 0.0915 M CH_3CO_2H, a weak acid
    (b) 1.4 M $(CH_3)_2NH$, a weak base
    (c) 0.0909 M $HC_2O_4^-$, a weak acid, $K_a = 6.4 \times 10^{-5}$
    (d) 0.66 M CH_3NH_2, a weak base
    (e) 0.24 M $V(H_2O)_6^{2+}$, a weak acid, $K_a = 1.4 \times 10^{-7}$

61. Calculate the concentrations of all solute species in each of the following solutions of salts of weak acids or weak bases. Assume that the ionization of water can be ne-

glected, and show that the changes in the initial concentrations can be neglected.

    (a) 0.125 M NH_4Cl, K_b for $NH_3 = 1.8 \times 10^{-5}$
    (b) 0.25 M KF, K_a for HF $= 7.2 \times 10^{-4}$
    (c) 0.311 M $(CH_3)_3NHBr$, K_b for $(CH_3)_3N = 7.4 \times 10^{-5}$
    (d) 0.11 M $Ba(CN)_2$, K_a for HCN $= 4.0 \times 10^{-10}$

62. Calculate the concentrations of all solute species in each of the following solutions of salts of weak acids or weak bases. Assume that the ionization of water can be neglected, and show that the changes in the initial concentrations can be neglected.

    (a) 0.12 M $[(CH_3)NH_2]NO_3$, K_b for $(CH_3)_2NH = 7.4 \times 10^{-4}$
    (b) 2.2 M CH_3NH_3Br, K_b for $CH_3NH_2 = 4.4 \times 10^{-4}$
    (c) 0.023 M $Ca(CNO)_2$, K_a for HCNO $= 3.46 \times 10^{-4}$
    (d) 1.1 M Na_2SO_4, K_a for $HSO_4^- = 1.2 \times 10^{-2}$

63. Calculate the concentrations of all solute species in each of the following solutions. Assume that the ionization of water can be neglected, and show that the changes in the initial concentrations cannot be neglected.

    (a) 0.0184 M HCNO, a weak acid
    (b) 0.11 M $(CH_3)_2NH$, a weak base
    (c) 0.100 M HF, a weak acid
    (d) 2.0×10^{-3} M $(CH_3)_3N$, a weak base
    (e) 0.050 M $Fe(H_2O)_6^{3+}$, a weak acid, $K_a = 6.5 \times 10^{-3}$

64. Calculate the concentrations of all solute species in each of the following solutions. Assume that the ionization of water can be neglected, and show that the changes in the initial concentrations cannot be neglected.

    (a) 0.12 M HSO_3NH_2, a weak acid, $K_a = 0.10$
    (b) 4.113×10^{-2} M $(CH_3)_2NH$, a weak base
    (c) 0.2173 M CH_2ClCO_2H, a weak acid, $K_a = 1.4 \times 10^{-3}$
    (d) 0.1050 M CH_3NH_2, a weak base
    (e) 0.017 M $Mn(H_2O)_5(OH)^+$, a weak base, $K_b = 3.9 \times 10^{-4}$

65. Calculate the concentrations of all solute species in each of the following solutions. Assume that the ionization of water can be neglected, and show that the changes in the initial concentrations cannot be neglected.

    (a) 0.00253 M LiC_6H_5O, K_a for $C_6H_5OH = 1.28 \times 10^{-10}$
    (b) 2.38×10^{-3} M $C_6H_5NH_3Cl$, K_b for $C_6H_5NH_2 = 4.6 \times 10^{-10}$
    (c) 0.0010 M $[C_5H_5NH]I$, K_b for $C_5H_5N = 1.7 \times 10^{-9}$
    (d) 0.010 M $Ca(OI)_2$, K_a for HOI $= 2.3 \times 10^{-11}$

66. Calculate the concentrations of all solute species in each of the following solutions. Assume that the ionization of water can be neglected, and show that the changes in the initial concentrations cannot be neglected.

    (a) 0.0101 M $CH_3OC_6H_4ONa$, K_a for $CH_3OC_6H_4OH = 6.7 \times 10^{-11}$
    (b) 0.079 M $[ClC_6H_4NH_3]Br$, K_b for $ClC_6H_4NH_2 = 3.7 \times 10^{-12}$

(c) 1.3×10^{-4} M, $[C_5H_5NH]Cl$, K_b for $C_5H_5N = 1.7 \times 10^{-9}$

(d) 7.5×10^{-4} M $Ca(OBr)_2$, K_a for $HOBr = 2.1 \times 10^{-9}$

67. Calculate the pH and pOH of the solutions in Exercise 59.

68. Calculate the pH and pOH of the solutions in Exercise 60.

69. Calculate the pH and pOH of the solutions in Exercise 61.

70. Calculate the pH and pOH of the solutions in Exercise 62.

71. Calculate the pH of a 0.19 M solution of HNO_2.

72. Calculate the pH of a 1.4×10^{-2} M solution of HCNO.

73. Propionic acid, $C_2H_5CO_2H$ ($K_a = 1.34 \times 10^{-5}$), is used in the manufacture of calcium propionate, a food preservative. What is the hydronium ion concentration in a 0.698 M solution of $C_2H_5CO_2H$?

74. White vinegar (Fig. 16.10) is a 5.0% by mass solution of acetic acid in water. If the density of white vinegar is 1.007 g/cm^3, what is the pH?

75. The ionization constant of lactic acid, $CH_3CH(OH)CO_2H$, an acid found in the blood after strenuous exercise, is 1.36 $\times 10^{-4}$. If 20.0 g of lactic acid is used to make a solution with a volume of 1.00 L, what is the concentration of hydronium ion in the solution?

76. Novocaine, $C_{13}H_{21}O_2N_2Cl$, is the salt of the base procaine and hydrochloric acid. The ionization constant for procaine is 7×10^{-6}. Is a solution of novocaine acidic or basic? Determine $[H_3O^+]$, $[OH^-]$, and the pH of a 2.0% solution by mass of novocaine, assuming that the density of the solution is 1.0 g mL^{-1}.

Polyprotic Acids and Bases

77. Even though both HF and HCN are weak acids, HF is a much stronger acid than HCN. Which of the following is correct at equilibrium for a solution that is initially 0.10 M in HF and 0.10 M in HCN? (a) $[HF] = [HCN]$. (b) $[F^-] = [CN^-]$. (c) $[H_3O^+] = [F^-]$. (d) $[H_3O^+] = [CN^-]$. (e) Both (a) and (b) are correct.

78. Even though both NH_3 and $C_6H_5NH_2$ are weak bases, NH_3 is a much stronger base than $C_6H_5NH_2$. Which of the following is correct at equilibrium for a solution that is initially 0.10 M in NH_3 and 0.10 M in $C_6H_5NH_2$? (a) $[OH^-] = [NH_4^+]$ (b) $[NH_4^+] = [C_6H_5NH_3^+]$. (c) $[OH^-] = [C_6H_5NH_3^+]$. (d) $[NH_3] = [C_6H_5NH_2]$. (e) Both (a) and (b) are correct.

79. Calculate the equilibrium concentrations of the nonionized acids and all ions in a solution that is 0.25 M in HCO_2H and 0.10 M in HClO.

80. Calculate the equilibrium concentrations of the nonionized acids and all ions in a solution that is 0.134 M in HNO_2 and 0.120 M in HBrO.

81. Calculate the equilibrium concentrations of the nonionized bases and all ions in a solution that is 0.25 M in CH_3NH_2 and 0.10 M in C_5H_5N ($K_b = 1.7 \times 10^{-9}$).

82. Calculate the equilibrium concentrations of the nonionized bases and all ions in a solution that is 0.115 M in NH_3 and 0.100 M in $C_6H_5NH_2$.

83. Tell which of the following concentrations would be equal in a calculation of the equilibrium concentrations in a 0.134 M solution of H_2CO_3, a diprotic acid: $[H_3O^+]$, $[OH^-]$, $[H_2CO_3]$, $[HCO_3^-]$, $[CO_3^{2-}]$. No calculations are needed to answer this question.

84. Tell which of the following concentrations would be equal in a calculation of the equilibrium concentrations in a 0.10 M solution of S^{2-}, a diprotic base: $[H_3O^+]$, $[OH^-]$, $[H_2S]$, $[HS^-]$, $[S^{2-}]$. No calculations are needed to answer this question.

85. Calculate the equilibrium concentrations of the nonionized acid and all ions in a 0.134 M solution of H_2CO_3, a diprotic acid.

86. Calculate the equilibrium concentrations of the nonionized acid and all ions in a 0.10 M solution of S^{2-}, a diprotic base.

87. One of the important components of proteins is glycine, $H_3NCH_2CO_2H^+$, which we can abbreviate as H_2gly^+. Calculate the equilibrium concentrations of H_3O^+, H_2gly^+, Hgly, and gly$^-$ in a 0.090 M solution of H_2gly^+. For glycine, $K_{a_1} = 4.5 \times 10^{-3}$ and $K_{a_2} = 2.5 \times 10^{-10}$.

88. Calculate the concentration of each species present in a 0.050 M solution of H_2S.

89. Calculate the concentration of each species present in a 0.010 M solution of phthalic acid, $C_6H_4(CO_2H)_2$.

$C_6H_4(CO_2H)_2 + H_2O \rightleftharpoons$
$\quad H_3O^+ + C_6H_4(CO_2H)(CO_2)^- \quad K_a = 1.1 \times 10^{-3}$
$C_6H_4(CO_2H)(CO_2)^- + H_2O \rightleftharpoons$
$\quad H_3O^+ + C_6H_4(CO_2)_2^{2-} \quad K_a = 3.9 \times 10^{-6}$

90. Calculate the concentration of each species present in a 0.050 M solution of Na_3PO_4.

$PO_4^{3-} + H_2O \rightleftharpoons HPO_4^{2-} + OH^- \quad K_b = 2.8 \times 10^{-2}$
$HPO_4^{2-} + H_2O \rightleftharpoons H_2PO_4^- + OH^- \quad K_b = 1.6 \times 10^{-7}$
$H_2PO_4^- + H_2O \rightleftharpoons$
$\quad\quad\quad\quad H_3PO_4 + OH^- \quad K_b = 1.3 \times 10^{-12}$

The Contribution of Water

91. The contribution of water to the concentrations of H_3O^+ in the solutions of acids in Exercise 59 can be neglected. Explain why we cannot neglect the contribution of water to the concentration of OH^-.

92. The contribution of water to the concentration of OH^- in a solution of the bases in Exercise 59 can be neglected. Explain why we cannot neglect the contribution of water to the concentration of H_3O^+.

93. Draw a curve similar to that shown in Fig. 16.2 for a series of solutions of HF. Plot $[H_3O^+]_{total}$ on the vertical axis and the total concentration of HF (the sum of the concentrations of both the ionized and the nonionized HF molecules) on the horizontal axis. Let the total concentration of HF vary from 1×10^{-10} M to 1×10^{-2} M.

94. Draw a curve similar to that shown in Fig. 16.3 for a series of solutions of NH_3. Plot $[OH^-]$ on the vertical axis and the total concentration of NH_3 (both ionized and nonionized NH_3 molecules) on the horizontal axis. Let the total concentration of NH_3 vary from 1×10^{-10} M to 1×10^{-2} M.

95. Show that the contribution of water to $[H_3O^+]$ from the acids and to $[OH^-]$ from the bases in Exercise 59 can be neglected.

96. Show that the contribution of water to $[H_3O^+]$ from the acids and to $[OH^-]$ from the bases in Exercise 61 can be neglected.

97. Calculate the hydrogen ion concentration in a 3.4×10^{-5} M solution of HBrO.

98. Calculate the hydrogen ion concentration in a 9.2×10^{-6} M solution of HCN.

99. Calculate the hydroxide concentration in a 0.0233M solution of acetamide, CH_3CONH_2, $K_b = 2.5 \times 10^{-13}$.

100. Calculate the hydroxide concentration in a 8.99×10^{-6} M solution of chloroanaline, $ClC_6H_4NH_2$, $K_b = 8.45 \times 10^{-11}$.

Common Ions and Buffers

101. What is $[H_3O^+]$ in a 0.25 M solution of CH_3CO_2H that is 0.030 M with respect to $NaCH_3CO_2$?

$$CH_3CO_2H + H_2O \rightleftharpoons$$
$$H_3O^+ + CH_3CO_2^- \quad K_a = 1.8 \times 10^{-5}$$

102. What is $[H_3O^+]$ in a 0.075 M solution of HNO_2 that is 0.030 M with respect to $NaNO_2$?

$$HNO_2 + H_2O \rightleftharpoons H_3O^+ + NO_2^- \quad K_a = 4.5 \times 10^{-5}$$

103. What is $[OH^-]$ in a 0.125 M solution of CH_3NH_2 that is 0.130 M with respect to CH_3NH_3Cl?

$$CH_3NH_2 + H_2O \rightleftharpoons$$
$$CH_3NH_3^+ + OH^- \quad K_b = 4.4 \times 10^{-4}$$

104. What is $[OH^-]$ in a 1.25 M solution of NH_3 that is 0.78 M with respect to NH_4NO_3?

$$NH_3 + H_2O \rightleftharpoons NH_4^+ + OH^- \quad K_b = 1.8 \times 10^{-5}$$

105. What concentration of NH_4NO_3 is required to make $[OH^-] = 1.0 \times 10^{-5}$ in a 0.200 M solution of NH_3?

106. What concentration of NaF is required to make $[H_3O^+] = 2.3 \times 10^{-4}$ in a 0.300 M solution of HF?

107. What is the effect on the concentrations of acetic acid, hydronium ion, and acetate ion when a little of each of the following is added to an acidic buffer solution of equal concentrations of acetic acid and sodium acetate? (a) HCl, (b) KCH_3CO_2, (c) NaCl, (d) KOH, and (e) CH_3CO_2H.

108. What is the effect on the concentrations of ammonia, hydroxide ion, and ammonium ion when a little of each of the following is added to a basic buffer solution of equal concentrations of ammonia and ammonium nitrate? (a) KI, (b) NH_3, (c) HI, (d) NaOH, and (e) NH_4Cl.

109. What will be the pH of a buffer solution prepared from 0.20 mol of NH_3, 0.40 mol of NH_4NO_3, and just enough water to give 1.00 L of solution?

110. Calculate the pH of a buffer solution prepared from 0.155 mol of phosphoric acid, 0.250 mol of KH_2PO_4, and enough water to make 0.500 L of solution.

111. How much solid $NaCH_3CO_2 \cdot 3H_2O$ must be added to 0.300 L of a 0.50 M acetic acid solution to give a buffer with a pH of 5.00? (*Hint:* Assume a negligible change in volume as the solid is added.)

112. What mass of NH_4Cl must be added to 0.750 L of a 0.100 M solution of NH_3 to give a buffer solution with a pH of 9.26? (*Hint:* Assume a negligible change in volume as the solid is added.)

113. A buffer solution is prepared from equal volumes of 0.200 M acetic acid and 0.600 M sodium acetate. Use 1.80×10^{-5} as K_a for acetic acid.

  (a) What is the pH of the solution?
  (b) Is the solution acidic or basic?
  (c) What is the pH of a solution that results when 3.00 mL of 0.034 M HCl is added to 0.2000 L of the original buffer?

114. A 5.36-g sample of NH_4Cl was added to 25.0 mL of 1.00 M NaOH, and the resulting solution is diluted to 0.100 L.

  (a) What is the pH of this buffer solution?
  (b) Is the solution acidic or basic?
  (c) What is the pH of a solution that results when 3.00 mL of 0.034 M HCl is added to the solution?

115. Which acid in Table 16.3 is most appropriate for preparation of a buffer solution with a pH of 3.1? Explain your choice.

116. Which acid in Table 16.3 is most appropriate for preparation of a buffer solution with a pH of 3.7? Explain your choice.

117. Which base in Table 16.4 is most appropriate for preparation of a buffer solution with a pH of 10.65? Explain your choice.

118. Which base in Table 16.4 is most appropriate for preparation of a buffer solution with a pH of 9.20? Explain your choice.

Titration Curves and Indicators

119. Calculate the pH at the following points in a titration of 40 mL (0.040 L) of 0.100 M barbituric acid ($K_a = 9.8 \times 10^{-5}$) with 0.100 M KOH.

    (a) No KOH added.
    (b) 20 mL of KOH solution added
    (c) 39 mL of KOH solution added
    (d) 40 mL of KOH solution added
    (e) 41 mL of KOH solution added

120. Calculate the pH at the following points in a titration of 50 mL (0.040 L) of 0.150 M dimethyl amine ($K_a = 7.4 \times 10^{-4}$) with 0.150 M HCl.

    (a) No HCl added.
    (b) 25 mL of HCl solution added.
    (c) 49 mL of HCl solution added.
    (d) 50 mL of HCl solution added.
    (e) 51 mL of HCl solution added.

121. Determine a theoretical titration curve for the titration of 20.0 mL of 0.100 M $HClO_4$, a strong acid, with 0.200 M KOH.

122. Determine a theoretical titration curve for the titration of 25.0 mL of 0.150 M NH_3 with 0.300 M HCl.

123. Which of the indicators listed in Fig. 16.22 would be appropriate for the titration described in Exercise 119?

124. Which of the indicators listed in Fig. 16.22 would be appropriate for the titration described in Exercise 120?

125. Which of the indicators listed in Fig. 16.22 would be appropriate for the titration described in Exercise 121?

126. Which of the indicators listed in Fig. 16.22 would be appropriate for the titration described in Exercise 122?

127. The indicator dinitrophenol is an acid with a K_a of 1.1×10^{-4}. In a 1.0×10^{-4} M solution, it is colorless in acid and yellow in base. Calculate the pH range over which it goes from 10% ionized (colorless) to 90% ionized (yellow).

128. Show that the hydronium ion concentration at the half-equivalence point (the point at which half of the amount of base necessary to reach the equivalence point has been added) in the titration of a weak acid with a strong base is equal to K_a for the weak acid.

Applications and Additional Exercises

129. In many detergents, silicates have replaced phosphates as water conditioners. If 125 g of a detergent that contains 8.00% Na_2SiO_3 by weight is used in 4.0 L of water, what are the pH and the concentrations of all ions in the wash water?

    $SiO_3{}^{2-} + H_2O \rightleftharpoons SiO_3H^- + OH^-$ $K_b = 1.6 \times 10^{-3}$
    $HSiO_3{}^- + H_2O \rightleftharpoons H_2SiO_3 + OH^-$ $K_b = 3.1 \times 10^{-5}$

130. How many moles of CH_3CO_2H were used to make the solution of acetic acid shown in Fig. 16.8 if its volume is 17 mL?

131. Saccharin, $C_7H_4NSO_3H$, is a weak acid ($K_a = 2.1 \times 10^{-12}$). Say 0.250 L of diet cola with a buffered pH of 5.48 was prepared from 2.00×10^{-3} g sodium saccharide, $Na(C_7H_4NSO_3)$. What are the final concentrations of saccharine and sodium saccharide in the solution?

132. What is the pH of 1.000 L of a solution of 100.0 g of glutamic acid ($C_5H_9NO_4$, a diprotic acid; $K_{a_1} = 8.5 \times 10^{-5}$, $K_{a_2} = 3.39 \times 10^{-10}$) to which has been added 20.0 g of NaOH during the preparation of monosodium glutamate, the flavoring agent? What is the pH when exactly 1 mol of NaOH per mole of acid has been added?

133. Salicylic acid, $HOC_6H_4CO_2H$, and its derivatives have been used as pain relievers for a long time. Salicylic acid occurs in small amounts in the leaves, bark, and roots of some vegetation. Extracts of these plants have been used as medications for centuries. The acid was first isolated in the laboratory in 1838.

    (a) Both functional groups of salicylic acid ionize in water, with $K_a = 1.0 \times 10^{-3}$ for the $-CO_2H$ group and 4.2×10^{-13} for the $-OH$ group. What is the pH of a saturated solution of the acid (solubility 1.8 g L^{-1})?
    (b) Aspirin was discovered as a result of efforts to produce a derivative of salicylic acid that would not be irritating to the stomach lining. Aspirin is acetyl-salicylic acid, $CH_3CO_2C_6H_4CO_2H$. The $-CO_2H$ functional group is still present, but its acidity is reduced, $K_a = 3.0 \times 10^{-4}$. What is the pH of a solution of aspirin with the same concentration as a saturated solution of salicylic acid [see part (a)]?
    (c) Under some conditions, aspirin reacts with water and forms a solution of salicylic acid and acetic acid.

    $CH_3CO_2C_6H_4CO_2H + H_2O \longrightarrow$
    $HOC_6H_4CO_2H + CH_3CO_2H$

    Does salicylic acid or acetic acid produce more hydronium ions in such a solution?

134. The label on a bottle of household bleach states that the active ingredient is 5.25% by mass sodium hypochlorite, NaClO. Calculate the pOH of the solution if the density is 1.12 g mL^{-1}.

135. Boric acid, H_3BO_3, a very weak acid, is widely used as a mild antiseptic. Can the self-ionization of water be neglected when calculating the pH of a 0.100 M solution of boric acid? What is the pH?

136. Determine the K_a of H_3PO_4, for $H_2PO_4{}^-$, and for $HPO_4{}^{2-}$ and the K_b for $H_2PO_4{}^-$, for $HPO_4{}^{2-}$, and for $PO_4{}^{3-}$ from the following concentrations present in an equilibrium mixture of H_3PO_4 in water. $[H_3PO_4] = 0.076$ M, $[H_3O^+] = [H_2PO_4{}^-] = 0.024$ M, $[HPO_4{}^{2-}] = 6.3 \times 10^{-8}$, $[PO_4{}^{3-}] = 9.4 \times 10^{-19}$ M, and $OH^- = 4.2 \times 10^{-13}$ M.

137. Is the self-ionization of water endothermic or exothermic? The ionization constant for water (K_w) is 2.9×10^{-14} at 40°C and 9.6×10^{-14} at 60°C.

138. Nicotine, $C_{10}H_{14}N_2$, is a base that will accept two protons ($K_{b_1} = 7 \times 10^{-7}$, $K_{b_2} = 1.4 \times 10^{-11}$). What is the concentration of each species present in a 0.050 M solution of nicotine?

139. Which of the following solutions in water has the lowest freezing temperature: a 0.10 M solution of acetic acid, boric acid, ammonia, or methyl amine? Assume the molality and molarity are equal for these solutions.

140. (a) The ion HTe^- is an amphiphrotic species; it can act as either an acid or a base. What is K_a for the acid reaction of HTe^- with H_2O? What is K_b for the reaction in which HTe^- functions as a base in water?
    (b) Demonstrate whether the second ionization of H_2Te can be neglected in the calculation of $[HTe^-]$ in a 0.10 M solution of H_2Te.

141. Using the K_a values in Appendix F, place $Al(H_2O)_6^{3+}$ ($K_a = 1.4 \times 10^{-5}$) in the correct location in Table 15.1, the table of relative strengths of conjugate acid–base pairs in Chapter 15.

142. The following question is taken from the 1991 Chemistry Advanced Placement Examination and is used with the permission of the Educational Testing Service.

    The acid ionization constant, K_a, for propanoic acid, C_2H_5COOH, is 1.3×10^{-5}.

    (a) Calculate the hydrogen ion concentration, $[H^+]$, in a 0.20 M solution of propanoic acid.
    (b) Calculate the percentage of propanoic acid molecules that are ionized in part (a).
    (c) What is the ratio of propanoate ion, $C_2H_5COO^-$, to that of propanoic acid in a buffer solution with a pH of 5.20?
    (d) In a 100.-milliliter sample of a different buffer solution, the propanoic acid concentration is 0.35-molar and the sodium propanoate concentration is 0.50-molar. To this buffer solution, 0.0040 mol of solid NaOH is added. Calculate the pH of the resulting solution.

CHAPTER OUTLINE

17.1 The Solubility Product

Precipitation and Dissolution

17.2 Solubilities and Solubility Products
17.3 Calculation of Solubilities from Solubility Products
17.4 The Precipitation of Slightly Soluble Solids
17.5 Concentrations Necessary to Form a Precipitate
17.6 Concentrations Following Precipitation
17.7 Fractional Precipitation

Multiple Equilibria Involving Solubility

17.8 Dissolution by Formation of a Weak Electrolyte
17.9 Dissolution by Formation of a Complex Ion

17

The Formation and Dissolution of Precipitates

Tufa beds near Mono Lake in eastern California are pillars of calcium carbonate.

Tufa beds in Mono Lake, preserving medical laboratory blood samples, mining sea water for magnesium, developing photographic film, milk of magnesia, tooth decay, hard water, stalactites and stalagmites, and the antacid Tums all involve a similar equilibrium: the equilibrium between a slightly soluble ionic solid and a solution of its ions. A slightly soluble ionic solid, such as apatite, $Ca_5(PO_4)_3OH$ (the principal inorganic component of our tooth enamel), dissolves to a very limited extent, and an equilibrium between the solid and a saturated solution of the ions of the solid results.

In some cases we want to prevent dissolution from occurring. Tooth decay, for example, occurs when the apatite in our teeth dissolves; preventing the dissolution prevents the decay. On the other hand, sometimes we want the reaction to occur. We want the calcium carbonate in a Tums tablet to dissolve, because this process can help soothe our upset stomach.

In this chapter we will find out how we can control the dissolution of a slightly soluble ionic solid by the application of Le Châtelier's principle. We will also see that we can use the equilibrium constant of the reaction to determine the concentration of ions necessary for controlling the reaction.

17.1 The Solubility Product

Silver chloride is a slightly soluble ionic solid (Fig. 17.1). When we add an excess of solid AgCl to water, it dissolves to a small extent and produces a mixture that includes a very dilute solution of Ag^+ and Cl^- ions in equilibrium with undissolved silver chloride.

$$AgCl(s) \rightleftharpoons Ag^+(aq) + Cl^-(aq)$$

This equilibrium, like other equilibria, is dynamic (Section 14.1); some of the solid AgCl continues to dissolve, but at the same time, Ag^+ and Cl^- ions in the solution combine to produce an equal amount of the solid. The state of equilibrium exists because the opposing processes have equal rates.

The equilibrium constant for the equilibrium between a slightly soluble ionic solid and a solution of its ions is called the **solubility product** of the solid. The solubility product is equal to the reaction quotient when the mixture is at equilibrium. Again, we use an ion's concentration as an approximation of its activity in a dilute solution (Section 14.2). For silver chloride, at equilibrium,

$$AgCl(s) \rightleftharpoons Ag^+(aq) + Cl^-(aq)$$

and
$$K_{sp} = Q = [Ag^+][Cl^-]$$

Figure 17.1

A precipitate of the slightly soluble ionic solid silver chloride, AgCl, forms when solutions of $AgNO_3$ and NaCl are mixed.

The solubility product is equal to the product of the concentrations of Ag^+ and Cl^- when a saturated solution is in equilibrium with undissolved AgCl. Although AgCl is a reactant, it is a solid, so [AgCl] does not appear in the expression for K_{sp} (Section 14.3). A table of solubility products appears in Appendix D. Each of these equilibrium constants is much smaller than 1, because the compounds listed are only slightly soluble.

Example 17.1	Writing Equations and Solubility Products

Write the ionic equation for the dissolution, and the equation for the solubility product, for each of the following slightly soluble ionic compounds:
(a) AgI, silver iodide, a light-sensitive solid used in photographic film.

(b) $CaCO_3$, calcium carbonate, the active ingredient in Tums.
(c) $Mg(OH)_2$, magnesium hydroxide, the active ingredient in milk of magnesia.
(d) $Mg(NH_4)PO_4$, magnesium ammonium phosphate, an essentially insoluble substance used in tests for magnesium.
(e) $Ca_5(PO_4)_3OH$, the mineral apatite, a component of teeth and a source of phosphate for fertilizers.

Solution:
(a) $AgI(s) \rightleftharpoons Ag^+(aq) + I^-(aq)$ $K_{sp} = [Ag^+][I^-]$
(b) $CaCO_3(s) \rightleftharpoons Ca^{2+}(aq) + CO_3^{2-}(aq)$ $K_{sp} = [Ca^{2+}][CO_3^{2-}]$
(c) $Mg(OH)_2(s) \rightleftharpoons Mg^{2+}(aq) + 2OH^-(aq)$ $K_{sp} = [Mg^{2+}][OH^-]^2$
(d) $Mg(NH_4)PO_4(s) \rightleftharpoons Mg^{2+}(aq) + NH_4^+(aq) + PO_4^{3-}(aq)$
$$K_{sp} = [Mg^{2+}][NH_4^+][PO_4^{3-}]$$
(e) $Ca_5(PO_4)_3OH(s) \rightleftharpoons 5Ca^{2+}(aq) + 3PO_4^{3-}(aq) + OH^-(aq)$
$$K_{sp} = [Ca^{2+}]^5[PO_4^{3-}]^3[OH^-]$$

Check your learning: Write the ionic equation for the dissolution, and the solubility product, for each of the following slightly soluble compounds: $BaSO_4$, Ag_2SO_4, $Al(OH)_3$, $Pb(OH)Cl$.

Answer: $BaSO_4(s) \rightleftharpoons Ba^{2+}(aq) + SO_4^{2-}(aq)$ $K_{sp} = [Ba^{2+}][SO_4^{2-}]$
$Ag_2SO_4(s) \rightleftharpoons 2Ag^+(aq) + SO_4^{2-}(aq)$ $K_{sp} = [Ag^+]^2[SO_4^{2-}]$
$Al(OH)_3(s) \rightleftharpoons Al^{3+}(aq) + 3OH^-(aq)$ $K_{sp} = [Al^{3+}][OH^-]^3$
$Pb(OH)Cl(s) \rightleftharpoons Pb^{2+}(aq) + OH^-(aq) + Cl^-(aq)$ $K_{sp} = [Pb^{2+}][OH^-][Cl^-]$

We can determine how to shift the concentration of ions in the equilibrium between a slightly soluble solid and a solution of its ions by applying Le Châtelier's principle. For example, one way to control the concentration of manganese(II) ion, Mn^{2+}, in a solution is to adjust the pH of the solution and, consequently, to manipulate the equilibrium between the slightly soluble solid manganese(II) hydroxide and a solution of manganese(II) ion and hydroxide ion.

$$Mn(OH)_2(s) \rightleftharpoons Mn^{2+}(aq) + 2OH^-(aq) \quad K_{sp} = [Mn^{2+}][OH^-]^2$$

This could be important to a laundry, because clothing washed in water that has a manganese concentration exceeding 0.1 mg per liter may be stained by the manganese. A laundry can reduce the concentration of manganese by increasing the concentration of hydroxide ion, for example, by adding a small amount of NaOH or some other base such as the silicates found in many laundry detergents. As the concentration of OH^- ion increases, the equilibrium responds by shifting to the left and reducing the concentration of Mn^{2+} ion while increasing the amount of solid $Mn(OH)_2$ in the equilibrium mixture, as predicted by Le Châtelier's principle (Fig. 17.2).

Figure 17.2

Manganese hydroxide, $Mn(OH)_2$, precipitates when a solution of a base is added to a solution of a manganese(II) salt.

Example 17.2   Solubility Equilibrium of a Slightly Soluble Solid

Determine the effect on the concentrations of Mg^{2+} and OH^- and on the amount of solid $Mg(OH)_2$ that dissolves when each of the following is added to a mixture of solid $Mg(OH)_2$ in water at equilibrium: (a) $MgCl_2$, (b) KOH, (c) an acid, (d) $NaNO_3$, and (e) $Mg(OH)_2$.

Solution: The equilibrium between solid $Mg(OH)_2$ and a solution of Mg^{2+} and OH^- is

$$Mg(OH)_2(s) \rightleftharpoons Mg^{2+}(aq) + 2OH^-(aq)$$

(a) The reaction shifts to the left to relieve the stress produced by the additional Mg^{2+} ion, in accordance with Le Châtelier's principle (Section 14.4). In quantitative terms, the added Mg^{2+} causes the reaction quotient to be larger than the solubility product ($Q > K_{sp}$), and $Mg(OH)_2$ forms until the reaction quotient again equals K_{sp}. At the new equilibrium, $[OH^-]$ is less and $[Mg^{2+}]$ is greater than in the solution of $Mg(OH)_2$ in pure water. More solid $Mg(OH)_2$ is present.

(b) The reaction shifts to the left to relieve the stress of the additional OH^- ion. $Mg(OH)_2$ forms until the reaction quotient again equals K_{sp}. At the new equilibrium, $[OH^-]$ is greater and $[Mg^{2+}]$ is less than in the solution of $Mg(OH)_2$ in pure water. More solid $Mg(OH)_2$ is present.

(c) The concentration of OH^- is reduced as the OH^- reacts with the acid. The reaction shifts to the right to relieve the stress of less OH^- ion. In quantitative terms, the reduction of $[OH^-]$ causes the reaction quotient to be smaller than the solubility product ($Q < K_{sp}$), and additional $Mg(OH)_2$ dissolves until the reaction quotient again equals K_{sp}. At the new equilibrium, $[OH^-]$ is less and $[Mg^{2+}]$ is greater than in the solution of $Mg(OH)_2$ in pure water. More $Mg(OH)_2$ is dissolved.

(d) $NaNO_3$ contains none of the species involved in the equilibrium, so we should expect it to have no appreciable effect on the concentrations of Mg^{2+} and OH^-. (As we saw in Section 16.8, dissolved salts change the activities of the ions of an electrolyte. However, the salt effect is generally small, and we shall neglect the slight errors that may result from it.)

(e) The addition of solid $Mg(OH)_2$ has no effect on the solubility of $Mg(OH)_2$ or on the concentrations of Mg^{2+} and OH^-. The concentration of $Mg(OH)_2$ does not appear in the equation for the reaction quotient.

$$Q = [Mg^{2+}][OH^-]^2$$

Thus changing the amount of solid magnesium hydroxide in the mixture has no effect on the value of Q, and no shift is required to restore Q to the value of the equilibrium constant.

Check your learning: Determine the effect on the concentrations of Ni^{2+} and CO_3^{2-} and on the amount of solid $NiCO_3$ that dissolves when each of the following is added to a mixture of the slightly soluble solid $NiCO_3$ and water at equilibrium: **(a)** $Ni(NO_3)_2$, **(b)** $KClO_4$, **(c)** $NiCO_3$, **(d)** K_2CO_3, and **(e)** HNO_3 (reacts with carbonate giving HCO_3^- or H_2O and CO_2).

Answer: **(a)** mass of $NiCO_3(s)$ increases, $[Ni^{2+}]$ increases, $[CO_3^{2-}]$ decreases; **(b)** no appreciable effect; **(c)** no effect except to increase the amount of solid $NiCO_3$; **(d)** mass of $NiCO_3(s)$ increases, $[Ni^{2+}]$ decreases, $[CO_3^{2-}]$ increases; **(e)** mass of $NiCO_3(s)$ decreases, $[Ni^{2+}]$ increases, $[CO_3^{2-}]$ decreases

We determine the value of the solubility product of a slightly soluble solid from the concentrations of ions in a solution that is in equilibrium with the solid. The value of the solubility product is equal to the reaction quotient at equilibrium.

Figure 17.3

These fluorite crystals are solid solutions containing traces of colored transition metal ions.

Example 17.3   Calculation of K_{sp} from Equilibrium Concentrations

Fluorite, CaF_2, a semiprecious stone used in jewelry (Fig. 17.3), is a slightly soluble solid that dissolves according to the equation

$$CaF_2(s) \rightleftharpoons Ca^{2+}(aq) + 2F^-(aq)$$

The concentration of Ca^{2+} in a saturated solution of CaF_2 is 2.1×10^{-4} M; that of F^- is 4.2×10^{-4} M. What is the solubility product of fluorite?

Solution: The value of the solubility product is equal to the reaction quotient for the reaction at equilibrium.

$$CaF_2(s) \rightleftharpoons Ca^{2+}(aq) + 2F^-(aq)$$

A saturated solution is a solution at equilibrium with the solid. Thus

$$K_{sp} = Q = [Ca^{2+}][F^-]^2 = (2.1 \times 10^{-4})(4.2 \times 10^{-4})^2 = 3.7 \times 10^{-11}$$

Check your learning: In a buffered solution with $[OH^-] = 1.0 \times 10^{-6}$ M that is in contact with solid $Fe(OH)_2$, the concentration of Fe^{2+} is 7.9×10^{-3} M. What is the solubility product for $Fe(OH)_2$?

$$Fe(OH)_2(s) \rightleftharpoons Fe^{2+}(aq) + 2OH^-(aq)$$

Answer: 7.9×10^{-15}

Precipitation and Dissolution

17.2   Solubilities and Solubility Products

We can determine the solubility product of a slightly soluble solid from its solubility, provided the only significant reaction that occurs when the solid dissolves is the formation of its ions; that is, the only equilibrium involved is

$$M_pX_q(s) \rightleftharpoons pM^{m+}(aq) + qX^{n-}(aq)$$

In this case, the solubility product can be calculated from the molar solubility (the solubility in moles per liter).

In this section and the next three sections, we will consider solubility phenomena that we can approximate as involving only a single equilibrium. Situations that involve two or more equilibria will be considered later in this chapter.

Figure 17.4

Silver chromate, Ag_2CrO_4, a slightly soluble ionic solid.

Example 17.4   Determination of K_{sp} from Molar Solubility

Determine the solubility product, K_{sp}, of silver chromate, Ag_2CrO_4 (Fig. 17.4). The solubility of silver chromate is 1.3×10^{-4} mol L^{-1}.

Solution: The reaction is

$$Ag_2CrO_4(s) \rightleftharpoons 2Ag^+(aq) + CrO_4^{2-}(aq)$$

and the solubility product is equal to the reaction quotient at equilibrium.

$$K_{sp} = Q = [Ag^+]^2[CrO_4^{2-}]$$

We can determine the equilibrium concentrations of Ag^+ and CrO_4^{2-} and the solubility product in two steps:

$$[Ag_2CrO_4] \xrightarrow{\;1\;} \begin{array}{c}[Ag^+] \,\&\, [CrO_4^{2-}] \\ \text{at equilibrium}\end{array} \xrightarrow{\;2\;} K_{sp}$$

Step 1.   From the equation for the dissolution of Ag_2CrO_4, we see that 2 mol of $Ag^+(aq)$ and 1 mol of $CrO_4^{2-}(aq)$ are formed for each mole of Ag_2CrO_4 that dissolves. Thus

$$[Ag^+] = \frac{1.3 \times 10^{-4} \text{ mol } Ag_2CrO_4}{1 \text{ L}} \times \frac{2 \text{ mol } Ag^+}{1 \text{ mol } Ag_2CrO_4} = 2.6 \times 10^{-4} \text{ M}$$

$$[CrO_4^{2-}] = \frac{1.3 \times 10^{-4} \text{ mol } Ag_2CrO_4}{1 \text{ L}} \times \frac{1 \text{ mol } CrO_4^{2-}}{1 \text{ mol } Ag_2CrO_4} = 1.3 \times 10^{-4} \text{ M}$$

Step 2.   At equilibrium,

$$\begin{aligned} K_{sp} = Q &= [Ag^+]^2[CrO_4^{2-}] \\ &= (2.6 \times 10^{-4})^2(1.3 \times 10^{-4}) \\ &= 8.8 \times 10^{-12} \end{aligned}$$

Check your learning: The solubility of BaF_2 is 7.5×10^{-3} M. What is its solubility product?

Answer: 1.7×10^{-6}

When the solubility of a compound is given in some unit other than moles per liter, we must convert the solubility to moles per liter.

Figure 17.5

Yellow lead chromate, $PbCrO_4$, the colored compound in this paint, is not appreciably soluble in water.

Example 17.5   Determination of K_{sp} from Gram Solubility

The solubility of the artist's pigment chrome yellow, $PbCrO_4$ (Fig. 17.5), is 4.3×10^{-5} g L^{-1}. Determine the solubility product for $PbCrO_4$.

Solution: We are given the solubility of $PbCrO_4$ in grams per liter. If we convert this solubility to moles per liter, we can find the equilibrium concentrations of Pb^{2+} and CrO_4^{2-} and, then, K_{sp}.

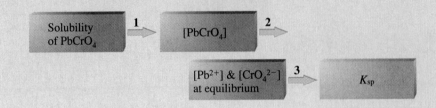

Step 1.   We use the molar mass of $PbCrO_4$ (323.2 g mol^{-1}) to convert the solubility of $PbCrO_4$ in grams per liter to moles per liter.

$$[PbCrO_4] = \frac{4.3 \times 10^{-5} \text{ g PbCrO}_4}{1 \text{ L}} \times \frac{1 \text{ mol PbCrO}_4}{323.2 \text{ g PbCrO}_4}$$

$$= \frac{1.3 \times 10^{-7} \text{ mol PbCrO}_4}{1 \text{ L}} = 1.3 \times 10^{-7} \text{ M}$$

Step 2.   The chemical equation for the dissolution,

$$PbCrO_4(s) \rightleftharpoons Pb^{2+}(aq) + CrO_4^{2-}(aq)$$

indicates that 1 mol of $PbCrO_4$ gives 1 mol of $Pb^{2+}(aq)$ and 1 mol of $CrO_4^{2-}(aq)$. Thus both $[Pb^{2+}]$ and $[CrO_4^{2-}]$ are equal to the molar solubility of $PbCrO_4$.

$$[Pb^{2+}] = [CrO_4^{2-}] = 1.3 \times 10^{-7} \text{ M}$$

Step 3. $K_{sp} = Q = [Pb^{2+}][CrO_4^{2-}]$

$$= (1.3 \times 10^{-7})(1.3 \times 10^{-7}) = 1.7 \times 10^{-14}$$

Check your learning: The solubility of TlCl is 2.9 g L^{-1}. What is its solubility product?

Answer: 1.5×10^{-4} (1.4×10^{-4} if we round the solubility to two digits)

17.3   Calculation of Solubilities from Solubility Products

If the only significant reaction that occurs when a slightly soluble ionic solid dissolves is the formation of its ions, we can determine its solubility from its solubility product. In the calculation, we determine the equilibrium concentrations from initial concentrations (which are zero) and the equilibrium constant for the dissolution, as explained in Section 14.10.

Example 17.6   **Calculating the Solubility of Hg_2Cl_2**

Calomel, Hg_2Cl_2, is a compound composed of the diatomic ion of mercury(I), Hg_2^{2+}, and chloride ions, Cl^-. Although most mercury compounds are very poisonous, six-teenth-century physicians used calomel as a medication. Their patients did not die of mercury poisoning (usually) because calomel is quite insoluble (Fig. 17.6).

$$Hg_2Cl_2(s) \rightleftharpoons Hg_2^{2+}(aq) + 2Cl^-(aq) \qquad K_{sp} = 1.1 \times 10^{-18}$$

Calculate the molar solubility of Hg_2Cl_2.

Solution: The molar solubility of Hg_2Cl_2 is equal to the concentration of Hg_2^{2+} ions, because for each mole of Hg_2Cl_2 that dissolves, 1 mol of Hg_2^{2+} forms. We can determine that concentration using the standard procedure.

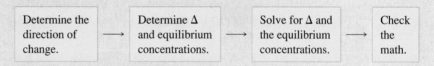

| Determine the direction of change. | Determine Δ and equilibrium concentrations. | Solve for Δ and the equilibrium concentrations. | Check the math. |

Step 1.   *Determine the direction of change.*

Before any Hg_2Cl_2 dissolves, Q is zero, and the reaction will shift to the right to reach equilibrium.

Step 2.   *Determine Δ and equilibrium concentrations.*

Concentrations and changes are given in the following table.

	Hg_2Cl_2 $\rightleftharpoons$	Hg_2^{2+}	+	$2Cl^-$
Initial Concentration, M		0		0
Change		Δ		2Δ
Equilibrium Concentration, M		$0 + \Delta = \Delta$		$0 + 2\Delta = 2\Delta$

Note that the change in the concentration of Cl^- (2Δ) is twice as large as the change in the concentration of Hg_2^{2+} (Δ), because 2 mol of Cl^- form for each mole of Hg_2^{2+} that forms. Hg_2Cl_2 is a solid, so it does not appear in the calculation.

Step 3.   *Solve for Δ and the equilibrium concentrations.*

We substitute the equilibrium concentrations into the expression for K_{sp} and calculate the value of Δ.

$$K_{sp} = [Hg_2^{2+}][Cl^-]^2$$
$$1.1 \times 10^{-18} = (\Delta)(2\Delta)^2$$
$$4\Delta^3 = 1.1 \times 10^{-18}$$
$$\Delta = \left(\frac{1.1 \times 10^{-18}}{4}\right)^{\frac{1}{3}} = 6.5 \times 10^{-7}$$

Figure 17.6

Mercury(I) chloride is essentially insoluble in water.

Now we determine the equilibrium concentrations.

$$[Hg_2^{2+}] = 0 + \Delta = 6.5 \times 10^{-7} \text{ M}$$

$$[Cl^-] = 0 + 2\Delta = 2(6.5 \times 10^{-7})$$

$$= 1.3 \times 10^{-6} \text{ M}$$

The molar solubility of Hg_2Cl_2 is equal to $[Hg_2^{2+}]$, or 6.5×10^{-7} M.

Step 4. *Check the work.*

At equilibrium, $Q = K_{sp}$.

$$Q = [Hg_2^{2+}][Cl^-]^2$$

$$= (6.5 \times 10^{-7})(1.3 \times 10^{-6})^2$$

$$= 1.1 \times 10^{-18}$$

The calculations check.

Check your learning: Determine the molar solubility of MgF_2 from its solubility product. $K_{sp} = 6.4 \times 10^{-9}$.

Answer: 1.2×10^{-3} M

17.4   The Precipitation of Slightly Soluble Solids

The equation describing the equilibrium between calcium carbonate and its ions is

$$CaCO_3(s) \rightleftharpoons Ca^{2+}(aq) + CO_3^{2-}(aq)$$

We can approach equilibrium either by adding solid calcium carbonate to water or by mixing a solution that contains calcium ions with a solution that contains carbonate ions. If we add solid calcium carbonate to water, it dissolves until the concentrations are such that the value of the reaction quotient ($Q = [Ca^{2+}][CO_3^{2-}]$) is equal to the solubility product ($K_{sp} = 4.8 \times 10^{-9}$). If we mix a solution of calcium nitrate, which contains Ca^{2+} ions, with a solution of sodium carbonate, which contains CO_3^{2-} ions, then the slightly soluble ionic solid $CaCO_3$ precipitates, provided the concentrations of Ca^{2+} and CO_3^{2-} ions are such that Q for the mixture is greater than K_{sp}. The reaction shifts to the left (Section 14.4), and the concentrations of the ions are reduced by formation of the solid until the value of Q equals K_{sp}. If the concentrations are such that Q is less than K_{sp}, then no precipitate forms.

The tufa beds around Mono Lake (seen in the photograph at the beginning of this chapter) are pillars of calcium carbonate that formed under water when springs at the bottom of the lake introduced water carrying Ca^{2+} ions into the waters of the lake, which had high concentrations of CO_3^{2-} ion. The tufa pillars became exposed when the lake level fell as it received less runoff from the nearby mountains.

We can compare Q with K_{sp} to predict whether any given mixture of ions will form a slightly soluble solid. If the reaction quotient for the mixture is greater than the solubility product of the solid ($Q > K_{sp}$), then the reaction will shift to the left, forming the solid.

Example 17.7   Precipitation of $Mg(OH)_2$

The first step in the preparation of magnesium metal is the precipitation of $Mg(OH)_2$ from sea water by the addition of lime, $Ca(OH)_2$, a readily available inexpensive source of OH^- ion (Section 8.9).

$$Mg(OH)_2(s) \rightleftharpoons Mg^{2+}(aq) + 2OH^-(aq) \quad K_{sp} = 1.5 \times 10^{-11}$$

The concentration of $Mg^{2+}(aq)$ in sea water is 0.0537 M. Will $Mg(OH)_2$ precipitate when enough $Ca(OH)_2$ is added to give $[OH^-]$ of 0.0010 M?

Solution: This problem asks whether the reaction

$$Mg(OH)_2(s) \rightleftharpoons Mg^{2+}(aq) + 2OH^-(aq)$$

shifts to the left and forms solid $Mg(OH)_2$ when $[Mg^{2+}]$ = 0.0537 M and $[OH^-]$ = 0.0010 M. The reaction shifts to the left if $Q > K_{sp}$. The reaction quotient for the solution *before any reaction occurs* is

$$Q = [Mg^{2+}][OH^-]^2 = (0.0537)(0.0010)^2 = 5.4 \times 10^{-8}$$

Because $Q > K_{sp}$ (5.4×10^{-8} is larger than 1.5×10^{-11}), we can expect the reaction to shift to the left and form solid magnesium hydroxide. $Mg(OH)_2(s)$ forms until the concentrations of magnesium ion and hydroxide ion are reduced sufficiently that the value of Q is equal to K_{sp}.

Check your learning: Use the solubility product in Appendix D to determine whether $CaHPO_4$ will precipitate from a solution with $[Ca^{2+}]$ = 0.0001 M and $[HPO_4^{2-}]$ = 0.001 M.

Answer: No; $Q = 1 \times 10^{-7}$, which is less than K_{sp}.

Example 17.8   Precipitation of AgCl upon Mixing Solutions

Does silver chloride precipitate when equal volumes of a 2×10^{-4} M solution of $AgNO_3$ and a 2×10^{-4} M solution of NaCl are mixed?

Solution: The equation for the equilibrium between solid silver chloride and a solution of silver ion and chloride ion is

$$AgCl(s) \rightleftharpoons Ag^+(aq) + Cl^-(aq)$$

The solubility product is 1.8×10^{-10} (Appendix D).

   AgCl precipitates if the reaction quotient calculated from the concentrations in the mixture of $AgNO_3$ and NaCl is greater than K_{sp}. The volume doubles when we mix equal volumes, so each concentration is reduced to half its initial value. Consequently, immediately upon mixing, $[Ag^+]$ and $[Cl^-]$ are both equal to

$$\tfrac{1}{2}(2 \times 10^{-4}) \text{ M} = 1 \times 10^{-4} \text{ M}$$

The reaction quotient, Q, is *momentarily* greater than K_{sp} for AgCl.

$$Q = [Ag^+][Cl^-] = (1 \times 10^{-4})(1 \times 10^{-4}) = 1 \times 10^{-8} > K_{sp}$$

AgCl will precipitate from the mixture.

Check your learning: Does $KClO_4$ precipitate when 20 mL of a 0.05 M solution of K^+ is added to 80 mL of a 0.50 M solution of ClO_4^-?

Answer: No: $Q = 4 \times 10^{-3}$, which is less than K_{sp}.

17.5 Concentrations Necessary to Form a Precipitate

In the previous two examples, we saw that $Mg(OH)_2$ or $AgCl$ precipitates when $Q > K_{sp}$. In general, when a solution of a soluble salt of the M^{m+} ion is mixed with a solution of a soluble salt of the X^{n-} ion, the solid, M_pX_q precipitates if the value of Q for the mixture of M^{m+} and X^{n-} is greater than K_{sp} for M_pX_q. Thus if we know the concentration of one of the ions of a slightly soluble ionic solid and the value for the solubility product of the solid, we can calculate the concentration that the other ion must exceed for precipitation to begin. To simplify the calculation, we will assume that precipitation begins when the reaction quotient becomes equal to the solubility product constant.

Example 17.9   Precipitation of Calcium Oxalate

Blood will not clot if calcium ion is removed from its plasma. Some blood collection tubes contain salts of the oxalate ion, $C_2O_4^{2-}$, for this purpose (Fig. 17.7). At sufficiently high concentrations, the calcium and oxalate ions form a solid, $CaC_2O_4 \cdot H_2O$ (which also contains water bound in the solid). The concentration of Ca^{2+} in a sample of blood serum is 2.2×10^{-3} M. What concentration of $C_2O_4^{2-}$ ion must be added before $CaC_2O_4 \cdot H_2O$ begins to precipitate?

Solution: The equilibrium is

$$CaC_2O_4 \cdot H_2O(s) \rightleftharpoons Ca^{2+}(aq) + C_2O_4^{2-}(aq) + H_2O(l)$$

For this reaction,

$$K_{sp} = [Ca^{2+}][C_2O_4^{2-}] = 2.27 \times 10^{-9} \text{ (Appendix D)}$$

$CaC_2O_4 \cdot H_2O$ does not appear in this expression because it is a solid. Water does not appear because it is the solvent.

Solid $CaC_2O_4 \cdot H_2O$ does not begin to form until $Q = K_{sp}$. Because we know K_{sp} and $[Ca^{2+}]$, we can solve for the concentration of $C_2O_4^{2-}$ that is necessary to produce the first trace of solid.

$$Q = K_{sp} = [Ca^{2+}][C_2O_4^{2-}] = 2.27 \times 10^{-9}$$

$$(2.2 \times 10^{-3})[C_2O_4^{2-}] = 2.27 \times 10^{-9}$$

$$[C_2O_4^{2-}] = \frac{2.27 \times 10^{-9}}{2.2 \times 10^{-3}} = 1.0 \times 10^{-6}$$

A concentration of $C_2O_4^{2-} = 1.0 \times 10^{-6}$ M is necessary to initiate the precipitation of $CaC_2O_4 \cdot H_2O$ under these conditions.

Figure 17.7

The white solid in this blood collection tube combines with the Ca^{2+} ion in blood plasma and prevents the blood from clotting.

Check your learning: If a solution contains 0.0020 mol of $CrO_4{}^{2-}$ per liter, what concentration of Ag^+ ion must be added as $AgNO_3$ before Ag_2CrO_4 begins to precipitate? Neglect any increase in volume upon adding the solid silver nitrate.

Answer: 7×10^{-5} M

17.6   Concentrations Following Precipitation

It is sometimes useful to know the concentration of an ion that remains in solution after precipitation. We can use the solubility product for this calculation, too. If we know the value of K_{sp} and the concentration of one ion in solution, we can calculate the concentration of the second ion remaining in solution. The calculation is the same type as that in Example 17.9—calculation of the concentration of a species in an equilibrium mixture from the concentrations of the other species and the equilibrium constant. However, the conditions are different; we are calculating concentrations that exist after precipitation is complete, rather than at the start of precipitation.

Example 17.10   Concentrations Following Precipitation

As we saw in Section 17.1, clothing washed in water that has a manganese concentration exceeding 0.1 mg L^{-1} (1.8×10^{-6} M) may be stained by the manganese, but the amount of Mn^{2+} in the water can be reduced by adding base. If a laundry wishes to add a buffer to keep the pH high enough to precipitate manganese as the hydroxide, $Mn(OH)_2$, what pH is required to keep $[Mn^{2+}]$ equal to 1.8×10^{-6} M?

Solution: The dissolution of $Mn(OH)_2$ is described by the equation

$$Mn(OH)_2(s) \rightleftharpoons Mn^{2+}(aq) + 2OH^-(aq) \qquad K_{sp} = 4.5 \times 10^{-14}$$

We need to calculate the concentration of OH^- when the concentration of Mn^{2+} is 1.8×10^{-6} M. From that, we can calculate the pH.
  At equilibrium,

$$Q = [Mn^{2+}][OH^-]^2 = K_{sp}$$

or
$$(1.8 \times 10^{-6})[OH^-]^2 = 4.5 \times 10^{-14}$$

so
$$[OH^-] = 1.6 \times 10^{-4}\ M$$

Now we calculate the pH from the pOH (Section 16.2).

$$pOH = -\log[OH^-] = -\log(1.6 \times 10^{-4}) = 3.80$$
$$pH = 14.00 - pOH = 14.00 - 3.80 = 10.20$$

  If the laundry adds a base, such as the sodium silicate in some detergents, to the wash water until the pH is raised to 10.20, the manganese ion will be reduced to a concentration of 1.8×10^{-6} M; at that concentration or less, the ion will not stain the clothing.

Check your learning: The first step in the preparation of magnesium metal is the precipitation of $Mg(OH)_2$ from sea water by the addition of $Ca(OH)_2$. The concentration of $Mg^{2+}(aq)$ in sea water is 5.37×10^{-2} M. Calculate the pH at which $[Mg^{2+}]$ is reduced to 1.0×10^{-5} M by the addition of $Ca(OH)_2$.

Answer: 11.09

17.7   Fractional Precipitation

Mixtures of silver halides are used in photographic film. Even though AgCl, AgBr, and AgI are all quite insoluble, we cannot prepare a homogeneous mixture of these solids (a solid solution) by adding Ag^+ to a solution of Cl^-, Br^-, and I^-; essentially all of the AgI will precipitate before any of the other solid halides form. However, we can prepare a homogeneous mixture of the solids by adding a solution of Cl^-, Br^-, and I^- to a solution of Ag^+.

Let us consider a simpler mixture—AgBr and AgI—to see what the problem is. Both silver iodide and silver bromide are slightly soluble ionic compounds, but silver iodide is less soluble than silver bromide. A saturated solution of silver iodide has a lower molar concentration than a saturated solution of silver bromide. If we add a solution of silver nitrate to a solution of a mixture of potassium bromide and potassium iodide, then the less soluble silver iodide begins to precipitate before the more soluble silver bromide. In fact, almost all of the iodide ion reacts and forms silver iodide before any silver bromide begins to form. However, if we add the halides to a solution of silver ion, then both halides precipitate at the same time.

When two anions form slightly soluble compounds with the same cation, or when two cations form slightly soluble compounds with the same anion, the less soluble compound (usually the compound with the smaller K_{sp}) generally precipitates first when we add a precipitating agent to a solution that contains both anions (or both cations). When the solubilities of the two compounds differ by a factor of about 20 or more, almost all of the less soluble compound precipitates before any of the more soluble one does. However, any remaining less soluble compound precipitates along with the more soluble one (coprecipitation) when enough precipitating agent is added to cause the more soluble compound to precipitate.

Figure 17.8

A precipitate of silver iodide, AgI (left), and a precipitate of silver chloride, AgCl (right).

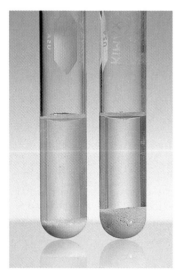

Example 17.11   Precipitation of Silver Halides

A solution contains 0.0010 mol of KI and 0.10 mol of KCl per liter. $AgNO_3$ is gradually added to this solution. Which forms first, solid AgI or solid AgCl (Fig. 17.8)?

Solution: The two equilibria involved (Fig. 17.8) are

$$AgCl(s) \rightleftharpoons Ag^+(aq) + Cl^-(aq) \qquad K_{sp} = 1.8 \times 10^{-10}$$
$$AgI(s) \rightleftharpoons Ag^+(aq) + I^-(aq) \qquad K_{sp} = 1.5 \times 10^{-16}$$

If the solution contained about *equal* concentrations of Cl^- and I^-, then the silver salt with the smallest K_{sp} (AgI) would precipitate first. The concentrations are not equal, however, so we should find the $[Ag^+]$ at which AgCl begins to precipitate and the $[Ag^+]$ at which AgI begins to precipitate. The salt that forms at the lower $[Ag^+]$ precipitates first.

For AgI. AgI precipitates when Q equals K_{sp} for AgI (1.5×10^{-16}). When $[I^-] = 0.0010$ M,

$$Q = [Ag^+][I^-] = [Ag^+](0.0010) = 1.5 \times 10^{-16}$$

$$[Ag^+] = \frac{1.5 \times 10^{-16}}{0.0010} = 1.5 \times 10^{-13} \text{ M}$$

AgI begins to precipitate when $[Ag^+]$ is 1.5×10^{-13} M.

For AgCl. AgCl precipitates when Q equals K_{sp} for AgCl (1.8×10^{-10}). When $[Cl^-] = 0.10$ M,

$$Q_{sp} = [Ag^+][Cl^-] = [Ag^+](0.10) = 1.8 \times 10^{-10}$$

$$[Ag^+] = \frac{1.8 \times 10^{-10}}{0.10} = 1.8 \times 10^{-9} \text{ M}$$

AgCl begins to precipitate when $[Ag^+]$ is 1.8×10^{-9} M.

AgI begins to precipitate at a lower $[Ag^+]$ than AgCl, so AgI begins to precipitate first.

Check your learning: What is the concentration of I^- in the solution described in this example when AgCl begins to precipitate, and what fraction of the original I^- remains in solution at this point? *Hint:* The fraction of I^- remaining is

$$\frac{[I^-] \text{ when precipitation of AgCl begins}}{[I^-] \text{ originally present}}$$

Answer: $[I^-] = 8.3 \times 10^{-8}$ M, fraction $= 8.3 \times 10^{-5}$

Multiple Equilibria Involving Solubility

Copper carbonate is essentially insoluble; its solubility product is small

$$CuCO_3(s) \rightleftharpoons Cu^{2+}(aq) + CO_3^{2-}(aq) \qquad K_{sp} = 1.37 \times 10^{-10}$$

However, copper carbonate dissolves in a solution of ammonia or in a solution of nitric acid. In both cases a second equilibrium reduces the concentration of either the copper(II) ion or the carbonate ion so that $Q < K_{sp}$. As a consequence, copper carbonate continues to dissolve as the system shifts to increase the concentrations of Cu^{2+} and CO_3^{2-} so that equilibrium is restored. In some cases, all of the solid dissolves before equilibrium can be restored.

Two of the equilibria that are established when ammonia is added to a mixture of copper carbonate and water are

$$CuCO_3(s) \rightleftharpoons Cu^{2+}(aq) + CO_3^{2-}(aq) \qquad (1)$$

$$Cu^{2+}(aq) + 4NH_3(aq) \rightleftharpoons Cu(NH_3)_4^{2+}(aq) \qquad (2)$$

In a mixture of $CuCO_3(s)$, Cu^{2+}, CO_3^{2-}, NH_3, and $Cu(NH_3)_4^{2+}$ at equilibrium in water, the same concentration of Cu^{2+} is found in both equilibria.

$$\text{For Reaction 1:} \qquad K_{sp} = [Cu^{2+}][CO_3^{2-}]$$

$$\text{For Reaction 2:} \qquad K = \frac{[Cu(NH_3)_4^{2+}]}{[Cu^{2+}][NH_3]^4}$$

It is possible to change the concentration of Cu^{2+} in the solution by adjusting the concentration of NH_3 in the second equilibrium (Reaction 2). As the concentration of NH_3 increases, the concentration of Cu^{2+} decreases as Reaction 2 comes to equilibrium. Additional $CuCO_3$ then dissolves to compensate for the decreased Cu^{2+} concentration.

Two of the equilibria that are established when nitric acid, a strong acid, is added to a mixture of copper carbonate and water are

$$CuCO_3(s) \rightleftharpoons Cu^{2+}(aq) + CO_3^{2-}(aq) \qquad (1)$$

$$CO_3^{2-}(aq) + 2H_3O^+(aq) \rightleftharpoons H_2CO_3(aq) + 2H_2O(l) \qquad (3)$$

In a mixture of $CuCO_3(s)$, Cu^{2+}, CO_3^{2-}, H_3O^+, and H_2CO_3 at equilibrium, the same concentration of CO_3^{2-} is found in both equilibria.

$$\text{For Reaction 1:} \qquad K_{sp} = [Cu^{2+}][CO_3^{2-}]$$

$$\text{For Reaction 2:} \qquad K = \frac{[H_2CO_3]}{[CO_3^{2-}][H_3O^+]^2}$$

We can change the concentration of CO_3^{2-} in the solution by adjusting the concentration of acid. As the concentration of H_3O^+ increases, the concentration of CO_3^{2-} decreases as Reaction 3 returns to equilibrium. This results in the dissolution of additional $CuCO_3$ to restore the equilibrium of Reaction 1.

In the following sections, we will see several ways in which we can control the concentrations of ions of a slightly soluble solid by using a second equilibrium.

17.8 Dissolution by Formation of a Weak Electrolyte

Slightly soluble solids derived from weak acids generally dissolve in strong acids (Fig. 17.9), unless their solubility products are extremely small (PbS, HgS, and CuS are examples). For instance, we can dissolve $CuCO_3$, FeS, and $Ca_3(PO_4)_2$ in HCl because

Figure 17.9

Zinc sulfide is insoluble in neutral solution (left) but dissolves in acidic solution (right). The pH paper shows that the pH of the solution is less than 1.

their anions react and form weak acids (H_2CO_3, H_2S, and $H_2PO_4^-$). The resulting decrease in the concentration of the anion causes a shift of the equilibrium concentrations to the right in accordance with Le Châtelier's principle.

Of particular relevance to us is the dissolution of apatite, $Ca_3(PO_4)_3OH$, in acid. Apatite is a mineral (Fig. 17.10), and it is also found as the principal mineral in the enamel of our teeth. A mixture of apatite and water (or saliva) contains an equilibrium mixture of solid $Ca_5(PO_4)_3OH$ and dissolved Ca^{2+}, PO_4^{3-}, and OH^- ions.

$$Ca_5(PO_4)_3OH(s) \rightleftharpoons 5Ca^{2+}(aq) + 3PO_4^{3-}(aq) + OH^-(aq)$$

When exposed to acid, phosphate ion reacts with the hydronium ion to form the dihydrogen phosphate ion.

$$PO_4^{3-}(aq) + 2H_3O^+ \rightleftharpoons H_2PO_4^- + 2H_2O$$

Hydroxide ion reacts to form water.

$$OH^-(aq) + H_3O^+ \rightleftharpoons 2H_2O$$

Figure 17.10

A crystal of the mineral apatite, $Ca_5(PO_4)_3OH$. Pure apatite is white, but like many other minerals, this sample is colored because of the presence of impurities.

These reactions reduce the phosphate and hydroxide ion concentrations, and additional apatite dissolves in an acidic solution in accordance with Le Châtelier's principle. Our teeth develop cavities when acid waste produced by bacteria growing on them causes the apatite of the enamel to dissolve. Fluoride toothpastes contain sodium fluoride, NaF, or stannous fluoride [more properly named tin(II) fluoride], SnF_2. They function by replacing the OH^- ion in apatite with F^- ion, producing fluorapatite, $Ca_5(PO_4)_3F$. The resulting $Ca_5(PO_4)_3F$ is less soluble than $Ca_5(PO_4)_3OH$, and F^- is a weaker base than OH^-. Both of these factors make the fluorapatite more resistant than apatite to attack by acids.

A similar reaction occurs when acid rain attacks limestone or marble, which are calcium carbonates. The hydronium ion from the acid combines with the carbonate ion and forms the hydrogen carbonate ion, a weak acid.

$$H_3O^+(aq) + CO_3^{2-}(aq) \rightleftharpoons HCO_3^-(aq) + H_2O(l)$$

Figure 17.11

Calcium carbonate in a Tums antacid tablet dissolves in an acid solution, with evolution of gaseous CO_2.

Calcium hydrogen carbonate, $Ca(HCO_3)_2$, is soluble, so limestone and marble objects slowly dissolve in acid rain.

When we add calcium carbonate to a concentrated acid, hydronium ion reacts with the carbonate ion according to the equation

$$2H_3O^+(aq) + CO_3^{2-}(aq) \rightleftharpoons H_2CO_3(aq) + 2H_2O(l)$$

(Acid rain is usually not acidic enough to cause this reaction, but laboratory acids are.) The solution may become saturated with the weak electrolyte carbonic acid, which is unstable, and carbon dioxide gas can be evolved (Fig. 17.11).

$$H_2CO_3(aq) \rightleftharpoons CO_2(g) + H_2O(l)$$

These reactions reduce the carbonate ion concentration, and additional calcium carbonate dissolves. If enough acid is present, all the calcium carbonate will dissolve.

The reaction of a metal carbonate with an acid (Fig. 17.11) also decreases the concentration of the acid. This is the reaction that occurs when the antacid Tums, whose active ingredient is calcium carbonate, reduces stomach acidity. An added benefit of the reaction is that the concentration of dissolved calcium ion in the stomach increases and the calcium ion becomes available for absorption by the body.

Other reactions can also lead to dissolution. Lead sulfate is only slightly soluble. However, we can dissolve it in a solution of ammonium acetate because the formation of lead acetate, a soluble weak electrolyte, reduces the concentration of lead ion,

and the reaction quotient for the dissolution of lead sulfate becomes smaller than the value of the solubility product for lead sulfate. Because $Q = [Pb^{2+}][SO_4^{2-}] < K_{sp}$, Reaction 4 shifts completely to the right.

$$PbSO_4(s) \rightleftharpoons Pb^{2+}(aq) + SO_4^{2-}(aq) \qquad (4)$$

$$Pb^{2+}(aq) + 2CH_3CO_2^-(aq) \rightleftharpoons Pb(CH_3CO_2)_2(aq, nonionized)$$

Most metal hydroxides—such as $Al(OH)_3$, $Mg(OH)_2$, and $Fe(OH)_3$,—dissolve in solutions of acids because the hydroxide ion reacts with the acid and forms water. For example,

$$Al(OH)_3(s) \rightleftharpoons Al^{3+}(aq) + 3OH^-(aq)$$

$$3H_3O^+(aq) + 3OH^-(aq) \rightleftharpoons 6H_2O(l)$$

Net: $Al(OH)_3(s) + 3H_3O^+(aq) \rightleftharpoons Al^{3+}(aq) + 6H_2O(l)$

When the hydroxide ion in a solution of a slightly soluble metal hydroxide reacts with acid, the reaction quotient for the dissolution becomes smaller than the value of the solubility product, and additional solid dissolves. If there is enough acid to react with all the hydroxide in the solid, the dissolution reaction will go to completion.

We can cause a solid to dissolve by adjusting the concentration of one of its component ions using a second equilibrium. We can also prevent a solid from forming by adjusting the concentration of one of its component ions (Fig. 17.12).

Example 17.12   Prevention of Precipitation of $Mg(OH)_2$

Calculate the concentration of ammonium ion that is required to prevent the precipitation of $Mg(OH)_2$ in a solution with $[Mg^{2+}] = 0.10$ M and $[NH_3] = 0.10$ M. Two equilibria are involved in this system.

$$Mg(OH)_2(s) \rightleftharpoons Mg^{2+}(aq) + 2OH^-(aq) \qquad K_{sp} = 1.5 \times 10^{-11} \quad (5)$$

$$NH_3(aq) + H_2O(l) \rightleftharpoons NH_4^+(aq) + OH^-(aq) \qquad K_b = 1.8 \times 10^{-5} \quad (6)$$

Solution: To prevent the formation of solid $Mg(OH)_2$, we must adjust the concentration of OH^- so that the reaction quotient for Equation 5, $Q = [Mg^{2+}][OH^-]^2$, is less than K_{sp} for $Mg(OH)_2$. (To simplify the calculation, we determine the concentrations when $Q = K_{sp}$.) $[OH^-]$ can be reduced by the addition of NH_4^+, which shifts Reaction 6 to the left and reduces $[OH^-]$.

Step 1.   We determine the $[OH^-]$ at which $Q = K_{sp}$ when $[Mg^{2+}] = 0.10$ M.

$$Q = [Mg^{2+}][OH^-]^2 = (0.10)[OH^-]^2 = 1.5 \times 10^{-11}$$

$$[OH^-] = 1.2 \times 10^{-5} \text{ M}$$

Solid $Mg(OH)_2$ does not form in this solution when $[OH^-]$ is less than 1.2×10^{-5} M.

Step 2.   We calculate the $[NH_4^+]$ needed to reduce $[OH^-]$ to 1.2×10^{-5} M when $[NH_3] = 0.10$.

$$K_b = \frac{[NH_4^+][OH^-]}{[NH_3]} = \frac{[NH_4^+](1.2 \times 10^{-5})}{(0.10)} = 1.8 \times 10^{-5}$$

$$[NH_4^+] = 0.15 \text{ M}$$

When $[NH_4^+]$ equals 0.15 M, $[OH^-]$ is 1.2×10^{-5} M. Any $[NH_4^+]$ greater than 0.15 M reduces $[OH^-]$ below 1.2×10^{-5} M and prevents the formation of $Mg(OH)_2$.

Check your learning: Consider the two equilibria

$$ZnS(s) \rightleftharpoons Zn^{2+}(aq) + S^{2-}(aq) \qquad K_{sp} = 1 \times 10^{-27}$$

$$2H_2O(l) + H_2S(aq) \rightleftharpoons 2H_3O^+(aq) + S^{2-}(aq) \qquad K = 1.0 \times 10^{-26}$$

and calculate the concentration of hydronium ion required to prevent the precipitation of ZnS in a solution that is 0.050 M in Zn^{2+} and saturated with H_2S (0.10 M H_2S).

Answer: $[H_3O^+] > 0.2$ M ($[S^{2-}]$ is less than 2×10^{-26} M, and precipitation of ZnS does not occur).

Now we can consider one reason why precise calculations of the solubility of solids from the solubility product are limited to cases in which the only significant reaction that occurs when the solid dissolves is the formation of its ions (as we said in Section 17.3).

In some cases the concentration of hydronium ion or hydroxide ion in pure water is sufficient to affect the solubility of some salts of weak acids, salts of weak bases, or hydroxides. The anions or cations of these compounds react with water when they dissolve, and a second equilibrium is established. For example, when we add a relatively insoluble sulfide such as lead sulfide, PbS, to water, some of the sulfide ion produced by its dissolution reacts with water and forms the hydrogen sulfide ion and, in some cases, hydrogen sulfide.

$$PbS(s) \rightleftharpoons Pb^{2+}(aq) + S^{2-}(aq)$$

$$S^{2-}(aq) + H_2O(l) \rightleftharpoons HS^-(aq) + OH^-(aq)$$

$$HS^-(aq) + H_2O(l) \rightleftharpoons H_2S(aq) + OH^-(aq)$$

This reduces the sulfide ion concentration and, in accordance with Le Châtelier's principle, more PbS dissolves to relieve the stress of the reduced S^{2-} concentration. Thus PbS is more soluble than we calculate by using its solubility product alone. On the basis of K_{sp}, we would expect the solubility of PbS to be 6×10^{-15} g L^{-1}; the experimental value is about 0.006 g L^{-1}.

The solubility of PbS in water is equal to the concentration of lead ion in the solution, not to the concentration of sulfide ion. Some of the sulfide ions produced by the dissolution of PbS react and form HS^- or H_2S, but one lead ion is produced for each formula unit of PbS that dissolves. This means that the concentration of lead ion equals the sum of the concentrations of the sulfide ion, S^{2-}, the hydrogen sulfide ion, HS^-, and the hydrogen sulfide, H_2S.

$$[Pb^{2+}] = [S^{2-}] + [HS^-] + [H_2S]$$

$[Pb^{2+}]$ is equal to the sum of the concentrations of all the sulfur-containing species. This means that in calculating the solubility of a slightly soluble sulfide from the solubility product, we must consider the reactions of the S^{2-} ion and the HS^- ion with water.

Slightly soluble hydroxides and salts of very weak acids, such as insoluble carbonates, behave similarly to the sulfides. When we add a relatively insoluble carbonate such as barium carbonate, $BaCO_3$, to water, its solubility is larger than we would calculate from the value of its solubility product, because the carbonate ion reacts with water.

$$H_2O(l) + CO_3^{2-}(aq) \rightleftharpoons HCO_3^-(aq) + OH^-(aq)$$

$$H_2O(l) + HCO_3^-(aq) \rightleftharpoons H_2CO_3(aq) + OH^-(aq)$$

17.9   Dissolution by Formation of a Complex Ion

Many slightly soluble ionic solids dissolve when the concentration of the metal ion in solution is reduced through the formation of **complex ions** in a Lewis acid–base reaction (Section 15.13). For example, silver chloride dissolves in a solution of ammonia because the silver ion reacts with ammonia to form the complex ion $Ag(NH_3)_2{}^+$. The Lewis structure of the $Ag(NH_3)_2{}^+$ ion is

$$\left[\begin{array}{ccc} & H & & H \\ & | & & | \\ H-&N&-Ag-&N&-H \\ & | & & | \\ & H & & H \end{array}\right]^+$$

The equations for the dissolution of AgCl in a solution of NH_3 are

$$AgCl(s) \rightleftharpoons Ag^+(aq) + Cl^-(aq)$$
$$Ag^+(aq) + 2NH_3(aq) \rightleftharpoons Ag(NH_3)_2{}^+(aq)$$
$$\textit{Net:}\quad AgCl(s) + 2NH_3(aq) \rightleftharpoons Ag(NH_3)_2{}^+(aq) + Cl^-(aq)$$

Aluminum hydroxide dissolves in a solution of sodium hydroxide or other strong base because of the formation of the complex ion $Al(OH)_4{}^-$. The Lewis structure of the $Al(OH)_4{}^-$ ion is

$$\left[\begin{array}{ccc} & :\ddot{O}-H & \\ & | & \\ H-\ddot{O}-&Al&-\ddot{O}-H \\ & | & \\ & :\ddot{O}-H & \end{array}\right]^-$$

The equations for the dissolution are

$$Al(OH)_3(s) \rightleftharpoons Al^{3+}(aq) + 3OH^-(aq)$$
$$Al^{3+}(aq) + 4OH^-(aq) \rightleftharpoons Al(OH)_4{}^-(aq)$$
$$\textit{Net:}\quad Al(OH)_3(s) + OH^-(aq) \rightleftharpoons Al(OH)_4{}^-(aq)$$

Mercury(II) sulfide dissolves in a solution of sodium sulfide because HgS reacts with the S^{2-} ion, forming $HgS_2{}^{2-}$, $[S-Hg-S]^{2-}$.

$$HgS(s) \rightleftharpoons Hg^{2+}(aq) + S^{2-}(aq)$$
$$Hg^{2+}(aq) + 2S^{2-}(aq) \rightleftharpoons HgS_2{}^{2-}(aq)$$
$$\textit{Net:}\quad HgS(s) + S^{2-}(aq) \rightleftharpoons HgS_2{}^{2-}(aq)$$

The equilibrium constant for the reaction of the components of a complex ion to form the complex ion in solution is called a **formation constant, K_f**. For example, the complex ion $Cu(CN)_2{}^-$, $[:N\equiv C-Cu-C\equiv N:]^-$, forms by the reaction

$$Cu^+(aq) + 2\,CN^-(aq) \rightleftharpoons Cu(CN)_2{}^-(aq)$$

At equilibrium,

$$K_f = Q = \frac{[Cu(CN)_2{}^-]}{[Cu^+][CN^-]^2}$$

A formation constant is sometimes called a **stability constant** or **association constant**. Appendix E contains a table of formation constants. The larger the formation constant, the more stable the complex.

Alternatively, we could describe the stability of a complex ion by its **dissociation constant,** K_d, the equilibrium constant for the decomposition of a complex ion into its components in solution. For $Cu(CN)_2^-$ the dissociation is

$$Cu(CN)_2^-(aq) \rightleftharpoons Cu^+(aq) + 2CN^-(aq)$$

At equilibrium,

$$K_f = Q = \frac{[Cu^+][CN^-]^2}{[Cu(CN)_2^-]}$$

It should be apparent that K_d is the inverse of K_f.

$$K_d = \frac{1}{K_f}$$

The smaller K_d, the more stable the complex.

As an example of dissolution by complex ion formation, consider what happens when we add aqueous ammonia to a mixture of silver chloride in water. Silver chloride dissolves in water, giving a small concentration of Ag^+ ($[Ag^+] = 1.3 \times 10^{-5}$ M).

$$AgCl(s) \rightleftharpoons Ag^+(aq) + Cl^-(aq)$$

However, if NH_3 is present in the water, the complex ion $Ag(NH_3)_2^+$ can form according to the equation

$$Ag^+(aq) + 2NH_3(aq) \rightleftharpoons Ag(NH_3)_2^+(aq)$$

with

$$K_f = \frac{[Ag(NH_3)_2^+]}{[Ag^+][NH_3]^2} = 1.6 \times 10^7$$

The large size of this formation constant indicates that most of the free silver ions produced by the dissolution of AgCl combine with NH_3 to form $Ag(NH_3)_2^+$. As a consequence, the concentration of silver ions, $[Ag^+]$, is reduced, and the reaction quotient for the dissolution of silver chloride, $[Ag^+][Cl^-]$, falls below the solubility product of AgCl.

$$Q = [Ag^+][Cl^-] < K_{sp}$$

More silver chloride then dissolves. If the concentration of ammonia is great enough, all the silver chloride dissolves.

Example 17.13 | Dissociation of a Complex Ion

Calculate the concentration of the silver ion in a solution that initially is 0.10 M with respect to $Ag(NH_3)_2^+$.

Solution: We use the familiar path to solve this problem.

Determine the direction of change.		Determine Δ and equilibrium concentrations.		Solve for Δ and the equilibrium concentrations.		Check the math.
	$\longrightarrow$		$\longrightarrow$		$\longrightarrow$	

Step 1.   *Determine the direction of change.*

The complex ion $Ag(NH_3)_2^+$ is in equilibrium with its components, as represented by the equation

$$Ag^+(aq) + 2NH_3(aq) \rightleftharpoons Ag(NH_3)_2^+(aq)$$

We write the equilibrium as a formation reaction because Appendix E lists formation constants for complex ions. Before equilibrium, the reaction quotient is larger than the equilibrium constant [$K_f = 1.6 \times 10^7$, and $Q = 0.10/(0 \times 0)$; a large, though undefined quantity], so the reaction shifts to the left to reach equilibrium.

Step 2.   *Determine Δ and equilibrium concentrations.*

We let the change in the increase in concentration of Ag^+ be Δ. Dissociation of 1 mol of $Ag(NH_3)_2^+$ gives 1 mol of Ag^+ and 2 mol of NH_3, so the change in [NH_3] is 2Δ and that of $Ag(NH_3)_2^+$ is $-Δ$. In summary,

	Ag^+	+	$2NH_3$	$\rightleftharpoons$	$Ag(NH_3)_2^+$
Initial Concentration, M	0		0		0.10
Change	Δ		2Δ		−Δ
Equilibrium Concentration, M	0 + Δ		0 + 2Δ		0.10 − Δ

Step 3.   *Solve for Δ and the equilibrium concentrations.*

At equilibrium,

$$K_f = \frac{[Ag(NH_3)_2^+]}{[Ag^+][NH_3]^2}$$

$$1.6 \times 10^7 = \frac{(0.10 - \Delta)}{(\Delta)(2\Delta)^2}$$

Both Q and K_f are much larger than 1, so let us assume that the changes in concentrations needed to reach equilibrium are small. Thus $0.10 - \Delta$ is approximated as 0.10.

$$1.6 \times 10^7 = \frac{(0.10)}{(\Delta)(2\Delta)^2}$$

$$\Delta^3 = \frac{0.10}{4(1.6 \times 10^7)} = 1.6 \times 10^{-9}$$

$$\Delta = (1.6 \times 10^{-9})^{1/3} = 1.2 \times 10^{-3}$$

Because only 1.2% of the $Ag(NH_3)_2^+$ dissociates to Ag^+ and NH_3, the assumption that Δ is small is justified.

Now we determine the equilibrium concentrations.

$$[Ag^+] = 0 + \Delta = 1.2 \times 10^{-3} \text{ M}$$

$$[NH_3] = 0 + 2\Delta = 2.4 \times 10^{-3} \text{ M}$$

$$[Ag(NH_3)_2^+] = 0.10 - \Delta = 0.10 - 0.0012 = 0.10$$

The concentration of uncomplexed silver ion in the solution is 0.0012 M.

Step 4.   *Check the work.*

The value of Q calculated by using the equilibrium concentrations is equal to K_f within the error associated with the significant figures in the calculation.

Check your learning: Calculate the silver ion concentration, $[Ag^+]$, of a solution prepared by dissolving 1.00 g of $AgNO_3$ and 10.0 g of KCN in enough water to make 1.00 L of solution. *Hint:* Because $Q < 1$ and $K_f > 1$, assume the reaction goes to completion and then calculate the $[Ag^+]$ produced by dissociation of the complex.

Answer: 3×10^{-21} M

Example 17.14   **Multiple Equilibria**

Unexposed silver halides are removed from photographic film (Fig. 17.13) when they react with sodium thiosulfate ($Na_2S_2O_3$, called hypo) to form the complex ion $Ag(S_2O_3)_2^{3-}$ ($K_f = 4.7 \times 10^{13}$). The reaction with silver bromide is

$$AgBr \rightleftharpoons Ag^+ + Br^-$$

What mass of $Na_2S_2O_3$ is needed to prepare 1.00 L of a solution that dissolves 1.00 g of AgBr by the formation of $Ag(S_2O_3)_2^{3-}$?

Solution: Two equilibria are involved when AgBr dissolves in a solution that contains the $S_2O_3^{2-}$ ion.

$$AgBr(s) \rightleftharpoons Ag^+(aq) + Br^-(aq) \qquad K_{sp} = 3.3 \times 10^{-13} \quad (7)$$
$$Ag^+(aq) + 2S_2O_3^{2-}(aq) \rightleftharpoons Ag(S_2O_3)_2^{3-}(aq) \qquad K_f = 4.7 \times 10^{13} \quad (8)$$

Figure 17.13

An exposed and developed negative. The transparent or light gray portions of the film were not exposed to light or were minimally exposed. During the developing process, the silver halide in these portions reacts with the hypo (sodium thiosulfate) and forms the soluble $Ag(S_2O_3)_2^{3-}$ ion. The darker portions of the film were exposed to light, reducing the metal halide to metallic silver (gray or black, depending on the degree of exposure). The metallic silver is not removed by the developing process.

In order for 1.00 g of AgBr to dissolve, the $[Ag^+]$ in the solution that results must be low enough for Q for Reaction 7 to be smaller than K_{sp} for this reaction. We reduce $[Ag^+]$ by adding $S_2O_3^{2-}$ and thus cause Reaction 8 to shift to the right. We need the following steps to determine what mass of $Na_2S_2O_3$ is needed to provide the necessary $S_2O_3^{2-}$.

Step 1.   We calculate the $[Br^-]$ produced by the complete dissolution of 1.00 g of AgBr (5.33×10^{-3} mol AgBr) in 1.00 L of solution. Result:

$$[Br^-] = 5.33 \times 10^{-3} \text{ M}$$

Step 2.   We use $[Br^-]$ and K_{sp} to determine the maximum possible concentration of Ag^+ that can be present without causing reprecipitation of AgBr. Result:

$$[Ag^+] = 6.2 \times 10^{-11} \text{ M}$$

Step 3.   We determine the $[S_2O_3^{2-}]$ required to make $[Ag^+] = 6.2 \times 10^{-11}$ M after the remaining Ag^+ ion has reacted with $S_2O_3^{2-}$ according to the equation

$$Ag^+ + 2S_2O_3^{2-} \rightleftharpoons Ag(S_2O_3)_2^{3-} \qquad K_f = 4.7 \times 10^{13}$$

Because 5.33×10^{-3} mol of AgBr dissolves,

$$(5.33 \times 10^{-3}) - (6.2 \times 10^{-11}) = 5.33 \times 10^{-3} \text{ mol } Ag(S_2O_3)_2^{3-}$$

must form. Thus at equilibrium, $[Ag(S_2O_3)_2^{3-}] = 5.33 \times 10^{-3}$ M, $[Ag^+] = 6.2 \times 10^{-11}$ M, and $Q = K_f = 4.7 \times 10^{13}$.

$$K_f = \frac{[Ag(S_2O_3)_2^{3-}]}{[Ag^+][S_2O_3^{2-}]^2} = 4.7 \times 10^{13}$$

$$[S_2O_3^{2-}] = 1.4 \times 10^{-3} \text{ M}$$

When $[S_2O_3^{2-}]$ is 1.4×10^{-3} M, $[Ag^+]$ is 6.2×10^{-11} M and all AgBr remains dissolved.

Step 4.   We determine the total number of moles of $S_2O_3^{2-}$ that must be added to the solution. This equals the amount that reacts with Ag^+ to form $Ag(S_2O_3)_2^{3-}$ plus the amount of free $S_2O_3^{2-}$ in solution at equilibrium. To form 5.33×10^{-3} mol of $Ag(S_2O_3)_2^{3-}$ requires $2 \times (5.33 \times 10^{-3})$ mol of $S_2O_3^{2-}$. In addition, 1.4×10^{-3} mole of unreacted $S_2O_3^{2-}$ is present (Step 3). Thus the total amount of $S_2O_3^{2-}$ that must be added is

$$2 \times (5.33 \times 10^{-3} \text{ mol } S_2O_3^{2-}) + 1.4 \times 10^{-3} \text{ mol } S_2O_3^{2-}$$
$$= 1.21 \times 10^{-2} \text{ mol } S_2O_3^{2-}$$

Step 5.   We determine the mass of $Na_2S_2O_3$ needed to give 1.21×10^{-2} mol $S_2O_3^{-2}$ by using the molar mass of $Na_2S_2O_3$.

$$1.21 \times 10^{-2} \text{ mol } S_2O_3^{2-} \times \frac{158.1 \text{ g } Na_2S_2O_3}{1 \text{ mol } Na_2S_2O_3} = 1.91 \text{ g } Na_2S_2O_3$$

Thus 1.00 L of a solution prepared from 1.9 g $Na_2S_2O_3$ dissolved 1.0 g of AgBr.

Check your learning: Check all the calculations in this example.

For Review   Summary

The value of the equilibrium constant for an equilibrium that involves the precipitation or dissolution of a slightly soluble ionic solid is called the **solubility product,** K_{sp}, of the solid. When we have a heterogeneous equilibrium involving the slightly soluble solid M_pX_q and its ions M^{m+} and X^{n-},

$$M_pX_q(s) \rightleftharpoons pM^{m+}(aq) + qX^{n-}(aq)$$

we write the solubility product equation as

$$K_{sp} = [M^{m+}]^p[X^{n-}]^q$$

The solubility product of a slightly soluble electrolyte can be calculated from its solubility (in moles per liter); conversely, its solubility can be calculated from its K_{sp}, provided the only significant reaction that occurs when the solid dissolves is the formation of its ions.

A slightly soluble electrolyte begins to precipitate when the magnitude of the reaction quotient for the dissolution reaction exceeds the magnitude of the solubility product. Precipitation continues until the reaction quotient equals the solubility product. Consequently, if we have a solution containing one of the ions of

a slightly soluble electrolyte, we can calculate what amounts of the other ions of the electrolyte must be added to cause precipitation to begin or to reduce the concentration of the first ion to any desired value.

Many slightly soluble ionic solids contain anions of weak acids. These anions are bases, and the solids dissolve in acidic solutions because reactions of the anions with hydronium ion reduce the concentration of the anion to the point where the reaction quotient is less than the solubility product. In some cases, the anion of a slightly soluble solid is a strong enough base that its reaction with water is sufficient to change the concentration of the anion, at equilibrium, to such an extent that the solubility of the solid is much greater than that predicted from the solubility product alone. The solubility of a slightly soluble solid can also be changed by formation of a complex ion, which reduces the concentration of the cation. The stability of a complex ion is described by its **formation constant,** K_f, or its **dissociation constant,** K_d.

Key Terms and Concepts

association constant   (17.9)
complex ion   (17.9)
dissociation constant   (17.9)
dissolution of precipitates   (17.1, 17.8, 17.9)

formation constant   (17.9)
molar solubility   (17.2)
precipitation   (17.4, 17.5)

solubility product   (17.1)
stability constant   (17.9)

Exercises

Questions

1. Which of the following symbols are used to represent the same thing? Which are used to represent different things? Explain your answers.

    (a) The symbol $[H_3O^+]$ in Chapter 13 and the symbol $[H_3O^+]$ in this chapter.
    (b) The symbol K in Chapter 10 and the symbol K in this chapter.
    (c) The symbol k in Chapter 13 and the symbol K in this chapter.
    (d) The symbol K in Chapter 14 and the symbol K in this chapter.

2. A saturated solution of a slightly soluble electrolyte in contact with some of the solid electrolyte is said to be a system

    in equilibrium. Explain. Why is such a system called a heterogeneous equilibrium?

3. Identify the microscopic species present in a solution of $BaSO_4$ and list these species in decreasing order of their concentrations. (*Hint:* Remember that the SO_4^{2-} ion is a weak base.)

4. Identify the microscopic species present in a solution of $Ca_3(PO_4)_2$ and list these species in decreasing order of their concentrations. (*Hint:* Remember that the PO_4^{3-} ion is a weak base.)

5. Give the effect on the amount of solid $Ca(OH)_2$ that dissolves and on the concentrations of Ca^{2+} and OH^- when each of the following is added to a mixture of

solid $Ca(OH)_2$ and water at equilibrium: (a) KOH, (b) $Ca(NO_3)_2$, (c) $HClO_4$, (d) $Ca(OH)_2$, and (e) $NaNO_3$.

6. Give the effect on the amount of $CaHPO_4$ that dissolves and on the concentrations of Ca^{2+} and $HPO_4{}^-$ when each of the following is added to a mixture of solid $CaHPO_4$ and water at equilibrium: (a) $CaCl_2$ (b) HCl, (c) $KClO_4$, (d) NaOH, and (e) $CaHPO_4$.

7. How do the concentrations of Ag^+ and $CrO_4{}^{2-}$ at equilibrium in a liter of water above 1.0 g of solid Ag_2CrO_4 change when 100 g of solid Ag_2CrO_4 is added to the system? Explain.

8. How do the concentrations of Pb^{2+} and S^{2-} change when K_2S is added to a saturated solution of PbS?

9. Under what circumstances, if any, does a sample of solid AgCl completely dissolve in pure water?

10. Explain what additional information we need to answer the following question: How is the equilibrium of solid silver bromide with a saturated solution of its ions affected when the temperature is raised?

11. Which of the following compounds, when dissolved in a 0.01 M solution of $HClO_4$, has a solubility greater than in pure water: CuCl, $CaCO_3$, MnS, $PbBr_2$, CaF_2? Explain your answer.

12. Which of the following compounds, when dissolved in a 0.01 M solution of $HClO_4$, has a solubility greater than in pure water: AgBr, BaF_2, $Ca_3(PO_4)_3$, ZnS, PbI_2? Explain your answer.

13. Which of the following slightly soluble compounds has a solubility greater than that calculated from its solubility product because the anion reacts with water: $CoSO_3$, CuI, $PbCO_3$, $PbCl_2$, Tl_2S, $KClO_4$?

14. Which of the following slightly soluble compounds has a solubility greater than that calculated from its solubility product because the anion reacts with water: AgCl, $BaSO_4$, CaF_2, Hg_2I_2, $MnCO_3$, ZnS?

15. Explain why adding NH_3 or HNO_3 to a saturated solution of Ag_2CO_3 in contact with solid Ag_2CO_3 increases the solubility of the solid.

16. Explain why adding NH_3 or HNO_3 to a saturated solution of $Cu(OH)_2$ in contact with solid $Cu(OH)_2$ increases the solubility of the solid.

Determination of Solubility Products

17. Write the ionic equation for the dissolution and the equation for the solubility product for each of the following slightly soluble ionic compounds.

(a) $PbCl_2$ (b) Ag_2S
(c) $Sr_3(PO_4)_2$ (d) $SrSO_4$

18. Write the ionic equation for the dissolution and the equation for the solubility product for each of the following slightly soluble ionic compounds.

(a) LaF_3
(b) $CaCO_3$
(c) Ag_2SO_4
(d) $Pb(OH)_2$

19. The following concentrations are found in mixtures of ions in equilibrium with slightly soluble solids. From the concentrations given, calculate K_{sp} for each of the slightly soluble solids indicated.

(a) AgBr: $[Ag^+] = 5.7 \times 10^{-7}$ M; $[Br^-] = 5.7 \times 10^{-7}$ M
(b) $CaCO_3$: $[Ca^{2+}] = 5.3 \times 10^{-3}$ M; $[CO_3{}^{2-}] = 9.0 \times 10^{-7}$ M
(c) PbF_2: $[Pb^{2+}] = 2.1 \times 10^{-3}$ M; $[F^-] = 4.2 \times 10^{-3}$ M
(d) Ag_2CrO_4: $[Ag^+] = 5.3 \times 10^{-5}$ M; $[CrO_4{}^{2-}] = 3.2 \times 10^{-3}$ M
(e) InF_3: $[In^{3+}] = 2.3 \times 10^{-3}$ M; $[F^-] = 7.0 \times 10^{-3}$ M

20. The following concentrations are found in mixtures of ions in equilibrium with slightly soluble solids. From the concentrations given, calculate K_{sp} for each of the slightly soluble solids indicated.

(a) TlCl: $[Tl^+] = 1.4 \times 10^{-2}$ M; $[Cl^-] = 1.4 \times 10^{-2}$ M
(b) $Ce(IO_3)_4$: $[Ce^{4+}] = 1.8 \times 10^{-4}$ M; $[IO_3{}^-] = 7.3 \times 10^{-4}$ M
(c) $Gd_2(SO_4)_3$: $[Gd^{3+}] = 0.132$ M; $[SO_4{}^{2-}] = 0.198$ M
(d) Ag_2SO_4: $[Ag^+] = 2.40 \times 10^{-2}$ M; $[SO_4{}^{2-}] = 2.05 \times 10^{-2}$ M
(e) $BaSO_4$: $[Ba^{2+}] = 0.500$ M; $[SO_4{}^{2-}] = 2.16 \times 10^{-10}$ M

21. The *Handbook of Chemistry and Physics* gives solubilities of the following compounds in grams per 100 mL of water. Because these compounds are only slightly soluble, assume that the volume does not change on dissolution, and calculate the solubility product for each.

(a) $BaSiF_6$, 0.026 g/100 mL (contains $SiF_6{}^{2-}$ ions)
(b) $Ce(IO_3)_4$, 1.5×10^{-2} g/100 mL
(c) $Gd_2(SO_4)_3$, 3.98 g/100 mL
(d) $(NH_4)_2PtBr_6$, 0.59 g/100 mL (contains $PtBr_6{}^{2-}$ ions)

22. The *Handbook of Chemistry and Physics* gives solubilities of the following compounds in grams per 100 mL of water. Because these compounds are only slightly soluble, assume that the volume does not change on dissolution, and calculate the solubility product for each.

(a) $BaSeO_4$, 0.0118 g/100 mL
(b) $Ba(BrO_3)_2 \cdot H_2O$, 0.30 g/100 mL
(c) $NH_4MgAsO_4 \cdot 6H_2O$, 0.38 g/100 mL
(d) $La_2(MoO_4)_3$, 0.00179 g/100 mL

Dissolution of Solids

23. Complete the changes in concentrations for each of the following reactions.

    (a) $AgI(s) \rightleftharpoons Ag^+(aq) + I^-(aq)$
        Δ _____

    (b) $CaCO_3(s) \rightleftharpoons Ca^{2+}(aq) + CO_3^{2-}(aq)$
        _____ Δ

    (c) $Mg(OH)_2(s) \rightleftharpoons Mg^{2+}(aq) + 2OH^-(aq)$
        Δ _____

    (d) $Mg_3(PO_4)_2(s) \rightleftharpoons 3Mg^{2+}(aq) + 2PO_4^{3-}(aq)$
        _____ 2Δ

    (e) $Ca_5(PO_4)_3OH(s) \rightleftharpoons$
        $5Ca^{2+}(aq) + 3PO_4^{3-}(aq) + OH^-(aq)$
        _____   _____ Δ

24. Complete the changes in concentrations for each of the following reactions.

    (a) $BaSO_4(s) \rightleftharpoons Ba^{2+}(aq) + SO_4^{2-}(aq)$
        Δ _____

    (b) $Ag_2SO_4(s) \rightleftharpoons 2Ag^+(aq) + SO_4^{2-}(aq)$
        _____ Δ

    (c) $Al(OH)_3(s) \rightleftharpoons Al^{3+}(aq) + 3OH^-(aq)$
        Δ _____

    (d) $Pb(OH)Cl(s) \rightleftharpoons Pb^{2+}(aq) + OH^-(aq) + Cl^-(aq)$
        _____ Δ _____

    (e) $Ca_3(AsO_4)_2(s) \rightleftharpoons 3Ca^{2+}(aq) + 2AsO_4^{3-}(aq)$
        3Δ _____

25. Assuming that no equilibria other than dissolution are involved, calculate the concentrations of ions in a saturated solution of each of the following (see Appendix D for solubility products).

    (a) AgI
    (b) Ag_2SO_4
    (c) $Mn(OH)_2$
    (d) $Sr(OH)_2 \cdot 8H_2O$
    (e) The mineral brucite, $Mg(OH)_2$

26. Assuming that no equilibria other than dissolution are involved, calculate the concentrations of ions in a saturated solution of each of the following (see Appendix D for solubility products).

    (a) $TlCl$
    (b) BaF_2
    (c) Ag_2CrO_4
    (d) $CaC_2O_4 \cdot H_2O$
    (e) The mineral anglesite, $PbSO_4$

27. Use solubility products and predict which of the following salts is the most soluble, in terms of moles per liter, in pure water: $AgBr$, CuI, HgS, or $SrSO_4$.

28. Use solubility products and predict which of the following salts is the most soluble, in terms of moles per liter, in pure water: CaF_2, Hg_2Cl_2, PbI_2, or $Sn(OH)_2$.

29. Assuming that no equilibria other than dissolution are involved, calculate the molar solubility of each of the following from its solubility product.

    (a) $KHC_4H_4O_6$, a salt containing the $HC_4H_4O_6^-$ ion
    (b) PbI_2
    (c) $Ag_4[Fe(CN)_6]$, a salt containing the $Fe(CN)_6^{4-}$ ion
    (d) Hg_2I_2

30. Assuming that no equilibria other than dissolution are involved, calculate the molar solubility of each of the following from its solubility product.

    (a) Ag_2SO_4
    (b) $PbBr_2$
    (c) AgI
    (d) $CaC_2O_4 \cdot H_2O$

31. Assuming that no equilibria other than dissolution are involved, calculate the concentration of all solute species in each of the following solutions of salts in contact with a solution containing a common ion. Show that changes in the initial concentrations of the common ions can be neglected.

    (a) $AgCl(s)$ in 0.025 M NaCl
    (b) $CaF_2(s)$ in 0.00133 M KF
    (c) $Ag_2SO_4(s)$ in 0.500 L of a solution containing 19.50 g of K_2SO_4
    (d) $Zn(OH)_2(s)$ in a solution buffered at a pH of 11.45

32. Assuming that no equilibria other than dissolution are involved, calculate the concentration of all solute species in each of the following solutions of salts in contact with a solution containing a common ion. Show that changes in the initial concentrations of the common ions can be neglected.

    (a) $TlCl(s)$ in 1.250 M HCl
    (b) $PbI_2(s)$ in 0.0355 M CaI_2
    (c) $Ag_2CrO_4(s)$ in 0.225 L of a solution containing 0.856 g of K_2CrO_4
    (d) $Cd(OH)_2(s)$ in a solution buffered at a pH of 10.995

33. Assuming that no equilibria other than dissolution are involved, calculate the concentration of all solute species in each of the following solutions of salts in contact with a solution containing a common ion. Show that it is not appropriate to neglect the changes in the initial concentrations of the common ions.

    (a) $TlCl(s)$ in 0.025 M $TlNO_3$
    (b) $BaF_2(s)$ in 0.0313 M KF
    (c) MgC_2O_4 in 2.250 L of a solution containing 8.156 g of $Mg(NO_3)_2$
    (d) $Ca(OH)_2(s)$ in an unbuffered solution initially with a pH of 12.700

34. Assuming that no equilibria other than dissolution are involved, calculate the concentration of all solute species in each of the following solutions of salts in contact with a solution containing a common ion. Show that it is not appropriate to neglect the changes in the initial concentrations of the common ions.

    (a) $Hg_2SO_4(s)$ in 9.94×10^{-4} M Na_2SO_4
    (b) $PbCl_2(s)$ in 0.0900 M $Pb(NO_3)_2$
    (c) $Ag_2SO_4(s)$ in 0.500 L of a solution containing 1.50 g of K_2SO_4
    (d) $Ni(OH)_2(s)$ in an unbuffered solution initially with a pH of 9.45

35. Explain why the changes in concentrations of the common ions in Exercise 31 can be neglected.

36. Explain why the changes in concentrations of the common ions in Exercise 33 cannot be neglected.

37. Calculate the solubility of aluminum hydroxide, $Al(OH)_3$, in a solution buffered at pH 11.00.

38. Refer to Appendix D for solubility products for calcium salts. Determine which of the calcium salts listed is most soluble in moles per liter and which is most soluble in grams per liter.

39. Most barium compounds are very poisonous; however, barium sulfate is often administered internally as an aid in the X-ray examination of the lower intestinal tract. This use of $BaSO_4$ is possible because of its insolubility. Calculate the molar solubility of $BaSO_4$ and the mass of barium present in 1.00 L of water saturated with $BaSO_4$.

40. Public Health Service standards for drinking water set a maximum of 250 mg L^{-1} of SO_4^{2-} ($[SO_4^{2-}] = 2.60 \times 10^{-3}$ M) because of its cathartic action (it is a laxative). Does natural water that is saturated with $CaSO_4$ ("gyp" water) as a result of passing through soil containing gypsum, $CaSO_4 \cdot 2H_2O$, meet these standards? What is $[SO_4^{2-}]$ in such water?

41. (a) Calculate $[Ag^+]$ in a saturated aqueous solution of AgBr.
    (b) What will $[Ag^+]$ be when enough KBr has been added to make $[Br^-] = 0.050$ M?
    (c) What will $[Br^-]$ be when enough $AgNO_3$ has been added to make $[Ag^+] = 0.020$ M?

42. The solubility product of $CaSO_4 \cdot 2H_2O$ is 2.4×10^{-5}. What mass of this salt will dissolve in 1.0 L of 0.010 M SO_4^{2-}?

Precipitation

43. Which of the following compounds precipitates from a solution that has the concentrations indicated? (See Appendix D for K_{sp} values.)

    (a) $KClO_4$: $[K^+] = 0.01$ M; $[ClO_4^-] = 0.01$ M
    (b) K_2PtCl_6: $[K^+] = 0.01$ M; $[PtCl_6^{2-}] = 0.01$ M

    (c) PbI_2: $[Pb^{2+}] = 0.003$ M; $[I^-] = 1.3 \times 10^{-3}$ M
    (d) Ag_2S: $[Ag^+] = 1 \times 10^{-10}$ M; $[S^{2-}] = 1 \times 10^{-13}$ M

44. Which of the following compounds precipitates from a solution that has the concentrations indicated? (See Appendix D for K_{sp} values.)

    (a) $CaCO_3$: $[Ca^{2+}] = 0.003$ M; $[CO_3^{2-}] = 0.003$ M
    (b) $Co(OH)_2$: $[Co^{2+}] = 0.01$ M; $[OH^-] = 1 \times 10^{-7}$ M
    (c) $CaHPO_4$: $[Ca^{2+}] = 0.01$ M; $[HPO_4^{2-}] = 2 \times 10^{-6}$ M
    (d) $Pb_3(PO_4)_2$: $[Pb^{2+}] = 0.01$ M; $[PO_4^{3-}] = 1 \times 10^{-13}$ M

45. Calculate the concentration of Tl^+ when TlCl just begins to precipitate from a solution that is 0.0250 M in Cl^-.

46. Calculate the concentration of sulfate ion when $BaSO_4$ just begins to precipitate from a solution that is 0.0758 M in Ba^{2+}.

47. Calculate the concentration of Sr^{2+} when SrF_2 ($K_{sp} = 3.7 \times 10^{-12}$) starts to precipitate from a solution that is 0.0025 M in F^-.

48. Calculate the concentration of PO_4^{3-} when Ag_3PO_4 starts to precipitate from a solution that is 0.0125 M in Ag^+.

49. Calculate the concentration of F^- required to begin precipitation of CaF_2 from a solution that is 0.010 M in Ca^{2+}.

50. Calculate the concentration of Ag^+ required to begin precipitation of Ag_2CO_3 from a solution that is 2.50×10^{-6} M in CO_3^{2-}.

51. What $[Ag^+]$ is required to reduce $[CO_3^{2-}]$ to 8.2×10^{-4} M by precipitation of Ag_2CO_3?

52. What $[F^-]$ is required to reduce $[Ca^{2+}]$ to 1.0×10^{-4} M by precipitation of CaF_2?

53. A volume of 0.800 L of a 2×10^{-4} M $Ba(NO_3)_2$ solution is added to 0.200 L of 5×10^{-4} M Li_2SO_4. Does $BaSO_4$ precipitate? Explain your answer.

54. (a) With what volume of water must a precipitate containing $NiCO_3$ be washed to dissolve 0.100 g of this compound? Assume that the wash water becomes saturated with $NiCO_3$ ($K_{sp} = 1.36 \times 10^{-7}$).
    (b) If the $NiCO_3$ were a contaminant in a sample of $CoCO_3$ ($K_{sp} = 1.0 \times 10^{-12}$), what mass of $CoCO_3$ would have been lost? Keep in mind that both $NiCO_3$ and $CoCO_3$ dissolve in the same solution.

55. Iron concentrations greater than 5.4×10^{-6} M in water used for laundry purposes can cause staining. What $[OH^-]$ is required to reduce $[Fe^{2+}]$ to this level by precipitation of $Fe(OH)_2$?

56. A solution is 0.010 M in both Cu^{2+} and Cd^{2+}. What percentage of Cd^{2+} remains in the solution when 99.9% of the Cu^{2+} has been precipitated as CuS by adding sulfide?

57. A solution is 0.15 M in both Pb^{2+} and Ag^+. If Cl^- is added to this solution, what is $[Ag^+]$ when $PbCl_2$ begins to precipitate?

58. What reagent might be used to separate the ions in each of the following mixtures, which are 0.1 M with respect to each ion? In some cases, it may be necessary to control the pH. (*Hint*: Consider the K_{sp} values given in Appendix D.)

(a) Hg_2^{2+} and Cu^{2+} (b) SO_4^{2-} and Cl^-
(c) Hg^{2+} and Co^{2+} (d) Zn^{2+} and Sr^{2+}
(e) Ba^{2+} and Mg^{2+} (f) CO_3^{2-} and OH^-

59. A solution contains 1.0×10^{-5} mol of KBr and 0.10 mol of KCl per liter. $AgNO_3$ is gradually added to this solution. Which forms first, solid AgBr or solid AgCl?

60. A solution contains 1.0×10^{-2} mol of KI and 0.10 mol of KCl per liter. $AgNO_3$ is gradually added to this solution. Which forms first, solid AgI or solid AgCl?

Complex Ion Equilibria

61. Calculate the equilibrium concentration of Ni^{2+} in a 1.0 M solution of $[Ni(NH_3)_6](NO_3)_2$.

62. Calculate the equilibrium concentration of Zn^{2+} in a 0.30 M solution of $Zn(CN)_4^{2-}$.

63. Calculate the equilibrium concentration of Cu^{2+} in a solution initially with 0.050 M Cu^{2+} and 1.00 M NH_3.

64. Calculate the equilibrium concentration of Zn^{2+} in a solution initially with 0.150 M Zn^{2+} and 2.50 M CN^-.

65. Calculate the Fe^{3+} equilibrium concentration when 0.0888 mol of $K_3[Fe(CN)_6]$ is added to a solution with 0.00010 M CN^-.

66. Calculate the Co^{2+} equilibrium concentration when 0.100 mol of $[Co(NH_3)_6](NO_3)_2$ is added to a solution with 0.025 M NH_3.

67. The equilibrium constant for the reaction

$$Hg^{2+}(aq) + 2Cl^-(aq) \rightleftharpoons HgCl_2(aq)$$

is 1.6×10^{13}. Is $HgCl_2$ a strong electrolyte or a weak electrolyte? What are the concentrations of Hg^{2+} and Cl^- in a 0.015 M solution of $HgCl_2$?

68. Calculate the cadmium ion concentration, $[Cd^{2+}]$, in a solution prepared by mixing 0.100 L of 0.0100 M $Cd(NO_3)_2$ with 1.50 L of 0.100 $NH_3(aq)$.

69. Sometimes equilibria for complex ions are described in terms of dissociation constants, K_d. For the complex ion AlF_6^{3-} the dissociation reaction is

$$AlF_6^{3-} \rightleftharpoons Al^{3+} + 6F^-$$

and $K_d = \dfrac{[Al^{3+}][F^-]^6}{[AlF_6^{3-}]} = 2 \times 10^{-24}$

Calculate the value of the formation constant, K_f, for AlF_6^{3-}.

70. Using the value of the formation constant for the complex ion $Co(NH_3)_6^{2+}$, calculate the dissociation constant.

71. Using the dissociation constant $K_d = 7.8 \times 10^{-18}$, calculate the equilibrium concentrations of Cd^{2+} and CN^- in a 0.250 M solution of $Cd(CN)_4^{2-}$.

72. Using the dissociation constant $K_d = 3.4 \times 10^{-15}$, calculate the equilibrium concentrations of Zn^{2+} and OH^- in a 0.0465 M solution of $Zn(OH)_4^{2-}$.

73. Using the dissociation constant $K_d = 2.2 \times 10^{-34}$, calculate the equilibrium concentrations of Co^{3+} and NH_3 in a 0.500 M solution of $Co(NH_3)_6^{3+}$.

74. Using the dissociation constant $K_d = 1 \times 10^{-44}$, calculate the equilibrium concentrations of Fe^{3+} and CN^- in a 0.333 M solution of $Fe(CN)_6^{3-}$.

Multiple Equilibria

75. Calculate the molar solubility of $Sn(OH)_2$ in a buffer solution containing equal concentrations of NH_3 and NH_4^+.

76. Calculate the molar solubility of $Al(OH)_3$ in a buffer solution with 0.100 M NH_3 and 0.400 M NH_4^+.

77. What is the molar solubility of CaF_2 in a 0.100 M solution of HF? K_a for HF is 7.2×10^{-4}.

78. What is the molar solubility of $BaSO_4$ in a 0.250 M solution of $NaHSO_4$? K_a for HSO_4^- is 1.2×10^{-2}.

79. What is the molar solubility of $Tl(OH)_3$ in a 0.10 M solution of NH_3?

80. What is the molar solubility of $Pb(OH)_2$ in a 0.138 M solution of CH_3NH_2?

81. A solution of 0.075 M $CoBr_2$ is saturated with H_2S ($[H_2S]$ = 0.10 M). What is the minimum pH at which CoS begins to precipitate?

$$CoS(s) \rightleftharpoons Co^{2+}(aq) + S^{2-}(aq) \qquad K_{sp} = 4.5 \times 10^{-27}$$
$$H_2S(aq) + 2H_2O(l) \rightleftharpoons 2H_3O^+(aq) + S^{2-}(aq)$$
$$K = 1.0 \times 10^{-26}$$

82. A 0.125 M solution of $Mn(NO_3)_2$ is saturated with H_2S ($[H_2S]$ = 0.10 M). At what pH does MnS begin to precipitate?

$$MnS(s) \rightleftharpoons Mn^{2+}(aq) + S^{2-}(aq) \qquad K_{sp} = 4.3 \times 10^{-22}$$
$$H_2S(aq) + 2H_2O(l) \rightleftharpoons 2H_3O^+(aq) + S^{2-}(aq)$$
$$K = 1.0 \times 10^{-26}$$

83. Calculate the molar solubility of BaF_2 in a buffer solution containing 0.20 M HF and 0.20 M NaF.

84. Calculate the molar solubility of $CdCO_3$ in a buffer solution containing 0.115 M Na_2CO_3 and 0.120 M $NaHCO_3$.

85. To a 0.10 M solution of $Pb(NO_3)_2$ enough HF(g) is added to make $[HF]$ = 0.10 M.

(a) Does PbF_2 precipitate from this solution? Show the calculations that support your conclusion.

(b) What is the minimum pH at which PbF_2 precipitates?

86. Calculate the concentration of Cd^{2+} resulting from the dissolution of $CdCO_3$ in a solution that is 0.250 M in CH_3CO_2H, 0.375 M in $NaCH_3CO_2$, and 0.010 M in H_2CO_3.

87. Both AgCl and AgI dissolve in NH_3.

    (a) What mass of AgI dissolves in 1.0 L of 1.0 M NH_3?
    (b) What mass of AgCl dissolves in 1.0 L of 1.0 M NH_3?

88. A volume of 50 mL of 1.8 M NH_3 is mixed with an equal volume of a solution containing 0.95 g of $MgCl_2$. What mass of NH_4Cl must be added to the resulting solution to prevent the precipitation of $Mg(OH)_2$?

89. Calculate the mass of potassium cyanide ion that must be added to 100 mL of solution to dissolve 2.0×10^{-2} mol of silver cyanide, AgCN.

90. Calculate the minimum concentration of ammonia needed in 1.0 L of solution to dissolve 3.0×10^{-3} mol of silver bromide.

91. A roll of 35-mm black and white film contains about 0.27 g of unexposed AgBr before developing. What mass of $Na_2S_2O_3 \cdot 5H_2O$ (hypo) in 1.0 L of developer is required to dissolve the AgBr as $Ag(S_2O_3)_2{}^{3-}$? ($K_f = 4.7 \times 10^{13}$)

92. Calculate the volume of 1.50 M CH_3CO_2H required to dissolve a precipitate that consists of 350 mg each of $CaCO_3$, $SrCO_3$, and $BaCO_3$.

93. The maximum allowable chloride ion concentration in drinking water is 0.25 g/L. A commercial kit for the analysis of chloride ion in water contains a solution of K_2CrO_4 as an indicator and a standard solution of $AgNO_3$, which is used as a titrant. As the $AgNO_3$ solution is added to the water sample drop by drop, insoluble white AgCl is formed. After "all" of the chloride ion has been precipitated, the next drop of $AgNO_3$ solution reacts with the K_2CrO_4 to form an orange-colored precipitate of Ag_2CrO_4, which indicates the end of the titration. What percentage of the initial chloride ion content remains unprecipitated when the first trace of Ag_2CrO_4 forms in a solution for which initial $[Cl^-] = 7.1 \times 10^{-3}$ M and $[CrO_4{}^{2-}] = 1.0 \times 10^{-4}$ M? Assume that the titration does not change the volume of the solution.

Applications and Additional Exercises

94. The calcium ions in human blood serum are necessary for coagulation. Potassium oxalate, $K_2C_2O_4$, is used as an anticoagulant when a blood sample is drawn for laboratory tests because it removes the calcium as a precipitate of $CaC_2O_4 \cdot H_2O$. It is necessary to remove all but 1.0% of the Ca^{2+} in serum in order to prevent coagulation. If normal blood serum with a buffered pH of 7.40 contains 9.5 mg of Ca^{2+} per 100 mL of serum, what mass of $K_2C_2O_4$ is required to prevent the coagulation of a 10-mL blood sample that is 55% serum by volume? [All volumes are accurate to

two significant figures. Note that the volume of fluid (serum) in a 10-mL blood sample is 5.5 mL. Assume that the K_{sp} value for CaC_2O_4 in serum is the same as in water.]

95. Even though $Ca(OH)_2$ is an inexpensive base, its limited solubility restricts its use. What is the pH of a saturated solution of $Ca(OH)_2$?

96. About 50% of urinary calculi (kidney stones) consist of calcium phosphate, $Ca_3(PO_4)_2$. The normal mid-range calcium content excreted in the urine is 0.10 g of Ca^{2+} per day. The normal mid-range amount of urine passed may be taken as 1.4 L per day. What is the maximum concentration of phosphate ion that urine can contain before a calculus begins to form?

97. The pH of normal urine is 6.30, and the total phosphate concentration ($[PO_4{}^{3-}] + [HPO_4{}^{2-}] + [H_2PO_4{}^-] + [H_3PO_4]$) is 0.020 M. What is the minimum concentration of Ca^{2+} necessary to induce calculus formation? (See Exercise 96 for additional information.)

98. Calculate $[HgCl_4{}^{2-}]$ in a solution prepared by adding 0.0200 mol of NaCl to 0.250 L of a 0.100 M $HgCl_2$ solution.

99. Magnesium metal (a component of alloys used in aircraft and a reducing agent used in the production of uranium, titanium, and other active metals) is isolated from sea water by the following sequence of reactions.

$$Mg^{2+}(aq) + Ca(OH)_2(aq) \longrightarrow Mg(OH)_2(s) + Ca^{2+}(aq)$$

$$Mg(OH)_2(s) + 2HCl(aq) \longrightarrow MgCl_2(s) + 2H_2O(l)$$

$$MgCl_2(l) \xrightarrow{\text{Electrolysis}} Mg(s) + Cl_2(g)$$

Sea water has a density of 1.026 g/cm^3 and contains 1272 parts per million of magnesium as $Mg^{2+}(aq)$ by mass. What mass, in kilograms, of $Ca(OH)_2$ is required to precipitate 99.9% of the magnesium in 1.00×10^3 L of sea water?

100. In a titration of cyanide ion, 28.72 mL of 0.0100 M $AgNO_3$ is added before precipitation begins. (The reaction of Ag^+ with CN^- goes to completion, producing the $Ag(CN)_2{}^-$ complex. Precipitation of solid AgCN takes place when excess Ag^+, above the amount needed to complete the formation of $Ag(CN)_2{}^-$ is added to the solution.) How many grams of NaCN were in the original sample?

101. Use the data from Figure 12.17 and determine whether the solubility product for $Ce_2(SO_4)_3 \cdot 9H_2O$ increases, decreases, or stays the same as the temperature increases from 10°C to 20°C.

102. A 0.010-mol sample of solid AgCN is rendered soluble in 1 L of solution by adding just enough cyanide ion to form $[Ag(CN)_2]^-$. When all of the solid silver cyanide has just dissolved, the concentration of free cyanide ion is 1.125×10^{-7} M. Neglecting hydrolysis of the cyanide ion, determine the concentration of free, uncomplexed silver ion in the solution. If more cyanide ion is added (without changing the volume) until the equilibrium concentration of free

cyanide ion is 1.0×10^{-6} M, what will be the equilibrium concentration of free silver ion?

103. What are the concentrations of Ag^+, CN^-, and $Ag(CN)_2^-$ in a saturated solution of AgCN?

104. Hydrogen sulfide is bubbled into a solution that is 0.10 M in both Pb^{2+} and Fe^{2+} and 0.30 M in HCl. After the solution has come to equilibrium, it is saturated with H_2S ($[H_2S] = 0.10$ M). What concentrations of Pb^{2+} and Fe^{2+} remain in the solution? For a saturated solution of H_2S, we can use the equilibrium

$$H_2S(aq) + 2H_2O(l) \rightleftharpoons 2H_3O^+(aq) + S^{2-}(aq)$$
$$K = 1.0 \times 10^{-26}$$

(*Hint:* $[H_3O^+]$ changes as metal sulfides precipitate.)

105. What mass of NaCN must be added to 1 L of 0.010 M $Mg(NO_3)_2$ in order to produce the first trace of $Mg(OH)_2$?

106. (a) Several salts of iodic acid are only slightly soluble. The iodate concentrations of saturated solutions of the salts may be determined by an oxidation–reduction titration. In a typical investigation, the iodate ion concentration of a saturated solution of $La(IO_3)_3$ was found to be 3.1×10^{-3} mol/L. Calculate K_{sp}.
    (b) Calculate the concentration of iodate ions in a saturated solution of $Cu(IO_3)_2$. ($K_{sp} = 7.4 \times 10^{-8}$)

107. Magnesium hydroxide and magnesium citrate function as mild laxatives when they reach the small intestine. Why do magnesium hydroxide and magnesium citrate, two very different substances, have the same effect in your small in-

testine? (*Hint:* The contents of the small intestine are basic.)

108. The following question is taken from the 1994 Chemistry Advanced Placement Examination and is used with the permission of the Educational Testing Service.

Solve the following problem.

$$MgF_2(s) \rightleftharpoons Mg^{2+}(aq) + 2F^-(aq)$$

In a saturated solution of MgF_2 at 18°C, the concentration of Mg^{2+} is 1.21×10^{-3} molar. The equilibrium is represented by the equation above.

(a) Write the expression for the solubility-product constant, K_{sp}, and calculate its value at 18°C.
(b) Calculate the equilibrium concentration of Mg^{2+} in 1.000 liter of saturated MgF_2 solution at 18°C to which 0.100 mole of solid KF has been added. The KF dissolves completely. Assume the volume change is negligible.
(c) Predict whether a precipitate of MgF_2 will form when 100.0 milliliters of a 3.00×10^{-3}-molar $Mg(NO_3)_2$ solution is mixed with 200.0 milliliters of a 2.00×10^{-3}-molar NaF solution at 18°C. Calculations to support your prediction must be shown.
(d) At 27°C the concentration of Mg^{2+} in a saturated solution of MgF_2 is 1.17×10^{-3} molar. Is the dissolving of MgF_2 in water an endothermic or an exothermic process? Give an explanation to support your conclusion.

CHAPTER OUTLINE

Work, Heat, and Changes in Internal Energy

18.1   Systems and Surroundings
18.2   Internal Energy, Heat, and Work
18.3   The First Law of Thermodynamics
18.4   State Functions

Enthalpy, Entropy, and Spontaneous Processes

18.5   Spontaneous Chemical Reactions
18.6   Minimization of Energy, Increase in Disorder, and Spontaneous Change
18.7   Changes in Energy: Enthalpy Changes
18.8   Changes in Disorder: Entropy and Entropy Changes
18.9   Calculation of Entropy Changes
18.10 The Third Law of Thermodynamics

Free Energy Changes

18.11 Free Energy Changes, Enthalpy Changes, and Entropy Changes
18.12 Free Energy of Formation
18.13 The Second Law of Thermodynamics
18.14 Free Energy Changes and Nonstandard States
18.15 The Relationship Between Free Energy Changes and Equilibrium Constants

18

Chemical Thermodynamics

Chemical reactions produce work and heat in the body.

In several of the preceding chapters we have seen that chemical processes are accompanied either by absorption or by evolution of energy. For example, removing the electron from a hydrogen atom, breaking a carbon–hydrogen bond, evaporating water, melting ice, and forming calcium oxide and carbon dioxide from calcium carbonate (Fig. 18.1) all proceed with the absorption of energy. On the other hand, burning of carbon (Fig. 18.2), digestion of food, reaction of hydrogen and oxygen, condensation of water vapor to a solid (frost) or a liquid (rain), reaction of hydrochloric acid with sodium hydroxide, and reaction of sodium with water all proceed with the evolution of energy.

When we want to understand the interplay of an energy transformation and the accompanying chemical or physical changes, we employ **chemical thermodynamics,** the topic of this chapter. In it we will see examples of the types of questions that chemical thermodynamics can answer. However, thermodynamics is an extensive subject and a far larger topic than we can cover in one chapter. Thus we will apply many simplifying assumptions and constraints on the conditions under which a chemical or physical change occurs. Under different conditions, the relations that we discuss in this chapter may not hold, or analyzing them may require more sophisticated mathematics than we will employ in our study.

Work, Heat, and Changes in Internal Energy

The changes in energy that accompany chemical reactions are of great importance, both in everyday situations and in the study of the behavior of matter. We use energy produced by chemical reactions to heat our homes, to power our automobiles, and to move our bodies, for example. We also use energy produced by chemical reactions to measure bond energies and lattice energies.

We can measure the energy change that accompanies a chemical reaction by determining the amount of work and heat that the reaction produces. For example, if we are interested in the energy changes that accompany the combustion of carbon, we can start by considering a mixture that consists of 1 mole of carbon and 2 moles of oxygen molecules. As the components of the mixture react and form carbon dioxide, we can determine the amount of heat and work produced under carefully controlled

Figure 18.1

The decomposition of $CaCO_3$ to CaO and CO_2 is an endothermic change; heating is required.

Figure 18.2

Carbon burns in an exothermic reaction; heat is produced.

conditions during the change. The total amount of heat and work produced is equal to the change in energy that accompanies the reaction.

18.1   Systems and Surroundings

We identify the part of the universe that undergoes a change and whose energy content interests us as the **system;** the rest of the universe is called the **surroundings.** Because energy can be neither created nor destroyed (Section 1.3), any gain or loss of energy by a system must be accompanied by an equivalent loss or gain of energy in the surroundings. In fact, the change in energy content of a system is often determined by measuring the energy gained by or withdrawn from the surroundings.

A system may be as complex as a human body or as simple as a mixture consisting of a drop of acid and a drop of base. We will restrict our attention to changes in systems that can be treated as **closed systems**—that is, systems that do not exchange matter with the surroundings during any exchange of energy. Generally we do not need to concern ourselves with the exact details of the surroundings. We can regard the surroundings either as a source of energy to put into our system or as a sink into which we can dump energy from the system.

We will be interested in how the amount of energy in a system changes as a chemical or physical change occurs in the system and heat or work is transferred to (or from) the surroundings. We might be interested, for example, in the amount of glycogen (the carbohydrate that stores energy in our body) that must be used to do the work necessary to swim a mile in cold water and to keep our body warm by replacing the heat lost to the water during the swim. This is an example of one of three types of situations that can arise (Fig. 18.3): (1) The amount of energy in a system

Figure 18.3

The internal energy of a system changes as energy is lost to or gained from the surroundings. If no energy is exchanged, then the internal energy cannot change.

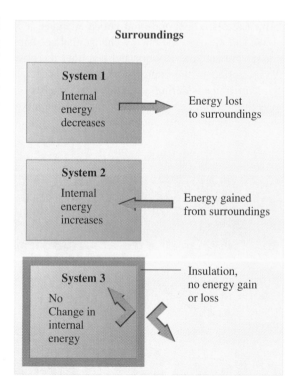

can decrease as energy is transferred from the system to the surroundings. (2) The amount of energy in a system can increase as energy is transferred into the system from the surroundings. (3) The amount of energy in a system may undergo no change because the system is insulated from the surroundings.

18.2   Internal Energy, Heat, and Work

Substances act as reservoirs of energy. The **internal energy** of a system, **E,** is simply the total of all the possible kinds of energy associated with the substances in that system—the total of kinetic energies, ionization energies of the electrons, bond energies, vibrational energies, lattice energies, and so on.

We cannot measure the internal energy in a system; we can only determine the **change in internal energy, ΔE,** that accompanies a change in the system. We have seen that the H—H covalent bond energy is 436 kilojoules per mole of bonds (Section 6.12). Thus a system that consists of 1 mole of hydrogen molecules can be converted into 2 moles of hydrogen atoms with the input of 436 kilojoules of heat or other energy (Fig. 18.4). During this process the internal energy of the system increases by 436 kilojoules. Although we cannot measure the initial internal energy, $E_{initial}$, of the system (when it exists as 1 mole of hydrogen molecules) or its final energy, E_{final} (when it exists as 2 moles of hydrogen atoms), we can determine the differences in internal energy between the two states, because 436 kilojoules of energy has been added to the system. The difference between E_{final} and $E_{initial}$ is equal to the change in internal energy.

$$\Delta E = E_{final} - E_{initial}$$

The value of ΔE is positive when energy is transferred from the surroundings to the system. The opposite is also true, when energy is transferred from the system to the surroundings, ΔE is negative.

Figure 18.4

When an initial state composed of 1 mole of H_2 molecules undergoes a change to a final state composed of 2 moles of H atoms, the internal energy of the system increases by 436 kJ, and $\Delta E = 436$ kJ. The system consists of 1.2×10^{24} H atoms in both states, but the arrangements of the atoms are different, as the magnified view of a small portion of the system shows.

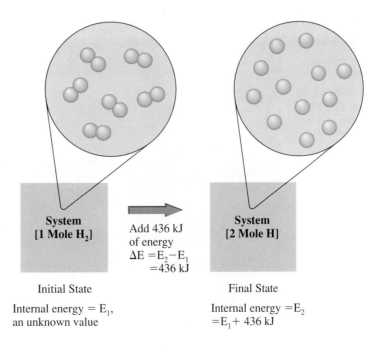

System
[1 Mole H_2]

Add 436 kJ
of energy
$\Delta E = E_2 - E_1$
$= 436$ kJ

System
[2 Mole H]

Initial State

Internal energy = E_1,
an unknown value

Final State

Internal energy $= E_2$
$= E_1 + 436$ kJ

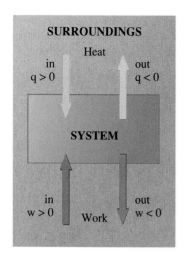

Figure 18.5

The signs of q and w as heat and work are transferred to or from a system.

We can identify two ways in which energy can be transferred into or out of a system: work and heat transfer. As we saw in Section 4.4, when we do work on a paper clip by rapidly bending it back and forth, it becomes hot. This temperature change does not occur because we have added heat but because we have done work and the work increased the thermal energy in the paper clip. This increase in the amount of thermal energy is reflected as an increase in the wire's temperature. We could also heat the paper clip by holding it in a flame where it would absorb heat and increase its thermal energy.

In Section 4.2 we discussed how to measure the amount of **heat, q,** transferred into or out of a system. We saw that when heat flows into the system from the surroundings (an **endothermic process**), the value of q is positive (Fig. 18.5). When heat flows from the system to the surroundings (an **exothermic process**), the value of q is negative. For example, if a system consisting of 1 mole of liquid water at 100°C were converted to gaseous water (steam) at 100°C and did no work, the internal energy of the system would increase, because 37.6 kilojoules of heat would flow into the system: $q = 37.6$ kilojoules.

We also have seen that when the surroundings do **work, w,** on a system, the internal energy of the system increases by the amount of work done on it and that when a system does work on the surroundings, the internal energy of the system decreases by the amount of work it does (Section 4.4). When energy is transferred from the system to the surroundings as work, work is done on the surroundings, and the value of w is negative (Fig. 18.5). When energy is transferred to the system from the surroundings as work, the surroundings do work on the system, and the value of w is positive.

Although there are several different forms of work, in this chapter we shall concentrate on one particular type, work that results from expansion against a constant pressure. This kind of work is called **expansion work** (Fig. 18.6). Expansion work occurs when a system pushes back the surroundings against a restraining pressure or when the surroundings compress the system against a pressure. When water freezes, it expands and pushes back the atmosphere. Pushing back the atmosphere is work,

Figure 18.6

The expansion of steam in the cylinders of this locomotive produces expansion work that is converted into kinetic energy.

Figure 18.7

When a system that consists of a gas expands, the system does work on the surroundings ($w < 0$). When the system is compressed, the surroundings do work on the system ($w > 0$).

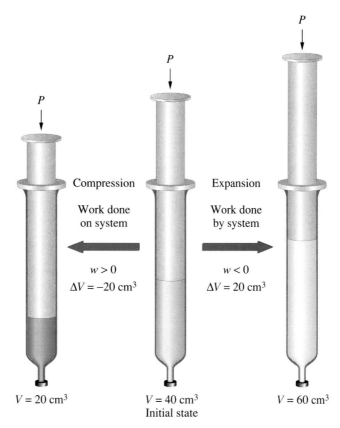

just as pumping air into a bicycle tire is work. As the water freezes, it uses some of its internal energy to do the work of pushing back the atmosphere.

We can calculate the amount of work done during expansion work by using the change in volume of the system and the restraining pressure. A system with an initial volume $V_{initial}$ that goes to a larger volume V_{final} will lose energy when it expands against a restraining pressure and does expansion work on the surroundings. The value of w is negative in this case ($w < 0$). When a system with an initial volume $V_{initial}$ is compressed to a smaller volume V_{final}, the system gains energy because the surroundings do work on the system ($w > 0$, Fig. 18.7).

When the restraining pressure, P, remains constant, the amount of work can be calculated from the expression

$$w = -P\Delta V = -P(V_{final} - V_{initial})$$

If the pressure is expressed in pascals and the volume change in cubic meters (m³), the resulting product has units of newton meters, which are the same as joules.*

---

*Pressure is defined as force per unit of area and therefore has units of force/length²; volume has units of length³. Thus the units for work are force × length.

$$w = -P(V_{final} - V_{initial}) = \frac{\text{force}}{\text{length}^2} \times \text{length}^3 = \text{force} \times \text{length}$$

Units of force × length are units of work, or energy. If the pressure is expressed in pascals (newtons/meter²) and the volume change in cubic meters (m³), the resulting product has units of newton meters, which are the same as joules or pascal meters³ (Section 4.1).

Figure 18.8

Water expands when it freezes.

Water expands when it freezes (Fig. 18.8). Determine the amount of work, in joules, done when a system consisting of 1.0 L of liquid water freezes under a constant pressure of 1.0 atm (101,325 Pa) and forms 1.1 L of ice.

Solution: As the water freezes, it expands against a restraining pressure of 1 atm and pushes back the atmosphere: The system does work on its surroundings. The amount of work done, w, can be calculated from the expression $w = -P\Delta V = -P(V_{final} - V_{initial})$. However, we must convert the volumes to cubic meters (1 m^3 = 1000 L) and express the pressure in pascals (1 atm = 101,325 Pa) in order to get units of joules from this question.

$$V_{initial} = 1.0\ L \times \frac{1\ m^3}{1 \times 10^3\ L} = 1.0 \times 10^{-3}\ m^3$$

$$V_{final} = 1.1\ L \times \frac{1\ m^3}{1 \times 10^3\ L} = 1.1 \times 10^{-3}\ m^3$$

$$
\begin{aligned}
w &= -P\Delta V = -P(V_{final} - V_{initial}) \\
&= -101{,}325\ Pa \times (1.1 \times 10^{-3}\ m^3 - 1.0 \times 10^{-3}\ m^3) \\
&= -101{,}325\ Pa \times (1 \times 10^{-4}\ m^3) \\
&= -10\ Pa\ m^3 \\
&= -10\ J
\end{aligned}
$$

The negative value of w indicates that work is done by the system. The magnitude of w indicates that very little work is done by this expansion. We will see that other changes may involve a great deal more work.

Check your learning: Calculate the amount of work done during the expansion and compression illustrated in Fig. 18.7 if the pressure is 250,000 Pa.

Answer: Expansion, $w = -5$ J; compression, $w = 5$ J

18.3   The First Law of Thermodynamics

When heat is added to a system and/or the surroundings do some work on the system, the internal energy of the system increases. The law of conservation of energy (Section 1.3), which is also known as the **first law of thermodynamics,** requires that the change in the internal energy of this system be equal to the energy added as heat from the surroundings plus the energy added as work from the surroundings:

$$\Delta E = q + w$$

In other words, the change in the internal energy of any system must equal the heat transferred into or out of the system plus the work done on or by the system.

Example 18.2   Calculation of ΔE

A system consisting of 1 mol of water vapor, $H_2O(g)$, at 100°C and 1 atm with a volume of 30.12 L (0.03012 m³) condenses at a constant pressure of 1 atm to liquid water, $H_2O(l)$, at 100°C and 1 atm with a volume of 18.8 cm³ (1.88 × 10^{-5} m³).

$$H_2O(g) \longrightarrow H_2O(l)$$

There is 40.7 kJ of heat released to the surroundings. Calculate ΔE for this process.

Solution: We need two values to determine ΔE: q, the amount of heat transferred to or from the system, and w, the amount of work done on or by the system. The problem gives us q, but we must calculate w.

Because 40.7 kJ of heat is transferred to the surroundings, q is −40.7 kJ. The sign of q is negative because heat leaves the system.

The amount of work (in joules) can be calculated from the expression

$$w = -P\Delta V = -P(V_{final} - V_{initial})$$

with P in pascals (1 atm = 101,325 Pa) and volumes in cubic meters. Thus

$$
\begin{aligned}
w &= -P(V_{final} - V_{initial}) \\
&= -101{,}325 \text{ Pa} \times (1.88 \times 10^{-5} \text{ m}^3 - 3012 \times 10^{-5} \text{ m}^3) \\
&= 3050 \text{ Pa m}^3 = 3050 \text{ J} = 3.050 \text{ kJ}
\end{aligned}
$$

As indicated in Section 18.2, units of Pa m³ are equivalent to joules. We can check our calculated sign of w by considering whether the volume of the system increases or decreases during a process. If the volume increases, the system does work and w is negative. If the volume of the system decreases, work is done on the system and the value of w must be positive. Here the volume of the system decreases and w is positive: The sign checks.

Now we can calculate ΔE.

$$
\begin{aligned}
\Delta E &= q + w \\
&= (-40.7 \text{ kJ}) + (3.050 \text{ kJ}) \\
&= -37.6 \text{ kJ}
\end{aligned}
$$

The internal energy of the system decreases by 37.6 kJ ($\Delta E < 0$), because the system gives up 40.7 kJ to the surroundings ($q < 0$), and the surroundings do 3.050 kJ of work on the system ($w > 0$). These changes are summarized in Fig. 18.9.

Figure 18.9

Work and heat changes accompanying a change in a system (composed of 1 mole of O atoms and 2 moles of H atoms combined as water molecules) from the gas phase to the liquid phase.

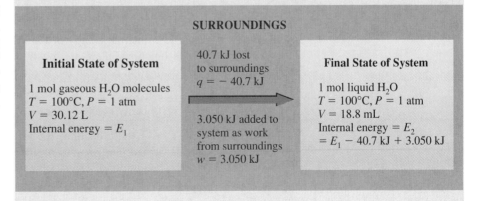

SURROUNDINGS

Initial State of System

1 mol gaseous H_2O molecules
$T = 100$°C, $P = 1$ atm
$V = 30.12$ L
Internal energy $= E_1$

40.7 kJ lost
to surroundings
$q = -40.7$ kJ

3.050 kJ added to
system as work
from surroundings
$w = 3.050$ kJ

Final State of System

1 mol liquid H_2O
$T = 100$°C, $P = 1$ atm
$V = 18.8$ mL
Internal energy $= E_2$
$= E_1 - 40.7$ kJ + 3.050 kJ

> **Check your learning:**   Lime is made commercially by the decomposition of limestone, $CaCO_3$. What is the change in internal energy when 1.00 mol of solid $CaCO_3$ (volume = 34.2 mL) absorbs 177.9 kJ of heat and decomposes at 25°C against a pressure of 1.00 atm to give solid CaO (volume = 16.9 mL) and $CO_2(g)$ (volume = 24.4 L)?
>
> *Answer: $q = 177.9$ kJ, $w = -2.5$ kJ, $\Delta E = 175.4$ kJ*

18.4   State Functions

We use the term **state function** to describe any property of a system that does not depend on how the system gets to the state exhibiting that property. The potential energy of a tennis ball on a third-floor window ledge is an example of a state function. (The potential energy is a measure of how hard the ball would hit the ground if it fell.) The potential energy of the ball on the window ledge would be the same whether we took the ball directly to the third floor or carried it to the roof and then back to the third floor. Any difference in the path by which it is moved to the third floor makes no difference in how hard the ball will hit the ground if it falls from the window ledge.

The change in the value of a state function equals its value in the final state minus its value in the initial state. The path by which a system changes does not matter to the change in a state function; the change in the state function will be the same no matter what path is used to make that change.

The increase in internal energy that accompanies the evaporation of a mole of water is a change in a state function—the internal energy of the water. A mole of water that changes from a liquid to a gas at 100°C and 1 atmosphere undergoes a change in internal energy, ΔE, that is simply the difference between the internal energies of 1 mole of gaseous H_2O and of 1 mole of liquid H_2O, both at 100°C and 1 atm. This difference does not depend on how we convert the liquid water to steam. We could convert it directly to gas at 100°C and 1 atmosphere:

$$H_2O(l) \longrightarrow H_2O(g)$$

or we could use a two-step process. We could first convert the $H_2O(l)$ to 1 mole of $H_2(g)$ and 1/2 mole of $O_2(g)$ and then reconvert the $H_2(g)$ and $O_2(g)$ to 1 mole of $H_2O(g)$ at 100°C and 1 atmosphere:

$$H_2O(l) \longrightarrow H_2(g) + \tfrac{1}{2}O_2(g) \longrightarrow H_2O(g)$$

Either way, the change in internal energy of the system, ΔE, is 37.6 kJ.

In contrast to state functions, there are functions whose values do depend on the paths followed. The distinction between the two types of functions can be illustrated by the variables in the expression

$$\Delta E = E_2 - E_1 = q + w$$

The internal energy, E, is a state function; ΔE for a reaction is the same regardless of how the reaction is carried out, because it reflects a change in a state function as the discussion of the evaporation of water in the previous paragraph indicates. However, heat, q, and work, w, do not reflect changes in state functions. For example, if

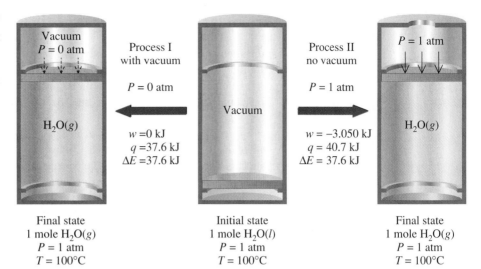

Figure 18.10

Water evaporating in two different ways. Process I: the evaporating water pushes a piston against a vacuum. Process II: the water pushes the piston against a pressure of 1 atmosphere. ΔE is 37.6 kilojoules in both cases, but q and w differ.

1 mole of liquid water at 100°C is allowed to expand against a vacuum to a volume of 30.12 liters (process I in Fig. 18.10), it does no work ($w = 0$) and absorbs only 37.5 kilojoules of heat from the surroundings. On the other hand, if the water is allowed to expand against a constant pressure of 1 atmosphere (process II in Fig. 18.10), it does 3.050 kilojoules of work and absorbs 40.7 kilojoules of heat from the surroundings. Note that ΔE is the same for the two processes.

As a second example, consider the reaction of glucose with oxygen.

$$C_6H_{12}O_6(s) + 6O_2(g) \longrightarrow 6CO_2(g) + 6H_2O(l)$$

When glucose combines with oxygen in air and produces CO_2 and H_2O, it produces about 5800 kilojoules of heat and almost no work. However, when glucose combines with oxygen and produces CO_2 and H_2O in the body, the chemical change can produce as little as 4100 kilojoules of heat and as much as 1700 kilojoules of work.

Now we can appreciate why E is capitalized and q and w are not. Quantities that are state functions are designated by capital letters, those that are not by lowercase letters.

Enthalpy, Entropy, and Spontaneous Processes

In the following sections we will consider the differences between two types of changes in a system: those that occur spontaneously and those that must be forced to occur. At first glance this may look like a very different topic from what we have talked about in preceding sections, but we will see that heat transfer plays a significant role. In addition, we will see that changes in the disorder in the system also play an important role.

18.5   Spontaneous Chemical Reactions

Figure 18.11

Solid MnS and a solution of $NaNO_3$ form spontaneously when a solution of Na_2S is added to a solution of $Mn(NO_3)_2$.

A **spontaneous change,** in a thermodynamic sense, is a change in a system that proceeds without any outside influence (Fig. 18.11). The following list includes several processes that are spontaneous and others that are not.

- You could slip and fall out of a tree, but you will never slip and fall up into a tree.
- An egg cooks in boiling water, but it never uncooks in boiling water.
- Our hair grows longer, but though it may fall out, it never grows shorter.
- Our cars may rust if the steel is exposed to air and moisture, but the iron oxide in the rust, by itself, never reconverts itself into steel.
- Gasoline burns and forms carbon dioxide and water, but carbon dioxide and water do not react and form gasoline and oxygen.
- When we blow up a balloon, the gas fills the balloon evenly, but all of the gas never shifts, by itself, into one end of the balloon.
- An ice cube melts on a warm day, but liquid water never forms a solid at the same temperature.

Note that in each case, *the reverse of a spontaneous process is not spontaneous.* This is an important idea. Also note that some of these processes are fast and others are slow. Thermodynamics helps us determine whether a process is possible, but it gives us no information about the time the process will take to occur. Changes may be spontaneous even though they are very slow. For those who own diamonds (which are essentially pure carbon), it is fortunate that the spontaneous reaction of carbon with oxygen is very slow at room temperature.

Some chemical reactions are so familiar that we need no help determining whether they are spontaneous. For example, we know that a strong acid will react spontaneously with a strong base and that wood will inflame spontaneously at high temperatures. However, other chemical reactions may be less familiar or even untried.

If we worked in an industry that made plastic bags of polyethylene, we might wish to find an inexpensive source of ethylene, C_2H_4, the monomer used to produce polyethylene (Section 9.15). Ethylene can be produced from ethane, C_2H_6, an inexpensive product of the petrochemical industry, at high temperatures according to the equation

$$C_2H_6(g) \longrightarrow H_2(g) + C_2H_4(g)$$

If we could get this reaction to go at room temperature (say, by using an appropriate catalyst), we could save the cost of heating the reaction and reduce the cost of our ethylene. But before we start looking for a catalyst, we should ask whether the reaction will go at room temperature under any conditions. That is, we ask, "Is the decomposition of ethane into hydrogen and ethylene spontaneous at room temperature?"

One way to answer this question is to consider a mixture of ethane, hydrogen, and ethylene at 25°C, each with a pressure of 1 atmosphere (reaction quotient, Q, = 1), and ask, "Which way will the reaction shift to reach equilibrium?"

$$C_2H_6(g) \rightleftharpoons H_2(g) + C_2H_4(g)$$
$$\text{1 atm} \qquad \text{1 atm} \qquad \text{1 atm}$$

This is another way of asking which reaction is spontaneous, the decomposition of ethane

$$C_2H_6(g) \longrightarrow H_2(g) + C_2H_4(g)$$

or the reaction of hydrogen and ethylene

$$H_2(g) + C_2H_4(g) \longrightarrow C_2H_6(g)$$

We know that the equilibrium constant tells us which way concentrations or pressures will shift to reach equilibrium (Section 14.5). If $Q = 1$ and $K > 1$, the pressures will shift so that the amount of products increases: The forward reaction is spontaneous. For a mixture with $Q = 1$ and $K < 1$, the shift will increase the pressures of reactants: The reverse reaction is spontaneous. For the equilibrium

$$C_2H_6(g) \rightleftharpoons H_2(g) + C_2H_4(g) \qquad K_p = 2.0 \times 10^{-18}$$

K_p is much less than 1, so the mixture will react in such a way that the concentrations of H_2 and C_2H_4 decrease. Thus the spontaneous reaction at 25°C is

$$H_2(g) + C_2H_4(g) \longrightarrow C_2H_6(g)$$

As we will see, ethylene will not form spontaneously at room temperature, so we cannot use this reaction to reduce the energy cost in the production of ethylene.

Example 18.3   Predicting the Spontaneity of a Chemical Reaction from an Equilibrium Constant

Which is the spontaneous reaction, the ionization of acetic acid or the reaction of hydronium ion and acetate ion?

Solution: This is another way of asking the question, "Which reaction is spontaneous when all of the concentrations are one molar, the ionization of acetic acid or the reaction of hydronium ion and acetate ion?" The equations for these two reactions are

$$CH_3CO_2H(aq) + H_2O(l) \longrightarrow H_3O^+(aq) + CH_3CO_2{}^-(aq)$$

and

$$H_3O^+(aq) + CH_3CO_2{}^-(aq) \longrightarrow CH_3CO_2H(aq) + H_2O(l)$$

For a mixture of CH_3CO_2H, H_3O^+, and $CH_3CO_2{}^-$ in an aqueous solution, the equilibrium constant, K_a, is much less than 1 (Appendix F).

$$CH_3CO_2H(aq) + H_2O(l) \longrightarrow H_3O^+(aq) + CH_3CO_2{}^-(aq) \qquad K_a = 1.8 \times 10^{-5}$$

A mixture with $[CH_3CO_2H]$, $[H_3O^+]$, and $[CH_3CO_2{}^-]$ each equal to 1 M will react in such a way that the concentrations of H_3O^+ and $CH_3CO_2{}^-$ decrease and the concentration of CH_3CO_2H increases. The spontaneous reaction in this case is

$$H_3O^+ + CH_3CO_2{}^- \longrightarrow CH_3CO_2H + H_2O$$

Check your learning: Which of the following reactions are spontaneous?
(a) $2Cl_2(g) + 2H_2O(g) \longrightarrow 4HCl(g) + O_2(g)$ $\qquad K_p = 3.2 \times 10^{-14}$
(b) $2NO(g) + Cl_2(g) \longrightarrow 2NOCl(g)$ $\qquad K_p = 4.6 \times 10^4$
(c) $H_2(g) + CO_2(g) \longrightarrow H_2O(g) + CO(g)$ $\qquad K_p = 1.6$
(d) $2SO_2(g) + O_2(g) \longrightarrow 2SO_3(g)$ $\qquad K_p = 0.06$

Answer: **(b)** and **(c)** are spontaneous.

As the preceding paragraphs indicate, we can use equilibrium constants to tell us which reaction is spontaneous when we have a mixture of reactants and products. However, note that we must use K_p for an equilibrium involving gases.

For now we will restrict our predictions to reactions that would occur if we had a mixture of reactants and products at the **standard state conditions** defined in Section 4.5: 25°C, 1 atm, and unit activity. Remember our discussion of activities in Section 14.2:

- The activity of a pure solid or a pure liquid is 1.
- The activity of a solvent in a dilute solution may be approximated as 1.
- The activity of a gas is approximated by its pressure in atmospheres (a pressure of 1 atm approximates unit activity).
- The activity of a dilute solute is approximated by its molar concentration (a 1 M solution approximates unit activity).

In a later section we will discuss how to predict the spontaneity of reactions under different conditions, and we will see that the decomposition of ethane becomes spontaneous only at temperatures above 860°C, which is the reason that it is necessary to produce it at elevated temperatures.

18.6   Minimization of Energy, Increase in Disorder, and Spontaneous Change

A physical or a chemical change that occurs spontaneously is accompanied by one or both of two shifts—a shift toward a lower energy and/or a shift to a more disordered state. This observation reflects two fundamental laws of nature that are particularly important in thermodynamics.

1. **Systems tend toward a state of lower energy.** For example, suppose we have a system that consists of a box containing an assembled jigsaw puzzle. If we drop the box, it falls to the floor. During this change in the box's position, part of its potential energy (energy due to its height above the floor) is converted to kinetic energy. When the box hits the floor, that kinetic energy is converted to heat. The loss of heat indicates that the system (the box and the puzzle) has gone to a lower energy state. It certainly has a lower potential energy, because it cannot fall as far as it could before the change in its position.

2. **Systems tend toward a state of maximum disorder.** In thermodynamics, **disorder** is a measure of the number of equivalent ways in which a system can be arranged. There are very few ways for an assembled jigsaw puzzle to be arranged in its box (for example, two arrangements are rightside up and upside down). There are many more arrangements of a disassembled jigsaw puzzle in its box (for example, there are many jumbled arrangements with one piece up and the remainder down, many others with two pieces up and the remainder down, still more with three pieces up and the remainder down, and so on). Thus the disorder of a puzzle (the number of ways it could be arranged) increases as it becomes more disassembled.

   If the jigsaw puzzle in the box were assembled before the fall, it would almost certainly be disassembled (more disordered) after the fall. No one would try to assemble a jigsaw puzzle by dropping the separated pieces on the floor, whereas the reverse process (making the system more disordered) is readily accomplished by dropping the assembled puzzle. A system tends to become less orderly because there are so many more ways to be disorderly than to be orderly. The probability of a system's becoming more disordered or more random is greater than that of its becoming more ordered.

The effects of a shift toward lower energy and of a shift toward a more disordered state are discussed in connection with the formation of solutions in Section 12.2. It could be beneficial for you to reread that section now.

18.7   Changes in Energy: Enthalpy Changes

Generally, the amount of heat, q, that a system absorbs or evolves as it changes is not a state function; rather, it varies with the way the process occurs (see Fig. 18.10). However, if the change occurs in such a way that the only work done is to effect a change in volume of the system at constant pressure, then q reflects a change in a state function of the system called **enthalpy, H.** We cannot determine the enthalpy of a system, but we can measure changes in enthalpy as the system changes. Having enthalpy as a state function is useful because it lets us use the value of q to determine how the internal energy of a system changes, and, hence, is a guide to whether or not a reaction is spontaneous.

The amount of heat, q, exchanged during a process that converts a system's reactants to products when the only work done by the system is expansion work at constant pressure is called the **enthalpy change, ΔH,** of the system ($\Delta H = q$).

Most chemical reactions of interest to us occur under the essentially constant pressure of the atmosphere. Consequently, we can determine the enthalpy changes of these or any other changes that occur at constant pressure by measuring the heat transferred during the process. We can calculate the change in internal energy for these reactions from the equation

$$\Delta E = q + w = \Delta H + w$$

Because the reaction occurs at constant pressure and the only work done results from the change in volume of the system, the work term in this case is equal to $-P(V_2 - V_1)$, or $-P\Delta V$.

We have previously seen enthalpy changes that include the standard enthalpy of formation, ΔH_f°, and enthalpies of combustion, vaporization, and fusion (Section 4.5). In addition, enthalpy changes are used to define ionization energy and electron affinity (Section 5.10), covalent bond energy (Section 6.12), and lattice energy (Section 6.3). Moreover, many chemical reactions occur under the essentially constant pressure of the atmosphere, so their enthalpy change is equal to the amount of heat they evolve or absorb (Fig. 18.12).

When ΔH for a chemical reaction is negative, the reaction is exothermic and the system evolves heat to the surroundings. Thus a negative value of ΔH indicates a decrease in energy of a system, a situation that favors a spontaneous reaction. When ΔH for a chemical reaction is positive, the reaction is endothermic and heat is absorbed by the system from the surroundings. A positive value of ΔH indicates an increase in energy of a system, a situation that favors a nonspontaneous reaction. When ΔH has a large negative or a large positive value (ΔH less than about -150 kilojoules or greater than about 150 kilojoules), it dominates the other factor that determines whether a reaction is spontaneous *at temperatures near room temperature.*

Figure 18.12

The heat released when 2 moles of Al and 1 mole of Fe_2O_3 at 25°C react and form 1 mole of Al_2O_3 and 2 moles of Fe at 25°C is the enthalpy change of the reaction $2Al + Fe_2O_3 \longrightarrow Al_2O_3 + 2Fe$. The heat is released in two stages: Heat is released as the reaction occurs, and then additional heat is released as the molten iron and hot Al_2O_3 cool back down to 25°C.

Table 18.1

Standard Enthalpies of Formation,
Standard Free Energies of Formation,
and Standard Entropies (298.15 K,
1 atm)^a

Substance	ΔH°_f, kJ mol^{-1}	ΔG°_f, kJ mol^{-1}	S°_{298}, J mol^{-1} K^{-1}
Carbon			
C(s) (*graphite*)	0	0	5.74
C (diamond)	1.897	2.900	2.38
C(g)	716.681	671.289	157.987
CO(g)	−110.52	−137.15	197.56
CO$_2$(g)	−393.51	−394.36	213.6
CH$_4$(g)	−74.81	−50.75	186.15
CH$_3$OH(g)	−200.7	−162.0	239.7
CH$_3$OH(l)	−238.7	−166.7	27
C$_2$H$_4$(g)	52.26	68.12	219.5
C$_2$H$_6$(g)	−84.68	−32.9	229.5
Chlorine			
Cl$_2$(g)	0	0	222.96
Cl(g)	121.7	105.7	165.09
Copper			
Cu(s)	0	0	33.15
CuS(s)	−53.1	−53.6	66.5
Hydrogen			
H$_2$(g)	0	0	130.57
H(g)	218.0	203.3	114.6
H$_2$O(g)	−241.82	−228.59	188.71
H$_2$O(l)	−285.8	−237.2	69.91
HCl(g)	−92.31	−95.3	186.8
H$_2$S(g)	−20.6	−33.6	205.7
Oxygen			
O$_2$(g)	0	0	205.03
Silver			
Ag$_2$O(s)	−31.0	−11.2	121.0
Ag$_2$S(s)	−32.6	−40.7	144.0

^{a}See Appendix I for additional values.

Thus an exothermic reaction that produces a moderate to large amount of heat will be spontaneous near room temperature. An endothermic reaction that absorbs a moderate to large amount of heat will be nonspontaneous near room temperature.

We can measure ΔH for a reaction by using calorimetry (Section 4.3) or we can calculate ΔH by using **Hess's law** (Section 4.6). The most common form of Hess's law is written as

$$\Delta H^{\circ} = \Sigma \Delta H^{\circ}_{f_{product}} - \Sigma \Delta H^{\circ}_{f_{reactant}}$$

and is used to determine an enthalpy change from enthalpies of formation. In this equation, Σ means "the sum of." Enthalpies of formation, ΔH°_f, can be found in Table 18.1 and Appendix I.

Figure 18.13

This chafing dish is warmed by the heat produced when methanol burns in air.

| Example 18.4 | Calculation of ΔH°_{298} from Enthalpies of Formation |

The heaters on many buffet counters burn methanol, CH_3OH (Fig. 18.13).

$$2CH_3OH(l) + 3O_2(g) \longrightarrow 2CO_2(g) + 4H_2O(g)$$

Calculate the standard enthalpy change, ΔH°_{298}, for the reaction.

Solution: We can determine ΔH°_{298} for this reaction by using Hess's law in the form

$$\Delta H^\circ_{298} = \Sigma \Delta H^\circ_{f_{product}} - \Sigma \Delta H^\circ_{f_{reactant}}$$

$$\Delta H^\circ_{298} = [2\Delta H^\circ_{f_{CO_2(g)}} + 4\Delta H^\circ_{f_{H_2O(g)}}] - [2\Delta H^\circ_{f_{CH_3OH(l)}} + 3\Delta H^\circ_{f_{O_2(g)}}]$$

The standard enthalpies of formation of the compounds involved can be found in Table 18.1. Remember that the standard enthalpy of formation of an element in its most stable state is zero (Section 4.5); hence $\Delta H^\circ_{f_{O_2(g)}} = 0$ kJ.

$$\begin{aligned}\Delta H^\circ_{298} = {} & 2 \text{ mol } CO_2(g) \times (-393.51 \text{ kJ mol}^{-1}) \\ & + 4 \text{ mol } H_2O(g) \times (-241.82 \text{ kJ mol}^{-1}) \\ & - 3 \text{ mol } O_2(g) \times (0 \text{ kJ mol}^{-1}) \\ & - 2 \text{ mol } CH_3OH(l) \times (-238.7 \text{ kJ mol}^{-1}) \\ = {} & -787.02 + (-967.28) - (-477.4) \text{ kJ} = -1276.9 \text{ kJ}\end{aligned}$$

$$2CH_3OH(l) + 3O_2(g) \longrightarrow 2CO_2(g) + 4H_2O(g) \qquad \Delta H^\circ_{298} = -1276.9 \text{ kJ}$$

Note that the reaction is exothermic and that ΔH°_{298} favors a spontaneous reaction.

Check your learning: Calculate ΔH°_{298} for the reaction

$$C_2H_6(g) \longrightarrow H_2(g) + C_2H_4(g)$$

and identify the reaction as endothermic or exothermic.

Answer: 136.94 kJ, endothermic

At this point we need to digress briefly and remind ourselves of the notation used to indicate enthalpy changes for processes that occur at various temperatures, pressures, and concentrations. We use ΔH°_{298} as the label for the enthalpy change of a process that occurs under standard state conditions (Section 18.5) and that converts a system's reactants, at unit activities, to products at unit activities. Thus the amount of heat, q, exchanged during a process at 25°C that converts a system's reactants, at unit activities, to products, at unit activities, when the only work done by a system is expansion work at a constant pressure of 1 atmosphere, is equal to the **standard enthalpy change, ΔH°_{298},** of the system; that is, $\Delta H^\circ_{298} = q$ when $w = -P\Delta V$ at 25°C with $P = 1$ atmosphere and all reactants and products at unit activities. We use the symbol ΔH° for the enthalpy change of a process that occurs at some temperature other than 25°C but for which all reactants and products have unit activities. The symbol ΔH is used to indicate an enthalpy change with no restrictions on the temperature or the activities of the components. The various symbols used to indicate enthalpy changes are summarized in Table 18.2 on page 684.

Table 18.2

Symbols for Various Thermodynamic Parameters

Enthalpy Change	Free Energy Change	Entropy and Entropy Change	Conditions
ΔH°_{298}	ΔG°_{298}	$S^\circ_{298}, \Delta S^\circ_{298}$	Standard state conditions, $T = 298.15$, unit activities (pure solids, pure liquids, $P = 1$ atm, 1 M concentrations)
ΔH°	ΔG°	$S^\circ, \Delta S^\circ$	Unit activities (pure solids, pure liquids, $P = 1$ atm, 1 M concentrations)
ΔH	ΔG	$S, \Delta S$	No restrictions

The difference between the enthalpy change for a process at 298.15 K and that for the same process at some other temperature is generally small and is often neglected. In this text we will make the approximation that ΔH°_{298} equals ΔH° for the same process at some other temperature. Thus we assume that the amount of heat produced by combustion of 2 moles of methanol at 450°C is the same as that calculated in Example 18.4 for the reaction at 25°C. Precise measurements show that this is not exactly correct, but the approximation is quite satisfactory for introductory study, particularly when it is used as a guide to whether a reaction is spontaneous or nonspontaneous.

18.8   Changes in Disorder: Entropy and Entropy Changes

We have just seen that we can use the enthalpy change as one guide to whether a process is spontaneous or nonspontaneous. As a second guide, we can use the change in the disorder in the system; an increase in disorder during a change favors a spontaneous change. Remember that disorder, as used in thermodynamics, is a measure of the number of equivalent ways in which a system can be arranged; that is, the number of arrangements that have the same internal energy.

Consider two changes that are spontaneous but endothermic: the melting of ice at room temperature,

$$H_2O(s) \longrightarrow H_2O(l) \qquad \Delta H^\circ = +6.0 \text{ kJ}$$

and the decomposition of calcium carbonate at high temperature,

$$CaCO_3(s) \longrightarrow CaO(s) + CO_2(g) \qquad \Delta H^\circ = +178 \text{ kJ}$$

The disorder of both systems increases when these changes occur, and the increasing disorder of the system is the reason these two reactions are spontaneous even though they are endothermic.

The disorder of the system increases when ice melts, because the water molecules become free to move about. Water molecules are fixed in a regular, repeating array in an ice crystal (Fig. 18.14 on page 685). Thus there are a limited number of equivalent ways in which the water molecules can arrange themselves, and they have a relatively low disorder. However, when the ice melts, the molecules are free to move as well as to change their orientations. Consequently, molecules in the liquid can adopt

Figure 18.14

The amount of disorder in ice is limited because the water molecules are fixed in position.

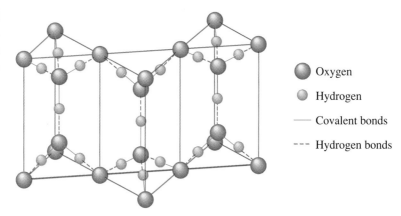

○ Oxygen

○ Hydrogen

— Covalent bonds

--- Hydrogen bonds

many more equivalent arrangements than those in the solid, and there is more disorder in the liquid than in the solid.

As calcium carbonate decomposes, the system changes from the ordered array of Ca^{2+} ions and CO_3^{2-} ions to an ordered array of Ca^{2+} ions and O^{2-} ions in solid calcium oxide plus a disordered collection of CO_2 molecules in the gas phase. The major increase in the disorder of the system is due to the formation of the gaseous CO_2 molecules. The disorder of these molecules is even greater than the disorder of the molecules in liquid water. There are more possible equivalent arrangements of the CO_2 molecules in the gas phase than of an equal number of water molecules in the liquid phase.

The amount of disorder of a substance (or a system) can be determined quantitatively and is called the **entropy, S,** of the substance (or the system). Every substance has an entropy as one of its characteristic properties. The larger the value of the entropy, the greater the amount of disorder present in the substance. The entropy of a system is equal to the sum of the entropies of its components. Like the internal energy of a system, the entropy, S, of a system is a state function.

The **standard entropy, S_{298}°,** of a substance is the entropy content of exactly 1 mole of the substance at standard state conditions (298.15 K, 1 atmosphere, unit activities). Values of standard entropies can be found in Table 18.1 and Appendix I. Note that solids (most ordered) tend to have lower entropies than liquids (less ordered) and that liquids tend to have lower entropies than gases (least ordered) (Fig. 18.15). That is,

$$S_{solid}^\circ < S_{liquid}^\circ < S_{gas}^\circ$$

Figure 18.15

The entropy of a substance increases $(\Delta S > 0)$ as it is transformed from an ordered solid to a less ordered liquid and then to a still less ordered gas. The entropy decreases $(\Delta S < 0)$ as the substance is transformed from a gas to a liquid and then to a solid.

Increasing entropy

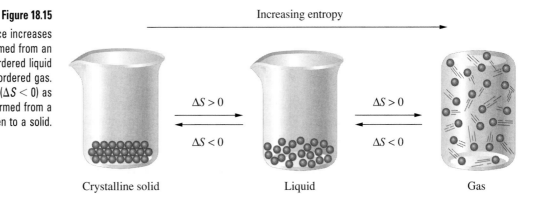

$\Delta S > 0$

$\Delta S < 0$

$\Delta S > 0$

$\Delta S < 0$

Crystalline solid   Liquid   Gas

In the same physical state, substances with simple molecules tend to have lower entropies than substances with more complicated molecules. Because larger molecules have more atoms per molecule to move about, they can exhibit greater disorder; thus they have higher entropies. Hard crystalline substances, such as diamond, tend to be more ordered and to have lower entropies than softer materials, such as graphite or sodium.

The notation used for entropies (and changes in entropies) at various temperatures, pressures, and concentrations is analogous to that used for enthalpy changes (Table 18.2). The entropy of a substance at standard state conditions is labeled S_{298}° (a change in entropy under these conditions is ΔS_{298}°). The entropy of a substance at some temperature other than 298.15 K but with unit activity is given the symbol S°. (A change is ΔS°.) The symbol S is used to indicate an entropy with no restrictions on the temperature or the activity of the substance. (A change is ΔS.)

The **entropy change, ΔS,** that accompanies a change in a system is the difference between the sum of the entropies of the products of that change and the sum of the entropies of the reactants. For the reaction

$$mA + nB \longrightarrow xC + yD$$

the entropy change is

$$\Delta S = \Sigma S_{\text{product}} - \Sigma S_{\text{reactant}}$$
$$= [(x \times S_C) + (y \times S_D)] - [(m \times S_A) + (n \times S_B)]$$

where S_A is the entropy of 1 mole of A, S_B is the entropy of 1 mole of B, and so on. A positive value for ΔS ($\Delta S > 0$) indicates an increase in disorder during the change; thus a positive value of ΔS indicates a situation that favors a spontaneous reaction. A negative value for ΔS ($\Delta S < 0$) indicates a decrease in disorder—that is, an increase in order—during a change; thus a negative value of ΔS indicates a situation that favors a nonspontaneous reaction.

18.9   Calculation of Entropy Changes

We can determine the **standard entropy change, ΔS_{298}°,** for a process that occurs under standard state conditions and that converts a system's reactants, at unit activities, to products, at unit activities, by using standard enthalpy values, S_{298}°, for the values of S° in the equation

$$\Delta S_{298}^{\circ} = \Sigma S_{\text{product}}^{\circ} - \Sigma S_{\text{reactant}}^{\circ}$$

We use an approximation to determine the value of an entropy change, ΔS°, at temperatures different from 298.15 K. Even though the entropy, S°, of a single substance varies with temperature, the difference between the entropy change for a process at 298.15 K and the entropy change for that same process at some other temperature is generally small. Thus we assume that ΔS° equals ΔS_{298}° for the same process.

Example 18.5   Determination of ΔS°_{298}

Does the entropy change that accompanies the condensation of 1 mol of water during a rain storm,

$$H_2O(g) \longrightarrow H_2O(l)$$

involve an increase or a decrease in disorder? Check the answer by using the assumption that the entropy change is approximately equal to the standard entropy change, ΔS°_{298}, for the reaction.

Solution: Liquids have less disorder (smaller entropies) than gases, so the condensation of water

$$H_2O(g) \longrightarrow H_2O(l)$$

proceeds with a decrease in disorder, which would be indicated with a negative value of ΔS. (The negative value of ΔS does not favor the spontaneous formation of rain. The formation of rain is a spontaneous process because condensation of water is exothermic.)

The value of the standard entropy change, ΔS°_{298}, is the difference between the standard entropy of the product, $H_2O(l)$, and the standard entropy of the reactant, $H_2O(g)$; see Table 18.1. We will omit the subscript 298 on S°_{298} in the following equation to simplify writing the equation.

$$\Delta S^{\circ}_{298} = \Sigma S^{\circ}_{product} - \Sigma S^{\circ}_{reactant}$$
$$\Delta S^{\circ}_{298} = S^{\circ}_{H_2O(l)} - S^{\circ}_{H_2O(g)}$$
$$= 1 \text{ mol } H_2O(l) \times (69.91 \text{ J mol}^{-1} \text{ K}^{-1})$$
$$- 1 \text{ mol } H_2O(g) \times (188.71 \text{ J mol}^{-1} \text{ K}^{-1})$$
$$= -118.80 \text{ J K}^{-1}$$

The value for ΔS°_{298} is negative, as predicted.

Check your learning: Predict the sign of the standard entropy change for the reaction

$$H_2(g) + C_2H_4(g) \longrightarrow C_2H_6(g)$$

and check your prediction by using standard entropy values from Table 18.1.

Answer: Negative, -120.6 J K^{-1}

Example 18.6   Determination of ΔS°_{298}

Predict the sign of ΔS°_{298} for the combustion of methanol, CH_3OH (Fig. 18.13 and Example 18.4), and check your prediction.

$$2CH_3OH(l) + 3O_2(g) \longrightarrow 2CO_2(g) + 4H_2O(g)$$

Solution: The products of this reaction contain more gas molecules than the reactants, so the products possess more disorder. An increase in disorder is indicated by a positive value of ΔS.

The value of entropy change is equal to the difference between the standard entropies of the products and the entropies of the reactants.

$$\Delta S^{\circ}_{298} = \Sigma S^{\circ}_{product} - \Sigma S^{\circ}_{reactant}$$

We are using S_{298}° values (Table 18.1) for each substance involved in this equation, but we have omitted the subscript 298 on S_{298}° to simplify writing the equation.

$$\Delta S_{298}^{\circ} = [2S_{CO_2(g)}^{\circ} + 4S_{H_2O(g)}^{\circ}] - [2S_{CH_3OH(l)}^{\circ} + 3S_{O_2(g)}^{\circ}]$$
$$= 2 \text{ mol } CO_2(g) \times (213.6 \text{ J mol}^{-1} \text{ K}^{-1})$$
$$+ 4 \text{ mol } H_2O(g) \times (188.71 \text{ J mol}^{-1} \text{ K}^{-1})$$
$$- 2 \text{ mol } CH_3OH(l) \times (127 \text{ J mol}^{-1} \text{ K}^{-1})$$
$$- 3 \text{ mol } O_2(g) \times (205.03 \text{ J mol}^{-1} \text{ K}^{-1})$$
$$= 313 \text{ J K}^{-1}$$

Note that standard entropies of the elements are not zero at 25°C. As this reaction proceeds, the entropy of the system increases.

$$2CH_3OH(l) + 3O_2(g) \longrightarrow 2CO_2(g) + 4H_2O(g) \qquad \Delta S_{298}^{\circ} = 313 \text{ J K}^{-1}$$

Check your learning: Predict the sign of the standard entropy change for the reaction

$$Ca(OH)_2(s) \longrightarrow CaO(s) + H_2O(l)$$

and check your prediction by using standard entropy values from Appendix I.

Answer: Positive, 34 J K^{-1}

The results of Examples 18.5 and 18.6 are consistent with the entropy changes of many other chemical reactions that involve gases. The behavior can be summarized as follows:

- If the products of a reaction contain more molecules of gas than do the reactants (Example 18.6), then the disorder of the system increases during the reaction (ΔS is positive).
- If the number of gas molecules decreases (Example 18.5), then the disorder decreases (ΔS is negative).
- If the number of gas molecules does not change, then the disorder can increase or decrease, but the magnitude of ΔS is relatively small.

18.10   The Third Law of Thermodynamics

Although it is not possible to measure the value of some state functions, such as enthalpy, H, and internal energy, E, it is possible to measure the entropy, S, of a pure substance at any given temperature. The reason for this is found in the **third law of thermodynamics:** *The entropy of any pure, perfect crystalline element or compound at absolute zero (0 K) is equal to zero.* At absolute zero all molecular motion is at a minimum, and at that temperature a pure perfect crystalline substance has no disorder; its entropy is zero (Fig. 18.16 on page 689). All molecular motion is also at a minimum in an impure substance or in a pure crystalline substance with defects at absolute zero, but the impurity or defects can be distributed in different ways, giving rise to disorder (and a nonzero value for entropy).

If we take a pure, perfect crystalline substance at absolute zero (with an entropy of zero) and measure the entropy change as its temperature is increased, then we have

Figure 18.16

(A) A representation of a perfect crystal of hydrogen chloride at 0 K. The smaller spheres represent the hydrogen atoms. The entropy of a perfect crystal is zero at 0 K. (B) As the temperature rises above 0 K, the vibration of molecules produces some disorder and the entropy increases.

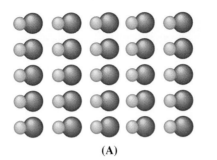

(A)

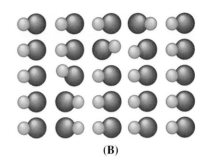

(B)

measured the absolute entropy of the substance at the higher temperature. Hence we can find entropies of pure substances. In contrast, for internal energy and enthalpy, we can find only differences between two values, not the actual values. Standard entropies enable us to compare the relative amounts of disorder present in different pure substances, and they can be used to determine entropy changes (Section 18.8). However, we must exercise caution when using entropies, because the entropy of a pure elemental substance at standard state conditions is not equal to zero. Table 18.1 and Appendix I give the standard entropies, S°_{298}, of some common substances.

Free Energy Changes

We have used the equilibrium constant or the values of ΔH and ΔS for a process to determine whether it is spontaneous. There remain two questions to be considered.

1. Is there a relationship between the equilibrium constant and the values of ΔH and ΔS?

2. What do we do when ΔH suggests that a reaction will be spontaneous and ΔS suggests that it will not be spontaneous (or vice versa)?

As we will see, the answer to both these questions lies in changes in a state function called the free energy.

18.11 Free Energy Changes, Enthalpy Changes, and Entropy Changes

We can combine the enthalpy change, ΔH, and the entropy change, ΔS, that accompany a reaction in a system to find the change in another state function that we call the **free energy, G,** of the system. We cannot determine the value of the free energy of a system, but we can determine changes in free energy from ΔH and ΔS and by several other ways. The relationship among the **free energy change, ΔG,** the enthalpy change, and the entropy change for a reaction is given by the expression

$$\Delta G = \Delta H - T\Delta S$$

where T is the temperature of the reaction on the Kelvin scale.

ΔH	ΔS	Comment	Example	ΔH°_{298}, kJ	ΔS°_{298}, J K^{-1}
−	+	Spontaneous at all temperatures	$H_2(g) + Cl_2(g) \longrightarrow 2HCl(g)$	−185	141
			$C(s) + O_2(g) \longrightarrow CO_2(g)$	−394	3
−	−	Spontaneous at lower temperatures, nonspontaneous at higher temperatures	$H_2O(g) \longrightarrow H_2O(l)$	−44	−119
			$2SO_2(g) + O_2(g) \longrightarrow 2SO_3(g)$	−198	−187
+	+	Nonspontaneous at lower temperatures, spontaneous at higher temperatures	$NH_4Cl(s) \longrightarrow NH_3(g) + HCl(g)$	176	284
			$N_2(g) + O_2(g) \longrightarrow 2NO(g)$	180	25
+	−	Nonspontaneous at all temperatures	$3O_2(g) \longrightarrow 2O_3(g)$	286	−137
			$2H_2O(l) + O_2(g) \longrightarrow 2H_2O_2(l)$	196	−126

Table 18.3

Examples of Spontaneous and Non-spontaneous Reactions

Unlike the value of ΔH or ΔS alone, we can use the value of the free energy change alone as an unambiguous predictor of the spontaneity of a chemical reaction run at constant temperature and pressure, because ΔG combines the effects of both ΔH and ΔS. Reactions for which the value of ΔG is negative ($\Delta G < 0$) are spontaneous, reactions with positive values of ΔG ($\Delta G > 0$) are nonspontaneous, and reactions with ΔG equal to zero are at equilibrium.

Values of T are always positive in the free energy equation because temperature is measured on the Kelvin scale; therefore, ΔS (which can be either positive or negative) determines the sign of the $T\Delta S$ term. As T increases, the value of $T\Delta S$ gets larger and ΔS plays an increasingly important role in the value of ΔG.

Table 18.3 and Figure 18.17 illustrate how the four combinations of positive and negative values of ΔH and ΔS affect the sign of ΔG. For example, the free energy

Figure 18.17

The dependence of both ΔG and the spontaneity of a process on the temperature for the four combinations of ΔH and ΔS.

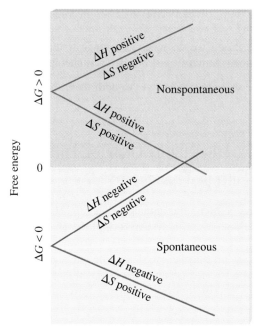

equation $\Delta G = \Delta H - T\Delta S$ tells us that a nonspontaneous process that is endothermic (ΔH positive) and produces a more disordered system (ΔS positive) can become spontaneous as the temperature increases. In this case, the initial value of ΔG is positive because of the magnitude of the positive value of ΔH. However, as the temperature increases, the negative value of the $T\Delta S$ term becomes great enough that the value of $\Delta H - T\Delta S$ becomes negative and the process becomes spontaneous. This is exactly the situation for the boiling of water and other phase changes. For example, we know that ice does not melt spontaneously at $-10°C$ but does melt at $10°C$ (See Example 18.7).

The notations ΔG°_{298} (**standard free energy change**), ΔG°, and ΔG for free energy changes (Table 18.2) are similar to those used for enthalpy changes and entropy changes at different temperatures, pressures, and concentrations. Because we will restrict ourselves to using standard enthalpy changes, ΔH°_{298}, and standard entropy changes, ΔS°_{298} in the free energy equation, we limit ourselves to calculating values of ΔG°_{298} or ΔG° from the equations.

$$\Delta G^{\circ}_{298} = \Delta H^{\circ}_{298} - T\Delta S^{\circ}_{298} \quad (T = 298.15 \text{ K})$$

$$\Delta G^{\circ} = \Delta H^{\circ}_{298} - T\Delta S^{\circ}_{298} \quad (T = \text{some other temperature})$$

The calculation of ΔG° from ΔH°_{298} and ΔS°_{298} values uses the approximation that ΔH° and ΔS° do not change with temperature. However, ΔG° does vary with temperature.

Example 18.7   Evaluation of a Free Energy Change from ΔH° and ΔS°

Use the values for ΔH°_{298} (6.00×10^3 J) and ΔS°_{298} (22.1 J K^{-1}) to show that ΔG° predicts that the melting of ice is not spontaneous at $-10°C$ (263 K) but is spontaneous at $10°C$ (283 K).

Solution: We can calculate ΔG° at 263 K and 283 K for the reaction

$$H_2O(s) \longrightarrow H_2O(l)$$

from the free energy equation

$$\Delta G^{\circ} = \Delta H^{\circ}_{298} - T\Delta S^{\circ}_{298}$$

with $T = 263$ K and 283 K, respectively.

For $T = 263$ K $\Delta G^{\circ} = \Delta H^{\circ}_{298} - T\Delta S^{\circ}_{298}$

$$= 6000 \text{ J} - (263 \text{ K} \times 22.1 \text{ J K}^{-1})$$

$$= 6000 \text{ J} - 5810 \text{ J}$$

$$= 1.9 \times 10^2 \text{ J}$$

At 263 K ($-10°C$), ΔG° is positive and melting is not predicted to be spontaneous.

For $T = 263$ K $\Delta G^{\circ} = \Delta H^{\circ}_{298} - T\Delta S^{\circ}_{298}$

$$= 6000 \text{ J} - (283 \text{ K} \times 22.1 \text{ J K}^{-1})$$

$$= 6000 \text{ J} - 6250 \text{ J}$$

$$= -2.5 \times 10^2 \text{ J}$$

At 283 K ($10°C$), ΔG° is negative and melting is predicted to be spontaneous.

Check your learning: Use the values for ΔH°_{298} (136.94 kJ; Example 18.4, Check your learning) and ΔS°_{298} (120.6 J K^{-1}; Example 18.5, Check your learning) to show that the following reaction is not spontaneous at 25°C (298.15 K).

$$C_2H_6(g) \longrightarrow H_2(g) + C_2H_4(g)$$

Answer: ΔG°_{298} = 101.0 kJ. Processes with positive ΔG°_{298} values are not spontaneous at 25°C.

Again, it should be emphasized that predicting whether a given process will proceed spontaneously says nothing about the rate of the reaction. A reaction with a negative ΔG value will proceed spontaneously, but it may do so at a very rapid rate or at an incredibly slow rate.

18.12   Free Energy of Formation

There are several ways to evaluate a change in free energy, ΔG, of a system. We have just seen one way: use the relationship

$$\Delta G^\circ = \Delta H^\circ_{298} - T\Delta S^\circ_{298}$$

We can also determine ΔG°_{298} from the equilibrium constant for a reaction (Section 18.15), from measurements of voltages of electrochemical cells (Chapter 19), or from standard free energies of formation.

The **standard free energy of formation, ΔG°_f,** of a substance is the free energy change that accompanies the formation of exactly 1 mole of the pure substance from free elements in their most stable states under standard state conditions. For example, the standard free energy of formation of $CO_2(g)$ is −394.36 kilojoules per mole, which is the standard free energy change for the reaction

$$C(s) + O_2(g) \longrightarrow CO_2(g) \qquad \Delta G^\circ_f = \Delta G^\circ_{298} = -394.36 \text{ kJ mol}^{-1}$$

starting with the reactants at a pressure of 1 atmosphere and 25°C with the carbon present as graphite (the most stable form of carbon under this set of standard conditions) and ending with the product, gaseous CO_2, also at 1 atmosphere and a temperature of 25°C.

The free energy of a system is a state function. Hence we calculate values of ΔG°_{298}, the free energy change for a reaction run under standard state conditions, from standard free energies of formation, ΔG°_f, in the same way in which we calculate ΔH°_{298} values from standard molar enthalpies of formation. We use the free energy equivalent of Hess's law (Sections 4.6 and 18.7). For the reaction

$$m\text{A} + n\text{B} \longrightarrow x\text{C} + y\text{D}$$

the standard free energy change may be found by using the following equation, in which $\Delta G^\circ_{\text{product}}$ represents the standard free energy of a product and $\Delta G^\circ_{\text{reactant}}$ the standard free energy of a reactant.

$$\Delta G^\circ_{298} = \Sigma\Delta G^\circ_{\text{product}} - \Sigma\Delta G^\circ_{\text{reactant}}$$
$$= [(x \times \Delta G^\circ_C) + (y \times \Delta G^\circ_D)] - [(m \times \Delta G^\circ_A) + (n \times \Delta G^\circ_B)]$$

Standard free energy of formation values for several substances are given in Table 18.1 and in Appendix I for the standard state conditions of 25°C, 1 atmosphere, and unit activity. Note that the standard free energy of formation of any free element in its most stable state is zero.

Example 18.8 Calculation of ΔG°_{298} from ΔG°_f

The reaction of calcium oxide with the pollutant sulfur trioxide,

$$CaO(s) + SO_3(g) \longrightarrow CaSO_4(s)$$

has been proposed as one way of removing SO_3 from the smoke resulting from burning coal that contains sulfur. Using the following ΔG°_f values, determine ΔG°_{298} for the reaction and determine whether it is spontaneous under standard state conditions.

$$CaO(s): \quad \Delta G^{\circ}_f = -604.2 \text{ kJ mol}^{-1}$$
$$SO_3(g): \quad \Delta G^{\circ}_f = -371.1 \text{ kJ mol}^{-1}$$
$$CaSO_4(s): \quad \Delta G^{\circ}_f = -1320.3 \text{ kJ mol}^{-1}$$

Solution:

$$\Delta G^{\circ}_{298} = \Sigma \Delta G^{\circ}_{\text{product}} - \Sigma \Delta G^{\circ}_{\text{reactant}}$$
$$\Delta G^{\circ}_{298} = \Delta G^{\circ}_{f\,CaSO_4(s)} - [\Delta G^{\circ}_{f\,CaO(s)} + \Delta G^{\circ}_{f\,SO_3(g)}]$$
$$= 1 \text{ mol } CaSO_4(s) \times (-1320.3 \text{ kJ mol}^{-1}) - [1 \text{ mol } CaO(s)$$
$$\times (-604.2 \text{ kJ mol}^{-1}) + 1 \text{ mol } SO_3(g) \times (-371.1 \text{ kJ mol}^{-1})]$$
$$= -345.0 \text{ kJ}$$

ΔG°_{298} is less than zero, so the reaction is spontaneous.

Check your learning: Using ΔG°_{298} determined from free energies of formation (Table 18.1), determine whether the following reaction is spontaneous.

$$C_2H_6(g) \longrightarrow H_2(g) + C_2H_4(g)$$

Answer: 101.0 kJ, nonspontaneous

18.13   The Second Law of Thermodynamics

When a process occurs in an isolated system—a system in which no energy or matter can get in or out—ΔH must be zero. When the process is spontaneous, the free energy change, ΔG, is negative, and because ΔH is zero, the entropy change, ΔS, must be positive. Thus the entropy increases in an isolated system when a spontaneous process occurs. The universe is an isolated system; the universe includes everything, so nothing can get in or out. Thus *any spontaneous change that occurs in the universe must be accompanied by an increase in the entropy of the universe:* This is the **second law of thermodynamics.** It is difficult to use the second law in its pure form to predict the spontaneity of chemical reactions. However, the second law does tell us that the disorder of the universe is increasing because spontaneous processes are occurring all the time.

We can understand how the order of a portion of the universe can increase (and its entropy decrease) if we divide the universe up into a system (the portion where the order increases) and surroundings. We can represent the change in entropy of the

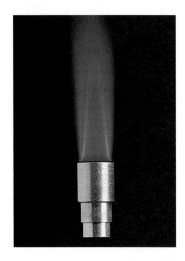

universe as the sum of the entropy changes in the system and in the surroundings.

$$\Delta S_{univ} = \Delta S_{sys} + \Delta S_{surr}$$

where ΔS_{sys} and ΔS_{surr} represent the entropy changes in the system and surroundings, respectively. This equation tells us that a spontaneous process with a decrease in disorder and a negative entropy change can occur in the system *if* another process that leads to a greater increase in disorder occurs in the surroundings. In such a situation, the overall entropy of the universe increases, even though the entropy in a small portion of the universe (the system) decreases.

18.14   Free Energy Changes and Nonstandard States

Many chemical reactions occur when the reactants and products are not in their standard states (see Fig. 18.18, for example). We can determine the free energy change, ΔG, of a reaction in a nonstandard state from the free energy change, ΔG°, of the same reaction when both the reactants and the products are at unit activities (pure solids or liquids, $P = 1$ atmosphere, concentrations $= 1$ M) by using the equation

$$\Delta G = \Delta G^\circ + RT \ln Q \tag{1}$$

In this equation T is the temperature on the Kelvin scale, R is a constant (8.314 J K^{-1}), and Q is the **reaction quotient** of the chemical reaction. We have seen (Section 14.2) that for the chemical reaction

$$mA + nB + \cdots \longrightarrow xC + yD + \cdots$$

the reaction quotient is

$$Q = \frac{[C]^x[D]^y \cdots}{[A]^m[B]^n \cdots}$$

Thus the reaction quotient Q has the same form as the equilibrium constant K for the reaction. However, Q has no fixed value; it is determined by the concentrations of reactants and products at whatever stage in the reaction we choose to evaluate Q. When A and B are first mixed, no products are present, and Q is equal to zero. The value of Q increases as the reaction proceeds: Q is equal to K only at equilibrium. Activities of the reactants and products should be used to evaluate Q. However, we will continue to use pressure (in atmospheres) for gases, concentrations (in moles per liter) for dissolved species, and unity for solvents and for pure solids and liquids as approximations of the activities of these species.

Example 18.9   Calculation of ΔG at 25°C

The reaction of calcium oxide with sulfur trioxide (Example 18.8) rarely occurs under standard state conditions. Calculate ΔG for this reaction at 25°C (298.15 K) when the pressure of SO_3 is 0.15 atm.

Solution: To determine ΔG, we use Equation 1 with $\Delta G^\circ = \Delta G^\circ_{298}$ because the temperature is 25°C (298.15 K).

$$\Delta G = \Delta G^\circ_{298} + RT \ln Q$$

CaO and CaSO$_4$ are solids, so their concentrations (activities) are 1. The activity of the gas SO$_3$ is taken as its pressure in atmospheres, 0.15 atm. In Example 18.8, ΔG°_{298} was shown to be -345.0 kJ. Thus

$$\Delta G = \Delta G^\circ_{298} + RT \ln Q$$

$$= \Delta G^\circ_{298} + RT \ln \frac{[CaSO_4]}{[CaO]P_{SO_3}}$$

$$= \Delta G^\circ_{298} + RT \ln \frac{1}{1 \times 0.15}$$

$$= -345.0 \text{ kJ} + (0.008314 \text{ kJ K}^{-1}) \times (298.15 \text{ K}) \times \ln 6.67$$

$$= -340.3 \text{ kJ}$$

The change of conditions caused a change in ΔG from -345.0 kJ to -340.3 kJ, but the reaction is still spontaneous under these conditions.

Check your learning: Is the decomposition of ammonia at 25°C spontaneous when $P_{N_2} = 0.870$ atm, $P_{H_2} = 0.250$ atm, and $P_{NH_3} = 12.9$ atm?

$$2NH_3(g) \longrightarrow 3H_2(g) + N_2(g) \qquad \Delta G^\circ_{298} = 33.2 \text{ kJ}$$

Answer: No, $\Delta G^\circ = 9.9$ kJ

Example 18.10   Calculation of ΔG at Elevated Temperatures

Over 17 million tons of lime are produced in the United States each year by thermal decomposition of limestone, CaCO$_3$. Calculate ΔG for the conversion of limestone (CaCO$_3$) to quicklime (CaO) in air at 1000°C.

$$CaCO_3(s) \longrightarrow CaO(s) + CO_2(g)$$

The partial pressure of carbon dioxide in the atmosphere is 0.0033 atm.

Solution: This process converts the initial state [1 mol CaCO$_3$(s)] to the final state [1 mol CaO(s) and 1 mol CO$_2$(g, $P = 0.0033$ atm)] under nonstandard conditions. We can determine ΔG with two steps.

1.   Calculate ΔG° for a CO$_2$ pressure of 1 atm at 1000°C from ΔH°_{298} and ΔS°_{298}, assuming that ΔH° and ΔS° for the reaction are equal to ΔH°_{298} and ΔS°_{298}, respectively. We cannot use ΔG°_{298} here because the temperature is not equal to 25°C (298 K).

2.   Calculate ΔG at 1000°C when the pressure of CO$_2$ is 0.0033 atm.

Step 1.   We evaluate ΔH°_{298} and ΔS°_{298} from the standard enthalpy and entropy values in Appendix I.

$$\Delta H^\circ_{298} = (1 \text{ mol CaO}(s) \times -635.5 \text{ kJ mol}^{-1}) + (1 \text{ mol CO}_2(g) \times -393.51 \text{ kJ mol}^{-1})$$
$$- (1 \text{ mol CaCO}_3(s) \times 1206.9 \text{ kJ mol}^{-1}) = 177.9 \text{ kJ}$$

$$S^\circ_{298} = (1 \text{ mol CaO}(s) \times 40 \text{ J mol}^{-1} \text{ K}^{-1}) + (1 \text{ mol CO}_2(g) \times 213.6 \text{ J mol}^{-1} \text{ K}^{-1})$$
$$- (1 \text{ mol CaCO}_3(s) \times 92.9 \text{ J mol}^{-1} \text{ K}^{-1}) = 161 \text{ J K}^{-1}$$

Next, assuming $\Delta H^\circ = \Delta H^\circ_{298}$ and $\Delta S^\circ = \Delta S^\circ_{298}$, we find ΔG° at 1000°C.

$$\Delta G^\circ = \Delta H^\circ - T\Delta S^\circ = 177.9 \text{ kJ} - 1273 \text{ K} \times 0.161 \text{ kJ K}^{-1} = -27 \text{ kJ}$$

(This value of ΔG° indicates that the decomposition of limestone is spontaneous at 1000°C when the pressure of CO_2 is 1 atm.)

Step 2. Now we can calculate ΔG.

$$\Delta G = \Delta G^\circ + RT \ln Q$$

$$Q = \frac{P_{CO_2}[\text{CaO}]}{[\text{CaCO}_3]} = \frac{0.0033 \times 1}{1} = 0.0033$$

$$\Delta G = 27 \text{ kJ} + (8.314 \text{ J K}^{-1})\left(\frac{1 \text{ kJ}}{1000 \text{ J}}\right)(1273 \text{ K})(\ln 0.0033)$$

$$= -27 \text{ kJ} + (-60.5 \text{ kJ}) = -87 \text{ kJ}$$

Changing the pressure of $CO_2(g)$ reduces the free energy change from -27 kJ (at a pressure of 1 atm) to -87 kJ (at a pressure of 0.0033 atm) and makes the reaction even more spontaneous at 1000°C.

Check your learning: Is the decomposition of ammonia spontaneous at 875°C? See Appendix I for values of ΔH° and S°.

$$2NH_3(g) \longrightarrow 3H_2(g) + N_2(g)$$

Answer: Yes, $\Delta G^\circ = -135.8$ kJ

Figure 18.19

The value of the equilibrium constant ($K_p = 0.282$ atmosphere) for the phase change $Br_2(l) \longrightarrow Br_2(g)$ at 25°C can be used to determine ΔG°_{298} (3.14 kilojoules) for the change.

18.15   The Relationship Between Free Energy Changes and Equilibrium Constants

When ΔG for any reaction is negative, the reaction occurs spontaneously as written, and the quantities of products increase. When ΔG is positive, the reverse reaction occurs spontaneously; that is, the species on the right react to give increased quantities of the species on the left. At equilibrium, ΔG can be neither positive nor negative; it can only be zero. At equilibrium, we have

$$\Delta G = 0 = \Delta G^\circ + RT \ln Q$$

where we use the value of ΔG° for unit activities of the reactants and products at temperature T. Because the reaction is at equilibrium, the value of the reaction quotient must be equal to that of the **equilibrium constant** for the reaction (Q is equal to K), so we have

$$\Delta G = 0 = \Delta G^\circ + RT \ln Q$$

and

$$\Delta G^\circ = -RT \ln K \tag{2}$$

Thus we can use the value of the standard free energy change, ΔG°, for a reaction to determine the equilibrium constant for that reaction, or we can determine values of ΔG° from equilibrium constants (Fig. 18.19). However, we should note that when a reactant or product is a gas, the equilibrium constant calculated from Equation 2 involves the pressure of that species in atmospheres.

Now we see one reason why calculating the value of $\Delta G°$ is useful. It enables us to predict the value of the equilibrium constant for a reaction and thus to determine whether the reaction will give significant amounts of products.

Example 18.11   Calculation of a Free Energy Change from K_a

Calculate the standard state free energy change for the ionization of acetic acid at 0°C.

$$CH_3CO_2H(aq) + H_2O(l) \longrightarrow H_3O^+(aq) + CH_3CO_2^-(aq)$$

The equilibrium constant for the reaction has been found to be 1.657×10^{-5} at 0°C.

Solution: At equilibrium,

$$\Delta G° = -RT \ln K$$
$$= -(8.314 \text{ J K}^{-1})(273.15 \text{ K})[\ln (1.657 \times 10^{-5})]$$
$$= -(8.314 \text{ J K}^{-1})(273.15 \text{ K})(-11.0079) = 2.500 \times 10^4 \text{ J}$$
$$= 25.00 \text{ kJ}$$

The free energy change for the *complete* transformation at 0°C of 1 L of 1 M acetic acid in water into 1 mol of hydrogen ion and 1 mol of acetate ion, both at 1 M concentrations in water, is 25.00 kJ. $\Delta G°$ is positive, so the reaction is not spontaneous. However, the reverse reaction,

$$H_3O^+(aq) + CH_3CO_2^-(aq) \longrightarrow CH_3CO_2H(aq) + H_2O(l)$$

is spontaneous ($\Delta G° = -25.00$ kJ).

Check your learning: Determine the value of the equilibrium constant at 298.15 K for the reaction

$$C_2H_6(g) \rightleftharpoons H_2(g) + C_2H_4(g) \qquad \Delta G°_{298} = 101.0 \text{ kJ}$$

Answer: 2.0×10^{-18}

Example 18.12   Calculation of an Equilibrium Constant from $\Delta H°_{298}$ and $\Delta S°_{298}$

Determine the value of K at 100°C for the reaction

$$I_2(g) + Cl_2(g) \longrightarrow 2ICl(g)$$

For this reaction, $\Delta H°_{298} = -26.9$ kJ and $\Delta S°_{298} = 11.3$ J K^{-1}.

Solution: We can solve this problem with two steps.

1.   Determine the value of $\Delta G°$ at 100°C (393.15 K) by using the following equation and the assumption that $\Delta H°_{298}$ and $\Delta S°_{298}$ do not vary with temperature.

$$\Delta G° = \Delta H°_{298} - T\Delta S°_{298}$$

2.   Determine the value of K by rearranging Equation 2.

$$\Delta G° = -RT \ln K$$

Step 1.

$$\Delta G° = \Delta H°_{298} - T\Delta S°_{298}$$
$$= -26.9 \text{ kJ} - (373.15 \text{ K} \times 0.0113 \text{ kJ K}^{-1})$$
$$= -31.1 \text{ kJ}$$

Step 2.

$$\Delta G° = -RT \ln K$$
$$\ln K = -\frac{\Delta G°}{RT} = -\frac{-31.1 \text{ kJ}}{(0.008314 \text{ kJ K}^{-1})(373.15 \text{ K})}$$
$$= 10.0$$
$$K = e^{10.0} = 2.2 \times 10^4$$

Check your learning: Determine the temperature at which $K = 1$ for the reaction

$$C_2H_6(g) \rightleftharpoons H_2(g) + C_2H_4(g)$$

Hint: Determine the value of $\Delta G°$ when $K = 1$. Then, using $\Delta H°_{298}$ and $\Delta S°_{298}$ for the reaction, determine the temperature at which $\Delta G°$ has this value.

Answer: 862°C

There is an alternative way to determine the equilibrium constant of a reaction at temperatures other than 25°C. If we combine the equation $\Delta G° = -RT \ln K$ and the equation $\Delta G° = \Delta H° - T\Delta S°$, we have

$$\Delta G° = -RT \ln K = \Delta H° - T\Delta S° \tag{3}$$

We can write two equations for an equilibrium reaction at two different temperatures, $T_1 < T_2$, with equilibrium constants K_{T_1} and K_{T_2}:

$$\Delta H° - T_1\Delta S° = -RT_1 \ln K_{T_1} \tag{4}$$
$$\Delta H° - T_2\Delta S° = -RT_2 \ln K_{T_2} \tag{5}$$

Now, if we make the approximation that $\Delta H°$ and $\Delta S°$ are independent of temperature, we can subtract Equation 4 from Equation 5 and rearrange the result to get an equation that relates change in temperature to changes in equilibrium constant.

$$\frac{\Delta H°(T_2 - T_1)}{RT_1T_2} = \ln \frac{K_{T_2}}{K_{T_1}} \tag{6}$$

Equations 3 and 6 prove to be quite useful. If we know $\Delta G°_{298}$, we can determine the equilibrium constant, K, at 298 K by using Equation 3. Then, if we know $\Delta H°_{298}$, we can obtain K for any other temperature by means of Equation 6 (within the limitation of the approximation that both $\Delta H°$ and $\Delta S°$ are independent of T). In addition, we can calculate the value of $\Delta G°$ at a temperature other than 298 K if $\Delta G°_{298}$ and $\Delta H°_{298}$ are known, because the value of $\Delta S°_{298}$ can be determined by using the equation

$$\Delta G°_{298} = \Delta H°_{298} - T\Delta S°_{298} \quad (T = 298 \text{ K})$$

Then, assuming both $\Delta H°$ and $\Delta S°$ are essentially independent of temperature, we can calculate $\Delta G°$ at other temperatures.

For Review   Summary

Chemical thermodynamics is the study of the energy transformations and transfers that accompany chemical and physical changes. From such a study we can determine whether a change will occur spontaneously and how far it will proceed—the equilibrium position of the change. The spontaneity of a chemical change is determined by the changes in energy and in disorder that accompany the change.

The **first law of thermodynamics** states that in any change that occurs in nature, the total energy of the universe remains constant. Thus by measuring the energy lost or gained by a system, we can determine the energy change within it. Energy can be lost from a system as the system does **work, w,** on the surroundings ($w < 0$) or as the surroundings do work on the system ($w > 0$). The amount of work due to expansion at constant pressure is given by the expression $-P(V_{final} - V_{initial})$. Energy can also be transferred as **heat, q.** In an **endothermic process,** heat is transferred into the system from the surroundings ($q > 0$). In an **exothermic process,** heat is transferred from the system to the surroundings ($q < 0$).

Although the amounts of heat and work that accompany a change may vary depending on how the change is carried out, the change in internal energy of the system is independent of how the change is accomplished. A property, such as internal energy, that does not depend on how the system gets from one state to another is called a **state function.** When a chemical change is carried out at constant pressure such that the only work is expansion work, q results from a change in a state function and is called the **enthalpy change, ΔH,** of the reaction. The enthalpy change that accompanies the formation of 1 mole of a substance from the elements in their most stable states, all at 298.15 K and 1 atmosphere with unit activity (a standard state), is called the **standard enthalpy of formation, ΔH_f°,** of the substance. Using **Hess's law,** we can describe the **standard enthalpy change** of any reaction, ΔH_{298}°, as the sum of the standard enthalpies of all the products minus those of all the reactants.

A **spontaneous change**—a change in a system that proceeds without any outside influence—is favored when the change is exothermic or when the change leads to an increase in the disorder of the system. The **second law of thermodynamics** states that any spontaneous change must be accompanied by an increase in the disorder of the universe.

The disorder of a substance can be determined quantitatively and is called the **entropy, S,** of the substance. The **third law of thermodynamics** states that the entropy of a pure crystalline substance at 0 K is equal to zero. Thus a measurement of the entropy change of such a substance as it is heated from 0 K to a higher temperature gives the entropy of the substance at the higher temperature. S is a state function, and the value of the **standard entropies, S_{298}°,** of many substances in the standard state at 298.15 K and 1 atmosphere have been measured and tabulated. The **standard entropy change, ΔS_{298}°,** for many reactions can be calculated as the sum of the standard entropies of all the products minus the sum of those of all the reactants. A positive value of an entropy change ($\Delta S > 0$) indicates that the disorder of the system has increased, whereas a negative value ($\Delta S < 0$) indicates that the disorder of the system has decreased.

The **free energy change, ΔG,** of a reaction is the difference between the enthalpy change of the reaction and the product of the temperature and entropy change of the reaction.

$$\Delta G = \Delta H - T\Delta S$$

For a reaction at constant temperature and pressure, a negative value of ΔG ($\Delta G < 0$) indicates that the reaction is spontaneous. Values of the **standard free energy change, ΔG_{298}°,** for a reaction under standard state conditions at 25°C can be calculated from **standard free energies of formation, ΔG_f°,** or from the values of ΔH_{298}° and ΔS_{298}° for the reaction.

The free energy change of a reaction that involves reactants or products at concentrations other than 1 M or pressures other than 1 atm, ΔG, can be determined using the equation

$$\Delta G = \Delta G^\circ + RT \ln Q$$

where R is equal to 8.314 J K^{-1} and Q is the reaction quotient. When a reaction has reached equilibrium, Q is equal to K, the equilibrium constant for the reaction, and ΔG is equal to zero. Thus, at equilibrium,

$$\Delta G^\circ = -RT \ln K$$

Key Terms and Concepts

change in internal energy, ΔE (18.2)
closed system   (18.1)
disorder   (18.6)
endothermic process   (18.2)
enthalpy change, ΔH (18.7)
entropy, ΔS (18.8, 18.10)
entropy change, S (18.8)

equilibrium constant   (18.15)
exothermic process   (18.2)
expansion work   (18.2)
first law of thermodynamics   (18.3)
free energy, G (18.11)
free energy change, ΔG (18.11)
heat, q (18.2)

Hess's law   (18.7)
internal energy, E (18.2)
nonstandard state   (18.14)
reaction quotient   (18.14)
second law of thermodynamics   (18.13)
spontaneous change   (18.5)
standard enthalpy change, ΔH_{298}° (18.7)

standard entropy, $S_{298}^°$ (18.8)
standard entropy change, $\Delta S_{298}^°$ (18.9)
standard free energy change, $\Delta G_{298}^°$
   (18.11)

standard free energy of formation, $\Delta G_f^°$
   (18.12)
standard state conditions   (18.5)
state functions   (18.4)

surroundings   (18.1)
system   (18.1)
third law of thermodynamics   (18.10)
work, w (18.2)

Exercises

Questions

1. What is an enthalpy change?

2. Define the following terms and describe what they have in common and in what ways they differ: heat, work, internal energy, enthalpy.

3. What is the first law of thermodynamics?

4. Explain the difference between q and ΔH. Under what conditions are they equal?

5. Explain the difference between a change in internal energy, ΔE, and an enthalpy change, ΔH.

6. Under what conditions are a change in internal energy and an enthalpy change equal (or very nearly equal)? Give an example of a chemical reaction where this is the case.

7. Explain why the heat produced by the combustion of 1 mol of carbon in Fig. 18.2 is not equal to $\Delta H_{298}^°$ for the reaction

$$C(s) + O_2(g) \longrightarrow CO_2(g)$$

8. What is the difference among ΔH, $\Delta H°$, and $\Delta H°_{298}$ for a chemical change?

9. In which of the following changes at constant pressure is work done by the surroundings on the system? By the system on the surroundings? In which is essentially no work done? What is the value of w in each case: $w > 0$, $w < 0$, or $w = 0$ or almost 0?

Initial State	Final State
(a) $H_2O(g)$	$H_2O(l)$
(b) $H_2O(s)$	$H_2O(g)$
(c) $2Na(s) + Cl_2(g)$	$2NaCl(s)$
(d) $H_2(g) + Cl_2(g)$	$2HCl(g)$
(e) $Na_2SO_4 \cdot 10H_2O(s)$	$Na_2SO_4(s) + 10H_2O(g)$
(f) $NO_2(g) + CO(g)$	$NO(g) + CO_2(g)$

10. In which of the following changes at constant pressure is work done by the surroundings on the system? By the system on the surroundings? In which is essentially no work done? What is the value of w in each case: $w > 0$, $w < 0$, or $w = 0$ or almost 0?

Initial State	Final State
(a) $H_2O(s)$	$H_2O(l)$
(b) $H_2O(g)$	$H_2O(s)$
(c) $2Na(s) + 2H_2O(l)$	$2NaOH(s) + H_2(g)$
(d) $3H_2(g) + N_2(g)$	$2NH_3(g)$
(e) $CaCO_3(s)$	$CaO(s) + CO_2(g)$
(f) $N_2(g) + O_2(g)$	$2NO(g)$

11. Define the term *entropy.*

12. Explain what happens when a process occurs with an increase in entropy.

13. What are the second and third laws of thermodynamics?

14. What is the difference among ΔS, $\Delta S°$, and $\Delta S_{298}^°$ for a chemical change?

15. Does the entropy of the system pictured in Fig. 18.1 increase, decrease, or remain essentially constant on going from the initial to the final state? If the entropy does change, give the sign of ΔS. Explain your answers.

16. Does the entropy of the system pictured in Fig. 18.2 increase, decrease, or remain essentially constant on going from the initial to the final state? If the entropy does change, give the sign of ΔS. Explain your answers.

17. Arrange the following systems, each of which consists of 1 mol of substance, in order of increasing entropy.

(a) $H_2(g)$, $HBrO_4(g)$, $HBr(g)$
(b) $H_2O(l)$, $H_2O(g)$, $H_2O(s)$
(c) $He(g)$, $Cl_2(g)$, $P_4(g)$

18. Arrange the following systems, each of which consists of 1 mol of substance, in order of increasing entropy.

(a) CCl_4, C_2Cl_4, C_2Cl_6
(b) $Ne(g)$, $Hg(l)$, $C(s)$
(c) SF_4, S_2F_{10}, SF_6

19. Arrange the following systems, each of which consists of 1 mol of substance, in order of increasing entropy: $H_2O(g)$ at 100°C, $N_2(s)$ at −215°C, $C_2H_5OH(g)$ at 100°C, $H_2O(l)$ at 25°C, $H_2O(s)$ at −215°C, $C_2H_5OH(s)$ at 0 K.

20. Does the entropy of each of the following systems increase, decrease, or not change in going from the initial to the final state? If the entropy does change, give the sign of ΔS and explain your answers.

Initial State	Final State
(a) $NaCl(s)$ at 298 K	$NaCl(s)$ at 0 K
(b) $H_2O(s)$ at 273 K and 1 atm	$H_2O(l)$ at 273 K and 1 atm
(c) 1 mol Si and 1 mol O_2	1 mol SiO_2
(d) 1 mol $CaCO_3$	1 mol CaO and 1 mol CO_2

21. What is a spontaneous reaction?

22. What is the difference among ΔG, $\Delta G°$, and $\Delta G_{298}^°$ for a chemical change?

23. Give an example of a reaction that is spontaneous under standard state conditions but that is slow.

24. Give an example of a reaction that is not spontaneous under standard state conditions but that is complete in a few minutes or less.

25. Explain under what conditions ΔG is equal to $\Delta G°$ for the reaction

$$2H_2O_2(l) \longrightarrow 2H_2O(g) + O_2(g)$$

26. Explain why the free energy change of a reaction varies with temperature.

27. As ammonium nitrate dissolves spontaneously in water at constant pressure, the solution gets cold. What is the sign of ΔH for this process? Is it possible to identify the sign of ΔS for this process from this information? Explain your answer.

28. As sulfuric acid dissolves spontaneously in water at constant pressure, heat is produced and the solution gets hot. What is the sign of ΔH for this process? Is it possible to identify the sign of ΔS for this process from this information? Explain your answer.

29. The reaction $3O_2(g) \longrightarrow 2O_3(g)$ is endothermic and proceeds with a decrease in the entropy of the system. Is this likely to be a spontaneous reaction? Explain.

30. No matter what their bond energy, all compounds decompose if heated to a sufficiently high temperature. Explain why the reaction $AB(g) \longrightarrow A(g) + B(g)$, where A and B represent atoms, will eventually become spontaneous with $K > 1$ as the temperature of the system is increased.

31. For the conversion $C(s, \text{ graphite}) \longrightarrow C(s, \text{ diamond})$, $\Delta H°_{298} = 1.897$ kJ, $\Delta S°_{298} = -3.36$ J K^{-1}, and $\Delta G°_{298} = 2.900$ kJ. The conversion is not spontaneous under standard state conditions. Will this conversion become spontaneous at any temperature, high or low? Explain your answer, assuming that $\Delta H°_{298}$ and $\Delta S°_{298}$ do not change with temperature.

32. Assuming that $\Delta H°_{298}$ and $\Delta S°_{298}$ do not change with temperature, is it possible for a reaction that is nonspontaneous under standard state conditions to become spontaneous both at some higher temperature and at some lower temperature? Explain your answer.

33. Describe the relative values of $\Delta H°_{298}$, $\Delta S°_{298}$, and $\Delta G°_{298}$ for a reaction that is not spontaneous under standard state conditions but that becomes spontaneous as the temperature is increased. Make up a set of values of $\Delta H°_{298}$, $\Delta S°_{298}$, and $\Delta G°_{298}$ for this reaction and indicate the temperature at which it becomes spontaneous, assuming that $\Delta H°_{298}$ and $\Delta S°_{298}$ do not change with temperature.

34. Describe the relative values of $\Delta H°_{298}$, $\Delta S°_{298}$, and $\Delta G°_{298}$ for a reaction that is not spontaneous under standard state conditions but that becomes spontaneous as the temperature is decreased. Make up a set of values of $\Delta H°_{298}$, $\Delta S°_{298}$, and $\Delta G°_{298}$ for this reaction and indicate the temperature at which it becomes spontaneous, assuming that $\Delta H°_{298}$ and $\Delta S°_{298}$ do not change with temperature.

35. Explain what happens as a reaction starts with $\Delta G < 0$ (negative) and reaches the point where $\Delta G = 0$.

36. A system at standard state conditions with a negative $\Delta G°_{298}$ for a particular reaction will undergo that reaction spontaneously. When the system reaches equilibrium for that reaction, the reaction is no longer spontaneous. Explain the changes in the system that shift the reaction from spontaneous to nonspontaneous.

Heat, Work, and Internal Energy

37. Describe the system in each of the reactions represented by the following equations. Describe the initial state and the final state at standard state conditions.

(a) $MnO_2(s) \longrightarrow Mn(s) + O_2(g)$
(b) $H_2(g) + Br_2(l) \longrightarrow 2HBr(g)$
(c) $Cu(s) + S(g) \longrightarrow CuS(s)$
(d) $2LiOH(s) + CO_2(g) \longrightarrow Li_2CO_3(s) + H_2O(g)$
(e) $CH_4(g) + O_2(g) \longrightarrow C(s) + 2H_2O(g)$
(f) $CS_2(g) + 3Cl_2(g) \longrightarrow CCl_4(g) + S_2Cl_2(g)$
(g) $SnCl_4(l) \longrightarrow SnCl_4(g)$
(h) $CS_2(g) \longrightarrow CS_2(l)$
(i) $Cu(s) \longrightarrow Cu(g)$

38. Describe the system in each of the reactions represented by the following equations. Describe the initial state and the final state at standard state conditions.

(a) $C(s) + O_2(g) \longrightarrow CO_2(g)$
(b) $O_2(g) + N_2(g) \longrightarrow 2NO(g)$
(c) $2Cu(s) + S(g) \longrightarrow Cu_2S(s)$
(d) $CaO(s) + H_2O(l) \longrightarrow Ca(OH)_2(s)$
(e) $Fe_2O_3(s) + 3CO(g) \longrightarrow 2Fe(s) + 3CO_2(g)$
(f) $CaSO_4 \cdot 2H_2O(s) \longrightarrow CaSO_4(s) + 2H_2O(g)$
(g) $C_6H_6(l) \longrightarrow C_6H_6(g)$
(h) $TiCl_4(g) \longrightarrow TiCl_4(l)$
(i) $Li(s) \longrightarrow Li(g)$

39. Calculate the missing value of ΔE, q, or w for each system, given the following data.

(a) $q = 570$ J; $w = 300$ J
(b) $\Delta E = -7500$ J; $w = -4500$ J
(c) $\Delta E = -250$ J; $q = 300$ J
(d) The system absorbs 2.000 kJ of heat and does 1425 J of work on the surroundings.

40. Calculate the missing value of ΔE, q, or w for each system, given the following data.

(a) $q = 570$ J; $w = -300$ J
(b) $\Delta E = -7500$ J; $w = 4500$ J
(c) $\Delta E = 250$ J; $q = 300$ J
(d) The system absorbs 1.000 kJ of heat and does 650 J of work on the surroundings.

41. During the process of expanding against a constant pressure of 0.50 atm from 10.0 L to 16.0 L, a gas absorbs 125 J of heat. What is the change in internal energy of the gas?

42. While expanding from 0.125 L to 10.000 L against a constant pressure of 4.25 atm, a gas evolves 25 J of heat. What is the change in internal energy of the gas?

43. Calculate the work involved in a system consisting of exactly 1 mol of H_2O as it changes from a liquid at 373 K (volume = 18.9 mL) to a gas at 373 K (volume = 30.6 L) under a constant pressure of 1 atm. Does this work increase or decrease the internal energy of the system?

44. Calculate the work involved as a system composed of exactly 1 mol of CH_3OH changes from a gas at 338 K (volume = 27.7 L) to a liquid at 338 K (volume = 25.3 mL) under a constant pressure of 1 atm. Does this work increase or decrease the internal energy of the system?

45. Assume that the only change in volume is due to the production of hydrogen and calculate w, the work done, when 2.00 mol of Zn dissolves in hydrochloric acid, giving H_2 at 35°C and 1.00 atm.

$$Zn(s) + 2HCl(aq) \longrightarrow ZnCl_2(aq) + H_2(g)$$

46. Assuming that the volume change in the solid and liquids can be neglected and that gases exhibit ideal behavior, calculate the work done when 17.0 g of $CaCO_3$ reacts at 75°C and 0.855 atm according to the following reaction.

$$CaCO_3(s) + 2HNO_3(aq) \longrightarrow$$
$$Ca(NO_3)_2 + CO_2(g) + H_2O(aq)$$

Enthalpy Change

47. Use the data in Appendix I to determine the enthalpy change for each of the following reactions, which are run under standard state conditions. Identify whether each is endothermic or exothermic.

(a) $MnO_2(s) \longrightarrow Mn(s) + O_2(g)$
(b) $H_2(g) + Br_2(l) \longrightarrow 2HBr(g)$
(c) $Cu(s) + S(g) \longrightarrow CuS(s)$
(d) $2LiOH(s) + CO_2(g) \longrightarrow Li_2CO_3(s) + H_2O(g)$
(e) $CH_4(g) + O_2(g) \longrightarrow C(s) + 2H_2O(g)$
(f) $CS_2(g) + 3Cl_2(g) \longrightarrow CCl_4(g) + S_2Cl_2(g)$

48. Use the data in Appendix I to determine the enthalpy change for each of the following reactions, which are run under standard state conditions. Identify whether each is endothermic or exothermic.

(a) $C(s) + O_2(g) \longrightarrow CO_2(g)$
(b) $O_2(g) + N_2(g) \longrightarrow 2NO(g)$
(c) $2Cu(s) + S(g) \longrightarrow Cu_2S(s)$
(d) $CaO(s) + H_2O(l) \longrightarrow Ca(OH)_2(s)$
(e) $Fe_2O_3(s) + 3CO(g) \longrightarrow 2\,Fe(s) + 3CO_2(g)$
(f) $CaSO_4 \cdot 2H_2O(s) \longrightarrow CaSO_4(s) + 2H_2O(g)$

49. Which of the reactions in Exercise 47 exhibit an enthalpy change that favors a spontaneous reaction under standard state conditions?

50. Which of the reactions in Exercise 48 exhibit an enthalpy change that favors a spontaneous reaction under standard state conditions?

51. Predict whether each of the following phase changes is endothermic or exothermic. Calculate ΔH_{298}° for each phase change and confirm your prediction.

(a) $SnCl_4(l) \longrightarrow SnCl_4(g)$ (b) $CS_2(g) \longrightarrow CS_2(l)$
(c) $Cu(s) \longrightarrow Cu(g)$

52. Predict whether each of the following phase changes is endothermic or exothermic. Calculate ΔH_{298}° for each phase change and confirm your prediction.

(a) $C_6H_6(l) \longrightarrow C_6H_6(g)$
(b) $TiCl_4(g) \longrightarrow TiCl_4(l)$
(c) $Li(s) \longrightarrow Li(g)$

53. Which of the phase changes in Exercise 51 exhibit an enthalpy change that favors a spontaneous reaction under standard state conditions?

54. Which of the phase changes in Exercise 52 exhibit an enthalpy change that favors a spontaneous reaction under standard state conditions?

55. Show that the standard enthalpy of formation of phosphoric acid is a change in a state function by evaluating ΔH_f° for $H_3PO_4(s)$ as it is prepared by two different stepwise processes involving the following reactions.

Process A. $P_4(s) + 5O_2(g) \longrightarrow P_4O_{10}(s)$
$$O_2(g) + 2H_2(g) \longrightarrow 2H_2O(g)$$
$$6H_2O(g) + P_4O_{10}(s) \longrightarrow 4H_3PO_4(s)$$

Process B. $P_4(s) + 6H_2(g) \longrightarrow 4PH_3(g)$
$$PH_3(g) + 2O_2(g) \longrightarrow H_3PO_4(s)$$

56. Show that the standard enthalpy of formation of sulfuric acid is a change in a state function by evaluating ΔH_f° for $H_2SO_4(l)$ as it is prepared by two different stepwise processes involving the following reactions.

Process A. $S_8(s) + 12O_2(g) \longrightarrow 8SO_3(g)$
$$2H_2(g) + O_2(g) \longrightarrow 2H_2O(g)$$
$$SO_3(g) + H_2O(l) \longrightarrow H_2SO_4(l)$$

Process B. $S_8(s) + 8H_2(g) \longrightarrow 8H_2S(g)$
$$H_2S(g) + 2O_2(g) \longrightarrow H_2SO_4(l)$$

57. Using the data in Appendix I, determine the enthalpy change for the combustion of liquid ethanol, C_2H_5OH, under standard state conditions to give gaseous carbon dioxide and liquid water.

58. Using the data in Appendix I, determine the enthalpy change for the combustion of gaseous propane, C_3H_8, under standard state conditions to give gaseous carbon dioxide and water vapor.

59. Calculate ΔH_{298}° for the reaction

$$2Na(s) + 2H_2O(l) \longrightarrow 2NaOH(s) + H_2(g)$$

The standard enthalpy of formation of $NaOH(s)$ is -426.8 kJ mol^{-1}; other enthalpy values are in Appendix I.

60. Calculate ΔH_{298}° for the reaction

$$Ca(s) + 2H_2O(l) \longrightarrow Ca(OH)_2(s) + H_2(g)$$

The standard enthalpies of formation for the compounds involved may be found in Appendix I.

61. The white pigment TiO_2 is prepared by the hydrolysis of titanium tetrachloride, $TiCl_4$, in the gas phase.

$$TiCl_4(g) + 2H_2O(g) \longrightarrow TiO_2(s) + 4HCl(g)$$

How much heat is evolved in the production of exactly 1 mol of $TiO_2(s)$ under standard state conditions of 25°C and 1 atm?

62. Consider the conversion of graphite to diamond.

$$C(s, graphite) \longrightarrow C(s, diamond) \qquad \Delta H° = 1.90 \text{ kJ}$$

How much do the enthalpy changes for the combustion of graphite and diamond differ? That is, how different are the standard enthalpies for the reactions represented by the following equations?

$$C(s, graphite) + O_2(g) \longrightarrow CO_2(g)$$
$$C(s, diamond) + O_2(g) \longrightarrow CO_2(g)$$

63. In 1774 Joseph Priestley prepared oxygen by heating red mercury(II) oxide with the light from the sun focused through a lens.

   (a) What is the standard enthalpy change for the decomposition under standard state conditions?

$$2HgO(s, red) \longrightarrow 2Hg(l) + O_2(g)$$

   (b) What is the work involved in this process if the decomposition were carried out under standard state conditions?

64. The decomposition of hydrogen peroxide, H_2O_2, has been used to provide thrust in the control jets of various space vehicles.

   (a) How much heat is produced by the decomposition of exactly 1 mol of H_2O_2 under standard state conditions?

$$H_2O_2(l) \longrightarrow H_2O(g) + \tfrac{1}{2}O_2(g)$$

   (b) How much work would be done if the decomposition were carried out under standard state conditions?

Entropy Change

65. Predict whether the entropy of each of the following systems should increase, decrease, or remain approximately constant during the following reactions, which occur under standard state conditions. If the entropy does change, give the sign of ΔS. Use the data in Appendix I to determine the entropy changes and verify your predictions.

   (a) $MnO_2(s) \longrightarrow Mn(s) + O_2(g)$
   (b) $H_2(g) + Br_2(l) \longrightarrow 2HBr(g)$
   (c) $Cu(s) + S(g) \longrightarrow CuS(s)$
   (d) $2LiOH(s) + CO_2(g) \longrightarrow Li_2CO_3(s) + H_2O(g)$
   (e) $CH_4(g) + O_2(g) \longrightarrow C(s) + 2H_2O(g)$
   (f) $CS_2(g) + 3Cl_2(g) \longrightarrow CCl_4(g) + S_2Cl_2(g)$

66. Predict whether the entropy of each of the following systems should increase, decrease, or remain approximately constant during the following reactions, which occur under standard state conditions. If the entropy does change, give the sign of ΔS. Use the data in Appendix I to determine the entropy changes and verify your predictions.

   (a) $C(s) + O_2(g) \longrightarrow CO_2(g)$
   (b) $O_2(g) + N_2(g) \longrightarrow 2NO(g)$
   (c) $2Cu(s) + S(g) \longrightarrow Cu_2S(s)$
   (d) $CaO(s) + H_2O(l) \longrightarrow Ca(OH)_2(s)$
   (e) $Fe_2O_3(s) + 3CO(g) \longrightarrow 2Fe(s) + 3CO_2(g)$
   (f) $CaSO_4 \cdot 2H_2O(s) \longrightarrow CaSO_4(s) + 2H_2O(g)$

67. Which of the reactions in Exercise 65 exhibit an entropy change that favors a spontaneous reaction under standard state conditions?

68. Which of the reactions in Exercise 66 exhibit an entropy change that favors a spontaneous reaction under standard state conditions?

69. Calculate $\Delta S°_{298}$ for the following phase changes.

   (a) $SnCl_4(l) \longrightarrow SnCl_4(g)$
   (b) $CS_2(g) \longrightarrow CS_2(l)$
   (c) $Cu(s) \longrightarrow Cu(g)$

70. Calculate $\Delta S°_{298}$ for the following phase changes.

   (a) $C_6H_6(l) \longrightarrow C_6H_6(g)$
   (b) $TiCl_4(g) \longrightarrow TiCl_4(l)$
   (c) $Li(s) \longrightarrow Li(g)$

71. Which of the phase changes in Exercise 69 exhibit an entropy change that favors a spontaneous reaction under standard state conditions?

72. Which of the phase changes in Exercise 70 exhibit an entropy change that favors a spontaneous reaction under standard state conditions?

73. (a) Explain the difference between the *standard entropy* of phosphoric acid and the *standard entropy of formation* of phosphoric acid, $\Delta S°_{f_{298}}$, the entropy change that accompanies the reaction

$$\tfrac{3}{2}H_2(g) + 2O_2(g) + \tfrac{1}{4}P_4(s) \longrightarrow H_3PO_4(s)$$

   (b) Evaluate $\Delta S°_{f_{298}}$ for solid phosphoric acid.

74. (a) Explain the difference between the *standard entropy* of sulfuric acid and the *standard entropy of formation* of sulfuric acid, $\Delta S°_{f_{298}}$, the entropy change that accompanies the reaction

$$H_2(g) + 2O_2(g) + \tfrac{1}{8}S_8(s) \longrightarrow H_2SO_4(l)$$

   (b) Evaluate $\Delta S°_{f_{298}}$ for sulfuric acid.

75. Using the data in Appendix I, determine the entropy change for the combustion of liquid ethanol, C_2H_5OH, under standard state conditions to give gaseous carbon dioxide and liquid water.

76. Using the data in Appendix I, determine the entropy change for the combustion of gaseous propane, C_3H_8, under standard state conditions to give gaseous carbon dioxide and water vapor.

77. Predict the sign of the entropy change and calculate ΔS°_{298} for the following reaction.

$$2NH_3(g) \longrightarrow N_2(g) + 3H_2(g)$$

78. Predict the sign of the entropy change and calculate ΔS°_{298} for the formation of ozone, $O_3(g)$, from oxygen, $O_2(g)$.

79. Calculate the standard entropy change for the decomposition of red mercury(II) oxide under standard state conditions.

$$2HgO(s, red) \longrightarrow 2Hg(l) + O_2(g)$$

Does ΔS°_{298} favor a spontaneous reaction?

80. Calculate the standard entropy change for the decomposition of hydrogen peroxide in a control jet of a space vehicle.

$$H_2O_2(l) \longrightarrow H_2O(g) + \tfrac{1}{2}O_2(g)$$

Does ΔS°_{298} favor a spontaneous reaction?

Free Energy Change

81. Use the standard free energy data in Appendix I to determine the free energy change for each of the following reactions, which are run under standard state conditions.

(a) $MnO_2(s) \longrightarrow Mn(s) + O_2(g)$
(b) $H_2(g) + Br_2(l) \longrightarrow 2HBr(g)$
(c) $Cu(s) + S(g) \longrightarrow CuS(s)$
(d) $2LiOH(s) + CO_2(g) \longrightarrow Li_2CO_3(s) + H_2O(g)$
(e) $CH_4(g) + O_2(g) \longrightarrow C(s) + 2H_2O(g)$
(f) $CS_2(g) + 3Cl_2(g) \longrightarrow CCl_4(g) + S_2Cl_2(g)$

82. Use the standard free energy data in Appendix I to determine the free energy change for each of the following reactions, which are run under standard state conditions.

(a) $C(s) + O_2(g) \longrightarrow CO_2(g)$
(b) $O_2(g) + N_2(g) \longrightarrow 2NO(g)$
(c) $2Cu(s) + S(g) \longrightarrow Cu_2S(s)$
(d) $CaO(s) + H_2O(l) \longrightarrow Ca(OH)_2(s)$
(e) $Fe_2O_3(s) + 3CO(g) \longrightarrow 2Fe(s) + 3CO_2(g)$
(f) $CaSO_4 \cdot 2H_2O(s) \longrightarrow CaSO_4(s) + 2H_2O(g)$

83. Use the ΔH and ΔS values determined in Exercises 47 and 65 to determine the standard free energy change for each of the reactions in Exercise 81.

84. Use the ΔH and ΔS values determined in Exercises 48 and 66 to determine the standard free energy change for each of the reactions in Exercise 82.

85. For which of the reactions in Exercise 81 does the free energy change indicate a spontaneous reaction under standard state conditions?

86. For which of the reactions in Exercise 82 does the free energy change indicate a spontaneous reaction under standard state conditions?

87. Calculate ΔG°_{298} for the following phase changes and determine which are spontaneous under standard state conditions.

(a) $SnCl_4(l) \longrightarrow SnCl_4(g)$
(b) $CS_2(g) \longrightarrow CS_2(l)$
(c) $Cu(s) \longrightarrow Cu(g)$

88. Calculate ΔG°_{298} for the following phase changes and determine which are spontaneous under standard state conditions.

(a) $C_6H_6(l) \longrightarrow C_6H_6(g)$
(b) $TiCl_4(g) \longrightarrow TiCl_4(l)$
(c) $Li(s) \longrightarrow Li(g)$

89. Show that the standard free energy of formation of phorphoric acid is a change in a state function by evaluating ΔG°_{298} for $H_3PO_4(s)$ as it is prepared by two different stepwise processes that involve the following reactions.

Process A. $P_4(s) + 5O_2(g) \longrightarrow P_4O_{10}(s)$
$$O_2(g) + 2H_2(g) \longrightarrow 2H_2O(g)$$
$$6H_2O(g) + P_4O_{10}(s) \longrightarrow 4H_3PO_4(s)$$

Process B. $P_4(s) + 6H_2(g) \longrightarrow 4PH_3(g)$
$$PH_3(g) + 2O_2(g) \longrightarrow H_3PO_4(s)$$

90. Show that the standard free energy of formation of sulfuric acid is a change in a state function by evaluating ΔG°_{298} for $H_2SO_4(l)$ as it is prepared by two different stepwise processes that involve the following reactions.

Process A. $S_8(s) + 12O_2(g) \longrightarrow 8SO_3(g)$
$$SO_3(g) + H_2O(l) \longrightarrow H_2SO_4(l)$$
Process B. $S_8(s) + 8H_2(g) \longrightarrow 8H_2S(g)$
$$H_2S(g) + 2O_2(g) \longrightarrow H_2SO_4(l)$$

91. Using the standard free energy data in Appendix I, determine the free energy change for the combustion of liquid ethanol, C_2H_5OH, under standard state conditions to give gaseous carbon dioxide and liquid water.

92. Using the standard free energy data in Appendix I, determine the free energy change for the combustion of gaseous propane, C_3H_8, under standard state conditions to give gaseous carbon dioxide and water vapor.

93. Calculate ΔG°_{298} for the formation of ozone, $O_3(g)$, from oxygen, $O_2(g)$. Is the formation spontaneous under standard state conditions?

94. (a) Using the data in Appendix I, calculate the standard free energy changes for the following reactions.
    i. $2Al(s) + 3F_2(g) \longrightarrow 2AlF_3(s)$
    ii. $3C_2H_2 \longrightarrow C_6H_6$
    iii. $TiO_2 + C + 2Cl_2 \longrightarrow TiCl_4 + CO_2$
    iv. $2NO_2 \longrightarrow N_2O_4$
    (b) Which of those reactions are spontaneous under standard state conditions?

95. For a certain process at 375 K, $\Delta S = 67.0 \text{ J K}^{-1}$ and $\Delta H = -56.9$ kJ. Find the free energy change for this process at this temperature.

96. Calculate ΔG°_{298} for the decomposition of red mercury(II) oxide under standard state conditions.

$$2HgO(s, \text{red}) \longrightarrow 2Hg(l) + O_2(g)$$

(a) Is the decomposition spontaneous under standard state conditions?

(b) Assuming ΔH and ΔS are independent of temperature, determine whether the reaction is spontaneous at 408°C.

(c) Determine at what Celsius temperature the decomposition will become spontaneous, assuming ΔH and ΔS are independent of temperature.

97. The standard molar enthalpies of formation of $NO(g)$, $NO_2(g)$, and $N_2O_3(g)$ are 90.25 kJ mol^{-1}, 33.2 kJ mol^{-1}, and 83.72 kJ mol^{-1}, respectively. Their standard molar entropies are 210.65 J mol^{-1} K^{-1}, 239.9 J mol^{-1} K^{-1}, and 312.2 J mol^{-1} K^{-1}, respectively.

(a) Use the foregoing data to calculate the free energy change for the following reaction at 25.0°C.

$$N_2O_3(g) \longrightarrow NO(g) + NO_2(g)$$

(b) Repeat the above calculation for 0.00°C and 100.0°C, assuming that the enthalpy and entropy changes do not vary with a change in temperature. Is the reaction spontaneous at 0.00°C? At 100.0°C?

98. Among other things, an ideal fuel for the control jet of a space vehicle should decompose in a spontaneous exothermic reaction when exposed to the appropriate catalyst. Evaluate the following substances under standard state conditions as suitable candidates for fuels.

(a) Ammonia: $2NH_3(g) \longrightarrow N_2(g) + 3H_2(g)$
(b) Diborane: $B_2H_6(g) \longrightarrow 2B(g) + 3H_2(g)$
(c) Hydrazine: $N_2H_4(g) \longrightarrow N_2(g) + 2H_2(g)$
(d) Hydrogen peroxide: $H_2O_2(l) \longrightarrow H_2O(g) + \frac{1}{2}O_2(g)$

Free Energy and Equilibrium Constants

99. Calculate $\Delta G°$ for each of the following reactions from the equilibrium constant at the temperature given.

(a) $N_2(g) + O_2(g) \longrightarrow 2NO(g)$
$T = 2000°C, K_p = 4.1 \times 10^{-4}$
(b) $H_2(g) + I_2(g) \rightleftharpoons 2HI(g)$
$T = 400°C, K_p = 50.0$
(c) $CO_2(g) + H_2(g) \longrightarrow CO(g) + H_2O(g)$
$T = 980°C, K_p = 1.67$
(d) $CaCO_3(s) \longrightarrow CaO(s) + CO_2(g)$
$T = 900°C, K_p = 1.04$
(e) $HF(aq) + H_2O(l) \longrightarrow H_3O^+(aq) + F^-(aq)$
$T = 25°C, K = 7.2 \times 10^{-4}$
(f) $AgBr(s) \longrightarrow Ag^+(aq) + Br^-(aq)$
$T = 25°C, K = 3.3 \times 10^{-13}$

100. Calculate $\Delta G°$ for each of the following reactions from the equilibrium constant at the temperature given.

(a) $Cl_2(g) + Br_2(g) \rightleftharpoons 2BrCl(g)$
$T = 25°C, K_p = 4.7 \times 10^{-2}$
(b) $2SO_2(g) + O_2(g) \rightleftharpoons 2SO_3(g)$
$T = 500°C, K_p = 48.2$
(c) $H_2O(l) \rightleftharpoons H_2O(g)$
$T = 60°C, K_p = 0.196$ atm
(d) $CoO(s) + CO(g) \rightleftharpoons Co(s) + CO_2(g)$
$T = 550°C, K_p = 4.90 \times 10^2$
(e) $CH_3NH_2(aq) + H_2O(l) \longrightarrow$
$$CH_3NH_3^+(aq) + OH^-(aq)$$
$T = 25°C, K = 4.4 \times 10^{-4}$
(f) $PbI_2(s) \longrightarrow Pb^{2+}(aq) + 2I^-(aq)$
$T = 25°C, K = 8.7 \times 10^{-9}$

101. Calculate the equilibrium constant at 25°C for each of the following reactions from the value of $\Delta G°_{298}$ given.

(a) $O_2(g) + 2F_2(g) \longrightarrow 2OF_2(g)$ $\Delta G°_{298} = -9.2$ kJ
(b) $I_2(s) + Br_2(l) \longrightarrow 2IBr(g)$ $\Delta G°_{298} = 7.3$ kJ
(c) $2LiOH(s) + CO_2(g) \longrightarrow Li_2CO_3(s) + H_2O(g)$
$$\Delta G°_{298} = -79 \text{ kJ}$$
(d) $N_2O_3(g) \longrightarrow NO(g) + NO_2(g)$ $\Delta G°_{298} = -1.6$ kJ
(e) $SnCl_4(l) \longrightarrow SnCl_4(g)$ $\Delta G°_{298} = 8.0$ kJ

102. Calculate the equilibrium constant at 25°C for each of the following reactions from the value of $\Delta G°_{298}$ given.

(a) $I_2(s) + Cl_2(g) \longrightarrow 2ICl(g)$ $\Delta G°_{298} = -10.88$ kJ
(b) $H_2(g) + I_2(s) \longrightarrow 2HI(g)$ $\Delta G°_{298} = 3.4$ kJ
(c) $CS_2(g) + 3Cl_2(g) \longrightarrow CCl_4(g) + S_2Cl_2(g)$
$$\Delta G°_{298} = -39 \text{ kJ}$$
(d) $2SO_2(g) + O_2(g) \longrightarrow 2SO_3(g)$
$$\Delta G°_{298} = -141.82 \text{ kJ}$$
(e) $CS_2(g) \longrightarrow CS_2(l)$ $\Delta G°_{298} = -1.88$ kJ

103. Using the data in Appendix I, calculate the equilibrium constant at the temperature given.

(a) $O_2(g) + 2F_2(g) \longrightarrow 2OF_2(g)$ $T = 100°C$
(b) $I_2(s) + Br_2(l) \longrightarrow 2IBr(g)$ $T = 0.0°C$
(c) $2LiOH(s) + CO_2(g) \longrightarrow Li_2CO_3(s) + H_2O(g)$
$$T = 575°C$$
(d) $N_2O_3(g) \longrightarrow NO(g) + NO_2(g)$ $T = -10.0°C$
(e) $SnCl_4(l) \longrightarrow SnCl_4(g)$ $T = 200°C$

104. Using the data in Appendix I, calculate the equilibrium constant at the temperature given.

(a) $I_2(s) + Cl_2(g) \longrightarrow 2ICl(g)$ $T = 100°C$
(b) $H_2(g) + I_2(s) \longrightarrow 2HI(g)$ $T = 0.0°C$
(c) $CS_2(g) + 3Cl_2(g) \longrightarrow CCl_4(g) + S_2Cl_2(g)$
$$T = 125°C$$
(d) $2SO_2(g) + O_2(g) \longrightarrow 2SO_3(g)$ $T = 675°C$
(e) $CS_2(g) \longrightarrow CS_2(l)$ $T = 90°C$

105. What is the assumption made in Exercises 103 and 104?

106. At 298 K the equilibrium constant, K_p, for the reaction $N_2O_4(g) \rightleftharpoons 2NO_2(g)$ is 0.142. What is $\Delta G°_{298}$ for the reaction?

107. If the standard enthalpy of vaporization of CH_2Cl_2 is 29.0 kJ mol^{-1} at 25.0°C and the entropy change accompanying vaporization is 92.5 J mol^{-1} K^{-1}, calculate the normal boiling temperature of CH_2Cl_2.

108. Consider the decomposition of dinitrogen trioxide described in Exercise 97.

$$N_2O_3(g) \longrightarrow NO(g) + NO_2(g)$$

At what temperature does this reaction become spontaneous?

109. Calculate ΔG°_{298} for the reaction of 1 mol of $H_3O^+(aq)$ with 1 mol of $OH^-(aq)$, using the equilibrium constant for the self-ionization of water at 298 K.

$$2H_2O(l) \rightleftharpoons H_3O^+(aq) + OH^-(aq) \quad K_w = 1.00 \times 10^{-14}$$

110. Hydrogen sulfide is a pollutant found in natural gas. Following its removal, it is converted to sulfur by the reaction

$$2H_2S(g) + SO_2(g) \longrightarrow 3S(s) + 2H_2O(g)$$

What is the equilibrium constant for this reaction? Is the reaction endothermic or exothermic?

111. (a) Using the ΔH and ΔS data in Appendix I, determine the temperature at which liquid water and gaseous water are in equilibrium with each other at 1 atm.
    (b) The boiling point of water at 1 atm is 373 K (100°C), but the calculation of the value in part (a) was based on the assumptions that ΔH° and ΔS° are independent of temperature. Recalculate the temperature at which liquid water and gaseous water are in equilibrium with each other at 1 atm, this time using the values of ΔH° and ΔS° at 100°C ($\Delta H^\circ = 40,656$ J and $\Delta S^\circ = 108.95$ J K^{-1}).

112. Consider the decomposition of $CaCO_3(s)$ into $CaO(s)$ and $CO_2(g)$ at 1 atm.
    (a) Estimate the minimum temperature at which you would conduct the reaction.
    (b) Calculate the equilibrium vapor pressure of $CO_2(g)$ above $CaCO_3(s)$ in a closed container at 298 K.

113. Hydrogen chloride, $HCl(g)$, and ammonia, $NH_3(g)$, escape from bottles of their solutions and react to form the white glaze often seen on glass in chemistry laboratories.

$$HCl(g) + NH_3(g) \longrightarrow NH_4Cl(s)$$

    (a) Calculate the free energy change, ΔG°_{298}, for this reaction.
    (b) At what temperature will ΔG° for the reaction be equal to zero?
    (c) Calculate the equilibrium constant for the decomposition of solid NH_4Cl to $HCl(g)$ and $NH_3(g)$ at 25°C.

114. Benzene can be prepared from acetylene.

$$3C_2H_2(g) \longrightarrow C_6H_6(g)$$

    (a) Calculate the standard free energy change, ΔG°_{298}, for this reaction.
    (b) Calculate the equilibrium constant for the reaction at room temperature.
    (c) Although the reaction is spontaneous at room temperature, it is very slow. The gas in a cylinder of acetylene does not convert to benzene. Heating increases the

rate of a reaction. Will the reaction be spontaneous at 850°C?
    (d) What is the equilibrium constant for this reaction at 850°C?

115. Sketch a graph for ΔG as a sample of N_2O_4 initially with a pressure of 0.2510 atm comes to equilibrium at 25°C according to the equation

$$N_2O_4(g) \longrightarrow 2NO_2(g)$$

Plot values of ΔG where $P_{NO_2} = 0.2500$ atm, 0.240 atm, 0.200 atm, and 0.172 atm, the equilibrium pressure.

116. Sketch a graph for ΔG as the mixture of NH_3, N_2, and H_2 described in Example 14.13 (and in Exercise 24 in Chapter 14) comes to equilibrium at 200°C according to the following equation

$$N_2(g) + 3H_2(g) \longrightarrow 2NH_3(g)$$

Start with $P_{NH_3} = 5.00$ atm, $P_{N_2} = 2.00$ atm, and $P_{H_2} = 6.00$ atm, and plot values of ΔG as P_{NH_3} changes from 5 atm to 5.5 atm, to 6 atm, to 6.5 atm, to 7 atm, to 7.5 atm, to 8 atm, and finally to the equilibrium pressure, 8.30 atm.

Applications and Additional Exercises

117. Alcohols can be produced by the reaction of water with olefins (Chapter 9). Judging by the spontaneity of the reaction, explain whether a chemical company should spend money on a search for a room temperature catalyst for the reaction represented by the equation

$$CH_2{=}CH_2(g) + H_2O(g) \longrightarrow CH_3CH_2OH(g)$$

If the reaction is spontaneous but proves to be slow, how will heating affect the yield? Explain your answer.

118. As discussed in Chapter 9, oxidation of alcohols can produce carboxylic acids. Would the oxidation of liquid ethanol, C_2H_5OH, by solid CrO_3 to a mixture of liquid acetic acid, CH_3CO_2H, liquid water, and solid Cr_2O_3 be spontaneous? Would the reaction produce heat and require cooling?

119. (a) Water gas, a mixture of H_2 and CO, is an important industrial fuel produced by the reaction of steam with red-hot coke (essentially pure carbon).

$$C(s) + H_2O(g) \longrightarrow CO(g) + H_2(g)$$

    Assuming that coke has the same enthalpy of formation as graphite, calculate ΔH°_{298} for this reaction. Does the coke heat or cool as the reaction proceeds?
    (b) Methanol, a liquid fuel that could possibly replace gasoline, can be prepared from water gas and additional hydrogen at high temperatures and pressures in the presence of a suitable catalyst.

$$2H_2(g) + CO(g) \longrightarrow CH_3OH(g)$$

    Under the conditions of the reaction, methanol forms as a gas. Calculate ΔG° for this reaction at 345°C.

(c) Calculate the equilibrium constant for the reaction

$$2H_2(g) + CO(g) \longrightarrow CH_3OH(g)$$

at 345°C.

(d) Calculate the heat of combustion of 1 mol of liquid methanol to $H_2O(g)$ and $CO_2(g)$.

120. (a) The effect of chlorofluorocarbons, which are used as aerosol propellants, on the ozone layer is now common knowledge. Another aerosol propellant, which was first used to dispense whipped cream and the safety of which for use in food is well established, is N_2O. This gas is made by the decomposition of NH_4NO_3 under controlled conditions; gaseous water is the other product. Using the data from Appendix I, calculate the standard enthalpy of the reaction for the preparation of $N_2O(g)$.

(b) Perhaps surprisingly because of its common uses, the stable gas N_2O explodes when subjected to the shock of a detonator. What are the standard enthalpy change and the free energy change for the decomposition of N_2O?

121. Carbon dioxide decomposes into CO and O_2 at elevated temperatures. What is the equilibrium partial pressure of oxygen in a sample at 1000°C for which the initial pressure of CO_2 was 1.15 atm?

122. Carbon tetrachloride, an important industrial solvent, is prepared by the chlorination of methane at 850 K.

$$CH_4(g) + 4Cl_2(g) \longrightarrow CCl_4(g) + 4HCl(g)$$

(a) What is the equilibrium constant for the reaction at 850 K?

(b) Will the reaction vessel need to be heated or cooled to keep the temperature of the reaction constant?

123. Acetic acid, CH_3CO_2H, can form a dimer, $(CH_3CO_2H)_2$, in the gas phase.

$$2CH_3CO_2H(g) \longrightarrow (CH_3CO_2H)_2(g)$$

The dimer is held together by two hydrogen bonds with a total strength of 66.5 kJ per mole of dimer.

At 25°C the equilibrium constant for the dimerization is 1.3×10^3 (pressure in atmospheres). What is $\Delta S°$ for the reaction?

124. Using the data in Appendix I, calculate the enthalpy change, the entropy change, and the free energy change for the vaporization under standard state conditions of exactly 1 mol of each of the following.

(a) $I_2(s)$

(b) $CHCl_3(l)$

(c) $C_2H_5OH(l)$

125. Nitric acid, HNO_3, can be prepared by the following sequence of reactions:

$$4NH_3(g) + 5O_2(g) \longrightarrow 4NO(g) + 6H_2O(g)$$

$$2NO(g) + O_2(g) \longrightarrow 2NO_2(g)$$

$$3NO_2(g) + H_2O(l) \longrightarrow 2HNO_3(l) + NO(g)$$

How much heat is evolved when 1 mol of $NH_3(g)$ is converted to $HNO_3(l)$? Assume that all reactants and products are in their standard states at 25°C and 1 atm.

126. Calculate the internal energy change (ΔE) as a system composed of exactly 1 mol of CH_2Cl_2 changes from a liquid at 313 K (density = 1.327 g cm^{-1}) to a gas at 313 K under a constant pressure of 1 atm. (*Hint:* ΔH values can be found in Appendix I. You need w and q to determine ΔE.)

127. At 1000 K the equilibrium constant for the decomposition of bromine molecules, $Br_2(g) \rightleftharpoons 2Br(g)$, is 2.8×10^4 (pressure in atmospheres). What is $\Delta G°$ for the reaction? Assume that the bond energy of Br_2 does not change between 298 K and 1000 K, and calculate the approximate value of $\Delta S°$ for the reaction at 1000 K.

128. Determine $\Delta G°$ for each of the following changes.

(a)

$$CH_3CH_2CH_2CH_3 \longrightarrow CH_3-\underset{\underset{CH_3}{|}}{\overset{\overset{CH_3}{|}}{C}}-H$$

$K_p = 2.5$ at 25°C

(b) Antimony pentachloride decomposes at 448°C according to the equation

$$SbCl_5(g) \longrightarrow SbCl_3(g) + Cl_2(g)$$

An equilibrium mixture in a 5.00-L flask at 448°C contains 3.85 g of $SbCl_5$, 9.14 g of $SbCl_3$, and 2.84 g of Cl_2.

(c) Chlorine molecules are 1.00% dissociated at 975 K at a pressure of 1.00 atm.

$$Cl_2(g) \longrightarrow 2Cl(g)$$

129. Consider the reaction represented by the equation

$$CuS(s) + H_2(g) \longrightarrow Cu(s) + H_2S(g)$$

(a) Calculate $\Delta G°_{298}$ and $\Delta H°_{298}$ from the data found in Appendix I.

(b) Calculate the value for the equilibrium constant, K, at 298.15 K and 1 atm.

(c) Estimate the value for K at 798 K and 1 atm.

(d) Calculate $\Delta S°_{298}$.

(e) Estimate $\Delta G°$ at 798 K and 1 atm.

(f) Estimate the temperature at which $\Delta G°$ is equal to zero, assuming that $\Delta H°$ and $\Delta S°$ do not change significantly as the temperature increases.

130. Using the data given in Appendix I, show that H, S, and G are state functions by calculating $\Delta H°_{298}$, $\Delta S°_{298}$, and $\Delta G°_{298}$ for the formation of $HCl(g)$ from $H_2(g)$ and $Cl_2(g)$ by two pathways, both at standard state conditions.

Path 1: $H_2(g) + Cl_2(g) \longrightarrow 2HCl(g)$

Path 2: $H_2(g) \longrightarrow 2H(g)$

$Cl_2(g) \longrightarrow 2Cl(g)$

$2H(g) + 2Cl(g) \longrightarrow 2HCl(g)$

131. Consider the vaporization of bromine liquid to bromine gas, $Br_2(l) \longrightarrow Br_2(g)$, at 25°C.

(a) Calculate the change in enthalpy and the change in entropy at standard state conditions.

(b) Discuss the relative disorder in bromine liquid compared to that in bromine gas. State what you can about the spontaneity of the vaporization.

(c) Estimate the value of ΔG°_{298} for the vaporization of bromine from the values of ΔH°_{298} and ΔS°_{298} you determined in part (a).

(d) State what you can about the spontaneity of the process from the value you obtained for ΔG°_{298} in part (c).

(e) Estimate the temperature at which liquid Br_2 and gaseous Br_2 with a pressure of 1 atm are in equilibrium. (Assume that ΔH° and ΔS° are independent of temperature.)

(f) State in which direction the process would be spontaneous at 298 K and at 398 K, using the temperature value you obtained in part (e).

132. The following question is taken from the 1991 Chemistry Advanced Placement Examination and is used with the permission of the Educational Testing Service.

$$BCl_3(g) + NH_3(g) \rightleftharpoons Cl_3BNH_3(s)$$

(a) Predict the sign of the entropy change, ΔS, as the reaction proceeds to the right. Explain your prediction.

(b) If the reaction proceeds spontaneously to the right, predict the sign of the enthalpy change, ΔH. Explain your prediction.

(c) The direction in which the reaction spontaneously proceeds changes as the temperature is increased above a specific temperature. Explain.

(d) What is the value of the equilibrium constant at the temperature referred to in (c); that is, the specific temperature at which the direction of the spontaneous reaction changes? Explain.

19

Electrochemistry and Oxidation–Reduction

CHAPTER OUTLINE

Galvanic Cells and Cell Potentials

19.1 Galvanic Cells
19.2 Cell Potentials
19.3 Standard Electrode Potentials
19.4 Calculation of Cell Potentials
19.5 Cell Potential, Electrical Work, and Free Energy
19.6 The Effect of Concentration on Cell Potential: The Nernst Equation
19.7 Relationship of the Cell Potential and the Equilibrium Constant

Batteries

19.8   Primary Cells
19.9   Secondary Cells
19.10 Fuel Cells
19.11 Corrosion

Electrolytic Cells

19.12 The Electrolysis of Molten Sodium Chloride
19.13 The Electrolysis of Aqueous Solutions
19.14 Electrolytic Deposition of Metals
19.15 Faraday's Law of Electrolysis

Oxidation–Reduction Reactions

19.16 Balancing Redox Equations

Corrosion of iron is an electrochemical process.

Every day, we use chemical reactions to provide electrical energy. The batteries in our portable radios, cassette players, toys, power tools, and cars are galvanic cells, which produce electrical energy by chemical reactions. The reverse situation is common too: Electrical energy is used to bring about chemical changes. The aluminum in our soft drink cans, the chlorine and chlorine salts used to purify swimming pools, and the sodium hydroxide in Drano and many other commercial products are manufactured in electrolytic cells, in which electrical energy is used to bring about a chemical change. When we recharge a battery, we use electrical energy to accomplish a chemical change that returns the battery to the state in which it delivers electricity.

This chapter deals with electrochemistry, the study of electric currents generated by chemical reactions and of chemical changes produced by electrical currents. In it we will see why a chemical reaction in a battery or other electrochemical cell can produce an electric current and why an electric current can produce a chemical change. We will find out how the voltage of a battery is calculated. We will see that electrochemical cells can be used to make precise measurements of the free energy change of a reaction (ΔG) and its equilibrium constant because the quantity of electrical energy produced or consumed by a chemical reaction can be measured very accurately. Finally, the reactions that take place at the electrodes of electrochemical cells will be used to clarify the processes of oxidation and reduction.

Galvanic Cells and Cell Potentials

Galvanic cells produce electrical energy from chemical reactions. Batteries are the most common forms of galvanic cells, but there are other types, such as fuel cells. In the following sections, we will consider how these devices function and will examine some of the factors that affect the voltage they produce.

19.1   Galvanic Cells

When we place magnesium metal in a solution of hydrochloric acid, the metal and hydronium ion from the acid react spontaneously: Hydrogen gas and a solution of magnesium ion are produced (Fig. 19.1) in an oxidation–reduction reaction (Section 8.5).

$$Mg(s) + 2H_3O^+(aq) \longrightarrow H_2(g) + Mg^{2+}(aq) + 2H_2O(l)$$

An oxidation–reduction reaction has two "halves." In one half of this reaction, the magnesium metal is oxidized, losing two electrons and giving the Mg^{2+} ion. The net ionic equation that describes the oxidation half of the reaction is

$$Mg(s) \longrightarrow Mg^{2+}(aq) + 2e^-$$

In the other half of the reaction, two hydronium ions are reduced. They combine with the two electrons, giving hydrogen gas and water. The net ionic equation that describes the reduction half of the equation is

$$2H_3O^+(aq) + 2e^- \longrightarrow H_2(g) + 2H_2O(l)$$

These reactions, the oxidation half and the reduction half of the overall reaction, are called **half-reactions.** The net ionic equation that describes the overall reaction is the

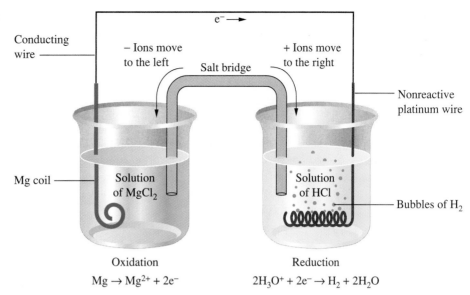

Figure 19.2

The oxidation of magnesium to magnesium ion occurs in the left-hand beaker in this apparatus; the reduction of hydronium ion to hydrogen and water occurs in the right-hand beaker. The wire conducts electrons from the left-hand beaker to the nonreactive electrode in the right-hand beaker, and the salt bridge allows ions to migrate from one side to the other.

sum of the two half-reactions, with the coefficients of the reactants and products in each half-reaction adjusted so that both half-reactions involve the same number of electrons. This is necessary because the electrons produced by the oxidation half-reaction are consumed in the reduction half-reaction. In this example the coefficients do not need adjusting, because both of the half-reactions involve two electrons.

Oxidation half-reaction: $\qquad Mg(s) \longrightarrow Mg^{2+}(aq) + 2e^-$

Reduction half-reaction: $\quad 2H_3O^+(aq) + 2e^- \longrightarrow H_2(g) + 2H_2O(l)$

Sum: $\qquad\qquad\qquad Mg(s) + 2H_3O^+(aq) \longrightarrow H_2(g) + Mg^{2+}(aq) + 2H_2O(l)$

We can run the oxidation half and the reduction half of the reaction in two separate containers without magnesium actually coming in contact with hydrochloric acid. Figure 19.2 shows a diagram of an apparatus for doing this. A magnesium strip is immersed in a solution of a magnesium salt (such as $MgCl_2$) in the beaker on the left-hand side of the diagram. The metal is oxidized according to the oxidation half-reaction.

$$Mg(s) \longrightarrow Mg^{2+}(aq) + 2e^-$$

The electrons produced by this oxidation are conducted by the wire to a nonreactive platinum wire immersed in a solution of hydrochloric acid in the beaker on the right-hand side of the diagram. The wire does not react with the acid but simply conducts the electrons into the acidic solution, where they combine with hydronium ions and produce hydrogen gas and water. The net change in this beaker is the reduction half-reaction

$$2H_3O^+(aq) + 2e^- \longrightarrow H_2(g) + 2H_2O(l)$$

The net ionic equation for the reaction occurring in the apparatus is the sum of the two halves of the reaction:

$$Mg(s) + 2H_3O^+(aq) \longrightarrow H_2(g) + Mg^{2+}(aq) + 2H_2O(l)$$

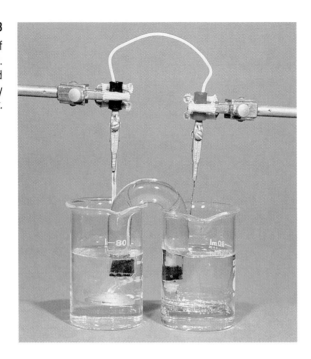

Figure 19.3 is a photograph of one such apparatus in action. Bubbles of hydrogen can be seen in the righthand beaker.

In addition to the wire, we need to connect the two parts of the apparatus by a **salt bridge,** a tube containing a concentrated solution of an electrolyte (Section 12.3) such as potassium nitrate. The salt bridge allows the ions of the electrolyte to migrate, but the solutions in the two beakers cannot mix. As positively charged magnesium ions are produced in the left beaker, enough negatively charged nitrate ions from the salt bridge migrate into that beaker for the solution to remain electrically neutral. That is, the total number of negative charges on the negative ions in the left-hand beaker remains equal to the total number of positive charges on the positive ions in the beaker. As positively charged hydronium ions are destroyed in the right-hand beaker, positively charged potassium ions from the salt bridge replace them and maintain equal numbers of positive and negative charges there.

The reaction shown in Fig. 19.3 is the same as that shown in Fig. 19.1; electrons are transferred from magnesium to hydronium ions, and hydrogen gas and magnesium ions are produced in both cases. However, in the apparatus shown in Fig. 19.3, the reaction takes place without any direct contact of the reactants. The electrons are transferred from the magnesium to the hydronium ions through a wire. This flow of electrons constitutes an electric current, and we can use the energy made available to do electrical work. For example, we could use the current to operate an electric motor, to light an electric lamp, or to produce some other form of energy. The apparatus shown in Figs. 19.2 and 19.3 is one type of **galvanic cell,** a device in which energy from a chemical reaction produces an electric current.

A galvanic cell consists of two **half-cells** with the oxidation half of the reaction occurring in one half-cell and the reduction half of the reaction occurring in the other half-cell. The sum of the two half-reactions is called the **cell reaction.** The half-reaction that produces electrons is the oxidation half-reaction; the half-cell where oxida-

Figure 19.4

Two half-cells connected by a porous disk.

Figure 19.5

No reaction occurs when two half-cells are connected by a wire but no salt bridge or porous disk.

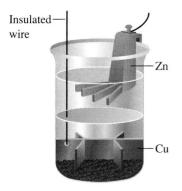

Figure 19.6

The Daniell cell. Copper metal (Cu) is surrounded by saturated copper(II) sulfate solution and crystals of copper sulfate; zinc metal (Zn) is surrounded by zinc sulfate solution.

tion occurs is called the **anode** of the cell. The half-reaction that accepts electrons is the reduction half-reaction; the half-cell where reduction occurs is called the **cathode** of the cell. (It will help us recall these definitions if we remember that *anode* and *oxidation* both start with a vowel and that *cathode* and *reduction* both start with a consonant.)

For a galvanic cell to deliver electrical energy, we must connect the two half-cells by an external wire through which the electrons pass and by a salt bridge or porous disk (Fig. 19.4) that allows ions to flow with no mixing of the solutions in the two half-cells. If the two solutions were allowed to mix, a direct reaction would take place, the half-cells would be shorted out, and no current would flow through the wire.

The importance of the salt bridge or porous disk becomes even more apparent when we consider what happens without one. When our two half-cells are connected by a wire but no salt bridge (Fig. 19.5), a few electrons pass from the magnesium strip through the wire into the acid solution, but then the current stops. The right-hand half-cell becomes negatively charged because of the transfer of electrons into it, and the left-hand half-cell becomes positively charged because electrons leave it. The attraction of electrons by the positive charge in the left-hand half-cell and the repulsion of electrons by the negative charge in the right-hand half-cell prevent any further transfer of electrons. Adding a salt bridge allows the two half-cells to lose their excess charge and permits more electrons to flow.

Example 19.1   Components of a Galvanic Cell

Identify the anode, the cathode, and the net ionic half-reactions in a Daniell cell, a battery first constructed in 1836 by the English chemist John Fredrick Daniell and used as the source of electricity in early telegraph systems. A Daniell cell (Fig. 19.6) consists of a piece of copper metal (Cu) with crystals of copper(II) sulfate in a saturated solution of copper(II) sulfate below a piece of zinc metal (Zn) suspended near the top of the cell in a dilute solution of zinc sulfate. (The crystals of copper sulfate simply keep the copper sulfate solution saturated.) The zinc sulfate solution floats on the denser solution of copper(II) sulfate. Because of the difference in density between the two solutions, they mix very slowly and the boundary between them acts like a porous disk. The cell reaction is

$$Zn(s) + Cu^{2+}(aq) \longrightarrow Cu(s) + Zn^{2+}(aq)$$

Solution: The anode is the half-cell where oxidation occurs. Zinc metal is oxidized, so the anode consists of the zinc metal and the solution of zinc sulfate. The net ionic oxidation half-reaction at the anode is

$$Zn(s) \longrightarrow Zn^{2+}(aq) + 2e^-$$

The cathode is the half-cell where reduction occurs. Copper(II) ion is reduced to copper metal during the reaction, so the cathode consists of the copper metal and the copper sulfate solution. The net ionic reduction reaction at the cathode is

$$Cu^{2+}(aq) + 2e^- \longrightarrow Cu(s)$$

When the two metals are connected by a wire, the zinc metal is oxidized and electrons from the zinc flow through the wire to the copper metal, where the copper(II) ions in the solution are reduced to metallic copper.

Check your learning: Sodium and chlorine are produced by the electrolysis of molten sodium chloride:

$$2Na^+ + 2Cl^- \longrightarrow 2Na + Cl_2$$

Write the equations for the anode and cathode half-reactions in this process.

Answer: Cathode: $2Na^+ + 2e^- \longrightarrow 2Na$ or $Na^+ + e^- \longrightarrow Na$
Anode: $2Cl^- \longrightarrow Cl_2 + 2e^-$

19.2   Cell Potentials

A galvanic cell produces a driving force that pushes electrons through an external circuit (for example, the wire in Fig. 19.3). This driving force is called the **cell potential,** E_{cell}, of the cell. We measure a cell potential in units of volts, V.

Let us consider the potential of a version of the Daniell cell described in Example 19.1. This cell (Fig. 19.7) is constructed from two half-cells, one a strip of metallic zinc immersed in a 1.0 M solution of zinc sulfate and the other a strip of metallic copper immersed in a 1.0 M solution of copper sulfate and operated at 25°C. The half-reaction at the anode is

$$Zn(s) \longrightarrow Zn^{2+}(aq) + 2e^-$$

The half-reaction at the cathode is

$$Cu^{2+}(aq) + 2e^- \longrightarrow Cu(s)$$

Figure 19.7

A cell constructed from a Zn^{2+}/Zn electrode (left) and a Cu^{2+}/Cu electrode (right) produces a potential of 1.10 V when the temperature is 25°C and the concentrations are 1 M.

The sulfate ion is unchanged during the reaction, so it does not appear in the net ionic equation for the cell reaction:

$$Zn(s) + Cu^{2+}(aq) \longrightarrow Cu(s) + Zn^{2+}(aq)$$

We measure the cell potential with a potentiometer placed in the circuit of the cell. The potential of the cell in Fig. 19.7 is 1.10 volts. Because we constructed the cell with the reactants and products at standard state conditions (Section 18.5), at the beginning of the reaction the cell potential is measured under standard state conditions. A cell potential measured under standard state conditions is called a **standard cell potential, E°_{cell}.** As in Chapter 18, we use a superscript to indicate that all reactants and products are at standard state conditions. For the net ionic reaction,

$$Zn(s) + Cu^{2+}(aq) \longrightarrow Cu(s) + Zn^{2+}(aq)$$

the standard cell potential, E°_{cell}, is 1.100 volts if measured with a more precise potentiometer than that used in Fig. 19.7.

Information related to cell potentials is usually tabulated at standard state conditions. However, the potential of a cell depends on the concentrations of its reactants and products and on the temperature, which we will take as 25°C unless otherwise noted.

A galvanic cell has a positive voltage when the cell reaction is spontaneous. A reaction that occurs in a galvanic cell with a positive potential also will be spontaneous when the reactants are mixed (Fig. 19.8).

As we will see in Section 19.3, each half-cell in a cell reaction can be assigned a half-cell potential. The sum of the half-cell potential E_{ox} of the anode and the half-cell potential E_{red} of the cathode is the cell potential, E_{cell}.

$$E_{cell} = E_{ox} + E_{red}$$

Figure 19.8

When a zinc strip is immersed in a solution of copper sulfate, copper metal and a colorless solution of zinc sulfate form. On the left are a zinc strip and a solution of copper sulfate. On the right, the reaction is almost complete; the blue color of the Cu^{2+} ion has almost disappeared, and copper metal can be seen.

For the cell illustrated in Fig. 19.7 operating under standard state conditions, the anode potential is taken as 0.763 volt and the cathode potential as 0.337 volt. Their sum is equal to E_{cell}°.

Anode:	$Zn(s) \longrightarrow Zn^{2+}(aq) + 2e^-$	$E_{ox}^{\circ} = 0.763$ V
Cathode:	$Cu^{2+}(aq) + 2e^- \longrightarrow Cu(s)$	$E_{red}^{\circ} = 0.337$ V
Sum:	$Zn(s) + Cu^{2+}(aq) \longrightarrow Cu(s) + Zn^{2+}(aq)$	$E_{cell}^{\circ} = E_{ox}^{\circ} + E_{red}^{\circ}$
		$= 1.100$ V

Example 19.2   Calculation of a Standard Cell Potential

The rechargeable cells used in calculators and battery-operated tools are based on nickel and cadmium electrodes. What are the cell reaction and the voltage of a rechargeable nickel–cadmium battery (a NICAD battery) if it is operated under standard state conditions? The standard state half-cell potentials are

Anode: $Cd(s) + 2OH^-(aq) \longrightarrow Cd(OH)_2(s) + 2e^-$ $\qquad E_{ox}^{\circ} = 0.40$ V

Cathode: $NiO_2(s) + 2H_2O(l) + 2e^- \longrightarrow Ni(OH)_2(s) + 2OH^-$ $\qquad E_{red}^{\circ} = 0.49$ V

Solution: A battery is a galvanic cell; under standard state conditions its voltage equals the standard cell potential. The cell reaction is the sum of the anode and the cathode half-reactions. The cell potential is the sum of the half-cell potentials.

Anode:	$Cd(s) + 2OH^-(aq) \longrightarrow Cd(OH)_2(s) + 2e^-$	$E_{ox}^{\circ} = 0.40$ V
Cathode:	$NiO_2(s) + 2H_2O(l) + 2e^- \longrightarrow Ni(OH)_2(s) + 2OH^-(aq)$	$E_{red}^{\circ} = 0.49$ V
Sum:	$Cd(s) + NiO_2(s) + 2H_2O(l) \longrightarrow Cd(OH)_2(s) + Ni(OH)_2(s)$	$E_{cell}^{\circ} = E_{ox}^{\circ} + E_{red}^{\circ}$
		$= 0.40$ V $+ 0.49$ V
		$= 0.89$ V

The standard state cell potential is 0.89 volt. Note that hydroxide ion does not appear in the cell reaction because the hydroxide ion consumed in the anode half-reaction is replaced by the hydroxide ion produced in the cathode half-reaction.

Check your learning: Determine the cell reaction and the cell potential for a cell based on the following half-reactions.

Anode:	$Sn^{2+}(aq) \longrightarrow Sn^{4+}(aq) + 2e^-$	$E_{ox}^{\circ} = -0.15$ V
Cathode:	$Hg_2Cl_2(s) + 2e^- \longrightarrow 2Hg(l) + 2Cl^-(aq)$	$E_{red}^{\circ} = 0.27$ V

Answer: $Sn^{2+}(aq) + Hg_2Cl_2(s) \longrightarrow Sn^{4+}(aq) + 2Hg(l) + 2Cl^-(aq)$ $\qquad E_{cell}^{\circ} = 0.12$ V

19.3   Standard Electrode Potentials

We need to be careful with the term *electrode*. It is used in two ways: (1) to refer to a wire or some other conductor that delivers electricity into a half-cell and (2) to refer to a complete half-cell. The platinum wire in Figs. 19.2 and 19.3 is an electrode in the first sense; it serves to carry electrons into the half-cell. The solution of zinc sulfate and the strip of zinc in Fig. 19.7 is an electrode in the second sense; together, they constitute a half-cell. The metal strip is both the conductor and the reduced

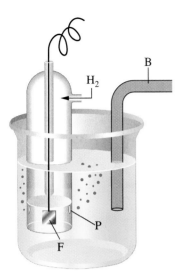

Figure 19.9

Diagram of a simple standard hydrogen electrode. F indicates the inert platinum foil, P a hole for escape of hydrogen gas, and B part of the salt bridge.

species; the metal ions are the oxidized species. In this section we will discuss electrodes that are half-cells.

Although we can measure the potential of a galvanic cell, there is no satisfactory method for measuring the actual potential of an individual electrode. We can only measure the sum of the potentials of two electrodes. However, if one electrode is *assigned* a standard potential, potentials of other electrodes can be reported by comparison to the assigned value of the standard. The electrode most commonly used as a reference is the **standard hydrogen electrode,** which is assigned a potential of zero volts. The potentials of all other electrodes are reported relative to the standard hydrogen electrode. (This approach is similar to arbitrarily establishing sea level as zero elevation and reporting all elevations in terms of how much higher or lower than this level they are.)

The standard hydrogen electrode is a **gas electrode,** a half-cell with a gas as one of the reactants. We can make a gas electrode by bubbling a reactive gas around an inert conductor—a conductor that carries electrons but does not enter into the electrode reaction. In a hydrogen electrode (Fig. 19.9), hydrogen gas is bubbled through a solution of hydrogen ions and around a platinum foil or wire covered with very finely divided platinum and immersed in the solution. The net ionic equation for reduction in the electrode is

$$2H_3O^+(aq) + 2e^- \longrightarrow H_2(g) + 2H_2O(l)$$

The potential of a gas electrode changes with changing pressure of the gas and with changing concentration of the other components, so we must be careful to keep the pressures and concentrations constant at standard state conditions when we measure standard potentials. A standard hydrogen electrode operates at a temperature of 25°C with hydrogen gas at a pressure of 1 atmosphere bubbling around platinum immersed in a solution containing a 1 M concentration of hydronium ions.

To measure the potential of an electrode relative to that of the standard hydrogen electrode, we can use a galvanic cell consisting of the electrode being measured and a standard hydrogen electrode. For example, we can measure the potential of a copper electrode by using a cell with the following cell reaction:

$$Cu^{2+}(aq) + H_2(g) + 2H_2O(l) \longrightarrow Cu(s) + 2H_3O^+(aq)$$

The cell is diagrammed in Fig. 19.10 on page 718. The Cu^{2+}/Cu electrode consists of a copper strip in contact with a 1 M solution of copper ion. The copper is involved in the reduction half-reaction:

$$Cu^{2+}(aq) + 2e^- \longrightarrow Cu(s) \qquad E^\circ_{red} = ? \text{ V}$$

The oxidation half-reaction is

$$H_2(g) + 2H_2O(l) \longrightarrow 2H_3O^+(aq) + 2e^- \qquad E^\circ_{ox} = 0.0 \text{ V}$$

We connect a potentiometer to the copper strip in the Cu^{2+} solution and the platinum wire of the hydrogen electrode to determine the potential difference—E_{cell}, the cell potential—between the two electrodes. For this cell we observe a potential of 0.337 volt. Because the cell potential is the sum of the potentials of the copper electrode and the hydrogen electrode (which has an assigned potential of zero volts), the cell potential is equal to the potential of the copper electrode; that is, it is 0.337 V.

$$E^\circ_{cell} = E^\circ_{ox} \text{ (for the hydrogen electrode)} + E^\circ_{red} \text{ (for the copper electrode)}$$
$$0.337 \text{ V} = 0 \text{ V} + E^\circ_{red}$$
$$E^\circ_{red} = 0.337 \text{ V}$$

Figure 19.10

A galvanic cell with a hydrogen electrode (left) and a Cu^{2+}/Cu electrode (right).

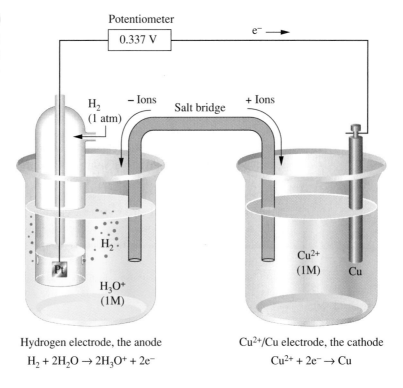

Hydrogen electrode, the anode

$$H_2 + 2H_2O \rightarrow 2H_3O^+ + 2e^-$$

Cu^{2+}/Cu electrode, the cathode

$$Cu^{2+} + 2e^- \rightarrow Cu$$

By international agreement, the values of electrode potentials are given for the reduction process. The potential for the reverse reaction, an oxidation potential, is equal to the negative of the reduction potential. The reduction potential and the oxidation potential of an electrode have the same absolute value but are of opposite sign.

Once a standard potential has been established relative to the standard hydrogen electrode, it can be used to determine other potentials.

Example 19.3   Determination of an Electrode Potential

A galvanic cell at 25°C has an anode consisting of an iron strip immersed in a 1 M solution of iron(II) perchlorate (a Fe^{2+}/Fe electrode) and a cathode consisting of a copper strip immersed in a 1 M solution of copper(II) perchlorate (a Cu^{2+}/Cu electrode, $E^\circ_{red} = 0.337$ V). A potentiometer shows the cell potential (E°_{cell}) to be 0.777 V (Fig. 19.11 on page 719). The cell reaction is

$$Fe(s) + Cu^{2+}(aq) \longrightarrow Fe^{2+}(aq) + Cu(s)$$

What is the standard reduction potential of the Fe^{2+}/Fe electrode?

Solution: The standard cell potential (0.777 V) is the sum of the standard electrode potentials of the anode and of the cathode. Iron metal is oxidized in the anode half-reaction; copper(II) ion is reduced in the cathode half-reaction.

Anode:	$Fe(s) \longrightarrow Fe^{2+}(aq) + 2e^-$	$E = E^\circ_{ox}$
Cathode:	$Cu^{2+}(aq) + 2e^- \longrightarrow Cu(s)$	$E = E^\circ_{red} = 0.337$ V
Sum:	$Fe(s) + Cu^{2+}(aq) \longrightarrow Fe^{2+}(aq) + Cu(s)$	$E^\circ_{cell} = E^\circ_{ox} + E^\circ_{red}$
		0.777 V $= E^\circ_{ox} + 0.337$ V

Figure 19.11

A galvanic cell with a Fe^{2+}/Fe electrode (left) and a Cu^{2+}/Cu electrode (right).

Thus

$$E^{\circ}_{ox} = E^{\circ}_{cell} - 0.337 \text{ V} = 0.440 \text{ V}$$

The iron electrode is the anode of the cell and is engaged in an oxidation half-reaction. In order to find the standard reduction potential for the electrode, we must reverse the sign of the oxidation potential.

$$E^{\circ}_{red} = -E^{\circ}_{ox} = -0.440 \text{ V}$$

The standard reduction potential of the Fe^{2+}/Fe electrode is -0.440 V.

$$Fe^{2+}(aq) + 2e^- \longrightarrow Fe(s) \qquad E^{\circ}_{red} = -0.440 \text{ V}$$

Check your learning: Determine the standard reduction potential for the half-reaction $Cl_2 + 2e^- \longrightarrow 2Cl^-$ from the following information:

$$Pt^{2+} + 2Cl^- \longrightarrow Pt + Cl_2 \qquad E^{\circ}_{cell} = -0.15 \text{ V}$$
$$Pt^{2+} + 2e^- \longrightarrow Pt \qquad E^{\circ}_{red} = 1.20 \text{ V}$$

Answer: 1.35 V

Table 19.1, on page 720, and Appendix H list the **standard reduction potentials** of several electrodes that involve aqueous solutions. *These standard reduction potentials refer only to reactions in the standard state (pure solids, 1 M concentrations in water, 1 atmosphere pressure, and a temperature of 25°C).* When the conditions change, the potentials change. The effect of the concentration on potential is illustrated by the reduction potentials for hydrogen electrodes with two different concentrations. Under standard state conditions, E°_{red} is defined as 0.0 volts.

Table 19.1

Standard Reduction Potentials

Half-reaction	$E°$, V
$K^+ + e^- \longrightarrow K$	-2.925
$Ba^{2+} + 2e^- \longrightarrow Ba$	-2.90
$Ca^{2+} + 2e^- \longrightarrow Ca$	-2.87
$Na^+ + e^- \longrightarrow Na$	-2.714
$Mg^{2+} + 2e^- \longrightarrow Mg$	-2.37
$Al^{3+} + 3e^- \longrightarrow Al$	-1.66
$Zn(OH)_2 + 2e^- \longrightarrow Zn + 2OH^-$	-1.245
$Mn^{2+} + 2e^- \longrightarrow Mn$	-1.18
$Fe(OH)_2 + 2e^- \longrightarrow Fe + 2OH^-$	-0.877
$Zn^{2+} + 2e^- \longrightarrow Zn$	-0.763
$Cr^{3+} + 3e^- \longrightarrow Cr$	-0.74
$Fe^{2+} + 2e^- \longrightarrow Fe$	-0.440
$Cd^{2+} + 2e^- \longrightarrow Cd$	-0.403
$PbSO_4 + 2e^- \longrightarrow Pb + SO_4^{2-}$	-0.356
$Co^{2+} + 2e^- \longrightarrow Co$	-0.277
$Ni^{2+} + 2e^- \longrightarrow Ni$	-0.257
$Sn^{2+} + 2e^- \longrightarrow Sn$	-0.136
$Pb^{2+} + 2e^- \longrightarrow Pb$	-0.126
$2H_3O^+ + 2e^- \longrightarrow H_2 + 2H_2O$	0.00
$Sn^{4+} + 2e^- \longrightarrow Sn^{2+}$	$+0.15$
$AgCl + e^- \longrightarrow Ag + Cl^-$	$+0.222$
$Hg_2Cl_2 + 2e^- \longrightarrow 2Hg + 2Cl^-$	$+0.27$
$Cu^{2+} + 2e^- \longrightarrow Cu$	$+0.337$
$NiO_2 + 2H_2O + 2e^- \longrightarrow Ni(OH)_2 + 2OH^-$	$+0.49$
$I_2 + 2e^- \longrightarrow 2I^-$	$+0.5355$
$MnO_4^- + 2H_2O + 3e^- \longrightarrow MnO_2 + 4OH^-$	$+0.588$
$Fe^{3+} + e^- \longrightarrow Fe^{2+}$	$+0.771$
$Hg_2^{2+} + 2e^- \longrightarrow 2Hg$	$+0.789$
$Ag^+ + e^- \longrightarrow Ag$	$+0.7991$
$Br_2(l) + 2e^- \longrightarrow 2Br^-$	$+1.0652$
$Pt^{2+} + 2e^- \longrightarrow Pt$	$\sim +1.20$
$O_2 + 4H_3O^+ + 4e^- \longrightarrow 6H_2O$	$+1.23$
$Cl_2 + 2e^- \longrightarrow 2Cl^-$	$+1.3595$
$Au^{3+} + 3e^- \longrightarrow Au$	$+1.50$
$MnO_4^- + 8H_3O^+ + 5e^- \longrightarrow Mn^{2+} + 12H_2O$	$+1.51$
$PbO_2 + SO_4^{2-} + 4H_3O^+ + 2e^- \longrightarrow PbSO_4 + 6H_2O$	$+1.685$
$F_2 + 2e^- \longrightarrow 2F^-$	$+2.87$

$$2H_3O^+(aq) + 2e^- \longrightarrow H_2(g) + 2H_2O(l) \qquad E°_{red} = 0.0 \text{ V}$$

When the concentration of hydronium ion is that in neutral water, 1×10^{-7} M, the electrode potential decreases to -0.41 volt.

The position of a metal in the table of reduction potentials, when the electrodes are arranged from the most negative to the most positive, is the same as in the activity series (Section 8.7).

19.4   Calculation of Cell Potentials

The values in Table 19.1 provide information that we can utilize in determining standard state cell potentials. However, we must remember two points when using these values.

1. The $E°$ values are for reduction half-reactions, and the sign of a reduction potential must be reversed when it is used as a potential for an oxidation half-reaction. As we will see, $E°$ values are related to $\Delta G°$ values, and like $\Delta G°$, $E°$ changes sign when the direction of a reaction is reversed.

2. Changing the stoichiometric coefficients of a half-cell equation does not change the value of $E°$, because electrode potentials are intensive properties (Section 1.4). For example, from Table 19.1,

$$Ag^+(aq) + e^- \longrightarrow Ag(s) \qquad E°_{red} = 0.7991 \text{ V}$$

$E°_{red}$ does not change when we change the quantities involved.

$$3Ag^+(aq) + 3e^- \longrightarrow 3Ag(s) \qquad E°_{red} = 0.7991 \text{ V}$$

This might seem unrealistic, but we see it in batteries all the time: The voltage (potential) of a battery does not change with its size. A larger, D-size alkaline battery and a small AAA alkaline battery both have the same voltage.

Example 19.4   **Determination of Standard State Cell Potentials**

Write the cell reaction and determine the standard state potential for the cell diagrammed in Fig. 19.12.

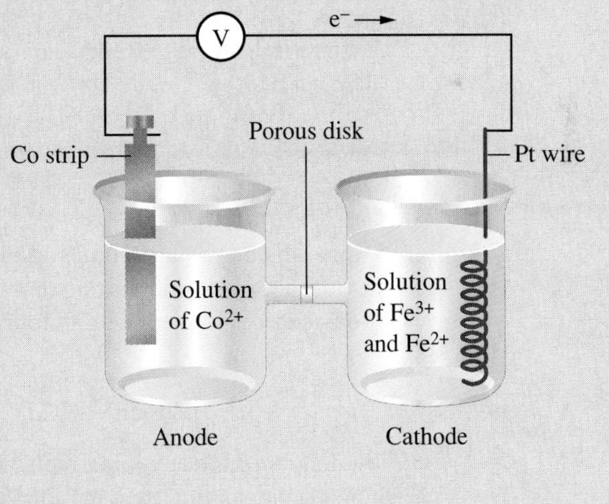

Figure 19.12

A galvanic cell with a Co^{2+}/Co anode (left) and a Fe^{3+}/Fe^{2+} cathode (right).

Solution: The anode half-reaction is

$$Co(s) \longrightarrow Co^{2+}(aq) + 2e^-$$

The cathode half-reaction is

$$Fe^{3+}(aq) + e^- \longrightarrow Fe^{2+}(aq)$$

The platinum wire in the cell is an inert electrode that delivers the electrons from the anode to the solution of iron(III) ions and iron(II) ions.

We cannot describe the cell reaction simply by adding the two half-reactions; the numbers of electrons in the two reactions differ. One cobalt atom produces two electrons, but one iron(III) ion reacts with one electron. There must be twice as many Fe^{3+} ions as Co atoms in order to make the number of electrons produced equal to the number of electrons consumed. Thus we double (multiply by 2) the number of electrons, Fe^{2+} ions, and Fe^{3+} ions in the cathode half-reaction in order to describe the correct ratio of Co and Fe^{3+}. This gives

Anode half-reaction: $\qquad\qquad\qquad Co(s) \longrightarrow Co^{2+}(aq) + 2e^-$

Cathode half-reaction: $\quad 2Fe^{3+}(aq) + 2e^- \longrightarrow 2Fe^{2+}(aq)$

Cell reaction: $\qquad\quad Co(s) + 2Fe^{3+}(aq) \longrightarrow Co^{2+}(aq) + 2Fe^{2+}(aq)$

The cell potential is equal to the sum of the potentials of the anode and cathode. The electrode potential of the cathode is equal to the potential given in Table 19.1; $E^\circ_{red} = 0.771$ V. The electrode potential of the anode is opposite in sign to that given in the table, because we reversed the direction of the half-reaction: $E^\circ_{ox} = -E^\circ_{red} = -(-0.277 \text{ V})$. Thus we have

Anode: $\qquad\qquad\qquad Co(s) \longrightarrow Co^{2+}(aq) + 2e^- \qquad\qquad E^\circ_{ox} = 0.277$ V

Cathode: $\quad 2Fe^{3+}(aq) + 2e^- \longrightarrow 2Fe^{2+}(aq) \qquad\qquad E^\circ_{red} = 0.771$ V

Sum: $\qquad Co(s) + 2Fe^{3+}(aq) \longrightarrow Co^{2+}(aq) + 2Fe^{2+}(aq) \qquad E^\circ_{cell} = E^\circ_{ox} + E^\circ_{red}$

$$= 0.227 \text{ V} + 0.771 \text{ V}$$
$$= 0.998 \text{ V}$$

Remember, changing the amounts of reactants and products in the cathode half-reaction does not change the half-cell potential.

Check your learning: Calculate the standard cell potential for a cell based on the reaction $Mn + 2AgCl \longrightarrow Mn^{2+} + 2Cl^- + 2Ag$.

Answer: 1.40 V

Line notation is sometimes used to represent a galvanic cell. The cell involving a reaction between zinc and a solution of hydrochloric acid under standard state conditions can be diagrammed as follows:

$$Zn \,|\, Zn^{2+} \text{ (1 M)} \xrightarrow{e^-} \| \text{ } H_3O^+ \text{ (1 M)} \,|\, H_2 \text{ (1 atm)} \,|\, Pt$$

The diagram indicates zinc metal in contact with a 1 M solution of zinc ions. The anion accompanying the zinc ion is not shown because it is not involved in the reaction, and water is usually not shown. The solution of zinc ion is connected by a salt bridge or porous disk (represented by ‖) to a 1 M solution of hydronium ions (in the hydrochloric acid) in a hydrogen electrode with a gaseous hydrogen pressure of 1 atmosphere. A single vertical line, |, is used to separate two different phases; two different species in the same phase are separated by a semicolon. The anode is always written on the left in such a diagram. The arrow and the symbol for the electron are not standard notation. We use them to show the direction of electron flow in the external circuit.

Example 19.5   **Using Line Notation to Describe a Cell**

Determine the standard cell potential, and write equations for the half-reactions and the cell reaction, for the cell described by the following line notation.

$$\text{Fe} \,|\, \text{Fe}^{2+} \,(1\ \text{M}) \xrightarrow{e^-} \text{MnO}_4^- \,(1\ \text{M});\ \text{Mn}^{2+}\,(1\ \text{M});\ \text{H}_3\text{O}^+\,(1\ \text{M}) \,|\, \text{Pt}$$

Solution: The species to the left of the double line are involved in the anode half-reaction,

$$\text{Fe}(s) \longrightarrow \text{Fe}^{2+}(aq) + 2e^- \qquad E^\circ_{\text{ox}} = -E^\circ_{\text{red}} = -(-0.440)\ \text{V} = 0.440\ \text{V}$$

The potential for this oxidation half-reaction is opposite in sign to its standard reduction potential (Table 19.1). The species to the right of the double line are involved in the cathode half-reaction (a reduction).

$$\text{MnO}_4^-(aq) + 8\text{H}_3\text{O}^+(aq) + 5e^- \longrightarrow \text{Mn}^{2+}(aq) + 12\text{H}_2\text{O}(l) \qquad E^\circ_{\text{red}} = 1.51\ \text{V}$$

We multiply the anode half-reaction by 5 and multiply the cathode half-reaction by 2 so that we have 10 electrons in each. Now we have

$$5\text{Fe} \longrightarrow 5\text{Fe}^{2+} + 10e^- \qquad\qquad\qquad E^\circ_{\text{ox}} = 0.440\ \text{V}$$

$$\underline{2\text{MnO}_4^- + 16\text{H}_3\text{O}^+ + 10e^- \longrightarrow 2\text{Mn}^{2+} + 24\text{H}_2\text{O} \qquad E^\circ_{\text{red}} = 1.51\ \ \text{V}}$$

$$5\text{Fe} + 2\text{MnO}_4^- + 16\text{H}_3\text{O}^+ \longrightarrow 5\text{Fe}^{2+} + 2\text{Mn}^{2+} + 24\text{H}_2\text{O} \quad E^\circ_{\text{cell}} = E^\circ_{\text{ox}} + E^\circ_{\text{red}}$$

$$= 0.440\ \text{V} + 1.51\ \text{V}$$

$$= 1.95\ \text{V}$$

The standard cell potential is 1.95 V. (*Note:* The phases of the reactants and products have been omitted in order to fit the equations on a single line.)

Check your learning: Determine the cell potential for the cell

$$\text{Pt} \,|\, \text{Br}^-(aq) \,|\, \text{Br}_2(l) \,\|\, \text{Cl}_2(g) \,|\, \text{Cl}^-(aq) \,|\, \text{Pt}$$

Answer: $E^\circ_{\text{cell}} = -0.2943\ \text{V}$

A positive cell potential indicates that a cell reaction proceeds spontaneously to the right and that the cell delivers an electric current. A negative potential indicates that the reaction does not proceed spontaneously to the right but rather proceeds spontaneously to the left. Thus the reaction of Fe with MnO_4^- ($E^\circ_{\text{cell}} = 1.95$ V, Example 19.5) is expected to proceed spontaneously. The reaction of Br_2 with Cl^- would be nonspontaneous as written but would proceed spontaneously in the opposite direction.

19.5   Cell Potential, Electrical Work, and Free Energy

We use batteries to do all sorts of useful things, many of which involve electric motors, because batteries are sources of electrical energy that can be used to do work. The amount of energy available depends on (1) the battery's potential, which pushes the electrons through a circuit and (2) the number of electrons involved.

We measure the potential of a battery (and other galvanic cells) in units of volts. One **volt (V)** is the potential required to impart 1 joule (J) of energy to a charge of 1 coulomb (C). Alternatively, one joule of energy is available when a charge of 1 coulomb passes through a potential of 1 volt.

$$1\ J = 1\ V \times 1\ C$$

A **coulomb (C)** of charge is the charge on 1/96,485 mole of electrons. One joule of work can be produced or consumed by a galvanic cell (depending on whether the charge moves with or against the potential) when 1 coulomb of charge passes between two electrodes with a potential difference of 1 volt. More work is available when the potential difference or the amount of charge moved is larger. Less work is available when the potential difference or the amount of charge moved is smaller.

When we use the current produced by a cell to do work—by running an electric motor, for example—the energy of the cell (the system) is reduced (because it does work on the surroundings). The sign of the work is negative. When a cell produces a current, the cell potential is positive. Thus the magnitude of the cell potential and the maximum amount of work, w_{max}, available from the cell have opposite signs. They are related by the equation

$$w_{max} = -nFE_{cell}$$

where the product nF is the charge, in coulombs, passed through the circuit and E_{cell} is the potential produced by the cell. The charge nF is determined by multiplying the number of moles of electrons n that pass through the circuit by the charge on 1 mole of electrons F. The charge on 1 mole of electrons is called a **faraday (F):** $1\ F = 96,485\ C\ mol^{-1}$, or, because $1\ C = 1\ J\ V^{-1}$, $1\ F = 96,485\ J\ V^{-1}\ mol^{-1}$. This equation gives the upper limit of available work. The actual amount of work a cell can do is less than w_{max}. For electrical work to be done, a current must flow, and when that current flows, some energy is converted to heat because of resistance in the circuit. The heat produced by the system reduces the total amount of work the system can actually do. (Remember that a reduction in energy in a system can result both from evolution of heat and from work; Section 18.2.)

The maximum amount of work, w_{max}, available from a process carried out at constant temperature and pressure is equal to the free energy change, ΔG, for the process.

$$w_{max} = \Delta G$$

We can write an equation that relates the potential of a cell and the free energy of the cell reaction, because the maximum amount of work available can be related both to the cell potential and to the free energy change of the cell reaction.

$$w_{max} = -nFE_{cell} = \Delta G \tag{1}$$

For our purposes, the important part of Equation 1 is the relationship between the free energy change and the cell potential:

$$\Delta G = -nFE_{cell}$$

or, for standard state conditions,

$$\Delta G^{\circ}_{298} = -nFE^{\circ}_{cell}$$

These equations relate the free energy change for a cell reaction, where the initial state is that of the reactants in the cell and the final state is that of the products in the cell. The equations indicate that we can find ΔG (or ΔG°) for a reaction if we know the potential of a galvanic cell with the same reaction. The equation also shows that a galvanic cell spontaneously produces current when it has a positive potential. A positive potential corresponds to a negative value of ΔG, which is the criterion for spontaneity of a reaction (Section 18.11).

Example 19.6   Determination of a Standard Free Energy Change

Calculate the standard free energy change at 25°C for the reaction

$$Cd(s) + Pb^{2+}(aq) \longrightarrow Cd^{2+}(aq) + Pb(s)$$

Solution: We can determine $\Delta G°$ with two steps:

1. Determine $E°_{cell}$ for the reaction.
2. Determine $\Delta G°_{298}$ from $E°_{cell}$.

Step 1.   We calculate $E°_{cell}$ as described in Section 19.4.

Anode:	$Cd(s) \longrightarrow Cd^{2+}(aq) + 2e^-$	$E°_{ox} = +0.403$ V
Cathode:	$Pb^{2+}(aq) + 2e^- \longrightarrow Pb(s)$	$E°_{red} = -0.126$ V
Sum:	$Cd(s) + Pb^{2+}(aq) \longrightarrow Cd^{2+}(aq) + Pb(s)$	$E°_{cell} = +0.277$ V

Step 2.   We calculate $\Delta G°_{298}$ for the change, using Equation 1 and taking into account that 2 moles of electrons are transferred when the ratio of reactants and products is that indicated by the cell reaction.

$$\Delta G° = -nFE°_{cell}$$
$$= -2 \text{ mol } (96.485 \text{ kJ V}^{-1} \text{ mol}^{-1})(0.277 \text{ V})$$
$$= -53.5 \text{ kJ}$$

The negative value for $\Delta G°$ indicates that the reaction is spontaneous both when it is run in a galvanic cell and when cadmium metal and Pb^{2+} ion are mixed.

Check your learning: Write the cell reaction, determine the cell potential, determine whether the cell reaction is spontaneous, and determine $\Delta G°_{298}$ for the cell reaction for the cell indicated by the following line diagram.

$$Pt|Fe^{2+} \text{ (1 M); } Fe^{3+} \text{ (1 M)} \| Sn^{2+} \text{ (1 M); } Sn^{4+} \text{ (1 M)}|Pt$$

Answer: $Sn^{4+}(aq) + 2Fe^{2+}(aq) \longrightarrow Sn^{2+}(aq) + 2Fe^{3+}(aq)$, $E°_{cell} = -0.62$ V, nonspontaneous, $\Delta G°_{298} = +120$ kJ

19.6   The Effect of Concentration on Cell Potential: The Nernst Equation

Up to this point, we have focused on cell potentials determined for cells with reactants and products in standard states ($E°_{cell}$); however, cells need not operate at standard state conditions.

We can use the equation for the free energy change of a reaction (Section 18.14) to derive a relationship that tells us how the cell potential changes as the concentrations and/or temperature change. The free energy change for a reaction is

$$\Delta G = \Delta G^\circ + RT \ln Q$$

where ΔG° is the free energy change for the reaction with reactants and products at standard state activities (1 molar concentrations, 1 atmosphere pressures, pure solids or liquids), ΔG is the free energy change under some other set of conditions, and Q is the reaction quotient for the reaction at the second set of conditions. Because $\Delta G = -nFE_{cell}$ and $\Delta G^\circ = -nFE_{cell}^\circ$, we can write

$$-nFE_{cell} = -nFE_{cell}^\circ + RT \ln Q$$

Dividing through by $-nF$ gives

$$E_{cell} = E_{cell}^\circ - \frac{RT}{nF} \ln Q \tag{2}$$

where

> E_{cell} = the cell potential under nonstandard state conditions
> E_{cell}° = the cell potential under standard state conditions
> R = the gas constant (8.314 J K^{-1})
> T = the Kelvin temperature
> F = the Faraday constant, 96,485 J V^{-1} mol^{-1}
> n = the number of moles of electrons exchanged in the cell reaction
> Q = the reaction quotient (Section 14.2)

Equation 2 is known as the **Nernst equation.** It was named after W. H. Nernst, a German chemist and physicist who was awarded the Nobel prize in 1920 for his contribution to thermodynamics.

For use with cells at 25°C, the Nernst equation is often written in a special form utilizing common (base-10) logarithms:

$$E_{cell} = E_{cell}^\circ - \frac{0.05916}{n} \log Q$$

The constant 0.05916 contains the value of RT/F and the conversion from natural to common logarithms.

Example 19.7   Calculating a Cell Potential Under Nonstandard State Conditions

Calculate the potential at 25°C for the cell

$$Cd|Cd^{2+} \ (2.00 \ M)\|Pb^{2+} \ (0.0010 \ M)|Pb$$

Solution: To determine the potential of a cell at nonstandard state conditions requires two steps:

1. Determine the potential of the cell under standard state conditions.
2. Determine the potential under nonstandard conditions using E_{cell}° and the Nernst equation.

Step 1. The cell reaction and standard state potential of the cell were determined in Example 19.6.

$$Cd(s) + Pb^{2+}(aq, \ 1 \ M) \longrightarrow Cd^{2+}(aq, \ 1 \ M) + Pb(s) \qquad E_{cell}^\circ = 0.277 \ V$$

Step 2. The line notation indicates that $[Pb^{2+}] = 0.0010$ M and $[Cd^{2+}] = 2.00$ M. At 25°C,

$$E_{cell} = E°_{cell} - \frac{0.05916}{n} \log Q$$

with $n = 2$, because 2 moles of electrons are exchanged in the cell reaction.

$$Q = \frac{[Cd^{2+}]}{[Pb^{2+}]} = \frac{2.00}{0.0010} = 2.0 \times 10^3$$

$$E_{cell} = E°_{cell} - \frac{0.05916}{n} \log Q$$

$$= 0.277 \text{ V} - \frac{0.05916}{2} \log 2.0 \times 10^3$$

$$= 0.277 \text{ V} - 0.098 \text{ V} = 0.179 \text{ V}$$

The cell potential decreases from 0.277 V at standard state concentrations to 0.179 V at the nonstandard concentrations.

Check your learning: Calculate the voltage produced by the cell

$$Co\,|\,Co^{2+}\ (0.00500 \text{ M})\,\|\,Al^{3+}\ (0.250 \text{ M})\,|\,Al$$

Answer: $E_{cell} = -1.36$ V

The magnitude of a cell potential is a measure of the spontaneity of a reaction: The more positive the cell potential, the more spontaneous the reaction. The change from 0.277 V to 0.179 V in Example 19.7 indicates that the spontaneity of the following reaction

$$Cd(s) + Pb^{2+}(aq) \longrightarrow Cd^{2+}(aq) + Pb(s)$$

decreases when the concentration of the reactant Pb^{2+} decreases and the concentration of the product Cd^{2+} increases. Reducing the amount of reactant relative to the amount of product (as described by Q, the reaction quotient) decreases the driving force of a reaction (as indicated by E_{cell}, the cell potential). Increasing the amount of reactant relative to the amount of product (as described by Q) increases the driving force of a reaction (as indicated by E_{cell}).

19.7   Relationship of the Cell Potential and the Equilibrium Constant

Electrochemical measurements provide data that we can use to determine thermodynamic parameters and equilibrium constants for a wide variety of chemical changes. Section 19.5 described the relationship between the standard free energy change of a cell reaction and its standard state potential.

$$\Delta G° = -nFE°_{cell}$$

In Section 18.15 we saw that the relationship between the standard state free energy change of a reaction and its equilibrium constant is

$$\Delta G° = -RT \ln K$$

Hence, if we know the standard state cell potential for a reaction, we can calculate its standard free energy change and, from that, the equilibrium constant for the reaction. Alternatively, we can use an equation that relates E°_{cell} and K. Both $-nFE^\circ_{cell}$ and $-RT \ln K$ are equal to ΔG°, so we can write

$$-nFE^\circ_{cell} = -RT \ln K = \Delta G^\circ$$

Dividing the first two terms by $-nF$ gives

$$E^\circ_{cell} = \frac{RT}{nF} \ln K \tag{3}$$

Thus we can use the standard potential of a cell reaction to calculate the equilibrium constant for the reaction, or we can use the equilibrium constant to calculate the standard potential of the cell.

Sometimes it is convenient to use a simpler version of Equation 3 for cells at 25°C (298.15 K). After conversion to the common logarithm, and with substitution of R, T, and F, the expression becomes

$$E^\circ_{cell} = \frac{0.05916}{n} \log K \tag{4}$$

Example 19.8   **Determination of an Equilibrium Constant**

Using electrochemical data, calculate the equilibrium constant at 25°C for the reaction

$$Cd(s) + Pb^{2+}(aq) \longrightarrow Cd^{2+}(aq) + Pb(s)$$

Solution: We use two steps to calculate the equilibrium constant of a reaction from electrochemical data:

1. Determine the standard state potential for the reaction.
2. Use Equation 3 or 4 to find the value of the equilibrium constant.

Step 1.   The standard state potential of the cell was determined in Example 19.7.

$$Cd(s) + Pb^{2+}(aq) \longrightarrow Cd^{2+}(aq) + Pb(s) \qquad E^\circ_{cell} = 0.277 \text{ V}$$

Step 2.   At 25° we can use Equation 4 to find K.

$$E^\circ_{cell} = \frac{0.05916}{n} \log K$$

Two electrons are involved in this cell reaction, so $n = 2$.

$$\log K = E^\circ_{cell} \times \frac{n}{0.05916}$$

$$= 0.277 \text{ V} \times \frac{2}{0.05916} = 9.36$$

$$K = 10^{9.36} = 2.3 \times 10^9$$

Check your learning: Determine the equilibrium constant at 25°C for the reaction

$$Br_2(l) + 2Cl^-(aq) \longrightarrow 2Br^-(aq) + Cl_2(g)$$

Answer: 1.12×10^{-10}

Table 19.2

The Relationships Among $E°_{cell}$, $\Delta G°$, and K

$E°_{cell}$	$\Delta G°$	K	Reaction Under Standard State Conditions
Positive	Negative	> 1	Spontaneous
0	0	$= 1$	At equilibrium
Negative	Positive	< 1	Nonspontaneous

The Nernst equation can be used to explain why a battery runs down—that is, why its voltage drops to zero as it is used. As the reaction in a cell proceeds, the cell reaction approaches equilibrium, and at equilibrium the cell will have a potential we can call E_{eq}. When the cell reaction reaches equilibrium, the reaction quotient is equal to the equilibrium constant; $Q = K$. At this point,

$$\frac{RT}{nF} \ln Q = \frac{RT}{nF} \ln K$$

and

$$E_{cell} = E_{eq}$$

Substituting this relationship into the Nernst equation (Section 19.6) gives

$$E_{cell} = E_{eq} = E°_{cell} - \frac{RT}{nF} \ln Q$$

or

$$E_{eq} = E°_{cell} - \frac{RT}{nF} \ln K$$

Earlier in this section we saw that

$$E°_{cell} = \frac{RT}{nF} \ln K$$

so we can write

$$E_{eq} = \frac{RT}{nF} \ln K - \frac{RT}{nF} \ln K$$

or

$$E_{eq} = 0$$

This tells us that when the concentrations in a cell reaction are equal to the equilibrium concentrations, the potential of the cell is zero. The relationships among E_{cell}, $\Delta G°$, and K are summarized in Table 19.2.

Batteries

Batteries are galvanic cells constructed so that they can deliver a large current and endure rough handling. Some batteries, such as flashlight batteries, consist of only one galvanic cell. Other batteries, such as the battery used to start a car, consist of several galvanic cells connected in series so that the potentials of the individual cells combine to give the total potential of the battery.

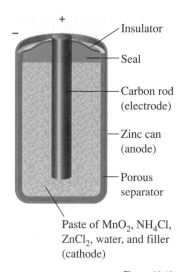

Paste of MnO_2, NH_4Cl, $ZnCl_2$, water, and filler (cathode)

Figure 19.13

Cross section of a flashlight battery, a dry cell.

19.8   Primary Cells

Primary cells are cells, or batteries, that cannot be recharged. Examples include many batteries for tape and CD players and the Daniell cell (Fig. 19.6), which was described in Example 19.1. These batteries cannot be recharged, because the electrodes and electrolytes cannot be restored to their original states by an external electrical potential.

Most inexpensive flashlight batteries are Leclanche cells (Fig. 19.13), a dry cell that is one form of a primary cell. A Leclanche cell consists of a zinc container, which serves as the anode; a carbon (graphite) rod, which serves as the electrode into the cathode; and a cathode consisting of a moist mixture of ammonium chloride, manganese dioxide, zinc chloride, and an inert filler such as sawdust. This mixture is separated from the zinc anode by a porous paper liner that serves as a salt bridge. When the cell delivers a current, the zinc anode is oxidized, forming zinc ions and electrons, which flow through the external circuit to the cathode and reduce the manganese(IV) oxide in the cathode. This reduction is not completely understood, but one possible reaction involves the reduction of the manganese from an oxidation state of +4 in MnO_2 to +3 in $MnO(OH)$:

$$e^- + NH_4^+ + MnO_2 \longrightarrow NH_3 + MnO(OH)$$

The ammonia that is produced combines with some of the zinc ions, forming the complex ion $[Zn(NH_3)_4]^{2+}$. This reaction helps to hold down the concentration of zinc ions, thereby keeping the potential of the zinc electrode more nearly constant. It also prevents the accumulation of an insulating layer of ammonia molecules on the carbon electrode, a condition called polarization, which would stop the action of the cell.

The more common battery, the alkaline cell, also uses zinc and manganese dioxide, but the electrolyte contains potassium hydroxide instead of ammonium chloride. Many of the small button-shaped batteries (Fig. 19.14) used in cameras, calculators, and watches are alkaline cells, as are the larger, longer-lived batteries used in radios and tape and CD players.

19.9   Secondary Cells

Secondary cells are galvanic cells that involve chemical reactions that are easily reversible. Such cells can be regenerated by passing a current of electricity through the cell in the reverse direction of that of discharge. This recharges the cell. The lead

Figure 19.14

Cross section of a small alkaline cell used in watches and calculators.

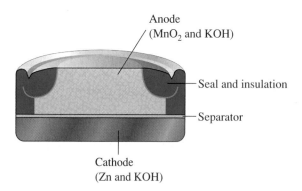

Anode (MnO_2 and KOH)

Seal and insulation

Separator

Cathode (Zn and KOH)

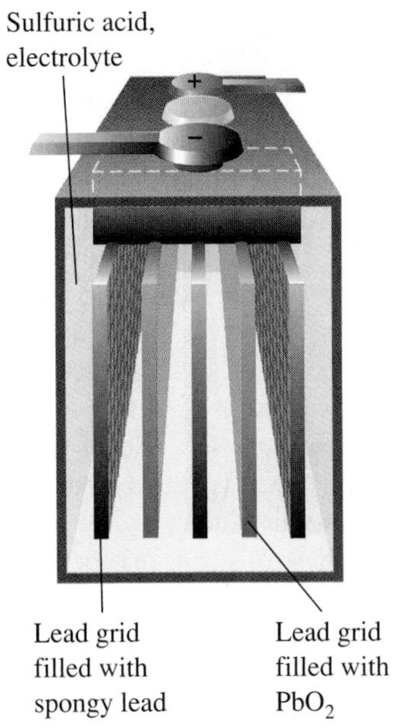

storage battery and the nickel–cadmium (NICAD) cell are examples of secondary cells.

The battery in an automobile is a lead storage battery (also called a lead–acid battery, Fig. 19.15). Its electrodes are lead alloy grids. One set of grids is filled with lead(IV) oxide and the other with spongy lead metal. Dilute sulfuric acid serves as the electrolyte. When the battery discharges (delivers a current), the spongy lead is oxidized to lead ions, which combine with sulfate ions of the electrolyte and coat the lead electrode with insoluble lead sulfate. The anode half-reaction is

$$Pb(s) + SO_4^{2-}(aq) \longrightarrow PbSO_4(s) + 2e^-$$

Electrons from the lead electrode flow through an external circuit and enter the lead(IV) oxide electrode. The lead(IV) oxide is reduced to lead(II) ions, and water is formed. Again lead ions combine with sulfate ions of the sulfuric acid electrolyte, and this plate also becomes coated with lead sulfate. The cathode half-reaction is

$$PbO_2(s) + 4H_3O^+(aq) + SO_4^{2-}(aq) + 2e^- \longrightarrow PbSO_4(s) + 6H_2O(l)$$

The alternator in a car recharges the battery by producing an external potential that pushes electrons in the reverse direction through each cell. The cell becomes an electrolytic cell during recharging. An **electrolytic cell** is a cell in which electrical energy is converted to chemical energy—the opposite of the process that occurs in a galvanic cell. The half-reactions are just the reverse of those that occur when the cell is producing a current. Electrons forced into the lead electrode reduce lead ions from the lead sulfate, so this electrode becomes the cathode during recharging. Electrons are withdrawn from the lead(IV) oxide electrode, and lead ions from the lead sulfate are oxidized.

The charge and discharge at the two plates may be summarized as follows. At the lead plate:

$$Pb(s) + SO_4^{2-}(aq) \underset{\text{Recharge}}{\overset{\text{Discharge}}{\rightleftharpoons}} PbSO_4(s) + 2e^-$$

At the lead oxide plate:

$$PbO_2(s) + 4H_3O^+(aq) + SO_4^{2-}(aq) + 2e^- \underset{\text{Recharge}}{\overset{\text{Discharge}}{\rightleftharpoons}} PbSO_4(s) + 6H_2O(l)$$

During recharging, hydronium ions and sulfate ions are regenerated.

The net cell reaction of the lead storage battery is obtained by adding the two electrode reactions together.

$$Pb + PbO_2 + 4H_3O^+ + 2SO_4^{2-} \underset{\text{Recharge}}{\overset{\text{Discharge}}{\rightleftharpoons}} 2PbSO_4 + 6H_2O$$

From the potentials in Table 19.1, we can determine that the potential of a single standard lead cell is 2.04 volts. The potential falls off slowly as the cell is used. A 12-volt automobile battery contains six lead storage cells.

The rechargeable NICAD batteries used in calculators and battery-operated tools are based on nickel and cadmium electrodes (Example 19.2). Cadmium metal serves as the anode, and nickel(IV) oxide is reduced to nickel(II) hydroxide at the cathode. The electrolyte is a hydroxide solution. When the cell delivers a current, cadmium is oxidized at the anode.

$$\textit{Anode:} \quad Cd + 2OH^- \longrightarrow Cd(OH)_2 + 2e^-$$

Sulfuric acid, electrolyte

Lead grid filled with spongy lead

Lead grid filled with PbO₂

Figure 19.15

One cell of a lead storage battery. A 12-volt automobile battery contains six of these cells.

Figure 19.16

A fuel cell used in the space shuttle.

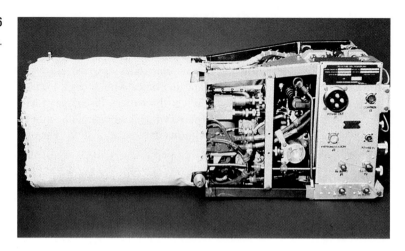

NiO_2 is reduced at the cathode.

$$\textit{Cathode:} \qquad NiO_2 + 2H_2O + 2e^- \longrightarrow Ni(OH)_2 + 2OH^-$$

These reactions are reversed during recharging. The net reaction is

$$Cd + NiO_2 + 2H_2O \underset{\text{Recharge}}{\overset{\text{Discharge}}{\rightleftharpoons}} Ni(OH)_2 + Cd(OH)_2$$

19.10   Fuel Cells

Fuel cells are galvanic cells in which electrode materials, usually in the form of gases, are supplied continuously and are consumed to produce electricity.

A typical fuel cell (Fig. 19.16), of the type used in the space shuttle, is based on the reaction of hydrogen and oxygen to form water. Hydrogen gas is diffused through the anode, a porous electrode with a catalyst such as finely divided platinum or palladium on its surface. Oxygen is diffused through the cathode, a porous electrode impregnated with cobalt oxide, platinum, or silver as a catalyst. The two electrodes are separated by a concentrated solution of sodium hydroxide or potassium hydroxide (Fig. 19.17 on page 733) as an electrolyte. The anode half-reaction is

$$2H_2 + 4OH^- \longrightarrow 4H_2O + 4e^-$$

The cathode half-reaction is

$$O_2 + 2H_2O + 4e^- \longrightarrow 4OH^-$$

The overall cell reaction is the combination of hydrogen and oxygen to produce water.

$$
\begin{array}{ll}
\textit{Anode:} & 2H_2 + 4OH^- \longrightarrow 4H_2O + 4e^- \\
\textit{Cathode:} & \underline{O_2 + 2H_2O + 4e^- \longrightarrow 4OH^-} \\
\textit{Cell reaction:} & 2H_2 + O_2 \longrightarrow 2H_2O
\end{array}
$$

Fuel cells have the potential for providing electricity from methane and other hydrocarbons much more efficiently than steam-driven generators. Thus a great deal of effort is being spent investigating fuels such as methane and electrode systems for these fuels.

Figure 19.17

Diagram of a hydrogen–oxygen fuel cell. It would actually take many such cells to light an ordinary 110-V bulb.

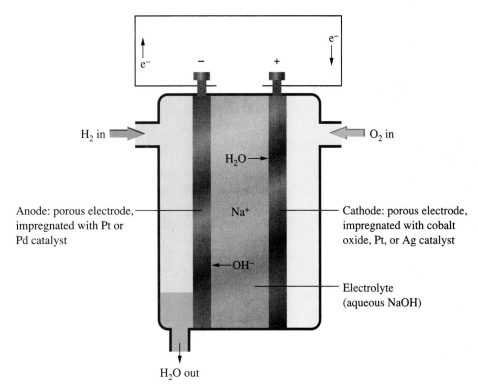

Anode: porous electrode, impregnated with Pt or Pd catalyst

Cathode: porous electrode, impregnated with cobalt oxide, Pt, or Ag catalyst

Electrolyte (aqueous NaOH)

19.11   Corrosion

Many metals, particularly iron, undergo corrosion when exposed to air and water. (See the photograph at the beginning of the chapter.) Losses caused by corrosion of metals total billions of dollars annually in the United States.

Iron will not rust in dry air or in water that is free of dissolved oxygen. Both air and water are involved in the corrosion process. The presence of an electrolyte in the water accelerates corrosion, particularly when the solution is acidic.

Corrosion is an electrochemical process (Fig. 19.18). When iron is in contact with water—with even a microscopic drop of water—the iron tends to oxidize and give up electrons.

Figure 19.18

Corrosion of iron. Iron dissolves, giving Fe^{2+} ions and electrons and causing the pit to form. The electrons travel to the edge of the drop, where they react with O_2, with the formation of OH^- ions. The Fe^{2+} and OH^- ions form $Fe(OH)_2$, which oxidizes to form rust.

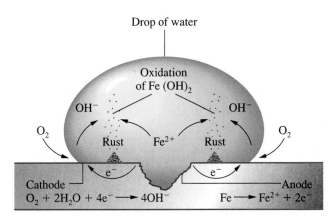

Drop of water

Oxidation of Fe $(OH)_2$

OH^- OH^-

O_2 Rust Fe^{2+} Rust O_2

Cathode

$O_2 + 2H_2O + 4e^- \longrightarrow 4OH^-$

Anode

$Fe \longrightarrow Fe^{2+} + 2e^-$

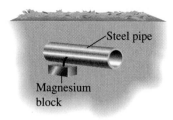

Figure 19.19

A pipeline is protected from corrosion by the active metal buried with it.

$$Anode: \quad Fe \longrightarrow Fe^{2+} + 2e^-$$

The electrons pass through the iron to the edge of the drop, where they reduce oxygen from the air to hydroxide ion.

$$Cathode: \quad O_2 + 2H_2O + 4e^- \longrightarrow 4OH^-$$

The iron(II) ions and hydroxide ions diffuse together and form insoluble iron(II) hydroxide.

$$Fe^{2+}(aq) + 2OH^-(aq) \longrightarrow Fe(OH)_2(s)$$

This precipitate is rapidly oxidized by oxygen to rust, an iron(III) compound with the approximate composition $Fe_2O_3 \cdot H_2O$.

Many methods and devices have been employed to prevent corrosion. We protect iron by coating it with paint, lacquer, grease, or asphalt; with another metal such as zinc, copper, nickel, chromium, or tin; with a ceramic enamel like that used on sinks, bathtubs, stoves, refrigerators, and washers; or with an adherent tough coating of Fe_3O_4, formed by exposing iron to superheated steam. Some alloys of iron are corrosion-resistant; examples include stainless steel (a solid solution of Fe, Cr, and Ni) and Duriron (a solution of Si in Fe).

Another means of preventing the corrosion of iron or steel involves an application of electrochemistry called cathodic protection. Cathodic protection is used on iron and steel, such as underground pipeline, that is in contact with soil (Fig. 19.19). If the iron is connected by a wire to a more active metal, such as zinc, aluminum, or magnesium, the iron becomes a cathode at which oxygen is reduced, rather than an anode where iron is oxidized. The difference in activity of the two metals causes a current to flow between them, producing corrosion of the more active metal and protecting the iron. With magnesium as the more active metal, the following reactions occur.

$$
\begin{aligned}
Anode: &\quad 2Mg \longrightarrow 2Mg^{2+} + 4e^- \\
Cathode: &\quad O_2 + 2H_2O + 4e^- \longrightarrow 4OH^- \\
\hline
Sum: &\quad 2Mg + O_2 + 2H_2O \longrightarrow 2Mg^{2+} + 4OH^-
\end{aligned}
$$

In this series of reactions no iron is oxidized. The active metal is slowly consumed and must be replaced periodically, but this is less expensive than replacing a pipeline.

Some of the more active metals such as aluminum and magnesium, which might be expected to corrode rapidly, are protected by a tightly adhering oxide coat of the metal oxide that forms when the metal is exposed to air. The metal is made passive by this coating.

Electrolytic Cells

Electrolytic cells are the opposite of galvanic cells. In an electrolytic cell, an electric current is used to produce a chemical reaction.

19.12   The Electrolysis of Molten Sodium Chloride

Lithium, sodium, magnesium, calcium, and aluminum are extracted from their compounds by **electrolysis**—the input of electrical energy as a direct current of electric-

Figure 19.20

The electrolysis of molten sodium chloride. The reduction of sodium ions at the cathode produces sodium metal, and the oxidation of chloride ions at the anode produces chlorine gas. The screen prevents the reaction of the sodium and chlorine produced.

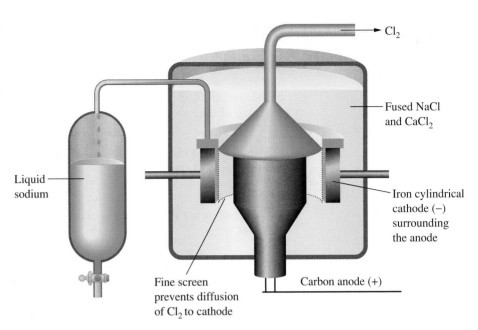

ity, which forces a nonspontaneous reaction to occur. As an example of this type of reaction, let us examine the electrolysis of molten sodium chloride.

The Downs cell used for the commercial electrolysis of molten sodium chloride (Fig. 19.20) is complex because of the need to prevent the product sodium from reacting with air or with the product chlorine. The essential elements of this cell are an air-tight container of molten sodium chloride, two inert electrodes, and a porous separator screen that permits diffusion of ions through the cell but prevents the molten sodium produced at one electrode from reacting with the chlorine gas produced at the other.

Molten sodium chloride contains equal numbers of sodium ions and chloride ions, which move about with considerable freedom. During electrolysis with a direct current, the sodium ions combine with electrons to form sodium atoms, which form metallic sodium at the cathode.

$$Na^+ + e^- \longrightarrow Na$$

The chloride ions give up one electron each and become chlorine atoms, which then combine to form molecules of chlorine gas at the anode.

$$2Cl^- \longrightarrow Cl_2 + 2e^-$$

The net reaction is

$$2Na^+ + 2Cl^- \longrightarrow 2Na + Cl_2$$

or
$$2NaCl(l) \longrightarrow 2Na(l) + Cl_2(g)$$

In order to cause the nonspontaneous decomposition of sodium chloride into sodium and chloride in a Downs cell, a potential of about 5 volts is applied with currents of thousands of amps.

The electrolysis of molten LiCl, $MgCl_2$, or $CaCl_2$ is very similar to that of NaCl. In each case, the metal is produced by reduction of the metal ion at the cathode, and chlorine is produced by oxidation of the chloride ion at the anode.

19.13   The Electrolysis of Aqueous Solutions

When an electrolyte is dissolved in water and an electrode current is passed through it, several reactions are possible. The ions of the electrolyte or the water itself can be oxidized or reduced. The half-reactions that occur depend on the ease of oxidation or reduction of the ions relative to that of water.

Electrolysis of Hydrochloric Acid.   When we electrolyze an aqueous solution of hydrochloric acid, hydronium ions are reduced to hydrogen gas at the cathode.

$$2H_3O^+ + 2e^- \longrightarrow H_2 + 2H_2O$$

Two oxidations are possible at the anode, the oxidation of chloride ion and the oxidation of water:

$$2Cl^- \longrightarrow Cl_2 + 2e^- \qquad\qquad E^\circ_{ox} = -1.36 \text{ V}$$
$$6H_2O \longrightarrow O_2 + 4H_3O^+ + 4e^- \qquad E^\circ_{ox} = -1.23 \text{ V}$$

Chloride ion and water are oxidized with almost equal ease, so the concentration of the chloride ion plays a significant role in determining the product. If the chloride ion concentration is high, chloride ion is oxidized and chlorine is formed at the anode. If the chloride concentration is low, water is oxidized as well, and oxygen is formed in addition to the chlorine. In a very dilute solution of hydrochloric acid, very little chlorine is formed, and the primary product is oxygen.

In a concentrated solution of hydrochloric acid, chlorine is produced at the anode, and the overall cell reaction can be obtained by adding the electrode reactions. Under standard state conditions,

Cathode:	$2H_3O^+ + 2e^- \longrightarrow H_2 + 2H_2O$	$E^\circ_{red} =$	0 V
Anode:	$2Cl^- \longrightarrow Cl_2 + 2e^-$	$E^\circ_{ox} =$	-1.36 V
Sum:	$2H_3O^+ + 2Cl^- \longrightarrow H_2 + Cl_2 + 2H_2O$	$E^\circ_{cell} =$	-1.36 V

Thus a potential greater than 1.36 volts must be applied to the electrolysis cell to cause this reaction to occur at standard state conditions.

Electrolysis of a Solution of Sodium Chloride.   Hydrogen gas, chlorine, and an aqueous solution of sodium hydroxide are produced when a concentrated solution of sodium chloride is electrolyzed (Fig. 19.21 on page 737). Chlorine is formed at the anode, and hydrogen gas and hydroxide ion are formed at the cathode. In order to account for these products, we need to consider all of the possible electrode reactions.

There are two species that might be reduced at the cathode: the sodium ion and water.

$$Na^+ + e^- \longrightarrow Na \qquad\qquad E^\circ_{red} = -2.714 \text{ V}$$
$$2H_2O + 2e^- \longrightarrow H_2 + 2OH^- \qquad E^\circ_{red} = -0.828 \text{ V}$$

Water is much more easily reduced than sodium ion, so hydroxide ion and hydrogen gas form. The hydroxide ions migrate toward the anode.

As we have just seen for the electrolysis of HCl, both chloride ion and water can be oxidized at the anode. In order to optimize the production of chlorine, a concentrated sodium chloride solution is used. The net reaction for the electrolysis of concentrated aqueous sodium chloride can be obtained by adding the electrode reactions.

Figure 19.21

The electrolysis of an aqueous solution of sodium chloride. The reduction of water produces hydrogen gas and hydroxide ions, and the oxidation of chloride ions produces chlorine gas. A solution of sodium hydroxide remains.

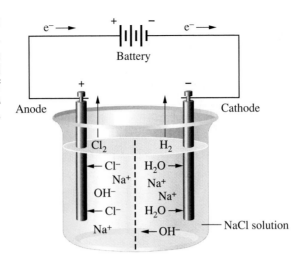

$$\text{Cathode:} \quad 2H_2O + 2e^- \longrightarrow H_2 + 2OH^- \qquad E^\circ_{red} = -0.828 \text{ V}$$
$$\text{Anode:} \quad \underline{2Cl^- \longrightarrow Cl_2 + 2e^-} \qquad \underline{E^\circ_{ox} = -1.36 \text{ V}}$$
$$\text{Sum:} \quad 2H_2O + 2Cl^- \longrightarrow H_2 + Cl_2 + 2OH^- \qquad E^\circ_{cell} = -2.19 \text{ V}$$

Hydroxide ions are formed during the electrolysis as chloride ions are removed from solution. Because the sodium ions remain unchanged, sodium hydroxide accumulates in the solution as electrolysis proceeds.

Cells for the industrial production of sodium hydroxide, chlorine, and hydrogen use currents of 10,000–60,000 amps at about 3.8 volts.

Electrolysis of a Solution of Sulfuric Acid.

Sulfuric acid ionizes in water in two steps:

$$H_2SO_4 + H_2O \longrightarrow H_3O^+ + HSO_4^- \qquad \text{(essentially complete)}$$
$$HSO_4^- + H_2O \rightleftharpoons H_3O^+ + SO_4^{2-} \qquad K_a = 1.2 \times 10^{-2}$$

As in a solution of hydrochloric acid, the reduction at the cathode is

$$2H_3O^+ + 2e^- \longrightarrow H_2 + 2H_2O$$

Two oxidations are possible at the anode: oxidation of the sulfate ion to the peroxydisulfate ion and oxidation of water.

$$2SO_4^{2-} \longrightarrow S_2O_8^{2-} + 2e^- \qquad E^\circ_{ox} = -1.96 \text{ V}$$
$$6H_2O \longrightarrow O_2 + 4H_3O^+ + 4e^- \qquad E^\circ_{red} = -1.23 \text{ V}$$

Oxygen and hydronium ions form because water is more easily oxidized than sulfate ion.

The net reaction is the sum of the anode and cathode reactions. (The cathode reaction is multiplied by 2 to balance the number of electrons.)

$$\text{Cathode:} \quad 4H_3O^+ + 4e^- \longrightarrow 2H_2 + 4H_2O$$
$$\text{Anode:} \quad \underline{6H_2O \longrightarrow O_2 + 4H_3O^+ + 4e^-}$$
$$\text{Sum:} \quad 2H_2O \longrightarrow 2H_2 + O_2$$

Figure 19.22

The electrolysis of a dilute solution of H_2SO_4.

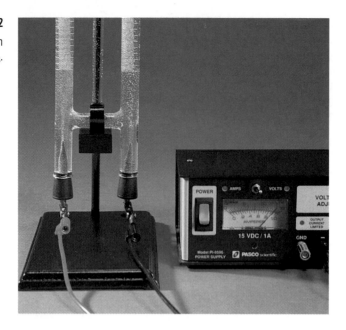

The electrolysis of an aqueous solution of sulfuric acid produces hydrogen and oxygen in a 2:1 ratio (Fig. 19.22). Although hydronium ion from the sulfuric acid is consumed by cathodic reduction, it is regenerated at the anode at the same rate at which it disappears at the cathode. Water is consumed, and the sulfuric acid becomes more concentrated as electrolysis proceeds.

19.14   Electrolytic Deposition of Metals

We use many objects with metal coatings deposited by passage of an electric current through an aqueous solution. Chrome plating, bronzing (Fig. 19.23), and the refining of copper all employ this process, which is another form of electrolysis. Bronzing and refining of copper are particularly interesting because copper is both the anode and the cathode in these processes.

In some electrolytic cells, the anode reaction may involve oxidation of the metal electrode. When a strip of metallic copper is used in an anode, the copper dissolves.

$$Cu \longrightarrow Cu^{2+} + 2e^-$$

Figure 19.23

A pair of bronzed baby shoes.

The copper(II) ions formed are reduced at the cathode, and copper metal plates out onto the cathode.

$$Cu^{2+} + 2e^- \longrightarrow Cu$$

If the cathode consists of a pair of baby shoes dusted with graphite powder to make them conducting, they become bronzed. If the cathode consists of a sheet of copper metal, refined copper is recovered.

Crude copper is refined electrochemically to improve its electrical conductivity. If impurities are not removed, they increase the electrical resistance of the metal, so impure copper wires may get dangerously hot when they conduct electricity. When impure copper is used as an anode, impurities such as gold and silver, which do not

Figure 19.24

The electrolytic refining of copper.

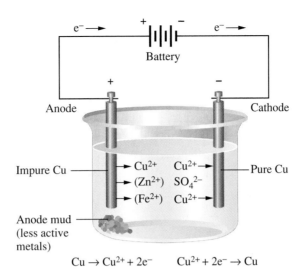

$$Cu \rightarrow Cu^{2+} + 2e^- \qquad Cu^{2+} + 2e^- \rightarrow Cu$$

oxidize so easily as copper, do not dissolve (are not oxidized) and fall to the bottom of the cell, forming a mud from which they are readily recovered (Fig. 19.24). Impurities such as zinc and iron, which oxidize more easily than copper, are oxidized and go into the solution as ions. When the electrical potential between the electrodes is carefully regulated, these ions are not reduced at the cathode; only the more easily reduced copper is deposited. The refined copper is deposited either upon a thin sheet of pure copper, which serves as a cathode, or upon some other metal from which the deposit can be stripped (Fig. 19.25).

A similar electrochemical technique is used to refine aluminum by the Hoopes process.

Figure 19.25

Copper anodes and stainless steel starter blanks (cathodes) are lured into an elecrolytic cell.

Figure 19.26

Amounts of various elements produced at the cathode by 1 faraday of
electricity (96,485 coulombs, or 1 mole
of electrons).

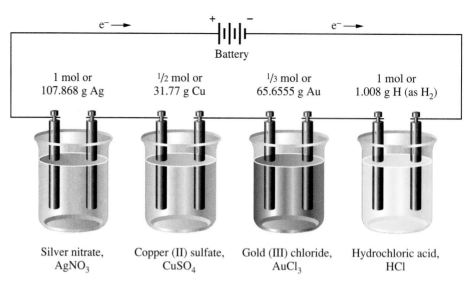

Silver nitrate,   Copper (II) sulfate,   Gold (III) chloride,   Hydrochloric acid,
$AgNO_3$ $CuSO_4$ $AuCl_3$ HCl

19.15   Faraday's Law of Electrolysis

In 1832–1833, Michael Faraday performed experiments demonstrating that the
amount of a substance undergoing a chemical change at each electrode during electrolysis is directly proportional to the quantity of electricity that passes through the
electrolytic cell (Fig. 19.26). We can express the quantity of electricity as the number of moles of electrons passing through a cell. The quantity of a substance undergoing chemical change is related to the number of electrons that are involved in its
half-reaction and can be expressed in terms of moles of the substance or in terms of
equivalents of the substance. An electrochemical **equivalent** is the amount of a substance that combines with or releases 1 mole of electrons. Thus 1 equivalent of an
oxidizing agent combines with 1 mole of electrons, and 1 equivalent of a reducing
agent releases 1 mole of electrons. The number of equivalents of solute in a liter of
solution (equivalents per liter) is called the **normality, N,** of the solution. A solution
that contains 1 equivalent of a substance per liter is called a 1 normal solution. One
liter of a 1 N solution will combine with or release 1 mole of electrons.

Experimental results show that 1 electron reduces 1 silver ion, whereas 2 electrons
are required to reduce 1 copper(II) ion.

$$Ag^+ + e^- \longrightarrow Ag$$

$$Cu^{2+} + 2e^- \longrightarrow Cu$$

Therefore 6.022×10^{23} electrons reduce 6.022×10^{23} silver ions (1 mole) to silver
atoms. The mass of silver produced is 107.9 grams; thus the mass of 1 equivalent of
silver is the same as the mass of 1 mole. Because 2 electrons are required to reduce
1 Cu^{2+} ion to atomic copper, 6.022×10^{23} electrons reduce only 1/2 of a mole of
copper. Thus the mass of 1 equivalent of copper is 1/2 the mass of 1 mole.

A faraday is the charge on 1 mole of electrons (Section 19.5). When 1 faraday is
passed through an electrochemical cell, 1 mole of electrons passes through the cell,
1 equivalent of a substance is reduced at the cathode, and 1 equivalent of a substance
is oxidized at the anode. Other units of electricity that you will encounter often are
the coulomb and the ampere. The relationships among these units are as follows:

$$1 \text{ faraday} = 6.022 \times 10^{23} \text{ electrons} = 1 \text{ mole of electrons}$$
$$= 96{,}485 \text{ coulombs (C)}$$

$$1 \text{ coulomb} = \text{quantity of electricity involved when a}$$
$$\text{current of 1 ampere (A) flows for 1 second}$$
$$= 1 \text{ A s}$$

$$1 \text{ ampere} = 1 \text{ coulomb per second} = 1 \text{ C s}^{-1}$$

We handle calculations that involve passage of an electric current (measured in amperes) through a cell just like any other stoichiometry problem: We convert the current in amperes to coulombs, then to faradays, and finally to moles of electrons. Use the moles of electrons as you would moles of any other reactant.

Example 19.9   Determining the Mass of Product

Calculate the mass of copper that is produced by the reduction of Cu^{2+} ion at the cathode during the passage of 1.600 amperes of current through a solution of copper(II) sulfate for 1.000 hour.

Solution: Copper(II) ion is reduced according to the equation

$$Cu^{2+} + 2e^- \longrightarrow Cu$$

We can determine the mass of copper produced from the number of moles of copper produced. We can calculate the moles of copper if we know the moles of electrons that pass through the cell. The moles of electrons can be determined from the amperage and time of the current flow. The chain of calculations is

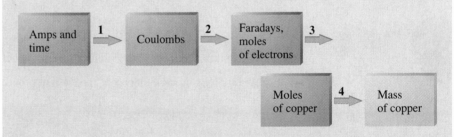

The numbers above the arrows refer to the steps below.

Step 1.   We calculate the number of coulombs used from the amperes and time.

$$1.600 \text{ A} \times 1.000 \text{ h} \times \frac{60 \text{ min}}{1 \text{ h}} \times \frac{60 \text{ s}}{1 \text{ min}} = 5760 \text{ A s} = 5760 \text{ C}$$

Step 2.   The number of faradays (F), or moles of electrons, involved is determined.

$$5760 \text{ C} \times \frac{1 \, F}{96{,}485 \text{ C}} = 5.970 \times 10^{-2} \, F = 5.970 \times 10^{-2} \text{ mol e}^-$$

Step 3.   We determine the number of moles of copper produced.

$$5.970 \times 10^{-2} \text{ mol e}^- \times \frac{1 \text{ mol Cu}}{2 \text{ mol e}^-} = 2.985 \times 10^{-2} \text{ mol Cu}$$

Step 4.   Finally we convert the moles of copper to the mass of copper.

$$2.985 \times 10^{-2} \text{ mol Cu} \times \frac{63.55 \text{ g Cu}}{1 \text{ mol Cu}} = 1.897 \text{ g Cu}$$

Check your learning: What mass of nickel is produced by passage of a 0.200-A current for 30.0 min through a solution of Ni^{2+} ion?

Answer: 0.109 g

Example 19.10   Determining the Length of an Electrolysis

Very large currents are used in many industrial electrolytic cells. How much time is required to produce exactly 1 metric ton (1000 kg) of magnesium by passage of a current of 150,000 amperes through molten $MgCl_2$? Assume a 100% yield of magnesium based on the current.

Solution: We can solve this problem using the following steps:

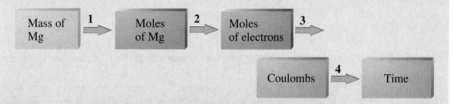

The numbers above the arrows refer to the steps below.

Step 1.   We convert the mass of magnesium produced to moles of magnesium.

$$1000 \text{ kg Mg} \times \frac{1000 \text{ g}}{1 \text{ kg}} \times \frac{1 \text{ mol}}{24.305 \text{ g}} = 4.114 \times 10^{4} \text{ mol Mg}$$

Step 2.   The moles of magnesium are converted to moles of electrons needed to produce it.

$$4.114 \times 10^{4} \text{ mol Mg} \times \frac{2 \text{ mol e}^{-}}{1 \text{ mol Mg}} = 8.228 \times 10^{4} \text{ mol e}^{-}$$

Step 3.   We find the coulombs required.

$$8.228 \times 10^{4} \text{ mol e}^{-} \times \frac{96,485 \text{ C}}{1 \text{ mol e}^{-}} = 7.939 \times 10^{9} \text{ C}$$

Step 4.   Finally we determine the length of time necessary to pass the coulombs with a current of 150,000 amps.

$$s = \frac{C}{A} = \frac{7.939 \times 10^{9} \text{ C}}{150,000 \text{ C s}^{-1}} = 5.293 \times 10^{4} \text{ s}$$

It requires 5.293×10^4 s, or 14.70 h, to produce 1 metric ton of magnesium with a current of 150,000 A.

Check your learning: How much time is required to form 3.02 g of Ag by electrolysis of a solution of Ag^+ with a current of 0.500 A?

Answer: 5.40×10^3 s, or 90.0 min

Oxidation–Reduction Reactions

Oxidation–reduction reactions (redox reactions) involve oxidation and reduction of the reactants (Section 8.5). The chemical changes that occur in electrochemical cells involve oxidation of one species at the anode and reduction of another species at the cathode. Thus the net reactions of most electrochemical cells are oxidation–reduction reactions.

19.16   Balancing Oxidation–Reduction Equations

Equations for oxidation–reduction reactions can be complicated and difficult to balance by trial and error. The half-reaction method for balancing redox equations provides a systematic approach using the same process that is used to write a cell reaction (Section 19.1). We use the following steps.

Step 1.   *Write the balanced oxidation half-reaction.*

Step 2.   *Write the balanced reduction half-reaction.*

Step 3.   *Balance the number of electrons lost and gained.*

Step 4.   *Add the balanced half-reactions and cancel species that appear on both sides of the arrow.*

Step 5.   *Check to see that both the number of atoms and the charges are balanced.*

Example 19.11   **Balancing an Oxidation–Reduction Reaction**

Balance the equation for the oxidation of iron(II) to iron(III) by chlorine.

$$Fe^{2+} + Cl_2 \longrightarrow Fe^{3+} + Cl^-$$

Solution:

Step 1.   *Write the balanced oxidation half-reaction.*

$$Fe^{2+} \longrightarrow Fe^{3+} + e^-$$

Step 2. *Write the balanced reduction half-reaction.*

$$Cl_2 + 2e^- \longrightarrow 2Cl^-$$

Step 3. *Balance the numbers of electrons lost and gained.*

In the oxidation half-reaction one electron is lost, whereas in the reduction half-reaction two electrons are gained. We multiply the oxidation half-reaction by 2 to balance the electrons.

$$2Fe^{2+} \longrightarrow 2Fe^{3+} + 2e^-$$

Step 4. *Add the balanced half-reactions and cancel species that appear on both sides of the arrow.*

When the balanced half-reactions are added together, the two electrons on either side of the equation cancel one another, and the balanced equation is obtained.

$$2Fe^{2+} \longrightarrow 2Fe^{3+} + 2e^-$$
$$\underline{Cl_2 + 2e^- \longrightarrow 2Cl^-}$$
$$2Fe^{2+} + Cl_2 \longrightarrow 2Fe^{3+} + 2Cl^-$$

Step 5. *Check to see that both the numbers of atoms and the charges are balanced.*

In this equation both sides have two Fe ions, two Cl atoms or ions, and a net charge of $+4$: $(2 \times +2)$ on the left and $(2 \times +3) + (2 \times -1)$ on the right.

Check your learning: Balance the equation $I_3^- + S_2O_3^{2-} \longrightarrow I^- + S_4O_6^{2-}$.

Answer: $I_3^- + 2S_2O_3^{2-} \longrightarrow 3I^- + S_4O_6^{2-}$

Many oxidation–reduction reactions involve the transfer of hydrogen or oxygen atoms without changes in their oxidation states. For example, when permanganate ion is reduced in acidic solution, the oxygen of the permanganate ion combines with hydronium ion to form water.

$$MnO_4^- + 8H_3O^+ + 5e^- \longrightarrow Mn^{2+} + 4H_2O$$

The manganese changes its oxidation state during this step, but the hydrogen and oxygen do not.

There are four general rules for the transfer of hydrogen and oxygen in such cases.

In acidic solution:

1. Excess oxide combines with hydronium ion to give water. Oxide is obtained from water, leaving hydronium ion.

$$O^{2-} + 2H_3O^+ \rightleftharpoons 3H_2O$$

2. Hydronium ion is the source of hydrogen ion. Excess hydrogen ion combines with water to form hydronium ion.

$$H_3O^+ \rightleftharpoons H^+ + H_2O$$

In basic solution:

3. Excess oxide combines with water to give hydroxide ion. Oxide is obtained from hydroxide ion, leaving water.

$$O^{2-} + H_2O \rightleftharpoons 2OH^-$$

4. Water is the source of hydrogen ion; hydroxide ion is also produced. Excess hydrogen ion combines with hydroxide ion to form water.

$$H_2O \rightleftharpoons H^+ + OH^-$$

Example 19.12   **Balancing an Oxidation–Reduction Reaction in Acidic Solution**

The dichromate ion oxidizes iron(II) to iron(III) in acid solution. Complete and balance the equation for the reaction

$$Cr_2O_7^{2-} + Fe^{2+} \longrightarrow Cr^{3+} + Fe^{3+}$$

Solution:

Step 1. *Write the balanced oxidation half-reaction.*

Iron(II) is oxidized, so the half-reaction is

$$Fe^{2+} \longrightarrow Fe^{3+} + e^-$$

This half-reaction is balanced.

Step 2. *Write the balanced reduction half-reaction.*

The reduction half-reaction involves the change

$$Cr_2O_7^{2-} + e^- \longrightarrow Cr^{3+}$$

In acid solution, the excess oxide combines with hydronium ion to give water: This half-reaction also involves H_3O^+ as a reactant and H_2O as a product.

$$Cr_2O_7^{2-} + H_3O^+ + e^- \longrightarrow Cr^{3+} + H_2O$$

To balance this half-reaction, we must balance both the numbers of atoms on either side of the arrow and the total charge. This gives

$$Cr_2O_7^{2-} + 14H_3O^+ + 6e^- \longrightarrow 2Cr^{3+} + 21H_2O$$

The charge is balanced in this equation. On the left side we have 14 positive charges (on the 14 H_3O^+ ions) and 8 negative charges (2 on the $Cr_2O_7^{2-}$ ion and 6 on the electrons) for an excess of 6 positive charges. On the right side we have 6 positive charges (on the 2 Cr^{3+} ions).

Step 3. *Balance the numbers of electrons lost and gained.*

We multiply the oxidation half-reaction by 6 so that both half-reactions will involve the same number of electrons.

$$6Fe^{2+} \longrightarrow 6Fe^{3+} + 6e^-$$

Step 4. *Add the balanced half-reactions and cancel species that appear on both sides of the arrow.*

Upon addition of the balanced half-reactions, the six electrons on either side of the equation cancel one another, and we have

$$6Fe^{2+} \longrightarrow 6Fe^{3+} + 6e^-$$
$$\underline{Cr_2O_7{}^{2-} + 14H_3O^+ + 6e^- \longrightarrow 2Cr^{3+} + 21H_2O}$$
$$Cr_2O_7{}^{2-} + 14H_3O^+ + 6Fe^{2+} \longrightarrow 2Cr^{3+} + 6Fe^{3+} + 21H_2O$$

Step 5.   *Check to see that both the numbers of atoms and the charges are balanced.*

In this equation both sides have 2 Cr ions, 6 Fe ions, 21 O atoms, 42 H atoms, and a net charge of $+24$.

Check your learning: In acidic solution, hydrogen peroxide oxidizes Fe^{2+} to Fe^{3+}. Complete and balance the equation

$$H_2O_2 + Fe^{2+} \longrightarrow H_2O + Fe^{3+}$$

Answer: $H_2O_2 + 2H_3O^+ + 2Fe^{2+} \longrightarrow 4H_2O + 2Fe^{3+}$

Example 19.13   **Balancing an Oxidation–Reduction Reaction in Basic Solution**

Hydrogen peroxide reduces permanganate, $MnO_4{}^-$, to manganese(IV) oxide, MnO_2, in basic solution. Write a balanced equation for the reaction

Solution: Initially we have

$$H_2O_2 + MnO_4{}^- \longrightarrow MnO_2$$

and we know that the reaction is in basic solution. We balance this equation one half-reaction at a time.

Step 1.   *Write the balanced oxidation half-reaction.*

Permanganate is reduced in this reaction. Thus the oxidation half-reaction must involve oxidation of hydrogen peroxide. Oxygen already has an oxidation state of -1 in H_2O_2, so it can be oxidized only to O_2. The oxidation half-reaction involves the change

$$H_2O_2 \longrightarrow O_2$$

In basic solution, excess hydrogen combines with hydroxide ion to give water.

$$H_2O_2 + 2OH^- \longrightarrow O_2 + 2H_2O$$

Electrons are also products of an oxidation half-reaction. Adding electrons to balance the charge gives us

$$H_2O_2 + 2OH^- \longrightarrow O_2 + 2H_2O + 2e^-$$

Step 2.   *Write the balanced reduction half-reaction.*

The reduction half-reaction involves the change

$$MnO_4{}^- \longrightarrow MnO_2$$

Because it is in basic solution, the excess oxide combines with water to give hydroxide ion. Thus H_2O is needed as a reactant, and OH^- is a product.

$$MnO_4^- + 2H_2O \longrightarrow MnO_2 + 4OH^-$$

Electrons are reactants in a reduction half-reaction. Adding electrons to balance the charge gives us the following half-reaction:

$$MnO_4^- + 2H_2O + 3e^- \longrightarrow MnO_2 + 4OH^-$$

Steps 3 and 4. *Balance the numbers of electrons lost and gained; then add the half-reactions.*

We multiply the reduction half-reaction by 2 and the oxidation half-reaction by 3 and then add them to give the overall equation.

$$2MnO_4^- + 4H_2O + 6e^- \longrightarrow 2MnO_2 + 8OH^-$$
$$3H_2O_2 + 6OH^- \longrightarrow 3O_2 + 6H_2O + 6e^-$$
$$\overline{2MnO_4^- + 4H_2O + 3H_2O_2 + 6OH^- \longrightarrow 2MnO_2 + 8OH^- + 3O_2 + 6H_2O}$$

Cancellation of identical species on the left and right sides of this equation yields

$$2MnO_4^- + 3H_2O_2 \longrightarrow 2MnO_2 + 3O_2 + 2OH^- + 2H_2O$$

Step 5. *Check to see that both the numbers of atoms and the charges are balanced.*

In this equation both sides have the same numbers of atoms and the same net charge, -2.

Check your learning: Complete and balance the equation for the following reaction that occurs in basic solution:

$$CrO + ClO^- \longrightarrow CrO_4^{2-} + Cl^-$$

Answer: $CrO + 2ClO^- + 2OH^- \longrightarrow CrO_4^{2-} + 2Cl^- + H_2O$

For Review   Summary

In a **galvanic cell,** chemical changes are used to produce electrical energy. In an **electrolytic cell,** electrical energy is used to produce a chemical change. In both types of cells, the electrode at which oxidation occurs is the **anode,** and the electrode at which reduction occurs is the **cathode.**

The potential that causes electrons from a galvanic cell to move from one electrode through an external circuit to the other electrode is called the **cell potential, E_{cell},** of the cell. A **standard cell potential, E°_{cell},** is the potential of a cell operating under standard state conditions (1 M concentrations, 1 atmosphere pressures, pure solids and liquids, 25°C). The potential of a cell is the sum of the electrode potential of the anode and the electrode potential of the cathode. A positive E_{cell} value indicates that the cell reaction occurs spontaneously as written. **Standard reduction potentials** are tabulated for a variety of reduction (cathode) **half-reactions** at standard state conditions. The potential of an oxida-

tion half-cell (the anode half-reaction) has the same absolute value as the reduction potential but is opposite in sign. The more positive the potential associated with a half-reaction, the greater the tendency of that reaction to occur as written. The standard free energy change, ΔG°_{298}, for a cell reaction can be calculated from the potential of a cell based on the reaction. At standard conditions,

$$\Delta G^\circ = -nFE^\circ_{cell}$$

The equilibrium constant for the reaction is related to the standard free energy change and thus to E°. At 25°C,

$$E^\circ_{cell} = \frac{RT}{nF} \ln K = \frac{0.05916}{n} \log K$$

Electrode potentials vary with the concentrations of the reactants and products involved in the half-reaction. We can calculate them from standard electrode potentials by using the **Nernst equation:**

$$E_{cell} = E°_{cell} - \frac{RT}{nF} \ln Q$$

or, at 25°C,

$$E_{cell} = E°_{cell} - \frac{0.05916}{n} \log Q$$

where n is the number of moles of electrons involved in the cell reaction.

Examples of batteries include the Daniell cell and the familiar flashlight battery, a dry cell. These are **primary cells,** which cannot be recharged. The lead storage battery is an example of a **secondary cell,** which is rechargeable.

Typical changes carried out in electrolytic cells include electrolysis of molten sodium chloride, producing sodium and chlorine; electrolysis of an aqueous solution of hydrogen chloride, producing hydrogen and chlorine; electrolysis of an aqueous solution of sodium chloride, producing hydrogen, chlorine, and a solution of sodium hydroxide; and electrolysis of a solution of sulfuric acid, producing hydrogen and oxygen. The extent of the chemical changes in these processes can be related to the amount of electricity that passes through the cell. A **faraday,** which corresponds to passage of 1 mole of electrons, reduces 1 **equivalent** of a substance at the cathode or is produced by the oxidation of 1 equivalent of a substance at the anode. The amount of electrical charge possessed by a mole of electrons is 96,485 coulombs; 1 **coulomb** is the quantity of electricity involved when a current of 1 ampere flows for 1 second.

Oxidation–reduction reactions can be considered to consist of two half-reactions. After the half-reactions are balanced, we can add them to obtain the overall balanced equation for the reaction.

Key Terms and Concepts

anode   (19.1)
balancing redox equations   (19.16)
cathode   (19.1)
cell potential, E_{cell} (19.2)
cell reaction   (19.1)
coulomb, C   (19.2, 19.15)
electrode   (19.3)
electrolysis   (19.12)
electrolytic cell   (19.9)
equilibrium constant   (19.7)
equivalent   (19.15)

faraday, F   (19.5)
free energy change   (19.5)
fuel cell   (19.10)
galvanic cell   (19.1)
gas electrode   (19.3)
half-cell   (19.1)
half-reaction   (19.1)
line notation   (19.4)
maximum work   (19.5)
Nernst equation   (19.6)

normality   (19.15)
primary cell   (19.8)
salt bridge   (19.1)
secondary cell   (19.9)
spontaneity of cell reactions   (19.5)
standard cell potential, $E°_{cell}$ (19.2)
standard electrode potential   (19.3)
standard hydrogen electrode   (19.3)
standard reduction potential   (19.3)
volt, V   (19.5)

Exercises

Questions

1. Define the terms *anode* and *cathode*.

2. What is the purpose of a salt bridge in an electrochemical cell?

3. An electrochemical cell is constructed as shown below.

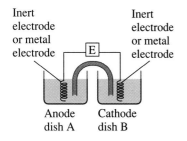

Inert
electrode
or metal
electrode

Inert
electrode
or metal
electrode

E

Anode   Cathode
dish A   dish B

(a) Indicate whether the electrode is a metal or an inert electrode and describe which ions or compounds are in each dish in such a cell with each of the following reactions or diagrams.

   i. $Cu^+ + Fe(CN)_6{}^{3-} \longrightarrow Cu^{2+} + Fe(CN)_6{}^{4-}$

  ii. $MnO_4{}^- + 8H_3O^+ + 5Au \longrightarrow$
$$Mn^{2+} + 12H_2O + 5Au^+$$

 iii. $2MnO_4{}^- + 3NO_2{}^- + H_2O \longrightarrow$
$$2MnO_2 + 3NO_3{}^- + 2OH^-$$

 iv. $Ni\,|\,Ni^{2+}\,\|\,Cu^{2+}\,|\,Cu$

  v. $Pt\,|\,SO_4{}^{2-};\ HSO_3{}^-,\ H_3O^+\,\|\,ClO_3{}^-;\ HClO^-;$
$H_3O^+\,|\,Pt$

(b) Write the net ionic equation for the half-reaction that occurs in each dish.

4. (a) Indicate whether the electrode is a metal or an inert electrode and describe which ions or compounds are

in each dish in a cell like that illustrated in Exercise 3 with each of the following reactions or diagrams.

i. $Sn^{2+} + 2Fe^{3+} \longrightarrow Sn^{4+} + 2Fe^{2+}$

ii. $2Mn^{2+} + 24H_2O + 5Fe^{2+} \longrightarrow$
$$2MnO_4^- + 16H_3O^+ + 5Fe$$

iii. $Cr_2O_7^{2-} + 5H_3O^+ + 3HNO_2 \longrightarrow$
$$2Cr^{3+} + 3NO_3^- + 9H_2O$$

iv. $Co\,|\,Co^{2+}\,\|\,Cr^{3+}\,|\,Cr$

v. $Pt\,|\,Mn^{2+};\ MnO_4^-;\ H_3O^+\,\|\,Fe^{3+};\ Fe^{2+}\,|\,Pt$

(b) Write the net ionic equation for the half-reaction that occurs in each dish.

5. Define the term *standard reduction potential*.

6. Why is the potential for the standard hydrogen electrode listed as 0.00 V?

7. Which is the better oxidizing agent in each of the following pairs at standard conditions?
(a) Cr^{3+} or Co^{2+}
(b) I_2 or Mn^{2+}
(c) MnO_4^- or $Cr_2O_7^{2-}$ (in acid solution)
(d) Pb^{2+} or Sn^{4+}

8. Which is the better reducing agent in each of the following pairs at standard conditions?
(a) F^- or Cu
(b) Sn^{2+} or Fe^{2+}
(c) H_2 or I^-
(d) Fe^{2+} or Cr

9. What is the cell with the highest potential that can be constructed from the metals iron, nickel, copper, and silver?

10. Describe the importance of the Nernst equation in electrochemistry.

11. How does the potential of a galvanic cell change as the concentrations of the species involved in the cell reaction change?

12. Write two different cell reactions that show that the electrode reaction for the zinc electrode is reversible.

13. How does a galvanic cell differ from an electrolytic cell?

14. By means of a description and chemical equations, explain the electrolytic purification of copper and its separation from impurities such as silver, zinc, and gold.

15. Describe two general methods that are commonly used to protect metals from corrosion.

16. In this and other chapters, we note commercial electrolytic processes for the preparation of several substances, some of which are hydrogen, oxygen, chlorine, and sodium hydroxide. Write equations for the net cell reaction and the individual electrode half-reactions that describe these commercial processes.

17. Tarnished silverware is coated with Ag_2S. The tarnish can be removed by placing the silverware in an aluminum pan and covering it with a solution of an inert electrolyte such as $NaCl$. Explain the electrochemical basis for this procedure.

18. What features determine the voltage of a battery? What determines the number of coulombs that a battery can deliver?

19. Soon after a copper metal rod is placed in a silver nitrate solution, copper ions are observed in the solution and silver metal has been deposited on the rod. Can the copper rod be considered an electrode? If so, is it an anode or a cathode?

20. Explain briefly how you could determine the solubility product of CuI ($K_{sp} \approx 10^{-7}$) by using an electrochemical measurement.

21. Explain why ClO_3^- is a poorer oxidizing agent in base than in acid.

22. Is H_2O_2 a better oxidizing agent in acid or in base?
$$H_2O_2 + 2e^- \longrightarrow 2OH^-$$

23. Which metals listed in Table 19.1 could be purified by an electrolysis similar to that used to purify copper?

24. Why are different products obtained at the cathode when molten $ZnCl_2$ is electrolyzed than when a solution of $ZnCl_2$ is electrolyzed? Why do a solution of $CuCl_2$ and molten $CuCl_2$ give the same products at the cathode upon electrolysis?

Standard Reduction Potentials and Cell Potentials

25. Complete and balance each of the following half-reactions. In each case, indicate whether oxidation or reduction occurs.
(a) $Sn^{4+} \longrightarrow Sn^{2+}$
(b) $[Ag(NH_3)_2]^+ \longrightarrow Ag$
(c) $Hg_2Cl_2 \longrightarrow Hg$
(d) $H_2O \longrightarrow O_2$ (in acid)
(e) $O_2 \longrightarrow OH^-$ (in base)
(f) $SO_3^{2-} \longrightarrow SO_4^{2-}$ (in acid)
(g) $MnO_4^- \longrightarrow Mn^{2+}$ (in acid)
(h) $Cl^- \longrightarrow ClO_3^-$ (in base)

26. Complete and balance each of the following half-reactions. In each case, indicate whether oxidation or reduction occurs.
(a) $Cr^{2+} \longrightarrow Cr^{3+}$
(b) $Hg \longrightarrow HgBr_4^{2-}$
(c) $ZnS \longrightarrow Zn$
(d) $H_2 \longrightarrow H_2O$ (in base)
(e) $H_2 \longrightarrow H_3O^+$ (in acid)
(f) $NO_3^- \longrightarrow HNO_2$ (in acid)
(g) $MnO_2 \longrightarrow MnO_4^-$ (in base)
(h) $Cl^- \longrightarrow ClO_3^-$ (in acid)

27. Write the half-reactions for cells that have the following net reactions. Using a line diagram of the type

Anode | anode solution ‖ cathode solution | cathode

diagram the electrolytic cell of each.

(a) $Mn + 2Ag^+ \longrightarrow Mn^{2+} + 2Ag$

(b) $MnO_4^- + 8H_3O^+ + 5Au \longrightarrow$
$$Mn^{2+} + 12H_2O + 5Au^+$$

(c) $4Cr + 3O_2 + 6H_2O + 4OH^- \longrightarrow$
$$4Cr(OH)_4^- \text{ (in base)}$$

(d) $Cr_2O_7^{2-} + 5H_3O^+ + 3HNO_2 \longrightarrow$
$$2Cr^{3+} + 3NO_3^- + 9H_2O$$

(e) $Cl^- + NO_3^- \longrightarrow ClO^- + NO_2^- \text{ (in base)}$

28. Write the half-reactions for cells that have the following net reactions. Using a line diagram of the type

Anode | anode solution ‖ cathode solution | cathode

diagram the electrolytic cell of each.

(a) $Cd + Pb^{2+} \longrightarrow Cd^{2+} + Pb$

(b) $Sn^{4+} + H_2 + 2H_2O \longrightarrow Sn^{2+} + 2H_3O^+$

(c) $Zn + 2Fe(CN)_6^{3-} + 4CN^- \longrightarrow$
$$Zn(CN)_4^{2-} + 2Fe(CN)_6^{4-}$$

(d) $2Mn^{2+} + 24H_2O + 5Fe^{2+} \longrightarrow$
$$2MnO_4^- + 16H_3O^+ + 5Fe$$

(e) $2MnO_4^- + 3NO_2^- + H_2O \longrightarrow$
$$2MnO_2 + 3NO_3^- + 2OH^-$$

29. Calculate the potential at standard conditions of a cell based on the reactions in Exercise 27.

30. Calculate the potential at standard conditions of a cell based on the reactions in Exercise 28.

31. Calculate the potential of a cell based on the following reactions at standard conditions.

(a) $Zn + I_2 \longrightarrow Zn^{2+} + 2I^-$

(b) $2Co(OH)_3 + Ni(OH)_2 \longrightarrow$
$$2Co(OH)_2 + NiO_2 + 2H_2O$$

32. Calculate the cell potential of a cell based on the following reactions at standard conditions.

(a) $Mn + 2AgCl \longrightarrow Mn^{2+} + 2Cl^- + 2Ag$

(b) $Fe + NiO_2 + 2H_2O \longrightarrow Fe(OH)_2 + Ni(OH)_2$

33. Determine the potential for each of the following cells.

(a) $Co|Co^{2+} (1 M)‖Cr^{3+} (1 M)|Cr$

(b) $Ni|Ni^{2+} (1 M)‖Br^- (1 M)|Br_2(l)|Pt$

34. Determine the potential for each of the following cells.

(a) $Pb; \ PbSO_4(s)|SO_4^{2-} \ (1 \ M)‖H_3O^+ \ (1 \ M)|H_2$ (1 atm)$|Pt$

(b) $Pt|Mn^{2+} \ (1 \ M); \ MnO_4^- \ (1 \ M); \ H_3O^+ \ (1 \ M)‖Fe^{3+}$ (1 M); $Fe^{2+} \ (1 \ M)|Pt$

35. Write the cell reaction for a galvanic cell based on each of the following pairs of half-reactions, and calculate the potential of the cell under standard conditions.

(a) $Sc^{3+} + 3e^- \longrightarrow Sc$
$Ag^+ + e^- \longrightarrow Ag$

(b) $S + 2e^- \longrightarrow S^{2-}$
$Cl_2 + 2e^- \longrightarrow 2Cl^-$

36. Write the cell reaction for a galvanic cell based on each of the following pairs of half-reactions, and calculate the potential of the cell under standard conditions.

(a) $ZnS + 2e^- \longrightarrow Zn + S^{2-}$
$CdS + 2e^- \longrightarrow Cd + S^{2-}$

(b) $Co(OH)_3 + e^- \longrightarrow Co(OH)_2 + OH^-$
$Cr(OH)_4^- + 3e^- \longrightarrow Cr + 4OH^-$

(c) $HClO_2 + 2H_3O^+ + 2e^- \longrightarrow HClO + 3H_2O$
$ClO_3^- + 3H_3O^+ + 2e^- \longrightarrow HClO_2 + 4H_2O$

37. Rechargeable nickel–cadmium cells are used in calculators and other battery-powered devices. Communication satellite also use these cells. The cell reaction is

$$NiO_2 + Cd + 2H_2O \longrightarrow Ni(OH)_2 + Cd(OH)_2$$

Calculate the cell potential, using the following half-cell potentials.

$NiO_2 + 2H_2O + 2e^- \longrightarrow Ni(OH)_2 + 2OH^- \quad E° = +0.49 \text{ V}$

$Cd(OH)_2 + 2e^- \longrightarrow Cd + 2OH^- \quad\quad\quad E° = -0.81 \text{ V}$

38. (a) Calculate the standard state potential of a single lead storage cell.

(b) A lead storage battery, such as that used in an automobile, contains six lead storage cells. What maximum voltage is expected of such a battery under standard state conditions?

39. Under standard conditions the potential of the following cell is +1.05 V. What is the metal M? Show your calculations.

$$M|M^{n+}‖Cu^+|Cu$$

40. Under standard conditions the potential of the following cell is +0.126 V. What is the metal M? Show your calculations.

$$Cd|Cd^{n+}‖M^{n+}|M$$

41. Should each of the following compounds be stable in a 1 M aqueous solutions? (*Hint:* Check the possibility of oxidation or reduction of the anion by the cation.)

(a) $Ba(MnO_4)_2$

(b) FeI_3

(c) $Pd[HgBr_4]$

(d) $[Co(NH_3)_6](ClO)_2$

(e) $Na_2[Cd(CN)_4]$

42. The zinc–air battery is commonly used in hearing aids. Air is admitted through tiny holes to form a cathode of O_2 adsorbed on porous carbon. The electrolyte is NaOH. Using the reduction potentials of Zn and O_2 in basic solution, write the cell reaction and calculate the cell potential and free energy change.

The Nernst Equation

43. Calculate the potential of a cell at 25°C with reactants and products that have the conditions indicated. If no pressure or concentration is indicated for a species, assume unit activity.

(a) $Mn(s) + 2Ag^+(0.0100 \text{ M}) \longrightarrow$
$$Mn^{2+}(0.0100 \text{ M}) + Ag(s)$$

(b) $MnO_4^-(0.155$ M$) + 8H_3O^+(1.200$ M$) + 5Au(s) \longrightarrow$
$Mn^{2+}(0.0100$ M$) + 12H_2O(l) + 5Au^+(0.0200$ M$)$

(c) $4Cr(s) + 3O_2(0.20$ atm$) + 6H_2O(l) + 4OH^-(1.0$ M$)$
$\longrightarrow 4Cr(OH)_4^-(0.15$ M$)$ (in base)

(d) $Cr_2O_7^{2-}(0.403$ M$) + 5H_3O^+(0.252$ M$) +$
$3HNO_2(0.100$ M$) \longrightarrow 2Cr^{3+}(0.000504$ M$) +$
$3NO_3^-(0.00102$ M$) + 9H_2O(l)$

(e) $Cl^-(0.250$ M$) + NO_3^-(0.250$ M$) \longrightarrow$
$ClO^-(0.250$ M$) + NO_2^-(0.250$ M$)$ (in base)

44. Calculate the potential of a cell at 25°C with reactants and products that have the conditions indicated. If no pressure or concentration is indicated for a species, assume unit activity.

(a) $Cd(s) + Pb^{2+}(1.55$ M$) \longrightarrow Cd^{2+}(1.55$ M$) + Pb(s)$

(b) $Sn^{4+}(0.0500$ M$) + H_2(0.0100$ atm$) + 2H_2O(l) \longrightarrow$
$Sn^{2+}(0.750$ M$) + 2H_3O^+(0.666$ M$)$

(c) $Zn(s) + 2Fe(CN)_6^{3-}(0.100$ M$) + 4CN^-(0.650$ M$)$
$\longrightarrow Zn(CN)_4^{2-}(0.250$ M$) + 2Fe(CN)_6^{4-}(0.150$ M$)$

(d) $2Mn^{2+}(0.00100$ M$) + 24H_2O(l) + 5Fe^{2+}(0.00100$ M$)$
$\longrightarrow 2MnO_4^-(0.00100$ M$) + 16H_3O^+(0.00100$ M$) +$
$5Fe(s)$

(e) $2MnO_4^-(2.00$ M$) + 3NO_2^-(2.00$ M$) + H_2O \longrightarrow$
$2MnO_2(s) + 3NO_3^-(0.100$ M$) + 2OH^-(1.00$ M$)$

45. Calculate the voltage produced by each of the following cells.

(a) $Zn|Zn^{2+}(0.0100$ M$)\|Cu^{2+}(1.00$ M$)|Cu$

(b) $Al|Al^{3+}(0.250$ M$)\|Co^{2+}(0.0500$ M$)|Co$

46. Calculate the potential of each of the following cells.

(a) $Pt|Br_2(l)|Br^-(0.450$ M$)\|Cl^-(0.0500$ M$)|Cl_2(g,$ 0.900 atm$)|Pt$

(b) $Pt|H_2(g, 0.790$ atm$)|H_3O^+(0.500$ M$)\|Cl^-(0.0500$ M$)|Cl_2(g, 0.100$ atm$)|Pt$

47. Sketch a graph for Q and E_{cell} as the following electrochemical cell runs down—that is, as it changes from standard state conditions to equilibrium.

$$Zn|Zn^{2+}\|Fe^{2+}|Fe$$

Plot values of Q and E_{cell} when $[Fe^{2+}] = 1.00$ M, 0.50 M, 0.10 M, 1.0×10^{-4} M, 1.0×10^{-8} M, 1.0×10^{-10} M, and the equilibrium concentration. Assume the volumes of the solution of Zn^{2+} and the solution of Fe^{2+} are equal, so the change in concentrations of Zn^{2+} and Fe^{2+} are equal but opposite in direction.

48. Sketch a graph for Q and E_{cell} as the following electrochemical cell runs down—that is, as it changes from standard state conditions to equilibrium.

$$Sn|Sn^{2+}\|Pb^{2+}|Pb$$

Plot values of Q and E_{cell} when $[Fe^{2+}] = 1.00$ M, 0.50 M, 0.10 M, 0.010 M, 0.0010 M, and the equilibrium concentration. Assume the volumes of the solution of Sn^{2+} and the solution of Pb^{2+} are equal, so the change in concentrations of Sn^{2+} and Pb^{2+} are equal but opposite in direction.

49. The Nernst equation can be used to determine the potential of a half-reaction. Calculate the potential for each of the following half-reactions. *Note:* e^- does not appear in Q.

(a) $Sn^{2+}(0.0100$ M$) + 2e^- \longrightarrow Sn$

(b) $Hg \longrightarrow Hg^{2+}(0.2500$ M$) + 2e^-$

(c) $O_2(0.0010$ atm$) + 4H_3O^+(0.1130$ M$) + 4e^- \longrightarrow 6H_2O(l)$

50. Calculate the potential for each of the following half-reactions, using the Nernst equation. *Note:* e^- does not appear in Q.

(a) $Cr_2O_7^{2-}(0.150$ M$) + 14H_3O^+(0.100$ M$) + 6e^- \longrightarrow 2Cr^{3+}(0.000100$ M$) + 21H_2O(l)$

(b) $Mn^{2+}(0.0125$ M$) + 12 H_2O(l) \longrightarrow MnO_4^-(0.0125$ M$) + 8H_3O^+(0.100$ M$) + 5e^-$

(c) $Sn^{4+}(0.000100$ M$) + 2e^- \longrightarrow Sn^{2+}(4.0$ M$)$

51. Hypochlorous acid, HOCl, is a stronger oxidizing agent in acidic solution than in neutral solution. Calculate the potential for the reduction of HOCl to Cl^- in a solution with a pH of 7.00 in which [HOCl] and $[Cl^-]$ are both 1.00 M.

52. The standard reduction potential of oxygen in acidic solution is 1.23 V ($O_2 + 4H_3O^+ + 4e^- \longrightarrow 6H_2O$). Calculate the standard reduction potential of oxygen in basic solution, and compare your result with the value in Appendix H. (*Hint:* What is $[H_3O^+]$ when $[OH^-]$ is 1 M?)

53. At what $[OH^-]$ does the following half-reaction have a potential of 0 V when concentrations except $[OH^-]$ are 1 M?

$NO_3^- + H_2O + 2e^- \longrightarrow NO_2^- + 2OH^-$ $\quad E_{cell}^\circ = +0.01$ V

54. At what $[I^-]$ does the following half-reaction have a potential of 0 V when $[HgI_4^-]$ is 1 M?

$HgI_4^- + 2e^- \longrightarrow Hg + 4I^-$ $\quad E_{cell}^\circ = -0.04$ V

55. What is the theoretical potential required to electrolyze a 0.0300 M solution of copper(II) chloride, producing metallic copper and chlorine at 0.300 atm of pressure?

56. A standard zinc electrode is combined with a hydrogen electrode with H_2 at 1 atm. If the cell potential is 0.46 V, what is the pH of the electrolyte in the hydrogen electrode?

Free Energy Changes and Equilibrium Constants

57. Calculate the standard free energy change at 25°C of the reactions in Exercise 27.

58. Calculate the standard free energy change at 25°C of the reactions in Exercise 28.

59. Calculate the free energy change at 25°C of the reactions in Exercise 43.

60. Calculate the free energy change at 25°C of the reactions in Exercise 44.

61. For a cell based on each of the following reactions run at standard conditions, calculate the cell potential, the standard free energy change of the reaction, and the equilibrium constant of the reaction.

(a) $Mn(s) + Cd^{2+}(aq) \longrightarrow Mn^{2+}(aq) + Cd(s)$

(b) $2Br^-(aq) + I_2(s) \longrightarrow Br_2(l) + 2I^-(aq)$

62. For a cell based on each of the following reactions run at standard conditions, calculate the cell potential, the standard free energy change of the reaction, and the equilibrium constant of the reaction.

(a) $Cr_2O_7^{2-} + 3Fe(s) + 14H_3O^+(aq) \longrightarrow$
$$2Cr^{3+}(aq) + 3Fe^{2+}(aq) + 21H_2O(l)$$

(b) $2Al(s) + 3Co^{2+}(aq) \longrightarrow 2Al^{3+}(aq) + 3Co(s)$

63. Calculate the standard free energy change and equilibrium constant for the reaction

$$2Br^- + F_2 \longrightarrow 2F^- + Br_2$$

64. Using the standard reduction potentials for the half-reactions in the hydrogen–oxygen fuel cell, calculate the standard free energy change and equilibrium constant for the combustion of hydrogen.

$$2H_2 + O_2 \longrightarrow 2H_2O$$

65. Copper(I) salts disproportionate in water to form copper(II) salts and copper metal.

$$2Cu^+ \longrightarrow Cu^{2+} + Cu$$

What is the equilibrium constant for this reaction and what concentration of Cu^+ remains at equilibrium in 1.00 L of a solution prepared from 1.00 mol of Cu_2SO_4?

66. Use the potentials of the following half-cells and show that hydrogen peroxide, H_2O_2, is unstable with respect to decomposition into oxygen and water.

$$H_2O_2 + 2H_3O^+ + 2e^- \longrightarrow 4H_2O \quad E° = +1.77 \text{ V}$$
$$O_2 + 2H_3O^+ + 2e^- \longrightarrow H_2O_2 + 2H_2O \quad E° = +0.68 \text{ V}$$

Electrolytic Cells and Faraday's Law of Electrolysis

67. Using a line diagram of the type

Anode | anode soln ‖ cathode soln | cathode

diagram the electrolytic cell for each of the following cell reactions.

(a) $MgCl_2 \longrightarrow Mg + Cl_2$ (using Mg and inert carbon as electrodes)

(b) $Fe + Cu^{2+} \longrightarrow Fe^{2+} + Cu$

68. Using a line diagram of the type

Anode | anode soln ‖ cathode soln | cathode

diagram the electrolytic cell for each of the following cell reactions.

(a) $2NaCl(aq) + 2H_2O \longrightarrow 2NaOH(aq) + H_2 + Cl_2$ (using inert electrodes of specially treated titanium, Ti)

(b) $Cl_2 + 2Br^- \longrightarrow Br_2 + 2Cl^-$ (using platinum electrodes)

69. Write the anode half-reaction, the cathode half-reaction, and the cell reaction for each of the following electrolytic cells.

(a) $C | NaCl(l) | Cl_2 \| NaCl(l) | Na(l) | Fe$

(b) $Pt | Cl_2 | Cl^-(aq) \| H^+(aq) | H_2 | Pt$

70. Write the anode half-reaction, the cathode half-reaction, and the cell reaction for each of the following electrolytic cells.

(a) $C | CO_2(g) | O^{2-} \| Al^{3+} | Al$ (The solvent for Al_2O_3 is molten Na_3AlF_6.)

(b) $Pb | PbSO_4; PbO_2 | H_2SO_4(aq) \| H_2SO_4(aq) | PbSO_4 | Pb$

71. How many moles of electrons are involved in each of the following electrochemical changes?

(a) 0.800 mol of I_2 is converted to I^-.

(b) 118.7 g of Sn^{2+} is converted to Sn^{4+}.

(c) 0.174 g of MnO_4^- is converted to Mn^{2+}.

(d) The MnO_4^- in 15.80 mL of 0.1145 M MnO_4^- is converted to Mn^{2+}.

72. How many moles of electrons are involved in each of the following electrochemical changes?

(a) 0.250 mol of Al^{3+} is converted to Al.

(b) 27.6 g of SO_3 is converted to SO_3^{2-}.

(c) 1.0 L of O_2 at STP is converted to H_2O in acid solution.

(d) The Cu^{2+} in 1100 mL of 0.50 M Cu^{2+} is converted to Cu.

73. How many faradays of electricity are involved in each of the electrochemical changes described in Exercise 71?

74. How many coulombs of electricity are involved in each of the electrochemical changes described in Exercise 72?

75. Aluminum is manufactured by the electrolysis of a molten mixture of Al_2O_3 and Na_3AlF_6. How many moles of electrons are required to convert 1.0 mol of Al^{3+} to Al? How many faradays? How many coulombs?

76. Ammonium perchlorate, NH_4ClO_4, which is used in the solid fuel in the booster rockets on the space shuttle, is prepared from sodium perchlorate, $NaClO_4$, which is produced commercially by the electrolysis of a hot, stirred solution of sodium chloride.

$$NaCl + 4H_2O \longrightarrow NaClO_4 + 4H_2$$

How many moles of electrons are required to produce 1.00 kg of sodium perchlorate? How many faradays? How many coulombs?

77. How many moles of electrons are required to prepare 1.000 metric ton (1000 kg) of chlorine gas by electrolysis of an aqueous solution of sodium chloride?

$$2NaCl(aq) + 2H_2O(l) \longrightarrow 2NaOH(aq) + H_2(g) + Cl_2(g)$$

How many faradays? How many coulombs? Assume that the efficiency of the electrochemical cell is 100%; that is, every electron involved results in the production of a chlorine atom. (In an actual commercial cell, the efficiency is about 65%.)

78. Electrolysis of a sulfuric acid solution with a certain amount of current produces 0.3718 g of hydrogen. What

mass of silver would be produced by the same amount of current? What mass of copper?

79. How many grams of zinc are deposited from a solution of zinc(II) sulfate by 3.40 faradays of electricity?

80. Say we conducted an experiment using the apparatus depicted in Fig. 19.26. How many grams of gold would be plated out of solution by the current required to plate out 4.97 g of copper? How many moles of hydrogen would simultaneously be released from the hydrochloric acid solution? How many moles of oxygen would be freed at each anode in the copper sulfate and silver nitrate solutions?

81. How many moles of electrons flow through a lamp that draws a current of 2.0 A for 1.0 min?

82. How many grams of cobalt are deposited from a solution of cobalt(II) chloride that is electrolyzed with a current of 20.0 A for 54.5 min?

83. How many faradays of electricity would be required to reduce 21.0 g of $Na_2[CdCl_4]$ to metallic cadmium? How long would this take (in minutes) with a current of 7.5 A?

84. Chromium metal can be plated electrochemically from an acidic aqueous solution of CrO_3.

   (a) What is the half-reaction for the process?
   (b) What mass of chromium, in grams, is deposited by a current of 2.50 A passing for 20.0 min?
   (c) How long does it take to deposit 1.0 g of chromium using a current of 10.0 A?

85. A single commercial electrolytic cell for the production of chlorine draws a current of 150,000 A. Assume that the efficiency of the cell is 100%, and calculate the mass of chlorine, in kilograms, produced by such a cell in 1.00 h.

86. A single commercial electrolytic cell for the production of aluminum draws a current of 125,000 A. Assume that the efficiency of the cell is 100%, and calculate the mass of aluminum, in kilograms, produced by such a cell in 1.00 h.

Oxidation–Reduction Reactions

87. Balance each of the following redox equations.

   (a) $P + Cl_2 \longrightarrow PCl_5$
   (b) $IF_5 + Fe \longrightarrow FeF_3 + IF_3$
   (c) $Sn^{2+} + Cu^{2+} \longrightarrow Sn^{4+} + Cu^+$
   (d) $H_2S + Hg_2^{2+} + H_2O \longrightarrow Hg + S + H_3O^+$
   (e) $CN^- + ClO_2 + H_2O \longrightarrow CNO^- + Cl^- + H_3O^+$
   (f) $Fe^{2+} + Ce^{4+} \longrightarrow Fe^{3+} + Ce^{3+}$
   (g) $HBrO + H_2O \longrightarrow H_3O^+ + Br^- + O_2$

88. Balance each of the following redox equations.

   (a) $Zn + BrO_4^- + OH^- + H_2O \longrightarrow$
   $$[Zn(OH)_4]^{2-} + Br^-$$
   (b) $H_2SO_4 + HBr \longrightarrow SO_2 + Br_2 + H_2O$
   (c) $MnO_4^- + S^{2-} + H_2O \longrightarrow MnO_2 + S + OH^-$
   (d) $NO_3^- + I_2 + H_3O^+ \longrightarrow IO_3^- + NO_2 + H_2O$
   (e) $Cu + H_3O^+ + NO_3^- \longrightarrow Cu^{2+} + NO_2 + H_2O$

   (f) $Zn + H_3O^+ + NO_3^- \longrightarrow Zn^{2+} + N_2O + H_2O$
   (g) $Cu + H_3O^+ + NO_3^- \longrightarrow Cu^{2+} + NO + H_2O$

89. Balance each of the following redox equations.

   (a) $Al + [Sn(OH)_4]^{2-} \longrightarrow [Al(OH)_4]^- + Sn + OH^-$
   (b) $H_2S + H_2O_2 \longrightarrow S + H_2O$
   (c) $MnO_4^{2-} + Cl_2 \longrightarrow MnO_4^- + Cl^-$
   (d) $Br_2 + CO_3^{2-} \longrightarrow Br^- + BrO_3^- + CO_2$
   (e) $C + HNO_3 \longrightarrow NO_2 + H_2O + CO_2$
   (f) $ClO_3^- + H_2O + I_2 \longrightarrow IO_3^- + Cl^- + H_3O^+$

90. Balance each of the following redox equations.

   (a) $Cl^- + H_3O^+ + NO_3^- \longrightarrow Cl_2 + NO_2 + H_2O$
   (b) $MnO_4^- + Se^{2-} + H_2O \longrightarrow MnO_2 + Se + OH^-$
   (c) $OH^- + NO_2 \longrightarrow NO_3^- + NO_2^- + H_2O$
   (d) $NH_3 + O_2 \longrightarrow NO + H_2O$
   (e) $HClO_3 \longrightarrow HClO_4 + ClO_2 + H_2O$
   (f) $Cr_2O_7^{2-} + HNO_2 + H_3O^+ \longrightarrow$
   $$Cr^{3+} + NO_3^- + H_2O$$

91. Complete and balance each of the following equations. (Note that when a reaction occurs in acidic solution, H_3O^+ and/or H_2O can be added on either side of the equation, as necessary, to balance the equation properly. When a reaction occurs in basic solution, OH^- and/or H_2O can be added, as necessary, on either side of the equation. No indication of the acidity of the solution is given if neither H_3O^+ nor OH^- is involved as a reactant or product.)

   (a) $Zn + NO_3^- \longrightarrow Zn^{2+} + N_2$ (acidic solution)
   (b) $Zn + NO_3^- \longrightarrow Zn^{2+} + NH_3$ (basic solution)
   (c) $CuS + NO_3^- \longrightarrow Cu^{2+} + S + NO$ (acidic solution)
   (d) $NH_3 + O_2 \longrightarrow NO_2$ (gas phase)
   (e) $H_2SO_4 + HI \longrightarrow I_2 + SO_2$ (acidic solution)
   (f) $Cl_2 + OH^- \longrightarrow Cl^- + ClO_3^-$ (basic solution)
   (g) $H_2O_2 + MnO_4^- \longrightarrow Mn^{2+} + O_2$ (acidic solution)
   (h) $NO_2 \longrightarrow NO_3^- + NO_2^-$ (basic solution)
   (i) $KClO_3 \longrightarrow KCl + O_2$ (no solvent)
   (j) $Fe^{3+} + I^- \longrightarrow Fe^{2+} + I_2$
   (k) $P_4 \longrightarrow PH_3 + HPO_3^{2-}$ (basic solution)
   (l) $P_4 \longrightarrow PH_3 + HPO_3^{2-}$ (acidic solution)

92. Complete and balance the following reactions. (See the instructions for Exercise 91.)

   (a) $Zn + H_3O^+ \longrightarrow$ (acidic solution)
   (b) $MnO_2 + Cl^- \longrightarrow$ (acidic solution)
   (c) $Pb^{4+} + Sn^{2+} \longrightarrow$
   (d) $Fe^{2+} + H_2O_2 \longrightarrow$ (acidic solution)
   (e) $Cl_2 + SO_2 \longrightarrow$ (acidic solution)
   (f) $ZnS + O_2 \xrightarrow{\Delta}$ (no solvent)
   (g) $ClO^- + Sn^{2+} \longrightarrow$ (acidic solution)
   (h) $PbO_2 + SeO_3^{2-} \longrightarrow$ (basic solution)
   (i) $Cr_2O_7^{2-} + Br^- \longrightarrow$ (basic solution)

Applications and Additional Exercises

93. Several battery companies, independent research laboratories, and government agencies involving at least a few

hundred chemists have worked for several years to develop a lithium battery. What prompts this interest?

94. When aluminum gutters were first installed on houses, iron nails were used. After a period of time, holes around the nails grew so large that the gutter fell off. Write the cell diagram for the reaction that occurred and the cell reaction.

95. Tell which of the following statements accurately describes the effect of adding CN^- to the cathode of a cell with the cell reaction

$$Cd + 2Ag^+ \longrightarrow 2Ag + Cd^{2+} \quad E° = 1.2 \text{ V}$$

(i) $E°$ increases because $Cd(CN)_4^{2-}$ forms.
(ii) $E°$ decreases because $Cd(CN)_4^{2-}$ forms.
(iii) $E°$ increases because $Ag(CN)_2^-$ forms.
(iv) $E°$ decreases because $Ag(CN)_2^-$ forms.
(v) Both $Cd(CN)_4^{2-}$ and $Ag(CN)_2^-$ form, so there is no change.

96. The cell potential for the *unbalanced* chemical reaction

$$Hg_2^{2+} + NO_3^- + H_3O^+ \longrightarrow Hg^{2+} + HNO_2 + H_2O$$

is measured under standard state conditions in the electrochemical cell shown in the accompanying diagram. The cell voltage is positive: $E° = 0.02$ V.

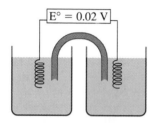

Dish A   Dish B
anode   cathode

(a) Which of the following statements must be true of the solutions in order for the cell to operate with the voltage indicated?
   i.   The solution in Dish A must be acidic.
   ii. The solution in Dish B must be acidic.
   iii. The solutions in both Dish A and Dish B must be acidic.
   iv. No acid may be in either Dish A or Dish B.
(b) Which ions and compounds are in each dish?
(c) Calculate the reduction potential for the half-cell

$$2Hg^{2+} + 2e^- \longrightarrow Hg_2^{2+}$$

given the following:

$$NO_3^- + 3H_3O^+ + 2e^- \longrightarrow HNO_2 + 4H_2O \quad E°_{red} = 0.94 \text{ V}$$

(d) What is the value of $\Delta G°_{298}$ for the reaction?
(e) What is the equilibrium constant for the reaction?
(f) At what pH will the cell potential be zero if all the other concentrations remain unchanged?
(g) How many moles of electrons pass through the circuit when 0.60 mol of Hg^{2+} and 0.30 mol of HNO_2 are

produced in a cell that contains 0.50 mol of Hg_2^{2+} and 0.40 mol of NO_3^- at the beginning of the reaction?
(h) How long will it take to produce 0.10 mol of HNO_2 by this reaction if a current of 10 A passes through the cell?
(i) Which of the following statements accurately expresses what will happen in a solution that is 1.0 M each in Hg_2^{2+}, NO_3^-, H_3O^+, Hg^{2+}, and HNO_2?
   i. Hg_2^{2+} will be oxidized and NO_3^- reduced.
   ii. Hg_2^{2+} will be reduced and NO_3^- oxidized.
   iii. Hg^{2+} will be oxidized and HNO_2 reduced.
   iv. Hg^{2+} will be reduced and HNO_2 oxidized.
   v.   There will be no change because the reaction is at standard state conditions.

97. When chlorine dissolves in water, it disproportionates, producing chloride ion and hypochlorous acid. Find at what hydronium ion concentration the potential for the disproportionation changes from a negative value to a positive value, assuming 1.00 atm of pressure and concentrations of 1.00 M for all species except hydrogen ion. (The standard reduction potential for chlorine to chloride ion is 1.36 V, and for hypochlorous acid to chlorine it is 1.63 V.) Could chlorine be produced from hypochlorite and chloride ions in solution, through the reverse of the disproportionation reaction, by acidifying the solution with strong acid? Explain.

98. The standard reduction potentials for the reactions

$$Ag^+ + e^- \longrightarrow Ag$$

and

$$AgCl + e^- \longrightarrow Ag + Cl^-$$

are +0.7991 V and +0.222 V, respectively. From these data and the Nernst equation, calculate a value for the solubility product (K_{sp}) for AgCl.

99. Calculate the standard reduction potential for the reaction $H_2O + e^- \longrightarrow \frac{1}{2}H_2 + OH^-$, using the Nernst equation and the fact that the standard reduction potential for the reaction $H_3O^+ + e^- \longrightarrow \frac{1}{2}H_2 + H_2O$ is by definition equal to 0.00 V.

100. The standard reduction potentials for the reactions

$$Ag^+ + e^- \longrightarrow Ag$$

and

$$[Ag(NH_3)_2]^+ + e^- \longrightarrow Ag + 2NH_3$$

are +0.7991 V and +0.373 V, respectively. From these values and the Nernst equation, determine K_f for the $[Ag(NH_3)_2]^+$ ion.

101. When gold is plated electrochemically from a basic solution of $[Au(CN)_4]^-$, O_2 forms at one electrode and Au is deposited at the other. Write the half-reactions that occur at the two electrodes and the net reaction for the electrolytic cell. (The cyanide ion, CN^-, is not oxidized or reduced under these conditions.)

102. A current of 9.0 A flowed for 45 min through water containing a small quantity of sodium hydroxide. How many

liters of gas were formed at the anode at 27.0°C and 750 torr of pressure?

103. A lead storage battery has initially 200 g of lead and 200 g of PbO_2, plus excess H_2SO_4. Theoretically, how long could this cell deliver a current of 10.0 A, without recharging, if it were possible to operate it so that the reaction goes to completion?

104. A total of 69,500 C of electricity was required to reduce 37.7 g of M^{3+} to the metal. What is M?

105. A current of 10.0 A is applied for 1.0 h to 1.0 L of a solution containing 1.0 mol of HCl. Calculate the pH of the solution at the end of this time.

106. A solution containing copper(I), nickel, and zinc cyanide complexes was electrolyzed, and a deposit of 0.175 g was obtained. The deposit contained 72.8% Cu, 4.3% Ni, and 22.9% Zn. No other element was released. Calculate the number of coulombs passed through the solution.

107. Balance each of the following redox equations.
(a) $MnO_4^- + NO_2^- + H_2O \longrightarrow$
$$MnO_2 + NO_3^- + OH^-$$
(b) $MnO_4^{2-} + H_2O \longrightarrow MnO_4^- + OH^- + MnO_2$
(c) $Br_2 + SO_2 + H_2O \longrightarrow H_3O^+ + Br^- + SO_4^{2-}$

108. The following question is taken from the 1992 Chemistry Advanced Placement Examination is used with the permission of the Educational Testing Service.

An unknown metal M forms a soluble compound $M(NO_3)_2$.

(a) A solution of $M(NO_3)_2$ is electrolyzed. When a current of 2.50 amperes is applied for 35.0 minutes, 3.06 grams of the metal is deposited. Calculate the molar mass of M and identify the metal.

(b) The metal identified in (a) is used with zinc to construct a galvanic cell, as shown below. Write the net ionic equation for the cell reaction and calculate the cell potential, $E°$.

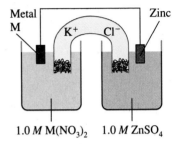

1.0 M $M(NO_3)_2$ 1.0 M $ZnSO_4$

(c) Calculate the free energy change, $\Delta G°$, at 25°C for the reaction in (b).

(d) Calculate the potential, E, for the cell shown in (b) if the initial concentration of $ZnSO_4$ is 0.10-molar, but the concentration of the $M(NO_3)_2$ solution remains unchanged.

20

Nuclear Chemistry

CHAPTER OUTLINE

The Stability of Nuclei

20.1 The Nucleus
20.2 Nuclear Binding Energy
20.3 Nuclear Stability
20.4 The Half-Life of Radioactive Materials

Nuclear Reactions

20.5 Equations for Nuclear Reactions
20.6 Radioactive Decay
20.7 Radioactive Dating
20.8 Synthesis of Nuclides

Nuclear Energy and Other Applications

20.9 Nuclear Fission and Nuclear Power
       Reactors
20.10 Nuclear Fusion and Fusion Reactors
20.11 Uses of Radioisotopes
20.12 Interaction of Radiation with Matter

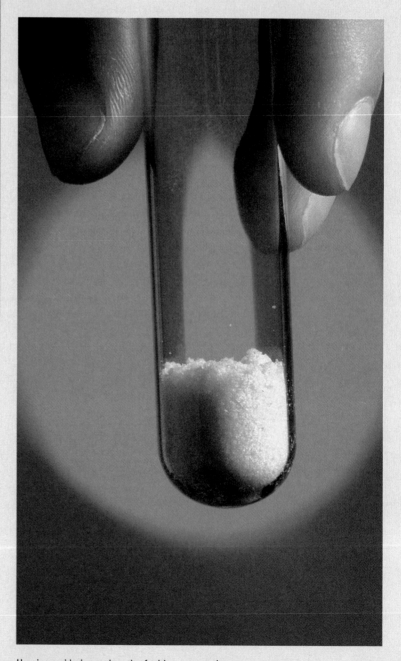

Uranium oxide is used as the fuel in many nuclear reactors.

The chemical changes that we considered in previous chapters involve changes in the arrangement of the electrons around atoms, ions, or molecules. The nuclei of the atoms involved in these chemical changes do not change; they simply provide the mass of the atom and the positive charges that determine the number of electrons involved. However, atomic nuclei do undergo changes. This chapter introduces the branch of chemistry, called **nuclear chemistry,** that considers changes and differences in atomic nuclei.

Nuclear chemistry is on the borderline between physics and chemistry. It began with the discovery of radioactivity in 1896 by Antoine Becquerel, a French physicist, and has become increasingly important during the twentieth century. A knowledge of nuclear chemistry is important when we consider nuclear power, nuclear medicine, and nuclear wastes, all of which developed during the past 50 years and all of which play an important role in our society.

The Stability of Nuclei

20.1 The Nucleus

We know that the nucleus of an atom is composed of protons and, with the exception of ^1_1H, neutrons (Section 2.2). The number of protons in the nucleus is called the **atomic number, Z,** of the element, and the sum of the number of protons and the number of neutrons is the **mass number, A.** Atoms with the same atomic number but different mass numbers are **isotopes** of the same element. When talking about a single type of nucleus, we often use the term **nuclide** and identify it by the notation ^A_ZX, where X is the symbol for the element, A is the mass number, and Z is the atomic number (for example, $^{14}_6C$). Sometimes a nuclide is referred to by the name of the element followed by a hyphen and the mass number; for example, $^{14}_6C$ may be called carbon-14.

Protons and neutrons, collectively called **nucleons,** are packed together tightly in a nucleus. With a radius of about 10^{-13} centimeters, a nucleus is quite small compared to the entire atom (which has a radius of about 10^{-8} centimeter). Nuclei are extremely dense; they average 1.8×10^{14} grams per cubic centimeter. This density is almost unimaginably large compared to the densities of more familiar materials. Water, for example, has a density of 1 gram per cubic centimeter. Iridium, one of the densest elements known, has a density of 22.6 grams per cubic centimeter. If the earth's density were equal to the average nuclear density, the earth's radius would be only about 200 meters. (The actual radius of the earth is approximately 6.4×10^6 meters, about 30,000 times larger.)

To hold positively charged protons together in the very small volume of a nucleus requires very strong attractive forces, because the positively charged protons repel one another strongly at such short distances. The force of attraction that holds the nucleus together is called the **nuclear force.** This force acts between protons, between neutrons, and between protons and neutrons. It is very different from, and much stronger than, the electrostatic force that holds negatively charged electrons around a positively charged nucleus; the nuclear force is about 30 to 40 times stronger than electrostatic repulsions between protons in a nucleus. The exact nature of the nuclear force is not known, but it is a short-range force, effective only within distances about the size of the nucleus (10^{-13} centimeter).

20.2   Nuclear Binding Energy

Although the nature of the nuclear force is unknown, we can determine the magnitude of the energy changes associated with its action. As an example, let us consider the formation of a helium nucleus, which consists of two protons and two neutrons. If a helium nucleus were to be formed by the combination of two protons and two neutrons without any change of mass, the mass of the helium atom (including the two electrons outside the nucleus) would be

$$\underset{\text{Protons}}{(2 \times 1.0073 \text{ amu})} + \underset{\text{Neutrons}}{(2 \times 1.0087 \text{ amu})} + \underset{\text{Electrons}}{(2 \times 0.00055 \text{ amu})} = 4.0331 \text{ amu}$$

However, mass spectrometric measurements show the mass of a $^4_2$He atom is only 4.0026 atomic mass units. This difference between the calculated and experimental masses, the **mass defect** of the atom, indicates a loss in mass of 0.0305 atomic mass unit. The loss in mass accompanying the formation of an atom from protons, neutrons, and electrons is due to conversion of that much mass to energy, which is lost as the atom forms. This energy is called the **nuclear binding energy.** The formation of an atom is extremely exothermic; thus the reverse reaction, decomposition of an atom to protons, neutrons, and electrons is extremely endothermic—and very difficult to accomplish.

Nearly all of us have seen the **Einstein equation,**

$$E = mc^2$$

This equation can be used to find the amount of energy that results when matter is converted to energy. The amount of energy, in joules, is determined by multiplying the mass of the matter converted into energy, m, in kilograms, by c, the speed of light in meters per second. We can use the Einstein equation to calculate the nuclear binding energy of a nucleus from its mass defect. The resulting binding energy is usually reported in units of millions of electron-volts per nuclide (1 MeV = 1.602×10^{-13} J).

Example 20.1   Calculation of Nuclear Binding Energy

What is the binding energy for the nuclide $^4_2$He in joules per nuclide, joules per mole, and MeV per nucleus? (1 MeV = 1.602×10^{-13} J)

Solution: We can answer such a question in four steps.

1. *Determine the mass defect of the nuclide.*
2. *Calculate the binding energy (in joules) for one nuclide from the mass defect using the Einstein equation.*
3. *Convert the binding energy per nuclide to a binding energy per mole.*
4. *Convert the binding energy per nuclide (in joules) to units of MeV per nuclide.*

Step 1.   As we have just seen, the difference in mass between a $^4_2$He nucleus and two protons plus two neutrons is 0.0305 amu.

Step 2. In order to use the Einstein equation to convert the mass defect into the equivalent energy in joules, we must express the mass defect in kilograms (1 amu $= 1.6605 \times 10^{-27}$ kg).

$$0.0305 \text{ amu} \times \frac{1.6605 \times 10^{-27} \text{ kg}}{1 \text{ amu}} = 5.06 \times 10^{-29} \text{ kg}$$

Now we use the Einstein equation.

$$E = mc^2$$
$$E = 5.06 \times 10^{-29} \text{ kg} \times (2.998 \times 10^8 \text{ m s}^{-1})^2$$
$$= 4.55 \times 10^{-12} \text{ kg m}^2 \text{ s}^{-2}$$
$$= 4.55 \times 10^{-12} \text{ J}$$

Units of kg m^2 s^{-2} are equivalent to J (see Section 4.2).

Step 3. The binding energy in 1 mol of helium nuclei is

$$4.55 \times 10^{-12} \text{ J nucleus}^{-1} \times 6.022 \times 10^{23} \text{ nuclei mol}^{-1} = 2.74 \times 10^{12} \text{ J mol}^{-1}$$
$$= 2.74 \times 10^9 \text{ kJ mol}^{-1}$$

A tremendous quantity of energy results from the conversion of even a very small quantity of matter to energy.

Step 4. Convert units.

$$4.55 \times 10^{-12} \text{ J nucleus}^{-1} \times \frac{1 \text{ MeV}}{1.602 \times 10^{-13} \text{ J}} = 28.4 \text{ MeV nucleus}^{-1}$$

Check your learning: What is the binding energy for the nuclide $^{19}_{9}$F (atomic mass, 18.9984 amu) in MeV per nucleus?

Answer: 148.4 MeV

The changes in mass in all ordinary chemical reactions are negligible, because only chemical bonds form or break. On the other hand, the changes in mass in nuclear reactions are very significant. If the nuclear reaction of 2 moles of neutrons with 2 moles of hydrogen atoms to give 1 mole of helium atoms could be made to occur, 2.74×10^9 kilojoules of energy would be released (Example 20.1). For comparison, burning 1 mole of methane (an ordinary chemical reaction) releases only 8.9×10^2 kilojoules, about three million times less energy.

The most stable nuclei are those with the greatest **binding energy per nucleon** (the total binding energy for the nucleus divided by the sum of the numbers of protons and neutrons present in the nucleus). The binding energy per nucleon is greatest for the nuclei of elements with mass numbers of approximately 56 (Fig. 20.1 on page 760). Thus the most stable nuclei are those in the vicinity of iron, cobalt, and nickel in the periodic table. These elements have the largest binding energies per nucleon. The binding energy per nucleon in helium is

$$\frac{28.4 \text{ MeV}}{4 \text{ nucleons}} = 7.10 \text{ MeV nucleon}^{-1}$$

Figure 20.1

Binding energy curve for the elements.

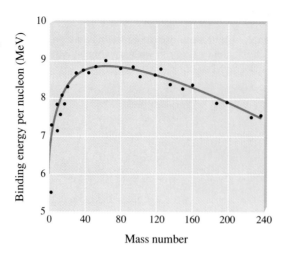

Mass number

The binding energy per nucleon in $^{56}_{26}Fe$ is almost 25% larger, as shown in the following example.

Example 20.2   **Calculation of Binding Energy per Nucleon**

The iron nuclide $^{56}_{26}Fe$ lies near the top of the binding energy curve shown in Fig. 20.1 and is one of the more stable. What is the binding energy per nucleon (in MeV) for the nuclide $^{56}_{26}Fe$ (atomic mass, 55.9349 amu)?

Solution: We can answer such a question in four steps.

1. *Determine the mass defect of the nuclide.*
2. *Calculate the binding energy (in joules) for one nucleus from the mass defect using the Einstein equation.*
3. *Convert the binding energy per nucleus (in joules) to units of MeV per nuclide.*
4. *Determine the binding energy per nucleon.*

Step 1.   The mass defect is the difference between the mass of 26 protons, 30 neutrons, and 26 electrons and the observed mass of a $^{56}_{26}Fe$ atom.

$$\text{Mass defect} = [(26 \times 1.0073) + (30 \times 1.0087) + (26 \times 0.00055)] - 55.9349$$

$$= 56.4651 - 55.9349$$

$$= 0.5302 \text{ amu}$$

Step 2.   Use the Einstein equation and calculate the binding energy in joules per nucleus.

$$E = 0.5302 \text{ amu} \times \frac{1.6605 \times 10^{-27} \text{ kg}}{1 \text{ amu}} \times (2.998 \times 10^8 \text{ m s}^{-1})^2$$

$$= 7.913 \times 10^{-11} \text{ J}$$

Step 3.   Convert from units of joules to units of MeV. The binding energy is 493.9 MeV.

Step 4.   The binding energy per nucleon is found by dividing the total nuclear binding energy by the number of nucleons in the atom.

$$\text{Binding energy per nucleon} = \frac{493.9 \text{ MeV}}{56 \text{ nucleons}}$$

$$= 8.820 \text{ MeV nucleon}^{-1}$$

Check your learning: What is the binding energy per nucleon in $^{19}_{9}F$ (atomic mass, 18.9984 amu)?

Answer: 7.810 MeV nucleon^{-1}

20.3   Nuclear Stability

A nucleus is stable if it cannot be transformed into another configuration without adding energy from the outside. A plot of the number of neutrons versus the number of protons for stable nuclei (the curve shown in Fig. 20.2) shows that the stable isotopes fall into a narrow band, which is called the **band of stability.** The straight line in Fig. 20.2 represents equal numbers of protons and neutrons for comparison. The figure indicates that the lighter stable nuclei, in general, have equal numbers of protons and neutrons. For example, nitrogen-14 has 7 protons and 7 neutrons. Heavier stable nuclei, however, have slightly more neutrons than protons, as the curve shows. For example, iron-56 has 30 neutrons and 26 protons, a neutron-to-proton ratio of

Figure 20.2

A plot of number of neutrons versus number of protons for stable nuclei. Each dot shows the number of protons and neutrons in a known stable nucleus. All isotopes of elements with atomic numbers greater than 83 are unstable.

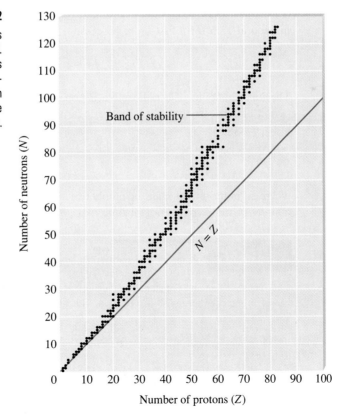

1.15. The stable nuclide lead-207 has 125 neutrons and 82 protons, thus it has a neutron-to-proton ratio of 1.52.

The nuclei that are to the left or to the right of the band of stability are radioactive. They change spontaneously (decay) to other nuclei that are either in or closer to the band of stability. All nuclei with atomic numbers above 83 are radioactive. We will discuss the nature and products of this radioactive decay in subsequent sections.

20.4   The Half-Life of Radioactive Materials

We often hear about the **half-life** of a radioactive substance: the time required for half of the atoms in a sample to decay. An isotope's half-life is important in several ways. It determines how long a sample of an undesirable or dangerous isotope must be stored before it decays to a level that does not represent a hazard. For example, cobalt-60, an isotope that emits gamma rays used to treat cancer, has a half-life of 5.2 years. The intensity of the radiation produced by a cobalt-60 source decreases by one-half every 5.2 years as half of the cobalt-60 decays. Thus a cobalt-60 source must be replaced regularly. The disposal of unwanted plutonium-239 presents a problem because it has a half-life of 24,000 years. One-half of the mass of plutonium in a sample of plutonium-239 remains after 24,000 years; at the end of another 24,000 years the mass of plutonium will be one-fourth of its initial mass; and so on.

The half-lives of radioactive nuclides vary widely. For example, the half-life of $^{142}_{58}Ce$ is 5×10^{15} years, that of $^{32}_{15}P$ is 14.28 days, that of $^{222}_{86}Rn$ is 3.82 days, and that of a recently reported element, element 111, is 1.5×10^{-3} seconds. The half-lives of a number of nuclides are listed in Appendix J.

Radioactive decay is a first-order process (see Sections 13.4 and 13.5). The rate constant k can be expressed in terms of the half-life, $t_{1/2}$, by the equation developed in Section 13.6 for a first-order process.

$$k = \frac{0.693}{t_{1/2}}$$

The initial amount of a radioactive isotope and the amount remaining after a given period of time are expressed by the logarithmic relationship that applies to all first-order reactions.

$$\ln \frac{c_0}{c_t} = kt$$

where c_0 is the initial mass or number of moles of the isotope and c_t is the mass or number of moles remaining at time t.

Example 20.3   Rates of Radioactive Decay

(a) What is the rate constant for the radioactive disintegration of cobalt-60, an isotope used in cancer therapy? $^{60}_{27}Co$ decays with a half-life of 5.2 years to produce $^{60}_{28}Ni$.

Solution:
$$k = \frac{0.693}{t_{1/2}}$$

$$= \frac{0.693}{5.2 \text{ y}} = 0.13 \text{ y}^{-1}$$

(b) Calculate the fraction of a sample of the $^{60}_{27}$Co isotope that will remain after 15 years.

Solution:

$$\ln \frac{c_0}{c_t} = kt$$

$$= (0.13 \text{ y}^{-1})(15 \text{ y}) = 1.95$$

$$\frac{c_0}{c_t} = \text{anti ln } 1.95 = 7.03 \quad \text{(Use the } e^x \text{ button on your calculator.)}$$

The fraction that will remain is equal to the amount at time t divided by the initial amount, or c_t/c_0.

$$\frac{c_0}{c_t} = \frac{1}{7.03} = 0.14$$

The fraction of $^{60}_{27}$Co remaining after 15 years is 0.14. Thus 14% of the $^{60}_{27}$Co that was originally present will remain after 15 years.

(c) How long does it take for a sample of $^{60}_{27}$Co to disintegrate to the extent that only 2.0% of the original amount remains?

Solution:

$$c_t = 0.020 \times c_0$$

Therefore

$$\frac{c_0}{c_t} = \frac{1}{0.020} = 50$$

$$\ln \frac{c_0}{c_t} = kt$$

$$\ln 50 = (0.13 \text{ y}^{-1})t$$

$$t = \frac{\ln 50}{0.13 \text{ y}^{-1}} = 30 \text{ y}$$

Check your learning: Radon-222, $^{222}_{86}$Rn, has a half-life of 3.823 days. How long will it take a sample of radon-222 with a mass of 0.750 g to decay to other elements, leaving only 0.100 g of radon-222?

Answer: 11.1 days

Nuclear Reactions

Changes in nuclei that result in changes in their atomic numbers, mass numbers, and/or energy states are called **nuclear reactions.**

20.5 Equations for Nuclear Reactions

To describe a nuclear reaction, we use an equation that identifies the nuclides involved in the reaction, their mass numbers and atomic numbers, and the other particles involved in the reaction. These particles include

1. **Alpha particles** ($_2^4$He, or α), helium nuclei, consisting of two neutrons and two protons

2. **Beta particles** ($_{-1}^0$e, or β), electrons

3. **Positrons** ($_{+1}^0$e, or β^+), particles with the same mass as an electron but with 1 unit of *positive* charge

4. **Protons** ($_1^1$H, or p), nuclei of hydrogen atoms

5. **Neutrons** ($_0^1$n, or n), particles with a mass approximately equal to that of a proton but with no charge

Nuclear reactions also often involve **gamma rays** (γ). Gamma rays are electromagnetic radiation, somewhat like X rays in character but with higher energies and shorter wavelengths.

The equations of several nuclear reactions that have played an important role in the history of nuclear chemistry follow.

1. The first naturally occurring radioactive element that was isolated, polonium, was discovered by the Polish scientist Marie Curie and her husband, Pierre, in 1898. It decays by alpha emission.

$$_{84}^{212}\text{Po} \longrightarrow {}_{82}^{208}\text{Pb} + {}_2^4\text{He}$$

2. Technetium was first prepared in 1937. This was the first element to be prepared that does not occur naturally on the earth. It decays by beta emission.

$$_{43}^{98}\text{Tc} \longrightarrow {}_{44}^{98}\text{Ru} + {}_{-1}^0\text{e}$$

3. The first controlled nuclear chain reaction was carried out in a reactor at the University of Chicago in 1942. One of the reactions involved was

$$_{92}^{235}\text{U} + {}_0^1\text{n} \longrightarrow {}_{35}^{87}\text{Br} + {}_{57}^{146}\text{La} + 3\,{}_0^1\text{n}$$

4. One of the reactions involved in the detonation of a thermonuclear weapon, and one proposed for the peaceful production of energy by nuclear fusion, is

$$_1^3\text{H} + {}_1^2\text{H} \longrightarrow {}_2^4\text{He} + {}_0^1\text{n}$$

5. The first element prepared by artificial means was prepared in 1919 by bombarding nitrogen atoms with α particles.

$$_7^{14}\text{N} + {}_2^4\text{He} \longrightarrow {}_8^{17}\text{O} + {}_1^1\text{H}$$

Each of these nuclear equations is balanced in two ways.

1. The sum of the mass numbers of the reactants equals the sum of the mass numbers of the products.

2. The sum of the atomic numbers of the reactants equals the sum of the atomic numbers of the products.

If the atomic number and the mass number of all but one of the particles in a nuclear reaction are known, we can identify the particle by balancing the reaction. For example, we could determine that $_8^{17}$O is a product of the reaction of $_7^{14}$N and $_2^4$He if we know that a proton, $_1^1$H, was one of the two products.

Example 20.4 Balancing Equations for Nuclear Reactions

The reaction of an α particle with magnesium-25 ($^{25}_{12}$Mg) produces a proton ($^{1}_{1}$H) and a nuclide of another element. Identify the new nuclide produced.

Solution: The nuclear reaction can be written as

$$^{25}_{12}\text{Mg} + {}^{4}_{2}\text{He} \longrightarrow {}^{A}_{Z}\text{X} + {}^{1}_{1}\text{H}$$

where A is the mass number and Z is the atomic number of the new nuclide, X. Because the sum of the mass numbers of the reactants must equal the sum of the mass numbers of the products,

$$25 + 4 = A + 1, \text{ or } A = 28$$

Similarly, the atomic numbers must balance, so

$$12 + 2 = Z + 1, \text{ and } Z = 13$$

Check the periodic table: The element with atomic number 13 is aluminum. Thus the product is $^{28}_{13}$Al. (This is an unstable nuclide with a half-life of 2.3 min that decays to $^{28}_{14}$Si and a β particle.)

Check your learning: The nuclide $^{125}_{53}$I captures an electron ($_{-1}^{0}$e) to produce a new nucleus and no other massive particles. What is the equation for this reaction?

Answer: $^{125}_{53}\text{I} + {}^{0}_{-1}\text{e} \longrightarrow {}^{125}_{52}\text{Te}$

20.6 Radioactive Decay

The spontaneous change of an unstable nuclide into another is called **radioactive decay.** The unstable nuclide is often called the **parent nuclide,** the nuclide that results from the decay the **daughter nuclide.** The daughter nuclide may be stable, or it may decay itself.

The radiation produced during decay is such that the daughter nuclide lies closer to the band of stability (Section 20.3) than does the parent nuclide, so the location of a nuclide relative to the band of stability can serve as a guide to the kind of decay it will undergo. We classify different types of radioactive decay by the radiation produced.

1. **α decay,** the loss of an α particle (which contains two neutrons and two protons), occurs primarily in heavy nuclei ($A > 200$, $Z > 83$). Because loss of an α particle gives a daughter nuclide with a mass number 4 units smaller and an atomic number 2 units smaller than those of the parent nuclide, the daughter nuclide has a larger neutron-to-proton ratio than the parent nuclide. If the parent nuclide undergoing α decay lies below the band of stability (Fig. 20.2), the daughter nuclide will lie closer to the band.

2. **β decay,** the loss of an electron *from a nucleus* by conversion of a neutron into a proton and a β particle, is observed in nuclides with a large neutron-to-proton ratio. Such nuclei lie above the band of stability. Emission of an electron does not change the mass number of the nuclide but does increase the number of its protons and

decrease the number of its neutrons. Consequently, the neutron-to-proton ratio is decreased, and the daughter nuclide lies closer to the band of stability than did the parent nuclide.

3. **β^+ decay,** the emission of a positron, is observed with nuclides in which the neutron-to-proton ratio is low. These nuclides lie below the band of stability. During the course of β^+ decay, a proton is converted into a neutron with the emission of a positron. The neutron-to-proton ratio increases, and the daughter nuclide lies closer to the band of stability than did the parent nuclide.

β^+ decay is usually accompanied by the subsequent appearance of two γ rays, each with an energy of 0.511 MeV. These γ rays result when the positron interacts with an electron and the two annihilate each other. All of their mass is converted into energy; two 0.511-MeV γ rays are produced.

$$\underset{\text{Electron}}{_{-1}^{0}e} \; + \; \underset{\text{Positron}}{_{+1}^{0}e} \; \longrightarrow \; 2\,\gamma \;(0.511 \text{ MeV each})$$

4. **Electron capture** occurs when one of the electrons in an atom is captured by the nucleus and a proton is converted to a neutron. Like β^+ decay, electron capture occurs when the neutron-to-proton ratio is low. (The nuclide lies below the band of stability.) Electron capture has the same effect on the nucleus as positron emission; the atomic number is decreased by 1 as a proton is converted into a neutron. This increases the neutron-to-proton ratio, and the daughter nuclide lies closer to the band of stability than did the parent nuclide.

5. **γ emission** is observed when a nuclide is formed in an excited state and then decays to its ground state with the emission of a γ ray, a quantum of high-energy electromagnetic radiation. Figure 20.3 illustrates the relationships for the decay of uranium-233 to thorium-229. A uranium-233 nuclide can decay into one of three states of thorium-229. If it decays into one of the two excited states, the excited state gives up its excess energy as a γ ray. There is no change of mass number or atomic number during emission of a γ ray unless the γ emission accompanies one of the other modes of decay.

The naturally occurring radioactive isotopes of the heaviest elements fall into chains of successive disintegrations, or decays, and all the species in one chain constitute a radioactive family, or series. Three of these series include most of the naturally radioactive elements of the periodic table. They are the uranium series, the actinide series, and the thorium series. Each series is characterized by a parent (first member) that has a long half-life and a series of daughter nuclides that ultimately lead to a stable end product—that is, a nuclide on the band of stability (Fig. 20.2). In all three series the end product is an isotope of lead.

Figure 20.3

The decay of uranium-233 to thorium-229 can occur by three paths. One path leads directly to the ground state of the $_{90}^{229}$Th nuclide. Two other paths give excited states of the $_{90}^{229}$Th nuclide, which decay to the ground state by the emission of γ rays.

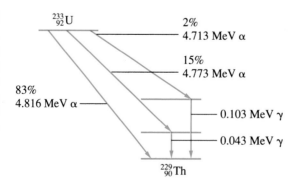

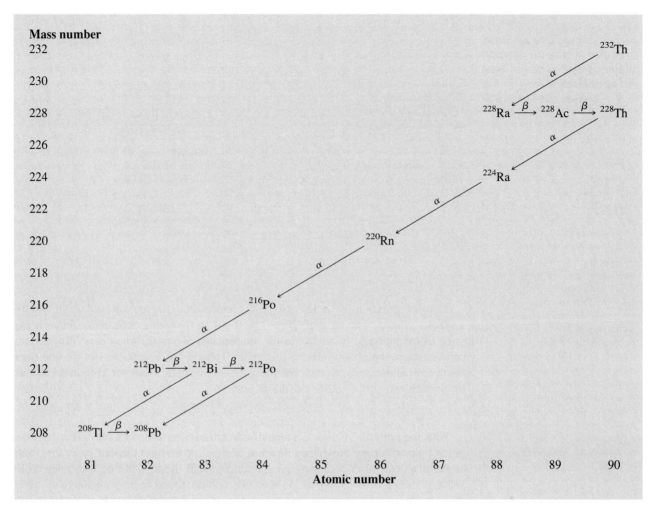

Table 20.1

The Thorium Decay Series

The steps in the thorium decay series are given in Table 20.1 as an illustration of one natural chain of successive decays.

20.7   Radioactive Dating

Radioactive Dating Using Carbon-14. The radioactivity of carbon-14 provides a method for dating objects (Fig. 20.4 on page 768) that have been living matter or have been produced from living matter within the past 10,000 to 30,000 years. Carbon-14 is produced in the upper atmosphere by the reaction of nitrogen atoms with neutrons from cosmic rays in space.

$$^{14}_{7}\text{N} + ^{1}_{0}\text{n} \longrightarrow ^{14}_{6}\text{C} + ^{1}_{1}\text{H}$$

The $^{14}_{6}\text{C}$ then reacts with oxygen molecules to form $^{14}_{6}\text{CO}_2$. Other isotopes of carbon (principally carbon-12) also react with oxygen to product CO_2 molecules. The ratio of $^{14}_{6}\text{CO}_2$ and $^{12}_{6}\text{CO}_2$ depends on the ratio of $^{14}_{6}\text{C}$ to $^{12}_{6}\text{C}$ in the atmosphere.

Figure 20.4

Pages from the Dead Sea Scrolls.
Carbon-14 dating has shown that
these scrolls were written or copied
on paper made from plants that died
between 100 B.C. and A.D. 50.

The incorporation of $^{14}_{6}CO_2$ and $^{12}_{6}CO_2$ into plants is a regular part of the photosynthesis process. The ratio of $^{14}_{6}C$ to $^{12}_{6}C$ found in the living portion of a plant is the same as the ratio of $^{14}_{6}C$ to $^{12}_{6}C$ in the atmosphere. When the plant dies, it no longer participates in the photosynthesis life cycle. Because $^{12}_{6}C$ is a stable isotope and does not decay radioactively, its concentration does not change. However, carbon-14 decays by β emission with a half-life of 5730 years.

$$^{14}_{6}C \longrightarrow {}^{14}_{7}N + {}^{0}_{-1}e$$

Thus the ratio of $^{14}_{6}C$ to $^{12}_{6}C$ gradually decreases after the plant dies. The decrease in the ratio with time provides a measure of the time that has elapsed since the death of the plant (or other organism). For example, with the half-life of $^{14}_{6}C$ being 5730 years, if the ratio of $^{14}_{6}C$ to $^{12}_{6}C$ in a wooden object found in an archeological dig is half what it is in a living tree, this indicates that the wooden object is 5730 years old. Highly accurate determinations of ratios of $^{14}_{6}C$ to $^{12}_{6}C$ can be obtained from very small samples (as little as a milligram) by the use of a mass spectrometer (see Section 2.4).

The accuracy of this technique depends on the ratio of $^{14}_{6}C$ to $^{12}_{6}C$ in a living plant being the same now as it was in an earlier era, but this assumption is not strictly valid. As a result of the increasing accumulation of CO_2 molecules (largely $^{12}_{6}CO_2$) in the atmosphere caused by combustion of the fossil fuels coal and oil (in which essentially all of the $^{14}_{6}C$ has decayed), the ratio of $^{14}_{6}C$ to $^{12}_{6}C$ in the atmosphere has changed significantly since 1945. This human-made increase in $^{12}_{6}CO_2$ in the atmosphere causes the ratio of $^{14}_{6}C$ to $^{12}_{6}C$ to decrease, and this in turn affects the ratio in organisms currently living on the earth. Fortunately, however, we can use other data, such as tree dating via examination of annual growth rings, to calculate correction factors. With these correction factors, accurate dates can be determined.

Radioactive Dating Using Nuclides Other Than Carbon-14.

Investigators can also utilize other radioactive nuclides with longer half-lives to date older events. Both uranium-238 (which decays in a series of steps to lead-206) and rubidium-87 (which decays with a half-life of 4.7×10^{10} years, to the stable isotope strontium-87) can be used for establishing the age of rocks and the approximate age of the earth from the ages of the oldest rocks.

To estimate the lower limit for the earth's age, scientists determine the age of various rocks and minerals, making the assumption that the earth must be at least as old as the rocks and minerals in its crust. One method, which relies on rubidium-87 dating, is to measure the relative amounts of rubidium-87 and strontium-87 in rock. One gram of rubidium-87 would product 0.50 gram of strontium-87 and leave 0.50 gram of rubidium-87 after decaying for 47 billion years. Such a radioactive dating method in a rock formation in southwestern Greenland has shown it to have an age of nearly 3.8 billion years. This formation is one of the oldest known rocks on earth (Fig. 20.5).

Figure 20.5

One of the oldest known rock formations on earth, found in Greenland, is about 3.8 billion years old. (The pocket knife is used to indicate scale.)

20.8   Synthesis of Nuclides

The first nucleus prepared in a laboratory was made in Lord Rutherford's laboratory in 1919 by a **transmutation reaction**—the bombardment of nuclei of one type with other nuclei or with neutrons. Rutherford bombarded nitrogen atoms with high-speed α particles from a natural radioactive isotope of radium and observed protons resulting from the reaction

$$^{14}_{7}\text{N} + {}^{4}_{2}\text{He} \longrightarrow {}^{17}_{8}\text{O} + {}^{1}_{1}\text{H}$$

The $^{17}_{8}\text{O}$ and $^{1}_{1}\text{H}$ nuclei are stable, so no further changes occur.

Prior to 1940 the heaviest known element was uranium, whose atomic number is 92. Now one heavy artificial element has been synthesized and isolated on such a large scale that it has had a profound effect on society. In 1940 McMillan and Abelson were able to make element 93, neptunium (Np), by bombarding uranium-238 with neutrons. The reaction gives unstable uranium-239, with a half-life of 23 minutes, which decays to neptunium-239. Neptunium-239 is also radioactive, with a half-life of 2.3 days, and decays to plutonium (Pu), element 94. The nuclear reactions are

$$^{238}_{92}\text{U} + {}^{1}_{0}\text{n} \longrightarrow {}^{239}_{92}\text{U}$$
$$^{239}_{92}\text{U} \longrightarrow {}^{239}_{93}\text{Np} + {}^{0}_{-1}\text{e}$$
$$^{239}_{93}\text{Np} \longrightarrow {}^{239}_{94}\text{Pu} + {}^{0}_{-1}\text{e}$$

Since 1940, hundreds of tons of plutonium have been prepared for military use and as a by-product of the nuclear power industry.

Plutonium forms by the capture of neutrons in nuclear reactors (Section 20.9). Initially, uranium-238 in the reactor's fuel reacts to give uranium-239, then neptunium-239 and plutonium-239 by the reactions represented by the preceding equations. As a reactor continues to operate, heavier isotopes of plutonium (Pu-240, Pu-241, and Pu-242) are produced by subsequent capture of neutrons by lighter plutonium nuclei. The longer the fuel stays in the reactor, the higher the concentration of heavier plutonium isotopes. It is estimated that about 1650 tons of plutonium will exist in the year 2000. Most of this will be a component of used nuclear reactor fuels, but worldwide, about 200 tons have been separated for use in military weapons.

Although they have not been prepared in the same quantity as plutonium, many other nuclei have been produced artificially. Nuclear medicine has developed from our ability to convert atoms of one type to other types of atoms. Radioactive isotopes of 35 different elements are currently produced and used for medical applications.

Table 20.2

Preparation of Some of the Transuranium Elements

Name	Symbol	Atomic Number	Reaction
Americium	Am	95	$^{239}_{94}Pu + ^{1}_{0}n \longrightarrow ^{240}_{95}Am + ^{0}_{-1}e$
Curium	Cm	96	$^{239}_{94}Pu + ^{4}_{2}He \longrightarrow ^{242}_{96}Cm + ^{1}_{0}n$
Berkelium	Bk	97	$^{241}_{95}Am + ^{4}_{2}He \longrightarrow ^{243}_{97}Bk + 2\,^{1}_{0}n$
Californium	Cf	98	$^{242}_{96}Cm + ^{4}_{2}He \longrightarrow ^{245}_{98}Cf + ^{1}_{0}n$
Einsteinium	Es	99	$^{238}_{92}U + 15\,^{1}_{0}n \longrightarrow ^{253}_{99}Es + 7\,^{0}_{-1}e$
Fermium	Fm	100	$^{239}_{94}Pu + 15\,^{1}_{0}n \longrightarrow ^{254}_{100}Fm + 6\,^{0}_{-1}e$
Mendelevium	Md	101	$^{253}_{99}Es + ^{4}_{2}He \longrightarrow ^{256}_{101}Md + ^{1}_{0}n$
Nobelium	No	102	$^{246}_{96}Cm + ^{12}_{6}C \longrightarrow ^{254}_{102}No + 4\,^{1}_{0}n$
Lawrencium	Lr	103	$^{250}_{98}Cf + ^{11}_{5}B \longrightarrow ^{257}_{103}Lr + 4\,^{1}_{0}n$
Rutherfordium	Rf	104	$^{249}_{98}Cf + ^{12}_{6}C \longrightarrow ^{257}_{104}Rf + 4\,^{1}_{0}n$
Hahnium	Ha	105	$^{249}_{98}Cf + ^{15}_{7}N \longrightarrow ^{260}_{105}Ha + 4\,^{1}_{0}n$
Seaborgium	Sg	106	$^{206}_{82}Pb + ^{54}_{24}Cr \longrightarrow ^{257}_{106}Sg + 3\,^{1}_{0}n$
			$^{249}_{98}Cf + ^{18}_{8}O \longrightarrow ^{263}_{106}Sg + 4\,^{1}_{0}n$
Meitnerium	Mt	109	$^{209}_{83}Bi + ^{58}_{26}Fe \longrightarrow ^{266}_{109}Mt + ^{1}_{0}n$

The radiation produced by their decay is used to image or to treat various organs or portions of the body (Section 20.11).

Elements 95 through 112 have also been prepared artificially. The elements beyond element 92 (uranium) are called **transuranium elements.** Elements 89 through 103 make up the actinide series.

In the previous edition of this book, published in 1991, we noted that "Elements 106, 107, 108, and 109 are very unstable. Whether or not additional elements can be made cannot be predicted at this time, but there is reason to hope that some can be, particularly within the range of elements 110–115." That hope was realized in 1994 and again in 1996. Elements 110 and 111 were prepared in 1994; element 112 was prepared in 1996. Element 112, $^{277}_{112}Uub$ (ununbium in the IUPAC system of nomenclature), was prepared in 1996 by a West German group who bombarded a target of lead with accelerated zinc nuclei. The researchers could detect single atoms of the new element. The decay products of the atom were characteristic of those expected for element 112. It began to decay about 0.28 second after its formation, with the emission of an α particle and the formation of an isotope of element 110, an element that the same research team had been the first to produce. Equations describing the preparation of several other isotopes of the transuranium elements are given in Table 20.2.

The kinetic energy of the nuclei used to cause transmutation reactions is increased in machines (**accelerators**) that use magnetic and electric fields. As these charged particles move in these fields, they reach the kinetic energies necessary to produce reactions. To avoid collisions with gas molecules, the particles move in a vacuum in all accelerators. When neutrons are required for transmutation reactions, they are usually obtained from radioactive decay reactions or from various nuclear reactions occurring in nuclear reactors (Section 20.9).

> # Nuclear Energy and Other Applications

20.9 Nuclear Fission and Nuclear Power Reactors

Many heavy elements with smaller binding energies per nucleon decompose into more stable elements that have intermediate mass numbers and larger binding energies per nucleon. Sometimes neutrons are also produced. We call this decomposition **fission.** The first report of such behavior appeared in 1939 when it was reported that uranium-235 atoms bombarded with slow-moving neutrons split into smaller fragments consisting of several neutrons and elements near the middle of the periodic table. Among the fission products were barium, krypton, lanthanum, and cerium, all of which have nuclei that are more stable than that of uranium-235. Subsequently, over 200 different isotopes of 35 elements have been observed among the fission products of uranium-235. One of the reactions involved is

$$^{235}_{92}U + ^1_0n \longrightarrow ^{87}_{35}Br + ^{146}_{57}La + 3\,^1_0n$$

Similar fission reactions have been observed with other uranium isotopes, as well as with a variety of additional isotopes, including those of plutonium.

A large amount of energy is produced by the fission of heavy elements. A loss of mass of about 0.2 gram per mole of uranium atoms occurs during fission of $^{235}_{92}U$. This mass is converted into a fantastic quantity of energy (Section 20.2). The fission of 1 pound of uranium-235, for example, yields about 2.5 million times as much energy as is produced by burning 1 pound of coal.

Fission of a uranium-235 nucleus produces, on the average, 2.5 neutrons as well as fission fragments. These neutrons may cause the fission of other uranium-235 atoms, which in turn provides more neutrons, setting up a **chain reaction.** Nuclear fission becomes self-sustaining when the number of neutrons produced by fission equals or exceeds the number of neutrons absorbed by splitting nuclei plus the number that escape into the surroundings. The amount of a fissionable material that will support a self-sustaining chain reaction is called a **critical mass.**

The critical mass of a fissionable material depends on the shape of the sample as well as on the type of material. An atomic bomb contains several pounds of fissionable material ($^{235}_{92}U$ or $^{239}_{94}Pu$), a source of neutrons, and an explosive device for compressing it quickly into a small volume. When fissionable material is in small pieces, the proportion of neutrons that escape through the relatively large surface area is great, and a chain reaction does not take place. When the small pieces of fissionable material are brought together quickly to form a body with a mass larger than the critical mass, the relative number of escaping neutrons decreases, and a chain reaction and explosion result.

Chain reactions of fissionable materials can be controlled without explosion in a nuclear reactor (Fig. 20.6). In a power reactor, used for the production of electricity, the energy released by fission reactions is trapped as thermal energy and used to boil water and produce steam. The steam is used to turn a turbine, which powers a generator for the production of electricity (Fig. 20.7).

Any nuclear reactor that produces power via the fission of uranium or plutonium by bombardment with neutrons must have at least five components (Fig. 20.8).

Figure 20.6

The core of a research reactor at Sandia National Laboratories, seen looking down through shielding water. The blue glow is Cerenkov radiation, which is light produced by rapidly moving charged particles.

Figure 20.7

A power-generating plant that employs a nuclear power reactor to produce steam. In a coal-fired power plant, the steam is generated in a boiler. Note that the reactor coolant is contained in a closed system and does not come in contact with outside cooling water. The reactor shielding has been omitted for clarity.

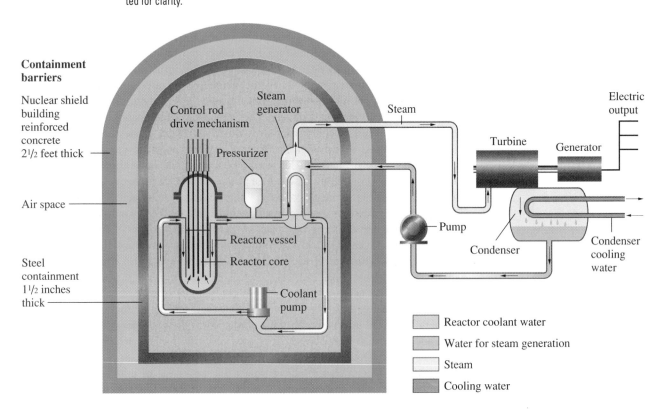

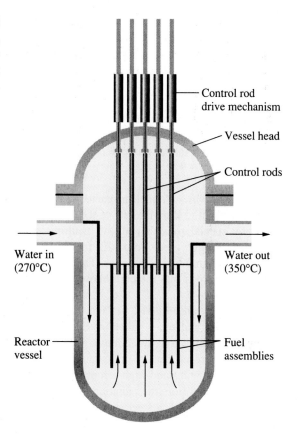

Figure 20.8

A light-water nuclear reactor. This reactor uses pressurized liquid water at 280°C and 150 atm as both a coolant and a moderator.

Control rod drive mechanism

Vessel head

Control rods

Water in (270°C)

Water out (350°C)

Reactor vessel

Fuel assemblies

A Nuclear Fuel.
Fissionable isotopes (commonly, provided by a mixture of uranium-238 with uranium-235) must be present in sufficient quantity to provide a self-sustaining chain reaction. Most reactors in the United States use pellets of the uranium oxide U_3O_8, in which the concentration of uranium-235 has been increased (enriched) from its 1% natural level to about 3%. The U_3O_8 pellets are contained in a tube (a fuel rod) of a protective material, usually a zirconium alloy. The core of a typical nuclear power reactor in the United States has about 40,000 kilograms of enriched U_3O_8 contained in several thousand fuel rods.

Naturally occurring uranium is a mixture of several isotopes; uranium-238 is the most abundant. About 1 in every 140 uranium atoms is the uranium-235 isotope. To obtain a higher concentration of uranium-235, it is necessary to separate it from the uranium-238. The most successful method uses fractional diffusion of gaseous UF_6 (uranium hexafluoride) at low pressure through porous barriers to separate $^{235}UF_6$ from $^{238}UF_6$. This method is based on the fact that the slightly lighter $^{235}UF_6$ molecules diffuse through a porous barrier slightly faster than the heavier $^{238}UF_6$ molecules (Graham's Law, Section 10.10). The enriched UF_6 is then chemically converted to U_3O_8.

A Moderator.
Neutrons produced by nuclear reactions move too fast to cause fission. They must be slowed down before they will be absorbed by the fuel and produce additional nuclear reactions. In a reactor, neutrons are slowed by collision with

the nuclei of a moderator such as heavy water, D_2O, graphite, carbon dioxide, or light (ordinary) water, H_2O. These materials are used because they do not react with or absorb neutrons. Most reactors in operation in the United States use light water as the moderator. Heavy water is used in Canada, and graphite is used in reactors in many other countries.

A Coolant.

The coolant carries the heat produced by the fission reaction to an external boiler and turbine, where it is transformed into electricity (Fig. 20.7). The coolant is a gas or liquid that is pumped through the reactor core. Some coolants also serve as moderators.

A Control System.

A nuclear reactor is controlled by adjusting the number of slow neutrons present to keep the rate of the chain reaction at a safe level. Control is maintained by special rods that absorb neutrons. Control rods containing cadmium or boron-10 are often used. Boron-10, for example, absorbs neutrons by a reaction that produces lithium-7 and α particles. The chain reaction can be stopped by inserting all of the control rods into the core between the fuel rods.

A Shield and Containment System.

During its operation, a nuclear reactor produces neutrons and other radiation. Even when the reactor is shut down, the decay products are radioactive. In addition, an operating reactor is (thermally) very hot, and high pressures result from the circulation of water or another coolant through it. Thus a reactor must withstand high temperatures and pressures, and operating personnel must be protected from the radiation. A reactor container often consists of three parts: (1) the reactor vessel, a steel shell that is 3–20 centimeters thick and that, with the moderator, absorbs much of the radiation produced by the reactor; (2) a main shield of 1–3 meters of high-density concrete; and (3) a personnel shield of lighter materials that protects operators from γ rays and X rays. In addition, reactors are often covered with a steel or concrete dome designed to contain any radioactive materials that might be released by a reactor accident (Fig. 20.9).

The importance of a containment vessel is amply illustrated by two accidents at nuclear reactors. In March 1979, the cooling system of one of the reactors at the Three

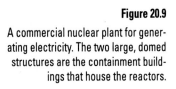

Figure 20.9

A commercial nuclear plant for generating electricity. The two large, domed structures are the containment buildings that house the reactors.

Mile Island plant in Pennsylvania failed, and the cooling water spilled from the reactor onto the floor of the containment building. The temperature of the core climbed to at least 2200°C, and the upper portion of the core began to melt. In addition, the zirconium alloy cladding of the fuel rods began to react with steam and produced hydrogen.

$$Zr(s) + 2H_2O(g) \longrightarrow ZrO_2(s) + 2H_2(g)$$

The hydrogen accumulated in the confinement building, and an explosion of the mixture of hydrogen and air in the building was feared. Consequently, hydrogen gas and radioactive gases (primarily krypton and xenon) were vented from the building. Within a week, cooling-water circulation was restored and the core began to cool. The plant was closed for nearly 10 years during the cleanup process.

Although zero discharge of radioactive material is desirable, the discharge of radioactive krypton and xenon, such as occurred at the Three Mile Island plant, is among the most tolerable. These gases readily disperse in the atmosphere and thus do not produce highly radioactive areas. Moreover, they are noble gases and are not incorporated into plant and animal matter in the food chain. Effectively none of the heavy elements of the core of the reactor were released into the environment, and no cleanup of the area outside the containment building was necessary.

A second major nuclear accident occurred in April 1986 near Chernobyl in the Ukraine. While operating at low power during an unauthorized experiment with some of its safety devices shut off, one of the reactors at the plant became unstable. Its chain reaction became uncontrollable and increased to a level far beyond that for which the reactor was designed. The steam pressure in the reactor went to between 100 and 500 times the full-power pressure and ruptured the reactor. Because the reactor was not enclosed in a containment building, the hot core was exposed to the atmosphere. In the initial burst, a large amount of radioactive material spewed out, and additional fission products were released as the graphite moderator of the core ignited and continued to burn. The fire was controlled, but over 200 plant workers and firemen developed acute radiation sickness, and at least 32 soon died from the effects of the radiation. The reactor has since been encapsulated in steel and concrete. After a year of decontamination of the reactor site and surrounding countryside, other reactors at the plant were restarted and residents began returning to some nearby towns and villages. However, significant radiation problems persist in the area.

It should be noted that the reactors used in some places in the former Soviet Union, including Chernobyl, are fundamentally different in design from the reactors used in other countries. They are the only reactors that are designed with potential low-power instability. These reactors have been modified since the accident to reduce the risk of a recurrence.

The energy produced by a reactor fueled with enriched uranium results from the fission of uranium as well as from the fission of plutonium produced as the reactor operates. As we have noted, the plutonium forms from the combination of neutrons with the uranium in the fuel. In any nuclear reactor, only about 0.1% of the mass of the fuel is converted into energy. The other 99.9% remains in the fuel rods as fission products and unused fuel. All of the fission products absorb neutrons, and after a period of several months to a few years, depending on the reactor, the fission products must be removed by changing the fuel rods. Otherwise, the concentration of these fission products would increase and they would absorb more neutrons until the reactor could no longer operate.

Spent fuel rods contain a variety of products: unstable nuclei ranging in atomic number from 25 to 60, some transuranium elements, including plutonium and americium, and unreacted uranium isotopes. The unstable nuclei and the transuranium isotopes give the spent fuel a dangerously high level of radioactivity. The long-lived isotopes require thousands of years to decay to a safe level. The ultimate fate of the nuclear reactor as a significant source of energy in the United States probably rests on whether a scientifically and politically satisfactory technique for processing and storing the components of spent fuel rods can be developed.

20.10   Nuclear Fusion and Fusion Reactors

The process of combining very light nuclei into heavier nuclei is also accompanied by the conversion of mass into large amounts of energy. The process is called **fusion** and is the focus of an intensive research effort to develop a practical thermonuclear reactor.

The principal source of energy in the sun is the net fusion reaction in which four hydrogen nuclei fuse and produce one helium nucleus and two positrons.

$$4\ {}_{1}^{1}\text{H} \longrightarrow {}_{2}^{4}\text{He} + 2\ {}_{+1}^{0}\text{e}\ (\beta^{+})$$

A helium nucleus has a mass that is 0.7% less than that of four hydrogen nuclei; this lost mass is converted into energy during the fusion.

It has been determined that deuterium, ${}_{1}^{2}\text{H}$, and tritium, ${}_{1}^{3}\text{H}$, the heavy isotopes of hydrogen, undergo fusion at extremely high temperatures (thermonuclear fusion). They form helium nuclei and neutrons.

$$ {}_{1}^{2}\text{H} + {}_{1}^{3}\text{H} \longrightarrow {}_{2}^{4}\text{He} + {}_{0}^{1}\text{n} $$

This change proceeds with a mass loss of 0.0188 atomic mass unit, corresponding to the release of 1.69×10^{9} kilojoules per mole of ${}_{2}^{4}\text{He}$ formed. The very high temperature is necessary to give the nuclei enough kinetic energy to overcome the very

Figure 20.10

The JET (Joint European Torus) nuclear fusion reactor in the United Kingdom. Inside JET's doughnut-shaped containment vessel, a plasma of hydrogen isotopes is confined by a complex magnetic field.

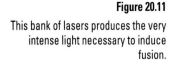

Figure 20.11

This bank of lasers produces the very intense light necessary to induce fusion.

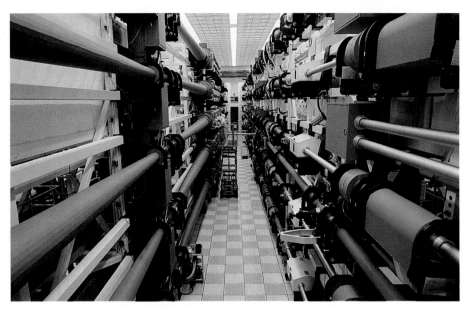

strong repulsive forces resulting from the positive charges on their nuclei so they can collide.

Uncontrolled fusion is the power source of the hydrogen bomb. However, if we could control fusion of heavy isotopes of hydrogen at reasonably low temperatures, hydrogen from the water of the oceans could supply immense amounts of energy for future generations.

A **fusion reactor** is a nuclear reactor in which fusion reactions of light nuclei are controlled. At the time of this writing, there are no self-sustaining fusion reactors operating in the world, although small-scale controlled fusion reactions have been run for very brief periods.

Useful fusion reactions appear to require very high temperatures for their initiation—about 10^8 K. At these temperatures all molecules dissociate into atoms, and the atoms ionize, forming a state of matter called a **plasma.** Because no solid materials are stable at 10^8 K, a plasma cannot be contained by mechanical devices. Two techniques to contain a plasma at the density and temperature necessary for a fusion reaction to occur have been studied. These involve containment by a magnetic field (Fig. 20.10 on page 776) and by the use of focused laser beams (Fig. 20.11).

20.11 Uses of Radioisotopes

Radioactive isotopes have the same chemical properties as stable isotopes of the same element, so we can track them in compounds by monitoring their radioactive emissions. When they are used in this way, we call them **radioactive tracers.**

Radioactive isotopes are also used in other applications when a source of radiation is needed. Many isotopes are used as diagnostic tracers in medicine (Fig. 20.12 on page 778). Four typical examples are technetium-99 ($^{99}_{43}\text{Tc}$), thallium-201 ($^{201}_{81}\text{Tl}$), iodine-131 ($^{131}_{53}\text{I}$), and sodium-24 ($^{24}_{11}\text{Na}$). After injection, certain compounds of technetium-99 are absorbed preferentially by any damaged tissues in the heart, liver, or

Figure 20.12

A radioisotopic tracer scan used in medicine.

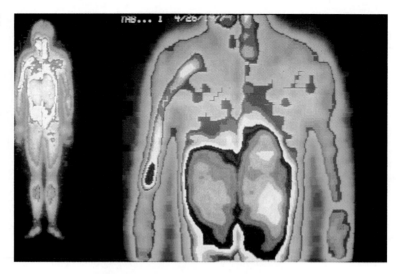

lungs. The location of the technetium compound, and hence of the damaged tissue, can be determined by detecting the γ rays emitted by the technetium isotope. Thallium-201 becomes concentrated in healthy heart tissue, so the two isotopes, Tc-99 and Tl-201, are used together to study heart tissue. Iodine-131 concentrates in the thyroid gland, the liver, and some parts of the brain. It can therefore be used to monitor goiter and other thyroid problems, as well as liver and brain tumors. Salt solutions containing compounds of sodium-24 are injected into the bloodstream to help locate obstructions to the flow of blood.

In some cases, an isotope used as a tracer can also be used, in higher doses, as treatment. For example, radiation from the decay of iodine-131 destroys cancer cells at a faster rate than healthy cells. The treatment of cancer by injected radioisotopes is one form of **chemotherapy.** An example of **radiation therapy** (wherein the isotope is not injected into the body but radiation from it is delivered externally) is the use of γ radiation from $^{60}_{27}$Co for the treatment of various forms of cancer, including leukemia (Fig. 20.13 on page 779).

Radioisotopes are used in diverse ways to study the mechanisms of chemical reactions in plants and animals. This research includes the use of labeling fertilizers in studies of nutrient uptake by plants and crop growth, investigations of digestive and milk-producing processes in cows, and studies on the growth and metabolism of animals and plants.

A radioisotope was helpful in the study of photosynthesis. The overall reaction is

$$6CO_2(g) + 6H_2O(l) \longrightarrow C_6H_{12}O_6(s) + 6O_2(g)$$

But the reaction mechanism is complex, proceeding through a series of steps in which various organic compounds are produced. In studies of the pathway of this reaction, plants were exposed to CO_2 containing a high concentration of $^{14}_6$C. At regular intervals, the plants were analyzed to determine which organic compounds contained carbon-14 and how much of each compound was present. From the time sequence in which the compounds appeared and the amount of each present at given time intervals, scientists learned more about the pathway of the reaction.

Industrial applications of radioactive materials are equally diverse. They include determining the thickness of films and thin metal sheets by exploiting the penetration

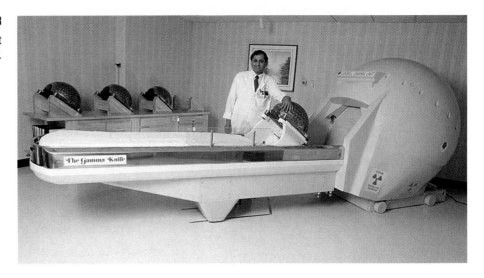

Figure 20.13

A cobalt-60 machine used in treatment of cancer.

power of various types of radiation. Flaws in metals used for structural purposes can be detected by using high-energy γ rays from cobalt-60 in a fashion similar to the way X rays are used to examine the human body.

In one form of pest control, flies are controlled by sterilizing male flies with γ radiation so that females breeding with them do not produce offspring.

Americium-241, an α emitter with a half-life of 458 years, is used in tiny amounts in ionization-type smoke detectors. The α emissions from the Am-241 ionize the air between two electrode plates in the ionizing chamber. A battery supplies a potential that causes movement of the ions, thus creating a small electric current. When smoke enters the chamber, the movement of the ions is impeded, reducing the conductivity of the air. This causes a marked drop in the current, triggering an alarm.

20.12   Interaction of Radiation with Matter

The increased use of radioisotopes has led to increased concern about the effects of these materials on biological systems. Some radioactive elements are sources of high-energy particles. Energy transferred to biological cells can break chemical bonds and ionize molecules; this can cause malfunctions in normal cell processes. The most serious biological damage from radioactive emissions results from the ability of some of these emissions to ionize and fragment molecules. For example, α and β particles emitted from nuclear decay reactions possess much higher energies than ordinary chemical bond energies. When these particles strike and penetrate matter, they produce ions and molecular fragments that are extremely reactive. In a biological system, this can seriously disrupt the normal operations of the cells.

The ability of various kinds of emissions to cause ionization varies greatly, and some particles have almost no tendency to produce ionization. Alpha particles have about twice the ionizing power of fast-moving neutrons, about 10 times that of β particles, and about 20 times that of γ rays and X rays.

There is a large difference in the magnitude of the biological effects of nonionizing radiation (for example, light and microwaves) and ionizing radiation (for example, α and β particles, γ rays, X rays, and high-energy ultraviolet radiation). Among the effects of energy absorbed from nonionizing radiation is an increase in the temperature of a system. Such radiation speeds up the movement of atoms and molecules, which is equivalent to heating the sample. Biological systems are sensitive to heat, as we all know from touching hot surfaces or spending a day at the beach in the sun. Nevertheless, a large amount of nonionizing radiation is necessary before dangerous levels are reached. Ionizing radiation is between 10 thousand and 10 billion times more dangerous than nonionizing radiation. Thus the most serious biological damage from radiation in a biological system comes from ionizing radiation.

It is impossible to avoid some exposure to ionizing radiation. Radiation is emitted both from natural sources and as a result of human activities. Two common units for measuring radiation doses are the **rad** (**r**adiation **a**bsorbed **d**ose) and the **rem** (**r**oentgen **e**quivalent in **m**an). The rad is the amount of radiation that results in absorption of 1×10^2 joules of energy per kilogram of tissue. The rem includes a biological factor referred to as the **RBE** (**r**elative **b**iological **e**ffectiveness).

$$\text{Number of rems} = \text{RBE} \times \text{number of rads}$$

Thus the rad indicates the specific energy of the radiation dose, whereas the rem takes into account both the energy and the biological effects of the type of radiation involved in the radiation dose.

We can compare the RBE values for particles when they are emitted by typical nuclear reactions. The approximate values are as follows:

	RBE
X rays (photons)	1
γ rays	1
β particles	1
Neutrons	2.0–2.3
Protons	2.0–2.3
α particles	10

Figure 20.14

Radon-222 seeps into houses and other buildings from rocks that contain uranium-238, a radon emitter. The radon enters through cracks in concrete foundations and basement floors, stone or porous cinderblock foundations, and openings for water and gas pipes.

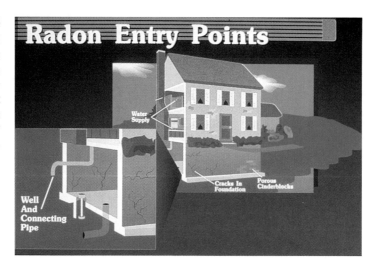

Figure 20.15

The aprons contain lead shielding that protects the operators of this medical X-ray machine from the cumulative effects of radiation exposure.

A short-term sudden dose of about 470 rems is estimated to have a 50% probability of causing the death of the victim within 30 days of exposure. Exposure to radioactive emissions has a cumulative effect in the body during a person's lifetime—another reason why it is important to avoid any unnecessary exposure to radiation.

In the United States, the average annual dose of radiation received by one man or woman is about 0.10 rem from natural sources; an additional, more variable amount ranging from about 0.05 to 0.10 rem per person is attributable to human activities. The average, taking into account all sources, is about 0.19 rem. Natural sources include cosmic rays from the sun and very small amounts of a variety of radionuclides that enter our bodies when we breathe (for example, carbon-14 and radon-222) or through the food chain (for example, potassium-40, strontium-90, and iodine-131). Radon-222 results from decay of traces of uranium present in soil. One product in the decay series is radon-222, which is an α emitter with a half-life of 3.82 days. Radon-222 continually seeps into many homes and other structures from rocks below that contain uranium (Fig. 20.14 on page 780). Sources of radiation derived from human activities include dental and medical X rays (Fig. 20.15), airplane flights (which are bombarded by increased numbers of cosmic rays in the upper atmosphere), and industrial and mining activities.

For Review   Summary

Protons and neutrons, collectively called **nucleons,** are held together by a short-range but very strong force called the nuclear force. The nuclear **binding energy** can be calculated from the **mass defect** (the difference in mass between a nucleus and the nucleons of which it is composed) by the **Einstein equation,**

$$E = mc^2$$

The binding energy per nucleon is largest for the elements with mass numbers near 56; these are the most stable nuclei.

Stable nuclei have equal numbers of neutrons and protons or a few more neutrons than protons. Nuclei that deviate from these stable neutron-to-proton ratios are radioactive, and they decay by losing one of several different kinds of particles. These include α **particles,** ^4_2He; β **particles,** $^0_{-1}\text{e}$; **positrons,** $^0_{+1}\text{e}$ (β^+); and **neutrons,** ^1_0n. Nuclear reactions also often involve γ **rays,** and some nuclei decay by **electron capture.** Each of these modes of decay leads to a new nucleus with a more stable neutron-to-proton ratio. The kinetics of the decay process are first-order. The **half-life** of a radioactive isotope is the time that is required for half of the atoms in a sample to decay. Each radioactive nuclide has its own characteristic half-life.

The ages of various objects from earlier times can be determined by a process called **radioactive dating.**

New atoms can be produced by bombarding other atoms with nuclei or high-speed particles. The products of these **transmutation reactions** can be stable or radioactive. A number of artificial elements, including technetium, astatine, and the transuranium elements, have been produced in this way.

Nuclear power can be generated through **fission** (reactions in which a heavy nucleus breaks up into two or more lighter nuclei and several neutrons). Because the neutrons may induce addi-

tional fission reactions when they combine with other heavy nuclei, a **chain reaction** can result. Useful power is obtained when the fission process is carried out in a nuclear reactor. The conversion of light nuclei into heavier nuclei (**fusion**) also produces energy. At present, this energy has not been contained adequately and is not feasible for commercial energy production.

Radioactive isotopes have many other practical uses in a wide variety of fields, including medicine, agriculture, industry, and research. In medicine, radioisotopes can be used as tracers in the diagnosis of disease or as therapy. The use, as treatment, of isotopes that are delivered from an external source is called radiation therapy; injecting isotopes into the body is one form of chemotherapy.

Key Terms and Concepts

alpha (α) decay   (20.6)
α particle   (20.5)
atomic number, Z   (20.1)
band of stability   (20.3)
beta (β) decay   (20.6)
β particle   (20.5)
binding energy per nucleon   (20.2)
chain reaction   (20.9)
critical mass   (20.9)
daughter nuclide   (20.6)
Einstein equation   (20.2)
electron capture   (20.6)
fission   (20.9)

fusion   (20.10)
fusion reactor   (20.10)
gamma (γ) emission   (20.6)
gamma (γ) ray   (20.5)
half-life   (20.4)
isotopes   (20.1)
mass defect   (20.2)
mass number, A   (20.1)
neutron   (20.5)
nuclear binding energy   (20.2)
nuclear force   (20.1)
nuclear reactor   (20.9)
nucleon   (20.1)

nuclide   (20.1)
parent nuclide   (20.6)
positron (β^+) emission   (20.5)
proton   (20.5)
rad   (20.12)
RBE   (20.12)
rem   (20.12)
radioactive decay   (20.6)
radioactive tracers   (20.11)
transmutation reaction   (20.8)
transuranium elements   (20.8)

Exercises

Questions

1. What are the two principal differences between nuclear reactions and ordinary chemical changes?

2. Each of the following is a term that can describe part of chemistry's macroscopic domain or part of chemistry's microscopic domain. Use each term in two sentences, one that applies to the macroscopic domain and one that applies to the microscopic domain. (a) α particle, (b) β particle, (c) chain reaction, (d) Einstein equation, (e) electron capture, (f) radioactive decay, (g) fission, (h) fusion, (i) mass defect.

3. Define and illustrate the term *half-life.*

4. Write a brief description or definition of each of the following terms. (a) nucleon, (b) α particle, (c) β particle, (d)

positron, (e) γ ray, (f) nuclide, (g) mass number, (h) atomic number, (i) electron capture.

5. What are the types of radiation emitted by the nuclei of radioactive elements?

6. The loss of an α particle by a nucleus causes what changes in the atomic number and the mass of the nucleus? What changes occur in the atomic number and mass when a β particle is emitted?

7. (a) What is the change in the nucleus that gives rise to a β particle? (b) To a β^+ particle?

8. Many nuclides with atomic numbers greater than 83 decay by processes such as electron emission. Explain why these unstable nuclides also commonly decay by paths that include α particles.

9. Which of the various particles (α particles, β particles, and so on) that may be produced in a nuclear reaction are actually nuclei?

10. Explain in terms of Fig. 20.2 how unstable heavy nuclides (atomic number greater than 83) may decompose to form nuclides of greater stability (a) if they are below the band of stability and (b) if they are above the band of stability.

11. Both fusion and fission are nuclear reactions. Why is a very high temperature required for fusion but not for fission?

12. How can a radioactive nuclide be used to show that the equilibrium $AgCl(s) \rightleftharpoons Ag^+(aq) + Cl^-(aq)$ is a dynamic equilibrium?

Nuclear Stability

13. Indicate the number of protons and neutrons in each of the following nuclei.

(a) $^{14}_{7}N$ (b) $^{1}_{1}H$ (c) $^{2}_{1}H$
(d) $^{3}_{1}H$ (e) neon-20 (f) lead-206
(g) uranium-235 (h) ^{12}C (i) ^{14}C

14. Indicate the number of protons and neutrons in each of the following nuclei.

(a) $^{60}_{28}Ni$ (b) $^{7}_{3}Li$ (c) $^{20}_{10}Ne$
(d) $^{3}_{2}He$ (e) argon-36 (f) lead-204
(g) thorium-232 (h) ^{13}C (i) ^{19}F

15. Which of the following nuclei lie within the band of stability shown in Fig. 20.2?

(a) chlorine-37 (b) calcium-40
(c) ^{204}Bi (d) ^{56}Fe
(e) ^{206}Pb (f) ^{211}Pb
(g) ^{222}Rn (h) carbon-14

16. Which of the following nuclei lie within the band of stability shown in Fig. 20.2?

(a) argon-40 (b) oxygen-16
(c) ^{122}Ba (d) ^{58}Ni
(e) ^{205}Tl (f) ^{210}Tl
(g) ^{226}Ra (h) magnesium-24

17. The following nuclei do not lie in the band of stability. How would they be expected to decay? Explain your answer.

(a) $^{34}_{15}P$ (b) $^{239}_{92}U$ (c) $^{38}_{20}Ca$
(d) $^{3}_{1}H$ (e) $^{245}_{91}Pu$

18. The following nuclei do not lie in the band of stability. How would they be expected to decay?

(a) $^{28}_{15}P$ (b) $^{235}_{92}U$ (c) $^{37}_{20}Ca$
(d) $^{9}_{3}Li$ (e) $^{245}_{96}Cm$

19. A $^{7}_{4}Be$ atom (mass = 7.0169 amu) decays to a $^{7}_{3}Li$ atom (mass = 7.0160 amu) by electron capture. How much energy (in millions of electron-volts) is produced by this reaction?

20. A $^{8}_{5}B$ atom (mass = 8.0246 amu) decays to a $^{8}_{4}Be$ atom (mass = 8.0053 amu) by loss of a β^+ particle (mass, 0.00055 amu). How much energy (in millions of electron-volts) is produced by this reaction?

21. The mass of the atom $^{20}_{10}Ne$ is 19.9924 amu.

(a) Calculate its binding energy per atom in millions of electron-volts.
(b) Calculate its binding energy per nucleon.

22. The mass of the atom $^{23}_{11}Na$ is 22.9898 amu.

(a) Calculate its binding energy per atom in millions of electron-volts.
(b) Calculate its binding energy per nucleon.

23. The mass of a hydrogen atom ($^{1}_{1}H$) is 1.007825 amu; that of a tritium atom ($^{3}_{1}H$) is 3.01605 amu; that of an α particle is 4.00150 amu and that of a neutron is 1.67493 amu. How much energy in kilojoules per mole of $^{4}_{2}He$ produced is released by the following fusion reaction?

$$^{1}_{1}H + {}^{3}_{1}H \longrightarrow {}^{4}_{2}He + {}^{1}_{0}n$$

24. Which of the following nuclei is most likely to decay by positron emission: chromium-53, manganese-51, or iron-59? Explain your choice.

Rates of Decay

25. Calculate the rate constant for the decay of each of the following radioactive isotopes.

(a) $^{240}_{94}Pu$ (half-life = 6.58×10^3 y)
(b) $^{13}_{5}B$ (half-life = 1.9×10^{-2} s)
(c) $^{233}_{92}U$ (half-life = 1.62×10^5 y)

26. Calculate the rate constant for the decay of each of the following radioactive isotopes.

(a) $^{202}_{81}Tl$ (half-life = 12.0 days)
(b) $^{61}_{30}Zn$ (half-life = 88 s)
(c) $^{230}_{90}Th$ (half-life = 8.0×10^4 y)

27. Calculate the time required for 99.999% of each of the radioactive isotopes in Exercise 25 to decay.

28. Calculate the time required for 99.999% of each of the radioactive isotopes in Exercise 26 to decay.

29. (a) What percentage of the nobelium in a 1.00×10^{-6}-g sample of $^{254}_{102}No$ remains 5.0 min after it is formed (half-life = 55 s)?
(b) What percentage remains 1.0 h after it is formed?

30. (a) What percentage of the radium in a 0.00100-g sample of $^{222}_{88}Ra$ remains 6.0 h after it is formed (half-life = 3.82 days)?
(b) What percentage remains 10.0 days after it is formed?

31. The half-life of ^{239}Pu is 2.41×10^4 y. What fraction of the ^{239}Pu present in nuclear wastes today will be present in 1000 y?

32. The first visible sample of plutonium, isolated in 1943, weighed about 0.0001 g. What fraction of that ^{239}Pu plutonium sample (half-life = 2.41×10^4 y) will be present in the year 2000?

33. The isotope ^{208}Tl undergoes β decay with a half-life of 3.1 min.

    (a) What isotope is produced by the decay?
    (b) How long will it take for 99.0% of a sample of pure ^{208}Tl to decay?
    (c) What percentage of a sample of pure ^{208}Tl remains un-decayed after 1.0 h?

34. If 1.000 g of $^{226}_{88}$Ra produces 0.0001 mL of the gas $^{222}_{86}$Rn at STP in 24 h, what is the half-life of ^{226}Ra in years?

35. The isotope $^{90}_{38}$Sr is one of the extremely hazardous species in the residues resulting from nuclear fission. The strontium in a 0.500-g sample diminishes to 0.393 g in 10.0 y. Calculate the half-life.

36. Technetium-99 (which is often used as a radioactive tracer for the assessment of heart, liver, and lung damage) has a half-life of 6.0 h. Calculate the rate constant for the decay of $^{99}_{43}$Tc.

37. (a) A sample of rock was found to contain 8.23 mg of rubidium-87 and 0.47 mg of strontium-87. Calculate the age of the rock if the half-life of the decay of rubidium by β emission is 4.7×10^{10} y.
    (b) If some $^{87}_{38}$Sr was initially present in the rock, would the age of the rock be less, more, or the same as the age calculated in part (a)? Explain your answer.

38. A laboratory investigation shows that a sample of uranium ore contains 5.37 mg of $^{238}_{92}$U and 2.52 mg of $^{206}_{82}$Pb. Calculate the age of the ore. The half-life of $^{238}_{92}$U is 4.5×10^9 y.

Nuclear Reactions

39. Write a balanced equation for each of the following nuclear reactions.

    (a) Uranium-230 undergoes α decay.
    (b) Bismuth-212 decays to polonium-212.
    (c) Beryllium-8 and a positron are produced by the decay of an unstable nucleus.
    (d) Neptunium-239 forms following the reaction of uranium-238 with a neutron and then spontaneously converts to plutonium-239.
    (e) Strontium-90 decays to yttrium-90.

40. Write a balanced equation for each of the following nuclear reactions.

    (a) Magnesium-27 undergoes β decay.
    (b) Mercury-180 decays to platinum-176.
    (c) Zirconium-90 and an electron are produced by the decay of an unstable nucleus.
    (d) Thorium-232 decays and produces an α particle and a radium-228 nucleus which decays into actinium-228 by β decay.
    (e) Neon-19 decays to fluorine-19.

41. Complete each of the following equations.

    (a) $^{27}_{13}\text{Al} + ^4_2\text{He} \longrightarrow ? + ^1_0\text{n}$
    (b) $^{239}_{94}\text{Pu} + ? \longrightarrow ^{242}_{96}\text{Cm} + ^1_0\text{n}$
    (c) $^{14}_7\text{N} + ^4_2\text{He} \longrightarrow ? + ^1_1\text{H}$
    (d) $^{235}_{92}\text{U} \longrightarrow ? + ^{135}_{55}\text{Cs} + 4\,^1_0\text{n}$

42. Complete each of the following equations.

    (a) $^7_3\text{Li} + ? \longrightarrow 2\,^4_2\text{He}$
    (b) $^{14}_6\text{C} \longrightarrow ^{14}_7\text{N} + ?$
    (c) $^{27}_{13}\text{Al} + ^4_2\text{He} \longrightarrow ? + ^1_0\text{n}$
    (d) $^{250}_{96}\text{Cm} \longrightarrow ? + ^{98}_{38}\text{Sr} + 4\,^1_0\text{n}$

43. Write a nuclear reaction for each step in the formation of $^{218}_{84}$Po from $^{238}_{92}$U, which proceeds by a series of decay reactions involving step-wise emission of $\alpha, \beta, \beta, \alpha, \alpha, \alpha,$ and α particles, in that order.

44. Write a nuclear reaction for each step in the formation of $^{208}_{82}$Pb from $^{228}_{90}$Th, which proceeds by a series of decay reactions involving step-wise emission of $\alpha, \alpha, \alpha, \alpha, \beta, \beta,$ and α particles, in that order.

45. Write equations to describe

    (a) The production of ^{17}O from ^{14}N by α-particle bombardment
    (b) The production of ^{14}C from ^{14}N by neutron bombardment
    (c) The production of ^{233}Th from ^{232}Th by neutron bombardment
    (d) The production of ^{239}U from ^{238}U by $^2_1$H bombardment

46. Technetium-99 is prepared from ^{98}Mo. Molybdenum-98 combines with a neutron to give molybdenum-99, an unstable isotope that decays by β emission to give an excited form of technetium-99 represented as ^{99}Tc*. This excited nucleus relaxes to the ground state, represented as ^{99}Tc, by emission of a γ ray. The ground state of ^{99}Tc decays by β emission. Write the equation for each of these nuclear reactions.

Nuclear Energy

47. How does nuclear fission differ from nuclear fusion? Why are both of these processes exothermic?

48. Cite the conditions necessary for a nuclear chain reaction to take place. Explain how it can be controlled to produce energy but not an explosion.

49. Describe the components of a nuclear reactor.

50. In usual practice, both a moderator and control rods are needed to operate a nuclear chain reaction safely for the purpose of energy production. Cite the function of each, and explain why both are necessary.

51. Describe how the potential energy of uranium is converted into electrical energy in a nuclear power plant.

52. List the advantages and disadvantages of nuclear energy (compared to energy from coal, fuel oil, natural gas, and water) as a source of electrical power.

Applications and Additional Exercises

53. Calculate the density of the $^{24}_{12}Mg$ nucleus in grams per cubic centimeter assuming that it has the typical nuclear diameter of 1×10^{-13} cm and is spherical in shape.

54. Plutonium was detected in trace amounts in natural uranium deposits by Glenn Seaborg and his associates in 1941. They proposed that the source of this ^{239}Pu was the capture of electrons by ^{238}U nuclei. Why is this plutonium not likely to be a primordial element trapped at the time the solar system formed 4.7×10^9 years ago?

55. What is the age of mummified primate skin that contains 8.25% of the original quantity of ^{14}C?

56. Why is electron capture accompanied by emission of an X ray?

57. Isotopes such as ^{26}Al (half-life $= 7.2 \times 10^5$ y) that are believed to have been present in our solar system as it formed but that have since decayed are called extinct nuclides.

    (a) ^{26}Al decays by β^+ emission or by electron capture. Write the equations for these two nuclear transformations.
    (b) The earth was formed about 4.7×10^9 (4.7 billion) years ago. How old was the earth when 99.999999% of the ^{26}Al originally present had decayed?

58. The iodine that enters the body is stored in the thyroid gland, from which it is released to control growth and metabolism. The thyroid can be imaged if iodine-131 is injected into the body. In larger doses, I-131 is also used as a means of treating cancer of the thyroid. I-131 has a half-life of 8.70 days and decays by β emission.

    (a) Write an equation for the decay.
    (b) How long will it take for 95.0% of a dose of I-131 to decay?

59. Predict by what mode(s) spontaneous radioactive decay will proceed for each of the following unstable isotopes.

    (a) ^6_2He (b) $^{60}_{30}Zn$
    (c) $^{235}_{91}Pa$ (d) $^{241}_{93}Np$
    (e) ^{18}F (f) ^{129}Ba
    (g) ^{237}Pu

60. Breeder reactors are nuclear reactors that produce more fuel than they consume by the conversion of less fissionable isotopes, such as thorium-232, into more fissionable isotopes, such as uranium-233. Write equations for the conversion of thorium-232 to thorium-233 followed by the decay of thorium-233 to protoactinium-233 and then to uranium-233.

CHAPTER OUTLINE

The Elemental Representative Metals

21.1 Periodic Relationships Among Groups
21.2 Preparation of Representative Metals

Compounds of the Representative Metals

21.3 Compounds with Oxygen
21.4 Hydroxides
21.5 Carbonates and Hydrogen Carbonates
21.6 Salts

21

The Representative Metals

Coral reefs are composed of calcium carbonate deposited by marine animals.

Metals can be divided into two large groups according to the valence electrons (Section 5.9) that are involved in forming compounds of these elements. Atoms of the 24 metallic elements of Groups 1A, 2A, 2B, 3A, 4A, 5A, and 6A (the elements shaded in yellow in Fig. 21.1) form ions by losing electrons from their outermost shells. We call these elements **representative metals.**

The majority of the representative metals are not found in the uncombined, elemental state because they react with water and oxygen in the air. However, elemental beryllium, magnesium, zinc, cadmium, mercury, aluminum, tin, and lead react very slowly with air, and we find several of these elements in everyday use. Elemental mercury appears in thermometers and dental amalgams. Elemental magnesium, aluminum, zinc, and tin are employed in the fabrication of many familiar items, including wire, cookware, foil, decorative castings, construction and automotive castings, and many household and personal objects. Although beryllium, cadmium, and lead are readily available, they are less commonly used because of their toxicity.

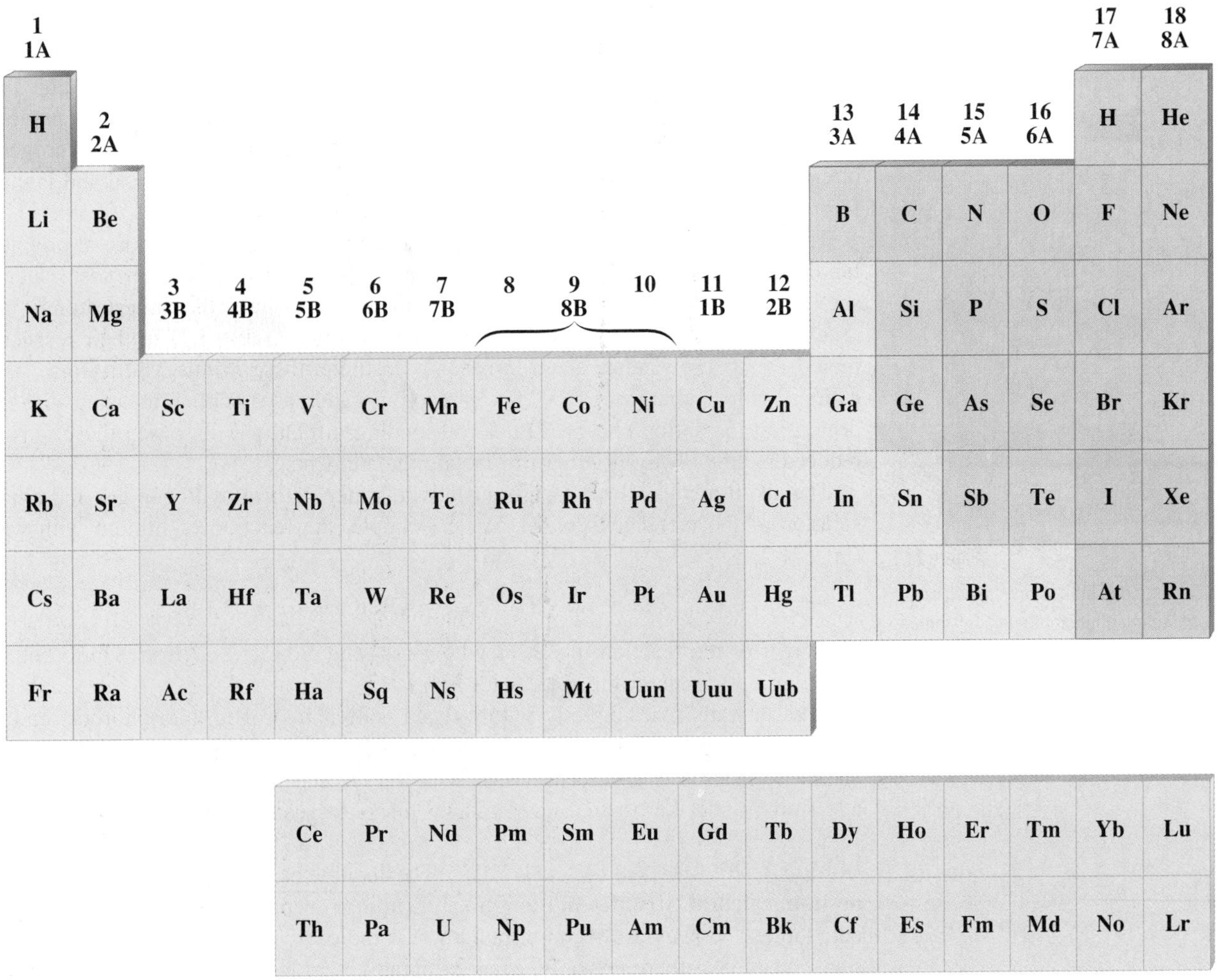

Figure 21.1 The location of the representative metals (shown in yellow) in the periodic table. Nonmetals are shown in green, semi-metals in blue, and the transition metals and inner transition metals in red.

Several representative metals form compounds that are essential to life. Examples include sodium chloride (salt), potassium salts, calcium salts, and chlorophyll (a magnesium compound). Many compounds of these elements appear as common components of rocks, soils, and minerals. Compounds of the representative metals play important roles in agriculture and in our technological society.

We begin this chapter by examining the behavior of the representative metals in terms of their positions in the periodic table. We will then see how these elements can be isolated and consider their more important uses. Finally, we will describe the formation and uses of a few of their more important compounds. The chemistry of the nonmetals and semi-metals will be considered in Chapter 22 and the chemistry of the transition metals in Chapter 23.

The Elemental Representative Metals

21.1   Periodic Relationships Among Groups

Group 1A, the Alkali Metals.
The alkali metals—**lithium, sodium, potassium, rubidium, cesium,** and **francium**—constitute Group 1A of the periodic table. The heaviest of these, francium, is a highly radioactive element that occurs in nature in very small quantities; accordingly, it is not well characterized. Although hydrogen is often shown in Group 1A (and in Group 7A, Fig. 21.1), it is a nonmetal and is therefore considered in Chapter 22.

The properties of the alkali metals are more closely related than those of any other family of elements in the periodic table. The single electron in the outermost (valence) shell of each is not tightly bound, so the alkali metals have the largest atomic radii (Section 5.10) and the lowest first ionization energies (Table 5.7) in their respective periods. Thus the valence electron is easily lost, and these metals readily form stable cations with a charge of +1. Their reactivity increases with increasing size and decreasing ionization energy. The corresponding difficulty with which these ions are reduced explains why it is hard to isolate the elements.

The alkali metals all react vigorously with water to form hydrogen gas and a basic solution of the metal hydroxide. As an example, the reaction of lithium with water (Fig. 21.2) is

Figure 21.2

A sample of lithium reacts with water containing phenolphthalein (an indicator that is pink in basic solution). Lithium floats because it is less dense than water.

$$2Li(s) + 2H_2O(l) \longrightarrow 2LiOH(aq) + H_2(g)$$

The vigor of the reaction increases with increasing atomic size (from top to bottom) in the group; compare Fig. 21.2 with Fig. 8.9.

Alkali metals react directly with oxygen, sulfur, hydrogen, phosphorous, and the halogens, giving binary ionic compounds containing the +1 metal ions. These metals are so reactive with moisture and with the oxygen in the air that they are generally stored under kerosene, under mineral oil, or in sealed containers (Fig. 21.3).

Group 2A, the Alkaline Earth Metals.
The alkaline earth metals—**beryllium, magnesium, calcium, strontium, barium,** and **radium**—constitute Group 2A of the periodic table. The heaviest of these, radium, is a radioactive element that occurs in very small quantities; there are only a few kilograms worldwide.

Because of the increasing nuclear charge and the addition of a second electron to the same principal quantum level (see Section 5.10) the atoms of the alkaline earth metals are smaller and have higher first ionization energies (Table 5.7) than those of the corresponding alkali metals within the same period. Because the valence shell electrons are more tightly bound, the alkaline earth metals are not as reactive as the corresponding alkali metals, but they are nevertheless very reactive elements. Their reactivity increases, as expected, with increasing size and decreasing ionization energy. Both valence electrons of the Group 2A elements are involved in chemical reactions, and these elements readily form compounds in which they exhibit the group oxidation state of +2. Because of their high reactivity, the alkaline earth metals, like the alkali metals, cannot be readily prepared by ordinary chemical techniques.

The lightest alkaline earth metal, beryllium, forms compounds in which it exhibits predominantly covalent character. Compounds of magnesium often show some covalent character, but the magnesium can be regarded as a +2 ion. Calcium, strontium, and barium form ionic compounds. The gradation from covalent to ionic character and the increase in reactivity from beryllium to barium and radium are due to the increasing atomic radius and the resulting lower ionization energy, leading to the increasing ease of loss of the valence electrons in these atoms.

The reactivity of magnesium metal with air at room temperature is often masked because the surface of the metal becomes covered with a tightly adhering layer of magnesium oxycarbonate that protects the metal and prevents additional reaction. Calcium, strontium, and barium react with water and air (Fig. 21.4) at room temperature; barium shows the most vigorous reaction. The products of the reactions with water are hydrogen and the metal hydroxide. These metals react directly with acids, sulfur, phosphorus, the halogens, and, with the exception of beryllium, with hydrogen to yield hydrides. Unlike most salts of the alkali metals, many of the common salts of the alkaline earth metals are insoluble in water (Section 12.5).

Group 2B.

Group 2B contains the four elements **zinc, cadmium, mercury** and **ununbium,** an artificial element prepared in 1996 but not chemically characterized. We will consider only zinc, cadmium, and mercury. Each of these elements has 2 electrons in its outer shell (ns^2) and 18 electrons in an underlying shell [$(n - 1)s^2$, $(n - 1)p^6$, $(n - 1)d^{10}$] (Section 5.9). When atoms of these metals form cations with a charge of +2, the two outer electrons are lost, giving pseudo-noble gas electron configurations (Section 6.2). These elements generally exhibit the group oxidation state of +2, although mercury also exhibits an oxidation state of +1 in compounds that contain the diatomic Hg_2^{2+} ion. Both cadmium and mercury, and their compounds, are toxic.

Of the three elements, zinc is the most reactive and mercury the least. This is a reversal of the reactivity found with the metals of Groups 1A and 2A, in which the metallic character and reactivity increase down a group. A decrease in reactivity with increasing atomic mass is also observed for the representative metals in Groups 3A through 5A. The decreasing reactivity is related to the formation of ions with a pseudo-noble gas configuration and to other factors that are beyond the scope of this discussion.

The chemical behaviors of zinc and cadmium are quite similar. Both elements lie above hydrogen in the activity series (see Section 8.7). They react with oxygen, sulfur, phosphorus, and the halogens to form compounds that contain the metals as cations with the group oxidation state of +2. They also react with solutions of acids,

Figure 21.3

To prevent contact with air and water, potassium for laboratory use is supplied as sticks stored under kerosene or mineral oil or in sealed containers.

Figure 21.4

A sample of calcium reacts with water containing phenolphthalein (an indicator that is pink in basic solution). The reaction is not so vigorous as that of sodium or of potassium with water. (See Fig. 8.9.)

Figure 21.5

Zinc is an active post-transition metal. It dissolves in hydrochloric acid, forming a solution of colorless Zn^{2+} ions, Cl^- ions, and hydrogen gas.

Figure 21.6

Mercury is an inactive post-transition metal. It does not react with hydrochloric acid (left). It dissolves in nitric acid (right), forming $Hg(NO_3)_2$ and NO_2 (the brown gas), because this acid is a strong oxidizing agent.

resulting in the liberation of hydrogen gas and the formation of salts that generally are soluble. The reaction of zinc with hydrochloric acid (Fig 21.5) is

$$Zn(s) + 2H_3O^+(aq) + 2Cl^-(aq) \longrightarrow H_2(g) + Zn^{2+}(aq) + 2Cl^-(aq) + 2H_2O(l)$$

Mercury is very different from zinc and cadmium. It is a liquid at 25°C, and many metals dissolve in it, forming solutions called amalgams. Mercury is a nonreactive element that lies well below hydrogen in the activity series. Thus it does not displace hydrogen from acids and it reacts only with oxidizing acids such as nitric acid (Fig. 21.6).

$$Hg(l) + HCl(aq) \longrightarrow \text{no reaction}$$
$$3Hg(l) + 8HNO_3(aq) \longrightarrow 3Hg(NO_3)_2(aq) + 4H_2O(l) + 2NO(g)$$

Mercury compounds decompose when they are heated, so elemental mercury reacts to form compounds only at relatively low temperatures—and then only with oxygen, sulfur, and the halogens—giving compounds that contain mercury atoms with a +2 oxidation state.

Group 3A.

Group 3A contains the semi-metal boron and the metals **aluminum, gallium, indium,** and **thallium.** The increase in metallic character moving down a group of the periodic table, which is only just apparent in Group 2A, is clearly illustrated by the elements of Group 3A. The lightest element, boron, is semiconducting, and its compounds, which are covalent, display distinctively nonmetallic properties (Chapter 22). The remaining elements of the group are metals, but their oxides and hydroxides change character. The oxides and hydroxides of aluminum and gallium exhibit both acidic and basic behavior, illustrative of both nonmetallic and metallic behavior in these two elements; that is, they are **amphoteric elements.** Indium and thallium oxides and hydroxides exhibit only basic behavior, in accordance with the clearly metallic character of these elements.

The Group 3A elements have a valence shell electron configuration of ns^2np^1. Aluminum uses all of its valence electrons when it reacts, giving compounds in which it has an oxidation state of +3. Although many of these compounds are covalent, others, such as AlF_3 and Al_2O_3, are ionic. Aqueous solutions of aluminum salts contain the cation $Al(H_2O)_6^{3+}$, abbreviated as $Al^{3+}(aq)$. Gallium, indium, and thallium also form ionic compounds containing the M^{3+} ions. A few compounds with a +1 oxidation state are known for gallium and indium, and both the +1 and +3 oxidation states are commonly observed with thallium. In aqueous solution, the $Tl^+(aq)$ ion is the stable ion.

The metals of Group 3A (Al, Ga, In, and Tl) are all reactive. However, these elements become coated by a hard, thin, tough film of the metal oxide when they are exposed to air, protecting them from chemical attack. When the film is broken, these elements react with water and oxygen (Fig. 21.7), giving compounds with oxidation states of +3 for the metals. Thallium also reacts with water and oxygen but gives thallium(I) derivatives. The metals of Group 3A all react directly with nonmetals such as sulfur, phosphorus, and the halogens, forming binary compounds.

Within Group 3A, gallium, indium, and thallium are progressively less reactive (unlike the elements of Groups 1A and 2A, in which reactivity increases down the group). The three Group 3A elements are also less reactive than aluminum.

All of the metals in Groups 1A and 2A exhibit only one oxidation state: +1 and +2, respectively. Group 3A differs in that gallium, indium, and thallium exhibit not only the oxidation state of +3 corresponding to their group number (3A) but also an oxidation state (in this case, +1) that is two below the group number. This phenom-

Figure 21.7

Aluminum oxide does not adhere to aluminum foil where the foil has been treated to produce a very thin film of mercury on its surface. Thus the white oxide film falls off, and the aluminum continues to "rust," producing additional aluminum oxide.

Figure 21.8

Tin(II) chloride is an ionic solid; tin(IV) chloride is a covalent liquid.

enon, the inert pair effect, is quite general in some other groups of representative elements as well (Section 6.2).

Group 4A. **Tin** and **lead** are the metallic members of Group 4A of the periodic table. The light elements in this group, carbon, silicon, and germanium, are primarily nonmetallic in character. Tin and lead form stable dipositive cations, Sn^{2+} and Pb^{2+}, with oxidation states two below the group oxidation state of $+4$. The stability of this oxidation state can also be attributed to the inert pair effect. The fact that the hydroxides of these ions are amphoteric is an indication of some nonmetallic character. Tin and lead also form covalent compounds in which the $+4$ oxidation state is exhibited; for example, $SnCl_4$ and $PbCl_4$ are low-boiling covalent liquids (Fig. 21.8).

Tin is a moderately active metal. It reacts with acids to form tin(II) compounds and with nonmetals to form either tin(II) or tin(IV) compounds, depending on the stoichiometry. Lead is less reactive. It lies just above hydrogen in the activity series and is attacked only by hot concentrated acids.

Group 5A. **Bismuth,** the heaviest member of Group 5A, is a less reactive metal. It readily gives up three of its five valence electrons to active nonmetals to form the tripositive ion, Bi^{3+}. It forms compounds with the group oxidation state of $+5$ only when treated with strong oxidizing agents.

Group 6A. The metal **polonium** is a member of Group 6A. It is an intensely radioactive element that results from the radioactive decay of uranium and thorium. We will not consider polonium further in this chapter.

21.2 Preparation of Representative Metals

Because of their reactivity, we do not find most representative metals as free elements in nature. However, compounds that contain ions of most representative metals are abundant. In this section we will consider the two common techniques used to isolate the metals from these compounds—electrolysis and chemical reduction. We will also consider several properties and uses of some of these metals.

Electrolysis. The metals of Groups 1A and 2A and aluminum are easily oxidized, making it very difficult to reduce the ions to neutral atoms. The largest quantities of these elements are prepared by electrolysis, in which the input of electrical energy forces reduction to occur (Section 19.12). Electrolysis is often used to carry out oxidation–reduction reactions of species that are oxidized or reduced with difficulty. The preparations of sodium and aluminum illustrate the process of electrolysis.

1. The Preparation of Sodium. The most important method for the production of sodium is the electrolysis of molten sodium chloride. The reaction involved in the process is

$$2NaCl(l) \xrightarrow[\text{600°C}]{\text{Electrolysis}} 2Na(l) + Cl_2(g)$$

The electrolysis is carried out in a cell (See Fig. 19.20) that contains molten sodium chloride (melting point 801°C), to which calcium chloride has been added to lower the melting point to 600°C (a colligative effect, see Section 12.16). When a direct current is passed through the cell, the sodium ions migrate to the negatively charged cathode, pick up electrons, and are thus reduced to sodium metal. Chloride ions migrate to the positively charged anode, lose electrons, and thus are oxidized to chlorine gas. The overall change is obtained by adding the following reactions, in which e^- represents an electron:

$$
\begin{array}{ll}
\textit{At the cathode:} & 2Na^+ + 2e^- \longrightarrow 2Na(l) \\
\textit{At the anode:} & \underline{2Cl^- \longrightarrow Cl_2(g) + 2e^-} \\
\textit{Overall change:} & 2Na^+ + 2Cl^- \longrightarrow 2Na(l) + Cl_2(g)
\end{array}
$$

Separation of the molten sodium and chlorine prevents recombination. The liquid sodium, which is less dense than molten sodium chloride, floats to the surface and flows into a collector. The gaseous chlorine is collected in tanks at high pressure. Chlorine is as valuable a product as the sodium.

Sodium and the other alkali metals, unlike magnesium and aluminum, are not suitable for structural applications because of their reactivity and softness. The major utility of these elements stems from their reactivity. Sodium is used as a reducing agent in the production of other metals (such as potassium, titanium, zirconium, and the heavier alkali metals) from their chlorides or oxides. Lithium and sodium are used as reducing agents in the manufacture of certain organic compounds, including dyes, drugs, and perfumes. Sodium and its compounds impart a yellow color to a flame (Fig. 21.9). Because the yellow light penetrates fog well, sodium is sometimes used in street lights. The synthetic rubber industry consumes large amounts of sodium, and the metal is used to prepare compounds such as sodium peroxide and sodium oxide that cannot be made from sodium chloride.

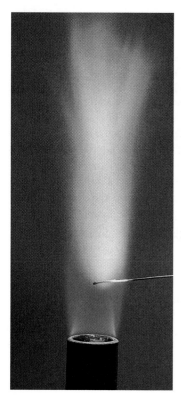

Figure 21.9

Heating sodium or sodium salts causes emission of a bright yellow light. This wire was dipped into a solution of a sodium salt.

2. The Preparation of Aluminum. Aluminum is prepared by a process invented in 1886 by Charles M. Hall, who began work on the problem while a student at Oberlin College. The process was discovered independently a month or two later by Paul L. T. Héroult in France.

The first step in the production of aluminum from the mineral bauxite, the most common source of aluminum, involves purification of the mineral. The reaction of bauxite, AlO(OH), with hot sodium hydroxide forms soluble sodium aluminate, while clay and other impurities remain undissolved.

$$AlO(OH)(s) + NaOH(aq) + H_2O(l) \longrightarrow Na[Al(OH)_4](aq)$$

After the impurities are removed by filtration, aluminum hydroxide is reprecipitated by adding acid to the aluminate.

$$Na[Al(OH)_4](aq) + H_3O^+(aq) \longrightarrow Al(OH)_3(s) + Na^+(aq) + 2H_2O(l)$$

The precipitated aluminum hydroxide is removed by filtration and heated, forming the oxide, Al_2O_3, which is then dissolved in a molten mixture of cryolite, Na_3AlF_6, and calcium fluoride, CaF_2. This solution is electrolyzed in a cell like that shown in Fig. 21.10 on page 793. Aluminum ions are reduced to the metal at the cathode, and oxygen, carbon monoxide, and carbon dioxide are liberated at the anode.

A tenacious oxide coating protects aluminum metal from corrosion (Section 21.1). The metal is very light, possesses high tensile strength, and, weight for weight, is twice as good an electrical conductor as copper.

The most important uses of aluminum are in the construction and transportation industries and in the manufacture of aluminum cans and aluminum foil. All of these uses depend on the lightness, toughness, and strength of the metal. About half of the aluminum produced in this country is converted to alloys for special uses. The fact

Figure 21.10

A cell for the production of aluminum.

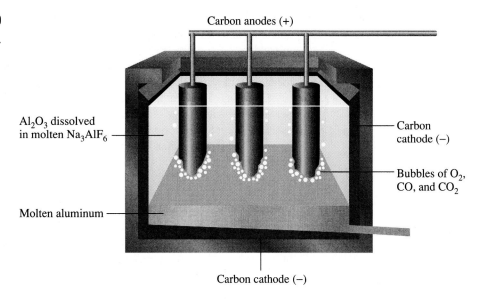

Carbon anodes (+)

Al$_2$O$_3$ dissolved in molten Na$_3$AlF$_6$

Carbon cathode (−)

Bubbles of O$_2$, CO, and CO$_2$

Molten aluminum

Carbon cathode (−)

that aluminum is an excellent conductor of heat and is lightweight and corrosion-resistant account for its use in the manufacture of cooking utensils.

When powdered aluminum and iron(III) oxide are mixed and ignited by means of a magnesium fuse, a vigorous and highly exothermic oxidation–reduction reaction occurs (Fig. 21.11).

$$2Al(s) + Fe_2O_3(s) \longrightarrow 2Fe(s) + Al_2O_3(s) \qquad \Delta H° = -851.4 \text{ kJ}$$

Figure 21.11

The thermite reaction, the reaction of aluminum with iron(III) oxide, produces a temperature of about 3000°C.

This is known as the **thermite reaction.** The temperature of the reaction mixture rises to about 3000°C, so the iron and aluminum oxide melt. The process is sometimes used in welding large pieces of iron or steel. Aluminum is used in the reduction of metallic oxides that are not readily reduced by carbon, such as MoO_3 and WO_3, and of metallic oxides that do not give pure metals when reduced by carbon, such as Cr_2O_3.

Magnesium is the other metal that is isolated in large quantities by electrolysis. Sea water, which contains approximately 0.5% magnesium chloride, serves as the usual source of magnesium. Addition of calcium hydroxide to sea water produces insoluble solid magnesium hydroxide. Magnesium chloride is produced from the magnesium hydroxide by addition of hydrochloric acid, isolated by evaporation of water from the solution, partially dried, and used in the production of metallic magnesium by electrolysis.

Magnesium is a silver-white metal that is malleable and ductile at high temperatures. Though very reactive, it does not undergo extensive reaction with air or water at room temperature because of the protective oxycarbonate film that forms on its surface (Section 21.1). Magnesium is the lightest of the widely used structural metals; most of the magnesium produced is used in making lightweight alloys.

The potent reducing power of hot magnesium is utilized in preparing many metals and nonmetals from their oxides. Indeed, the affinity of magnesium for oxygen is so great that burning magnesium reacts with carbon dioxide, producing elemental carbon.

$$2Mg + CO_2 \longrightarrow 2MgO + C$$

(A CO_2 fire extinguisher cannot be used to put out a magnesium fire.) The brilliant white light emitted by burning magnesium makes it useful in flashbulbs, flares, and fireworks.

Chemical Reduction.

With the exception of aluminum, the representative metals of Groups 2B–5A can be isolated from their compounds by **chemical reduction** using elemental carbon as the reducing agent. Because chemical reduction generally is much less expensive than electrolysis, this process is the method of choice for the isolation of these elements. Potassium, rubidium, and cesium (metals of Group 1A that are used in only small quantities) are also produced by chemical reduction. Their molten chlorides are reduced with very powerful reducing agents such as sodium metal.

Now we will consider the production of zinc and tin as additional examples of chemical reduction.

1. The Preparation of Zinc. Zinc ores usually contain zinc sulfide, zinc oxide, or zinc carbonate. After separation of these compounds from the ores, the sulfide is roasted (heated in air) to convert it to zinc oxide. Carbonate ores that contain zinc are converted to the oxide simply by heating.

$$2ZnS + 3O_2 \xrightarrow{\Delta} 2ZnO + 2SO_2$$
$$ZnCO_3 \xrightarrow{\Delta} ZnO + CO_2$$

Zinc oxide is reduced by heating it with coal in a fire-clay vessel.

$$ZnO + C \xrightarrow{\Delta} Zn + CO$$

As rapidly as zinc is produced, it distills out and is condensed. It contains impurities of cadmium, iron, lead, and arsenic, but it can be purified by careful redistillation.

Zinc is a silvery metal that quickly tarnishes to a blue-gray appearance. This color is due to an adherent coating of a basic carbonate, $Zn_2(OH)_2CO_3$, which protects the underlying metal from further corrosion.

A large amount of zinc is used in the manufacture of batteries for flashlights and portable radios; in the manufacture of the small button-shaped alkaline dry cell batteries used in some cameras, calculators, and watches, and in the production of alloys such as brass (Cu and Zn) and bronze (Cu, Sn, and Zn). About half of the zinc metal produced is used to protect iron and other metals from corrosion by air and water. The zinc coating on iron may be applied in several ways, and the product is called *galvanized iron.*

2. The Preparation of Tin. The ready reduction of tin(IV) oxide by the hot coals of a campfire accounts for the knowledge of tin the ancient world. In the modern process, ores containing SnO_2 are roasted to remove the contaminants arsenic and sulfur as volatile oxides; oxides of other metals are then removed by dissolving them in hydrochloric acid. The purified ore is reduced by carbon at temperatures above 1000°C.

$$SnO_2 + 2C \xrightarrow{\Delta} Sn + 2CO$$

Figure 21.12

The disintegration of this old tin can is due to its exposure to cold temperatures.

The molten tin collects at the bottom of the furnace and is drawn off and cast into blocks.

Tin exists in three forms: gray tin, white tin, and brittle tin. The white form is malleable. When white tin is heated, it changes to the brittle form. At temperatures below 13.2°C, white tin slowly changes into gray tin, which is powdery. Consequently, articles made of tin are likely to disintegrate in cold weather, particularly if the cold spell is lengthy (Fig. 21.12). The change progresses slowly from the spot of origin, and the gray tin that is first formed catalyzes further change. In a way, this effect is similar to the spread of an infection in a plant or animal body. For this reason, it is called tin disease, or tin pest.

The principal use of tin is in the electrolytic production of tin plate—sheet iron coated with tin to protect it from corrosion. Tin is also used in making alloys such as bronze (Cu, Sn, and Zn) and solder (Sn and Pb).

Lead is the other representative metal that is produced in large quantities by chemical reduction. It is a soft metal that has little tensile strength and the greatest density of the common metals (except for gold and mercury). Lead has a metallic luster when freshly cut but quickly acquires a dull gray color when exposed to moist air. It becomes oxidized on its surface, forming a protective layer that is both compact and adherent; this film is probably lead hydroxycarbonate, $Pb_3(OH)_2(CO_3)_2$.

The major uses of lead exploit the ease with which it is worked, its low melting point, its great density, and its resistance to corrosion. Lead is a principal constituent of the lead storage battery (see Chapter 19) and an important compound of solder (Sn and Pb).

Compounds of the Representative Metals

With a few exceptions, simple compounds of the representative metals are ionic. They contain cations formed from atoms of the metals and anions composed of atoms of the nonmetals. The anions may be monatomic, such as Cl^-, O^{2-}, and S^{2-}, or polyatomic, such as OH^-, CO_3^{2-}, NO_3^-, and SO_4^{2-}. Much of the chemical behavior of these compounds reflects the chemical properties of the anions.

The heavier metals of Groups 2B–5A sometimes form covalent compounds in which the metal exhibits its highest oxidation state. For example, with the exception of the fluorides, the halides of mercury(II), thallium(III), tin(IV) (Fig 21.8), lead(IV) and bismuth(V) are covalent molecules, although the halides of thallium(I), tin(II), lead(II), and bismuth(III) are ionic. The formation of covalent compounds is a characteristic of nonmetallic behavior, and the behavior of these elements when they exhibit their highest oxidation state is an example of the observation discussed in Section 8.6: The metallic behavior of an element decreases and its nonmetallic behavior increases as the positive oxidation state of the element in its compounds increases.

21.3   Compounds with Oxygen

Compounds of the representative metals with oxygen fall into three categories: (1) **oxides,** containing oxide ions, O^{2-}; (2) **peroxides,** containing peroxide ions, O_2^{2-}, with oxygen–oxygen covalent single bonds; and (3) a very limited number of **superoxides,** containing superoxide ions, O_2^-, with oxygen–oxygen covalent bonds that have a bond order of $1\frac{1}{2}$. All representative metals form oxides. The metals of Group 2A also form peroxides, MO_2, whereas the metals of Group 1A also form peroxides, M_2O_2, and superoxides, MO_2.

Oxides.
Oxides of most representative metals can be produced by heating the corresponding hydroxides (forming the oxide and gaseous water) or carbonates (forming the oxide and gaseous CO_2). Equations for sample reactions are

$$2Al(OH)_3 \xrightarrow{\Delta} Al_2O_3 + 3H_2O$$

$$CaCO_3 \xrightarrow{\Delta} CaO + CO_2$$

However, alkali metal salts generally are very stable and do not decompose when heated. Alkali metal oxides result from the oxidation–reduction reactions created by heating nitrates or hydroxides with the metals. Equations for sample reactions are

$$2KNO_3 + 10K \xrightarrow{\Delta} 6K_2O + N_2$$

$$2LiOH + 2Li \xrightarrow{\Delta} 2Li_2O + H_2$$

With the exception of mercury(II) oxide, oxides of the metals of Groups 2B–5A can be prepared by burning the corresponding metal in air. The heaviest member of each group—the member for which the inert pair effect (Section 21.1) is most pronounced—forms an oxide in which the oxidation state of the metal ion is two less than the group oxidation state. Thus Tl_2O, PbO, and Bi_2O_3 form when thallium, lead, and bismuth, respectively, are burned. The oxides of the lighter members of each group exhibit the group oxidation state. For example, SnO_2 if formed when tin is burned. Mercury(II) oxide, HgO, forms slowly when mercury is warmed below 500°C; it decomposes at higher temperatures.

Burning the members of Groups 1A and 2A in air is not a suitable way to form the oxides of these elements. These metals are reactive enough to combine with nitrogen in the air, so they form mixtures of oxides and ionic nitrides. Several also form peroxides or superoxides when heated in air.

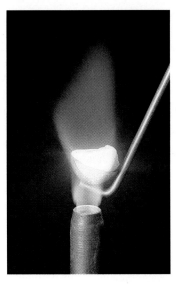

Figure 21.13
When heated, calcium oxide produces an intense white light.

Ionic oxides all contain the oxide ion, a very powerful hydrogen ion acceptor. With the exception of the very insoluble α-alumina and the oxides of tin(IV) and lead(IV), the oxides of the representative metals react with acids to form salts. Equations for sample reactions are

$$Na_2O + 2HNO_3 \longrightarrow 2NaNO_3 + H_2O$$
$$CaO + 2HCl \longrightarrow CaCl_2 + H_2O$$
$$\gamma\text{-}Al_2O_3 + 3H_2SO_4 \longrightarrow Al_2(SO_4)_3 + 3H_2O$$
$$SnO + 2HClO_4 \longrightarrow Sn(ClO_4)_2 + H_2O$$

The oxides of the metals of Groups 1A and 2A and of thallium(I) oxide react with water and form hydroxides. Example of such reactions are

$$Na_2O + H_2O \longrightarrow 2NaOH$$
$$CaO + H_2O \longrightarrow Ca(OH)_2$$
$$Tl_2O + H_2O \longrightarrow 2TlOH$$

Several different modifications of aluminum oxide exist. Heating aluminum hydroxide, $Al(OH)_3$, or aluminum oxyhydroxide, $AlO(OH)$, below 450°C drives off water and produces gamma-alumina (γ-Al_2O_3), a form of aluminum oxide, Al_2O_3, that reacts with water and dissolves in acids. Above 1000°C the reaction of aluminum with oxygen or loss of water from $Al(OH)_3$ or $AlO(OH)$ produces alpha-alumina (α-Al_2O_3), a very hard form of Al_2O_3 that does not react with water and is not attacked by acids.

The oxides of the alkali metals are used in the laboratory as sources of the metal ions and of the oxide ion, but they have little industrial utility, unlike magnesium oxide, calcium oxide, and aluminum oxide. Magnesium oxide is used widely in making fire brick, crucibles, furnace linings, and thermal insulation—applications that require chemical and thermal stability. Calcium oxide, sometimes called *quicklime* or *lime* in the industrial market, is very reactive, and its principal uses reflect its reactivity. Pure calcium oxide emits an intense white light when heated to a high temperature (Fig. 21.13). Blocks of calcium oxide heated by gas flames were used as stage lights in theaters before electricity was available. This is the source of the phrase "in the limelight."

Alpha-alumina occurs in nature as the mineral corundum, a very hard substance used as an abrasive for grinding and polishing. Several precious stones consist of α-Al_2O_3 with metal ion impurities that impart color. Artificial rubies and sapphires are now manufactured by melting aluminum oxide (mp = 2050°C) with small amounts of oxides to produce the desired colors and then cooling the melt in such a way as to produce large crystals. Ruby lasers use synthetic ruby crystals.

Zinc oxide, ZnO, is used as a white paint pigment. It is also used in the manufacture of automobile tires and other rubber goods and in the preparation of medicinal ointments. It is used in zinc-oxide-based sun screens to prevent sunburn (Fig. 21.14). Lead dioxide is a constituent of the charged lead storage battery. Lead(IV) tends to revert to the more stable lead(II) ion by gaining two electrons, so lead dioxide is a powerful oxidizing agent.

Figure 21.14
Zinc oxide protects exposed skin from sunburn.

Peroxides and Superoxides.

Peroxides and superoxides form when the metals or metal oxides of Groups 1A and 2A react with pure oxygen at elevated temperatures. Sodium peroxide and the peroxides of calcium, strontium, and barium are formed when the corresponding metal or metal oxide is heated in pure oxygen.

$$2Na + O_2 \xrightarrow{\Delta} Na_2O_2$$

$$2Na_2O + O_2 \xrightarrow{\Delta} 2Na_2O_2$$

$$Ca + O_2 \xrightarrow{\Delta} CaO_2$$

$$2SrO + O_2 \xrightarrow{\Delta} 2SrO_2$$

The peroxides of potassium, rubidium, and cesium can be prepared by heating the metal or its oxide in a carefully controlled amount of oxygen.

$$2K + O_2 \longrightarrow K_2O_2 \qquad (2 \text{ mol K per mol } O_2)$$

With an excess of oxygen, the superoxides KO_2, RbO_2, and CsO_2 form. For example,

$$K + O_2 \longrightarrow KO_2 \qquad (1 \text{ mol K per mol } O_2)$$

The stability of the peroxides and superoxides of the alkali metals increases as the size of the cation increases. Both peroxides and superoxides are strong oxidizing agents.

21.4   Hydroxides

Hydroxides, compounds that contain the OH^- ion, are prepared by two general types of reactions. Soluble metal hydroxides are generally prepared by reaction of the metal or metal oxide with water. Insoluble metal hydroxides form when a solution of a soluble salt of the metal combines with a solution containing hydroxide ions.

With the exception of beryllium and magnesium, the metals of Groups 1A and 2A form hydroxides and hydrogen gas when the metal is added to water. Examples of such reactions include

$$2Li + 2H_2O \longrightarrow 2LiOH + H_2$$

$$Ca + 2H_2O \longrightarrow Ca(OH)_2 + H_2$$

However, these reactions can be violent and dangerous, so soluble metal hydroxides are generally prepared by the reaction of the respective oxide with water.

$$Li_2O + H_2O \longrightarrow 2LiOH$$

$$CaO + H_2O \longrightarrow Ca(OH)_2$$

The insoluble hydroxides of beryllium, magnesium, and the metals of Groups 2B–5A can be prepared by addition of sodium hydroxide to a solution of a salt of the respective metal (Fig. 21.15 A on page 799). The net ionic equations for the reactions involving a magnesium salt, an aluminum salt, and a zinc salt are

$$Mg^{2+}(aq) + 2OH^-(aq) \longrightarrow Mg(OH)_2(s)$$

$$Al^{3+}(aq) + 3OH^-(aq) \longrightarrow Al(OH)_3(s)$$

$$Zn^{2+}(aq) + 2OH^-(aq) \longrightarrow Zn(OH)_2(s)$$

An excess of hydroxide must be avoided when preparing aluminum, gallium, zinc, and tin(II) hydroxides, or the hydroxides will dissolve (Fig. 21.15 B on page 799) with the formation of the corresponding complex ions: $Al(OH)_4^-$, $Ga(OH)_4^-$, $Zn(OH)_4^{2-}$, and $Sn(OH)_3^-$.

Figure 21.15

(A) Mixing solutions of NaOH and Zn(NO₃)₂ produces a white precipitate of Zn(OH)₂. (B) Addition of an excess of NaOH results in dissolution of the precipitate.

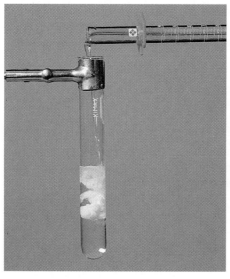

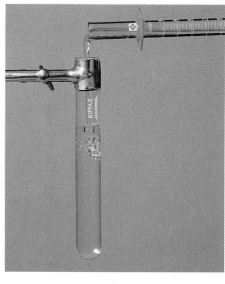

(A)   (B)

Figure 21.16

Calcium hydroxide, Ca(OH)₂, is added to a lake to neutralize the effects of acid rain.

Large quantities of sodium hydroxide are used industrially and are prepared from sodium chloride, because it is a less expensive starting material than the oxide. Sodium hydroxide was among the top ten chemicals in production in the United States in 1995 (more than 13 million tons), and this production was almost entirely by electrolysis of solutions of sodium chloride (Section 8.9).

$$2Na^+(aq) + 2Cl^-(aq) + 2H_2O(l) \xrightarrow{\text{Electrolysis}} 2Na^+(aq) + 2OH^-(aq) + H_2(g) + Cl_2(g)$$

Sodium hydroxide is an ionic compound, but it melts and boils without decomposition. It is very soluble in water, giving off a great deal of heat and forming very basic solutions; 40 grams of sodium hydroxide dissolves in only 60 grams of water at 25°C. Sodium hydroxide is employed in the production of other sodium compounds and is used to neutralize acidic solutions during the production of other chemicals such as petrochemicals and polymers.

Hydroxides are used to neutralize acids in a wide variety of applications (Fig. 21.16) and to prepare oxides by thermal decomposition (Section 21.3). An aqueous suspension of magnesium hydroxide constitutes the antacid *milk of magnesia*. Because of its ready availability (from the reaction of water with calcium oxide prepared by the decomposition of limestone, CaCO₃), cheapness, and activity, calcium hydroxide is used more extensively in commercial applications than any other base. The reaction of hydroxides with appropriate acids is also used to prepare salts.

21.5 Carbonates and Hydrogen Carbonates

The metals of Groups 1A and 2A, as well as zinc, cadmium, mercury, and lead(II), form ionic **carbonates**—compounds that contain the carbonate anion, CO_3^{2-}. The metals of Group 1A and magnesium, calcium, strontium, and barium also form **hydrogen carbonates**—compounds that contain the hydrogen carbonate anion, HCO_3^-.

With the exception of magnesium carbonate, the carbonates of the metals of Groups 1A and 2A can be prepared by the reaction of carbon dioxide with the respective oxide or hydroxide. Examples of such reactions include

$$Na_2O + CO_2 \longrightarrow Na_2CO_3$$
$$BaO + CO_2 \longrightarrow BaCO_3$$
$$2LiOH + CO_2 \longrightarrow Li_2CO_3 + H_2O$$
$$Ca(OH)_2 + CO_2 \longrightarrow CaCO_3 + H_2O$$

The carbonates of the alkaline earth metals (Group 2A), of Group 2B, and of thallium(I) and lead(II) are not soluble. They precipitate when solutions of soluble alkali metal carbonates (Group 1A) are added to solutions of soluble salts of these metals. The tufa beds shown in the introductory photograph to Chapter 17 were formed in this way. Examples of net ionic equations for the reactions are

$$Ca^{2+}(aq) + CO_3{}^{2-}(aq) \longrightarrow CaCO_3(s)$$
$$2Tl^+(aq) + CO_3{}^{2-}(aq) \longrightarrow Tl_2CO_3(s)$$
$$Pb^{2+}(aq) + CO_3{}^{2-}(aq) \longrightarrow PbCO_3(s)$$

If tin(II) or one of the trivalent or tetravalent ions such as Al^{3+} or Sn^{4+} is used in this reaction, then carbon dioxide and the corresponding oxide form.

Alkali metal hydrogen carbonates such as $NaHCO_3$ and $CsHCO_3$ form in solution when solutions of the hydroxides are saturated with carbon dioxide. The net ionic reaction involves hydroxide ion and carbon dioxide.

$$OH^-(aq) + CO_2(aq) \longrightarrow HCO_3{}^-(aq)$$

Figure 21.17

Stalactites and stalagmites are cave formations of calcium carbonate.

The solids can be isolated by evaporation of the water from the solution.

Although they are insoluble in pure water, the alkaline earth carbonates dissolve readily in water containing carbon dioxide because hydrogen carbonate salts are formed. For example, caves form in limestone when calcium carbonate, $CaCO_3$, dissolves in water containing dissolved carbon dioxide.

$$CaCO_3(s) + CO_2(aq) + H_2O(l) \longrightarrow Ca^{2+}(aq) + 2HCO_3{}^-(aq)$$

Hydrogen carbonates of the alkali metals remain stable only in solution, evaporation of the solution produces the carbonate. Stalactites and stalagmites form in caves when drops of water containing dissolved calcium hydrogen carbonate evaporate and slowly deposit calcium carbonate (Fig. 21.17).

The two carbonates used commercially in the largest quantities are sodium carbonate and calcium carbonate. Sodium carbonate is prepared industrially in the United States by extraction from the mineral trona, $Na_3(CO_3)(HCO_3)(H_2O)_2$. Following recrystallization to remove clay and other impurities, trona is roasted to produce Na_2CO_3.

$$2Na_3(CO_3)(HCO_3)(H_2O)_2 \longrightarrow 3Na_2CO_3 + 5H_2O + CO_2$$

Carbonates are moderately strong bases. Aqueous solutions are basic because the carbonate ion accepts a hydrogen ion from water in a reversible reaction, as shown by the equation

$$CO_3{}^{2-} + H_2O \rightleftharpoons HCO_3{}^- + OH^-$$

Carbonates react with acids to form salts of the metal, gaseous carbon dioxide, and water (Fig. 21.18 on page 801). The reaction of calcium carbonate, the active ingredient of the antacid Tums, with hydrochloric acid (stomach acid) illustrates the reaction.

$$CaCO_3 + 2HCl \longrightarrow CaCl_2 + CO_2 + H_2O$$

Many uses of carbonates are based on their conversion to salts by reaction with acids. Other applications include glass making—where carbonate ions serve as a source of oxide ions—and synthesis of oxides.

Hydrogen carbonates can act as both weak acids and weak bases. Hydrogen carbonate ions act as acids and react with solutions of soluble hydroxides to form a carbonate and water.

$$KHCO_3 + KOH \longrightarrow K_2CO_3 + H_2O$$

With acids, hydrogen carbonates form a salt, carbon dioxide, and water. Baking soda (bicarbonate of soda) is sodium hydrogen carbonate. Baking powders contain baking soda and a solid acid such as potassium hydrogen tartrate (cream of tartar), $KHC_4H_4O_6$. As long as the powder is dry, no reaction occurs; as soon as water is added, the acid reacts with the hydrogen carbonate ions to form carbon dioxide.

$$HC_4H_4O_6{}^- + HCO_3{}^- \longrightarrow C_4H_4O_6{}^{2-} + CO_2 + H_2O$$

If the carbon dioxide is trapped in the dough, it will expand as the dough is baked, producing the characteristic texture of baked goods.

Figure 21.18

The reaction of an acid and a carbonate produces gaseous carbon dioxide.

21.6 Salts

Thousands of **salts** of the representative metals have been prepared. Generally these salts are formed from the metals or from oxides, hydroxides, or carbonates. We will illustrate the general types of reactions involved with reactions used to prepare binary halides.

The binary compounds of a metal with the halogens are called **halides.** Most binary halides are ionic. However, mercury, the elements of Group 3A with oxidation states of +3, tin(IV), lead(IV), and bismuth(V) form covalent binary halides. Binary halides (and other salts) can be prepared by a variety of methods.

The direct reaction of a metal and a halogen produce the halide of the metal. Examples of these oxidation–reduction reactions include

$$Cd + Cl_2 \longrightarrow CdCl_2$$
$$2Ga + 3Br_2 \longrightarrow 2GaBr_3$$

Because of the extreme reactivity of the Group 1A and 2A metals, this method is not generally used to prepare their halides.

If a metal can exhibit two oxidation states, it may be necessary to control the stoichiometry in order to obtain the halide with the lower oxidation state. For example, preparation of tin(II) chloride requires a one-to-one ratio of Sn to Cl_2, whereas preparation of tin(IV) chloride requires a one-to-two ratio.

$$Sn + Cl_2 \longrightarrow SnCl_2$$
$$Sn + 2Cl_2 \longrightarrow SnCl_4$$

Other salts of the representative metals may be prepared by heating the elemental metals with other elements. For example, most representative metals form sulfides when heated with sulfur and selenides when heated with selenium.

The active representative metals—those that lie above hydrogen in the activity series (Section 8.7)—react with gaseous hydrogen halides to produce metal halides

Figure 21.19

Solid HgI_2 forms when solutions of KI and $Hg(NO_3)_2$ are mixed.

and hydrogen. The reaction of zinc with hydrogen fluoride is

$$Zn(s) + 2HF(g) \longrightarrow ZnF_2(s) + H_2(g)$$

The active representative metals also react with solutions of hydrogen halides to form hydrogen and solutions of the corresponding halides. Examples of such reactions include

$$Cd(s) + 2HBr(aq) \longrightarrow CdBr_2(aq) + H_2(g)$$
$$Sn(s) + 2HBr(aq) \longrightarrow SnBr_2(aq) + H_2(g)$$

Solutions of other acids also react with these metals, producing a wide variety of other salts. For example,

$$Zn(s) + H_2SO_4(aq) \longrightarrow ZnSO_4(aq) + H_2(g)$$
$$2Al(s) + 6HClO_4(aq) \longrightarrow 2Al(ClO_4)_3(aq) + 3H_2(g)$$

Hydroxides, carbonates, and some oxides react with solutions of the hydrogen halides to form solutions of halide salts. Additional salts can be prepared by the reaction of these hydroxides, carbonates, and oxides with aqueous solutions of other acids.

$$CaCO_3(s) + 2HCl(aq) \longrightarrow CaCl_2(aq) + CO_2(g) + H_2O(l)$$
$$TlOH(aq) + HF(aq) \longrightarrow TlF(aq) + H_2O(l)$$
$$Sn(OH)_2(s) + 2HNO_3(aq) \longrightarrow Sn(NO_3)_2(aq) + 2H_2O(l)$$

A few halides and many of the other salts of the representative metals are insoluble (see Section 8.5). These insoluble salts can be prepared by **metathesis reactions** (Section 2.10) that occur when solutions of soluble salts are mixed (Fig. 21.19).

Several halides occur in large quantities in nature. The ocean and underground brines contain many halides. For example, magnesium chloride in the ocean is the source of magnesium ions used in the production of magnesium (Section 21.2). Large underground deposits of sodium chloride (Fig. 21.20) are found in many parts of the world. These deposits serve as the source of sodium and chlorine in almost all other compounds containing these elements. Sodium chloride is used in the manufacture of the large industrial quantities of chlorine, sodium hydroxide, and hydrochloric acid reported in Table 8.8.

Figure 21.20

Salt is mined from beds up to 500 feet thick. This is the Akzo Nobel salt mine in Cleveland, Ohio.

For Review   Summary

The representative metals include the metals of Groups 1A, 2A, 2B, 3A, 4A, 5A, and 6A of the periodic table.

The **alkali metals** (Group 1A) are more alike in their properties than the members of any other group of elements. The single electron in the outermost shell of a Group 1A atom is loosely bound and easily lost. These metals readily form stable positive ions with a charge of +1 in ionic compounds that are usually soluble. The alkali metals all react vigorously with water to form hydrogen gas and a basic solution of the metal hydroxide. The metals are so reactive with moisture and air that they must be stored in kerosene, in mineral oil, or in sealed containers.

Each of the two electrons in the outermost shell of an atom of an **alkaline earth metal** (Group 2A) is more tightly bound than the single electron in the preceding alkali metal, so each alkaline earth atom is smaller than, and not so reactive as, the preceding alkali metal atom. These elements easily form compounds in which the metals exhibit an oxidation state of +2. Compounds of magnesium often show some covalent character, but the magnesium in them can generally be regarded as a +2 ion. Like the alkali metals, the alkaline earth elements are hard to prepare by chemical methods; the pure elements are isolated by **electrolysis.**

Zinc, cadmium, and **mercury** (Group 2B) commonly exhibit the group oxidation state of +2, although mercury also exhibits an oxidation state of +1 in compounds that contain the Hg_2^{2+} group. The chemical behaviors of zinc and cadmium are similar. Both elements lie above hydrogen in the **activity series.** Mercury is very different from zinc and cadmium. It is a nonreactive element that lies well below hydrogen in the activity series. These metals and, with the exception of aluminum, the representative metals of Groups 3A, 4A, 5A, and 6A can be prepared by **chemical reduction.**

Aluminum, gallium, indium and **thallium** (Group 3A) lie above hydrogen in the activity series and react with acids, liberating hydrogen. Aluminum, gallium, and indium are commonly found with an oxidation state of +3, but thallium is also commonly found as the Tl^+ ion. Aluminum forms both ionic and covalent compounds, in which it has an oxidation state of +3. Its amphoteric oxide and hydroxide react with both acids and bases. Aluminum is also manufactured by electrolysis.

Tin and **lead** (Group 4A) form stable dipositive cations, Sn^{2+} and Pb^{2+}, with oxidation states two below the group oxidation state of +4. Tin and lead also form covalent compounds in which the +4 oxidation state is exhibited. Tin is a moderately active metal. Lead is less reactive; it lies just above hydrogen in the activity series and is attacked only by hot concentrated acids.

Bismuth (Group 5A) readily gives up three of its five valence electrons to form the tripositive ion, Bi^{3+}. It forms compounds with the group oxidation state of +5 only when treated with strong oxidizing agents.

Polonium (Group 6A) is an intensely radioactive element that results from the radioactive decay of uranium and thorium.

Because of their reactivity, few of the representative metals are found as free elements in nature. They occur extensively as simple compounds such as chlorides, oxides or hydroxides, carbonates, sulfides, and nitrates. Aluminum and calcium are two of the most abundant metals in the earth's crust.

Compounds of the representative metals with oxygen fall into the three categories of **oxides, peroxides,** and **superoxides.** The oxides are usually produced by heating the corresponding hydroxides, nitrates, or carbonates. Peroxides and superoxides are formed by heating the metal or metal oxide in oxygen. The soluble oxides dissolve in water to form solutions of **hydroxides.** The hydroxides of the representative metals react with acids in acid–base reactions to form salts and water. The hydroxides have many commercial uses. Sodium hydroxide, NaOH, magnesium hydroxide [milk of magnesia, $Mg(OH)_2$], and calcium hydroxide, $Ca(OH)_2$, are all important metal hydroxides.

Carbonates of the alkali and alkaline earth metals are usually prepared by reaction of an oxide or hydroxide with carbon dioxide. Other carbonates form by precipitation. Metal carbonates or hydrogen carbonates such as limestone ($CaCO_3$), the antacid Tums, ($CaCO_3$), and baking soda ($NaHCO_3$) are well known to us.

All of the representative metals form **halides.** Several halides can be prepared when the metals react directly with elemental halogens or with solutions of the hydrohalic acids (HF, HCl, HBr, HI). Other laboratory preparations involve the addition of aqueous hydrohalic acids to compounds that contain such basic anions as hydroxides, oxides, or carbonates. Sodium chloride is one of the most abundant minerals, and it is used extensively in the chemical industry. Other **salts** of the representative metals can be prepared by acid–base reactions, oxidation–reduction reactions, or metathesis reactions.

Key Terms and Concepts

alkali metals   (21.1)
alkaline earth metals   (21.1)
amphoteric elements   (21.1)
carbonates   (21.5)
chemical reduction   (21.2)
halides   (21.1)

hydrogen carbonates   (21.5)
hydroxides   (21.4)
metathesis reaction   (21.6)
oxides   (21.3)
periodic relationships   (21.1)
peroxides   (21.3)

preparation of pure metals   (21.2)
representative metals   (21.1)
salts   (21.6)
superoxides   (21.3)
thermite reaction   (21.2)

Exercises

The Alkali Metals

1. Why does the reactivity of the alkali metals decrease from cesium to lithium?

2. How do the alkali metals differ from the alkaline earth metals in atomic structure and general properties?

3. Is the reaction of rubidium with water more or less vigorous than that of sodium? Than that of magnesium?

4. Write an equation for the reduction of cesium chloride by elemental calcium at high temperature.

5. Predict the product of burning francium in air.

6. Why must the chlorine and sodium resulting from the electrolysis of sodium chloride be kept separate during the production of sodium metal?

7. Using equations, describe the reaction of water with potassium and with potassium oxide.

8. Physiological saline concentration—that is, the sodium chloride concentration in our bodies—is approximately 0.16 M. Saline solution for contact lenses is prepared to match the physiological concentration. If you purchase 25 mL of contact lens saline solution, how many grams of sodium chloride have you bought?

9. A 25.00-mL sample of CsOH solution is exactly neutralized with 35.27 mL of 0.1062 M HNO_3. What is the concentration of the CsOH solution?

10. Give balanced equations for the overall reaction in the electrolysis of molten lithium chloride and for the reactions occurring at the electrodes.

11. Sodium chloride and strontium chloride are both white solids. How could you distinguish one from the other? ("Taste them" is not an acceptable answer.)

The Alkaline Earth Metals

12. Suppose you discovered a diamond completely encased in limestone. How would you chemically free the diamond without harming it?

13. What weight of Epsom salts ($MgSO_4 \cdot 7H_2O$) can be prepared from 5.0 kg of magnesium?

14. What weight, in grams, of hydrogen gas is produced by the complete reaction of 10.01 g of calcium with water?

15. How many grams of oxygen gas are required to react completely with 3.01×10^{21} atoms of magnesium to yield magnesium oxide?

16. Magnesium is an active metal; it is burned in the form of powder, ribbons, and filaments to provide flashes of brilliant light. Why is it possible to use magnesium in construction and even for the fabrication of cooking grills?

17. Predict the formulas for the nine compounds that are formed when each species in column 1 reacts with each species in column 2.

1	2
Na	I
Sr	Se
Al	O

18. Select:
    (a) the most metallic of the elements Al, Be, and Ba
    (b) the most covalent of the compounds NaCl, $CaCl_2$, and $BeCl_2$
    (c) the lowest first ionization energy among the elements Rb, Mg, and Sr
    (d) the smallest among Al, Al^+, and Al^{3+}
    (e) the largest among Cs^+, Ba^{2+}, and Xe

19. Solely on the basis of the lattice energy of the product, should we expect potassium metal or calcium metal to be more reactive with oxygen?

20. Write a balanced chemical equation describing the reaction of calcium metal with water.

21. The reaction of quicklime, CaO, with water produces slaked lime, $Ca(OH)_2$, which is widely used in the construction industry to make mortar and plaster. The reaction of quicklime and water is highly exothermic.

    $$CaO(s) + H_2O(l) \longrightarrow Ca(OH)_2(s) \quad \Delta H = -350 \text{ kJ mol}^{-1}$$

    (a) What is the enthalpy of reaction per gram of quicklime that reacts?
    (b) How much heat, in kilojoules, is associated with the production of 1 ton of slaked lime?

22. Write a balanced equation for the reaction of elemental strontium with each of the following: oxygen, hydrogen bromide, hydrogen, nitrogen, phosphorus, and water.

23. When $MgNH_4PO_4$ is heated to 1000°C, it is converted to $Mg_2P_2O_7$. A 1.203-g sample containing magnesium yielded 0.5275 g of $Mg_2P_2O_7$ after precipitation of $MgNH_4PO_4$ and heating. What percent by mass of magnesium was present in the original sample?

24. A sample of $MgSO_4 \cdot xH_2O$ weighing 5.018 g is heated until the water of hydration is completely driven off. The resulting anhydrous $MgSO_4$ weighs 2.449 g. What is the formula of the hydrated compound?

Group 2B

25. Write balanced chemical equations for the following reactions.

(a) Zinc metal is heated in a stream of oxygen gas.

(b) Zinc metal is added to a solution of lead(II) nitrate.

(c) Elemental zinc is added to a solution of cadmium nitrate.

(d) Zinc carbonate is heated until loss of mass stops.

(e) Zinc carbonate is added to a solution of acetic acid, CH_3CO_2H.

(f) Zinc is added to a solution of hydrobromic acid.

(g) Zinc is heated with sulfur.

26. Write balanced chemical equations for the following reactions.

(a) Cadmium is burned in air.

(b) Cadmium is heated with sulfur.

(c) Elemental cadmium is added to a solution of hydrochloric acid.

(d) Cadmium hydroxide is added to a solution of acetic acid, CH_3CO_2H.

(e) Cadmium hydroxide is heated until loss of mass stops.

(f) Cadmium metal is added to a solution of mercury(II) nitrate.

(g) Elemental cadmium metal is added to a solution of lead(II) nitrate.

27. A dilute solution of perchloric acid is dripped into a solution of sodium zincate, $Na_2[Zn(OH)_4]$. A white gelatinous precipitate is formed; analysis reveals it to be a hydroxide. Upon addition of more acid, a clear solution results. Use chemical equations to explain these observations.

28. The roasting of an ore of a metal usually results in conversion of the metal to the oxide. Why does the roasting of cinnabar, HgS, produce metallic mercury rather than an oxide of mercury?

29. Write balanced chemical equations for the following reactions.

(a) Mercury(II) oxide is added to a solution of nitric acid.

(b) Elemental mercury is warmed with sulfur.

(c) Cadmium metal is added to a solution of mercury(II) nitrate.

(d) An excess of elemental zinc is added to a solution of mercury(II) nitrate.

30. What does it mean to say that mercury(II) halides are weak electrolytes?

31. How many moles of ionic species are present in 1.0 L of a solution marked 1.0 M mercury(I) nitrate?

32. What is the mass of fish, in kilograms, that one would have to consume to obtain a fatal dose of mercury, if the fish contains 30 parts per million of mercury by weight? (Assume that all the mercury from the fish ends up as mercury(II) chloride in the body and that a fatal dose is 0.20 g of $HgCl_2$.) How many pounds of fish is this?

Group 3A

33. Illustrate the amphoteric nature (Section 15.5) of aluminum hydroxide by citing suitable equations.

34. The elements sodium, aluminum, and chlorine belong to the same period.

(a) Which has the greatest electronegativity?

(b) Which of the atoms is smallest?

(c) Write the Lewis structure for the covalent compound formed between aluminum and chlorine.

(d) Will the oxide of each element be acidic, basic, or amphoteric?

35. Why can aluminum, which is an active metal, be used so successfully as a structural metal?

36. Write balanced chemical equations for the following reactions.

(a) Gaseous hydrogen fluoride is bubbled through a suspension of bauxite in molten sodium fluoride.

(b) Metallic aluminum is burned in air.

(c) Elemental aluminum is heated in an atmosphere of chlorine.

(d) Aluminum is heated in hydrogen bromide gas.

(e) Aluminum hydroxide is added to a solution of nitric acid.

37. Describe the production of metallic aluminum by electrolytic reduction.

38. Write balanced chemical equations for the following reactions.

(a) Gallium metal is heated in air.

(b) An aqueous solution of sodium hydroxide is added dropwise to a solution of gallium chloride in water until $Ga(OH)_3$ has precipitated and then dissolved and the solution becomes clear again.

(c) Elemental indium is heated with an excess of sulfur.

(d) Indium metal is added to a solution of hydrobromic acid.

(e) Gallium(III) hydroxide is added to a solution of nitric acid.

(f) Indium(III) hydroxide is heated until loss of mass stops.

(g) Gallium metal is added to a solution of indium nitrate.

(h) Indium metal is added to a solution of lead(II) nitrate.

Group 4A

39. Why is $SnCl_4$ not classified as a salt?

40. Write balanced chemical equations describing:

(a) the burning of tin metal

(b) the dissolution of tin in a solution of hydrochloric acid

(c) the reaction of dry bromine with tin (write equations for both possible products)

(d) the reactions involved when an excess of aqueous NaOH is slowly added to a solution of tin(II) chloride

(e) the thermal decomposition of tin(II) hydroxide

(f) the reaction of tin with sulfur (write equations for both possible products)

41. What is the common ore of tin, and how is tin separated from it?

42. A 1.497-g sample of type metal (an alloy of Sn, Pb, Sb, and Cu) is dissolved in nitric acid, and metastannic acid, H_2SnO_3, precipitates. This is dehydrated by heating to tin(IV) oxide, which is found to weigh 0.4909 g. What percentage of tin was in the original type metal sample?

43. Does metallic tin react with HCl?

44. What is tin pest, also known as tin disease?

45. Compare the nature of the bonds in $PbCl_2$ to that of the bonds in $PbCl_4$.

46. Why should water to be used for human consumption not be conveyed in lead pipes?

Group 5A

47. Write balanced chemical equations for the following reactions.

(a) Bismuth is heated in air.

(b) An aqueous solution of sodium hydroxide is added dropwise to a solution of bismuth(III) chloride.

(c) Bismuth is heated with an excess of sulfur.

(d) Bismuth is added to a solution of hydrobromic acid.

(e) Bismuth(III) oxide is added to a solution of nitric acid.

(f) Bismuth(III) hydroxide is heated until loss of mass stops.

(g) Gallium metal is added to a solution of bismuth(III) nitrate.

(h) Bismuth metal is added to a solution of lead(II) nitrate.

48. (a) How long will it take to produce 100 kg of sodium metal when a current of 50,000 A is passed through a cell like that shown in Fig. 19.20 if the yield of sodium is 100% of the theoretical yield?

(b) What volume of chlorine at 25°C and 1.00 atm is produced?

Applications and Additional Exercises

49. In 1774 Joseph Priestley prepared O_2 by heating red HgO with sunlight focused through a lens. How much heat is required to decompose exactly 1 mol of red HgO(s) to Hg(l) and O_2(g) under standard state conditions?

50. Which of the following is the most basic: LiOH, $P(OH)_3$, $Sr(OH)_2$, CsOH, or $As(OH)_3$?

51. What mass of magnesium is produced when 100,000 A is passed through a $MgCl_2$ melt for 1.00 h if the yield of magnesium is 85% of the theoretical yield?

52. What volume of oxygen at 25°C and 1.00 atm pressure is required to prepare 1.00 kg of potassium peroxide? To prepare 1.00 kg of potassium superoxide?

53. Write balanced chemical equations for the following reactions.

(a) Sodium oxide is added to water.

(b) Cesium carbonate is added to an excess of an aqueous solution of HF.

(c) An electric current is passed through a sample of molten $CaBr_2$.

(d) γ-Alumina is added to an aqueous solution of $HClO_4$.

(e) Barium and barium hydroxide are heated together.

(f) Aluminum is heated with calcium oxide in a vacuum.

(g) A solution of sodium carbonate is added to a solution of barium nitrate.

(h) A solution of sodium carbonate is added to a solution of magnesium nitrate.

(i) Titanium metal is produced from the reaction of titanium tetrachloride with elemental sodium.

54. Determine the oxidation state of oxygen in lithium oxide, in sodium peroxide, and in cesium superoxide.

55. Peroxides, like oxides, are basic; they form H_2O_2 upon treatment with an acid. What volume of 0.250 M H_2SO_4 solution is required to neutralize a solution that contains 5.00 g of CaO_2?

CHAPTER OUTLINE

The Semi-Metals

22.1   The Chemical Behavior of the Semi-Metals
22.2   Structures of the Semi-Metals
22.3   Isolation of Boron and Silicon
22.4   Boron and Silicon Hydrides
22.5   Boron and Silicon Halides
22.6   Boron and Silicon Oxides and Derivatives

The Nonmetals

22.7   Periodic Trends and the General Behavior of the Nonmetals
22.8   Structures of the Nonmetals
22.9   Chemical Behavior of the Nonmetals

Isolation and Uses of the Nonmetals

22.10   Hydrogen
22.11   Oxygen
22.12   Nitrogen
22.13   Phosphorus
22.14   Sulfur
22.15   The Halogens
22.16   The Noble Gases

Properties of the Nonmetals

22.17   Hydrogen
22.18   Oxygen
22.19   Nitrogen
22.20   Phosphorus
22.21   Sulfur
22.22   The Halogens
22.23   The Noble Gases

Compounds of Selected Nonmetals

22.24   Hydrogen Compounds
22.25   Nonmetal Halides
22.26   Nonmetal Oxides
22.27   Nonmetal Oxyacids and Their Salts

22

The Semi-Metals and the Nonmetals

Comet Hyakutake, which was visible in the spring of 1996, is essentially an evaporating ball of ice.

Representative elements consist of metals, semi-metals, and nonmetals (Fig. 22.1). In Chapter 21 we discussed the chemistry of the representative metals, and in this chapter we will examine the chemistry of the semi-metals and the nonmetals.

A series of five elements called the **semi-metals,** sometimes referred to as the **metalloids,** separate the metals from the nonmetals in the periodic table. The semi-metals are boron, silicon, germanium, antimony, and tellurium (Fig. 22.1). These elements look metallic, but they conduct electricity poorly: They are semiconductors. Their chemical behavior falls between that of the metals and that of the nonmetals; for example, the pure semi-metal elements form covalent crystals like the nonmetals, but like the metals, they generally do not form monatomic anions. Semi-metals are fairly nonreactive elements, having electronegativities slightly lower than that of hydrogen. In this chapter we will briefly discuss the chemical behavior of semi-metals and deal with two of these elements—boron and silicon—in more detail.

The nonmetals are the elements located in the upper right portion of the periodic table. Under normal conditions, more than half of the nonmetals are gases, one is a liquid, and the rest include some of the softest and hardest of solids. The nonmetals exhibit a rich variety of chemical behavior. They include the most reactive and the most nonreactive of the elements, and they form many different ionic and covalent

Figure 22.1

The location in the periodic table of the semi-metals, shown in yellow, with the nonmetals shown in green, and the metals in red.

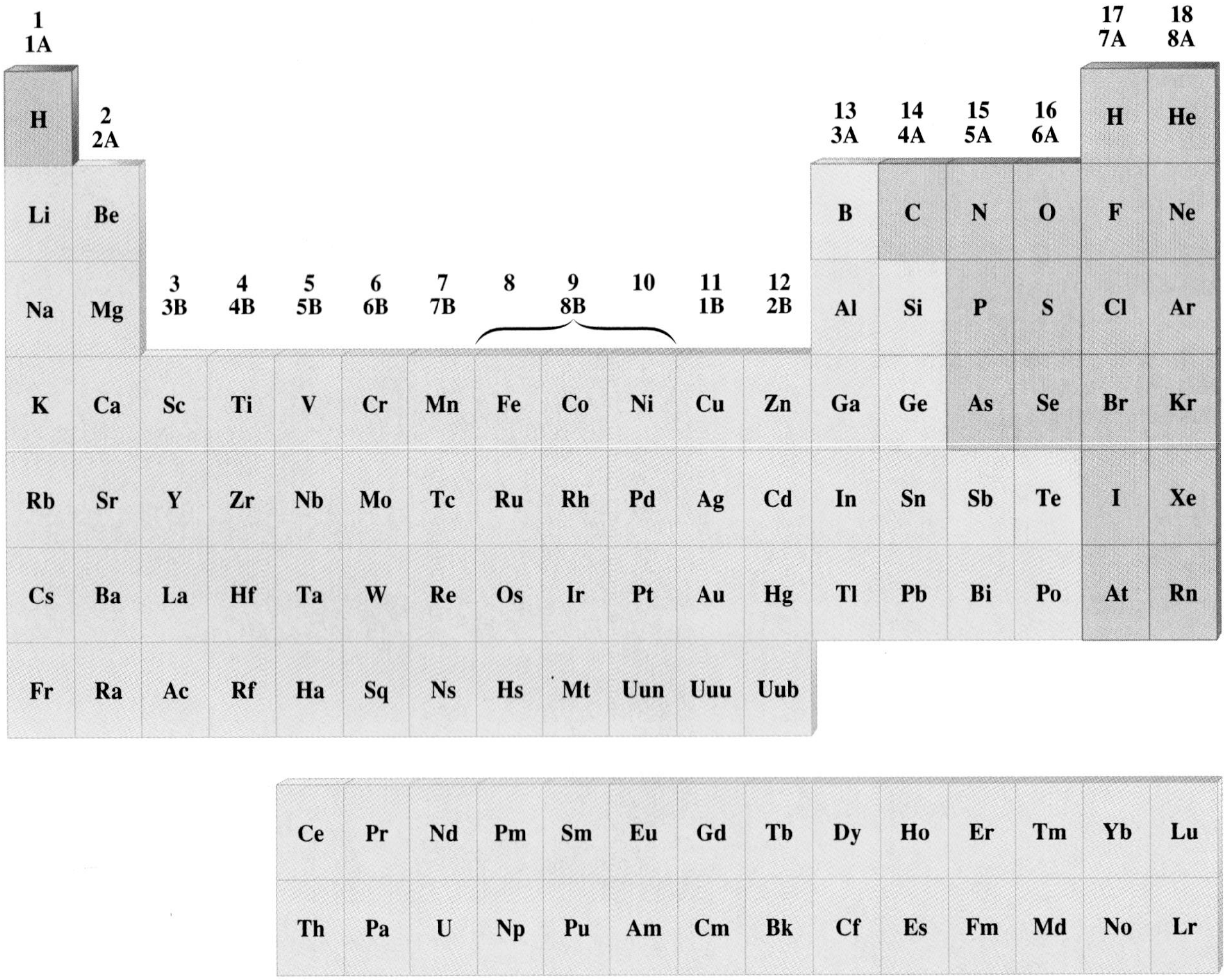

compounds. We have already discussed the varied chemistry of the nonmetal carbon in Chapter 9, and in the present chapter we present an overview of the properties and chemical behavior of the nonmetals as well as the chemistry of specific elements.

The Semi-Metals

22.1   The Chemical Behavior of the Semi-Metals

Elemental **boron** is chemically inert at room temperature, reacting with only fluorine and oxygen to form boron trifluoride, BF_3, and boric oxide, B_2O_3, respectively. At higher temperatures, boron reacts with all of the nonmetals except tellurium, and with nearly all metals. It is oxidized to B_2O_3 when heated with concentrated nitric or sulfuric acid. Boron does not react with nonoxidizing acids, with boiling concentrated aqueous sodium hydroxide, or with molten sodium hydroxide below 500°C. Many boron compounds react readily with water to give boric acid, $B(OH)_3$ (sometimes written as H_3BO_3), and with air to give boric oxide.

The semi-metal boron exhibits many similarities to its neighbor carbon (Chapter 9) and its diagonal neighbor silicon. All three elements form covalent compounds. However, boron has one distinct difference in that its $2s^2 2p^1$ outer electron structure gives it one less valence electron than it has valence orbitals. Although boron exhibits an oxidation state of $+3$ in most of its stable compounds, this "electron deficiency" provides boron with the ability to form other, sometimes fractional, oxidation states. These are found, for example, in the boron hydrides (Section 22.4).

The name **silicon** is derived from the Latin word for flint, *silex*. The semi-metal silicon readily forms compounds containing Si—O—Si bonds, which are of prime importance in the mineral world. This bonding capability is in contrast to the nonmetal carbon in the same group, whose ability to form carbon–carbon bonds gives it prime importance in the plant and animal worlds.

Silicon has the valence shell electron configuration $3s^2 3p^2 3d^0$, and it commonly forms tetrahedral compounds in which it is sp^3-hybridized with an oxidation state of $+4$. The major differences between the chemistry of carbon and that of silicon result from the relative strength of the carbon–carbon bond, the relative weakness of the silicon–silicon bond, carbon's ability to form stable bonds to itself, and the presence of the empty $3d$ valence shell orbitals in silicon. Silicon's empty d orbitals and boron's empty p orbital enable tetrahedral silicon compounds and triagonal planar boron compounds to act as Lewis acids (Section 15.13). Because they possess no available valence shell orbitals, tetrahedral carbon compounds cannot act as Lewis acids.

Silicon, like boron, is unreactive at low temperatures and resists attack by air, water, and acids. A very thin film of silicon dioxide, SiO_2, protects the surface from significant attack. This film dissolves in base and exposes the surface, so silicon dissolves in hot sodium hydroxide or potassium hydroxide solutions, forming hydrogen gas and anions of silicon and oxygen called silicates (Section 22.6). For example,

$$Si(s) + 2OH^-(aq) + H_2O(l) \longrightarrow SiO_3{}^{2-}(aq) + 2H_2(g)$$

Silicon reacts with the halogens at high temperatures, forming volatile tetrahalides, such as SiF_4. Silicon oxidizes in air at elevated temperatures to give silicon dioxide, SiO_2.

Unlike carbon, silicon does not readily form double or triple bonds. Silicon compounds of the general formula SiX_4, where X is a highly electronegative group, can act as Lewis acids and form six-coordinate silicon. For example, silicon tetrafluoride, SiF_4, reacts with sodium fluoride to give $Na_2[SiF_6]$, which contains the octahedral $[SiF_6]^{2-}$ ion in which silicon is sp^3d^2-hybridized.

$$2NaF(s) + SiF_4(g) \longrightarrow Na_2SiF_6(s)$$

Except for silicon tetrafluoride, silicon halides are extremely sensitive to water. For example, when $SiCl_4$ is exposed to water, it reacts rapidly and all four chlorine atoms are replaced by hydroxide groups. However, orthosilicic acid, $Si(OH)_4$ or H_4SiO_4 is unstable and slowly decomposes to SiO_2.

Germanium is very similar to silicon in its chemical behavior. However, it also forms a series of compounds, such as GeO and $GeCl_2$, in which it exhibits an oxidation state of +2. These compounds are very reactive and react with air or water.

Antimony generally forms compounds in which it exhibits an oxidation state of +3 or +5. The element tarnishes only slightly in dry air but is readily oxidized when warmed, giving antimony(III) oxide, Sb_4O_6. The element reacts readily with stoichiometric amounts of fluorine, chlorine, bromine, or iodine, giving trihalides or, with excess fluorine or chlorine, forming the pentahalides SbF_3 and $SbCl_5$. Depending on the stoichiometry, it forms antimony(III) sulfide, Sb_2S_3, or antimony(V) sulfide when heated with sulfur. It is oxidized by hot nitric acid, forming Sb_4O_6. It reacts slowly with hot concentrated sulfuric acid, forming $Sb_2(SO_4)_3$ and evolving SO_2. As would be expected, the metallic nature of the element is more pronounced than that of arsenic, which lies immediately above it in Group 5A.

Tellurium combines directly with most elements. The most stable tellurium compounds are (1) the tellurides, salts of Te^{2-} formed with the active metals and the lanthanides, and (2) compounds with oxygen, fluorine, and chlorine in which tellurium normally exhibits an oxidation state of +2 or +4. Although tellurium(VI) compounds are known (for example, TeF_6), there is a marked resistance to oxidation to this maximum group oxidation state.

22.2   Structures of the Semi-Metals

The crystal structures of the semi-metals are characterized by covalent bonding. In this regard, these elements resemble nonmetals in their behavior.

Elemental silicon, germanium, antimony, and tellurium are lustrous, metallic-looking solids. Silicon and germanium crystallize with a diamondlike structure (Fig. 9.2 A). Each atom within the crystal is covalently bonded to four neighboring atoms at the corners of a regular tetrahedron. Single crystals of silicon and germanium are giant three-dimensional molecules. The crystal structure of antimony is layerlike and contains puckered sheets of antimony atoms in which each antimony atom forms covalent bonds to three adjacent atoms in the sheet. Tellurium forms crystals that contain infinite spiral chains of tellurium atoms. Each atom in the chain is bonded to two other atoms.

Pure crystalline boron is transparent. The crystals consist of icosahedra (Fig. 22.2) with a boron atom at each corner. In the most common form of boron, the icosahedra are packed together in a manner similar to cubic closest packing of spheres (Fig. 22.3 on page 811; also see Section 11.14). All boron–boron bonds within each icosahedron are identical and are approximately 1.76 Å in length. However, there are two kinds of boron–boron bonds between the icosahedra. Each of six of the 12 boron

Figure 22.2

An icosahedron, a symmetrical solid shape with 20 faces, each of which is an equilateral triangle. The faces meet at 12 corners.

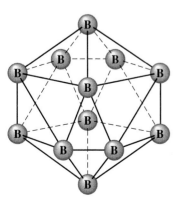

Figure 22.3

The structure of boron. Each icosahedron contains 12 boron atoms. Each of six of these is bonded to a boron atom in another icosahedron by a two-center bond (1.71Å), and each of the other six is bonded to two boron atoms, each in a separate icosahedron, by a three-center bond (2.03Å). Only the three-center bonds between icosahedra are shown here.

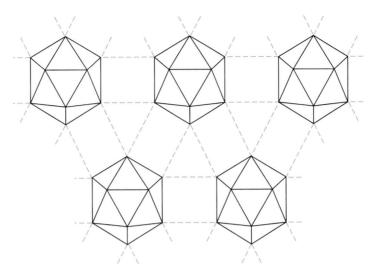

Figure 22.4

A three-center B—B—B bond involving the overlap of three sp^3 hybrid orbitals, which share one pair of electrons. (A) The overlap diagram for the hybrid orbitals. (B) The overall outline of the bonding molecular orbital.

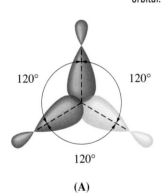

120°   120°

120°

(A)

(B)

atoms of an icosahedron is joined to a boron atom in adjacent icosahedra by a regular covalent bond between the two atoms, which is 1.71Å long. (These bonds are not shown in Fig. 22.3). Each of the other six atoms is bonded to *two* other atoms, one in each of two other icosahedra, by a bond in which two electrons bond three atoms, forming a bond that is 2.03Å long.

In Chapters 6 and 7, we discussed bonds in which two atoms are bonded together by one pair of electrons in a bond that results from the overlap of two atomic orbitals. However, in some situations three atoms are bonded together by one pair of electrons in a bond formed from the overlap of three atomic orbitals. These bonds are called **three-center two-electron bonds.** As was pointed out in Chapter 7, in a two-center two-electron bond, one bonding molecular orbital and one antibonding molecular orbital result from the overlap of two atomic orbitals (see Section 7.12). In a three-center two-electron bond, one bonding molecular orbital and two antibonding molecular orbitals (or one bonding orbital, one nonbonding orbital, and one antibonding orbital) result from the overlap of three atomic orbitals. The two electrons occupy the bonding molecular orbital. The formation of the bonding orbital in elemental boron is shown in Fig. 22.4.

The three-center two-electron bond is found with atoms that do not have enough electrons to satisfy ordinary two-center bond requirements. Boron makes the greatest use of the three-center bond; it exhibits this type of bond not only in elemental boron but also in boron hydrides (Section 22.4).

22.3   Isolation of Boron and Silicon

Boron constitutes less than 0.001% by weight of the earth's crust. It does not occur in the free state in nature but rather is found in compounds with oxygen. Boron is widely distributed in volcanic regions as boric acid, $B(OH)_3$, and in dry lake regions, including the desert areas of California, as **borates,** salts of boron oxyacids, such as **borax,** $Na_2B_4O_7 \cdot 10H_2O$, and **kernite,** $Na_2B_4O_7 \cdot 4H_2O$.

Reduction of boric oxide with powdered magnesium forms boron (95–98% pure) as a brown amorphous powder.

$$B_2O_3 + 3Mg \longrightarrow 2B + 3MgO$$

The magnesium oxide is removed by dissolving it in hydrochloric acid. Pure boron can be obtained by passing a mixture of boron trichloride and hydrogen either through an electric arc or over a hot tungsten filament.

$$2BCl_3(g) + 3H_2(g) \xrightarrow{1500°C} 2B(s) + 6HCl(g) \qquad \Delta H° = 253.7 \text{ kJ}$$

Silicon makes up nearly one-fourth of the mass of the earth's crust—second in abundance there only to oxygen. The crust is composed almost entirely of minerals in which silicon atoms are connected by oxygen atoms in complex structures involving chains, layers, and three-dimensional frameworks. These minerals constitute the bulk of most common rocks, soils, clays, and sands. Sand and sandstone are forms of impure silicon dioxide, as are quartz, amethyst, agate, and flint. Most rocks are built up of the common metal cations and silicate anions. Materials such as granite, bricks, cement, mortar, ceramics, and glasses are composed of silicon compounds.

Silicon can be obtained by the reduction of silicon dioxide with strong reducing agents at high temperatures. With carbon and magnesium as the reducing agents, the equations are

$$SiO_2(s) + 2C(s) \longrightarrow Si(s) + 2CO(g)$$
$$SiO_2(s) + 2Mg(s) \longrightarrow Si(s) + 2MgO(s)$$

Extremely pure silicon, such as that required for the manufacture of semiconductor electronic devices, is prepared by the decomposition of silicon tetrahalides, or silane, SiH_4, at high temperatures, followed by a purification method known as **zone refining.** In this method a rod of silicon is heated at one end by a heat source that produces a thin cross section of molten silicon. The heat source is moved slowly from one end of the rod to the other. As this thin, molten region moves, impurities in the silicon dissolve in the liquid silicon and move with the molten region. Ultimately they are concentrated in the other end of the silicon rod, which can be removed (Fig. 22.5 on page 813). This highly purified silicon, containing no more than one part impurity per million parts of silicon, is the most important element in the computer industry (Fig. 22.6 on page 813). It is used in semiconductor electronic devices such as transistors, computer chips, and solar cells. Elemental silicon is also used as a deoxidizer in the production of steel, copper, and bronze and in the manufacture of acid-resistant iron alloys.

22.4   Boron and Silicon Hydrides

Boron and silicon form a series of volatile hydrides, although boron hydrides are quite different from those of either carbon or silicon. The hydride BH_3 does not exist as the monomer at room temperature, although Lewis acid–base adducts, such as H_3B—CO and H_3B—PH_3, are known. BH_3 dimerizes instead, and the simplest stable boron hydride is **diborane,** $(BH_3)_2$, more commonly written as B_2H_6. The structure of diborane contains two kinds of hydrogen atoms (Fig. 22.7 on page 813).

The bonding in the B_2H_6 molecule is unusual. Four of the six hydrogen atoms are involved in two-center two-electron bonds, but the two bridging hydrogen atoms are bonded to the boron atoms in three-center two-electron bonds. The four regular two-center B—H bonds are all in the same plane and use eight of the 12 valence electrons. A bridging hydrogen above and below the plane is connected to the two boron atoms by a three-center two-electron B—H—B bond. The three atoms are bonded by

Figure 22.5

Zone refining apparatus that results in ultrapure elemental silicon.

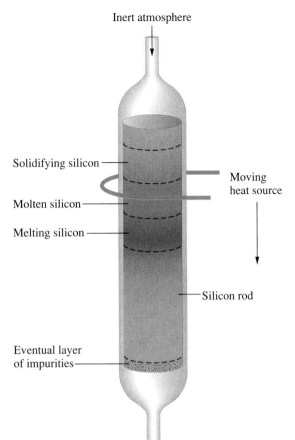

Inert atmosphere

Solidifying silicon

Molten silicon

Melting silicon

Moving heat source

Silicon rod

Eventual layer of impurities

Figure 22.6

Silicon computer chips are vital to the computer industry. Their small size aids in the manufacture of small, lightweight computers.

a bond that contains one pair of electrons, analogous to the B—B—B three-center bond in elemental boron (Section 22.2). The two three-center bonds hold the remaining four electrons.

The reaction of lithium aluminum hydride with boron trifluoride in diethyl ether solution yields diborane.

$$4BF_3 + 3LiAlH_4 \longrightarrow 2B_2H_6(g) + 3LiF + 3AlF_3$$

Figure 22.7

The structure of the diborane molecule, B_2H_6. (A) The molecular structure. (B) The spatial arrangement of the bonding orbitals.

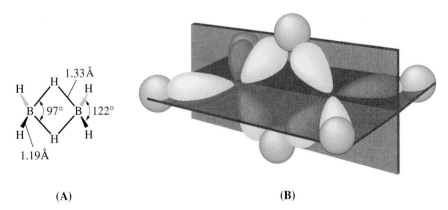

(A)

(B)

Diborane is not acidic; it reacts with water, forming boric acid and hydrogen.

$$B_2H_6(g) + 6H_2O(l) \longrightarrow 2B(OH)_3(s) + 6H_2(g) \qquad \Delta H° = -509 \text{ kJ}$$

Heating diborane above 100°C in a sealed vessel (a pyrolysis reaction) results in the formation of boron hydrides of higher molecular mass, as do the pyrolysis reactions of other boron hydrides. Over 25 neutral boron hydrides and an even larger number of boron hydride anions have been characterized. Examples include B_4H_{10}, $B_{10}H_{14}$, $B_3H_8^-$, and $B_{10}H_{10}^{2-}$. In all these compounds, boron uses all four of its valence orbitals and its three valence electrons, necessitating three-center two-electron bonding.

The silicon hydrides have formulas analogous to those of the alkanes (Section 9.3); SiH_4, Si_2H_6, Si_3H_8, and S_4H_{10} are examples. These compounds contain Si—H and Si—Si single bonds. Because of the presence of the silicon empty d orbitals, the chemical behavior of these silicon hydrides is decidedly different from that of the hydrocarbons of similar formulas. For example, the silicon hydrides inflame spontaneously in air, whereas the hydrocarbons do not.

Acids react with magnesium silicide to form **silane**, SiH_4, which has a structure analogous to that of methane, CH_4.

$$Mg_2Si(s) + 4H_3O^+(aq) \longrightarrow 2Mg^{2+}(aq) + SiH_4(g) + 4H_2O(l)$$

Silane, a colorless gas, is thermally stable at ordinary temperatures but spontaneously inflames when exposed to air. The products of its oxidation are silicon dioxide and water.

$$SiH_4(g) + 2O_2(g) \longrightarrow SiO_2(s) + 2H_2O(g) \qquad \Delta H° = -1429 \text{ kJ}$$

The hydrogen atoms in silane can be replaced one at a time by halogen atoms via reaction of silane with hydrogen halides, using the corresponding aluminum halide as a catalyst. With HBr, using $AlBr_3$ as the catalyst, the overall reaction to produce silicon tetrabromide is

$$SiH_4(g) + 4HBr(g) \longrightarrow SiBr_4(l) + 4H_2(g)$$

Silane is extremely sensitive to hydroxides, reacting readily as a Lewis acid to give silicates and hydrogen.

$$SiH_4(aq) + 2OH^-(aq) + H_2O(l) \longrightarrow SiO_3^{2-}(aq) + 4H_2(g)$$

In contrast, methane, CH_4, is unreactive to hydroxides.

Perhaps the most important reaction of compounds containing an Si—H bond, at least from a commercial standpoint, is the reaction with hydrocarbons such as propene, $CH_3CH=CH_2$, that contain carbon–carbon double bonds.

$$CH_3CH=CH_2 + H_2SiCl_2 \longrightarrow CH_3CH_2CH_2SiHCl_2$$
$$CH_3CH=CH_2 + CH_3CH_2CH_2SiHCl_2 \longrightarrow (CH_3CH_2CH_2)_2SiCl_2$$

These reactions are used in the preparation of **silicones,** which are polymeric organosilicon compounds containing Si—O—Si and Si—C bonds and which are stable toward heat, chemicals, and water.

22.5   Boron and Silicon Halides

Boron trihalides—BF_3, BCl_3, BBr_3, and BI_3—can be prepared by direct reaction of the elements. These nonpolar molecules contain boron that exhibits sp^2 hybridization and have a trigonal planar molecular geometry. The fluoride and chloride compounds are colorless gases, the bromide is liquid, and the iodide is a white crystalline solid.

The heavier boron trihalides readily hydrolyze in water to form boric acid and the corresponding hydrohalic acid. Boron trichloride reacts according to the equation

$$BCl_3(g) + 3H_2O(l) \longrightarrow B(OH)_3(aq) + 3HCl(aq)$$

When boron trifluoride is added to hydrofluoric acid, it reacts to give **fluoroboric acid,** HBF_4.

$$BF_3(aq) + HF(aq) + H_2O(l) \longrightarrow H_3O^+(aq) + BF_4^-(aq)$$

In the latter reaction, the BF_3 molecule acts as a Lewis acid (electron-pair acceptor) and accepts a pair of electrons from a fluoride ion, as shown by the equation

All the tetrahalides of silicon, SiX_4, have been prepared. **Silicon tetrachloride** can be prepared by direct chlorination at elevated temperatures or by heating silicon dioxide with chlorine and carbon.

$$SiO_2(s) + 2C(s) + 2Cl_2(g) \xrightarrow{\Delta} SiCl_4(g) + 2CO(g)$$

Silicon tetrachloride is a covalent tetrahedral molecule containing four covalent Si—Cl bonds. It is a nonpolar, low-boiling (57°C) colorless liquid that evaporates readily. Consequently it fumes strongly in moist air to produce a dense smoke of finely divided silica as the Si—Cl bonds in the gaseous $SiCl_4$ molecules are replaced by Si—O bonds.

$$SiCl_4(g) + 2H_2O(g) \longrightarrow SiO_2(s) + 4HCl(g)$$

Elemental silicon ignites spontaneously in an atmosphere of fluorine, forming gaseous **silicon tetrafluoride,** SiF_4. The reaction of hydrofluoric acid with silica or a silicate also produces SiF_4.

$$SiO_2(s) + 4HF(g) \longrightarrow SiF_4(g) + 2H_2O(l) \qquad \Delta H° = -191.2 \text{ kJ}$$
$$CaSiO_3(s) + 6HF(g) \longrightarrow SiF_4(g) + CaF_2(s) + 3H_2O(l)$$

Silicon tetrafluoride hydrolyzes in water, producing **fluorosilicic acid,** H_2SiF_6, as well as **orthosilicic acid.**

$$3SiF_4(g) + 8H_2O(l) \longrightarrow H_4SiO_4(s) + 4H_3O^+(aq) + 2SiF_6^{2-}(aq)$$

Orthosilicic acid   Fluorosilicic acid

The difference in the reactivity of SiF_4 and $SiCl_4$ with water can be attributed to the great strengths of Si—O, Si—C, and Si—F bonds. Exposure of $SiCl_4$ and most other silicon compounds to water or oxygen results in their decomposition to compounds containing Si—O bonds, unless the compounds are stabilized by the presence of Si—O, Si—C, or Si—F bonds.

22.6   Boron and Silicon Oxides and Derivatives

Boron burns at 700°C in oxygen, forming **boric oxide,** B_2O_3. Boric oxide is used in the production of heat-resistant borosilicate glass (Fig. 22.8) and certain optical glasses. It dissolves in hot water to form **boric acid,** $B(OH)_3$.

$$B_2O_3(s) + 3H_2O(l) \longrightarrow 2B(OH)_3(aq)$$

The boron atom in $B(OH)_3$ is sp^2-hybridized and is located at the center of an equilateral triangle with oxygen atoms at the corners.

$$
\begin{array}{c}
\text{H} \\
| \\
\text{O} \\
120° \nearrow | \nwarrow 120° \\
\text{H} \diagdown \quad \text{B} \quad \diagup \\
\text{O} \rightleftarrows \text{O} \\
\quad 120° \quad | \\
\text{H}
\end{array}
$$

In solid $B(OH)_3$ these triangular units are held together by hydrogen bonding. Boric acid is a very weak acid that does not act as a proton donor but rather as a Lewis acid accepting an unshared pair of electrons from the Lewis base OH^-.

$$B(OH)_3 + 2H_2O \rightleftharpoons B(OH)_4^- + H_3O^+ \qquad K_a = 5.8 \times 10^{-10}$$

When boric acid is heated to 100°C, molecules of water are split out between pairs of adjacent OH groups to form **metaboric acid,** HBO_2. With further heating at about 150°, additional B—O—B linkages form, connecting the BO_3 groups together with shared oxygen atoms to form **tetraboric acid,** $H_2B_4O_7$. At still higher temperatures, boric oxide is formed.

Borates are salts of the oxyacids of boron. Borates result from the reaction of a base with an oxyacid or from the fusion of boric acid or boric oxide with a metal oxide or hydroxide. Borate anions range from the simple trigonal planar BO_3^{3-} ion

Figure 22.8

Laboratory glassware, such as Pyrex and Kimax, is made of borosilicate glass because it does not break when heated.

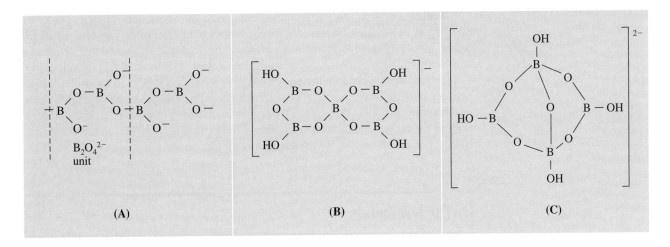

(A)   (B)   (C)

Figure 22.9

The borate anions found in (A) CaB_2O_4, (B) $KB_5O_8 \cdot 4H_2O$, and (C) $Na_2B_4O_7 \cdot 10H_2O$. The anion in CaB_2O_4 is an infinite chain.

to complex species containing chains and rings of three- and four-coordinated boron atoms. The structures of the anions found in CaB_2O_4, $K[B_5O_6(OH)_4] \cdot 2H_2O$ (commonly written $KB_5O_8 \cdot 4H_2O$), and $Na_2[B_4O_5(OH)_4] \cdot 8H_2O$ (commonly written $Na_2B_4O_7 \cdot 10H_2O$) are shown in Fig. 22.9. Commercially, the most important borate is **borax**, $Na_2[B_4O_5(OH)_4] \cdot 8H_2O$. Most of the supply of borax comes directly from dry lakes, such as Searles Lake in California, or is prepared from kernite, $Na_2B_4O_7 \cdot 4H_2O$.

Silicon dioxide, silica, is found in both crystalline and amorphous forms. The usual crystalline form of silicon dioxide is **quartz,** a hard, brittle, clear, colorless solid. It is used in many ways—for architectural decorations, semiprecious jewels, and frequency control in radio transmitters. Amorphous silicon dioxide occurs in nature as obsidian. The term *quartz* is also used for articles such as tubing and lenses that are manufactured from amorphous silica.

The contrast in structure and physical properties between silicon dioxide and its carbon analog, carbon dioxide, is interesting. Solid carbon dioxide (dry ice) contains single CO_2 molecules with each of the two oxygen atoms attached to the central carbon atom by double bonds. Very weak intermolecular forces hold the molecules together in the crystal. The low melting point and volatility of dry ice reflect these weak forces between molecules. In contrast, silicon dioxide is a covalent network solid (Section 11.12). Silicon does not form double bonds. In silicon dioxide, each silicon atom is linked to four oxygen atoms by single bonds directed toward the corners of a regular tetrahedron, and SiO_4 tetrahedra share oxygen atoms. This arrangement gives a three-dimensional, continuous silicon–oxygen network. A quartz crystal is a macromolecule of silicon dioxide.

At 1600°C, quartz melts to give a viscous liquid with a random internal structure. When the liquid is cooled, it does not crystallize readily but usually supercools and forms a glass, also called *silica.* The SiO_4 tetrahedra in glassy silica have the random arrangement characteristic of supercooled liquids, and the glass has some very useful properties. Silica is highly transparent to both visible and ultraviolet light. It is used in lamps that give radiation rich in ultraviolet light and in certain optical instruments that operate with ultraviolet light. The coefficient of expansion of silica glass is very low, so it is not easily fractured by sudden changes in temperature. It is used in Corning Ware and other ceramic cookware (Section 24.11).

Silicates are salts containing anions composed of silicon and oxygen. There are many types of silicates, because the silicon-to-oxygen ratio can vary widely. In all silicates, however, sp^3-hybridized silicon atoms are found at the centers of tetrahedra with oxygen at the corners, and silicon is tetravalent. The variation in the silicon-to-oxygen ratio occurs because the silicon–oxygen tetrahedra may exist as discrete, independent units or may share oxygen atoms at corners, edges, or (more rarely) faces, in a variety of ways. The silicon-to-oxygen ratio varies according to the extent of sharing of oxygen atoms by silicon atoms in the linking together of the tetrahedra. Many ceramics (Chapter 24) are composed of silicates.

The Nonmetals

The properties and behavior of the nonmetals in the upper right portion of the periodic table are quite different from those of metals on the left side. However, there are no distinct points at which changes from metallic behavior occur in periods or groups in the periodic table.

Chemists have slightly different ideas about which elements constitute the semi-metal group and which constitute the nonmetals. For our purposes, we will consider hydrogen, the noble gases, and those elements with electronegativities equal to or larger than that of hydrogen as nonmetals (Fig. 22.1).

22.7   Periodic Trends and the General Behavior of the Nonmetals

Acquaintance with periodic properties and trends can enhance our understanding and recall of the chemistry of the nonmetals (Table 8.2). Nonmetals exhibit widely differing chemical behaviors, but we can understand and predict similarities and trends if we are familiar with properties such as electronegativity, ionization energy, and size. Many reactions of nonmetals are Lewis acid–Lewis base, Brønsted acid–Brønsted base, or oxidation–reduction reactions. (To refresh your memory, you may want to review material on these types of reactions in Chapters 2, 8, and 15.)

In many cases, trends in electronegativity enable us to predict the type of bonding in compounds involving the nonmetals, as well as the physical states of the compounds. We know that electronegativity decreases as we move down a given group and increases as we move from left to right across a period. The nonmetals have higher electronegativities than metals, and compounds formed between metals and nonmetals are generally ionic in nature because of the large differences in electronegativity between the metals and the nonmetals. The metals form cations, the nonmetals form anions, and the resulting compounds are solids at room temperature and pressure. On the other hand, compounds formed between two or more nonmetals have small differences in electronegativity between the atoms, and covalent bonding—sharing of electrons—results. These substances thus tend to be molecular in nature and are gases, liquids, or volatile solids at room temperature and pressure.

In normal chemical processes, nonmetals do not form monatomic positive ions because their ionization energies are too high. All monatomic nonmetal ions are

Table 22.1

Common Oxidation States of the
Nonmetals in Compounds, by Group

1A	4A	5A	6A	7A	8A
H	C	N	O	F	
+1	+4	+5	−1	−1	
−1	to	to	−2		
	−4	−3			
		P–As	S–Se	Cl–I	Xe
		+5	+6	+7	+8
		+3	+4	+5	+6
		−3	−2	+3	+4
				+1	+2
				−1	

anions; examples include the chloride ion, Cl^-, the nitride ion, N^{3-}, and the selenide ion, Se^{2-}.

The common oxidation states that the nonmetals exhibit in their ionic and covalent compounds are shown in Table 22.1. Remember that an element exhibits a positive oxidation state when it is combined with a more electronegative element and that it exhibits a negative oxidation state when it is combined with a less electronegative element.

The first member of each nonmetal group exhibits different behavior, in many respects, from the other group members. The several reasons for this include smaller size, greater ionization energy, and (most important) the fact that the first member of each group has only four valence orbitals (one $2s$ and three $2p$) available for bonding, whereas other group members have empty d orbitals in their valence shells, making possible five, six, or even more bonds around the central atom. For example, nitrogen forms only NF_3, whereas phosphorus forms both PF_3 and PF_5.

Another difference between the first group member and subsequent members is the greater ability of the first member to form π bonds. This is primarily a function of the smaller size of the first member of each group, which allows better overlap of atomic orbitals. Nonmetals other than the first member of each group form π bonds, but not to the same extent, and they form their most stable π bonds to elements that are the first members of a group. For example, sulfur–oxygen π bonds are well known, whereas sulfur does not form stable π bonds to itself.

The variety of oxidation states displayed by most of the nonmetals means that many of their chemical reactions involve changes in these oxidation states in oxidation–reduction reactions. Four general aspects of the oxidation–reduction chemistry of the nonmetals are listed below.

Figure 22.10

Hot iron powder is vigorously oxidized by the oxygen in air, forming iron oxides.

1. Nonmetals oxidize most metals (Fig. 22.10). The oxidation state of the metal becomes positive as it is oxidized, and that of the nonmetal becomes negative as it is reduced. For example,

$$\overset{0}{4Fe} + \overset{0}{3O_2} \longrightarrow \overset{+3\ -2}{2Fe_2O_3}$$

2. With the exception of nitrogen and carbon, which are poor oxidizing agents, a more electronegative nonmetal oxidizes a less electronegative nonmetal or the anion of the nonmetal.

$$\overset{0}{S} + \overset{0}{O_2} \longrightarrow \overset{+4 \ -2}{SO_2}$$

$$\overset{0}{Cl_2} + 2I^- \longrightarrow \overset{0}{I_2} + 2Cl^-$$

Fluorine and oxygen are the strongest oxidizing agents within their respective groups; each oxidizes all the elements that lie below it in the group. Within any period, the strongest oxidizing agent is found in Group 7A. A nonmetal often oxidizes an element that lies to its left in the same period. For example,

$$\overset{0}{2As} + \overset{0}{3Br_2} \longrightarrow \overset{+3 \ -1}{2AsBr_3}$$

3. The stronger a nonmetal is as an oxidizing agent, the more difficult it is to remove electrons from the anion formed by the nonmetal. This means that the most stable negative ions are formed by elements at the top of the group or in Group 7A of the period.

4. Fluorine and oxygen are the strongest oxidizing elements known. Fluorine does not form compounds in which it exhibits positive oxidation states; oxygen exhibits a positive oxidation state only when combined with fluorine. For example,

$$\overset{0}{2F_2} + 2OH^- \longrightarrow \overset{+2 \ -1}{OF_2(g)} + 2F^- + H_2O$$

With the exception of the noble gases, all nonmetals form compounds with oxygen giving covalent oxides. Most of these oxides are acidic; that is, they react with water to form oxyacids that yield hydronium ions in aqueous solution. Notable exceptions are carbon monoxide, CO, nitrous oxide, N_2O, and nitric oxide, NO. We can summarize the general behavior of acidic oxides with the following three statements.

1. Oxides such as SO_2 and N_2O_5, in which the nonmetal exhibits one of its common oxidation states, are called **acid anhydrides** and react with water to form acids with no change in oxidation state. The product is an oxyacid. For example,

$$SO_2 + H_2O \longrightarrow H_2SO_3$$
$$N_2O_5 + H_2O \longrightarrow 2HNO_3$$

2. Those oxides such as NO_2 and ClO_2, in which the nonmetal does not exhibit one of its common oxidation states, also react with water. In these reactions the nonmetal is both oxidized and reduced. For example,

$$\overset{+4}{3NO_2} + H_2O \longrightarrow \overset{+5}{2HNO_3} + \overset{+2}{NO}$$
$$\overset{+4}{6ClO_2} + 3H_2O \longrightarrow \overset{+5}{5HClO_3} + \overset{-1}{HCl}$$

Reactions in which the same element is both oxidized and reduced are called **disproportionation reactions.**

3. The relative acid strength of oxyacids can be predicted on the basis of the electronegativity and oxidation state of the central atom. The acid strength increases as the electronegativity of the central atom increases (see Section 15.5). For example, perchloric acid, $HClO_4$, is stronger than sulfuric acid, H_2SO_4, which in turn

is stronger than phosphoric acid, H_3PO_4. For the same central atom, the acid strength increases as the oxidation state of the central atom increases. For example, nitric acid, HNO_3, is a stronger acid than nitrous acid, HNO_2; and sulfuric acid, H_2SO_4, is a stronger acid than sulfurous acid, H_2SO_3.

The binary hydrides of the nonmetals also exhibit acidic behavior in water, though only HCl, HBr, and HI are strong acids. The acid strength of the nonmetal hydrides increases from left to right across a period and down a group. For example, ammonia, NH_3, is a weaker acid than water, H_2O, which is weaker than hydrogen fluoride, HF. Water, H_2O, is also a weaker acid than hydrogen sulfide, H_2S, which is weaker than hydrogen selenide, H_2Se.

22.8 Structures of the Nonmetals

The structures of the nonmetals differ dramatically from those of metals. Metals crystallize in closely packed arrays that do not contain molecules or covalent bonds (see Section 11.14). Nonmetal structures contain covalent bonds, and many nonmetals consist of individual molecules.

The noble gases are all monatomic, whereas the other nonmetal gases—hydrogen, nitrogen, oxygen, fluorine, and chlorine—exist as the diatomic molecules H_2, N_2, O_2, F_2, and Cl_2. The other halogens are also diatomic; Br_2 is a liquid, and I_2 and At_2 exist as solids. The changes in state as one moves down the halogen family offer an excellent example of the increasing strength of intermolecular London forces with increasing molecular mass and increasing polarizability (see Section 11.3).

Allotropes are two or more forms of the same element that exhibit different structures in the same physical state. Several nonmetals exist as allotropes. In addition to diatomic O_2, elemental oxygen also occurs as the unstable allotrope O_3, which is called ozone. Carbon, phosphorus, and sulfur are other examples of nonmetals that occur as allotropes.

Descriptions of the physical properties of two nonmetals that are characteristic of molecular solids follow.

Phosphorus. The name *phosphorus* is derived from Greek words meaning "light-bringing." (When phosphorus was first isolated, scientists noted that it glowed in the dark and burned when exposed to air.) Although phosphorus is the only member of its group that is not found in the uncombined state in nature, it exists in many allotropic forms. We will consider two of those forms: **white phosphorus** and **red phosphorus.**

White phosphorus (Fig. 22.11) is a white, waxy solid that melts at 44.2°C and boils at 280°C. It is insoluble in water (in which it is stored), is very soluble in carbon disulfide, and bursts into flame in air. As a solid, as a liquid, as a gas, and in solution, white phosphorus exists as P_4 molecules with the four phosphorus atoms at the corners of a regular tetrahedron. Each phosphorus atom is covalently bonded to the other three atoms in the molecule by covalent, single bonds (Fig. 22.12 A on page 822). White phosphorus is the most reactive allotrope and is also very toxic.

Heating white phosphorus to 270–300°C in the absence of air yields red phosphorus (Fig 22.11). Red phosphorus is more dense, has a higher melting point (~600°C), is much less reactive, is essentially nontoxic, and is easier and safer to handle than white phosphorus. Its structure is highly polymeric and appears to contain three-dimensional networks of P_4 tetrahedra joined by P—P single bonds (Fig.

Figure 22.11

White phosphorus stored under water (left) and red phosphorus (right).

Figure 22.12

(A) The P_4 molecule found in white phosphorus. (B) The chain structure of red phosphorus.

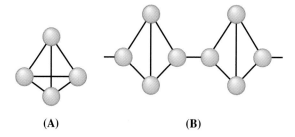

(A)   (B)

22.12 B). Red phosphorus is insoluble in solvents that dissolve white phosphorus. When red phosphorus is heated, P_4 molecules sublime from the solid. This accounts for the fact that the products of the reactions of red phosphorus are usually the same as those produced when the white form reacts.

Sulfur.   The allotropy of sulfur is far greater and more complex than that of any other element. Sulfur is the "brimstone" referred to in the Bible, and references to sulfur occur throughout recorded history—right up to the relatively recent discovery that it is a component of the atmospheres of Venus and of Io, a moon of Jupiter. The most common and most stable allotrope of sulfur is yellow, **rhombic sulfur,** so named because of the shape of its crystals (Fig. 22.13 A). Rhombic sulfur is the form to which all other allotropes revert at room temperature. Crystals of rhombic sulfur melt at 113°C and form a straw-colored liquid. When this liquid cools and crystallizes, long needles of **monoclinic sulfur** are formed (Fig. 22.13 B). This is the stable form above 96°C. It melts at 119°C. At room temperature it gradually reverts to the rhombic form.

Solid rhombic sulfur, rhombic sulfur at its melting point, solutions of rhombic and monoclinic sulfur in carbon disulfide, and monoclinic sulfur all contain S_8 molecules in which the atoms form eight-membered puckered rings that resemble crowns (Fig. 22.14 A on page 822). Each sulfur atom is bonded to each of its two neighbors in the ring by covalent S—S single bonds.

Figure 22.13

(A) Crystals of rhombic sulfur.
(B) Crystals of monoclinic sulfur.

(A)   (B)

Figure 22.14

(A) An S_8 molecule and (B) a chain of sulfur atoms.

(A)   (B)

When rhombic sulfur melts, the straw-colored liquid is quite mobile (Fig. 22.15 A); its viscosity is low because S_8 molecules are essentially spherical and offer relatively little resistance as they move past each other. As the temperature rises, S—S bonds in the rings break, and polymeric chains of sulfur atoms result (Fig. 22.14 B). These chains combine end to end, forming still longer chains that become entangled with one another. The liquid gradually darkens in color and becomes so viscous that finally (at about 230°C) it does not pour easily. The unpaired elements at the ends of the chains of sulfur atoms are responsible for the dark red color. When the liquid is cooled rapidly, a rubberlike amorphous mass, called **plastic sulfur,** results (Fig. 22.15 B).

Sulfur boils at 445°C and forms a vapor consisting of S_2, S_6, and S_8 molecules; at about 1000°C the vapor density corresponds to the formula S_2, which is a paramagnetic molecule like O_2 with a similar electronic structure and a weak sulfur–sulfur double bond.

An important feature of the structural behavior of the nonmetals is that the elements are usually found naturally with eight electrons in their valence shells. If necessary, the elements form enough covalent bonds to supplement the electrons already present in order to possess an octet. For example, the members of Group 5A have five valence elements and require only three additional electrons to fill their valence shells. These elements form three covalent bonds in their free state: triple bonds in the N_2 molecule and single bonds to three different atoms in arsenic and phosphorus. The elements of Group 6A require only two additional electrons. Oxygen forms a double bond in the O_2 molecule, and sulfur, selenium, and tellurium form two single

Figure 22.15

(A) Liquid rhombic sulfur. (B) Formation of plastic sulfur.

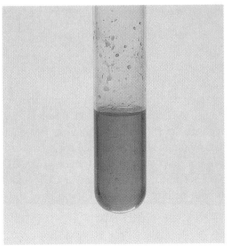

(A)   (B)

bonds in various rings and chains. The halogens form diatomic molecules in which each atom is involved in only one bond. This provides the one electron required to supplement the seven electrons in the valence shells of the individual atoms. An atom of a noble gas contains eight electrons in its valence shell, and noble gases do not form covalent bonds to other noble gas atoms.

Finally, note again that only elements that are filling their second principal quantum shell ($n = 2$) consistently form strong multiple bonds. Larger elements have covalent radii that are not conducive to effective overlap of their p orbitals.

22.9   Chemical Behavior of the Nonmetals

Before we examine the chemical behavior of specific nonmetals, it will be helpful to review Lewis structures (originally discussed in Section 6.6) and the Lewis concept of acids and bases (originally discussed in Section 15.13). Briefly, a Lewis acid is any species that uses a valence shell orbital to accept a pair of electrons, and a Lewis base is any species that donates an unshared pair of electrons. We can understand many chemical reactions if we examine the electronic structures of the reactants to ascertain their Lewis acid–base characteristics. For example, the simplest Lewis acid is the proton, H^+, which contains an empty valence orbital ($1s$), and the simplest Lewis base is the hydride ion, H^-, which has an unshared pair of electrons in its valence shell (also $1s$). Anytime a source of protons and a hydride ion are in proximity to one another, hydrogen gas will be produced by a Lewis acid-base reaction.

$$H^+ + H^- \longrightarrow H_2 + H_2O$$

Understanding this Lewis acid–base reaction enables us to predict that when a metal hydride is added to an acid or to water, both of which are sources of H^+, hydrogen gas should be evolved.

$$CaH_2 + H_2O \longrightarrow Ca(OH)_2 + H_2(g)$$

This concept also makes it easy to understand why water is formed when a proton source encounters the hydroxide ion, another Lewis base.

$$H_3O^+ + :\overset{..}{\underset{..}{O}}-H^- \longrightarrow 2\ \overset{..}{\underset{H\quad\ \ H}{O}}$$

We also recognize this reaction as a Brønsted acid–Brønsted base reaction (see Section 8.5).

Many molecules of the nonmetals and the semi-metals contain empty valence orbitals or unshared electron pairs. Here again, we can use their Lewis structures to explain much of their chemical behavior. For example, using Lewis acid–base concepts we can explain why water, a Lewis base with two unshared electron pairs on oxygen, does not react with carbon tetrachloride, CCl_4, but does react with silicon tetrachloride, $SiCl_4$. The Lewis structures of CCl_4 and $SiCl_4$ are identical, because carbon and silicon both belong to Group 4A and have the same number of valence electrons.

$$
\begin{array}{cc}
\ddot{\ddot{C}l} & \ddot{\ddot{C}l} \\
| & | \\
:\ddot{C}l-\overset{\displaystyle}{C}-\ddot{C}l: \qquad & :\ddot{C}l-\overset{\displaystyle}{Si}-\ddot{C}l: \\
| & | \\
:\ddot{C}l: & :\ddot{C}l:
\end{array}
$$

However, these two molecules differ in that carbon uses all of its valence orbitals (a 2*s* orbital and three 2*p* orbitals) for bonding, whereas silicon uses its 3*s* orbital and its three 3*p* orbitals but still has five empty 3*d* orbitals. Thus, the carbon atom in CCl_4 is neither a Lewis acid nor a Lewis base, whereas the silicon atom in $SiCl_4$ (with empty valence orbitals) is a Lewis acid and reacts with the Lewis base H_2O.

The initial step in the reaction of $SiCl_4$ with H_2O is the formation of an unstable Lewis acid–Lewis base adduct (a compound that contains a coordinate covalent bond) to form $SiCl_4 \cdot H_2O$. A molecule of hydrogen chloride then leaves.

Unstable

This happens four times, producing $Si(OH)_4$, which is unstable and decomposes to SiO_2 and two molecules of water.

Isolation and Uses of the Nonmetals

22.10 Hydrogen

Hydrogen is the most abundant element in the universe, and it is the third most abundant element (after oxygen and silicon) on the earth's surface. The sun and other stars are composed largely of hydrogen, as are the gases found in interstellar space. It is estimated that 90% of the atoms in the universe are hydrogen atoms.

Hydrogen accounts for nearly 11% of the mass of water, its most abundant compound on earth. It is an important part of the tissues of all plants and animals, petroleum, many minerals, cellulose and starch, sugar, fats, oils, alcohols, acids, and thousands of other substances. Hydrogen is a component of more compounds than any other element.

At ordinary temperatures, hydrogen is a colorless, odorless, tasteless, and nonpoisonous gas consisting of diatomic molecules, H_2.

Hydrogen is composed of three isotopes: (1) ordinary hydrogen, ^1_1H; (2) **deuterium,** ^2_1D or ^2_1H; and (3) **tritium,** ^3_1T or ^3_1H. In a naturally occurring sample of hydrogen, there is one atom of deuterium for every 7000 ^1_1H atoms and one atom of tritium for every 10^{18} ^1_1H atoms. The chemical properties of the different isotopes are very similar, because they have identical electron structures, but they differ in some physical properties because of their differing atomic masses. Elemental deuterium and tritium have lower vapor pressures than ordinary hydrogen. Consequently, when liquid hydrogen evaporates, the heavier isotopes are somewhat concentrated in the last portions to evaporate. Deuterium is isolated by the electrolysis of heavy water, D_2O. Tritium is made by a nuclear reaction (Chapter 20).

Hydrogen must be prepared from compounds by breaking chemical bonds. The most common methods of preparing hydrogen follow.

Figure 22.16

The electrolysis of water produces
hydrogen and oxygen.

1. **From Steam and Carbon or Hydrocarbons.** Water is the cheapest and most abundant source of hydrogen. Passing steam over coke (an impure form of carbon) at 1000°C produces a mixture of carbon monoxide and hydrogen.

$$C(s) + H_2O(g) \xrightarrow{1000°C} CO(g) + H_2(g)$$
Water gas

This gaseous mixture is known as **water gas** and is used as an industrial fuel. Additional hydrogen can be obtained by mixing the water gas with steam in the presence of a catalyst to convert the CO to CO_2.

$$CO(g) + H_2O(g) \underset{\text{Catalyst}}{\xrightleftharpoons{}} H_2(g) + CO_2(g)$$

This reaction is sometimes called the **water gas shift reaction.**

Hydrocarbons from natural gas or petroleum can be mixed with steam and passed over a nickel-based catalyst to produce carbon monoxide and hydrogen. Propane is an example of a hydrocarbon reactant.

$$C_3H_8(g) + 3H_2O(g) \xrightarrow[\text{Catalyst}]{900°C} 3CO(g) + 7H_2(g)$$

2. **Electrolysis.** Hydrogen is liberated when a direct current of electricity is passed through water containing a small amount of an electrolyte such as H_2SO_4, NaOH, or Na_2SO_4 (Fig. 22.16). Bubbles of hydrogen are formed at the cathode, and oxygen is evolved at the anode. The net reaction can be summarized by the equation

$$2H_2O(l) + \text{electrical energy} \longrightarrow 2H_2(g) + O_2(g)$$

3. **Reaction of Metals with Acids.** This is the most convenient laboratory method of producing hydrogen. Electropositive metals reduce the hydrogen ion in dilute acids to produce hydrogen gas and metal salts. For example, iron in dilute hydrochloric acid produces hydrogen gas and iron(II) chloride (Fig. 22.17).

$$Fe(s) + 2H_3O^+(aq) + 2Cl^-(aq) \longrightarrow Fe^{2+}(aq) + 2Cl^-(aq) + H_2(g) + 2H_2O(l)$$

4. **Reaction of Ionic Metal Hydrides with Water.** Hydrogen can also be produced by the reaction of hydrides of the active metals, which contain the very strongly basic H^- anion, with water.

$$CaH_2(s) + 2H_2O(l) \longrightarrow Ca^{2+}(aq) + 2OH^-(aq) + 2H_2(g)$$

Metal hydrides are expensive but convenient sources of hydrogen, especially where space and weight are important factors, as they are in the inflation of life jackets, life rafts, and military balloons.

Two-thirds of the world's hydrogen production is devoted to the manufacture of ammonia, which is used primarily as a fertilizer and in the manufacture of nitric acid. Hydrogen is also used extensively in the process of **hydrogenation,** in which vegetable oils are changed from liquids to solids. Crisco is an example of a hydrogenated oil. The change results from the addition of H_2 to carbon–carbon double bonds to give single bonds (see Section 9.5). Methyl alcohol, an important industrial solvent and raw material, is produced synthetically by the catalyzed reaction of hydrogen with carbon monoxide.

$$2H_2 + CO \xrightarrow{\text{Catalyst}} CH_3OH$$

Figure 22.17

The reaction of iron with an acid produces hydrogen. Here iron reacts with hydrochloric acid.

Hydrogen is currently used in several ways as a fuel, and its potential uses may come to be even more important. The reaction of hydrogen with oxygen is a very exothermic reaction, releasing 286 kJ of energy per mole of water formed. Hydrogen burns without explosion under controlled conditions. As we noted before, water gas (a mixture of hydrogen and carbon monoxide) is an important industrial fuel. The oxygen–hydrogen torch, because of the high heat of combustion of hydrogen, can achieve temperatures up to 2800°C. The hot flame of this torch is used in "cutting" thick sheets of many metals. Hydrogen is also an important rocket fuel (Fig. 22.18).

22.11 Oxygen

Oxygen is the most abundant element on the earth's surface, and it forms compounds with most of the other elements. It is essential to combustion and to respiration in most plants and animals. Oxygen occurs as O_2 molecules and, to a limited extent, as O_3 (ozone) molecules in air. It forms about 23% of the mass of the air. About 89% of water by mass consists of combined oxygen. About 50% of the mass and 90% of the volume of the earth's crust consist of oxygen (combined with other elements, principally silicon). In combination with carbon, hydrogen, and nitrogen, oxygen is a large part of the weight of the bodies of plants and animals.

Approximately 97% of the oxygen isolated commercially comes from air and 3% from the electrolysis of water (Fig. 22.16), because air and water are abundant, cheap, and easy to process. Oxygen (boiling point, 90 K) is separated from the air in commercial quantities by cooling and compressing air until it liquefies and then distilling off the lower-boiling nitrogen (boiling point, 77 K) and some other elements (see Section 11.7 for a discussion of distillation). The electrolysis of water to produce hydrogen and oxygen is described in Section 22.10.

Oxygen is essential in combustion processes such as the burning of fuels. Oxygen is also required for the decay of organic matter. Plants and animals use the oxygen from the air in respiration. Oxygen-enriched air is administered in medical practice when a patient is receiving an inadequate supply of oxygen as a result of shock, pneumonia, or some other illness. Health services use about 13% of the oxygen produced commercially.

Approximately 30% of all oxygen produced commercially is used to remove carbon from iron during steel production. Large quantities of pure oxygen are also consumed in metal fabrication and in the cutting and welding of metals with oxyhydrogen and oxyacetylene torches (see Fig. 4.1). The chemical industry employs oxygen for oxidizing many substances.

Liquid oxygen is used as an oxidizing agent in the space shuttle and other rocket engines. It is also used to provide gaseous oxygen for life support in space.

As we know, oxygen is very important to life. The energy required for maintenance of normal body functions in human beings and in other organisms is derived from the slow oxidation of chemical compounds in the body. Oxygen is the final oxidizing agent in these reactions. In humans, oxygen passes from the lungs into the blood, where it combines with hemoglobin, producing oxyhemoglobin. In this form, oxygen is carried by the blood to tissues, where it is released and consumed. The ultimate products are carbon dioxide and water. The blood carries the carbon dioxide through the veins to the lungs, where it gives up the carbon dioxide and collects another supply of oxygen. The digestion and assimilation of food regenerate the materials consumed by oxidation in the body; in fact, the same amount of energy is liberated as if the food had been burned outside the body.

The oxygen in the atmosphere is continually replenished through the action of green plants by a process called **photosynthesis.** The products of photosynthesis may vary, but the process can be generalized as the conversion of carbon dioxide and water to glucose (a sugar) and oxygen using the energy of light.

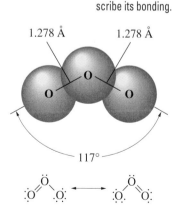

Figure 22.19

The bent O_3 molecule and the resonance structures necessary to describe its bonding.

1.278 Å   1.278 Å

117°

$$6CO_2 + 6H_2O \xrightarrow[\text{Light}]{\text{Chlorophyll}} C_6H_{12}O_6 + 6O_2$$

Carbon   Water   Glucose   Oxygen
dioxide

Thus the oxygen that is converted to carbon dioxide and water by the metabolic processes in plants and animals is returned to the atmosphere by these photosynthetic reactions.

Ozone.
When dry oxygen is passed between two electrically charged plates, a decrease in the volume of the gas occurs, and **ozone** (O_3, Fig. 22.19), an allotrope of oxygen possessing a distinctive odor, is formed. The formation of ozone from oxygen is an endothermic reaction, in which the energy is furnished in the form of an electrical discharge, heat, or ultraviolet light.

$$3O_2 \xrightarrow{\substack{\text{Electric} \\ \text{discharge}}} 2O_3 \qquad \Delta H° = 287 \text{ kJ}$$

The sharp odor associated with sparking electrical equipment is due in part to ozone.

Ozone is formed naturally in the upper atmosphere by the action of ultraviolet light from the sun on the oxygen there. Most atmospheric ozone is found in the stratosphere, a layer of the atmosphere extending from about 10 to 50 kilometers above the earth's surface. This ozone acts as a barrier to harmful ultraviolet light from the sun by absorbing it via a chemical decomposition reaction.

$$O_3(g) \xrightarrow{\substack{\text{Ultraviolet} \\ \text{light}}} O(g) + O_2(g)$$

The reactive oxygen atom recombines with molecular oxygen to complete the ozone cycle. The frequency of skin cancer and other damaging effects of ultraviolet radiation are decreased by the presence of stratospheric ozone. Concern has arisen that chlorofluorocarbons, CFCs (known commercially as Freons), which have been used as aerosol propellants in spray cans and as refrigerants, are causing depletion of ozone in the stratosphere (see photograph on page 473). This occurs because ultraviolet light also causes CFCs to decompose, producing atomic chlorine. The chlorine atoms react with ozone molecules, resulting in a net removal of O_3 molecules from the stratosphere.

The uses of the allotrope ozone depend on its reactivity with other substances. As a bleaching agent for oils, waxes, fabrics, and starch, it oxidizes the colored compounds in these substances to colorless compounds. It is sometimes used instead of chlorine to disinfect water.

22.12   Nitrogen

Nitrogen is obtained industrially by the fractional distillation of liquid air. The atmosphere consists of 78% nitrogen by volume and 75% nitrogen by mass. These are more than 20 million tons of nitrogen over every square mile of the earth's surface. Nitrogen is a component of proteins and of the genetic material of all plants and animals.

Large volumes of atmospheric nitrogen are used for making ammonia (Section 8.9), the principal starting material used for preparation of large quantities of other nitrogen-containing compounds. Most other uses of elemental nitrogen are based on its inactivity. It is used when a chemical process requires an inert atmosphere. Canned foods and luncheon meats cannot oxidize in a pure nitrogen atmosphere, so they retain a better flavor and color and spoil less rapidly when sealed in nitrogen instead of air.

22.13   Phosphorus

Phosphorus is produced commercially by heating calcium phosphate, obtained from phosphate rock (Section 8.9), with sand and coke.

$$2Ca_3(PO_4)_2 + 6SiO_2 + 10C \xrightarrow{\Delta} 6CaSiO_3 + 10CO + P_4$$

The phosphorus distills out of the furnace and is condensed to a solid or burned to form P_4O_{10}, from which other phosphorus compounds are manufactured. Elemental phosphorus is shipped to plants where it is converted into phosphoric acid and phosphates. This stratagem makes it unnecessary to pay freight costs for transporting the oxygen and water used in the manufacture of phosphorus compounds.

Large quantities of phosphorus compounds are converted into acids and salts to be used in fertilizers and in the chemical industries. Other uses are in the manufacture of special alloys such as ferrophosphorus and phosphor bronze. Phosphorus is also used in making pesticides, matches, and some plastics.

22.14   Sulfur

Sulfur exists in nature as elemental deposits as well as sulfides of iron, zinc, lead, and copper and sulfates of sodium, calcium, barium, and magnesium. Hydrogen sulfide is often a component of natural gas and occurs in many volcanic gases (Fig. 22.20). Sulfur compounds are also found in coal. Sulfur is a constituent of many proteins and therefore exists in the combined state in living matter.

Free sulfur is mined by the **Frasch process** (Fig. 22.21) from enormous underground deposits in Texas and Louisiana. Superheated water (170°C and 10 atm pressure) is forced down the outermost of three concentric pipes to the underground deposit. When the hot water melts the sulfur, compressed air is forced down the innermost pipe. The liquid sulfur mixed with air forms a foam that flows up through the outlet pipe. The emulsified sulfur is conveyed to large settling vats, where it solidi-

Figure 22.20

Volcanoes are thought to be the source of two-thirds of the sulfur compounds in the atmosphere through their emission of H_2S. At the high temperatures involved, H_2S reacts with oxygen to produce SO_2 or sulfur.

Figure 22.21

Diagram of the Frasch process for mining sulfur.

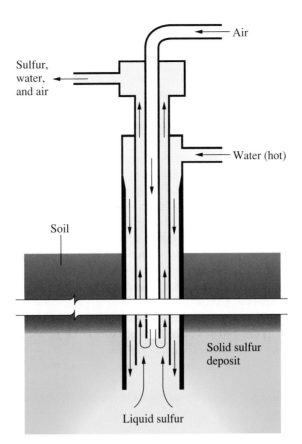

fies upon cooling. Sulfur produced by this method is 99.5 to 99.9% pure and requires no purification for most uses.

Sulfur is also obtained, in quantities exceeding that extracted by means of the Frasch process, from hydrogen sulfide recovered during the purification of natural gas.

22.15 The Halogens

The halogens are too reactive to occur free in nature, but their compounds are widely distributed. Chlorides are the most abundant, and although fluorides, bromides, and iodides are less common, they are reasonably available.

All of the halogens occur in sea water as halide ions. The concentration of the chloride ion is 0.54 M; that of the other halides is less than 10^{-4} M. Fluorine is also found in the minerals CaF_2, $Ca_5(PO_4)_3F$, and Na_3AlF_6, and in small amounts in teeth, bone, and the blood. Chlorine compounds are found in high concentrations in the Great Salt Lake and the Dead Sea and in extensive salt beds that contain $NaCl$, $MgCl_2$, or $CaCl_2$. Chlorine is a component of stomach acid, which is hydrochloric acid. Bromine compounds occur in the Dead Sea and in underground brines. Iodine compounds are found in small quantities in Chile saltpeter, in underground brines, and in sea kelp. Iodine is an essential component of the thyroid gland.

The best sources of the halogens (except iodine) are halide salts. The halide ions can be oxidized to free diatomic halogen molecules by various methods, depending on the ease of oxidation of the halide ion. This increases with increasing atomic size, in the order $F^- < Cl^- < Br^- < I^-$.

Fluorine is the most powerful oxidizing agent of the known elements; it spontaneously oxidizes most other elements. The reverse reaction, the oxidation of fluorides, is very difficult to accomplish; electrolytic oxidation is needed to prepare elemental fluorine. Electrolysis is often used to accomplish oxidation–reduction reactions with species that are oxidized or reduced with difficulty. The electrolysis is commonly performed in a molten mixture of potassium hydrogen fluoride, KHF_2, and anhydrous hydrogen fluoride (melting point 72°C). When electrolysis begins, HF is decomposed to form fluorine gas at the anode and hydrogen at the cathode. The two gases are kept separated to prevent their recombination to form hydrogen fluoride.

Most commercial **chlorine** is produced by electrolysis of the chloride ion in aqueous solutions of sodium chloride. Other products of the electrolysis are hydrogen and sodium hydroxide (see Section 8.9). Chlorine is also a product when metals such as sodium, calcium, and magnesium are produced by the electrolysis of their fused chlorides (see Section 21.23).

Chloride, bromide, and iodide ions are easier to oxidize than fluoride ions. Thus chlorine, bromine, and iodine also can be prepared by the chemical oxidation of the respective halides. Small quantities of chlorine are sometimes prepared by oxidation of the chloride ion in acid solution with strong oxidizing agents such as manganese dioxide (MnO_2), potassium permanganate ($KMnO_4$), or sodium dichromate ($Na_2Cr_2O_7$). The reaction with manganese dioxide is

$$MnO_2(s) + 2Cl^-(aq) + 4H_3O^+(aq) \longrightarrow Mn^{2+}(aq) + Cl_2(g) + 6H_2O(l)$$

Bromine is prepared commercially by the oxidation of bromide ion by chlorine.

$$2Br^-(aq) + Cl_2(g) \longrightarrow Br_2(l) + 2Cl^-(aq)$$

Chlorine is a stronger oxidizing agent than bromine, and the equilibrium for this reaction lies well to the right. Essentially all domestic bromine is produced by chlorine oxidation of bromide ions obtained from underground brines found in Arkansas.

Elemental **iodine** is sometimes produced by the oxidation of iodide ion with chlorine. An excess of chlorine must be avoided; it forms iodine monochloride, ICl, and iodic acid, HIO_3. Iodine is produced commercially by the reduction of sodium iodate, $NaIO_3$, an impurity in deposits of Chile saltpeter, with sodium hydrogen sulfite.

$$2IO_3^- + 5HSO_3^- \longrightarrow 3HSO_4^- + 2SO_4^{2-} + H_2O + I_2(s)$$

Fluorine gas has been used to fluorinate organic compounds (to replace hydrogen with fluorine) since its initial discovery by Moissan in 1886. The resulting fluorocarbon compounds are quite stable and nonflammable. Compounds of carbon, hydrogen, and fluorine are replacing Freons, compounds of carbon, chlorine, and fluorine, as refrigerants. Teflon is a polymer composed of $—CF_2CF_2—$ units. Perfluorodecalin, $C_{10}F_{18}$, is useful as a blood substitute, in part because oxygen is very soluble in this chemically inert substance. Fluorine gas is used in the production of uranium hexafluoride, UF_6, which is used in the separation of isotopes for the production of atomic energy. Fluoride ion is added to water supplies and to some toothpastes as SnF_2 or NaF to fight tooth decay.

Chlorine is used to bleach wood pulp and cotton cloth. The chlorine reacts with water to form hypochlorous acid, which oxidizes colored substances to colorless ones.

Large quantities of chlorine are used in chlorinating hydrocarbons (replacing hydrogen with chlorine) to produce compounds such as carbon tetrachloride (CCl_4), chloroform ($CHCl_3$), and ethyl chloride (C_2H_5Cl) and in the production of polyvinyl chloride (PVC) and other polymers. Chlorine is also used to kill the bacteria in community water supplies.

Bromine is used to produce certain dyes, light-sensitive silver bromide for photographic film, and sodium and potassium bromides for sedatives.

Iodine in alcohol solution with potassium iodide is used as an antiseptic (tincture of iodine). Iodide salts are essential for the proper functioning of the thyroid gland; an iodine deficiency may lead to the development of goiter. Iodized table salt contains 0.023% potassium iodide. Silver iodide is used in photographic film and in the seeding of clouds to induce rain. Iodoform, CHI_3, is an antiseptic.

22.16   The Noble Gases

All of the noble gases are present in the atmosphere in small amounts. Some natural gas contains 1–2% helium by mass. Helium is isolated from these natural gases by liquefying the condensable components, leaving only helium as a gas. The United States possesses most of the world's commercial supply of this element in its helium-bearing gas fields. Argon, neon, krypton, and xenon are produced by the fractional distillation of liquid air. Radon is collected from radium salts. More recently it has been observed that this radioactive gas is present in very small amounts in many different soils and minerals. Its accumulation in well-insulated, tightly sealed buildings constitutes a health hazard.

Helium is used for filling balloons and lighter-than-air craft; because it does not burn, it is safer to use than hydrogen. Helium at high pressures is not a narcotic; nitrogen is. Thus mixtures of oxygen and helium are used by divers working under high pressures in order to avoid the disoriented mental state known as nitrogen narcosis, the so-called rapture of the deep, that can result from breathing air. The tightness and rapid diffusion of helium decrease the muscular effort involved in breathing. Helium is used as an inert atmosphere for the melting and welding of easily oxidizable metals and for many chemical processes that are sensitive to air.

Liquid helium (boiling point, 4.2 K) is used to reach low temperatures for cryogenic research, and it is essential for achieving the low temperatures necessary to produce superconduction in traditional superconducting materials used in powerful magnets and other devices. Thus helium is required for magnetic resonance imaging (MRI), a common medical diagnostic procedure.

Neon is used in neon lamps and signs. When an electric spark is passed through a tube containing neon at low pressure, the familiar red glow of neon is emitted. The color of the light given off by a neon tube can be changed by mixing argon or mercury vapor with the neon and by utilizing tubes made of glasses of special color.

Argon is used in gas-filled electric light bulbs, where its lower heat conductivity and chemical inertness make it preferable to nitrogen for inhibiting vaporization of the tungsten filament and prolonging the life of the bulb. Fluorescent tubes commonly contain a mixture of argon and mercury vapor. Many Geiger-counter tubes are filled with argon.

Krypton–xenon flash tubes are used for taking high-speed photographs. An electric discharge through such a tube gives a very intense light that lasts only 1/50,000 of a second.

Properties of the Nonmetals

22.17   Hydrogen

An uncombined hydrogen atom consists of one proton in the nucleus and one valence electron in the $1s$ orbital. The $n = 1$ valence shell has a capacity of two electrons, and hydrogen can rightfully occupy two locations in the periodic table. It can be considered a Group 1A element because it has only one valence electron, and it can be considered a Group 7A element because it needs only one additional electron to fill its valence orbital. Thus hydrogen can lose an electron to form the proton, H^+; it can gain an electron to form a hydride ion, H^-; or it can share an electron to form a single, covalent bond. In reality, hydrogen is a unique element that almost deserves its own location in the periodic table.

Under normal conditions hydrogen is relatively inactive chemically, but when heated, it enters into many chemical reactions.

1. **Reaction with Elements.** When heated, hydrogen reacts with the metals of Group 1A and with Ca, Sr, and Ba (the more active metals in Group 2A). The compounds formed are crystalline **ionic hydrides** that contain the hydride anion (H^-), a strong reducing agent and a strong base, which reacts vigorously with water and other acids to form hydrogen gas (Fig. 22.22).

    The reactions of hydrogen with the nonmetals generally produce *acidic* hydrogen compounds with hydrogen in the $+1$ oxidation state. The reactions become more exothermic and vigorous as the electronegativity of the nonmetal increases. Hydrogen reacts with nitrogen and sulfur only when heated, but it reacts explosively with fluorine (producing gaseous HF) and, under some conditions, with chlorine (producing gaseous HCl). A mixture of hydrogen and oxygen explodes if ignited. Because of the explosive nature of the reaction, caution must be exercised in handling hydrogen (or any other combustible gas) in order to avoid the

Figure 22.22

Calcium hydride reacts rapidly with water, producing $H_2(g)$ and $Ca(OH)_2$.

Table 22.2

Chemical Reactions of Hydrogen with Other Elements

General Equation	Comments
$\frac{n}{2}H_2 + M \longrightarrow MH_n$	Ionic hydrides with Group 1A and Ca, Sr, Ba; metallic hydrides with transition metals
$H_2 + C \longrightarrow$ (no reaction)	
$3H_2 + N_2 \longrightarrow 2NH_3$	Requires high pressure and temperature; low yield
$2H_2 + O_2 \longrightarrow 2H_2O$	Exothermic and potentially explosive
$H_2 + S \longrightarrow H_2S$	Requires heating; low yield
$H_2 + X_2 \longrightarrow 2HX$	X = F, Cl, Br, I; explosive with F_2; low yield with I_2

formation of an explosive mixture in a confined space. Although most hydrides of the nonmetals are acidic (see Section 15.5), ammonia and phosphine (PH_3) are very, very weak acids and generally function as bases. The reactions of hydrogen with the elements are summarized in Table 22.2.

2. **Reaction with Compounds.** Hydrogen reduces the heated oxides of many metals, with the formation of the metal and water (Fig. 22.23). For example, when hydrogen is passed over heated CuO, copper and water are formed.

$$H_2(g) + CuO(s) \xrightarrow{\Delta} Cu(s) + H_2O(g)$$

Hydrogen may also reduce the metal ions in some metal oxides to lower oxidation states.

$$H_2(g) + MnO_2(s) \xrightarrow{\Delta} MnO(s) + H_2O(g)$$

22.18 Oxygen

Oxygen is a colorless, odorless, and tasteless gas at ordinary temperatures. It is slightly more dense than air. Although it is only slightly soluble in water (49 mL of gas dissolves in 1 L at STP), oxygen's solubility is very important to aquatic life.

Elemental oxygen is a strong oxidizing agent. It reacts with most other elements and with many compounds.

1. **Reaction with Elements.** Oxygen reacts directly at room temperature or at elevated temperatures with all other elements (Fig. 22.10) except the noble gases, the halogens, and a few second- and third-row transition metals of low reactivity [those below copper in the activity series, the oxides of which decompose upon heating (see Section 8.7)]. The more active metals form peroxides or superoxides (see Section 21.3). Less active metals and the nonmetals give oxides. Oxides of the halogens, of at least one of the noble gases, and of the metals at the bottom of the activity series can be prepared, but not by the direct action of the elements with oxygen. The reactions of oxygen with many of the elements are summarized in Table 22.3 on page 836.

2. **Reaction with Compounds.** Elemental oxygen also reacts with some compounds. A compound made up of elements that combine with oxygen when free may be expected to react with oxygen to form oxides of the constituent elements, when

Figure 22.23

Copper forms when CuO is heated in an atmosphere of H_2.

Table 22.3

Chemical Properties of Elemental Oxygen

General Equation	Comments
Reactions with Elements	
$nM + \dfrac{m}{2}O_2 \longrightarrow M_nO_m$	Oxygen reacts with most metals, M, except those at the bottom of the activity series (Section 8.7)
$2Na + O_2 \longrightarrow Na_2O_2$	A peroxide forms with Na (Section 21.3)
$M + O_2 \longrightarrow MO_2$	Peroxides form with M = Ca, Sr, Ba (Section 21.3)
$M + O_2 \longrightarrow MO_2$	M = K, Rb, Cs form superoxides (Section 21.3)
$2H_2 + O_2 \longrightarrow 2H_2O$	Potentially explosive reaction
$2C + O_2 \longrightarrow 2CO$	With a stoichiometric amount of O_2
$E + O_2 \longrightarrow EO_2$	With heating in excess O_2; E = lighter members of Group 4A: C, Si, Ge, Sn
$2Pb + O_2 \longrightarrow 2PbO$	Pb is the heaviest member of Group 4A
$N_2 + O_2 \longrightarrow 2NO$	High temperature required; low yield of product
$E_4 + 3O_2 \longrightarrow E_4O_6$	Requires a stoichiometric amount of O_2; E = P, As, Sb
$P_4 + 5O_2 \longrightarrow P_4O_{10}$	Phosphorus burns in air
$E + O_2 \longrightarrow EO_2$	E = heavier members of Group 6A: S, Se, Te
Reactions with Compounds	
$2CO + O_2 \longrightarrow 2CO_2$	Requires ignition
$2NO + O_2 \longrightarrow 2NO_2$	Spontaneous reaction
$P_4O_6 + 2O_2 \longrightarrow P_4O_{10}$	
$2SO_2 + O_2 \longrightarrow 2SO_3$	Requires heat and Pt catalyst; occurs very slowly in the atmosphere
$C_nH_m + O_2 \longrightarrow CO_2 + H_2O$ (For example, $CH_4 + 2O_2 \longrightarrow CO_2 + 2H_2O$)	This is the general unbalanced reaction for the combustion of a hydrocarbon
$2H_2S + 3O_2 \longrightarrow 2SO_2 + 2H_2O$	
$CS_2 + 3O_2 \longrightarrow CO_2 + 2SO_2$	

one or more of the atoms in the compound do not exhibit their maximum oxidation state. For example, hydrogen sulfide, H_2S, contains sulfur with an oxidation state of -2. Because the sulfur does not exhibit its maximum oxidation state, and because free sulfur reacts with oxygen, we would expect H_2S to react with oxygen. It does, giving water and sulfur dioxide. Oxides such as CO and P_4O_6 that contain an element with a lower oxidation state than is usually formed when the element combines with an excess of oxygen also react with additional oxygen. Examples are given in Table 22.3.

The ease with which elemental oxygen picks up electrons is mirrored by the difficulty of removing electrons from oxygen in most oxides. Of the elements, only the very reactive fluorine molecule oxidizes oxides to form oxygen gas.

The oxygen allotrope ozone is pale blue as a gas and deep blue as a liquid. It has a sharp, irritating odor, produces headaches, and is poisonous. Energy is absorbed when ozone is formed from molecular oxygen, so ozone is more active chemically than oxygen and decomposes readily into oxygen by the exothermic reaction

$$2O_3 \longrightarrow 3O_2 \qquad \Delta H° = -286 \text{ kJ}$$

As a consequence, it is less stable and more reactive than oxygen. Ozone is a powerful oxidizing agent and forms oxides with many elements at temperatures at which O_2 will not react.

22.19   Nitrogen

Under ordinary conditions, nitrogen is a colorless, odorless, and tasteless gas. It boils at $-195.8°C$ (77.4 K) and freezes at $-210.0°C$ (63 K). It is slightly less dense than air because air contains the heavier molecules of oxygen and argon as well as molecules of nitrogen. Nitrogen is very unreactive. The only common reactions of N_2 at room temperature occur with lithium to give Li_3N, with certain transition metal complexes, and with hydrogen or oxygen in nitrogen-fixing bacteria. Nitrogen forms nitrides upon heating with active metals and gives low yields of ammonia upon heating with hydrogen. Heating with oxygen followed by rapid cooling (quenching) produces nitric oxide, NO. The general unreactivity of nitrogen makes the remarkable ability of some bacteria to synthesize nitrogen compounds, using atmospheric nitrogen gas as the source, one of the exciting chemical events on our planet.

Compounds with nitrogen in all of its oxidation states from -3 to $+5$ are known. Much of the chemistry of nitrogen involves oxidation–reduction reactions. Some of the reactions of nitrogen and its compounds are shown in Table 22.4 on page 838.

Figure 22.24

The molecular structure of P_4S_3 (top), which is used in the heads of "strike-anywhere" matches (bottom).

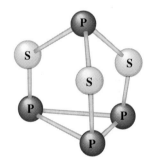

22.20   Phosphorus

Phosphorus is an active nonmetal. In compounds, phosphorus is most commonly observed with oxidation states of -3, $+3$, and $+5$. Phosphorus exhibits oxidation numbers that are unusual for a Group 5A element in compounds that contain phosphorus–phosphorus bonds; examples include diphosphorus tetrahydride, H_2P—PH_2, and tetraphosphorus trisulfide, P_4S_3 (Fig. 22.24).

The most important chemical property of elemental phosphorus is its reactivity with oxygen. Slow oxidation of white phosphorus causes it to get warm, and it spontaneously inflames when it reaches 35–45°C. Because of this, white phosphorus must be stored under water (see Fig. 22.11). Red phosphorus is less active than the white form; it does not ignite in air unless it is heated to about 250°C. However, the products of the reactions of red phosphorus are the same as those of the white form.

Phosphorus is one of the least electronegative of the nonmetals. It reacts with active metals, forming salts that contain the very basic **phosphide ion,** P^{3-}. With transition metals it forms phosphides that are not ionic. Phosphorus is oxidized by the nonmetals oxygen, sulfur, and the halogens. Some of the reactions of phosphorus and its compounds are summarized in Table 22.5 on page 839.

Table 22.4

Chemical Properties of Nitrogen and
Its Compounds

General Equation	Comments
Reactions with Elements	
$N_2 + 6Li \longrightarrow 2Li_3N$	N^{3-} forms with active metals
$N_2 + 3Mg \overset{\Delta}{\longrightarrow} Mg_3N_2$	
$N_2 + 3H_2 \rightleftharpoons 2NH_3$	Reversible, slow reaction; low yield at elevated temperatures
$N_2 + O_2 \overset{\Delta}{\rightleftharpoons} 2NO$	Low-yield; endothermic reaction; requires heating
Reactions of Compounds	
$NH_3 + H^+ \longrightarrow NH_4^+$	NH_3 acts both as a weak base and as a very weak acid
$NH_3 + LiCH_3 \longrightarrow LiNH_2 + CH_4$	
$NH_4^+ + OH^- \longrightarrow NH_3 + H_2O$	NH_4^+ is a weak acid
$4NH_3 + 5O_2 \overset{\Delta}{\longrightarrow} 4NO + 6H_2O$	
$NH_3 + OCl^- \longrightarrow NH_2Cl + OH^-$	
$NH_2Cl + NH_3 + OH^- \longrightarrow N_2H_4 + Cl^- + H_2O$	
$Cu + HNO_3 \longrightarrow NO_2$ or NO, $Cu(NO_3)_2$, and H_2O	Concentrated HNO_3 gives NO_2; dilute HNO_3 gives NO
$2NO + O_2 \longrightarrow 2NO_2$	
$NO + NO_2 \rightleftharpoons N_2O_3$	
$2NO_2 \rightleftharpoons N_2O_4$	
$2NO_2 + H_2O \longrightarrow HNO_3 + HNO_2$	NO_2 disproportionates in water
$3NO_2 + H_2O \longrightarrow 2HNO_3 + NO$	
$4HNO_3 + P_4O_{10} \longrightarrow 2N_2O_5 + 4HPO_3$	

22.21   Sulfur

Sulfur exists in several allotropic forms (see Section 22.8). The stable form at room temperature contains eight-membered rings and should technically be written as S_8. However, chemists commonly use the symbol S to simplify the coefficients in chemical equations; we will follow this practice in the rest of this book.

Like oxygen, which is also a member of Group 6A, sulfur exhibits distinctly nonmetallic behavior. It oxidizes metals, giving a variety of binary sulfides in which sulfur exhibits a negative oxidation state. Elemental sulfur oxidizes less electronegative nonmetals and is oxidized by more electronegative nonmetals, such as oxygen and the halogens. Sulfur is also oxidized by other strong oxidizing agents. For example, concentrated nitric acid oxidizes sulfur to the sulfate ion, with the concurrent formation of nitrogen(IV) oxide.

$$S + 6HNO_3 \longrightarrow 2H_3O^+ + SO_4^{2-} + 6NO_2$$

The chemistry of sulfur with an oxidation state of -2 is similar to that of oxygen. Unlike oxygen, however, sulfur forms a variety of compounds in which it exhibits positive oxidation states. Some of the general reactions of sulfur and its compounds are summarized in Table 22.6 on page 840.

Table 22.5

Chemical Properties of Phosphorus and Its Compounds

General Equation	Comments
Reactions of Phosphorus	
$P_4 + 12Na \longrightarrow 4Na_3P$	Active metals reduce P to P^{3-}
$P_4 + 6Mg \longrightarrow 2Mg_3P_2$	
$P_4 + 3O_2 \longrightarrow P_4O_6$	In about 50% yield; requires a stoichiometric amount of O_2
$P_4 + 5O_2 \longrightarrow P_4O_{10}$	Very exothermic reaction
$P_4 + S \longrightarrow P_4S_{10}, P_4S_3,$ and others	A variety of phosphorus sulfides form, depending on the stoichiometry
$P_4 + 6X_2 \longrightarrow 4PX_3$	X = F, Cl, Br, I
$P_4 + 10X_2 \longrightarrow 4PX_5$	X = F, Cl, Br
$P_4 + 3NaOH + 3H_2O \longrightarrow$ $3NaH_2PO_2 + PH_3$	P disproportionates in base
Reactions of Compounds	
$P^{3-} + 3H_3O^+ \longrightarrow PH_3 + 3H_2O$	
$PX_3 + 3H_2O \longrightarrow H_3PO_3 + 3HX$	Forms phosphorous acid; X = halogen
$PX_5 + 4H_2O \longrightarrow H_3PO_4 + 5HX$	Forms phosphoric acid; X = halogen
$P_4O_6 + 6H_2O \longrightarrow 4H_3PO_3$ $P_4O_{10} + 6H_2O \longrightarrow 4H_3PO_4$	The oxides of phosphorus are acidic
$H_3PO_4 + OH^- \longrightarrow$ $H_2PO_4^-, HPO_4^{2-},$ or PO_4^{3-}	Product depends on stoichiometry

22.22 The Halogens

Fluorine is a pale yellow gas, chlorine is a greenish-yellow gas, bromine is a deep reddish-brown liquid three times as dense as water, and iodine is a grayish-black crystalline solid with a low melting point (Fig. 22.25). Liquid bromine has a high vapor pressure, and the reddish vapor can easily be seen in Fig. 22.25. Iodine crystals have

Figure 22.25

Chlorine is a pale yellow-green gas, gaseous bromine is deep orange, and gaseous iodine is purple. (Fluorine is so reactive that it is too dangerous for the photographer to handle.)

Table 22.6

Chemical Properties of Sulfur and Its Compounds

General Equation	Comments
Reactions with Elements	
$nM + mS \longrightarrow M_nS_m$	Most metals combine with sulfur, giving sulfides that contain the S^{2-} ion
$H_2 + S \longrightarrow H_2S$	In low yield; H_2S decomposes at elevated temperatures
$E + 2S \longrightarrow ES_2$	E = C, Si, Ge, the lighter members of Group 4A
$E + S \longrightarrow ES$	E = Sn, Pb, the heavier members of Group 4A
$P_4 + 10S \longrightarrow P_4S_{10}$	With less S, many other phosphorus sulfides are possible
$2E + 3S \longrightarrow E_2S_3$	E = As, Sb, Bi, the heavier members of Group 5A
$S + O_2 \longrightarrow SO_2$	Traces of SO_3 also form when S burns in air
$S + 3F_2 \longrightarrow SF_6$	Other sulfur fluorides can be prepared indirectly
$S + nCl_2 \longrightarrow S_2Cl_2$ or SCl_2	Product depends on the reaction stoichiometry
$2S + Br_2 \longrightarrow S_2Br_2$	I_2 does not react with S
Reactions of Compounds	
$S^{2-} + 2H^+ \longrightarrow H_2S$	S^{2-} is a strongly basic anion
$S^{2-} + \text{oxidant} \longrightarrow S$	S^{2-} can be oxidized to S or to higher oxidation states
$CS_2 + 3O_2 \longrightarrow CO_2 + 2SO_2$	
$2SO_2 + O_2 \longrightarrow 2SO_3$	Catalyst required for satisfactory rate
$SO_2 + H_2O \longrightarrow H_2SO_3$	Sulfurous acid
$SO_2 + O^{2-} \longrightarrow SO_3^{2-}$	Sulfite ion
$SO_3 + H_2O \longrightarrow H_2SO_4$	Sulfuric acid
$SO_3 + O^{2-} \longrightarrow SO_4^{2-}$	Sulfate ion

a noticeable vapor pressure. When gently heated, these crystals sublime and form a beautiful deep violet vapor.

Bromine is only slightly soluble in water, but it is miscible in all proportions in less polar (or nonpolar) solvents such as alcohol, ether, chloroform, carbon tetrachloride, and carbon disulfide, forming solutions that vary from yellow to reddish-brown, depending on the concentration.

Iodine is soluble in chloroform, carbon tetrachloride, carbon disulfide, and many hydrocarbons, giving violet solutions of I_2 molecules. The solutions have the same color as I_2 molecules in the gas phase. Iodine dissolves only slightly in water, giving brown solutions. It is quite soluble in alcohol, ether, and aqueous solutions of iodides, with which it also forms brown solutions. These brown solutions result because iodine molecules have empty valence d orbitals and can act as weak Lewis acids (see Section 15.13). They can form Lewis acid–base complexes with solvent molecules that can function as Lewis bases or with the iodide ion, which can also act as a Lewis base. The equation for the reversible reaction of iodine with the iodide ion to give the triiodide ion, I_3^-, is

$$:\!\ddot{I}\!:^- + :\!\ddot{I}\!-\!\ddot{I}\!: \rightleftharpoons \left[:\!\ddot{I}\!-\!\ddot{I}\!-\!\ddot{I}\!:\right]^-$$

Table 22.7
Chemical Properties of the Elemental Halogens

General Equation	Comments
Reactions with Elements	
$2M + nX_2 \longrightarrow 2MX_n$	With almost all metals
$H_2 + X_2 \longrightarrow 2HX$	With decreasing reactivity in the order $F_2 > Cl_2 > Br_2 > I_2$
$Xe + \dfrac{n}{2}F_2 \longrightarrow XeF_n$	$n = 2, 4,$ or 6
$nX_2 + X'_2 \longrightarrow 2X'X_n$	X' heavier than X; n an odd integer
$S + 3F_2 \longrightarrow SF_6$	Se and Te can replace S
$S + Cl_2 \longrightarrow SCl_2$	
$2S + X_2 \longrightarrow S_2X_2$	With Cl_2 or Br_2
$2P + 3X_2 \longrightarrow 2PX_3$	With excess P; also with As, Sb, or Bi
$2P + 5X_2 \longrightarrow 2PX_5$	With excess halogen, except I_2
Reactions with Compounds	
$X_2 + 2X'^- \longrightarrow 2X^- + X'_2$	X' heavier than X
$X_2 + 2H_2O \longrightarrow H_3O^+ + X^- + HOX$	Not with F_2
$X_2 + CO \longrightarrow COX_2$	With Cl_2 or Br_2
$X_2 + SO_2 \longrightarrow SO_2X_2$	With F_2 or Cl_2
$X_2 + PX_3 \longrightarrow PX_5$	Not with I_2
$-\overset{\mid}{\underset{\mid}{C}}-H + X_2 \longrightarrow -\overset{\mid}{\underset{\mid}{C}}-X + HX$	With F_2, Cl_2 or Br_2 and many hydrocarbons
$\overset{}{\underset{}{>}}C{=}C\overset{}{\underset{}{<} } + X_2 \longrightarrow -\overset{\overset{X}{\mid}}{\underset{\mid}{C}}-\overset{\overset{X}{\mid}}{\underset{\mid}{C}}-$	With Cl_2, Br_2, or I_2 and many hydrocarbons containing C=C double bonds

The elemental (free) halogens are oxidizing agents with strengths decreasing in the order $F_2 > Cl_2 > Br_2 > I_2$. The easier it is to oxidize the halide ion, the more difficult it is for the halogen to act as an oxidizing agent. Fluorine generally oxidizes an element to its highest oxidation state, whereas the heavier halogens may not. For example, when fluorine reacts with sulfur, SF_6 is formed. Chlorine gives SCl_2 and bromine, S_2Br_2. Iodine does not react with sulfur.

The reactions of elemental halogens with a variety of substances are summarized in Table 22.7.

Fluorine reacts directly and forms binary fluorides with all of the elements except the lighter noble gases (He, Ne, and Ar). Fluorine is such a strong oxidizing agent that many substances ignite on contact with it. Drops of water inflame in fluorine and form O_2, OF_2, H_2O_2, O_3, and HF. Wood and asbestos ignite and burn in fluorine gas. Most hot metals burn vigorously in fluorine. However, fluorine can be handled in copper, iron, magnesium, or nickel containers, because an adherent film of the fluoride salt protects their surfaces from further attack.

Fluorine readily displaces chlorine and the other halogens from solid metal halides; in an excess of fluorine, halogen fluorides are formed. Fluorine and hydrogen react explosively. Fluorine is the only element that reacts directly with the noble gas xenon.

Figure 22.26

Molten sodium (mp 97.8°C) inflames in
an atmosphere of chlorine.

Although it is a strong oxidizing agent, chlorine is less active than fluorine. For example, fluorine and hydrogen react explosively, but when chlorine and hydrogen are mixed in the dark, the reaction between them is so slow as to be imperceptible. When the mixture is exposed to light, the reaction is explosive. Chlorine is less active toward metals than fluorine, and oxidation reactions usually require higher temperatures. Molten sodium ignites in chlorine (Fig. 22.26). Chlorine attacks most nonmetals (C, N_2, and O_2 are notable exceptions), forming covalent molecular compounds. Chlorine generally reacts with compounds that contain only carbon and hydrogen (hydrocarbons) by adding to multiple bonds or by substitution.

When chlorine is added to water, it is both oxidized and reduced in a **disproportionation** reaction (Section 22.7).

$$Cl_2 + 2H_2O \rightleftharpoons HOCl + H_3O^+ + Cl^-$$

Half the chlorine atoms oxidize to the $+1$ oxidation state (in hypochlorous acid), and the other half reduce to the -1 oxidation state (in chloride ion). This disproportionation is incomplete, so chlorine water is an equilibrium mixture of chlorine molecules, hypochlorous acid molecules, hydronium ions, and chloride ions. When exposed to light, this solution undergoes a photochemical decomposition.

$$2HOCl + 2H_2O \xrightarrow{\text{Sunlight}} 2H_3O^+ + 2Cl^- + O_2(g)$$

The chemical properties of bromine are similar to those of chlorine, although bromine is the weaker oxidizing agent and its reactivity is less than that of chlorine.

Iodine is the least reactive of the four naturally occurring halogens. It is the weakest oxidizing agent, and the iodide ion is the most easily oxidized halide ion. Iodine reacts with metals, but heating is often required. It does not oxidize other halide ions.

Compared with the other halogens, iodine reacts only slightly with water. Traces of iodine in water react with a mixture of starch and iodide ion, forming a deep blue color. This reaction is used as a very sensitive test for the presence of iodine in water.

22.23   The Noble Gases

The boiling points and melting points of the noble gases are extremely low compared to those of other substances of comparable atomic or molecular masses. This is because no strong chemical bonds hold the atoms together in the liquid or solid states. Only weak London dispersion forces (see Section 11.3) are present, and these forces can hold the atoms together only when molecular motion is very slight, as it is at very low temperatures. Helium is the only substance known that does not solidify on cooling at normal pressure. It remains liquid close to absolute zero (0.001 K) at ordinary pressures, but it solidifies under elevated pressure.

Stable compounds of xenon form when xenon reacts with fluorine. **Xenon difluoride,** XeF_2, is prepared by heating an excess of xenon gas with fluorine gas at 400°C and cooling. The material forms colorless crystals, which are stable at room temperature in a dry atmosphere. **Xenon tetrafluoride,** XeF_4 (Fig. 22.27), and **xenon hexafluoride,** XeF_6, are prepared in an analogous manner, with a stoichiometric amount of fluorine and an excess of fluorine, respectively. Compounds with oxygen are prepared by replacing fluorine atoms in the xenon fluorides with oxygen.

Xenon compounds are very strong oxidizing agents. They may disproportionate in water, and they react with strong Lewis acids by donating a fluoride ion. The reaction with AsF_5 is

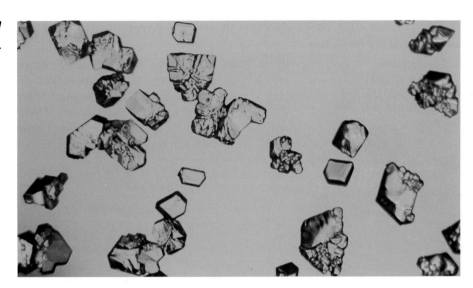

Figure 22.27
Crystals of xenon tetrafluoride, XeF_4.

$$XeF_2(s) + AsF_5(s) \longrightarrow [XeF]^+[AsF_6]^-$$

Dry, solid **xenon trioxide,** XeO_3, is extremely explosive and must be handled with care. When XeF_6 reacts with water, a solution of XeO_3 results and the $+6$ oxidation state for the xenon is retained.

$$XeF_6(s) + 3H_2O(l) \longrightarrow XeO_3(aq) + 6HF(aq)$$

Both XeF_6 and XeO_3 disproportionate in basic solution, producing xenon, oxygen, and salts of the **perxenate ion,** XeO_6^{4-}, in which xenon reaches its maximum oxidation state of $+8$.

Xenon difluoride reacts with acids such as HSO_3F, F_5TeOH, and $HClO_4$, which are resistant to oxidation, with evolution of hydrogen fluoride and formation of one or two $Xe-O$ bonds, depending on the stoichiometry of the reaction. Compounds formed this way include $FXeOSO_2F$, $FXeOTeF_5$, and $Xe(OTeF_5)_2$.

Although most of the chemistry of xenon involves $Xe-F$ and $Xe-O$ bonds, chemists have also been successful in preparing compounds containing $Xe-N$ and $Xe-C$ bonds. Compounds of this type that have been reported include $FXeN(SO_2F)_2$ and $Xe(CF_3)_2$.

Krypton forms a difluoride, KrF_2, which is thermally unstable at room temperature. Radon apparently forms RnF_2, but the evidence of the compound is based on radiochemical tracer techniques. Stable compounds of helium, neon, and argon are not known.

Compounds of Selected Nonmetals

As we saw earlier, the nonmetals exhibit widely differing chemical behaviors. However, there are common threads running through their chemistry. For example, the nonmetals form covalent hydrogen compounds and halides, as well as covalent, acidic oxides.

Figure 22.28
Structure of an ammonia molecule.

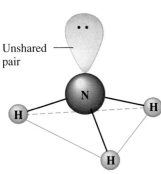

Unshared pair

22.24   Hydrogen Compounds

Of Nitrogen.   **Ammonia,** NH_3, is produced in nature when any nitrogen-containing organic material decomposes in the absence of air. Its odor is common in decaying organic matter. Ammonia is usually prepared in the laboratory by the reaction of an ammonium salt with a strong base such as sodium hydroxide. The acid–base reaction with the weakly acidic ammonium ion gives ammonia (Fig. 22.28). Ammonia also forms when ionic nitrides react with water. The nitride ion is a much stronger base than the hydroxide ion.

$$Mg_3N_2(s) + 6H_2O(l) \longrightarrow 3Mg(OH)_2(s) + 2NH_3(g)$$

Ammonia is produced commercially by the direct combination of the elements in the **Haber process.**

$$N_2(g) + 3H_2(g) \xrightleftharpoons{\text{Catalyst}} 2NH_3(g) \qquad \Delta H^\circ = -92 \text{ kJ}$$

See Section 14.4 for a discussion of this reaction.

Ammonia is a colorless gas with a sharp, pungent odor. Smelling salts utilize this powerful odor. Gaseous ammonia is readily liquefied, giving a colorless liquid that boils at $-33°C$. Liquid ammonia has a vapor pressure of only about 10 atmospheres at 25°C; it is readily handled in steel cylinders. Because of intermolecular hydrogen bonding (see Section 11.4), the enthalpy of vaporization of liquid ammonia is higher than that of any other liquid except water, so ammonia is used as a refrigerant. Ammonia is quite soluble in water (1180 L at STP dissolves in 1 L of H_2O).

The chemical properties of ammonia are as follows:

1. Ammonia acts as a Brønsted base, because it readily accepts protons, and as a Lewis base, in that it can be an electron-pair donor (Section 15.13). When ammonia dissolves in water, only about 1% reacts to form ammonium and hydroxide ions; the remainder is present as unreacted NH_3 molecules. Although it is a weak base, ammonia readily accepts protons from acids and hydronium ions, forming salts of the **ammonium ion,** NH_4^+. The ammonium ion is similar in size to the potassium ion, and ionic compounds of the two ions exhibit many similarities in their structures and solubilities.

    Ammonia forms **amines** by sharing electrons with metal ions, which act as Lewis acids. The diammine silver ion forms by the reaction

    $$Ag^+ + 2NH_3 \longrightarrow [H_3N-Ag-NH_3]^+$$

    In these species, ammonia functions as a Lewis base.

2. Ammonia can display acidic behavior, although it is a much weaker acid than water. It reacts with very strong bases, such as the CH_3^- ion. Like other acids, ammonia reacts with metals, although it is so weak that high temperatures are often required. Hydrogen and (depending on the stoichiometry) **amides** (salts of NH_2^-), **imides** (salts of NH^{2-}), or **nitrides** (salts of N^{3-}) are formed.

3. The nitrogen atom in ammonia has its lowest possible oxidation state (-3) and thus is not susceptible to reduction. However, it can be oxidized. Ammonia burns in air, giving NO and water. Hot ammonia and the ammonium ion are active reducing agents. Of particular interest are the oxidations of ammonium ion by nitrite ion, NO_2^-, to give pure nitrogen and by nitrate ion to give nitrous oxide, N_2O.

There are a number of compounds that can be considered derivatives of ammonia in that one or more of the hydrogen atoms in the ammonia molecule have been replaced by some other atom or group of atoms, bonded to the nitrogen atom by a single bond. Inorganic derivations include chloramine, NH_2Cl, and hydrazine, N_2H_4.

Ammonia   Chloramine   Hydrazine

Chloramine, NH_2Cl, results from the reaction of sodium hypochlorite, $NaOCl$, with ammonia in basic solution. In the presence of a large excess of ammonia at low temperature, the chloramine reacts further to produce **hydrazine,** N_2H_4.

$$NH_3(aq) + OCl^-(aq) \longrightarrow NH_2Cl(aq) + OH^-(aq)$$

$$NH_2Cl(aq) + NH_3(aq) + OH^-(aq) \longrightarrow N_2H_4(aq) + Cl^-(aq) + H_2O(l)$$

Anhydrous hydrazine is relatively stable, both thermally and kinetically, despite its positive enthalpy and free energy of formation.

$$N_2(g) + 2H_2(g) \longrightarrow N_2H_4(l) \qquad \Delta H_f^\circ = 50.6 \text{ kJ mol}^{-1}; \ \Delta G_f^\circ = 149.2 \text{ kJ mol}^{-1}$$

It is a fuming, colorless liquid that has some physical properties remarkably similar to those of H_2O (it melts at 2°C, boils at 113.5°C, and has a density at 25°C of 1.00 g/mL). It burns rapidly and completely in air with substantial evolution of heat.

$$N_2H_4(l) + O_2(g) \longrightarrow N_2(g) + 2H_2O(l) \qquad \Delta H^\circ = -621.5 \text{ kJ mol}^{-1}$$

Like ammonia, hydrazine is both a Brønsted base and a Lewis base, although it is weaker than ammonia. It reacts with strong acids and forms two series of salts that contain the $N_2H_5^+$ and $N_2H_6^{2+}$ ions, respectively.

Of Phosphorus.

The most important hydride of phosphorus is **phosphine,** PH_3, a gaseous analog of ammonia in formula and structure. Unlike ammonia, phosphine cannot be made by the direct union of the elements. It is prepared by the reaction of an ionic phosphide with acid or is one product of the disproportionation of white phosphorus in a hot concentrated solution of sodium hydroxide.

$$AlP + 3H_3O^+ \longrightarrow PH_3 + Al^{3+}$$

$$P_4 + 4OH^- + 2H_2O \longrightarrow 2HPO_3^{2-} + 2PH_3$$

The disproportionation reaction produces phosphine and sodium hydrogen phosphite, Na_2HPO_3.

Phosphine is a colorless, very poisonous gas, which has an odor like that of decaying fish. It is easily decomposed by heat ($4PH_3 \longrightarrow P_4 + 6H_2$) and burns in air. Like ammonia, gaseous phosphine unites with gaseous hydrogen halides, forming the phosphonium compounds, PH_4Cl, PH_4Br, and PH_4I. However, phosphine is a much weaker base than ammonia; these compound decompose in water, and the insoluble PH_3 escapes from solution.

Of Sulfur.

Hydrogen sulfide, H_2S, is a colorless gas that is responsible for the offensive odor of rotten eggs and of many hot springs. Hydrogen sulfide is as toxic as hydrogen cyanide. Great care must be exercised in handling it. Hydrogen sulfide is particularly deceptive because it paralyzes the olfactory nerves; after a short exposure, one does not smell it!

Figure 22.29

The reaction of FeS with H_2SO_4 produces gaseous H_2S.

The production of hydrogen sulfide by the direct reaction of the elements (H_2 + S) is unsatisfactory, because the reaction is reversible and hydrogen sulfide decomposes upon heating. A more effective preparation method is the reaction of a metal sulfide with a dilute strong acid (Fig. 22.29). For example,

$$FeS(s) + 2H_3O^+(aq) \longrightarrow Fe^{2+}(aq) + H_2S(g) + 2H_2O(l)$$

The sulfur in metal sulfides and in hydrogen sulfide is readily oxidized, making metal sulfides and H_2S good reducing agents (Fig. 22.30). In acidic solutions, hydrogen sulfide reduces Fe^{3+} to Fe^{2+}, Br_2 to Br^-, MnO_4^- to Mn^{2+}, $Cr_2O_7^{2-}$ to Cr^{3+}, and HNO_3 to NO_2. The sulfur in the H_2S is usually oxidized to elemental sulfur, unless a large excess of the oxidizing agent is present. In this case the sulfide may be oxidized to SO_3^{2-} or SO_4^{2-} (or to SO_2 or SO_3 in the absence of water).

Hydrogen sulfide burns in air, forming water and sulfur dioxide. When heated with a limited supply of air or with sulfur dioxide (which can be produced by the combustion of H_2S), elemental sulfur is formed.

$$2H_2S(g) + O_2(g) \longrightarrow 2S(s) + 2H_2O(l)$$

In this way sulfur is recovered from the hydrogen sulfide found in many sources of natural gas. The deposits of sulfur in volcanic regions may be the result of the oxidation of H_2S because it is a constituent of volcanic gases.

Hydrogen sulfide is a weak diprotic acid. An aqueous solution of hydrogen sulfide is called **hydrosulfuric acid.** The acid ionizes in two stages, giving hydrogen sulfide ions, HS^-, in the first stage and sulfide ions, S^{2-}, in the second (see Section 15.8). Salts of both the HS^- and S^{2-} ions are known. Because hydrogen sulfide is a weak acid, the sulfide ion and the hydrogen sulfide ion are strong bases. Aqueous solutions of soluble sulfides and hydrogen sulfides are basic.

$$S^{2-} + H_2O \rightleftharpoons HS^- + OH^-$$
$$HS^- + H_2O \rightleftharpoons H_2S + OH^-$$

The sulfide ion is slowly oxidized by oxygen, so sulfur precipitates when a solution of hydrogen sulfide or a sulfide salt is exposed to the air for a time.

Figure 22.30

When an orange solution of sodium dichromate, $Na_2Cr_2O_7$, is added to a colorless solution of sodium sulfide, Na_2S, the sulfide ions are oxidized to elemental sulfur, which is insoluble in aqueous solution.

Of the Halogens.

Binary compounds containing only hydrogen and a halogen are called **hydrogen halides.** At room temperature the pure hydrogen halides HF, HCl, HBr, and HI are gases.

The anhydrous hydrogen halides are rather inactive chemically and do not attack dry metals at room temperature. However, they react with many metals at elevated temperatures, forming metal halides and hydrogen. These reactions are sometimes used to prepare anhydrous metal halides.

$$Fe(s) + 2HCl(g) \xrightarrow{300°C} FeCl_2(s) + H_2(g)$$

The hydrogen halides can be prepared by the general techniques used to prepare other acids (see Section 15.7), although each technique is not suitable for every hydrogen halide. As indicated in Table 22.7, fluorine, chlorine, and bromine react directly with hydrogen to form the respective hydrogen halide. This reaction is used to prepare hydrogen chloride and hydrogen bromide commercially. Bromine reacts much less vigorously with hydrogen than does chlorine.

Hydrogen halides can be prepared by acid–base reactions between a nonvolatile strong acid and a metal halide. The escape of the gaseous hydrogen halide drives the

Figure 22.31

(left) The reaction of sulfuric acid with NaCl produces gaseous HCl. (right) Sulfuric acid oxidizes NaI and produces I_2.

reaction to completion. For example, hydrogen fluoride is usually prepared by heating a mixture of calcium fluoride, CaF_2, and concentrated sulfuric acid.

$$CaF_2(s) + H_2SO_4(aq) \longrightarrow CaSO_4(s) + 2HF(g)$$

Gaseous hydrogen fluoride is also a by-product in the preparation of phosphate fertilizers by the reaction of fluoroapatite, $Ca_5(PO_4)_3F$, with sulfuric acid (see Section 8.9). Hydrogen chloride is prepared, both in the laboratory and commercially, by the reaction of concentrated sulfuric acid with a chloride salt (Fig. 22.31 A). Sodium chloride is generally used, because it is the least expensive chloride. Hydrogen bromide and hydrogen iodide cannot be prepared in similar reactions, because sulfuric acid is such a strong oxidizing agent that it oxidizes both bromide and iodide ions (Fig. 22.32 B). However, both hydrogen bromide and hydrogen iodide can be prepared by the reaction of a covalent nonmetal bromide or iodide with water. For example, note the following equation.

$$PBr_3(l) + 3H_2O(l) \longrightarrow 3HBr(g) + H_3PO_3(aq)$$

All of the hydrogen halides are very soluble in water. With the exception of hydrogen fluoride, which has a strong hydrogen–fluorine bond, they are strong acids. Reactions of the hydrohalic acids with metals or with metal hydroxides, oxides, and carbonates are often used to prepare soluble salts of the halides. Most chloride salts are soluble ($AgCl$, $PbCl_2$, and Hg_2Cl_2 are the commonly encountered exceptions).

The halide ions in hydrohalic acids give these substances the properties associated with $X^-(aq)$. The heavier halide ions (Cl^-, Br^-, I^-) can act as reducing agents and are oxidized by lighter halogens or other oxidizing agents. They also serve as precipitating agents for insoluble metal halides.

Hydrofluoric acid is unique in its reactions with sand (silicon dioxide) and with glass, which is a mixture of silicates (mainly calcium silicate).

$$SiO_2 + 4HF \longrightarrow SiF_4(g) + 2H_2O$$
$$CaSiO_3 + 6HF \longrightarrow CaF_2 + SiF_4(g) + 3H_2O$$

The silicon escapes from these reactions as silicon tetrafluoride, a volatile compound. Because hydrogen fluoride attacks glass, it is used to frost or etch glass. Light bulbs are frosted with hydrogen fluoride, and markings on thermometers, burets, and other glassware are made with it.

The largest use for hydrogen fluoride is in the production of fluorocarbons for refrigerants such as the Freons, in plastics, and in propellants. The second largest use is in the manufacture of cryolite, Na_3AlF_6, which is important in the production of

Figure 22.32

The molecular structure of PCl₃ and that of PCl₅ in the gas phase.

Figure 22.32

The molecular structure of PCl_3 and that of PCl_5 in the gas phase.

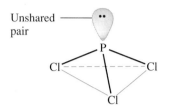

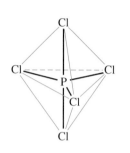

aluminum (Section 21.2). The acid is also used in the production of other inorganic fluorides (such as BF_3), which serve as catalysts in the industrial synthesis of certain organic compounds.

Hydrochloric acid is inexpensive. It is an important and versatile acid in industry and is used in the manufacture of metal chlorides, dyes, glue, glucose, and various other chemicals. A considerable amount is also used in the activation of oil wells and as a pickle liquor—an acid used to remove oxide coatings from iron or steel that is to be galvanized, tinned, or enameled. The amounts of hydrobromic acid and hydroiodic acid used commercially are insignificant by comparison.

22.25   Nonmetal Halides

We have already discussed the hydrogen halides (Section 22.24) and the relatively limited number of noble gas fluorides. Almost all of the other nonmetals also form halides. Those of nitrogen and sulfur are either unstable or of little practical importance. In this section, we will examine the halides of phosphorus and oxygen as well as the interesting class of compounds known as the interhalogens.

Of Phosphorus.   Phosphorus will react directly with the halogens, forming **trihalides,** PX_3, and **pentahalides,** PX_5. The trihalides are much more stable than the corresponding nitrogen trihalides; nitrogen pentahalides do not form because of nitrogen's inability to form more than four bonds.

The chlorides PCl_3 and PCl_5 (Fig. 22.32) are the most important halides of phosphorus. The colorless, liquid **phosphorus trichloride** is prepared by passing chlorine over molten phosphorus. Solid **phosphorus pentachloride** is prepared by oxidizing the trichloride with excess chlorine. The pentachloride is an off-white solid that sublimes when warmed and forms an equilibrium with the trichloride and chlorine when heated.

Like most other nonmetal halides, both phosphorus chlorides react with an excess of water and give hydrogen chloride and an oxyacid; PCl_3 gives phosphorous acid, H_3PO_3, and PCl_5 gives phosphoric acid, H_3PO_4. Partial reaction of phosphorus(V) halides produces **phosphorus(V) oxyhalides,** POX_3, which have tetrahedral structures.

The pentahalides of phosphorus are Lewis acids because of the empty valence d orbitals of phosphorus. They readily react with halide ions (Lewis bases) to give the anion PX_6^-. Whereas phosphorus pentafluoride is a molecular compound in all states, X-ray studies show that solid phosphorus pentachloride is an ionic compound, $[PCl_4^+][PCl_6^-]$, as are phosphorus pentabromide, $[PBr_4^+][Br^-]$, and phosphorus pentaiodide, $[PI_4^+][I^-]$.

Of Oxygen.   The halogens do not react directly with oxygen, but binary oxygen—halogen compounds can be prepared by the reactions of the halogens with oxygen-containing compounds. Oxygen compounds with chlorine, bromine, and iodine are called oxides because oxygen is the more electronegative element in these compounds. On the other hand, fluorine compounds with oxygen are called fluorides because fluorine is the more electronegative element.

As a class, the oxides are extremely reactive and unstable, and their chemistry has little practical importance. Dichlorine monoxide (Fig. 22.33 A) and chlorine dioxide (Fig. 22.33 B) are the only commercially important compounds. They are employed

Figure 22.33

Structures of the (A) Cl_2O and (B) ClO_2 molecules.

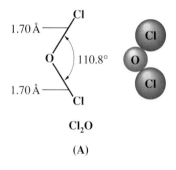

1.70 Å   110.8°

Cl_2O

(A)

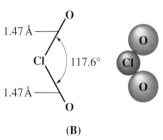

1.47 Å   117.6°

(B)

Figure 22.34

Structures of IF₃ (T-shaped), IF₅ (square pyramidal), and IF₇ (pentagonal bipyramidal).

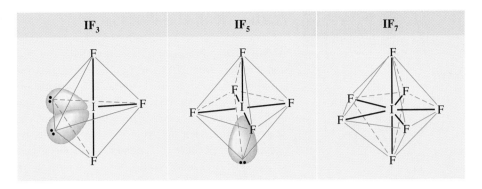

as bleaching agents (for use with wood pulp and flour) and for water treatment. The importance of and interest in these compounds derive from the fact that a rather extensive chemistry is associated with the oxyacids of the halogens and their salts. These will be discussed in a later section.

Of the Halogens.
Compounds formed from two different halogens are called **interhalogens**. Interhalogen molecules consist of one atom of the heavier halogen bonded by single bonds to an odd number of atoms of the lighter halogen. The structures of IF₃, IF₅, and IF₇ are shown in Fig. 22.34. Formulas for other interhalogens, each of which can be prepared by the reaction of the respective halogens, are given in Table 22.8.

Note from Table 22.8 that fluorine is able to oxidize iodine to its maximum oxidation state, +7, whereas bromine and chlorine, which are more difficult to oxidize, achieve only the +5 oxidation state. Because smaller halogens are grouped about a larger one, the maximum number of smaller atoms possible increases as the radius of the larger atom increases. Many of these compounds are unstable, and most are extremely reactive. The interhalogens react like their component halides; halogen fluorides, for example, are stronger oxidizing agents than halogen chlorides.

The ionic **polyhalides** of the alkali metals, compounds such as KI_3, $KICl_2$, $KICl_4$, $CsIBr_2$, and $CsBrCl_2$ that contain an anion composed of at least three halogen atoms, are closely related to the interhalogens. The formation of the polyhalide anion I_3^- is responsible for the solubility of iodine in aqueous solutions containing iodide ion.

$$I_2(s) + I^-(aq) \longrightarrow I_3^-(aq)$$

Table 22.8

Interhalogens[a]

XX′	XX′₃	XX′₅	XX′₇
ClF(g)	ClF₃(g)	ClF₅(g)	
BrF(g)	BrF₃(l)	BrF₅(l)	
BrCl(g)			
IF(s)	IF₃(s)	IF₅(l)	IF₇(g)
ICl(l)	ICl₃(s)		
IBr(s)			

[a]The physical states shown are those at STP.

Formula	Oxidation State of Nitrogen
N_2O	+1
NO	+2
N_2O_3	+3
NO_2	+4
N_2O_4	+4
N_2O_5	+5

Table 22.9

The Oxides of Nitrogen

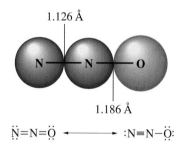

1.126 Å

1.186 Å

$\ddot{N}=N=\ddot{O} \longleftrightarrow :N\equiv N-\ddot{O}:$

Figure 22.35

The molecular and resonance structures of a molecule of nitrous oxide, N_2O.

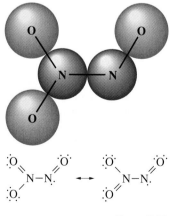

$\ddot{O}=N-\ddot{N} \longleftrightarrow \ddot{O}-\ddot{N}$

Figure 22.36

The molecular and resonance structures of a molecule of dinitrogen trioxide, N_2O_3.

22.26   Nonmetal Oxides

Of Nitrogen.   Nitrogen oxides in which nitrogen exhibits each of its positive oxidation numbers from +1 to +5 are well characterized. They are listed in Table 22.9.

When ammonium nitrate is heated, **nitrous oxide** (dinitrogen oxide), N_2O (Fig. 22.35), is formed. In this oxidation–reduction reaction, the nitrogen in the ammonium ion is oxidized by the nitrogen in the nitrate ion. Nitrous oxide is a colorless gas possessing a mild, pleasing odor and a sweet taste. It is used as an anesthetic for minor operations, especially in dentistry, under the name of laughing gas.

Low yields of **nitric oxide,** NO, are produced when nitrogen and oxygen are heated together. It also forms by direct union of nitrogen and oxygen in the air brought about by lightning during thunderstorms. Nitric oxide is produced commercially by burning ammonia. In the laboratory, nitric oxide is produced by reduction of nitric acid. When copper reacts with dilute nitric acid, nitric oxide is the principal reduction product.

$$3Cu + 8HNO_3 \longrightarrow 2NO(g) + 3Cu(NO_3)_2 + 4H_2O$$

Pure nitric oxide can be obtained by reducing nitric acid in dilute solution with an iron(II) salt.

$$3Fe^{2+} + NO_3^- + 4H_3O^+ \longrightarrow 3Fe^{3+} + NO(g) + 6H_2O$$

Gaseous nitric oxide is the most thermally stable of the nitrogen oxides and is also the simplest known thermally stable molecule with an unpaired electron. It is one of the air pollutants generated by internal combustion engines, resulting from the reaction of atmospheric nitrogen and oxygen during the combustion process.

At room temperature nitric oxide is a slightly soluble, colorless gas consisting of diatomic molecules. As is often the case with molecules that contain an unpaired electron, two molecules combine to form a dimer by pairing their unpaired electrons. Liquid NO is partially dimerized, and solid NO contains dimers.

$$2 \cdot\ddot{N}=\ddot{O}: \rightleftharpoons \ \ddot{O}-N-N \overset{O}{\underset{O}{}}$$

When a mixture of equal parts of nitric oxide and nitrogen dioxide is cooled to −21°C, the gases form **dinitrogen trioxide,** a blue liquid consisting of N_2O_3 molecules (Fig. 22.36). Dinitrogen trioxide exists only in liquid and solid states. When heated, it forms a mixture of NO and NO_2.

Nitrogen dioxide is made in the laboratory by heating the nitrate of a heavy metal, or by the reduction of concentrated nitric acid with copper metal (Fig. 22.37 on page 851). Nitrogen dioxide is made commercially by oxidizing nitric oxide with air.

The nitrogen dioxide molecule (Fig. 22.38) contains an unpaired electron, which is responsible for its color and paramagnetism. It is also responsible for the dimerization of NO_2. At low pressures or at high temperatures, nitrogen dioxide has a deep brown color that is due to the presence of the NO_2 molecule. At low temperatures the color almost entirely disappears as **dinitrogen tetraoxide,** N_2O_4, is formed (see Fig. 14.7). At room temperature an equilibrium exists.

$$2NO_2(g) \rightleftharpoons N_2O_4(g) \qquad K_p = 6.86$$

Dinitrogen pentaoxide, N_2O_5 (Fig. 22.39 on page 851), is a white solid formed by the dehydration of nitric acid by phosphorus(V) oxide.

Figure 22.37

The reaction of copper metal with concentrated HNO_3 produces a solution of $Cu(NO_3)_2$ and brown fumes of NO_2.

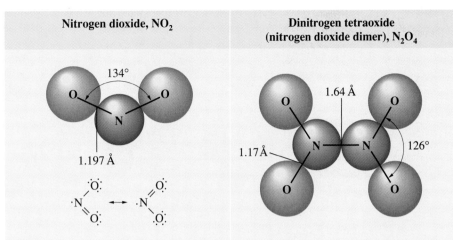

Figure 22.38 The molecular structures of molecules of nitrogen dioxide, NO_2, and dinitrogen tetraoxide, N_2O_4, and the resonance structures of nitrogen dioxide.

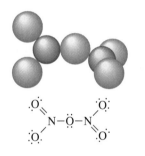

Figure 22.39

The molecular structure and one resonance structure of a molecule of dinitrogen pentaoxide, N_2O_5.

$$P_4O_{10} + 4HNO_3 \longrightarrow 4HPO_3 + 2N_2O_5$$

It is unstable above room temperature, decomposing to N_2O_4 and O_2.

$$2N_2O_5(g) \longrightarrow 2N_2O_4(g) + O_2(g)$$

The oxides of nitrogen(III), nitrogen(IV), and nitrogen(V) react with water and form nitrogen-containing oxyacids. Nitrogen(III) oxide, N_2O_3, is the anhydride of nitrous acid; HNO_2 forms when N_2O_3 reacts with water. There are no stable oxyacids containing nitrogen with an oxidation state of +4. Nitrogen(IV) oxide, NO_2, disproportionates in one of two ways when it reacts with water. In cold water a mixture of HNO_2 and HNO_3 is formed. At higher temperatures HNO_3 and NO form. Nitrogen(V) oxide, N_2O_5, is the anhydride of nitric acid; HNO_3 is produced when N_2O_5 reacts with water.

$$N_2O_5 + H_2O \longrightarrow 2HNO_3$$

The nitrogen oxides exhibit extensive oxidation–reduction behavior. Nitrous oxide, nitrogen(I) oxide, resembles oxygen in its behavior when heated with combustible substances. It is a strong oxidizing agent that decomposes when heated to form nitrogen and oxygen. Because one-third of the gas liberated is oxygen, nitrous oxide supports combustion better than air. A glowing splinter bursts into flame when thrust into a bottle of this gas. Nitric oxide acts both as an oxidizing agent and as a reducing agent. For example,

Oxidizing agent: $P_4 + 6NO \longrightarrow P_4O_6 + 3N_2$

Reducing agent: $Cl_2 + 2NO \longrightarrow 2ClNO$

Nitrogen dioxide (or dinitrogen tetraoxide) is a good oxidizing agent. For example,

$$NO_2 + CO \longrightarrow NO + CO_2$$
$$NO_2 + 2HCl \longrightarrow NO + Cl_2 + H_2O$$

Figure 22.40

The molecular structures of P_4O_6 and P_4O_{10}.

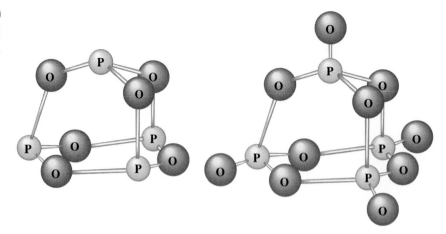

Of Phosphorus.

Phosphorus forms two common oxides, **phosphorus(III) oxide** (or tetraphosphorus hexaoxide), P_4O_6, and **phosphorus(V) oxide** (or tetraphosphorus decaoxide), P_4O_{10} (Fig. 22.40). Phosphorus(III) oxide is a white crystalline solid with a garliclike odor. Its vapor is very poisonous. It oxidizes slowly in air and inflames when heated to 70°C, forming P_4O_{10}. Phosphorus(III) oxide dissolves slowly in cold water to form phosphorus acid, H_3PO_3.

Phosphorus(V) oxide, P_4O_{10}, is a white flocculent powder that is prepared by burning phosphorus in excess oxygen. Its enthalpy of formation is very high (-2984 kJ), and it is quite stable and a very poor oxidizing agent. When P_4O_{10} is dropped into water, it reacts with a hissing sound, and heat is liberated as orthophosphoric acid is formed.

$$P_4O_{10}(s) + 6H_2O(l) \longrightarrow 4H_3PO_4(aq)$$

Because of its great affinity for water, phosphorus(V) oxide is used extensively for drying gases and removing water from many compounds.

Of Sulfur.

The two common oxides of sulfur are **sulfur dioxide,** SO_2, and **sulfur trioxide,** SO_3. The odor of burning sulfur comes from sulfur dioxide. Sulfur dioxide occurs in volcanic gases and in the atmosphere near industrial plants that burn coal or oil that contains sulfur compounds. The oxide forms when these sulfur compounds react with oxygen during combustion.

Sulfur dioxide is produced commercially by burning free sulfur and by heating sulfide ores such as ZnS, FeS_2, and Cu_2S in air (see Section 23.2). (Roasting, which forms the metal oxide, is the first step in the separation of the metals from the ores.) Sulfur dioxide is prepared conveniently in the laboratory by the action of sulfuric acid on either sulfite salts, containing the SO_3^{2-} ion, or hydrogen sulfite salts, containing HSO_3^-. Sulfurous acid, H_2SO_3, is formed first, but it quickly decomposes into sulfur dioxide and water. Sulfur dioxide is also formed when many reducing agents react with hot concentrated sulfuric acid. **Sulfur trioxide** forms slowly when sulfur dioxide and oxygen are heated together; the reaction is exothermic.

$$2SO_2 + O_2 \longrightarrow 2SO_3 \qquad \Delta H° = -197.8 \text{ kJ}$$

Sulfur dioxide is a gas at room temperature, and the SO_2 molecule is bent (Fig. 22.41). Sulfur trioxide melts at 17°C and boils at 43°C. In the vapor state its mole-

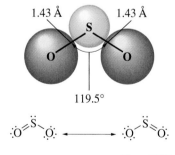

Figure 22.41

The molecular structure and resonance forms of sulfur dioxide.

Figure 22.42

The structure of sulfur trioxide in the gas phase and its resonance forms.

cules are single SO_3 units (Fig. 22.42), but in the solid state SO_3 exists in several polymeric forms.

The sulfur oxides react as Lewis acids with many oxides and hydroxides in Lewis acid–base reactions (Section 15.13), with the formation of **sulfites** or **hydrogen sulfites** and **sulfates** or **hydrogen sulfates,** respectively.

$$BaO + SO_2 \longrightarrow BaSO_3 \text{ (a sulfite)}$$
$$KOH + SO_2 \longrightarrow KHSO_3 \text{ (a hydrogen sulfite)}$$
$$BaO + SO_3 \longrightarrow BaSO_4 \text{ (a sulfate)}$$
$$KOH + SO_3 \longrightarrow KHSO_4 \text{ (a hydrogen sulfate)}$$

22.27 Nonmetal Oxyacids and Their Salts

Of Nitrogen. Both N_2O_5 and NO_2 react with water to form **nitric acid,** HNO_3. Nitric acid (Fig. 22.43) was known to the alchemists of the eighth century as *aqua fortis* (meaning "strong water"). It was prepared from KNO_3 and was used in the separation of gold from silver; it dissolves silver but not gold. Traces of nitric acid occur in the atmosphere after thunderstorms, and its salts are widely distributed in nature. Chile saltpeter, $NaNO_3$, is found in tremendous deposits (3 kilometers wide by 300 kilometers long, and as much as 2 meters thick) in the desert region near the boundary of Chile and Peru. Bengal saltpeter, KNO_3, is found in India and in other countries of the Far East.

Nitric acid can be prepared in the laboratory by heating a nitrate salt (such as sodium or potassium nitrate) with concentrated sulfuric acid.

$$NaNO_3(s) + H_2SO_4(l) \xrightarrow{\Delta} NaHSO_4(s) + HNO_3(g)$$

Nitric acid is produced commercially by the **Ostwald process:** oxidation of ammonia to nitric oxide, NO; oxidation of nitric oxide to nitrogen dioxide, NO_2; and conversion of nitrogen dioxide to nitric acid. Most of the 18 billion pounds of nitric acid produced in the United States in 1995 came from the Ostwald process.

Pure nitric acid is a colorless liquid. However, it is often yellow or brown in color because of the NO_2 formed as it decomposes. Nitric acid is stable in aqueous solution; solutions containing 68% of the acid are sold as concentrated nitric acid. It is both a strong oxidizing agent and a strong acid.

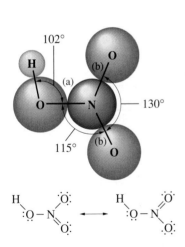

Figure 22.43

The molecular and resonance structures of a molecule of nitric acid, HNO_3. N—O distances are (a) 1.41Å and (b) 1.22Å.

The action of nitric acid on a metal rarely produces H_2 (by reduction of H^+) in more than small amounts. Instead, nitrogen is reduced. The products formed depend on the concentration of the acid, the activity of the metal, and the temperature. A mixture of nitrogen oxides, nitrates, and other reduction products is usually produced. Less active metals such as copper, silver, and lead reduce concentrated nitric acid primarily to nitrogen dioxide (Fig. 22.38). The reaction of dilute nitric acid with copper gives NO. The more active metals, such as zinc and iron, give nitrous oxide with dilute nitric acid. In each case, the nitrate salts of the metals crystallize when the resulting solutions are evaporated.

Nonmetallic elements, such as sulfur, carbon, iodine, and phosphorus, are oxidized by concentrated nitric acid to their oxides or oxyacids, with the formation of NO_2.

$$S(s) + 6HNO_3(aq) \longrightarrow H_2SO_4(aq) + 6NO_2(g) + 2H_2O(l)$$
$$C(s) + 4HNO_3(aq) \longrightarrow CO_2(g) + 4NO_2(g) + 2H_2O(l)$$

Many compounds are oxidized by nitric acid. Hydrochloric acid is readily oxidized by concentrated nitric acid to chlorine and chlorine dioxide. A mixture of one part concentrated nitric acid and three parts concentrated hydrochloric acid (called *aqua regia,* which means "royal water") reacts vigorously with metals. This mixture is particularly useful in dissolving gold and platinum and other metals that lie below hydrogen in the activity series. The action of aqua regia on gold can be represented, in a somewhat simplified form, by the equation

$$Au(s) + 4HCl(aq) + 3HNO_3(aq) \longrightarrow HAuCl_4(aq) + 3NO_2(g) + 3H_2O(l)$$

Nitric acid reacts with proteins, such as those in the skin, to give a yellow material called xanthoprotein. You may have noticed that if you get nitric acid on your fingers, they turn yellow.

Nitrates, salts of nitric acid, form when metals or their oxides, hydroxides, or carbonates react with nitric acid. Most nitrates are soluble in water; indeed, one of the significant uses of nitric acid is to prepare soluble metal nitrates.

Nitric acid is used extensively in the laboratory and in chemical industries as a strong acid and as an active oxidizing agent. It is used in the manufacture of explosives, dyes, plastics, and drugs. Salts of nitric acid (nitrates) are valuable as fertilizers. Gunpowder is a mixture of potassium nitrate, sulfur, and charcoal.

The reaction of N_2O_3 with water gives a pale blue solution of **nitrous acid,** HNO_2 (Fig. 22.44). However, HNO_2 is easier to prepare by addition of an acid to a solution of a nitrite; nitrous acid is a weak acid, so the nitrite ion is basic.

$$NO_2^- + H_3O^+ \longrightarrow HNO_2 + H_2O$$

Nitrous acid is very unstable and exists only in solution. It disproportionates slowly at room temperature (rapidly when heated) into nitric acid and nitric oxide. Nitrous acid is an active oxidizing agent with strong reducing agents, and it is oxidized to nitric acid by active oxidizing agents.

Sodium nitrite, $NaNO_2$, is the most important salt of nitrous acid. It is usually made by reducing molten sodium nitrate with lead.

$$NaNO_3 + Pb \longrightarrow NaNO_2 + PbO$$

Sodium nitrite is added to meats such as hot dogs and cold cuts. The nitrite ion has two functions. It limits the growth of bacteria that can cause food poisoning, and it prolongs the meat's retention of its red color. The addition of sodium nitrite to meat products is now controversial, because nitrous acid is known to react with certain

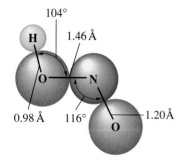

Figure 22.44

The molecular structure of a molecule of nitrous acid, HNO_2.

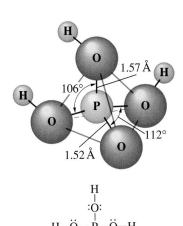

$$H-\overset{\displaystyle H}{\underset{\displaystyle :\ddot{O}:}{\overset{\displaystyle :O:}{\ddot{O}}}}-P-\overset{}{\ddot{O}}-H$$

Figure 22.45

The molecular and electronic structures of orthophosphoric acid, H_3PO_4.

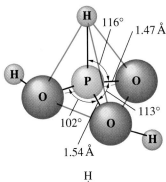

$$H-\overset{\displaystyle H}{\underset{\displaystyle :\ddot{O}:}{\ddot{O}}}-P-\overset{}{\ddot{O}}-H$$

Figure 22.46

Phosphorous acid, H_3PO_3. Only the two hydrogen atoms that are bonded to an oxygen atom are acidic.

organic compounds to form a class of compounds known as nitrosoamines. Nitrosoamines produce cancer in laboratory animals. This has prompted the U.S. Food and Drug Administration (FDA) to limit the amount of $NaNO_2$ that can legally be added to foods.

The nitrites are much more stable than the acid (the salts of all oxyacids are more stable than the acids themselves), but nitrites, like nitrates, can explode. Nitrites are soluble in water ($AgNO_2$ is only slightly soluble).

Of Phosphorus.

Pure **orthophosphoric acid**, H_3PO_4 (Fig. 22.45), forms colorless, deliquescent crystals that melt at 42°C. It is commonly called phosphoric acid, and is commercially available as a viscous, 82% solution known as "syrupy phosphoric acid."

One commercial method of preparing orthophosphoric acid is to treat calcium phosphate rock with concentrated sulfuric acid.

$$Ca_3(PO_4)_2(s) + 3H_2SO_4(aq) \longrightarrow 2H_3PO_4(aq) + 3CaSO_4(s)$$

The products are diluted with water, and the calcium sulfate is removed by filtration. This method gives a dilute acid that is contaminated with calcium dihydrogen phosphate, $Ca(H_2PO_4)_2$, and other compounds associated with calcium phosphate rock. Pure orthophosphoric acid is manufactured by dissolving P_4O_{10} in water.

The action of water on P_4O_6, PCl_3, PBr_3, or PI_3 forms **phosphorous acid**, H_3PO_3 (Fig. 22.46). Pure phosphorous acid is most readily obtained by hydrolyzing phosphorus trichloride,

$$PCl_3(l) + 3H_2O(l) \longrightarrow H_3PO_3(aq) + 3HCl(g)$$

heating the resulting solution to expel the hydrogen chloride, and evaporating the water until white crystals of phosphorous acid appear upon cooling. The crystals are deliquescent, very soluble in water, and have an odor like that of garlic. The solid melts at 70.1°C and decomposes at about 200°C by disproportionation into phosphine and orthophosphoric acid.

$$4H_3PO_3(l) \longrightarrow PH_3(g) + 3H_3PO_4(l)$$

Phosphorous acid and its salts are active reducing agents, because they are readily oxidized to phosphoric acid and phosphates, respectively. Phosphorous acid reduces the silver ion to free silver, mercury(II) salts to mercury(I) salts, and sulfurous acid to sulfur.

Phosphorous acid forms only two series of salts, which contain the **dihydrogen phosphite ion**, $H_2PO_3^-$, and the **hydrogen phosphite ion**, HPO_3^{2-}, respectively. The third atom of hydrogen cannot be replaced; it is bonded to the phosphorus atom rather than to an oxygen atom, and like the hydrogen atoms in phosphine, it is not very acidic.

The solution remaining from the preparation of phosphine from white phosphorus and a base contains the **hypophosphite ion**, $H_2PO_2^-$. The barium salt can be obtained by using barium hydroxide in the preparation. When barium hypophosphite is treated with sulfuric acid, barium sulfate precipitates and **hypophosphorous acid**, H_3PO_2 (Fig. 22.47 on page 856), forms in solution.

$$Ba^{2+} + 2H_2PO_2^- + 2H_3O^+ + SO_4^{2-} \longrightarrow BaSO_4(s) + 2H_3PO_2 + 2H_2O$$

The acid is weak and monoprotic, forming only one series of salts. Two nonacidic hydrogen atoms are bonded directly to the phosphorus atom. Hypophosphorous acid

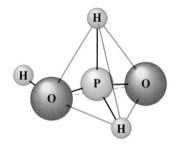

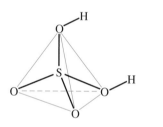

Figure 22.47

Hypophosphorous acid, H_3PO_2. Only the hydrogen atom bonded to an oxygen atom is acidic.

Figure 22.48

The tetrahedral molecular structure of sulfuric acid.

and its salts are strong reducing agents because phosphorus is in the unusually low oxidation state of $+1$.

Of Sulfur.
Sulfuric acid, H_2SO_4 (Fig. 22.49), is prepared by oxidizing sulfur to sulfur trioxide and then converting the trioxide to sulfuric acid (see Section 8.9). Pure sulfuric acid is a colorless, oily liquid that freezes at $10.5°C$. It fumes when heated because the acid decomposes to water and sulfur trioxide. More sulfur trioxide than water is lost during the heating, until a concentration of 98.33% acid is reached. Acid of this concentration boils at $338°C$ without further change in concentration (a constant boiling solution, Section 12.16) and is sold as concentrated H_2SO_4.

The strong affinity of concentrated sulfuric acid for water makes it a good dehydrating agent. Gases and immiscible liquids that do not react with the acid can be dried by passing them through it.

Sulfuric acid is a strong diprotic acid that ionizes in two stages. In aqueous solution the first stage is essentially complete. The secondary ionization is less nearly complete, but even so HSO_4^- is a moderately strong acid (about 25% ionized in a solution of a HSO_4^- salt: $K_a = 1.2 \times 10^{-2}$).

Being a diprotic acid, sulfuric acid forms both **sulfates,** such as Na_2SO_4, and **hydrogen sulfates,** such as $NaHSO_4$. The sulfates of barium, strontium, calcium, and lead are only slightly soluble in water. Their insolubility is the basis of a qualitative and quantitative test for the sulfate ion and the barium ion.

Among the important soluble sulfates are Glauber's salt, $Na_2SO_4\cdot10H_2O$, and Epsom salts, $MgSO_4\cdot7H_2O$. Because the HSO_4^- ion is an acid, hydrogen sulfates, such as $NaHSO_4$, exhibit acidic behavior. Sodium hydrogen sulfate is the primary ingredient in some household cleansers.

Hot, concentrated sulfuric acid is an oxidizing agent. Depending on its concentration, the temperature, and the strength of the reducing agent, sulfuric acid oxidizes many compounds and, in the process, undergoes reduction to either SO_2, HSO_3^-, SO_3^{2-}, S, H_2S, or S^{2-}. The displacement of volatile acids from their salts by concentrated sulfuric acid is described in Section 15.7. The amount of sulfuric acid used in industry exceeds that of any other manufactured compound (Section 8.9).

Sulfur dioxide dissolves in water to form a solution of sulfurous acid, as expected for the oxide of a nonmetal. **Sulfurous acid** is unstable, and anhydrous H_2SO_3 cannot be isolated. Boiling a solution of sulfurous acid expels the sulfur dioxide. Like other diprotic acids, sulfurous acid ionizes in two steps; the **hydrogen sulfite ion,** HSO_3^-, and the **sulfite ion,** SO_3^{2-}, are formed. Sulfurous acid is a moderately strong acid. Ionization is about 25% in the first stage, but it is much less in the second ($K_{a_1} = 1.2 \times 10^{-2}$ and $K_{a_2} = 6.2 \times 10^{-8}$).

Solid sulfite and hydrogen sulfite salts can be prepared by adding a stoichiometric amount of a base to a sulfurous acid solution and then evaporating the water. These salts are also formed by the reaction of SO_2 with oxides and hydroxides. Solid sodium hydrogen sulfite forms sodium sulfite, sulfur dioxide, and water when heated.

$$2NaHSO_3(s) \xrightarrow{\Delta} Na_2SO_3(s) + SO_2(g) + H_2O(l)$$

Sulfurous acid can be oxidized by strong oxidizing agents. Oxygen in the air oxidizes it slowly to the more stable sulfuric acid.

$$2H_2SO_3 + O_2 + 2H_2O \longrightarrow 2H_3O^+ + 2HSO_4^-$$

Solutions of sulfites are also very susceptible to air oxidation, whereby sulfates are formed. Thus solutions of sulfites always contain sulfates after being exposed to air.

Table 22.10

Oxyacids of the Halogens

Name	Fluorine	Chlorine	Bromine	Iodine
Hypohalous	HOF	HOCl	HOBr	HOI
Halous		$HClO_2$		
Halic		$HClO_3$	$HBrO_3$	HIO_3
Perhalic		$HClO_4$	$HBrO_4$	HIO_4
Paraperhalic				H_5IO_6

Of the Halogens.

The compounds HXO, HXO_2, HXO_3, and HXO_4, where X represents Cl, Br, or I, are called **hypohalous, halous, halic,** and **perhalic** acids, respectively. The strengths of these acids increase from the hypohalous acids, which are very weak acids, to the perhalic acids, which are very strong. The known acids are listed in Table 22.10.

The only known oxyacid of fluorine is the very unstable **hypofluorous acid,** HOF, which is prepared by the reaction of gaseous fluorine with ice.

$$F_2(g) + H_2O(s) \longrightarrow HOF(g) + HF(g)$$

This compound does not ionize in water, and no salts are known.

The reactions of chlorine and bromine with water are analogous to that of fluorine with ice, but these reactions do not go to completion, and mixtures of the halogen and the respective hypohalous and hydrohalic acids result. The reaction of the halogen with mercury(II) oxide is used to prepare a solution of the pure **hypohalous acid** (X = Cl, Br, I).

$$2X_2 + 3HgO + H_2O \longrightarrow HgX_2 \cdot 2HgO + 2HOX(aq)$$

None of the hypohalous acids except HOF has been isolated in the free state. They are stable only in solution. The hypohalous acids are all very weak acids; however, HOCl is a stronger acid than HOBr, which in turn is stronger than HOI.

Solutions of salts containing the basic **hypohalite** ions, OX^-, can be prepared by adding base to solutions of hypohalous acids. The salts have been isolated as solids. All of the hypohalites are unstable with respect to disproportionation in solution, but the reaction is slow for hypochlorite. Hypobromite and hypoiodite disproportionate rapidly, even in the cold.

$$3XO^- \longrightarrow 2X^- + XO_3^-$$

Sodium hypochlorite is used as an inexpensive bleach (Clorox) and germicide. It is produced commercially by the electrolysis of cold, dilute aqueous sodium chloride solutions under conditions where the resulting chlorine and hydroxide ion can react. The net reaction is

$$Cl^- + H_2O \xrightarrow{\text{Electrical energy}} ClO^- + H_2$$

The only definitely known halous acid is **chlorous acid,** $HClO_2$, obtained by the reaction of barium chlorite with dilute sulfuric acid.

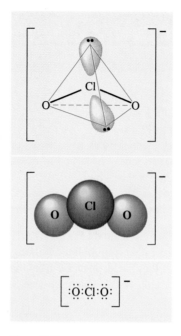

Figure 22.49
Structure of the chlorite ion, ClO_2^-.

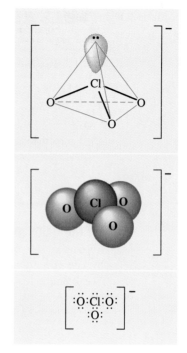

Figure 22.50
Structure of the chlorate ion, ClO_3^-.

$$Ba(ClO_2)_2(aq) + H_2SO_4(aq) \longrightarrow BaSO_4(s) + 2HClO_2(aq)$$

A solution of $HClO_2$ is obtained by filtering off the barium sulfate. Chlorous acid is not stable; it slowly decomposes in solution to give chlorine dioxide, hydrochloric acid, and water. Chlorous acid reacts with bases to give salts containing the **chlorite ion** (Fig. 22.49). Metal chlorite salts can also be prepared by the action of chlorine dioxide on a metal peroxide. For example, sodium peroxide reacts as follows:

$$2ClO_2(g) + Na_2O_2(s) \longrightarrow 2NaClO_2(s) + O_2(g)$$

Sodium chlorite is used extensively in the bleaching of paper, because it is a strong oxidizing agent that does not damage the paper.

Chloric acid, $HClO_3$, and **bromic acid,** $HBrO_3$, are stable only in solution, but **iodic acid,** HIO_3, can be isolated as a stable white solid from the reaction of iodine with concentrated nitric acid.

$$I_2(s) + 10HNO_3(aq) \longrightarrow 2HIO_3(s) + 10NO_2(g) + 4H_2O(l)$$

The lighter halic acids can be obtained from their barium salts by reaction with dilute sulfuric acid. The reaction is analogous to that used to prepare chlorous acid. All of the halic acids are strong acids and very active oxidizing agents. Salts containing **halate** ions (Fig. 22.50) can be prepared by reaction of the acids with bases. Metal chlorates are also prepared by electrochemical oxidation of a hot solution of a metal halide. Bromates can be produced by the oxidation of bromides with hypochlorite ion; oxidation of iodides with chlorates gives iodates. Sodium chlorate is used as a weed killer, potassium chlorate is used in some matches.

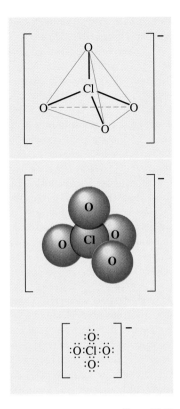

Figure 22.51
Structure of the perchlorate ion,
ClO_4^-.

Perchloric acid, $HClO_4$, can be obtained by treating a perchlorate, such as potassium perchlorate, with sulfuric acid under reduced pressure. The $HClO_4$ distills from the mixture.

$$KClO_4(s) + H_2SO_4(aq) \longrightarrow HClO_4(l) + KHSO_4(s)$$

Perchloric acid explodes above 92°C, but it distills at temperatures below 92°C at reduced pressures. Dilute aqueous solutions of perchloric acid are quite stable thermally, but concentrations above 60% are unstable and dangerous. Perchloric acid and its salts are powerful oxidizing agents. Serious explosions have occurred when concentrated solutions were heated with easily oxidized substances. However, its reactions as an oxidizing agent are slow when perchloric acid is cold and dilute. The acid is among the strongest of all acids. Most salts containing the **perchlorate ion** (Fig. 22.51) are soluble. They are prepared by reactions of bases with perchloric acid and, commercially, by the electrolysis of hot solutions of their chlorides.

Perbromate salts are different to prepare, and currently the best syntheses involve the oxidation of bromates in basic solution with fluorine gas. Perbromic acid, $HBrO_4$, is prepared by the acidification of perbromate salts.

Several different acids containing iodine in the +7 oxidation state are known; they include **metaperiodic acid,** HIO_4, and **paraperiodic acid,** H_5IO_6. Salts of these acids can be readily prepared by reactions with bases.

For Review   Summary

The elements boron, silicon, germanium, antimony, and tellurium separate the metals from the nonmetals in the periodic table. These elements, called **semi-metals** or sometimes **metalloids,** exhibit properties characteristic of both metals and nonmetals. The structures of these elements are similar to those of nonmetals, but the elements are electrical semiconductors.

Boron exhibits some metallic characteristics, but the great majority of its chemical behavior is that of a nonmetal that exhibits an oxidation state of +3 in its compounds (although other oxidation states are known). Its valence shell configuration is $2s^2 2p^1$, so it forms trigonal planar compounds with three single covalent bonds. The resulting compounds are Lewis acids, because the unhybridized p orbital does not contain an electron pair. The most stable boron compounds are those containing oxygen and fluorine. Diborane, the simplest stable boron hydride, contains **three-center two-electron bonds,** as does elemental boron.

Silicon is a semi-metal with a valence shell configuration of $3s^2 3p^2 3d^0$. It commonly forms tetrahedral compounds in which silicon exhibits an oxidation state of +4. Although the d orbital is unfilled in four-coordinate silicon compounds, its presence makes silicon compounds much more reactive than the corresponding carbon compounds. Silicon forms strong single bonds with carbon and oxygen, giving rise to the stability of **silicones. Silicates** contain oxyanions of silicon and are important components of minerals and glass.

The nonmetals are hydrogen (Group 1A or 7A); carbon (Group 4A); nitrogen and phosphorus (Group 5A); oxygen, sulfur, and selenium (Group 6A); fluorine, chlorine, bromine, iodine, and astatine (Group 7A); and helium, neon, argon, krypton, xenon, and radon (Group 8A). Compounds of the nonmetals with active metals, or with most other metals with oxidation states of +1, +2, and +3, are ionic and contain the nonmetal as a negative ion. Nonmetals form covalent compounds with other nonmetals and with metals with high oxidation states. With the exception of the hydrides of C, N, and P, the hydrides of the nonmetals generally are acidic in water, as are the oxides. The nonmetals become progressively better oxidizing agents as their electronegativity increases (except for nitrogen, which is an anomalously weak oxidizing agent).

Hydrogen has the chemical properties of a nonmetal with a relatively low electronegativity. It forms ionic hydrides with active metals, covalent compounds in which it has an oxidation state of -1 with less electronegative elements, and covalent compounds in which it has an oxidation state of $+1$ with more electronegative nonmetals. It reacts explosively with oxygen, fluorine, and chlorine; less readily with bromine; and much less readily with iodine, sulfur, and nitrogen. Hydrogen reduces the oxides of those metals lying below chromium in the activity series to form the metal and water.

Oxygen (Group 6A) forms compounds with almost all of the elements. Except in a few compounds with fluorine, oxygen exhibits negative oxidation states in its compounds. The most common oxidation state is -2, in **oxides.** The -1 oxidation state is found in compounds that contain O—O bonds, **peroxides.** Oxygen is a strong oxidizing agent. **Ozone,** O_3, is an allotrope of oxygen, O_2. Ozone forms from oxygen in an endothermic reaction and is a stronger oxidizing agent than O_2.

Nitrogen (Group 5A) is not very reactive. At room temperature it reacts with lithium. At elevated temperatures it reacts with active metals, forming **nitrides,** N^{3-}, and with hydrogen and with oxygen in reversible low-yield reactions, forming ammonia and nitric oxide, respectively.

Ammonia is prepared by the direct reaction of hydrogen and nitrogen. It exhibits both Brønsted and Lewis base behavior, and it undergoes oxidation. **Hydrazine, nitric oxide,** and **ammonium salts** are prepared from ammonia.

Nitrogen exhibits oxidation states ranging from -3 to $+5$. The most important oxides are **nitric oxide** and **nitrogen dioxide,** because they are intermediates in the preparation of nitric acid by oxidation of ammonia. These two oxides can also be prepared by reduction of nitric acid. Dinitrogen trioxide is the anhydride of nitrous acid, dinitrogen pentaoxide is the anhydride of nitric acid.

Nitric acid is one of the important strong acids of commerce. It is also a strong oxidizing agent. Metals dissolve in nitric acid by reduction of the nitrate ion, generally with formation of nitrogen dioxide or nitric oxide, rather than by reduction of hydrogen ion with the production of hydrogen gas. **Nitrous acid,** a weak acid, can be prepared by the acidification of a solution of a nitrite salt.

Phosphorus (Group 5A) commonly exhibits oxidation states of -3 with active metals and of $+3$ and $+5$ with more electronegative nonmetals. Hydrolysis of **phosphides** or disproportionation of phosphorus in base produces **phosphine,** PH_3. Phosphorus is oxidized by halogens and by oxygen. The oxides are phosphorus(V) oxide, P_4O_{10}, and phosphorus(III) oxide, P_4O_6.

Orthophosphoric acid, H_3PO_4, is prepared by the reaction of phosphates with sulfuric acid or of water with phosphorus(V) oxide. Orthophosphoric acid is a triprotic acid that forms three types of salts. Upon heating, orthophosphoric acid loses water and forms condensed phosphoric acids.

The reaction of PCl_3 with water produces **phosphorous acid,** H_3PO_3. One hydrogen atom is bonded directly to the phosphorus atom, so this acid contains only two acidic hydrogen atoms. **Hypophosphorous acid,** H_3PO_2, contains two hydrogen atoms bonded directly to the phosphorus atom and one acidic hydrogen.

Sulfur (Group 6A) reacts with almost all metals and readily forms the **sulfide ion,** S^{2-}, in which it has an oxidation state of -2. Sulfur also reacts with most nonmetals. It exhibits an oxidation state of -2 in covalent compounds with less electronegative elements. With more electronegative elements it exhibits oxidation states of $+4$ and $+6$ (commonly) and $+1$ and $+2$ (occasionally).

Sulfur burns in air and forms sulfur dioxide, which reacts with water to form the weak, unstable **sulfurous acid.** Neutralization of sulfurous acid and reactions of sulfur dioxide with metal oxides produce **sulfites.** The slow reaction of sulfur dioxide with oxygen produces sulfur trioxide. Sulfur trioxide is used to prepare **sulfuric acid,** a strong acid and an oxidizing agent. Neutralization of sulfuric acid and reactions of sulfur trioxide with metal oxides produce **sulfates.**

The **halogens** are members of Group 7A. The heavier halogens form compounds in which they have oxidation states of -1, $+1$, $+3$, $+5$, $+7$, although they sometimes exhibit other oxidation states. Fluorine always exhibits an oxidation state of -1 in compounds.

The oxidizing ability of the elemental halogens and the resistance of the halides to oxidation decrease as halogen size increases. This is reflected in the ease of preparation of the elements from their halide salts. The gaseous **hydrogen halides** form from the direct reaction of halogens and hydrogen and from the other general techniques used to make acids. They dissolve in water to give solutions of **hydrohalic acids.**

The halogens form **halides** with less electronegative elements. Halides of the metals vary from ionic to covalent; halides of nonmetals are covalent. **Interhalogens** are formed by the combination of two different halogens. Binary oxygen–halogen compounds, generally of low stability, are known, as are **hypohalous, halous, halic,** and **perhalic** acids and their salts.

The noble gas elements exhibit a very limited chemistry. Xenon reacts directly with fluorine, forming the xenon fluorides XeF_2, XeF_4, and XeF_6. Xenon oxides can be prepared by the reaction of water with the fluorides, and the reaction of certain acids with XeF_2 forms derivatives containing Xe—O bonds.

Key Terms and Concepts

acid anhydride (22.7)
allotrope (22.8)
borates (22.6)
deuterium (22.10)
disproportionation reaction (22.7)
Frasch process (22.14)
halides (22.24)
hydrogen halide (22.24)
hypohalous acid (22.27)

interhalogen (22.25)
metalloid (22.1)
nitrate (22.27)
nitride (22.24)
nitrite (22.27)
Ostwald process (22.27)
perhalic acid (22.27)
phosphide (22.20)
photosynthesis (22.11)

semi-metal (22.1)
silicates (22.6)
silicones (22.4)
sulfate (22.7)
sulfite (22.7)
three-center bond (22.4)
tritium (22.10)
zone-refining (22.3)

Exercises

Questions

1. Give the hybridization of the semi-metal and the molecular geometry for each of the following compounds.

   (a) GeH_4 (b) SbF_3 (c) $Te(OH)_6$
   (d) H_2Te (e) GeF_2 (f) $TeCl_4$
   (g) SiF_6^{2-} (h) $SbCl_5$ (i) TeF_6

2. Name each of the following compounds.

   (a) TeO_2 (b) Sb_2S_3 (c) GeF_4
   (d) SiH_4 (e) GeH_4 (f) $NaBH_4$

Boron

3. Why are the compounds known as boron hydrides said to exhibit electron-deficient bonding?

4. Write a Lewis structure for each of the following molecules or ions.

   (a) H_3BPH_3 (b) BF_4^- (c) BBr_3
   (d) $B(CH_3)_3$ (e) $B(OH)_3$ (f) H_3BCN^-

5. Describe the hybridization of boron and the molecular structure of each molecule and ion listed in Exercise 4.

6. Write a balanced equation for the reaction of elemental boron with F_2, O_2, S, Se, and Br_2, respectively. (Most of these reactions require high temperature.)

7. From the data given in Appendix I, determine the standard enthalpy change and the standard free energy change for each of the following reactions.

   (a) $BF_3(g) + 3H_2O(l) \longrightarrow B(OH)_3(s) + 3HF(g)$
   (b) $BCl_3(g) + 3H_2O(l) \longrightarrow B(OH)_3(s) + 3HCl(g)$
   (c) $B_2H_6(g) + 6H_2O(l) \longrightarrow 2B(OH)_3(s) + 6H_2(g)$

8. Why is boron limited to a maximum coordination number of 4 in its compounds?

Silicon

9. Write a formula for each of the following compounds.

   (a) silicon dioxide (b) silicon tetraiodide
   (c) silane (d) silicon carbide
   (e) magnesium silicide (f) fluorosilicic acid

10. Using only the periodic table, write the complete electron configuration for silicon, including any empty orbitals in the valence shell.

11. Write a Lewis structure for each of the following molecules and ions.

    (a) $(CH_3)_3SiH$ (b) SiO_4^{4-} (c) Si_2H_6
    (d) $Si(OH)_4$ (e) SiH_2F_2 (f) SiF_6^{2-}

12. Describe the hybridization of silicon and the molecular structure of the molecules and ions listed in Exercise 11.

13. Write two equations in which the semi-metal silicon acts as a metal and two equations in which it acts as a non-metal.

14. Describe the hybridization and the bonding of a silicon atom in elemental silicon.

15. Classify each of the following molecules as polar or non-polar.

    (a) SiH_4 (b) Si_2H_6 (c) $SiCl_3H$
    (d) $(CH_3)_2SiH_2$ (e) SiF_4 (f) $SiCl_2F_2$

16. Silicon reacts with sulfur at elevated temperatures. If 0.0923 g of silicon reacts with sulfur to give 0.3030 g of silicon sulfide, determine the empirical formula of silicon sulfide.

17. A hydride of silicon prepared by the reaction of Mg_2Si with acid exerted a pressure of 306 torr at 26°C in a bulb with a volume of 57.0 mL. If the mass of the hydride was 0.0861 g,

what is its molecular mass? What is the molecular formula for the hydride?

Hydrogen

18. Why does hydrogen not exhibit an oxidation state of -1 when bonded to nonmetals?

19. The reaction of calcium hydride, CaH_2, with water can be characterized as a Lewis acid–base reaction.

$$CaH_2 + 2H_2O \longrightarrow Ca(OH)_2 + 2H_2$$

Identify the Lewis acid and the Lewis base among the reactants. The reaction is also an oxidation–reduction reaction. Identify the oxidizing agent, the reducing agent, and the changes in oxidation number that occur in the reaction.

20. In drawing Lewis structures, we learn that a hydrogen atom forms only one two-center two-electron bond in a covalent compound. Why?

21. What mass of CaH_2 is required to react with water to provide enough hydrogen gas to fill a balloon at 20°C and 0.8 atm pressure with a volume of 4.5 L? (For the balanced equation, see Exercise 19.)

22. What mass of hydrogen gas results from the reaction of 8.5 g of KH with water?

$$KH + H_2O \longrightarrow KOH + H_2$$

Oxygen

23. How many moles of oxygen are contained in a 50.0-L cylinder at 22.0°C if the pressure gauge indicates 2000 lb/in^2?

24. Write a balanced chemical equation for the reaction of an excess of oxygen with each of the following. Remember that oxygen is a strong oxidizing agent and tends to oxidize an element to its maximum oxidation state.

(a) Mg   (b) Rb   (c) Bi
(d) Ga   (e) Na_2SO_3 (f) AlN
(g) C_2H_2 (h) CO

25. How many liters of oxygen gas are produced at 20°C and 750 torr when 10.0 g of sodium peroxide reacts with water to produce sodium hydroxide and oxygen?

26. Which is the stronger acid, H_2SO_4 or H_2SeO_4? Why?

Nitrogen

27. Write the Lewis structure for each of the following.

(a) NH^{2-} (b) N^{3-} (c) N_2F_4
(d) NH_2^- (e) NF_3 (f) N_3^-

28. For species (c) through (f) listed in Exercise 27, indicate the hybridization of the nitrogen atom (for N_3^-, the central nitrogen).

29. Explain how ammonia can function both as a Brønsted base and as a Lewis base.

30. Determine the oxidation state of nitrogen in each of the following.

(a) NCl_3 (b) ClNO   (c) N_2O_5
(d) N_2O_3 (e) NO_2^- (f) N_2O_4
(g) N_2O (h) NO_3^- (i) HNO_2
(j) HNO_3

31. Draw the Lewis structures of NO_2, NO_2^-, and NO_2^+. Predict the ONO bond angle in each species, and give the hybridization of the nitrogen in each species.

32. What mass of gaseous ammonia can be produced from the reaction of 3.0 g of hydrogen gas and 3.0 g of nitrogen gas?

33. Although PF_5 and AsF_5 are stable, nitrogen does not form NF_5 molecules. Explain this difference among members of the same group.

Phosphorus

34. Write the Lewis structure for each of the following.

(a) HCP   (b) PH_3 (c) PH_4^+
(d) P_2H_4 (e) PO_4^{3-} (f) PF_5

35. Describe the molecular structure of each of the molecules or ions listed in Exercise 34.

36. Complete and balance each of the following chemical equations. (In some cases there may be more than one correct answer.)

(a) $P_4 + Al \longrightarrow$ (b) $P_4 + Na \longrightarrow$
(c) $P_4 + F_2 \longrightarrow$ (d) $P_4 + xsCl_2 \longrightarrow$
(e) $P_4 + xsSe \longrightarrow$ (f) $P_4 + O_2 \longrightarrow$
(g) $P_4O_6 + O_2 \longrightarrow$

37. Describe the hybridization of phosphorus in each of the following compounds.

(a) P_4O_{10} (b) P_4O_6
(c) PH_4I (an ionic compound)   (d) PBr_3
(e) H_3PO_4 (f) H_3PO_3
(g) PH_3 (h) P_2H_4

38. What volume of 0.200 M NaOH is required to neutralize the solution produced by dissolving 2.00 g of PCl_3 is an excess of water? Note that when H_3PO_3 is titrated under these conditions, only one proton of the acid molecule reacts.

39. How much $POCl_3$ can be produced from 25.0 g of PCl_5 and the appropriate amount of H_2O?

40. How many tons of $Ca_3(PO_4)_2$ are needed to prepare 5.0 tons of phosphorus if a yield of 90% is obtained?

41. Write equations showing the stepwise ionization of phosphorous acid.

42. Compare the structures of PF_4^+, PF_5, PF_6^-, and POF_3.

43. Why does phosphorous acid form only two series of salts, even though the molecule contains three hydrogen atoms?

44. Assign an oxidation state to phosphorus in each of the following.

(a) NaH_2PO_4 (b) PF_5 (c) P_4O_{10}
(d) K_3PO_4 (e) Na_3P (f) $Na_4P_2O_7$

Sulfur

45. Explain why hydrogen sulfide is a gas at room temperature, whereas water, which has a lower molecular mass, is a liquid.

46. Give the hybridization and oxidation state for sulfur in SO_2, in SO_3, and in H_2SO_4.

47. Which is the stronger acid, $NaHSO_3$ or $NaHSO_4$?

48. Determine the oxidation state of sulfur in SF_6, SO_2F_2, and KHS.

49. Which is a stronger acid, sulfurous acid or sulfuric acid? Why?

50. Oxygen forms double bonds in O_2, but sulfur forms single bonds in S_8. Why?

51. Give the Lewis structure of each of the following.
(a) SF_4 (b) K_2SO_4 (c) HSSSH
(d) SO_2Cl_2 (e) H_2SO_3 (f) SO_3
(g) $O_3SSSSO_3^{2-}$ $(S_4O_6)^{2-}$

52. Write two balanced chemical equations in which sulfuric acid acts as an oxidizing agent.

Halogens

53. Which is the stronger acid, $HClO_3$ or $HBrO_3$? Why?

54. Which is the stronger acid, $HClO_4$ or $HBrO_4$? Why?

55. What is the hybridization of iodine in IF_3 and IF_5?

56. Predict the molecular geometries of each of the following.
(a) IF_5 (b) I_3^- (c) PCl_5
(d) $SiBr_4$ (e) SeF_4 (f) ClF_3

57. Which halogen has the highest ionization energy? Is this what you would predict on the basis of what you have learned about periodic properties?

58. Name each of the following compounds.

(a) $Fe(BrO_4)_3$ (b) BrF_3 (c) $NaBrO_3$
(d) PBr_5 (e) $NaClO_4$ (f) KClO

59. Explain why at room temperature fluorine and clorine are gases, bromine is a liquid, and iodine is a solid. (You may need to refer to Section 11.3.)

60. What is the oxidation state of the halogen in each of the following?
(a) H_5IO_6 (b) IO_4^- (c) ClO_2
(d) ICl_2 (e) ICl_3 (f) F_2

The Noble Gases

61. Give the hybridization of xenon in each of the following.
(a) XeF_2 (b) XeF_4 (c) XeO_2F_2
(d) XeO_3 (e) XeO_4 (f) $XeOF_4$

62. What is the molecular structure of each of the molecules listed in Exercise 61?

63. Indicate whether each molecule listed in Exercise 61 is polar or nonpolar.

64. What is the oxidation state of xenon in each of the following?
(a) XeO_2F_2 (b) XeF^+ (c) XeF_3^+
(d) XeO_6^{4-} (e) XeO_3 (f) XeO_3F_2

65. A mixture of xenon and fluorine was heated. A sample of the white solid that formed reacted with hydrogen to give 81 mL of xenon at STP and hydrogen fluoride, which was collected in water, giving a solution of hydrofluoric acid. The hydrofluoric acid solution was titrated, and 68.43 mL of 0.3172 M sodium hydroxide was required to reach the equivalence point. Determine the empirical formula for the white solid, and write balanced chemical equations for the reactions involving xenon.

66. Basic solutions of Na_4XeO_6 are powerful oxidants. What mass of $Mn(NO_3)_2 \cdot 6H_2O$ reacts with 125.0 mL of a 0.1717 M basic solution of Na_4XeO_6 that contains an excess of sodium hydroxide if the products include Xe and a solution of sodium permanganate?

CHAPTER OUTLINE

Periodic Relationships of the Transition Elements

23.1   The Transition Elements
23.2   Properties of the Transition Elements
23.3   Preparation of the Transition Elements
23.4   Compounds of the Transition Elements
23.5   Copper Oxide Superconductors

Coordination Compounds

23.6   Basic Concepts
23.7   The Naming of Complexes
23.8   The Structures of Complexes
23.9   Isomerism in Complexes
23.10 Uses of Complexes

Bonding and Electron Behavior in Coordination Compounds

23.11 Valence Bond Theory
23.12 Crystal Field Theory
23.13 Magnetic Moments of Molecules and Ions
23.14 Colors of Transition Metal Complexes

23

The Transition Elements and Coordination Compounds

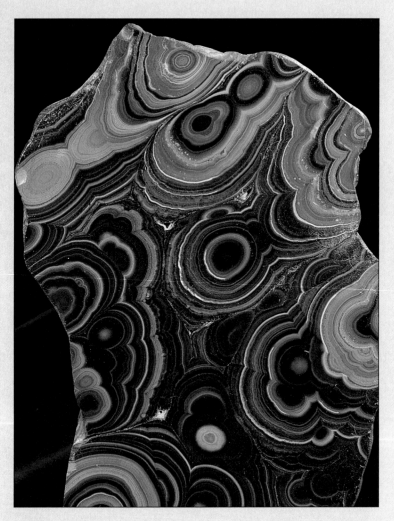

Malachite, a copper-bearing mineral, is one of many colorful transition metal compounds.

We have daily contact with transition elements: Iron, copper, chromium, silver, and gold are often encountered. Iron is used in many metal items, ranging from the rings in your notebook and the cutlery in your kitchen to automobiles, ships, and buildings. Copper is used in electrical wiring and coins, chromium as a protective plating on plumbing fixtures, and silver and gold in jewelry.

Compounds of the transition elements also play a familiar role in our daily lives. Silver iodide, AgI, is a component of photographic film, zirconium silicate, $ZrSiO_4$, is used in artificial gemstones, and chromium(IV) oxide, CrO_2, is used in magnetic recording tape. More complex compounds are also common; the hemoglobin in your blood is a compound that contains atoms of iron.

The variety of behaviors exhibited by most transition elements is due to their complex valence shells. The outermost s orbital, the first inner d orbitals, and, in some cases, the first inner f orbitals counting from the outside [the ns, $(n - 1)d$, and $(n - 2)f$ orbitals] make up the valence shells of atoms of the transition elements. Some or all of the electrons in these two subshells are used when these elements form compounds. Because a given element can often exhibit several different oxidation states, and because the presence of a partially filled d subshell leads to colored compounds, these elements exhibit a rich and fascinating chemistry.

Periodic Relationships of the Transition Elements

23.1 The Transition Elements

Two sets of metallic elements may be identified as transition elements (Fig. 23.1 on page 866): the **d-block elements,** which are usually called the *transition elements* (yellow in Fig. 23.1), and the **f-block elements,** usually called the *inner transition elements* (orange in the figure). The d-block elements are those elements in which the second shell, counting in from the outside, is filling from 8 to 18 electrons as its $(n - 1)d$ orbitals fill. The f-block elements are those elements in which the third shell, counting in from the outside, is filling from 18 to 32 electrons as its $(n - 2)f$ orbitals fill. Table 5.3 lists the electron configuration of the elements.

The d-block elements are divided into the **first transition series** (the elements Sc through Cu), the **second transition series** (the elements Y through Ag), and the **third transition series** (the element La and the elements Hf through Au). Actinium, Ac, is the first member of the **fourth transition series,** which also includes Rf through Uuu. The f-block elements are the elements Ce through Lu, which constitute the **lanthanide series** or **rare earth elements,** and the elements Th through Lr, which constitute the **actinide series.**

Because lanthanum behaves very much like the lanthanide elements, it is often considered a lanthanide element, even though its electron configuration makes it the first member of the third transition series. Similarly, the behavior of actinium often causes it to be considered with the actinide series, although its electron configuration makes it the first member of the fourth transition series.

The transition elements have common properties. They are almost all hard, strong, high-melting metals (Fig. 23.2 on page 867) that conduct heat and electricity well. They readily form alloys with one another and with other metallic elements. The

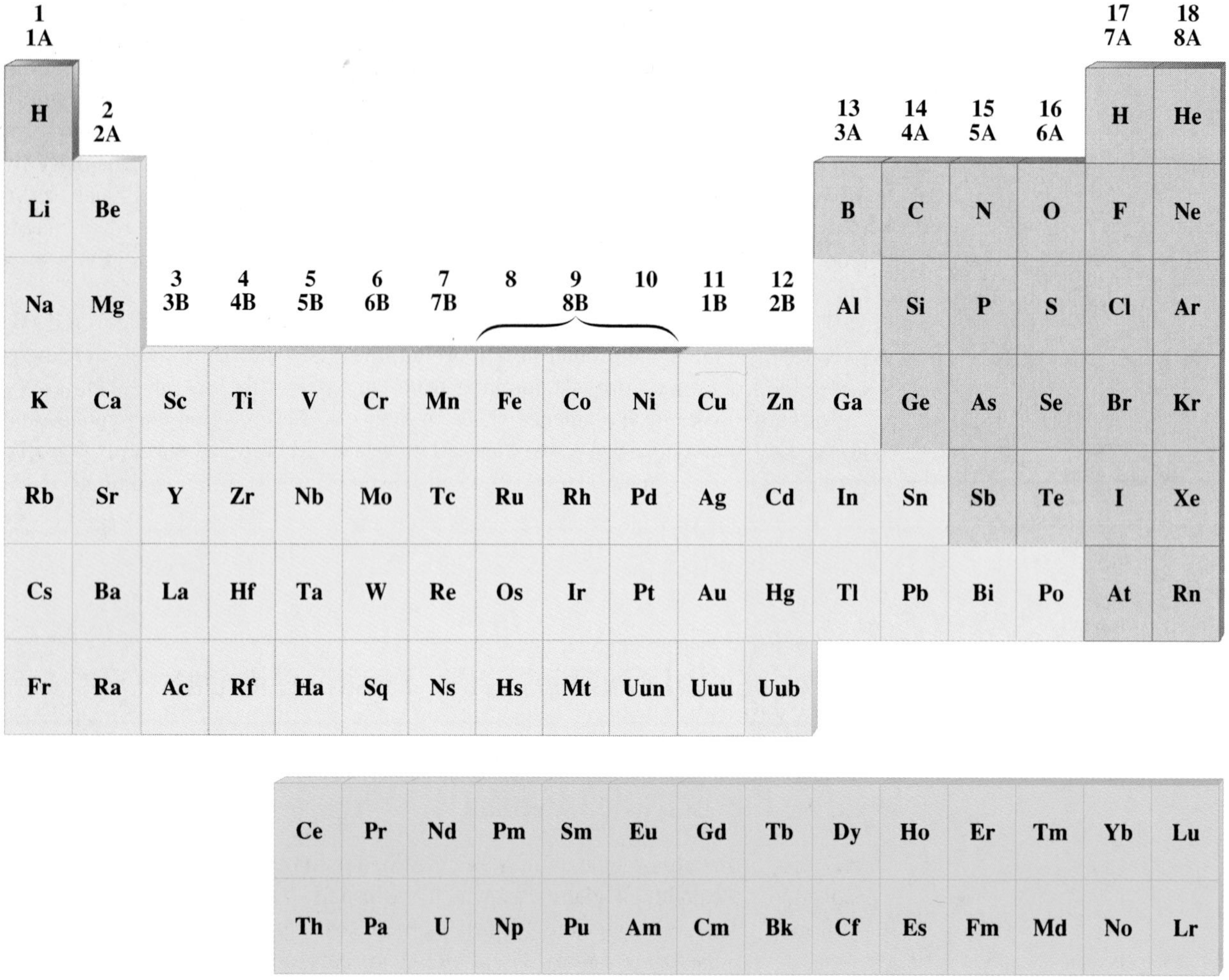

Figure 23.1

The location of the *d*-block elements (highlighted in yellow) and the *f*-block elements (highlighted in orange), in the periodic table. Nonmetals are shown in green, semi-metals in blue, and the representative metals in red.

elements of the *f*-block, of Group 3B, and of the first transition series are sufficiently active that they react with acids. Both the *d*-block and the *f*-block elements form a vast array of coordination compounds (see Section 23.6).

In this chapter we shall focus primarily on the chemical behavior of the elements of the first transition series.

23.2   Properties of the Transition Elements

Both the transition elements (the *d*-block elements) and the inner transition elements (the *f*-block elements) are metals. However, as may be seen from their positions in the activity series (Section 8.7) and their reduction potentials (Appendix H), these elements vary from active to very inactive metals. The *f*-block elements are more reactive than aluminum and are the most reactive of the transition elements.

The elements of the lanthanide series (La through Lu) and of the first transition series (Sc through Cu) form ionic compounds that dissolve in water giving stable solvated cations. The heavier *d*-block elements usually do not form simple positive ions

(A)   (B)

that are stable in water. Thus the Cr^{3+}, Fe^{3+}, and Co^{2+} ions are stable in aqueous solutions (Fig. 23.3), whereas the Mo^{3+}, Ru^{2+}, and Ir^{2+} ions are not. The majority of simple water-stable ions formed by the heavier d-block elements are oxyanions such as MoO_4^{2-} and ReO_4^{-}.

The heavier elements of Group 8B (ruthenium, osmium, rhodium, iridium, palladium, and platinum) are sometimes called the **platinum metals.** These elements and gold are particularly nonreactive. They do not form simple cations that are stable in water, and, unlike the earlier elements in the second and third transition series, they do not form stable oxyanions.

Both the d-block elements and the f-block elements react with nonmetals to form binary compounds; heating is usually required. These elements react with halogens to form a variety of halides ranging in oxidation state from $+1$ to $+6$. With the exception of palladium, platinum, and gold, they react with sulfur to form sulfides. On heating, oxygen reacts with all of the transition elements except palladium, platinum, silver, and gold; the oxides of these, even if formed, decompose upon heating (Section 8.7). The f-block elements, the elements of Group 3B, and the elements of the first transition series (except copper) are sufficiently active to react with aqueous solutions of strong acids, forming hydrogen and solutions of the corresponding salts.

As we cross the three series of the d-block elements from left to right in the periodic table, we find that each successive element through the manganese family can

Table 23.1

Common Oxidation States of the
Elements of the First Transition Series

Sc	Ti	V	Cr	Mn	Fe	Co	Ni	Cu
								+1
			+2	+2	+2	+2	+2	+2
+3	+3	+3	+3	+3	+3	+3		+3
	+4	+4	+4	+4				
		+5						
			+6					
				+7				

form compounds with a wider range of oxidation states. The oxidation states of the elements of the first transition series are listed in Table 23.1.

The elements at the beginning of the first transition series exhibit a highest oxidation state that corresponds to the loss of all of the electrons in both the s and d orbitals of their valence shell. The titanium(IV) ion, for example, is formed when the titanium atom loses its two $3d$ and two $4s$ electrons. These group oxidation states are the most stable oxidation states for scandium, titanium, and vanadium. Moving across the series, it becomes progressively more difficult to form these highest oxidation states. Compounds of titanium(IV) are harder to reduce than those of vanadium(V), which in turn are harder to reduce than those of chromium(VI) or manganese(VII). No compounds of iron(VIII) are known. Beyond manganese, the elements are stable with oxidation states of $+1$, $+2$, or $+3$. All the elements of the first transition series form ions with a charge of $+2$ or $+3$ that are stable in water, although those of the early members of the series can be readily oxidized by air.

The elements of the second and third transition series generally are more stable with higher oxidation states than are the elements of the first series. For example, the simple chemistry of molybdenum and tungsten, members of Group 6B, is limited to an oxidation state of $+6$ in aqueous solution. Chromium, the lightest member of the group, forms stable Cr^{3+} ions in water and, in the absence of air, stable Cr^{2+} ions in water. The sulfide with the highest oxidation state for chromium is Cr_2S_3, which contains the Cr^{3+} ion. Molybdenum and tungsten form sulfides in which the metals exhibit oxidation states of $+4$ and $+6$.

23.3   Preparation of the Transition Elements

Iron, copper, silver, and gold were known to the ancients because these metals occur freely in nature (Fig. 23.4 on page 869). Elemental iron resulting from meteorite falls is found in small quantities. Copper, silver, and gold occur in large amounts.

Generally, the transition elements are extracted from compounds found in a variety of ores. However, the ease of their recovery varies widely, depending on the concentration of the element in the ore, the extent to which it is mixed with other transition elements, and the difficulty of reducing the element to the free metal.

In general, it is not difficult to reduce ions of the d-block elements to the free element. Carbon is a sufficiently strong reducing agent in most cases. However, like the ions of the more active representative metals (Section 21.2), ions of the f-block ele-

(A)   (B)   (C)

Figure 23.4

Transition metals occur in nature in various forms. Examples include (A) a nugget of copper, (B) iron in the form of red hematite (Fe_2O_3), and (C) gold.

ments must be isolated by electrolysis or by reduction with an active metal such as calcium.

We shall discuss the processes used for the isolation of iron, copper, and silver because these three processes illustrate the principal means of isolating most of the *d*-block metals. In general, each of these processes involves three principal steps: preliminary treatment, smelting, and refining.

1. **Preliminary Treatment.** Ores are generally processed to make them suitable for extraction of the metals. This usually involves crushing or grinding the ore, concentrating the metal-bearing compounds, and sometimes treating these compounds chemically to convert them into substances that are more readily reduced to the metal.

2. **Smelting.** The next step is the extraction of the metal in the molten state, a process called **smelting,** which includes reduction of the metallic compound to the metal. Impurities may be removed by the addition of a compound that forms a slag—a substance with a low melting point that can be readily separated from the molten metal.

3. **Refining.** The final step in the recovery of a metal is refining the metal. Low-boiling metals such as zinc and mercury can be refined by distillation. When fused on an inclined table, low-melting metals like tin flow away from higher-melting impurities. Electrolysis is a common method for refining metals. Section 19.14 discussed the electrolytic refining of copper.

Isolation of Iron.

The early application of iron to the manufacture of tools and weapons was possible because of the wide distribution of iron ores and the ease with which the iron compounds in the ores could be reduced by carbon. For a long time charcoal was the form of carbon used in the reduction process. The production and use of iron on a large scale began about 1620, when coal was introduced as the reducing agent.

The first step in the metallurgy of iron is usually roasting the ore (heating the ore in air) to remove water, decompose carbonates to the oxides, and convert sulfides to the oxides. The oxides are then reduced in a blast furnace 80–100 feet high and about 25 feet in diameter (Fig. 23.5 on page 870). The roasted ore, coke, and limestone are introduced continuously into the top. Molten iron and slag are withdrawn at the bottom. The entire stock in a furnace may weigh several hundred tons.

Near the bottom of a furnace are nozzles through which preheated air is blown into the furnace. As soon as the air enters, the coke in the region of the nozzles is

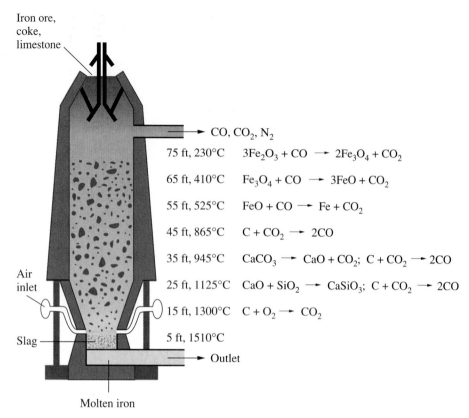

Figure 23.5

Reactions that occur at different levels within a blast furnace.

Iron ore, coke, limestone

CO, CO$_2$, N$_2$

75 ft, 230°C	$3Fe_2O_3 + CO \longrightarrow 2Fe_3O_4 + CO_2$
65 ft, 410°C	$Fe_3O_4 + CO \longrightarrow 3FeO + CO_2$
55 ft, 525°C	$FeO + CO \longrightarrow Fe + CO_2$
45 ft, 865°C	$C + CO_2 \longrightarrow 2CO$
35 ft, 945°C	$CaCO_3 \longrightarrow CaO + CO_2;\ C + CO_2 \longrightarrow 2CO$
25 ft, 1125°C	$CaO + SiO_2 \longrightarrow CaSiO_3;\ C + CO_2 \longrightarrow 2CO$
15 ft, 1300°C	$C + O_2 \longrightarrow CO_2$
5 ft, 1510°C	

Air inlet

Slag

Outlet

Molten iron

oxidized to carbon dioxide with the liberation of a great deal of heat. As the hot carbon dioxide passes upward through the overlying layer of white-hot coke, it is reduced to carbon monoxide.

$$CO_2 + C \longrightarrow 2CO$$

The carbon monoxide serves as the reducing agent in the upper regions of the furnace. The individual reactions are indicated in Fig. 23.5.

The iron oxides are reduced in the upper region of the furnace. In the middle region, limestone (calcium carbonate) decomposes, and the resulting calcium oxide combines with silica and silicates in the ore to form slag.

$$CaO + SiO_2 \longrightarrow CaSiO_3$$

Just below the middle of the furnace, the temperature is high enough to melt both the iron and the slag. They collect in layers at the bottom of the furnace; the less dense slag floats on the iron and protects it from oxidation. Several times a day, the slag and molten iron are withdrawn from the furnace. The iron is transferred to casting machines or to a steelmaking plant (Fig. 23.6 on page 871).

Most iron is refined by converting it into steel. **Steel** is made from iron by removing impurities and adding substances such as manganese, chromium, nickel, tungsten, molybdenum, and vanadium to produce alloys with properties that make the material suitable for specific uses. Most steels also contain small but definite percentages of carbon (0.04–2.5%). Thus a large part of the carbon contained in iron must be removed in the manufacture of steel.

Figure 23.6
Casting molten iron.

The principal process used in the production of steel utilizes a cylindrical furnace with a charge of about 300 tons of iron and 18 tons of limestone (to form slag). A jet of high-purity oxygen is directed into the white-hot molten charge through a water-cooled lance. The oxygen produces a vigorous reaction that oxidizes the impurities in the charge, and the oxidized impurities separate into the slag leaving the purified steel.

Isolation of Copper.

The most important ores of copper contain copper sulfides, although copper oxides and copper hydroxycarbonates are sometimes found. In the production of copper metal, the concentrated ore is roasted to remove part of the sulfur as sulfur dioxide. The remaining mixture, which consists of Cu_2S, FeS, FeO, and SiO_2, is mixed with limestone, which serves as a flux, and heated. Molten slag forms as the iron and silica are removed by the Lewis acid–base reactions

$$CaCO_3 + SiO_2 \longrightarrow CaSiO_3 + CO_2$$
$$FeO + SiO_2 \longrightarrow FeSiO_3$$

Reduction of the Cu_2S that remains after smelting is accomplished by blowing air through the molten material. The air converts part of the Cu_2S to Cu_2O. As soon as copper(I) oxide is formed, it is reduced by the remaining copper(I) sulfide to metallic copper.

$$2Cu_2S + 3O_2 \longrightarrow 2Cu_2O + 2SO_2$$
$$2Cu_2O + Cu_2S \longrightarrow 6Cu + SO_2$$

The copper obtained in this way is called blister copper because of its characteristic appearance, which is due to the air blisters it contains (Fig. 23.7 on page 872). The impure copper is cast into large plates, which are used as anodes in the electrolytic refining of the metal (see Section 19.14).

Figure 23.7

Blister copper.

Isolation of Silver.

Silver is found sometimes in large nuggets (Fig. 23.8) but more frequently in veins and related deposits. The extraction of silver from its ores is often accomplished by a process called **hydrometallurgy,** the general separation of a metal from other metals by converting it to a soluble ion and then precipitating it as the free metal via a suitable reducing agent. In the presence of air, alkali metal cyanides readily form the soluble dicyanoargentate(I) ion, $[Ag(CN)_2]^-$, from silver metal and all of its compounds. Representative equations are

$$4Ag + 8CN^- + O_2 + 2H_2O \longrightarrow 4[Ag(CN)_2]^- + 4OH^-$$
$$2Ag_2S + 8CN^- + O_2 + 2H_2O \longrightarrow 4[Ag(CN)_2]^- + 2S + 4OH^-$$
$$AgCl + 2CN^- \rightleftharpoons [Ag(CN)_2]^- + Cl^-$$

The silver is precipitated from the cyanide solution by addition of either zinc or aluminum, which serves as a reducing agent.

$$2[Ag(CN)_2]^- + Zn \longrightarrow 2Ag + [Zn(CN)_4]^{2-}$$

23.4   Compounds of the Transition Elements

The simple compounds of the transition elements range from ionic to covalent. In their lower oxidation states, the transition elements form ionic compounds; in their higher oxidation states, they form covalent compounds. The variation in oxidation states exhibited by the transition elements gives these compounds a metal-based oxidation–reduction chemistry. The chemistry of several classes of compounds containing elements of the first transition series follows.

Figure 23.8

Naturally occurring free silver may be found as nuggets or in veins.

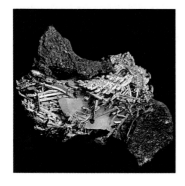

Halides.

Anhydrous halides of each of the transition elements can be prepared by the direct reaction of the metal with halogens. For example,

$$2Fe + 3Cl_2 \longrightarrow 2FeCl_3$$

Heating a metal halide with additional metal can be used to form a halide of the metal with a lower oxidation state.

$$Fe + 2FeCl_3 \longrightarrow 3FeCl_2$$

The stoichiometry of the metal halide that results from the reaction of a metal with a halogen is determined by the relative amounts of metal and halogen and by the strength of the halogen as an oxidizing agent. Generally fluorine forms fluorides containing metals in their highest oxidation states. The other halogens may not form analogous compounds.

Stable water solutions of the halides of the metals of the first transition series are generally prepared by addition of a hydrohalic acid to carbonates, hydroxides, oxides, or other compounds containing basic anions. Sample reactions are

$$NiCO_3(s) + 2H_3O^+(aq) + 2F^-(aq) \longrightarrow Ni^{2+}(aq) + 2F^-(aq) + 3H_2O(l) + CO_2(g)$$
$$Co(OH)_2(s) + 2H_3O^+(aq) + 2Br^-(aq) \longrightarrow Co^{2+}(aq) + 2Br^-(aq) + 4H_2O(l)$$

Many of these metals also dissolve in acids, forming a solution of the salt and hydrogen gas.

The nature of the bonding in the anhydrous halides of the elements of the first transition series varies with the oxidation state of the metal. Halides of metals with lower oxidation numbers are ionic; halides of metals with higher oxidation states are covalent. For example, the titanium chlorides $TiCl_2$ and $TiCl_3$ are ionic compounds with high melting points, but $TiCl_4$ is a volatile liquid with covalent titanium–chlorine bonds. All halides of the heavier d-block elements have significant covalent character.

The covalent behavior of the transition metals with higher oxidation states is exemplified by the reaction of the metal tetrahalides with water. Like covalent silicon tetrachloride (Section 22.5) both the covalent titanium and vanadium tetrahalides react with water to give solutions containing the corresponding hydrohalic acids and the covalent oxyions TiO^{2+} and VO^{2+}, respectively.

$$TiCl_4(l) + 3H_2O(l) \longrightarrow TiO^{2+}(aq) + 2H_3O^+(aq) + 4Cl^-(aq)$$
$$VBr_4(l) + 3H_2O(l) \longrightarrow VO^{2+}(aq) + 2H_3O^+(aq) + 4Br^-(aq)$$

Oxides.   Oxides of the transition elements with oxidation states of +1, +2, and +3 behave as ionic compounds that contain metal ions and basic oxide ions, whereas those with oxidation states of +4, +5, +6, and +7 contain covalent metal-oxygen bonds.

The oxides of the first transition series shown in Table 23.2 can be prepared by heating the metal in air. Alternatively, these oxides and oxides with other oxidation states can be produced by heating the corresponding hydroxides, carbonates, or oxalates in an inert atmosphere. Iron(II) oxide can be prepared by heating iron(II) oxalate, and cobalt(II) oxide by heating cobalt(II) hydroxide.

$$FeC_2O_4 \longrightarrow FeO + CO + CO_2$$
$$Co(OH)_2 \longrightarrow CoO + H_2O(g)$$

Table 23.2

Oxides Formed by Heating a First Transition Series Element in Air

Sc_2O_3
TiO_2
V_2O_5
Cr_2O_3
M_3O_4 M = Mn, Fe, Co
MO M = Ni, Cu

Oxides of the transition metals with higher oxidation states are formed by precipitation from acidic solution or by oxidation under special conditions. For example, chromium(VI) oxide, CrO_3, is produced as scarlet crystals when concentrated sulfuric acid is added to a concentrated solution of potassium dichromate. Cobalt(III) oxide, Co_2O_3, can be produced by gently heating cobalt(II) nitrate.

With the exception of CrO_3 and Mn_2O_7, transition metal oxides are not soluble in water. They exhibit their acid–base properties by reacting with acids or bases. Overall, oxides of transition metals with the lowest oxidation states are basic, the intermediate ones are amphoteric, and the highest ones are primarily acidic. The oxides of metals with oxidation states of +1, +2, and +3 are basic; they react with aqueous acids to form solutions of salts and water. Examples include

$$CoO(s) + 2H_3O^+(aq) + 2NO_3{}^-(aq) \longrightarrow Co^{2+}(aq) + 2NO_3{}^-(aq) + 3H_2O(l)$$
$$Sc_2O_3(s) + 6H_3O^+(aq) + 6Cl^-(aq) \longrightarrow 2Sc^{3+}(aq) + 6Cl^-(aq) + 9H_2O(l)$$

The oxides of metals with oxidation states of +4 are amphoteric, and most are not soluble in either acids or bases. Vanadium(V) oxide, chromium(VI) oxide, and manganese(VII) oxide are acidic. They react with solutions of hydroxides to form salts of the oxyanions $VO_4{}^{3-}$, $CrO_4{}^{2-}$, and $MnO_4{}^-$. For example,

$$CrO_3(s) + 2Na^+(aq) + 2OH^-(aq) \longrightarrow 2Na^+(aq) + CrO_4{}^{2-}(aq) + H_2O(l)$$

Chromium(VI) oxide and manganese(VII) oxide react with water to form the acids H_2CrO_4 and $HMnO_4$, respectively.

Hydroxides.
When a soluble hydroxide is added to an aqueous solution of a salt of a transition metal of the first transition series, a gelatinous precipitate forms. For example, adding a solution of sodium hydroxide to a solution of cobalt sulfate produces a gelatinous blue precipitate of cobalt(II) hydroxide. The net ionic equation is

$$Co^{2+}(aq) + 2OH^-(aq) \longrightarrow Co(OH)_2(s)$$

In this and many other cases these precipitates are hydroxides containing the transition metal ion, hydroxide ions, and water coordinated to the transition metal. In other cases the precipitates are hydrated oxides composed of the metal ion, oxide ions, and water of hydration. These substances do not contain hydroxide ions. However, both the hydroxides and the hydrated oxides react with acids to form salts and water.

Carbonates.
Many of the elements of the first transition series form insoluble carbonates. Thus these carbonates can be prepared by the addition of a soluble carbonate salt to a solution of a transition metal salt. For example, nickel carbonate can be prepared from solutions of nickel nitrate and sodium carbonate according to the following net ionic equation.

$$Ni^{2+}(aq) + CO_3{}^{2-}(aq) \longrightarrow NiCO_3(s)$$

The reactions of the transition metal carbonates are similar to those of the active metal carbonates (see Section 21.5). They react with acids to form metal salts, carbon dioxide, and water. Upon heating, they decompose, forming the transition metal oxides.

Other Salts.
In many respects, the chemical behavior of the elements of the first transition series is very similar to that of the representative metals. In particular, simple ionic salts of these elements can be prepared by the same types of reactions that are used to prepare salts of the representative metals (Section 21.6).

A variety of salts can be prepared from those metals that lie above hydrogen in the activity series (Section 8.7) by reaction with the corresponding acids: Scandium metal reacts with hydrobromic acid to form a solution of scandium bromide.

$$2Sc(s) + 6H_3O^+(aq) + 6Br^-(aq) \longrightarrow$$
$$2Sc^{3+}(aq) + 6Br^-(aq) + 3H_2(g) + 6H_2O(l)$$

The common compounds that we have just discussed can also be used to prepare salts. The reactions involved include the reactions of oxides, hydroxides, or carbonates with acids; for example,

$$Ni(OH)_2(s) + 2H_3O^+(aq) + 2ClO_4{}^-(aq) \longrightarrow Ni^{2+}(aq) + 2ClO_4{}^-(aq) + 4H_2O(l)$$

Metathetical reactions involving soluble salts may be used to prepare insoluble salts, for example,

$$Ba^{2+}(aq) + 2Cl^-(aq) + 2K^+(aq) + CrO_4{}^{2-}(aq) \longrightarrow$$
$$BaCrO_4(s) + 2K^+(aq) + 2Cl^-(aq)$$

In our discussion of oxides in this section, we have seen that reactions of the covalent oxides of the transition elements with hydroxides form salts that contain oxyanions of the transition elements.

23.5   Copper Oxide Superconductors

One of the most exciting scientific discoveries of the 1980s was the characterization of compounds that exhibit superconductivity at temperatures above 90 K. **Superconductors** conduct electricity with no resistance. They also possess unusual magnetic properties.

Typical among the high-temperature superconducting materials are ternary oxides containing yttrium (or one of several rare earth elements), barium, and copper in a 1:2:3 ratio. The formula of the yttrium compound is $YBa_2Cu_3O_7$, in which one-third of the copper is oxidized to copper(III).

The superconducting copper oxides are ionic compounds. The structure of a unit cell of $YBa_2Cu_3O_7$ is illustrated in Fig. 23.9. The copper(II) and copper(III) ions cannot be distinguished in the structure because the compound is a solid solution (Section 12.8) containing copper(II) and copper(III) ions.

Because these oxides are ionic compounds and are ceramics (glasslike in character), they are brittle and fragile and cannot be drawn out to form wires, as can a metal such as copper itself. However, ceramic wires and other shapes can be formed by mixing the ceramic powder or its ingredients with an organic binder (a glue) and then extruding a thin strip through a press. (This works just like squeezing a fine strip of toothpaste out of a tube.) The sample is then heated to burn off the binder and annealed, giving a current-carrying ceramic strip.

Most currently used commercial superconducting materials, niobium alloys such as NbTi and Nb_3Sn, do not become superconducting until they are cooled below 23 K ($-250°C$). This requires the use of liquid helium, which has a boiling temperature of 4 K and is expensive and difficult to handle. The new materials become superconducting at temperatures close to 90 K (Fig. 23.10 on page 876), temperatures that can be reached by cooling with liquid nitrogen (boiling temperature 77 K). A sample immersed in boiling liquid nitrogen is cooled to 77 K. Not only are liquid-nitrogen-cooled materials easier to handle, but the cooling costs are also about 1000 times less than for liquid helium.

Liquid-nitrogen-cooled superconductors could revolutionize society. Power companies, for example, could use underground superconducting transmission lines that would carry current for hundreds of miles with no loss of power due to resistance in the wires. This would allow generating stations to be located in areas remote from population centers and near to the natural resources necessary for power production.

Superconducting microchips could lead to more powerful supercomputers. Present chips produce waste heat as a result of the resistance of the electric current as it flows through the chip, so the chips have to be spaced in such a way as to allow a coolant to flow through them. Superconducting chips would produce no waste heat because

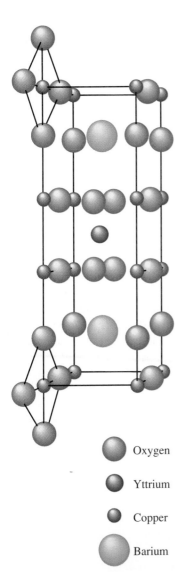

Oxygen

Yttrium

Copper

Barium

Figure 23.9

The unit cell of the superconductor $YBa_2Cu_3O_7$.

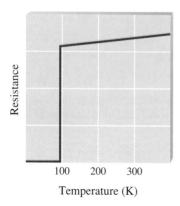

Figure 23.10

A graph of the relationship between resistance and temperature for the high-temperature superconductor $YBa_2Cu_3O_7$. This substance is a superconductor below 92 K. Note how the resistance falls to zero upon cooling.

they would have no resistance. Thus chips could be packed closer together and the signals would require less time to travel from chip to chip. This would make possible smaller computers that work faster.

Superconducting materials are particularly useful for generating very strong magnetic fields. Because they have no resistance, superconducting coils can carry very large currents without melting, and these large currents can be used to generate very strong magnetic fields.

The medical application of magnetic resonance imaging (MRI) devices requires the use of superconducting magnets to image tissues inside the body. The new superconductors could lower the price of these instruments and increase their power. The development of nuclear fusion (Section 20.12) would be enhanced if these materials could be used to produce magnetic fields strong enough to contain the hot plasma in which the nuclear reactions take place.

If high-temperature superconductors can be converted from laboratory to practical technology, it will require a tremendous amount of research and development. In particular, shaping the superconducting ceramics into useful shapes with the right physical and electrical properties presents a large challenge. Work in this area is under way, and experimental superconducting rods, rings, wires, tapes, and thin films have been fabricated.

Coordination Compounds

The hemoglobin in your blood, the blue dye in your ballpoint pen and in your blue jeans, chlorophyll, vitamin B-12, and the catalyst used in the manufacture of polyethylene all contain coordination compounds, or complexes. Ions of the transition elements, of the inner transition elements, and of a few representative metals are especially likely to form complexes; many of these are highly colored (Fig. 23.11). In the remainder of this chapter we will consider the structure and bonding of these remarkable and generally very colorful compounds.

Figure 23.11

Metal ions that contain partially filled d subshells usually form colored complex ions; ions with empty d subshells (d^0) or with filled d subshells (d^{10}) usually form colorless complexes. This figure shows, from left to right, solutions containing $[M(H_2O)_6]^{n+}$ ions with M = Sc^{3+} (d^0), Cr^{3+} (d^3), Co^{2+} (d^7), Ni^{2+} (d^8), Cu^{2+} (d^9), and Zn^{2+} (d^{10}).

23.6   Basic Concepts

A **coordination compound,** or **complex,** is a Lewis acid–base adduct (Section 15.13) in which neutral molecules or anions (called **ligands**) are bonded to a **central metal ion** or atom by coordinate covalent bonds (Section 6.8). The ligands are Lewis bases, each having at least one pair of electrons to donate to the metal ion or atom. The central metal ion or atom is a Lewis acid, which can accept the pairs of electrons from the Lewis bases. Within a ligand, the atom attached directly to the metal through the coordinate covalent bond is called the **donor atom.**

The **coordination sphere** consists of the central metal ion plus its attached ligands. It is usually enclosed in brackets when written in a formula. The **coordination number** of the central metal ion is the number of donor atoms bonded to it. The coordination number for the silver ion in $[Ag(NH_3)_2]^+$ is 2; that for the copper(II) ion in $[Cu(NH_3)_4]^{2+}$ is 4; and that for the cobalt(III) ion in $[Co(NH_3)_6]^{3+}$ is 6 (Fig. 23.12 A). In each of these examples, the coordination number is also equal to the number of ligands in the coordination sphere, but that is not always the case. Some ligands, such as ethylenediamine, $H_2NCH_2CH_2NH_2$,

$$
\begin{array}{c}
\quad\;\; H \quad\; H \quad\; H \quad\; H \\
\quad\;\; | \quad\quad | \quad\quad | \quad\quad | \\
H-N-C-C-N-H \\
\quad\;\; \cdot\cdot \;\; | \quad\quad | \;\; \cdot\cdot \\
\quad\quad\;\; H \quad\; H
\end{array}
$$

contain two donor atoms (shown in color). The coordination number for cobalt in $[Co(H_2NCH_2CH_2NH_2)_3]^{3+}$ is 6 (Fig. 23.12 B). The coordination sphere of this complex contains only three ligands, but six nitrogen atoms are bonded to the cobalt. The most common coordination numbers are 2, 4, and 6, but examples of all coordination numbers from 2 to 12 are known.

When a ligand bonds to a central metal ion by several donor atoms, it is called a **polydentate ligand,** or a **chelating ligand.** These names are based on the Greek words for *tooth* and *claw,* respectively. A polydentate ligand "bites" the metal ion with more than one tooth (hence *poly,* for *many*). A chelate holds the metal ion rather like a crab's claw would hold a marble. A complex with a chelating ligand is called a **chelate;** examples include $[Co(H_2NCH_2CH_2NH_2)_3]^{3+}$ (Fig. 23.12 B) and the heme

Figure 23.12

Two complexes with a coordination number of 6. The six donor atoms are shown in color. (A) $[Co(NH_3)_6]^{3+}$ is formed from a Co^{3+} ion and six NH_3 molecules as ligands. (B) $[Co(en)_3]^{3+}$ is formed from a Co^{3+} ion and three $H_2NCH_2CH_2NH_2$ (ethylenediamine) molecules as ligands (the latter is abbreviated as en in the formula). Two N atoms from each en molecule bond to the Co^{3+} ion, giving it a coordination number of 6.

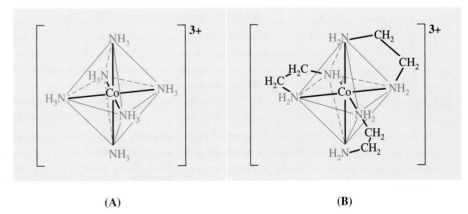

(A)   (B)

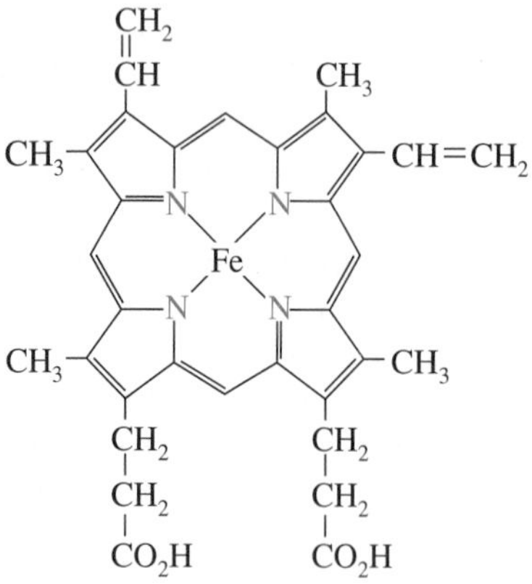

Figure 23.13

Heme, the square planar complex of iron found in hemoglobin. The donor atoms are shown in color.

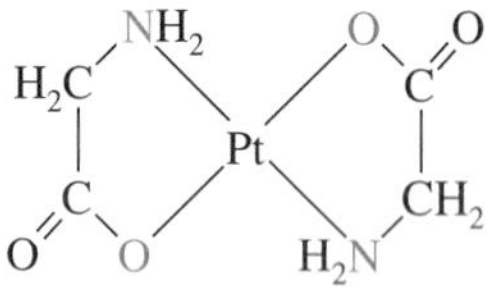

Figure 23.14

A complex composed of a Pt^{2+} ion and two anions of the amino acid glycine. The Pt^{2+} ion and the donor oxygen and nitrogen atoms (in color) are planar.

complex in hemoglobin (Fig. 23.13), which contains a polydentate ligand with four donor atoms.

Ligands are sometimes identified with prefixes that indicate the number of donor atoms in the ligand. Ligands with one donor atom are called **monodentate ligands** ("one-tooth" ligands); examples are Cl^-, OH^-, and NH_3. Ligands with two donor groups are called **bidentate ligands** ("two-tooth" ligands); examples are ethylenediamine, $H_2NCH_2CH_2NH_2$, and the anion of the amino acid glycine, $NH_2CH_2CO_2^-$ (Fig. 23.14). **Tridentate ligands, tetradentate ligands, pentadentate ligands,** and **hexadentate ligands** contain three, four, five, and six donor atoms, respectively. The ligand in heme (Fig. 23.13) is a tetradentate ligand.

23.7   The Naming of Complexes

The nomenclature of complexes is patterned after a system suggested by Alfred Werner, a Swiss chemist and Nobel laureate, whose outstanding work more than 90 years ago laid the foundation for a clearer understanding of these compounds. The following five rules are used for naming complexes.

1. If a coordination compound is ionic, name the cation first and the anion second, in accordance with usual nomenclature.

2. Name the ligands first, followed by the central metal.

3. Name the ligands alphabetically. Negative ligands (anions) have names formed by adding $-o$ to the stem name of the group; for example,

F^-	fluoro	NO_3^-	nitrato
Cl^-	chloro	OH^-	hydroxo
Br^-	bromo	O^{2-}	oxo
I^-	iodo	$C_2O_4^-$	oxalato
CN^-	cyano	CO_2^{2-}	carbonato

For most neutral ligands the name of the molecule is used. The four common exceptions are *aqua* (H_2O), *ammine* (NH_3), *carbonyl* (CO), and *nitrosyl* (NO).

4. If more than one ligand of a given type is present, the number is indicated by the prefixes *di-* (for two), *tri-* (for three), *tetra-* (for four), *penta-* (for five), and *hexa-*

(for six). Sometimes the prefixes *bis-* (for two), *tris-* (for three), and *tetrakis-* (for four) are used when the name of the ligand already includes *di-, tri-,* or *tetra-,* contains numbers, begins with a vowel, or is for a polydentate ligand.

5. When the complex is either a cation or a neutral molecule, the name of the central metal atom is spelled exactly like the name of the element and is followed by a Roman numeral in parentheses to indicate its oxidation state. When the complex is an anion, the suffix *-ate* is added to the stem for the name of the metal, followed by the Roman numeral designation of its oxidation state. Sometimes the Latin name of the metal is used when the English name is clumsy. For example, *ferrate* is used instead of *ironate, plumbate* instead of *leadate,* and *stannate* instead of *tinate.* Examples in which the complex is a cation (shown in color) are as follows:

$[Co(NH_3)_6]Cl_3$	Hexaamminecobalt(III) chloride
$[Pt(NH_3)_4Cl_2]^{2+}$	Tetraamminedichloroplatinum(IV) ion
$[Ag(NH_3)_2]^+$	Diamminesilver(I) ion
$[Cr(H_2O)_4Cl_2]Cl$	Tetraaquadichlorochromium(III) chloride
$[Co(H_2NCH_2CH_2NH_2)_3]_2(SO_4)_3$	Tris(ethylenediamine)cobalt(III) sulfate

Examples in which the complex is neutral:

$[Pt(NH_3)_2Cl_4]$	Diamminetetrachloroplatinum(IV)
$[Ni(H_2NCH_2CH_2NH_2)_2Cl_2]$	Dichlorobis(ethylenediamine)nickel(II)

Examples in which the complex is an anion (shown in color):

$[PtCl_6]^{2-}$	Hexachloroplatinate(IV) ion
$Na_2[SnCl_6]$	Sodium hexachlorostannate(IV)

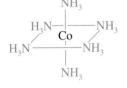

Figure 23.15

An abbreviated drawing of an octahedral complex ion.

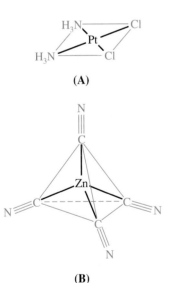

(A)

(B)

Figure 23.16

(A) The square planar configuration of $[Pt(NH_3)_2Cl_2]$. (B) The tetrahedral configuration of the $[Zn(CN)_4]^{2-}$ ion.

23.8 The Structures of Complexes

The structures of many compounds and ions were discussed in Chapter 7. Now we extend our study to coordination compounds with many of these same structures.

The most common structures of the complexes in coordination compounds are octahedral, tetrahedral, and square planar. Octahedral complexes have a coordination number of 6, and the six donor atoms are arranged at the corners of an octahedron around the central metal ion. Examples are shown in Fig. 23.12. Ions outside the brackets in the formula of a coordination compound do not form coordinate covalent bonds to the central metal atom but instead are bonded by ionic bonds. Thus the chloride and nitrate anions in $[Co(H_2O)_6]Cl_2$ and $[Cr(en)_3](NO_3)_3$ and the potassium cations in $K_2[PtCl_6]$ are not bonded to the metal ion.

Many chemists use an abbreviated drawing of an octahedron (Fig. 23.15) when describing the geometry about a metal ion.

Complexes in which the metal shows a coordination number of 4 exist in one of two different geometric arrangements: square planar or tetrahedral. Examples of four-coordinate complexes with a square planar geometry include $[Pt(NH_3)_2Cl_2]$ (Fig. 23.16 A) and $[Cu(NH_3)_4]^{2+}$; examples of four-coordinate complexes with a tetrahedral geometry include $[Zn(CN)_4]^{2-}$ (Fig. 23.16 B) and $[NiBr_4]^{2-}$.

Figure 23.17

The *cis* and *trans* isomers of
[Co(NH$_3$)$_4$Cl$_2$]$^+$.

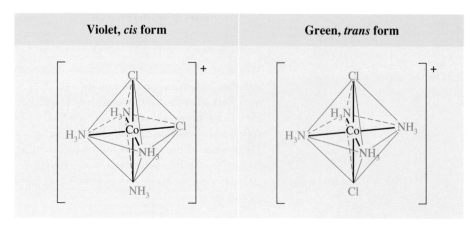

Violet, *cis* form	Green, *trans* form

23.9   Isomerism in Complexes

Certain complexes, such as the [Co(NH$_3$)$_4$Cl$_2$]$^+$ ion in [Co(NH$_3$)$_4$Cl$_2$]NO$_3$ and the Pt(NH$_3$)$_2$Cl$_2$ molecule, form **isomers,** different chemical species that have the same chemical formula (Section 9.5). The octahedral [Co(NH$_3$)$_4$Cl$_2$]$^+$ ion has two isomers. One form has a *cis* **configuration** (chloride ions occupy adjacent corners of the octahedron) and the other form has a *trans* **configuration** (chloride ions occupy opposite corners), as shown in Fig. 23.17. The square planar [Pt(NH$_3$)$_2$Cl$_2$] molecule also exists in *cis* and *trans* forms (Fig. 23.18). Isomers such as these, which differ only in the way in which the atoms are oriented in space relative to each other, are called **geometric isomers.**

Different geometric isomers of a substance are different chemical compounds. They exhibit different properties even though they have the same formula. For example, the two isomers of [Co(NH$_3$)$_4$Cl$_2$]NO$_3$ differ in color; the *cis* form is violet and the *trans* form is green. Furthermore, these isomers have different dipole moments, spectra, solubilities, and reactivities. The isomers of [Pt(NH$_3$)$_2$Cl$_2$] also differ in behavior. Perhaps the most interesting difference is that *cis*-[Pt(NH$_3$)$_2$Cl$_2$] exhibits antitumor activity whereas *trans*-[Pt(NH$_3$)$_2$Cl$_2$] does not.

Figure 23.18

The *cis* and *trans* isomers of
Pt(NH$_3$)$_2$Cl$_2$.

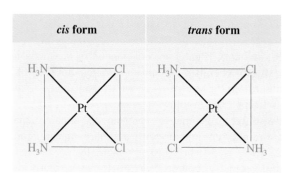

cis form	*trans* form

Figure 23.19

Optical isomers of [Co(H₂NCH₂CH₂NH₂)₃]³⁺. In these abbreviated formulas, N⌒N stands for H₂NCH₂CH₂NH₂.

Figure 23.19

Optical isomers of $[Co(H_2NCH_2CH_2NH_2)_3]^{3+}$. In these abbreviated formulas, N⌒N stands for $H_2NCH_2CH_2NH_2$.

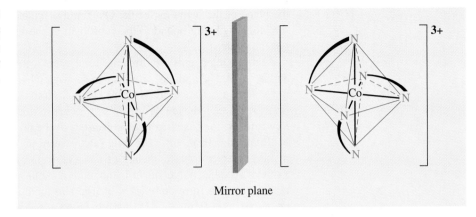

Mirror plane

Figure 23.20

The three isomeric forms of $[Co(en)_2Cl_2]^+$. In these abbreviated formulas, N⌒N stands for $H_2NCH_2CH_2NH_2$.

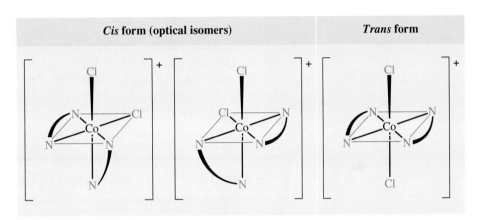

Cis form (optical isomers)   *Trans* form

Isomers that are mirror images of each other but not identical are called **optical isomers.** The tris(ethylenediamine)cobalt(III) ion, $[Co(H_2NCH_2CH_2NH_2)_3]^{3+}$, has two optical isomers, as shown in Fig. 23.19.

The $[Co(en)_2Cl_2]^+$ ion has two *cis* isomers, which are a pair of optical isomers, and one *trans* isomer (Fig. 23.20). The *trans* configuration has no optical isomerism. Its mirror images are superimposable, and therefore identical.

23.10   Uses of Complexes

Chlorophyll, the green pigment in plants, is a complex that contains magnesium. Plants appear green because chlorophyll absorbs red and purple light; the reflected light consequently appears green (Section 23.14). The energy resulting from the absorption of light is used in photosynthesis. The square planar copper(II) complex phthalocyanine blue (Fig. 23.21) is one of many complexes used as pigments or dyes. This complex is used in blue ink, blue jeans, and certain blue paints.

The structure of heme (Fig. 23.13), the iron-containing complex in hemoglobin, is very similar to that of chlorophyll. In hemoglobin the red heme complex is bonded to a large protein molecule (globin) by coordination of the protein to a position above

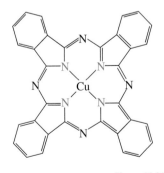

Figure 23.21

Copper phthalocyanine blue, a square planar copper complex found in some blue dyes.

the plane of the heme molecule. Oxygen molecules are transported by hemoglobin in the blood by being bound to the coordination site opposite the binding site of the globin molecule.

Complexing agents are often used for water softening because they tie up such ions as Ca^{2+}, Mg^{2+}, and Fe^{2+}, which make water hard. Complexing agents that tie up metal ions are also used as drugs. British Anti-Lewisite, $HSCH_2CH(SH)CH_2OH$, a drug developed during World War I as an antidote for the arsenic-based war gas Lewisite, is now used to treat poisoning by heavy metals such as arsenic, mercury, thallium, and chromium. The drug (abbreviated as BAL) is a ligand and functions by making a water-soluble chelate of the metal; this metal chelate is eliminated by the kidneys. Another polydentate ligand, enterobactin, which is isolated from certain bacteria, is used to form complexes of iron and thereby to control the severe iron build-up found in patients suffering from blood diseases such as Cooley's anemia, who require frequent transfusions. As the transfused blood breaks down, the usual metabolic processes that remove iron are overloaded, and excess iron can build up to fatal levels. Enterobactin forms a water-soluble complex with the excess iron, and this complex can be safely eliminated by the body.

In the electroplating industry it has been found that many metals plate out as a smoother, more uniform, better-looking, and more adherent surface when plated from a bath containing the metal as a complex ion. Thus complexes such as $[Ag(CN)_2]^-$ and $[Au(CN)_2]^-$ are used extensively in the electroplating industry.

Bonding and Electron Behavior in Coordination Compounds

Any theory of the arrangement of electrons in coordination compounds must explain four important properties of the compounds: their stabilities, their structures, their colors, and their magnetic properties. The first modern attempt to explain the properties of complex compounds invoked the concepts of valence bond theory (Section 7.4) and hybridization (Section 7.5) to describe the formation of bonds in these compounds. A later model relies on crystal field theory; it focuses on the electrostatic repulsions between the electrons of the central metal ion and the ligands to describe the behavior of the electrons that give rise to the color and magnetic properties of the compounds.

23.11   Valence Bond Theory

Valence bond theory treats a metal–ligand bond as a coordinate covalent bond (Section 6.8) formed when a filled orbital on the donor atom overlaps a hybrid orbital on the central metal atom. The electron pair from the ligand is shared with the metal, and this electron pair occupies both an atomic orbital on the ligand and one of several equivalent hybrid orbitals on the metal. The ligand is a Lewis base; the metal, a Lewis acid.

In a tetrahedral complex such as $[FeCl_4]^-$, four electron pairs from the four ligands also occupy hybrid orbitals on the metal ion. As we noted in Section 7.8, four equivalent hybrid orbitals on the metal result from the hybridization of one s and three *p* orbitals, giving sp^3 hybridization. Tetrahedral complexes can be described as sp^3 hybridized.

In an octahedral complex such as $[Co(NH_3)_6]^{3+}$, six electron pairs from the six ligands also occupy hybrid orbitals on the metal ion. As we noted in Section 7.9, six equivalent hybrid orbitals on the metal result from the hybridization of two *d*, one *s*, and three *p* orbitals. However, two different types of hybridization must be invoked to describe the bonding in octahedral complexes: inner-shell (d^2sp^3) hybridization and outer-shell (sp^3d^2) hybridization. The formation of a set of hybrid orbitals from two *d* orbitals of the inner *d* subshell and orbitals from the outer *s* and *p* subshells is called **inner-shell hybridization.** Hybridization of two 3*d* orbitals, a 4*s* orbital, and three 4*p* orbitals to give a set of d^2sp^3 hybrid orbitals on a cobalt ion is an inner-shell hybridization. The *d* orbitals come from an inner shell (the shell with $n = 3$), and the *s* and *p* orbitals come from the outer shell (the shell with $n = 4$). Hybridization of a 4*s* orbital, 4*p* orbitals, and 4*d* orbitals to give a set of sp^3d^2 hybrid orbitals on a cobalt ion is an **outer-shell hybridization.** The *d* orbitals come from the outer shell of the atom, the same shell that contains the *s* and *p* orbitals.

Whether an atom in an octahedral complex undergoes inner-shell hybridization or outer-shell hybridization must be determined experimentally. This is done by measuring the atom's magnetic moment (Section 23.13)—or by using some other experimental measurement—in order to determine the number of unpaired electrons on the atom. Let us consider as examples the hybridization in $[Fe(H_2O)_6]^{2+}$, which contains four unpaired electrons, and that in $[Fe(CN)_6]^{4-}$, which contains no unpaired electrons. Both of these ions are complexes of a Fe^{2+} ion. The electron configuration of a free Fe^{2+} ion is

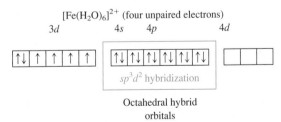

The Fe^{2+} ion contains four unpaired electrons. If the hybridization in the octahedral $[Fe(H_2O)_6]^{2+}$ ion is described as a sp^3d^2 outer-shell hybridization that involves the 4*d* orbitals, the four unpaired electrons of the free Fe^{2+} ion can be accommodated in the unhybridized 3*d* orbitals in the complex and remain unpaired, as is observed. The hybrid orbitals used in bonding must be formed from the 4*s*, the three 4*p*, and two of the 4*d* orbitals of the iron(II) ion. The distribution of electrons on Fe^{2+} in the $[Fe(H_2O)_6]^{2+}$ ion is

$[Fe(H_2O)_6]^{2+}$ (four unpaired electrons)

3*d* 4*s* 4*p* 4*d*

sp^3d^2 hybridization

Octahedral hybrid orbitals

The magnetic moment of $[Fe(CN)_6]^{4-}$ indicates that it has no unpaired electrons. The hybridization in this complex is an inner-shell hybridization. The 3*d* orbitals (rather than the 4*d* orbitals) appear to be involved in the inner-shell d^2sp^3 hybridization of the Fe^{2+} ion in this octahedral complex ion. This must force the unpaired electrons from the 3*d* orbitals used in the hybridization into the unhybridized 3*d* orbitals and results in the pairing of all electrons.

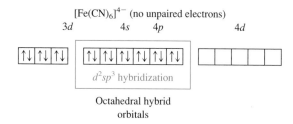

$[Fe(CN)_6]^{4-}$ (no unpaired electrons)

Octahedral hybrid
orbitals

In the case of four-coordinate structures with a square planar configuration, the four bonds of the central atom arise from dsp^2 hybridization, as illustrated here for $[Ni(CN)_4]^{2-}$.

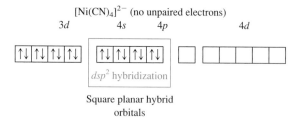

$[Ni(CN)_4]^{2-}$ (no unpaired electrons)

Square planar hybrid
orbitals

23.12   Crystal Field Theory

The colors of complexes (Fig. 23.22) and their magnetic properties can be considered to result from the electrons that are not involved in the covalent bonding between the metal ion and the ligands. It is possible to use a model involving simple electrostatic interactions between the ligands and the electrons in the unhybridized orbitals of the central metal atom to understand, interpret, and predict the colors, magnetic behavior, and some structures of coordination compounds of transition metals. This electrostatic model of the properties of complexes is called **crystal field theory.**

Crystal field theory does not explain the bonding in complexes; it describes only behavior that can be attributed to electrons on the metal atom in the complex. Like valence bond theory, crystal field theory tells only part of the story of the behavior of complexes. However, it tells the part that valence bond theory does not. In its pure form, crystal field theory ignores any covalent bonding between ligands and metal ions. Both the ligand and the metal are treated as infinitesimally small point charges.

Figure 23.22

Colors of various compounds of chromium, with different oxidation states and ligands. From left to right, solutions of K_2CrO_4, $K[Cr(H_2O)_2 (OH)_4]$, $[Cr(H_2O)_6](NO_3)_3$, $K_2Cr_2O_7$, and $[Cr(H_2O)_4Cl_2]Cl$.

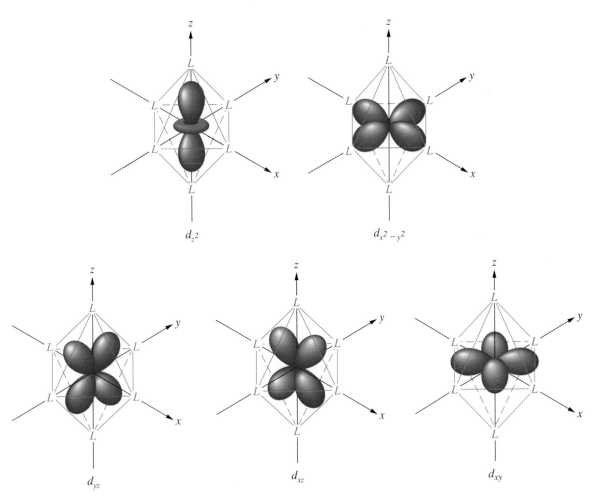

d_{z^2} $d_{x^2-y^2}$

d_{yz} d_{xz} d_{xy}

Figure 23.23

Diagrams showing the directional characteristics of the five *d* orbitals. *L* indicates a ligand at each corner of the octahedron.

The effect of ligands on the electrons of a metal results from the electrostatic repulsions between the negative charge on the ligands and the electrons of the central metal.

Let us consider the behavior of the electrons in the unhybridized *d* orbitals in an octahedral complex. The *d* orbitals, which occur in sets of five, consist of lobe-shaped regions and are arranged in space as shown in Fig. 23.23, reproduced within an octahedral structure.

The lobes in two of the five *d* orbitals, the d_{z^2} and $d_{x^2-y^2}$ orbitals, point toward the ligands on the corners of the octahedron around the metal (Fig. 23.23). These two orbitals are called the e_g **orbitals** (the symbol actually refers to the symmetry of the orbitals, but we will use it as a convenient name for these two orbitals in an octahedral complex). The other three orbitals, the d_{xy}, d_{xz}, and d_{yz} orbitals, the lobes of which point between the ligands on the corners of the octahedron, are called the t_{2g} **orbitals** (again the symbol really refers to the symmetry of the orbitals). As six ligands approach the metal ion along the axes of the octahedron, their point charges repel the electrons in the *d* orbitals of the metal ion. However, the repulsions between the electrons in the e_g orbitals (the d_{z^2} and $d_{x^2-y^2}$ orbitals) and the ligands are greater than the repulsions between the electrons in the t_{2g} orbitals (the d_{xy}, d_{xz}, and d_{yz} orbitals) and the ligands because the lobes of the e_g orbitals point directly at the ligands, whereas the lobes of the t_{2g} orbitals point between them. Thus electrons in e_g orbitals of a metal ion in an octahedral complex have higher potential energies than those of electrons in t_{2g} orbitals. The difference in energy may be represented as follows:

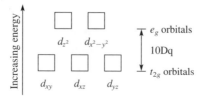

The difference in energy between the e_g and the t_{2g} orbitals is called the **crystal field splitting** and is symbolized by **10Dq.**

The size of the crystal field splitting (10Dq) depends on the nature of the six ligands located around the central metal ion. Different ligands produce different crystal field splittings. The increasing crystal field splitting produced by ligands is expressed in the **spectrochemical series,** a short version of which is given here.

$$I^- < Br^- < Cl^- < F^- < H_2O < C_2O_4^{2-} < NH_3 < en < NO_2^- < CN^-$$

A few ligands of the spectrochemical series, in order of increasing field strength of the ligand

In this series the ligands on the left have a low field strength, and those on the right have a high field strength. Thus the crystal field splitting produced by an iodide ion (I^-) as a ligand is much smaller than that produced by a cyanide ion (CN^-).

In a simple metal ion in the gas phase, the electrons are distributed among the five 3d orbitals in accord with Hund's rule (Section 5.8), because the orbitals all have the same energy. However, if the metal ion lies in an octahedron formed by six ligands, the energies of the d orbitals are no longer the same, and two opposing forces are set up. One force tends to keep the electrons of the metal ion distributed with unpaired spins within all of the d orbitals. It requires energy to pair up electrons in an orbital; this energy is called the **pairing energy, P.** The other force tends to reduce the average energy of the d electrons by placing as many of them as possible in the lower-energy t_{2g} orbitals. The d electrons end up with the lowest possible total energy. If it requires less energy for the d electrons to be excited to the upper e_g orbitals than to pair in the lower t_{2g} orbitals (10Dq $< P$), then they will remain unpaired. If it requires less energy to pair d electrons in the lower t_{2g} orbitals than to put them in the upper e_g orbitals (10Dq $> P$), then they will pair.

In $[Fe(CN)_6]^{4-}$ the strong field of six cyanide ions produces a large crystal field splitting. Under these conditions the electrons require less energy to pair than to be excited to the e_g orbitals (10Dq $> P$). The six 3d electrons of the Fe^{2+} ion pair in the three t_{2g} orbitals. The result is in agreement with the experimentally measured magnetic moment of $[Fe(CN)_6]^{4-}$ (no unpaired electrons).

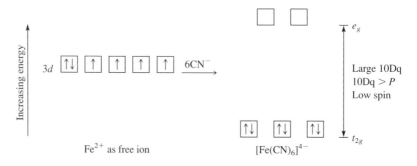

Complexes such as the $[Fe(CN)_6]^{4-}$ ion, in which the electrons are paired because of the large crystal field splitting, are called **low-spin complexes** because the number of unpaired electrons (spins) is a minimum.

In $[Fe(H_2O)_6]^{2+}$, on the other hand, the weak field of the water molecules produces only a small crystal field splitting. Because it requires less energy to excite electrons to the e_g orbitals than to pair them ($10Dq < P$), they remain distributed in all five $3d$ orbitals.

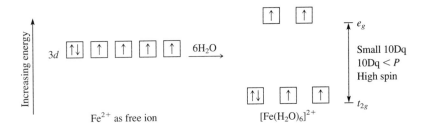

Fe^{2+} as free ion $[Fe(H_2O)_6]^{2+}$

Four unpaired electrons should be present, as is verified by measuring the magnetic moment of $[Fe(H_2O)_6]^{2+}$. Complexes such as the $[Fe(H_2O)_6]^{2+}$ ion, in which the electrons are unpaired because the crystal field splitting is not large enough to cause them to pair, are called **high-spin complexes** because the number of unpaired electrons (spins) is a maximum.

A similar line of reasoning shows why the $[Fe(CN)_6]^{3-}$ ion is a low-spin complex with only one unpaired electron, whereas the $[Fe(H_2O)_6]^{3+}$ ion and the $[FeF_6]^{3-}$ ion are high-spin complexes with five unpaired electrons.

23.13   Magnetic Moments of Molecules and Ions

Molecules such as O_2 (Section 7.16) and ions such as $[Fe(H_2O)_6]^{2+}$ (Section 23.12) that contain unpaired electrons are **paramagnetic.** As shown in the photograph of liquid oxygen on page 243, paramagnetic substances tend to be attracted into a magnetic field, such as that between the poles of a magnet. Many transition metal complexes have unpaired electrons and hence are paramagnetic. Molecules such as N_2 (Section 7.16) and ions such as Na^+ and $[Fe(CN)_6]^{4-}$ (Section 23.12) that contain no unpaired electrons are **diamagnetic.** Diamagnetic substances have a slight tendency to be repelled by a magnetic field.

An electron in an atom spins about its own axis. Because the electron is electrically charged, this spin gives it the properties of a small magnet, with north and south poles. Two electrons in the same orbital spin in opposite directions, and their magnetic moments cancel because their north and south poles are opposed. When an electron in an atom or ion is unpaired, the magnetic moment due to its spin makes the entire atom or ion paramagnetic. Thus a sample containing such atoms or ions is paramagnetic. The size of the magnetic moment of a system containing unpaired electrons is related directly to the number of such electrons; the greater the number of unpaired electrons, the larger the magnetic moment. Therefore, the observed magnetic moment is used to determine the number of unpaired electrons present.

Figure 23.24

Passing white light through a prism shows that it is actually a mixture of all colors of visible light (red, orange, yellow, green, blue, indigo, and violet).

23.14   Colors of Transition Metal Complexes

When atoms absorb light of the proper frequency, their electrons are excited to higher energy levels (Chapter 5). The same thing can happen in coordination compounds. Electrons can be excited from the lower-energy t_{2g} to the next higher-energy e_g orbitals, provided that the latter are not already filled with paired electrons.

The human eye perceives a mixture of all the colors, in the proportions present in sunlight, as white light (Fig. 23.24). The eye also utilizes complementary colors in color vision and perceives a mixture of two complementary colors, in the proper proportions, as white light. Likewise, when a color is missing from white light, the eye sees its complement (Fig. 23.25). For example, as shown in Table 23.3 on page 889, when red light is removed from white light, the eyes sees the color blue-green; when violet is removed from white light, the eye sees lemon yellow; when green light is removed, the eye sees purple. The blue color of the $[Cu(NH_3)_4]^{2+}$ ion (Fig. 23.26 on page 889) results because this ion absorbs orange and red light, leaving the complementary colors of blue and blue-green (Fig. 23.27 on page 889).

Consider $[Fe(CN)_6]^{4-}$ (Section 23.12). The electrons in the t_{2g} orbitals can absorb energy and be excited to the next higher energy level. The necessary energy corresponds to photons of violet light. When white light impinges on $[Fe(CN)_6]^{4-}$, violet light is absorbed (to accomplish the excitation), and the eye sees the unabsorbed complement, lemon yellow. $K_4[Fe(CN)_6]$ is lemon yellow (Fig. 23.28 on page 889). In contrast, when white light strikes $[Fe(H_2O)_6]^{2+}$, red light (longer wavelength, lower energy) is absorbed. The eye sees its complement, blue-green. $[Fe(H_2O)_6]SO_4$, for example, is therefore blue-green.

A coordination compound of the Cu^+ ion has a d^{10} configuration, and all the e_g orbitals are filled. In order to excite an electron to a higher level, such as the $4p$ orbital, photons of very high energy are needed. This energy corresponds to very short wavelengths in the ultraviolet region of the spectrum. No visible light is absorbed, so the eye sees no change and the compound appears white or colorless. A solution con-

Figure 23.25

(A) An object is black if it absorbs all colors of light. (B) If it reflects all colors of light, it is white. (C) An object (such as this yellow strip) has a color if it absorbs all colors except one (yellow in this case). (D) The strip also appears yellow if it absorbs the complementary color from white light (the complementary color of yellow is indigo).

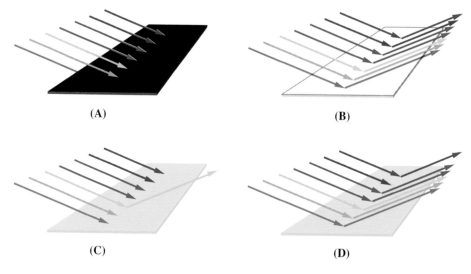

(A)   (B)

(C)   (D)

Table 23.3
Complementary Colors

Wavelength, Å	Spectral Color	Complementary Color[a]
4100	Violet	Lemon yellow
4300	Indigo	Yellow
4800	Blue	Orange
5000	Blue-green	Red
5300	Green	Purple
5600	Lemon yellow	Violet
5800	Yellow	Indigo
6100	Orange	Blue
6800	Red	Blue-green

[a]The complementary color is seen when the spectral color is removed from white light.

taining $[Cu(CN)_2]^-$, for example, is colorless. On the other hand, Cu^{2+} complexes have a vacancy in the e_g orbitals, and electrons can be excited to this level. The wavelength (energy) of the light absorbed corresponds to the visible part of the spectrum, and Cu^{2+} complexes are almost always colored—blue, blue-green, violet, or yellow (Fig. 23.25).

As we noted earlier, strong-field ligands cause a large split in the energies of the d orbitals of the central metal atom (significantly negative 10Dq value). Transition metal coordination compounds with these ligands are yellow, orange, or red because they absorb higher-energy violet or blue light. On the other hand, coordination compounds of transition metals with weak-field ligands are blue-green, blue, or indigo because they absorb lower-energy yellow, orange, or red light.

Figure 23.26
A solution of $[Cu(NH_3)_4]^{2+}$.

Figure 23.27   A solution of $[Cu(NH_3)_4]^{2+}$ is blue because the ion absorbs orange light and red light, the complementary colors of blue and blue-green, respectively.

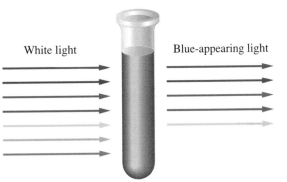

White light   Blue-appearing light

Figure 23.28
A solution of $K_4[Fe(CN)_6]$.

For Review | Summary

The **transition elements** are the elements with partially filled *d* orbitals (***d*-block elements**) or *f* orbitals (***f*-block elements**) in their valence shells. The *d*-block elements are then classified into the first, second, third, and fourth **transition series**; the *f*-block elements, into the **lanthanide series** and the **actinide series.**

The reactivity of the transition elements varies widely. The lanthanide elements are very active metals, while the *d*-block elements range from very active metals such as scandium and iron to almost inert elements such as the **platinum metals.** The type of chemistry used in the isolation of the elements from their ores depends upon the concentration of the element in its ore and the difficulty of reducing ions of the elements to the metal. More active metals are more difficult to reduce.

Compounds of the elements of the first transition series (Sc–Cu) exhibit chemical behavior typical of metals. For example, they oxidize in air upon heating and react with elemental halogens to form halides. Those elements that lie above hydrogen in the activity series react with acids producing salts and hydrogen gas. Simple compounds of these elements with oxidation states of +1, +2, and +3 are ionic. Oxides, hydroxides, and carbonates of such compounds are basic. Halides and other salts are generally stable in water, although oxygen must be excluded in some cases. With higher oxidation states, these elements form covalent compounds. The covalent oxides exhibit acidic behavior.

The transition elements and some representative metals can form **coordination compounds,** or **complexes,** in which a **central metal atom** or **ion** is bonded to two or more **ligands** by co-ordinate covalent bonds. Ligands with more than one donor atom are called **polydentate ligands** and form **chelates.** The common geometries found in complexes are tetrahedral and square planar (both with a **coordination number** of 4) and octahedral (with a coordination number of 6). *Cis* and *trans* **configurations** are possible in some octahedral and square planar complexes. In addition to these **geometrical isomers, optical isomers** (molecules or ions that are mirror images but not identical) are possible in certain octahedral complexes.

The bonding in coordination compounds can be described in terms of valence bond theory, although this model of bonding does not explain the magnetic properties and colors very well. In the valence bond model, the bonds are regarded as normal coordinate covalent bonds between a ligand, which acts as a Lewis base, and an empty hybrid metal atomic orbital.

Crystal field theory treats interactions between the electrons on the metal and the ligands as a simple electrostatic effect. The presence of the ligands near the metal ion changes the energies of the metal *d* orbitals relative to their energies in the free ion. Both the color and the magnetic properties of a complex can be attributed to this **crystal field splitting.** The magnitude of the splitting (**10Dq**) depends on the nature of the ligands bonded to the metal. Strong-field ligands produce large splittings and favor formation of **low-spin complexes,** in which the t_{2g} orbitals are completely filled before any electrons occupy the e_g orbitals. Weak-field ligands favor formation of **high-spin complexes.** The t_{2g} and the e_g orbitals are singly occupied before any are doubly occupied.

Key Terms and Concepts

central metal ion (23.6)
chelate (23.6)
chelating ligand (23.6)
cis configuration (23.9)
color (23.14)
complex (23.6)
coordination compound (23.6)
coordination number (23.6)
coordination sphere (23.6)

crystal field theory (23.12)
donor atom (23.6)
e_g orbitals (23.12)
geometry of complexes (23.9)
geometric isomer (23.9)
high-spin complex (23.12)
hydrometallurgy (23.3)
ligand (23.6)
low-spin complex (23.12)

optical isomer (23.9)
pairing energy (23.12)
polydentate ligand (23.6)
rare earth element (23.1)
spectrochemical series (23.12)
superconductor (23.5)
t_{2g} orbitals (23.12)
trans configuration (23.9)

Exercises

Chemistry of Transition Elements

1. Write the electron configurations for the following elements: Sc, Ti, Cr, Fe, Mo, Ru.

2. Write the electron configurations for Ti and the Ti^{2+}, Ti^{3+}, and Ti^{4+} ions.

3. Write the electron configurations for the following elements and their 3+ ions: La, Sm, and Lu.

4. Why are the elemental (free) rare earth elements not found in nature?

5. Which of the following elements is most likely to be used to prepare La by the reduction of La_2O_3: Al, C, or Fe? Why?

6. Which of the following is the strongest oxidizing agent: VO_4^{3-}, CrO_4^{2-}, or MnO_4^-?

7. Which of the following elements is most likely to form an oxide with the formula MO_3: Zr, Nb, Mo?

8. Predict the products of the following reactions. *Note:* In addition to using the information in this chapter, also use the knowledge you have accumulated at this stage of your study, including information on the prediction of reaction products (see Section 8.8).

   (a) $CuCO_3 + HI(aq) \longrightarrow$

   (b) $CoO + O_2 \xrightarrow{\Delta}$

   (c) $La + O_2 \xrightarrow{\Delta}$

   (d) $V + VCl_4 \xrightarrow{\Delta}$

   (e) $Co + xsF_2 \longrightarrow$

   (f) $CrO_3 + CsOH(aq) \longrightarrow$

   (g) $Fe + H_2SO_4 \longrightarrow$

   (h) $FeCl_3(aq) + NaOH(aq) \longrightarrow$

   (i) $Mn(OH)_2 + HBr(aq) \longrightarrow$

   (j) $Cr + O_2 \xrightarrow{\Delta}$

   (k) $Mn_2O_3 + HCl(aq) \longrightarrow$

   (l) $Ti + xsF_2 \xrightarrow{\Delta}$

9. Describe the electrolytic process for refining copper.

10. What reactions occur in a blast furnace? Which of these are oxidation–reduction reactions?

11. Predict the products of the following reactions. (See the note in Exercise 8.)

   (a) Zn is added to a solution of $Cr_2(SO_4)_3$ in acid.

   (b) $TiCl_3$ is added to a solution containing an excess of CrO_4^{2-} ion.

   (c) $Cr^{2+} + CrO_4^{2-}$ in acid solution.

   (d) Mn is heated with CrO_3.

   (e) $CrO + 2HNO_3$ in water.

   (f) $CrCl_3$ is added to an aqueous solution of NaOH.

12. Would you expect a manganese(VII) oxide solution to have a pH greater or less than 7.0? Justify your answer.

13. What is the gas produced when iron(II) sulfide is treated with a nonoxidizing acid?

14. Iron(II) can be titrated to iron(III) by dichromate ion, which is reduced to chromium(III) in acid solution. A 4.600-g sample of iron ore is dissolved and the iron converted to iron(II). Exactly 23.52 mL of 0.0150 M $Na_2Cr_2O_7$ is required in the titration. What percentage of the ore sample was iron?

15. Predict the products of the following reactions. (See the note in Exercise 8.)

   (a) Fe is heated in an atmosphere of steam.

   (b) NaOH is added to a solution of $Fe(NO_3)_3$.

   (c) $FeSO_4$ is added to an acidic solution of K_2CrO_4.

   (d) Fe is added to a dilute solution of H_2SO_4.

   (e) A solution of $Fe(NO_3)_2$ and HNO_3 is allowed to stand in air.

   (f) $FeCO_3$ is added to a solution of $HClO_4$.

   (g) Fe is heated in air.

16. Balance the following equation by oxidation–reduction methods, noting that *three* elements change oxidation state.

   $$Co(NO_3)_2 \longrightarrow Co_2O_3 + NO_2 + O_2$$

17. How many cubic feet of air at a pressure of 674 torr and 35°C is required per ton of Fe_2O_3 to convert that Fe_2O_3 into iron in a blast furnace? For this exercise assume air is 19% oxygen by volume.

18. What is the percent by mass of cobalt in sodium hexanitrocobaltate(III), $Na_3[Co(NO_2)_6]$?

19. Find the potential of the following electrochemical cell.

   $$Cd\,|\,Cd^{2+}, M = 0.10\,\|\,Ni^{2+}, M = 0.50\,|\,Ni$$

20. How many grams of $CuCl_2$ contain the same mass of copper as 100 g of CuCl?

21. Dilute sodium cyanide solution is slowly dripped into a slowly stirred silver nitrate solution. A white precipitate forms temporarily but dissolves as the addition of sodium cyanide continues. Use chemical equations to explain this observation. Silver cyanide is similar to silver chloride in its solubility.

22. The formation constant of $[Cu(NH_3)_4]^{2+}$ is 1.2×10^{12}. What will be the equilibrium concentration of Cu^{2+} if 1.0 g of Cu is oxidized and put into 1.0 L of 0.25 M NH_3 solution?

23. The formation constant of $[Ag(CN)_2]^-$ is 1.0×10^{20}. What will be the equilibrium concentration of Ag^+ if 1.0 g of Ag is oxidized and put into 1 L of 1.0×10^{-1} M CN^- solution?

24. A 2.5624-g sample of a pure solid alkali metal chloride is dissolved in water and treated with excess silver nitrate. The resulting precipitate, filtered and dried, weighs 3.03707 g. What was the percent by mass of chloride ion in the original compound? What is the identity of the salt?

25. Would you expect salts of the gold(I) ion, Au^+, to be colored? Explain.

Structure and Nomenclature of Coordination Compounds

26. Indicate the coordination number for the central metal atom in each of the following coordination compounds.

   (a) $[Pt(H_2O)_2Br_2]$

   (b) $[Pt(NH_3)(py)(Cl)(Br)]$ (py = pyridine, C_5H_5N)

(c) $[Zn(NH_3)_2Cl_2]$

(d) $[Zn(NH_3)(py)(Cl)(Br)]$

(e) $[Ni(H_2O)_4Cl_2]$

(f) $[Fe(en)_2(CN)_2]^+$ (en is ethylenediamine)

27. Give the coordination numbers and write the formulas for the following, including all isomers where appropriate.

   (a) tetrahydroxozincate(II) ion (tetrahedral)

   (b) hexacyanopalladate(IV) ion

   (c) dichloroaurate(I) ion (*aurum* is Latin for gold)

   (d) diamminedichloroplatinum(II)

   (e) potassium diamminetetrachlorochromate(III)

   (f) hexaamminecobalt(III) hexacyanochromate(III)

   (g) dibromobis(ethylenediamine)cobalt(III) nitrate

28. Give the coordination number for each coordinated metal ion in the following compounds.

   (a) $[Co(CO_3)_3]^{3-}$ (CO_3^{2-} is bidentate (in this complex.)

   (b) $[Cu(NH_3)_4]^{2+}$

   (c) $[Co(NH_3)_4Br_2]_2SO_4$

   (d) $[Pt(NH_3)_4][PtCl_4]$

   (e) $[Cr(en)_3](NO_3)_3$

   (f) $[Pd(NH_3)_2Br_2]$ (square planar)

   (g) $K_3[Fe(CN)_6]$

   (h) $[Zn(NH_3)_2Cl_2]$

29. Sketch the structures of the following complexes. Indicate any *cis*, *trans*, and optical isomers.

   (a) $[Pt(H_2O)_2Br_2]$ (square planar)

   (b) $[Pt(NH_3)(py)(Cl)(Br)]$ (square planar, py = pyridine, C_5H_5N)

   (c) $[Zn(NH_3)_3Cl]^+$ (tetrahedral)

   (d) $[Pt(NH_3)_3Cl]^+$ (square planar)

   (e) $[Ni(H_2O)_4Cl_2]$

   (f) $[Co(C_2O_4)_2Cl_2]^{3-}$ ($C_2O_4^{2-}$ is the bidentate oxalate ion, $^-O_2CCO_2^-$)

30. Draw diagrams for any *cis*, *trans*, and optical isomers that could exist for the following (en is ethylenediamine).

   (a) $[Co(en)_2(NO_2)Cl]^+$

   (b) $[Co(en)_2Cl_2]^+$

   (c) $[Cr(NH_3)_2(H_2O)_2Br_2]^+$

   (d) $[Pt(NH_3)_2Cl_4]$

   (e) $[Cr(en)_3]^{3+}$

   (f) $[Pt(NH_3)_2Cl_2]$

31. Name each of the compounds or ions given in Exercise 28.

32. Name each of the compounds or ions given in Exercise 30.

Bonding in Complexes

33. Draw orbital diagrams and indicate what type of hybridization you would expect for each of the following.

   (a) $[Co(NH_3)_6]^{3+}$

   (b) $[Zn(NH_3)_4]^{2+}$

   (c) $[Cd(CN)_4]^{2-}$

   (d) $[Fe(en)_3]^{3+}$

   (e) $[Fe(CN)_6]^{4-}$

34. Show by means of orbital diagrams the hybridization for each of the following complexes.

   (a) $[Cu(CN)_2]^-$ (linear)

   (b) $[HgCl_3]^-$ (trigonal planar)

   (c) $[CoBr_4]^{2-}$ (tetrahedral)

   (d) $[Ni(CN)_4]^{2-}$ (square planar)

   (e) $[Co(NH_3)_6]^{3+}$ (octahedral)

35. Determine the number of unpaired electrons expected for $[Fe(CN)_6]^{3-}$ and for $[Fe(H_2O)_6]^{3+}$ in terms of crystal field theory.

36. Is it possible for a complex of a metal of the first transition series to have six or seven unpaired electrons? Explain.

37. How many unpaired electrons are present in each of the following?

   (a) $[CoF_6]^{3-}$ (high-spin)

   (b) $[Co(en)_3]^{3+}$ (low-spin)

   (c) $[Mn(CN)_6]^{3-}$ (low-spin)

   (d) $[Mn(CN)_6]^{4-}$ (low-spin)

   (e) $[MnCl_6]^{4-}$ (high-spin)

   (f) $[RhCl_6]^{3-}$ (low-spin)

38. Explain how the diphosphate ion, $O_3POPO_3^{4-}$, can function as a water softener by complexing Fe^{2+} and preventing the precipitation of an insoluble iron salt.

39. For complexes of the same metal ion with no change in oxidation number, the stability increases as the number of electrons in the $\pm_{2g}$ orbitals increases. Which complex in each of the following pairs of complexes is the more stable? $[Fe(H_2O)_6]^{2+}$ or $[Fe(CN)_6]^{4-}$; $[Co(NH_3)_6]^{3+}$ or $[CoF_6]^{3-}$; $[Mn(CN)_6]^{4-}$ or $[MnCl_6]^{4-}$.

Applications and Additional Exercises

40. Determine the crystal field splitting of the *d* orbitals in a two-coordinated complex as the two ligands approach the metal along the *z* axis.

41. What is the crystal field splitting of the nickel ion in NiO? Of the Mn^{2+} ion in MnF_2? (See Section 11.15.)

42. Trimethylphosphine, $:P(CH_3)_3$, can act as a ligand by donating the lone pair of electrons on the phosphorus atom. If trimethylphosphine is added to a solution of nickel(II) chloride in acetone, a blue compound that has a molecular mass of approximately 270 and contains 21.5% Ni, 26.0% Cl, and 52.5% $P(CH_3)_3$ can be isolated. This blue compound does not have any isomeric forms. What are the geometry and molecular formula of the blue compound?

43. Calculate the concentration of free copper ion that is present in equilibrium with 1.0×10^{-3} M $[Cu(NH_3)_4]^{2+}$ and 1.0×10^{-1} M NH_3.

44. Using the radius ratio rule (Section 11.16) and the ionic radii given in Fig. 5.24, predict whether complexes of Ni^{2+}, Mn^{2+}, and Sc^{3+} with Cl^- will be tetrahedral or octahedral. Write the formulas for these complexes.

45. The standard reduction potential for the reaction

$$[Co(H_2O)_6]^{3+} + e^- \longrightarrow [Co(H_2O)_6]^{2+}$$

is about 1.8 V. The reduction potential for the reaction

$$[Co(NH_3)_6]^{3+} + e^- \longrightarrow [Co(NH_3)_6]^{2+}$$

is +0.1 V. Calculate the cell potentials to show which of the complex ions, $[Co(H_2O)_6]^{2+}$ or $[Co(NH_3)_6]^{2+}$, can be oxidized to the corresponding cobalt(III) complex by oxygen.

46. The complex ion $[Co(en)_3]^{3+}$ is diamagnetic. Would you expect the $[Co(en)_3]^{2+}$ ion to be diamagnetic or paramagnetic? The $[Co(CN)_6]^{3-}$ ion? Explain your reasoning in each case.

CHAPTER OUTLINE

Polymers

24.1   Factors That Affect the Properties of Polymers
24.2   Polymer Properties
24.3   Elastomers, Thermoplastics, and Thermosetting Polymers
24.4   Recycling of Polymers
24.5   The Future of Polymers

Metals, Insulators, and Semiconductors

24.6   Band Theory
24.7   Metals
24.8   Semiconductors and Insulators
24.9   The Solar Cell

Ceramics

24.10 Catalytic Converters and Space Shuttle Tiles
24.11 Glass
24.12 Superconducting Ceramics

Thin Films

24.13 Films on Windows and Lenses
24.14 Diamond Thin Films
24.15 Synthesis of Thin Films

Materials in Medicine

24.16 Artificial Hips
24.17 Nitinol

24

The Chemistry of Materials

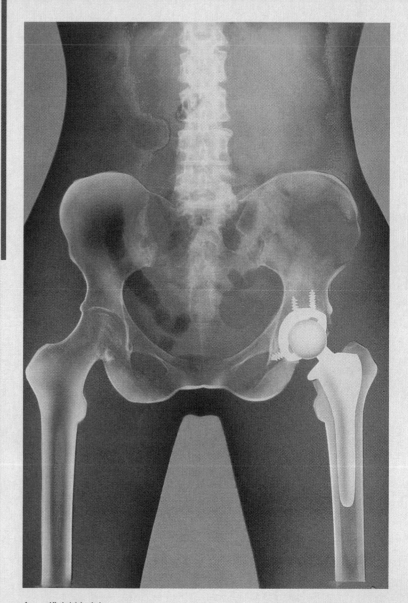

An artificial hip joint.

Progress in human development is often recorded as transformations from one "age" to another. As improved tools became available, the Stone Age gave way to the Bronze Age, the Bronze Age became the Iron Age, and so on. The materials available shaped the lives of the people who lived in those times. The same is true today, modern materials shape our lives.

The field of study that is concerned with the relationships between the structure and composition of a solid and its properties is called *materials science*. This is an interdisciplinary field that often involves teams of scientists, including chemists, engineers, biologists, and physicists, working together to solve problems. Chemists contribute to materials science in two main ways. First, they tend to be skilled at the preparation of materials. Second, they are able to visualize and understand the behavior of substances at the microscopic (atomic/molecular) level. In this chapter, we draw on what we have learned about these areas in previous chapters to understand the applications of materials science presented here.

Polymers

The modern era has been called the Plastic Age. As we look at our present surroundings, we see many examples of plastics—materials that can be shaped by applying heat or pressure. Most are synthetic polymers, though some are based on natural substances. (See Section 9.15 for an introduction to polymers.) Our clothing may contain one or more of the polymers cotton, nylon, orlon, and dacron. So might the carpet or chair covering. We may be seated near a phone, computer, or calculator that has a plastic case. Eyeglasses and contact lenses may be plastic, as may be our cup or soft drink bottle. Plastics can be stronger than steel, less dense than water, and corrosion resistant. Some can be implanted in the human body as replacements for blood vessels or tissue. The substitution of plastic for metal has made possible an increase in the mileage of the modern car and the production of military aircraft that are invisible to radar.

Though we may think of plastics and polymers as a recent development, natural polymers have existed for millions of years. Amber, starch, cellulose, proteins, and nucleic acids are all natural polymers. Only quite recently, however, have we been able to create human-made, or synthetic, polymers. For example, the first synthetic polymer (cellulose nitrate, or celluloid) was developed in the 1860s as the result of a contest to produce an alternative to ivory billiard balls made from elephant tusks. Currently, there are about 60,000 kinds of synthetic polymers known, and they have a wide variety of properties. Almost none of these synthetic polymers were known 50 years ago.

Chemists strive to understand the properties of matter at the atomic and molecular scale. They concern themselves, for example, with what white foam cups, clear plastic cups, and fishing line have in common and what makes them different or why some objects made out of polyethylene are "dishwasher safe," whereas others melt in a dishwasher. In the next few sections, we will see the outcomes of such studies, that is, how the properties of a polymer are related to its composition and structure.

24.1 Factors That Affect the Properties of Polymers

In Section 9.15 we saw that polymers are large molecules assembled via the reaction of smaller molecules called monomers. We can think of most individual polymer chains as being like a train with thousands of identical boxcars. A collection of polymer chains has been likened to a plate of very long spaghetti with flexible, coiled, and intertwined pieces. However, the strands of a polymer are relatively much longer than spaghetti. If the thickness of a polymer backbone were increased to the thickness of a piece of spaghetti, the corresponding length of even a moderate-size polymer chain would be about 40 feet.

Though thousands of polymers are known, each owes its physical properties to five main factors. These factors are the chemical composition of the polymer, the average molecular weight of the polymer chains, and the stereochemistry, topology, and morphology of the polymer. The job of someone trying to prepare a polymer with new properties is to understand how these five parameters are related to the desired properties of the product.

Chemical Composition.
Chemical composition consists of the number and identity of monomers that have been used to form the polymer (Table 24.1). A **homopolymer,** such as polyethylene or starch, is formed from only one type of monomer. **Copolymers** are formed from two monomer units, and the properties of a copolymer often are intermediate between those of the parent homopolymers. The monomer units in a copolymer may arrange themselves in a regular, alternating pattern (ABABABAB . . .), or they may attach randomly (ABBABAABABBBA . . .) or in blocks (AAAAABBBBBBBBAAAAAA . . .). Proteins can be considered polymers that are built out of amino acids.

The properties of a polymer are related to its chemical composition. For example, the inertness and non-stick surface of Teflon are related to the unusual strength of the carbon–fluorine bonds on the polymer backbone. On the other hand, polyethylene, a polymer with a backbone identical to Teflon but with H atoms replacing F atoms, will burn. For another example, substitution of one amino acid (replacement of a glutamic acid residue with a valine) at one point in the sequence of 146 amino acids in human hemoglobin produces profound changes in the properties of this polymer. Persons with this type of hemoglobin have sickle-shaped red blood cells (see Fig. 25.17 on page 956) that are less able to bind oxygen, resulting in conditions called sickle-cell trait and sickle-cell anemia.

Average Molar Mass.
With the exception of enzymes (Section 25.7), not all the chains in a polymer sample have the same mass. Rather, the average chain length and average molar mass are used to characterize a polymer. In addition, samples of the same polymer prepared under different conditions may have different average chain lengths. This is important because chain length is related to the ease of processing the polymer; in general, the longer the chains, the more difficult the material is to process. In most applications, the polymer must be able to flow so that it can be flattened into sheets, molded into bottles, and so on. However, as the molecules get larger, the polymer chains become more entangled and thereby more resistant to flow. As an analogy, imagine a plate of very long spaghetti. If we cut the spaghetti, it is easier to

Polymer (abbreviations, trade names)	Structure	Uses
Polyethylene (HDPE, LDPE, PE, polythene)	$\text{+CH}_2\text{—CH}_2\text{+}_n$	Packaging, containers, bags
Teflon	$\text{+CF}_2\text{—CF}_2\text{+}_n$	Non-stick coatings
Polypropylene (PP, Herculon)	$\text{+CH}_2\text{—CH+}_n$ | CH_3	Kitchenware, bottles, battery cases, indoor–outdoor carpets
Polyvinyl chloride (PVC, vinyl)	$\text{+CH}_2\text{—CH+}_n$ | Cl	Pipes, tile, car tops, hoses, imitation leather
Polystyrene (PS, styrofoam)	$\text{+CH}_2\text{—CH+}_n$ (with phenyl ring)	Packing material, food containers, insulation, cups
Nylon 6,6	$\text{+NH—(CH}_2)_6\text{—NH—C—(CH}_2)_4\text{—C+}_n$ (with two C=O)	Clothing, carpeting
Polyethylene terephthalate (PETE, Dacron)	$\text{+CH}_2\text{—CH}_2\text{—O—C—C} \cdots \text{C—C—O+}_n$ (with aromatic ring and C=O groups)	Backing for magnetic tape, clothing, soft drink bottles

Table 24.1

Some Common Synthetic Polymers and Their Uses

wind a mouth-sized portion onto a fork. The longer the strands, the harder they are to control. The situation is the same on the molecular level.

Foam cups, clear plastic cups, and strong, ultrathin fishing line are made of polystyrene with different average molar masses. White styrofoam cups are made from beads of polystyrene with an average molar mass of about 15,000 grams per mole plus about 5% of a volatile compound such as pentane. When the beads are heated inside a cup-shaped mold, the pentane boils, expanding the bead in much the same way as a kernel of popcorn pops. Clear, rigid polystyrene cups are made by melting polystyrene with an average molar mass of about 250,000 grams per mole. Some ultrathin fishing line is composed of polystyrene with an average molar mass of about 1,000,000 grams per mole.

Stereochemistry. The **stereochemistry** of a polymer is the spatial arrangement of the atoms in the polymer, or its molecular structure. For examples, the stereochemistries of starch and cellulose differ even though both are polymers of the same monomer (glucose). The monomers are connected in such a way that the shapes of

Figure 24.1

Starch (A) and cellulose (B) have the
same monomer unit, but these units
differ in the way they are arranged in
space. In starch, all the monomers
have the same relative orientation,
whereas in cellulose, every other
monomer unit is turned.

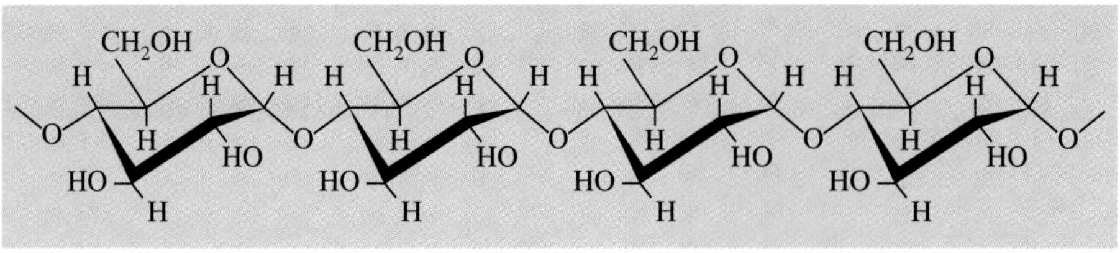

(A)

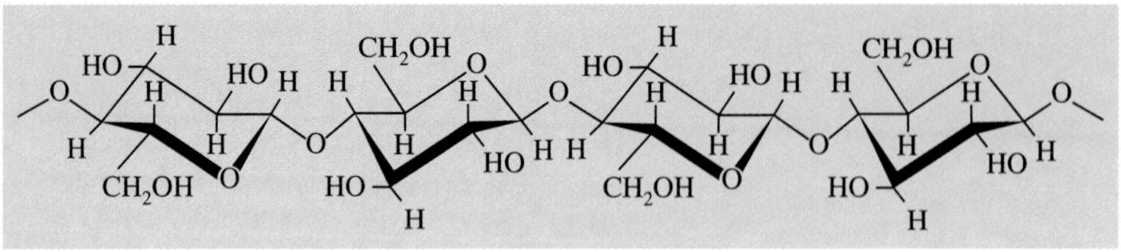

(B)

the two polymers are different (Fig. 24.1). These different shapes confer different properties: Even though the monomers in the two polymers are identical, starch is digestible by humans but cellulose is not.

Topology.   The **topology** of a polymer identifies the polymer chains as linear, branched, or cross-linked (Fig. 24.2). **Linear polymers** are those in which the monomer units form a continuous chain. **Branched polymers** have smaller chains extending out from the main polymer backbone. The branches extend outward like the branches on a tree and prevent the chains from packing closely. This tends to make branched polymers softer, lower-melting, and less dense than their linear counterparts. **Cross-linked polymers** are those in which the backbones of different polymer chains are connected. This creates a web-like structure. In general, the greater the number of cross-links, the more rigid the polymer.

As an example, linear polyethylene has a melting point of about 132°C and a density of 0.92–0.97 g cm^{-3}, and branched polyethylene has a melting point of about 106°C and a density of about 0.88 g cm^{-3}. The former material, high-density poly-

Figure 24.2

Linear (A), branched (B), and cross-linked (C) polymer chains.

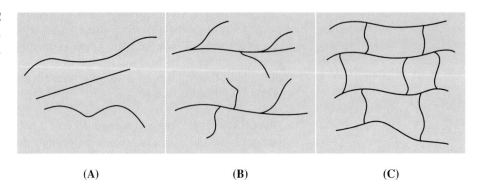

(A)   **(B)**   **(C)**

Figure 24.3

Polymer chains may arrange themselves in a regular, ordered array (A), may be amorphous (B), or may have ordered and amorphous regions (C).

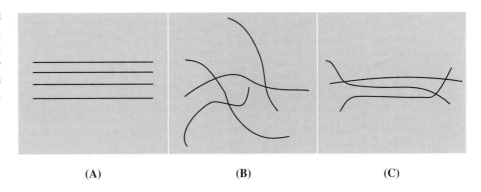

(A)   (B)   (C)

ethylene (HDPE), is used in plastic bottles and pipe. The latter, low-density polyethylene (LDPE), is used in plastic bags, wrappings, and squeeze bottles. Materials made of HDPE are "dishwasher safe," whereas those of LDPE tend to melt under the same conditions.

Morphology.   The way polymer chains pack in a solid is called the morphology of the polymer. A linear polyethylene molecule, for example, can wind itself around other molecules or can align itself with its neighbors so that the polymer backbones are approximately parallel and maximize interchain attractive forces (Fig. 24.3). The parallel arrangement makes regions of the polymer nearly crystalline; that is, the packing of the chains in these regions displays a regularly repeating order. However, the flexibility of polymers and the variations in chain lengths in a polymer sample make it very unlikely that the sample will be completely crystalline. Rather, it will have amorphous regions in which order is lacking. Branched polymers or linear polymers with bulky sidegroups are mostly amorphous. In general, high percentages of crystallinity in a polymer sample are associated with higher density, higher melting points, and higher transparency (lower opacity) than a less crystalline sample exhibits.

24.2   Polymer Properties

Changes in any of the five characteristics mentioned in the previous section change the properties of a polymer. Two important properties illustrate the structure–property relationship: The polymer backbone flexibility and the strengths of the intermolecular forces that hold adjacent polymer chains together.

The backbone flexibility is the ease with which a polymer chain can twist and turn. It is related to a measurable quantity called the **glass transition temperature** (T_g), the temperature at which a polymer transforms from a rigid, glassy state to a flexible solid. Above the glass transition temperature, portions of the polymer chains can rotate, and the sample behaves like a viscous liquid. The chains have the ability to slide past each other, imparting a flexibility to the polymer sample as a whole. Below T_g, the chains are more rigid, and the polymer is much more brittle and glass-like.

Many chemistry shows contain a demonstration of the changes in properties that accompany the cooling of a polymer through the glass transition temperature. The demonstrator takes a bouncy rubber ball, cools it in liquid nitrogen to below its glass transition temperature, and then shatters the cold ball by throwing it against a wall.

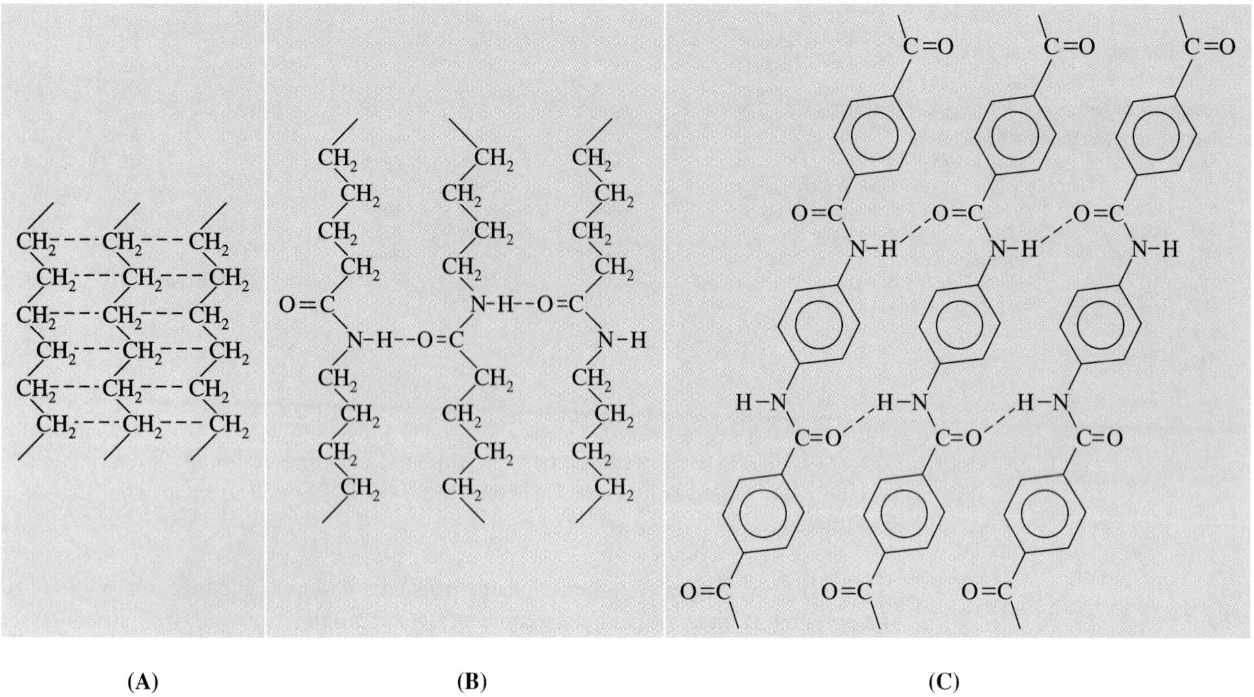

(A)   (B)   (C)

Figure 24.4

Polymer chains may be held together by London dispersion forces, as is the case in polyethylene (A), or by inter-chain hydrogen bonding, as is the case in nylon (B) and Kevlar (C).

Polymer flexibility depends on the monomer unit used to produce the polymer. By increasing the size of the side groups, a polymer chain can be rendered more rigid, thus increasing the glass transition temperature. For example, T_g for polyethylene is often about $-100°C$. This means that polyethylene will be flexible at room temperature, making it suitable for use in plastic bags. Polystyrene, on the other hand, has relatively large C_6H_5 groups replacing one-quarter of the hydrogen atoms on the polymer backbone. Polystyrene has a T_g in the vicinity of $+100°C$, which makes it rigid at room temperature, so it is used in such applications as CD cases. The glass transition temperature thus determines the potential uses of the polymer.

Properties of many polymers depend on the intermolecular forces between chains. In Chapter 11 we noted that the attractive forces between molecules can involve dipole–dipole interactions between polar molecules, whereas weaker London dispersion forces hold nonpolar molecules together. Certainly we consider polyethylene, which has only C—H and C—C bonds, to be nonpolar. However, the cumulative effect of having so many of these bonds in the backbone is that the London forces (Section 11.3) hold the chains together quite strongly (Fig. 24.4 A). This prevents polyethylene from being a gas (like many small nonpolar molecules) and from being dissolved by most solvents.

Hydrogen bonding between polymer chains provides even stronger intrachain bonding. This is what holds chains of nylon together (Fig. 24.4 B) and allows it to be pulled into threads. Because the hydrogen bonding sites on any given chain are already occupied in the formation of hydrogen bonds to other chains, there is little tendency for nylon to attract water. This makes nylon fabrics easy to dry. Kevlar, a polyamide like nylon, is used in bulletproof vests. The intrastrand hydrogen bonding

Figure 24.5

Cross-links connect the chains in rubber. Upon stretching, the chains may untangle and straighten, but the cross-links prevent any further motion of one chain with respect to another.

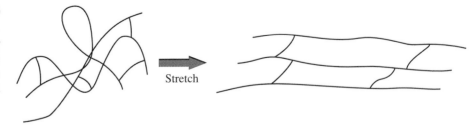

Stretch

causes the polymer to form sheets (Fig. 24.4 C). Layers of these sheets form a tough material that can slow or even stop bullets.

24.3 Elastomers, Thermoplastics, and Thermosetting Polymers

Synthetic polymers can be separated into three major categories: elastomers, thermoplastics, and thermosetting polymers. **Elastomers** are polymers that will stretch when a force is applied and then return to their original shape when the force is released. Rubber is an elastomer. On the molecular level, a relaxed rubber band has coiled or kinked polyisoprene chains held together by sulfur cross-links in its "relaxed" state (Fig. 24.5). As the band is stretched, the polymer chains straighten out to the extent the cross-links allow. There are many kinked orientations for the polymer chain, but only one orientation makes it straight. Thus releasing the rubber band causes it to return spontaneously to its original shape because of an increase in entropy (Section 18.8).

Thermoplastics soften upon heating and harden when cooled. They may be reheated and reshaped if desired. Thermoplastics can also change shape upon the application of force and can retain that shape when the force is removed. That is, they can be molded by pressure. Polyethylene and polystyrene are thermoplastics, as are some of the materials that form fibers, including nylon.

Thermosetting polymers become permanently hard upon heating to some particular temperature. This is often due to the forming of cross-links in the polymer. Examples include bakelite (Fig. 24.6 on page 902), which is used to form heat-resistant handles for cooking utensils, and epoxy resins, which are used as adhesives.

24.4 Recycling of Polymers

It is estimated that by the year 2000, consumers will discard 38 billion pounds of the 55 billion pounds of plastic produced annually. Today, plastics make up about 7% of solid waste by weight, but they are about 30% of the waste by volume. This causes a problem in landfills because most plastics biodegrade to only a small degree; few bacteria or other decomposers digest plastic. Responding to concerns that our landfills will soon be overflowing with plastic, many communities collect and recycle plastic containers.

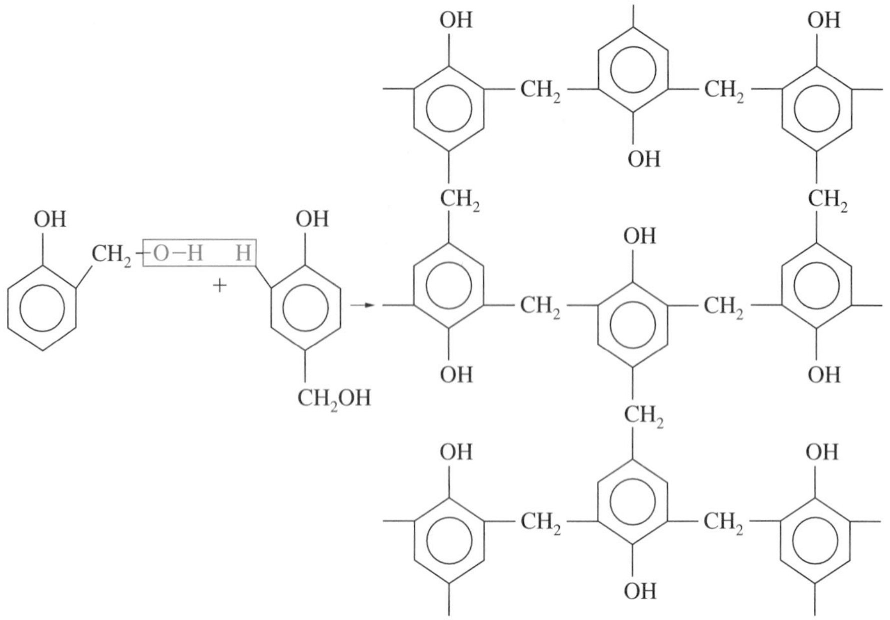

Figure 24.6

Bakelite is a thermosetting polymer. Covalent bonds, cross-linked in three dimensions, make this polymer hard.

Currently, only about 1% of all plastic is being recycled, compared to 60% of all aluminum cans and 20% of all paper. The reasons for the low percentage of recycled plastics are chemical as well as civic. Not all polymers are candidates for recycling. Thermosetting materials are difficult to recycle almost by definition, because they can't be remelted to form a similar material. Mixtures of polymers are difficult to recycle. Soft drink bottles, for example, contain several polymers. The body may be polyethylene terephthalate (PETE), the cap HDPE or aluminum. The base is likely to be HDPE, and an adhesive holds on the paper label. In order to obtain PETE pure enough to reuse, all of these other materials, as well as other plastics, must be removed. One PVC (polyvinyl chloride) bottle in with 1000 PETE bottles can ruin the batch for reuse as bottles.

To facilitate sorting of plastic, the objects are marked with codes such as HDPE and PVC. Numbers inside the recycling triangle indicate the composition, too. For example, a 2 indicates HDPE (Table 24.2 on page 903). At many recycling facilities, initial sorting of plastic containers is done by hand. Then the plastic is shredded and ground into chips. Next, washing removes the adhesive, and the light-weight paper is removed from the heavier plastic chips. The chips may be further separated by their density; separation occurs because polymers with different densities sink or float in different solvents. The chips are then dried and sold.

The most commonly recycled plastic is PETE, or polyethylene terephthalate, which is used in soft drink bottles. Approximately one-third of the annual U.S. production of PETE is recycled. Some is converted into new bottles; the rest is used for carpet fibers or insulating fill in jackets and sleeping bags.

Impure plastics also have uses. Mixtures of plastics and batches contaminated with paper or metal are converted into flower pots, "plastic lumber," park benches, and speed bumps.

Table 24.2
The Polymer Recycling Code

Code	Abbreviation	Polymer
♳ 1	PETE	Polyethylene terephthalate
♴ 2	HDPE	High-density polyethylene
♵ 3	V	Polyvinyl chloride
♶ 4	LDPE	Low-density polyethylene
♷ 5	PP	Polypropylene
♸ 6	PS	Polystyrene
♹ 7		Others

24.5   The Future of Polymers

Current "high-tech" uses for polymers may provide glimpses into future consumer uses of polymers. Polymers can be lighter and more resistant to corrosion than metals. The outer skin of some advanced aircraft is made not of metal but of a mixture of graphite fibers in an epoxy matrix. Such a mixture of two materials that produces a new material is called a **composite.** The density of these graphite composites is about two-thirds that of aluminum, yet the material is stronger than steel. Lighter aircraft can either carry more people or use less fuel to carry the same number of people. One type of passenger plane with composite substituted for aluminum is reported to have a full ton less mass, resulting in a savings of a quarter of a million gallons of fuel per year. In the future, both planes and automobiles are likely to be made more of plastic.

A second potential use for polymers is in electronics and energy storage. Though most polymers are electrical insulators, certain polymers can conduct electricity. A polymer chain or bundle of chains might be able to act as a molecular wire, continuing the trend of miniaturization in electronic circuitry. Similarly, the car batteries of the future may have electrodes of conducting polymers. Using the much lighter polymer-based batteries could overcome a major difficulty that currently prevents the widespread use of electric cars. Some current designs for these vehicles require a bank of batteries that weighs as much as a thousand pounds. This severely reduces speed, acceleration, and carrying capacity.

Improvement is also possible in the biodegradation of plastics. Some plastics, such as those in six-pack rings, diapers, and garbage bags, are not conveniently recycled and do not degrade at any appreciable rate under any circumstances. However, blending a small amount of starch with a polymer such as polyethylene can render the polymer biodegradable. Microorganisms digest the starch and begin the process of polymer breakdown. Another approach is to include polymer components that undergo

photodegradation; that is, they decompose upon exposure to light. Unfortunately, none of these methods works very rapidly buried in a landfill, where light, moisture, and oxygen are low. In fact, recent excavation of a landfill found hot dogs that were still recognizable ten years after burial.

Metals, Insulators, and Semiconductors

24.6   Band Theory

Metals and some compounds (such as $YBa_2Cu_3O_7$, ReO_3, and TiS_2) are **electrical conductors;** electrons can move through them with little resistance. Other elements (for example, sulfur and iodine) and most compounds are **insulators;** electrons cannot move through them, so they do not conduct electricity. A few elements and their compounds, including silicon and gallium arsenide (GaAs), are **semiconductors;** they conduct electricity better than insulators but not so well as metals.

The differences in conductivity among metals, insulators, and semiconductors result from differences in their electronic structures. These differences can be described by **band theory,** a type of molecular orbital theory (Section 7.12). In some solids, atomic orbitals of the atoms present combine to yield molecular orbitals that extend throughout the solid. As the number of atoms (and thus the number of atomic orbitals) in a molecule increases, the number of molecular orbitals increases (Fig. 24.7). As the number of molecular orbitals increases, the difference in energy between them becomes smaller until there is very little difference in energy between adjacent

Figure 24.7

Atomic orbitals in a Li atom; molecular orbitals in Li_2, Li_3, and Li_4; and bands in a lithium crystal, Li_n.

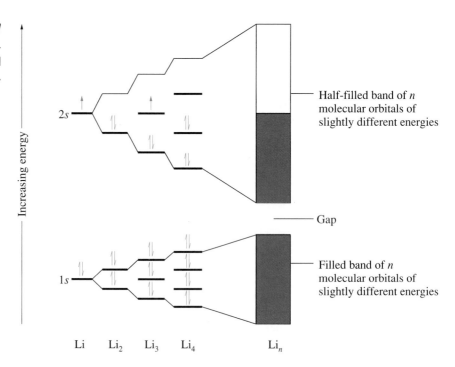

Figure 24.8

(A) Partially filled bands are found in metals. (B) Completely filled and completely empty bands separated by a large gap are found in insulators. (C) A completely filled band and a completely empty band separated by a small gap are found in semiconductors at 0 K. When the semiconductor is warmed to room temperature, a few electrons are thermally promoted from the previously filled band to the previously empty band, as shown.

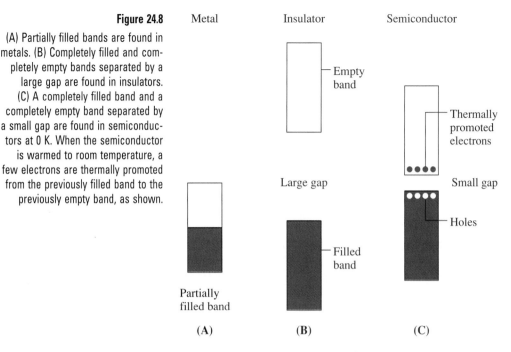

molecular orbitals. The result is a continuous **band** of molecular orbitals—or **energy levels,** as they are sometimes called—that extends throughout the entire crystal. There is one energy level in the band for each atomic orbital that participates in forming it. Each of the energy levels in the band can contain two electrons. However, only the higher-energy electrons in a band are sufficiently free, or mobile, to cause electrical conductivity.

To gain a better understanding of band theory, consider the case of lithium, which has an electron configuration of $1s^2 2s^1$. In a lithium crystal, one band arises from the $1s$ atomic orbitals and one from the $2s$ orbitals. A very small crystal of lithium contains about 10^{18} atoms. Thus there are about 10^{18} $1s$ and 10^{18} $2s$ atomic orbitals, each giving rise to a set of 10^{18} energy levels. For this number, the difference between energy levels is so small that the levels are essentially continuous, and they constitute a band. Because each atom contributes one $1s$ orbital and two electrons from the $1s$ orbital to the band, this s **band** is completely full. However, the second band, arising from $2s$ atomic orbitals, is only half filled with electrons. Between the bands is a **gap**—a range of energy in which no energy levels or molecular orbitals are located. No electrons can be found with these energies, because there are no energy levels for them to occupy.

The spacing of the bands in a substance and their filling determine whether the substance is a conductor, a nonconductor (insulator), or a semiconductor (a poor conductor). A substance such as lithium, which contains partially filled bands (Fig. 24.8 A) exhibits **metallic conduction.** If the bands are completely filled or completely empty and the energy gap between bands is large (Fig. 24.8 B), then the substance is an **insulator.** Diamond is an example of an insulator. A substance that contains a completely filled band and a completely empty band can behave as a **semiconductor** if the energy gap between the filled and empty bands is small enough that electrons from the filled band can be promoted to energy levels in the empty band by thermal energy

(Fig. 24.8 C). Promotion of electrons leaves empty levels called **holes** in the previously filled band. The previously empty band then contains a few electrons that can conduct an electric current.

24.7 Metals

A common property that we associate with metals is the ability to conduct electricity. But how is it that current can be conducted through a metal but not through an insulator? For a solid to be a good conductor, two requirements must be met: First, there must be a band in the solid. The molecular orbitals in the band act as a highway for electrons and enables them to move throughout the solid. Second, the band should be only partially filled, typically from 10% to 90% of its capacity.

When a small voltage is applied to a piece of metal, one end of the metal becomes positively charged and the other end negatively charged. Some electrons gain energy, move to previously unoccupied levels of the partially filled band, and migrate through these levels toward the positive end. If the electron-containing band is completely full, there is electron gridlock because there are no empty orbitals into which to promote the electrons. Electron migration does not occur. Electron flow in a band can be likened to a bottle that is half filled with sand (representing electrons). When the bottle is tilted (representing an applied voltage), some of the sand moves to new, low-energy positions. If the bottle is completely filled (or completely empty), then no sand can move. A band in an electrical insulator resembles a filled or empty bottle.

So far, we have cited lithium as a typical conductor, because it has a half-filled band constructed out of $2s$ orbitals. However, the half-filled band is also a general characteristic of each of the Group 1A alkali metals, which all have ns^1 configurations (Section 5.9). Looking at the alkaline earths (Group 2A), we might predict that they would be insulators, because ns^2 configurations should completely fill the s band. But remember that the alkaline earths also have empty p orbitals. These p orbitals also form a band, and this **p band** overlaps the s band (Fig. 24.9), creating the empty, accessible energy levels through which electron migration can occur. Thus band theory correctly predicts that the alkaline earth elements are metals. In general, the metallic elements in the periodic table have partially filled bands, derived from $s, p, d,$ and f orbitals as necessary, that allow migration of electrons.

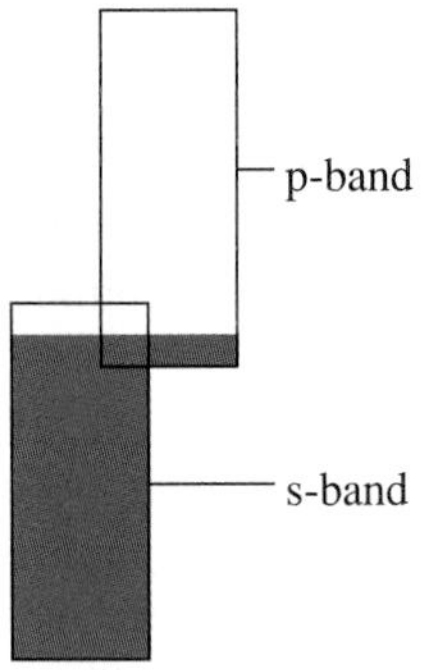

Another characteristic of a metal is that the conductivity decreases with increasing temperature. As the temperature is increased, the nuclei of the metal atoms vibrate with increasing energy about their positions in the lattice. This **thermal motion** increases the likelihood that the mobile electrons will collide with the nuclei, which slows their progress through the crystal. At temperatures above 0 K, the nuclei are always vibrating to some extent and creating the interference with free electron motion that we know as **resistance.**

Figure 24.9

The s band and the p band overlap in an alkaline earth element.

24.8 Semiconductors and Insulators

As we noted in Section 24.6, an insulator is a material that has a filled band, an empty band at higher energy, and a large band gap between them. A semiconductor has this same arrangement but the band gap is not so large. At room temperature, some electrons in a semiconductor pick up enough thermal energy to jump the band gap and enter the formerly empty band. These electrons are now mobile. As the temperature is raised, more and more electrons can jump the band gap, increasing the conductivity of the semiconductor.

As electrons are promoted in a semiconductor, empty energy levels (positively charged holes) form in the previously filled band. This breaks the electron gridlock, and electrons can move through this band as well. Sometimes it is convenient to think of current in a nearly filled band as resulting from the motion of the positively charged holes, though it is really the electrons that are moving. By analogy, when we invert a nearly full bottle of syrup or shampoo, we can see the air bubble slowly rise to the top. The syrup or shampoo moves too as it flows around the bubble to the bottom. The return to the stable state can be described by the movement of the liquid (analogous to electrons in a band) *or* by the movement of the air (analogous to the holes in a band).

It is possible to alter the inherent conductivity of a semiconductor by adding impurity atoms, or **dopants,** to the semiconductor during its preparation. Only very small amounts of dopants, on the order of parts per million, are needed to alter the conductivity of a semiconductor such as silicon. Suppose, for example, that one of every million silicon atoms in a solid is replaced by an arsenic atom. (Each silicon atom has four valence electrons; arsenic has five.) Because there are relatively few arsenic atoms, the basic crystal structure and band structure of the silicon are unaffected. However, the extra electron possessed by the arsenic atom cannot be contained in the already full lower band and must go into the nearly empty upper band. These extra electrons can migrate in an electrical field, causing the arsenic-doped semiconductor (an **extrinsic semiconductor**) to be a better conductor than an undoped semiconductor (an **intrinsic semiconductor**) at the same temperature. Semiconductors that are doped with comparatively electron-rich atoms such as arsenic are called **n-type semiconductors.** (The negative electrons are considered the current-carrying species.)

Another type of extrinsic semiconductor is formed when a few silicon atoms in the crystal are replaced by an atom such as aluminum. The presence of aluminum atoms, which have only three valence electrons, creates holes in the lower band, rendering it only "almost" full. This also allows for the possibility of electron motion through the solid. Doping with comparatively electron-poor atoms creates **p-type semiconductors.** (The positive holes are considered the current-carrying species.) In this case, conduction occurs by motion of the holes in the lower band.

Figure 24.10

A bank of solar cells is used to power this radio telephone in a remote location in Australia.

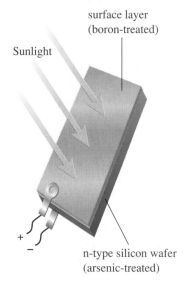

Figure 24.11

Diagram of a solar cell.

24.9   The Solar Cell

An enormous amount of energy is given off by the sun. Earth receives more energy from sunlight in two days than is stored in all known reserves of fossil fuels. Solar cells (Fig. 24.10) are currently available that generate electrical power from sunlight at the rate of 90 watts per square yard of illuminated surface.

The basic unit of a solar cell is a wafer of very pure silicon that is n-doped with small amounts of arsenic. On the surface of this wafer, a thin layer of p-type silicon (doped with boron) has been deposited. The area where these two layers meet is called a **p–n junction** (Fig. 24.11). Within this junction, some electrons diffuse from the electron-rich side (the n-type side) to the vacancies on the p-type side. This diffusion results in a net positive charge on the body of the wafer (which was neutral before the diffusion of electrons) and a net negative charge on the surface layer. Thus an electrostatic potential develops between the two regions, until it is just large enough to prevent any further charge migration. However, this redistribution of charge has

Figure 24.12

(A) A p–n junction is formed where a p-doped semiconductor is in contact with an n-doped semiconductor. Some electrons in the interface region move from the n-doped side to the p-doped side. This creates a net positive charge on the n-doped side and a negative charge on the p-doped side. (B) Absorption of light in the p–n interface region results in the promotion of electrons from the lower band to the upper band, as shown by arrow A. The electron moves into the n-type material toward the positive charge (arrow B), and the holes move into the p-type material (arrow C). This creates an electric current.

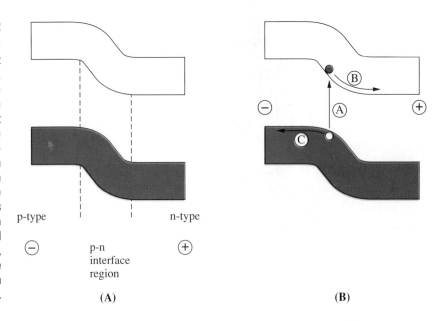

the effect of changing the relative energies of the electrons in the two regions of the solar cell. Because it is now negatively charged, the energy of the electrons in the p-type region increases while the energy of the electrons in the n-doped region decreases. This produces a bending of the band energy levels at the p–n interface.

In a solar cell, one electrical lead (a wire) is attached to the body of the wafer, and one lead to the surface. When the cell is exposed to sunlight, energy from the sunlight causes the electrons near the p–n junction to be excited into the nearly empty high-energy band. An empty hole is left behind. The promoted electrons move "downhill" to lower energy in the upper band, which pushes them toward the positive side of the junction (the n-doped portion of the semiconductor), where they can be collected by the electrical lead. As long as light shines on the cell, the electrons continue to move in this manner. This device is an electric cell with the positive terminal at the p-contact and the negative terminal at the n-contact (Fig. 24.12). In actual practice, a series of wafers containing p–n junctions make up a solar cell.

The magnitude of the current in a solar cell is determined by the amount of light absorbed; in theory, each photon creates one electron-hole pair, providing one electron for the current flow. When one mobile electron is produced for each photon that strikes the cell, a solar cell is said to be operating with 100% efficiency. However, several factors limit the efficiency of solar cells. First, some of the sunlight that strikes the solar cell may be reflected instead of absorbed. In order to minimize this effect, solar cells are usually given an antireflective coating (Section 24.13). Second, not all photons create electron-hole pairs. At a minimum, the photons must have energies in excess of the band gap in order to promote the electron to the upper band and create the electron-hole pair. Thus some low-energy photons do not create current. Third, defects in the solid may promote the recombination of the light-generated electron-hole pairs. This results in heat instead of providing current. Finally, if the solar cell is thin, some photons may pass through it without being absorbed at all. The net result of these effects is that most solar cells in use today operate at 10% to 20% efficiency, although under special conditions in the laboratory, efficiencies of 30% have been reported.

The first practical application of the solar cell was as a power source for eight telephones on a rural line in Georgia in 1955. Now they are common sources of small amounts of electrical power in many applications. Roughly 100 million solar-powered calculators are sold worldwide each year. Solar cells are used to power communication devices in satellites and spacecraft, especially those designed to remain in space for long periods. Solar cells are also used in experimental solar-powered cars.

Interest in large-scale solar power plants is growing, because the price of electricity from solar cells has dropped dramatically in recent years. Currently, the cost of electricity in many parts of the United States is about $0.07 per kilowatt-hour. (This means that it costs seven cents to light a 100-watt light bulb for 10 hours or to run a 1000-watt appliance for 1 hour.) At present, most of this electricity comes from coal-fired and nuclear power plants. If solar cells were used to produce electricity, the cost would rise to about $0.22 per kilowatt-hour employing today's technology. However, this represents a dramatic decrease in cost from $1 per kilowatt-hour in 1980 and $60 per kilowatt-hour in 1970. Given continued improvements in solar cell technology and the increasing costs of pollution control devices for coal-fired power plants, energy from solar cells could become less expensive than energy from fossil fuel plants in the not too distant future.

Ceramics

Each day we encounter numerous ceramic objects: Pottery, china plates and mugs, porcelain sinks, window glass, and drinking glasses are a few examples. In general, **ceramics** are inorganic, nonmetallic, covalent-network solids (Section 11.12). Many are silicates (Section 22.6); others are ionic compounds with highly charged cations and oxide anions. Ceramics may have very ordered lattice structures or they may lack clear structural patterns. Aluminum oxide, Al_2O_3, silicon nitride, Si_3N_4, silicon carbide, SiC, and crystalline and amorphous silicon dioxide, SiO_2, are examples of these classes of materials.

Most ceramics are electrical insulators. The bonds in covalent-network solids are strongly directional, because the electrons in the bonds are shared tightly between the two atoms participating in the bond. As a result, there are few mobile electrons in the solid to carry current. These strong bonds also tend to hold the atoms in their positions in the solid. This makes them have very high melting points; very high temperatures are required to disrupt the bonding. Silicon dioxide, for example, has a melting point in excess of 1400°C. Silicon carbide is stable up to about 3000°C. Compared to metals, ceramics are lighter, stiffer, and much more resistant to corrosion. (Ceramic oxides are already fully oxidized.)

The strong, directional bonding also makes ceramics brittle. When a sharp force is applied to a piece of metal, the relatively nondirectional bonding allows the atoms to reposition themselves. The metal may bend or dent but does not generally break. By contrast, the bonding in a ceramic tends to hold the atoms in place more rigidly. When sufficient force is applied, the bonds will break and the ceramic object will shatter rather than deform.

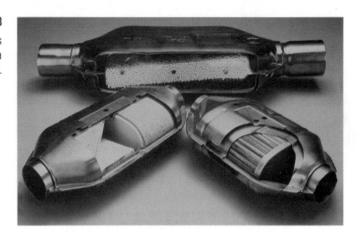

24.10   Catalytic Converters and Space Shuttle Tiles

Ceramic materials serve as the support for the precious metal catalyst that operates in a catalytic converter (Fig. 24.13 and Fig. 13.12). This device takes unburned hydrocarbons, nitrogen oxides, and carbon monoxide from a car's engine and catalyzes their conversion into nitrogen gas, water, and carbon dioxide. The reaction is heterogeneous—the exhaust gases come in contact with the surface of a solid catalyst that includes the metals rhodium and platinum. Because only the atoms near the surface of the catalyst interact with the exhaust gases, it is only these few atoms that are actually performing the catalysis. In order to maximize the efficiency of a heterogeneous catalyst, the surface area of the catalytic species must be high. One way to ensure this is to deposit atoms of the catalyst (palladium, rhodium, platinum or other metals) on the surface of a honeycomb of an oxide-based ceramic. This has several advantages: The honeycomb structure provides a high surface area for contact between the gases and the metal atoms, the ceramic is able to withstand the high temperatures of the exhaust gases, and only a thin film of the metal is required, which saves a great deal of money because rhodium and platinum are very expensive.

Figure 24.14

A Space Shuttle tile is an excellent thermal insulator. After a piece of tile is removed from a furnace and the outside has cooled, it can be handled even though the inside is still hot enough to glow.

Space Shuttle tiles are also made out of ceramics. These tiles are designed to insulate the ship against the heat generated upon reentry into the earth's atmosphere. Each tile is composed of a mat of fibers of silicon dioxide (Fig. 24.14). Not only is this material a good insulator itself, but it also traps air within the network of fibers, which improves the insulating properties even further. Though the tiles have a density of only 0.2 g cm^{-3}, they are capable of withstanding temperatures up to 1250°C.

24.11   Glass

Glass is a supercooled liquid consisting of a complex mixture of silicates (Section 22.6). It is transparent, brittle, and entirely lacking the ordered structure of a crystalline solid (Section 11.12). When heated, glass does not melt sharply. Instead, it gradually softens until it flows.

The glass used for window panes, bottles, and dishes is a mixture of sodium silicates and calcium silicates with an excess of silica (silicon dioxide). It is made by melting together sand (which is silicon dioxide), sodium carbonate (or sodium sulfate), and calcium carbonate.

$$Na_2CO_3 + SiO_2 \longrightarrow Na_2SiO_3 + CO_2(g)$$
$$Na_2SO_4 + SiO_2 \longrightarrow Na_2SiO_3 + SO_3(g)$$
$$CaCO_3 + SiO_2 \longrightarrow CaSiO_3 + CO_2(g)$$

After the bubbles of gas have been expelled, a clear viscous melt results. This is poured into molds or stamped with dies to produce pressed glassware. Bottles, flasks, and beakers are formed by taking a lump of molten glass on a hollow tube, inserting it into a mold, and blowing with compressed air until the glass assumes the outline of the mold. Plate glass (float glass) is made by pouring molten glass on a layer of very pure molten tin. Because the molten tin surface is perfectly smooth, the glass floating on it is also perfectly smooth and does not need to be ground or polished after hardening.

New glassware is annealed by heating it for a time just below its softening temperature and then cooling it slowly. Annealing lessens internal strains and thereby reduces the chances of breakage from temperature change or shock. Additional resistance to shattering can be achieved by combining glass and plastic. One form of safety glass contains a thin layer of plastic held between two pieces of thin plate glass. If the glass is broken, then adhesion to the flexible plastic reduces the danger from flying glass and jagged edges.

When sodium is replaced by potassium in a glass melt, a higher-melting and harder glass is obtained. If part of the calcium is replaced by lead, then a glass of high density and high refractive index is formed. This **flint glass** is used in making lenses and cut-glass articles. **Borosilicate glass** (in which some silicon atoms are replaced by boron atoms; brand names include Pyrex and Kimax) is used for test tubes, flasks, and other laboratory glassware because it is resistant to sudden changes in temperature and chemical action.

Still another way to make ceramics resistant to breaking upon sudden changes in temperature is exemplified in the cooking utensils called Corning Ware (Fig. 24.15).

Figure 24.15

A Corning Ware pan is resistant to extreme temperature changes. One end of this pan is frozen in a block of ice while the other end is heated.

Normal glass tends to break when subjected to thermal stress, because the strong directional bonds tend to reduce flexibility. When a cold piece of glass such as a glass pan is set on a hot surface, the bottom of the glass expands while the top of the glass does not. This has the effect of expanding any microcracks that are unavoidably present in the bottom of the pan and may cause it to shatter. A Corning Ware pan is made of regular glass that contains extremely small particles of TiO_2. These particles act as seed crystals, causing a huge number of very small SiO_2 crystals to form. The resulting crystals are less than 1 micrometer in diameter and pack together with very little empty space between them. The SiO_2 that did not crystallize returns to its glassy state and acts as a glue that holds the microcrystals together. The result is that any crack that might form on the bottom cannot easily pass through the many boundaries between the bottom and the top. This makes Corning Ware especially resistant to changes in temperature. (Because of the many small particles that make up a Corning Ware object, light cannot pass through the pan and the object is opaque. This is the same effect that makes an ice cube clear and transparent but makes a snowball, which is composed of many small ice crystals, white and opaque.)

Optical fibers also are made out of glass. In general, an optical fiber has a *core* of amorphous silica, SiO_2, surrounded by a second material called the *cladding*. When the core has a larger refractive index than the cladding, a condition called total internal reflection results: Light that is directed down the core stays within the core and is not lost as a result of reflection into the cladding (Fig. 24.16). In order to transmit signals over relatively long distances, the glass must be very pure. Even trace amounts of impurities can cause a significant decrease in the signal.

AT&T Bell Laboratories developed a process called Modified Chemical Vapor Deposition, or MCVD (Fig. 24.17 on page 913), to prepare glass with the purity required and to produce a core and cladding with precise refractive indices. In this technique, the optical fiber is prepared from the outside to the inside. First, $SiCl_4$, O_2, and a small amount of fluorine-containing gas such as SiF_4 are passed into a silica tube. $SiCl_4$ is a volatile liquid that can be obtained in high purity, and upon heating, it reacts with O_2 to form a cladding of amorphous SiO_2 and gaseous Cl_2 that escapes from the tube. Silicon tetrafluoride, SiF_4, is used as a source of fluorine atoms, which lower the refractive index of the cladding. In the second step, the core material is formed inside the cladding by passing a mixture of $SiCl_4$, O_2, and a small amount of $GeCl_4$ into the tube. The $GeCl_4$ behaves chemically like the $SiCl_4$ but serves to increase the refractive index of the core relative to the SiO_2 of the cladding. The tube is heated to fuse the components into glass, drawn into a fiber that may be as small as 8–9 microns in diameter, and coated with a plastic for protection.

Figure 24.16

End (left) and side (right) views of an optical fiber. Total internal reflection occurs when the refractive index of the cladding is less than the refractive index of the core material. This keeps the light within the fiber.

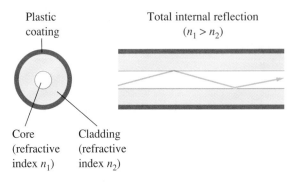

Plastic coating

Total internal reflection $(n_1 > n_2)$

Core (refractive index n_1)

Cladding (refractive index n_2)

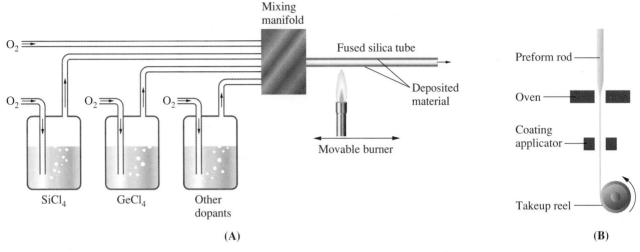

Figure 24.17

(A) A diagram of the MCVD process for making optical fibers. When heated, $SiCl_4$ and O_2 react to form SiO_2 and deposit on the inside of a fused silica tube. Other materials may be added to change the refractive index of the resulting glass. (B) After completion of the deposition process, the resulting rod is placed into an oven and stretched to form a thin glass fiber.

Optical fibers can carry many times more information than a metal wire. Literally thousands of phone conversations can be transmitted on the same optical fiber simultaneously. This is accomplished by dividing every second into many segments, each of which can be viewed as a channel for a phone call. The sender's message is converted into digital code by a laser that is rapidly turned on and off, corresponding to the 1's and 0's of digital information. The sender and the receiver are carefully coordinated so that they are tuned to the same time slot; thus the conversation occurs without scrambling of information.

24.12 Superconducting Ceramics

Most ceramic materials are electrical insulators. However, in 1986, J. G. Bednorz and K. A. Muller, working at the IBM research laboratories in Zurich, Switzerland, made the surprising discovery that at low temperatures a ceramic oxide containing lanthanum, copper, and barium ions was superconducting—that is, it had no electrical resistance. Within only a few years, Bednorz and Muller had won a Nobel Prize for their discovery, and worldwide research activity had produced several such superconducting ceramic oxides. These discoveries have led to predictions about dramatic breakthroughs in levitated trains, high-strength magnets, and high-efficiency power lines.

The phenomenon of superconductivity is not new. In 1911, the Dutch physicist H. Kamerlingh Onnes observed that elemental mercury, when cooled to around 4 K, loses all resistance to electric current. This total lack of resistance below a particular temperature, called the **critical temperature,** is known as **superconductivity.** In the 70 or so years that followed, a number of other metallic elements and alloys were found to exhibit the same behavior. Until 1986, the critical temperature for all of these materials was a frosty 23 K or below. Because of the difficulty and expense of maintaining these low temperatures, superconductivity was practical in only a few high-tech applications. Magnetic resonance imaging, or MRI, a powerful medical diagnostic tool employing a superconducting magnet, is one such application that has found widespread use.

Figure 24.18

A powerful magnet can be levitated over a superconducting material. This is known as the Meissner effect.

In the 1930s, another unusual characteristic of superconductors was discovered. In Germany, W. Meissner and R. Ochsenfeld found that a superconducting material will not permit a magnetic field to penetrate the sample. That is, a superconductor repels a magnetic field, acting like a magnetic mirror. This property became known as the **Meissner effect** (Fig. 24.18).

The ceramic superconductors are sometimes called "high-temperature" superconductors, because they become superconducting at temperature well above those of the metals and alloys first investigated. Because of the relatively high critical temperatures of these materials, liquid nitrogen (boiling point 77 K) can be used as the coolant instead of liquid helium (Boiling point 4 K). Not only are liquid nitrogen-cooled materials easier to handle, but the cooling costs are also about 1000 times less than for liquid helium. Probably the best-studied ceramic superconductor is $YBa_2Cu_3O_7$, sometimes called YBCO or the "1-2-3 superconductor" (after the stoichiometry of the metals in the compound). Discovered in 1987, $YBa_2Cu_3O_7$ has a critical temperature of 93 K. As of this writing, the material with the highest critical temperature is $HgBa_2Ca_2Cu_3O_{8+x}$, (where x represents a slight excess of oxygen), which is superconducting below 133 K.

The crystal structure of the 1-2-3 superconductor is shown in Section 23.5. Extensive research has indicated that the superconductivity takes place in the copper–oxygen planes located above and below the yttrium atom. Note also that the copper is in an unusual oxidation state. Assuming that the other elements adopt their usual oxidation states (+2 for barium, +3 for yttrium, and −2 for oxygen), this gives each copper atom an average oxidation state of +2.33. Another way of expressing this is to say that two-thirds of the copper ions are Cu(II) and one-third are Cu(III). The copper(II) and copper(III) sites cannot be distinguished because the compound is a solid solution (see Section 12.8). However, despite a great deal of research, the mechanism by which superconductivity occurs is still unknown.

The 1-2-3 superconductor was first prepared in what has been called a "heat and beat" synthesis. In general, this involves grinding the solid reactants together to make an intimate mixture and then heating them in a high-temperature oven to promote reaction. Often these steps must be repeated to get a uniform sample. To make $YBa_2Cu_3O_7$, stoichiometric amounts of Y_2O_3, $BaCO_3$, and CuO are ground together and heated to 950°C. The resulting solid is reground, pressed into a pellet, and reheated in an oxygen atmosphere at 500°C–600°C.

Figure 24.19

Experimental magnetic levitation train on the test track in Miyazaki, Japan. Repulsion forces between superconducting electromagnets in the train and ferromagnets attached to the guideway provide levitation, guidance, and propulsion or deceleration. The superconducting magnets on the train raise it about 4 inches above the guideway, eliminating essentially all friction and enabling it to achieve speeds of approximately 300 miles per hour.

There are two main difficulties with heat and beat syntheses. First, unlike reactions that take place in solution, it is difficult to get uniform mixing of the reactants. Extensive grinding of the reactants in the synthesis of $YBa_2Cu_3O_7$ typically produces particles that have a radius of 10^{-3} cm, but each of these contains only one type of metal. Copper atoms, for example, are separated by about 4×10^{-8} cm in the crystal structure of CuO; this means that in order to get into the center of this particle, the yttrium and barium atoms move past 25,000 copper atoms. This process is slow but is accelerated by heating; regrinding and reheating also speed the process. The second difficulty that can be encountered in these syntheses is that it is very difficult to remove impurities. Unlike syntheses performed in solution, where it is possible to separate side products or leftover reactants by filtration or distillation, the insolubility and nonvolatility of the ceramic prevent this. In the case of superconductors, significant amounts of impurities may lower the critical temperature of the material or render it nonsuperconducting.

The new superconductors may make technological breakthroughs possible in a number of areas. At present, a significant amount of energy is lost to electrical resistance in the wires that connect the power plants to their customers. If conventional power lines could be replaced with superconducting wires, then it might be possible to save this amount of energy, which would conserve fuel and result in less atmospheric pollution.

Superconducting microchips for computers and other electronic devices could also be more powerful. Present-day chips produce waste heat as a result of resistance to the electric current as it flows through the chip, so the chips have to be spaced in such a way as to dissipate heat. Superconducting chips would produce no waste heat because they would have no resistance. The chips could be placed closer together, and the signals would require less time to pass from chip to chip. This would make for smaller computers that work faster.

The Japanese National Railways has already built an experimental train that is levitated above the tracks via frictionless magnetic force generated by superconducting magnets cooled by liquid helium (Fig. 24.19). This train can travel up to 300 miles

Figure 24.20

This superconducting tape is composed of filaments of a high-temperature superconductor encased in a metal sheath. The tape is flexible because the filaments are very fine.

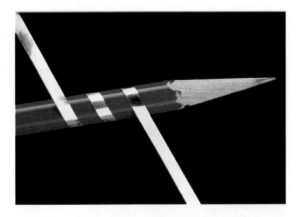

an hour, but the expense and difficulty of carrying liquid helium have so far prevented any practical routine use. Other such projects are under serious consideration. If the new high-temperature superconductors become adaptable for practical use, rendering cooling by liquid nitrogen feasible, then such trains could become commonplace.

Two primary difficulties have prevented widespread implementation of the new superconductors. First, because they are ceramic materials, they tend to be very brittle. This has limited their applications where flexibility is important, such as in superconducting wires. Second, the superconducting state of the ceramic materials in many cases is destroyed by the application of strong magnetic fields or high currents. However, prototype flexible superconducting cables have recently been prepared that can carry large amounts of current. One cable (Fig. 24.20) is a tape consisting of bundles of fine superconducting filaments prepared by packing a metal tube with the starting materials used to prepare the superconductor, stretching the tube into a fine wire, and then heating the wire to form the superconducting material inside. A second method involves depositing the superconducting material on a flexible metal alloy tape. In either case, some way of cooling the material must be provided so that it remains in the superconducting state, a major difficulty in a power transmission line that might be many miles long.

Thin Films

Very thin layers of metals, metal oxides, or polymers that have been deposited on the surface of another material, which is called the substrate, are classified as **thin films.** Though there are no precise limits, thin films are usually considered to have thicknesses of 0.1–300 microns. Paints, for example, are usually much thicker than this and are not considered thin films. Useful thin films are uniform, resist chemical attack and wear, and adhere well to the material covered. Such films are used to make materials tougher, to reflect light, or to modify the electrical properties of the substrate.

24.13   Films on Windows and Lenses

Many modern office buildings have windows that have a thin film of metal plated on them (Fig. 24.21 on page 917). This gives an attractive appearance and a reflection of external daylight keeps the interior cooler and reduces the need for air condition-

Figure 24.21

A thin film makes the windows in this skyscraper reflective.

Figure 24.21

A thin film makes the windows in this skyscraper reflective.

ing. Why do these windows appear mirrored during the day but transparent at night? It is a matter of which side of the window is the more brightly lit. A fraction of the light passing through the window (either the internal office lights or the external sunlight) is reflected while the rest is transmitted. Thus a person in the office receives external light transmitted through the window and office light reflected from the window. The external light is the brighter during daylight hours, and the glass appears transparent from the interior. A person outside sees a reflection of the brighter outdoor light. At night the situation is reversed; a person in the office sees mostly a reflection of the interior of the office, while the outdoor observer sees the interior lights of the office.

Many sunglasses and camera lenses are also treated with antireflective coatings. On camera lenses, a thin film of MgF_2 or Na_3AlF_6 prevents reflections from both the front and the back faces of the lens that might cause fogging and lack of contrast. A film of uniform thickness can be optimized to cause destructive interference for only one wavelength. Often 500 nm is chosen, because it is in the middle of the visible spectrum, and the eye is most sensitive in the region of this green light. As the green light is removed, some of the red and blue/violet light is reflected. This gives such lenses their oily purple color.

Metals may also be oxidized to produce an array of colors, a technique that has been used in the creation of jewelry. In this case, either an applied voltage or heat treatment in an oxygen atmosphere is used to form a thin layer of metal oxide on the surface of the metal itself. The color that is observed varies with the thickness of the oxide layer. In addition, the color may vary with the angle of viewing, so the appearance of the jewelry can change when it is viewed from different perspectives.

24.14   Diamond Thin Films

Diamond has a number of unusual and valuable properties. Not only is it one of the hardest materials known, but it also is electrically insulating, is an excellent conductor of heat, and is resistant to chemical attack. Recently, scientists have developed techniques to prepare diamond films for cutting tools. Like crystals of diamond, these thin films are very tough, and cutting tools so coated are more resistant to wear. Diamond "heat sinks" also are used to remove heat from circuit boards in electronic applications.

Thin films of diamond are prepared by exposing a dilute, low-pressure mixture of methane, CH_4, in hydrogen gas to microwave radiation at about 1000°C in the presence of the object to be coated. Under these conditions, free hydrogen atoms and carbon atoms are generated, and the presence of the hydrogen atoms seems to prevent the formation of graphite. Unfortunately, these temperatures are still too high for some applications; for example, steel loses its hardness under these conditions.

24.15   Synthesis of Thin Films

The preparation of thin films can be as simple as heating a metal in the presence of oxygen in order to prepare an oxide thin film on the surface of the metal. In this case, the film is formed as a result of a chemical reaction with the surface atoms. When the thin film is not related chemically to the substrate, more sophisticated techniques are needed.

One such technique is called **chemical vapor deposition,** or CVD (Fig. 24.22). In this method, the substrate is placed in an evacuated chamber. The thin-film precursor is evaporated by heating it or bombarding it with an electron beam. Atoms of the precursor condense on the substrate to form the film. The thickness of the film and its resulting properties can be controlled by controlling the temperature or voltage and the length of exposure.

Circuit components on silicon chips can be manufactured in a layer-by-layer manner. Each chip has insulating, conducting, and doped regions. Imagine the preparation of a small p–n junction on a silicon chip. In the first step, a p-doped region on the chip can be prepared by placing the silicon wafer in a CVD chamber and evaporating boron atoms onto the surface. A protective screen called a *template* or *mask* is used to cover all but the area to be doped. Second, heating the wafer allows the boron atoms to diffuse into the body of the silicon crystal. Third, phosphorus is vaporized onto the area, and heating produces a p-type region on top of the n-type region. Connections between circuit components are made by evaporating metal atoms to form "wires" and protecting other areas with masks.

Figure 24.22

A schematic diagram of the CVD process.

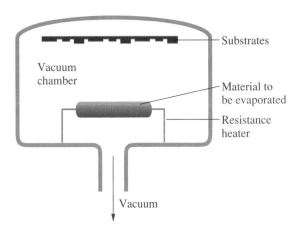

Circuits constructed in this manner can be extremely small. Currently, hundreds of thousands of circuit components fit on a chip the size of a pencil eraser. In the last 50 years, advances in microelectronics have enabled us to own hand-held calculators that are as powerful as the early computers that occupied entire rooms.

Materials in Medicine

Advances in medicine have made possible the replacement of certain body parts with artificial parts called *prostheses*. These advances have saved the lives of many victims of accidents and illness and have improved the quality of life of those who have been stricken with crippling pain, physical deformities, or diseases.

Not just any material can be implanted in the human body. Foreign objects provoke the immune response, which protects us from the germs that are constantly present. Implants not attacked by the immune system are *biocompatible*. In addition, desirable implants should be chemically inert and not decompose upon exposure to body fluids, which have corrosive properties similar to those of warm sea water. Implants should not cause the blood to clot, and they should be flexible (as are artificial blood vessels) or strong (as are artificial hips).

Because many polymers possess these properties, they have found numerous applications as prostheses. Teflon is used in replacement veins and arteries, eustachian tubes, bladders, and intestinal walls. Nylon is utilized for sutures and as the lining for the power cables to pacemakers. Dacron, a type of polyester, substitutes for blood vessels and serves as artificial skin for burn victims.

24.16 Artificial Hips

Figure 24.23

An artificial hip joint is visible in this X ray.

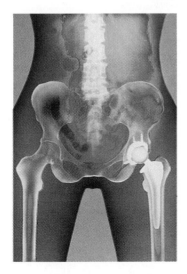

According to a 1994 estimate, 123,000 hip replacements are performed each year. Most of them involve pushing a stainless steel shaft into the femur (thigh bone) and gluing it there. The top of this shaft is ball-shaped and fits into a polyethylene socket that is glued to the patient's pelvis. These joints (Fig. 24.23) last 10–15 years and are very effective for older, less active patients. Because of the difficulty of securely fastening the socket and the shaft to the bone, younger, more active patients may need a second hip replacement after only 5 or so years. In addition, the implant slightly changes the way the weight is distributed from the hip to the femur. Generally, this means the femur is shielded from stress, and bone is lost as a result. Consequently, the search for new ways to fasten implants and new implant materials has continued.

One way to get better adhesion between the implant and the bone is to promote new bone growth in the area. Research indicates that titanium joints support the growth of new bone. The surface of a titanium prosthesis is not actually titanium metal but titanium oxide. This layer is relatively rough and lets new bone grow into pores on the oxide surface. This effectively fastens the titanium to the original bone.

A second method is to apply a thin layer of hydroxyapatite, $Ca_5(PO_4)_3OH$, to a stainless steel implant. A technique called thermal spraying melts the mineral and sprays it onto the metal. Hydroxyapatite is the inorganic component of bone, so the combination promotes a fusing of the natural bone with the implant.

Still another way to improve the useful length of implants is to reduce the amount of strain that tends to loosen the implant. Stainless steel, in particular, is very rigid and unbending. As a result, exercise causes the shaft slowly to loosen from the glue holding it inside the femur, perhaps creating the need for a second hip replacement. Natural bone flexes slightly to accommodate such stress, so metals and alloys with properties matching those of bone are sought. Research has indicated that a cobalt–chromium alloy and titanium-based alloys may be more effective than stainless steel.

24.17   Nitinol

Imagine a material that has the ability to remember its shape. Nitinol, an alloy of nickel and titanium, is just such a material. Its name stands for Nickel Titanium Naval Ordnance Laboratory, where it was originally discovered in 1965. No matter how it has been bent or twisted, application of gentle heat to about 50°C causes a sample of nitinol to return to its original shape. Bending most other metals generates defects in the structure that prevent the object from being bent back to its original shape. This is why it is very difficult to straighten a bent nail completely or remove all traces of a dent in a car fender.

Alloys such as nitinol can be "trained" to a shape they will return to upon heating. Temperatures of about 500°C (far above its transition temperature of about 50°C) are necessary to retrain nitinol. As the sample is held in the desired shape while it is heated, the atoms in the sample move slightly so as to relax into new low-energy configurations.

Nitinol is also called "memory metal" and has many applications, a number in medicine. The Simon vena cava filter (Fig. 24.24) is a medical device that takes

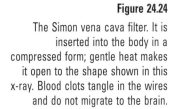

Figure 24.24

The Simon vena cava filter. It is inserted into the body in a compressed form; gentle heat makes it open to the shape shown in this x-ray. Blood clots tangle in the wires and do not migrate to the brain.

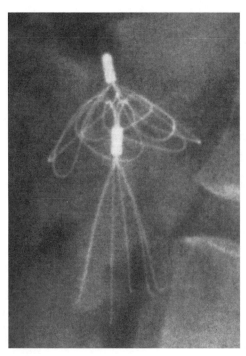

advantage of this property of nitinol. In a blood vessel, the filter can trap any blood clots that could cause a stroke in the brain. This device is trained to remember a shape that resembles an open umbrella. When cooled below body temperature, it can be shaped into a thin bundle of wire (very much like a closed umbrella) and pushed through a catheter into the patient's vein. Body heat or a warmed solution then causes it to return to its original shape, that of an open umbrella, which is ideal for trapping blood clots.

Nitinol has been used to expand or strengthen blood vessels that have become weakened or clogged. A compressed woven cylinder of nitinol is inserted into a blood vessel; slight warming causes it to return to the shape it "remembers"—that of an open tube.

But how does the shape-memory effect work? Nitinol undergoes a phase change without a change in state. Unlike the melting of ice, which involves a change of state *and* a change of phase, nitinol undergoes a transition from a solid phase to a second solid phase in which the atoms are arranged differently. The transition temperature is around 50°C; this temperature can be raised or lowered by making slight changes in the ratio of titanium to nickel.

The high-temperature phase of nitinol is called **austenite.** Like the CsCl structure we saw in Chapter 11, this phase consists of a cubic unit cell with a nickel atom in the center and eight titanium atoms at the corners of the cube, or vice versa. This is a very ordered structure that imparts rigidity to the material. In the low-temperature phase, which is called **martensite,** the atoms have a different, less ordered arrangement. As the austenite phase cools to below the transition temperature, it becomes unstable relative to martensite, in the same way that liquid water cooled below 0°C becomes unstable relative to ice. Though the details are somewhat complicated, the atoms move slightly, but without any visible change to the object, as the austenite is converted to martensite. The atomic movement can occur in 24 different ways, and all the resulting structures can interconvert. This makes the alloy very easy to bend when in the martensite form.

Imagine a piece of wire that is in the martensite form but that has been trained to return to a straight shape when it is in the austenite form. As the wire is bent, the new shape is accommodated by a repositioning of the atoms to match the new shape. When gentle heating is applied and the transition temperature is exceeded, the martensite is unstable with respect to austenite. (Again, imagine ice being unstable with respect to liquid water, above 0°C.) The atoms snap back to the more orderly austenite structure. Because there is only one pathway that restores the original structure of one Ni atom surrounded by eight Ti atoms (and vice versa), the original straight macroscopic shape of the object is restored.

Pressure can also be used to interconvert the austenite and martensite forms of nitinol. The martensite phase is slightly more dense; applying pressure to the austenite phase causes it to convert to martensite. (This is analogous to causing a gas to liquefy by applying pressure.) When the stress is removed, the martensite transforms back to the austenite phase. This property is called **superelasticity.** Eyeglass frames made out of nitinol have this property. When subjected to impact or applied force, they bend a great deal but always return to their original shape.

Design of the wires in dental braces also takes advantage of the superelastic properties of nitinol. As the wires are tightened, the pressure or stress that is created causes a phase change to martensite. As the sample slowly returns to austenite, the teeth are moved in an even and continuous manner. This treatment requires fewer trips to the orthodontist to retighten the braces, and the use of nitinol often cuts in half the time a patient must wear braces.

Guidewires in arthroscopic surgery are also made of nitinol. In this type of procedure, a surgeon manipulates tools inside the body by moving guidewires outside. The ideal guidewire should match every twist and turn of the surgeon's hand with an equal reaction in the patient's joint. The superelastic properties of nitinol are such that surgeons say using such a guidewire is like having their hand inside the patient.

For Review   Summary

Materials science involves the study of the relationship between the structure of a solid and its properties. For example, the properties of a polymer depend on its chemical composition, average molecular weight, stereochemistry, topology, and morphology. **Topology** indicates whether the polymer chains are **linear,** have side-chains extending out from the backbone (**branched polymers**), or are interconnected to other polymer chains (**cross-linked polymers**). **Morphology** refers to whether the polymer sample is crystalline or amorphous. Changes in any of these characteristics affect the properties of the polymer. One important property is the **glass transition temperature** (T_g), the temperature at which the polymer changes from a rigid, inflexible material to a more flexible solid. Polymers may also be classified as being **elastomers,** which can be stretched and will return to their original shape when the pressure is released, **thermoplastics,** which soften when heated but return to their original texture when cooled, or **thermosetting polymers,** which become permanently hard upon heating. Many plastics, especially the thermoplastic ones, can now be successfully recycled.

Molecular orbitals that extend throughout a solid form **bands** of energy levels. The distribution of electrons in bands results in **electrical conductors, semiconductors,** or **insulators.** Substances with partially filled bands are electrical conductors (metals). Substances with a filled band and an empty band separated by a large **band gap** are insulators. At low temperatures, semiconductors have a filled band and an empty band separated by a small band gap. As the temperature is raised, conductivity increases as some electrons are promoted to energy levels that were formerly empty, leaving behind **holes.** A material that is inherently a semiconductor is called an **intrinsic semiconductor.**

Adding more atoms to increase conductivity is called **doping,** which creates an **extrinsic semiconductor.** When electron-deficient atoms are added, a **p-type semiconductor** is created; adding electron-rich atoms creates an **n-type semiconductor.** When p-type and n-type materials are joined together, a **p–n junction,** the basis for solar cells, is formed.

Ceramics are nonmolecular, nonmetallic, covalent-network or ionic solids that are brittle and high-melting. **Glass** is an amorphous silicate-based ceramic that does not melt when heated but gradually softens. By adding other metal ions to a glass, properties such as refractive index and inertness to chemical attack can be changed. **Flint glass** and **borosilicate glass** are examples of this. Optical fibers are also silica-based. The outer part of the fiber, called the **cladding,** serves to contain the light in the **core,** allowing the light to pass through the fiber with little loss. Below their **critical temperature, superconducting** ceramics carry current with zero resistance. They may also exhibit the **Meissner effect.**

Thin films may be used to alter the optical, electronic, or physical properties of a material. They are often prepared by **chemical vapor deposition** of the film precursor on the substrate.

Materials science has made many contributions to the field of medicine, including the replacing of body parts with artificial ones called **prostheses.** In order to be successful, the prosthesis must be **biocompatible.** The durability of artificial hip implants has benefited from new materials. Nitinol, or memory metal, is an alloy that has many medical applications. It has the ability to return, when gentle heat is applied, to a shape it has been trained to remember. The basis for this is a solid-state phase transition between martensite and austenite forms of the alloy.

Key Terms and Concepts

band   (24.6)

band gap   (24.6)

band theory   (24.6)

branched polymer   (24.1)

chemical vapor deposition   (24.15)

composite   (24.5)

critical temperature   (24.12)

cross-linked polymer   (24.1)

dopant   (24.8)

elastomer   (24.3)

electrical conductor   (24.6)

energy levels   (24.6)

extrinsic semiconductor   (24.8)

glass   (24.11)

glass transition temperature   (24.2)

hole   (24.6)
intrinsic semiconductor   (24.8)
linear polymer   (24.1)
Meissner effect   (24.12)
metallic conduction   (24.6)
morphology   (24.1)

n-type semiconductor   (24.8)
p–n junction   (24.9)
p-type semiconductor   (24.8)
photodegradation   (24.5)
superconductor   (24.12)
thermal motion   (24.7)

thermoplastic   (24.3)
thermosetting   (24.3)
thin films   (24.13)
topology   (24.1)

Exercises

Polymers

1. Which would you expect to be more heavily cross-linked, the rubber in a rubber band or the rubber in an automobile tire? Explain.

2. How many ethylene monomer units add together to make a polyethylene chain that has a molecular weight of 1.00×10^8 g mol^{-1} (1 million g mol^{-1})?

3. PVC (polyvinyl chloride) is the polymer in some kinds of rigid plastic pipes. Would you expect the glass transition temperature for this polymer to be above or below room temperature? Explain.

4. PETE is made by the condensation polymerization of terephthalic acid and ethylene glycol.

(a) How would the production of the reaction differ if

   i. The ethylene glycol were contaminated with 5% ethyl alcohol, CH_3CH_2OH?

   ii. The ethylene glycol were contaminated with 5% propylene glycol, $CH_2(OH)CH(OH)CH_2OH$?

(b) In which case would the product of the reaction be most likely to be a thermosetting polymer?

5. Suppose you have a mixture of polypropylene, high-density polyethylene (HDPE), polystyrene, and polyethylene terephthalate (PETE). Given the densities of these polymers and the densities of certain liquids, design a procedure that would enable you to separate the polymer mixtures into batches of pure polymer.

Polymer	Density, g cm^{-3}
Polypropylene	0.90–0.91
HDPE	0.95–0.97
Polystyrene	1.05–1.07
PETE	1.39

Liquid	Density, g cm^{-3}
Ethanol	0.79
1:1 Ethanol/water	0.92
Water	1.0
10% NaCl in water	1.10
$CHCl_3$	1.49

Metals, Insulators, and Semiconductors

6. Explain why the partially filled band in lithium metal is exactly half-filled.

7. Explain why a semiconductor no longer conducts when it is cooled to 0 kelvins.

8. How are metals and semiconductors alike at room temperature? How do they differ?

9. How are semiconductors and insulators alike at 0 kelvins? How do they differ at 0 kelvins? How do their similarities and differences change as they warm from 0 K?

10. Which of the following elements would form p-type semiconductors with silicon? Which would form n-type semiconductors? Al, As, In, P, Se

11. Explain why some substances that are insulators at room temperature conduct electricity at higher temperatures.

12. The conductivity of extrinsic semiconductors is less temperature-dependent than that of intrinsic semiconductors. Explain why this is true.

13. Graphite and diamond are two forms of elemental carbon. They differ in their electrical properties: Graphite is a semi-conductor, whereas diamond is an insulator. How do the band structures of these two substances differ?

14. Which of the following solid materials contain partially filled bands? Mg, S, Cu, NaCl

15. A material has the following band structure. Is it a metal, a semiconductor, or an insulator? What would happen to the electrical properties of the material if all of the electrons from the upper band were removed? Would an oxidizing agent or a reducing agent have to be used to accomplish this?

Ceramics and Thin Films

16. Identify two ceramics that can be classified as ionic compounds.

17. Identify two ceramics that can be classified as network solids.

18. In the synthesis of a 1-2-3 superconductor, no solvents are used; all the reactants and products are solids (or gases). By contrast, solid PbI_2 can be prepared by mixing aqueous solutions of KI and $Pb(NO_3)_2$ and collecting the insoluble precipitate. Compare these two syntheses in terms of the ease of getting rid of unreacted starting material and the ease of achieving uniform mixing of the reactants.

19. During the preparation of a 1-2-3 superconductor, the material is pressed into a pellet using a very high pressure and then heated in an oven. Why does pressing the material into a pellet improve the synthesis?

20. How is CVD like the preparation of optical fiber?

21. How is the preparation of diamond thin films like the preparation of optical fiber?

Materials in Medicine

22. Several of the areas of materials science discussed in this chapter are involved in artificial hips. Describe the applications of polymer chemistry and thin films in this area of medicine.

23. What characteristics are desirable for the metal components of hip replacements?

24. Suggest a use for Nitinol not mentioned in this chapter. The use may be medical or nonmedical.

25. What is the coordination number for Ni in the high-temperature form of Nitinol? What is the coordination number for Ti in the high-temperature form of nitinol?

Additional Problems and Applications

26. Diamond is unusual in that it is an excellent conductor of heat yet is an electrical insulator. Compare and contrast this behavior with that of metals and with that of ceramics such as SiO_2.

27. Draw the Lewis structure of the monomer units that could be used to produce the following polymers.
    (a) Polyvinyl alcohol, $+CH_2CHCH_2CHCH_2CH+_n$

    $$\qquad\qquad\qquad\qquad\qquad OH \quad\ OH \quad\ OH$$

    (b) Saran, $+CH_2CCl_2CH_2CCl_2CH_2CCl_2+_n$
    (c) Orlon, $+CH_2CHCH_2CHCH_2CH+_n$

    $$\qquad\qquad\qquad\qquad CN \quad\ CN \quad\ CN$$

28. Nylon 44 is formed by the reaction of $H_2N(CH_2)_4NH_2$ with $HO_2C(CH_2)_2CO_2H$. Draw a portion of a nylon 44 chain.

29. The following is the structure of a portion of a condensation polymer called Kodel. Draw the monomers that are used to make Kodel.

30. (a) Define the term *linear polyethylene*. In reality, even if a chain of so-called linear polyethylene were stretched as much as possible, the molecule would not be linear. On the basis of what you know about molecular geometries, explain why not. What shape would a stretched linear polyethylene chain exhibit?
    (b) Assume that a polyethylene chain is truly linear. If a polymer chain had a molecular weight of 1.00×10^6 g mol^{-1}, how long, in centimeters, would the chain be? A carbon–carbon single bond has a length of 154 pm.

31. Like most polymers, polyethylene is insoluble in water. However, polyvinyl alcohol is water-soluble. Explain. (The structure of polyvinyl alcohol is shown in Exercise 27.)

32. A sample of a noncrystalline polymer is converted into a more nearly crystalline form. What is the sign of the entropy change of the system for the process? Explain.

33. What are the empirical formulas for ethylene and polyethylene? What are the percentages (use four significant figures) of C and H for these two species?

34. (a) Draw the Lewis dot structure for ethylene and a six-carbon unit of a polyethylene chain.
    (b) How do the bonding and the hybridization in an ethylene molecule change as it is incorporated into a polyethylene chain?

(c) Use bond enthalpies to estimate the enthalpy change per mole of ethylene incorporated into the chain during the polymerization reaction. Is it exothermic or endothermic?

(d) What effect would the net enthalpy change have on the way the reaction is performed on an industrial (large) scale?

35. What is the sign of the entropy change of the system when a sample of ethylene is converted into polyethylene? Explain.

36. Given your answers to Exercises 34 and 35, under what temperature conditions is the formation of polyethylene from ethylene spontaneous (above a certain temperature, below a certain temperature, at all temperatures, or never)?

37. How do the hybridization and geometry about the carbon atoms change when styrene is converted into polystyrene?

38. The interconversion of the martensite and austenite forms of nitinol can be viewed as an application of Le Châtelier's principle. If austenite is converted into martensite upon the application of pressure, which phase would you expect to be more dense?

39. How many grams of phosphorus would have to be added to a 1-cm^3 crystal of silicon for there to be 1.0×10^{18} charge carriers (phosphorus atoms with their free electrons) per cubic centimeter, a typical doping level for extrinsic semiconductors?

40. The residents of a particular house use 900 kilowatt-hours of energy a month. Using the information in the chapter, calculate the cost of that electricity (a) if it is provided by conventional sources; (b) if it were provided by photovoltaic cells.

41. One way to improve the efficiency of a solar cell is to stack semiconductors with different band gaps on top of each other. Suppose solar cell A has a band gap of 1.4 eV and solar cell B has a band gap of 2.2 eV. Which should be on top (nearest the sun) and why?

42. Suppose you want to prepare a 100-g sample of $YBa_2Cu_3O_7$ from Y_2O_3, $BaCO_3$, and CuO.

(a) Write a balanced equation for the reaction. *Hint:* In this reaction, O_2 is a reactant in addition to the metal-containing species. CO_2 is the carbon-containing product.

(b) How much of each metal-containing reactant should be mixed together to prepare the desired quantity of material?

43. The following compounds have also been reported to be superconducting. Calculate the average oxidation state of the copper ions in each compound.

(a) $YBa_2Cu_3O_{6.7}$

(b) $Bi_2Ca_3Sr_2Cu_4O_{10}$ (Assume the oxidation state of the bismuth is $+3$.)

44. A 1-2-3 superconductor can also be prepared by using Y_2O_3, BaO, and CuO.

(a) Write a balanced equation for this reaction.

(b) Do you see any advantages to using these reactants instead of the reactants given in Exercise 42? Explain.

45. A manufacturer of stylish sunglasses advertised that its sunglass lenses were made with thin films of iridium. Iridium is a rare transition metal with an atomic radius of 135 pm and a density of 22.42 g cm^{-3}. It sells for about $100 per gram, making it much more expensive than gold. An average lens has the area of a 2.5 cm $\times$ 5.0 cm rectangle, and one side is plated with iridium to a thickness of 125 nm.

(a) How many atoms, stacked on a line, are required to give a line 125 nm in length?

(b) Iridium crystallizes in a face-centered cubic unit cell with a 3.839-Å cell edge. How many lenses can be plated with $100 worth of iridium?

46. The surface atoms of a heterogeneous catalyst actually perform the catalysis. Assume that a gold-plated brick is to be used as a catalyst for a reaction and that a layer of gold that is one atom thick and covers the entire brick will be sufficient. Use the radius of a gold atom to determine the cost of gold-plating a brick that measures 2.0 in. $\times$ 4.0 in. $\times$ 10.0 in. with gold that sells for $300 an ounce. What would be the cost of a solid gold brick (density 19.3 g cm^{-3})?

CHAPTER OUTLINE

The Cell
25.1   Components of a Cell

Metabolism
25.2   Energy Sources and Energy Utilization
25.3   ATP

Proteins
25.4   Amino Acids
25.5   Peptides and Proteins
25.6   Classes and Structures of Proteins
25.7   Enzymes

Carbohydrates
25.8   Monosaccharides
25.9   Disaccharides and Polysaccharides

Lipids
25.10   Classes of Lipids

Nucleic Acids and the Genetic Code
25.11   Nucleic Acids
25.12   The Genetic Code and Protein Synthesis
25.13   Recombinant DNA Technology
25.14   Gene Mutations

25

Biochemistry

Most of the biochemistry of plants and animals is identical.

Biochemistry is the study of the composition and functions of chemical substances found in living organisms. Considering the vast diversity of those organisms and the complexity and organization of their cells, it is not surprising that biological molecules, both large and small, are among the most complicated substances known. However, it may be surprising that many of the kinds of molecules active in our bodies play similar roles in nearly all other living organisms. Moreover, most organisms use the same chemical reactions to make their biological molecules.

In this chapter we will explore some of the molecules and reactions that make life possible. In addition, we will see that several of the unique properties of living organisms can be classified in terms of the following themes.

- **Organization.** How do molecules fold up and interact with each other? How are cellular structures assembled? How do cells organize into tissues, organs, and whole organisms? (Sections 25.1, 25.6, 25.10, and 25.11)
- **Energy.** How do cells obtain energy to function? Where does this energy come from? (Sections 25.2, 25.3, 25.8, and 25.10)
- **Catalysis.** How can chemical reactions in cells happen so fast? How are complicated molecules made? (Section 25.7)
- **Information transfer.** How do cells obtain information and pass it on to other cells, including their daughter cells? In what form is the information found? (Sections 25.6 and 25.11–25.14)

Biochemistry is at the heart of the biotechnology revolution of the last two decades. This is because powerful new tools and research techniques have been developed to study and influence life processes, and consequently, a very large amount of basic information has been accumulated. Nearly all life scientists and health care professionals in the many disciplines that deal with living organisms look to biochemistry for basic understanding of the roles biomolecules play in whatever problem they are trying to understand.

The Cell

Despite the wide variety of cells found in living organisms, all cells share striking similarities. They all contain the same four classes of biomolecules: proteins, carbohydrates, nucleic acids, and lipids—molecules that will be discussed later in this chapter. All cells use DNA in the same way to store and retrieve genetic information, they employ closely related proteins such as enzymes for catalyzing reactions, they use similar lipids for the construction of membranes, and they obtain and utilize energy by one of two possible mechanisms.

25.1 Components of a Cell

Most cells are between 60% and 90% water. In fact, the principal constituent of all living matter is water. Large molecules are suspended in water as colloidal particles, and both organic and inorganic substances are dissolved in it and transported through the organism and to the cells. In addition to its role as a solvent, water has a large heat capacity and acts as a heat sink, thereby helping organisms resist drastic changes in temperature. Finally, water participates in many biological reactions. For example, we will see that a molecule of water is produced when a peptide bond is formed between two amino acids during protein synthesis (Section 25.5) and that water is decomposed during photosynthesis (Section 25.2).

To understand another role played by water in biological systems, we distinguish between substances that are **hydrophilic** (literally, "water-loving") and those that are **hydrophobic** ("water-fearing"). Biomolecules fall into both classes. Remember that water is a very polar molecule with the ability to hydrogen-bond (Section 11.4) and that it readily dissolves both polar and ionic substances (Chapter 12). Sugars and amino acids contain polar and/or ionic substituents, which impart hydrophilic character, and these substances may be very water-soluble. On the other hand, nonpolar substances such as hydrocarbons and lipids are hydrophobic. They dissolve very poorly in water and have a tendency to avoid aqueous environments. For example, oil molecules float on water, and lipid membranes surround water-filled cells: neither dissolves in water. Some biomolecules are hydrophilic in one part of the molecule and hydrophobic in another part. The polar part tends to dissolve in water while, at the same time, the hydrophobic part avoids association with water. This behavior is called **amphipathic;** it helps explain why some molecules self-organize, associate, or fold in a particular manner.

Organisms are commonly divided into two classes. Eukaryotic organisms are composed of cells in which most of the genetic material is contained in chromosomes within a nucleus (Fig. 25.1 A). Most multicellular organisms are eukaryotic. Prokaryotic organisms have cells with a single chromosome and no nucleus (Fig. 25.1 B). Microorganisms such as blue-green algae and bacteria are prokaryotic.

A chromosome in either kind of cell stores genetic information that enables the cell to replicate itself and to synthesize various proteins needed for a multitude of functions. Chromosomes are composed primarily of two substances: **deoxyribonucleic acid (DNA)** and proteins. The genetic information itself is stored in DNA. The proteins perform functions such as helping to replicate DNA, making **ribonucleic acid (RNA)** transcripts of DNA that are used in the actual protein synthesis, and ensuring DNA stabilization and repair.

All cells are surrounded with a cell membrane, which is the "skin" or envelope that contains the cellular contents. The cell membrane insulates the contents of the

Figure 25.1

(A) A normal human cell (eukaryotic). (B) A single cell bacterium, *clostridium tetani* (prokaryotic).

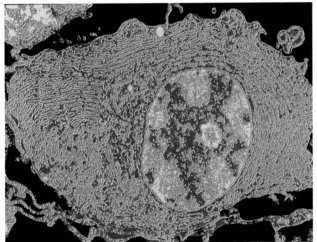

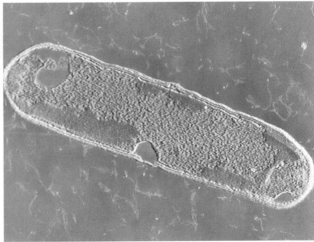

(A)   (B)

cell from the surrounding environment and actively controls what enters and what leaves the cell. The basic structure of membranes is assembled from lipids. It is studded with proteins and carbohydrates that provide entry and exit ports for the cell and that serve as markers of its identity. Bacteria, fungi, and plant cells are also surrounded by cell walls—complex structures containing lipids, carbohydrates, and proteins that impart additional strength, as well as other properties, to the cell.

The cytoplasm (the contents of a cell apart from any nucleus) contains subcellular structures called organelles. Specific steps of cell metabolism occur in these structures. Some organelles are found in the cytoplasm of the cell; others, on its surface.

All cells make proteins in the **ribosomes,** the site of directed polymerization of amino acids into proteins. Ribosomes are roughly spherical bodies composed of a number of RNAs and proteins.

Mitochondria are the "power plants" of eukaryotic cells. Oxidation of a variety of biomolecular fuels, including carbohydrates, amino acids, and fatty acids, is the chief path by which most nonphotosynthetic organisms obtain energy from their environment. Some of the energy released during these oxidations is converted to usable chemical energy through the synthesis of **nucleotide triphosphates;** adenosine triphosphate (**ATP**) and guanosine triphosphate (**GTP**) are examples. The remainder of the energy is converted to heat, which may be conserved or lost. An average human uses his or her own mass in ATP in a day, but only a small fraction of that amount is found in tissues at any one time. ATP is made only on demand and is not stored.

Chloroplasts are present in all eukaryotic plants that are capable of converting light energy into chemical energy through photosynthesis. Light energy is collected by the chloroplasts and used both to synthesize ATP and to convert water and CO_2 photochemically to O_2 and carbohydrates or other reduced metabolites. Nearly all nonphotosynthetic organisms, including humans, owe their lives to plants, which produce both food and oxygen.

Metabolism

The term *metabolism* refers to all the chemical reactions that take place in living organisms. Metabolism includes the reactions in the cell that produce energy, secure nutrients for the synthesis of biomolecules, degrade macromolecules to produce the basic building blocks for future construction, and shuffle intermediates between those cells or organelles that have plenty of them to those that are wanting.

25.2   Energy Sources and Energy Utilization

Metabolism results in either the synthesis or the degradation of a compound. **Anabolic pathways** result in the synthesis of a substance and require the input of chemical energy, and **catabolic pathways** lead to the degradation of a substance and release chemical energy. In general, anabolic and catabolic pathways are separated in cells so that the pathways are controlled independently, but they share substances. In both cases, the thousands of substances used or produced are called **metabolites.**

The energy required for the growth and maintenance of organisms is provided by metabolic reactions of two general types: photosynthesis and chemotrophy. Green plants and algae carry out **photosynthesis,** an endothermic, multistep process by which chloroplasts convert light into chemical energy. Photosynthetic reactions lead to the overall reaction of carbon dioxide and water, producing oxygen and reduced chemical substances, such as glucose.

$$6CO_2 + 6H_2O \xrightarrow{\text{Light}} \underset{\text{Glucose}}{C_6H_{12}O_6} + 6O_2 \qquad \Delta G° = +2870 \text{ kJ mol}^{-1}$$

Chemotrophy is the release of chemical energy through a multistep oxidation of reduced organic compounds, often compounds produced by photosynthetic organisms. Chemotropic organisms, which include humans, other animals, fungi, and the majority of bacteria, must consume an elaborate array of organic molecules and minerals in order to maintain their lives.

Chemotropic energy is provided by metabolic reactions of two general types: respiration and fermentation. The oxidation–reduction reactions in **respiration** involve O_2 as the ultimate electron acceptor (hence they are **aerobic** reactions), generating water and carbon dioxide as products. Respiration very efficiently generates usable chemical energy. **Fermentation** is much less efficient; the reactions end up producing organic molecules, which may be discarded as waste. Oxygen is not required for fermentation; it is an **anaerobic** reaction. Many organisms, including mammals, use both fermentation and respiration reactions to generate energy.

Glucose (Section 25.8) is a central molecule in energy metabolism. The sequence of reactions that anaerobically degrades a single molecule of glucose into two pyruvate ions is known as **glycolysis.**

Glucose   Pyruvate ions

The reaction also involves formation of two ATP molecules from two phosphate ions and two ADP molecules (Section 25.3), neither of which is shown in this unbalanced equation.

Anaerobic organisms (such as yeast cells used for ethanol production by the fermentation of sugars) in a few additional steps convert the pyruvate produced in glycolysis to carbon dioxide and ethanol. The reactions produce about 230 kilojoules per mole of glucose fermented.

Aerobic organisms, in many more steps, convert the pyruvate ions into carbon dioxide and water. The net reaction accomplishes the opposite of the photosynthetic reaction and therefore liberates approximately 2870 kilojoules per mole of glucose consumed. Note that there is more than 12 times as much energy available from oxidation of glucose in the aerobic pathway as is available from ethanol production via the anaerobic pathway.

25.3 ATP

About 50% of the energy released during the degradation of glucose appears as heat. The remainder is converted to chemical energy in nucleoside triphosphates such as **adenosine triphosphate, ATP** (see also Section 25.11).

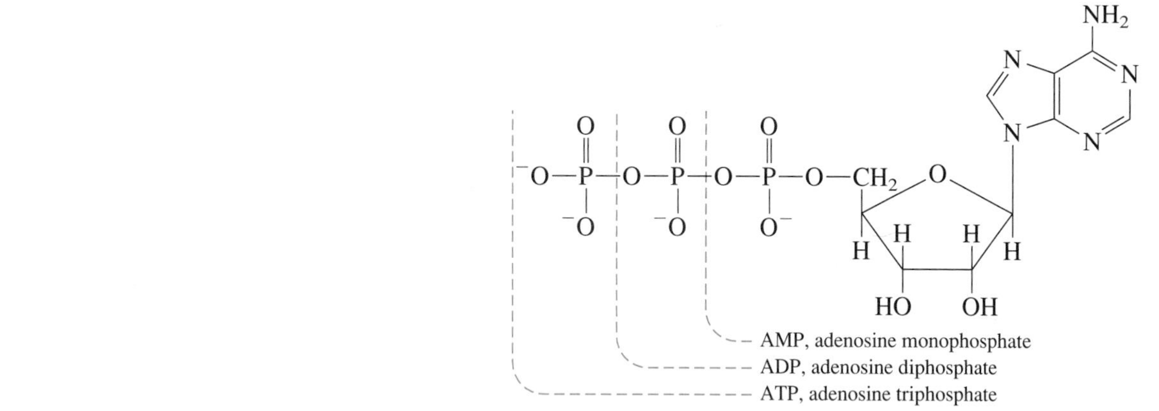

AMP, adenosine monophosphate
ADP, adenosine diphosphate
ATP, adenosine triphosphate

Nucleoside triphosphates do not accumulate and do not function as long-term stores of chemical energy. They form as needed.

ATP and the related triphosphates GTP, CTP, and UTP (Section 25.11) can react with water to produce a phosphate ion and adenosine diphosphate, ADP (or GDP, CDP, or UDP), with the release of about 30 kilojoules per mole of triphosphate. The nucleoside diphosphates, such as ADP, can undergo further reaction to form nucleoside monophosphates, such as AMP, with the release of an additional 30 kilojoules per mole. These reactions are essentially irreversible. If the hydrolysis of ATP or ADP can be coupled to another chemical reaction, such as the synthesis of proteins from amino acids, or can be coupled to a mechanical process, such as the contraction of muscle, then the process can be driven to completion as ATP is consumed.

During protein synthesis, chemical energy in ATP derived from catabolic processes (degradation) accomplishes anabolic processes (synthesis). Degradative and synthetic pathways can be linked in other ways as well. There are many points in catabolic pathways where intermediates can be removed to be used in synthetic (anabolic) pathways. For instance, pyruvate (Section 25.2) formed in the degradation of glucose can be converted in one step to alanine, an amino acid (see Fig. 25.3), by addition of an amino group from a donor molecule. Pyruvate can also be converted with the appropriate enzymes in times of plenty (abundant ATP and reducing equivalents) to polysaccharides (such as starch and glycogen) or fats. These represent medium- and long-term energy storage molecules, respectively.

Proteins

Proteins are the most abundant cellular components. Among them are enzymes, antibodies, storage components, structural elements, transport agents, and hormones. The word *protein* is derived from the Greek *proteios,* meaning "of first importance." All proteins are linear condensation polymers of amino acids. We will see that a staggering number of different polymers can be made. Because there are so many structural possibilities, proteins can have myriad functions.

25.4 Amino Acids

All proteins in the biological world are assembled on a ribosome from 20 amino acids, the **common amino acids.** Each amino acid has four substituents around a tetrahedral carbon atom called the alpha (α) carbon. Bonded covalently to it are an amino ($-NH_2$) group (or in one case, an amine), a carboxyl ($-CO_2H$) group, and a hydrogen atom ($-H$). The fourth bond is to a side group (R). The structures of the α-amino acids differ only in the nature of the side group and can be represented by the general structure

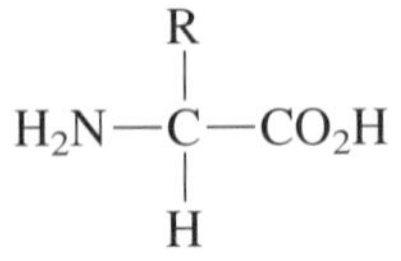

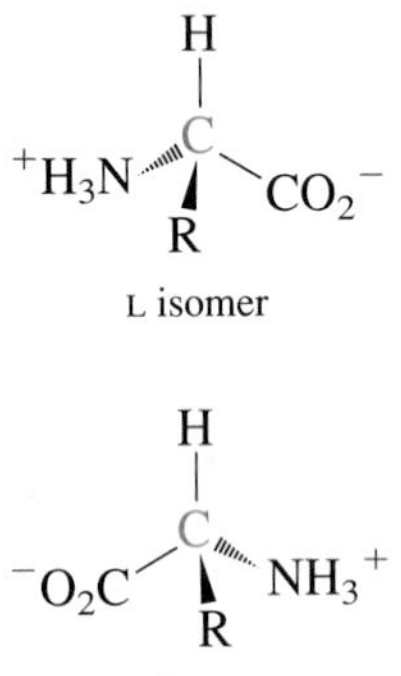

$$H_2N-\underset{\underset{H}{|}}{\overset{\overset{R}{|}}{C}}-CO_2H$$

Because the α-carbon is tetrahedral with four different groups bonded to it (with the exception of glycine, in which R is also a H atom), the α-amino acids are optically active (Section 25.8). Nearly all amino acids isolated from naturally occurring proteins have a structure called an L structure. Looking down the H—C bond of an L-amino acid, we see that the $-CO_2H$, $-R$, and $-NH_2$ substituents are arranged in a clockwise fashion. In the other arrangement, that in a D-amino acid, these substituents are arranged in a counterclockwise manner (Fig. 25.2). Though they are not common, D-amino acids are found occasionally in nature, especially in bacterial cell walls.

An amino acid molecule contains a weak base, the $-NH_2$ group, which acts as a proton acceptor. The $-CO_2H$ group functions as a weak acid, a proton donor. Proton transfer within the molecule forms a species called a **zwitterion,** a molecule that contains both a positive and a negative electrical charge.

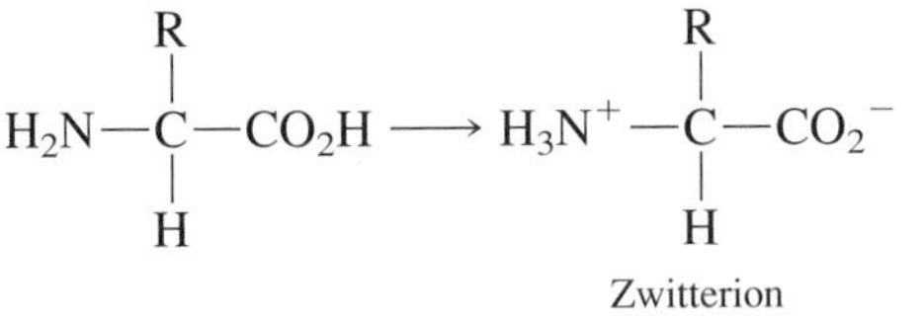

$$H_2N-\underset{\underset{H}{|}}{\overset{\overset{R}{|}}{C}}-CO_2H \longrightarrow H_3N^+-\underset{\underset{H}{|}}{\overset{\overset{R}{|}}{C}}-CO_2^-$$

Zwitterion

A zwitterion is the predominant form of an amino acid in aqueous solution at or near neutral pH (pH $=$ 7). In an acidic solution, however, the net charge on the molecule becomes positive as the carboxylate ion is protonated. Conversely, in a basic solution, the amino acid develops a net negative charge as the amino group loses its proton. Alanine is an example.

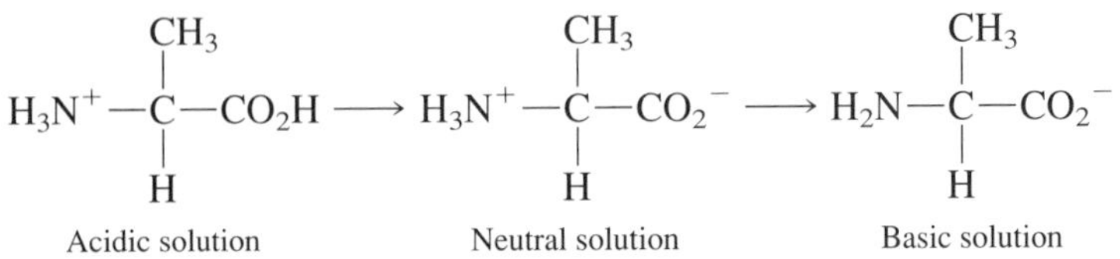

$$H_3N^+-\underset{\underset{H}{|}}{\overset{\overset{CH_3}{|}}{C}}-CO_2H \longrightarrow H_3N^+-\underset{\underset{H}{|}}{\overset{\overset{CH_3}{|}}{C}}-CO_2^- \longrightarrow H_2N-\underset{\underset{H}{|}}{\overset{\overset{CH_3}{|}}{C}}-CO_2^-$$

Acidic solution   Neutral solution   Basic solution

Figure 25.2

The molecular structures of the L and D isomers of an amino acid. These optical isomers are mirror images. The α-carbon is highlighted in color.

L isomer

D isomer

Amino acids are usually grouped on the basis of similarities in the structures of the side (R) groups. The names and structures of the 20 common α-amino acids are shown in Fig. 25.3. To emphasize their uncharged form in a protein, we have left the α-amino and α-carboxyl groups in their uncharged states in this figure, whereas any acidic or basic groups in the side chains are shown in the ionized form found at neutral pH.

Figure 25.3

The 20 common α-amino acids classified in terms of their functional groups (in color). The formal name of each is followed by its three-letter abbreviation.

Aliphatic R Group

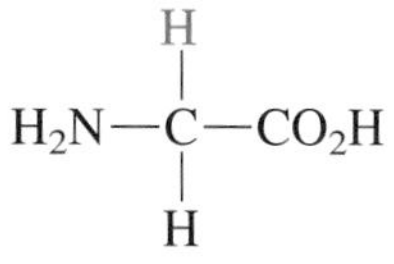

Glycine (Gly) 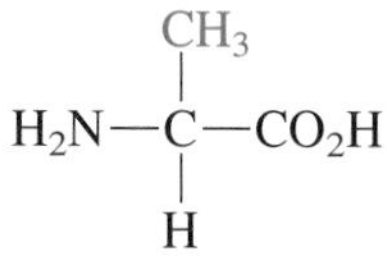   Alanine (Ala) 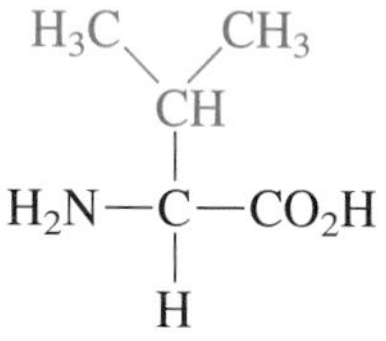   Valine (Val)

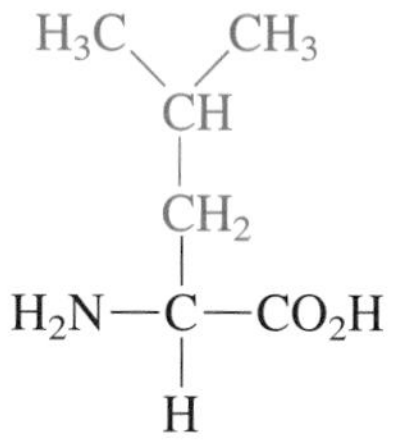

Leucine (Leu) 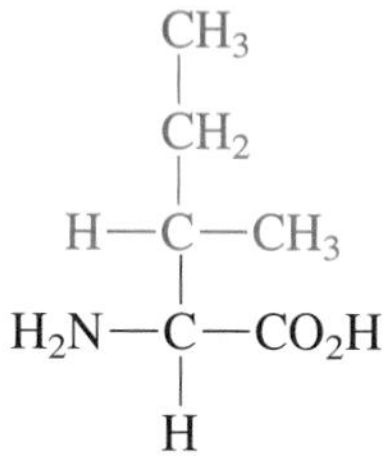   Isoleucine (Ile) 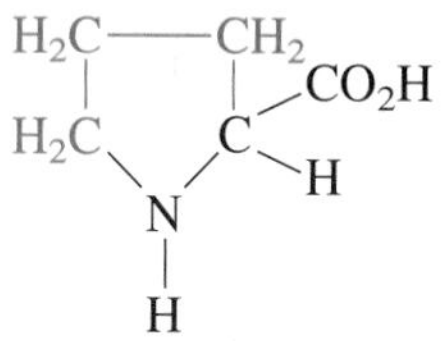   Proline (Pro)

Aromatic R Group

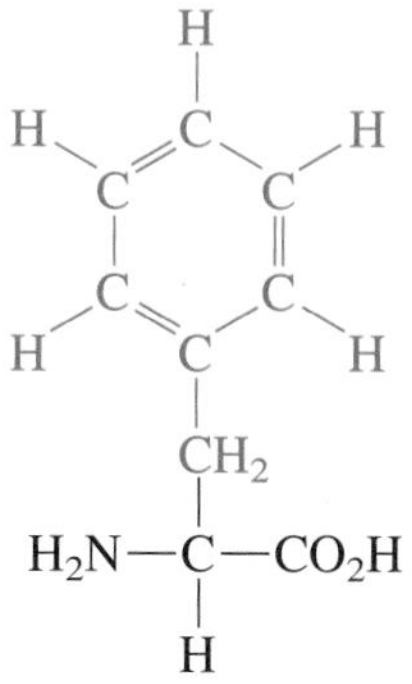

Phenylalanine (Phe) 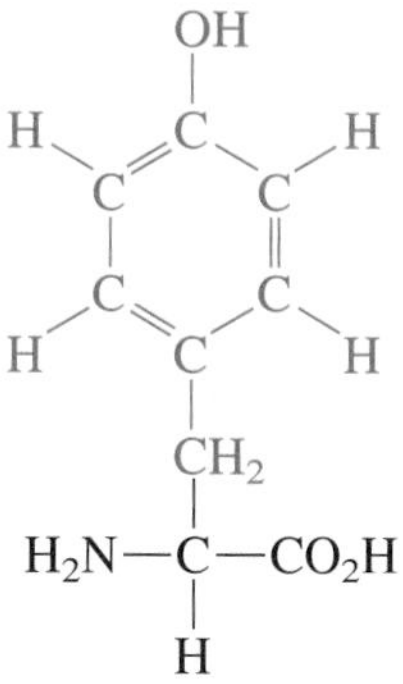   Tyrosine (Tyr) 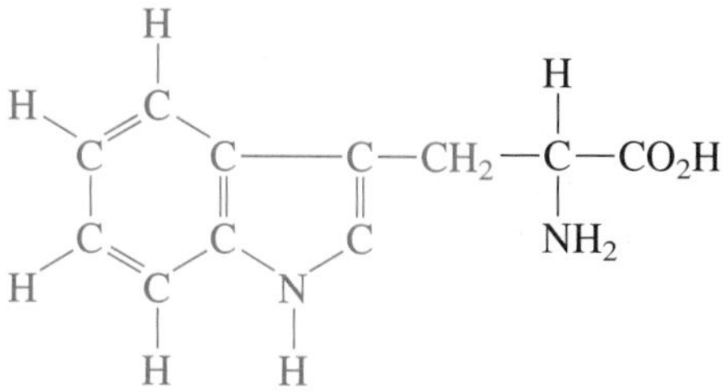   Tryptophan (Trp)

Carboxylic Acid R Group

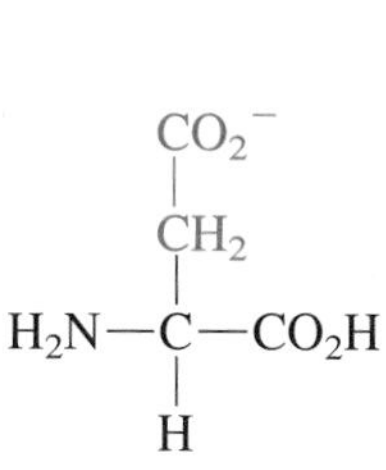

Aspartic Acid (Asp) 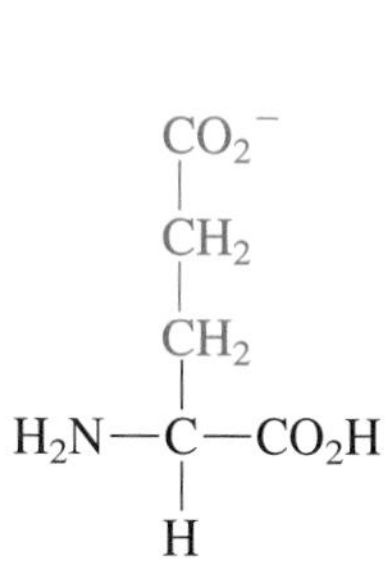   Glutamic Acid (Glu)

Figure 25.3   *Amide R Group*
(continued)

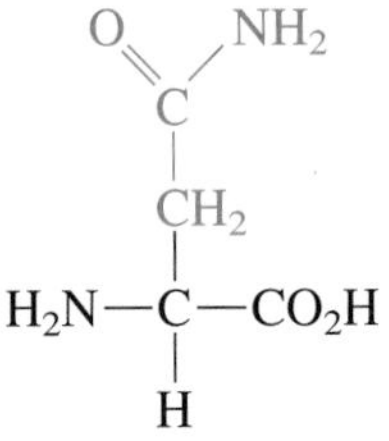

Asparagine (Asn) 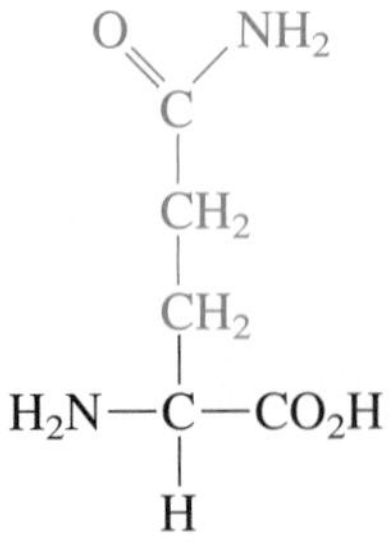   Glutamine (Gln)

Alcohol R Group

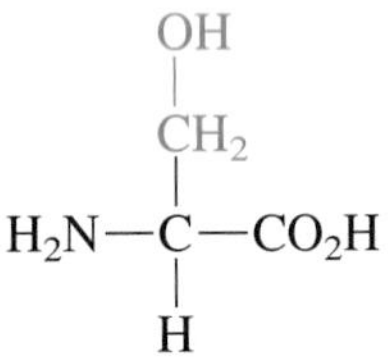

Serine (Ser) 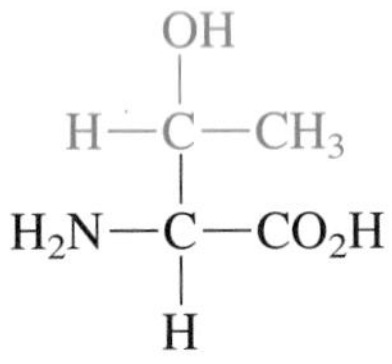   Threonine (Thr)

Basic R Group

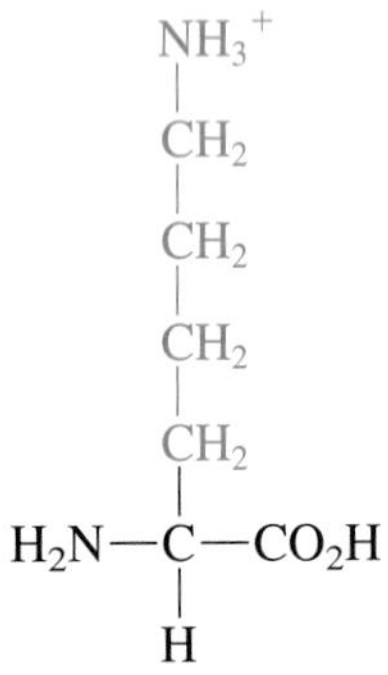

Lysine (Lys) 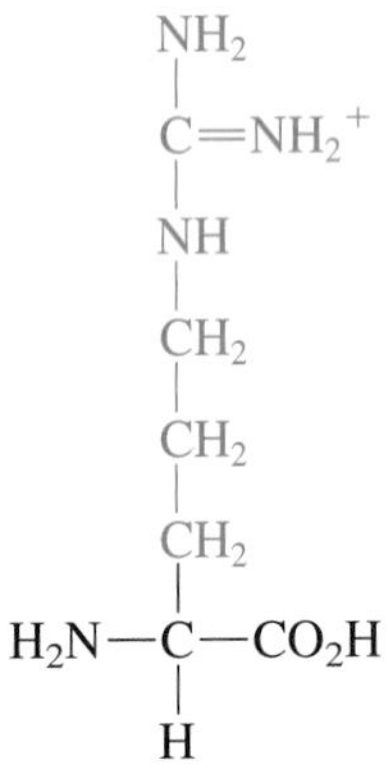   Arginine (Arg) 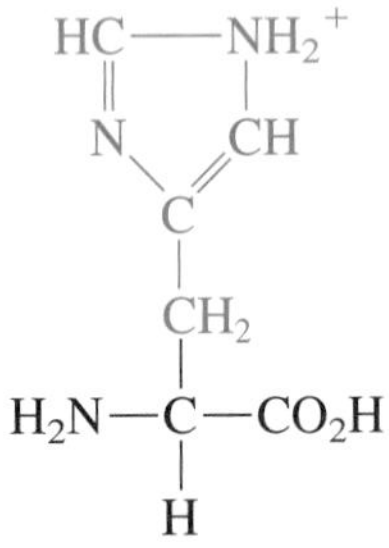   Histidine (His)

Sulfur-Containing R Group

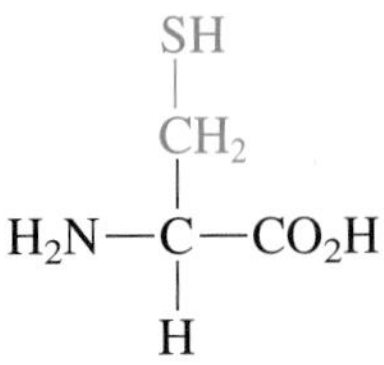

Cysteine (Cys) 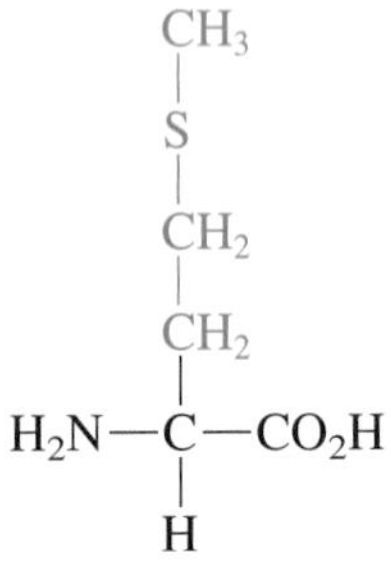   Methionine (Met)

The amino acid cysteine (Fig. 25.3) has a thiol-containing side group. Cysteine often links to a second cysteine side chain through oxidation to a covalent disulfide bond ($-S-S-$). The combination of the two amino acids is given the name cystine. Reduction cleaves the bond back to two cysteine molecules.

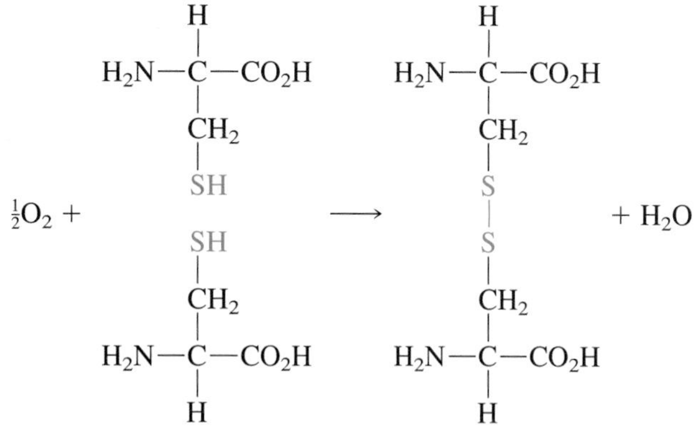

$$\tfrac{1}{2}O_2 + \quad \begin{array}{c} \text{H} \\ | \\ \text{H}_2\text{N}-\text{C}-\text{CO}_2\text{H} \\ | \\ \text{CH}_2 \\ | \\ \text{SH} \\ \\ \text{SH} \\ | \\ \text{CH}_2 \\ | \\ \text{H}_2\text{N}-\text{C}-\text{CO}_2\text{H} \\ | \\ \text{H} \end{array} \quad \longrightarrow \quad \begin{array}{c} \text{H} \\ | \\ \text{H}_2\text{N}-\text{C}-\text{CO}_2\text{H} \\ | \\ \text{CH}_2 \\ | \\ \text{S} \\ | \\ \text{S} \\ | \\ \text{CH}_2 \\ | \\ \text{H}_2\text{N}-\text{C}-\text{CO}_2\text{H} \\ | \\ \text{H} \end{array} \quad + H_2O$$

Two cysteine molecules   Cystine

25.5 Peptides and Proteins

The structures of peptides and proteins were first determined by the German chemist Emil Fischer between 1900 and 1910. **Peptides** are polymers formed from two or more amino acids. The link is an amide given the special name **peptide bond.** The nature of a peptide bond (shown in color) can be seen in the following structure for a dipeptide (two amino acids joined by a peptide bond).

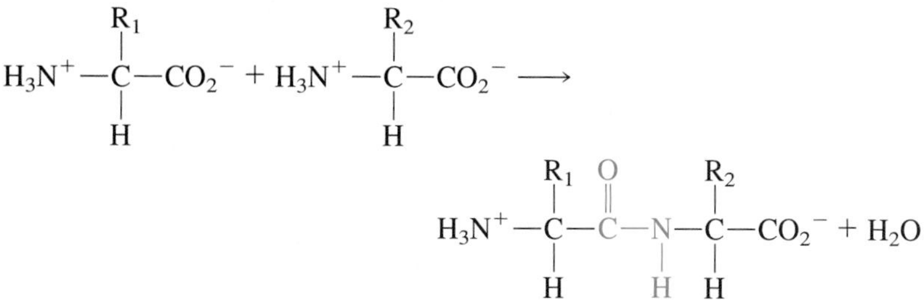

$$\begin{array}{c} \text{R}_1 \\ | \\ \text{H}_3\text{N}^+-\text{C}-\text{CO}_2{}^- \\ | \\ \text{H} \end{array} + \begin{array}{c} \text{R}_2 \\ | \\ \text{H}_3\text{N}^+-\text{C}-\text{CO}_2{}^- \\ | \\ \text{H} \end{array} \longrightarrow$$

$$\begin{array}{c} \text{R}_1 \quad \text{O} \quad \text{R}_2 \\ | \quad\ \parallel \quad | \\ \text{H}_3\text{N}^+-\text{C}-\text{C}-\text{N}-\text{C}-\text{CO}_2{}^- + H_2O \quad (1)\\ | \qquad\ \ | \ \ | \\ \text{H} \qquad\ \text{H}\ \ \text{H} \end{array}$$

Peptide bond in dipeptide

The peptide bond results from removal of the elements of H_2O from the α-carboxyl group of one amino acid and the α-amino group of another. The amino acids that make up a peptide are called **residues.** The amino acid residue that has a free α-amino group is the **amino-terminal** (or **N-terminal**) residue; the residue at the opposite end, which has a free carboxyl group, is the **carboxyl-terminal** (or **C-terminal**) residue. The residues in peptides and proteins are written beginning from the N-terminal residue. For example, two dipeptides can be formed containing the two amino acids glycine (Gly) and alanine (Ala); they are Gly-Ala and Ala-Gly.

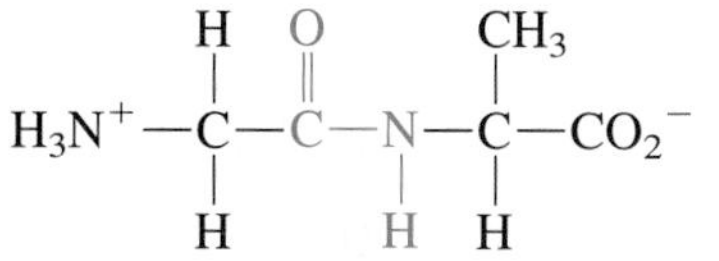

$$\begin{array}{c} \text{H} \quad \text{O} \qquad \text{CH}_3 \\ | \quad\ \parallel \qquad\ | \\ \text{H}_3\text{N}^+-\text{C}-\text{C}-\text{N}-\text{C}-\text{CO}_2{}^- \\ | \qquad\ | \ \ | \\ \text{H} \qquad\ \text{H}\ \ \text{H} \end{array} \qquad \begin{array}{c} \text{CH}_3 \quad \text{O} \qquad \text{H} \\ | \quad\ \ \parallel \qquad | \\ \text{H}_3\text{N}^+-\text{C}-\text{C}-\text{N}-\text{C}-\text{CO}_2{}^- \\ | \qquad\ \ | \ \ | \\ \text{H} \qquad\ \ \text{H}\ \ \text{H} \end{array}$$

Gly-Ala 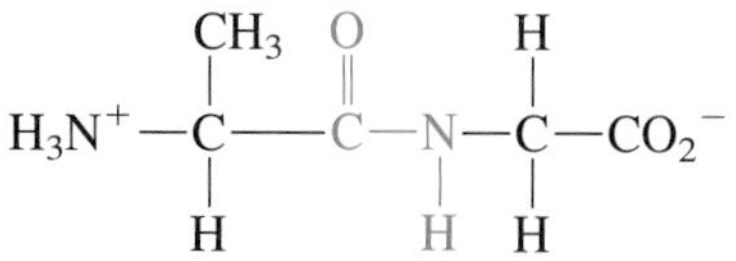   Ala-Gly

Aspartame, the artificial sweetener sold under the name NutraSweet, is the methyl ester of the dipeptide aspartylphenylalanine.

Aspartic acid residue   Phenylalanine methyl ester residue

Aspartylphenylalanine methyl ester (aspartame)

We should note several other important features of peptides. First, the C-terminal carboxyl group is available for condensation with yet another amino acid, so a linear polymer can be made of any length. Second, the backbone of the polymer is invariant, $(-CO-NH-CHR-)_n$, with only the R groups differing along the polymer. Third, the peptide bond neither donates nor accepts protons at neutral pH, although it is capable of hydrogen bonding.

For every particular protein found in nature, the number of amino acid residues and their order are fixed. The order of the residues in a peptide from the amino terminus to the carboxyl terminus is called the **amino acid sequence.** Any change in the amino acid sequence of a protein may lead to profound differences in its chemical and biological properties.

The name *peptide* is usually restricted to polymers made up of only a few amino acids. Such polymers are often called oligopeptides (from the Greek *oligo,* which means "few"). Polymers that contain more than a small number of residues are called either **polypeptides** or **proteins.** Some proteins contain 10,000 or more residues.

The free energy for hydrolysis of a peptide bond, the reaction of a peptide bond with water (the reverse of Reaction 1), is negative, so the reaction is spontaneous. Fortunately for a cell, the activation energy (Section 13.8) is high, so the reaction is slow unless catalysts such as strong acid, strong base, or enzymes called **proteases** are present. During enzymatic digestion in the small intestine, proteins in our diet are hydrolyzed completely to their amino acid constituents. As we age, cellular proteins are slowly degraded to amino acids and are replaced with new ones.

25.6   Classes and Structures of Proteins

The extraordinary variety of protein structures allows for remarkable differences in their types and functions. In addition, proteins often associate with other, smaller organic and inorganic substances called **prosthetic groups,** which enhance their functioning. For example, the heme groups in hemoglobin and myoglobin (see Fig. 25.6 on page 939) are prosthetic groups. Proteins play many different biological roles, because there are many different amino acid sequences and thus many possible structures. For example, from the 20 common amino acids there are 20^2 (400) possible dipeptides, 20^3 (8000) tripeptides, 20^4 (160,000) tetrapeptides, and so on. For an average-size protein of 300 amino acids, there are more than 10^{390} possible sequences, an astronomically large number.

Among the biologically important proteins are the following seven classes.

1. **Enzymes.** Enzymes are biological catalysts. They catalyze biochemical reactions with a high degree of efficiency and specificity. This is the most varied and most highly specialized group of proteins.

2. **Transport proteins.** These proteins are involved in the movement of both organic and inorganic nutrients across cell membranes. They are also contained in blood plasma and red cells, where they transport specific molecules or ions from one organ to another.

3. **Nutrient and storage proteins.** Many plant seeds store nutrient proteins necessary for growth of the embryonic plant. Casein, the major protein of milk, is a nutrient protein, and ferritin, which stores iron in mammalian liver, is a storage protein.

4. **Contractile proteins.** Muscles are composed of fibrous bundles of the proteins actin and myosin, which are involved in the contraction and relaxation of muscle tissue. Tubulin makes up microtubules, which move chromosomes during cell division.

5. **Structural proteins.** Many proteins serve as supports, giving biological structures strength or protection. The protein collagen is the most abundant protein in vertebrates. It has high tensile strength and is the major organic component of skin, tendons, cartilage, and bones (excepting the minerals). Hair, fingernails, and feathers consist mainly of the protein keratin.

6. **Defense proteins.** The antibodies produced by the immune response system are proteins that complex with and neutralize viruses, bacteria, and substances recognized as foreign. Interferons inhibit the replication of many kinds of RNA viruses in animals.

7. **Regulatory proteins.** Proteins bind to specific sites on the cell membrane or on DNA to regulate cellular or physiological activity. Among them are the hormone insulin, which helps regulate sugar metabolism, and repressors, which regulate gene expression.

For a protein to function properly, it must assume its own unique three-dimensional structure. There can be four components to that structure.

1. The **primary structure** of a protein is its covalent structure—that is, the sequence of amino acid residues, starting from the amino terminus.

2. A polypeptide is flexible and can fold upon itself to produce a **secondary structure,** in which the number of hydrogen bonds between peptide linkages is maximized.

$$\diagdown \!\!\!\! \underset{\diagup}{N}\!-\!H\cdots O\!=\!\underset{\diagdown}{\overset{\diagup}{C}}$$

Only a limited number of secondary structures allow a polypeptide chain to form the maximum number of hydrogen bonds. Two of the most common forms of secondary structure are the **α-helix** (Fig. 25.4 on page 938) and the **β-pleated**

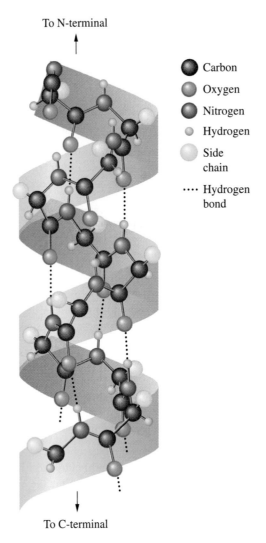

Figure 25.4

The α-helical structure of a peptide chain. Hydrogen bonds (dotted lines) within the chain stabilize this structure.

To N-terminal

- Carbon
- Oxygen
- Nitrogen
- Hydrogen
- Side chain
- Hydrogen bond

To C-terminal

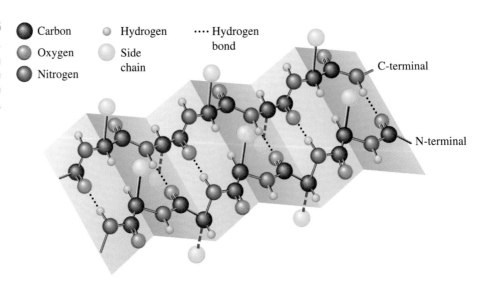

Figure 25.5

The β-pleated sheet secondary structure of protein molecules. Hydrogen bonds (dotted lines) between the chains stabilize parallel peptide sections.

- Carbon
- Oxygen
- Nitrogen
- Hydrogen
- Side chain
- Hydrogen bond

C-terminal

N-terminal

sheet (Fig. 25.5 on page 938), both of which were described by Linus Pauling and coworkers. In the α-helix, the polypeptide backbone forms a helix or coil. The α-helix is the dominant secondary structure in many proteins and was originally discovered in the α-keratins of hair and wool. The β-pleated sheet structure results from hydrogen bonds between adjacent strands of a polypeptide backbone. Extended arrays of the β-pleated sheet secondary structure are found in fibroin, the principal protein of silk (where they were first discovered), as well as in many other proteins.

3. Helixes and sheets in a protein may be folded into a complex conformation known as the **tertiary structure** (Fig. 25.6). The principal driving force for folding is the movement of hydrophobic side groups to the interior of the structure away from the surrounding polar water solvent. Interactions, such as hydrogen bonds between amino acid side groups, the attraction of opposite charges, and covalent disulfide bonds, also stabilize these structures.

4. Larger proteins may contain more than one folded peptide chain or **subunit.** The **quaternary structure** of a protein is determined by the way these subunits pack into the protein molecule. The forces that hold proteins in folded structures are usually noncovalent and therefore weak. Factors such as heat and extremes of pH can lead to loss of secondary, tertiary, and quaternary structure (denaturing) and, consequently, to loss of biological activity. It is believed that a fever is a defensive response that inactivates disease-causing viruses or microorganisms.

Figure 25.6

The tertiary structure of myoglobin, an oxygen-storing protein found in muscle. The folded structure is shown, as well as the α-helical secondary structures (cylindrical regions). Only the atoms of the backbone of the protein are shown, as a series of colored dots. A prosthetic group, heme, can be seen near the top.

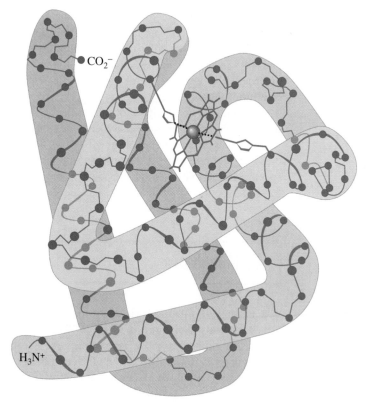

25.7   Enzymes

Enzymes are proteins that catalyze chemical reactions with remarkable efficiency and selectivity. The enzyme fumarase, for example, catalyzes the addition reaction

| Fumarate ion | | L-Malate ion |

Fumarase is such a good catalyst that a single enzyme molecule can transform over 48,000 fumarate ions per minute into L-malate ions. This is 100 million times faster than the rate of the uncatalyzed reaction. One explanation for an enzyme's efficiency is that the binding of the **substrate** (the molecule undergoing the reaction) to the enzyme may strain the substrate's bonds until their conformation more closely resembles the transition state for the reaction (Section 13.7), thereby reducing the activation energy.

Enzymes are not only efficient, they can also be remarkably selective. For example, only the *trans* isomer of the fumarate ion participates in the fumarase reaction. The *cis* isomer is not transformed, nor is any D-malate produced. The ability of most enzymes to be specific for a very few substrates suggests that the enzymes and substrates fit together somewhat like a lock and key. Thus the enzyme activates only those substrates that fit into a specific crevice called an **active site,** just as a lock is opened only by keys that fit.

Several substances may bind to an enzyme, but only a limited number of them can react. If a molecule resembling the substrate's structure competes with the substrate for the available active sites of the enzyme and slows down the rate of reaction, then the activity of the enzyme is inhibited. This decrease in activity is called **competitive inhibition.** There are other types of inhibitors, such as the heavy-metal poisons Hg^{2+} and Pb^{2+}, that bear no resemblance to the substrate and bind enzymes either in the presence or in the absence of the substrate. These inhibitors are called **noncompetitive inhibitors.**

The study of competitive inhibition is crucial to the design of some pharmacologically active substances, because enzymes may play a major role in a disease. For example, the multiplication of the HIV virus that causes AIDS (acquired immune deficiency syndrome) is dependent on the action of many enzymes. A substance that could specifically inhibit an enzyme critical to the virus's replication would be a candidate for a therapeutic drug.

Carbohydrates

The name *carbohydrate* originally meant a compound with the empirical formula CH_2O, literally a hydrate of carbon. This definition has since been broadened to include all aldehydes and ketones that have two or more hydroxyl groups and substances derived from them. Among the compounds classified as carbohydrates are sugars, starches, cellulose, glycogen, and chitin. Carbohydrates are an important source of energy for many organisms, and they form the supporting tissue of plants and many animals. Carbohydrates exist as monomers, dimers, and various polymers.

25.8 Monosaccharides

The simplest carbohydrates contain only a single aldehyde or ketone functional group (Section 9.12) and are known as **monosaccharides.** They are colorless, are crystalline as solids, and frequently have a sweet taste. A monosaccharide is classified as either an **aldose** or a **ketose,** depending on whether it contains an aldehyde or a ketone functional group. With the exception of the simplest ketose (the three-carbon monosaccharide dihydroxyacetone), all the monosaccharides are optically active. Although both D and L isomers are possible, the dominant form of carbohydrates isolated from natural sources is the D isomer. The structures of the D and L forms of the simplest aldose, glyceraldehyde, are

D-Glyceraldehyde   L-Glyceraldehyde

Some monosaccharides can undergo a reversible intramolecular reaction changing among linear and cyclic structures. One important aldose, glucose, forms a linear molecule and two isomeric six-membered pyranose rings. The two isomers differ only in orientation of the —OH group at the C(1) atom.

α isomer   Linear form of   β isomer
                    D-Glucose

Glucose, also called dextrose or blood sugar, $C_6H_{12}O_6$, is one of the most abundant organic compounds in nature. It is the monomer from which the polysaccharides starch, glycogen, and cellulose form. Metabolism of glucose is central to biological energy production. It is the most plentiful carbohydrate in the blood stream. Human blood normally contains slightly less than 1 gram of glucose per liter. Individuals who suffer from diabetes are unable to assimilate glucose into their cells easily, and their blood glucose concentration may rise to 2–4 grams per liter. They eliminate glucose through the kidneys, a symptom of the disease.

Fructose, also called levulose or fruit sugar, has the same molecular formula as glucose but contains a ketone rather than an aldehyde functional group. The two sugars are readily interconverted during metabolism. Fructose occurs naturally in both fruits and honey and is found combined with glucose is the sugar sucrose. Fructose is the sweetest of all sugars; a given mass of fructose is perceived as nearly twice as sweet as the same amount of sucrose. Like glucose, fructose undergoes isomerization, forming a linear molecule and two cyclic isomers that differ in the orientation of the —OH group at the C(2) atom.

α isomer   Linear form of   β isomer
                  D-Fructose

25.9   Disaccharides and Polysaccharides

Monosaccharides can polymerize, with the elimination of the elements of water, to form polysaccharides. However, the polymerization reaction can be reversed only in the presence of acid or a suitable enzyme catalyst, so polysaccharides are stable in aqueous solution.

Sucrose, also called cane sugar or table sugar, is the most common disaccharide. Sucrose can be hydrolyzed in the presence of acid or the enzyme sucrase to an equimolar mixture of fructose and glucose, commonly called invert sugar, the major component of honey. Sucrose contains a glucose ring and a fructose ring. In representations of ring structures such as sugars, the symbols for the carbon atoms are sometimes omitted, as in the following representations.

Sucrose

Lactose (milk sugar) is a disaccharide formed from glucose and another sugar, galactose.

Lactose

Some infants and many adults cannot digest lactose because they lack the enzyme lactase that hydrolyzes the ether linkage (the —O— group) between the rings. Because this causes intestinal discomfort, lactase has been made available at low cost for pretreatment of milk, one example of the therapeutic use of enzymes.

Two major classes of **polysaccharides** are formed by the polymerization of glucose rings: starch (and glycogen, animal starch) and cellulose. **Starch** (see Fig. 24.1 A) serves primarily as a long-term energy storage medium in plants. It accumulates in seeds, tubers, and fruits. Starch is one of the energy supplies for a young plant until the development of a leaf system enables it to capture energy through photosynthesis. **Glycogen** is a similar polysaccharide that serves as a source of glucose in animals; glycogen, however, is a short-term storage medium. Glycogen is found in muscle and in the liver.

Cellulose is a linear polysaccharide of glucose units (see Fig. 24.1 B) found in plants. As the main component of the rigid cell wall of plant cells, it provides structure rather than energy. Humans lack an enzyme capable of cleaving the β-linkage in cellulose, and this prevents us from digesting cotton and paper (which are mostly cellulose). Some animals, including termites and ruminants such as cattle, sheep, and goats, are able to digest cellulose because microorganisms in their digestive tracts have the required enzymes. Termites can be controlled by killing these bacteria; they still consume the cellulose, but they starve to death anyway.

Monosaccharides can be extensively modified. Aldehyde and alcohol functional groups can be oxidized to carboxylic acids. Amino groups can be substituted for hydroxyl groups and can then be acylated. Hydroxyl groups may be esterified with organic and inorganic acids or replaced with hydrogen atoms. All these transformations and more are found in linear and branched polysaccharides throughout biological systems. Complex polysaccharides, because of their dazzling variety, are used by cells to serve both structural and information functions.

Glycoproteins are protein molecules that have carbohydrate units attached to side groups of certain amino acid residues. Glycoproteins include blood-clotting proteins, collagen, and cell membrane proteins. Glycoproteins in membranes and oligosaccharides on the surfaces of cells provide identity so that cells can be recognized as "self" by the immune system. The blood groups A, B, and O, for instance, result from different glycolipids attached to the surface of red cell membranes.

> # Lipids

The **lipids** include widely different compounds with no obvious structural relationships. These biomolecules are defined instead by their behavior: They are soluble in moderately polar organic solvents such as chloroform, $CHCl_3$, and diethyl ether, $(C_2H_5)_2O$. Lipids are oily organic substances that do not form solutions in water.

25.10   Classes of Lipids

Lipids can be divided into five classes.

Fatty Acids and Triglycerides.   **Fatty acids** are long-chain carboxylic acids (Section 9.13). One class of lipids consists of derivatives of fatty acids such as stearic acid, $C_{17}H_{35}CO_2H$.

$$CH_3CH_2CH_2CH_2CH_2CH_2CH_2CH_2CH_2CH_2CH_2CH_2CH_2CH_2CH_2CH_2CH_2CO_2H$$
Stearic acid

Although fatty acids are available to a cell from a wide variety of sources, the concentration of free fatty acids in the cell is negligible. Most fatty acids occur in plant and animal cells in the form of triesters (Section 9.13) of the alcohol 1,2,3-propanetriol or **glycerol.**

HOCH$_2$ $CH_3CH_2CH_2CH_2CH_2CH_2CH_2CH_2CH_2CH_2CH_2CH_2CH_2CH_2CH_2CH_2CH_2\overset{\displaystyle O}{\overset{\|}{C}}OCH_2$

HOCH $CH_3CH_2CH_2CH_2CH_2CH_2CH_2CH_2CH_2CH_2CH_2CH_2CH_2CH_2CH_2CH_2CH_2\overset{\displaystyle O}{\overset{\|}{C}}OCH$

HOCH$_2$ $CH_3CH_2CH_2CH_2CH_2CH_2CH_2CH_2CH_2CH_2CH_2CH_2CH_2CH_2CH_2CH_2CH_2\overset{\displaystyle O}{\overset{\|}{C}}OCH_2$
Glycerol   Tristearylglycerol

Lipids such as tristearylglycerol are commonly known as **triglycerides.** Often the fatty acid groups in a triglyceride molecule are derived from different fatty acids. The primary function of triglycerides is the long-term storage of energy. Triglycerides are either liquids (oils) or solids (fats) at room temperature. Oils tend to have hydrocarbon chains containing one or more double bonds (derived from unsaturated fatty acids) or to have short chains. Many plants produce oils; these are of interest because polyunsaturated fatty acids tend to protect humans from cardiovascular disease.

Before a fat can be used as an energy source, hydrolysis of the ester linkage must take place. Like most esters, triglycerides can be hydrolyzed in the presence of strong acids, strong bases, or enzymes called lipases (esterases), yielding the alcohol, glycerol, and the fatty acids. When a base is used, the process is called saponification, and it yields the salts of the fatty acids, which are called **soaps** (Section 12.23).

The synthetic fat substitute olestra, which has been approved by the Food and Drug Administration for use in snack foods, is a polyester of sucrose rather than of glycerol. Olestra consists of a sucrose fragment bonded to seven or eight fatty acid groups by ester linkages. It is noncaloric; digestive enzymes that hydrolize the ester linkages in fats cannot get to these units in the larger olestra molecules, so it passes through the digestive tract intact.

Eicosanoids.

The hormone-like **eicosanoids** make up the second class of lipids and consist of the prostaglandins, the thromboxanes, and the leukotrienes. Hundreds of these closely related compounds control and influence a bewildering array of cellular and physiological activities. All are derived from a polyunsaturated fatty acid, **arachidonic acid,** which humans must obtain from their diet. Arachidonic acid has 20 carbon atoms and 4 double bonds. It may be released from membranes, stimulated by either hormones or some pathology, and can ultimately cause inflammation. For hundreds of years it has been known that plant extracts containing salicylic acid, and later aspirin (acetylsalicylic acid), relieve inflammation. It is now known that aspirin slows the production of prostaglandins by inhibition of the complex initial enzymatic steps in the oxidation of arachidonic acid to prostaglandins.

Arachidonic acid   Prostaglandin PGH$_2$

Structural Lipids.

The **structural lipids** make up the third class of lipids. Important examples are the phosphoglycerides (or glycerophosphatides), which are major components of cell membranes. Phosphoglycerides can be thought of as derivatives of phosphoric acid and glycerol. They consist of a hydrophilic polar charged head and two long hydrophobic acyl tails. Thus they are amphipathic. Among the most important of the phosphoglycerides are the **phosphatidyl cholines** or **lecithins,** of which the following is a representative example. In addition to the glycerol and the phosphate, it contains choline, a quaternary aminoalcohol and two acyl groups derived from fatty acids.

Charged hydrophilic head   Nonpolar hydrophobic tails

Phosphatidyl choline (a lecithin)

Other structural lipids are made of different components but have the same overall architecture and also are components of membranes. All these lipids can spontaneously form spherical **liposomes** in aqueous media; ionic polar heads face the water, both inside and outside, and the nonpolar tails form a bilayer hydrophobic membrane (Fig. 25.7 on page 946). Liposomal membranes are models for cellular membranes, which separate the cell from its surroundings. Liposomes are being studied as vehicles for drug delivery to tumors.

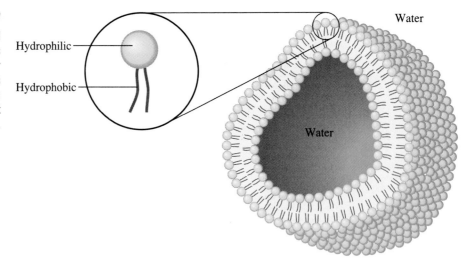

Figure 25.7

Structural lipids such as phosphatidyl choline associate to form liposomes whose bilayers serve as a model for cell membranes. The nonpolar tails associate to form a hydrophobic bilayer, and the polar heads point toward the internal and external water.

Waxes.

Waxes are rather simple both in structure and in function. They consist of esters of long-chain fatty acids and long-chain fatty alcohols. Beeswax, for instance, is primarily myricyl palmitate. Waxes are very hydrophobic and serve to protect plants and animals from desiccation.

$$CH_3(CH_2)_{14}-\overset{\overset{\displaystyle O}{\parallel}}{C}-O-(CH_2)_{29}CH_3$$
Myricyl palmitate

Isoprenoids.

Isoprenoids are assembled from five-carbon units related to isoprene, $CH_2{=}C(CH_3){-}CH{=}CH_2$. This large and very heterogeneous group of compounds includes the vitamin A precursor and pigment β-carotene, insect juvenile hormones, and steroids.

Steroids (Fig. 25.8) are a remarkable group of lipids that share the same carbon framework, which is synthesized from isoprenoid derivatives by a complex set of reactions. In humans, cholesterol is the steroid from which all other steroids are derived. Cholesterol is an important component of cell membranes, but at the same time it is a source of dietary concern because of the role it plays in atherosclerosis (hardening of the arteries). Among other activities, steroids in animals help regular reproduction,

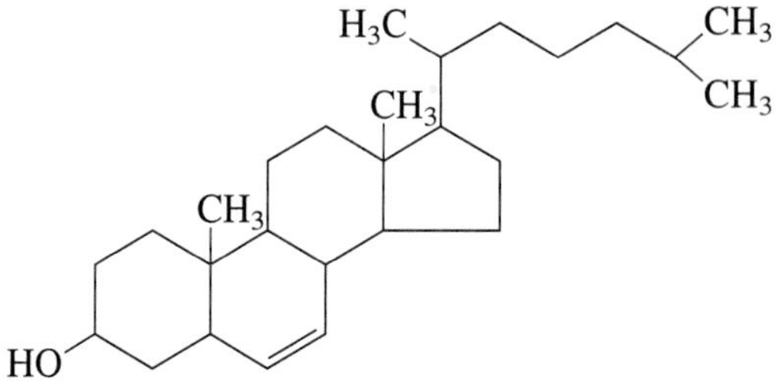

Figure 25.8

The steroids are a class of compounds that all exhibit the same essential fused-ring structure. Many functional groups can be attached to it. (C and H atoms in the rings and in the alkane tail are not shown.)

Cholesterol

sexual development, energy metabolism, and ion balance in the blood plasma. The complex carbon skeleton of the steroids allows a very large number of possible synthetic derivatives to be made. Chemists have made thousands of steroids in addition to the many naturally occurring ones, often finding useful anti-inflammatory agents or hormone mimics.

Nucleic Acids and the Genetic Code

Understanding the chemical nature of nucleic acids is the basis for comprehending molecular genetics; both are at the core of the sweeping biotechnological advances underway in human culture. There are only three major steps in the processing of the genetic information: replication, transcription, and translation. But before we can consider these processes, we need to look at the structure of the nucleic acids.

25.11 Nucleic Acids

There are two kinds of nucleic acids: *deoxyribonucleic acids (DNA)* and *ribonucleic acids (RNA)*. DNA molecules are linear polymers with molar masses of 10^{12} or more. Their function is the storage of genetic information in an organism's chromosome(s). RNA molecules also are polymers, but they vary in size and are much smaller than DNA. RNAs have a number of critical functions in processing the genetic information stored in DNA. Both types of nucleic acids are synthesized in the nuclei of eukaryotic cells (or on chromosomes, in prokaryotic cells). The DNA remains in the nucleus of a eukaryotic cell, whereas the RNAs function primarily in the cytoplasm or in certain organelles.

The ribonucleic acids are built on the framework of the ring of the sugar ribose, and the deoxyribonucleic acids contain a modified ribose ring in which the —OH group on the second, or 2', carbon is replaced by a hydrogen atom.

β-D-Ribofuranose   2-Deoxy-β-D-ribofuranose

A **nucleoside** (spelled with an s) forms when a sugar combines with a molecule of one of two types of cyclic nitrogen-containing bases: a pyrimidine or a purine. **Pyrimidines** are six-membered nitrogen-containing cyclic compounds (Fig. 25.9 on page 948); **purines** are pyrimidines with an additional five-membered nitrogen-containing ring (Fig. 25.9). A **nucleotide** (spelled with a t) is a phosphate, diphosphate, or triphosphate ester of a nucleoside. So, when the base **uracil** joins to a ribofuranose ring

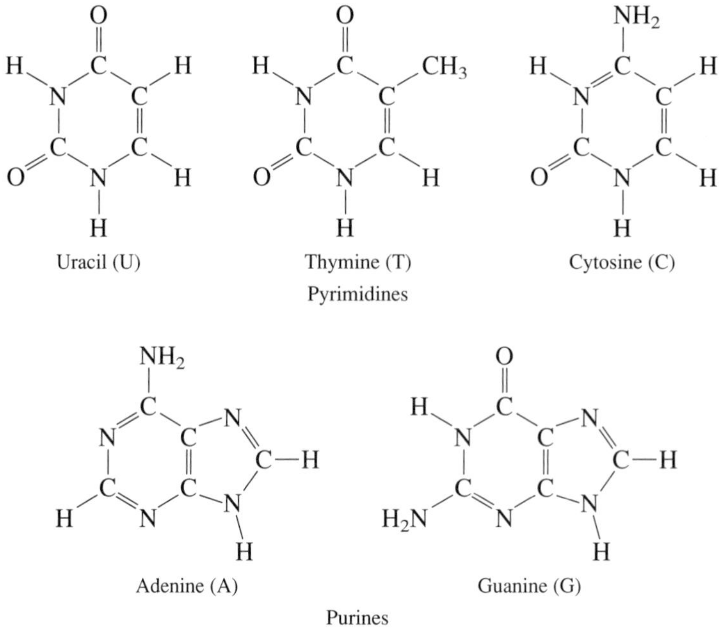

Figure 25.9

The five most common nitrogen-containing bases found in nucleic acids are the pyrimidines uracil (U), thymine (T), and cytosine (C) and the purines adenine (A) and guanine (G).

Uracil (U)   Thymine (T)   Cytosine (C)

Pyrimidines

Adenine (A)   Guanine (G)

Purines

at the $1'$ carbon, the nucleoside **uridine** forms. This nucleoside can then combine with the phosphate ion, PO_4^{3-}, to form the nucleotide **uridine monophosphate** (Fig. 25.10).

Nucleic acids form by polymerization of nucleotides. Polymerization occurs by reaction of the $3'$ hydroxyl group of one nucleotide with the phosphate attached to the $5'$ carbon of another nucleotide, forming an ester. Figure 25.11 on page 949 shows a segment of a DNA chain containing the four deoxy-nucleotides that are found in deoxyribonucleic acids.

Note that the polymeric backbone of nucleic acids is an invariant structure of sugar molecules alternating with phosphates (Fig. 25.11). Thus the backbone can be ignored, and the sequence of nucleotides in Fig. 25.11, reading from the $5'$ end of this chain to the $3'$ end, can be symbolized as T-A-C-G. This is the **base sequence.** The phos-

Figure 25.10

The progression from base to nucleoside to nucleotide. (Carbon atoms in the ribofuranose rings are not shown.)

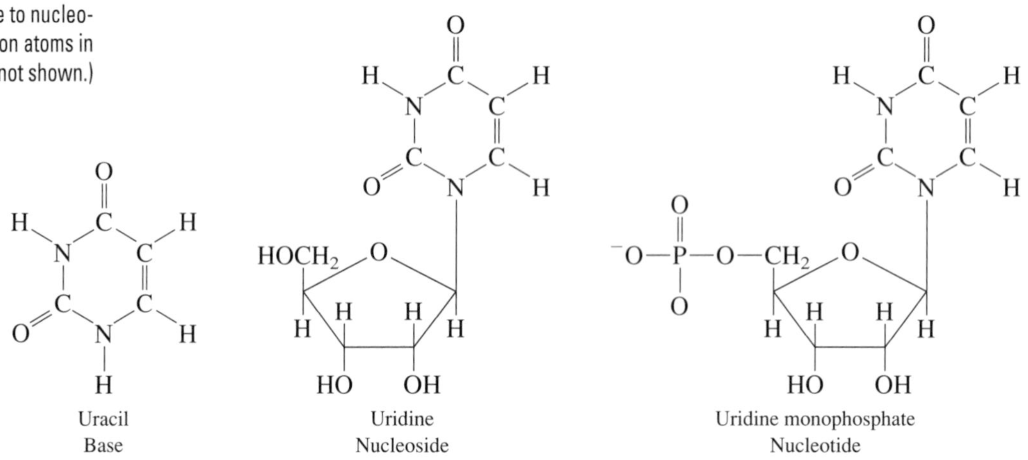

Uracil   Uridine   Uridine monophosphate
Base   Nucleoside   Nucleotide

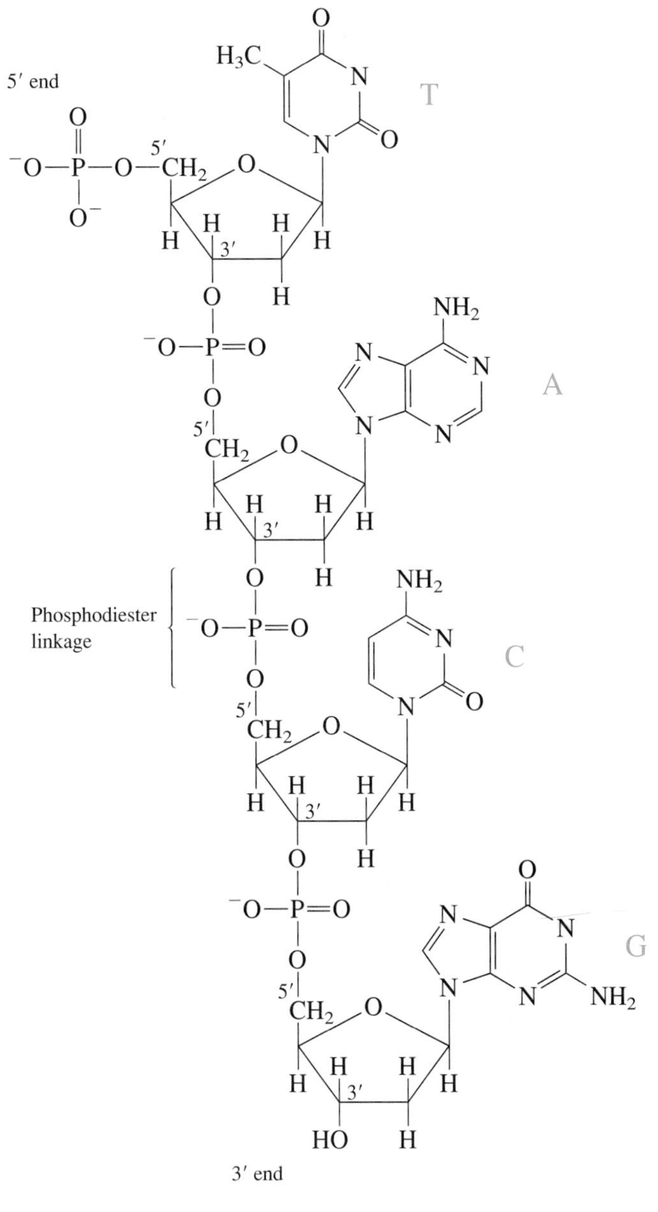

Figure 25.11

The structure of a tetranucleotide segment of a single strand of deoxyribonucleic acid (DNA). Reading from the 5′ end to the 3′ end, this tetranucleotide is symbolized as T-A-C-G. (Carbon atoms in the furanose rings and bases are not shown.)

phodiester linkages between adjacent nucleosides in nucleic acids are strong acids (hence their name) and are ionized at neutral pH. Figure 25.11 shows the tetranucleotide as it would exist at or near neutral pH—that is, carrying a negative charge on every phosphate. Thus nucleic acids are normally found as salts, associated with positively charged proteins such as histones or with cations such as Mg^{2+} ions.

There are important structural differences between DNA and RNA. First, RNA contains ribose, and DNA contains 2′-deoxyribose. Furthermore, they contain different bases; DNA contains exclusively the two purines adenine (A) and guanine (G) and the two pyrimidines thymine (T) and cytosine (C). RNA contains the pyrimidine uracil (U) instead of thymine and sometimes a number of modified bases. However, these

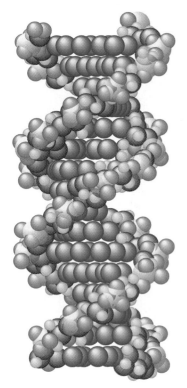

Figure 25.12

Model of the DNA double helix. Hydrogen atoms are omitted, and other atoms are represented as spheres. The sugar–phosphate backbones are outside the helix; purine and pyrimidine bases fill the interior.

are formed by modification of adenine, guanine, cytosine, or uracil after the RNA polymer is produced.

In 1953, J. D. Watson and F. H. C. Crick proposed a model for the structure of DNA that immediately helped explain how the genetic information stored in DNA is accurately passed on to subsequent generations of cells. For this work they received the Nobel prize. According to their model, DNA is composed of two strands, running in opposite directions, that are bridged by hydrogen bonds between specific pyrimidines in one strand and purines in the other (Fig. 25.12). These two strands are twisted into a **double helix.**

In the Watson–Crick model, two hydrogen bonds form between adenine and thymine and three between guanine and cytosine (Fig. 25.13). This ensures that wherever adenine is found on one strand, thymine is found on the other (A-T **base pairs**) and that wherever guanine is found on one strand, cytosine is found on the other (G-C base pairs). The sequence of nucleotides in one strand of the double helix determines the sequence in the other, which is the biological guarantee that nearly perfect copies can be made when DNA is replicated. It is the base sequence that stores information in the DNA molecule.

25.12 The Genetic Code and Protein Synthesis

Genes contain the information for the synthesis of specific RNAs and proteins. The double helix of DNA in a gene stores the code used to designate the sequence of amino acids in a polypeptide. To transmit this information during cell division, the DNA molecule must be capable of faultless replication so that each generation of new cells gets an exact copy of the information (Fig. 25.14 on page 951).

To maintain the code while passing on directions for protein synthesis, a portion of one strand in the DNA helix is transcribed to form another nucleic acid called messenger RNA. Finally, the mRNA are translated into the amino acid sequences of proteins.

We can summarize the three processes of replication, transcription, and translation (referred to as the "central dogma") as follows. In this case the arrows mean transfer of information and do not indicate a direct reaction product.

$$\text{DNA} \xrightarrow{\text{Replication}} \text{DNA} \xrightarrow{\text{Transcription}} \text{mRNA} \xrightarrow{\text{Translation}} \text{Protein}$$

Figure 25.13

A-T and G-C base pairs. The hydrogen bonds are represented as dots. (Carbon atoms in the bases are not shown.)

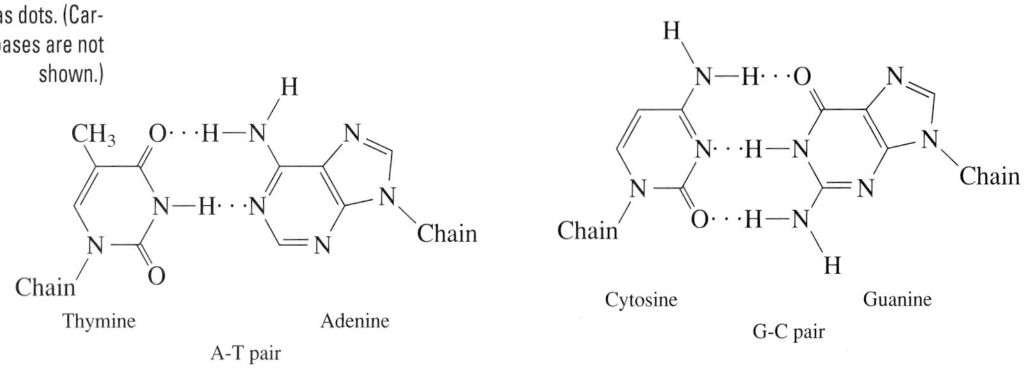

Thymine — Adenine

A-T pair

Cytosine — Guanine

G-C pair

Figure 25.14

Autoradiograph of replicating DNA from *Escherichia coli*. Replication started in the bacterium at the 1:30 o'clock position and has copied in both directions as far as the 12 o'clock and 3 o'clock positions, producing two loops.

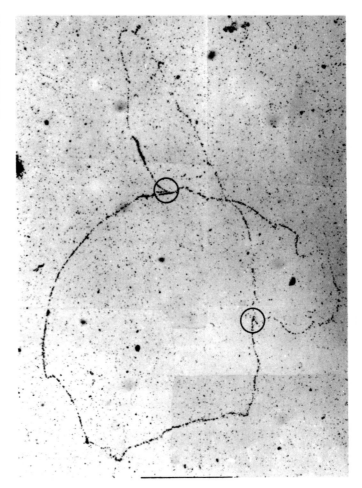

The precise replication of the DNA double helix is based on the specificity of A-T and G-C base pairing in the **complementary strands** of DNA. The first step in the replication of DNA during cell division involves the binding of certain proteins that separate a portion of the double helix into two single strands of DNA. Each single strand is then copied, making two new complementary strands, with A bonded to T in the old strands, G to C, T to A, and C to G. Replication proceeds until two identical double-strand copies of the original DNA molecules are formed. Replication is a semiconservative process: Each resulting double strand contains one strand from the parent DNA molecule and one completely new strand (Fig. 25.15 on page 952).

DNA stores the information that enables a cell to synthesize many thousands of different proteins in its sequences of bases. This information is processed in a two-step mechanism. In transcription, an *RNA complement* of a portion of one strand of the DNA double helix is synthesized under the control of certain enzymes. This segment of RNA carries the message contained in the DNA molecule, so it is called **messenger RNA,** or **mRNA.** Other kinds of RNA with different functions are made in the same way. Just as the two strands of the DNA double helix run in opposite directions, the mRNA segment synthesized during transcription in a complementary copy

Figure 25.15

The replication of the DNA double helix. Both strands are copied simultaneously. One new strand is a continuous strand (the leading strand), and one new strand is fragmented and subsequently joined together. Each new helix contains one new strand and retains one strand from the parent molecule.

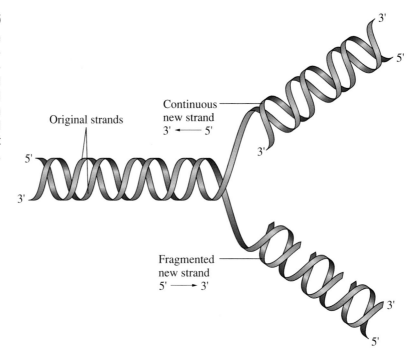

DNA double helix

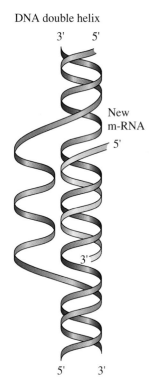

Figure 25.16

The transcription of a sequence of messenger RNA on the template of one strand of a DNA double helix. The mRNA chain is the complement of the DNA strand being copied.

of the DNA strand being copied (Fig. 25.16), with one major difference—RNA contains the base uracil, whereas DNA contains thymine. Thus the base pairs used during transcription are A-U and G-C.

After DNA information has been transcribed to an mRNA segment, that information is translated into the amino acid sequence of a protein. During **translation,** the mRNA segment acts as a template on which the protein is synthesized. The synthesis of a protein takes place on a ribosome (Section 25.1), which contains both proteins and additional forms of RNA known collectively as **ribosomal RNA,** or **rRNA.** The first step in translation, then, is binding of the end of an mRNA to a ribosome. The amino acids are transported to the ribosome by a third form of RNA known as **transfer RNA,** or **tRNA.** The tRNA molecules are among the smallest of all the nucleic acid structures, and there are at least 20 kinds of them, each carrying a different common amino acid. Individual tRNA molecules, each carrying an amino acid, attach to the ribosomes and match the mRNA chain through hydrogen bonds like those linking the base pairs in the DNA double helix. The amino acids are then joined sequentially (polymerized) to form a growing polypeptide chain. The elongation of the chain continues until a "termination" group of bases on the mRNA is read by the ribosome, whereupon the ribosome releases both the finished polypeptide chain and the mRNA segment for reuse.

The sequence of nucleotides in an mRNA strand is the basis for coding the sequence of amino acids in proteins. Each sequence of three adjacent nucleotides is specific for a single amino acid and is known as a **codon.** There are 64 possible codons (4 bases used in triplets, or 4^3) and only 20 amino acids to be encoded; therefore, some amino acids have more than one codon. Three codons are used to signal chain **termination** (stop translation): UAA, UGA, and UAG.

Table 25.1

Genetic Code Dictionary: The Amino Acids and Termination Signals Coded by Triplet Base Sequences (Codons) of Nucleotides in Messenger RNAs

First Position (5' end)	Second Position					Third Position (3' end)
	U	**C**	**A**	**G**		
U	Phe	Ser	Tyr	Cys		U
	Phe	Ser	Tyr	Cys		C
	Leu	Ser	Term[a]	Term[a]		A
	Leu	Ser	Term[a]	Trp		G
C	Leu	Pro	His	Arg		U
	Leu	Pro	His	Arg		C
	Leu	Pro	Gln	Arg		A
	Leu	Pro	Gln	Arg		G
A	Ile	Thr	Asn	Ser		U
	Ile	Thr	Asn	Ser		C
	Ile	Thr	Lys	Arg		A
	Met	Thr	Lys	Arg		G
G	Val	Ala	Asp	Gly		U
	Val	Ala	Asp	Gly		C
	Val	Ala	Glu	Gly		A
	Val	Ala	Glu	Gly		G

[a]The three codons UAA, UGA, and UAG code for the termination of the polypeptide chain.

The identity of the amino acid coded by any three-nucleotide sequence, or codon, in an mRNA can be determined from the **genetic code** dictionary in Table 25.1. The structure of an mRNA is always translated from the 5' end to the 3' end—that is, in the 5'-to-3' direction. This corresponds to the amino acid sequence of the protein encoded, starting with the amino terminus and ending with the carboxyl terminus.

Example 25.1   Predicting an Amino Acid Sequence from DNA

What is the sequence of amino acids coded by the following sequence of nucleotides on a strand of DNA, reading from the 5' end to the 3' end?

$$^{5'}\text{T-T-A-G-C-T-A-C-G-A-A-T}^{3'}$$

Solution:

DNA	$^{5'}$T-T-A-G-C-T-A-C-G-A-A-T$^{3'}$
mRNA	$^{3'}$A-A-U-C-G-A-U-G-C-U-U-A$^{5'}$
Protein	(Stop)HO$_2$C-Ser—Arg—Ile-NH$_2$

The mRNA sequence that forms on this DNA template is its complement and runs in the opposite direction. Thus the mRNA chain, reading from the 3' end to the 5' end, would be A-A-U-C-G-A-U-G-C-U-U-A. But the mRNA is translated beginning from its 5' end. Thus the codons are, in order, AUU, CGU, AGC, and UAA, which encode the amino acid sequence H$_2$N-Ile-Arg-Ser-CO$_2$H (termination), reading from the N-terminal residue to the C-terminal residue.

Check your learning: What is the sequence of amino acids coded by the following sequence of DNA nucleotides?

$$^{5'}\text{A-T-C-G-C-T-A-C-G-A-A-T}^{3'}$$

Answer: Ile-Arg-Ser-Asp

All the information necessary for the synthesis of the approximately 2000–5000 different proteins found in even the most simple bacterial cell is contained in the DNA molecules of the cell's chromosome(s). That portion of the DNA molecule that is transcribed to any kind of RNA is known as a **gene.** Each gene is usually specific for the synthesis of a single kind of polypeptide, although multiple copies of a gene may be found on a chromosome.

To put the vast amount of information carried by DNA into perspective, consider that the entire genetic material of humans, called the human **genome,** consists of 3 billion nucleotides. These 3 billion units are contained in the 23 pairs of human chromosomes, which are strands of DNA (and the associated histones or other cations) varying in length from 50 million to 500 million nucleotides. This length is in the centimeter range and contains the information for approximately 100,000 genes, which is packaged into microscopic cells. Even so, the function of about 95% of human DNA is not known; it does not code for any known proteins.

A number of controversies have arisen concerning DNA because of rapidly emerging biotechnologies: the human genome project, DNA "fingerprinting," and genetic engineering. The aim of the human genome project is to determine the sequence of all the base pairs in the human genome, a time-consuming and extremely costly task. In 1995 it was announced that the first complete genome (that of a bacterium) had been sequenced. Investigators are also studying the genomes of many other organisms, including plants and animals of economic interest and model organisms such as mice. Some scientists feel that the benefits of genome projects will not justify the cost, but advocates argue that the resulting information will have enormous impact on our ability to control major human diseases, understand basic biological processes, and improve food sources.

DNA fingerprinting is based on the fact that considerable variation is found in the DNA of every person (except identical siblings). Thus DNA fingerprinting, precisely cutting DNA into smaller pieces and comparing them, has great potential to be useful in forensic procedures (legal settings where identification of individuals is crucial). DNA fingerprinting can be admitted as evidence in court only when stringent standards for sample storage, analytical techniques, and analysis of results have been developed and codified. These kinds of studies are also showing the relationships between various human population groups, as they have migrated to populate the planet.

Genetic engineering facilitates the movement of genes from one individual to another—and even from one species to another. This happens in natural setting from time to time, but new technologies have made genetic engineering rather straightforward. Therefore, profound moral and ethical questions have arisen about which genes should be transferred to what other organisms and who has legal control of what genes.

25.13   Recombinant DNA Technology

In the mid-1970s, scientists discovered how to cut DNA strands precisely and to transfer pieces containing specific genes from one organism to another, thereby changing the characteristics of the recipient organism. This procedure has come to be known as **recombinant DNA technology.**

The recipient organisms are most often bacteria. As the bacteria reproduce, the fragment of transferred DNA and the recipient organism containing it are both duplicated (cloned) many times, producing millions of identical cells called a **clone** of cells. Thus it is possible to multiply a particular gene millions of times by placing it in a bacterial cell and allowing the cell to multiply. The gene cloning is part of a process known as **genetic engineering** whereby the gene may be subsequently transferred to cells of a third organism, where altering the characteristics may prove predictable and useful. Alternatively, the recipient bacteria may produce the protein for which its new DNA provides the code. There have been many successes in recombinant DNA technology, resulting in the production of valuable proteins for human (and animal) pharmacology, such as human insulin, growth hormone, and blood-clotting factors. These products are identical to those produced in the human body, even though they may have been synthesized in a bacterium.

Another approach, termed **gene therapy,** is the transfer of a gene to an animal or plant. This can impart defense against or resistance to a disease. Resistance to an insect pest or a herbicide can be developed in a plant. As we will see in the next section, both plant and animal diseases often result from defective proteins or missing proteins, a situation that may be remedied by gene transfer. An entire industry known as biotechnology has developed to promote genetic modification of plants and domestic animals and to provide new pharmacological products for treatment of human and animal diseases.

25.14   Gene Mutations

At the molecular level, **gene mutations** are substitutions, deletions, or additions of one or more nucleotides in a DNA molecule. In a medical or social context, we see harmful mutations more often than effects that are beneficial to a living species. For instance, many kinds of cancer arise from mutations in proteins that control cellular growth and division.

There are many varieties of hemoglobin mutations in humans. One such example of gene mutation results in sickle-cell anemia, which affects approximately three of every thousand Americans of African descent. These individuals carry the altered gene on both pairs of chromosomes. Sickle-cell trait results from carrying only one copy of the gene on one chromosome. Hemoglobin, a medium-size red protein, is the oxygen carrier found in red blood cells. Its quaternary structure consists of two alpha peptide chains and two beta peptide chains. In sickle-cell anemia, the red blood cells contain a modified hemoglobin in which just one glutamic acid residue, with a negatively charged R group, at the sixth position of the beta chain, is replaced by a valine residue, with its hydrophobic R group.

Adult hemoglobin β chain:	Val-His-Leu-Thr-Pro-Glu-Glu-Lys...
Sickle-cell adult hemoglobin β chain:	Val-His-Leu-Thr-Pro-Val-Glu-Lys...
Residue:	1 2 3 4 5 6 7 8

Figure 25.17

Photomicrograph of nonsickled (rounded and oval) red blood cells and sickled (crescent-shaped) cells, both from a person with sickle-cell anemia.

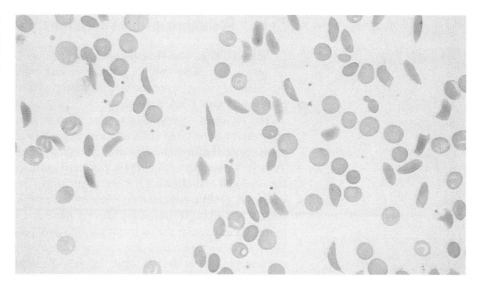

The difference in hydrophobicity of the two amino acids, which are found on the protein surface, causes the sickle-cell hemoglobin, in its deoxygenated form, to complex with other sickle-cell hemoglobin molecules and form rods inside the red blood cells. In turn, the cells change shape, or sickle (Fig. 25.17); are marked for removal by natural processes, which causes anemia; and obstruct blood flow in capillaries, resulting in pain and multiple organ failure.

On the other hand, persons with sickle-cell trait have resistance to malaria, which illustrates that the same gene mutation may have both beneficial and harmful effects. It also explains why the mutated gene has persisted in human populations in tropical areas where malaria is endemic. The resistance to malaria is partially due to the malarial parasite's inability, as it persists in red blood cells, to use sickle-cell hemoglobin as a source of the amino acids it needs to reproduce.

It is striking that the profound effects of the sickle-cell mutation arise from substitution of a *single* nucleotide base in the gene that controls the synthesis of the beta chain of a hemoglobin molecule containing a total of 574 amino acid residues. The mutation in the sickle-cell hemoglobin gene results in a valine residue replacing a glutamic acid residue in nonsickling hemoglobin. The genetic code dictionary for mRNA (Table 25.1) shows that Glu has two codons, GAA and GAG, whereas Val has four, GUU, GUC, GUA, and GUG. Two Val codons, GUA and GUG, differ from the Glu codons, GAA and GAG, by one nucleotide. The change of A to U in mRNA results from a single mutation (from T to A) in a DNA sequence at some point in the past.

For Review   Summary

Biochemistry attempts to understand the properties of living organisms in terms of organization, catalysis, energy, and information. Molecules present in organisms are often divided into four classes—**proteins, carbohydrates, nucleic acids,** and **lipids**—each of which has its own set of structures and functions. Cells consist primarily of water; biomolecules may be **hydrophilic** and dissolve in water, or **hydrophobic** and dissolve very poorly.

Metabolism comprises all the chemical reactions in organisms. It is divided into **anabolic pathways,** wherein molecules are synthesized and chemical energy is spent, and **catabolic**

pathways, wherein substances are degraded and chemical energy is released. **Metabolites** are the molecules that are shuttled within and between the pathways. **Aerobic** organisms are more efficient generators of free energy than **anaerobic** ones. Some energy released during the catabolism of **glucose** is transformed into heat, and the rest is transmitted as chemical energy to **ATP.** The hydrolysis of ATP can be coupled to other chemical reactions to accomplish chemical work.

Proteins are polymers composed of **amino acids.** Twenty common amino acids are genetically coded into proteins and are distinguished by the nature of the side group on the α-carbon. All amino acids except glycine are optically active. The combination of two or more amino acids by formation of **peptide bonds** produces a **peptide.** Proteins are very large peptides that perform a huge variety of functions because of the vast number of structures that can be made. The **primary structure** of a protein is the order in which the amino acid **residues** are connected. The **secondary structure** is produced by the local folding of the peptide to maximize hydrogen bonding between peptide bonds: The α-**helix** and β-**pleated sheet** are common secondary structures. **Tertiary structure** results when an entire polypeptide is folded into a unique conformation that is stabilized by interactions between amino acid side groups and by the exclusion of water from hydrophobic amino acid side groups. The **quaternary structure** describes how separate **subunits** within a large protein are related to each other. One of the most remarkable classes of proteins consist of the **enzymes,** which are catalysts of biochemical reactions. Some enzymes require helper **prosthetic groups** to accomplish catalysis.

Carbohydrates are linear aldehydes or ketones that have two or more —OH groups. They can be energy sources or stores in plants and animals, as well as playing structural roles. The simplest carbohydrates are **monosaccharides,** such as the sugars glucose and fructose. Glucose plays a central role in energy metabolism in humans. **Disaccharides** are dimers composed of two monosaccharides. **Starch** and **glycogen** are similar large polymers of glucose used for energy reserves in plants and animals, respectively. **Cellulose** is a structural polysaccharide of glucose in plants.

Lipids are classified by their behavior: solubility in polar organic solvents and relative insolubility in water. They consist of **triglycerides,** such as fats and oils used for long-term energy storage; **eicosanoids,** hormone-like substances derived from arachidonic acid; **structural lipids,** such as phosphatidyl choline, the chief components of cell membranes; **waxes,** which prevent desiccation; and **isoprenoids,** which include the steroids.

Nucleic acids are polymers of **nucleotides** ranging from trimers to polymers of very high mass that store and transfer genetic information. The **base sequence** encodes the information in the nucleic acids. The DNA polymer is a **double helix** held together by hydrogen bonds between pairs of **purines** and **pyrimidines,** one on each strand of the helix. A-T and C-G base pairing makes each strand the complement of the other. The sequence of three adjacent nucleotides (**codons**) in **mRNA,** copied from a DNA strand, is the basis for coding the sequence of amino acids in proteins. The flow of information from DNA to DNA, and from DNA to RNA to protein, entails **replication, transcription,** and **translation.**

Recombinant DNA technology makes **gene transfer** from one organism to another possible. It seeks to alter the characteristics of the recipient by promoting genetic modification of microorganisms, plants, and animals.

Key Terms and Concepts

active site   (25.7)
adenosine triphosphate, ATP   (25.1, 25.3)
aerobic   (25.2)
α-helix   (25.6)
amino acid sequence   (25.5)
anabolic pathway   (25.2)
anaerobic   (25.2)
base sequence   (25.11)
base pair   (25.11)
β-pleated sheet   (25.6)
carbohydrate   (25.8, 25.9)
catabolic pathway   (25.2)
clone   (25.13)
codon   (25.12)
common amino acids   (25.4)

competitive inhibition   (25.7)
complementary strands   (25.12)
deoxyribonucleic acid, DNA   (25.1, 25.11)
double helix   (25.11)
enzyme   (25.6, 25.7)
fatty acid   (25.10)
fermentation   (25.2)
gene   (25.12)
gene mutation   (25.14)
genetic code   (25.12)
genome   (25.12)
glycolysis   (25.2)
hydrophilic   (25.1)
hydrophobic   (25.1)
lipid   (25.10)

messenger RNA, mRNA   (25.12)
metabolism   (25.2)
monosaccharide   (25.8)
nucleic acid   (25.11)
nucleoside   (25.3, 25.11)
nucleotide   (25.11)
peptide   (25.5)
photosynthesis   (25.2)
polypeptide   (25.5)
polysaccharide   (25.9)
primary structure   (25.6)
proteins   (25.5, 25.6)
purine base   (25.11)
pyrimidine base   (25.11)
quaternary structure   (25.6)
respiration   (25.2)

ribonucleic acid, RNA   (25.1, 25.11)   substrates   (25.7)   translation   (25.12)
ribosomal RNA, rRNA   (25.12)   tertiary structure   (25.6)   triglycerides   (25.10)
secondary structure   (25.6)   transfer RNA, tRNA   (25.12)

Exercises

The Cell

1. What are some of the properties of living organisms that distinguish them from inanimate matter?

2. List the four principal organic constituents of living cells. Identify the specific functions of each of these constituents.

3. Define the terms *hydrophilic* and *hydrophobic*. What factors determine whether a compound is hydrophilic or hydrophobic? Give an example of each class of compound.

4. How do nonphotosynthetic organisms obtain energy from their environment?

Metabolism

5. Contrast anabolic pathways and catabolic pathways.

6. Distinguish between photosynthesis and chemotrophy.

7. Why are organisms that utilize fermentations much less energetically efficient than those that respire?

8. Can all the energy released during the degradation of glucose be recovered for work? Why?

9. Calculate the equilibrium constant for the reaction of water and ATP to produce ADP and phosphate ion. Where does the equilibrium lie?

Proteins

10. Define the terms *amino acid, peptide,* and *protein.*

11. Draw all possible ionic forms for the amino acids glutamic acid and lysine.

12. Are aliphatic R groups on amino acids polar or nonpolar? What about those with amides and alcohol R groups? Aromatic R groups?

13. Distinguish among the primary, secondary, tertiary, and quaternary structures of a protein.

14. How many possible tripeptides can be made using all the amino acids with acidic and basic R groups? Do you think they will be very water-soluble?

15. Contrast the denaturation of a protein with its hydrolysis.

16. What is an amino acid residue?

17. Draw symbolic structures of all four possible dipeptides that can be formed from the amino acids Gly and/or Ala.

18. Fungal laccase, a blue protein found in wood-rotting fungi, contains approximately 0.397% copper (a prosthetic group) by mass. If the molecular mass of fungal laccase is approximately 64,000 amu, how many copper atoms are there in each protein molecule?

19. What is an enzyme's active site?

Carbohydrates

20. What is the literal meaning of the term *carbohydrate,* and how did this name arise?

21. Which carbon atoms in glucose have four different functional groups on them and therefore are centers of optical activity?

22. How does a glucose ring differ from a fructose ring?

23. What is the difference between the α and β isomers of glucose?

24. How many grams of invert sugar can be produced by the hydrolysis of 1.00 g of sucrose, $C_{12}H_{22}O_{11}$?

25. Why is starch, and not cellulose, digestible by humans?

26. D-Glucose is an aldehyde that can reduce Cu(II) to Cu(I) in strongly basic aqueous solutions. However, nearly all glucose in aqueous solutions is in cyclic forms, which are not aldehydes. Suggest a reason why *all* glucose rapidly disappears from solution in the presence of excess Cu(II).

Lipids

27. What is the difference between a fatty acid and a fat? How do fats and oils differ in composition?

28. Define the term *saponification* and write an equation for the saponification of tripalmitylglycerol, the triester utilizing palmitic acid, $CH_3(CH_2)_{14}CO_2H$.

29. The molecular mass of an unknown triglyceride can be estimated by determining the mass of KOH required to saponify a known mass of lipid. What is the molecular mass of a triglyceride if 3.12 mg of the triglyceride requires 0.590 mg of KOH for saponification?

30. How many molecules of a drug are contained inside a spherical liposome with an outside diameter of 500 nm if the membrane bilayer is 7 nm thick and the drug concentration is 0.10 M?

31. Identify the functional groups in cholesterol.

Nucleic Acids and the Genetic Code

32. Explain how a purine and/or pyrimidine, a nucleoside, and a nucleotide differ. Explain how a nucleotide monophosphate, a diphosphate, and a triphosphate differ.

33. What structural features distinguish the 3′ end of a polynucleotide from the 5′ end?

34. Where is the difference in the structures of the nucleotides adenosine monophosphate and deoxyadenosine monophosphate?

35. If a single strand of DNA contains the nucleotide sequence A-G-G-C-T-C-A-G-C-T-A-G, reading from the 5′ end to the 3′ end, what would be the replication product reading in the same direction? What would be the transcription product reading in the same direction?

36. Which three codons do not code for any amino acid? Which amino acids have only one codon?

37. Human beings have assisted in the transfer of plant genes by hybridization from one related species to another for at least 5000 years. List some possible advantages and disadvantages of recombinant DNA technology in plants.

38. Complete the following table:

	DNA	RNA
Sugar unit		
Purine bases		
Pyrimidine bases		

Applications and Additional Exercises

39. Nonphotosynthetic microorganisms have been found that obtain energy from the oxidation of H_2S. From where might these organisms get carbon compounds to build the organic molecules for life processes?

40. Explain the molecular basis for the change in the shape of red blood cells in sickle-cell anemia.

41. List, in order of increasing strength, the kinds of weak forces that hold macromolecules in stable folded configurations.

42. Discuss the manner in which the energy requirements of a plant might be met in the dark.

43. How would you explain the observation that proteins can contain additional amino acids that are structurally related to the 20 genetically coded amino acids.

44. What evidence do we have that a twenty-first genetically coded amino acid will not be found?

45. Would alanine in a solution of pH 2.0 migrate toward a positive electrode, migrate toward a negative electrode, or show no preference for either upon electrolysis of the solution? What about at pH 10?

46. Discuss the ethical and moral issues related to the use of recombinant gene technology in humans to relieve disease symptoms.

47. If the degradation of 1.00 mol of glucose and that of 1.00 mol of palmitic acid yield 38 and 129 mol of ATP, respectively, calculate the moles of ATP produced per gram of lipid or carbohydrate consumed. The formula of palmitic acid is $CH_3(CH_2)_{14}COOH$.

48. (a) The degradation of fatty acids, such as palmitic acid, and carbohydrates, such as glucose, leads to the formation of carbon dioxide and water. However, the degradation of palmitic acid produces 2 1/2 times as much energy as that of glucose for each gram of substance consumed. Explain this observation in terms of the average oxidation states of the carbon atoms in fatty acids, carbohydrates, and carbon dioxide.

    (b) Ethanol, CH_3CH_2OH, is also oxidized completely by humans to CO_2 and water. Using the methods above, predict how much energy per gram of ethanol consumed will be produced compared to glucose.

Appendixes

Appendix A: Chemical Arithmetic

A.1 Exponential Arithmetic

Exponential notation is used to express very large and very small numbers as a product of two numbers. The first number of the product, the *digit term,* is usually a number not less than 1 and not greater than 10. The second number of the product, the *exponential term,* is written as 10 with an exponent. Some examples of exponential notation are

$$1000 = 1 \times 10^3 \qquad 0.01 = 1 \times 10^{-2}$$
$$100 = 1 \times 10^2 \qquad 0.001 = 1 \times 10^{-3}$$
$$10 = 1 \times 10^1 \qquad 2386 = 2.386 \times 1000 = 2.386 \times 10^3$$
$$1 = 1 \times 10^0 \qquad 0.123 = 1.23 \times 0.1 = 1.23 \times 10^{-1}$$
$$0.1 = 1 \times 10^{-1}$$

The power (exponent) of 10 is equal to the number of places the decimal is shifted to give the digit number. The exponential method is a particularly useful notation for very large and very small numbers. For example, $1,230,000,000 = 1.23 \times 10^9$ and $0.00000000036 = 3.6 \times 10^{-10}$.

Addition of Exponentials. Convert all numbers to the same power of 10, add the digit terms of the numbers, and, if appropriate, convert the digit term back to a number between 1 and 10 by adjusting the exponential term.

Example A.1

Add 5.00×10^{-5} and 3.00×10^{-3}.

$$3.00 \times 10^{-3} = 300 \times 10^{-5}$$
$$(5.00 \times 10^{-5}) + (300 \times 10^{-5}) = 305 \times 10^{-5} = 3.05 \times 10^{-3}$$

Subtraction of Exponentials. Convert all numbers to the same power of 10, take the difference of the digit terms, and, if appropriate, convert the digit term back to a number between 1 and 10 by adjusting the exponential term.

Example A.2

Subtract 4.0×10^{-7} from 5.0×10^{-6}.

$$4.0 \times 10^{-7} = 0.40 \times 10^{-6}$$
$$(5.0 \times 10^{-6}) - (0.40 \times 10^{-6}) = 4.6 \times 10^{-6}$$

Multiplication of Exponentials. Multiply the digit terms in the usual way and add the exponents of the exponential terms.

Example A.3

Multiply 4.2×10^{-8} by 2.0×10^3.

$$(4.2 \times 10^{-8}) \times (2.0 \times 10^3) = (4.2 \times 2.0) \times 10^{(-8)+(+3)} = 8.4 \times 10^{-5}$$

Division of Exponentials. Divide the digit term of the numerator by the digit term of the denominator and subtract the exponents of the exponential terms.

Example A.4

Divide 3.6×10^{-5} by 6.0×10^{-4}.

$$\frac{3.6 \times 10^{-5}}{6.0 \times 10^{-4}} = \left(\frac{3.6}{6.0}\right) \times 10^{(-5)-(-4)} = 0.60 \times 10^{-1} = 6.0 \times 10^{-2}$$

Squaring of Exponentials. Square the digit term in the usual way and multiply the exponent of the exponential term by 2.

Example A.5

Square the number 4.0×10^{-6}.

$$(4.0 \times 10^{-6})^2 = 4 \times 4 \times 10^{2 \times (-6)} = 16 \times 10^{-12} = 1.6 \times 10^{-11}$$

Cubing of Exponentials. Cube the digit term in the usual way and multiply the exponent of the exponential term by 3.

Example A.6

Cube the number 2×10^4.

$$(2 \times 10^4)^3 = 2 \times 2 \times 2 \times 10^{3 \times 4} = 8 \times 10^{12}$$

Taking Square Roots of Exponentials. If necessary, decrease or increase the exponential term so that the power of 10 is evenly divisible by 2. Extract the square root of the digit term and divide the exponential term by 2.

Example A.7

Find the square root of 1.6×10^{-7}.

$$1.6 \times 10^{-7} = 16 \times 10^{-8}$$

$$\sqrt{16 \times 10^{-8}} = \sqrt{16} \times \sqrt{10^{-8}} = \sqrt{16} \times 10^{-8/2} = 4.0 \times 10^{-4}$$

A.2   Significant Figures

A beekeeper reports that he has 525,341 bees. The last three figures of the number are obviously inaccurate, for during the time the keeper was counting the bees, some of them died and others hatched; this makes it quite difficult to determine the exact number of bees. It would have been more accurate if the beekeeper had reported the number 525,000. In other words, the last three figures are not significant, except to set the position of the decimal point. Their exact values have no meaning. In reporting any information as numbers, use only as many significant figures as the accuracy of the measurement warrants.

The importance of significant figures lies in their application to fundamental computation. In addition and subtraction, the last digit that is retained in the sum or difference should correspond to the first doubtful decimal place (indicated by underscoring in the following example).

Example A.8

Add 4.383 g and 0.0023 g.

$$4.38\underline{3} \text{ g}$$
$$0.002\underline{3}$$
$$4.38\underline{5} \text{ g}$$

In multiplication and division, the product or quotient should contain no more digits than the least number of significant figures in the numbers involved in the computation.

Example A.9

Multiply 0.6238 by 6.6.

$$0.623\underline{8} \times 6.\underline{6} = 4.\underline{1}$$

In rounding off numbers, increase the last digit retained by 1 if it is followed by a number larger than 5 or by a 5 followed by other nonzero digits. Do not change the last digit retained if the digits that follow are less than 5. If the last digit retained is followed by 5, increase the last digit retained by 1 if it is odd and leave the last digit retained unchanged if it is even.

A.3 The Use of Logarithms and Exponential Numbers

The common logarithm of a number (log) is the power to which 10 must be raised to equal that number. For example, the common logarithm of 100 is 2, because 10 must be raised to the second power to equal 100. Additional examples follow.

Number	Number Expressed Exponentially	Common Logarithm
1000	10^3	3
10	10^1	1
1	10^0	0
0.1	10^{-1}	−1
0.001	10^{-3}	−3

What is the common logarithm of 60? Because 60 lies between 10 and 100, which have logarithms of 1 and 2, respectively, the logarithm of 60 must lie between 1 and 2. The logarithm of 60 is 1.7782; that is,

$$60 = 10^{1.7782}$$

The common logarithm of a number less than 1 has a negative value. The logarithm of 0.03918 is -1.4069, or

$$0.03918 = 10^{-1.4069} = \frac{1}{10^{1.4069}}$$

To obtain the common logarithm of a number, use the *log* button on your calculator. To calculate a number from its logarithm, enter the logarithm into your calculator and push the *antilog* button, take the inverse log of the logarithm, or calculate 10^x (where x is the logarithm of the number).

The natural logarithm of a number (ln) is the power to which e must be raised to equal the number; e is the constant 2.7182818. For example, the natural logarithm of 10 is 2.303; that is,

$$10 = e^{2.303} = 2.7182818^{2.303}$$

To obtain the natural logarithm of a number, use the *ln* button on your calculator. To calculate a number from its natural logarithm, enter the natural logarithm and take the inverse ln of the natural logarithm, or calculate e^x (where x is the natural logarithm of the number).

Logarithms are exponents; thus operations involving logarithms follow the same rules as operations involving exponents.

1. The logarithm of a product of two numbers is the sum of the logarithms of the two numbers: $\log xy = \log x + \log y$, and $\ln xy = \ln x + \ln y$.
2. The logarithm of the number resulting from the division of two numbers is the difference between the logarithms of the two numbers: $\log x/y = \log x - \log y$, and $\ln x/y = \ln x - \ln y$.
3. The logarithm of the square root of a number is one-half of the logarithm of the number: $\log x^{1/2} = 1/2 \log x$, and $\ln x^{1/2} = 1/2 \ln x$.
4. The logarithm of the cube root of a number is one-third of the logarithm of the number: $\log x^{1/3} = 1/3 \log x$, and $\ln x^{1/3} = 1/3 \ln x$.

A.4 The Solution of Quadratic Equations

Any quadratic equation can be expressed in the following form:

$$ax^2 + bx + c = 0$$

In order to solve a quadratic equation, use the following formula:

$$x = \frac{-b \pm \sqrt{b^2 - 4ac}}{2a}$$

Example A.10

Solve the quadratic equation $3x^2 + 13x - 10 = 0$.

Substituting the values $a = 3$, $b = 13$, and $c = -10$ in the formula, we obtain

$$x = \frac{-13 \pm \sqrt{(13)^2 - 4 \times 3 \times (-10)}}{2 \times 3}$$

$$= \frac{-13 \pm \sqrt{169 + 120}}{6} = \frac{-13 \pm \sqrt{289}}{6} = \frac{-13 \pm 17}{6}$$

The two roots are therefore

$$x = \frac{-13 + 17}{6} = 0.67 \quad \text{and} \quad x = \frac{-13 - 17}{6} = -5$$

Equations constructed on physical data always have real roots, and of these real roots, usually only those having positive values are of any significance.

Appendix B: Units and Conversion Factors

Units of Length

Meter (m) = 39.37 inches (in.)
      = 1.094 yards (yd)
Centimeter (cm) = 0.01 m
Millimeter (mm) = 0.001 m
Kilometer (km) = 1000 m
Angstrom unit (Å) = 10^{-8} cm
         = 10^{-10} m

Yard = 0.9144 m
Inch = 2.54 cm (definition)
Mile (U.S.) = 1.60934 km

Units of Volume

Liter (L) = 0.001 m^3 = 1000 cm^3
     = 1.057 (U.S.) quarts
Milliliter (mL) = 0.001 L = 1 cm^3

Liquid quart (U.S.) = 0.9463 L
          = 32 (U.S.) liquid
           ounces
          = $\frac{1}{4}$ (U.S.) gallon
Dry quart = 1.1012 L
Cubic foot (U.S.) = 28.316 L

Units of Mass

Gram (g) = 0.001 kg
Milligram (mg) = 0.001 g
Kilogram (kg) = 1000 g = 2.205 lb
Ton (metric) = 1000 kg = 2204.62 lb

Ounce (oz) (avoirdupois) = 28.35 g
Pound (lb) (avoirdupois) = 0.4535924 kg
Ton (short) = 2000 lb = 907.185 kg
Ton (long) = 2240 lb = 1.016 metric ton

Units of Energy

4.184 joule (J) = 1 thermochemical calorie (cal) = 4.184×10^7 erg
Erg = 10^{-7} J
Electron-volt (eV) = $1.60217733 \times 10^{-19}$ J = 23.061 kcal mol^{-1}
Liter atmosphere = 24.217 cal = 101.325 J

Units of Pressure

Torr = 1 mm Hg
Atmosphere (atm) = 760 mm Hg = 760 torr = 101,325 N m^{-2} = 101,325 Pa
Pascal (Pa) = kg m^{-1} s^{-2} = N m^{-2}

Appendix C: General Physical Constants

Avogadro's number	6.0221367×10^{23} mol^{-1}
Electron charge, e	$1.60217733 \times 10^{-19}$ coulomb (C)
Electron rest mass, m_e	9.109390×10^{-31} kg
Proton rest mass, m_p	$1.6726231 \times 10^{-27}$ kg
Neutron rest mass, m_n	$1.6749286 \times 10^{-27}$ kg
Charge-to-mass ratio for electron, e/m_e	$1.75881962 \times 10^{11}$ coulomb kg^{-1}
Faraday constant, F	9.6485309×10^4 coulomb/equivalent
Planck constant, h	$6.6260755 \times 10^{-34}$ J s
Boltzmann constant, k	1.380658×10^{-23} J K^{-1}
Gas constant, R	8.205784×10^{-2} L atm mol^{-1} K^{-1}
	= 8.314510 J mol^{-1} K^{-1}
Speed of light (in vacuum), c	2.99792458×10^8 m s^{-1}
Atomic mass unit (= $\frac{1}{12}$ the mass of an atom of the ^{12}C nuclide), amu	1.660542×10^{-27} kg
Rydberg constant, R_∞	1.0973731534×10^7 m^{-1}

Appendix D: Solubility Products

Substance	K_{sp} at 25°C	Substance	K_{sp} at 25°C
Aluminum		BaF$_2$	1.7×10^{-6}
Al(OH)$_3$	1.9×10^{-33}	Ba(OH)$_2$·8H$_2$O	5.0×10^{-3}
Barium		Ba$_3$(PO$_4$)$_2$	1.3×10^{-29}
BaCO$_3$	8.1×10^{-9}	Ba$_3$(AsO$_4$)$_2$	1.1×10^{-13}
BaC$_2$O$_4$·2H$_2$O	1.1×10^{-7}	Bismuth	
BaSO$_4$	1.08×10^{-10}	BiO(OH)	1×10^{-12}
BaCrO$_4$	2×10^{-10}	BiOCl	7×10^{-9}

Solubility Products (Continued)

Substance	K_{sp} at 25°C	Substance	K_{sp} at 25°C
Bi_2S_3	7.3×10^{-91}	$MgCO_3 \cdot 3H_2O$	$ca\ 1 \times 10^{-5}$
Cadmium		$MgNH_4PO_4$	2.5×10^{-13}
$Cd(OH)_2$	1.2×10^{-14}	MgF_2	6.4×10^{-9}
CdS	2.8×10^{-35}	MgC_2O_4	8.6×10^{-5}
$CdCO_3$	2.5×10^{-14}	Manganese	
Calcium		$Mn(OH)_2$	4.5×10^{-14}
$Ca(OH)_2$	7.9×10^{-6}	$MnCO_3$	8.8×10^{-11}
$CaCO_3$	4.8×10^{-9}	MnS	4.3×10^{-22}
$CaSO_4 \cdot 2H_2O$	2.4×10^{-5}	Mercury	
$CaC_2O_4 \cdot H_2O$	2.27×10^{-9}	$Hg_2O \cdot H_2O$	1.6×10^{-23}
$Ca_3(PO_4)_2$	1×10^{-25}	Hg_2Cl_2	1.1×10^{-18}
$CaHPO_4$	5×10^{-6}	Hg_2Br_2	1.26×10^{-22}
CaF_2	3.9×10^{-11}	Hg_2I_2	4.5×10^{-29}
Chromium		Hg_2CO_3	9×10^{-17}
$Cr(OH)_3$	6.7×10^{-31}	Hg_2SO_4	6.2×10^{-7}
Cobalt		Hg_2S	8×10^{-52}
$Co(OH)_2$	2×10^{-16}	Hg_2CrO_4	2×10^{-9}
$CoS(\alpha)$	4.5×10^{-27}	HgS	2×10^{-59}
$CoS(\beta)$	6.7×10^{-29}	Nickel	
$CoCO_3$	1.0×10^{-12}	$Ni(OH)_2$	1.6×10^{-14}
$Co(OH)_3$	2.5×10^{-43}	$NiCO_3$	1.36×10^{-7}
Copper		$NiS(\alpha)$	2×10^{-27}
CuCl	1.85×10^{-7}	$NiS(\beta)$	8×10^{-33}
CuBr	5.3×10^{-9}	Potassium	
CuI	5.1×10^{-12}	$KClO_4$	1.07×10^{-2}
CuSCN	4×10^{-14}	K_2PtCl_6	1.1×10^{-5}
Cu_2S	1.2×10^{-54}	$KHC_4H_4O_6$	3×10^{-4}
$Cu(OH)_2$	5.6×10^{-20}	Silver	
CuS	6.7×10^{-42}	$\frac{1}{2}Ag_2O\ (Ag^+ + OH^-)$	2×10^{-8}
$CuCO_3$	1.37×10^{-10}	AgCl	1.8×10^{-10}
Iron		AgBr	3.3×10^{-13}
$Fe(OH)_2$	7.9×10^{-15}	AgI	1.5×10^{-16}
$FeCO_3$	2.11×10^{-11}	AgCN	1.2×10^{-16}
FeS	8×10^{-26}	AgSCN	1.0×10^{-12}
$Fe(OH)_3$	1.1×10^{-36}	Ag_2S	8×10^{-58}
Lead		Ag_2CO_3	8.2×10^{-12}
$Pb(OH)_2$	2.8×10^{-16}	Ag_2CrO_4	9×10^{-12}
PbF_2	3.7×10^{-8}	$Ag_4Fe(CN)_6$	1.55×10^{-41}
$PbCl_2$	1.7×10^{-5}	Ag_2SO_4	1.18×10^{-5}
$PbBr_2$	6.3×10^{-6}	Ag_3PO_4	1.8×10^{-18}
PbI_2	8.7×10^{-9}	Strontium	
$PbCO_3$	1.5×10^{-13}	$Sr(OH)_2 \cdot 8H_2O$	3.2×10^{-4}
PbS	6.5×10^{-34}	$SrCO_3$	9.42×10^{-10}
$PbCrO_4$	1.8×10^{-14}	$SrCrO_4$	3.6×10^{-5}
$PbSO_4$	1.8×10^{-8}	$SrSO_4$	2.8×10^{-7}
$Pb_3(PO_4)_2$	3×10^{-44}	$SrC_2O_4 \cdot H_2O$	5.61×10^{-8}
Magnesium		Thallium	
$Mg(OH)_2$	1.5×10^{-11}	TlCl	1.9×10^{-4}

Solubility Products (Continued)

Substance	K_{sp} at 25°C	Substance	K_{sp} at 25°C
TlSCN	5.8×10^{-4}	Sn(OH)$_4$	1×10^{-56}
Tl$_2$S	9.2×10^{-31}	Zinc	
Tl(OH)$_3$	1.5×10^{-44}	ZnCO$_3$	6×10^{-11}
Tin		Zn(OH)$_2$	4.5×10^{-17}
Sn(OH)$_2$	5×10^{-26}	ZnS	1×10^{-27}
SnS	6×10^{-35}		

Appendix E: Formation Constants for Complex Ions

Equilibrium	K_f
$Al^{3+} + 6F^- \rightleftharpoons [AlF_6]^{3-}$	5×10^{23}
$Cd^{2+} + 4NH_3 \rightleftharpoons [Cd(NH_3)_4]^{2+}$	4.0×10^6
$Cd^{2+} + 4CN^- \rightleftharpoons [Cd(CN)_4]^{2-}$	1.3×10^{17}
$Co^{2+} + 6NH_3 \rightleftharpoons [Co(NH_3)_6]^{2+}$	8.3×10^4
$Co^{3+} + 6NH_3 \rightleftharpoons [Co(NH_3)_6]^{3+}$	4.5×10^{33}
$Cu^+ + 2CN^- \rightleftharpoons [Cu(CN)_2]^-$	1×10^{16}
$Cu^{2+} + 4NH_3 \rightleftharpoons [Cu(NH_3)_4]^{2+}$	1.2×10^{12}
$Fe^{2+} + 6CN^- \rightleftharpoons [Fe(CN)_6]^{4-}$	1×10^{37}
$Fe^{3+} + 6CN^- \rightleftharpoons [Fe(CN)_6]^{3-}$	1×10^{44}
$Fe^{3+} + 6SCN^- \rightleftharpoons [Fe(NCS)_6]^{3-}$	3.2×10^3
$Hg^{2+} + 4Cl^- \rightleftharpoons [HgCl_4]^{2-}$	1.2×10^{15}
$Ni^{2+} + 6NH_3 \rightleftharpoons [Ni(NH_3)_6]^{2+}$	1.8×10^8
$Ag^+ + 2Cl^- \rightleftharpoons [AgCl_2]^-$	2.5×10^5
$Ag^+ + 2CN^- \rightleftharpoons [Ag(CN)_2]^-$	1×10^{20}
$Ag^+ + 2NH_3 \rightleftharpoons [Ag(NH_3)_2]^+$	1.6×10^7
$Zn^{2+} + 4CN^- \rightleftharpoons [Zn(CN)_4]^{2-}$	1×10^{19}
$Zn^{2+} + 4OH^- \rightleftharpoons [Zn(OH)_4]^{2-}$	2.9×10^{15}

Appendix F: Ionization Constants of Weak Acids

Acid	Formula	K_a at 25°C
Acetic	CH$_3$CO$_2$H	1.8×10^{-5}
Arsenic	H$_3$AsO$_4$	4.8×10^{-3}
	H$_2$AsO$_4^-$	1×10^{-7}

Ionization Constants of Weak Acids (Continued)

Acid	Formula	K_a at 25°C
	$HAsO_4^{2-}$	1×10^{-13}
Arsenous	H_3AsO_3	5.8×10^{-10}
Boric	H_3BO_3	5.8×10^{-10}
Carbonic	H_2CO_3	4.3×10^{-7}
	HCO_3^-	7×10^{-11}
Cyanic	HCNO	3.46×10^{-4}
Formic	HCO_2H	1.8×10^{-4}
Hydrazoic	HN_3	1×10^{-4}
Hydrocyanic	HCN	4×10^{-10}
Hydrofluoric	HF	7.2×10^{-4}
Hydrogen peroxide	H_2O_2	2.4×10^{-12}
Hydrogen selenide	H_2Se	1.7×10^{-4}
	HSe^-	1×10^{-10}
Hydrogen sulfate ion	HSO_4^-	1.2×10^{-2}
Hydrogen sulfide	H_2S	1.0×10^{-7}
	HS^-	1.0×10^{-19}
Hydrogen telluride	H_2Te	2.3×10^{-3}
	HTe^-	1×10^{-5}
Hypobromous	HBrO	2×10^{-9}
Hypochlorous	HClO	3.5×10^{-8}
Nitrous	HNO_2	4.5×10^{-4}
Oxalic	$H_2C_2O_4$	5.9×10^{-2}
	$HC_2O_4^-$	6.4×10^{-5}
Phosphoric	H_3PO_4	7.5×10^{-3}
	$H_2PO_4^-$	6.3×10^{-8}
	HPO_4^{2-}	3.6×10^{-13}
Phosphorous	H_3PO_3	1.6×10^{-2}
	$H_2PO_3^-$	7×10^{-7}
Sulfurous	H_2SO_3	1.2×10^{-2}
	HSO_3^-	6.2×10^{-8}

Appendix G: Ionization Constants of Weak Bases

Base	Ionization Equation	K_b at 25°C
Ammonia	$NH_3 + H_2O \rightleftharpoons NH_4^+ + OH^-$	1.8×10^{-5}
Dimethylamine	$(CH_3)_2NH + H_2O \rightleftharpoons (CH_3)_2NH_2^+ + OH^-$	7.4×10^{-4}
Methylamine	$CH_3NH_2 + H_2O \rightleftharpoons CH_3NH_3^+ + OH^-$	4.4×10^{-4}
Phenylamine (aniline)	$C_6H_5NH_2 + H_2O \rightleftharpoons C_6H_5NH_3^+ + OH^-$	4.6×10^{-10}
Trimethylamine	$(CH_3)_3N + H_2O \rightleftharpoons (CH_3)_3NH^+ + OH^-$	7.4×10^{-5}

Appendix H: Standard Electrode (Reduction) Potentials

Half-reaction	$E°$, V	Half-reaction	$E°$, V
$Li^+ + e^- \longrightarrow Li$	-3.09	$Fe(OH)_2 + 2e^- \longrightarrow Fe + 2OH^-$	-0.877
$K^+ + e^- \longrightarrow K$	-2.925	$SiO_2 + 4H_3O^+ + 4e^- \longrightarrow Si + 6H_2O$	-0.86
$Rb^+ + e^- \longrightarrow Rb$	-2.925	$NiS + 2e^- \longrightarrow Ni + S^{2-}$	-0.83
$Ra^{2+} + 2e^- \longrightarrow Ra$	-2.92	$2H_2O + 2e^- \longrightarrow H_2 + 2OH^-$	-0.828
$Ba^{2+} + 2e^- \longrightarrow Ba$	-2.90	$Zn^{2+} + 2e^- \longrightarrow Zn$	-0.763
$Sr^{2+} + 2e^- \longrightarrow Sr$	-2.89	$Cr^{3+} + 3e^- \longrightarrow Cr$	-0.74
$Ca^{2+} + 2e^- \longrightarrow Ca$	-2.87	$HgS + 2e^- \longrightarrow Hg + S^{2-}$	-0.72
$Na^+ + e^- \longrightarrow Na$	-2.714	$[Cd(NH_3)_4]^{2+} + 2e^- \longrightarrow Cd + 4NH_3$	-0.597
$La^{3+} + 3e^- \longrightarrow La$	-2.52	$Ga^{3+} + 3e^- \longrightarrow Ga$	-0.53
$Ce^{3+} + 3e^- \longrightarrow Ce$	-2.48	$S + 2e^- \longrightarrow S^{2-}$	-0.48
$Nd^{3+} + 3e^- \longrightarrow Nd$	-2.44	$[Ni(NH_3)_6]^{2+} + 2e^- \longrightarrow Ni + 6NH_3$	-0.47
$Sm^{3+} + 3e^- \longrightarrow Sm$	-2.41	$Fe^{2+} + 2e^- \longrightarrow Fe$	-0.440
$Gd^{3+} + 3e^- \longrightarrow Gd$	-2.40	$[Cu(CN)_2]^- + e^- \longrightarrow Cu + 2CN^-$	-0.43
$Mg^{2+} + 2e^- \longrightarrow Mg$	-2.37	$Cr^{3+} + e^- \longrightarrow Cr^{2+}$	-0.41
$Y^{3+} + 3e^- \longrightarrow Y$	-2.37	$Cd^{2+} + 2e^- \longrightarrow Cd$	-0.403
$Am^{3+} + 3e^- \longrightarrow Am$	-2.32	$Se + 2H_3O^+ + 2e^- \longrightarrow H_2Se + 2H_2O$	-0.40
$Lu^{3+} + 3e^- \longrightarrow Lu$	-2.25	$[Hg(CN)_4]^{2-} + 2e^- \longrightarrow Hg + 4CN^-$	-0.37
$\frac{1}{2}H_2 + e^- \longrightarrow H^-$	-2.25	$ClO_4^- + H_2O + 2e^- \longrightarrow ClO_3^- + 2OH^-$	-0.36
$Sc^{3+} + 3e^- \longrightarrow Sc$	-2.08	$PbSO_4 + 2e^- \longrightarrow Pb + SO_4^{2-}$	-0.356
$[AlF_6]^{3-} + 3e^- \longrightarrow Al + 6F^-$	-2.07	$In^{3+} + 3e^- \longrightarrow In$	-0.342
$Pu^{3+} + 3e^- \longrightarrow Pu$	-2.07	$[Ag(CN)_2]^- + e^- \longrightarrow Ag + 2CN^-$	-0.31
$Th^{4+} + 4e^- \longrightarrow Th$	-1.90	$Co^{2+} + 2e^- \longrightarrow Co$	-0.277
$Np^{3+} + 3e^- \longrightarrow Np$	-1.86	$Ni^{2+} + 2e^- \longrightarrow Ni$	-0.257
$Be^{2+} + 2e^- \longrightarrow Be$	-1.85	$[SnF_6]^{2-} + 4e^- \longrightarrow Sn + 6F^-$	-0.25
$U^{3+} + 3e^- \longrightarrow U$	-1.80	$Sn^{2+} + 2e^- \longrightarrow Sn$	-0.136
$Hf^{4+} + 4e^- \longrightarrow Hf$	-1.70	$CrO_4^{2-} + 4H_2O + 3e^- \longrightarrow Cr(OH)_3 + 5OH^-$	-0.13
$SiO_3^{2-} + 3H_2O + 4e^- \longrightarrow Si + 6OH^-$	-1.70	$Pb^{2+} + 2e^- \longrightarrow Pb$	-0.126
$Al^{3+} + 3e^- \longrightarrow Al$	-1.66	$MnO_2 + 2H_2O + 2e^- \longrightarrow Mn(OH)_2 + 2OH^-$	-0.05
$Ti^{2+} + 2e^- \longrightarrow Ti$	-1.63	$[HgI_4]^{2-} + 2e^- \longrightarrow Hg + 4I^-$	-0.04
$Zr^{4+} + 4e^- \longrightarrow Zr$	-1.53	$2H_3O^+ + 2e^- \longrightarrow H_2 + 2H_2O$	0.00
$ZnS + 2e^- \longrightarrow Zn + S^{2-}$	-1.44	$NO_3^- + H_2O + 2e^- \longrightarrow NO_2^- + 2OH^-$	$+0.01$
$Cr(OH)_3 + 3e^- \longrightarrow Cr + 3OH^-$	-1.3	$[Ag(S_2O_3)_2]^{3-} + e^- \longrightarrow Ag + 2S_2O_3^{2-}$	$+0.01$
$[Zn(CN)_4]^{2-} + 2e^- \longrightarrow Zn + 4CN^-$	-1.26	$[Co(NH_3)_6]^{3+} + e^- \longrightarrow [Co(NH_3)_6]^{2+}$	$+0.1$
$Zn(OH)_2 + 2e^- \longrightarrow Zn + 2OH^-$	-1.245	$S + 2H_3O^+ + 2e^- \longrightarrow H_2S + 2H_2O$	$+0.141$
$[Zn(OH)_4]^{2-} + 2e^- \longrightarrow Zn + 4OH^-$	-1.216	$Sn^{4+} + 2e^- \longrightarrow Sn^{2+}$	$+0.15$
$CdS + 2e^- \longrightarrow Cd + S^{2-}$	-1.21	$Cu^{2+} + e^- \longrightarrow Cu^+$	$+0.153$
$[Cr(OH)_4]^- + 3e^- \longrightarrow Cr + 4OH^-$	-1.2	$Co(OH)_3 + e^- \longrightarrow Co(OH)_2 + OH^-$	$+0.17$
$[SiF_6]^{2-} + 4e^- \longrightarrow Si + 6F^-$	-1.2	$[HgBr_4]^{2-} + 2e^- \longrightarrow Hg + 4Br^-$	$+0.21$
$V^{2+} + 2e^- \longrightarrow V$	$ca\ -1.18$	$AgCl + e^- \longrightarrow Ag + Cl^-$	$+0.222$
$Mn^{2+} + 2e^- \longrightarrow Mn$	-1.18	$Hg_2Cl_2 + 2e^- \longrightarrow 2Hg + 2Cl^-$	$+0.27$
$[Cd(CN)_4]^{2-} + 2e^- \longrightarrow Cd + 4CN^-$	-1.03	$ClO_3^- + H_2O + 2e^- \longrightarrow ClO_2^- + 2OH^-$	$+0.33$
$[Zn(NH_3)_4]^{2+} + 2e^- \longrightarrow Zn + 4NH_3$	-1.03	$Cu^{2+} + 2e^- \longrightarrow Cu$	$+0.337$
$FeS + 2e^- \longrightarrow Fe + S^{2-}$	-1.01	$[Fe(CN)_6]^{3-} + e^- \longrightarrow [Fe(CN)_6]^{4-}$	$+0.36$
$PbS + 2e^- \longrightarrow Pb + S^{2-}$	-0.95	$[Ag(NH_3)_2]^+ + e^- \longrightarrow Ag + 2NH_3$	$+0.373$
$SnS + 2e^- \longrightarrow Sn + S^{2-}$	-0.94	$O_2 + 2H_2O + 4e^- \longrightarrow 4OH^-$	$+0.401$
$Cr^{2+} + 2e^- \longrightarrow Cr$	-0.91	$[RhCl_6]^{3-} + 3e^- \longrightarrow Rh + 6Cl^-$	$+0.44$

Standard Electrode (Reduction) Potentials (Continued)

Half-reaction	$E°$, V	Half-reaction	$E°$, V
$Ag_2CrO_4 + 2e^- \longrightarrow 2Ag + CrO_4^{2-}$	+0.446	$NO_3^- + 4H_3O^+ + 3e^- \longrightarrow NO + 6H_2O$	+0.96
$NiO_2 + 2H_2O + 2e^- \longrightarrow$		$Pd^{2+} + 2e^- \longrightarrow Pd$	+0.987
$\qquad\qquad\qquad Ni(OH)_2 + 2OH^-$	+0.49	$Br_2(l) + 2e^- \longrightarrow 2Br^-$	+1.0652
$Cu^+ + e^- \longrightarrow Cu$	+0.521	$ClO_4^- + 2H_3O^+ + 2e^- \longrightarrow ClO_3^- + 3H_2O$	+1.19
$TeO_2 + 4H_3O^+ + 4e^- \longrightarrow Te + 6H_2O$	+0.529	$Pt^{2+} + 2e^- \longrightarrow Pt$	ca +1.2
$I_2 + 2e^- \longrightarrow 2I^-$	+0.5355	$ClO_3^- + 3H_3O^+ + 2e^- \longrightarrow HClO_2 + 4H_2O$	+1.21
$[PtBr_4]^{2-} + 2e^- \longrightarrow Pt + 4Br^-$	+0.58	$O_2 + 4H_3O^+ + 4e^- \longrightarrow 6H_2O$	+1.23
$MnO_4^- + 2H_2O + 3e^- \longrightarrow$		$MnO_2 + 4H_3O^+ + 2e^- \longrightarrow Mn^{2+} + 6H_2O$	+1.23
$\qquad\qquad\qquad MnO_2 + 4OH^-$	+0.588	$Cr_2O_7^{2-} + 14H_3O^+ + 6e^- \longrightarrow 2Cr^{3+} + 21H_2O$	+1.33
$[PdCl_4]^{2-} + 2e^- \longrightarrow Pd + 4Cl^-$	+0.62	$Cl_2 + 2e^- \longrightarrow 2Cl^-$	+1.3595
$ClO_2^- + H_2O + 2e^- \longrightarrow ClO^- + 2OH^-$	+0.66	$HClO + H_3O^+ + 2e^- \longrightarrow Cl^- + 2H_2O$	+1.49
$[PtCl_6]^{2-} + 2e^- \longrightarrow [PtCl_4]^{2-} + 2Cl^-$	+0.68	$Au^{3+} + 3e^- \longrightarrow Au$	+1.50
$O_2 + 2H_3O^+ + 2e^- \longrightarrow H_2O_2 + 2H_2O$	+0.682	$MnO_4^- + 8H_3O^+ + 5e^- \longrightarrow Mn^{2+} + 12H_2O$	+1.51
$[PtCl_4]^{2-} + 2e^- \longrightarrow Pt + 4Cl^-$	+0.73	$Ce^{4+} + e^- \longrightarrow Ce^{3+}$	+1.61
$Fe^{3+} + e^- \longrightarrow Fe^{2+}$	+0.771	$HClO + H_3O^+ + e^- \longrightarrow \frac{1}{2}Cl_2 + 2H_2O$	+1.63
$Hg_2^{2+} + 2e^- \longrightarrow 2Hg$	+0.789	$HClO_2 + 2H_3O^+ + 2e^- \longrightarrow HClO + 3H_2O$	+1.64
$Ag^+ + e^- \longrightarrow Ag$	+0.7991	$Au^+ + e^- \longrightarrow Au$	ca +1.68
$Hg^{2+} + 2e^- \longrightarrow Hg$	+0.854	$NiO_2 + 4H_3O^+ + 2e^- \longrightarrow Ni^{2+} + 6H_2O$	+1.68
$HO_2^- + H_2O + 2e^- \longrightarrow 3OH^-$	+0.88	$PbO_2 + SO_4^{2-} + 4H_3O^+ + 2e^- \longrightarrow$	
$ClO^- + H_2O + 2e^- \longrightarrow Cl^- + 2OH^-$	+0.89	$\qquad\qquad\qquad PbSO_4 + 6H_2O$	+1.685
$2Hg^{2+} + 2e^- \longrightarrow Hg_2^{2+}$	+0.920	$H_2O_2 + 2H_3O^+ + 2e^- \longrightarrow 4H_2O$	+1.77
$NO_3^- + 3H_3O^+ + 2e^- \longrightarrow$		$Co^{3+} + e^- \longrightarrow Co^{2+}$	+1.82
$\qquad\qquad\qquad HNO_2 + 4H_2O$	+0.94	$F_2 + 2e^- \longrightarrow 2F^-$	+2.87

Appendix I: Standard Enthalpies of Formation, Standard Free Energies of Formation, and Absolute Standard Entropies [298.15 K (25°C), 1 atm]

Substance	$\Delta H_f°$, kJ mol^{-1}	$\Delta G_f°$, kJ mol^{-1}	$S_{298}°$, J K^{-1} mol^{-1}
Aluminum			
$Al(s)$	0	0	28.3
$Al(g)$	326	286	164.4
$Al_2O_3(s)$	−1676	−1582	50.92
$AlF_3(s)$	−1504	−1425	66.44
$AlCl_3(s)$	−704.2	−628.9	110.7
$AlCl_3 \cdot 6H_2O(s)$	−2692	—	—
$Al_2S_3(s)$	−724	−492.4	—
$Al_2(SO_4)_3(s)$	−3440.8	−3100.1	239
Antimony			
$Sb(s)$	0	0	45.69

Standard Enthalpies of Formation, Standard Free Energies of Formation, and Absolute Standard Entropies [298.15 K (25°C), 1 atm] (Continued)

Substance	ΔH_f°, kJ mol^{-1}	ΔG_f°, kJ mol^{-1}	S_{298}°, J K^{-1} mol^{-1}
Sb(g)	262	222	180.2
Sb$_4$O$_6$(s)	−1441	−1268	221
SbCl$_3$(g)	−314	−301	337.7
SbCl$_5$(g)	−394.3	−334.3	401.8
Sb$_2$S$_3$(s)	−175	−174	182
SbCl$_3$(s)	−382.2	−323.7	184
SbOCl(s)	−374	—	—
Arsenic			
As(s)	0	0	35
As(g)	303	261	174.1
As$_4$(g)	144	92.5	314
As$_4$O$_6$(s)	−1313.9	−1152.5	214
As$_2$O$_5$(s)	−924.87	−782.4	105
AsCl$_3$(g)	−258.6	−245.9	327.1
As$_2$S$_3$(s)	−169	−169	164
AsH$_3$(g)	66.44	68.91	222.7
H$_3$AsO$_4$(s)	−906.3	—	—
Barium			
Ba(s)	0	0	66.9
Ba(g)	175.6	144.8	170.3
BaO(s)	−558.1	−528.4	70.3
BaCl$_2$(s)	−860.06	−810.9	126
BaSO$_4$(s)	−1465	−1353	132
Beryllium			
Be(s)	0	0	9.54
Be(g)	320.6	282.8	136.17
BeO(s)	−610.9	−581.6	14.1
Bismuth			
Bi(s)	0	0	56.74
Bi(g)	207	168	186.90
Bi$_2$O$_3$(s)	−573.88	−493.7	151
BiCl$_3$(s)	−379	−315	177
Bi$_2$S$_3$(s)	−143	−141	200
Boron			
B(s)	0	0	5.86
B(g)	562.7	518.8	153.3
B$_2$O$_3$(s)	−1272.8	−1193.7	53.97
B$_2$H$_6$(g)	36	86.6	232.0
B(OH)$_3$(s)	−1094.3	−969.01	88.83
BF$_3$(g)	−1137.3	−1120.3	254.0
BCl$_3$(g)	−403.8	−388.7	290.0
B$_3$N$_3$H$_6$(l)	−541.0	−392.8	200
HBO$_2$(s)	−794.25	−723.4	40
Bromine			
Br$_2$(l)	0	0	152.23
Br$_2$(g)	30.91	3.142	245.35
Br(g)	111.88	82.429	174.91
BrF$_3$(g)	−255.6	−229.5	292.4

Standard Enthalpies of Formation, Standard Free Energies of Formation, and Absolute Standard Entropies [298.15 K (25°C), 1 atm] (Continued)

Substance	ΔH_f°, kJ mol^{-1}	ΔG_f°, kJ mol^{-1}	S_{298}°, J K^{-1} mol^{-1}
HBr(g)	−36.4	−53.43	198.59
Cadmium			
Cd(s)	0	0	51.76
Cd(g)	112.0	77.45	167.64
CdO(s)	−258	−228	54.8
CdCl$_2$(s)	−391.5	−344.0	115.3
CdSO$_4$(s)	−933.28	−822.78	123.04
CdS(s)	−162	−156	64.9
Calcium			
Ca(s)	0	0	41.6
Ca(g)	192.6	158.9	154.78
CaO(s)	−635.5	−604.2	40
Ca(OH)$_2$(s)	−986.59	−896.76	76.1
CaSO$_4$(s)	−1432.7	−1320.3	107
CaSO$_4$·2H$_2$O(s)	−2021.1	−1795.7	194.0
CaCO$_3$(s) (calcite)	−1206.9	−1128.8	92.9
CaSO$_3$·2H$_2$O(s)	−1762	−1565	184
Carbon			
C(s) (graphite)	0	0	5.740
C(s) (diamond)	1.897	2.900	2.38
C(g)	716.681	671.289	157.987
CO(g)	−110.52	−137.15	197.56
CO$_2$(g)	−393.51	−394.36	213.6
CH$_4$(g)	−74.81	−50.75	186.15
CH$_3$OH(l)	−238.7	−166.4	127
CH$_3$OH(g)	−200.7	−162.0	239.7
CCl$_4$(l)	−135.4	−65.27	216.4
CCl$_4$(g)	−102.9	−60.63	309.7
CHCl$_3$(l)	−134.5	−73.72	202
CHCl$_3$(g)	−103.1	−70.37	295.6
CS$_2$(l)	89.70	65.27	151.3
CS$_2$(g)	117.4	67.15	237.7
C$_2$H$_2$(g)	226.7	209.2	200.8
C$_2$H$_4$(g)	52.26	68.12	219.5
C$_2$H$_6$(g)	−84.68	−32.9	229.5
CH$_3$CO$_2$H(l)	−484.5	−390	160
CH$_3$CO$_2$H(g)	−432.25	−374	282
C$_2$H$_5$OH(l)	−277.7	−174.9	161
C$_2$H$_5$OH(g)	−235.1	−168.6	282.6
C$_3$H$_8$(g)	−103.85	−23.49	269.9
C$_6$H$_6$(g)	82.927	129.66	269.2
C$_6$H$_6$(l)	49.028	124.50	172.8
CH$_2$Cl$_2$(l)	−121.5	−67.32	178
CH$_2$Cl$_2$(g)	−92.47	−65.90	270.1
CH$_3$Cl(g)	−80.83	−57.40	234.5
C$_2$H$_5$Cl(l)	−136.5	−59.41	190.8
C$_2$H$_5$Cl(g)	−112.2	−60.46	275.9
C$_2$N$_2$(g)	308.9	297.4	241.8

Standard Enthalpies of Formation, Standard Free Energies of Formation, and Absolute Standard Entropies [298.15 K (25°C), 1 atm] (Continued)

Substance	ΔH_f°, kJ mol^{-1}	ΔG_f°, kJ mol^{-1}	S_{298}°, J K^{-1} mol^{-1}
HCN(l)	108.9	124.9	112.8
HCN(g)	135	124.7	201.7
Chlorine			
Cl$_2$(g)	0	0	222.96
Cl(g)	121.68	105.70	165.09
ClF(g)	−54.48	−55.94	217.8
ClF$_3$(g)	−163	−123	281.5
Cl$_2$O(g)	80.3	97.9	266.1
Cl$_2$O$_7$(l)	238	—	—
Cl$_2$O$_7$(g)	272	—	—
HCl(g)	−92.307	−95.299	186.80
HClO$_4$(l)	−40.6	—	—
Chromium			
Cr(s)	0	0	23.8
Cr(g)	397	352	174.4
Cr$_2$O$_3$(s)	−1140	−1058	81.2
CrO$_3$(s)	−589.5	—	—
(NH$_4$)$_2$Cr$_2$O$_7$(s)	−1807	—	—
Cobalt			
Co(s)	0	0	30.0
CoO(s)	−237.9	−214.2	52.97
Co$_3$O$_4$(s)	−891.2	−774.0	103
Co(NO$_3$)$_2$(s)	−420.5	—	—
Copper			
Cu(s)	0	0	33.15
Cu(g)	338.3	298.5	166.3
CuO(s)	−157	−130	42.63
Cu$_2$O(s)	−169	−146	93.14
CuS(s)	−53.1	−53.6	66.5
Cu$_2$S(s)	−79.5	−86.2	121
CuSO$_4$(s)	−771.36	−661.9	109
Cu(NO$_3$)$_2$(s)	−303	—	—
Fluorine			
F$_2$(g)	0	0	202.7
F(g)	78.99	61.92	158.64
F$_2$O(g)	−22	−4.6	247.3
HF(g)	−271	−273	173.67
Hydrogen			
H$_2$(g)	0	0	130.57
H(g)	217.97	203.26	114.60
H$_2$O(l)	−285.83	−237.18	69.91
H$_2$O(g)	−241.82	−228.59	188.71
H$_2$O$_2$(l)	−187.8	−120.4	110
H$_2$O$_2$(g)	−136.3	−105.6	233
HF(g)	−271	−273	173.67
HCl(g)	−92.307	−95.299	186.80
HBr(g)	−36.4	−53.43	198.59
HI(g)	26.5	1.7	206.48

Standard Enthalpies of Formation, Standard Free Energies of Formation, and Absolute Standard Entropies [298.15 K (25°C), 1 atm] (Continued)

Substance	ΔH_f°, kJ mol^{-1}	ΔG_f°, kJ mol^{-1}	S_{298}°, J K^{-1} mol^{-1}
$H_2S(g)$	−20.6	−33.6	205.7
$H_2Se(g)$	30	16	218.9
Iodine			
$I_2(s)$	0	0	116.14
$I_2(g)$	62.438	19.36	260.6
$I(g)$	106.84	70.283	180.68
$IF(g)$	95.65	−118.5	236.1
$ICl(g)$	17.8	−5.44	247.44
$IBr(g)$	40.8	3.7	258.66
$IF_7(g)$	−943.9	−818.4	346
$HI(g)$	26.5	1.7	206.48
Iron			
$Fe(s)$	0	0	27.3
$Fe(g)$	416	371	180.38
$Fe_2O_3(s)$	−824.2	−742.2	87.40
$Fe_3O_4(s)$	−1118	−1015	146
$Fe(CO)_5(l)$	−774.0	−705.4	338
$Fe(CO)_5(g)$	−733.9	−697.26	445.2
$FeCl_2(s)$	−341.79	−302.30	117.95
$FeCl_3(s)$	−399.49	−334.00	142.3
$FeO(s)$	−272	—	—
$Fe(OH)_2(s)$	−569.0	−486.6	88
$Fe(OH)_3(s)$	−823.0	−696.6	107
$FeS(s)$	−100	−100	60.29
$Fe_3C(s)$	25	20	105
Lead			
$Pb(s)$	0	0	64.81
$Pb(g)$	195	162	175.26
$PbO(s)$ (yellow)	−217.3	−187.9	68.70
$PbO(s)$ (red)	−219.0	−188.9	66.5
$Pb(OH)_2(s)$	−515.9	—	—
$PbS(s)$	−100	−98.7	91.2
$Pb(NO_3)_2(s)$	−451.9	—	—
$PbO_2(s)$	−277	−217.4	68.6
$PbCl_2(s)$	−359.4	−314.1	136
Lithium			
$Li(s)$	0	0	28.0
$Li(g)$	155.1	122.1	138.67
$LiH(s)$	−90.42	−69.96	25
$Li(OH)(s)$	−487.23	−443.9	50.2
$LiF(s)$	−612.1	−584.1	35.9
$Li_2CO_3(s)$	−1215.6	−1132.4	90.4
Manganese			
$Mn(s)$	0	0	32.0
$Mn(g)$	281	238	173.6
$MnO(s)$	−385.2	−362.9	59.71
$MnO_2(s)$	−520.03	−465.18	53.05
$Mn_2O_3(s)$	−959.0	−881.2	110

Standard Enthalpies of Formation, Standard Free Energies of Formation, and Absolute Standard Entropies [298.15 K (25°C), 1 atm] (Continued)

Substance	ΔH_f°, kJ mol^{-1}	ΔG_f°, kJ mol^{-1}	S_{298}°, J K^{-1} mol^{-1}
$Mn_3O_4(s)$	−1388	−1283	156
Mercury			
$Hg(l)$	0	0	76.02
$Hg(g)$	61.317	31.85	174.8
$HgO(s)$ (red)	−90.83	−58.555	70.29
$HgO(s)$ (yellow)	−90.46	−57.296	71.1
$HgCl_2(s)$	−224	−179	146
$Hg_2Cl_2(s)$	−265.2	−210.78	192
$HgS(s)$ (red)	−58.16	−50.6	82.4
$HgS(s)$ (black)	−53.6	−47.7	88.3
$HgSO_4(s)$	−707.5	—	—
Nitrogen			
$N_2(g)$	0	0	191.5
$N(g)$	472.704	455.579	153.19
$NO(g)$	90.25	86.57	210.65
$NO_2(g)$	33.2	51.30	239.9
$N_2O(g)$	82.05	104.2	219.7
$N_2O_3(g)$	83.72	139.4	312.2
$N_2O_4(g)$	9.16	97.82	304.2
$N_2O_5(g)$	11	115	356
$NH_3(g)$	−46.11	−16.5	192.3
$N_2H_4(l)$	50.63	149.2	121.2
$N_2H_4(g)$	95.4	159.3	238.4
$NH_4NO_3(s)$	−365.6	−184.0	151.1
$NH_4Cl(s)$	−314.4	−201.5	94.6
$NH_4Br(s)$	−270.8	−175	113
$NH_4I(s)$	−201.4	−113	117
$NH_4NO_2(s)$	−256	—	—
$HNO_3(l)$	−174.1	−80.79	155.6
$HNO_3(g)$	−135.1	−74.77	266.2
Oxygen			
$O_2(g)$	0	0	205.03
$O(g)$	249.17	231.75	160.95
$O_3(g)$	143	163	238.8
Phosphorus			
$P_4(s)$	0	0	164
$P_4(g)$	58.91	24.5	280.0
$P(g)$	314.6	278.3	163.08
$PH_3(g)$	5.4	13	210.1
$PCl_3(g)$	−287	−268	311.7
$PCl_5(g)$	−375	−305	364.5
$P_4O_6(s)$	−1640	—	—
$P_4O_{10}(s)$	−2984	−2698	228.9
$HPO_3(s)$	−948.5	—	—
$H_3PO_2(s)$	−604.6	—	—
$H_3PO_3(s)$	−964.4	—	—
$H_3PO_4(s)$	−1279	−1119	110.5
$H_3PO_4(l)$	−1267	—	—

Standard Enthalpies of Formation, Standard Free Energies of Formation, and Absolute Standard Entropies [298.15 K (25°C), 1 atm] (Continued)

Substance	ΔH_f°, kJ mol^{-1}	ΔG_f°, kJ mol^{-1}	S_{298}°, J K^{-1} mol^{-1}
$H_4P_2O_7(s)$	−2241	—	—
$POCl_3(l)$	−597.1	−520.9	222.5
$POCl_3(g)$	−558.48	−512.96	325.3
Potassium			
$K(s)$	0	0	63.6
$K(g)$	90.00	61.17	160.23
$KF(s)$	−562.58	−533.12	66.57
$KCl(s)$	−435.868	−408.32	82.68
Silicon			
$Si(s)$	0	0	18.8
$Si(g)$	455.6	411	167.9
$SiO_2(s)$	−910.94	−856.67	41.84
$SiH_4(g)$	34	56.9	204.5
$H_2SiO_3(s)$	−1189	−1092	130
$H_4SiO_4(s)$	−1481	−1333	190
$SiF_4(g)$	−1614.9	−1572.7	282.4
$SiCl_4(l)$	−687.0	−619.90	240
$SiCl_4(g)$	−657.01	−617.01	330.6
$SiC(s)$	−65.3	−62.8	16.6
Silver			
$Ag(s)$	0	0	42.55
$Ag(g)$	284.6	245.7	172.89
$Ag_2O(s)$	−31.0	−11.2	121
$AgCl(s)$	−127.1	−109.8	96.2
$Ag_2S(s)$	−32.6	−40.7	144.0
Sodium			
$Na(s)$	0	0	51.0
$Na(g)$	108.7	78.11	153.62
$Na_2O(s)$	−415.9	−377	72.8
$NaCl(s)$	−411.00	−384.03	72.38
Sulfur			
$S_8(s)$ (rhombic)	0	0	254
$S(g)$	278.80	238.27	167.75
$SO_2(g)$	−296.83	−300.19	248.1
$SO_3(g)$	−395.7	−371.1	256.6
$H_2S(g)$	−20.6	−33.6	205.7
$H_2SO_4(l)$	−813.989	690.101	156.90
$H_2S_2O_7(s)$	−1274	—	—
$SF_4(g)$	−774.9	−731.4	291.9
$SF_6(g)$	−1210	−1105	291.7
$SCl_2(l)$	−50	—	—
$SCl_2(g)$	−20	—	—
$S_2Cl_2(l)$	−59.4	—	—
$S_2Cl_2(g)$	−18	−32	331.4
$SOCl_2(l)$	−246	—	—
$SOCl_2(g)$	−213	−198	309.7
$SO_2Cl_2(l)$	−394	—	—
$SO_2Cl_2(g)$	−364	−320	311.8

Standard Enthalpies of Formation, Standard Free Energies of Formation, and Absolute Standard Entropies [298.15 K (25°C), 1 atm] (Continued)

Substance	ΔH_f°, kJ mol^{-1}	ΔG_f°, kJ mol^{-1}	S_{298}°, J K^{-1} mol^{-1}
Tin			
Sn(s)	0	0	51.55
Sn(g)	302	267	168.38
SnO(s)	−286	−257	56.5
SnO$_2$(s)	−580.7	−519.7	52.3
SnCl$_4$(l)	−511.2	−440.2	259
SnCl$_4$(g)	−471.5	−432.2	366
Titanium			
Ti(s)	0	0	30.6
Ti(g)	469.9	425.1	180.19
TiO$_2$(s)	−944.7	−889.5	50.33
TiCl$_4$(l)	−804.2	−737.2	252.3
TiCl$_4$(g)	−763.2	−726.8	354.8
Tungsten			
W(s)	0	0	32.6
W(g)	849.4	807.1	173.84
WO$_3$(s)	−842.87	−764.08	75.90
Zinc			
Zn(s)	0	0	41.6
Zn(g)	130.73	95.178	160.87
ZnO(s)	−348.3	−318.3	43.64
ZnCl$_2$(s)	−415.1	−369.43	111.5
ZnS(s)	−206.0	−201.3	57.7
ZnSO$_4$(s)	−982.8	−874.5	120
ZnCO$_3$(s)	−812.78	−731.57	82.4
Complexes			
[Co(NH$_3$)$_4$(NO$_2$)$_2$]NO$_3$, *cis*	−898.7	—	—
[Co(NH$_3$)$_4$(NO$_2$)$_2$]NO$_3$, *trans*	−896.2	—	—
NH$_4$[Co(NH$_3$)$_2$(NO$_2$)$_4$]	−837.6	—	—
[Co(NH$_3$)$_6$][Co(NH$_3$)$_2$(NO$_2$)$_4$]$_3$	−2733	—	—
[Co(NH$_3$)$_4$Cl$_2$]Cl, *cis*	−997.0	—	—
[Co(NH$_3$)$_4$Cl$_2$]Cl, *trans*	−999.6	—	—
[Co(en)$_2$(NO$_2$)$_2$]NO$_3$, *cis*	−689.5	—	—
[Co(en)$_2$Cl$_2$]Cl, *cis*	−681.1	—	—
[Co(en)$_2$Cl$_2$]Cl, *trans*	−677.4	—	—
[Co(en)$_3$](ClO$_4$)$_3$	−762.7	—	—
[Co(en)$_3$]Br$_2$	−595.8	—	—
[Co(en)$_3$]I$_2$	−475.3	—	—
[Co(en)$_3$]I$_3$	−519.2	—	—
[Co(NH$_3$)$_6$](ClO$_4$)$_3$	−1035	−227	636
[Co(NH$_3$)$_5$NO$_2$](NO$_3$)$_2$	−1089	−418.4	350
[Co(NH$_3$)$_6$](NO$_3$)$_3$	−1282	−530.5	469
[Co(NH$_3$)$_5$Cl]Cl$_2$	−1017	−582.8	366
[Pt(NH$_3$)$_4$]Cl$_2$	−728.0	—	—
[Ni(NH$_3$)$_6$]Cl$_2$	−994.1	—	—
[Ni(NH$_3$)$_6$]Br$_2$	−923.8	—	—
[Ni(NH$_3$)$_6$]I$_2$	−808.3	—	—

Appendix J: Composition of Commercial Acids and Bases

Acid or Base	Density, g/mL	Percentage by Mass	Molarity	Normality
Hydrochloric acid	1.19	38%	12.4	12.4
Nitric acid	1.42	70%	15.8	15.8
Sulfuric acid	1.84	95%	17.8	35.6
Acetic acid	1.05	99%	17.3	17.3
Aqueous ammonia	0.90	28%	14.8	14.8

Appendix K: Half-Life Times for Several Radioactive Isotopes

Isotope	Half-Life	Type of Emission[a]	Isotope	Half-Life	Type of Emission[a]
$^{14}_{6}C$	5730 y	(β^-)	$^{210}_{83}Bi$	5.01 d	(β^-)
$^{13}_{7}N$	9.97 m	(β^+)	$^{212}_{83}Bi$	60.5 m	$(\alpha \text{ or } \beta^-)$
$^{15}_{9}F$	5×10^{-22} s	(p)	$^{210}_{84}Po$	138.4 d	(α)
$^{24}_{11}Na$	14.97 h	(β^-)	$^{212}_{84}Po$	3×10^{-7} s	(α)
$^{32}_{15}P$	14.28 d	(β^-)	$^{216}_{84}Po$	0.16 s	(α)
$^{40}_{19}K$	1.26×10^9 y	$(\beta^- \text{ or } E.C.)$	$^{218}_{84}Po$	3.11 m	(α)
$^{49}_{26}Fe$	0.08 s	(β^+)	$^{215}_{85}At$	1.0×10^{-4} s	(α)
$^{60}_{26}Fe$	1.5×10^6 y	(β^-)	$^{218}_{85}At$	1.6 s	(α)
$^{60}_{27}Co$	5.2 y	(β^-)	$^{220}_{86}Rn$	55.6 s	(α)
$^{87}_{37}Rb$	4.7×10^{10} y	(β^-)	$^{222}_{86}Rn$	3.82 d	(α)
$^{90}_{38}Sr$	29 y	(β^-)	$^{224}_{88}Ra$	3.66 d	(α)
$^{115}_{49}In$	4.4×10^{14} y	(β^-)	$^{226}_{88}Ra$	1590 y	(α)
$^{131}_{53}I$	8.040 d	(β^-)	$^{228}_{88}Ra$	5.75 y	(β^-)
$^{142}_{58}Ce$	5×10^{15} y	(α)	$^{228}_{89}Ac$	6.13 h	(β^-)
$^{208}_{81}Tl$	3.052 m	(β^-)	$^{228}_{90}Th$	1.912 y	(α)
$^{210}_{82}Pb$	22.6 y	(β^-)	$^{232}_{90}Th$	1.4×10^{10} y	(α)
$^{212}_{82}Pb$	10.6 h	(β^-)	$^{233}_{90}Th$	23 m	(β^-)
$^{214}_{82}Pb$	26.8 m	(β^-)	$^{234}_{90}Th$	24.10 d	(β^-)
$^{206}_{83}Bi$	6.243 d	$(E.C.)$	$^{233}_{91}Pa$	27 d	(β^-)

Half-Life Times for Several Radioactive Isotopes (Continued)

Isotope	Half-Life	Type of Emission[a]	Isotope	Half-Life	Type of Emission[a]
$^{233}_{92}U$	1.62×10^5 y	(α)	$^{242}_{96}Cm$	162.8 d	(α)
$^{234}_{92}U$	2.45×10^5 y	(α)	$^{243}_{97}Bk$	4.5 h	$(\alpha$ or $E.C.)$
$^{235}_{92}U$	7.04×10^8 y	(α)	$^{253}_{99}Es$	20.47 d	(α)
$^{238}_{92}U$	4.51×10^9 y	(α)	$^{254}_{100}Fm$	3.24 h	$(\alpha$ or $S.F.)$
$^{239}_{92}U$	23.54 m	(β^-)	$^{255}_{100}Fm$	20.1 h	(α)
$^{239}_{93}Np$	2.3 d	(β^-)	$^{256}_{101}Md$	76 m	$(\alpha$ or $E.C.)$
$^{239}_{94}Pu$	2.411×10^4 y	(α)	$^{254}_{102}No$	55 s	(α)
$^{240}_{94}Pu$	6.58×10^3 y	(α)	$^{257}_{103}Lr$	0.65 s	(α)
$^{241}_{94}Pu$	14.4 y	$(\alpha$ or $\beta^-)$	$^{260}_{105}Ha$	1.5 s	$(\alpha$ or $S.F.)$
$^{241}_{95}Am$	458 y	(α)	$^{263}_{106}Sg$	0.8 s	$(\alpha$ or $S.F.)$

[a]$E.C.$ = electron capture, $S.F.$ = spontaneous fission; y = years, d = days, h = hours, m = minutes, s = seconds.

Photo Credits

Glossary

Absolute entropy The entropy change of a substance taken from absolute zero to a given temperature (T). (18.10)

Absolute zero The temperature at which all possible heat has been removed from an object. (10.4)

Acid A compound that donates a proton (hydrogen ion, H^+) to another compound. (2.10)

Acid–base reaction Reaction occurring when a proton is transferred from a Brønsted acid to a Brønsted base. (8.5)

Activated complex An unstable combination of reacting molecules that is intermediate between reactants and products. (13.7)

Activation energy (E_a) The minimum energy necessary to form an activated complex in a reaction. (13.8)

Addition polymer A polymer formed by an addition reaction. (9.15)

Addition reaction Reaction of two or more substances to give another substance. (2.10, 9.5)

Adhesive force Force of attraction between two separate phases. (11.11)

Alkyl group A substituent that contains one less hydrogen than the corresponding alkane. (9.4)

Alpha decay The loss of an alpha particle during radioactive decay. (20.6)

Alpha (α) particle A helium nucleus; that is, a helium atom that has lost two electrons. (20.5)

Amino acid A substance containing both an amine group and a carboxylic acid group. Proteins are composed of amino acids. (25.4)

Amorphous solid A solid that lacks a crystalline structure. (11.12)

Amphiprotic A species that may either gain or lose a proton in reaction. (15.2)

Amphoteric compound A compound that can exhibit the properties of either an acid or a base. (8.6)

Angular momentum number (l) A quantum number distinguishing the different shapes of orbitals. (5.6)

Anion A negative ion. (6.1)

Anode The electrode at which oxidation takes place in an electrochemical cell. (19.1)

Antibonding orbital Molecular orbital located outside of the region between two nuclei. Electrons in an antibonding orbital destablize the molecule. (7.12)

Arrhenius equation ($k = A \times e^{-E_a/RT}$) Expresses the relationship between the rate constant and the activation energy of a reaction. (13.8)

Atom The smallest particle of an element that can enter into a chemical combination. (1.6)

Atomic mass (atomic weight) The average mass of an atom expressed in amu. (2.2)

Atomic mass unit (amu) A unit of mass equal to $\frac{1}{12}$ of the mass of a ^{12}C atom. (2.2)

Atomic number (Z) The number of protons in the nucleus of an atom. (2.2)

Aufbau process Process by which chemists illustrate the electronic structures of the elements by "building" them in atomic order, adding one proton to the nucleus and one electron to the proper subshell at a time. (5.8)

Avogadro's law Equal volumes of all gases, measured under the same conditions of temperature and pressure, contain the same number of molecules. (10.5)

Avogadro's number The number of atoms contained in exactly 12 grams of ^{12}C, equal to 6.022×10^{23} atoms. (3.2)

Azeotropic mixture A solution that forms a vapor with the same concentration as the solution, distilling without a change in concentration. (12.15)

Band The orbitals, or energy levels, that extend through a crystal. The way in which these bands are filled or not filled with electrons determines whether the substance is a metal, a semiconductor, or an insulator. (24.6)

Barometer A device used to measure air pressure. (10.1)

Base A compound that accepts a proton (hydrogen ion, H^+). (2.10)

Beta decay The breakdown of a neutron into a proton, which remains in the nucleus, and an electron, which is emitted as a beta particle. (20.6)

Beta (β) particle An electron emitted during radioactive decay. (20.5)

Bimolecular reaction The collision and combination of two reactants to give an activated complex in an elementary reaction. (13.9)

Binary compound A compound containing two different elements. (2.11)

Body-centered cubic structure   A crystalline structure that has a cubic unit cell with lattice points at the corners and in the center of the cell. (11.14)

Boiling point   The temperature at which the vapor pressure of a liquid equals the pressure of the gas above it. (11.6)

Boiling-point elevation $(\Delta T = K_b m)$ The elevation of the boiling point of a liquid by addition of a solute. (12.14)

Bond angle   The angle between any two covalent bonds that share a common atom. (7.1)

Bond distance   The distance between the nuclei of two bonded atoms. (7.1)

Bond energy   The energy required to break a covalent bond in a gaseous substance. (6.12)

Bond order   In a Lewis formula, the number of bonding pairs of electrons between two atoms. The molecular-orbital bond order is the net number of pairs of bonding electrons, or the difference between the number of bonding and antibonding electrons divided by two. (7.14)

Bonding orbital   Molecular orbital located between two nuclei. Electrons in a bonding orbital stabilize a molecule. (7.12)

Born–Haber cycle   Cyclic process used to relate the enthalpy of formation (ΔH_f) of a compound to its lattice energy (U), the ionization energy (I), the electron affinity $(E.A.)$, the enthalpy of sublimation (ΔH_s), and the bond dissociation energy (D) of its constituents. (11.19)

Boyle's law   The volume of a given mass of gas held at constant temperature is inversely proportional to the pressure under which it is measured. $PV = k$. (10.2)

Bragg equation $(n\lambda = 2d \sin\theta)$ An equation that relates the angles (θ) at which X rays of wavelength λ are scattered by planes with a separation d. (11.20)

Brønsted acid   A compound that donates a proton to another compound. (15.1)

Brønsted base   A compound that accepts a proton. (15.1)

Buffer capacity   The amount of an acid or base that can be added to a volume of a buffer solution before its pH changes significantly. (16.9)

Buffer solution   A mixture of a weak acid or a weak base and its salt. The pH of a buffer resists change when small amounts of acid or base are added. (16.9)

Calorie   A non-SI unit representing the amount of heat or other energy necessary to raise the temperature of 1 gram of water 1 degree Celsius. 1 cal = 4.184 J. (4.2)

Calorimetry   The process of measuring the amount of heat involved in a chemical or physical change. (4.3)

Catalysis   The effect of a catalyst in increasing the speed of a chemical reaction. (14.4)

Catalyst   A substance that changes the speed of a chemical reaction without affecting the yield or undergoing permanent chemical change. (14.4)

Cathode   The electrode at which reduction takes place in an electrochemical cell. (19.1)

Cathodic protection   Method of preventing corrosion of iron or steel by connecting a metal that is more active than iron to it. (19.11)

Cation   A positive ion. (6.1)

Cell potential   The difference in potential between the two electrodes of a cell. (19.2)

Chain mechanism   A series of elementary reactions that repeat over and over to produce a product. (13.10)

Chain reaction   Repeated fission caused when the neutrons released in fission bombard other atoms. (20.9)

Charles's law   The volume of a given mass of gas is directly proportional to its Kelvin temperature when the pressure is held constant. $V/T = k$. (10.3)

Chelating ligand   A ligand that is attached to a central metal ion by bonds from two or more donor atoms. (23.6)

Chemical change   Change producing a different kind of matter from the original kind of matter. (1.4)

Chemical property   Behavior that is related to the change of one kind of matter into another kind. Examples include metallic versus nonmetallic and acid versus base behavior. (1.4)

Chemical thermodynamics   The chemical science that deals with the energy transfers and transformations that accompany chemical and physical changes. (18.1)

***Cis* configuration**   Configuration of a geometrical isomer in which two groups are on the same side of an imaginary reference line on the molecule. (9.5, 23.9)

Cloning   The exact duplication of a cell's DNA and, therefore, of an organism's characteristics. (25.13)

Colligative properties   Properties of a solution that depend only on the concentration of a solute species. (12.13)

Colloid   Insoluble particles (larger than simple molecules) in a stable suspension. (12.21)

Common ion effect   The shift in equilibrium caused by the addition of a substance that has an ion in common with the substances in equilibrium. (16.8)

Compound   A pure substance with an invariant composition that can be decomposed, producing either elements or other compounds, by chemical change. (1.5)

Concentration   The relative amounts of solute and solvent present in a solution. (3.7)

Condensation   The change from a vapor to a condensed state (solid or liquid). (11.5)

Condensation polymer   A polymer formed by linking together molecules in a reaction that eliminates small molecules such as water. (9.15)

Conjugate acid   Substance formed when a base gains a proton (hydrogen ion). Considered an acid because it can lose a proton (hydrogen ion) to re-form the base. (15.1)

Conjugate base   Substance formed when an acid loses a proton (hydrogen ion). Considered a base because it can gain a proton (hydrogen ion) to re-form the acid. (15.1)

Constant boiling solution   A solution that forms a vapor with the same concentration as the solution, distilling without a change in concentration. (12.15)

Coordinate covalent bond   A bond formed when one atom provides both electrons in a shared pair. (6.8)

Coordination compound (complex)   A molecule or ion formed by the bonding of a metal atom or ion to two or more ligands by coordinate covalent bonds. (23.6)

Coordination number   The number of atoms closest to any given atom in a crystal or to the central metal atom in a complex (11.14, 23.6)

Coordination sphere   The central metal atom or ion plus the attached ligands of a complex. (23.6)

Coulomb   The quantity of electricity involved when a current of 1 ampere flows for 1 second. (19.2)

Covalent bond   Bond formed when pairs of electrons are shared between atoms. (6.4)

Covalent radius   Half the distance between the nuclei of two identical atoms when they are joined by a single covalent bond. (5.10)

Critical pressure   The pressure required to liquefy a gas at its critical temperature. (11.9)

Critical temperature   The temperature above which a gas cannot be liquefied, no matter how much pressure is applied. (11.9)

Crystal defect   A variation in the regular arrangement of the atoms or molecules of a crystal. (11.13)

Crystal field splitting (10Dq)   The difference in energy between a metal's d orbitals in a coordination complex. (23.12)

Crystal field stabilization energy (CFSE)   A measure of the increased stability of a complex that shows crystal field splitting. (23.12)

Crystalline solid   A homogeneous solid in which the atoms, ions, or molecules assume ordered positions. (11.12)

Cubic closest packed structure   A crystalline structure in which planes of closest packed atoms or ions are stacked ABCABC. (11.14)

Dalton's law   The total pressure of a mixture of ideal gases is equal to the sum of the partial pressures of the component gases. (10.9)

Daughter nuclide   A nuclide produced by the radioactive decay of another nuclide. May be stable or may decay further. (20.6)

Decomposition reaction   Reaction in which one compound breaks down into two or more substances. (2.10)

Degenerate orbitals   Orbitals that have the same energy. (5.7)

Density   Mass of a unit volume of a substance. (1.11)

Diamagnet substance   A substance that contains no unpaired electrons. Diamagnetic substances are repelled by a magnetic field. (6.8)

Diffusion   The movement of gas molecules through the gas. (10.10)

Dipole   The separation of charge in a bond or a molecule with a positively charged end and a negatively charged end. (6.5)

Dipole–dipole attraction   The intermolecular attraction of two permanent dipoles. (11.2)

Diprotic acid   An acid containing two ionizable hydrogen atoms per molecule. A diprotic acid ionizes in two steps. (15.8)

Disorder   A measure of the number of equivalent ways in which a system can be arranged. (18.6)

Dissociation constant (K_d)   The equilibrium constant for the decomposition of a complex ion into its components in solution. (17.9)

Double bond   A covalent bond in which two pairs of electrons are shared between two atoms. (6.6)

Einstein equation   Equation for determining the amount of energy resulting from the conversion of matter to energy, $E = mc^2$. (20.2)

Electrode   A system in which a conductor is in contact with a mixture of oxidized and reduced forms of some chemical species. (19.3)

Electrode potential   The difference between the charge on an electrode and the charge in the solution. (19.3)

Electrolysis   The input of electrical energy as a direct current to force a nonspontaneous reaction to (19.12)

Electrolyte   An ionic or covalent compound that melts to give a liquid that contains ions or that dissolves to give a solution that contains ions. (12.3)

Electrolytic conduction   The movement of ions through a molten substance, a solution, or (occasionally) a solid. (12.3)

Electron   A relatively light, negatively charged subatomic particle. (2.2)

Electron affinity   A measure of the energy involved when an electron is added to a gaseous atom to form a negative ion. (5.10)

Electron capture   Capture of an electron by an unstable nucleus. The electron converts a proton to a neutron in the nucleus. (20.6)

Electron configuration   Electronic structure of an atom. (5.7)

Electronegativity   The relative attraction of an atom for the electrons in a covalent bond. (6.5)

Element   A substance that is composed of a single type of atom; a substance that cannot be decomposed by a chemical change. (1.5)

Elementary reaction   A reaction that cannot be broken down into smaller steps. (13.9)

Empirical formula   A formula showing the composition of a compound given as the simplest whole-number ratio of atoms. (3.5)

End point   The point during a titration when an indicator shows that the amount of reactant necessary for a complete reaction has been added to a solution. (3.13)

Endothermic process   A chemical reaction or physical change that occurs with the absorption of heat. (4.1)

Energy   The capacity to do work. (1.3)

Enthalpy change (ΔH)   The heat lost or absorbed by a system under constant pressure during a reaction or other change. (4.5, 18.7)

Enthalpy of fusion (ΔH_{fus})   The energy needed to change a given quantity of a substance from the solid state to the liquid state at a constant temperature. (4.5)

Enthalpy of reaction   The heat lost or absorbed by a system in a reaction. (4.5)

Enthalpy of vaporization (ΔH_{vap})   The energy needed to evaporate a given quantity of liquid at a constant specified temperature. (4.5)

Entropy (S)   The randomness, or amount of disorder, of a system. (18.8)

Entropy change (ΔS)   The change in entropy that accompanies a chemical or physical change. ΔS is given by the sum of the entropies of the products of a chemical change minus the sum of the entropies of the reactants. $\Delta S = \Sigma S_{products} - \Sigma S_{reactants}$. (18.8)

Equilibrium   The state at which the conversion of reactants into products and the conversion of products back into reactants occur simultaneously at the same rate. (14.1)

Equilibrium constant (K)   The value of the reaction quotient for a system at equilibrium. (14.2)

Equivalent   An equivalent of an acid is the amount of acid required to provide 1 mole of protons (hydrogen ions) in a reaction; an equivalent of a base is the amount of base required to react with 1 mole of protons (hydrogen ions). (15.12) An equivalent of an oxidizing agent or of a reducing agent is the amount of the agent that combines with or releases 1 mole of electrons, respectively. (19.15)

Evaporation   The change of a liquid into a gas. (11.5)

Excited state   State in which an atom or molecule picks up outside energy, causing an electron to move into a higher-energy orbital. (5.3)

Exothermic process   A chemical reaction or physical change that produces heat. (4.1)

Expansion work   Work transferred between a system and its surroundings as the system expands or contracts against a constant pressure. (4.4, 18.2)

Extensive property   A property of a substance that depends on the amount of the substance. (1.4)

Face-centered cubic structure   A crystalline structure consisting of a cubic unit cell with lattice points on the corners and in the center of each face. (11.14)

Faraday (F)   The charge on 1 mole of electrons. 1 F = 96,485 coulombs. (19.5)

Faraday's law   The amount of a substance undergoing a chemical change at each electrode during electrolysis is directly proportional to the quantity of electricity that passes through the electrolytic cell. (19.15)

First ionization energy   The energy required to remove the most loosely bound electron from a gaseous atom. (5.10)

First law of thermodynamics   The total amount of energy in the universe is constant. (18.3)

Fission   The splitting of a heavier nucleus into two or more lighter nuclei, usually accompanied by the conversion of mass into large amounts of energy. (20.9)

Formal charge   The charge on a species that would result if the electrons in a covalently bonded atom were shared evenly. (6.9)

Formation constant (K_f)   The equilibrium constant for the formation of a complex ion from its components in solution. (17.9)

Formula mass   The sum of the atomic masses of the atoms found in one formula unit of a compound. (3.1)

Free energy change (ΔG)   A predictor of the spontaneity of a chemical reaction at constant temperature. $\Delta G = \Delta H - T\Delta S$. (18.11)

Freezing point   The temperature at which the solid and liquid phases of a substance are in equilibrium. (11.8)

Freezing-point depression   ($\Delta T = K_f m$) The lowering of the freezing point of a liquid by addition of a solute. (12.16)

Frequency factor (A)   In the Arrhenius equation, a constant indicating how many collisions have the correct orientation to lead to products. (13.8)

Functional group   A part of an organic molecule responsible for chemical behavior of the molecule. (9.3)

Fusion   The combining of very light nuclei into heavier nuclei, accompanied by the conversion of mass into large amounts of energy. (20.10)

Galvanic cell   A device in which chemical energy is converted into electrical energy. Also called a voltaic cell. (19.1)

Gamma (γ) ray   High-energy electromagnetic radiation. (20.5)

Gas   The state in which matter has neither definite volume nor shape. (1.2)

Gas constant (R)   Constant derived from the ideal gas equation, $PV = nRT$. $R = 0.08206$ L atm mol^{-1} K^{-1} or 8.314 L kPa mol^{-1} K^{-1}. (10.6)

Gene cloning   A procedure for transferring small pieces of genetic information from one organism to another, thereby producing many copies of the second organism as it replicates. (25.13)

Genetic code   The information coded in a DNA molecule that is replicated when new cells are produced. (25.12)

Geometrical isomers   Isomers that differ only in the way in which atoms are oriented in space relative to each other. (9.3, 23.9)

Glass   An amorphous solid. (11.12)

Graham's law   The rates of diffusion of gases are inversely proportional to the square roots of their densities (or their molecular masses). (10.10)

Gram (g)   A unit of measure for mass. 1 g $= 1 \times 10^{-3}$ kg. (1.10)

Ground state   State in which the electrons in an atom, ion, or molecule are in the lowest-energy orbitals possible. (5.3)

Half-cell   An electrode containing both an oxidized and a reduced species. (19.1)

Half-life ($t_{1/2}$)   The time required for half of the atoms in a radioactive sample to decay. (20.4)

Half-life of a reaction ($t_{1/2}$)   The time required for half of the original concentration of the limiting reactant to be consumed. (13.6)

Half-reaction   One of the two parts (oxidation or reduction) of an oxidation–reduction reaction. (19.1)

Heat   Energy that is transferred between two bodies that are at different temperatures. (4.1)

Heat capacity   A property of a body of matter that represents the quantity of heat required to increase its temperature by 1 degree Celsius (or 1 kelvin). (4.2)

Heisenberg uncertainty principle   It is impossible to determine accurately both the momentum and the position of a particle simultaneously. (5.4)

Henry's law   The mass of a gas that dissolves in a definite volume of liquid is directly proportional to the pressure of the gas, provided that the gas does not react with the solvent. (12.5)

Hess's law   If a process can be written as the sum of several stepwise processes, then the enthalpy change of the total process equals the sum of the enthalpy changes of the various steps. (4.6)

Heterogeneous catalyst   A catalyst present in a different phase from the reactants, furnishing a surface at which a reaction can occur. (13.11)

Heterogeneous equilibrium   An equilibrium between two or more different phases, involving a boundary surface between the two phases. (14.3)

Hexagonal closest packed structure   A crystalline structure in which close packed layers of atoms or ions are stacked ABABAB; the unit cell is hexagonal. (11.14)

Homogeneous catalyst   A catalyst present in the same phase as the reactants. (13.11)

Homogeneous equilibrium   An equilibrium within a single phase. (14.3)

Homonuclear diatomic molecule   A molecule composed of two identical atoms. (7.12)

Hund's rule   Every orbital in a subshell is singly occupied with one electron before any one orbital is doubly occupied, and all electrons in singly occupied orbitals have the same spin. (5.8)

Hybridization   A model that describes the changes in the atomic orbitals of an atom when it forms a covalent compound. (7.5)

Hydrocarbon   A compound composed only of hydrogen and carbon. The major component of fossil fuels. (4.7, 9.3)

Hydrogen bond   The strong electrostatic attraction that exists between molecules in which hydrogen is in a covalent bond with a highly electronegative element (fluorine, oxygen, or nitrogen). (11.4)

Hydronium ion (H_3O^+)   A water molecule with an added proton (hydrogen ion). (2.10)

Ideal gas   A gas that follows Boyle's law, Charles's law, and Avogadro's law perfectly. (10.2, 10.13)

Ideal gas equation $PV = nRT$. (10.6)

Ideal solution   A solution formed with no accompanying energy change, when the intermolecular attractive forces between the molecules of the solvent are the same as those between the molecules in the separate components. (12.2, 12.13)

Insulator   A crystal that has bands that are completely filled or completely empty, with large energy gaps between them. An insulator does not conduct electricity. (24.6)

Intensive property   A property of a substance that is independent of the amount of substance. (1.4)

Intermolecular force   The attractive force between two molecules. (11.2)

Internal energy (E)   The total of all possible kinds of energy present in a substance or substances. (4.4)

International System of Units (SI Units)   An updated version of the metric system used by scientists, adopted by the Institute of Standards and Technology in 1964. (1.9)

Interstitial site   A position between the regular positions in an array of atoms or ions that can be occupied by other atoms or ions. (11.13)

Intramolecular force   The attractive force between the atoms that make up a molecule. (11.2)

Ion   Charged particle resulting from the loss or gain of one or more electrons from an atom or a molecule. (2.6)

Ion activity   The effective concentration of any particular kind of ion in solution. It is less than indicated by the actual concentration of a solution. (12.20)

Ion product for water (K_w)   The equilibrium constant for the autoionization of water. At 25°C, $K_w = 1.00 \times 10^{-14}$. (15.1)

Ion–dipole attraction   The electrostatic attraction between an ion and the dipole of a molecule. (12.3)

Ionic bond   Bond due to the electrostatic attraction between positive and negative ions in an ionic compound. (6.1)

Ionic compound   A compound composed of ions. (6.1)

Ionic radius   The radius of an ion. (11.16)

Ionization constant   The equilibrium constant for the ionization of a weak acid or base. (15.3)

Ionization energy   The amount of energy required to remove an electron from a gaseous atom. (5.10)

Isoelectronic species   A group of ions, atoms, or molecules that have the same number of electrons. (5.10)

Isomorphous structures   Two elements or compounds that crystallize with the same structure. (11.14)

Isotopes   Atoms with the same atomic number and different numbers of neutrons. (2.4)

Joule   The SI unit of energy. One joule is the kinetic energy of an object with a mass of 2 kilograms moving with a velocity of 1 meter per second. $1 \text{ J} = 1 \text{ kg m}^2 \text{ s}^{-2}$ and $4.184 \text{ J} = 1 \text{ cal}$. (4.2)

K_a   The equilibrium constant for the reaction of an acid with water. (15.3)

K_b   The equilibrium constant for the reaction of a base with water. (15.3)

Kelvin (K)   The SI unit of temperature. 273.15 K = 0° Celsius. (1.10)

Kilogram   Standard SI unit of mass. Approximately 2.2 pounds. (1.10)

Kinetic energy (KE)   The kinetic energy of a moving body, in joules, is equal to $\frac{1}{2}mu^2$ (m = mass in kilograms; u = speed in meters per second). (10.14)

Kinetic-molecular theory   Theory that explains the properties of an ideal gas and assumes that such a gas consists of continuously moving molecules of negligible size. (10.13)

Lattice energy (U)   The energy required to separate the ions or molecules in a mole of a compound by infinite distances. (6.3, 11.19)

Lattice point   A point in a space lattice. (11.17)

Law of definite proportion   All samples of a pure compound contain the same elements in the same proportion by mass. (2.1)

Law of mass action   When a reversible reaction has attained equilibrium at a given temperature, the reaction quotient remains constant. (14.2)

Le Châtelier's principle   When a stress is applied to a system in equilibrium, the equilibrium shifts in a way that tends to minimize the effect of the stress. (14.4)

Lewis acid   Any species that can accept a pair of electrons and form a coordinate covalent bond. (15.13)

Lewis base   Any species that can donate a pair of electrons and form a coordinate covalent bond. (15.3)

Lewis structure   Diagram showing shared and unshared pairs of electrons in an atom, a molecule, or an ion. (6.6)

Lewis symbol   The symbol for an element or monatomic ion that uses a dot to represent each valence electron in the element or ion. (6.6)

Ligand   An ion or a neutral molecule attached to the central metal ion in a coordination compound. (23.6)

Limiting reagent   The reactant that is completely consumed by a chemical reaction. The amount of the limiting reagent limits the amount of product that can be formed. (3.12)

Liquid   The state in which matter takes the shape of its container, assumes a horizontal upper surface, and has a fairly definite volume. (1.2)

Liter (L)   Unit of volume. $1 \text{ L} = 1000 \text{ cm}^3$. (1.11)

London dispersion force   The attraction between two rapidly fluctuating, temporary dipoles. Significant only if atoms are very close together. (11.3)

Magnetic quantum number (m)   Quantum number signifying the orientation of an orbital around the nucleus. (5.6)

Mass   The quantity of matter contained by an object. Mass is measured in terms of the force required to change the speed or direction of its movement. (1.2)

Mass defect   The difference between the calculated mass and the experimental mass of a nucleus. (20.2)

Mass number (A)   The sum of the numbers of neutrons and protons in the nucleus of an atom. (2.4, 20.1)

Matter   Anything that occupies space and has mass. (1.2)

Melting point   The temperature at which the solid and liquid phases of a substance are in equilibrium. (11.8)

Metal   A substance that is malleable and ductile, has a characteristic luster, and is generally a good conductor of heat and electricity. (1.7) The bands of a metal are partially filled. (24.6)

Metallic conduction   The movement of electrons through a metal, with no changes in the metal and no movement of the metal atoms. (24.6)

Metathesis reaction   A reaction in which two or more compounds exchange parts. (2.10)

Meter   Standard metric and SI unit of length. Approximately 1.094 yards (1.10)

Miscibility   The ability of a liquid to mix with another liquid. (12.6)

Mixture   Matter that can be separated into its components by physical means. (1.5)

Molality (m)   The number of moles of solute dissolved in exactly 1 kilogram of solvent. (12.11)

Molar mass   Mass in grams of 1 mole of an element or compound. Numerically equal to the molecular mass of a molecule or the atomic mass of an atom. (3.2, 3.3)

Molarity (M)   The number of moles of solute dissolved in 1 liter of solution. (3.7)

Mole (mol)   The number of atoms contained in exactly 12 grams of ^{12}C (Avogadro's number). $1 \text{ mol} = 6.022 \times 10^{23}$ atoms, molecules, or ions. (3.2, 3.3)

Mole fraction (X)   The number of moles of a component of a solution divided by the total number of moles of all components. (12.2)

Molecular formula   A formula indicating the composition of a molecule of a compound and giving the actual number of atoms of each element in a molecule of the compound. (2.3)

Molecular mass (molecular weight)   The average mass of a molecule expressed in amu. (3.1)

Molecular orbital   The discrete energy and region of space in which an electron can be found around the nuclei of the atoms in a molecule. (7.12)

Molecular structure   The three-dimensional, geometrical arrangement of the atoms in a molecule. (7.1)

Molecule   A bonded collection of two or more atoms of the same or different elements. (1.6)

Monoprotic acid   An acid containing one ionizable hydrogen atom per molecule. (15.8)

Nernst equation   Used to calculate cell potentials at other than standard conditions. At $25°C$, $E = E° - RT/nF \ln Q$. (19.6)

Neutralization   A reaction that occurs when stoichiometrically equivalent quantities of an acid and a base are mixed. (2.10, 15.4)

Neutron   An uncharged subatomic particle with a mass of 1.0087 amu. (2.2)

Nonelectrolyte   A compound that does not ionize when dissolved in water. (12.3)

Nonmetal   An element that is a gas, is a liquid, or is brittle and nonductile as a solid; has no luster; and is a poor conductor of heat and electricity. (1.7)

Normal boiling point The temperature at which a liquid's vapor pressure equals 1 atm (760 torr). (11.6)

Normality (N) The number of equivalents of solute dissolved in 1 liter of solution. (15.12, 19.15)

Nuclear binding energy The energy produced by the loss of mass accompanying the formation of an atom from protons, electrons, and neutrons. (20.2)

Nuclear force The force of attraction between nucleons that holds a nucleus together. (20.1)

Nucleon Collective term for protons and neutrons. (20.1)

Nucleus The very heavy, positively charged body located at the center of an atom. (2.2)

Nuclide The nucleus of a particular isotope. (20.1)

Octahedral hole An octahedral space between six atoms or ions in a crystal. (11.15)

Optical isomers Molecules that are nonsuperimposable mirror images and that are optically active. (23.9)

Optically active molecules Molecules that rotate the plane of vibration of plane-polarized light. (23.9)

Orbital A three-dimensional region around the nucleus in which an electron moves; can hold up to two electrons. (5.5)

Order of a reaction With respect to one of the reactants, the order of a reaction is equal to the power to which the concentration of that reactant is raised in the rate equation. (13.4)

Osmosis The tendency of a solvent to diffuse through a semipermeable membrane from the less concentrated to the more concentrated solution. (12.18)

Osmotic pressure (π) The pressure required to stop the osmosis from a pure solvent into a solution. $\pi = MRT$. (12.18)

Oxidation The loss of electrons or an increase in oxidation state. (8.5)

Oxidation state The charge on the species that would result if the electrons in a bond were assigned to the more electronegative atom. (6.10)

Oxidation–reduction reaction Reaction in which oxidation numbers change as electrons are lost by one atom and gained by another. (8.5)

Oxidizing agent The substance in an oxidation–reduction reaction that gains electrons and the oxidation state of which is reduced. (8.5)

Oxyacid A hydroxide of a nonmetal. (2.11)

Paramagnetic substance A substance that contains unpaired electrons. Paramagnetic substances are attracted to a magnetic field. (6.8)

Parent nuclide An unstable nuclide that changes spontaneously into another (daughter) nuclide. (20.6)

Partial pressure The pressure exerted by an individual gas in a mixture of gases. (10.9)

Pauli exclusion principle No two electrons in the same atom can have the same set of four quantum numbers. (5.8)

Percent yield The actual yield of an experiment divided by the theoretical yield and multiplied by 100. (3.11)

Periodic law The properties of the elements are periodic functions of their atomic numbers. (2.5)

pH The negative logarithm of the concentration of hydronium ions in a solution. (16.2)

Photon A quantum of light or other electromagnetic radiation. The energy of a photon equals the product of Planck's constant and its frequency. (5.2)

Physical change A change in the state or properties of a particular kind of matter that does not involve a chemical change. (1.4)

Physical property A characteristic of a substance that can change without signaling a change of one kind of matter into another. Examples include color, hardness, and physical state. (1.4)

Pi (π) bond A covalent bond formed by side-by-side overlap of atomic orbitals. The electron density is found above and below the internuclear axis. (7.4)

Pi (π) orbital A molecular orbital formed by side-by-side overlap of atomic orbitals, in which the electron density is found above and below the internuclear axis. (7.14)

pOH The negative logarithm of the concentration of hydroxide ions in a solution. (16.2)

Polar covalent bond Covalent bond between atoms of different electronegativities; a covalent bond with a positive end and a negative end. (6.5)

Polydentate ligand A ligand that is attached to a central metal ion by bonds from several donor atoms. (23.6)

Polymer A compound of high molecular mass that is built up of a large number of simple molecules, or monomers. (9.15)

Polymorphism The assumption of two or more crystal structures by the same substance. (11.17)

Positron An atomic particle with the same mass as an electron but with one unit of positive charge. (20.5)

Pressure Force exerted on a unit area. The SI unit of pressure is the pascal (Pa). Pa = 1 newton m^{-2}. (10.1)

Principal quantum number (n)   Quantum number specifying the shell of an electron in an atom or a monatomic ion. (5.6)

Proton   A nuclear particle that has a mass of 1.0073 amu and carries a charge of $+1$. (2.2)

Quantized   A description of the discrete, or individual, values by which the energy of an electron can vary. (5.3)

Rad (radiation absorbed dose)   A unit of radiation dosage; the amount of radiation that deposits 1×10^2 J of energy per kilogram of tissue. (20.12)

Radioactive decay   The spontaneous change of an unstable nuclide (parent) into another nuclide (daughter) with the simultaneous emission of rays or particles. (20.6)

Radius ratio   In an ionic compound, the radius of the positive ion, $r+$, divided by the radius of the negative ion, $r-$. (11.16)

Raoult's law   The vapor pressure of the solvent in an ideal solution (P_{solv}) is equal to the mole fraction of the solvent (X_{solv}) times the vapor pressure of the pure solvent ($P°_{solv}$). (12.13)

Rate constant   The proportionality constant in the relationship between reaction rate and concentrations of reactants. (13.4)

Rate equations   Equations giving the relationship between reaction rate and concentrations of reactants. (13.4)

Rate-determining step   The slowest elementary reaction in a reaction path, which determines the maximum rate of the overall reaction. (13.10)

Reaction mechanism   The stepwise sequence of elementary reactions in an overall reaction. (13.9)

Reaction quotient (Q)   A ratio of the product of molar concentrations (or pressures) of the products to that of the reactants, each concentration (or pressure) being raised to the power equal to the coefficient in the equation. For the reaction $a\text{A} + b\text{B} + \cdots \rightarrow c\text{C} + d\text{D} + \cdots$, $Q = [\text{C}]^c[\text{D}]^d \ldots /[\text{A}]^a[\text{B}]^b \ldots$ (14.2)

Recombinant DNA   DNA produced by cutting a DNA chain and inserting particular pieces from another organism, changing the characteristics of the original DNA. (25.13)

Reducing agent   The substance in an oxidation–reduction reaction that gives up electrons and the oxidation state of which is increased. (8.5)

Reduction   The gain of electrons or a decrease in oxidation state. (8.5)

Rem (roentgen equivalent in man)   A unit of radiation dosage that includes a biological factor referred to as the RBE (relative biological effectiveness). rems = RBE $\times$ rads. (20.12)

Resonance forms   Two or more Lewis structures that have the same arrangement of atoms but different arrangements of electrons. (6.11)

Resonance hybrid   The average of the resonance forms shown by the individual Lewis structures. (6.11)

Reversible reaction   A chemical reaction that can proceed in either direction. (14.1)

Salt   An ionic compound composed of cations and anions other than hydroxide or oxide ions. (8.5)

Saturated solution   A solution in which no more solute can be dissolved. (12.1)

Second law of thermodynamics   Any spontaneous change that occurs in the universe must be accompanied by an increase in the entropy of the universe. (18.13)

Semiconductor   A substance that contains a full band and an empty band, with small energy gaps between the bands. It is a poor conductor. (24.6)

Semi-metals   Substances that possess some of the properties of both metals and nonmetals. (1.7)

Shell   All of the orbitals in an atom or monatomic ion with the same value of n. (5.6)

Sigma (σ) bond   A covalent bond formed by overlap of atomic orbitals along the internuclear axis. The electron density is found along the axis of the bond. (7.4)

Sigma (σ) orbital   A molecular orbital in which the electron density is found along the axis of the bond. (7.14)

Simple cubic structure   A crystalline structure with a cubic unit cell with lattice points only on the corners. (11.14)

Single bond   A bond in which a single pair of electrons is shared between two atoms. (6.6)

Solid   The state in which matter is rigid, has a definite shape, and has a fairly constant volume. (1.2)

Solid solution   A homogeneous and stable solution of one solid substance in another. (12.8)

Solubility product (K_{sp})   The equilibrium constant for the dissolution of a slightly soluble electrolyte. (17.1)

Solution   A homogeneous mixture of a solute in a solvent. (3.7)

Space lattice   All points within a crystal that have identical environments. (11.17)

Specific heat   A property of a substance that represents the quantity of heat required to raise the temperature of 1 gram of the substance 1 degree Celsius (or 1 kelvin). (4.2)

Spectrum   The component colors, or wavelengths, of light or other forms of electromagnetic radiation. (5.2)

Spin quantum number (s)   Number specifying the direction of the spin of an electron around its own axis. (5.6)

Spontaneous process   A physical or chemical change that occurs without the addition of energy. (12.2) $\Delta G < 0$ for a spontaneous process. (18.5)

Stability curve   A plot of the number of neutrons versus the number of protons for stable nuclei. (20.3)

Standard conditions (STP)   273.15 K (0°C) and 1 atmosphere of pressure (760 torr or 101.325 kilopascals). (10.7)

Standard electrode potential ($E°$)   Potential measured with respect to a standard hydrogen electrode at 25°C with 1 M concentration of each ion in solution and 1 atmosphere of pressure of each gas involved. (20.3)

Standard enthalpy of formation ($\Delta H_f°$)   The enthalpy change of a chemical reaction in which 1 mole of a pure substance is formed from the free elements in their most stable states under standard state conditions. (4.5)

Standard entropy ($S_{298}°$)   Actual entropy content of 1 mole of a substance in a standard state. (18.8)

Standard hydrogen electrode   Assigned an electrode potential of exactly zero. The potential of all other electrodes is reported relative to that of the standard hydrogen electrode. (13.3)

Standard state   298.15 K (25°C) and 1 atmosphere of pressure. (4.5)

State function   A property of a system that is not dependent on the way the system gets to the state in which it exhibits that property. (18.4)

Strong acid   An acid that gives a 100% yield of hydronium ions when dissolved in water. (15.3)

Strong base   A base that gives a 100% yield of hydroxide ions when dissolved in water. (15.3)

Strong electrolyte   An electrolyte that gives a 100% yield of ions when dissolved in water. (12.3)

Structural isomers   Two substances that have the same molecular formula but have different physical and chemical properties because their component atoms are arranged differently. (9.3)

Sublimation   The passing of a solid directly to the vapor state without first melting. (11.5)

Subshell   A set of degenerate orbitals with the same values of n and l. (5.6)

Substitution reaction   A reaction in which one atom replaces another in a molecule. (9.3)

Supersaturated solution   A solution that contains more solute than it would if the dissolved solute were in equilibrium with the undissolved solute. (12.1)

Surface tension   The force that causes the surface of a liquid to contract, reducing its surface area to a minimum. (11.11)

Surroundings   The universe outside a thermodynamic system. (4.4)

System   The substance or substances involved in a reaction or change that is being studied. (4.4, 18.1)

Termolecular reaction   An elementary reaction involving the simultaneous collision of any combination of three molecules, ions, or atoms. (13.9)

Ternary compound   A compound containing three different elements. (2.11)

Tetrahedral hole   A tetrahedral space formed by four atoms or ions in a crystal. (11.15)

Theoretical yield   The calculated yield of a reaction based on the assumptions that there is only one reaction involved, that all the reactant is converted into product, and that all the product is collected. (3.11)

Third law of thermodynamics   The entropy of any pure, perfect crystalline element or compound at absolute zero (0 K) is equal to zero. (18.10)

Three-center bond   The bonding of three atoms by one pair of electrons in a molecular orbital formed from the overlap of three atomic orbitals. (22.4)

Titration   Method of determining concentration by adding a solution of a reactant of known concentration to a solution of sample until an indicator changes color. (3.13)

***Trans* configuration**   Configuration of a geometrical isomer in which two groups are on opposite sides of an imaginary reference line on the molecule. (9.3, 23.9)

Transition state   A combination of reacting molecules that is intermediate between reactants and products. (13.7)

Triple bond   A bond in which three pairs of electrons are shared between two atoms. (6.6)

Triple point   The point at which an equilibrium exists among the vapor, liquid, and solid phases of a substance. (11.10)

Triprotic acid   An acid that contains three ionizable hydrogen atoms per molecule. Ionization of triprotic acids occurs in three stages. (15.8)

Unimolecular reaction   An elementary reaction in which the rearrangement of a single molecule produces one or more molecules of product. (13.9)

Unit cell   The portion of a space lattice that is repeated in order to form the entire lattice. (11.17)

Unsaturated solution   A solution in which more solute can be dissolved. (12.1)

Unshared pair   Electrons not used to form a covalent bond. (6.6)

Valence electrons   Electrons in the valence shell of an atom. The number of valence electrons determines how an element reacts. (5.9)

Valence shell   The outermost shell of electrons of a representative element; the outermost shell of electrons and the d electrons in the next inner shell of a d-block element; or the outermost shell of electrons, the d electrons in the next inner shell and the f electrons in the next inner shell of an f-block element. (5.9)

Valence shell electron-pair repulsion (VSEPR) theory   A theory used to predict the bond angles in a molecule, based on the positioning of regions of high electron density as far apart as possible to minimize electrostatic repulsion. (7.1)

Van der Waals equation   A quantitative expression of the deviations of real gases from the ideal gas laws. (10.17)

Van der Waals force   Intermolecular attractive force. (11.2)

Vapor pressure   The pressure exerted by a vapor in equilibrium with a solid or a liquid at a given temperature. (11.5)

Vapor-pressure lowering   The lowering of the vapor pressure of a liquid by addition of a solute. (12.13)

Volt (V)   Difference in electrical potential when 1 joule of energy is required to move 1/96,485 mole of electrons (1 coulomb of charge) from a lower potential to a higher potential. (19.2)

Wave function (ψ)   A mathematical function that describes the shape of the orbital that an electron occupies, the energy of the electron in the orbital, and the probability of finding the electron at any given location in the orbital. (5.5)

Weak acid   An acid that gives about 10% or less yield of hydronium ions when dissolved in water. (15.3)

Weak base   A base that gives about 10% or less yield of hydroxide ions when dissolved in water. (15.3)

Work (w)   One process for removing energy from a system or adding energy to it. (4.4, 18.2)

Yield   The quantity of a product of a chemical reaction. (3.11)

Answers to Odd-Numbered Exercises

Chapter 1

1. Study of the composition, structure, and properties of matter and the changes from one form of matter to another. **3.** Liquids can change their shape (flow); solids don't. Gases can undergo large volume changes as the pressure changes; liquids don't. Gases flow and change volume; solids don't. **5.** The mixture can have a variety of compositions, a pure substance has a definite composition. Both have the same composition from point to point. **7.** Molecules of elements contain only one type of atom; molecules of compounds contain two or more types of atoms. **9.** Place a glass of water outside. It will freeze if the temperature is below 0°C. **11.** There are many possible answers. Some include a penny (solid), milk (liquid), air (gas), gold or iron (element), water or salt (compound), cereal with milk (heterogeneous mixture), Coca Cola or Pepsi (homogeneous mixture), gold, sugar, aluminum foil (pure substance). **13.** Gasoline (a mixture of compounds), oxygen, and to a lesser extent, nitrogen are consumed. Carbon dioxide and water are the principal products. Carbon monoxide and nitrogen oxides are produced in lesser amounts. **15.** physical, chemical, physical, physical, chemical, chemical **17.** a. element; b. compound; c. element; d. element; e. compound; f. element **19.** a. physical; b. chemical; c. chemical; d. physical; e. physical **21.** physical **23.** A sulfur molecule is composed of eight sulfur atoms. **25.** a. macroscopic; b. microscopic; c. symbolic, macroscopic; d. symbolic, microscopic **27.** microscopic. The behavior is explained in terms of the behavior of microscopic particles (molecules). **29.** 3A, 4A, 5A, 6A **31.** Mg **33.** Ca, Fe, K **35.** F, Br, I, At **37.** beryllium, magnesium, calcium, strontium, barium, radium **39.** scandium, yttrium, lanthanum, actinium **41.** a yard **43.** a. meters; b. kilograms; c. cubic meters; d. meters/second; e. kilograms/cubic meter; f. square meters; g. kelvins **45.** a. 1.3×10^{-4} kg; b. 2.32×10^{8} kg; c. 5.23×10^{-12} m; d. 8.63×10^{-5} kg; e. 3.76×10^{-1} m; f. 5.4×10^{-5} m; g. 1×10^{12} s; h. 2.7×10^{-11} s; i. 1.5×10^{-4} K **47.** kilo, k; centi, c; deci, d; milli, m; mega, M; micro, μ **49.** 5.0×10^{-3} mL **51.** a. 7.110×10^{2}; b. 2.39×10^{-1}; c. 9.0743×10^{4}; d. 1.342×10^{2}; e. 5.499×10^{-2}; f. 1.00000×10^{4}; g. 7.38592×10^{-7} **53.** Exact: a, d. Uncertain: b, c, e, f **55.** a. 3; b. 1; c. 7; d. 6; e. 3; f. 3; g. 4 **57.** a. 0.44; b. 9.0; c. 27; d. 140; e. 1.5×10^{-3}; f. 0.44 **59.** a. 2.15×10^{5}; b. 4.2×10^{6}; c. 2.08; d. 0.19; e. 27440; f. 43.0 **61.** 1.094 yd/1 m; 0.9463 L/1 qt; 2.205 lb/1 kg **63.** 69–71 cm; 400–450 g **65.** 355 mL **67.** 8×10^{-4} cm **69.** Yes. weight = 89.4 kg **71.** 45.4 L **73.** 2.2×10^{8} kg **75.** a. 8.903 m; b. 10 km; c. 2.5118×10^{5} km^{2}; d. 118.3 mL; e. 1.4×10^{9} km^{3}; f. 1599 kg; g. 65 g **77.** 5.0 km, 3.1 mi **79.** 2.70 g/cm^{3} **81.** a. 81.6 g; b. 17.6 g **83.** a. 5.1 mL; b. 37 L **85.** 5371°F; 3239 K **87.** −23°C; 250 K **89.** −33.4°C; 239.8 K **91.** a. 2; b. exact; c. 1; d. 2; e. 3; f. 3; g. 1; h. 4 **93.** 28.3 g = (exactly) 1 oz, 3 significant figures **95.** Yes, the acid's volume is 123 mL. **97.** 113°F **99.** 113.61 mi/h; 50.786 m/s; 166.62 ft/s **101.** 1.0×10^{-5} g

Chapter 2

1. a **3.** The symbol represents one atom of oxygen, a molecule of oxygen contains two oxygen atoms **5.** d **7.** d, e **9.** a **11.** Atoms are neither created nor destroyed during a chemical change; they rearrange. **13.** Atoms are neither created nor destroyed during a chemical change; they rearrange. (This explains why the mass of the reactant equals the mass of the products.) **15.** a. 5 protons, 5 electrons, 5 neutrons; b. 80 protons, 80 electrons, 119 neutrons; c. 29 protons, 29 electrons, 34 neutrons; d. 6 protons, 6 electrons, 7 neutrons; e. 34 protons, 34 electrons, 43 neutrons **17.** 35.46 amu **19.** Sodium, potassium, iron, mercury, silver, copper, lead, antimony, tin, gold **21.** Ca, Cl, I, Mg, P, Na, Zn **23.** CO_2, CO_2; C_2H_2, CH; C_2H_4, CH_2; H_2SO_4, H_2SO_4 **25.** a. $C_4H_5N_2O$; b. $C_{12}H_{22}O_{11}$; c. HO; d. CH_2O; e. $C_3H_4O_3$ **27.** a. $^{23}_{11}Na$; b. $^{129}_{54}Xe$; c. $^{73}_{33}As$; d. $^{226}_{88}Ra$ **29.** Ionic: KCl, $MgCl_2$. Covalent: NCl_3, ICl, PCl_5, CCl_4 **31.** a. He; b. Be; c. Li; d. O **33.** Metal: a, c, e, f, g, j; nonmetal: b, d, h, i; representative element: b, c, d, h, i, j; transition metal, e, g; inner transition metal: a, f **35.** a. Kr; b. Ca; c. F; d. Te **37.** Ionic: BaO (Ba^{2+}, O^{2-}), $(NH_4)_2CO_3$ (NH_4^+, CO_3^{2-}), $Sr(H_2PO_4)_2$ (Sr^{2+}, $H_2PO_4^-$), Na_2O (Na^+, O^{2-}); covalent: NF_3, IBr **39.** a. RbBr; b. MgSe; c. Na_2O; d. $CaCl_2$; e. HF; f. GaP; g. $AlBr_3$; h. $(NH_4)_2SO_4$ **41.** a. Cs^+; b. I^-; c. P^{3-}; d. Co^{3+} **43.** A reaction is part of the microscopic or macroscopic world. An equation is a symbol. **45.** a. 1,1,1,2; b. 3,8,3,4,2; c. 1,1,2; d. 4,3,2; e. 2,2,2,1; f. 1,1,1,4; g. 1,6,4; h. 1,1,2 (Coefficients of 1 are usually not written.) **47.** a. $CaCO_3(s) \xrightarrow{\Delta} CaO(s) + CO_2(g)$, b. $2C_4H_{10}(g) + 13O_2(g) \longrightarrow 8CO_2(g) + 10H_2O(g)$, c. $MgCl_2(aq) + 2NaOH(aq) \longrightarrow Mg(OH)_2(s) + 2NaCl(aq)$, d. $2H_2O(g) + 2Na(s) \longrightarrow 2NaOH(s) + H_2(g)$ **49.** a. addition; b. neutralization; c. combustion; d. decomposition; e. addition **51.** Cesium chloride, barium oxide, potassium sulfide, beryllium chloride, hydrogen bromide, aluminum fluoride **53.** ClO_2, N_2O_4, K_3P, Ag_2S, AlN, SiO_2 **55.** a. chromium(III) oxide; b. iron(II) chloride; c. chromium(VI) oxide; d. titanium(IV) chloride; e. cobalt(II) oxide; f. molybdenum(IV) sulfide **57.** a. K_3PO_4; b. $CuSO_4$; c. $CaCl_2$; d. TiO_2; e. NH_4NO_3; f. $NaHSO_3$ **59.** a. Iron, 26 protons, 26 electrons, and 32 neutrons; b. iodine, 53 protons, 53 electrons, and 74 neutrons **61.** manganese(IV) oxide, mercury(I) chloride, iron(III) nitrate, titanium(IV) chloride, copper(II) bromide **63.** 79.90 **65.** mineral, 6.94; military, 6.978 **67.** $(NH_4)_3PO_4$, KNO_3, $CuSO_4$, $ZnSO_4$ **69.** CH_2O, C_2H_4O **71.** a. $4HF(aq) + SiO_2(s) \longrightarrow SiF_4(g) + 2H_2O(l)$, b. $2Na^+(aq) + 2F^-(aq) + Ca^{2+}(aq) + 2Cl^-(aq) \longrightarrow CaF_2(s) + 2Na^+(aq) + 2Cl^-(aq)$, c. $2F^-(aq) + Ca^{2+}(aq) \longrightarrow CaF_2(s)$ **73.** a. $CaCO_3 \longrightarrow CaO + CO_2$, b. $CaO + H_2O \longrightarrow Ca(OH)_2$, c. $MgCl_2 + Ca(OH)_2 \longrightarrow Mg(OH)_2 + CaCl_2$, d. $Mg(OH)_2 + 2HCl \longrightarrow MgCl_2 + 2H_2O$, e. $MgCl_2 \longrightarrow Mg + Cl_2$

Chapter 3

1. The two masses have the same numerical value, but the units are different: the molecular mass is the mass of 1 molecule while the molar mass is the mass of 6.022×10^{23} molecules.   **3.** The mole. One mole of a substance (atoms, molecules, ions, or formula units) is the amount that contains a number of atoms, molecules, ions, or formula units that is equal to the number of atoms in exactly 12 g of the isotope carbon-12.   **5.** Determine the mass of one mole of the compound from its formula and the molar masses of its atoms, then determine the mass of the sample of the compound from the number of moles of compound in the sample.   **7.** Its empirical formula and its molar mass (or molecular mass).   **9.** The number of moles of sulfuric acid dissolved in the solution and the volume of the solution.   **11.** The mass and number of moles of $Cu(NO_3)_2$ stay the same. The volume of the solution and the concentration change.   **13.** a. The initial mass of sulfadiazine, molecular formulas, and molar masses of both sulfadiazine and silver sulfadiazine, and the chemical equation relating the mole ratio between sulfadiazine and silver sulfadiazine. b. The theoretical yield and the experimentally determined actual yield. c. Atomic masses, the chemical formulas of sulfadiazine and silver oxide, the masses of both reactants, and the mole ratio of these reactants in the equation for the reaction.   **15.** 100%   **17.** $CaSO_4$ **19.** In a given compound, the number of atoms of each type always has the same ratio. Atoms are neither created nor destroyed during a chemical change; they are redistributed.   **21.** a. 123.896 amu, b. 18.015 amu, c. 164.086 amu, d. 60.052 amu, e. 342.297 amu   **23.** a. 100.086 amu, 100.086 g/mol; b. 58.443 amu, 58.443 g/mol; c. 537.498 amu, 537.498 g/mol, d. 221.114 amu, 221.114 g/mol, e. 813.434 amu, 813.434 g/mol   **25.** a. 98.916 g/mol, b. 26.038 g/mol, c. 185.846 g/mol, d. 98.078 g/mol   **27.** 1.327×10^{-22} g   **29.** C_2H_4 **31.** 1 g H_2O (0.06 mol) contains the greatest mass of O atoms. It contains 3×10^{22} O atoms; one molecule of C_2H_5OH contains one O atom; one molecule of CO_2, two O atoms   **33.** a. F (largest atomic mass), b. O_3 (largest molecular mass)   **35.** i. 20.006 g, 1.205×10^{25} amu; ii. 17.031 g, 1.026×10^{25} amu; iii. 63.012 g, 3.795×10^{25} amu; iv. 311.798 g, 1.878×10^{26} amu; v. 61.832 g, 3.724×10^{25} amu   **37.** i. 0.594 mol C_3H_6, 1.78 mol C, 3.56 mol H; ii. 4.08×10^{-5} mol $C_2H_5NO_2$, 8.16×10^{-5} mol C and mol O, 2.04×10^{-4} mol H, 4.08×10^{-5} mol N; iii. 40.1 mol $C_{13}H_{16}N_2O_4F$, 521 mol C, 642 mol H, 80.2 mol N, 160 mol O, 40.1 mol F; iv. 0.202 mol $Cu_4(AsO_3)_2(CH_3CO_2)_2$, 0.808 mol Cu and mol C, 0.404 mol As, 1.21 mol H, 2.02 mol O; v. 1.80×10^{-3} mol $C_6H_4(CO_2H)(CO_2CH_3)$, 1.62×10^{-2} mol C and 1.44×10^{-2} mol H, 7.20×10^{-3} mol O   **39.** a. 0.819 g, b. 307 g, c. 0.23 g, d. 1.235×10^6 g (1235 kg), e. 765 g   **41.** 8.4×10^{-3} mol   **43.** 4.5×10^{21} molecules, 4.8×10^2 g, 1.1 lb   **45.** 1 mol CH_4 **47.** 0.10 mol of glucose   **49.** 3.113×10^{25} C atoms   **51.** 2.38×10^{-4} mol   **53.** 0.865 servings   **55.** a. 82.24% N, 17.76% H; b. 29.08% Na, 40.56% S, 30.36% O; c. 38.76% Ca^{2+} **57.** 38.2% NH_3 **59.** a. CS_2, CH_2O **61.** C_6H_6 **63.** $Na_2S_2O_3(H_2O)_5$ **65.** 55.0%   **67.** a. 0.679 M, b. 1.00 M, c. 0.06998 M, d. 1.75 M, e. 0.070 M, f. 6.6 M   **69.** a. 4.0 g, b. 3.4×10^{-2} g, c. 0.39 g   **71.** a. 1.00 mol, 164 g; b. 0.118 mol, 7.11 g; c. 0.012 mol, 1.5 g; d. 0.012 mol, 4.2 g   **73.** a. 0.250 L, b. 0.1718 L, c. 4.12 L, d. 1.9 mL, e. 0.794 M   **75.** 0.148 M   **77.** 11.9 M   **79.** i. a. $2Na + Cl_2 \longrightarrow 2NaCl$, Mass of Na $\longrightarrow$ Moles of Na $\longrightarrow$ Moles of Cl_2 $\longrightarrow$ Mass of Cl_2, b. 0.217 mol Cl_2, 15.4 g Cl_2; ii. a. $2HgO \longrightarrow 2Hg + O_2$, Mass of HgO $\longrightarrow$ Moles of HgO $\longrightarrow$ Moles of O_2 $\longrightarrow$ Mass of O_2, b. 2.890×10^{-3} mol

O_2, 9.248×10^{-2} g O_2; iii. a. $2NaNO_3 \longrightarrow 2NaNO_2 + O_2$, Mass of O_2 $\longrightarrow$ Moles of O_2 $\longrightarrow$ Moles of $NaNO_3$ $\longrightarrow$ Mass of $NaNO_3$, b. 8.00 mol $NaNO_3$, 6.80×10^2 g $NaNO_3$; iv. a. $C + O_2$ $\longrightarrow CO_2$, Mass of C $\longrightarrow$ Moles of C $\longrightarrow$ Moles of CO_2 $\longrightarrow$ Mass of CO_2, b. 1.67×10^3 mol CO_2, 73.3 kg CO_2; v. a. $CuCO_3$ $\longrightarrow CuO + CO_2$, Mass of CuO $\longrightarrow$ Moles of CuO $\longrightarrow$ Moles of $CuCO_3$ $\longrightarrow$ Mass of $CuCO_3$, b. 18.86 mol $CuCO_3$, 2.329 kg $CuCO_3$; vi. a. $C_2H_4 + Br_2 \longrightarrow C_2H_4Br_2$, Mass of C_2H_4 $\longrightarrow$ Moles of C_2H_4 $\longrightarrow$ Moles of $C_2H_4Br_2$ $\longrightarrow$ Mass of $C_2H_4Br_2$, b. 0.4580 mol $C_2H_4Br_2$, 86.05 g $C_2H_4Br_2$ **81.** a. Volume of H_3PO_4 solution $\longrightarrow$ Moles of H_3PO_4 $\longrightarrow$ Moles of H_2 $\longrightarrow$ Mass of H_2; b. 0.1560 mol H_2, 0.3144 g H_2 **83.** 1.275×10^{23} molecules, 53.74 g   **85.** 8.11 g   **87.** 712 g   **89.** 92%   **91.** 1036 kg   **93.** 5.1×10^2 g, 25%   **95.** Determine the moles of CO_2 produced by 30.0 g of propane. Determine the moles of CO_2 produced by 75.0 g of oxygen. The limiting reagent is the one that produces the smaller amount of CO_2; that is, O_2.   **97.** Li   **99.** 8   **101.** 4   **103.** a. O_2, b. 88.1%   **105.** 238 mL   **107.** 0.3374 M   **109.** Yes, concentration is 4.21 M   **111.** Determine the mass of Cl atoms in the AgCl, determine the mass of I atoms in the 0.6548 g sample by subtracting the mass of Cl atoms, determine the numbers of moles of Cl and moles of I in the 0.6548 g sample, determine the simplest mole ratio; ICl_3 **113.** BH_3, B_2H_6 **115.** a, b, d   **117.** a. 1.0×10^2 mg, b. 3.2×10^{21} F atoms   **119.** 4.8×10^{-6} M   **121.** No, 2.1 kg is required   **123.** 0.0055 M   **125.** a. 1.2×10^2 g, b. 9.0 mol, c. S, d. 24.5 g   **127.** a. 0.155 M; b. i. 68.8 mL, ii. 64.5 mL, iii. 59.9 mL, iv. 66 mL

Chapter 4

1. The 2 L of water lost more heat than the 500 mL.   **3.** Smaller, more heat would be lost to the environment and the temperature increase would be less.   **5.** The enthalpy change is for exactly 1 mol HCl and 1 mol NaOH; the heat in the example is produced by 0.0500 mol HCl and 0.0500 mol NaOH.   **7.** The heat converts solid water to liquid water, it does not increase the kinetic energy of the liquid water (and does not increase the temperature of the liquid).   **9.** Separate the nitrogen, adjust its pressure to 1 atm, and adjust its temperature to 25°C.   **11.** Improve the measurement of the amount of heat by determining the amount of heat absorbed by the calorimeter.   **13.** 10.9°C   **15.** 5.11×10^4 J, 1.22×10^4 cal.   **17.** a. 119 J, 28.4 cal; b. 3.66 J, 0.876 cal   **19.** 170 mL   **21.** 5.7×10^2 kJ   **23.** 2.2 kJ, exothermic   **25.** 1.4 kJ   **27.** 6.9°C. Twice as much heat would be produced, but it would be absorbed by twice as much solution.   **29.** 1.1 kJ, 0.28 J $g^{-1} °C^{-1}$ **31.** 5.02×10^3 kJ   **33.** 1.83×10^{-2} mol   **35.** 796 kJ mol^{-1} **37.** 26 kJ mol^{-1} (This is a positive heat of solution because the process is endothermic)   **39.** 140 kJ mol^{-1} **41.** 462.7 kJ   **43.** 9.5 kJ   **45.** 367 g   **47.** Formation of the solid   **49.** −982.8 kJ   **51.** 891.2 kJ   **53.** 44.01 kJ   **55.** a. −1614.9 kJ, b. −484.5 kJ, c. 164 kJ, d. −238 kJ   **57.** a. −2044 kJ, b. 2657 kJ mol^{-1}, c. propane   **59.** −124.9 kJ   **61.** 3.7 kg   **63.** 1.2×10^2 g, 0.27 lb   **65.** 30%   **67.** 3.9 g   **69.** a. 82°C, b. The temperature of the coffee cannot increase when a colder spoon is placed in it.   **71.** 7.5 kWh, assuming that the density of water is a 1.0 g/cm³ and that it takes as much energy to keep the water at 85°F as to heat it from 72°F to 85°F.   **73.** −15.6 kJ   **75.** B_2H_6 **77.** isooctane 7.3×10^4 kJ/$, cereal, 4.1×10^3 kJ/$. Isooctane is the least expensive source.   **79.** 1.4 kJ assuming that

ΔH°_{298} is not different from ΔH under the conditions of the reaction. 1.2°C, assuming no heat escapes.   **81.** −88.1 kJ **83.** a. $C_3H_8(g) + 5O_2(g) \longrightarrow 3CO_3(g) + 4H_2O(l)$, b. 133 L, c. −101.6 kJ, d. 44.2°C

Chapter 5

1. Both are charged particles that are components of an atom. Although charges are the same size, the signs of the charges are opposite. Protons are 1836 times heavier than electrons.   **3.** The spectrum consists of colored lines, one of which (probably the brightest) is red.   **5.** On average the electron with $n = 3$ is farther from the nucleus and the average electrostatic attraction between it and the nucleus is less than for an electron with $n = 1$.   **7.** Wave-like: a, d; particle-like: b, c; both: e   **9.** A microscopic property; it describes a property of an atom.   **11.** The charge on the ion.   **13.** Identified a nucleus as part of the atomic structure.   **15.** Both involve a relatively heavy nucleus with electrons around it. The Rutherford model says nothing about the behavior of the electrons. The Bohr model describes electron behavior in terms of quantized energies and orbits.   **17.** 2.908 m **19.** 4.558×10^{-19} J, 2.854 eV   **21.** 4.469×10^{14} s^{-1}, 6.709×10^{-7} m, 178 kJ mol^{-1}, red   **23.** a. 3.8×10^{14} s^{-1}; b. 7.1×10^{14} s^{-1}, blue to violet   **25.** a. 13.595 eV, b. 13.60 eV (difference is due to the number of significant figures in the equation)   **27.** 122 eV   **29.** 0.3778 eV   **31.** 8.46 Å   **33.** 2.856 eV, 4.576×10^{-19} J **35.** a. 6; b. low: 0.660 eV, high: 12.745 eV; c. low: 1.60×10^{14} s^{-1}, 1.88×10^{-6} m, high: 3.082×10^{15} s^{-1}, 9.728×10^{-8} m **37.** A shell consists of one or more subshells; each subshell contains a different type of orbital. A subshell consists of one or more orbitals of the same type with the same energy. An orbital is a region of space that can be occupied by a maximum of two electrons.   **39.** $n = 1 \ldots \infty$, $l = 0 \ldots n - 1$, $m_l = -l \ldots l$, $s = \pm\frac{1}{2}$ **41.** a. $2p$, b. $4d$, c. $6s$ **43.** a. 3, b. 5 **45.** $n = 3$, $l = 0$, $m_l = 0$, $s = \frac{1}{2}$
$n = 3$, $l = 0$, $m_l = 0$, $s = -\frac{1}{2}$
$n = 3$, $l = 1$, $m_l = -1$, $s = \frac{1}{2}$
$n = 3$, $l = 1$, $m_l = -1$, $s = -\frac{1}{2}$
$n = 3$, $l = 1$, $m_l = 0$, $s = \frac{1}{2}$
$n = 3$, $l = 1$, $m_l = 0$, $s = -\frac{1}{2}$
$n = 3$, $l = 1$, $m_l = 1$, $s = \frac{1}{2}$
$n = 3$, $l = 1$, $m_l = 1$, $s = -\frac{1}{2}$
$n = 3$, $l = 2$, $m_l = -2, -1, 0, 1,$ or 2, $s = \pm\frac{1}{2}$
47.

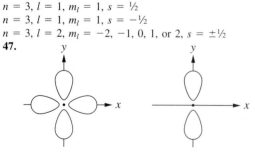

49. a. i. 2, ii. 2, iii. 2; b. i. 1, ii. 3, iii. 0; c. i. 4 0 0 ½, ii. 2 1 0 ½, iii. 3 2 0 ½; d. i. 1, ii. 2, iii. 3; e. i. $l = 0$, $m_l = 0$, ii. $l = 1$, $m_l = -1, 0,$ or +1, iii. $l = 2$, $m_l = -2 -1 0 +1 +2$ **51.** a. $1s^2 2s^2 2p^2$
b. $1s^2 2s^2 2p^6 3s^2 3p^3$
c. $1s^2 2s^2 2p^6 3s^2 3p^6 4s^2 3d^3$
d. $1s^2 2s^2 2p^6 3s^2 3p^6 3d^{10} 4s^2 4p^6 4d^{10} 5s^2 5p^3$
e. $1s^2 2s^2 2p^6 3s^2 3p^6 3d^{10} 4s^2 4p^6 4d^{10} 5s^2 5p^6 6s^2 4f^6$

53. a.   ↑↓ | ↑ ↑ |
         $2s$ $2p$

b.   ↑↓ | ↑ ↑ ↑ |
         $3s$ $3p$

c.   ↑↓ | ↑ ↑ ↑ | |
         $4s$ $3d$

d.   ↑↓ | ↑ ↑ ↑ |
         $5s$ $5p$

e.   ↑↓ | ↑↓ ↑ ↑ ↑ ↑ |
         $5s$ $4d$

55. a. $1s^2 2s^2 2p^6$
b. $1s^2 2s^2 2p^6 3s^2 3p^6$
c. $1s^2 2s^2 2p^6 3s^1$
d. $1s^2 2s^2 2p^4$
e. $1s^2 2s^2 2p^6 3s^2 3p^6 3d^5$
f. $1s^2 2s^2 2p^6 3s^2 3p^6 3d^{10} 4s^2 4p^6 4d^{10} 5s^2 5p^6 4f^5$
57. Zr   **59.** Tc^{2+}, Ru^{3+} **61.** B   **63.** Bi   **65.** Co^{2+}
$1s^2 2s^2 2p^6 3s^2 3p^6 3d^7$, Co^{3+} $1s^2 2s^2 2p^6 3s^2 3p^6 3d^6$ **67.** Cl   **69.** O **71.** Rb < Li < N < F   **73.** 5A   **75.** Mg < Ca < Rb < Cs **77.** Si^{4+} < Al^{3+} < Ca^{2+} < K$^+$ **79.** Se, As$^-$ **81.** Mg^{2+} < K$^+$ < Br$^-$ < As^{3-} **83.** Examples include Na$^+$ $1s^2 2s^2 2p^6$
Ca^{2+} $1s^2 2s^2 2p^6 3s^2 3p^6$
Sn^{2+} $1s^2 2s^2 2p^6 3s^2 3p^6 3d^{10} 4s^2 4p^6 4d^{10} 5s^2$
F$^-$ $1s^2 2s^2 2p^6$
O^{2-} $1s^2 2s^2 2p^6$
Cl$^-$ $1s^2 2s^2 2p^6 3s^2 3p^6$
85. 9.502×10^{-15} J, 5.931×10^4 eV, 1.434×10^{19} s^{-1} **87.** See Fig. 5.4   **89.** 11236 times larger (106^2) **91.** Co $1s^2 2s^2 2p^6 3s^2 3p^6 4s^2 3d^7$;
I $1s^2 2s^2 2p^6 3s^2 3p^6 3d^{10} 4s^2 4p^6 4d^{10} 5s^2 5p^5$
93. 12   **95.** O   **97.** The S^{2-} ion has the smallest nuclear charge and attracts the electrons least strongly.   **99.** Ra   **101.** a. The laser is in the infra red with a frequency of 3.84×10^{14} s^{-1}, b. 7.09×10^{19} photons, c. 1.0×10^{-6} m, d. 4.4×10^9 bits, a page requires 1.8×10^4 bits.

Chapter 6

1. Macroscopic: a, c, g, i. Microscopic: b, f, i. Symbolic: d, e, h; d and h symbolize microscopic properties; e, neither—formal charge is a hypothetical property.   **3.** There is one electron more than protons (32 electrons and 31 protons) in the ion.   **5.** NaCl consists of ions and contains no molecules.   **7.** Ionic b, d, e, g, i; covalent: a, c, f, h, j, k   **9.** Two valence electrons per Pb atom are transferred to Cl atoms; the resulting Pb^{2+} ion has a $6s^2$ valence shell configuration. Two of the valence electrons in the HCl molecule are shared and the other six are located on the Cl atom as lone pairs of electrons.   **11.** The atoms are not in the same arrangement.   **13.** The difference in electronegativity of the atoms involved.   **15.** Elements   **17.** H_2O_2 contains an O—O bond and the pair of electrons in the bond are distributed evenly. **19.** Anions: P, I, Cl, O; cations: Mg, In, Cs, Pb, Co   **21.** P^{3-}, Mg^{2+}, Al^{3+}, O^{2-}, Cl$^-$, Cs$^+$.
23. a. $1s^2 2s^2 2p^6 3s^2 3p^6 4s^2 3d^{10} 4p^6$, noble gas
b. $1s^2 2s^2 2p^6 3s^2 3p^6 3d^{10} 4s^2 4p^6 4d^{10} 5s^2 5p^6$, noble gas
c. $1s^2$, noble gas

d. $1s^2 2s^2 2p^6 3s^2 3p^6 3d^{10} 4s^2 4p^6 4d^{10}$, pseudo-noble gas

e. $1s^2 2s^2 2p^6$, noble gas

f. $1s^2 2s^2 2p^6 3s^2 3p^6 3d^{10}$, pseudo-noble gas

g. $1s^2$, noble gas

h. $1s^2 2s^2 2p^6$, noble gas

i. $1s^2 2s^2 2p^6 3s^2 3p^6 3d^{10} 4s^2 4p^6 4d^{10} 5s^2$, neither

j. $1s^2 2s^2 2p^6 3s^2 3p^6 3d^7$, neither

25. $:\overset{..}{\underset{..}{As}}:^{3-}$, $:\overset{..}{\underset{..}{I}}:^-$, Be^{2+}, $:\overset{..}{\underset{..}{O}}:^{2-}$, Ga^{3+}, Li^+, $:\overset{..}{\underset{..}{N}}:^{3-}$, $Sn:^{2+}$.

27.

a. $\cdot Al\cdot$ $1s^2 2s^2 2p^6 3s^2 3p^1$
   Al^{3+} $1s^2 2s^2 2p^6$

b. $:\overset{..}{\underset{..}{Br}}\cdot$ $1s^2 2s^2 2p^6 3s^2 3p^6 3d^{10} 4s^2 4p^5$
   $:\overset{..}{\underset{..}{Br}}:^-$ $1s^2 2s^2 2p^6 3s^2 3p^6 3d^{10} 4s^2 4p^6$

c. $Sr\cdot$ $1s^2 2s^2 2p^6 3s^2 3p^6 3d^{10} 4p^6 5s^2$
   Sr^{2+} $1s^2 2s^2 2p^6 3s^2 3p^6 3d^{10} 4s^2 4p^6$

d. $Li\cdot$ $1s^2 2s^1$ Li^+ $1s^2$

e. $\cdot\overset{..}{As}\cdot$ $1s^2 2s^2 2p^6 3s^2 3p^6 3d^{10} 4s^2 4p^3$
   $:\overset{..}{\underset{..}{As}}:^{3-}$ $1s^2 2s^2 2p^6 3s^2 3p^6 3d^{10} 4s^2 4p^6$

f. $\cdot\overset{..}{S}\cdot$ $1s^2 2s^2 2p^6 3s^2 3p^4$
   $:\overset{..}{\underset{..}{S}}:^{2-}$ $1s^2 2s^2 2p^6 3s^2 3p^6$

29. a. MgS, b. $AlCl_3$, c. Na_2S, d. Al_2S_3 **31.** a. MgO, b. MgO, c. Li_2O, d. MgO   **33.** MgO   **35.** 924 kJ mol^{-1}

37. a. H—H b. $H—\overset{..}{\underset{..}{Br}}:$, c. $:\overset{..}{\underset{..}{Cl}}—\overset{..}{P}—\overset{..}{\underset{..}{Cl}}:$ with $:\overset{..}{\underset{..}{Cl}}:$, d. $:\overset{..}{S}:$, e. $\underset{H}{\overset{H}{>}}C=C\underset{H}{\overset{H}{<}}$,

f. $H—\overset{..}{N}=\overset{..}{N}—H$, g. $H—C≡\overset{..}{N}—H$, h. $\overset{..}{N}=\overset{..}{\underset{..}{O}}:^-$,

i. $:N≡N:$, j. $:C≡O:$, k. $:C≡N:^-$

39. a. $:\overset{..}{\underset{..}{Cl}}—\overset{..}{\underset{..}{F}}:$, b. $:\overset{..}{\underset{..}{Cl}}—\underset{:\overset{..}{\underset{..}{Cl}}:}{\overset{:\overset{..}{\underset{..}{Cl}}:}{P}}—\overset{..}{\underset{..}{Cl}}:$, c. $\underset{:\overset{..}{\underset{..}{F}}:}{\overset{:\overset{..}{\underset{..}{F}}:}{B}}—\overset{..}{\underset{..}{F}}:$, d. $:\overset{..}{\underset{..}{F}}—\underset{:\overset{..}{\underset{..}{F}}:}{\overset{:\overset{..}{F}:}{P}}—\overset{..}{\underset{..}{F}}:^-$

41. a. $:\overset{..}{\underset{..}{O}}—\underset{:\overset{..}{\underset{..}{O}}:}{\overset{:\overset{..}{O}:}{P}}—\overset{..}{\underset{..}{O}}:^{3-}$ b. $:\overset{..}{\underset{..}{Cl}}—\underset{:\overset{..}{\underset{..}{Cl}}}{I}—\overset{..}{\underset{..}{Cl}}:^-$

c. $:\overset{..}{\underset{..}{S}}—\overset{:\overset{..}{O}:}{\underset{:\overset{..}{\underset{..}{O}}:}{}}\overset{..}{\underset{..}{O}}:^{2-}$ d. $H—\overset{..}{\underset{..}{O}}—\overset{..}{N}=\overset{..}{\underset{..}{O}}:$

43. Form coordinate-covalent bonds: BF_3, CO, H_2O, O^{2-}, SiF_4

45. a. $\underset{F}{\overset{F}{B}}—F + \underset{H}{\overset{H}{N}}—H \longrightarrow \underset{F}{\overset{F}{F—B}}—\underset{H}{\overset{H}{N}}—H$

b. $Pb^{4+} + 6:\overset{..}{\underset{..}{Cl}}:^- \longrightarrow :\overset{..}{\underset{..}{Cl}}—\underset{:\overset{..}{\underset{..}{Cl}}:}{\overset{:\overset{..}{\underset{..}{Cl}}:\ :\overset{..}{\underset{..}{Cl}}:}{Pb}}—\overset{..}{\underset{..}{Cl}}:^{2-}$

c. $:\overset{..}{\underset{..}{Cl}}—\overset{..}{Sn}—\overset{..}{\underset{..}{Cl}}: + :\overset{..}{\underset{..}{Br}}:^- \longrightarrow :\overset{..}{\underset{..}{Cl}}—\underset{:\overset{..}{\underset{..}{Br}}:}{Sn}—\overset{..}{\underset{..}{Cl}}:^-$

d. $:\overset{:\overset{..}{O}:}{\underset{:\overset{..}{\underset{..}{O}}:}{S}}—\overset{..}{\underset{..}{O}}:^{2-} + :\overset{..}{S}\cdot \longrightarrow :\overset{:\overset{..}{O}:}{\underset{:\overset{..}{\underset{..}{O}}:}{S}}—\overset{..}{S}—\overset{..}{\underset{..}{O}}:^{2-}$

47. The donor is listed first in each case. a. H_2O, H^+; b. OH^-, H^+; c. NH_3, Ag^+; d. Cl^-, PCl_5; e. O^{2-}, SO_3; f. O^{2-}, CO_2

49. $H—\overset{..}{\underset{..}{O}}—\overset{\overset{..}{O}—H}{\underset{:\overset{..}{O}:}{S}}—\overset{..}{\underset{..}{O}}:$

51. $:\overset{:\overset{..}{\underset{..}{Cl}}:}{\underset{:\overset{..}{\underset{..}{Cl}}:}{Cl—C—Cl}}:$ $\overset{:\overset{..}{\underset{..}{Cl}}:}{\underset{:\overset{..}{\underset{..}{Cl}}:}{C}}=\overset{..}{\underset{..}{O}}:$

53. a. $:\overset{..}{O}=Se—\overset{..}{\underset{..}{O}}: \longleftrightarrow :\overset{..}{\underset{..}{O}}—Se=\overset{..}{O}:$

b. $\left[\underset{:\overset{..}{\underset{..}{O}}:}{\overset{:\overset{..}{\underset{..}{O}}:\ \overset{..}{O}:}{N}} \right]^- \longleftrightarrow \left[\underset{:\overset{..}{\underset{..}{O}}:}{\overset{:\overset{..}{O}:\ :\overset{..}{\underset{..}{O}}:}{N}} \right]^- \longleftrightarrow \left[\underset{\overset{..}{\underset{..}{O}}:}{\overset{:\overset{..}{\underset{..}{O}}:\ :\overset{..}{\underset{..}{O}}:}{N}} \right]^-$

c. $H—\overset{..}{\underset{..}{O}}—\underset{\overset{..}{\underset{..}{O}}:}{N}—\overset{..}{\underset{..}{O}}: \longleftrightarrow H—\overset{..}{\underset{..}{O}}—\underset{:\overset{..}{\underset{..}{O}}:}{N}=\overset{..}{O}:$

d.

e. $H—\underset{:\overset{..}{\underset{..}{O}}:}{\overset{:\overset{..}{O}:}{C}}=\overset{..}{\underset{..}{O}}:^- \longleftrightarrow H—\underset{:\overset{..}{\underset{..}{O}}:}{\overset{..}{O}}—\overset{..}{\underset{..}{O}}:^-$

55. $:\overset{..}{O}=\overset{..}{O}—\overset{..}{\underset{..}{O}}: \longleftrightarrow :\overset{..}{\underset{..}{O}}—\overset{..}{O}=\overset{..}{O}:$

57. The acetic acid molecule contains a C—O double bond and a C—O single bond. The acetate ion is described by two resonance structures which average the two C—O bonds.   **59.** The bond order changes from $1\frac{1}{2}$ in NO_2^- to $1\frac{1}{3}$ in NO_3^-.   **61.** a. H 0, Cl 0; b. C 0, F 0; c. Cl +2, O −1; d. P 0, Cl 0; e. P 0, F 0   **63.** 0,

0, 0 **65.** a. O 0; b. S 0; O 0; c. N 0, O $-\frac{1}{2}$; d. N +1, $-\frac{2}{3}$; e. H 0, O, 0, N 0, O 0 **67.** HOCl **69.** a. Cl, b. O, c. O, d. S, e. N, f. P, g. N **71.** a. H, C, N, O, F; b. H, I, Br, Cl, F; c. H, P, S, O, F; d. Na, Al, H, P, O; e. Ba, H, As, N, O **73.** N, O, F, Cl

75. H—C̈l:, :Ö=C=Ö:, :C≡O:

77. O—H **79.** NO_3^-, CO_2, H_2S, BH_4^- **81.** a. -114 kJ, b. 30 kJ, c. -1055 kJ **83.** a **85.** F_2 158.0 per mole of bonds, O_2 498.34 kJ per mole of bonds, N_2 945.408 kJ per mole of bonds **87.** The S—F bond in SF_4 **89.** The bond energy involves breaking HCl into H and Cl atoms. The enthalpy of formation involves making HCl from H_2 and Cl_2 molecules. **91.** F, Cl, O **93.** a. H +1, Cl -1; b. C +4, F -1; c. N -3, H +1; d. P +3, Cl -1; e. P +5, Cl -1; f. H +1, O -2; g. Cl +5, O -2; h. O -1; i. C +2, O -2; j. H +1, O -2, N +3; k. P +2, Cl -1; l. Mg +2, F -1; m. Na +1, P -3 **95.** Mg, Al, Na, Fe **97.** Examples include sodium chloride (salt), NaCl; sodium hydrogen carbonate (baking soda), $NaHCO_3$; sodium hydroxide, NaOH; potassium chloride (a salt substitute), KCl; stanous fluoride (in toothpaste), SnF_2; calcium carbonate (antacid), $CaCO_3$; ammonium nitrate (fertilizer) NH_4NO_3.

99. a.

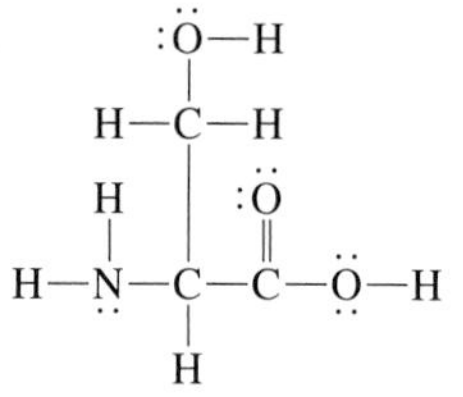

b.

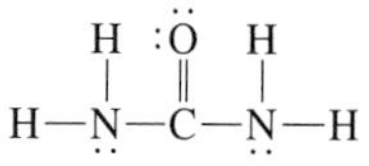

c.

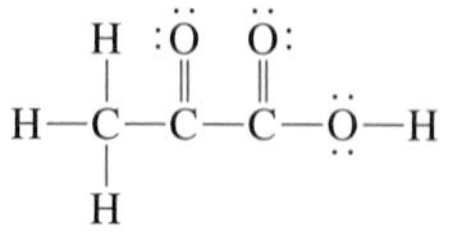

d.

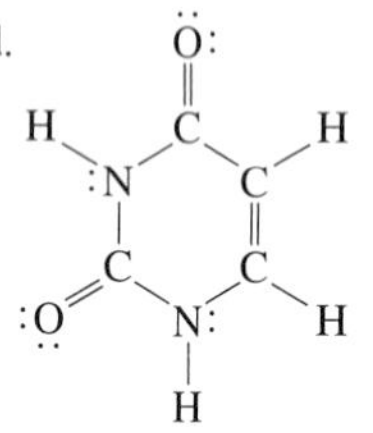

e.

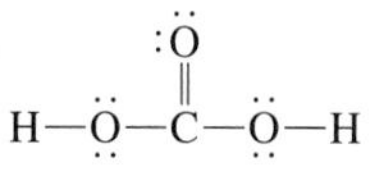

101. a.

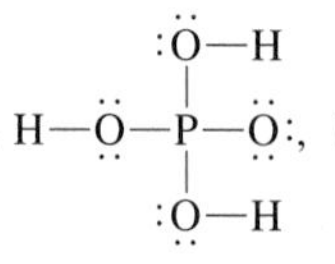

, P +5

b.

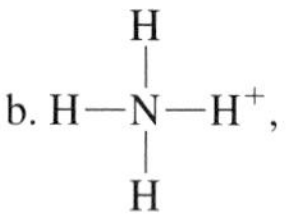

, N -3, :C̈l:$^-$

c. :C̈l—S̈—S̈—C̈l:, S +1

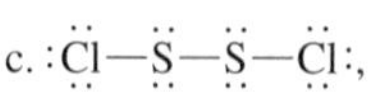

d.

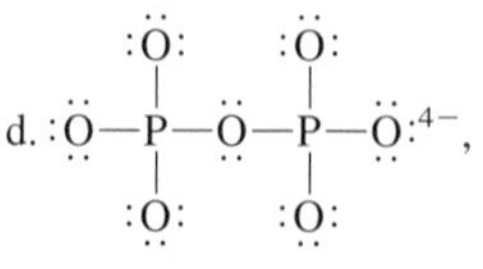

, P +5

103. :Ï—F̈:, Ox. State +1, formal charge 0.

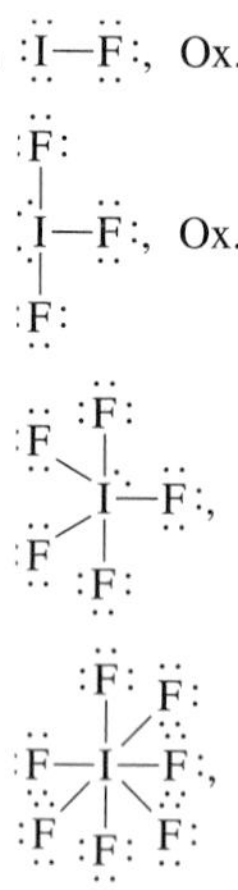

, Ox. State +3, formal charge 0.

, Ox. State +5, formal charge 0.

, Ox. State +7, formal charge 0.

105.

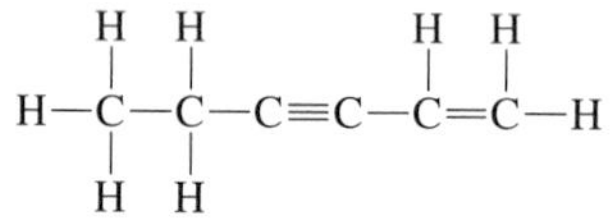

The C—C single bonds are longest.

107. yes

109.

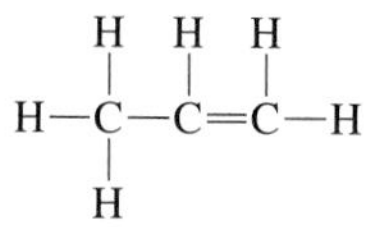

111.

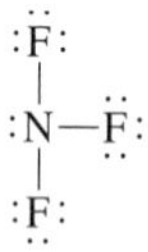

; NF_3; Ox. States: N +3, F -1; Formal Charge: N 0, F 0. **113.** K_2PtCl_6

Chapter 7

1. All are part of the microscopic domain. **3.** The presence or absence of unshared pairs on the central atom. **5.** NH_3 contains three bonding pairs of electrons and an unshared pair of electrons, which is bigger than a bonding pair and takes up more room. Both CH_4 and NH_4^+ contain 4 bonding pairs and no unshared pairs. **7.** MX_2E_3 and MX_4E_2, where E represents an unshared pair of electrons. XeF_2 and XeF_4 are examples. **9.** Similarities: Both types of bonds result from overlap of atomic orbitals on adjacent atoms and contain a maximum of two electrons. Differences: sigma bonds are stronger because they result from end-to-end overlap; pi bonds between the same two atoms are weaker because they result from side-by-side overlap. **11.** There are no d orbitals in the valence shell of carbon. **13.** A triple bond consists of 1 σ bond and 2 π bonds. A σ bond is stronger than a π bond. **15.** See Figs. 7.37 and 7.38 **17.** a. octahedral, octahedral; b. trigonal bipyramidal, trigonal bipyramidal; c. tetrahedral, tetrahedral; d. linear, linear; e. trigonal planar, trigonal planar **19.** a. octahedral, square pyramidal; b. tetrahedral, bent (109°); c. octahedral, square planar; d. tetrahedral, trigonal pyramidal; e. trigonal bipyramidal, seesaw; f. tetrahedral, bent (109°); g. trigonal bipyramidal, linear **21.** a. trigonal planar, bent (120°); b. linear, linear; c. trigonal planar, trigonal planar; d. tetrahedral, trigonal pyramidal; e. tetrahedral, tetrahedral; f. trigonal bipyramidal, seesaw; g. tetrahedral, trigonal pyramidal **23.** All of the molecules and ions in Exercises 17 and 19 contain

polar bonds. Only ClF_5, ClO_2^-, PCl_3, SeF_4, and PH_2^- have dipole moments.   **25.** b, c, d, e   **27.** P   **29.** nonpolar
31. a. tetrahedral; b. trigonal pyramidal; c. bent (109°); d. trigonal planar; e. bent (109°); f. bent (120°); g. $C\underline{H_3}CCH$ tetrahedral, $CH_3\underline{CC}H$, linear; h. tetrahedral, i. $H_2\underline{C}CCH_2$ linear; $H_2C\underline{CC}H_2$ trigonal planar   **33.** The single bond present in each molecule results from overlap of $1s$ orbitals in H_2, overlap of the H $1s$ orbital and Cl $3p$ orbital in HCl, and overlap of $3p$ orbitals in Cl_2.
35. a. sp^3d^2, b. sp^3d, c. sp^3, d. sp, e. sp^2 **37.** a. sp^3d^2, b. sp^3, c. sp^3d^2, d. sp^3 e. sp^3d, f. sp^3, g. sp^3d **39.** a. sp^2, b. sp, c. sp^2, d. sp^3, e. sp^3, f. sp^3d, g. sp^3 **41.** a. sp^3; b. sp^3; c. sp^3; d. sp^2; e. sp^3; f. sp^2; g. $C\underline{H_3}CCH$ sp^3, $CH_3\underline{CC}H$, sp; h. sp^3; i. $H_2\underline{C}CCH_2$ sp, $H_2C\underline{CC}H_2$ sp^2 **43.** Na_2^{2+}: $(\sigma_{3s})^0$, unstable; Mg_2^{2+} $(\sigma_{3s})^2$, stable; Al_2^{2+} $(\sigma_{3s})^2(\sigma_{3s}*)^2$, unstable; Si_2^{2+} $(\sigma_{3s})^2(\sigma_{3s}*)^2(\sigma_{3p_x})^2$, stable; P_2^{2+} $(\sigma_{3s})^2(\sigma_{3s}*)^2(\sigma_{3p_x})^2(\pi_{3p_y}, \pi_{3p_z})^2$, stable; S_2^{2+} $(\sigma_{3s})^2(\sigma_{3s}*)^2(\sigma_{3p_x})^2(\pi_{3p_y}, \pi_{3p_z})^4$, stable; Cl_2^{2+} $(\sigma_{3s})^2(\sigma_{3s}*)^2(\sigma_{3p_x})^2(\pi_{3p_y}, \pi_{3p_z})^4(\pi_{3p_y}*, \pi_{3p_z}*)^2$, stable; Ar_2^{2+} $(\sigma_{3s})^2(\sigma_{3s}*)^2(\sigma_{3p_x})^2(\pi_{3p_y}, \pi_{3p_z})^4(\pi_{3p_y}*, \pi_{3p_z}*)^4$, stable
45. $(\sigma_{2s})^2(\sigma_{2s}*)^2(\sigma_{2p_x})^2(\pi_{2p_y}, \pi_{2p_z})^4(\pi_{2p_y}*, \pi_{2p_z}*)^3$ See Fig. 7.41 for the order of energies in the energy diagram.   **47.** a. H_2, b. N_2, c. O, d. C_2, e. B_2 **49.** linear, CO_2; bent with an approximately 109° angle, H_2O; bent with an approximately 120° angle, SO_2

51. a.

$:\ddot{S}=C=\ddot{S}:$; $:C\equiv S:$

b. CS_3^{2-}, trigonal planar; CS_2, linear

53.

Phosphorus and nitrogen can form sp^3 hybrids to form 3 bonds and hold one lone pair in PF_3 and NF_3, respectively. However, nitrogen has no valence d orbitals, so it cannot form a set of sp^3d hybrid orbitals to bind 5 F atoms in NF_5. Phosphorus has d orbitals and can bind 5 F atoms with dsp^3 hybrid orbitals in PF_5.

55. S_8

Each S has a bent (109°) geometry, sp^3

SO_2 $:\ddot{O}=S-\ddot{O}: \longleftrightarrow :\ddot{O}-S=\ddot{O}:$, bent (120°), sp^2

SO_3

$S=\ddot{O}:$, trigonal planar, sp^2

H_2SO_4

tetrahedral, sp^3

57. XeF_2, $:\ddot{F}-Xe-\ddot{F}:$, linear, sp^3d
59. $:\ddot{F}-N-N-\ddot{F}:$, trigonal pyramidal at each N, sp^3;

$:\ddot{F}-N=N-\ddot{F}:$, bent (120°) at each N, sp^2.

61. a. $H-\ddot{O}-N=\ddot{O}:$; b. O tetrahedral, bent (109°); N trigonal planar, trigonal planar;   c. O sp^3, N sp^2 **63.** All of the C—O bonds are equivalent because the actual electronic structure is the average of three resonance forms.   **65.** $H-C\equiv N:$; linear; C atom has sp hybridization. The H—C single bond results from the end-to-end overlap of the H $1s$ orbital with one of the sp hybrid orbitals on the C atom. The sigma bond in the $C\equiv N$ triple bond results from the end-to-end overlap of the second sp hybrid orbital on the C atom with a p orbital on the N atom. The second and third bonds in the $C\equiv N$ triple bond, which are pi bonds, result from side-by-side overlap of unhybridized p orbitals of the C atom with p orbitals of the N atom.   **67.** i. tetrahedral, sp^3; ii. NH_4^+ tetrahedral, sp^3: NO_3^- trigonal planar, sp^2; iii. bent (109°), sp^3; iv. tetrahedral, sp^3 **69.** Left to right: sp^3, sp^3, sp, sp, sp^2, sp^2 **71.** trigonal planar, sp^2 **73.** trigonal pyramidal; sp^3; formal charge—N, 0; F, 0; oxidation state—N +3, F −1

Chapter 8

1. Metal hydroxides are ionic; they contain a metal ion and hydroxide ion, and exhibit basic behavior due to the hydroxide ion. Nonmetal hydroxides are covalent and exhibit acidic behavior.   **3.** Look for the positions of Ca, Si, and S in the periodic table   **5.** Like xenon, +2, +4, +6, and +8
7. Determine whether the compound contains a metal combined with the hydroxide group (the hydroxide would be a base) or a nonmetal combined with the hydroxide group (an acid).   **9.** SO_2. N_2 does not dissolve in water and form an acidic solution; the S atom in SO_3 is in its maximum oxidation state so SO_3 will not react with oxygen.   **11.** Of the choices, only oxygen is more electronegative than chlorine and cannot be oxidized by it.
13. Atoms of Group 2A form positive ions by loss of electrons. The most electrons an atom of Group 2A can lose in a chemical reaction are the electrons in its valence shell.   **15.** An oxidation-reduction reaction because the oxidation state of the silver changes during the reaction.   **17.** a. Ca, b. Cl, c. N_2O_3, d. Al, e. N, f. Br   **19.** a. CsOH, b. $B(OH)_3$, c. S, d. RbCl, e. O
21. a. KF, b. $Al(OH)_3$, c. O_2, d. Mg, e. F_2, f. KNO_2, g. P_4O_6
23. a. Na, b. Sr, c. In, d. Si, e. N   **25.** b, d, f, h, i   **27.** Examples include: a. $CaCl_2$, CaO; b. Li_3N, NH_3; c. N_2O_3, NCl_3; d. H_2S, TiS_2; e. HCl, $FeCl_3$; f. BrF_3, $BrCl_3$; g. PCl_3, P_4O_6; h. PbS, $PbCl_2$
29. a. Na, +1, I −1; b. Gd, +3 Cl −1; c. Li +1, N +5, O −2; d. H +1, Se −2; Mg +2, Si −4; Rb +1, O −½; H +1, F −1
31. H +1, P +5, O −2; b. Al +3, H +1, O −2; c. Se +4, O −2; d. K +1, N +3, O −2; In +3, S −2; P +3, O −2
33. a. +3; b. −1, +1, +3, +5, +7; c. +2; d. +1; e. −3, +3, +5; f. −2, +4, +6; g. +4; h. −1, +1   **35.** a. iron(III) chloride, b. nitrogen(IV) oxide, c. sulfur(VI) fluoride, d. arsenic(III) oxide, e. cobalt(II) nitrate, f. vanadium(III) sulfate   **37.** sulfur(VI) oxide, b. nitrogen(V) oxide, c. phosphorus(III) oxide, d. indium(I) iodide, e. carbon(IV) oxide, f. tin(IV) chloride   **39.** $Ca_3(PO_4)_2$: Ca +2, P +5, O −2; C: C 0; SiO_2: Si +4, O −2; $CaSiO_3$: Ca +2, Si +4, O −2; CO: C +2, O −2; P_4: P 0. C is oxidized, P is reduced.
41. a. $2HF \longrightarrow H_2 + F_2$, $2K_2MnF_6 + 4SbF_5 \longrightarrow F_2 + 2MnF_3$

+ $4KSbF_6$; b. HF: H +1, F −1; H_2: H 0; F_2: F 0; K_2MnF_6: K +1, Mn +4, F −1; SbF_5: Sb +5, F −1; MnF_3: Mn +3, F −1; $KSbF_6$: K +1, Sb +5, F −1; c. Electrolysis of HF: H is reduced (+1 ⟶ 0), F is oxidized (−1 ⟶ 0). Second reaction: Mn is reduced (+4 ⟶ +3), F is oxidized (−1 ⟶ 0); d. Electrolysis of HF: H in HF is oxidizing agent, F in HF is reducing agent. Second reaction: Mn is oxidizing agent, F is reducing agent
43. Acid: b, c, d, g, h. Base: a, e, f, i
45. a. $2HCl(g) + Ca(OH)_2(s) \longrightarrow CaCl_2(s) + 2H_2O(l)$
b. $Sr(OH)_2(aq) + 2HNO_3(aq) \longrightarrow Sr(NO_3)_2(aq) + 2H_2O(l)$
c. $SO_2(g) + KOH(s) \longrightarrow KHSO_3(s)$ or $SO_2(g) + 2KOH(s) \longrightarrow K_2SO_3(s) + H_2O(l)$
d. $2HBr(g) + CoO(s) \longrightarrow CoBr_2(s) + H_2O(l)$
e. $Na_2O(s) + CO_2(g) \longrightarrow Na_2CO_3(s)$
47. a. $2Al + 3F_2 \longrightarrow 2AlF_3$
b. $2Al + 3CuBr_2 \longrightarrow 3Cu + 2AlBr_3$
c. $P_4 + 5O_2 \longrightarrow P_4O_{10}$ (or $P_4 + 5O_2 \longrightarrow 2P_2O_5$)
d. $Ca + 2H_2O \longrightarrow Ca(OH)_2 + H_2$
e. $2In + 6HCl \longrightarrow 2InCl_3 + 3H_2$
49. c, e, f, i, k
51. a. $6HBr(g) + In_2O_3(s) \longrightarrow 2InBr_3(s) + 3H_2O(l)$
b. $Mg(OH)_2(s) + 2HClO_4(aq) \longrightarrow Mg^{2+}(aq) + 2ClO_4^-(aq) + 2H_2O(l)$ c. $SO_3(g) + 2H_2O(l) \longrightarrow H_3O^+(aq) + HSO_4^-(aq)$ (a solution of H_2SO_4) d. $Na_2O(s) + H_2O(l) \longrightarrow 2NaOH(s)$
e. $Li_2O(s) + 2CH_3CO_2H(l) \longrightarrow 2LiCH_3CO_2(s) + H_2O(l)$
f. $SrO(s) + H_2SO_4(l) \longrightarrow SrSO_4(s) + H_2O(l)$
53. a. $Li[CH_3CO_2] + H_3PO_4 \longrightarrow CH_3CO_2H + LiH_2PO_4$
b. $SrF_2 + H_2SO_4 \longrightarrow 2HF + SrSO_4$
c. $NaCN + HCl \longrightarrow HCN + NaCl$
d. $CaCO_3 + 2HCl \longrightarrow CaCl_2 + CO_2 + H_2O$
55. $C + O_2 \longrightarrow CO_2$ **57.** $H_2 + F_2 \longrightarrow 2HF$ **59.** $2NaBr + Cl_2 \longrightarrow 2NaCl + Br_2$ **61.** $2Mg + O_2 \longrightarrow 2MgO$, $3Mg + N_2 \longrightarrow Mg_3N_2$ **63.** Co   **65.** a. yes, b. no, c. no, d. yes, e. yes
67. Li above Mg. $2Li + 2H_2O \longrightarrow H_2 + 2LiOH$; same reaction for steam; $4Li + O_2 \longrightarrow 2Li_2O$; $2Li + 2HCl \longrightarrow 2LiCl + H_2$. Be near Mg and Al. $Be + H_2O$, no reaction; $Be + H_2O(steam) \longrightarrow BeO + H_2$; $2Be + O_2 \longrightarrow 2BeO$; $Be + 2HCl \longrightarrow BeCl_2 + H_2$ **69.** Examples include: calcium as $CaCO_3$; iron as ferrous fumarate, an Fe^{2+} salt; zinc as ZnO; and magnesium as magnesium stearate, a Mg^{2+} salt   **71.** $CaCO_3 + 2C_2H_5CO_2H \longrightarrow Ca(C_2H_5CO_2)_2 + CO_2 + H_2O$ **73.** $CaCl_2$: a. $CaSO_4 + SrCl_2 \longrightarrow CaCl_2 + SrSO_4$ b. $Ca + Cl_2 \longrightarrow CaCl_2$ c. $Ca(OH)_2 + 2HCl \longrightarrow CaCl_2 + H_2O$ K_2SO_4: a. $2KCl + Ag_2SO_4 \longrightarrow K_2SO_4 + 2AgCl$. b. $2K + H_2SO_4 \longrightarrow K_2SO_4 + H_2$ c. $2KOH + H_2SO_4 \rightarrow K_2SO_4 + 2H_2O$ $Mg(H_2PO_4)_2$: $MgCl_2 + 2KH_2PO_4 \longrightarrow Mg(H_2PO_4)_2 + 2KCl$ b. $Mg + 2H_3PO_4 \longrightarrow Mg(H_2PO_4)_2 + H_2$ c. $Mg(OH)_2 + 2H_3PO_4 \longrightarrow Mg(H_2PO_4)_2 + 2H_2O$
75. $CN^-(aq) + H_2SO_4(aq) \longrightarrow HCN(g) + HSO_4^-(aq)$; $NaOH(aq) + HCN(g) \longrightarrow NaCN(aq) + H_2O(l)$ **77.** a. $SO_2 + NaOH \longrightarrow NaHSO_3$; b. The H atoms in $NaBH_4$ are oxidized (−1 ⟶ +1), S atoms are reduced (+4 ⟶ +3);
c.

$$\left[\begin{array}{c} H \\ H-B-H \\ H \end{array} \right]^- \quad \left[\begin{array}{c} H-\overset{..}{\underset{..}{O}}-\overset{..}{\underset{..}{S}}-\overset{..}{\underset{..}{O}}: \\ \overset{..}{\underset{..}{O}}: \end{array} \right]^- \quad \left[\begin{array}{c} :\overset{..}{\underset{..}{O}}-\overset{..}{\underset{..}{S}}-\overset{..}{\underset{..}{S}}-\overset{..}{\underset{..}{O}}: \\ :\overset{..}{O}: \; :\overset{..}{O}: \end{array} \right]^{2-}$$

d. BH_4^- tetrahedral, tetrahedral; HSO_3^- tetrahedral, trigonal pyramidal; $S_2O_4^{2-}$ tetrahedral, trigonal pyramidal   **79.** 45.93%
81. 0.0583 M   **83.** linear   **85.** from sp^2 to sp^3 **87.** a. P_2O_5; b. P_4O_{10}; c. $P_4 + 5O_2 \longrightarrow P_4O_{10}$, $P_4O_{10} + 6H_2O \longrightarrow 4H_3PO_4$;

d. 3.23 tons   **89.** a. Sulfur: SO_2 +4, H_2SO_4 +6; nitrogen: NO +2, NO_2 + 4; b. NO_2 nitrogen dioxide, nitrogen(IV) oxide; NO nitrogen monoxide, nitrogen(II) oxide; SO_2 sulfur dioxide, sulfur(IV) oxide;
c.

$$:\overset{..}{O}=\overset{..}{S}-\overset{..}{\underset{..}{O}}: \longleftrightarrow :\overset{..}{\underset{..}{O}}-\overset{..}{S}=\overset{..}{O}:, \quad H-\overset{..}{\underset{..}{O}}-\overset{\displaystyle :\overset{..}{O}-H}{\underset{\displaystyle :\overset{..}{\underset{..}{O}}:}{S}}-\overset{..}{\underset{..}{O}}:,$$

$$:\overset{..}{O}=\overset{..}{N}-\overset{..}{\underset{..}{O}}: \longleftrightarrow :\overset{..}{\underset{..}{O}}-\overset{..}{N}=\overset{..}{O}:, \quad :N=\overset{..}{O}:$$

SO_2 and NO_2 bent (120°), sp^2; H_2SO_4 S tetrahedral, sp^3, internal O sp^3, bent (109°); d. 3.54×10^5 g (354 metric tons); e. 10.5 M; f. 33 L; g. i. −296.83 kJ (this is an enthalpy of formation, ii. −114.1 kJ. iii −296 kJ; each reaction is exothermic.

Chapter 9

1. A single underline indicates a macroscopic property; a double underline indicates a microscopic property. Symbols are in italics. (a) Buckminsterfullerene, C_{60}, is a soluble form of the element carbon that is composed of soccer-ball-shaped molecules. (b) Graphite is slippery because the planes of carbon atoms in the crystal slide past each other very easily. (c) Even though diamond crystals are very hard, they will burn. (d) CO_2 and H_2O are produced when hexane burns in air. (e) When a Br_2 molecule adds to a $H_2C=CH_2$ molecule, the two C-Br sigma bonds formed are stronger than the pi bond between the carbon atoms. (f) Nylon is more stable at higher temperatures than is polyethylene because nylon contains polar functional groups and polyethylene does not. (g) Vinegar tastes sour because molecules of acetic acid in vinegar react with water to give hydronium ions.
3. There are several sets of answers; one is:

alkane, C_5H_{12},
$$H-\overset{\overset{H}{|}}{\underset{\underset{H}{|}}{C}}-\overset{\overset{H}{|}}{\underset{\underset{H}{|}}{C}}-\overset{\overset{H}{|}}{\underset{\underset{H}{|}}{C}}-\overset{\overset{H}{|}}{\underset{\underset{H}{|}}{C}}-\overset{\overset{H}{|}}{\underset{\underset{H}{|}}{C}}-H$$

alkene, C_5H_{10},
$$H-\overset{\overset{H}{|}}{\underset{\underset{H}{|}}{C}}-\overset{\overset{H}{|}}{\underset{\underset{H}{|}}{C}}-\overset{\overset{H}{|}}{C}=\overset{\overset{H}{|}}{C}-\overset{\overset{H}{|}}{\underset{\underset{H}{|}}{C}}-H$$

alkyne, C_5H_8,
$$H-\overset{\overset{H}{|}}{\underset{\underset{H}{|}}{C}}-\overset{\overset{H}{|}}{\underset{\underset{H}{|}}{C}}-C\equiv C-\overset{\overset{H}{|}}{\underset{\underset{H}{|}}{C}}-H$$

5. When bromine reacts with a saturated hydrocarbon, a hydrogen atom from a molecule of the unsaturated hydrocarbon is replaced by a bromine atom and a molecule of hydrogen bromide is also produced. When bromine reacts with an unsaturated hydrocarbon, the electrons in the pi bond of the alkene functional group are used to form carbon-bromine sigma bonds on the carbon atoms that were formerly bonded by the double bond.   **7.** Each contains a chain of six carbon atoms.   **9.** The two compounds have different formulas;

C_6H_{12} and C_6H_{14}.   **11.** A ketone contains a $-\overset{\overset{\displaystyle O}{\|}}{C}-$ group bonded to two additional carbon atoms; thus a minimum of three carbon atoms are needed.   **13.** The hydrocarbon chain in a saturated fatty acid contains no double or triple bonds. The hydrocarbon chain in an unsaturated fatty acid contains one or more multiple bonds.
15. polar: a, b, c, e; nonpolar: d, f   **17.** sp, sp^2, sp **19.** CO. The

triple bond in CO is stronger than the double bond in CO_2, and the C—O bond in CO_3^{2-} has a bond order of $1\frac{1}{3}$ as the resonance description of the ion's bonding shows.   **21.** $:C\equiv C:^{2-}, :C\equiv O:$; both have a triple bond and one lone pair of electrons on each atom.
23. CO +2, CO_2 +4, CN_2^{2-} +4, CO_3^{2-} +4, CaC_2 −1, HCN +2
25. $CaC_2 + 2H_2O \rightarrow C_2H_2 + Ca(OH)_2$; $2C_2H_2 + 5O_2 \longrightarrow 4CO_2 + 2H_2O$

27. a. C_6H_{14},

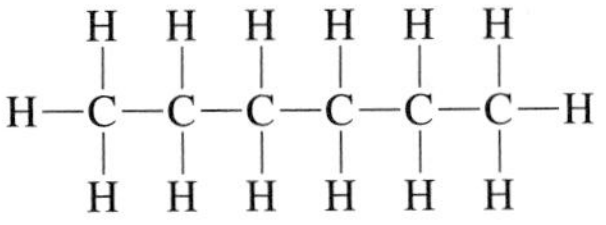

b. C_6H_{14},

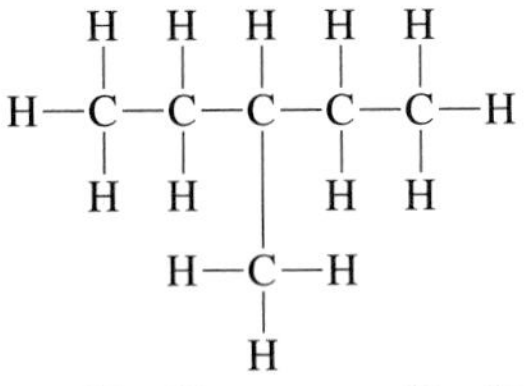

c. C_6H_{12},

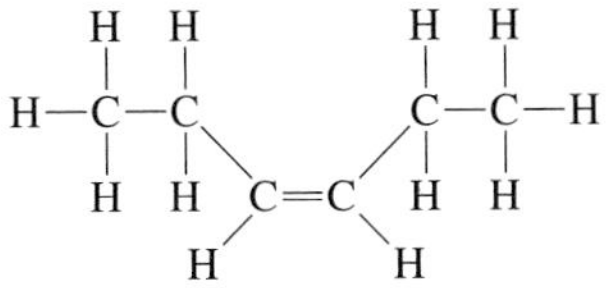

d. C_6H_{12},

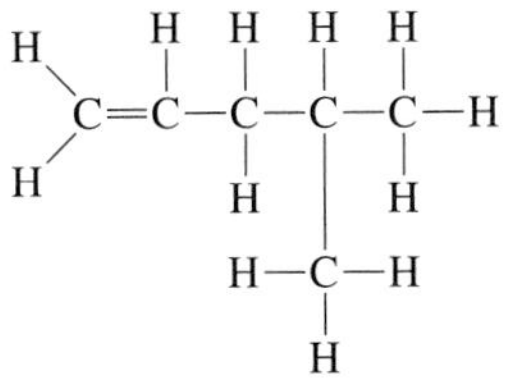

e. C_6H_{10},

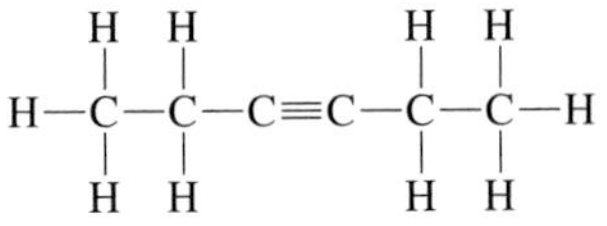

f. C_6H_{10},

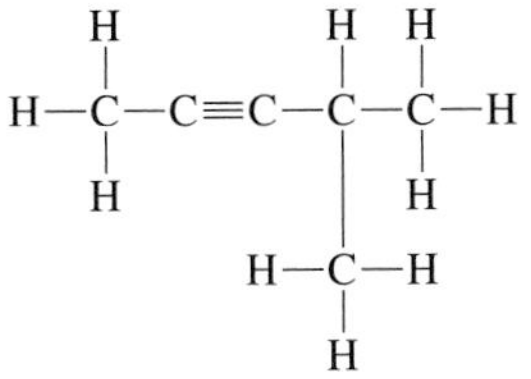

29. a. 2,2-dibromobutane, b. 2-chloro-2-methylpropane,
c. 2-methylbutane, d. 1-butyne, e. 4-fluoro-4-methyl-1-octyne,
f. *trans*-1-chloropropene, g. 4-methyl-1-pentene   **31.** a. dichlorodifluoromethane, b. 1,1,2-trichloro-1,2,2-trifluoroethane,
c. 1,1,1,2-tetrafluoroethane, d. 1,1-dichloro-1-fluoroethane

33. *n*-hexane

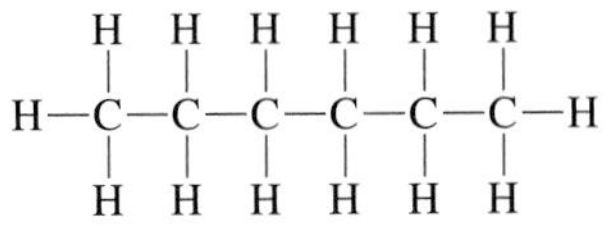

2-methylpentane

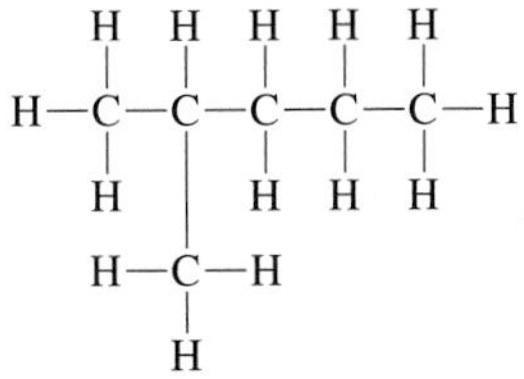

3-methylpentane

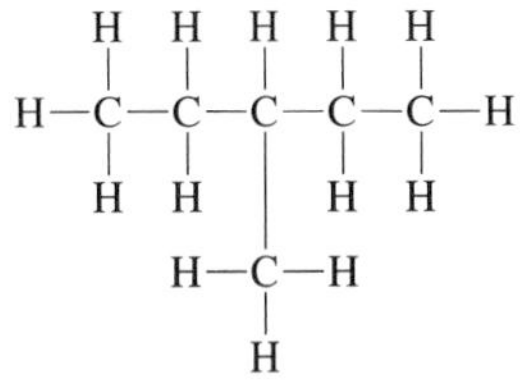

2,3-dimethylbutane

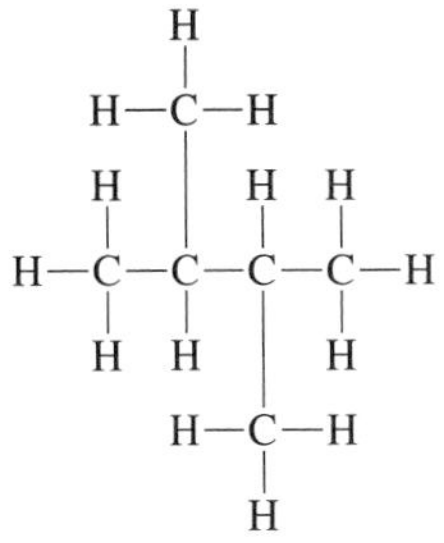

2,2-dimethylbutane

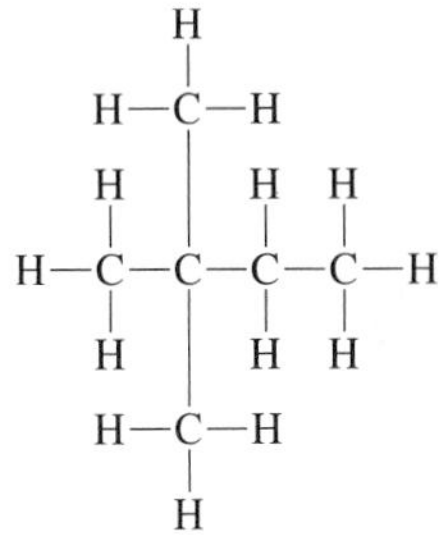

35.

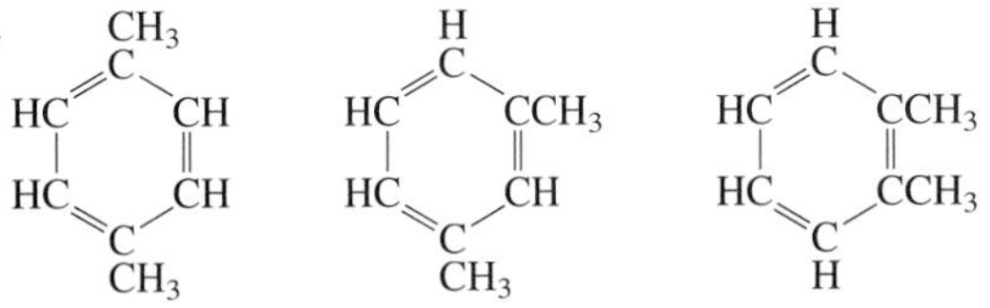

Like benzene, each isomer has two resonance forms.

37. 1-butyne

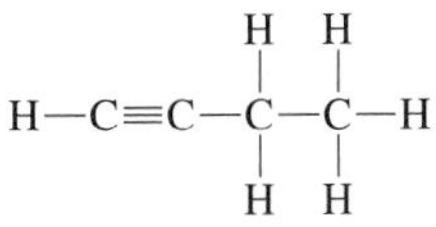

2-butyne

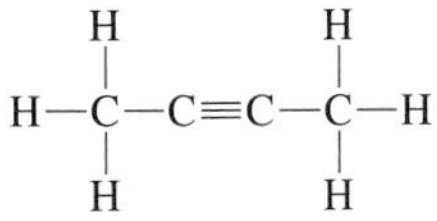

39. See Table 9.2 for the two propyl and four butyl isomers

41. a.

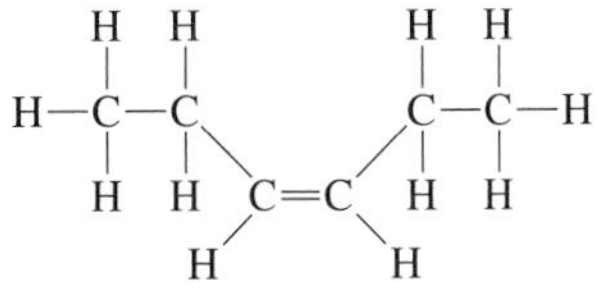

Left-to-right: tetrahedral, tetrahedral, trigonal planar, trigonal planar, tetrahedral, tetrahedral

b.

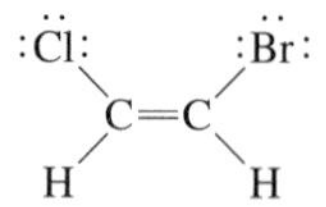

$$:\ddot{C}l: \quad :\ddot{B}r:$$
C=C Both are trigonal planar
(with H, H)

c. 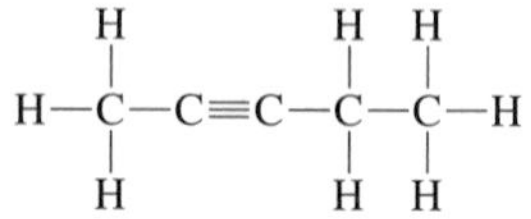 H—C—C≡C—C—C—H

Left-to-right: tetrahedral, linear, linear, tetrahedral, tetrahedral

d. 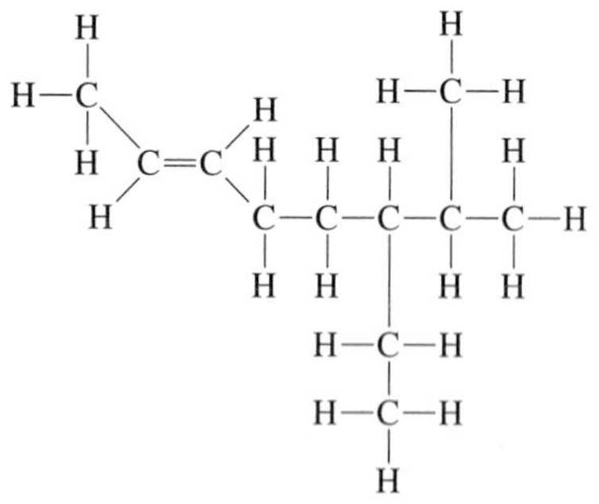 H—C ... C=C ... C—C—C—C—C—H (with H—C—H and H—C—H branches)

All carbons tetrahedral except the two trigonal planar C atoms in the double bond. **43.** a. CH_3CH=CH_2: tetrahedral, trigonal planar, trigonal planar; b. $CH_3CH_2CH_2CH_2OH$: all tetrahedral; c. $CH_3CH_2OCH_2CH_2CH_3$: all tetrahedral; d. CH_3CH=$CHCHBrCH_2CH_2CH_3$: all tetrahedral except the two trigonal planar C atoms in the double bond; e. $CH_3C(CH_3)_2CH(CH_3)CH_2CH_2CH_3$ all tetrahedral; f. H_2CO: trigonal planar **45.** a. —CO_2H carboxylic acid, —NH_2 amine, —OH alcohol; b. two —OH alcohol groups, two —CO_2H carboxylic acid groups; c. three —C(O)O— ester groups; d. —CO_2H carboxylic acid, —OC(O)— ester, C_6 ring aromatic hydrocarbon; e. —NH_2 amine, —C(O)O— ester, —CO_2H carboxylic acid groups, —C(O)NH— amide, C_6 ring aromatic hydrocarbon; f. —CO_2H carboxylic acid, C_6 ring aromatic hydrocarbon; g. —O— ether; h. five —OH alcohol groups, —O— ether; i. two —C(O)— ketone groups

47.

a. 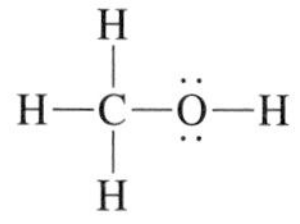 H—C—Ö—H

b. H—C—C—C—Ö—H (with H—C—H branch)

c. H—C—C—Ö—C—H

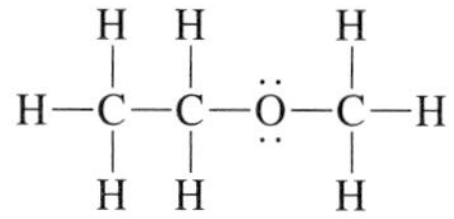

d. 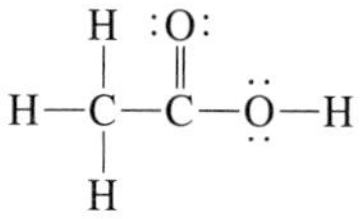 H—C—C—Ö—H

e. 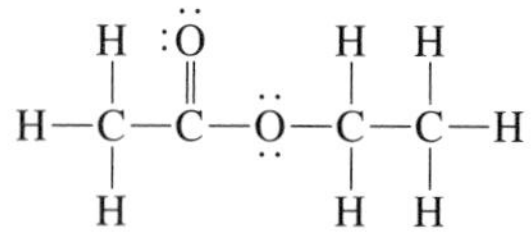 H—C—C—Ö—C—C—H

f. 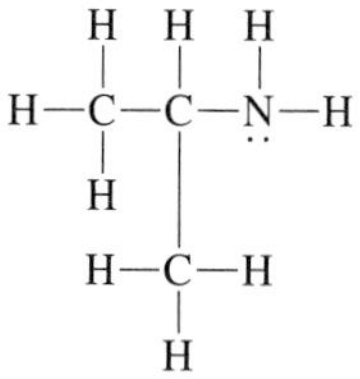 H—C—C—N—H (with H—C—H branch)

49.

a. 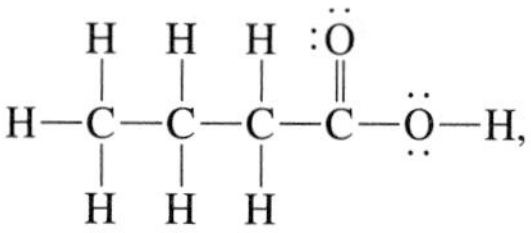 H—C—C—C—C—Ö—H,

Oxidation states: −3, −2, −2, +3
Hybridization: $sp^3\ sp^3\ sp^3\ sp^2$

b. 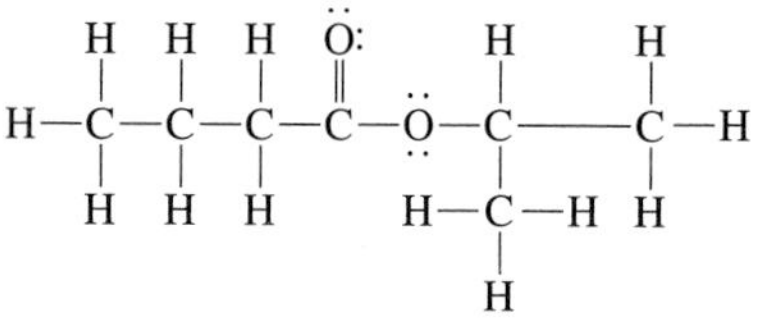 H—C—C—C—C—Ö—C ... C—H (with H—C—H branch)

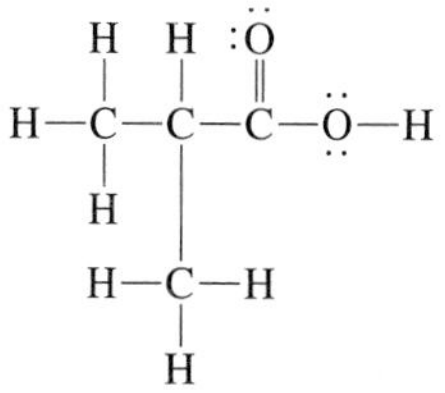

51. CH_3OCH_3, ether; CH_3CH_2OH, alcohol

53. 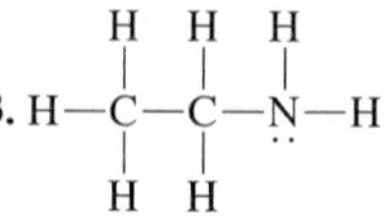 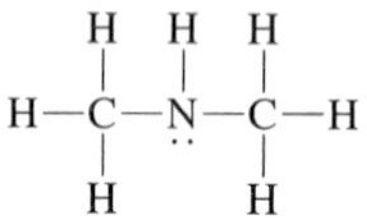 H—C—C—N—H   H—C—N—C—H

55. a. $CH_3CH_2OH + CH_3CH_2CO_2H \longrightarrow$
$$CH_3CH_2CO_2CH_2CH_3 + H_2O$$

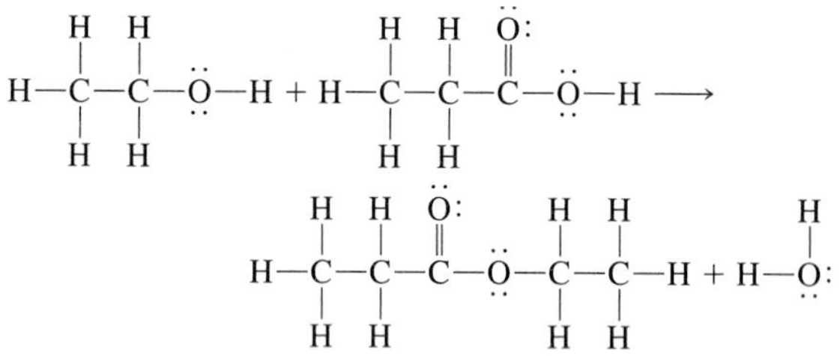

H—C—C—Ö—H + H—C—C—C—Ö—H ⟶

H—C—C—C—Ö—C—C—H + H—Ö:

b. CH≡$CCH_2CH_3 + 2I_2 \longrightarrow CHI_2CI_2CH_2CH_3$

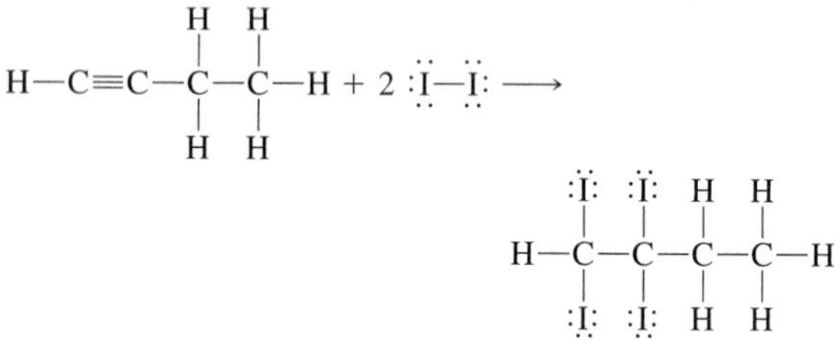

H—C≡C—C—C—H + 2 :Ï—Ï: ⟶

H—C—C—C—C—H (with :Ï: :Ï: and H H)

c. $CH_3CH_2CH_2CH_2CH_3 + 8O_2 \longrightarrow 5CO_2 + 6H_2O$

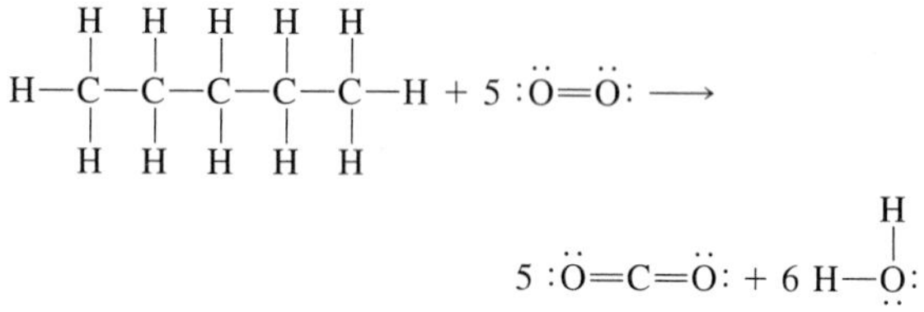

H—C—C—C—C—C—H + 5 :Ö=Ö: ⟶

5 :Ö=C=Ö: + 6 H—Ö:

d. $2CH_3CH_2CH_2OH + \longrightarrow CH_3CH_2CH_2OCH_2CH_2CH_3 + H_2O$

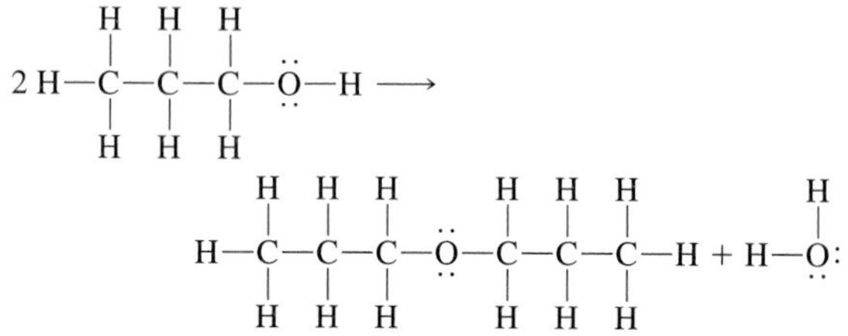

2 H—C—C—C—Ö—H ⟶

H—C—C—C—Ö—C—C—C—H + H—Ö:

e. $CH_3C \equiv CH + H_2O \longrightarrow CH_3CH(OH)CH_3$ or $CH_3CH_2CH_2OH$
One of these isomers of propanol will form, but we do not have enough information to determine which one.

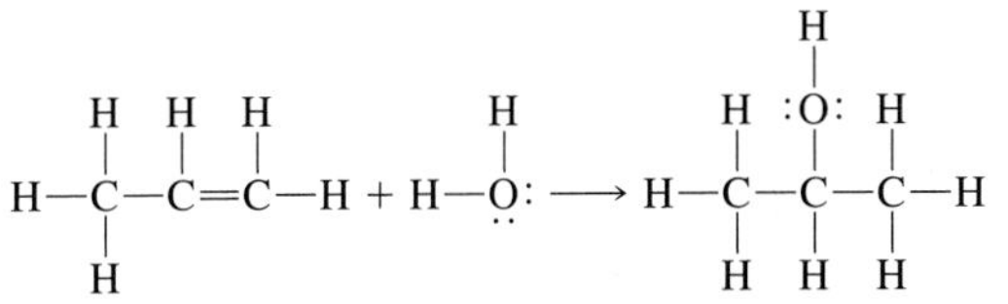

f. $CH_3NH_2 + H_3O^+ + Cl^- \longrightarrow CH_3NH_3^+ + Cl^- + H_2O$

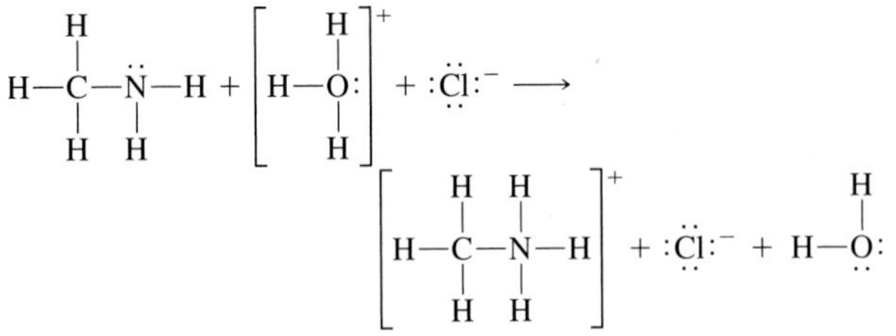

g. $C_6H_5CO_2H + Na^+ + OH^- \longrightarrow Na^+ + C_6H_5CO_2^+ + H_2O$

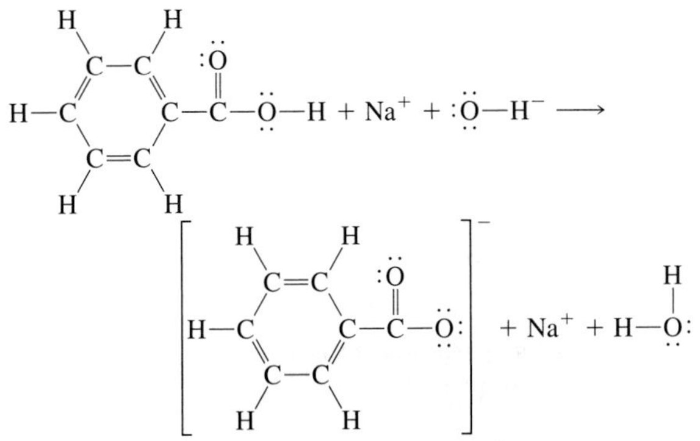

57. b. $\underline{C}H \equiv \underline{C}CH_2CH_3 \ (sp) \longrightarrow CHI_2CI_2CH_2CH_3 \ (sp^3)$
c. $CH_3\underline{C}H_2CH_2\underline{C}H_2\underline{C}H_3 \ (sp^3) \longrightarrow CO_2 \ (sp)$
e. $CH_3\underline{C} = \underline{C}H_2 \ (sp^2) \longrightarrow CH_3CH(OH)CH_3 \ (sp^3)$

59. Trimethyl amine: trigonal pyramidal, sp^3; trimethyl ammonium ion: tetrahedral, sp^3

61.

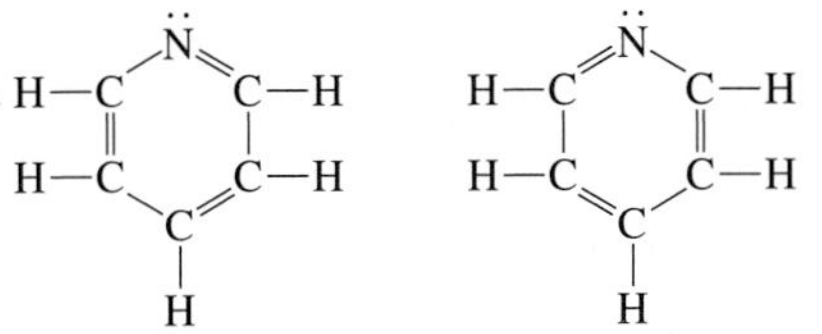

63.

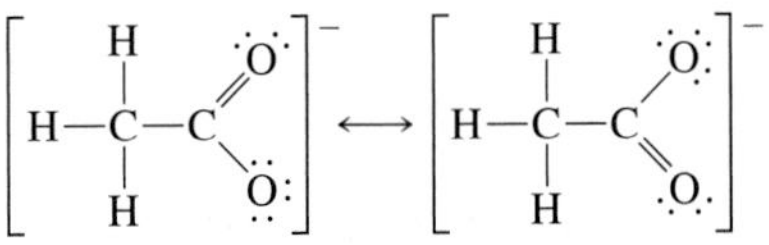

65. A. $CH_3CH_2CH_2OH$, alcohol; B. $CH_3CH_2OCH_3$, ether; C. $CH_3CH(OH)CH_3$, alcohol **67.** Morphine: 2 alcohol groups, an amine, an ether, and an aromatic C_6 ring; codeine: an alcohol, an amine, two ether groups, and an aromatic C_6 ring; heroin: 2 ester groups, an amine, an ether, and an aromatic C_6 ring

69. n

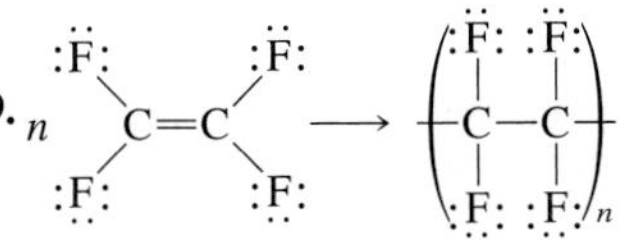

71.

a.

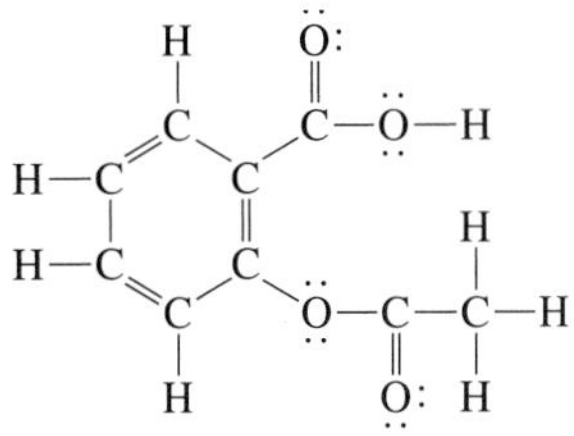

b.

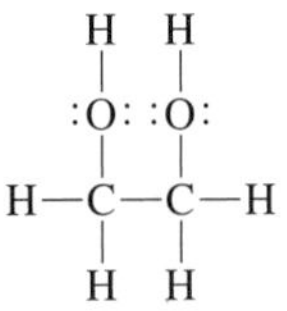

c.

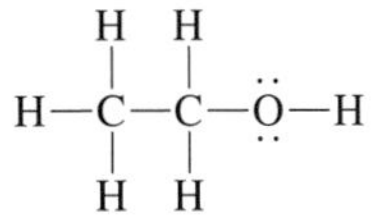

d.

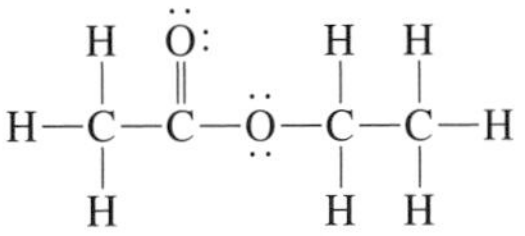

e.

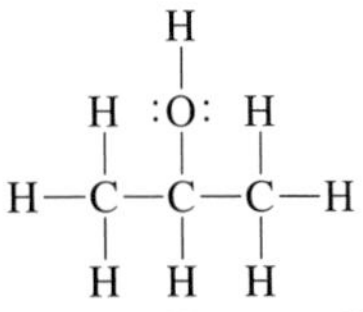

f.

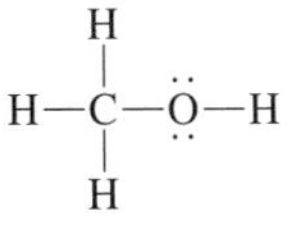

g.

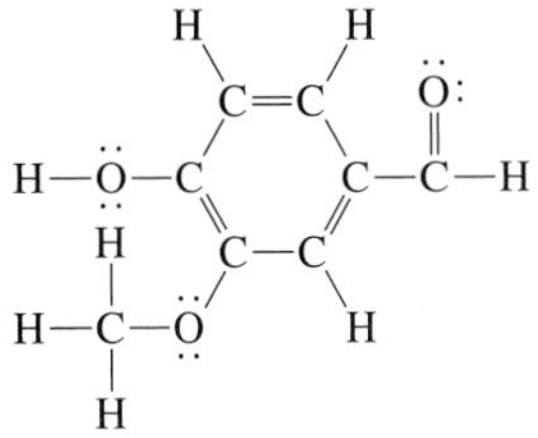

h.

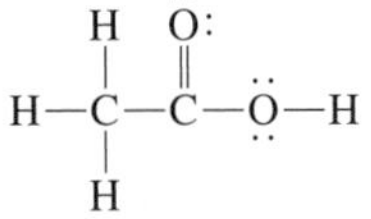

73. Lewis Structures.

a.

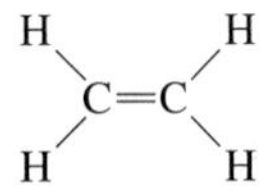

b.

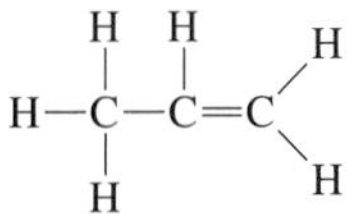

c.

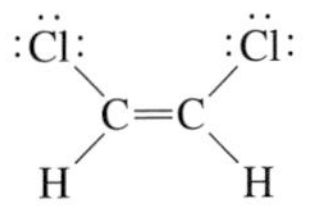

d.

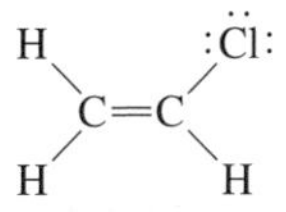

e.

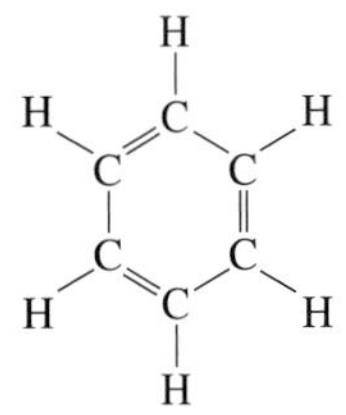

f.

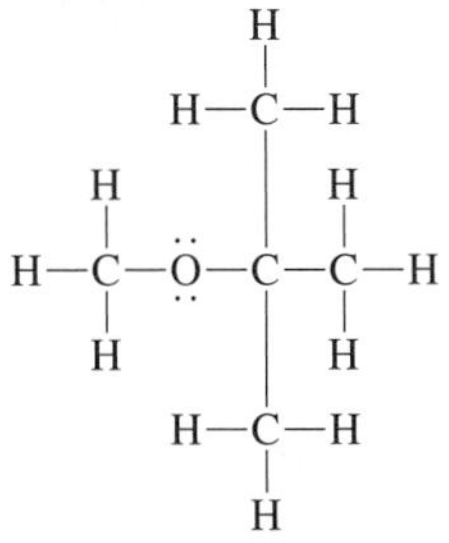

g.

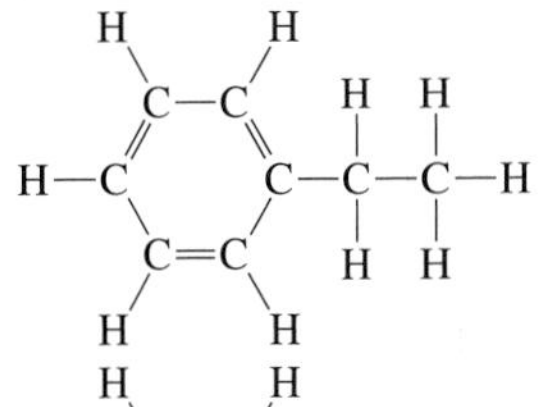

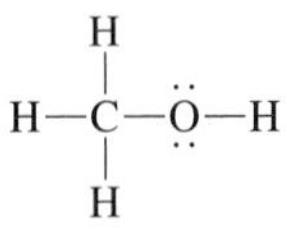

h.

i.

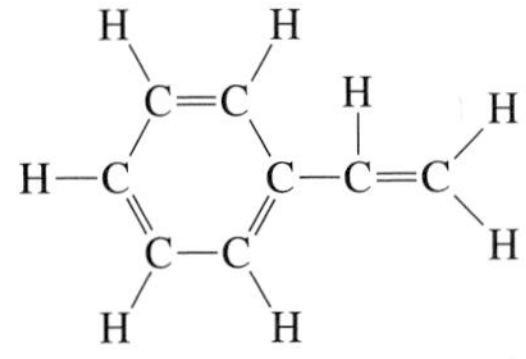

k. (structure)

l. H—C—H with O above

n. (structure)

o. (structure)

j. one of three isomers, see Exercise 35

m. (structure)

p. (structure)

q. (structure)

r. (structure)

s. (structure)

t. (structure)

u. (structure)

w. (structure)

v. H—C=C—C≡N:

x. (structure)

y. (structure)

z. (structure)

Functional Groups. a. C=C double bond; b. C=C double bond; c. C=C double bond, halogens; d. C=C double bond, halogen; e. aromatic ring; f. ether; g. aromatic ring, ethyl group; h. aromatic ring, C=C double bond; i. alcohol; j. aromatic ring, methyl groups; k. aromatic ring, carboxylic acid groups; l. aldehyde; m. ether; n. aromatic ring, methyl group; o. aromatic ring, methyl groups; p. alcohols; q. aromatic ring, isopropyl group; r. aromatic ring, alcohol; s. carboxylic acid; t. ether; u. C=C double bonds; v. C=C double bond, C≡N triple bond (cyanide group); w. C=C double bond, ester; x. ketone; y. carboxylic acid groups; z. alcohol.
75. Hybridization: a. sp^3, sp^2, sp^2; b. sp^2, sp; c. both sp; d. sp^2; e. sp^3, sp^2; f. sp^3, sp^2, sp^3. Geometry: a. tetrahedral, trigonal planar, trigonal planar; b. trigonal planar, linear; c. both linear; d. trigonal planar; e. tetrahedral, trigonal planar; f. tetrahedral, trigonal planar, tetrahedral.
77. :S=C=S:, linear, sp

79. 2.400 mol   **81.** 88.4%   **83.** 145 mL   **85.** 9.330×10^2 kg
87. −1.897 kJ, exothermic   **89.** 4.82×10^4 kJ
91. a. (reaction structure)

b. 959.2 L

Chapter 10

1. Cooling a balloon cools the gas inside. As the gas cools, its molecules move more slowly and the pressure in the balloon decreases. The atmospheric pressure outside the balloon collapses the balloon. As the volume of the balloon decreases, the pressure of the gas inside increases until it just balances the pressure of the atmosphere and the balloon stops shrinking.   **3.** a. The pressure increases as the volume decreases, and vice-versa. Mathematically this can be expressed $P \times V =$ a constant, or $P =$ constant $\times 1/V$. b. Amount of gas and temperature.   **5.** A balloon filled with H_2 will sink first. H_2 has the smaller molar mass and will leak out (diffuse through pinholes) 1.4 times faster than the heavier He.   **7.** H_2O. Cooling shows the velocities of the He atoms.
9. The slope would increase by a factor of 2.   **11.** Yes. As Fig. 10.30 shows, at any given instant there are a range of values of molecular speeds in a sample of gas. Any single molecule can speed up or slow down as it collides with other molecules. The *average* velocity of all the molecules is constant at constant temperature.   **13.** CO_2 molecules attract each other and have a finite volume. Thus they only approximate ideal behavior.
15. Strictly microscopic: d; strictly macroscopic: a, b, c, e; both: f
17. 0.809 atm; 82.0 kPa   **19.** 2.2×10^2 kPa   **21.** 0.987 atm, 100 kPa   **23.** 4.88 torr, 0.650 kPa   **25.** $CO_2 < N_2O <$ He
27. 755 mm   **29.** 16.3 to 16.5 L   **31.** 944 mL   **33.** 11.2 L
35. 1.27×10^5 L   **37.** 3.40×10^3 torr   **39.** 9.74×10^{-5} mol
41. 8.190×10^{-2} mol, 5.553 g   **43.** 5.49 L   **45.** 420 K or 147°C   **47.** 68.4 atm   **49.** 1.84 g L^{-1} **51.** 215 mol   **53.** a. 7.25 $\times 10^{-2}$ g, b. 23.1 g, c. 1.5×10^{-4} g   **55.** 46.4 g   **57.** 6 atm
59. 1.136 atm, 863.1 torr, 115.1 kPa   **61.** CO_2, 7.1 atm; O_2, 17.5 atm; N_2 121 atm   **63.** 740 torr   **65.** 503 mL   **67.** Dry air is denser; it has a larger average molecular mass.   **69.** a. Determine the moles of HgO that decompose; using the chemical equation, determine the moles of O_2 produced by decomposition of this

amount of HgO; determine the volume of O_2 from the moles of O_2, temperature, and pressure. b. 0.308 L   **71.** a. Determine the moles of water in 15.0 g; using the chemical equation, determine the moles of H_2 produced from this amount of water; determine the volume of H_2 from the number of moles of H_2, its temperature, and its pressure. b. 20.4 L   **73.** 4.0 L, 8.0 L   **75.** a. Look up the moles of O_2 consumed (Example 10.9); using the chemical equation, determine the moles of glucose that react with this amount of O_2; determine the molar mass of glucose; use the molar mass and the number of moles of glucose consumed to determine the mass required. b. 17 g   **77.** 30.0 g mol^{-1}
79. 3:1   **81.** N_2O **83.** 71.4 L   **85.** 88.0 g mol^{-1}, PF_3
87. B_2Cl_4 **89.** Average kinetic energies are equal (1:1 ratio). $u_{SO_2}/u_{O_2} = 0.70673$ **91.** 0.43 km/s   **93.** H_2O diffuses 1.055 times faster than D_2O.   **95.** 1.4 to 1; 1.2 to 1   **97.** 51.7 cm
99. Most like ideal gas, b. Deviations occur with a and c.
101. SF_6 **103.** One liter of CH_4 and one liter of H_2 contain the same number of molecules at STP (Avogadro's law), but each CH_4 molecule contains 4 H atoms while each H_2 molecule contains 2 H atoms.   **105.** a. Determine the empirical formula of the compound from the percent composition; determine the molar mass from the mass of the sample and its volume, temperature, and pressure; determine the molecular formula from the molar mass and the empirical formula; b. CH_5N or CH_3NH_2
107. a. Determine the empirical formula from the percent composition; determine the molecular formula from the empirical formula and the molar mass; draw the Lewis structure of the compound; identify the geometry from the electron pair geometry; b. All 3 carbon atoms are tetrahedral.   **109.** 2×10^{-10} atm or 2×10^{-7} torr   **111.** 18.0 L, 0.533 atm   **113.** 6.8×10^4 g mol^{-1}
115. 17.2%

Chapter 11

1. Liquids have no fixed shape (they flow), are incompressible, and have larger densities than gases. Solids are rigid, are incompressible, and have larger densities than gases.   **3.** Macroscopic: critical temperature, boiling point, phase diagram, surface tension, vapor pressure, variable shape; microscopic: kinetic-molecular description, a dipole-dipole attraction, a hydrogen bond   **5.** Macroscopic: 11.8, 11.11, 11.15, 11.18, 11.21; microscopic: 11.1, 11.3, 11.5, 11.7, 11.24
7. Macroscopic: θ, the angle of diffraction; microscopic: λ, the wave length and d, the interplanar spacing.   **9.** The heat increases the kinetic energy of a limited number of molecules in the liquid water so that they have enough energy to escape into the gas phase. The escape of these molecules removes the additional energy so the temperature of the remaining liquid does not increase. The amount of liquid water changes as the additional heat causes some to change to a gas.   **11.** Intramolecular: covalent bonds in the HCO_3^- ion. Intermolecular: attractions between Na^+ and HCO_3^- ions, $O—H \bullet \bullet \bullet O$ hydrogen bonds, dipole-dipole attractions between CO bonds, London forces among ions.   **13.** Similarities: both are composed of atoms, ions, or molecules; there is very little empty space between these particles; motion of the particles changes with changing temperature. Differences: particles in the solid are fixed in a regular arrangement; although they jiggle, they do not move about; particles in the liquid have no regular arrangement and are free to move past each other.   **15.** The motion changes from vibration to a mixture of vibration and translation.   **17.** The molecules are forced closer together and their attractive forces become strong relative to their kinetic energy.   **19.** Molecules in a liquid are free to move

past each other while molecules in a solid are held in fixed positions. **21.** The London forces increase as the number of electrons increases.   **23.** Both have approximately equal London forces, but because it contains an OH group, 1-propanol forms hydrogen bonds which are stronger intermolecular forces than the dipole-dipole attractions in acetone that result from the C=O bond.   **25.** One mole of C_3H_8 molecules contains more H atoms and more $H \bullet \bullet \bullet H$ intermolecular forces than C_2H_6, which contains more than CH_4. The larger the number of $H \bullet \bullet \bullet H$ attractions to be overcome, the greater the enthalpy of vaporization.   **27.** $-85°C$ **29.** The molecules become more polar and the intermolecular forces increase as the melting temperatures increase from $-185°C$ to $-85°C$ **31.** Cu, metallic bonds; O_2 London forces; NO, dipole-dipole forces; HF, hydrogen bonding; Si, covalent bonds; CaO, ionic bonds   **33.** d
35. a. Br_2, b. LiCl, c. Kr, d. Cl_2 **37.** Molecules of gases are widely separated while molecules in a liquid touch; thus a given volume of a gas contains a much smaller number of molecules and a much smaller mass than the same volume of a liquid.   **39.** The thermal energy (heat) needed to evaporate the liquid is removed from the skin.   **41.** The attractions in diethyl ether (CH_3OCH_3) are due to London forces and dipole-dipole attraction. Ethanol (CH_3CH_2OH) has hydrogen bonding, in addition.   **43.** We can see the amount of liquid in an open container decrease and we can smell the vapor of some liquids.   **45.** The vapor pressure of a series of liquids decreases as the strength of the intermolecular forces increases.
47. As the temperature increases, the average kinetic energy of the molecules increases and more molecules have sufficient energy to escape from the liquid.   **49.** See the answer to exercise 47.
51. When the pressure on the liquid is exactly 1 atm.   **53.** a. $SiH_4 <$ $H_2S <$ HCl; b. $F_2 < Cl_2 < Br_2$; c. $CH_4 < C_2H_6 < C_3H_8$; d. $O_2 <$ NO < CaO   **55.** a. About 450 torr; b. decrease   **57.** Yes, ice will sublime although it may take it several days.   **59.** a. solid, b. liquid, c. liquid, d. gas, e. solid, f. gas   **61.** At 2 torr: s $\longrightarrow$ g at about $-10°C$. At 400 torr: s $\longrightarrow$ l at 0°C, l $\longrightarrow$ g at about 82°C.
63. a. liquid, b. solid, c. gas, d. gas, e. gas, f. gas   **65.** At 15 atm: s $\longrightarrow$ l at about $-45°C$, l $\longrightarrow$ g at about $-20°C$. At 5 atm: s $\longrightarrow$ g at about $-58°C$.
67.

69. The temperature and pressure at which a substance can exist as a mixture of solid, liquid, and gas in equilibrium.   **71.** 647.1K and 217.7 atm, the critical temperature and pressure of water.
73. Metallic or covalent network, atoms; ionic, ions; molecular, molecules   **75.** e   **77.** It is not crystalline; the B—O network has a random arrangement.   **79.** metallic: Cu; network: SiO_2, C; ionic: KCl, $BaSO_4$, NH_4F; molecular: NH_3, C_2H_5OH, CO   **81.** 8   **83.** 12
85. Face-centered cubic, or cubic close packed.   **87.** 2.176 Å, 3.595 g cm^{-3} **89.** 3.90 Å, 21.8 g cm^{-3} **91.** CdS   **93.** Co_3O_4
95. TlI   **97.** NaCl: a, c; CsCl: d; ZnS: b, e; CaF_2: f   **99.** RbI
101. a. 1.49 Å, b. 1.372 g cm^{-3} **103.** MnF_3, 6, 4.02 Å, 2.86 g cm^{-3} **105.** 890 kJ mol^{-1} **107.** Each would make the reaction less exothermic.   **109.** 19.3 g cm^{-3} **111.** a   **113.** a. The crystal structure of Si shows that it is less tightly packed (coordination number 4) in the solid than Al (C.N. = 12). b. Si   **115.** When water melts, only a fraction of the hydrogen bonds in the solid are

broken. When it evaporates all of the hydrogen bonds are broken.
117. 2.79×10^{-15} J or 1.74×10^4 eV   **119.** 59.95%, +4

Chapter 12

1. A solution can vary in composition, a compound cannot vary in composition. Solutions are homogeneous, other mixtures are heterogeneous.   **3. a.** The solutions are the same throughout (the color is constant throughout), and the composition of these solutions of $K_2Cr_2O_7$ in water can vary.   **5.** Macroscopic: boiling-point elevation, Henry's law, molarity, nonelectrolyte, nonstoichiometric compound, osmosis; microscopic: hydrogen bond, ion-dipole attraction, solvated ion   **7.** Macroscopic: 12.6, 12.12, 12.17; microscopic: 12.3, 12.8, 12.24   **9.** Molecules of sugar do not form ions in solution. Solid potassium chloride consists of ions that separate when they dissolve in water, move about independently, and carry a charge. A small fraction of the molecules of acetic acid present in solution react with water to form hydronium ions and acetate ions; the relative small number of these ions makes the solution only weakly conducting.   **11.** The solubility of solids usually decreases upon cooling a solution, while the solubility of gases usually decreases upon warming.   **13.** It decreased.   **15. a.** the more dilute, **b.** the more concentrated, **c.** the more concentrated   **17. a.** A solution that forms a vapor with the same composition (mole fractions) of components as the solution. **b.** Boil a solution of HNO_3 with a little more HNO_3 than in a constant boiling solution or with a little less HNO_3 than in a constant boiling solution.   **19.** Add a small crystal of $Na_2S_2O_3$. It will dissolve in an unsaturated solution, remain unchanged in a saturated solution, or cause a supersaturated solution to crystallize.   **21. a.** Ion-dipole, **b.** hydrogen bonds, **c.** London forces, **d.** dipole-dipole attractions, **e.** London dispersion forces   **23.** HBr molecules react with water molecules and form H_3O^+ and Br^- ions.   **25.** Heat is released when the total intermolecular forces (IMFs) between the solute and solvent molecules are stronger than the total IMFs in the pure solute and in the pure solvent: replacing weaker IMFs with stronger IMFs releases heat. Heat is absorbed when the total IMFs in the solution are weaker than the total of those in the pure solute and in the pure solvent: replacing stronger IMFs with weaker IMFs absorbs heat.   **27.** The hydrogen bonds between water and C_2H_5OH are much stronger than the hydrogen bonds between water and C_2H_5SH.   **29.** The ions in the solids are fixed in place while those in the molten compounds are free to move around.   **31. a.** Calculate the mass of pure sulfuric acid needed to make 200 g of a 20% solution; calculate the mass of the 95% solution that contains this amount of pure sulfuric acid. **b.** 42 g   **33.** 375 g   **35.** 19.9 g   **37. a.** determine the number of moles of glucose in 0.500 L of solution; determine the molar mass of glucose; determine the mass of glucose from the number of moles and its molar mass. **b.** 27 g   **39. a.** 37.0 mol, 3.63 kg; **b.** 3.8×10^{-6} mol, 1.9×10^{-4} g; **c.** 73.1 mol, 2.20 kg **d.** 5.8×10^{-7} mol, 8.9×10^{-5} g   **41.** Determine the molar mass of $KMnO_4$; determine the number of moles of $KMnO_4$ in the solution; from the number of moles and the volume of solution, determine the molarity. **b.** 1.15×10^{-3} M   **43. a.** 5.04×10^{-3} M, **b.** 0.499 M, **c.** 9.92 M, **d.** 1.1×10^{-3} M   **45.** 0.025 M   **47.** 1.75×10^{-3} M   **49. a.** Determine the molar mass of H_3PO_4; determine the number of moles of acid in the solution; from the number of moles and the mass of solvent, determine the molality. **b.** 1.18 m   **51. a.** 3.96 m, **b.** 0.15 m, **c.** 1.247 m, **d.** 0.79 m   **53.** 4.2×10^{-3} m   **55. a.** Determine the molar masses of H_3PO_4 and of H_2O; determine the number of moles of H_3PO_4; determine the number of moles of H_2O; determine the

mole fraction of H_3PO_4 and H_2O. **b.** 0.0209, 0.979.   **57. a.** H_2SO_4 0.0666, H_2O 0.9334; **b.** NaCl 0.0026, H_2O 0.9974; **c.** $C_{18}H_{21}NO_3$ 0.05433, C_2H_5OH 0.94567; **d.** I_2 0.035, C_2H_5OH 0.965   **59.** Methanol 0.39, ethanol 0.27, water 0.34   **61. a.** 0.95 g, **b.** 5.3×10^{-3} M   **63.** 1.6 L   **65.** 49.6 mL   **67.** 0.91 L   **69.** H_2 0.78, CO 0.038, CO_2 0.18   **71.** 54%, 14 m   **73.** Distilled water. **75. a.** Determine the molar mass of sucrose; determine the number of moles of sucrose in the solution; from the number of moles and the mass of solvent, determine the molality; determine the difference between the boiling point of water and the boiling point of the solution; determine the new boiling point. **b.** 100.491°C.   **77. a.** Determine the molar mass of sucrose; determine the number of moles of sucrose in the solution; from the number of moles and the mass of solvent, determine the molality; determine the difference between the freezing temperature of water and the freezing temperature of the solution; determine the new freezing temperature. **b.** −1.8°C **79. a.** Determine the molar mass of $Ca(NO_3)_2$; determine the number of moles of $Ca(NO_3)_2$ in the solution; determine the number of moles of ions in the solution; determine the molarity of ions from the moles of ions and the volume of the solution, then determine the osmotic pressure. **b.** 2.67 atm   **81. a.** Determine the molal concentration from the change in boiling point and K_b; determine the moles of solute in the solution from the molal concentration and mass of solvent; determine the molar mass from the number of moles and the mass of solute. **b.** 2.1×10^2 g mol^{-1} **83.** No. If HCl were ionized the concentration of ions would be 2 m and the freezing point would be −5°C.   **85.** 144 g mol^{-1} **87.** Dissolve 284 g of glycerin in 1.00 kg of water, −5.73°C   **89.** S_8 **91.** 1.39×10^4 amu **93.** 54 g   **95.** 100.25°C   **97.** Colloidal dispersions consist of particles that are much bigger than the average small molecule such as H_2O or C_8H_{18}. Colloidal particles are either very large molecules or aggregates of smaller species that usually are big enough to scatter light. Suspensions are homogeneous on a macroscopic scale. **99.** See Section 12.23.   **101.** Colloid: milk, ink, and some shampoos; solution: concentrated $KMnO_4$; pure substance, mercury, bromine   **103.** As a solute, dissolved air will change the freezing point.   **105.** 43%   **107.** 47 M, 4.3 m, 0.44%   **109.** 243 amu **111.** 6.59%, 0.442 m   **113.** 15.5 mL   **115.** No   **117.** $C_{12}H_{10}$ **119.** $C_6H_{12}O_6$ **121.** 0.282 M

Chapter 13

1. A rate of reaction is a change in concentration per unit of time; the rate of a reaction usually varies with concentration of reactants and products. A rate constant is a number (that is usually multiplied by concentrations) used to calculate a rate of reaction. The rate constant for a given reaction does not change unless the temperature changes. **3.** Macroscopic: first order reaction, half-life of a reaction, homogeneous catalysts, integrated rate laws, rate of reaction; microscopic: activated complex, elementary reaction, reaction mechanism, unimolecular elementary reaction   **5.** Macroscopic: 13.1, 13.4, 13.9, 13.16, 13.20; microscopic: 13.19, 13.23   **7.** A linear plot of [A] *versus* t indicates a 0 order reaction with slope = −k. A linear plot of ln[A] *versus* t indicates a first order reaction with k = −slope. A linear plot of 1/[A] *versus* t indicates a second order reaction with k = slope.   **9.** The reactants may be moving too slowly to have enough kinetic energy to exceed the activation energy for the reaction.   **11.** Rate = $k[A][B]^2$; Rate = $k[A]^3$ **13.** The instantaneous rate is the rate of a reaction at any particular point in time, a period of time that is so short that the concentrations of reactants and products change by a negligible amount. The initial rate is the instantaneous rate of a reaction as it starts (as product just begins to form). Average rate is the average of the instantaneous rates over a time period.   **15.** The temperature of boiling water at higher elevations is lower than at sea level. The lower cooking temperature at Denver makes the rate of cooking slower and the time to completion, longer.   **17.** $\nu = -\frac{1}{2}\,\Delta[O_3]/\Delta t = \frac{1}{3}\,\Delta[O_2]/\Delta t$ **19.** a. 3.10×10^{-6} M s^{-1}, 1.04×10^{-6} M s^{-1}; b. about 7×10^{-7} M s^{-1} c. 1.55×10^{-6} M s^{-1}, about 3.5×10^{-7} M s^{-1} **21.** 2.1×10^{-4} M s^{-1} **23.** a. 2nd order, b. 1st order   **25.** a. Reduce the rate by a factor of 4; b. No effect; c. Increase the rate, but by an unknown amount.   **27.** 1.6×10^{-4} mol d^{-1} **29.** 4.3×10^{-5} mol L^{-1} s^{-1} **31.** Rate $= k$; $k = 2.0 \times 10^{-2}$ mol L^{-1} h^{-1} (about 0.9 g L^{-1} h^{-1} for the average male); 0 order   **33.** Rate $= k[NOCl]^2$; $k = 8.0 \times 10^{-8}$ L mol^{-1} s^{-1}; 2nd order   **35.** Rate $= k[NO]^2[Cl_2]$; $k = 9.12$ L^2 mol^{-2} h^{-1}; 2nd order in NO; 1st order in Cl$_2$ **37.** a. Second, Rate $= k[A]^2$ b. 7.88×10^{-3} L mol^{-1} s^{-1} **39.** 1st order; $k = 2.2 \times 10^{-5}$ s^{-1} **41.** 2nd order, $k = 50.4$ L mol^{-1} h^{-1} **43.** 14.3 day   **45.** 8.3×10^7 s   **47.** a. 2.5×10^{-4} mol L^{-1} min^{-1} b. 5.0×10^{-4} mol L^{-1} min^{-1}; 1.2×10^{-4} mol L^{-1} min^{-1} **49.** Rate $= k[I^-][OCl^-]$, $k = 6.1 \times 10^{-2}$ L mol^{-1} s^{-1} **51.** 0.826 s   **53.** The activation energy is the minimum amount of energy necessary to form the activated complex in a reaction. It is usually expressed as the energy necessary to form one mole of activated complex.   **55.** It increases the rate by a factor of 2 because the average kinetic energy of the molecules involved increases so more collisions are effective in producing activated complexes.   **57.** 12 min   **59.** 191 kJ   **61.** c   **63.** No, yes   **65.** b, d, e   **67.** Both change the mechanism to one with a lower activation energy, thus producing a faster reaction. Homogeneous catalysts work in the same phase as the reactants; heterogeneous catalysts work in a different phase than the reactants.   **69.** a. Rate$_1 = k[O_3]$, Rate$_2 = k[O_3][Cl]$, Rate$_3 = k[ClO][O]$; b. Rate$_1 = k[O_3]$, Rate$_2 = k[O_3][NO]$, Rate$_3 = k[NO_2][O]$ **71.** Rate $= k[\text{penicillin}][\text{penicillinase}]$, $k = 1.0 \times 10^7$ L mol^{-1} min^{-1} **73.** The rate of the first reaction (breaking a C—C bond) will be largest because the C—C bond energy (its activation energy) is smaller than the C—H bond energy.   **75.** Tc, 0.4%; Tl, 63%   **77.** 1.5×10^{-3} M, 38 torr   **79.** a. 6.319×10^{-3} min, b. 12.0%, c. 1458 min (24.29 hr)

Chapter 14

1. Macroscopic behavior. The study is concerned with concentrations of compounds; these are macroscopic properties.   **3.** The system is not closed; one of the components of the equilibrium, the Br$_2$ vapor, would escape from the bottle. Thus more liquid would evaporate than gas condense.   **5.** There is a significant effect only when the number of moles of gaseous reactants in the equation that describes the equilibrium reaction differs from the number of moles of gaseous products.   **7.** The amount of CaCO$_3$ must be so small that P_{CO_2} is less than Q_P when all of the CaCO$_3$ has decomposed.   **9.** a. $K = [Ag^+][Cl^-]$ is less than 1. AgCl is insoluble thus the concentrations of ions are much less than 1 M. b. $K = 1/[Pb^{2+}][Cl^-]^2$ is greater than one because PbCl$_2$ is insoluble and formation of the solid will reduce the concentration of ions to a low level.   **11.** K about 10.   **13.** It would become lighter as Fe^{3+} was removed from the solution and the FeNCS^{2+} ion decomposed to compensate.   **15.** No, the system is not confined so products escape continuously and reactants are added continuously.   **17.** $K > 1$.   **19.** a. $[CH_3Cl][HCl]/[CH_4][Cl_2]$, b. $[NO]^2/[N_2][O_2]$, c. $[SO_3]^2/[SO_2]^2[O_2]$, d. $[SO_2]$, e. $1/[P_4][O_2]^5$, f. $[Br]^2/[Br_2]$, g. $[CO_2]/[CH_4][O_2]^2$, h. $[H_2O]^5$ **21.** a. 25, shifts left; b. 0.22, shifts

right; c. ∞, shifts left; d. 1, shifts right; e. 0, shifts right; f. 4, shifts left   **23.** Values for [NO$_2$], [N$_2$O$_4$] and Q are 0.10 M, 0 M, 0; 0.080 M, 0.01 M, 1.6; 0.060 M, 0.020 M, 5.6; 0.040 M, 0.030 M, 19; 0.020 M, 0.040 M, 100; 0.016 M, 0.042 M, 1.6×10^2 **25.** Homogeneous: a, b, c, f; heterogeneous: d, e, g, h.   **27.** a, b.   **29.** a. 1.6×10^{-4} atm^{-2}; b. 50.2; c. 5.33×10^{-39}; d. 4.60×10^{-3} **31.** Add N$_2$, add H$_2$, increase the pressure, heat the reaction.   **33.** a. shift right, shift left; b. shift right, no effect; c. shift left, shift left; d. shift left, shift right.   **35.** a. $K = [CH_3OH]/[H_2]^2[CO]$; b. 1, [H$_2$] increase, [CO] decrease, [CH$_3$OH] increase; 2, [H$_2$] increase, [CO] decrease, [CH$_3$OH] decrease; 3, [H$_2$] increase, [CO] increase, [CH$_3$OH] increase; 4, [H$_2$] increase, [CO] increase, [CH$_3$OH] increase; 5, [H$_2$] increase, [CO] increase, [CH$_3$OH] decrease; 6. no changes.   **37.** a. $K = [CO][H_2]/[H_2O]$; b. In each of the following cases the mass of carbon will change, but its concentration (activity) will not change. 1. [H$_2$O] no change, [CO] no change, [H$_2$] no change; 2. [H$_2$O] decrease, [CO] decrease, [H$_2$] decrease; 3. [H$_2$O] increase, [CO] increase, [H$_2$] decrease; 4. [H$_2$O] increase, [CO] increase, [H$_2$] increase; 5. [H$_2$O] decrease, [CO] increase, [H$_2$] increase.   **39.** b.   **41.** Add NaCl or some other salt that produces Cl$^-$ in the solution. Cool the solution.   **43.** a.   **45.** 7.1×10^2 **47.** a. rate $= k[CO][Cl]$; b. $K = [Cl]^2/[Cl_2]$; c. rate $= kK^{1/2}[CO][Cl_2]^{1/2} = k'[CO][Cl_2]^{1/2}$ **49.** Rate $= k[A]$, first order; b. k $= 0.294$ h^{-1}; c. k$_2 = 0.588$ h^{-1} **51.** $K = [C]^2/[A][B]^2$; There are many different sets of equilibrium concentrations; two are [A] $= 0.1$ M, [B] $= 0.1$ M, [C] $= 1$ M, and [A] $= 0.01$, [B] $= 0.250$, [C] $= 0.625$.   **53.** f.   **55.** 6.00×10^{-2} **57.** 0.50 mol L^{-1} **59.** 1.9×10^3 atm^{-1} **61.** 3.06 atm^2 **63.** a. -2Δ, 2Δ, -0.250 M, 0.250 M; b. 4Δ, -2Δ, -6Δ, 0.32 M, -0.16 M, -0.48 M; c. -2Δ, 3Δ, -50 torr, 75 torr; d. Δ, $-\Delta$, -3Δ, 5 atm, -5 atm, -15 atm; e. Δ, 1.03×10^{-4} M; f. -4Δ, -1.6 atm   **65.** Activities of pure crystalline solids equal 1 and are constant; however, the mass of Ni does change.   **67.** [NH$_3$] $= 9.1 \times 10^{-2}$ M   **69.** 4.9×10^{-2} atm   **71.** [CO] $= 2.0 \times 10^{-4}$ M   **73.** 3.6×10^{-3} atm   **75.** Calculate Q based on the calculated concentrations and see if it is equal to K.   **77.** a. [NO$_2$] $= 1.17 \times 10^{-3}$ M, [N$_2$O$_4$] $= 0.128$ M; b. The change in concentration of N$_2$O$_4$ is less than 1%.   **79.** Pressure of S$_2$ $= 7.2 \times 10^{-3}$ atm, pressure of H$_2$ $= 1.4 \times 10^{-2}$ atm, pressure of H$_2$S $= 0.810$ atm; b. The change in the pressure of H$_2$S is 0.9%.   **81.** [H$_2$O] $=$ [Cl$_2$O] $= 0.869$ M, [HOCl] $= 0.262$ M   **83.** Pressure of Cl$_2$ $= 0.028$ atm; of NO $= 0.056$ atm; of NOCl $= 0.47$ atm   **85.** [O$_2$] $= 0.0035$ M, [NO] $= 0.0070$ M, [NO$_2$] $= 0.19$ M   **87.** Pressure of O$_3$ $= 4.9 \times 10^{-26}$ atm.   **89.** 507 g   **91.** 33 g. With less than 33 g the CaCO$_3$ will decompose completely.   **93.** H$_2$ 0.612 mol, CO$_2$ 1.61 mol, H$_2$O 1.14 mol, CO 1.39 mol.   **95.** a. H$_2$, decrease; N$_2$, increase; NH$_3$, decrease; b. pressure of N$_2$ $= 250$ torr, pressure of NH$_3$ $= 880$ torr, pressure of H$_2$ $= 266$ torr   **97.** SbCl$_5$ 6.22 g, SbCl$_3$ 7.35 g, Cl$_2$ 2.28 g.   **99.** a.   **101.** $Q = [H_3O^+][OCN^-]/[HOCN]$ **103.** Approximately 450 torr.   **105.** a. $K = [CO_2\text{—}Hb\text{—}H^+][O_2]/[HbO_2][H_3O^+][CO_2]$; b. Lactic acid releases H$_3$O$^+$. This coupled with the increased concentration of CO$_2$ shifts the equilibrium to the right, releasing O$_2$.   **107.** a. [sucrose] $= 1.65 \times 10^{-7}$ M, [glucose] $=$ [fructose] $= 0.15$ M; b. 6.5×10^{11} s.   **109.** 1.01×10^{-4} atm, 1.26×10^{-6} mol   **111.** a. HB is stronger. b. For HA, K $= 5 \times 10^{-4}$; for HB K $= 3 \times 10^{-3}$ **113.** 0.63 mol

Chapter 15

1. The introductory definition is a macroscopic definition in that it involves amounts of hydronium ion; amounts, such as concentrations, are macroscopic concepts. The other two definitions are microscopic because they are presented in terms of transfer of microscopic entities—a hydrogen ion or a pair of electrons. **3.** a. base and solvent, acidic; b. acid and solvent, basic; c. acid and solvent, basic; d. acid and solvent, basic; e. solvent, acidic; f. solvent, basic; g. solvent, neutral.   **5.** One example for NH_3 as a conjugate acid: $NH_2^- + H^+ \longrightarrow NH_3$; as a conjugate base: $NH_4^+ + OH^- \longrightarrow NH_3 + H_2O$.   **7.** $[H_2O] > [CH_3CO_2H] > [H_3O^+] = [CH_3CO_2^-] > [OH^-]$.   **9.** At a given concentration, stronger acids produce greater concentrations of hydronium ion. A K_a is a kind of ratio of hydronium ion concentration to acid concentration, and K_as increase in proportion to the increasing ability of an acid to produce more hydronium ion.   **11.** a. Both give a 100% yield of OH^- when dissolved in water because both are stronger bases than OH^-. b. Measure the percent yields of product in the reactions of NH_2^- with C_2H_5OH and of $C_2H_5O^-$ with NH_3; the stronger base will give the higher yield.   **13.** The oxidation state of the sulfur in H_2SO_4 is greater than the oxidation state of the sulfur in H_2SO_3. See Section 15.5 for an explanation of how oxidation state affects acid strength.   **15.** Examine the reactions of samples of the hydroxide with a strong acid and a strong base. An amphoteric hydroxide can react both as an acid and as a base.   **17.** $H_3O^+ + OH^- \longrightarrow 2H_2O$ **19.** a. $H_3O^+ \longrightarrow H^+ + H_2O$, b. $HCl \longrightarrow H^+ + Cl^-$, c. $NH_3 \longrightarrow H^+ + NH_2^-$, d. $CH_3CO_2H \longrightarrow H^+ + CH_3CO_2^-$, e. $NH_4^+ \longrightarrow H^+ + NH_3$, f. $HSO_4^- \longrightarrow H^+ + SO_4^{2-}$.   **21.** a. $H_2O + H^+ \longrightarrow H_3O^+$, b. $OH^- + H^+ \longrightarrow H_2O$, c. $NH_3 + H^+ \longrightarrow NH_4^+$, d. $CN^- + H^+ \longrightarrow HCN$, e. $S^{2-} + H^+ \longrightarrow HS^-$, f. $H_2PO_4^- + H^+ \longrightarrow H_3PO_4$.   **23.** Acid, base: a. H_2O, O^{2-}; b. H_3O^+, OH^-; c. H_2CO_3, CO_3^{2-}; d. NH_4^+, NH_2^-; e. H_2SO_4, SO_4^{2-}; f. $H_3O_2^+$, HO_2^-; g. H_2S, S^{2-}; h. $H_6N_2^{2+}$, H_4N_2.   **25.** The labels are Brønsted acid BA; its conjugate base CB; Brønsted base BB; its conjugate acid CA. a. HNO_3(BA), H_2O(BB), H_3O^+(CA) NO_3^-(CB); b. CN^-(BB), H_2O(BA), HCN(CA), OH^-(CB); c. H_2SO_4(BA), Cl^-(BB), HCl(CA), HSO_4^-(CB); d. HSO_4^-(BA), OH^-(BB), SO_4^{2-}(CB), H_2O(CA); e. O^{2-}(BB), H_2O(BA) OH^-(CB and CA); f. $[Cu(H_2O)_3(OH)]^+$(BB), $[Al(H_2O)_6]^{3+}$(BA), $[Cu(H_2O)_4]^{2+}$(CA), $[Al(H_2O)_5(OH)]^{2+}$(CB); g. H_2S(BA) NH_2^-(BB), HS^-(CB), NH_3(CA).   **27.** a. $HCl + H_2O \longrightarrow H_3O^+ + Cl^-$ b. $HNO_3 + H_2O \longrightarrow H_3O^+ + NO_3^-$ c. $NH_3 + H_2O \longrightarrow NH_4^+ + OH^-$ d. $NH_2^- + H_2O \longrightarrow NH_3 + OH^-$ e. $HClO_4 + H_2O \longrightarrow H_3O^+ + ClO_4^-$ f. $F^- + H_2O \longrightarrow HF + OH^-$ g. $NH_4^+ + H_2O \longrightarrow H_3O^+ + NH_3$ **29.** $CuO + 2HNO_3 \longrightarrow Cu(NO_3)_2 + H_2O$ **31.** Amphiprotic: a. $H_2O + HBr \longrightarrow H_3O^+ + Br^-$, $H_2O + CN^- \longrightarrow HCN + OH^-$; b. $H_2PO_4^- + HBr \longrightarrow H_3PO_4 + Br^-$, $H_2PO_4^- + OH^- \longrightarrow HPO_4^{2-} + H_2O$; e. $HSO_4^- + HClO_4 \longrightarrow H_2SO_4 + ClO_4^-$, $HSO_4^- + OH^- \longrightarrow SO_4^{2-} + H_2O$. Not amphiprotic: S^{2-}, CH_4, $Al(H_2O)_6^{3+}$ **33.** 2.3×10^{-11} **35.** 1×10^{-8} **37.** $(CH_3)_2NH$, $CH_3NH_3^+$ **39.** $HBrO$ **41.** a. HF; b. $B(OH)_3$; c. HSO_4^-; d. H_2S; e. H_2Te.   **43.** a. $HCl < HBr < HI$; b. $Cl^- < H_2O < OH^- < H^-$, c. $ClO_3(OH) < Si(OH)_4 < Mg(OH)_2$; d. $CH_4 < NH_3 < H_2O < HF$; e. $ClO_4^- < ClO_3^- < ClO_2^- < ClO^-$; f. $HOI < HOIO < HOIO_2 < HOIO_3$.   **45.** a. $[H_2O] > [Na^+] = [NO_3^-] > [H_3O^+] = [OH^-]$; b. $[H_2O] > [Br^-] > [NH_4^+] > [NH_3] = [H_3O^+] > [OH^-]$; c. $[H_2O] > [Na^+] = [NO_3^-] > [NH_3] > [NH_4^+] = [OH^-] > [H_3O^+]$; d. $[H_2O] > [K^+] = [Cl^-] > [CH_3CO_2H] > [H_3O^+] = [CH_3CO_2^-] > [OH^-]$ **47.** CN^- **49.** Evolution of H_2 with active metals, $Mg + 2HCl \longrightarrow MgCl_2 + H_2$; Neutralize oxides or hydroxides, $K_2O + 2HBr \longrightarrow 2KBr + H_2O$; React with salts of weaker acids, $NiS +$

$H_2SO_4 \longrightarrow NiSO_4 + H_2S$.   **51.** $H_2S + H_2O \longrightarrow H_3O^+ + HS^-$, $HS^- + H_2O \longrightarrow H_3O^+ + S^{2-}$; $H_3PO_4 + H_2O \longrightarrow H_3O^+ + H_2PO_4^-$, $H_2PO_4^- + H_2O \longrightarrow H_3O^+ + HPO_4^{2-}$, $HPO_4^{2-} + H_2O \longrightarrow H_3O^+ + PO_4^{3-}$.   **53.** a. $KOH(aq) + HClO_4(aq) \longrightarrow KClO_4(aq) + H_2O(l)$; b. $NH_3(g) + HClO_4(aq) \longrightarrow NH_4ClO_4(aq)$; c. $2Al(s) + 6HClO_4(aq) \longrightarrow 2Al(ClO_4)_3(aq) + 3H_2(g)$; d. $Al_2S_3(s) + 6HClO_4(aq) \longrightarrow 2Al(ClO_4)_3(aq) + 3H_2S(aq)$, or $2Al(ClO_4)_3(aq) + 3H_2S(g)$; e. $Na_2O(s) + 2HClO_4(aq) \longrightarrow 2NaClO_4(aq) + H_2O(l)$; f. $CaCO_3(s) + 2HClO_4(aq) \longrightarrow Ca(ClO_4)_2(aq) + CO_2(g) + H_2O(l)$.   **55.** Acidic; the ammonium ion is a weak acid.   **57.** Basic; the fluoride ion is a weak base.   **59.** Strongly acidic; the ClO_4^- ion is such a weak base that no H_3O^+ is removed from the solution.   **61.** See Section 15.9 for methods and examples.   **63.** Convert the mass of Na_2CO_3 to moles of Na_2CO_3, find the moles of HCl required to react with this number of moles of Na_2CO_3, find the volume of the solution of HCl that contains the required number of moles of HCl; 78.6 mL   **65.** Find the number of moles of oxalic acid contained in 15.00 mL of its solution, find the moles of NaOH required to react with this number of moles of oxalic acid, find the volume of the solution of NaOH that contains the required number of moles of NaOH; 13.64 mL   **67.** a. LiOH 23.9 g, H_2SO_4 49.0 g; b. HBr 80.9 g, NH_3, 17.0 g; c. KOH 56.1 g, H_2SO_3, 82.1 g; d. $Al(OH)_3$ 26.0 g, HNO_3, 63.0 g.   **69.** a. 2.0 N, b. 0.050 N, c. 0.75 N, d. 1.50 N   **71.** a. 0.20 M; b. 0.20 N   **73.** a. 9.213×10^{-4} M. b. 1.843×10^{-3} N   **75.** Write a balanced equation for the reaction. Determine the equivalents of KOH in 32.00 mL of KOH solution, determine the volume of 0.366 N H_2SO_4 required given that 1 equivalent of H_2SO_4 reacts with 1 equivalent of KOH. From the balanced equation determine the number of moles of H_2SO_4 that will react with 1 mol of KOH (= 1 equivalent of KOH), determine the moles of H_2SO_4 that will react with the KOH in 32.00 mL of KOH solution, determine the volume of 0.366 M solution required. 16.0 mL of 0.366 M H_2SO_4; 32.0 mL of 0.366 N H_2SO_4 **77.** a. 63.77 g. b. HNO_3 **79.** 266 g   **81.** The Lewis acid is listed first in each equation.

a. $\ddot{O}=C=\ddot{O}: + :\ddot{O}-H^- \longrightarrow H-\ddot{O}-C \begin{array}{c} \ddot{O} \\ \ddot{O}:^- \end{array}$

b. $H-\ddot{O}-B\begin{array}{c} \ddot{O}-H \\ \ddot{O}-H \end{array} + :\ddot{O}-H^- \longrightarrow \left[H-\ddot{O}-B\begin{array}{c} :\ddot{O}-H \\ \ddot{O}-H \end{array}-\ddot{O}-H \right]^-$

c. $:\ddot{I}-\ddot{I}: + :\ddot{I}:^- \longrightarrow :\ddot{I}-\ddot{I}-\ddot{I}:^-$

d. $\begin{array}{c} :\ddot{Cl}: \\ :\ddot{Cl}-Al \\ :\ddot{Cl}: \end{array} + :\ddot{Cl}:^- \longrightarrow \left[\begin{array}{c} :\ddot{Cl}: \\ :\ddot{Cl}-Al-\ddot{Cl}: \\ :\ddot{Cl}: \end{array} \right]^-$

e. $\begin{array}{c} :\ddot{O}: \\ S=\ddot{O}: \\ :\ddot{O}: \end{array} + :\ddot{O}:^{2-} \longrightarrow \left[\begin{array}{c} :\ddot{O}: \\ :\ddot{O}-S-\ddot{O}: \\ :\ddot{O}: \end{array} \right]^{2-}$

83. Lewis acid-base reaction: $Al^{3+} + 6H_2O \longrightarrow Al(H_2O)_6^{3+}$; Brønsted acid-base reaction: $Al(H_2O)_6^{3+} + H_2O \longrightarrow Al(H_2O)_5(OH)^{2+} + H_3O^+$

85.

$$H-\ddot{O}-\overset{\ddot{O}}{\underset{H-\ddot{O}}{S}}-\ddot{O}: + \ddot{S}=\ddot{O} \longrightarrow H-\ddot{O}-\overset{\ddot{O}}{\underset{\ddot{O}}{S}}-\ddot{O}-\overset{\ddot{O}}{\underset{\ddot{O}}{S}}-\ddot{O}-H$$

  Lewis base   Lewis acid

87. a. HCl, b. HBr, c. H_3O^+, d. H_2O, e. HSO_4^-.   **89.** $HNO_3 + HF \longrightarrow H_2NO_3^+ + F^-$; $HF + BF_3 \longrightarrow H^+ + BF_4^-$.

$$H-\ddot{O}-N\overset{\overset{\ddot{O}}{\|}}{\underset{\ddot{O}:}{}} + H-\ddot{F}: \longrightarrow \left[H-\ddot{O}-N\overset{\overset{\ddot{O}}{\|}}{\underset{\ddot{O}-H}{}} \right]^+ + :\ddot{F}:^-$$

$$H-\ddot{F}: + :\ddot{F}-B\overset{\ddot{F}:}{\underset{\ddot{F}:}{}} \longrightarrow H^+ + \left[:\ddot{F}-\overset{:\ddot{F}:}{\underset{:\ddot{F}:}{B}}-\ddot{F}: \right]^-$$

91.

a. in HCl,
$$\left[H-\overset{H}{\underset{H}{N}}-CH_2-\overset{\overset{\ddot{O}:}{\|}}{C}-\ddot{O}-H \right]^+$$

in NaOH,
$$\left[H-\overset{}{\underset{H}{N}}-CH_2-\overset{\overset{\ddot{O}:}{\|}}{C}-\ddot{O}: \right]^-$$

b.
$$H-\overset{H}{\underset{H}{N}}-CH_2-\overset{\overset{\ddot{O}:}{\|}}{C}-\ddot{O}:$$

93. Inversely proportional.   **95.** 35.0 mL   **97.** $[H_2O] > [Na^+] > [C_6H_5CO_2^-] > [C_6H_5CO_2H] = [OH^-] > [H_3O^+]$ **99.** 0.05312 M   **101.** $2Na + 2NH_3 \longrightarrow 2NaNH_2 + H_2$ **103.** 3.03 g
105. The strong acid HCl is 100% ionized giving a total concentration of solute species of 2 m. The weak acid HF is only partly ionized giving a lower concentration of solute species.
107. Faster in 0.1 M HCl, a strong acid that is completely ionized giving $[H_3O^+] = 0.1$ M. The weak acid produces a much lower concentration of H_3O^+.

Chapter 16

1. a, d.   **3.** $[H_2O] > [Na^+] = [OH^-] > [H_3O^+]$ **5.** $[H_2O] > [NH_3] > [NH_4^+] = [OH^-] > [H_3O^+]$ **7.** $[H_2O] > [NO_2^-] > [Ba^+] > [HNO_2] = [OH^-] > [H_3O^+]$ **9.** We assume that strong electrolytes are 100% ionized and as long as the component ions are neither weak acids nor weak bases, the ionic species present result from the dissociation of the strong electrolyte. Equilibrium calculations are necessary when one or more of the ions is a weak acid or a weak base.   **11.** In a neutral solution $[H_3O^+] = [OH^-]$. At 40°C, $[H_3O^+] = [OH^-] = (2.9 \times 10^{-14})^{1/2} = 1.7 \times 10^{-7}$.

13. (1) Assume the change in initial concentration of the acid does not change as equilibrium is established. (2) Assume we can neglect the contribution of water to the equilibrium concentration of H_3O^+.
15. $[NH_3]$, $[NH_4^+]$, $[OH^-]$, Q: 0.100 M, 0 M, 0 M, 0; 0.0998 M, 0.0002 M, 0.0002 M, 4.0×10^{-7}; 0.0996 M, 0.0004 M, 0.0004 M, 1.6×10^{-6}; 0.0994 M, 0.0006 M, 0.0006 M, 3.6×10^{-6}; 0.0992 M, 0.0008 M, 0.0008 M, 6.4×10^{-6}; 0.099 M, 0.001 M, 0.001 M, 1×10^{-5}.   **17.** b.   **19.** a. $[NO_2^-]$ decreases, $[HNO_2]$ increases, $[OH^-]$ decreases. b. $[NO_2^-]$ increases, $[HNO_2]$ increases, $[OH^-]$ decreases. c. $[NO_2^-]$ increases, $[HNO_2]$ decreases, $[OH^-]$ increases. d. no effect expected. e. $[NO_2^-]$ increases, $[HNO_2]$ increases, $[OH^-]$ increases.   **21.** The reaction forms the conjugate acid of the weak base. The conjugate acid of a weak base is a weak acid and reacts with water to a slight extent giving an acidic solution.
23. The 0.1 M $[H_3O^+]$ produced by the complete ionization of the strong acid HCl suppresses the ionization of the weak acid HCO_2H (an example of Le Châtelier's principle).   **25.** The base is neutralized by the H_3PO_4 in the buffer; the acid by the $H_2PO_4^-$. Since either reaction changes the relative concentrations of H_3PO_4 and $H_2PO_4^-$ only slightly, the concentration of H_3O^+ changes only slightly.   **27.** Select an indicator that changes color over a pH range that includes the pH of the titration at its equivalence point.
29. a. $[K^+]$ 0.10 M, $[NO_3^-]$ 0.10 M; b. $[Ba^{2+}]$ 0.050 M, $[OH^-]$ 0.10 M; c. $[Na^+]$ 2.6 M, $[SO_4^{2-}]$ 1.3 M; d. $[H_3O^+]$ 0.45 M, $[Br^-]$ 0.45 M.   **31.** $[H_3O^+] = [OH^-] = 1.7 \times 10^{-7}$ M, pH = pOH = 6.77.
33. pH, pOH: a. 0.699, 13.30; b. 12.16, 1.845; c. −0.48, 14.48; d. 11.79, 2.21.   **35.** pH, pOH: a. 2.89, 11.11; b. 8.18, 5.82; c. 1.54, 12.46.   **37.** pH −0.30, pOH 14.30.   **39.** $[H_3O^+]$ 3×10^{-4}, $[OH^-]$ 3×10^{-11}.   **41.** a, d, e.   **43.** a. $K_a = 1.82 \times 10^{-5}$, b. $K_b = 3.5 \times 10^{-8}$, c. $K_a = 1.8 \times 10^{-4}$, d. $K_a = 2.3 \times 10^{-5}$.   **45.** K_a: $K_{H_2CO_3} = 4.4 \times 10^{-7}$, $K_{HCO_3^-} = 7 \times 10^{-11}$; K_b: $K_{HCO_3^-} = 2.3 \times 10^{-8}$, $K_{CO_3^{2-}} = 1.4 \times 10^{-4}$.   **47.** 1.7×10^{-5}.   **49.** 1.7×10^{-5}.
51. 1.2×10^{-2}.   **53.** 7.7×10^{-4}.   **55.** 2.1×10^{-5}.   **57.** a. K_b 1.4×10^{-11}, b. K_a 5.6×10^{-10}, c. K_b 0.1, d. K_a 1.4×10^{-11}, e. K_b 2.2×10^{-11}, f. K_b 1.7×10^{-13}.   **59.** a. $[HClO]$ 0.0092 M, $[ClO^-] = [H_3O^+]$ 1.8×10^{-5} M, $[OH^-]$ 5.6×10^{-10} M; b. $[C_6H_5NH_2]$ 0.0784 M, $[C_6H_5NH_3^+] = [OH^-]$ 6.0×10^{-6} M, $[H_3O^+]$ 1.7×10^{-9} M; c. $[HCN]$ 0.0810 M, $[CN^-] = [H_3O^+]$ 6×10^{-6} M, $[OH^-]$ 2×10^{-9} M; d. $[(CH_3)_3N]$ 0.11 M, $[(CH_3)_3NH^+] = [OH^-]$ 2.8×10^{-3} M, $[H_3O^+]$ 3.6×10^{-12} M; e. $[Fe(H_2O)_6^{2+}]$ 0.120 M, $[Fe(H_2O)_5(OH)^+] = [H_3O^+]$ 1.4×10^{-4} M, $[OH^-]$ 7.1×10^{-11} M.
61. a. $[Cl^-] = [NH_4^+]$ 0.125 M, $[NH_3] = [H_3O^+]$ 8.3×10^{-6} M, $[OH^-] = 1.2 \times 10^{-9}$ M; b. $[K^+] = [F^-]$ 0.25 M, $[HF] = [OH^-]$ 1.9×10^{-6} M, $[H_3O^+]$ 5.4×10^{-9} M; c. $[Br^-] = [(CH_3)_3NH^+]$ 0.311 M, $[(CH_3)_3N] = [H_3O^+]$ 6.5×10^{-6} M, $[OH^-]$ 1.5×10^{-9} M, d. $[Ba^{2+}]$ 0.11 M, $[CN^-]$ 0.22 M, $[HCN] = [OH^-]$ 2.3×10^{-3} M, $[H_3O^+]$ 4.3×10^{-12} M.   **63.** a. $[HCNO]$ 0.0160 M, $[CNO^-] = [H_3O^+]$ 2.36×10^{-3} M, $[OH^-]$ 4.24×10^{-12} M; b. $[(CH_3)_2NH]$ 0.10 M, $[(CH_3)_2NH_2^+] = [OH^-]$ 8.6×10^{-3} M, $[H_3O^+]$ 1.2×10^{-12} M; c. $[HF]$ 0.092 M, $[F^-] = [H_3O^+]$ 8.1×10^{-3} M, $[OH^-]$ 1.2×10^{-12} M; d. $[(CH_3)_3N]$ 1.7×10^{-3} M, $[(CH_3)_3NH^+] = [OH^-]$ 3.5×10^{-4} M, $[H_3O^+]$ 2.9×10^{-11} M. e. $[Fe(H_2O)_6^{3+}]$ 0.035 M, $[Fe(H_2O)_5(OH)^{2+}] = [H_3O^+]$ 0.015 M, $[OH^-]$ 6.7×10^{-13} M.   **65.** a. $[Li^+]$ 2.53×10^{-3} M, $[C_6H_5O^-]$ 2.12×10^{-3} M, $[C_6H_5OH] = [OH^-]$ 4.07×10^{-4} M, $[H_3O^+]$ 2.5×10^{-11} M; b. $[Cl^-]$ 2.38×10^{-3} M, $[C_6H_5NH_3^+]$ 2.16×10^{-3} M, $[C_6H_5NH_2] = [H_3O^+]$ 2.2×10^{-4} M, $[OH^-] = 4.6 \times 10^{-11}$ M; c. $[I^-]$ 1.0×10^{-3} M, $[C_5H_5NH^+]$ 9×10^{-4} M, $[C_5H_5N] = [H_3O^+]$ 7.4×10^{-5} M, $[OH^-]$ 1.4×10^{-10}; d. $[Ca^{2+}]$ = 0.010 M, $[OI^-]$ = 0.017 M, $[HOI] = [OH^-]$ 2.7×10^{-3} M, $[H_3O^+]$ 3.7×10^{-12} M.   **67.** pH, pOH: a. 4.74, 9.26; b. 8.77, 5.23; c. 5.2, 8.8; d. 11.44, 2.56, e. 3.85,

10.15. **69.** pH, pOH: a. 5.08, 8.92; b. 8.23, 5.77; c. 5.19, 8.81; d. 11.37, 2.63. **71.** pH = 2.04; pOH = 11.96. **73.** 3.06×10^{-3} M. **75.** 5.5×10^{-3} M. **77.** c. **79.** [HClO] 0.10 M, [HCO_2H] 0.24 M, [ClO^-] 5.2×10^{-7} M, [HCO_2^-] = [H_3O^+] 0.0067 M, [OH^-] 1.5×10^{-12} M. **81.** [C_5H_5N] 0.10 M, [CH_3NH_2] 0.24 M, [$CH_3NH_3^+$] = [OH^-] 0.010 M, [$C_5H_5NH^+$] 1.7×10^{-8} M, [H_3O^+] 1.0×10^{-12} M. **83.** [H_3O^+] and [HCO_3^-]. **85.** [H_2CO_3] 0.134 M, [H_3O^+] = [HCO_3^-] 2.4×10^{-4} M, [CO_3^{2-}] 7×10^{-11} M, [OH^-] 4.2×10^{-11} M. **87.** [H_2gly^+] = 0.072 M, [H_3O^+] = [Hgly] = 0.018 M, [gly^-] = 2.5×10^{-10} M. **89.** [$C_6H_4(CO_2H)_2$] 7.2×10^{-3} M, [$C_6H_4(CO_2H)(CO_2)^-$] = [H_3O^+] 2.8×10^{-3} M, [$C_6H_4(CO_2)_2^{2-}$] 3.9×10^{-6} M, [OH^-] 3.6×10^{-12} M. **91.** Although water is a negligible source of H_3O^+, it is the only source of OH^- in a solution of an acid. **93.** Sample points for [HF]$_{total}$ and [H_3O^+]: 1×10^{-2} M, 2.4×10^{-3} M; 1.0×10^{-6} M, 1.1×10^{-6} M; 1×10^{-7} M, 2×10^{-7} M; 1×10^{-10} M, 1×10^{-7} M. **95.** In each case [H_3O^+] in an acid solution or [OH^-] in a basic solution is greater than about 4.5×10^{-7} M. **97.** 3×10^{-7} M. **99.** 1.3×10^{-7} M. **101.** 1.5×10^{-4} M. **103.** 4.2×10^{-4} M. **105.** 0.36 M. **107.** [CH_3CO_2H], [H_3O^+], [$CH_3CO_2^-$]: a. increase, increase slightly, decrease; b. increase, decrease slightly, increase; c. no significant effect expected; d. decrease, decrease slightly, increase; e. increase, increase slightly, increase. **109.** 8.95. **111.** 37 g. **113.** a. 5.220, b. acidic, c. 5.220. **115.** HF. **117.** CH_3NH_2. **119.** a. 2.50, b. 4.01, c. 5.60, d. 8.35, e. 11.09. **121.** For selected points, mL KOH added, pH: 0 mL, 1.00; 2 mL, 1.14; 5 mL, 1.40; 9 mL, 2.16; 9.5 mL, 2.47; 9.9 mL, 3.17; 10 mL, 7.00; 10.1 mL, 10.82, 10.5 mL, 11.52; 15 mL, 12.46. **123.** Cresol red (second change), thymol blue (second change), o-cresolphthalein, or phenolphthalein **125.** Bromthymol blue, although phenolphthalein would also work. **127.** 3.0—4.9. **129.** 11.70, [SiO_3^{2-}] 0.0155 M, [SiO_3H^-] = [OH^-] 5.0×10^{-3} M, [H_2SiO_3] 3.1×10^{-5} M, [H_3O^+] 2.0×10^{-12} M. **131.** Saccharine 3.9×10^{-5} M; sodium saccharine 2.5×10^{-11} M. **133.** a. 2.50; b. 2.74; c. Salicylic acid. **135.** Yes, 5.12. **137.** Endothermic. **139.** Methyl amine. **141.** Between acetic acid and carbonic acid.

Chapter 17

1. a and d are the same. **3.** [H_2O] > [Ba^{2+}] > [SO_4^{2-}] > [HSO_4^-] = [OH^-] > [H_3O^+]. **5.** Effect on amount of solid $Ca(OH)_2$, [Ca^{2+}], [OH^-]: a. increase, decrease, increase; b. increase, increase, decrease; c. decrease, increase, decrease; d. increase, no effect, no effect; e. no effect predicted. **7.** There is no change. A solid has an activity of 1 whether there is a little or a lot. **9.** When the amount of solid is so small that a saturated solution is not produced. **11.** $CaCO_3$, MnS, CaF_2. Each is a salt of a weak acid and the [H_3O^+] from perchloric acid reduces the equilibrium concentration of the anion. **13.** $CoSO_3$, $PbCO_3$, Tl_2S. **15.** NH_3 forms a complex ion with Ag^+, reducing its concentration. HNO_3 reacts with CO_3^{2-}, reducing its concentration. **17.** a. $PbCl_2(s) \rightleftharpoons Pb^{2+}(aq) + 2Cl^-(aq)$ K_{sp} = [Pb^{2+}][Cl^-]^2; b. $Ag_2S(s) \rightleftharpoons 2Ag^+(aq) + S^{2-}(aq)$ K_{sp} = [Ag^+]^2[S^{2-}]; c. $Sr_3(PO_4)_2(s) \rightleftharpoons 3Sr^{2+}(aq) + 2PO_4^{3-}(aq)$ K_{sp} = [Sr^{2+}]^3[PO_4^{3-}]^2; d. $SrSO_4(s) \rightleftharpoons Sr^{2+}(aq) + SO_4^{2-}(aq)$ K_{sp} = [Sr^{2+}][SO_4^{2-}] **19.** a. 3.2×10^{-13}; b. 4.8×10^{-9}; c. 3.7×10^{-8}; d. 9.0×10^{-12}; e. 7.9×10^{-10} **21.** a. 8.7×10^{-7}, b. 4.6×10^{-17}, c. 1.36×10^{-4}, d. 2.3×10^{-6} **23.** a. Δ; b. Δ; c. 2Δ; d. 3Δ; e. 5Δ, 3Δ. **25.** a. [Ag^+] = [I^-] 1.2×10^{-8} M; b. [Ag^+] 2.86×10^{-2} M, [SO_4^{2-}] 1.43×10^{-2} M; c. [Mn^{2+}] 2.2×10^{-5} M, [OH^-] 4.5×10^{-5} M; d. [Sr^{2+}] 4.3×10^{-2} M, [OH^-] 8.6×10^{-2} M; e. [Mg^{2+}] 1.6×10^{-4} M; [OH^-]

3.2×10^{-4} M **27.** $SrSO_4$ **29.** a. 2×10^{-2} M; b. 1.3×10^{-3} M; c. 2.27×10^{-9} M; d. 2.2×10^{-10} M **31.** a. [Ag^+] 7.2×10^{-9} M, [Cl^-] 0.025 M; b. [Ca^{2+}] 2.2×10^{-5} M, [F^-] 0.00133 M; c. [Ag^+] 7.26×10^{-3} M, SO_4^{2-} 0.2239 M; d. [Zn^{2+}] 5.7×10^{-12} M; [OH^-] 2.8×10^{-3} M. In each case, the changes in initial concentration of the common ion is less than 5%. **33.** a. [Tl^+] 3.1×10^{-2}, [Cl^-] 6.1×10^{-3}; b. [Ba^{2+}] 1.6×10^{-3} M, [F^-] 0.0329 M; c. [Mg^{2+}] 2.67×10^{-2} M, [$C_2O_4^{2-}$] 3.1×10^{-3} M; d. [Ca^{2+}] 2.6×10^{-3}, [OH^-] 5.3×10^{-2} M. In each case, the initial concentration of the common ion changes by more than 5%. **35.** The changes in concentration are less than 5%. **37.** 1.0×10^{-24} M **39.** 1.04×10^{-5} M, 1.43×10^{-3} g **41.** a. 5.7×10^{-7} M; b. 6.6×10^{-12} M; c. 1.6×10^{-11} M. **43.** d. **45.** 7.6×10^{-3} M **47.** 5.9×10^{-7} M **49.** 6.2×10^{-5} M **51.** 1.0×10^{-4} M **53.** Yes, $Q > K_{sp}$. **55.** 3.8×10^{-5} M **57.** 1.7×10^{-8} M **59.** AgCl **61.** 0.014 M **63.** 1.0×10^{-13} M. **65.** 8.8×10^{-22} M **67.** [Hg^{2+}] 6.2×10^{-6} M, [Cl^-] 1.2×10^{-5} M. **69.** 5×10^{23} **71.** [Cd^{2+}] 9.5×10^{-5} M, [CN^-] 3.8×10^{-4} M. **73.** [Co^{3+}] 3.0×10^{-6} M, [NH_3] 1.8×10^{-5} M **75.** 2×10^{-16} M **77.** 5.9×10^{-7} M **79.** 6.2×10^{-36} M **81.** 0.89 **83.** 4.2×10^{-5} M **85.** a. yes, b. 0.93 **87.** a. 0.012 g, b. 7.0 g AgCl **89.** 1.3 g **91.** 0.80 g **93.** 0.008% **95.** 12.40 **97.** 3×10^{-3} M **99.** 3.99 kg **101.** decreases **103.** [Ag^+] 1×10^{-6} M, [CN^-] 1×10^{-10} M, [$Ag(CN)_2^-$] 1×10^{-6} M **105.** 5×10^{-3} g **107.** The active agent is $Mg(OH)_2$.

Chapter 18

1. The amount of heat released or absorbed by a system when it undergoes a change at constant temperature and pressure, the only work being due to a volume change at that pressure. **3.** The total amount of energy in the universe is constant. **5.** An enthalpy change is a heat transfer under the conditions described in the answer to Exercise 1. An internal energy change is a loss or gain of energy by both heat transfer and work. **7.** The temperature is not 298 K, and the pressures of O_2 and CO_2 are not 1 atm. **9.** a. Work done on system, $w > 0$; b. work done on surroundings, $w < 0$; c. work done on system, $w > 0$; d. essentially no work done, $w = 0$; e. work done on surroundings, $w < 0$; f. essentially no work done, $w = 0$ **11.** The amount of disorder in a substance—that is, the number of equivalent ways that a system can be arranged. **13.** Second law: Any spontaneous change occurring in the universe is accompanied by an increase in the entropy of the universe. Third law: The entropy of any pure, perfect crystalline element or compound is zero at 0 K. **15.** Increases, $\Delta S > 0$. **17.** a. $H_2 <$ HBr < $HBrO_4$; b. s < l < g; c. He < $Cl_2 < P_4$. **19.** $C_2H_5OH(s)$ at 0 K < $N_2(s)$ at $-215°C < H_2O(s)$ at $-215°C < H_2O(l)$ at 25°C < $H_2O(g)$ at 100°C < $C_2H_5OH(g)$ at 100°C. **21.** A reaction that occurs without outside influence. For a spontaneous reaction, $\Delta G < 0$. **23.** Examples include $C + O_2 \longrightarrow CO_2$; iron rusting; wood oxidizing (rotting). **25.** When H_2O_2 is a pure liquid (activity = 1) and the pressures of $H_2O(g)$ and $O_2(g)$ are both 1 atm. **27.** $\Delta H > 0$ (an endothermic process), $\Delta S > 0$. ΔS must be positive for $\Delta H - T\Delta S$ to be negative. **29.** No. **31.** No, it is not possible for $\Delta H - T\Delta S$ (ΔG) to be negative at any temperature **33.** $\Delta H_{298}°$ positive (endothermic reaction), $\Delta S_{298}°$ positive, and $\Delta H_{298}° > 298 \times \Delta S_{298}°$. A reaction with $\Delta H_{298} = 100$ kJ and $\Delta S_{298}° = 250$ J K^{-1} will become spontaneous above 400 K. **35.** The reaction starts with $Q < K$ and proceeds to equilibrium. **37.** In each case, temperature is 25°C and pressure is 1 atm. a. system: 1 mol Mn atoms, 2 mol O atoms; initial state: 1 mol solid MnO_2; final state: 1 mol solid Mn, 1 mol gaseous O_2. b. system: 2 mol H atoms, 2 mol Br atoms; initial state: 1 mol

gaseous H_2, 1 mol liquid Br_2; final state: 2 mol gaseous HBr.
c. system: 1 mol Cu atoms, 1 mol S atoms; initial state: 1 mol solid Cu, 1 mol gaseous S; final state: 1 mol solid CuS. d. system: 2 mol Li atoms, 4 mol O atoms, 2 mol H atoms, 1 mol C atoms; initial state: 2 mol solid LiOH, 1 mol gaseous CO_2; final state: 1 mol solid Li_2CO_3, 1 mol gaseous H_2O. e. system: 1 mol C atoms, 4 mol H atoms, 2 mol O atoms; initial state: 1 mol gaseous CH_4, 1 mol gaseous O_2; final state: 1 mol solid C, 2 mol gaseous H_2O. f. system: 1 mol C atoms, 2 mol S atoms, 6 mol Cl atoms; initial state: 1 mol gaseous CS_2, 3 mol gaseous Cl_2; final state: 1 mol gaseous CCl_4 1 mol gaseous S_2Cl_2. g. system: 1 mol Sn atoms, 4 mol Cl atoms; initial state: 1 mol liquid $SnCl_4$; final state: 1 mol gaseous $SnCl_4$. h. system: 1 mol C atoms, 2 mol S atoms; initial state: 1 mol gaseous CS_2; final state: 1 mol liquid CS_2. i. system: 1 mol Cu atoms; initial state: 1 mol solid Cu; final state: 1 mol gaseous Cu.
39. a. 870 J, b. −3000 J, c. −550 J, d. 575 J. **41.** -1.8×10^2 J
43. −3.10 kJ, decreases the internal energy of the system.
45. -5.12×10^3 J **47.** a. 520.03 kJ, b. −72.8 kJ, c. −331.9 kJ, d. −89.4 kJ, e. −408.83 kJ, f. −238 kJ. Exothermic: b, c, d, e, f.
49. b, c, d, e, f. **51.** a. Endothermic, 39.7 kJ; b. exothermic, −27.7 kJ; c. endothermic, 338.3 kJ **53.** b.
55. A. i. $\frac{1}{4}P_4(s) + \frac{5}{4}O_2(g) \longrightarrow \frac{1}{4}P_4O_{10}(s)$ $\Delta H_{298}^\circ = -746$ kJ
ii. $\frac{3}{4}O_2(g) + \frac{3}{2}H_2(g) \longrightarrow \frac{3}{2}H_2O(g)$ $\Delta H_{298}^\circ = -362.73$ kJ
iii. $\frac{3}{2}H_2O(g) + \frac{1}{4}P_4O_{10}(s) \longrightarrow H_3PO_4(s)$ $\Delta H_{298}^\circ = -170$ kJ
Total A = i + ii + iii
$\frac{1}{4}P_4(s) + 2O_2(g) + \frac{3}{2}H_2(g) \longrightarrow H_3PO_4(s)$ $\Delta H_f^\circ = -1279$ kJ
B. iv. $\frac{1}{4}P_4(s) + \frac{3}{2}H_2(g) \longrightarrow PH_3(g)$ $\Delta H_{298}^\circ = 5.4$ kJ
v. $PH_3(g) + 2O_2(g) \longrightarrow H_3PO_4(s)$ $\Delta H_{298}^\circ = -1284$ kJ
Total B = iv + v
$\frac{1}{4}P_4(s) + 2O_2(g) + \frac{3}{2}H_2(g) \longrightarrow H_3PO_4(s)$ $\Delta H_f^\circ = -1279$ kJ **57.** −1366.8 kJ **59.** $\Delta H_{298}^\circ = -281.9$ kJ
61. $\Delta H_{298}^\circ = -67.1$ kJ; 67.1 kJ is evolved as heat.
63. a. 181.66 kJ, b. $w = -2.48$ kJ. **65.** a. increase, 184.0 J K^{-1}, b. increase, 114.38 J K^{-1}, c. decrease −134.4 J K^{-1}, d. remain constant, −34.9 J K^{-1}, e. remain constant, −8.0 J K^{-1}, f. decrease, −265.5 J K^{-1}. **67.** a, b. **69.** a. 107 J K^{-1}, −86.4 J K^{-1}, 133.2 J K^{-1} **71.** a, c. **73.** a. S_{298}° is the entropy change for the process $H_3PO_4(s, T = 0$ K$) \longrightarrow H_3PO_4(s, T = 298$ K$)$; ΔS_{298}° is the entropy change for the process $\frac{3}{2}H_2(g) + 2O_2(g) + \frac{1}{4}P_4(s) \longrightarrow H_3PO_4(s)$ at 298 K. b. −536 J K^{-1} **75.** −139 J K^{-1} **77.** positive, 198.6 J K^{-1} **79.** 216.49 J K^{-1}, yes. **81.** a. 465.18 kJ, b. −106.86 kJ, c. −291.9 kJ, d. −78.8 kJ, e. −406.43 kJ, f. −160 kJ.
83. a. 465.17 kJ, b. −106.9 kJ, c. −291.8 kJ, d. −79.0 kJ, e. −406.44 kJ, f. −159 kJ **85.** Spontaneous: b, c, d, e, f.
87. a. 8.0 kJ, nonspontaneous; b. −1.88 kJ, spontaneous; c. 298.5 kJ, nonspontaneous. **89.** See the answer to Exercise 55 for the identity of equations i-v, Total A and Total B. For i, $\Delta G_{298}^\circ =$ −674 kJ; for ii, $\Delta G_{298}^\circ = -342.88$ kJ; for iii, $\Delta G_{298}^\circ = -102$ kJ; for Total A, $\Delta G_{298}^\circ = -1119$ kJ; for iv, $\Delta G_{298}^\circ = 13$ kJ; for v, $\Delta G_{298}^\circ =$ −1132 kJ; for Total B, $\Delta G_{298}^\circ = -1119$ kJ; **91.** −1325.4 kJ
93. 326 kJ, nonspontaneous. **95.** −82.0 kJ **97.** a. −1.6 kJ; b. At 0°C $\Delta G^\circ = 1.9$ kJ, not spontaneous, at 100°C $\Delta G^\circ = -11.9$, spontaneous. **99.** a. 1.5×10^2 kJ; b. −21.9 kJ; c. −5.34 kJ; d. −0.383 kJ; e. 18 kJ; f. 71 kJ. **101.** a. 41, b. 0.053, c. 6.9×10^{13}, d. 1.9, e. 0.040. **103.** a. 1, b. 2.51×10^{-3}, c. 4.83×10^3, d. 0.219, e. 16.1 **105.** The value of ΔH° and ΔS° do not change with temperature. **107.** 40°C **109.** −79.9 kJ **111.** a. 97.2°C; b. 100.01°C. **113.** a. −89.7 kJ, b. 345.4°C, c. 1.90×10^{-16}
115. Values for Q and ΔG are 1.6×10^{-5}, −23 kJ; 2.0×10^{-3}, −11 kJ; 5.20×10^{-2}, −2.55 kJ; 0.145, 0. **117.** It might be possi-

ble to find a catalyst because the reaction is spontaneous; $\Delta G_{298}^\circ =$ −8.1 kJ. Heating would reduce the yield; ΔS_{298}° is negative (−125.6 J K^{-1}). **119.** a. 131.30 kJ, cool; b. 45 kJ; c. 1.6×10^{-4}; d. −638.4 kJ. **121.** 1.3×10^{-5} atm **123.** −0.16 kJ K^{-1}
125. $\Delta H_{298}^\circ = -307.3$ kJ, so 307.3 kJ of heat is evolved.
127. ΔG°, −85 kJ; ΔS°, 278 J K^{-1} **129.** a. ΔG_{298}° 20.0 kJ, ΔH_{298}° 32.5 kJ; b. 3.13×10^{-4}; c. 1.16; d. 41.9 J K^{-1}; e. −0.9 kJ; f. 776 K. **131.** a. ΔH_{298}° 30.91 kJ, ΔS_{298}° 93.12 J K^{-1}; c. ΔG_{298}° 3.17 kJ; d. nonspontaneous; e. 331.9 K; f. Vapor to liquid is spontaneous at 298 K. Liquid to vapor is spontaneous at 398 K.

Chapter 19

1. Anode: an electrode at which oxidation occurs. Cathode: an electrode at which reduction occurs. **3.** a. Water is present in each dish in all cells. i. Dish A: inert electrode, Cu^+, Cu^{2+}, Dish B: inert electrode, $Fe(CN)_6^{3-}$, $Fe(CN)_6^{4-}$; ii. Dish A: Au electrode, Au^+, Dish B: inert electrode, MnO_4^-, Mn^{2+}, H_3O^+; iii. Dish A: inert electrode, NO_2^-, NO_3^-, OH^-, Dish B: inert electrode, MnO_4^-, MnO_2, OH^-; iv. Dish A: Ni electrode, Ni^{2+}, Dish B: Cu electrode, Cu^{2+}; v. Dish A: inert electrode (Pt), SO_4^{2-}, HSO_3^-, H_3O^+, Dish B: inert electrode (Pt), ClO_3^-, $HClO$, H_3O^+. b. i. Dish A: $Cu^+ \longrightarrow Cu^{2+} + e^-$ Dish B: $Fe(CN)_6^{3-} + e^- \longrightarrow Fe(CN)_6^{4-}$ ii. Dish A: $Au \longrightarrow Au^+ + e^-$ Dish B: $MnO_4^- + 8H_3O^+ + 5e^- \longrightarrow Mn^{2+} + 12H_2O$ iii. Dish A: $NO_2^- + 2OH^- \longrightarrow NO_3^- + H_2O + 2e^-$ Dish B: $MnO_4^- + 2H_2O + 3e^- \longrightarrow MnO_2 + 4OH^-$ iv. Dish A: $Ni \longrightarrow Ni^{2+} + 2e^-$ Dish B: $Cu^{2+} + 2e^- \longrightarrow Cu$ v. Dish A: $HSO_3^- + 4H_2O \longrightarrow SO_4^{2-} + 3H_3O^+ + 2e^-$ Dish B: $ClO_3^- + 5H_3O^+ + 4e^- \longrightarrow HClO + 7H_2O$. **5.** The potential, measured relative to the standard hydrogen electrode, for an electrode with a reduction half-reaction at 25°C in which all reactants and products have an activity of 1 (1 M concentrations, 1 atm pressures, pure liquids and solids). **7.** a. Co^{2+}, b. I_2, c. MnO_4^-, d. Sn^{4+}.
9. $Fe|Fe^{2+}\|Ag^+|Ag$ **11.** As the relative concentrations of the reactants decrease (as described by Q), the cell voltage decreases. As the relative concentrations of the reactants increase (as described by Q), the cell voltage increases. **13.** A galvanic cell produces an electric current from a chemical change. An electric current is used to produce a chemical change in an electrolytic cell. **15.** Use of a protective coating and cathodic protection. **17.** Al and Ag salts form an electrochemical cell in which silver salts are reduced to solid Ag as the Al is oxidized. The NaCl acts as the electrolyte and salt bridge. **19.** The copper rod is part of a Cu/Cu^{2+} electrode, an anode. **21.** Hydronium ion is a reactant in the reduction process in an acidic solution, whereas hydroxide is a product in basic solution.
23. Zn, Cr, Fe, Cd, Co, Ni, Sn, Pb, Ag, Hg, Pt, Au **25.** a. $Sn^{4+} + 2e^- \longrightarrow Sn^{2+}$, reduction; b. $[Ag(NH_3)_2]^+ + e^- \longrightarrow Ag + 2NH_3$, reduction; c. $Hg_2Cl_2 + 2e^- \longrightarrow 2Hg + 2Cl^-$, reduction; d. $6H_2O \longrightarrow O_2 + 4H_3O^+ + 4e^-$, oxidation; e. $O_2 + 2H_2O + 4e^- \longrightarrow 4OH^-$, reduction; f. $SO_3^{2-} + 3H_2O \longrightarrow SO_4^{2-} + 2H_3O^+ + 2e^-$; oxidation; g. $MnO_4^- + 8H_3O^+ + 5e^- \longrightarrow Mn^{2+} + 12 H_2O$, reduction; h. $Cl^- + 6OH^- \longrightarrow ClO_3^- + 6e^- + 3H_2O$, oxidation.
27. a. $Mn \longrightarrow Mn^{2+} + 2e^-$, $Ag^+ + e^- \rightarrow Ag$, $Mn|Mn^{2+}\|Ag^+|Ag$; b. $Au \longrightarrow Au^+ + e^-$, $MnO_4^- + 8H_3O^+ + 5e^- \longrightarrow Mn^{2+} + 12H_2O$, $Au|Au^+\|MnO_4^-$; Mn^{2+}; $H_3O^+|Pt$; c. $Cr + 4OH^- \longrightarrow Cr(OH)_4^- + 3e^-$, $O_2 + 2H_2O + 4e^- \longrightarrow 4OH^-$, $Cr|Cr(OH)_4^-$; $OH^-\|OH^-|O_2|Pt$: d. $HNO_2 + 4H_2O \longrightarrow NO_3^- + 3H_3O^+ + 2e^-$, $Cr_2O_7^{2-} + 14H_3O^+ + 6e^- \longrightarrow 2Cr^{3+} + 21H_2O$, $Pt|HNO_2$; NO_3^-; $H_3O^+\|Cr_2O_7^{2-}$; Cr^{3+}, $H_3O^+|Pt$;
e. $Cl^- + 2OH^- \longrightarrow ClO^- + H_2O + 2e^-$, NO_3^-; $H_2O + 2e^-$

$\longrightarrow NO_2^- + 2OH^-$, $Pt|ClO^-$; Cl^-; $OH^-\|NO_3^-$; NO_2^-; $OH^-|Pt$.   **29.** a. $+1.98$ V; b. -0.17 V; c. $+1.6$ V; d. $+0.39$ V; e. -0.88 V.   **31.** a. 1.298 V; b. -0.32 V.   **33.** a. -0.46 V, b. $+1.315$ V.   **35.** a. $Sc + 3Ag^+ \longrightarrow 3Ag + Sc^{3+}$ $E_{cell}^\circ = +2.88$ V; b. $Cl_2 + S^{2-} \longrightarrow S + 2Cl^-$ $E_{cell}^\circ = +1.84$ V   **37.** 1.30 V.   **39.** Ga   **41.** a, c, e are stable.   **43.** a. 1.92 V; b. -0.05 V; c. 1.6 V; d. 0.48 V, e. -0.88 V   **45.** a. 1.159 V, b. 1.36 V.   **47.** $[Fe^{2+}]$, Q, E_{cell}: 1.00 M, 1.00, 0.323 V: 0.50 M, 3.00, 0.309 V; 0.10 M, 19.0, 0.285 V; 1.0×10^{-4} M, 2.00×10^4, 0.196 V; 1.0×10^{-8} M, 2.00×10^8, 0.077 V; 1.0×10^{-10} M, 2×10^{10}, 0.018 V; 2.4×10^{-11} M, 8.3×10^{10}, 0 V.   **49.** a. -0.195 V, b. -0.836 V, c. 1.13 V.   **51.** 1.28 V.   **53.** 1.48 M   **55.** -1.124 V   **57.** a. -382 kJ, b. 82 kJ, c. -1.9×10^3 kJ, d. -2.3×10^2 kJ, e. 1.7×10^2 kJ   **59.** a. -371 kJ, b. 20 kJ, c. -1.9×10^3 kJ, d. -2.8×10^2 kJ, e. 1.7×10^2 kJ   **61.** a. 0.78 V, -150 kJ, 2.4×10^{26}, b. -0.5297 V, 102.2 kJ, 1.238×10^{-18}.   **63.** -348 kJ, 1.05×10^{61} **65.** 1.7×10^6, 7.7×10^{-4} M   **67.** $C|MgCl_2(l)$; $Cl_2(g)\|MgCl_2(l)|Mg$; b. $Fe|Fe^{2+}\|Cu^{2+}|Cu$ **69.** a. Anode: $2Cl^- \longrightarrow Cl_2 + 2e^-$, cathode: $2Na^+ + 2e^- \longrightarrow 2Na$, $2Na^+ + 2Cl^- \longrightarrow 2Na + Cl_2$; b. Anode: $2Cl^- \longrightarrow Cl_2 + 2e^-$, cathode: $2H_3O^+ + 2e^- \longrightarrow H_2 + 2H_2O$, $2H_3O^+ + 2Cl^- \longrightarrow H_2 + Cl_2 + 2H_2O$.   **71.** a. 1.60 mol, b. 2.000 mol, c. 7.31×10^{-3} mol, d. 9.046×10^{-3} mol.   **73.** a. 1.60 F, b. 2.000 F, c. 7.31×10^{-3} F, d. 9.046×10^{-3} F.   **75.** 3.0 mol, 3.0 F, 2.9×10^5 C.   **77.** 2.820×10^4 mol e^-, 2.820×10^4 F, 2.721×10^9 C   **79.** 111 g   **81.** 1.2×10^{-3} mol   **83.** 0.140 F, 30 min   **85.** 198 kg   **87.** a. 2, 5, 2; b. 3, 2, 2, 3; c. 1, 2, 1, 2; d. 1, 1, 2, 2, 1, 2; e. 5, 2, 3, 5, 2, 2; f. 1, 1, 1, 1; g. 2, 2, 2, 2, 1.   **89.** a. $2Al + 3Sn(OH)_4^{2-} \longrightarrow 2Al(OH)_4^- + 3Sn + 4OH^-$; b. $H_2S + H_2O_2 \longrightarrow S + 2H_2O$; c. $2MnO_4^{2-} + Cl_2 \longrightarrow 2MnO_4^- + 2Cl^-$; d. $3Br_2 + 3CO_3^{2-} \longrightarrow 5Br^- + BrO_3^- + 3CO_2$; e. $C + 4HNO_3 \longrightarrow 4NO_2 + 2H_2O + CO_2$; f. $5ClO_3^- + 9H_2O + 3I_2 \longrightarrow 6IO_3^- + 5Cl^- + 6H_3O^+$ **91.** a. $5Zn + 12H_3O^+ + 2NO_3^- \longrightarrow 5Zn^{2+} + N_2 + 18H_2O$; b. $4Zn + NO_3^- + 6H_2O \longrightarrow 4Zn^{2+} + NH_3 + 9OH^-$; c. $3CuS + 8H_3O^+ + 2NO_3^- \longrightarrow 3Cu^{2+} + 3S + 2NO + 12H_2O$; d. $4NH_3 + 7O_2 \longrightarrow 4NO_2 + 6H_2O$; e. $H_2SO_4 + 2HI \longrightarrow I_2 + SO_2 + 2H_2O$; f. $3Cl_2 + 6OH^- \longrightarrow 5Cl^- + ClO_3^- + 3H_2O$; g. $5H_2O_2 + 2MnO_4^- + 6H_3O^+ \longrightarrow 2Mn^{2+} + 5O_2 + 14H_2O$; h. $2NO_2 + 2OH^- \longrightarrow NO_3^- + NO_2^- + H_2O$; i. $2KClO_3 \longrightarrow 2KCl + 3O_2$; j. $2Fe^{3+} + 2I^- \longrightarrow 2Fe^{2+} + I_2$; k. $P_4 + 4OH^- + 2H_2O \longrightarrow 2PH_3 + 2HPO_3^{2-}$; l. $P_4 + 10H_2O \longrightarrow 2PH_3 + 2HPO_4^{2-} + 4H_3O^+$ **93.** Li is a powerful reducing agent and is light. The half-reaction $Li \longrightarrow Li^+ + e^-$ has an oxidation potential of 3.09 V. 6.9 g of Li produces 1 mole of electrons.   **95.** iv.   **97.** $[H_3O^+]$ 2.7×10^{-5}, yes.   **99.** -0.83 V.   **101.** Anode: $4OH^- \longrightarrow O_2 + 2H_2O + 4e^-$; cathode: $Au(CN)_4^- + 3e^- \longrightarrow Au + 4CN^-$; $4Au(CN)_4^- + 12OH^- \longrightarrow 4Au + 3O_2 + 6H_2O + 16CN^-$.   **103.** 4.48 h.   **105.** 0.20.   **107.** a. 2, 3, 1, 2, 3, 2; b. 3, 2, 2, 4, 1; c. 1, 1, 6, 4, 2, 1.

Chapter 20

1. Nuclear reactions usually change one type of nucleus into another, chemical changes rearrange atoms. Nuclear reactions involve much larger energies than chemical reactions.   **3.** Time required for half the atoms in a sample to decay. For C-14 the half-life is 5730 years. A 10 g sample of C-14 would contain 5 g of C-14 after 5730 years; a 0.20 g sample of C-14, 0.10 g.   **5.** α (helium nuclei), β (electrons), β^+ (positrons), γ rays, neutrons.   **7.** a. Conversion of a neutron to a proton. b. Conversion of a proton to a neutron.   **9.** Alpha particles are He-4 nuclei; protons are H-1 nuclei.   **11.** Two

nuclei must collide for fusion to occur. High temperatures are required to give the nuclei enough kinetic energy to overcome the very strong repulsion resulting from their positive charges. Fission is induced by a collision between a neutron (a neutral particle) and a nucleus.   **13.** a. 7 $^1_1$p, 7 $^1_0$n; b. 1 $^1_1$p, 0 $^1_0$n; c. 1 $^1_1$p, 1 $^1_0$n; d. 1 $^1_1$p, 2 $^1_0$n; e. 10 $^1_1$p, 10 $^1_0$n; f. 82 $^1_1$p, 124 $^1_0$n; g. 92 $^1_1$p, 143 $^1_0$n; h. 6 $^1_1$p, 6 $^1_0$n; i. 6 $^1_1$p, 8 $^1_0$n.   **15.** a, b, c, d, e.   **17.** a. Too many neutrons, β decay; b. Atomic number greater than 82, α or β decay; c. Too few neutrons, electron capture or β^+ decay; d. Too many neutrons, β decay; e. Atomic number greater than 82, α or β decay.   **19.** 0.8 MeV   **21.** a. 161.3 MeV per nucleus, b. 8.063 MeV per nucleon   **23.** 2.011×10^9 kJ mol^{-1} **25.** a. 1.05×10^{-4} y^{-1}; b. 36 s^{-1}; c. 4.28×10^{-6} y^{-1}.   **27.** a. 1.09×10^5 y; b. 0.32 s; c. 2.69×10^6 y.   **29.** a. 2.3%; b. $2.0 \times 10^{-18}\%$.   **31.** 97.2% **33.** a. $^{208}_{82}Pb$, b. 21 min, c. $1.5 \times 10^{-4}\%$.   **35.** 28.8 y.   **37.** a. 3.8×10^9 y; b. Younger.   **39.** a. $^{230}_{92}U \rightarrow {}^{226}_{90}Th + {}^4_2He$; b. $^{212}_{83}Bi \longrightarrow {}^{212}_{84}Po + {}^0_{-1}e$; c. $^8_5B \longrightarrow {}^8_4Be + {}^0_{+1}e$; d. $^{239}_{92}U + {}^1_0n \longrightarrow {}^{239}_{93}Np + {}^0_{-1}e$, $^{239}_{93}Np \longrightarrow {}^{239}_{94}Pu + {}^0_{-1}e$; e. $^{90}_{38}Sr \longrightarrow {}^{90}_{39}Y + {}^0_{-1}e$.   **41.** a. $^{30}_{15}P$, b. ^4_2He, c. $^{17}_8O$, d. $^{96}_{37}Rb$.   **43.** $^{238}_{92}U \longrightarrow {}^{234}_{90}Th + {}^4_2He$; $^{234}_{90}Th \longrightarrow {}^{234}_{91}Pa + {}^0_{-1}e$; $^{234}_{91}Pa \longrightarrow {}^{234}_{92}U + {}^0_{-1}e$; $^{234}_{92}U \longrightarrow {}^{230}_{90}Th + {}^4_2He$; $^{230}_{90}Th \longrightarrow {}^{226}_{88}Ra + {}^4_2He$; $^{226}_{88}Ra \longrightarrow {}^{222}_{86}Rn + {}^4_2He$; $^{222}_{86}Rn \longrightarrow {}^{218}_{84}Po + {}^4_2He$ **45.** a. $^{14}_7N + {}^4_2He \longrightarrow {}^{17}_8O + {}^1_1H$; b. $^{14}_7N + {}^1_0n \longrightarrow {}^{14}_6C + {}^1_1H$; c. $^{232}_{90}Th + {}^1_0n \longrightarrow {}^{233}_{90}Th$; d. $^{238}_{92}U + {}^2_1H \longrightarrow {}^{239}_{92}U + {}^1_1H$.   **47.** Fission is the conversion of heavier nuclei into two or more lighter nuclei and other fragments. Fusion is the combination of two smaller nuclei into a heavier nucleus. In both cases the mass of the products is less than the mass of the reactants and the mass lost is converted into energy.   **49.** Components are fuel, moderator, coolant, control system, and a shield and containment system. See Section 20.9 for a description of each.   **51.** The fission of uranium generated heat which is carried to an external steam generator (boiler). The resulting steam turns a turbine that powers an electrical generator.   **53.** 8×10^{16} g mL^{-1}.   **55.** 2.06×10^4 yr.   **57.** a. $^{26}_{13}Al \longrightarrow {}^{26}_{12}Mg + {}^0_{+1}e$, $^{26}_{13}Al + {}^0_{-1}e \longrightarrow {}^{26}_{12}Mg$; b. 1.9×10^7 yr (19 million years)   **59.** a. n/p ratio high, β decay; b. n/p ratio low, β^+ decay or electron capture; c. n/p ratio high and Z > 82, α or β decay; d. n/p ratio high and Z > 82, α or β decay; e. n/p ratio low, β^+ decay or electron capture; f. n/p ratio low, β^+ decay or electron capture; g. n/p ratio high and Z > 82, α or β decay.

Chapter 21

1. The ns^1 valence electron configuration is more tightly bound in Li than in Cs.   **3.** Rb is more reactive than Na or Mg.   **5.** $Fr(s) + O_2(g) \longrightarrow FrO_2(s)$ **7.** $2K(s) + 2H_2O(l) \longrightarrow 2KOH(aq) + H_2(g)$; $K_2O_2(s) + 2H_2O(l) \longrightarrow 2KOH(aq) + H_2O_2(aq)$ **9.** 0.1498 M   **11.** Flame test: Na is yellow and Sr is red; also use solubility and density differences.   **13.** 5.1×10^4 g   **15.** 8.00×10^{-2} g   **17.** NaI, Na_2Se, Na_2O; SrI_2, $SrSe$, SrO; AlI_3, Al_2Se_3, Al_2O_3 **19.** Ca   **21.** a. -6.24 kJ/g; b. -4.29×10^6 kJ   **23.** 9.577% **25.** a. $2Zn(s) + O_2(g) \longrightarrow 2ZnO(s)$; b. $Zn(s) + Pb(NO_3)_2(aq) \longrightarrow Zn(NO_3)_2(aq) + Pb(s)$; c. $Zn(s) + Cd(NO_3)_2(aq) \longrightarrow Zn(NO_3)_2(aq) + Cd(s)$; d. $ZnCO_3(s) \longrightarrow ZnO(s) + CO_2(g)$; e. $ZnCO_3(s) + 2CH_3COOH(aq) \longrightarrow Zn(CH_3COO)_2(aq) + CO_2(g) + H_2O(l)$; f. $Zn(s) + 2HBr(aq) \longrightarrow ZnBr_2(aq) + H_2(g)$; g. $Zn(s) + S(s) \longrightarrow ZnS(s)$ **27.** $2HClO_4(aq) + Na_2[Zn(OH)_4](aq) \longrightarrow Zn(OH)_2(s) + 2NaClO_4(aq) + 2H_2O(l)$; $2HClO_4(aq) + Zn(OH)_2(s) \longrightarrow Zn(ClO_4)_2(aq) + 2H_2O(l)$ **29.** a. $HgO(s) + 2HNO_3(aq) \longrightarrow Hg(NO_3)_2(aq) + H_2O(l)$; b. $Hg(l) + S(s) \longrightarrow HgS(s)$; c. $Cd(s) + Hg(NO_3)_2(aq) \longrightarrow Cd(NO_3)_2(aq) + Hg(l)$;

d. $Zn(s) + Hg(NO_3)_2(aq) \longrightarrow Zn(NO_3)_2(aq) + Hg(l)$ **31.** 3 mol of ionic species **33.** $Al(OH)_3(s) + 3H^+(aq) \longrightarrow Al^{3+} + 3H_2O(l)$; $Al(OH)_3(s) + OH^- + 2H_2O(l) \longrightarrow [Al(H_2O)_2(OH)_4]^-(aq)$ **35.** $Al(s)$ forms a protective oxide coating. **37.** Extract from ore: $AlO(OH)(s) + NaOH(aq) + H_2O(l) \longrightarrow Na[Al(OH)_4](aq)$; recover: $2Na[Al(OH)_4](s) + H_2SO_4(aq) \longrightarrow 2Al(OH)_3(s) + Na_2SO_4(aq) + 2H_2O(l)$; sinter: $2Al(OH)_3(s) \longrightarrow Al_2O_3(s) + 3H_2O(g)$; dissolve in $Na_3AlF_3(l)$ and electrolyze: $Al^{3+} + 3e^- \longrightarrow Al(s)$ **39.** It is a covalently bonded compound. **41.** Cassiterite or tinstone, SnO_2; a three step process; roasting, acid to remove impurities, reduction by carbon **43.** Yes **45.** $PbCl_2$: ionic solid; $PbCl_4$: volatile covalent liquid **47.** a. $4Bi(s) + 3O_2(g) \longrightarrow 2Bi_2O_3(s)$; b. $BiCl_3(aq) + 3NaOH(aq) \longrightarrow Bi(OH)_3(s) + 3NaCl(aq)$; c. $2Bi(s) + 3S(s) \longrightarrow Bi_2S_3(s)$; d. $2Bi(s) + 6HBr(aq) \longrightarrow 2BiBr_3(aq) + 3H_2(g)$; e. $Bi_2O_3(s) + 6HNO_3(aq) \longrightarrow 2Bi(NO_3)_3(aq) + 3H_2O(l)$; f. $Bi(OH)_3(s) \longrightarrow BiO(OH)(s) + H_2O(l)$; g. $Ga(s) + Bi(NO_3)_3(aq) \longrightarrow Ga(NO_3)_3(aq) + Bi(s)$; h. $Bi(s) + Pb(NO_3)_2(aq) \longrightarrow$ No Reaction **49.** +90.83 kJ/mol **51.** 39 kg **53.** a. $Na_2O(s) + H_2O(l) \longrightarrow 2NaOH(aq)$; b. $Cs_2CO_3(s) + 2HF(aq) \longrightarrow 2CsF(aq) + CO_2(g) + H_2O(l)$; c. $CaBr_2(l) \longrightarrow Ca(s) + Br_2(l)$; d. $Al_2O_3(s) + 6HClO_4(aq) \longrightarrow 2Al(ClO_4)_3(aq) + 3H_2O(l)$; e. $Ba(s) + Ba(OH)_2(s) \longrightarrow 2BaO(s) + H_2(g)$; f. $2Al(s) + 3CaO(s) \longrightarrow Al_2O_3(s) + 3Ca(s)$; g. $Na_2CO_3(aq) + Ba(NO_3)_2(aq) \longrightarrow 2NaNO_3(aq) + BaCO_3(s)$; h. $Na_2CO_3(aq) + Mg(NO_3)_2(aq) \longrightarrow 2NaNO_3(aq) + MgCO_3(s)$; i. $TiCl_4(l) + 4Na(s) \longrightarrow Ti(s) + 4NaCl(s)$ **55.** 277 mL

Chapter 22

1. a. sp^3, tetrahedral; b. sp^3, trigonal pyramid; c. sp^3d^2, octahedral; d. sp^3, bent; e. sp^2, bent; f. sp^3d, seesaw-shaped; g. sp^3d^2, octahedral; h. sp^3d, trigonal bipyramid; i. sp^3d^2, octahedral **3.** Boron hydrides are said to have electron-deficient bonding because they lack two electrons in comparison with similar carbon-containing compounds. **5.** a. sp^3, linear; b. sp^3, tetrahedral; c. sp^2, trigonal planar; d. sp^2, trigonal planar; e. sp^2, trigonal planar; f. sp^3, linear along $B-C-N$ **7.** a. 87 kJ, 44 kJ; b. -109.9 kJ, -154.7 kJ; c. -510 kJ, -601.5 kJ **9.** a. SiO_2; b. SiI_4; c. SiH_4; d. SiC; e. Mg_2Si; f. H_2SiF_6

11.

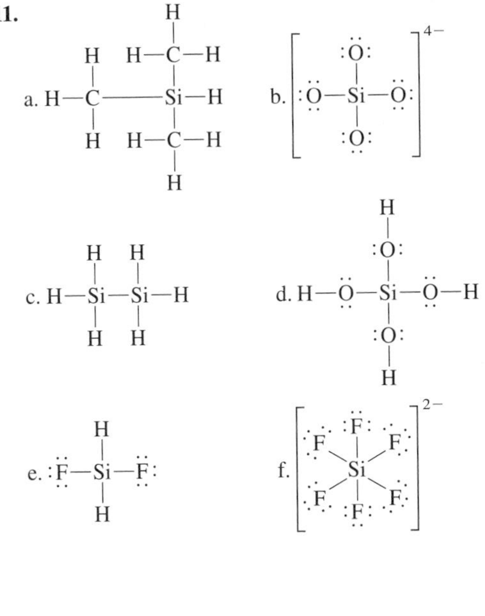

13. Metallic: $Si(s) + 2OH^-(aq) + H_2O(l) \longrightarrow SiO_3{}^{2-}(aq) + 2H_2(g)$, $Si(s) + 2F_2(g) \longrightarrow SiF_4(g)$; nonmetallic: $SiF_4(g) + 2NaF(s) \longrightarrow Na_2SiF_6(s)$, $SiCl_4(l) + 4H_2O(l) \longrightarrow Si(OH)_4(s) + 4HCl(g)$ **15.** a. nonpolar; b. nonpolar; c. polar; d. polar; e. nonpolar; f. polar **17.** 92 g/mol; Si_3H_8 **19.** Lewis acid: H_2O; Lewis base: CaH_2; oxidizing agent: H_2O, H from +1 to 0; reducing agent: CaH_2, H from -1 to 0 **21.** 3 g **23.** 283 mol **25.** 1.56 L

27. a. NH^{2-}: b. N^{3-}: c. N_2F_4:

d. $NH_2{}^-$: e. NF_3: f. $N_3{}^-$:

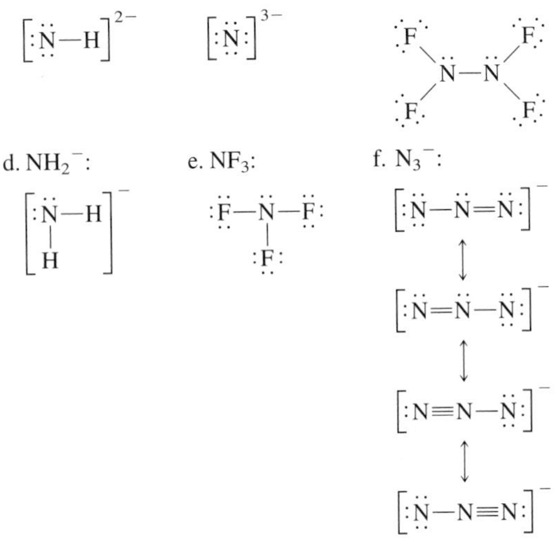

29. Ammonia acts as a Brønsted base because it readily accepts protons and a Lewis base in that it has an electron pair to donate.

31. NO_2:

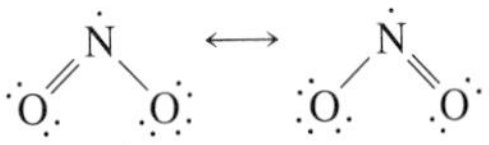

Nitrogen is sp^2 hybridized. The molecule has a bent geometry with an ONO bond angle of approximately 120°.

$NO_2{}^-$:

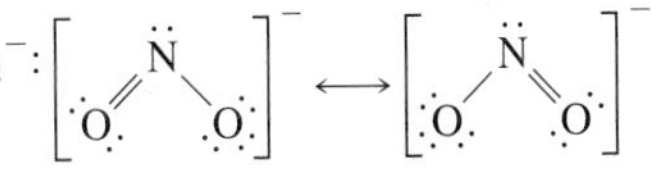

Nitrogen is sp^2 hybridized. The molecule has a bent geometry with an ONO bond angle of 120°.

$NO_2{}^+$: $\left[\ddot{O}=N=\ddot{O}\right]^+$

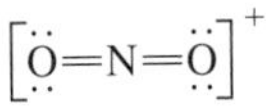

Nitrogen is sp hybridized. The molecule has a linear geometry with an ONO bond angle of 180°.
33. The valence shell of nitrogen does not contain d orbitals, as do the valence shells of P and As. **35.** a. linear; b. trigonal pyramid; c. tetrahedral; d. each P is trigonal pyramid; e. tetrahedral; f. trigonal bipyramidal **37.** P is sp^3 in all given compounds. **39.** 18.4 g **41.** $H_3PO_3(aq) + H_2O(l) \rightleftharpoons H_3O^+(aq) + H_2PO_3{}^-(aq)$; $H_2PO_3{}^- + H_2O(l) \rightleftharpoons H_3O^+(aq) + HPO_3{}^{2-}(aq)$ **43.** Phosphorous acid has one hydrogen atom bonded directly to a phosphorus atom; this hydrogen is not acidic. **45.** H_2O has very strong hydrogen bonding although H_2S has almost twice the molar mass. **47.** $NaHSO_4 > NaHSO_3$; oxidation number of S is greater in H_2SO_4 **49.** $H_2SO_4 > H_2SO_3$ due to higher oxidation number of S.

51.

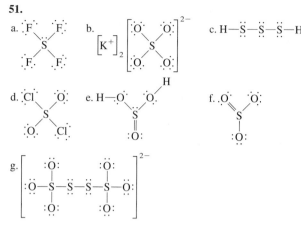

a. [F₄SF₂ structure] b. [K⁺]₂[SO₄]²⁻ structure c. H—S—S—S—H

d. Cl₂SO₂ structure e. H—O—O (with S) f. O=S=O structure

g. [O₃S—S—S—SO₃]²⁻ structure

53. $HClO_3 > HBrO_3$; Cl is more electronegative than Br.   **55.** IF_3: sp^3d; IF_5: sp^3d^2 **57.** Fluorine has the highest ionization energy; yes.   **59.** All have London dispersive forces; physical state is then determined by increasing size and molar mass.   **61.** a. sp^3d; b. sp^3d^2; c. sp^3d; d. sp^3; e. sp^3; f. sp^3d^2 **63.** a. nonpolar; b. nonpolar; c. polar; d. polar; e. nonpolar; f. polar   **65.** XeF_6; $Xe(g) + 3F_2(g) \longrightarrow XeF_6(s)$; $XeF_6(s) + 3H_2(g) \longrightarrow 6HF(g) + Xe(g)$

Chapter 23

1. Sc: $1s^22s^22p^63s^23p^64s^23d^1$; Ti: $1s^22s^22p^63s^23p^64s^23d^2$; Cr: $1s^22s^22p^63s^23p^64s^13d^5$; Fe: $1s^22s^22p^63s^23p^64s^23d^6$; Mo: $1s^22s^22p^63s^23p^64s^23d^{10}4p^65s^14d^5$; Ru: $1s^22s^22p^63s^23p^64s^23p^{10}4p^65s^24d^6$
3. La: $1s^22s^22p^63s^23p^64s^23d^{10}4p^65s^24d^{10}5p^66s^25d^1$; La³⁺: $1s^22s^22p^63s^23p^64s^23d^{10}4p^65s^24d^{10}5p^6$; Sm: $1s^22s^22p^63s^23p^64s^23d^{10}4p^65s^24d^{10}5p^66s^24f^6$; Sm³⁺: $1s^22s^22p^63s^23p^64s^23d^{10}4p^65s^24d^{10}5p^64f^5$; Lu: $1s^22s^22p^63s^23p^64s^23d^{10}4p^65s^24d^{10}5p^66s^24f^{14}5d^1$; Lu³⁺: $1s^22s^22p^63s^23p^64s^23d^{10}4p^65s^24d^{10}5p^64f^{14}$
5. Al is used because Al_2O_3 has the highest negative free energy per oxygen atom of any compound.   **7.** MoO_3 **9.** See text.
11. a. $Cr_2(SO_4)_3(aq) + 2Zn(s) + 2H_3O^+(aq) \longrightarrow 2Zn^{2+}(aq) + H_2(g) + 2H_2O(l) + 2Cr^{2+}(aq) + 3SO_4^{2-}(aq)$; b. $3TiCl_3(s) + CrO_4^{2-}(aq) + 5H^+(aq) \longrightarrow 3Ti^{4+}(aq) + Cr(OH)_3(s) + 9Cl^-(aq) + H_2O(l)$; c. $3Cr^{2+}(aq) + CrO_4^{2-}(aq) + 8H_3O^+(aq) \longrightarrow 4Cr^{3+}(aq) + 12H_2O(l)$; d. $16CrO_3(s) + 18Mn(s) \longrightarrow 8Cr_2O_3(s) + 6Mn_3O_4(s)$; e. $CrO(s) + 2H_3O^+(aq) + 2NO_3^-(aq) \longrightarrow Cr^{2+}(aq) + 2NO_3^-(aq) + 3H_2O(l)$; f. $CrCl_3(s) + 3NaOH(aq) \longrightarrow Cr(OH)_3(s) + 3Na^+(aq) + 3Cl^-(aq)$ **13.** $H_2S(g)$ **15.** a. $3Fe(s) + 4H_2O(l) \longrightarrow Fe_3O_4(s) + 4H_2(g)$; b. $3NaOH(aq) + Fe(NO_3)_3(aq) \longrightarrow Fe(OH)_3(s) + 3Na^+(aq) + 3NO_3^-(aq)$; c. $6Fe^{2+}(aq) + Cr_2O_7^{2-}(aq) + 14H_3O^+(aq) \longrightarrow 6Fe^{3+}(aq) + 2Cr^{3+}(aq) + 21H_2O(l)$; d. $Fe(s) + 2H_3O^+(aq) + SO_4^{2-}(aq) \longrightarrow Fe^{2+}(aq) + SO_4^{2-}(aq) + H_2(g) + 2H_2O(l)$; e. $4Fe^{2+}(aq) + O_2(g) + 4HNO_3(aq) \longrightarrow 4Fe^{3+}(aq) + 2H_2O(l) + 4NO_3^-(aq)$; f. $FeCO_3(s) + 2HClO_4(aq) \longrightarrow Fe(ClO_4)_2(aq) + H_2O(l) + CO_2(g)$; g. $3Fe(s) + 2O_2(g) \longrightarrow Fe_3O_4(s)$ **17.** 3.8×10^4 ft³/ton   **19.** 0.17 V
21. $Ag^+(aq) + CN^-(aq) \longrightarrow AgCN(s)$; $AgCN(s) + CN^-(aq) \longrightarrow Ag(CN)_2^-(aq)$ **23.** 1.4×10^{-20} M   **25.** No, Au^+ has a complete 5d sublevel.   **27.** a. 4, $Zn(OH)_4^{2-}$; b. 6, $Pd(CN)_6^{2-}$; c. 2, $AuCl_2^-$; d. 4, $Pt(NH_3)_2Cl_2$; e. 6, $K[Cr(NH_3)_2Cl_4]$; f. 6, 6, $[Co(NH_3)_6][Cr(CN)_6]$; g. 6, $[Co(en)_2Br_2]NO_3$

29. a. $[Pt(H_2O)_2Br_2]$:

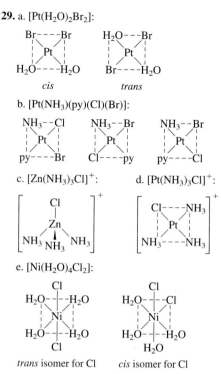

cis   *trans*

b. $[Pt(NH_3)(py)(Cl)(Br)]$:

c. $[Zn(NH_3)_3Cl]^+$:   d. $[Pt(NH_3)_3Cl]^+$:

e. $[Ni(H_2O)_4Cl_2]$:

trans isomer for Cl   *cis* isomer for Cl

f. $[Co(C_2O_4)_2Cl_2]^{3-}$; dichlorobisoxalatocobaltate(III) ion; this is octahedral and has *cis* and *trans* positions for the Cl ligands. The *cis* form has an optical isomer.

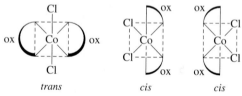

trans   *cis*   *cis*

31. a. tricarbonatocobaltate(III) ion b. tetraamminecopper(II) ion c. tetraamminedibromocobalt(III) sulfate d. tetraammineplatinum(II) tetrachloroplatinate(II) e. tris(ethylenediamine)chromium(III) nitrate f. diamminedibromopalladium(II) g. potassium hexacyanoferrate(III) h. diamminedichlorozinc(II)
33.

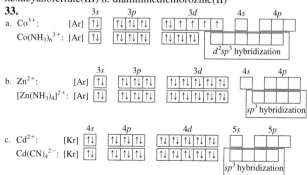

a. Co^{3+}: [Ar] ... ; $Co(NH_3)_6^{3+}$: [Ar] ... d^2sp^3 hybridization

b. Zn^{2+}: [Ar] ... ; $[Zn(NH_3)_4]^{2+}$: [Ar] ... sp^3 hybridization

c. Cd^{2+}: [Kr] ... ; $Cd(CN)_4^{2-}$: [Kr] ... sp^3 hybridization

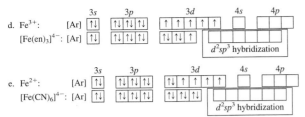

d. Fe^{3+}: [Ar]

$[Fe(en)_3]^{4-}$: [Ar]

d^2sp^3 hybridization

e. Fe^{2+}: [Ar]

$[Fe(CN)_6]^{4-}$: [Ar]

d^2sp^3 hybridization

35. $Fe(CN)_6^{3-}$ is a d^5 strong field complex: one unpaired electron; $Fe(H_2O)_6^{3+}$ is a d^5 weak field complex: five unpaired electrons. **37.** a. 4; b. 0; c. 2; d. 1; e. 5; f. 0 **39.** $CN^- > H_2O$, $Fe(CN)_6^{4-}$ more stable; $NH_3 > F^-$, $Co(NH_3)_6^{3+}$ more stable; $H_2O > Cl^-$, $Mn(CN)_6^{4-}$ more stable **41.** NiO: $-12Dq$; Mn^{2+}: 0 **43.** 8.3×10^{-12} M **45.** $Co(H_2O)_6^{2+}$, $E° = -0.6$ V; $Co(NH_3)_6^{2+}$, $E° = +1.1$ V; $Co(NH_3)_6^{2+}$ can be oxidized.

Chapter 24

1. The rubber in a tire; it is stiffer. **3.** Above room temperature, because the pipe is not flexible at room temperature. **5.** Float the poproplyene in a 1:1 ethanol-water mixture, float the HDPE in water, float the polystyrene in 10% NaCl in water, leaving PETE. **7.** Its bands are completely filled or completely empty. **9.** At 0 K, both have bands that are completely filled or completely empty. Upon warming, electrons are promoted across the gap in the semi-conductor. **11.** Electrons are thermally promoted from filled bands into empty bands. **13.** The band gap in diamond is much larger than the band gap in graphite. **15.** This metal would become an insulator or a semiconductor depending upon the size of the band gap. An oxidizing agent would remove the electrons from the upper band. **17.** SiC and Si_3N_4 **19.** Pressing increases the surface contact between particles of reactants and facilitates diffusion of ions. **21.** In both cases a gaseous precursor is decomposed at elevated temperatures and the product precipitates from the gas phase. **23.** They should not corrode in biological fluids, they should have some flexibility, and bone should adhere to their surface. **25.** 8, 8.
27.

a. H—C=C—O—H b. H—C=C—Cl: c. H—C=C—C≡N:

29.

HOCH₂—C⟨⟩C—CH₂OH and

H—O—C(=O)—C⟨⟩C—C(=O)—O—H

31. Polyvinyl alcohol chains can form hydrogen bonds to water that are about the same strength as the hydrogen bonds in the polymer. This results in solubility. (See Section 12.2) **33.** CH_2 and C 85.63%, H 14.37%, for both **35.** Negative; the entropy decreases as gaseous ethylene molecules are incorporated into solid polyethylene chains. **37.** The geometry and hybridization of the C atoms in the C_6 ring do not change. The hybridization of the carbon atoms in

the —CH=CH₂ substituent changes from sp^2 to sp^3 and the geometry changes from trigonal planar to tetrahedral. **39.** 5.1×10^{-5} g. **41.** The cell with the 2.2 eV gap should be on top. The next cell will trap lower energy light as well as light with an energy greater than 2.2 eV that may have passed through the top cell. **43.** a. $+2.1$, b. $+1$ **45.** a. 463 atoms, b. 2.8×10^2 lenses

Chapter 25

1. Living organisms can extract and utilize energy from their environments, have metabolisms based on complex carbon chemistry, are highly organized, are dependent on polymers, and have the ability to reproduce and pass on information to their descendants. **3.** Determining factors: hydrogen bonding and non-hydrogen bonding or polarity and lack of polarity

description	example
hydrophilic water-liking	glycine, sugar
hydrophobic water-fearing	myristyl palmitate, oil

5. *Anabolic pathways* are series of linked synthetic reactions which consume chemical energy. *Catabolic pathways* are a series of linked degradative reactions which release chemical energy. **7.** Because they are not able to *completely* oxidize organic compounds to CO_2 and H_2O. **9.** Assuming the temperature is 25°C, for the reaction $ATP + H_2O \longrightarrow ADP + $ phosphate, $K_{eq} = 1.6 \times 10^5$, and the equilibrium lies far to the right; that is, towards hydrolysis. **11.** *Glutamic acid:* there are 3 functional groups each of which can exist in 2 forms uncharged or charged; that is, $-CO_2H$ or $-CO_2^-$ and $-NH_2$ or $-NH_3^+$. Thus there are 8 possible forms; one of them is not ionic. *Lysine:* the same situation exists for lysine as glutamic acid. **13.** *Primary structure:* the linear amino acid polymer of a protein. *Secondary structure:* localized folding of a peptide chain maximizing hydrogen bonds. *Tertiary structure:* completely folded conformation of a peptide chain. *Quaternary structure:* the association of two or more folded subunits of a protein having more than one peptide chain. **15.** *Denaturation* refers to the unfolding of a protein caused by harsh treatment such as high temperature or acid or base; denaturation does not involve breaking covalent bonds. *Hydrolysis* involves the cleavage of the peptide bonds with water to produce the amino acids from which the protein was polymerized. **17.** Gly-Gly, Gly-Ala, Ala-Gly, and Ala-Ala. **19.** The active site of an enzyme is the crevice into which the substrates must bind to be transformed into products. **21.** In the linear structure, carbons 2,3,4, and 5 each have four different functional groups. In the cyclic structure, carbons 2,3,4,5, and 6 have four different groups. **23.** The hydroxyl group at carbon 1 can assume two positions, down α and up β. **25.** Humans have no hydrolytic digestive enzyme for the ether linkages between the rings in cellulose. **27.** Fatty acids are long-chain carboxylic acids. Fats and oils are fatty acid triesters of glycerol. Fats are solids at room temperature and oils are liquids. The fatty acids from which fats are made tend to be longer and have fewer double bonds than those of oils. **29.** The triglyceride has a molar mass of 890. **31.** There is a double bond and a hydroxyl group. **33.** The 3' end of a polynucleotide has a free 3' hydroxyl group on the sugar. The 5' end has a terminal phosphate ester at the 5' carbon of the sugar. **35.** The replication product would read, from its 5' end: C-T-A-G-C-T-G-A-G-C-C-T. The transcription product would read from its 5' end: C-U-A-G-C-U-G-A-G-C-C-U. **37.** Advantages: confer specific disease, herbicide, and insect resistance; increase crop yield; improve flavor and nutritional value; utilize plants for producing non-plant products; create thrifty plants.

Disadvantages; decrease diversity of crop varieties; increase costs of seed; plant genomes controlled by agribusinesses; foreign proteins in foods derived from plants may cause human health problems. **39.** Absorb carbon dioxide (as plants do) and synthesize reduced organic compounds from it.   **41.** London forces, polar bonds, hydrogen bonds, and ionic interactions.   **43.** Proteins undergo chemical modification after they are synthesized.   **45.** At pH 2, alanine would have a net charge of $+1$ and would move toward the cathode $(-)$. At pH 10, alanine would have a net charge of -1 and would move toward the anode $(+)$.   **47.** 0.50 mol ATP per g of palmitic acid.

Index

Abelson, P. H., 769
Absolute zero, 336
Accelerators, 770
Acetic acid, 313, 596
Acetone, 312
Acid(s), 57–58, 267
    amino, 932–935
    anhydrides, 820
    arachidonic, 945
    Brønsted, 267, 554
        preparation of, 566–567
        properties of, 566
        strengths of, 559–563
    carboxylic, 312–314
    conjugate, 555
    deoxyribonucleic, 928, 947–950
    diprotic, 567, 605–607
    equivalents of, 570–572
    fatty, 944–945
    ionization constant, 559
    Lewis, 572–574
    mixtures of, 604–609
    monoprotic, 567
    nucleic, 947–950
    polyprotic, 604–609
    preparation of, 566–567
    properties of, 566
    protonic concept of, 554–558
    quantitative reactions of, 569–570
    reactions with metals, 826
    ribonucleic, 928, 947–952
    strength of, 559–563, 564–566
    strong, 268, 559
    triprotic, 567–568, 607–609
    water as, 556–557
    weak, 269, 559
        ionization of, 590–595,
          609–612
Acid-base
    adduct, Lewis, 572
    concepts
        Brønsted-Lowry, 554–572
        Lewis, 572–574
    indicators, 625–626
    neutralization, 563–564
        titration, 626–629
    properties of solutions of salts,
        602–604
    reaction, 267–270, 554
Acidic oxides, 271
    reaction of basic oxides and, 271
Acid rain, 355
Acquired immune deficiency syndrome
    (AIDS), 940
Actinide series, 13, 171, 766, 770, 865

Activated complex, 487
Activation energy, 488–492
Activation substrate, 940
Active site, 940
Activity, 513
    ion, 448
    series, 273–274
Actual yield, 98
Addition polymers, 316
Addition reaction, 56, 305
Adduct, Lewis acid-base, 572
Adenine, 949–950
Adenosine diphosphate (ADP), 931
Adenosine monophosphate (AMP), 931
Adenosine triphosphate (ATP), 929, 931
Adhesive forces, 388
Aerobic reactions, 930
Aerosols, 829
Affinities, electron, 176–177
AIDS, 940
Alanine, 931, 932
Alcohol(s), 309, 943
    ethyl (ethanol), 309
    naming, 309
Aldehydes, 311–312, 943
Aldose, 941
Alkali metals, 45, 788
Alkaline cell, 730
Alkaline earth metals, 45, 788–789
Alkaloids, 314
Alkanes, 300–302
    nomenclature of, 302–304
Alkenes, 304–306
    isomers of, 306
    naming, 305
Alkyl group, 303
Alkynes, 304, 306–307
    naming, 307
Allotropes, 821
    of phosphorus, 821–822
    of sulfur, 822–824
Alloys, 427
Alpha
    decay, 765
    helix, 937
    particles, 146, 764
Alpha-alumina, 797
Aluminum, 272, 790
    hydroxide, 272, 797
    oxide, 272, 797
    preparation of, 792–794
Americium, 779
Amide linkage, 316
Amides, 844
Amines, 310–311

Amino acid(s), 932–935
    common, 932
    sequence, 936
Amino-terminal residue, 935
Ammines, 844
Ammonia, 826
    chemical properties of, 844–845
    Haber process, 844
    preparation of, 844
    uses of, 282–283
Ammonium ion, 844
Ammonium nitrate, 418
Amorphous solids, 389
Ampere, 740–741
Amphipathic behavior, 928
Amphiprotic species, 558–559
Amphoteric compounds, 272, 565–566
Amphoteric elements, 790
Anabolic pathways, 929
Anaerobic reactions, 930
Analysis
    combustion, 104–106
    gravimetric, 102–104
Anemia
    Cooley's, 882
    sickle-cell, 896, 955–956
Angular momentum quantum number, 159
Anhydrides, 820
Anions, defined, 46, 184
Anode, 713
Antibodies, 937
Antibonding orbitals, 244
Anti-Lewisite, British, 882
Antimony, 810, 811
Approximations, successive, 537–539
Aqua regia, 854
Aqueous solutions, 443
    electrolysis of, 736–738
Arachidonic acid, 945
Argon, 833, 843
    uses of, 833
Aromatic hydrocarbons, 307
Arrhenius, Svante, 420, 554
Arrhenius equation, 488–492
Artificial hips, 919–920
Aspartame, 936
Aspirin, 945
Association constants, 656
Assumption that Δ is small, 536–537
Atmosphere, 352–355
    pressure of, 327
Atom(s)
    assigning hybrid orbitals to central,
        238–240
    Bohr model of, 151–155

Atom(s) (continued)
  central metal, 877
  composition of, 35–37
  defined, 10–11, 34
  donor, 877
  isoelectronic, 173
  moles of, 73–76
    mass of, 73
  nuclear, 146–147
  quantum mechanical model of, 157–162
Atomic
  fission, 771
  mass, 36, 71
  mass unit (amu), 36
  nucleus, 757
  numbers, 36, 757
  orbitals, 157
    d, 159, 885
    f, 160
    hybridization of, 232–233
    overlap of, 230–231
    p, 159
    s, 159
  properties, electron configuration and,
    164–177
  radii, inside back cover
  structure, 144–177
    orbital energies and, 163–164
  theory
    Dalton's, 34–35
    historical development of, 145–155
  weight. See Atomic mass
ATP, 929
Attraction
  dipole-dipole, 374–375
  ion-dipole, 419
Aufbau process, 164–169
Austenite, 921
Autoionization, 557
Avogadro, Amedeo, 73, 337
Avogadro's law, 337–338, 339
Avogadro's number, 73
Axial position, 220
Azeotropic mixtures, 440
Azimuthal quantum number, 159

Baking soda, 801
Balancing equations, 53–54
  oxidation-reduction, 743–747
Band, 905
  gap, 905
  of stability, 761
  theory, 904–906
Barium, 788, 789
  peroxide, 79
Barometer, 328
Base(s), 58
  Brønsted, 267, 554
  conjugate, 555
  equivalents of, 570–572
  Lewis, 572–574

mixtures of, 604–609
pairs, 950
preparation of, 568–569
properties of, 568
protonic concept of, 554–558
quantitative reactions of, 569–570
sequence, 948
strengths of, 559–563, 564–566
strong, 270, 560
water as, 556–557
weak, 270, 560
  ionization of, 590–595
Basic oxides, 270
  reactions of acidic oxides and, 271
Battery. See Cell(s)
Becquerel, Antoine, 757
Bednorz, J. G., 913
Bengal saltpeter, 843
Benzene, 307
Beryllium, 788, 789
Beta decay, 765–766
Beta+ decay, 766
Beta particles, 764
Beta-pleated sheet, 937–939
Bicarbonate of soda, 801
Bidentate ligands, 878
Bimolecular elementary reactions, 493–494
Binary compounds, naming of, 59–63
Binding energy, 758–761
  per nucleon, 759–760
Biochemistry, 927–956
Biocompatibility, 919
Bismuth, 791
Blast furnace, 869–870
Blister copper, 871
Blood sugar, 941
Body-centered cubic structure, 393
Body-centered cubic unit cell, 399–400
Bohr, Niels, 146, 151
Bohr model of the atom, 151–155
Boiling point
  elevation of, 437–438
  normal, 379
Bond(s)
  angle, 217
  chemical, 184–209
  coordinate covalent, 199
  covalent, 49, 51, 188–189
    coordinate, 199
    polar, 189–191
    strengths of, 206–209
  distance, 217
  double, 194, 240–242
  energies (dissociation), 206–209
  hydrogen, 376–377
  ionic, 49, 184–185
  order, 247
  peptide, 935
  pi, 232
  polar covalent, 189–191
  sigma, 231, 232

single, 193
  three-center two-electron, 811
  triple, 194, 240–242
Bonding orbitals, 244
Borates, 811, 816–817
Borax, 811, 817
Boric acid, 811, 816
Boric oxide, 816
Born-Haber cycle, 403–404
Boron, 790
  chemical behavior of, 809
  halides, 815–816
  hydrides, 812–814
  isolation of, 811–812
  oxide, 816–817
  structure of, 810–811
Borosilicate glass, 911
Boyle, Robert, 329, 331, 554
Boyle's law, 329–333, 339, 359
Bragg, William, 405
Bragg equation, 404–405
Branched polymers, 898
British anti-Lewisite (BAL), 882
Brittle tin, 795
Broglie, Louis de, 155
Bromates, 858
Bromic acid, 858
Bromine, 831, 832, 833
  preparation of, 832
  properties of, 839–842
Brønsted, Johannes, 267, 554
Brønsted acids, 267
Brønsted bases, 267, 554
Brønsted-Lowry acid-base concept,
  554–572
Buckminsterfullerene, 296
Buffer, 617
  capacity, 621
  solutions, 617–623
Buret, 101
Butane, 301
Butene, 306

Cadmium, 789–790
  cell, with nickel, 731–732
Calcium, 788, 789
  carbide, 299
  carbonate, 800
  cyanamide, 29
  hydroxide, 283, 799
  oxide, 283, 797
  peroxide, 7
Calculations
  boiling point, 437–438
  bond energy, 206–209
  cell potentials, 721–723
  complex ions, 656–660
  concentrations, 427–434, 469
  enthalpy, 124–130, 681–684
  entropy, 684–686
  equilibrium, 527–543, 652–660

Faraday's law, 740–743
free energy, 689–698
Hess's law, 131–134
ideal gas, 338–341
ionic radii, 402–403
lattice energies, 403–404
mass, 71–72, 445–447
of molarity, 85
molecular mass, 71–72
Nernst equation, 726
nuclear binding energy, 758–761
percent yield, 98–99
solubility product, 644–646
successive approximation, 537–539
titration curves, 626–629
yield, 98–99
Calorie, 119
Calorimeter, 120
Calorimetry, 120–123
Cane sugar, 942
Capillary action, 388
Carbohydrates, 940–943
defined, 940
disaccharides, 942–943
monosaccharides, 941–942
polysaccharides, 931, 942–943
starch, 943
Carbon
dioxide, 297–299
elemental, 294–296
inorganic compounds of, 297–299
monoxide, 297
organic compounds of, 299–307
tetrachloride, 833
Carbon-14 dating, 767–769
Carbonate(s), 799–801
ion, 299
of transition metals, 874
Carbon black, 296
Carbonyl group, 311
Carboxylic acids, 312–314
Carboxyl-terminal residue, 935
Casein, 937
Catabolic pathways, 929
Catalysts, 280
in catalytic converters, 470, 499
effect of
on activation energy, 499
on equilibrium, 520–521
on reaction rate, 470, 497–499
heterogeneous, 498–499
homogeneous, 497–498
Catalytic converters, 470, 499, 910
Cathode, 713
Cations, 46, 184
Cell(s)
alkaline, 730
components of, 927–929
Daniell, 713, 714, 730
Downs, 735
dry, 730

electrolytic, 731, 734–743
eukaryotic, 947
fuel, 732
galvanic, 710–714
lead-acid, 731
Leclanche, 730
nickel-cadmium, 731–732
potential, 714–716
calculation of, 721–723
concentration and, 725–727
equilibrium constant and, 727–729
free energy, electrical work, and, 723–725
standard, 715
primary, 730
prokaryotic, 947
reaction, 712
spontaneity of, 724–725
ribosomes of, 929
secondary, 730–732
solar, 907–909
unit, 397–401
Cellulose, 943
Celsius temperature scale, 16
Central metal atom or ion, 877
Ceramics, 909–916
superconducting, 913–916
Cesium, 788
Chain mechanism, 496
Chain reaction, 771
Chalcogens, 45
Change
chemical vs. physical, 7
in concentration or pressure, 527–529
in internal energy, 671
spontaneous, 678–681
Charcoal, 296
Charge, formal, 200–203
Charles, S. A. C., 334
Charles's law, 333–336, 339, 359–360
Chelate, 877
Chelating ligand, 877
Chemical
bonding, 184–209
change, 7
stoichiometry and, 90–100
equations, 52–55
mole relationships and, 90–92
equilibrium, 508–543
kinetics, 461–499
properties, 7
reactions
classification of, 56–59
spontaneous, 678–680
reduction, of representative metals, 794–795
stoichiometry, 71–106
synthesis, 274
thermodynamics, 123–124, 669–698
vapor deposition, 918

Chemistry, 3–4
biochemistry, 927–956
colloid, 448–453
domains of, 13–14
nuclear, 757–781
Chemotherapy, 778
Chemotrophy, 930
Chernobyl nuclear disaster, 775
Chile saltpeter, 831, 832, 853
Chloramine, 845
Chlorates, 858
Chloric acid, 852
Chlorine, 831, 832–833
dioxide, 848–849
preparation of, 832
properties of, 839, 841, 842
uses of, 284–285, 832–833
Chlorite ion, 858
Chlorofluorocarbons (CFCs), 829
Chloroform, 833
Chlorophyll, 881
Chloroplasts, 929
Chlorous acid, 857–858
Cholesterol, 946
Cholines, phosphatidyl, 945
cis (configuration) isomer, 306, 880–881
Cladding, 912
Clone, 955
Closed systems, 670
Closest packing, 391–393
cubic, 393
hexagonal, 392
Coal, 135
Codons, 952–953
Coefficients, 53
Cohesive forces, 387–388
Coke, 296
Collagen, 937
Colligative properties, 435–448
Collision theory of reaction rate, 486–488
Colloid(s), 448–449
electrical properties of, 452–453
gels, 453
preparation of, 450
Colloidal dispersion, 448–449
Colors, of transition metal complexes, 888–889
Combination reaction, 56
Combustion
analysis, 104–106
enthalpy of, 126–127
reaction, 57
Common amino acids, 932
Common ion effect, 613–617
Competitive inhibition, 940
Complementary strands, 951
Complex(es), 877
activated, 487
bonding and electron behavior in, 882–889
calculations involving, 656–660

Complex(es) (*continued*)
  enzyme-substrate, 940
  high-spin, 887
  isomerism in, 880–881
  low-spin, 887
  naming of, 878–879
  structures of, 879
  transition metal, colors, of, 888–889
  uses of, 881–882
Composite, 903
Compounds
  amphoteric, 272, 565–566
  binary, 59–63
  coordination. *See* Coordination
    compounds
  covalent, 51
  defined, 9, 34
  hydrogen, 844–848
  ionic, 49–51, 184
  nonstoichiometric, 426
  with oxygen, 796–798
  of representative metals, 795–802
  ternary, 59
Concentrated solution, 85, 415
Concentration(s)
  effect on cell potentials, 725–727
  equilibrium, 529–535
  expressing, 427–434
  molality, 430–431
  molarity, 429–430
  mole fraction, 431–434
  percent by mass, 428–429
  and precipitation, 649–650
  and pressure changes, equilibrium and,
    527–529
  and reaction rate, 469
  of solutions, 85, 427–434
  of weak acids or weak bases,
    596–602
Condensation, 378
  method of preparing colloids, 450
  polymers, 316
Conduction, metallic, 905
Conductors, electrical, 904
Conjugate acids and bases, 555
Conservation laws
  energy, 6
  matter, 5
Constant(s)
  association, 656
  dissociation, 657
  equilibrium, 511
    calculation of, 530–531
    cell potential and, 727–729
    free energy changes and, 696–698
    reaction quotients and, 510–513
  formation, 656
  gas, 338
  ionization, 559–563, 590
    of acid, 559
    of base, 560

molal boiling-point elevation, 437
molal freezing-point depression,
    441
rate, 470
stability, 656
Constant boiling solution, 440
Constant composition, law of, 35
Containment system, 774–775
Continuous spectrum, 149
Control system, 774
Conversion factors, 21–23
Coolant, 774
Cooley's anemia, 882
Coordinate covalent bond, 199
Coordination
  number, 877
    for crystals, 393
  sphere, 877
Coordination compounds, 876–889
  association constants, 656
  bonding and electron behavior in,
    882–889
  definitions of terms, 877–879
  formation constants, 565
  isomerism in, 880–881
  naming of, 878–879
  structures of, 879
  transition metal, 874
  uses of, 881–882
Copolymers, 896
Copper
  blister, 871
  isolation of, 871
  oxide superconductors, 875–876
Core, 912
Corrosion, 733–734
Corundum, 797
Cottrell, Frederick, 452–453
Coulomb, 724, 740–741
Covalent
  compounds, 51
  network solids, 390
  radii, 171–172
Covalent bonds, 49, 51, 188–189
  coordinate, 199
  polar, 189–191
  strengths of, 206–209
Crick, Francis, 950
Critical
  mass, 771
  pressure and temperature, 383–384
  temperature, 913
Cross-linked polymers, 898
Cryolite, 847–848
Crystal(s)
  body-centered cubic, 393
  defects in, 391
  face-centered cubic, 393
  field splitting, 886
  field theory, 884–887
  hexagonal, 392

ionic
  lattice energies of, 187–188,
    403–404
  structures of, 394–395
  simple cubic, 393–394
Crystalline solids, 389
C-terminal residue, 935
Cubic
  centimeter, 17
  closest packing, 393
  holes, 394
  meter, 17
  structure
    body-centered, 393
    face-centered, 393
    simple, 393–394
Curie, Marie and Pierre, 764
Cysteine, 935
Cytoplasm, 929, 947
Cytosine, 949–950

Dacron, 316, 919
Dalton, John, 10, 34, 146
Dalton's atomic theory, 34–35
Dalton's law, 345–347, 360
Daniell cell, 713, 714, 730
Daughter nuclide, 765
Davisson, C. J., 155
Davy, Sir Humphry, 554
d-block elements, 865
Debye, Peter, 448
Decay, radioactive, 765–767
Decomposition reaction, 57
Defects, crystal, 391
Definite proportion, law of, 35
Degenerate orbitals, 161, 245
Degree Celsius, 16
d electron, 159
Density
  calculations, 17–18
  defined, 17
  electron, 158
  of gases, 343–344
Deoxyribonucleic acids (DNA), 928,
    947–950
  genetic information in, 954
  recombinant technology, 955
  replication, transcription, and trans-
    lation of, 950–952
  structure of, 949–950
Derivatives, of hydrocarbon, 308–317
Detergents, 451–452
Deuterium, 825
Dextrose, 941
Diamagnetic molecules, 198
Diamagnetism, 887
Diamond, 295, 905
  thin films, 917–918
Diatomic molecules, of second period,
    248–251
Diborane, 812–814

Dichlorine monoxide, 848–849
Differential rate law, 470
Diffraction, X-ray, 404–405
Diffusion
    of gases, 348–349
    reactions controlled by, 486
Dihelium molecule, 248
Dihydrogen
    molecule, 247
    phosphite ion, 855
Dilute solutions, 85, 415
Dilution of solutions, 87–90
Dimethyl ketone, 312
Dinitrogen
    oxides, 850
    pentaoxide, 850–851
    tetraoxide, 850, 851
    trioxide, 850
Dipeptide, 935
Diphosphorus tetrahydride, 837
Dipole-dipole attractions, 374–375
Dipole moment, 227
Diprotic acids, 567
    ionization of, 605–607
Direction of shift, 522–523
Direct proportionality, 334
Disaccharides, 942–943
Disorder
    changes in, 684–686
    increase in, 680–681
Dispersed phase, 449
Dispersion
    colloidal, 448–449
    forces, 375–376
    medium, 449
    method of preparing colloids, 450
Disproportionation reactions, 820, 842
Dissociation constant, 657
Dissolution
    by complex ion formation, 656–660
    of ionic compounds, 419–421
    of molecular electrolytes, 421
    process of, 416–421
    by weak electrolyte formation,
        652–655
Distillation, 380–381, 439–440
    fractional, 440
DNA. See Deoxyribonucleic acids (DNA)
Domains of chemistry
    macroscopic, 13
    microscopic, 13
    symbolic, 14
Donor atom, 877
Dopants, 907
d orbitals, 159, 885
Double
    bond, 194
        hybridization involving, 240–242
    displacement reaction, 57
    helix, in DNA, 950
Downs cell, 735

Dry cell, 730
d^2sp^3 hybridization, 883
Dual nature of matter, 155
Dynamic equilibrium, 378, 415

Effusion of gases, 348–349
e_g orbitals, 885
Eicosanoids, 945
Einstein, Albert, 149
Einstein equation, 758
Elastomers, 901
Electrical conductors, 904
Electrical work, cell potential, and free
        energy, 723–725
Electrode
    defined, 716–717
    gas, 717
    hydrogen, standard, 717
    potential of, 716–720
Electrolysis, 734–735
    of aqueous solutions, 736–738
    of hydrochloric acid, 736
    of molten sodium chloride, 734–735
    preparation of representative metals
        by, 791–794
    of sodium chloride solution, 736–737
    of sulfuric acid solution, 737
    of water, 826
Electrolyte(s), 420
    colligative properties of, 447–448
    dissolution
        of ionic, 419–421
        of molecular, 421
    Faraday's law of, 740–743
    strong, 421
        ion concentrations in solutions of,
            583–585
    weak, 421
Electrolytic cells, 731, 734–743
Electrolytic deposition of metals,
        738–739
Electromagnetic radiation, 147
Electron(s)
    affinities, 176–177
    capture, 766
    charge of, 36
    configuration(s), 163
        and atomic properties, 164–177
        aufbau process, 164–169
        of elements, 167
        noble gas, 186
        and periodic table, 169–171
        pseudo-noble gas, 186
    d, 159
    defined, 36
    density, 158
    f, 159–160
    mass of, 36
    molecules with extra, 199
    p, 159
    paired, 162

    s, 159
    sharing, chemical bonding by, 188–189
    valence, 169
Electron-deficient molecules, 198–199
Electronegativity, 189–191
Electron-pair geometry, 220
    rules for predicting, 222–227
Electron-pair repulsion theory (VSEPR),
        217–229
Electroplating, 882
Elementary reactions, 492–495
Elements
    actinides, 13, 171, 766, 770, 865
    amphoteric, 790
    classifying, 11–13
    d-block, 865
    defined, 9, 34
    electron configurations of, 167
    electronegativity of, 189–191
    f-block, 865
    inner transition, 45, 171, 865
    lanthanides, 13, 171, 865
    main group, 45, 170
    new, 37–39
    noble gases, 45–46, 833
    periods of, 12
    rare earth, 865
    representative, 45, 170
    transition, 45, 171, 865–866
    transuranium, 770
Emission spectra, 150
Empirical formula, 39–40, 81–84
Emulsifying agents, 450
Emulsions, 450
Endothermic process, 118, 672
End point of titration, 101
Energy(ies)
    activation, 488–492
    of antibonding and nonbonding
        orbitals, 244
    binding, 758–761
    bond (dissociation), 206–209
    changes, 681–684
    conservation of, 6
    defined, 5
    diagrams, 245–246
    free, 689–698
    internal, 123, 671–672
    kinetic, 6, 357–359
    lattice, 187–188
    levels, 905
    minimization of, 680–681
    molecular orbital, 245–246
    nuclear, 771–781
    nuclear binding, 758–761
    orbital, 163–164
    pairing, 886
    potential, 6
    quantized, 151
    sources, 929–930
    thermal, 118

Energy(ies) (*continued*)
    thermodynamics, 123–124, 669–698
    utilization, 929–930
Enterobactin, 882
Enthalpy, 681
    changes, 124–130, 681–684
    of combustion, 126–127
    of formation, standard, 129–130
    of fusion, 127–128
    of vaporization, 128–129
Entropy, 685
    changes in, 684–686
    standard, 685
Enzyme(s), 940
    inhibition, 940
    substrate activation of, 940
Epsom salts, 856
Equations
    balancing, 53–54
    chemical, 52–55
        mole relationships and, 90–92
    for ionic reactions, 55–56
    rate. *See* Rate laws
    redox, balancing of, 743–747
    See also names of specific equations
Equatorial position, 220
Equilibrium, 378
    calculations, 527–543
    catalyst effect on, 520–521
    chemical, 508–543
    concentration(s)
        calculating, 529–535
        changes in, 517–521
        and pressure changes, 527–529
    constant(s), 511
        calculation of, 530–531
        cell potential and, 727–729
        free energy changes and, 696–698
        reaction quotients and, 510–513
    dynamic, 378
    heterogeneous, 516–517
    homogeneous, 513–516
    kinetics and, 523–527
    multiple, involving solubility,
        651–660
    position of, 517
    pressure effects on, 518–519
    problems, techniques for solving,
        535–543
    reaction mechanisms involving,
        525–527
    relationship of reaction rates and,
        523–525
    of saturated solutions, 415
    state of, 508–509
    temperature effects on, 519–520
Equivalence point, 626–629
Equivalent
    of acids and bases, 570–572
    of electrons, 740
Esters, 312–314

Ethane, 237, 308
Ethanol, 308, 309
Ethene. *See* Ethylene
Ethers, 310
Ethyl alcohol, 309
Ethyl chloride, 833
Ethylene, 194, 304–305
Eukaryotic cells, 947
Evaporation, 378–379
Excited state, 153
Exothermic process, 118, 672
Expansion work, 124, 672–673
Extensive property, 7
Extrinsic semiconductor, 907

Face-centered cubic structure, 393
Face-centered cubic unit cell, 400
Fahrenheit temperature scale, 25
Faraday, 724, 740–741
Faraday, Michael, 740
Faraday's law, 740–743
Fats, 314, 944
Fatty acids, 944–945
f-block elements, 865
f electron, 159–160
Fermentation, 930
Ferritin, 937
First
    ionization energy, 174
    law of thermodynamics, 674–676
    transition series, 171, 865
First-order reactions, 472, 477–478
    half-life of, 483–484
Fischer, Emil, 935
Fission, nuclear, 771
Flint glass, 911
Fluorides, 832
Fluorine, 831, 832
    occurrence and preparation of, 832
    properties of, 839–842
    uses of, 832
Fluoroapatite, 847
Fluoroboric acid, 815
Fluorocarbons, 832, 847
Fluorosilicic acid, 815
Food, fuel and, 134–136
f orbitals, 160
Force(s)
    adhesive, 388
    cohesive, 387–388
    dispersion, 375–376
    intermolecular, 374–377
    London dispersion, 375
    nuclear, 757–758
    van der Waals, 374
Formal charge, 200–203
Formaldehyde, 194, 312
Formation, standard enthalpy of,
    129–130
Formation constant, 656
Formic acid, 313

Formula(s)
    defined, 39–41
    empirical, 39–40
        derivation of, 81–84
    mass, 72
    molecular, 39, 40
        derivation of, 84–85
    percent composition from, 80–81
    simplest, 39
    structural, 41
Fourth transition series, 171, 865
Fractional distillation, 440
Fractional precipitation, 650–651
Francium, 788
Frasch process, 830–831
Free energy
    changes in, 689–698
        and enthalpy and entropy changes,
        689–692
        equilibrium constants and,
        696–698
        and nonstandard states, 694–696
        standard, 691
    of formation, standard, 692–693
Freezing point, 382
    depression of, 440–443
Freons, 829
Frequency, 147–148
Frequency factor, 489
Fructose (fruit sugar), 941–942
Fuel
    cells, 732
    food and, 134–136
    nuclear, 773
Fuller, Buckminster, 184, 296
Functional groups, 302
Fusion
    enthalpy of, 127–128
    nuclear, 776–777
    reactor, 777

Gallium, 790
Galvanic cell(s), 710–714
    batteries, 710–714
    potential of, 714–716
Galvanized iron, 795
Gamma-alumina, 797
Gamma rays, 764
    emission of, 766
Gap, band, 905
Gas(es)
    behavior of, 359–361
    constant, 338
    densities of, 343–344
    diffusion of, 348–349
    effusion of, 348–349
    electrode, 717
    equation, 338–341, 361–362
    homogeneous equilibria among,
        514–516
    ideal, 331

kinetic-molecular theory for, 356–357
macroscopic behavior of, 326–355
microscopic behavior of, 356–364
molar mass of, 344
molecular velocities of, 357–359
natural, 134
pressure of, 326–329
    of mixture of, 345–347
properties of, 4
real, 332–333
solutions of, 422
    in liquids, 422–424
standard conditions for, 341–343
stoichiometry involving, 349–352
volume
    related to pressure, 329–333
    related to temperature, 333–336
water, 826, 827
    shift reaction, 826
Gay-Lussac, J. L., 337–338, 554
Geiger, Hans, 146
Gels, 453
Gene
    defined, 954
    mutations, 955–956
    therapy, 955
Genetic code, 950–954
Genetic engineering, 955
Genome, 954
Geometrical isomers, 306, 880
Geometry
    electron-pair, 220
    molecular, 220
Germanium, 810
Germer, L. H., 155
Glass, 389, 910–913
    borosilicate, 911
    flint, 911
    transition temperature, 899–900
Glauber's salt, 856
Glucose, 941, 942
Glutamic acid, 955, 956
Glyceraldehyde, 941
Glycerol, 944
Glycine, 932
Glycogen, 943
Glycolysis, 930
Glycoproteins, 943
Graham, Thomas, 348, 449
Graham's law, 348–349, 360
Graphite, 295–296
Gravimetric analysis, 102–104
Gray tin, 795
Ground state, 153
Groups
    of elements, 12
    functional, 302
    See also Periodic relationships among
        groups
Guanine, 949–950
Guanosine triphosphate (GTP), 929

Haber, Fritz, 520
Haber process, 844
Half-cell, 712
Half-life
    of radioactive materials, 762–763
    of reaction, 483–486
Half-reactions, 710–711
Halic acid, 857
Halides, 801–802
    boron, 815–816
    hydrogen, 846–847
    nonmetal, 848–849
    silicon, 815–816
    of transition metals, 872–873
    See also names of specific halides
Hall, Charles M., 792
Halogens, 45, 831–833
    halides of, 849
    hydrogen compounds of, 846–848
    interhalogens, 849
    oxyacids of, 857–859
    properties of, 839–842
    See also Bromine; Chlorine; Fluorine;
        Iodine
Halous acid, 857
Heat, 118, 672
    capacity, 119
    measurement of, 119–120
    specific, 119
    See also Enthalpy; Temperature
Heisenberg, Werner, 156
Heisenberg uncertainty principle, 156
Helium, 833
    properties of, 842, 843
    uses of, 833
Helix
    alpha, 937
    double, 950
Heme, 881–882, 936
Hemoglobin, in sickle-cell anemia,
    955–956
Henderson-Hasselbalch equation, 622–623
Henry's law, 422
Héroult, P. L. T., 792
Hertz, 148
Hess's law, 131–134, 682
Heterogeneous
    catalysts, 498–499
    equilibrium, 516–517
    mixture, 8
Hexadentate ligands, 878
Hexagonal closest packing, 392
High-spin complex, 887
Holes, in band, 906
Homogeneous
    catalysts, 497–498
    equilibrium, 513–516
    mixture, 8
    substance, 8
Homopolymer, 896
Huckel, Erich, 448

Hund's rule, 165
Hybridization
    of atomic orbitals, 232–233
    d^2sp^3, 883
    involving double and triple bonds,
        240–242
    linear, 233
    octahedral, 883
    sp, 233–234
    sp^2, 235
    sp^3, 235–237
    sp^3d, 237–238
    sp^3d^2, 237–238, 883
    square planar, 884
    tetrahedral, 235–237
    trigonal bipyramidal, 237
    trigonal planar, 235
Hybrid orbitals, 232–238
    assigning, to central atoms, 238–240
Hybrid resonance, 205
Hydrated metal ions, 623–625
Hydrazine, 845
Hydrides
    boron, 812–814
    ionic, 826, 834
    silicon, 812–814
Hydrobromic acid, 848
Hydrocarbons, 134
    alkanes, 300–302
    alkenes, 304–306
    alkynes, 306–307
    aromatic, 307
    derivatives of, 308–317
    nomenclature of, 302–304
    saturated, 300–302
Hydrochloric acid, 848, 854
    electrolysis of, 736
Hydrofluoric acid, 847
Hydrogen, 825–827
    bonding, 376–377
    bromide, 847
    carbonate ion, 299
    chemical properties of, 834–835
    chloride, 847
    compounds, 844–848
    cyanide, 299
    electrode, standard, 717
    fluoride, 847–848
    halides of, 846–847
    iodide, 847
    isotopes of, 825
    occurrence of, 825
    phosphite ion, 855
    preparation of, 825–826
    properties of, 825
    sulfates, 853, 856
    sulfide, 845–846
    sulfite ion, 856
    sulfites, 853
    uses of, 826–827
Hydrogenation, 826

Hydroiodic acid, 848
Hydrometallurgy, 872
Hydronium ion, 57–58, 268
Hydrophilic substances, 928
Hydrophobic substances, 928
Hydrosulfuric acid, 846
Hydroxide(s), 798–799
  ion, 58
  preparation of, 568–569
Hypertonic solution, 445
Hypofluorous acid, 857
Hypohalite ions, 857
Hypohalous acids, 857
Hypophosphite ion, 855
Hypophosphorous acid, 855–856
Hypothesis, 4
Hypotonic solution, 445

Ideal gas, 331
  deviations from, 362–364
  equation, 338–341
    kinetic-molecular theory and,
      359–362
Ideal solution, 416
Imides, 844
Immiscibility, 425
Indicators, 101
  acid-base, 625–626
Indium, 790
Industrial chemicals, chemical properties
    of, 279–285
Inert-pair effect, 185–186
Inhibitors, 499
Initial rate of reaction, 464
Inner-shell hybridization, 883
Inner transition metals (or elements), 45,
    171, 865
Inorganic compounds
  of carbon, 297–299
  naming, 59–63
Instantaneous rate, 463–464
Insulators, 904, 905, 906–907
Intensive property, 7
Interhalogens, 849
Intermediates, 492
Intermolecular attractions, 373
Internal energy, 123, 671–672
International System of Units (SI), 15,
    16–17
Interstitial sites, 391
Interstitial solid solutions, 427
Intrinsic semiconductor, 907
Inverse proportionality, 329–331
Invert sugar, 942
Iodates, 858
Iodic acid, 8
Iodine, 831, 832
  occurrence and preparation of, 831, 832
  properties of, 839–842
  uses of, 833
Iodoform, 833

Ion(s), 184
  activity, 448
  amphiprotic, 558–559
  central metal, 877
  common-ion effect, 613–617
  complex, 656–660
  concentration of, in solutions of strong
      electrolytes, 583–585
  -dipole attraction, 419
  electronic structures of, 185–187
  electrostatic attractions between,
      184–185
  formation of, 46–49
  hydrated metal, 623–625
  hydronium, 57–58, 268
  hydroxide, 58
  isoelectronic, 173
  magnetic moment of, 887
  monatomic, 48
  pair, 448
  polyatomic, 48–49
  product for water, 557
  spectator, 55
Ionic
  bonds, 49, 184–185
  compounds, 49–51, 184
    dissolution of, 419–421
    unit cells of, 400–401
  crystals
    lattice energies of, 187–188
    structures of, 394–395
  equation, 55
    net, 55
  hydrides, 826, 834
  radii
    calculation of, 402–403
    variation in, 173–174
  reactions, equations for, 55–56
  solids, 389
Ionization
  constants, 559–563, 590
    of acid, 559
    of base, 560
  energy(ies)
    first, 174
    second, 174
    third, 174
    variation in, 174–176
  of hydrated metal ions, 623–625
  of polyprotic acids, 604–609
  of water, 557, 585–586
  of weak acids, 590–595, 609–612
  of weak bases, 590–595
Ionizing, 556
Iron
  galvanized, 795
  isolation of, 869–871
Isoelectronic species, 173
Isomers
  cis and trans, 306, 880–881
  geometrical, 306, 880

  optical, 880
  structural, 301
Isomorphous structures, 393
Isoprenoids, 946–947
Isotonic solutions, 445
Isotopes, 41–44, 757
IUPAC
  names of new elements, 37–38, 39
  nomenclature for alkanes, 302–304
  periodic table, 13

Joule, 119

K_a, 590
K_b, 591
Kelvin temperature scale, 16, 25, 336
Keratin, 937
Kernite, 811
Kerosene, 302
Ketones, 311–312
Ketose, 941
Kevlar, 900–901
Kilocalorie, 119
Kilogram, 16
Kilojoule, 119
Kilopascal, 327
Kinetic energy, 6
  molecular velocities and, 357–359
Kinetic-molecular theory, 356–357
  gas behavior and, 359–361
  ideal gas equation derived from,
      361–362
  and liquids and solids, 372–374
Kinetics
  chemical, 461–499
  and equilibria, 523–527
  of reaction, 497
K_p, 515
Krypton, 833, 843
  uses of, 833

Lactose, 942–943
Lanthanide series, 13, 171, 865
Lattice
  energies, 187–188, 403–404
  space, 397–398
Laws, 4
  See also names of specific laws
Lead, 791, 795
  -acid cell, 731
  storage battery, 731
Le Châtelier's principle, 517
Lecithins, 945
Leclanche cell, 730
Length, 16
Leveling effect, 564
Levulose, 941
Lewis, G. N., 554, 572
Lewis
  acid-base adduct, 572
  acid-base concept, 572–574

structures, 192–194
    writing, 194–200
symbols, 192, 193
Ligands, 877–878
    bidentate, 878
    chelating, 877
    hexadentate, 878
    monodentate, 878
    pentadentate, 878
    polydentate, 877
    tetradentate, 878
    tridentate, 878
Lime, 797
Limiting reactant, 99–100
Linear polymers, 898
Line notation, 722–723
Lipids, 944–947
    classes of, 944–947
    structural, 945
Liposomes, 945
Liquid(s)
    boiling of, 379–380
    distillation of, 380–381
    evaporation of, 378–379
    freezing point of, 382
    kinetic-molecular theory for, 372–374
    miscibility of, 424–425
    properties of, 4, 378–388
    solutions
        of gases in, 422–424
        of liquids in, 424–425
    surface tension of, 388
    temperature effect on solubilities,
        425–426
    viscosity of, 387
Liter, 17
Lithium, 788
London, Fritz, 375
London dispersion force, 375
Lone pairs, 193
Lowry, Thomas, 554
Low-spin complex, 887

Macroscopic domain, 13
Magnesia, milk of, 799
Magnesium, 788, 789
    chloride, 802
    oxide, 797
    preparation of, 794
Magnetic
    moment, 887
    quantum number, 160–161
    resonance imaging (MRI), 913
Main group elements, 45, 170
Manometer, 328
Marsden, Ernest, 146
Martensite, 921
Mass
    action, law of, 511
    atomic, 36, 71
    critical, 771

defect, 758
defined, 5
formula, 72
molar, 73, 77, 344, 896–897
molecular, 71–72, 445–447
number, 36, 757
percent by, 428–429
unit, 16
    atomic (amu), 36
Material balances, 92–97
Materials science, 895
Matter
    classification of, 8–10
    conservation of, 5
    defined, 4
    dual nature of, 155
    interaction of radiation with, 779–781
    states of, 4
McMillan, E. M., 769
Measurement, 14
    density, 17
    of heat, 119–120
    International System of Units (SI), 15,
        16–17
    length, 16
    mass, 5
    metric system of, 16–17
    temperature, 24–26
    uncertainty in, 18–20
    units and, 14–15
    volume, 17
    weight, 5
Mechanism(s)
    chain, 496
    reaction, 492, 495–497
        involving equilibria, 525–527
Medicine, materials in, 919–922
    artificial hips, 919–920
    nitinol, 920–922
Meissner, W., 914
Meissner effect, 914
Melting point, 382–383
Membrane, semipermeable, 444
Mendeleev, Dimitri, 44, 258–259
Mercury, 789, 790
    (II) oxide, 796
Messenger RNA (mRNA), 950, 951
Metabolism, 929–931
Metabolites, 929
Metaboric acid, 816
Metal(s), 11, 259–261
    activity series of, 273–274
    alkali, 45, 788
    alkaline earth, 45, 788–789
    chelates, 877
    electrolytic deposition of, 738–739
    of Group 1A, 788
    of Group 2A, 788–789
    of Group 2B, 789–790
    of Group 3A, 790–791
    of Group 4A, 791

of Group 5A, 791
of Group 6A, 791
hydrides, 826
inner transition, 45, 171, 865
ions, 623–625
platinum, 867
properties of, 906
reactions with acids or water, 826
representative, 787–802
    compounds of, 795–802
    preparation of, 791–795
structures of, 391–394
transition, 45, 171, 865–866
    complexes of, 888–889
    preparation of, 868–872
    properties of, 866–868
    series of, 865
unit cells of, 399–400
Metallic
    behavior of representative elements,
        variation in, 272–273
    conduction, 905
    solids, 389–390
Metalloids, 259, 808
    chemical behavior of, 809–810
    structures of, 810–811
Metaperiodic acid, 859
Metathesis reaction, 57, 802
Meter, 16
Methane, 235–237, 308
Methanol, 308
Methyl orange, 629
Metric system, 16–17
Meyer, Lothar, 44
Microscopic domain, 13
Microscopic world, behavior in, 155–157
Milk
    of magnesia, 799
    sugar, 942–943
Milliliter, 17
Millimeters of mercury (mm Hg), 328
Miscibility, 424–425
Mitochondria, 929
Mixture
    azeotropic, 440
    defined, 8
    heterogeneous, 8
    homogeneous, 8
mm Hg, 328
Moderator, nuclear, 773–774
Modified Chemical Vapor Deposition
        (MCVD), 912
Moissan, Henri, 832
Molal boiling-point elevation constant, 437
Molal freezing-point depression constant,
        441
Molality, 430–431
Molar
    mass, 73, 77, 344, 896–897
    solubilities, 643
    volume, standard, 341

Molarity, 85, 429–430
Mole(s), 73, 77
    of atoms, 73–76
        mass of, 73
    fraction, 431–434
    of gas and volume, 337–338
    of molecules, 77–80
        mass of, 71–72
    relationships and chemical equations,
        90–92
Molecular
    electrolytes, 421
    formulas, 39, 40, 84–85
    geometry, 220
        rules for predicting, 222–227
    mass, 71–72, 445–447
    orbital(s)
        antibonding, 244
        bonding, 244
        energy diagrams, 245–246
        theory, 243–251
    polarity, 227–229
    solids, 390–391
    structures
        defined, 202, 217
        prediction of, 217–220
    velocities, 357–359
Molecule(s)
    defined, 11
    diamagnetic, 198
    diatomic, of second period, 248–251
    dihydrogen and dihelium, 247–248
    electron-deficient, 198–199
    with extra electrons, 199
    forces between, 374–377
    magnetic moments of, 887
    moles of, 77–80
    odd-electron, 198
    paramagnetic, 198
    polar, 227–229
Moment
    dipole, 227
    magnetic, 887
Monatomic ion, 48
Monoclinic sulfur, 822
Monodentate ligand, 878
Monomers, 315
Monoprotic acids, 567
Monosaccharides, 941–942
Morphine, 315
Morphology, 899
Moseley, Henry, 147
Motion, thermal, 906
Muller, K. A., 913
Multiple proportions, law of, 35
Muscle tissue, 937
Mutations, 955–956

Names of new elements, 37–39
Natural gas, 134
Natural products, 314–315

Neon, 833, 843
    uses of, 833
Neptunium, 769
Nernst, W. H., 726
Nernst equation, 726
Net ionic equation, 55
Neutralization, acid-base, 563–564
Neutralization reaction, 57, 563
Neutral solution, 563
Neutron-proton ratio, 761–762
Neutrons, 764
    charge of, 36
    defined, 36
    mass of, 36
New elements, 37–39
Newton, Sir Isaac, 149
Nickel-cadmium cell, 731–732
Nitinol, 920–922
Nitrates, 854
Nitric acid, 850, 851, 853–854
Nitric oxide, 850, 851
Nitrides, 844
Nitrites, 855
Nitrogen, 829
    dioxide, 850, 851
    hydrogen compounds of, 844–845
    oxides of, 850–851
    oxyacids of, 853–855
    properties of, 837
    uses of, 283, 829
Nitrosoamines, 855
Nitrous acid, 851, 854–855
Nitrous oxide, 850, 851
Noble gas(es), 45–46, 833
    electron configuration, 186
        pseudo-noble gas, 186
    properties of, 842–843
    uses of, 833
Nobel prize, 520, 878, 950
Nomenclature
    of alcohols, 309
    of alkanes, 302–304
    of alkenes, 305
    of alkynes, 307
    of binary compounmds, 59–63
    of coordination compounds,
        878–879
    of inorganic compounds, 59–63
    of new elements, 37–39
Noncompetitive inhibitors, 940
Nonelectrolytes, 420
    solution of, 443
        phase diagram of, 443
Nonmetallic behavior, of representative
        elements, variation in, 272–273
Nonmetals, 11, 259–261, 808–809, 818
    chemical behavior of, 824–825
    compounds of selected, 843–859
    electronegativity of, 818
    isolation and uses of, 825–833
    oxidation states of, 819

periodic trends and general behavior
    of, 818–821
properties of, 834–843
structures of, 821–824
Nonstandard states, free energy changes
    and, 694–696
Nonstoichiometric compounds, 426
Normal boiling point, 379
Normality, 571–572, 740
N-terminal residue, 935
n-type semiconductors, 907
Nuclear
    accidents, 774–775
    atom, 146–147
    binding energy, 758–761
    chemistry, 757–781
    energy, 771–781
    fission, 771
    force, 757–758
    fuel, 773
    fusion, 776–777
    power reactors, 771–776
        components of, 771–775
        fusion, 777
    reactions, 763–770
        equations for, 763–765
        transmutation, 769
    stability, 761–762
        band of, 761
        neutron-proton ratio and,
            761–762
Nucleic acids, 947–950
Nucleon(s), 757
    binding energy per, 759–760
Nucleosides, 947–948
Nucleotides, 947–948
Nucleotide triphosphates, 929
Nucleus
    atomic, 757
    composition of, 757
    diameter of, 36
    stability of, 761–762
Nuclides, 757
    parent and daughter, 765
    synthesis of, 769–770
Number
    atomic, 36, 757
    quantum, 152, 159–162
Nutritional calorie, 119
Nylon, 316, 919

Ochsenfeld, R., 914
Octahedral hole, 394
Octahedral hybridization, 883
Odd-electron molecules, 198
Oils, 944
    hydrogenation of, 826
Oleic acid, 314
Olestra, 944–945
Onnes, H. Kamerlingh, 913
Optical isomers, 880

Orbital(s)
  antibonding, 244
  atomic, 157
    hybridization of, 232–233
  bonding, 244
  d, 159, 885
  degenerate, 161, 245
  diagram, 164
  e_g, 885
  energies, 163–164
  f, 160
  hybrid, 232–233
  molecular, 243
    defined, 243
    energy diagrams, 245–246
    pi, 244
    sigma, 244
    theory, 243–251
  p, 159
  s, 159
  t_{2g}, 885
Order of reaction, 472
Organic compounds, of carbon, 299–307
Orthophosphoric acid, 855
Orthosilicic acid, 815
Osmosis, 444–445
  reverse, 445
Ostwald process, 853
Outer-shell hybridization, 883
Overall order of a reaction, 472
Overlap of atomic orbitals, 230–231
Oxidation numbers, 203
Oxidation-reduction reactions, 261, 271–272
  balancing, 743–747
Oxidation state, 203–205, 261–265
  periodic variations in, 265–266
Oxide(s)
  boron, 816–817
  nonmetal, 850–853
  reactions of, with water, 270–271
  of representative metals, 796–797
  silicon, 817–818
  of transition elements, 873–874
Oxidized atom, 271
Oxidizing agents, 272
Oxyacids, 62–63, 271, 565
  nonmetal, and their salts, 853–859
Oxygen, 827–829
  compounds with, 796–798
  halides of, 848–849
  importance of, 828
  occurrence and preparation of,
    827–828
  properties of, 835–837
  uses of, 283–284, 828
Ozone, 828–829, 837

Paired electrons, 162
Pairing energy, 886
Palmitic acid, 314
Parallel spins, 162

Paramagnetic molecules, 198
Paramagnetism, 887
Paraperiodic acid, 859
Parent nuclide, 765
Partial miscibility, 425
Partial pressure, 345–347
Particles, nuclear
  alpha, 146, 764
  beta, 764
  neutrons, 36, 764
  positrons, 764, 766
  protons, 36, 764
Pascal, 327
Path, reaction, 492, 495
Pathways, anabolic and catabolic, 929
Pauli exclusion principle, 165
Pauling, Linus, 189, 939
p band, 906
p electron, 159
Pentadentate ligands, 878
Pentahalides, 848
Peptide(s)
  bonds, 935
  polypeptides, 936
  and proteins, 935–936
Perbromic acid, 859
Percent
  composition, 428–429
    from formulas, 80–81
  by mass, 428–429
  yield, 98–99
Perchlorate ion, 859
Perchloric acid, 859
Perfluorodecatin, 832
Perhalic acids, 857
Period, 12
Periodic
  law, 44
  relationships among groups
    Group 1A, 788
    Group 2A, 788–789
    Group 2B, 789–790
    Group 3A, 790–791
    Group 4A, 791
    Group 5A, 791
    Group 6A, 791
  relationships of transition elements,
    865–876
  table, *inside front cover,* 164–177
    chemical behavior and, 258–259
    discussed, 11–13, 44–46
    electron configuration and, 169–171
    groups (families) and periods
      (series) in, 12
    inner transition metals in, 171
    nonmetals in, 818
    representative elements in, 170
    semi-metals in, 11, 259–261, 808
    transition metals in, 171, 856–866
    variation of atomic properties within
      periods and groups, 171–177

Peroxides, 796, 797–798
Perxenate ion, 843
Petroleum, 134–135
pH, 585–590
Phase diagrams, 384–387, 443
Phenolphthalein, 628, 629
Pheromones, 314
Phosphatidyl cholines, 945
Phosphide ion, 837
Phosphine, 845
Phosphoglycerides, 945
Phosphorous acid, 855
Phosphorus, 829–830
  allotropes of, 821–822
  halides of, 848
  hydrogen compounds of, 845
  occurrence and preparation of, 837
  (III) oxide, 852
  (V) oxide, 852
  oxides of, 852
  oxyacids of, 855–856
  oxyhalides, 848
  pentachloride, 848
  properties of, 837
  red, 821–822, 837
  trichloride, 848
  uses of, 830
  white, 821–822, 837
Photochemical smog, 354
Photodegradation, 904
Photons, 149
Photosynthesis, 828, 930
Physical change, 7
Physical properties, 7
Pi bond, 232
Pi molecular orbital, 244
Pipet, 101
pK, 622
Planck, Max, 149
Plasma, 4, 777
Plastics, 315, 895
Plastic sulfur, 823
Platinum metals, 867
Pleated sheet, beta-, 937–939
Plutonium, 769
p-n junction, 907
pOH, 586
Polar covalent bonds, 189–191
Polar molecules, 227–229
Polonium, 764, 791
Polyatomic ions, 48–49
Polydentate ligand, 877
Polyethylene, 316, 896, 898–899, 900
Polyethylene terephthalate (PETE), 902
Polyhalides, 849
Polymers, 315–317, 895–904
  addition, 316
  average molar mass of, 896–897
  branched, 898
  chemical composition of, 896
  condensation, 316

Polymers (*continued*)
cross-linked, 898
factors affecting properties of, 896–899
future of, 903–904
linear, 898
morphology of, 899
properties of, 899–901
recycling of, 901–902
stereochemistry of, 897–898
thermosetting, 901
topology of, 898–899
Polypeptides, 936
Polyprotic acids, 604–609
Polysaccharides, 931, 942–943
Polystyrene, 897, 900
Polyvinyl chloride, 833, 902
p orbitals, 159
Positron(s), 764
emission, 766
Potassium, 788
chlorate, 858
hydrogen tartrate, 801
Potential(s)
cell, 714–716
calculation of, 721–723
concentration and, 725–727
equilibrium constant and, 727–729
free energy, electrical work, and,
723–725
standard, 715
electrode, 716–720
energy, 6
reduction, standard, 719
Precipitation
concentrations following, 649–650
fractional, 650–651
reactions, 267
of slightly soluble solids, 646–649
Precipitates, 57
Prediction of reaction products, 274–279
Pre-equilibrium, 526
Preparation of representative metals,
792–795
Pressure(s)
and concentration changes,
equilibrium and, 527–529
critical, 383–384
effects on equilibrium, 518–519
gas, 326–329
osmotic, 444–445
partial, 345–347
standard conditions of, 341–343
vapor, 378, 435–437
volume related to, 329–333
Primary galvanic cell, 730
Primary structure of protein, 937
Principal quantum number, 159
Principles. *See names of specific principles*
Procedures for analysis. *See* Qualitative
analysis

Products
defined, 53
natural, 314–315
Prokaryotic cells, 947
Properties
chemical, 7
colligative, 435–448
extensive, 7
intensive, 7
physical, 7
Proportionality
direct, 334
inverse, 329–331
Prostheses, 919
Prosthetic groups, 936
Proteases, 936
Protein(s), 931–940
alpha-helix, 937
beta-pleated sheet, 937–939
classes and structures of, 936–939
peptides and, 935–936
primary structure of, 937
quaternary structure of, 939
secondary structure of, 937
synthesis, genetic code and, 950–954
tertiary structure of, 939
types of
contractile, 937
defense, 937
enzymes, 937
nutrient and storage, 937
regulatory, 937
structural, 937
transport, 937
Proton, 764
charge of, 36
defined, 36
mass of, 36
Proust, Joseph, 35
Pseudo-noble gas electron configuration,
186
p-type semiconductors, 907
Pure substances, 8
Purines, 947
Pyrex glass, 911
Pyrimidines, 947

Qp, 515
Quadratic formula, 534
Quanta, 149
Quantitative analysis, 101–106
Quantitative reactions, of acids and
bases, 569–570
Quantized energy, 151
Quantum-mechanical model of the atom,
157–162
Quantum mechanics, 155–164
Quantum number, 152
angular momentum, 159
azimuthal, 159

magnetic, 160–161
principal, 159
spin, 161–162
Quartz, 817
Quaternary structure of proteins, 939
Quicklime, 797

rad, 780
Radiation
electromagnetic, 147
interaction of, with matter, 779–781
therapy, 778
Radii
atomic, *inside back cover*
covalent, 171–172
ionic, 402–403
variation in, 173–174
Radioactive
dating, 767–769
decay, 765–767
materials, half-life of, 762–763
tracers, 777
Radioisotopes, uses of, 777–779
Radium, 788, 789
Radius
covalent, 171–172
ionic, 402–403
ratio rule, 396–397
Radon, 833, 843
Randomness. *See* Entropy
Raoult's law, 435
Rare earth elements, 865
Rare gases. *See* Noble gases
Rate(s)
constant, 470
-determining step, 496
of diffusion, 348–349
of effusion, 348–349
equations. *See* Rate laws
instantaneous, 463–464
of reaction, 461–486
catalyst and, 470
collision theory of, 486–488
and concentration, 469
and equilibria, 523–525
factors affecting, 468–470
initial, 464
microscopic explanation of, 486–499
and nature of reactants, 468
relative, and reaction velocity,
464–468
and state of subdivision of reactants,
468
and temperature, 468–469
Rate laws, 470–476, 525–527
differential, 470
integrated, 477–482
Ratio, 21
using, to convert units, 21–22
Rays, gamma, 764, 766

RBE, 780
Reactant(s), 53
  limiting, 99–100
Reaction(s)
  acid-base, 267–270, 554
  addition, 56, 305
  aerobic, 930
  anaerobic, 930
  bimolecular elementary, 493–494
  cell, 712
    spontaneity of, 724–725
  chain, 771
  classification of chemical, 56–59
  combination, 56
  combustion, 57
  decomposition, 57
  diffusion-controlled, 486
  disproportionation, 820, 842
  elementary, 492–495
  endothermic, 118, 672
  energy changes, 681–684
  enthalpy of, 124–130, 681–684
  entropy of, 684–686
  equations for, 52–55
  exothermic, 672
  first-order, 472, 477–478, 483–484
  free energy of, 692–693, 696–698
  half-life of, 483–486
  ionic equations for, 55–56
  kinetics of, 497
  mechanism(s), 492, 495–497
    involving equilibria, 525–527
  metathesis, 57, 802
  neutralization, 57, 563
  nuclear, 763–770
  order of, 472
  oxidation-reduction, 261, 271–272,
    743–747
  path, 492, 495
  precipitation, 267
  products, prediction of, 274–279
  quotient, 510–513, 694
  rates of, 461–486
    catalyst and, 470
    collision theory of, 486–488
    and concentration, 469
    and equilibria, 523–525
    factors affecting, 468–470
    initial, 464
    microscopic explanation of,
      486–499
    and nature of reactants, 468
    relative, and reaction velocity,
      464–468
    and state of subdivision of reactants,
      468
    and temperature, 468–469
  redox, 261, 271–272, 743–747
  reversible, 508–509
    predicting direction of, 521–523

second-order, 472, 480–481, 484–485
spontaneous, 416
substitution, 302
termolecular elementary, 494–495
thermite, 794
transmutation, 769
unimolecular, 492–493
velocity, 464–468
water gas shift, 826
zero-order, 472, 482, 485
Real gas(es), 332–333
  descriptions of, 362–364
Recombinant DNA technology, 955
Recycling of polymers, 901–902
Redox reactions, 261, 271–272
  balancing equations for, 743–747
Red phosphorus, 821–822, 837
Reduced atom, 271
Reducing agent, 272
Reduction
  chemical, of representative metals,
    794–795
  potentials, standard, 719
Refining, 869
rem, 780
Replication, 950–951
Representative elements, 45, 170, 272–273
Representative metals, 787–802
Residues, 935
Resistance, 906
Resonance, 205–206
  forms, 205
  hybrid, 205
Respiration, 930
Reverse osmosis, 445
Reversible reaction, 508–509
  predicting direction of, 521–523
Rhombic sulfur, 822
Ribonucleic acids (RNA), 928, 947–950
  messenger (mRNA), 950, 951
  ribosomal, 952
  transfer (tRNA), 952
Ribosomal RNA, 952
Ribosomes, 929
RNA. See Ribonucleic acids (RNA)
Rounding numbers, 19–20
Rubber, 315–316
Rubidium, 788
  dating, 768–769
Rules. See names of specific rules
Rutherford, Lord Ernest, 146–147, 769

Salt(s), 268, 801–802
  acid-base properties of solutions of,
    602–604
  bridge, 712
  defined, 58
  effect, 613–617
  nonmetal oxyacids and their, 853–859
  of transition metals, 874–875

Saltpeter
  Bengal, 853
  Chile, 831, 832, 853
Saponification, 944
Saturated hydrocarbons, 300–302
Saturated solutions, 415
s band, 905
Scientific method, 4
Seaborg, Glenn T., 38–39
Second, as unit of time, 16
Second
  ionization energy, 174
  law of thermodynamics, 693–694
  period, diatomic molecules of,
    248–251
  transition series, 171, 865
Secondary galvanic cell, 730–732
Secondary structure of proteins, 937
Second-order reactions, 472, 480–481
  half-life of, 484–485
s electron, 159
Self-ionization, 557
Semiconductors, 904, 905, 906–907
  extrinsic, 907
  intrinsic, 907
  n-type, 907
  p-type, 907
Semi-metals, 11, 259–261, 808
  chemical behavior of, 809–810
  structures of, 810–811
Semipermeable membrane, 444
Series, periodic, 12
Shell, 159
  valence, 169
Shield and containment system,
    nuclear, 774–775
Sickle-cell anemia, 896, 955–956
Sigma bond, 231, 232
Sigma orbitals, 244
Significant figures, 18–20
Silane, 814
Silica, 817
Silicates, 818
Silicon
  chemical behavior of, 809–810
  dioxide, 817
  halides of, 815–816
  hydrides of, 812–814
  isolation of, 812
  oxides, 817–818
  structure of, 810
  tetrachloride, 815
  tetrafluoride, 815
Silicones, 814
Silver
  iodide, 833
  isolation of, 872
Simple cubic structure, 393–394
Simple cubic unit cell, 399
Single bonds, 193

SI units, 15
Slag, 870, 871
Smelting, 869
Smog, photochemical, 354
Soaps, 450–452, 944
Sodium, 788
    carbonate, 800
    chlorate, 858
    chloride, 802, 847
        electrolysis of molten, 734–735
        electrolysis of solution of, 736–737
    cyanide, 299
    hydrogen carbonate, 801
    hydroxide, 284–285, 798, 799
    nitrite, 854
    peroxide, 797
    preparation of, 791–792
Solar cell, 907–909
Solid(s)
    amorphous, 389
    covalent network, 390
    crystalline, 389
        structures of, 389–405
    evaporation of, 378–379
    ionic, 389
    kinetic-molecular theory and,
        372–374
    melting point of, 382–383
    metallic, 389–390
    molecular, 390–391
    precipitation of slightly soluble,
        646–649
    properties of, 4, 378–388
    solutions of, 422
        in solids, 426–427
        in water, 425–426
    types of, 389–391
Solid solutions, 426–427
    interstitial, 427
    substitutional, 427
Solubility(ies)
    defined, 415
    of gases in liquids, 422–424
    of liquids in liquids, 424–425
    molar, 643
    multiple equilibria involving, 651–660
    pressure effects on, 422–423
    product, 639–644
        calculations of, 644–646
    of solids in water, 425–426
    temperature effects on, 425–426
Solute, 85, 414
Solution(s), 85–87
    aqueous, 443
    buffer, 617–623
    calculations involving, 644–646
    colligative properties of, 435–448
    concentrated, 85, 415
    concentration of, 85, 427–434
    constant boiling, 440

defined, 85
dilute, 85, 415
dilution of, 87–90
distillation of, 439–440
electrolysis of, 736–738
formation of, 416–418
    of gases in liquids, 422–424
    homogeneous equilibria in, 513–514
    homogeneous mixtures as, 8
    hypertonic, 445
    hypotonic, 445
    ideal, 416
    isotonic, 445
    of liquids in liquid, 424–425
    macroscopic properties of, 422–427
    molality of, 430–431
    molarity of, 429–430
    nature of, 414–416
    neutral, 563
    of nonelectrolytes, 443
    normality of, 571–572
    osmosis and osmotic pressure of,
        444–445
    percent by mass, 428–429
    phase diagrams for, 443
    saturated, 415
    standard, 569
    supersaturated, 415
    unsaturated, 415
    of weak acids or weak bases,
        concentrations in, 596–602
Solvent, 85, 414
    boiling point elevation of, 437–438
    freezing point depression of,
        440–443
    lowering vapor pressure of, 435–437
s orbitals, 159
sp hybridization, 233–234
sp^2 hybridization, 235
sp^3 hybridization, 235–237
sp^3d hybridization, 237–238
sp^3d^2 hybridization, 237–238, 883
Space lattice, 397–398
Space Shuttle tiles, 910
Specific heat, 119
Spectator ions, 55
Spectrum(s)
    continuous, 149
    emission, 150
Spin(s)
    parallel, 162
    quantum number, 161–162
Splitting, crystal field, 886
Spontaneous change or process, 416,
    678–681
Square planar structure, 884
Stability
    band of, 761
    constant, 656
    nuclear, 761–762

Standard
    cell potential, 715
    conditions of temperature and pressure
        (STP), 341–343
    electrode potentials, 716–720
    enthalpy
        change, 683
        of formation, 129–130
    entropy, 685
        change, 686–688
    free energy
        change, 691
        of formation, 692–693
    hydrogen electrode, 717
    molar volume, 341
    reduction potentials, 719
    solution, 569
    state, 125–126
        conditions, 680
State(s)
    excited, 153
    functions, 676–677
    ground, 153
    of matter, 4
    nonstandard, free energy changes
        and, 694–696
    standard, 125–126
    transition, 487
Stearic acid, 314
Steel, 870–871
Stereochemistry, 897
Steroids, 946–947
Stoichiometry, 71–106
    and chemical change, 90–100
    defined, 71
    involving gases, 349–352
Strong
    acids, 268, 559
    bases, 270, 560
    electrolytes, 421
Strontium, 788, 789
    peroxide, 797
Structural
    formula, 41
    isomers, 301
    lipids, 945
    protein, 937
Styrene, 307
Sublimation, 378, 379
Subshell, 159
Substance
    homogeneous, 8
    pure, 8
Substituents, 302–303
Substitutional solid solutions, 427
Substitution reaction, 302
Substrate activation, 940
Subunit, 939
Successive approximations, 537–539
Sucrose, 942

Sugars
    disaccharides, 942–943
    monosaccharides, 941–942
    polysaccharides, 931, 942–943
Sulfates, 853, 856
Sulfite(s), 853, 856
    ion, 856
Sulfur, 830–831
    allotropes of, 822–824
    dioxide, 852
    extraction of, 830–831
    Frasch process, 830–831
    hydrogen compounds of, 845–846
    monoclinic, 822
    oxides of, 852–853
    oxyacids of, 856
    plastic, 823
    properties of, 838
    rhombic, 822
    trioxide, 852–853
Sulfuric acid, 856
    as dehydrating agent, 856
    electrolysis of, 737
    uses of, 279–282
Sulfurous acid, 852, 856
Superconductivity, 913–916
Superconductors, 875–876
Supercritical fluids, 383–384
Superelasticity, 921
Superoxides, 796, 797–798
Supersaturated solution, 415
Surface tension, 388
Surroundings, thermodynamic, 123, 670–671
Symbolic domain, 14
Symbols
    for elements, 37
    Lewis, 192, 193
Synthesis
    chemical, 274
    of nuclides, 769–770
    photosynthesis, 828, 930
    of proteins, genetic code and, 950–954
    of thin films, 918–919
Synthetic rubber, 316
System, thermodynamic, 123–124, 670–671

Table salt. See Sodium chloride
Technetium, 764
Teflon, 832, 896, 919
Tellurium, 810
Temperature(s)
    Celsius, 24
    critical, 383–384, 913
    defined, 24
    effect on equilibrium, 519–520
    effect on solubility in water, 425–426
    Fahrenheit, 25

glass transition, 899–900
heat and, 118
Kelvin, 16, 25, 336
rate of reaction and, 468–469
and solubility, 425–426
standard conditions of, 341–343
units of, 16
volume and, 333–336
10Dq, 886
Termination, 952
Termolecular elementary reactions, 494–495
Ternary compounds, 59
Tertiary structure of proteins, 939
Tetraboric acid, 816
Tetradentate ligands, 878
Tetrahedral hole, 394
Tetrahedral hybridization, 235–237
Tetraphosphorus trisulfide, 837
t_{2g} orbitals, 885
Thallium, 790
Theoretical yield, 98
Theories, 4
    See also names of specific theories
Thermal energy, 118
Thermal motion, 906
Thermite reaction, 794
Thermochemistry, 117–136
    defined, 117
    and thermodynamics, 123–124
Thermodynamic(s), 123–124, 669–698
    enthalpy changes, 124–130
    first law, 674–676
    second law, 693–694
    surroundings, 123, 670–671
    system, 123–124, 670–671
    thermochemistry and, 123–124
    third law, 688–689
Thermoplastics, 901
Thermosetting polymers, 901
Thin films, 916–919
    diamond, 917–918
    synthesis of, 918–919
    on windows and lenses, 916–917
Third
    ionization energy, 174
    law of thermodynamics, 688–689
    transition series, 171, 865
Thomson, Sir J. J., 147
Thorium
    decay series, 767
    series, 766
Three-center two-electron bonds, 811
Three Mile Island, 774–775
Thymine, 949
Tiles, Space Shuttle, 910
Time, measurement of, 16
Tin, 791
    brittle, 795
    gray, 795

preparation of, 795
uses of, 795
white, 795
Titration, 101–102
    curves, 626–629
    equivalence point of, 626–629
Toluene, 307
Topology, of polymers, 898–899
Torr, 328
Torricelli, Evangelista, 328
trans (configuration) isomer, 306, 880–881
Transcription, 950, 951–952
Transfer RNA (tRNA), 952
Transition metals (or elements), 45, 171, 865–866
    carbonates of, 874
    complexes, colors of, 888–889
    compounds of, 872–875
    halides of, 872–873
    hydroxides of, 874
    oxides of, 873–874
    periodic relationships of, 856–876
    preparation of, 868–872
    properties of, 866–868
    salts of, 874–875
    series of, 865
Transition state, 487
Translation, 950, 952
Transmutation reactions, 769
Transport proteins, 937
Transuranium elements, 770
Tridentate ligands, 878
Triglycerides, 944–945
Trigonal bipyramidal structure, 237
Trigonal planar structure, 235
Trihalides, 848
Triple bond, 194
    hybridization involving, 240–242
Triple point, 387
Triprotic acids, 567–568
    ionization of, 607–609
Tritium, 825
Trona, 800
Tyndall effect, 448

Ultraviolet radiation, 829
Uncertainty
    of measurement, 18–20
    principle, Heisenberg, 156
Unimolecular elementary reactions, 492–493
Unit(s)
    cells, 397–401
        of ionic compounds, 400–401
        of metals, 399–400
    conversion factor, 23
    International System of (SI), 18–20
    of measurement, 14–15
    metric system of, 16–17
Unsaturated hydrocarbons, 304

Unsaturated solutions, 415
Unshared pairs, 193
Ununbium, 770, 789
Uracil, 947–948, 949–950
Uranium
    dating, 768–769
    fission of, 771
    series, 766
Uridine, 948
    monophosphate, 948

Vacancies, 391
Valence
    bond theory, 230–232, 882–884
    electrons, 169
    shell, 169
        electron-pair repulsion (VSEPR)
            theory, 217–229
Valine, 955, 956
van der Waals, Johannes, 363
van der Waals equation, 363
van der Waals forces, 374
Vaporization, enthalpy of, 128–129
Vapor pressure, 378
    lowering of, 435–437
Variation(s)
    in metallic character, 272–273
    in nonmetallic character, 272–273
    in oxidation state, 265–266
Viscosity, 387
Volts, 724
Volume
    defined, 17
    molar, standard, 341

pressure related to, 329–333
temperature related to, 333–336
units of, 17
Vulcanization, 315–316

Water
    as acid, 556–557
    as base, 556–557
    distillation of, 381
    electrolysis of, 826
    gas, 826, 827
        shift reaction, 826
    hard, 380–381
    ionization of, 557, 585–586
    ion product for, 557
    leveling effect of, 564
    reaction of ionic metal hydrides
        with, 826
    reactions of oxides with, 270–271
    temperature's effect on solubility of
        solids in, 425–426
Watson, J. D., 950
Wave function, 158
Wavelength, 147
Wave mechanics, 157
Waxes, 946
Weak acid(s), 269, 559
    concentrations in solutions of, 596–602
    ionization of, 590–595, 609–612
Weak base(s), 270, 560
    concentrations in solutions of,
        596–602
    ionization of, 590–595
Weak electrolytes, 421

Weight, 5
    See also Mass
Werner, Alfred, 878
White phosphorus, 821–822, 837
White tin, 795
Work, 5, 124, 672
    electrical, 723–725
    expansion, 124, 672–673
    See also Thermodynamic(s)

Xanthoprotein, 854
Xenon, 833
    difluoride, 842
    hexafluoride, 842
    properties of, 842–843
    tetrafluoride, 842
    trioxide, 843
    uses of, 833
X-ray diffraction, 404–405
Xylene, 307

Yield, 98–99

Zero-order reaction, 472, 482, 485
Zinc, 789–790
    oxide, 797
    preparation of, 794–795
Zone refining, 812
Zwitterion, 932